ORGANIC FUNCTIONAL GROUP PREPARATIONS

Second Edition

VOLUME I

This is Volume 12 of
ORGANIC CHEMISTRY
A series of monographs
Editor: HARRY H. WASSERMAN

A complete list of the books in this series appears at the end of the volume.

PREFACE

Since 1968, when the first edition of "Organic Functional Group Preparations" was published, this work has grown to a series of three volumes, followed by another three-volume series dealing with the synthesis of various polymers.

The purpose of this book, as with the first edition, is to provide for the organic chemist a convenient and updated source of reliable preparative procedures for the most common functional groups.

The unique features of this work have been continued and expanded. The preparations of each functional group are subdivided into the various reaction types such as condensation, elimination, and oxidation and reduction reactions. The miscellaneous section has also been expanded to include a variety of other procedures for information purposes.

In this revised volume, several of the chapters have been extensively rewritten, and the others contain substantial new information and references to the recent patent and journal literature. In other cases, older procedures have been kept intact when the recent literature has made no major improvement in them.

Information on safety is given where available. However, we particularly urge the use of great care in the handling of all materials because of unknown long- or short-term toxicity. Reactions should not be scaled up unless extensive tests are first run to ensure that scale-up reactions can be run safely. Since many of the preparations have not been checked either by us or by an independent laboratory for hazards or toxicity, we do not warrant the preparations against any safety or toxic hazards and assume no liability with respect to the use of the preparations.

We express our gratitude to our wives and our families for their patience, understanding, and encouragement at all stages of the preparation of this manuscript. We also express our appreciation to the production staff at Academic Press for their help in all phases of publication of this volume. Finally, we are certainly fortunate again for having Ms. Emma Moesta, who typed the manuscript of the first edition, provide us with another highly professional typing of the manuscript of this edition.

STANLEY R. SANDLER
WOLF KARO

FROM THE PREFACE TO THE FIRST EDITION

This work is an outgrowth of an attempt by us to ease the burden of those in our laboratory engaged in organic syntheses. Since suitable directions for the preparation of a functional group attached to an invariant carbon chain are scattered throughout the literature, we undertook to compile procedures which we believe to be reliable as to yield and certainty of the structures produced. Some attention was paid to the problem of presenting specific laboratory direction for the many name reactions which are often presented only as a set of equations in textbooks. Thus, it is the purpose of this book to provide in a single volume a convenient source of modern procedures for the preparation of functional groups.

The unique feature of this work is that each chapter deals with the preparation of a given functional group by various reaction types (condensation, elimination, oxidation, reduction) and a variety of starting materials and reagents. Detailed laboratory directions are given which are also representative of a general class of procedures. To a limited extent indications of the scope of the reactions are presented.

Since the primary synthetic object in most cases is to convert one functional group to another, the book serves as a guide to effect these changes. This work is not intended to be encyclopedic in nature nor to mention every available procedure ever published up to the date of this printing. Rather, it is a brief critical review; and because of the limitations of space, it was necessary to be selective in the choice of procedures cited.

The procedures, for inclusion in this text, had to meet the following requirements: (1) The procedures should be generally used for a wide range of organic structures. (2) The yield of product should be high. (3) The preparation should be relatively uncomplicated and should be able to be carried out in most laboratories. (4) The laboratory operations should be safe and free from the danger of explosion.

Most of the preparations were taken from the recent literature; however, since several well-known classic reactions are included, the older literature in some cases was deemed satisfactory in light of recent developments. No attempt was made to include preparations with a 1966–1967 date just for the sake of having an up-to-date bibliography. The preparations were chosen solely for their ability to meet the four criteria mentioned above. The recent literature is cited; and, where possible, methods showing wide applicability have been included. Otherwise, they are briefly mentioned as being worthy of further scrutiny.

While information on the safety of various procedures is not always available, an effort was made to point out hazards, particularly toxicity and explosion hazards, of many of the reactions discussed. Unfortunately, we are not in a position to state that where no hazards are mentioned in connection with a given reaction, none exist. The literature abounds with examples of reactions that have been used for many years which, for reasons frequently not at all clear, suddenly lead to violent explosions. Also, the toxicity hazards of many reactants which have been used with impunity for years have only recently come to light. The extent to which this may be true of other materials cannot be stated at this time.

In several cases the preparations have been repeated in our laboratory and supplementary information is given. In some cases unpublished information is cited where insufficient data are available in the literature.

It should be emphasized that where possible chromatography should be used to check the purity and isomeric nature of the product. In some cases minor variations of a procedure will cause changes in the above factors.

This book is also written with the hope that it will stimulate interest in the reader to reinvestigate several of the methods cited either to improve them or to aid in further extending their general applicability.

STANLEY R. SANDLER
WOLF KARO

CHAPTER 1 / HYDROCARBONS (PARAFFINIC AND AROMATIC)

1. Introduction		2
2. Reduction Reactions		3
A. Reduction of Unsaturated Compounds (Olefins)		3
2-1a.	Conversion of 1-Hexene to n-Hexane by the Hydroboration Method	4
2-1b.	Hydrogenation of Diethyl Maleate to Diethyl Succinate	5
2-1c.	Hydrogenation of β-Pinene	6
2-1d.	Hydrogenation of Ethyl Oleate to Ethyl Stearate	6
B. Reduction of Aromatic Compounds		9
C. Reduction of Carbonyl Compounds		9
a. The Wolff–Kishner Method		9
2-2.	Preparation of 2-(n-Octyl)naphthalene by the Huang–Minlon Method	10
b. The Clemmensen Method		10
2-3.	Preparation of 2-(n-Octyl)naphthalene	11
c. The Reduction of Thioketals or Thioacetals		11
2-4.	Preparation of 7,7,10-Trimethyl-$\Delta^{1(9)}$-octalin	11
D. Reduction of Alcohols		12
2-5.	Catalytic High-Pressure Hydrogenolysis of 2,2,3-Trimethyl-1-butanol to 2,2,3-Trimethylbutane	12
2-6.	Conversion of Ketones to Alcohols and Subsequent Reduction to Alkanes—Reduction of Benzophenone to 1,1-Diphenylethane	12
E. Reduction to Halides		13
2-7.	Lithium Aluminum Hydride Reaction of 1-Bromooctane	14
2-8.	Sodium Borohydride Reduction of Methyl Chloride	14
2-9.	Magnesium in Isopropanol Reduction of Chlorobenzene	15
3. Condensation Reactions		15
A. Friedel–Crafts Alkylation of Hydrocarbons		15
3-1.	1,3,5-Triethylbenzene by the Friedel–Crafts Ethylation of Benzene	17
3-2.	Jacobsen Reaction of 6,7-Dimethyltetralin	18
B. Hydrocarbon Polymers		19
3-3.	Synthesis of Polystyrene by Thermal Activation	19
3-4.	Emulsion Polymerization of Styrene	20
C. Small Ring Hydrocarbon Syntheses		20
3-5.	Preparation of n-Hexylcyclopropane	21
D. The Diels–Alder Reaction		22
E. Coupling Reactions		23
3-6.	Grignard Coupling—Synthesis of Neohexane	24
3-7.	Ullmann Synthesis of 2,2'-Diethylbiphenyl	25
3-8.	Preparation of Hexamethylbenzene	26
4. Elimination Reactions		27
4-1.	Conversion of p-Tolyldiazonium Chloride to Toluene	28

4-2. Elbs Reaction—Synthesis of 1,2-Benzanthracene	29
5. Dehydrogenations	29
6. Miscellaneous Methods	31
References	32

1. INTRODUCTION

Hydrocarbons are conveniently prepared in the laboratory by reduction, condensation, elimination, or hydrolysis reactions. Isomerization, oxidation, and photochemical reactions are less common on a preparative scale.

The reduction methods depend on converting a given functional group to a methylene group. For example, olefins, aromatic rings, alcohols, aldehydes, ketones, and halides give hydrocarbons on reduction. These methods allow the preparation of hydrocarbons of known structure. The Clemmensen (zinc amalgam and hydrochloric acid) and Wolff–Kishner (hydrazine and base) methods can be used to reduce aldehydes and ketones. Catalytic hydrogenation methods can be used to reduce olefins and aromatic compounds. The catalytic hydrogenation method can also be used for ketones, provided that a high-pressure apparatus is available. Nickel and platinum are the most commonly used catalysts.

The use of sodium borohydride and palladium chloride has been described for the reduction of olefins in excellent yields. The method is quite reliable and it has been applied as an analytical technique for the quantitative estimation of the degree of unsaturation of a compound or of a mixture.

$$RCH=CH_2 + NaBH_4 \xrightarrow[H^+]{PdCl_2} RCH_2-CH_3 \qquad (1)$$

Condensation reactions are used to synthesize a hydrocarbon from two or more compounds which may or may not be the same as described in Eq. (2)

$$R-Y + RX \longrightarrow R-R + XY \qquad (2)$$

where X or Y may be a hydrogen, halogen, diazo, or organometallic group. The Friedel–Crafts, the Wurtz, the Wurtz–Fittig, organometallic coupling, Ullmann, and Pschorr syntheses are some representative condensation reactions. The Friedel–Crafts reaction and the coupling of organometallics with halides are the most useful laboratory syntheses of hydrocarbons, especially branched-chain hydrocarbons such as neohexane.

The hydrolysis of the Grignard reagent is a useful method of preparing hydrocarbons from halides [Eq. (3)].

$$ROH \xrightarrow{HX} RX \xrightarrow{Mg} RMgX \xrightarrow{H_2O} RH \qquad (3)$$

The yields by this method are usually excellent and pure hydrocarbons are obtained. The chloromethylation reaction can thus be used as a step in the addition of a methyl group to an aromatic nucleus.

Carboxylic acids eliminate carbon dioxide when heated with soda lime or electrolyzed to produce paraffins (Kolbe reaction). The former is a more useful method in the laboratory.

Aldehyde, diazo, and sulfuric acid groups are a few of the other groups that can be eliminated and replaced by hydrogen to give hydrocarbons.

Cyclodehydration of aromatic alcohols and ketones gives tetralins, anthracenes, phenanthrenes, and other ring systems.

The Jacobsen reaction involves the isomerization by sulfuric acid of an aromatic system containing several alkyl halo groups to give vicinal derivatives.

Other procedures such as photochemical, oxidative, electrochemical, will be mentioned only briefly unless they have been shown to be of general synthetic applicability in the laboratory. (Several of the methods in the earlier literature are mentioned where appropriate since they are of a classical nature.) However, one should be cautious of reactions prior to the 1950s since vapor phase chromatography (VPC) and nuclear magnetic resonance (NMR) techniques were not available for the determination of the purity and structure of the products. Several research problems today are based on a reexamination of earlier reported research findings by VPC.

2. REDUCTION REACTIONS

A. Reduction of Unsaturated Compounds (Olefins)

Olefins can be reduced by one of the following methods: Adkins catalytic hydrogenation [1], sodium borohydride and palladium chloride [2], triphenylsilane [3], triphenyltin hydride [4], rhodium on alumina in the presence of hydrogen for halogenated olefins [5], and sodium hydride [6].

The platinum oxide catalyst is useful for hydrogenations of olefins at room temperature and low pressure [7]. The use of tris(triphenylphosphine)–chlororhodium complexes have been reported to be extremely efficient catalysts for the homogeneous hydrogenation of nonconjugated olefins and acetylenes at atmospheric pressure and at room temperature [8]. On the other hand, Raney nickel requires high-pressure equipment [9].

Diimide [10–15] has also been used as a reducing agent for activated double bonds. The diimide is unstable and, therefore, is produced *in situ* by first oxidizing hydrazine or its derivatives in the presence of olefins [Eq. (4)]

$$\diagup_{C=C}\diagdown + NH_2-NH_2 \xrightarrow{Base} \diagup_{C=C}\diagdown + [NH=NH] \longrightarrow$$

$$\left[\begin{array}{c} \diagup_{C\text{-----}C}\diagdown \\ H \quad\quad H \\ N=\!=\!=N \end{array} \right] \longrightarrow \begin{array}{c} \diagup_{C-C}\diagdown \\ | \ \ | \\ H \ H \end{array} + N_2 \quad (4)$$

Olefins can also be hydroborated to organoboranes which are then converted to the hydrocarbon by refluxing with propionic acid [16]. This procedure is a convenient noncatalytic laboratory method for the hydrogenation of olefins.

$$3 \text{ RCH}=CH_2 + NaBH_4 + BF_3 \longrightarrow (RCH_2CH_2)_3B \xrightarrow{C_2H_5COOH} 3 \text{ RCH}_2CH_3 \quad (5)$$

Terminal olefins are readily hydroborated but internal olefins require additional reaction time and heating prior to refluxing with propionic acid. Substituents such as active sulfur, chlorine, and nitrogen are not affected by this hydrogenation procedure.

2-1a. Conversion of 1-Hexene to n-Hexane by Hydroboration Method [16]

$$3 \text{ CH}_3(CH_2)_3-CH=CH_2 + NaBH_4 + BF_3 \longrightarrow [CH_3(CH_2)_3CH_2CH_2]B \longrightarrow$$
$$CH_3(CH_2)_4CH_3 \quad (6)$$

To a three-necked flask equipped with a mechanical stirrer, dropping funnel, and reflux condenser with attached drying tube are added 16.8 gm (0.20 mole) of 1-hexene and 2.0 gm of sodium borohydride (0.055 mole) in 55 ml of diglyme. While stirring under nitrogen 10.0 gm (0.075 mole) of boron trifluoride etherate in 25 ml of diglyme is added during a period of 1.5 hr. Then 222 gm (0.3 mole) of propionic acid is added and the mixture is refluxed for 2 hr while ether and the product distill over. The product is washed with sodium bicarbonate solution, then water, dried, and fractionally distilled to yield 15.6 gm (91%) of n-hexane, bp 68°–69°C (738 mm), n_D^{20} 1.3747.

Another reduction technique utilizing sodium borohydride is the *in situ* preparation of platinum black while hydrogen is evolved from the borohydride [17a,b]. An active nickel catalyst [18] in the latter procedure has been found to be a very selective hydrogenation catalyst for the conversion of acetylenes to yield cis olefins [19]. In addition there has been developed a simple, pressure-activated device for controlling the rate of addition of the borohydride solution (Fig. 1). In this device a syringe barrel or a buret fitted to a hypodermic needle is inserted through a rubber serum cap into a mercury well to a depth adequate to support the column of borohydride solution. As hydrogen is utilized in the hydrogenation flask, the pressure drops 10 to 20 mm below atmospheric, drawing a small quantity of the borohydride solution through the mercury seal where it rises to the top of the mercury and runs into the flask through the small vent

§ 2. Reduction Reactions

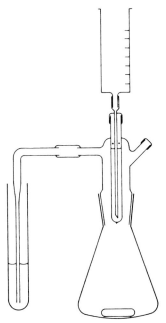

FIG. 1. Hydrogenation apparatus with automatic valve. [Reprinted from C. A. Brown and H. C. Brown, *J. Am. Chem. Soc.* **84**, 2829 (1962). Copyright 1962 by the American Chemical Society. Reprinted by permission of the copyright owner.]

holes located just above the mercury interface. The acidic solution in the flask hydrolyzes the borohydride and the resulting increase in pressure seals the valve. The addition proceeds smoothly to the completion of the hydrogenation, with the amount of the borohydride solution corresponding quantitatively to the amount of unsaturated compound contained in the flask.

2-1b. Hydrogenation of Diethyl Maleate to Diethyl Succinate [17b]

$$\begin{array}{c} HC-COOC_2H_5 \\ \| \\ HC-COOC_2H_5 \end{array} + NaBH_4 + \text{Chloroplatinic acid} \longrightarrow \begin{array}{c} H_2C-COOC_2H_5 \\ | \\ H_2C-COOC_2H_5 \end{array} \quad (7)$$

The following procedure involving the hydrogenation of diethyl maleate is representative. In a 500-ml Erlenmeyer flask are placed 5 gm of Darco K-B carbon, 100 ml of anhydrous ethanol, and 5.0 ml of 0.20 M chloroplatinic acid solution. The system is assembled (Fig. 1) and the solution stirred vigorously by a magnetic stirrer as 20 ml of 1.0 M solution of sodium borohydride in ethanol is injected to reduce the catalyst. This is followed in approximately 1 min by 25 ml of concentrated hydrochloric acid to decompose the borohydride and provide a hydrogen atmosphere. The reaction is initiated by injecting 81 ml, 86.0 gm, of diethyl maleate. The reaction is complete in 60 to 70 min. The reaction solution is filtered to remove catalyst, treated with 5% sodium bicarbonate, and extracted with methylene chloride. Distillation of the extract yields 77.6 gm (90% yield) of diethyl succinate, bp 103°–104.5°C (15 mm), n_D^{20} 1.4201.

In some cases the presence of a strong acid, such as hydrochloric acid, may be undesirable. In such cases, acetic acid may be utilized. Moreover, by maintaining essentially anhydrous conditions in the hydrogenation flask, one can minimize solubility problems encountered with higher hydrocarbons, terpenes, and steroids. This procedure is illustrated by the hydrogenation of β-pinene.

2-1c. Hydrogenation of β-Pinene [17b]

$$ \text{(8)} $$

In the 250-ml flask are placed 5 gm of Darco K-B carbon, 100 ml of anhydrous ethanol, and 5.0 ml of a 0.20 M solution of chloroplatinic acid in ethanol. After addition of 20 ml of 1.0 M sodium borohydride in ethanol to reduce the catalyst, 10.0 ml of acetic acid is added. The hydrogenation is initiated by the addition of 78.5 ml, 68 gm, of (−)-β-pinene. There is isolated 60.0 gm (87%) of pinane, bp 164°–165.5°C (740 mm), n_D^{20} 1.4618, $[\alpha]_D^{25}$ −21.3°.

In some cases it might be desirable to generate hydrogen in one flask as the hydrogen is being absorbed in the hydrogenation flask. For large-scale hydrogenations this has the advantage of reducing the amounts of solvent that must be handled. The following procedure is representative.

2-1d. Hydrogenation of Ethyl Oleate to Ethyl Stearate [17b]

$$CH_3(CH_2)_7\diagdown(CH_2)_7COOC_2H_5$$
$$C=C$$
$$HH$$
$$\longrightarrow CH_3(CH_2)_{16}COOC_2H_5 \qquad (9)$$

Hydrogen is generated by adding a 2.5 M solution of sodium borohydride in water to aqueous acetic acid in a generator fitted with the valve previously described. In the hydrogenation flask, a 500-ml Erlenmeyer with a stirrer, are placed 5 gm of carbon, 100 ml of ethanol, 5.0 ml of 0.20 M chloroplatinic acid in ethanol, and 20 ml of 1.0 M sodium borohydride in ethanol. Ethyl oleate, 179 ml (155 gm), is added, and the flask connected to the generator. The system is flushed with hydrogen from the generator, acetic acid (10.0 ml) is injected into the hydrogenation flask, and the hydrogenation allowed to proceed. The reaction is complete in 2 hr. The solution is filtered and added slowly to ice water to recover ethyl stearate, mp 32°–33°C, in 91% yield.

FIG. 2. Commercial form of the H. C. Brown hydrogenator. (Reproduced by permission of the Delmar Scientific Laboratories, Inc., Maywood, Illinois.)

1. Hydrocarbons (Paraffinic and Aromatic)

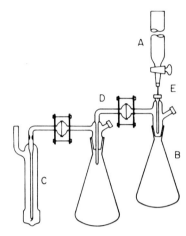

FIG. 3. Diagram of the H. C. Brown hydrogenator. A, Buret with Luer–Lok attachment provides addition of sodium borohydride solution; B, modified Erlenmeyer flask is used for hydrogenation and for hydrogen generation; C, Brown pressure control bubbler prevents pressure buildup within the system; D, Brown atuomatic hydrogen inlet adapted supplies hydrogen to the reaction flask where hydrogenation takes place; E, Brown automatic hydrogenator valve delivers sodium borohydride only as the system requires hydrogen for hydrogenation of the organic compound.

The procedure has been applied successfully to the hydrogenation of 500 gm of ethyl oleate, using a 1-liter flask, with a larger quantity of catalyst [17b].

The controlled generation of hydrogen should be very helpful even in cases where it is desirable to follow literature procedures for hydrogenations. For example, the hydrogenation of cholesterol is a capricious reaction. However, Hershberg *et al.* reported that the erratic tendencies of this reaction could be overcome by performing the hydrogenation with platinic oxide in ethyl acetate in the presence of a small quantity of perchloric acid. The reaction was carried out (on a scale 1/150 of that described) utilizing the automatic hydrogen generator. The reaction required approximately 1 hr for completion. If the amount of platinic oxide was doubled, the hydrogenation was complete in 20 min.

These new procedures should greatly facilitate laboratory-scale hydrogenations. The technique has also proven valuable for chemical analysis and for following the rates of hydrogenation.

Other compounds that can easily be hydrogenated under similar conditions are 1-octene, 4-methylcyclohexene, 1,5,9-cyclododecatriene, and many others.

Other forms of the apparatus shown in Fig. 1 are shown in Figs. 2 and 3.*

---

*Procedures 2-1a to 2-1d are reprinted from C. A. Brown and H. C. Brown, *J. Am. Chem. Soc.* **84**, 2829–2830 (1962). Copyright 1962 by the American Chemical Society. Reprinted by permission of the copyright owner.

B. Reduction of Aromatic Compounds

Aromatic compounds are quantitatively converted to cyclohexanes by catalytic hydrogenation. Platinum catalysts effect reduction at room temperature [20,21], whereas nickel catalysts require 100°–200°C [22,23]. Isomerization occurs when nickel catalysts at 170°C are used [24]. Aromatic amines have been reported to be reduced using hydrogen and a ruthenium catalyst and lithium hydroxide [25]. The Birch reduction of aromatic compounds to 1,4-dihydroaromatic compounds is discussed in Chapter 2, Section 4.

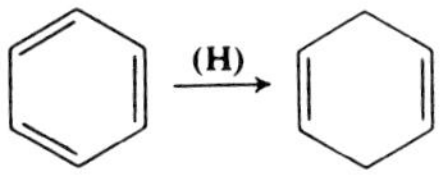

$$\text{C}_6\text{H}_6 \xrightarrow{(H)} \text{C}_6\text{H}_{10} \tag{10}$$

C. Reduction of Carbonyl Compounds

$$\underset{\underset{R}{\overset{\overset{O}{\|}}{-}}\!\!\!C\!\!-\!\!R}{} \xrightarrow{(H)} R_2CH_2 \tag{11}$$

The most commonly employed laboratory methods for the reduction of aldehydes and ketones to hydrocarbons are (1) the Wolff–Kishner method utilizing hydrazine in the presence of base [22–24], (2) the Clemmensen method utilizing zinc and hydrochloric acid [25], (3) catalytic hydrogenation utilizing a metal catalyst and hydrogen [26,27], (4) metal hydrides such as lithium aluminum hydride in the presence of aluminum chloride [32,33], and (5) formation of thioketal or thioacetal and reaction of it with Raney nickel in ethanol [34a].

Aromatic carboxylic acids have been reported to be conveniently reduced to the corresponding methyl compound by means of trichlorosilane/tri-*n*-propylamine [34b].

Since the Friedel–Crafts reaction usually yields a mixture of isomers on alkylation of benzene, the reduction of the pure alkyl aryl ketone is a more practical and reliable method for the preparation of pure di- and polyalkylbenzenes.

a. THE WOLFF–KISHNER METHOD

The Wolff–Kishner [27,28,35] procedure depends on reacting an aldehyde or ketone with hydrazine in the presence of base to yield the corresponding hydrocarbon.

$$RR'C=O + H_2N-NH_2 + KOH \longrightarrow RR'-CH_2 \tag{12}$$

Semicarbazones or azines undergo the same reaction upon being heated with base.

Huang–Minlon [36–38] modified and improved the original procedure by using diethylene glycol as a reaction medium at 180°–200°C at atmospheric pressure for 2–4 hr to obtain 60–90% yields of hydrocarbon products [42]. It should be mentioned that aromatic nitro compounds yield amines in this procedure.

Steric effects influence the efficiency of reaction in many cases. For example 2,3,5,6-tetramethylacetophenone is not reduced, whereas the 2,3,4,5-isomer gives 75% reduction to the corresponding ethylbenzene [39]. A special procedure for sterically hindered ketones has recently been published which is especially useful for 11-oxo steroids [40].

Cram and co-workers [41] added the pure aldehyde or ketone hydrazones slowly to a potassium hydroxide solution in dimethyl sulfoxide over a period of 8 hr at 25°C and found that yields of the methylene compound varied from 65% to 90% with some azine as a by-product. The procedure has the advantage that a reaction temperature of only 25°C is required. However, the disadvantage is that the pure hydrazones have first to be isolated.

2-2. *Preparation of 2-(n-Octyl)naphthalene by the Huang–Minlon Method* [42]

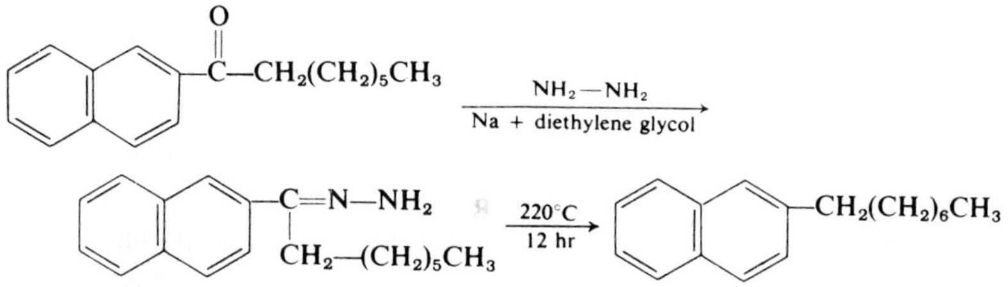

$$\text{Naphthyl-CO-CH}_2(CH_2)_5CH_3 \xrightarrow[\text{Na + diethylene glycol}]{NH_2-NH_2} \text{Naphthyl-C(=N-NH}_2)\text{-CH}_2-(CH_2)_5CH_3 \xrightarrow[\text{12 hr}]{220°C} \text{Naphthyl-CH}_2(CH_2)_6CH_3 \quad (13)$$

To a reaction flask containing a solution of 25 gm of sodium in 700 ml of diethylene glycol are added 100 gm of 2-naphthyl heptyl ketone and 50 ml of 90% hydrazine hydrate. The mixture is heated for 3 hr under reflux to form the hydrazone and then for an additional 12 hr at 220°C. During this period, the upper layer changes color from orange-red to almost colorless. The mixture is cooled, acidified with dilute hydrochloric acid, and extracted with benzene. The benzene layer is concentrated and the residue is crystallized from acetone at −60°C to yield 75% of 2-(*n*-octyl)naphthalene, mp 13°C. In order to obtain purer products, it is recommended that the crude product be distilled under reduced pressure first and then recrystallized.

b. The Clemmensen Method

The Clemmensen reaction [29] of carbonyl compounds, which requires refluxing for 1–2 days with amalgamated zinc (from 100 gm of mossy zinc, 5 gm mercuric chloride, 5 ml of concentrated hydrochloric acid, and 100–150 ml water) and hydrochloric acid, yields hydrocarbons.

$$RR'C{=}O + Zn(Hg) + HCl \longrightarrow RR'CH_2 \quad (14)$$

§ 2. Reduction Reactions

The yields of paraffins [43–46] and alicyclic hydrocarbons are poor, and the products are frequently contaminated with olefins [47]. Aromatic ketones are reduced in much better yields [48].

2-3. Preparation of 2-(n-Octyl)naphthalene [42]

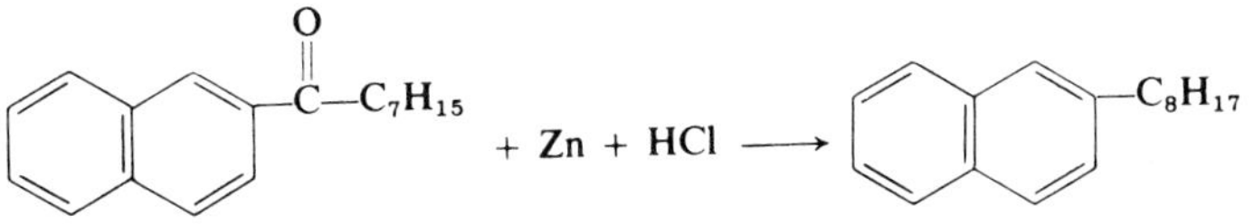

$$\text{Naphthyl-}\underset{\underset{O}{\|}}{C}-C_7H_{15} + Zn + HCl \longrightarrow \text{Naphthyl-}C_8H_{17} \quad (15)$$

Naphthyl heptyl ketone (20 gm) is added to a mixture of 100 gm of granulated amalgamated zinc, 100 ml of concentrated hydrochloric acid, and 75 ml of water. The mixture is refluxed for 18 hr. During this time an additional 90 ml of acid are added every 6 hr. The reaction mixture is cooled, decanted from the zinc, and the residue is washed with benzene. The reaction mixture is extracted with benzene, and the combined extracts are washed with water, dried, and concentrated. The residue is then recrystallized from acetone at $-60°C$ to yield 58% of 2-(n-octyl)naphthalene, mp 13°C. Purer products are obtained by a distillation of the product prior to recrystallization.

c. The Reduction of Thioketals or Thioacetals

This method involves the reaction of aldehydes or ketones with ethanedithiol in the presence of BF_3 etherate to give the thioacetal or thioketal, respectively. The latter is reacted with Raney nickel in absolute ethanol to give the hydrocarbon [34a].

2-4. Preparation of 7,7,10-Trimethyl-$\Delta^{1(9)}$-octalin [34a]

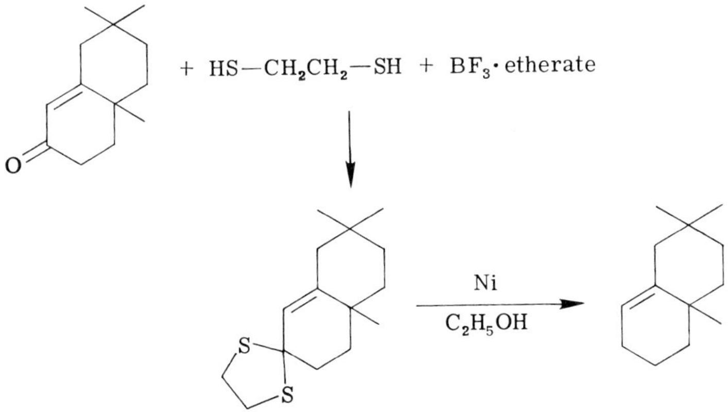

$$\text{(ketone)} + HS-CH_2CH_2-SH + BF_3\cdot\text{etherate} \longrightarrow \text{(thioketal)} \xrightarrow[C_2H_5OH]{Ni} \text{(octalin)} \quad (16)$$

To a flask are added 2 ml of boron trifluoride etherate and 5 gm (0.035 mole) of 7,7,10-trimethyl-$\Delta^{1(9)}$-octal-2-one and 4.25 ml of ethanedithiol at 0°C. The mixture is stirred to 0°C for ½ hr, diluted with 25 ml methanol, cooled, and filtered to give 6.38 gm (91%) of the dithioketal, mp 67°–69°C (mp 70.5°–71°C from methanol).

To 6.3 gm of the dithioketal are added 50 gm W-2 Raney nickel and 450 ml of absolute ethanol. The mixture is refluxed for 14 hr, the nickel filtered, and the solvent removed by evaporation. Then pentane is added to the solid residue, and the filtrate is added to the alcohol filtrate. The combined filtrate is concentrated under reduced pressure (30 mm Hg) to yield 3.40 gm (81%) of the product, n_D^{22} 1.4951, having IR bands at 6.00 and 12.44 μm.

D. Reduction of Alcohols

$$ROH \xrightarrow{(H)} RH \qquad (17)$$

Alcohols are ordinarily reduced by hydrogenolysis over a metallic catalyst such as cobalt on alumina [49] or vanadium pentoxide on aluminum oxide at about 300°C [50]. Sodium or lithium in liquid ammonia can reduce alcohol groups alpha to an aromatic ring [51]. Recently, chloroaluminum hydrides [52a] have been used in the absence of hydrogen gas to reduce alcohols. This procedure is only of limited value since wholly aliphatic secondary and tertiary alcohols give some reduction but large amounts of olefins are formed. Migration of phenyl substituents is also common with this latter reducing system.

2-5. Catalytic High-Pressure Hydrogenolysis of 2,2,3-Trimethyl-1-butanol to 2,2,3-Trimethylbutane [49]

$$\underset{\underset{CH_3}{|}}{\overset{\overset{CH_3}{|}\ \overset{CH_3}{|}}{CH_3CH-C-CH_2OH}} + H_2 + Co-Al_2O_3 \xrightarrow[300°C]{225\ atm} \underset{\underset{CH_3}{|}}{\overset{\overset{CH_3}{|}\ \overset{CH_3}{|}}{CH_3-CH-C-CH_3}} \qquad (18)$$

To a stainless steel shaker bomb are added 100 gm of 2,2,3-trimethyl-1-butanol and 20 gm of cobalt-on-alumina catalyst. The bomb is closed, pressurized with hydrogen, and kept at 298°–308°C for 18 hr, during which time the pressure drops from the maximum of 965 atm to 740 atm. The bomb is cooled, vented of hydrogen, and the contents are distilled to yield 27 gm (31%) of 2,2,3-trimethylbutane, bp 81°–82°C. Approximately 32 gm of starting alcohol is recovered.

2-6. Conversion of Ketones to Alcohols and Subsequent Reduction to Alkanes—Reduction of Benzophenone to 1,1-Diphenylethane [52b]

$$(C_6H_5)_2C=O \xrightarrow{CH_3Li} (C_6H_5)_2C\begin{smallmatrix}CH_3\\OLi\end{smallmatrix} \xrightarrow{Li/NH_3} (C_6H_5)_2CH-CH_3 \qquad (19)$$

A dry (all glassware is oven-dried) 500-ml, 3-necked flask is equipped with a Dry Ice condenser, glass-coated magnetic stirring bar, pressure-equalizing dropping funnel, and rubber septum on one neck. Then the flask is flushed with argon and the argon is kept throughout the reaction (controlled via a gas bubbler). Through the septum neck are injected 30 ml of anhydrous ether and 19.5 ml of 1.89 M (0.037 mole) of methyllithium in ether (Foote Mineral Co.). From the dropping funnel is added over a 20-min period a solution of 4.54 gm (0.025 mole) of benzophenone in 35 ml of anhydrous ether. The mixture is stirred for ½ hr and Dry Ice and isopropanol are added to the condenser. In place of the septum is attached a side arm attached with Tygon tubing to a tank of anhydrous ammonia. Then 75 ml of ammonia is condensed into the flask, and this is followed by the addition of 0.53 gm of lithium wire segments (0.5 cm). After 15 min the dark blue color is discharged by adding portionwise 5 gm of ammonium chloride over a 15-min period. Then the ammonia is allowed to evaporate in the hood and to the residue is added 100 ml of sat. aq. NaCl and then 100 ml of ether. The ether extract is separated and the sodium chloride solution is further extracted twice with 50 ml of ether. The combined ether extracts are dried and concentrated to yield 4.4–4.5 gm of crude product. This is then dissolved in 150 ml of petroleum ether and filtered through 60 gm of Woelm alumina (grade III) to yield after evaporation 4.2–4.3 gm (92–95%) of 1,1-diphenylethane, bp 100°C. (0.25 mm), n_D^{28} 1.5691.

E. Reduction of Halides

Halides are converted to the corresponding hydrocarbons by one of several methods. The Grignard reaction of a halide and subsequent reactions, with water produce a hydrocarbon [53]. Magnesium, lithium, zinc, and acetic acid [54], lithium aluminum hydride [55], lithium hydride [56a], sodium hydride [56b], sodium borohydride [57], sodium in alcohol [58], magnesium in methanol [59a] or isopropanol [59b], and nickel aluminum alloy in aqueous alkali [60a] can be used for the reduction of halides. The last three methods have been used specifically for aryl halogen atoms. More recently Pd/C has been used to reduce aryl halogen atoms in the presence of triethylammonium formate [60b].

Sodium borohydride efficiently reduces organic halides, especially if they can form stable carbonium ions. For example, *tert*-cumyl chloride (2-phenyl-2-chloropropane) and benzhydryl chloride are converted into cumene (82%) and diphenylmethane (72%), respectively. The following conditions are used for the reduction of these compounds: 50°C reaction temperature, 1–2 hr reaction time, 65 vol% diglyme, 0.5 mole of the organic halide, 4.0 mole of sodium borohydride, and 1.0 mole of sodium hydroxide (added in order to minimize the hydrolysis of sodium borohydride) [57]. The use of DMSO as a solvent has also been reported [60c].

Sodium borohydride is more convenient to handle than lithium aluminum hydride because it is not sensitive to water. In fact, sodium borohydride reduction can be carried out in aqueous solution, whereas lithium aluminum hydride requires strictly anhydrous conditions. Saturated aqueous solutions of sodium borohydride at 30°–40°C are stable in the presence of 0.2% sodium hydroxide.

More recently sodium cyanoborohydride $NaBH_3CN$ has been reported to be be effective in reducing primary, secondary, and tertiary halides and tosylates [61a].

Tri-n-butyltin hydride has also been reported selectively to reduce polyhalides such as benzotrichloride or dihalocyclopropanes.

$$R_3SnH + R'X \longrightarrow R_3SnX + R'H \qquad (20)$$

CAUTION: Lithium aluminum hydride may also spontaneously ignite when rubbed or ground vigorously in air. A nitrogen or argon atmosphere is recommended for safe grinding operations in a hood. A safer product is reported to be available from Metallgesellschaft A.G. Frankfurt, West Germany (See *Chem. & Eng. News,* May 11, 1981, p. 3)

Catalytic trialkyltin hydride reductions have been reported using sodium borohydride [61c].

2-7. *Lithium Aluminum Hydride Reduction of 1-Bromooctane* [56]

$$LiAlH_4 + LiH + 5\ C_8H_{17}Br \longrightarrow 5\ C_8H_{18} + 2\ LiBr + AlBr_3 \qquad (21)$$

To a flask equipped with a stirrer, dropping funnel, reflux condenser, and drying tube, and a thermometer are added 5 gm lithium aluminum hydride (0.13 mole) and 12 gm of lithium hydride (1.5 moles), and the flask is cooled. Tetrahydrofuran (THF) (300 ml) is added with stirring and then the contents are heated to reflux. 1-Bromooctane (193 gm, 1 mole) is added dropwise at such a rate that a moderate reflux is maintained without external heating. The mixture is then refluxed for an additional hour, cooled to 10°C, and cautiously hydrolyzed with 100 ml of a 60/40 mixture of THF/water so that the temperature is kept below 20°C. The mixture is then transferred to a 2-liter beaker containing 80 ml of sulfuric acid in crushed ice water. The organic layer is separated, dried over potassium carbonate, and distilled to yield 96% of n-octane, bp 125°C, n_D^{20} 1.3975.

2-8. *Sodium Borohydride Reduction of Methyl Chloride* [62]

$$NaBH_4 + 4\ CH_3Cl \xrightarrow{\text{Diglyme}} 4\ CH_4 + NaBCl_4 \qquad (22)$$

To a flask are added 25 ml of diglyme, 0.0874 gm (23 mmole) of sodium borohydride, and 85 ml (3.8 mmole) of methyl chloride. The contents are allowed to stand for 1 hr. Recovery of the volatile products using a liquid nitrogen trap yields 78 ml of methane (3.5 mmole, 92%) and 7 ml (0.3 mmole) of recovered methyl chloride. (At the start flask is connected to 1 aq. N_2 trap.)

2-9. *Magnesium in Isopropanol Reduction of Chlorobenzene* [59b]

$$C_6H_5Cl + Mg + CH_3-\underset{OH}{CH}-CH_3 \longrightarrow C_6H_6 + (CH_3)_2CHOMgCl \quad (23)$$

To a 250-ml round bottom flask equipped with a mechanical stirrer, thermometer, pressure equalizing dropping funnel, and a condenser, with a nitrogen bubbler are added 6.0 gm (0.25 gm atom) of freshly ground magnesium powder, 50 ml of freshly distilled decahydronaphthalene, and a crystal of iodine. The nitrogen is swept through the flask and gently maintained throughout the reaction. The reaction mixture is heated to reflux without stirring and then ⅕ of the solution of 11.3 gm (0.1 mole) of freshly distilled chlorobenzene in 9.0 gm (0.15 mole) of dry isopropanol is added. The reaction should start almost immediately in the presence of the iodine crystal and after about 15 min the stirrer is started while reducing the external heating. If the reaction is not sustained, then another crystal of iodine and a drop or two of methyl iodide (or *n*-butyl bromide) are added. Once the reaction has been initiated, the remainder of the chlorobenzene solution is added over a period of ½ hr or at such a rate to keep the reaction gently refluxing without external heating. After the addition 25 ml of additional decahydronaphthalene is added and the mixture is refluxed for 1 hour. The reaction mixture is then cooled, and 6 *N* HCl solution is added to dissolve the solids. The organic layer is separated, washed 4 times with 30-ml portions of water, dried, and distilled through a packed column (or spinning band column) to yield 5.5–6.5 gm (70–83%) benzene, bp 80°–82°C, n_D^{20} 1.5007.

The above procedure is also applicable to other compounds such as 1-chlorobutane, 2-chloro-2-methylbutane, carbon tetrachloride, monochloroacetone, bromobenzene, iodobenzene, 1-bromonaphthalene, β-bromostyrene, *p*-bromaniline, and *p*-bromophenol.

3. CONDENSATION REACTIONS

A. Friedel–Crafts Alkylation of Hydrocarbons

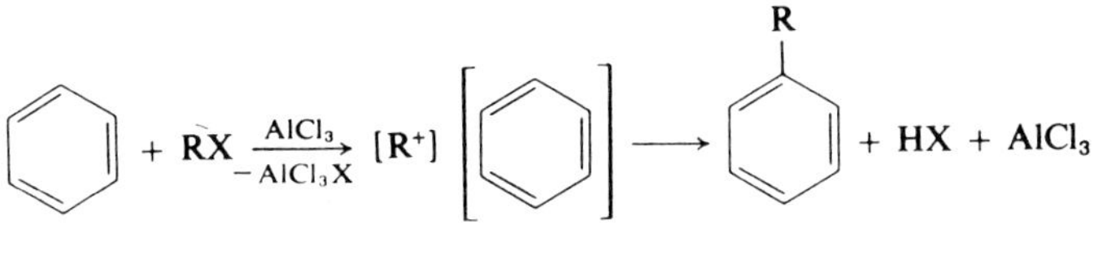

$$C_6H_6 + RX \xrightarrow[-AlCl_3X]{AlCl_3} [R^+] \; [C_6H_6] \longrightarrow C_6H_5R + HX + AlCl_3 \quad (24)$$

The Friedel–Crafts reaction is a popular laboratory method for the alkylation of benzene [63], naphthalene [64], and other aromatic hydrocarbons to give monoalkyl, dialkyl, and polyalkyl aromatics [65]. Since isomerization occurs when the alkylating agent is greater than two carbon atoms in length, the Friedel–Crafts reaction cannot be used for the preparation of long-chain n-alkyl aromatic derivatives [66].

The catalyst used is usually anhydrous aluminum chloride [67] but it may also be boron trifluoride [68], hydrogen chloride [69], hydrogen fluoride [70], ferric chloride [71], beryllium chloride [72], silicophosphoric acid [73], and sulfuric acid [74]. A trace amount of water is sometimes necessary as a cocatalyst [68]. Russell [75] reported on a series of catalysts and their activity to alkylate benzene. They had the following order of activity: $Al_2Br_6 > Ga_2Br_6 > Ga_2Cl_6$ $Fe_2Cl_6 > SbCl_5 > ZrCl_4 > BF_3 > SnCl_4 > SbCl_3$.

The alkylating agents are not limited to alkyl halides [76], olefins [77], acetylenes [78], alcohols, ethers, esters [79–81], alkyl sulfate [82], cyclopropane [83], or other active hydrocarbons (isoparaffins [67] which may produce carbonium ions have also been used [84]). The activity of alkyl halides is in the sequence RF > RCl > RBr > RI. The fluoro group in 1-chloro-2-fluoroethane is predominantly used in the alkylation of benzene as shown below [85].

$$C_6H_6 + FCH_2CH_2Cl \xrightarrow[-10°C]{BBr_3} C_6H_5-CH_2CH_2-Cl \qquad (25)$$

The Haworth synthesis [86] of polynuclear compounds initially involves the Friedel–Crafts reactions of aromatics with succinic anhydride followed by Clemmensen reduction, acid-catalyzed cyclization to a ketone, reduction, and selenium metal dehydrogenation to yield the aromatic system in 50% yields [Eq. (26)].

The Friedel–Crafts reaction is also discussed in Chapters 8 and 9 in connection with the synthesis of ketones and carboxylic acids. Several reviews [87–89] and a monograph [90] are available on the Friedel–Crafts reaction.

The effect of experimental conditions on the orientation and production compositions during alkylation can not always be generalized. However, it is known that an excess of the aromatic compound favors monoalkyl derivatives. Long reaction times and/or elevated temperatures cause isomerization and fragmentation to occur. For example, o-, m-, and p-diethylbenzenes are isomerized at room temperature in the presence of 0.2 mole of aluminum chloride per mole of diethylbenzene in the presence of 1 ml of water [91]. An equilibrium is reached after 20 hr to give about 3% ortho, 69% meta, and 28% para isomer. Vapor phase chromatographic analysis is used to determine the purity and isomer distribution. Similar isomerizations are reported for the *tert*-butyltoluenes [92] and the diisopropylbenzenes [93].

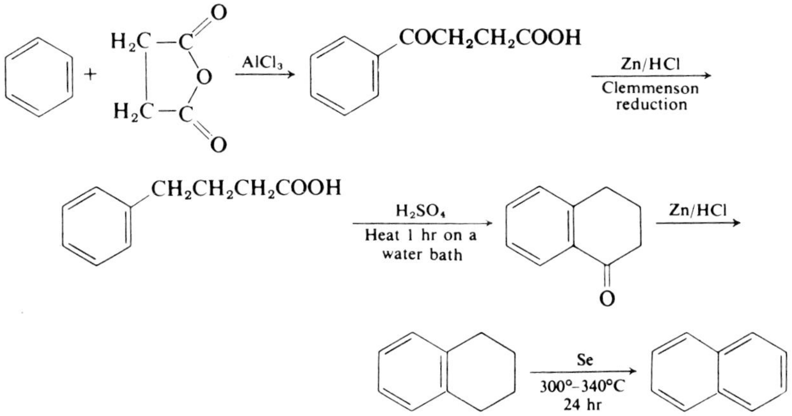

§ 3. Condensation Reactions

$$\text{C}_6\text{H}_6 + \begin{matrix}\text{H}_2\text{C}-\text{C}\\ | \\ \text{H}_2\text{C}-\text{C}\end{matrix}\overset{O}{\underset{O}{\diagdown}} \overset{O}{\diagup} \xrightarrow{\text{AlCl}_3} \text{C}_6\text{H}_5\text{COCH}_2\text{CH}_2\text{COOH} \xrightarrow[\text{Clemmenson reduction}]{\text{Zn/HCl}}$$

$$\text{C}_6\text{H}_5\text{CH}_2\text{CH}_2\text{CH}_2\text{COOH} \xrightarrow[\text{Heat 1 hr on a water bath}]{\text{H}_2\text{SO}_4} \text{[tetralone]} \xrightarrow{\text{Zn/HCl}}$$

$$\text{[tetralin]} \xrightarrow[\substack{300°-340°\text{C} \\ 24 \text{ hr}}]{\text{Se}} \text{[naphthalene]} \qquad (26)$$

Mild catalysts such as boron trifluoride (with an alcohol), hydrogen fluoride (with an olefin), or ferric chloride (with an alkyl halide) may produce almost pure para dialkylation products or 1,2,4-trialkylation compound. An excess of aluminum chloride at elevated temperatures favors meta-dialkyl or symmetrical trialkyl derivatives.

The catalyst quantity varies with the alkylating agent. Trace amounts of aluminum chloride are required only for the alkylations involving alkyl halides or olefins. However, with alcohols or their derivatives large amounts of catalyst are required to offset the deactivating effect on the catalyst by hydroxyl groups from alcohol or water.

3-1. 1,3,5-Triethylbenzene by the Friedel–Crafts Ethylation of Benzene [94]

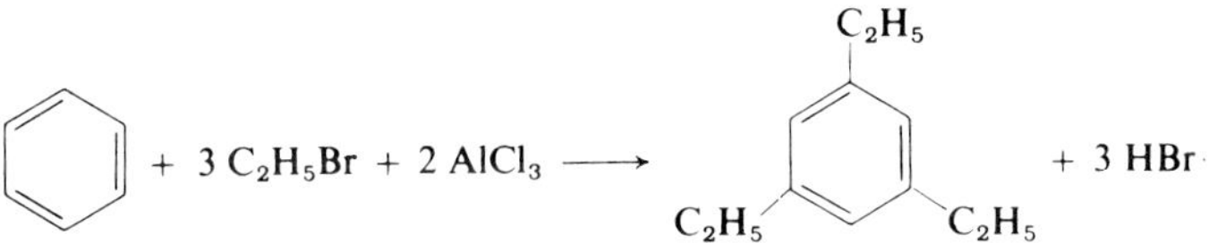

$$\text{C}_6\text{H}_6 + 3\,\text{C}_2\text{H}_5\text{Br} + 2\,\text{AlCl}_3 \longrightarrow \text{1,3,5-(C}_2\text{H}_5)_3\text{C}_6\text{H}_3 + 3\,\text{HBr} \qquad (27)$$

To an ice-cooled three-necked flask equipped with a stirrer, dropping funnel, condenser, and ice-water trap is added 267 gm (2 moles) of anhydrous aluminum chloride. Then ½ to ⅔ of the total ethyl bromide (335 gm, 3 moles) is added to moisten the aluminum chloride. The benzene (78 gm, 1 mole) is added dropwise over a period of ½ hr at 0° to −5°C. The remaining ethyl bromide is added cautiously over a period of ½ hr. The reaction mixture is stirred and allowed to come to room temperature slowly. After 24 hr the yellow intermediate is decomposed by pouring it, with stirring, into 500 ml of crushed ice and 50 ml of concentrated hydrochloric acid contained in a 4-liter beaker. More ice is added and the organic layer is separated, washed with sodium hydroxide solution, dried, and distilled through a Vigreaux column to yield 85–90% of 1,3,5-triethylbenzene, bp 215°–216°C.

The ethyl bromide that is carried over with the hydrogen bromide is caught in an ice-water trap and an amount equal to it is added later in the synthesis.

1,2,3-Trimethylbenzene, which cannot be made by the Friedel–Crafts method, can be synthesized using the Tiffeneau rearrangement which occurs with benzyl Grignard reagents in the presence of formaldehyde [87].

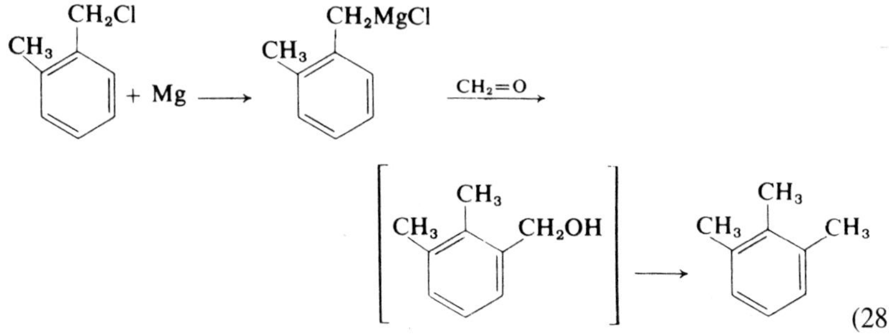

(28)

The Jacobsen reaction [95] which is described below can be considered a Friedel–Crafts rearrangement to give vicinal-substituted aromatic compounds. The example cited is for a 6,7-dimethyltetralin which may be considered a 1,2,4,5-tetraalkylbenzene being converted to a 5,6-dimethyltetralin (1,2,3,4-tetraalkylbenzene).

3-2. *Jacobsen Reaction of 6,7-Dimethyltetralin* [96]

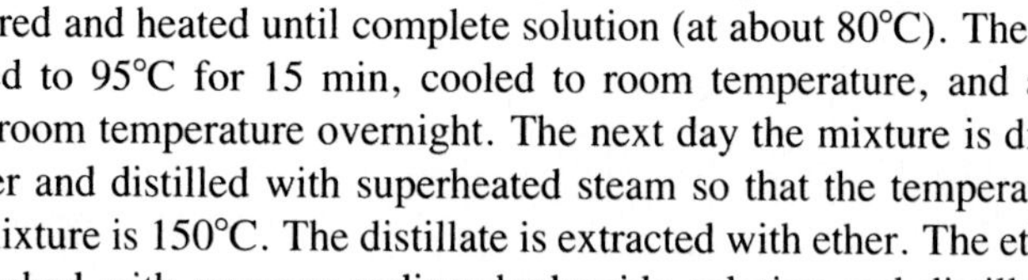

(29)

A mixture of 6,7-dimethyltetralin (18 gm) and 50 gm of concentrated sulfuric acid is stirred and heated until complete solution (at about 80°C). The mixture is then heated to 95°C for 15 min, cooled to room temperature, and allowed to remain at room temperature overnight. The next day the mixture is diluted with some water and distilled with superheated steam so that the temperature of the reaction mixture is 150°C. The distillate is extracted with ether. The ether extract is then washed with aqueous sodium hydroxide solution and distilled to yield 5,6-dimethyltetralin, 4.3 gm (24%), bp 110°–115°C (7 mm), n_D^{20} 1.5530.

Dehydrogenation of the above dialkyltetralin gives the corresponding dialkylnaphthalene in good purity. In this reaction, if the substituents are *n*-propyl groups there is no rearrangement of them to isopropyl groups.

The preparation of adamantane also involves an internal condensation–rearrangement using aluminum halides [96b].

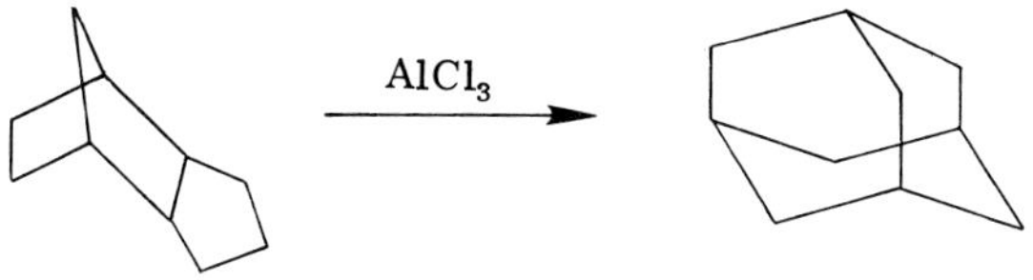

$$\text{(structure)} \xrightarrow{\text{AlCl}_3} \text{(adamantane)} \qquad (30)$$

B. Hydrocarbon Polymers

As a part of a discussion of the methods of synthesis of hydrocarbons, it is pertinent to mention hydrocarbons produced by polymerization reactions (addition reactions).

$$\underset{R}{CH=CH_2} \xrightarrow[\text{or ionic catalysis}]{\text{Free radical}} \left[\underset{R}{-CH-CH_2-} \right]_n \qquad (31)$$

Polyethylene, polystyrene, polypropylene, polyisobutylene, and polybutadiene are a few of the hydrocarbon polymers that are of great commercial importance today. Copolymers of mixtures of two or more monomers have also been prepared.

The polymerization reactions are effected either thermally, free radically, cationically, or anionically depending on the monomers involved. Ziegler–Natta type catalysts are used to give stereospecific polymers of either the isotactic or syndiotactic conformation.

The references will lead the reader to more theoretical discussions and sources for more synthetic methods [97–100].

The polymerization of styrene by two different techniques is described below.

3-3. Synthesis of Polystyrene by Thermal Activation [101a]

$$C_6H_5CH=CH_2 \longrightarrow \left[\underset{C_6H_5}{-CH-CH_2-} \right]_n \qquad (32)$$

To a test tube is added 25 gm of styrene monomer; the tube is flushed with nitrogen and stoppered. The test tube is immersed in an oil bath at 125°–130°C for 24 hr, cooled, and broken open to recover a clear glasslike mold of polystyrene. If all the oxygen has been excluded the polystyrene will be free of yellow stains, especially on the surface. The polystyrene can be purified and freed of residual monomer by dissolving it in benzene and reprecipitating it in a stirred solution of methanol. The solids are filtered and dried in a vacuum oven at 50°–60°C to give a 90% yield (22.5 gm). The molecular weight is about 150,000 to 300,000 as determined by viscometry in benzene at 25°C.

3-4. Emulsion Polymerization of Styrene [101b]

To a standard resin kettle equipped with a stirrer, condenser, and dropping funnel are added 100 gm of styrene and 128.2 gm of distilled water. To the water the following are added separately: 0.68% potassium persulfate solution and 100 ml of a 3.56% soap solution. Nitrogen is used to purge the system of dissolved air, and then the temperature is raised to 50°C and kept at this temperature for 24 hr. A 90% conversion to polystyrene emulsion will have occurred at this point.

Other polymerization techniques are briefly illustrated below [97]:

$$C_6H_5-CH=CH_2 \longrightarrow \left[\begin{array}{c} -CH-CH_2- \\ | \\ C_6H_5 \end{array} \right]_n \qquad (33)$$

Catalysts: Thermal
Peroxide, or other free radical source
Emulsifier + free radical source
Suspending agent + free radical source

$$CH_3-\underset{\underset{CH_3}{|}}{C}=CH_2 + BF_3 \xrightarrow{\text{Ionic polymerization}} \left[\begin{array}{c} CH_3 \\ | \\ -C-CH_2- \\ | \\ CH_3 \end{array} \right]_n \qquad (34)$$

$$CH_3-\underset{\underset{C_6H_5}{|}}{C}=CH_2 \xrightarrow[\text{n-Butyllithium}]{\text{Anionic catalyst}} \left[\begin{array}{c} CH_3 \\ | \\ -C-CH_2- \\ | \\ C_6H_5 \end{array} \right]_n \qquad (35)$$

$$CH_2=\underset{\underset{CH_3}{|}}{C}-CH=CH_2 \xrightarrow[\text{Catalyst from Al(C_2H_5)_3 + VCl_3}]{\text{Stereospecific [92]}} \text{Synthetic Balata rubber} \qquad (36)$$

$$CH_2=CH-CH=CH_2 \xrightarrow[\text{from (C_2H_5)_3Al_2Cl_3- cobalt octoate}]{\text{Stereospecific Catalyst [93]}} \text{cis-1,4-Polybutadiene} \qquad (37)$$

C. Small Ring Hydrocarbon Syntheses

Cyclobutane [104] and cyclopropane [105,106] are two popular small ring compounds that have attracted a great deal of research effort. Cyclobutane syntheses have recently been reviewed and will not be discussed here.

Carbenes [105,106] and specifically methylene have been shown to play an important role in the synthesis of hydrocarbons either by insertion or, more importantly, by their addition to olefins to yield cyclopropanes. The proof of whether certain species are truly carbenes, i.e., singlet state divalent carbon species, is still an active research area. Proof exists to indicate that the photolysis of certain diazoalkanes yields divalent carbon species [107]. Recent investigations have indicated, however, that certain organometallics react with alkyl

§ 3. Condensation Reactions

halides not to give free carbenes but α-halolithium compounds which add to olefins to give cyclopropanes [108]. The Simmons–Smith [109] reaction is an example in which an organozinc halide reacts with an olefin to give products identical to those which might be obtained from free carbenes.

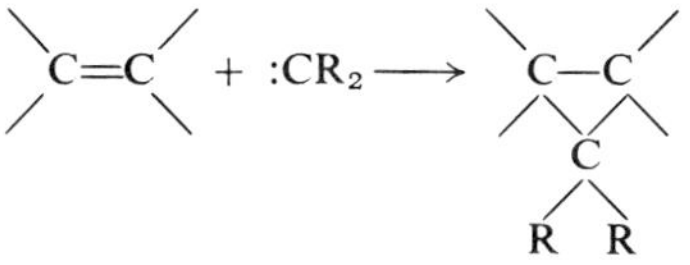

$$\diagup C=C\diagdown + :CR_2 \longrightarrow \diagup C-C\diagdown \atop \underset{R\ R}{C} \qquad (38)$$

The photolysis of diazomethane (see Chapter 15) has been used for the synthesis of methylene, but the hazardous nature of the reagent has hindered the widespread use of this reaction [110] in the synthesis of cyclopropanes.

$$CH_2N_2 \longrightarrow :CH_2 + N_2 \qquad (39)$$

A novel method for the generation of methylene is via iodomethylzinc iodide [109] as shown in Eq. (40).

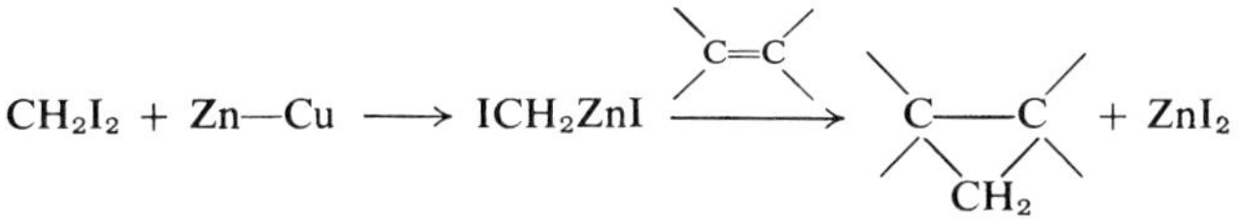

$$CH_2I_2 + Zn\text{—}Cu \longrightarrow ICH_2ZnI \xrightarrow{\diagup C=C\diagdown} \diagup C\text{—}C\diagdown \atop CH_2 + ZnI_2 \qquad (40)$$

The organozinc compound is stable for several hours but decomposes rapidly in the presence of olefins. The methylene thus generated as an intermediate reacts stereospecifically [111] with olefins, as does methylene generated from diazomethane.

The newly reported synthesis of free C_1 and C_3 carbon species offers the possibility of preparing spiropentane and allene structures by vaporization of carbon in the presence of olefins [112–114]. However, more development in this area is necessary before this technique can be applied for the synthesis of compounds in good yields.

3-5. Preparation of n-Hexylcyclopropane [115]

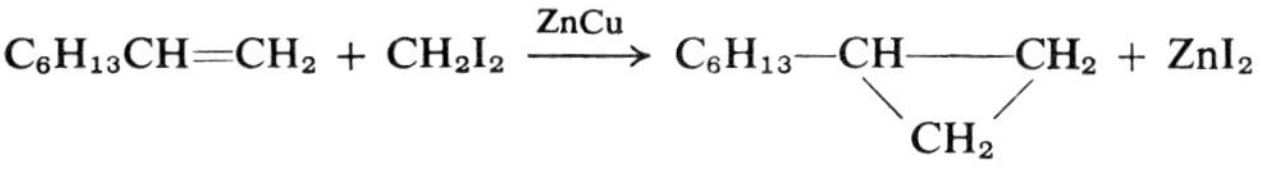

$$C_6H_{13}CH=CH_2 + CH_2I_2 \xrightarrow{ZnCu} C_6H_{13}\text{—}CH\text{———}CH_2 \atop \diagdown CH_2 \diagup + ZnI_2 \qquad (41)$$

(a) *Synthesis of the zinc-copper complex.* Zinc powder Analytical Reagent (Mallinckrodt), 65.6 gm, 1.0 mole, is washed rapidly in a beaker with hydrochloric acid (3%, 4 × 50 ml), decanted, and then washed with distilled water (4 × 60 ml), aqueous copper sulfate (2%, 2 × 100 ml), distilled water (4 × 60 ml), absolute ethanol (4 × 60 ml), and absolute ether (5 × 50 ml). The ethanol and ether washings are decanted onto a Büchner funnel to prevent loss of the couple. Additional ether is used to wash the couple in the Büchner funnel, which is then covered with a rubber dam and dried under suction until the couple reaches room temperature. The couple is used immediately or stored in a vacuum desiccator over phosphorus pentoxide.

(b) Reaction of methylene iodide with 1-octene in the presence of the zinc-copper couple. To a flask containing the zinc-copper couple (16.3 gm of zinc, 0.25 mole) prepared as above and 165 ml of anhydrous ether are added, with stirring, 53.6 gm (0.20 mole) of methylene iodide and 0.15 gm of iodine (0.006 mole). The iodine color disappears instantly and the gray-colored mixture is refluxed for approximately ½ hr. The color of the mixture becomes darker and a gentle exothermic reaction is evident. External heating is then stopped and 44.8 gm (0.4 mole) of 1-octene in 25 ml of anhydrous ether is added dropwise over a period of ½ hr. The mixture continues to reflux during the addition. Heating is then continued so that the mixture refluxes for another 4–6 hr. The flask is cooled, the contents filtered, and the residue is washed with ether. The ether solutions are combined and washed with three 50-ml portions of 5% hydrochloric acid, aqueous sodium bicarbonate, and with three 50-ml portions of a saturated aqueous sodium chloride solution. The aqueous washings are extracted with ether and the combined ether solutions are dried. The ether is removed by distillation and the residue is distilled through a Nester–Faust (Perkin-Elmer®, Norwalk, Connecticut 06856) spinning band column to give *n*-hexylcyclopropane, 24 gm (48%), bp 148°–150°C, n_D^{25} 1.4173, showing no olefinic absorption at 6.08 μm in the infrared spectrum.

Other methods of forming small rings via carbenes involves first the formation of the dihalocarbene and *in situ* reduction to the cyclopropane. For example:

$$\text{cyclohexene} \xrightarrow[\text{[116, 117]}]{\text{CHCl}_3} \text{bicyclic product} \tag{42}$$

$$\text{naphthalene-like} \xrightarrow[\substack{\text{2. Na, NH}_3 \\ \text{[118]}}]{\substack{\text{1. CHCl}_3, \\ t\text{-BuOK}}} \text{product} \tag{43}$$

D. The Diels–Alder Reaction

The Diels–Alder reaction is discussed in Chapter 2, Olefins, but it is also relevant to mention it here, since six-membered rings can be produced by it, which, in turn, may be aromatized.

$$\underset{\substack{\text{CX} \\ \text{CY}}}{\overset{\text{CH}_2}{\underset{\text{CH}_2}{\|}}} + \underset{\text{HC—CO}}{\overset{\text{HC—CO}}{\underset{\text{O}}{\diagdown}}} \longrightarrow \underset{Y}{\overset{X}{\diagup}}\!\!\!\underset{\text{CO}}{\overset{\text{CO}}{\diagdown\!\!\!O}} \longrightarrow \underset{Y}{\overset{X}{\diagup}}\!\!\!\underset{\text{CO}}{\overset{\text{CO}}{\diagdown\!\!\!O}} \longrightarrow \underset{Y}{\overset{X}{\diagup}}$$

(44)

§ 3. Condensation Reactions

It has been reported that *trans,trans*-1,4-diacetoxybutadiene was added to dienophiles and the resulting products were thermally aromatized by the facile elimination of the methyl acetate groups [119].

$$\begin{array}{c}\text{HC}-\text{COOCH}_3\\ \text{HC}\\ |\\ \text{HC}\\ \text{HC}-\text{COOCH}_3\end{array} + \begin{array}{c}\text{CH}-\text{X}\\ \|\\ \text{CH}-\text{Y}\end{array} \longrightarrow \begin{array}{c}\text{COOCH}_3\\ \diagup\text{X}\\ \diagdown\text{Y}\\ \text{COOCH}_3\end{array} \xrightarrow{\text{Heat}} \begin{array}{c}\diagup\text{X}\\ \diagdown\text{Y}\end{array} \quad (45)$$

E. Coupling Reactions

Organometallic compounds can be coupled with certain halides to give unrearranged hydrocarbons. Neopentane [120], neohexane [121], hexamethylethane [122], and other branched hydrocarbons are formed in good yields by coupling a primary Grignard reagent with tertiary alkyl halides. The organometallic compounds used are mainly the Grignard [123] reagents, organozinc [124], lithium [125], and sodium [126] derivatives. The halides may be primary [127], secondary, and tertiary [120–122].

Benzyl chlorides [128] and α-phenylethyl chloride [129] can be converted to the corresponding Grignard reagents but these also react rapidly with the starting halides to give diphenylethanes.

$$\text{RM} + \text{R}'\text{X} \longrightarrow \text{R}-\text{R} + \text{MX} \quad (46)$$

The Wurtz reaction [130] is a poor method for the laboratory preparation of pure hydrocarbons since mixtures are obtained which are also contaminated by some olefins formed by some dehydrohalogenation.

$$2\,\text{RX} + 2\,\text{Na} \longrightarrow \text{R}-\text{R} + 2\,\text{NaX} \quad (47)$$

Cyclopropanes [131,132] and spiropentanes [133] are formed by the reaction of zinc dust on 1,3-dichloropropane and pentaerythrityl tetrabromide, respectively. Spiropentane is isolated only in 24–28% yields, in addition to 54–58% methylenecyclobutane, 13–18% 2-methyl-1-butene, and 1–3% 1,1-dimethylcyclopropane. Cyclobutane is made by the specific coupling with lithium amalgam [134].

The use of lithium dialkylcopper has shown utility to give specific alkyl compounds by displacement of a halide. For example [135,136a], see eq. (48) below.

$$2\,\text{LiCH}_3 + \text{CuI} \longrightarrow \text{LiCu(CH}_3)_2$$
$$\text{LiCu(CH}_3) + \text{RBr} \longrightarrow \text{R}-\text{CH}_3 \quad (48)$$

Photocyclization of activated olefins can also be considered a photolytic coupling reaction. For example photocyclization of *o-*, *p-*, or α-substituted stilbenes results in the corresponding 1-, 3-, or 9-substituted phenanthrenes [136b].

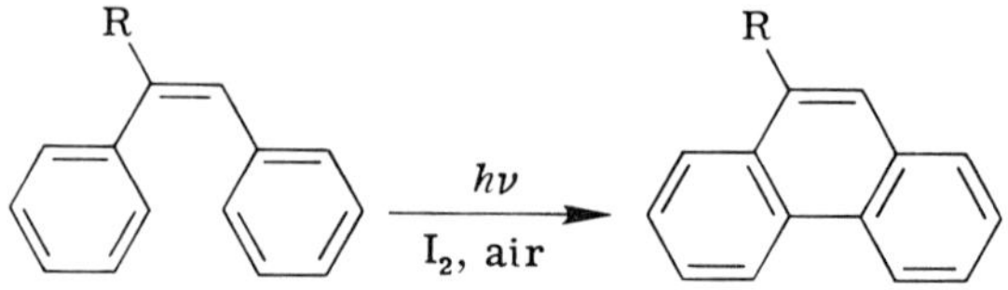

24　　　　　　　**1. Hydrocarbons (Paraffinic and Aromatic)**

$$\text{Ar-CH=CH-Ar} \xrightarrow[I_2, \text{ air}]{h\nu} \text{phenanthrene derivative} \quad (49)$$

3-6. Grignard Coupling—Synthesis of Neohexane [121]

$$\text{CH}_3\text{MgCl} + \text{Cl}-\underset{\underset{\text{CH}_3}{|}}{\overset{\overset{\text{CH}_3}{|}}{\text{C}}}-\text{CH}_2\text{CH}_3 \longrightarrow \text{CH}_3-\underset{\underset{\text{CH}_3}{|}}{\overset{\overset{\text{CH}_3}{|}}{\text{C}}}-\text{CH}_2-\text{CH}_3 + \text{MgCl}_2 \quad (50)$$

To methylmagnesium chloride [from the action of gaseous methyl chloride on 121.5 gm (5 moles) of magnesium in 1600 ml of dry di-n-butyl ether] is added 610 ml (5 moles) of *tert*-amyl chloride in 1 liter of di-n-butyl ether at 5°C over a period of 8 hr. Decomposition with ice and distillation of the ether yielded the crude product at bp 37°–50°C. Washing with sulfuric acid and redistillation yielded pure neohexane, bp 49°–50°C (740 mm), n_D^{20} 1.3688, in 36–39% yield.

Methylmagnesium iodide can be prepared more conveniently from liquid methyl iodide and used in place of methylmagnesium chloride in the above coupling reaction.

Organolithium reagents can also react with hydrocarbons to give 15–50% yields of alkylated products. For example, *tert*-butyllithium is reported to react with naphthalene in decalin solution at 165°C for 41 hr to give a 30% yield of mono-*tert*-butylnaphthalene and a 50% yield of di-*tert*-butylnaphthalene [137a]. The position of substitution was not reported. The generality of this reaction has not been explored beyond benzene, phenanthrene and perylene [137b].

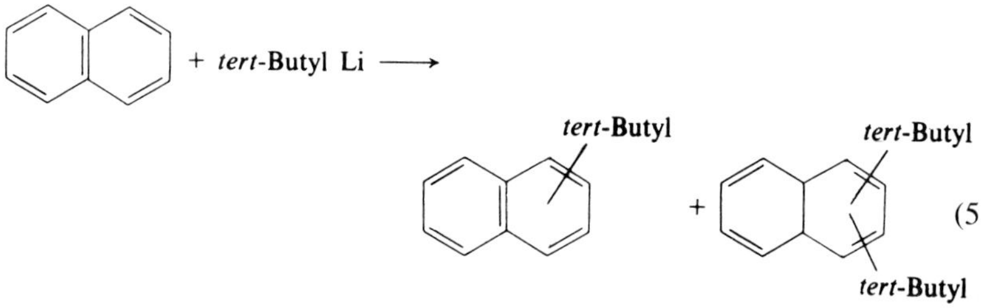

$$\text{naphthalene} + \textit{tert}\text{-Butyl Li} \longrightarrow \textit{tert}\text{-butylnaphthalene} + \text{di-}\textit{tert}\text{-butylnaphthalene} \quad (51)$$

Grignard reagents also react with inorganic halides to give coupled products. Examples of such halides are cupric chloride [138], lead chloride [139], silver bromide [140], silver cyanide [141], nickel chloride [142], palladium chloride [143], chromic chloride [144], iron halide [145], ruthenium halide [145], and rhodium halide [145].

$$2\,\text{RMgX} + \text{MX}_2 \longrightarrow \text{R—R} + 2\,\text{MgX}_2 + \text{M} \quad (52)$$

§ 3. Condensation Reactions

The Ullmann reaction [146] is related to the Wurtz reaction as a method of coupling aryl halides. In this reaction activated copper bronze is used in place of sodium. The most reactive halides are the iodides.

3-7. Ullmann Synthesis of 2,2'-Diethylbiphenyl [147]

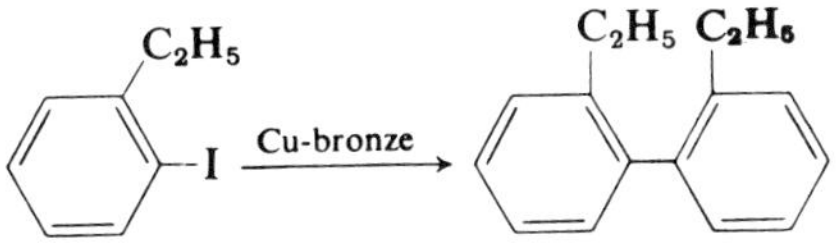

$$\text{o-C}_2\text{H}_5\text{C}_6\text{H}_4\text{-I} \xrightarrow{\text{Cu-bronze}} (\text{o-C}_2\text{H}_5\text{C}_6\text{H}_4)_2 \quad (53)$$

To a Pyrex flask are added 150 gm (0.647 mole) of o-ethyliodobenzene and 150 gm of copper bronze. The mixture is heated for 3 hr at approximately 240°C. The product 2,2'-diethylbiphenyl is isolated by extraction with boiling chlorobenzene and then by concentration under reduced pressure. The residue is distilled from sodium. The product is obtained in 60% yield (42.5 gm), bp 142°–143°C (14–15 mm), n_D^{25} 1.5620.

Grignard reagents also couple in good yields with dimethyl and diethyl sulfate [148] to give alkyl aromatics. The alkyl esters of arylsulfonic acids [149] also couple, as do the sulfates.

Diazonium salts react with aromatic compounds in the presence of sodium hydroxide to give low yields of coupled products [150].

The Pschorr synthesis [151] involves the diazotization and coupling of the resulting diazonium halides with an aromatic hydrocarbon in the presence of copper.

$$\text{ArN}_2^+\text{X}^- + \text{Ar'H} \xrightarrow{\text{Cu}} \text{Ar—Ar'} + \text{N}_2 + \text{HX} \quad (54)$$

The reductive coupling of alcohols has recently been reported to occur in 38–51% yields with some alcohols [152].

$$\text{ROH} \longrightarrow [(\text{RO})_2\text{Ti}] \longrightarrow \text{R—R} + \text{TiO}_2 \quad (55)$$

The scope and generality of this reaction still remain to be determined.

Substituted acetylenes condense with each other to give polyalkylated aromatics in 49–80% yields at 23°–36°C [153a]. This reaction offers a convenient laboratory preparation at room temperature of highly alkylated aromatics difficult to obtain by alternate routes. Several examples of this reaction are presented in Table I.

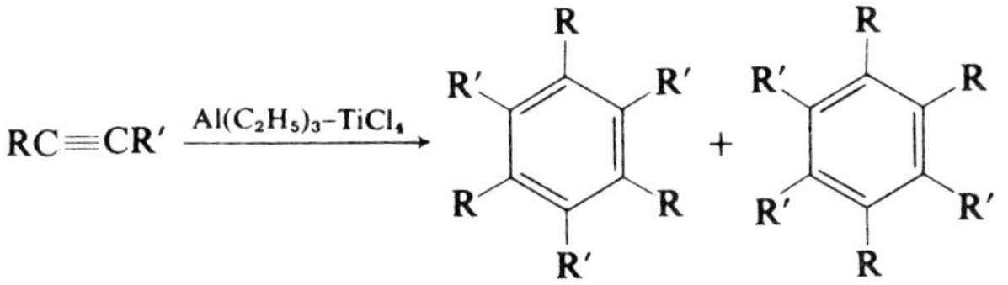

$$\text{RC}\equiv\text{CR'} \xrightarrow{\text{Al(C}_2\text{H}_5)_3\text{–TiCl}_4} \text{1,2,4-(R)}_3\text{(R')}_3\text{C}_6 + \text{1,3,5-(R)}_3\text{(R')}_3\text{C}_6 \quad (56)$$

TABLE I
Acetylenes Used and Aromatic Products Obtained by Cyclization[a]

Substitution on acetylene	Reaction temp. (°C)	Trimer	Yield (%)	High polymer (%)[d]
None	23°–39°[b,c]	Benzene	49.1	24.1
Methyl	23°–36°[b]	Mesitylene	40.4	9.3
		Pseudocumene	21.1	
Ethyl	23°–30°[b]	1,3,5-Triethylbenzene	35.5	7.3
		1,2,4-Triethylbenzene	17.0	
Butyl	24°–32°[b]	1,3,5-Tributylbenzene	59.8	—
Dimethyl	23°–34°[b]	Hexamethylbenzene	80.2	2.2[e]
Diethyl	24°–32°[b]	Hexaethylbenzene	76.5	—
Dibutyl	~86°	Hexabutylbenzene[g]	52.2[f]	4.6[e]

[a] Table I is reprinted from E. F. Lutz, *J. Am. Chem. Soc.* **83**, 2551 (1961). Copyright 1961 by the American Chemical Society and reprinted by permission of the copyright owner.

[b] Reactions were exothermic, producing an 11°–16° rise in temperature.

[c] The temperature gradually dropped toward the end of the reaction, indicating a loss of catalyst activity.

[d] Estimated by subtracting the weight of the organometallic catalyst from the total weight of the solids obtained from reaction.

[e] Because of the reaction work-up used, this represents the upper limit of high polymer formation.

[f] A small amount of another product, perhaps a dimer, also was obtained.

[g] This appears to be the first synthesis of this compound.

This reaction appears to be related to the method used by Schäfer to trimerize 2-butyne to hexamethyl–Dewar benzene with the aid of aluminum chloride. Heating hexamethyl–Dewar benzene yielded hexamethylbenzene [153b]. (See Chapter 2.)

3-8. Preparation of Hexamethylbenzene [153a]

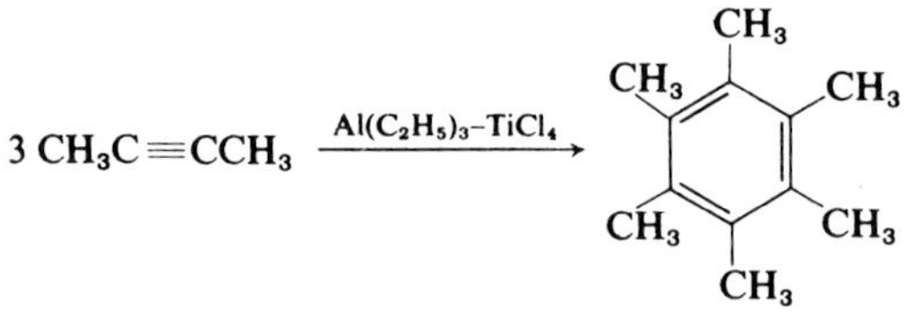

$$3\ CH_3C{\equiv}CCH_3 \xrightarrow{Al(C_2H_5)_3-TiCl_4} \text{hexamethylbenzene} \tag{57}$$

(a) General procedure for catalyst preparation. To a three-necked 1-liter Morton flask under anhydrous conditions (in a hood) is added 3.4 gm (0.03 mole) of aluminum triethyl in 200 ml of pure dry *n*-heptane. The solution is

stirred and 1.9 gm (0.01 mole) of titanium tetrachloride in 5 ml of *n*-heptane is added. A black, insoluble organometallic complex immediately precipitates. The catalyst solution is stirred very fast to reduce the particle size of the precipitated catalyst. Under a nitrogen atmosphere 400 ml of additional pure dry *n*-heptane is added.

(b) Reaction of 2-butyne with catalyst solution. To the stirred catalyst solution is added 16.3 gm (0.3 mole) of 2-butyne in 60 ml of dry *n*-heptane over a period of 2 hr. The reaction temperature rises from 23° to 34°C during this time. After the reaction the black precipitate is centrifuged and the *n*-heptane is concentrated to dryness using a water aspirator. The resulting white solid is washed several times with ether to extract the product. The black solids from the catalyst are also washed with ether to recover any product. Concentration of the combined ether extracts yields 13.1 gm (80.2%) of crude hexamethylbenzene, mp 155°–160°C. Recrystallization from absolute alcohol raises the melting point to 162°–163°C.

4. ELIMINATION REACTIONS

Hydrocarbons can be prepared from substituted compounds by elimination of the substituent which is other than hydrocarbon in nature

$$RX \longrightarrow R + X \tag{58}$$

where X may be halogen, amino, alcohol, aldehyde, ketone, carboxyl, sulfonic acid, etc. Ring closure with elimination of water or some other group will also be mentioned. Several of the earlier mentioned coupling reactions and condensation reactions can also be considered elimination reactions.

Aromatization can be considered an elimination where hydrogen is the functional group that is eliminated.

Halogenated compounds can be converted to hydrocarbons by conversion to the Grignard reagent and hydrolysis with aqueous ammonium chloride [138]. This method has been quite useful in the introduction of a methyl group into aromatic systems by first chloromethylating and then converting to the Grignard prior to hydrolysis [154a].

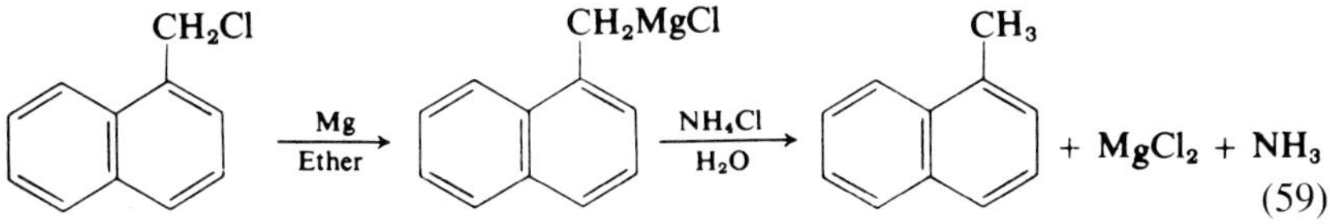

$$\text{Naphthyl-CH}_2\text{Cl} \xrightarrow[\text{Ether}]{\text{Mg}} \text{Naphthyl-CH}_2\text{MgCl} \xrightarrow[\text{H}_2\text{O}]{\text{NH}_4\text{Cl}} \text{Naphthyl-CH}_3 + \text{MgCl}_2 + \text{NH}_3 \tag{59}$$

The titanium-catalyzed isomerization of the Grignard reagent derived from secondary halides can also be utilized in preparation of primary (linear) alkanes by hydrolysis [154b].

Conversion of halides to lithium or sodium derivatives followed by hydrolysis also gives the hydrocarbons. In some cases dehydrohalogenation to the olefin occurs, which may contaminate the product.

A better procedure for converting halides to hydrocarbons has already been discussed in Section 2,E, Reduction of Halides.

Hydrocarbons can be acetylated by the Friedel–Crafts method and then converted to the amide via the Willgerodt method. Hydrolysis to the acetic acid derivative and subsequent heating with soda lime yields the corresponding methyl derivative.

$$ArH + CH_3COCl \xrightarrow{AlCl_3} ArCOCH_3 \xrightarrow{NH_4(S)_x} ArCH_2CONH_2 \longrightarrow$$

$$ArCH_2COOH \xrightarrow[\text{Heat}]{\text{Soda lime}} ArCH_3 + CO_3 \qquad (60)$$

When Ar is phenanthrene, an 84% yield of methylphenanthrene is obtained by this method [155].

Carbonyl groups may be eliminated by the Kolbe electrolysis to give coupled products [156].

$$2\ RCOO^- \longrightarrow R-R + 2\ CO_2 + 2\ e^- \qquad (61)$$

The diazonium group may be replaced by hydrogen by reaction of the diazonium salt with a 5 M excess of hypophosphorous acid at 0°–5°C [157]. The yields are usually good and the method is suitable for laboratory synthesis problems. Hydrochloric acid is the recommended acid for diazotization unless nuclear halogenation is a competing reaction. Substituents such as halo, nitro, and carboxyl groups do not interfere.

4-1. Conversion of p-Tolyldiazonium Chloride to Toluene [158]

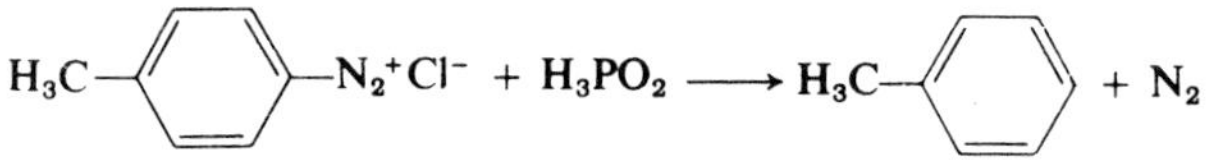

$$H_3C\text{-}C_6H_4\text{-}N_2^+Cl^- + H_3PO_2 \longrightarrow H_3C\text{-}C_6H_5 + N_2 \qquad (62)$$

The diazonium solution is prepared (see Chapter 15) at 0°C from 0.4 mole of the p-toluidine, 1.2 moles of hydrochloric acid, and 0.4 moles of sodium nitrite, with the final volume of the mixture being approximately 750 ml. The p-toluidine is added dropwise as a solution in 200 ml of aqueous hydrochloric acid to the aqueous sodium nitrite at 0°C. The solution is then poured with vigorous stirring into a mixture of 660 gm (5 moles) of 50% hydrophosphorous acid and 100 gm of ice. (CAUTION: foaming may occur.) After the ether has been extracted, the mixture is held at 0°C for 24 hr and for 5 days at 25°C. An additional extraction is done and the combined ether layer is distilled to yield toluene (72%).

§ 5. Dehydrogenations

Ring closure reactions such as the Elbs reaction of *o*-diaryl ketones yields anthracene on pyrolysis at 400°–500°C.

4-2. Elbs Reaction—Synthesis of 1,2-Benzanthracene [159]

To a Pyrex tube are added 6 gm of α-naphthyl *o*-tolyl ketone and 2 gm of zinc dust. The tube is heated at 400°–410°C for 3 hr, cooled, and the contents distilled and then passed through a column of alumina in benzene to yield 2.75 gm of 1,2-benzanthracene, mp 159.5°–160.5°C. In addition, a yellow product is collected, mp 159°–160°C [probably impure (?)]. The total yield is 61%.

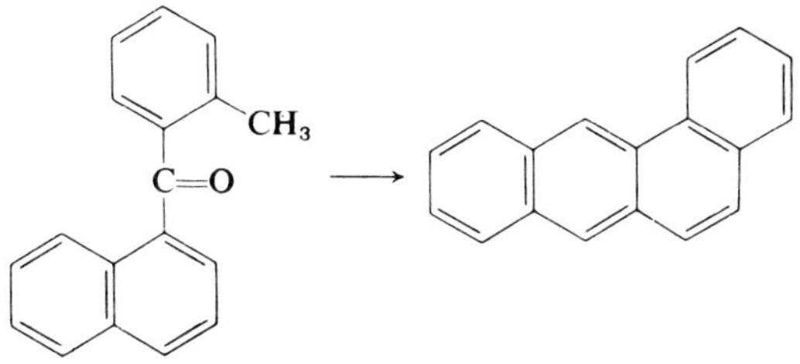

(63)

Other ring closures are the cyclization of 2-, 3-, 4-, or 5-hydroxy-1-phenylpentane as well as the 5-phenyl-1-pentane using 85–90% sulfuric acid [160].

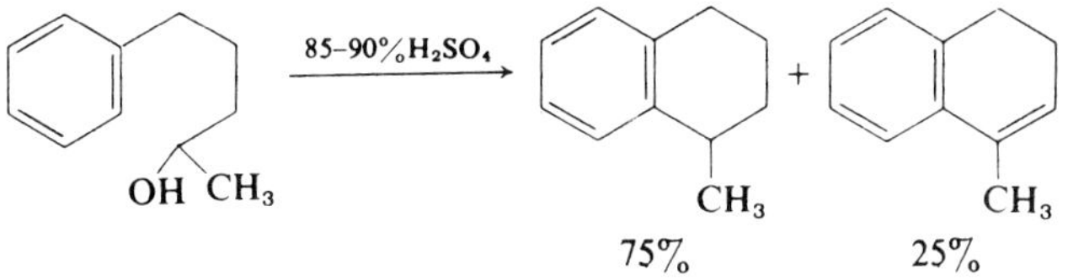

(64)

In addition, β-styrylacetaldehyde yields naphthalene upon refluxing with a mixture of hydrobromic acid and acetic acid [161a].

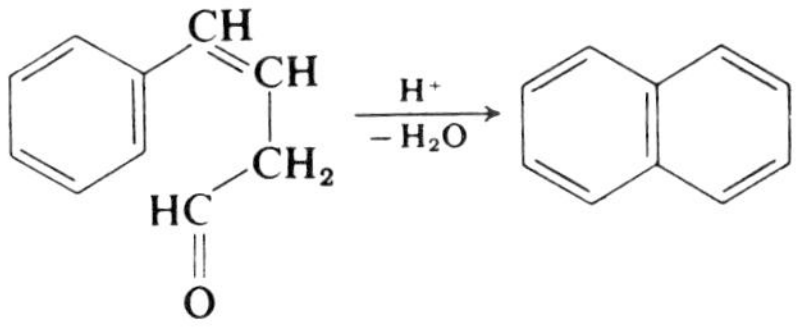

(65)

Polyphosphoric acid also has been reported to be effective in causing the cyclodehydration of pinacols and pinacolones to alkylindenes [161b].

Both the latter two reactions can be explained by electrophilic attack on the aromatic ring by a carbonium ion.

5. DEHYDROGENATIONS

Dehydrogenations from saturated or partially saturated six-membered ring hydrocarbons yield aromatic compounds.

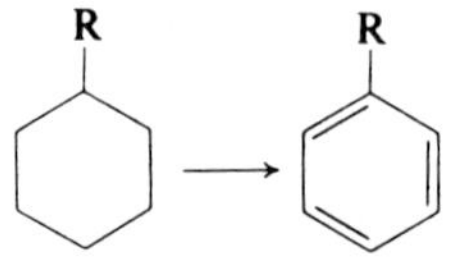

$$\underset{\text{R}}{\bigcirc} \longrightarrow \underset{\text{R}}{\bigcirc} \tag{66}$$

The common dehydrogenation catalysts are platinum and palladium [162, 163], which are also used for hydrogenation reactions; using the latter catalysts, cyclohexanes yield alkylbenzenes. Other catalysts are nickel [164], nickel or chromium oxide [165], metallic oxides [166], and chromium-alumina [167]. The latter catalyst has found widespread use in preparation of large quantities of polyalkylbenzenes from the corresponding cyclohexenes at 450°–470°C [167]. At 600°–650°C alkylbenzenes are dehydrogenated to styrenes [168].

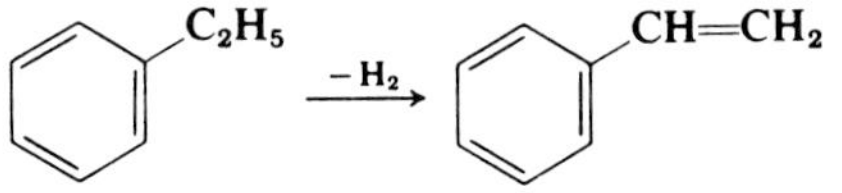

$$\text{C}_6\text{H}_5\text{-C}_2\text{H}_5 \xrightarrow{-\text{H}_2} \text{C}_6\text{H}_5\text{-CH=CH}_2 \tag{67}$$

Dehydrogenations are also effected using chemical reactants such as sulfur [169], selenium [170], or chloranil [171] in refluxing xylene. Chloranil has been used to produce biphenyl, terphenyls, and quaterphenyls [171, 172]. For example, 4-(2,5-dimethyl-1-cyclohexen)-*p*-terphenyl was refluxed for 4 hr in the presence of slightly more than the equimolar amounts of chloranil in xylene. The reaction mixture was cooled and the chloranil hydroquinone was separated by filtration. The xylene solution was evaporated off under reduced pressure to leave 2,5-dimethyl-*p*-quaterphenyl in 42% yield, mp 183° [172].

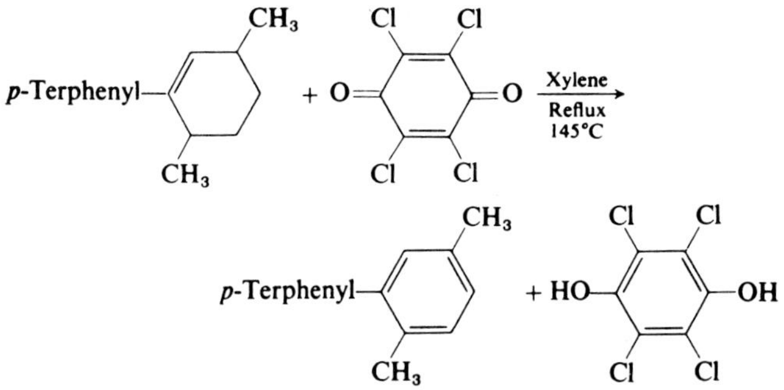

$$p\text{-Terphenyl-(dimethylcyclohexene)} + \text{chloranil} \xrightarrow[\text{145°C}]{\text{Xylene Reflux}} p\text{-Terphenyl-(dimethylbenzene)} + \text{tetrachlorohydroquinone} \tag{68}$$

Substituents such as primary hydroxyl, carboxyl, ester, alkoxy, or keto groups do not interfere [164, 165, 173–175]. Secondary and tertiary hydroxyl groups are eliminated as water and yield the corresponding hydrocarbon [176, 177].

It has been reported [178] that pyrolyzed polyacrylonitrile can be used as a chemical dehydrogenation agent. Using this latter reagent, cyclohexene is converted in 50% yields per pass to benzene.

6. MISCELLANEOUS METHODS

(1) Preparation of paracyclophanes [179–180].

(2) Cyclization of alkenyl arenes catalyzed by potassium metal [181].

(3) Dealkylation and fragmentation reactions of aryl alkyls induced by aluminum chloride and water [182–185].

(4) Photolysis of stilbene to yield phenanthrene [186].

(5) Photolytic coupling of aryl iodides [187, 188].

(6) Cyclobutane formation by the mercury-photosensitized reactions of ethylene [189].

(7) Reaction of hydrocarbons with iodine at 500°C, 2,5-dimethylhexane yields p-xylene in 98% yield [190].

(8) Light-induced decarboxylation of aldehydes [191, 192].

(9) Aromatization of hydrocarbons on chromic aluminum at 531°C [193].

(10) Electrolytic reductive coupling [194].

(11) Hydrogenolysis of the Grignard reagent [195].

(12) The addition of phenyllithium to allylic chlorides to form phenylcyclopropanes [196].

(13) Aromatic alkylation via diazotization [197].

(14) Phenylation with nitrosoacetanilides [198].

(15) Benzene polymerized to p-polyphenyl by ferric chloride/H_2O [199].

(16) The synthesis of p-sexiphenyl from biphenyl or p-terphenyl using a Lewis acid catalyst-oxidant [200].

(17) A general synthesis for the preparation of macrocyclic compounds [201].

(18) A novel synthesis of [2.2] paracyclophanes [202].

(19) Preparation of diamantane [203].

(20) Preparation of 4,6,8-trimethylazulene [204].

(21) Conversion of methanol into hydrocarbons [205].

(22) Cleavage of sulfonic esters with Raney nickel catalysts [206].

(23) Conversion of lower alkyl ester group of an aromatic acid to an alkyl aromatic compound [207].

(24) Reductive cyclization of α,ω-dihalides with chromium(II) complexes [208].

(25) Preparation of bicyclo[1,1,0]butane [209].

(26) Reduction of acetylenes to alkanes with ferrous chloride–sodium hydride [210].

(27) Preparation of aryl aklyls via photosubstitution reactions [211].

(28) Conversion of aldehydes to substituted cyclopropyl compounds [212].

(29) Preparation of benzocyclobutene by n-butyllithium coupling of dibromide [213].

1. Hydrocarbons (Paraffinic and Aromatic)

REFERENCES

1. H. Adkins, "Reactions of Hydrogen." Univ. of Wisconsin Press, Madison, 1937; also H. Adkins and R. L. Shriner, in "Organic Chemistry" (H. Gilman, ed.), Vol. 1, p. 779. Wiley, New York, 1943.
2. H. C. Brown and C. A. Brown, *J. Am. Chem. Soc.* **84**, 1495 (1962).
3. H. Merton and H. Gilman, *J. Am. Chem. Soc.* **76**, 5798 (1954).
4. G. J. M. Van der Kerk, J. G. Noltes, and J. G. A. Luijten, *J. Appl. Chem.* **7**, 356 (1959).
5. G. E. Ham and W. P. Coker, *J. Org. Chem.* **29**, 194 (1964).
6. J. J. Brunet, L. Mordenti, B. Loubinoux and P. Caubere, *Tetrahedron Lett.* 1069 (1977); J. J. Brunet and P. Caubere, *ibid.* p. 3947.
7. G. Crane, C. E. Boord, and A. L. Henne, *J. Am. Chem. Soc.* **67**, 1237 (1965).
8. R. S. Monson "Advanced Organic Synthesis," pp. 43–44. Academic Press, New York, 1971.
9. C. E. Boord, A. L. Henne, K. W. Greenlee, W. L. Perilstein, and J. M. Derfer, *Ind. Eng. Chem.* **41**, 609, 613 (1949).
10. S. Hunig, H. R. Miller, and W. Thier, *Angew. Chem., Int. Ed. Engl.* **4**, No. 4, 271 (1965).
11. C. E. Miller, *J. Chem. Educ.* **42**, No. 5, 254 (1965).
12. E. J. Corey and W. L. Mock, *J. Am. Chem. Soc.* **84**, 685 (1962).
13. E. J. Corey, D. J. Pasto, and W. L. Mock, *J. Am. Chem. Soc.* **83**, 2957 (1961).
14. E. J. Garbisch, Jr., S. M. Schildorant, D. B. Patterson, and C. M. Sprecher, *J. Am. Chem. Soc.* **87**, 2932 (1965).
15. K. Kondo, S. Murai, and N. Sonodo, *Tetrahedron Lett.* p. 3727 (1977).
16. H. C. Brown and K. Murray, *J. Am. Chem. Soc.* **81**, 4108 (1959).
17a. H. C. Brown and C. A. Brown, *J. Am. Chem. Soc.* **84**, 1493–1495 (1962).
17b. C. A. Brown and H. C. Brown, *J. Am. Chem. Soc.* **84**, 2829 (1962).
18. H. C. Brown and C. A. Brown, *J. Am. Chem. Soc.* **85**, 1003 (1963).
19. H. C. Brown and C. A. Brown, *J. Am. Chem. Soc.* **85**, 1004 (1963).
20. R. H. Baker and R. D. Schultz, *J. Am. Chem. Soc.* **69**, 1250 (1947).
21. R. Adams and J. R. Marshall, *J. Am. Chem. Soc.* **50**, 1970 (1928).
22. H. Adkins and H. I. Cramer, *J. Am. Chem. Soc.* **52**, 4348 (1930).
23. K. T. Serijan, P. H. Wise, and L. C. Gibbons, *J. Am. Chem. Soc.* **71**, 2205 (1949).
24. W. F. Seyer, M. M. Wright, and R. C. Bell, *Ind. Eng. Chem.* **31**, 759 (1939).
25. S. Nishimura, Y. Kono, Y. Otsuki, and Y. Fukaya, *Bull. Chem. Soc. Jpn.* **44**, 240 (1981).
26. Huang-Minlon, *J. Am. Chem. Soc.* **68**, 2487 (1946).
27. D. Todd, *Org. React.* **4**, 378 (1948).
28. C. H. Herr, F. C. Whitmore, and R. W. Schiessler, *J. Am. Chem. Soc.* **67**, 2061 (1945).
29. F. L. Martin, *Org. React.* **1**, 155 (1942).
30. H. Adkins and R. Connor, *J. Am. Chem. Soc.* **53**, 1091 (1931).
31. H. Pines, D. R. Strehlau, and V. N. Ipatieff, *J. Am. Chem. Soc.* **72**, 1563 (1950).
32. B. R. Brown and A. M. S. White, *J. Chem. Soc.* p. 3755 (1957).
33. R. F. Nystrom and C. R. A. Berger, *J. Am. Chem. Soc.* **80**, 2896 (1958).
34a. F. Sondheimer and S. Wolfe, *Can. J. Chem.* **37**, 1870 (1959).
34b. G. S. Li, D. F. Ehler, and R. A. Benkeser, *Org. Synth.* **56**, 83 (1977).
35. H. H. Szmant, H. F. Hannsberger, T. J. Butler, and W. P. Barie, *J. Am. Chem. Soc.* **74**, 2724 (1952).
36. Huang-Minlon, *J. Am. Chem. Soc.* **71**, 3301 (1949).
37. Huang-Minlon, *J. Am. Chem. Soc.* **70**, 2802 (1948).
38. Huang-Minlon, *Sci. Sin. Chin. Ed.* **10**, 711 (1961); *Chem. Abstr.* **56**, 11407h (1962).
39. N. P. Buu-Hoi, M. Sy, and J. Riche, *Bull. Soc. Chim. Fr.* p. 1493 (1960).
40. W. Nagata and H. Itazaki, *Chem. Ind. (London)* p. 1194 (1964).

§ References

41. D. L. Cram, M. R. V. Sahyun, and G. R. Knox, *J. Am. Chem. Soc.* **84,** 1734 (1962).
42. B. Bannister and B. B. Elsner, *J. Chem. Soc.* p. 1055 (1951).
43. M. A. Dolliver, T. L. Gresham, G. B. Kistiakowsky, and W. E. Vaughan, *J. Am. Chem. Soc.* **59,** 831 (1937).
44. E. L. Martin, *J. Am. Chem. Soc.* **58,** 1438 (1936).
45. L. I. Smith and C. P. Lo, *J. Am. Chem. Soc.* **70,** 2209 (1948).
46. O. L. Brady and J. N. E. Day, *J. Chem. Soc.* p. 116 (1934).
47. D. M. Cowan, G. H. Jeffery, and A. I. Vogel, *J. Chem. Soc.* p. 1862 (1939).
48. W. E. Bachmann and G. D. Cortes, *J. Am. Chem. Soc.* **65,** 1329 (1943).
49. T. A. Ford, H. W. Jacobson, and F. C. McGrew, *J. Am. Chem. Soc.* **70,** 3793 (1948).
50. V. I. Komarewsky, C. F. Price, and J. R. Coley, *J. Am. Chem. Soc.* **69,** 238 (1947).
51. A. J. Birch, *J. Chem. Soc.* p. 809 (1945).
52a. J. H. Brewster, S. F. Osman, H. O. Bayer, and H. B. Hopps, *J. Org. Chem.* **29,** 121 (1964).
52b. S. D. Lipsky and S. S. Hall, *Org. Synth.* **55,** 7 (1976).
53. A. Grummitt and A. C. Buck, *J. Am. Chem. Soc.* **65,** 295 (1943).
54. P. A. Levene, *Org. Synth. Collect.* **2,** 320 (1943).
55. W. G. Brown, *Org. React.* **6,** 469 (1941); E. C. Ashby and J. J. Lin, *Tetrahedron Lett.* p. 4481 (1977).
56a. J. E. Johnson, R. H. Blizzard, and H. W. Carhart, *J. Am. Chem. Soc.* **79,** 3664 (1948).
56b. B. Loubinoux, R. Vanderesse, and P. Canbere, *Tetrahedron Lett.* p. 3951 (1977).
57. H. C. Brown and H. M. Bell, *J. Org. Chem.* **27,** 1928 (1962).
58. C. A. Buehler, D. E. Cooper, and E. O. Scrudder, *J. Org. Chem.* **8,** 316 (1943).
59a. L. Zechmeister and P. Rom, *Justus Liebigs Ann. Chem.* **468,** 127 (1929).
59b. D. Bryce-Smith and B. J. Wakefield, *Org. Synth. Collect. Vol.* **5,** 998 (1973).
60a. H. Hart, *J. Am. Chem. Soc.* **71,** 1966 (1949).
60b. N. A. Cortese and R. F. Heck, *J. Org. Chem.* **42,** 3491 (1977).
60c. J. Jacobus, *Chem. Commun.* p. 338 (1970).
61a. R. O. Hutchins, D. Kandasamy, C. A. Maryanoff, D. Masilamani, and B. E. Maryanoff, *J. Org. Chem.* **42,** 82 (1977); R. O. Hutchins, C. A. Milewski and B. E. Maryanoff, *Org. Syn.* **53,** 107 (1973).
61b. H. G. Kuivila, L. W. Monapace, and C. R. Warner, *J. Am. Chem. Soc.* **54,** 3584 (1962); D. Seyforth, H. Yamazaki, and D. L. Alleston, *J. Org. Chem.* **28,** 703 (1963).
61c. E. J. Corey and S. W. Suggs, *J. Org. Chem.* **40,** 2554 (1975).
62. H. C. Brown and P. A. Tierney, *J. Am. Chem. Soc.* **80,** 1552 (1958).
63. L. I. Smith, *Org. Syn. Collect. Vol.* **2,** 248 (1943).
64. F. C. Whitmore and W. H. James, *J. Am. Chem. Soc.* **65,** 2088 (1943).
65. A. Newton, *J. Am. Chem. Soc.* **65,** 320 (1943).
66. H. Gilman and R. H. Meals, *J. Org. Chem.* **8,** 126 (1943).
67. L. Schmerling, *Ind. Eng. Chem.* **40,** 2072 (1948).
68. G. F. Hennion and R. A. Kurtz, *J. Am. Chem. Soc.* **65,** 1001 (1943).
69. J. H. Simons and H. Hart, *J. Am. Chem. Soc.* **66,** 1309 (1944).
70. J. H. Simons and S. Archer, *J. Am. Chem. Soc.* **60,** 2952, 2953 (1938).
71. E. Wertyporoch, *Ber. Dtsch. Chem. Ges. B* **66B,** 1232 (1933).
72. H. Bredereck, G. Lehmann, C. Schonfeld, and E. Fritzsche, *Ber. Dtsch. Chem. Ges. B* **72B,** 1414 (1939).
73. H. Pines, J. D. La Zerte, and V. N. Ipatieff, *J. Am. Chem. Soc.* **72,** 2850 (1950).
74. B. B. Corson and V. N. Ipatieff, *Org. Synth. Collect. Vol.* **2,** 151 (1943).
75. G. A. Russell, *J. Am. Chem. Soc.* **81,** 4834 (1959).
76. E. F. Pratt, R. K. Preston, and J. N. Draper, *J. Am. Chem. Soc.* **72,** 1367 (1950).
77. W. S. Calcott, J. M. Tinker, and V. Weinmayer, *J. Am. Chem. Soc.* **61,** 1012 (1939).

78. J. S. Reichert and J. A. Nieuwland, *J. Am. Chem. Soc.* **45**, 3090 (1923).
79. W. M. Potts and R. D. Dodson, *J. Am. Chem. Soc.* **61**, 2553 (1939).
80. J. F. Norris and B. M. Sturgis, *J. Am. Chem. Soc.* **61**, 1413 (1939).
81. L. Spiegler and J. M. Tinker, *J. Am. Chem. Soc.* **61**, 1002 (1939).
82. J. Epelberg and A. Lowry, *J. Am. Chem. Soc.* **63**, 101 (1941).
83. A. V. Grosse and V. N. Ipatieff, *J. Org. Chem.* **2**, 447 (1937).
84. C. C. Price, *Chem. Rev.* **29**, 44 (1941).
85. G. A. Olah and J. J. Kuhn, *J. Org. Chem.* **29**, 2317 (1964).
86. R. D. Haworth, *J. Chem. Soc.* p. 1125 (1932).
87. C. C. Price, *Org. React.* **3**, 1 (1946).
88. A. W. Francis, *Chem. Rev.* **43**, 257 (1948).
89. N. O. Calloway, *Chem. Rev.* **17**, 327 (1935).
90. G. A. Olah, Ed., "Friedel–Crafts and Related Reactions," Vols. I and II. Wiley, New York, 1963, 1964.
91. G. A. Olah, M. W. Meyer, and N. A. Overchuck, *J. Org. Chem.* **29**, 2313 (1964).
92. G. A. Olah, M. W. Meyer, and N. A. Overchuck, *J. Org. Chem.* **29**, 2310 (1964).
93. G. A. Olah, M. W. Meyer, and N. A. Overchuck, *J. Org. Chem.* **29**, 2315 (1964).
94. J. F. Norris and D. Rubinstein, *J. Am. Chem. Soc.* **61**, 1163 (1939).
95. L. I. Smith and L. J. Spillane, *J. Am. Chem. Soc.* **62**, 2369 (1940).
96a. L. I. Smith and C. P. Lo, *J. Am. Chem. Soc.* **70**, 2210 (1948).
96b. P. von R. Schleyer, M. M. Donaldson, R. D. Nicholas, and C. Cupas, *Org. Synth. Collect. Vol.* **5**, 16 (1973).
97. P. J. Flory, "Principles of Polymer Chemistry," Cornell Univ. Press, Ithaca, New York, 1953.
98. N. G. Gaylord and H. Mark, "Linear and Stereospecific Addition Polymers." Wiley (Interscience), New York, 1959.
99. W. R. Sorenson, *J. Chem. Educ.* **42**, No. 1, 8 (1965).
100. S. R. Sandler and W. Karo "Polymer Syntheses," Vol. 1. Academic Press, New York, 1974.
101a. Authors' Laboratory.
101b. I. M. Kolthoff and W. J. Dale, *J. Am. Chem. Soc.* **69**, 441 (1947).
102. J. S. Lasky, H. K. Garner, and R. H. Ewart, *Ind. Eng. Chem., Prod. Res. Dev.* **1**, 82 (1962).
103. C. W. Childers, *J. Am. Chem. Soc.* **85**, 229 (1963).
104. J. D. Roberts, *Org. React.* **12**, 1 (1962).
105. J. Hine, "Divalent Carbon." Ronald Press, New York, 1964.
106. W. Kirmse, *Angew. Chem.* **73**, 161 (1961).
107. A. M. Trozzolo, R. W. Murray, and E. Wasserman, *J. Am. Chem. Soc.* **84**, 4990 (1962).
108. C. L. Closs and R. A. Moss, *J. Am. Chem. Soc.* **86**, 4042 (1964).
109. H. E. Simmons and R. D. Smith, *J. Am. Chem. Soc.* **81**, 4256 (1959).
110. T. J. De Boer and H. T. Backer, *Org. Synth. Collect. Vol.* **4**, 250 (1963).
111. P. S. Skell and R. C. Woodward, *J. Am. Chem. Soc.* **78**, 449 (1956).
112. P. S. Skell and R. R. Engel, *J. Am. Chem. Soc.* **87**, 1135 (1965).
113. P. S. Skell, L. D. Wescott, Jr., J. P. Goldstein, and R. R. Engel, *J. Am. Chem. Soc.* **87**, 2829 (1965).
114. C. Mackay and R. Wolfgang, *Science* **148**, 899 (1965); P. S. Skell and R. R. Engel, *Abstr. Pap., 150th Meet., Am. Chem. Soc., 1965* p. 13S (1965).
115. R. S. Shank and H. Shechter, *J. Org. Chem.* **24**, 1825 (1959).
116. W. E. Billups, A. J. Blakeney, and W. Y. Chow, *Org. Synth.* **55**, 12 (1976).
117. W. E. Billups, A. J. Blakeney, and W. Y. Chow, *Chem. Commun.* p. 1461 (1971).
118. E. Vogel, W. Klag, and A. Brever, *Org. Synth.* **54**, 11 (1974).
119. R. K. Hill and R. M. Carlson, *J. Org. Chem.* **30**, 2414 (1965).

120. F. C. Whitmore and G. H. Fleming, *J. Am. Chem. Soc.* **55**, 3804 (1933).
121. F. C. Whitmore, H. I. Bernstein, and L. W. Nixon, *J. Am. Chem. Soc.* **60**, 2539 (1938).
122. G. Calingaert, H. Soroos, V. Hnizda, and H. Shapiro, *J. Am. Chem. Soc.* **66**, 1389 (1944).
123. H. Soroos and H. B. Willis, *J. Am. Chem. Soc.* **63**, 881 (1941).
124. C. R. Noller, *J. Am. Chem. Soc.* **51**, 594 (1929).
125. D. M. Hull, M. S. Lesslie, and E. E. Turner, *J. Chem. Soc.* p. 711 (1950).
126. A. A. Morton and G. M. Richardson, *J. Am. Chem. Soc.* **62**, 123 (1940).
127. G. B. Bachman and R. I. Hoaglin, *J. Am. Chem. Soc.* **63**, 621 (1941).
128. T. Reichstein and R. Oppenauer, *Helv. Chim. Acta* **16**, 1377 (1933).
129. H. J. Barber, R. Slack, and A. M. Woolman, *J. Chem. Soc.* p. 99 (1943).
130. A. A. Morton, J. B. Davidson, T. R. P. Gibbs, Jr., E. L. Little, E. F. Clarke, and A. G. Green, *J. Am. Chem. Soc.* **64**, 2250 (1942).
131. H. B. Haas, E. T. McBee, G. E. Hinds, and E. N. Glissenkamp, *Ind. Eng. Chem.* **28**, 1178 (1936).
132. R. W. Shortridge, R. A. Craig, K. W. Greenlee, J. M. Derfer, and C. E. Boord, *J. Am. Chem. Soc.* **70**, 946 (1948).
133. V. A. Slabey, *J. Am. Chem. Soc.* **68**, 1335 (1946).
134. D. S. Cobnor and E. R. Wilson, *Tetrahedron Lett.* p. 4925 (1967).
135. E. J. Corey and G. H. Posner, *J. Am. Chem. Soc.* **90**, 5615 (1968).
136a. W. Hertz and R. Ullrich, *Makromol. Chem.* **98**, 29 (1966).
136b. F. B. Mallory and C. S. Wood, *Org. Synth. Collect. Vol.* **5**, 952 (1973).
137a. J. A. Dixon and D. H. Fishman, *J. Am. Chem. Soc.* **85**, 1356 (1963).
137b. H. E. Zieger and E. Ellis, *Abstr. Pap., 150th Meet., Am. Chem. Soc., 1965* p. 46S (1965).
138. W. Krizewsky and E. E. Turner, *J. Chem. Soc.* **115**, 559 (1919).
139. S. R. Sandler, *Int. J. Appl. Radiat. Isot.* **16**, 425 (1965).
140. J. H. Gardner and P. Bergstrom, *J. Am. Chem. Soc.* **51**, 3375 (1929).
141. H. Gilman and J. E. Kirby, *Recl. Trav. Chim. Pays-Bas* **48**, 155 (1929).
142. M. S. Kharasch, W. Nudenberg, and S. Archer, *J. Am. Chem. Soc.* **65**, 49 (1943).
143. M. S. Kharasch, R. Morrison, and W. H. Urry, *J. Am. Chem. Soc.* **66**, 368 (1944).
144. G. M. Bennett and E. E. Turner, *J. Chem. Soc.* **105**, 1057 (1914).
145. H. Gilman and M. Lichtenwalter, *J. Am. Chem. Soc.* **61**, 957 (1939).
146. P. E. Ranta, *Chem. Rev.* **64**, 613 (1964).
147. P. M. Everitt, D. M. Hall, and E. E. Turner, *J. Chem. Soc.* p. 2286 (1956).
148. H. Gilman and R. E. Hoyle, *J. Am. Chem. Soc.* **44**, 2621 (1922).
149. H. Gilman and L. L. Heck, *J. Am. Chem. Soc.* **50**, 2223 (1928).
150. M. Gomberg and J. C. Pernert, *J. Am. Chem. Soc.* **48**, 1372 (1928).
151. J. R. Johnson, *Org. React.* **1**, 246 (1942).
152. E. E. Van Tamelen and M. A. Schwartz, *J. Am. Chem. Soc.* **87**, 3277 (1965).
153a. E. F. Lutz, *J. Am. Chem. Soc.* **83**, 2551 (1961).
153b. W. Schäfer, *Angew. Chem., Int. Ed. Engl.* **7**, 699 (1966).
154a. O. Grumitt and A. C. Buck, *J. Am. Chem. Soc.* **65**, 295 (1943).
154b. G. D. Cooper and H. L. Finkbeiner, *J. Org. Chem.* **27**, 1493 (1962).
155. W. E. Bachmann and G. D. Cortes, *J. Am. Chem. Soc.* **65**, 1332 (1943).
156. F. Fichter, *Helv. Chim. Acta* **22**, 970 (1939).
157. N. Kornblum, *Org. React.* **2**, 262 (1944); M. M. Robison and B. L. Robison, *Org. Synth. Collect. Vol.* **4**, 947 (1963); J. F. Bunnett and H. Takayama, *J. Org. Chem.* **33**, 1924 (1968); N. Kornblum, *Org. Synth. Collect. Vol.* **3**, 295 (1955); A. H. Lowin and T. Cohen, *J. Org. Chem.* **32**, 3844 (1967); R. A. Henry and W. G. Finnigan, *J. Am. Chem. Soc.* **76**, 290 (1954).
158. N. Kornblum, A. E. Kelley, and G. D. Cooper, *J. Am. Chem. Soc.* **74**, 3074 (1952).
159. L. F. Fieser and E. Hershberg, *J. Am. Chem. Soc.* **59**, 2502 (1937).

160. R. O. Roblin, Jr., D. Davidson, and M. T. Bogert, *J. Am. Chem. Soc.* **57,** 151 (1935).
161a. C. K. Bradsher, *J. Am. Chem. Soc.* **64,** 1007 (1942).
161b. N. D. Heindel and S. W. McNeil, *Abstr. Pap., 150th Meet., Am. Chem. Soc., 1965* p. 20P (1965).
162. N. D. Zelinskii, *Ber. Dtsch. Chem. Ges.* **56,** 787 (1923).
163. M. C. Kloetzel and H. L. Herzog, *J. Am. Chem. Soc.* **72,** 1993 (1950).
164. H. Adkins and N. A. Reid, *J. Am. Chem. Soc.* **63,** 741 (1941).
165. H. Adkins, L. M. Richards, and J. M. Davis, *J. Am. Chem. Soc.* **63,** 1320 (1941).
166. R. G. Flowers and H. E. Miller, *J. Am. Chem. Soc.* **69,** 1388 (1947).
167. T. W. Reynolds, E. R. Ebersole, J. M. Lamberti, H. H. Chanan, and P. M. Ordin, *Ind. Eng. Chem.* **40,** 1751 (1948).
168. J. E. Nickels, G. A. Webb, W. Heintzelman, and B. B. Corson, *Ind. Eng. Chem.* **41,** 563 (1949).
169. R. Weiss, *Org. Synth.* **24,** 84 (1944).
170. L. F. Fieser and E. B. Hershberg, *J. Am. Chem. Soc.* **57,** 2192 (1935).
171. R. T. Arnold, C. Collins, and W. Zenk, *J. Am. Chem. Soc.* **62,** 983 (1940).
172. S. R. Sandler, *Int. J. Appl. Radiat. Isot.* **16,** 426 (1965).
173. E. E. J. Marler and E. E. Turner, *J. Chem. Soc.* p. 266 (1937).
174. M. S. Newman and F. T. J. O'Leary, *J. Am. Chem. Soc.* **68,** 258 (1946).
175. R. P. Linstead and K. O. A. Michaelis, *J. Chem. Soc.* p. 1134 (1940).
176. W. E. Bachmann and R. C. Edgerton, *J. Am. Chem. Soc.* **62,** 2220 (1940).
177. W. S. Johnson, A. Goldman, and W. P. Schneider, *J. Am. Chem. Soc.* **67,** 1357 (1945).
178. J. Manassen and J. Wallach, *J. Am. Chem. Soc.* **87,** 2671 (1965).
179. D. T. Longone and C. L. Warren, *J. Am. Chem. Soc.* **84,** 1507 (1962).
180. D. T. Longone and F. P. Boettscher, *J. Am. Chem. Soc.* **85,** 3436 (1963); F. Voegtte and J. Gruetze, *Angew. Chem.* **87**(15), 543 (1975).
181. E. Lewicki, H. Pines, and N. C. Sih, *Chem. Ind. (London)* p. 154 (1964).
182. R. Anschutz, *Justus Liebigs Ann. Chem.* **235,** 177 (1886).
183. A. W. Schorger, *J. Am. Chem. Soc.* **39,** 2671 (1917).
184. R. Heise and A. Tohl, *Justus Liebigs Ann. Chem.* **270,** 155 (1892).
185. R. M. Roberts, E. E. Bayliss, and G. J. Fonkin, *J. Am. Chem. Soc.* **85,** 3454 (1963).
186. R. Srinivasan and J. C. Powers, Jr., *J. Am. Chem. Soc.* **85,** 1355 (1963).
187. W. Wolf and N. Kharasch, *J. Org. Chem.* **26,** 283 (1961).
188. W. M. Hutchinson, P. S. Hudson, and R. C. Doss, *J. Am. Chem. Soc.* **85,** 3358 (1963); N. Kharasch and R. K. Sharma, *Abstr. Pap., 150th Meet., Am. Chem. Soc., 1965* p 10S (1965).
189. J. P. Chesnick, *J. Am. Chem. Soc.* **85,** 3718 (1963).
190. R. D. Mullineaux and J. H. Raley, *J. Am. Chem. Soc.* **85,** 3178 (1963).
191. J. D. Berman, J. H. Stanley, W. V. Sherman, and S. G. Cohen, *J. Am. Chem. Soc.* **85,** 4010 (1963).
192. S. Winstein and F. H. Seubold, Jr., *J. Am. Chem. Soc.* **69,** 2917 (1947).
193. C. T. Goetschel and H. Pines, *J. Org. Chem.* **29,** 399 (1964).
194. M. M. Baizer and J. D. Anderson, *J. Org. Chem.* **30,** 1348 (1965).
195. W. E. Becker and E. C. Ashby, *J. Org. Chem.* **29,** 954 (1964).
196. S. Wawzonek, B. Studnicka, H. J. Bluhm, and R. F. Kallio, *J. Am. Chem. Soc.* **87,** 2069 (1965).
197. D. E. Pearson, C. V. Breder, and J. C. Craig, *J. Am. Chem. Soc.* **86,** 5054 (1964).
198. E. L. Elliel and J. G. Saha, *J. Am. Chem. Soc.* **86,** 3581 (1964).
199. P. Kovacic, F. W. Koch, and C. E. Stephen, *J. Polym. Sci.* **2,** 1193 (1964).
200. P. Kovacic and R. M. Lange, *J. Org. Chem.* **29,** 2416 (1964).

201. P. R. Story, D. D. Denson, C. E. Bishop, B. C. Clark, Jr., and J. C. Farine, *J. Am. Chem. Soc.* **90,** 817 (1968).
202. G. W. Brown and F. Sondheimer, *J. Am. Chem. Soc.* **89,** 7116 (1967).
203. T. M. Gund, W. Thielecka, and P. V. R. Schleyer, *Org. Synth.* **53,** 30 (1973).
204. K. Hafner and H. Kaiser, *Org. Synth. Collect. Vol.* **5,** 1088 (1973).
205. D. E. Pearson, *J. Chem. Soc., Chem. Commun.* No. 10, p. 397 (1974).
206. G. W. Kenner and M. A. Murray, *J. Chem. Soc.*, No. 5, p. 5178 (1949).
207. R. V. Norton, U.S. Patent 3,880,905 (1975).
208. J. K. Kochi and D. M. Singleton, *J. Org. Chem.* **33,** 1027 (1968).
209. G. M. Lampman and J. C. Aumiller, *Org. Synth.* **51,** 55 (1976).
210. T. Fujisawa and K. Sugimoto, *Chem. Lett.* p. 581 (1976).
211. J. Cornelisse and E. Havinga, *Chem. Rev.* **75,** 353 (1975).
212. J. Elthimoff-Felkin and P. Sarda, *Tetrahedron* **31,** 2785 (1975).
213. W. E. Parham, L. D. Jones, and Y. A. Sayed, *J. Org. Chem.* **41,** 1184 (1976).

CHAPTER 2 / **OLEFINS**

1. Introduction   39
2. Elimination Reactions   41
   A. Dehydration of Alcohols   41
   2-1. Preparation of Cyclohexene   46
   2-2. Preparation of Dimethylstyrenes   46
   B. Pyrolysis Reactions   49
   2-3. Pyrolysis of the Acetate of p-*Cyanophenylmethylcarbionol to* p-*Cyanostyrene*   51
   *2-4. Chugaev Method—Preparation of 3-*tert-*Butyl-1-cyclohexene*   51
   2-5. Decarboxylation of p-*Chlorocinnamic Acid to* p-*Chlorostyrene*   52
   C. Dehydrohalogenation Reactions   52
   2-6. Preparation of 3-Chloro-2-methyl- and 3-Chloro-4-methyl-α-methylstyrene   53
   2-7. Boord Method—Preparation of 1,4-Hexadiene   54
      a. Reactions of 1,1-Dihalocyclopropanes—Insertion of a Carbon Atom between the Atoms of a Double Bond   56
   2-8. Preparation of 2-Chloro-3-hydroxycyclohexene from Cyclopentene   57
   2-9. Preparation of 1,1-Diphenyl-2-bromo-3-acetoxy-1-propene   57
   D. Dehalogenation of Dihalides   58
3. Condensation Reactions   59
   A. The Wittig Synthesis of Olefins   59
   3-1. Wittig Method—Preparation of Methylenecyclohexane in Dimethyl Sulfoxide Solvent   60
   3-2. Preparation of 1,2-Distyrylbenzene   60
   3-3. Preparation of trans-*Stilbene with Phosphonate Carbanions*   61
   B. Condensations Involving Acetylenes (Vinylation Reactions)   62
   3-4. Preparation of Vinyl Chloroacetate   63
   3-5. Preparation of Hexamethyl–Dewar Benzene (Hexamethylbicyclo[2.2.0]-2,5-hexadiene) from 2-Butyne   63
   C. Condensation of Aldehydes and Ketones with Themselves or with Other Active Methylene Compounds   64
   3-6. Aldol Condensation—Preparation of α-*Ethylcinnamaldehyde*   65
   D. Coupling and Grignard Reactions   65
   3-7. Preparation of Diallyl Isophthalate   67
   E. The Diels–Alder Reaction   67
   3-8. Preparation of cis-*4-Cyclohexene-1,2-dicarboxylic Anhydride (*cis-*Tetrahydrophthalic Anhydride)*   67
4. Reduction Reactions   69
   4-1. Hydroboration of 1-Hexyne to 1-Hexene   69
   4-2. Modified Birch Reduction with Lithium in an Amine–Alcohol System—Preparation of 2,5-Dihydroethylbenzene   70
5. Isomerization Reactions   71
   5-1. Claisen Rearrangement—Preparation of Allyl Phenyl Ether and Its Rearrangement to 2-Allylphenol   72

6. Miscellaneous Methods 72
  A. Elimination Reactions 72
  B. Condensation Reactions 73
  C. Oxidation Reactions 74
  D. Reduction Reactions 74
  E. Isomerization and Rearrangement Reactions 74
  References 75

1. INTRODUCTION

Olefins are commonly prepared by elimination reactions (loss of water, hydrogen halides, acids, etc.) and condensation reactions. Methods utilizing oxidation, reduction, isomerization or rearrangement, free radical, photolytic, and enzyme reactions are less commonly used in the laboratory to prepare a center of unsaturation.

The elimination reactions are summarized by Eq. (1)

$$R-\underset{Z}{CH}-CH_3 \xrightarrow{\text{Acids, bases, or heat}} RCH=CH_2 + ZH \quad (1)$$

where Z may be a hydroxyl, halogen, ester, ether, methyl xanthate (Chugaev reaction), carbamate, carbonate, sulfite, amine, quaternary ammonium hydroxide (Hofmann degradation), amine oxide, or one of many other labile groups.

Another elimination reaction involves disubstituted derivatives as in Eq. (2)

$$R-\underset{Z}{CH}-\underset{Z}{CH_2} \longrightarrow RCH=CH_2 \quad (2)$$

where Z may be hydroxyl or halogen.

Condensation reactions, such as the Boord synthesis, are good methods for converting aliphatic aldehydes [Eq. (3)] to substituted olefins via the preparation of dibromoethyl ethers, Grignard coupling, and elimination of bromoethoxy zinc. (See the text that follows for more detailed equations.)

$$RCH_2CH=O \longrightarrow RCH=CR'R'' \quad (3)$$

Aldehydes can also be converted to olefins by reaction with active methylene compounds [Eq. (4)] by the Knoevenagel, Perkin, Claisen, and aldol condensation reactions

$$RCH_2CH=O + CH_2(R')_2 \longrightarrow RCH_2CH=C(R')_2 \quad (4)$$

where R's are carboxylic acid or ester groups, nitro, nitrile, carboxylic anhydride groups, aldehydes, and ketones or any other strongly electron-withdrawing substituents.

The Wittig reaction [Eq. (5)] is a convenient laboratory method useful for the conversion of aldehydes and ketones to olefins via the reaction of triphenylphosphinemethylenes

$$RCH=O + (C_6H_5)_3P=CX_2 \longrightarrow RCH=CX_2 \qquad (5)$$

where X is hydrogen, alkyl, alkyl carboxylate, and halogen.

A convenient modification of the Wittig method uses phosphonate carbanions, $(RO)_2P^-$ OCX_2, in place of $(C_6H_5)_3P=CX_2$.

The condensation of acetylenes with carbon monoxide, hydrogen halides, alcohols, acids, amines, mercaptans, halogens, etc., gives extremely useful unsaturated compounds as generalized by Eq. (6).

$$HC{\equiv}CH + HZ \longrightarrow CH_2{=}CHZ \qquad (6)$$

However, as a result of the hazards involved in handling acetylene, a great many laboratory workers avoid its use. Nevertheless, Reppe's pioneering experimental work has shown that under the proper conditions the hazards associated with acetylene may be minimized and that acetylene is quite useful for vinylation reactions. Some industrial processes today utilize acetylene reactions to prepare vinyl ethers, acrylic acid, vinyl fluoride, 2-butene-1,4-diol, and some other olefins used for the preparation of plastics.

Recently Dewar benzene (bicyclo[2.2.0]-2,5-hexadiene) derivatives have been prepared from acetylene trimerization or by a Diels–Alder reaction of cyclobutadieneiron tricarbonyl complex with activated acetylenes.

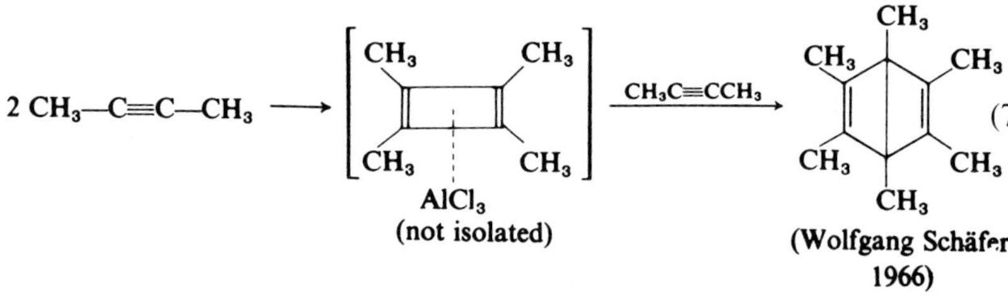

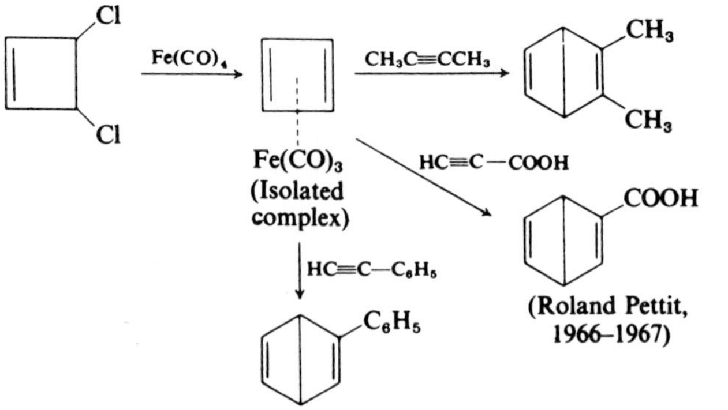

§ 2. Elimination Reactions

It is interesting to note that both reactions proceed via a cyclobutadiene-metal complex.

The above-mentioned methods depend on conversion of an existing functional group to an olefin. However, olefinic groups can also be added on to an existing molecule to give a site of unsaturation.

The reaction of vinylsodium, allylmagnesium halide, and other vinyl metallics with aldehydes, ketones, and halogens are synthetically useful methods.

Allyl alcohol, allyl halide, and other unsaturated compounds may condense with reactive compounds to give a substituted olefin.

The reduction of acetylenic compounds to olefins by hydroboration procedures is of importance. However, the acetylenes are not common starting materials. Therefore this type of method will have only limited value.

The Birch reduction of aromatic compounds by sodium and liquid ammonia in the presence of ethanol gives 1,4-cyclohexadienes and cyclohexenes. This method is quite useful since aromatic compounds are plentiful.

Oxidation methods are less commonly used for the preparation of olefins.

Computer-assisted analysis for the synthesis of olefins, especially the stereoselective aspects, has been an active area of interest [1]. A data base of C=C transforms (retroreactives) has been computed to enable one to synthesize a wide variety of compounds. The data base information is shown in Table 1. For example, transform 431 shows that the preparation of the unsaturated ester can be seen as being

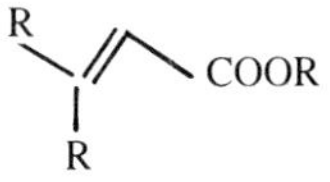

obtained by a Wittig reaction involving ketone (A) and bromoacetate ester (B).

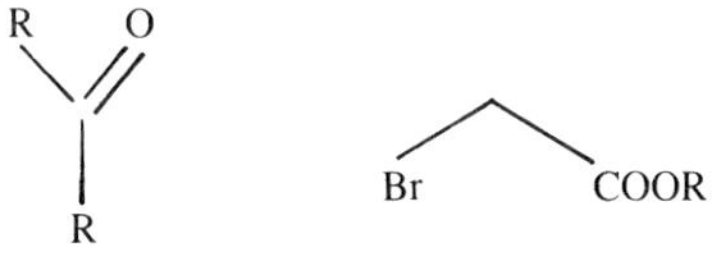

An earlier review on stereoselective and stereospecific olefin synthesis is worth consulting [2].

2. ELIMINATION REACTIONS

A. Dehydration of Alcohols

The acid-catalyzed or thermal elimination of water from alcohols is a favorite laboratory method for the preparation of olefins. Isomeric mixtures usually arise

TABLE I

The Following Transforms Form the Data Base for the Olefin Package in LHASA[a]

TRANSFORM 410
CLAISEN REARRANGEMENT
TL 3243(1969), JACS 92 741,4461,4463(1970); 95 553(1973)

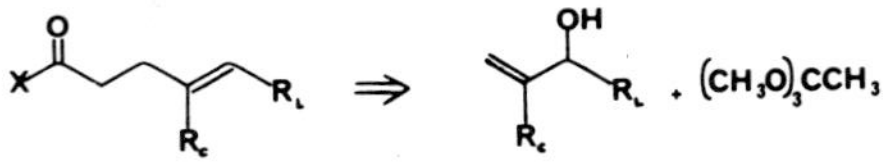

TRANSFORM 411
ACETAL CLAISEN REARRANGEMENT
TL 3469(75)

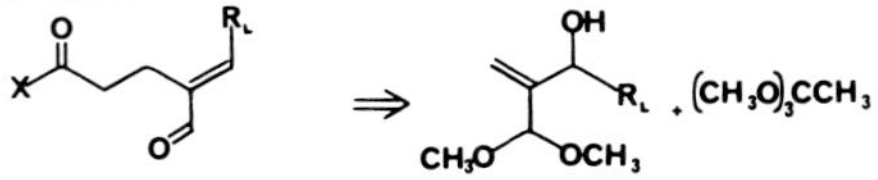

TRANSFORM 412
TRICHLOROACETIMIDATE REARRANGEMENT
JACS 96 597(74), 98 2901(76)

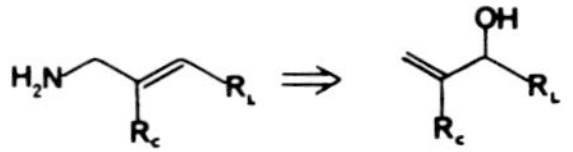

TRANSFORM 413
1,4 ADDITION-ELIMINATION
CHEM LETT 1097(1973), TL 925(1974)

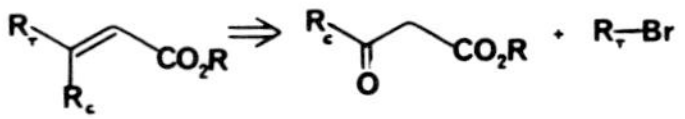

TRANSFORM 414
ALPHA-CHLORO ALDEHYDE ELIMINATION
TL 2465(73)

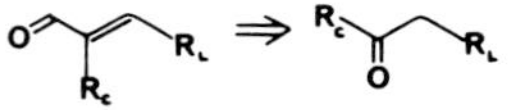

TRANSFORM 415
DIRECTED ALDOL CONDENSATION
NEUERE METHODEN DER PRAP. ORG CHEM, BAND VI, P. 42 (70)
ACIE 7 7(68)

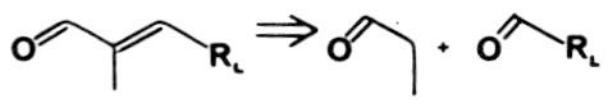

TRANSFORM 420
ORGANOCUPRATE ADDITION TO PROPARGYLIC ESTER
JACS 94 4395(1972), TL 1277,1281(1973); JOC 38 2733(1973)
JACS 97 1197(75), JOC 41 3629(76)

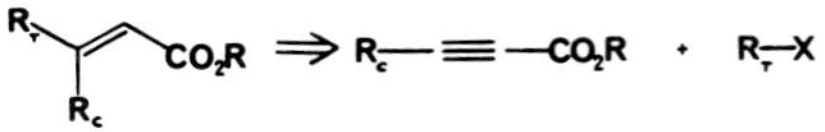

TRANSFORM 421
ELECTROPHILIC ATTACK BY VINYL COPPER COMPLEX
SYN 245(76), TET 32 1675(76), TL 2023(77)
J ORGMET CHEM 40 C49(72), 77 269,281(74)

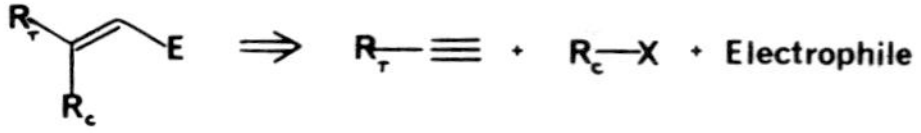

TRANSFORM 422
CONJUGATE ADDITION BY VINYL COPPER REAGENT
JACS 99 253 (77)

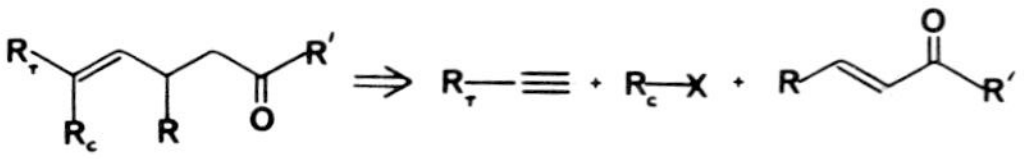

TRANSFORM 423
ORGANOCOPPER ADDITION TO PROPARGYLIC ACETAL
TL 2313 (76)

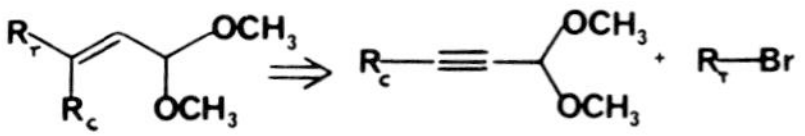

TRANSFORM 424
LAH REDUCTION OF PROPARGYLIC ALCOHOL
JACS 89 4245(1967), JOC 38 2733(1973), TL 1983(1973)

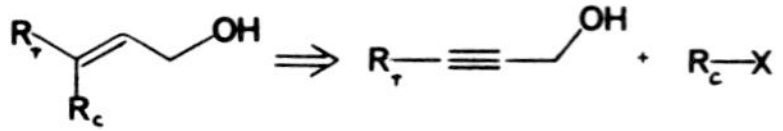

TRANSFORM 425
R2AlH REDUCTION OF PROPARGYLIC ALKOXIDE
JACS 92 6314(1970)

TRANSFORM 426
EPOXIDE OPENING BY ALKYNYL BORATE
TL 2741(1973), TET 30 3037(74), BUT CHEM LETT 397(75)

TRANSFORM 427
ALKYLATION OF ACETYLENIC BORATE
TL 795,4491(73); 3327(75)

TRANSFORM 428
DOUBLE WITTIG WITH FORMALDEHYDE
JACS 92 226,6635-7(1970) ; TL 3231(77)

TRANSFORM 429
DOUBLE WITTIG WITH TWO ALDEHYDES
JACS 92 226,6635-7(1970) ; TL 3231(77)

TRANSFORM 416
JULIA SYNTHESIS
JACS 90 2882(1968), TL 111(1973)

TRANSFORM 417
JULIA SYNTHESIS ON TERTIARY CARBINOL
TL 3445 (1974)

TRANSFORM 418
JULIA SYNTHESIS ON TERTIARY CARBINOL
TET. 28 3739(1972)

TRANSFORM 419
JULIA SYNTHESIS ON TERTIARY CARBINOL
TL 745(76)

(*continued*)

TABLE I (*Continued*)

TRANSFORM 430
EMMONS-WADSWORTH-HORNER WITH ALDEHYDE
BCSJ 40 2968(1967), JACS 83 1733(1961), CHEM REV 87(74)

TRANSFORM 431
EMMONS-WADSWORTH-HORNER WITH KETONE
JACS 90 3769(1968), CHEM REV 87(74)

TRANSFORM 432
WITTIG REACTION TO GAMMA-HALO TIGLATE
TL 1679(75), 167(77)

TRANSFORM 433
ALLYLIC REARRANGEMENT WITH SOCL2
BSCF 3568(1969), PNAS 68 1294(1971),
JACS 92 737, 4461(1970), 95 7067(1973)

TRANSFORM 434
ALLYLIC SULFOXIDE REARRANGEMENT
CHEM COMM 702 (72), TL 1389(73), ACR 7 147 (74)
JOC 38 2245, 2572(73)

TRANSFORM 440
CLAISEN REARRANGEMENT OF ACETYLENIC ALCOHOL
TL 2607 (76)

TRANSFORM 441
ALPHA-KETO SULFOXIDE ELIMINATION
JACS 98 4887 (76)

TRANSFORM 442
BETA-DIKETONE ENOL ETHER TRANSPOSITION
JACS 98 2351 (76)

TRANSFORM 443
ALKYLATION OF DIMETHYLTHIO-ALLYLLITHIUM
JACS 93 1724 (1971)

TRANSFORM 444
MODIFIED FAVORSKI REARRANGEMENT
PROC CHEM SOC 148 (64)

TRANSFORM 435
ALLYLIC SULFOXIDE REARRANGEMENT OF UNSAT. ESTER
TL 4215 (76)

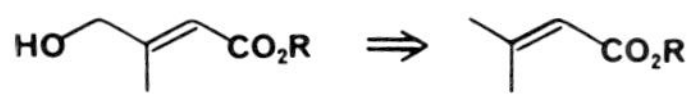

TRANSFORM 436
ALLYLIC REARRANGEMENT OF VINYL-WITHDRAWING GROUP
TL 2751,2755 (74); JACS 98 3384 (76)

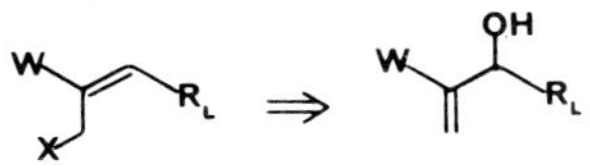

TRANSFORM 437
SELENIUM DIOXIDE OXIDATION
JACS 94 7154(72), 93 4835(71)

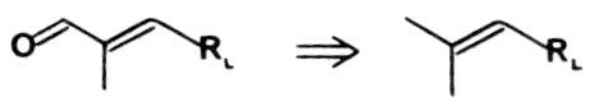

TRANSFORM 438
THIO-CLAISEN REARRANGEMENT
JACS 95 2693 (1973)

TRANSFORM 439
CYCLIC ORTHOESTER CLAISEN REARRANGEMENT
TL 847 (74)

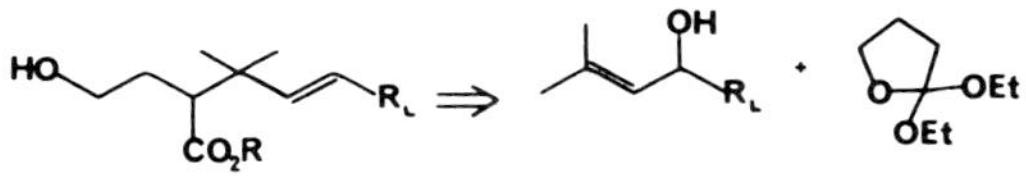

TRANSFORM 445
ACETYLENE REDUCTION
MARCH 361,593; HOUSE 19,91,124,172,205,252; B+P 106,162
ORGANOMET. CHEM SYN 1 249(71),CHEM COMM 17,452(76)
TL 1815(72), 1927(76); JACS 96 316(74)
JOC 41 221,2215, 3484(76), 42 579(77)
SYN 525,816(76)

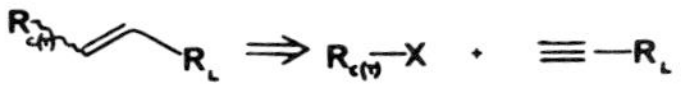

TRANSFORM 446
HYDROBORATION OF TERMINAL ACETYLENE
JACS 95 5786,6456 (1973)

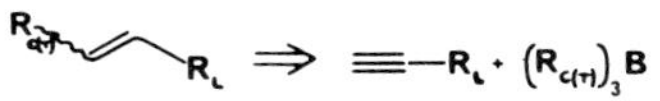

TRANSFORM 447
CONJUGATE ADDITION OF VINYL ALUMINUM COMPLEX
CJC 51 2098 (1973)

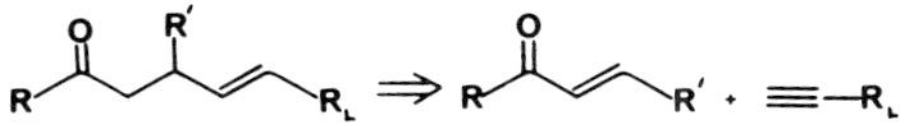

TRANSFORM 448
ALKYLATION OF VINYL CUPRATE
TL 2583 (71), 3461 (75)
J ORGMET CHEM 40 C49 (72), 77 269 (74)

TRANSFORM 449
WITTIG REACTION WITH REACTIVE YLIDE
MARCH 702; HOUSE 682-709; B+P 141
JACS 98 5653(66), 89 2758(67), 91 5675(69)
ACIE 4 689(65), 5 126(65); ANN 708 1(67)

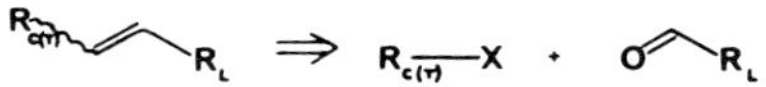

---

[a] Reprinted from E. J. Corey and A. K. Long, *J. Org. Chem.* **43**, 2208 (1978). Copyright (1978) by the American Chemical Society. Reprinted by permission of the copyright owner.

with the acid-catalyzed method. The order of reactivity in dehydration usually follows the order of stability of the intermediate (transient) carbonium ion, i.e., tertiary > secondary > primary. The acid-catalyzed procedure is illustrated below, where 79–87% yield of cyclohexene is obtained [1–3].

Tertiary arylcarbinols have been reported to be converted within 30 sec to the corresponding alkenes (70% yield) with warm 20% sulfuric–acetic acid (by volume) [4]. The yields are much lower with aliphatic tertiary or secondary arylcarbinols. Some tertiary alcohols, such as those obtained from tetralone and the Grignard reagent, dehydrate on simple distillation and in the presence of anhydrous cupric sulfate as a catalyst [5].

Secondary and tertiary alcohols can be dehydrated in dimethyl sulfoxide when heated to 160°–185°C for 14–16 hr to give olefins in yields of 70–85% [6]. The solution is diluted with water, extracted with petroleum ether (30°–60°C), dried, and then distilled. Other acid catalysts that have been reported for dehydration of alcohols are: anhydrous or aqueous oxalic [7, 8], phosphoric [9], and potassium acid sulfate [10, 11]. In addition, acidic oxides such as phosphorus pentoxide [10–12] and acidic chlorides such as phosphorus oxychloride or thionyl chloride [13] have been reported to be effective as catalysts for the dehydration reaction.

2-1. Preparation of Cyclohexene [3]

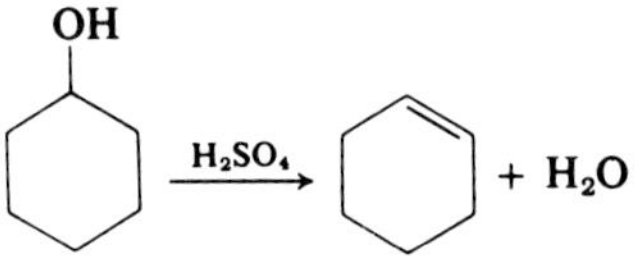

$$\text{cyclohexanol} \xrightarrow{H_2SO_4} \text{cyclohexene} + H_2O \qquad (9)$$

To 400 gm (4 moles) of cyclohexanol, in a flask set up for distillation of the contents into an ice-cooled receiver, is added 12 ml of concentrated sulfuric acid. The flask is heated to 130°–150°C by means of an oil bath for 5–6 hr. The cyclohexene is salted out of the distillate, dried, and fractionated to give 260–285 gm (78–87%) bp 80°–82°C.

Aluminum oxide vapor phase dehydrations at 300°–400°C have the advantage that isomerization is reduced as a result of the short contact time of the alcohol and olefin with the catalyst. The main by-products are ethers [14, 15]. The dehydration of alcohols on an alumina catalyst may involve the preliminary formation of a surface alkoxide which then thermally decomposes to an olefin [16]. By this procedure 1-butene is prepared from n-butyl alcohol [17] and styrenes are obtained from arylmethylcarbinols [18].

2-2. Preparation of Dimethylstyrenes [19]

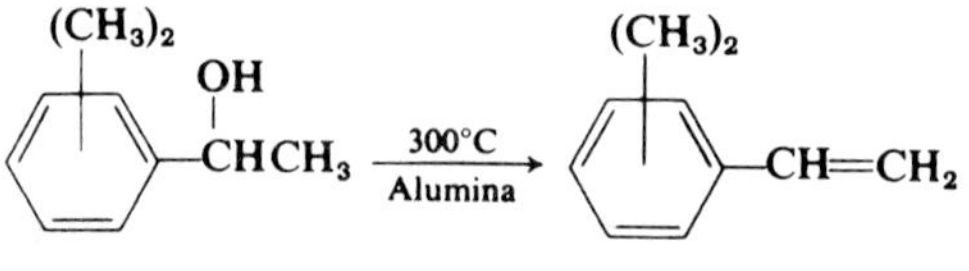

$$(CH_3)_2\text{-Ar-CH(OH)CH}_3 \xrightarrow[\text{Alumina}]{300°C} (CH_3)_2\text{-Ar-CH=CH}_2 \qquad (10)$$

§ 2. Elimination Reactions

(a) Preparation of the dehydration column [19]. A 2½ ft Pyrex column, 1 inch in diameter, having appropriate 24/40 ground glass joints at either end, is wound with two layers of moist asbestos paper. The column is then carefully wound with 30–35 turns of nichrome wire $\frac{5}{32} \times 0.0063$ B & S 22, 0.899 ohm/ft, supplied by the Driver and Harris Co., Harrison, New Jersey 07029) as tightly as possible. The nichrome wire is then covered with two layers of asbestos paper. The asbestos is then covered with aluminum foil and the latter is covered with glass fiber cloth. The electrical connections are made by connecting alligator clamps to the terminal nichrome wires and these in turn are connected to a standard two-prong plug. The current is controlled by means of a variable transformer (5–6 amp capacity); such a column gives an internal temperature, after being packed with alumina, of over 300°C. The assembled dehydration apparatus is shown in Fig. 1. (CAUTION: Asbestos is a known carcinogen.)

(b) Dehydration. The 1-(dimethylphenyl)ethanols (obtained by the sodium borohydride reduction of the dimethyl acetophenones) are dehydrated by dropping the organic liquid into a flask heated to 450°–500°C. The vapors are then led through the 2½ ft column packed with activated alumina (alumina Catalyst AL-0104T $\frac{5}{32}$ inch, The Harshaw Chemical Co.) and heated to 300°C under vacuum. The crude material is dried and vacuum-distilled through a 3-ft Vigreaux column. The physical constants of the products are as follows: 3,4-dimethylstyrene, bp 63°–64°C (5 mm), n_D^{27} 1.5405; 2,5-dimethylstyrene, bp 47°–48°C (3 mm), n_D^{27} 1.5382; 2,4-dimethylstyrene, bp 47°–51°C (3 mm), n_D^{27} 1.5435.

More recently α-alkylbenzyl alcohols have been reported to be converted in high yield and purity to styrene and substituted styrenes by contacting the alcohol in the vapor phase with silica gel and steam [20]

$$H_3C-CH(OH)-C_6H_5 \xrightarrow[\text{silica gel}]{\substack{\text{toluene,}\\\text{steam}\\300°C}} H_2C=CH-C_6H_5 \quad (11)$$

95% Yield, 95% Purity

The selective dehydration of secondary alcohols to 1-olefins can be effected using thorium oxide or lanthanide metal oxides in yields seldom below 95% at 350°–450°C [21]. These catalysts differ in behavior from the alumina or chromium oxide catalysts which give 2-olefins from secondary alcohols. For example, 4-methyl-2-pentanol gives 4-methyl-1-pentene in 98% yield using a thorium oxide catalyst. Therefore acetaldehyde may be reacted with a Grignard reagent to yield secondary alcohols which may then be dehydrated to the desired 1-olefin.

Dehydrations of alcohols can also occur under basic conditions as in the case

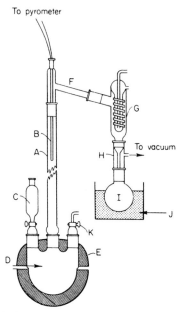

Fig. 1. Assembled dehydration apparatus. A, dehydrating column; B; thermocouple well; C, dropping funnel; D, Pyrex flask, E, heating mantles; F, Claisen head; G, Friedrichs condenser; H, connecting tube; I, flask; J, ice bath; K, stopcock.

of β-phenylethyl alcohols. Molten sodium or potassium hydroxide and the alcohol are heated to 140°C to give the styrenes in good yields [22]. However, the phenylmethylcarbinols require acidic catalysts for successful dehydration to the styrene. Reaction of ethylene oxide with xylene in the presence of aluminum chloride gives good yields of β-xylylethyl alcohols, which dehydrate in good yields to dimethylstyrenes using molten potassium hydroxide [19].

Aliphatic alcohols can also be dehydrated in low yields under basic conditions in the presence of bromoform [23]. The fact that the same olefin mixtures are obtained as those produced by the reaction of nitrous acids on amines indicates that this reaction may proceed via a carbonium ion mechanism [24]. Whether this reaction can be further developed into a useful synthetic method is yet to be determined.

$$ROH + CHBr_3 + Base \longrightarrow [R^+] \longrightarrow Olefins \qquad (12)$$

α-Cyclopropylcarbinols have been reported by Julia [25, 26] to be cleaved by acid in a stereoselective *trans*-manner.

§ 2. Elimination Reactions

$$R^2 \underset{OH}{\overset{}{\underset{|}{\overset{|}{C}}}} R^1 \longrightarrow \underset{R^2}{\overset{Br}{\diagdown}} \underset{R^1}{\diagup} \quad (13)$$

In the case where $R^1 = nC_4H_9$ and $R^2 = CH_3$, the use of 48% HBr gives a mixture of trans/cis isomers in the ratio of 3:1 [27]. However, in the case of R^1 = alkyl and $R^2 = H$, the trans-olefin is obtained in 90–95% stereoselectivity [25].

Recently Ashby, Willard, and Goel reported [28] a new, convenient, and stereospecific method for the dehydration of alcohols by the thermal decomposition of its magnesium, zinc, and aluminum alkoxides. The reaction proceeds in a manner similar to the pyrolysis of esters or the Chugaev reaction of xanthates (syn elimination to give the olefin) as described in Section B.

B. Pyrolysis Reactions

Although the vapor phase dehydration of alcohols over alumina or other solid catalysts is superior in some ways to the acid dehydration in solution, the high temperatures of 300°–500°C limit its usefulness to olefins that are stable at these temperatures for the relatively brief time during which the product is exposed.

The pyrolysis of acetate esters [29–31] is usually carried out at 300°–600°C, but that of the xanthates employed in the Chugaev [32] method requires only 100°–250°C. Esters of boric acid appear to be pyrolyzed easily at 260°–270°C

$$RCH_2CH_2OH + CH_3COCl \longrightarrow RCH_2CH_2O\overset{O}{\overset{\|}{C}}CH_3 \longrightarrow RCH=CH_2 \quad (14)$$

$$RCH_2CH_2OH \xrightarrow[CS_2]{NaOH} RCH_2CH_2OCS_2Na \xrightarrow{CH_3I}$$

$$RCH_2CH_2OCS_2CH_3 \xrightarrow[\text{Reaction}]{\text{Chugaev}} RCH=CH_2 \quad (15)$$

$$3\, RCH_2CH_2OH + H_3BO_3 \longrightarrow (RCH_2CH_2O)_3B \xrightarrow{\text{Heat}} 3\, RCH=CH_2 \quad (16)$$

[33–35]. An improved procedure for the pyrolysis of esters has recently been described using a recycling apparatus [36].

The pyrolysis of xanthates by the Chugaev reaction is limited mainly to secondary alcohols. Primary alcohols give only low yields and the use of this method for tertiary alcohols is less common.

The Chugaev reaction suffers from the disadvantage that the preparation of the xanthates is more difficult than that of the esters, especially during purification. The pyrolysis of xanthates also yields sulfur-containing contaminants which are sometimes difficult to separate. The main advantages are the use of low temperatures for the pyrolysis under basic conditions and the absence of rearrangements.

The potassium xanthate salts of some tertiary alcohols have been reported to be easily prepared and pyrolyzed at 200°–250°C to yield olefins [37].

In a related process Ashby and co-workers [38] reported a new convenient and stereospecific method for the dehydration of metal alkoxides to olefins.

Carbonate and carbamate [39–41] esters are pyrolyzed at temperatures between those used for esters and xanthates. The carbonates yield carbon dioxide and an alcohol on pyrolysis, which offers the advantage that these do not contaminate the olefin product. The carbamates give the olefin, an amine, and carbon dioxide.

Sulfites [42] decompose at temperatures similar to xanthates.

Ethers [43] eliminate a molecule of alcohol when heated to about 300°C in the presence of alumina, or at 60°–100°C over phosphorus pentoxide. Acetals [43, 44] lose alcohols when heated to 147°–170°C in the presence of phosphorus pentoxide, quinoline, or with phthalic anhydride.

The pyrolysis of quaternary ammonium salts (Hofmann exhaustive methylation) [45] is a useful method mainly for proof of structure since carbon skeleton rearrangement does not occur. However, as a preparative method it affords good yields when three of the alkyl groups are methyl radicals.

$$\underset{\underset{CH_3}{|}}{RCH}-\overset{+}{N}(CH_3)_3 OH^- \longrightarrow RCH=CH_2 + (CH_3)_3N + H_2O \qquad (17)$$

where $R = C_3H_7$ the yield is 63% [46] from the iodide salt and using sodium ethoxide. Substitution of t-butoxide increases the yield to 84% (97% pure olefin) [47]. The elimination is only possible when the quaternary ammonium hydroxide contains a β-hydrogen atom [48, 49].

A reaction related to the pyrolysis of quaternary ammonium salts is the pyrolysis of Mannich bases or their hydrochloride salts at 120°C [50–56].

$$RCOCH_2R \xrightarrow[R_2NH, HCl]{CH_2=O} \underset{\underset{CH_2NR_2 \cdot HCl}{|}}{RCOCHR} \xrightarrow{Heat} \underset{\underset{CH_2}{\|}}{RCOC-R} + R_2NH \cdot HCl \qquad (18)$$

The Cope elimination of amine oxides has the advantage over the Hofmann elimination in that there is little or no isomerization of the olefin produced [57]. The reaction occurs usually at 120°–150°C but it also has been reported to take place at 25°C in dimethyl sulfoxide and tetrahydrofuran [58].

$$R_2CH-CR_2 \longrightarrow R_2C=CR_2 + (CH_3)_2NOH \qquad (19)$$
$$\underset{H_3C\overset{N}{\underset{O}{\downarrow}}CH_3}{}$$

Corey and Winter reported a novel olefin synthesis from 1,2-diols that involves conversion of the diol to the cyclic thionocarbonate, followed by treatment with trimethyl phosphite [59, 60].

§ 2. Elimination Reactions   51

$$\underset{\underset{OH}{|}\;\underset{OH}{|}}{>\!C\!-\!C\!<} \longrightarrow \underset{\underset{O\diagdown_{\underset{\|}{C}}\diagup O}{}}{>\!C\!-\!C\!<} \xrightarrow{(CH_3O)_3P} >\!C\!=\!C\!< \;+\; CO_2 \;+\; (CH_3O)_3P\!=\!S \quad (20)$$

The decarboxylation of olefinic acids (such as those obtained from the Perkin and related reactions) at 220°C in the presence of quinoline and copper powder gives olefins in yields up to 86% [61]. Thermal decomposition of cinnamic acids (without catalysts) gives styrenes in 41% yield [62].

$$ArCH\!=\!CH\!-\!COOH \xrightarrow[\substack{Cu\ powder\\220°C}]{Quinoline} ArCH\!=\!CH_2 \quad (21)$$

Nuclear substituents effect the ease of decarboxylation but halo, methoxy, cyano, and nitro styrenes have been prepared in yields ranging from 30% to 76% [61, 63, 64]. Unsaturated aliphatic acids also decarboxylate thermally, as is true for β-ethoxycrotonic acid [62]. *cis* and *trans*-Stilbenes have recently been obtained in good yield by the pyrolysis of α-phenylcinnamic acids [65].

2-3. Pyrolysis of the Acetate of p-Cyanophenylmethylcarbinol to p-Cyanostyrene [66].

$$\underset{CN}{\underset{|}{C_6H_4}}\!-\!\underset{\underset{}{}}{CH(CH_3)}\!-\!O\!-\!\underset{\underset{\|}{O}}{C}\!-\!CH_3 \xrightarrow{575°-600°C} \underset{CN}{\underset{|}{C_6H_4}}\!-\!CH\!=\!CH_2 \;+\; CH_3COOH \quad (22)$$

To the acetate of *p*-cyanophenylmethylcarbinol, 58 gm (0.307 mole), is added 1 gm of *p-tert*-butylcatechol. The material is dropped through a vertical 40 cm by 20-mm Pyrex tube packed with glass beads heated to 575°–600°C by means of an electric furnace. The addition rate is 1 drop per second. The product is collected in a chilled receiver, washed twice with 100-ml portions of water, and similarly with 100 ml of 10% sodium bicarbonate solution. The organic layer is dried, inhibited with a small amount of *p-tert*-butylcatechol, and distilled to yield 30 gm (76%) of *p*-cyanostyrene, bp 92°–93°C (3 mm); n_D^{20} 1.5772.

2-4. Chugaev Method—Preparation of tert-Butyl-1-cyclohexene [42]

To a solution of 78 gm (0.56 moles) of *cis*-2-*tert*-butylcyclohexanol in 400 ml of benzene is added 11.5 gm (0.5 gm atom) of sodium in small pieces. At the end of 4 hr, the sodium salt has formed. It has a gel-like consistency. To the gel is added 400 ml of benzene and then 40 gm (0.53 moles) of carbon bisulfide. After refluxing for 8 hr, 75 gm (0.53 mole) of methyl iodide is added and the refluxing

2. Olefins

$$\text{OH-cyclohexyl-C(CH}_3)_3 + \text{Na} \longrightarrow \text{ONa-cyclohexyl-C(CH}_3)_3 \xrightarrow{\text{CS}_2}$$

$$\text{OCS}_2\text{Na-cyclohexyl-C(CH}_3)_3 \xrightarrow{\text{CH}_3\text{I}} \text{OCS}_2\text{CH}_3\text{-cyclohexyl-C(CH}_3)_3 \longrightarrow \text{cyclohexenyl-C(CH}_3)_3 \quad (23)$$

is continued overnight. The next day the sodium iodide is filtered off and the solvent is evaporated to yield 91 gm of a yellow solid, mp 41°–43°C. Heating the solid under gentle reflux at 200°–205°C for 3 hr using a nitrogen stream over the liquid yields 30.0 gm (60%) of the product, bp 165°–170°C, n_D^{25} 1.4568.

2-5. Decarboxylation of p-Chlorocinnamic Acid to p-Chlorostyrene [61]

$$\text{Cl-C}_6\text{H}_4\text{-CH=CH-COOH} \xrightarrow[\text{Cu powder}]{\text{Quinoline}} \text{Cl-C}_6\text{H}_4\text{-CH=CH}_2 + \text{CO}_2 \quad (24)$$

To a Claisen or distilling flask are added 200 gm (1.1 mole) of p-chlorocinnamic acid, 400 ml of quinoline, and 20 gm of copper powder. The flask is heated with a Glass-Col heater so that the vapors remain below 220°C. One-third to two-thirds of the reaction mixture is distilled in about 1 hr. The end of the reaction is evidenced in the rise of temperature of the vapors to the boiling point of quinoline, p-chlorostyrene is obtained, 126 gm (83%), bp 60°–62°C (6.5 mm), n_D^{20} 1.5650. An excess of quinoline may be used if convenient. Other bases such as lepidine may also be used.

C. Dehydrohalogenation Reactions

The dehydrohalogenation reaction is complex because the nucleophile B, can remove the β-proton to produce elimination, it can attack the α-carbon to give the SN$_2$ product or give α-elimination. Normally α-elimination leads to the same product as obtained by β-elimination.

With simple unbranched alkyl halides, the use of alcoholic bases gives the Saytzeff olefin. Steric effects on the β-position usually increases Hofmann elimination.

$$2\ \text{CH}_3\text{CH}_2\text{CH}_2\text{CH}_2\text{CH}_2\text{Br} + \text{C}_2\text{H}_5\text{OK} \xrightarrow{\text{C}_2\text{H}_5\text{OH}} \text{CH}_3\text{CH}_2\text{CH=CH-CH}_3$$

69% Saytzeff olefin
(18% *cis*, 51% *trans*)

$$+\ \text{CH}_3\text{CH}_2\text{CH}_2\text{CH=CH}_2$$

31% Hofmann olefin (25)

§ 2. Elimination Reactions

$$(CH_3)_3C-CH_2-\underset{Br}{\underset{|}{\overset{CH_3}{\overset{|}{C}}}}-CH_3 + C_2H_5-OK \xrightarrow{C_2H_5OH} (CH_3)_3C-CH_2-\overset{CH_3}{\underset{|}{C}}=CH_2 \quad (26)$$

81%

2-6. Preparation of 3-Chloro-2-methyl- and 3-Chloro-4-methyl-α-methylstyrene [67]

$$\text{Cl-}\underset{}{\text{Ar}}(\text{CH}_3)\text{CH-CH}_2\text{Cl} \xrightarrow{\text{KOH}}_{\text{CH}_3\text{OH}} \text{H}_3\text{C-}\underset{}{\text{Ar}}(\text{Cl, CH}_3)\text{C}=\text{CH}_2 + \underset{}{\text{Ar}}(\text{Cl, CH}_3, \text{CH}_3)\text{C}=\text{CH}_2 \quad (27)$$

Four hundred and eight grams (2.0 moles) of chloropropylated o-chlorotoluene (from propylene chlorhydrin, o-chlorotoluene, boron trifluoride, and phosphorous pentoxide) is refluxed with an 85% solution of potassium hydroxide in methanol [392 gm (7 moles) of KOH in 1850 ml of methanol]. The methanol is removed by distillation and the remaining liquid is washed with water, dried with calcium chloride, and distilled through an efficient column under reduced pressure to give 90 gm (26%) of 3-chloro-2-methyl-α-methylstyrene, bp 64°–65°C (4 mm), n_D^{25} 1.5340 and 152 gm (48%) of 3-chloro-4-methyl-α-methylstyrene, bp 73°–74°C (4 mm), n_D^{25} 1.5520. The purity of these materials should be checked by vapor phase chromatography.

Dehydrohalogenation can also occur during decarboxylation reactions to give olefins [68]

$$\underset{R'}{\underset{|}{RCH=C}}-COOH + HX \longrightarrow \underset{X\ \ R'}{\underset{|\ \ \ |}{RCH-CH}}-COOH \xrightarrow{NaCO_3} RCH=CHR' \quad (28)$$

Although alcoholic potassium hydroxide is the most common base for the dehydrohalogenation reaction, the following other bases have been reported to be effective: triethylamine [69], pyridine [70], sodium acetate [71], silver oxide [72], potassium t-butoxide [73] and lithium N-alkylamides [74].

Trimethyl phosphite has been used to dehydrohalogenate 8 α- and 7 β-bromocholesteryl benzoate to give 56% of 7-dehydrocholesteryl benzoate [75].

The rate of elimination can be greatly accelerated in aqueous systems using phase transfer catalysts such as benzyltriethylammonium chloride [76, 77].

The Boord synthesis [78] is an interesting method for converting an aldehyde of the type RCH_2CHO into an olefin of the type $RCH=CR'R''$.

$$RCH_2CH=O \xrightarrow[HCl]{C_2H_5OH} \underset{Cl}{\underset{|}{RCH_2-CH}}-OC_2H_5 \xrightarrow{Br_2} \underset{Br\ \ Br}{\underset{|\ \ \ |}{RCH-CH}}-OC_2H_5 \quad (29)$$

2. Olefins

$$RCHCH-OC_2H_5 \xrightarrow{R'MgBr} \underset{\underset{Br}{|}\;\underset{R'}{|}}{RCH-CH-OC_2H_5} \xrightarrow[C_2H_5OH]{KOH} \underset{\underset{R'}{|}}{RCH=C-OC_2H_5} \quad (30)$$
(with Br, Br below first structure)

$$\underset{\underset{R'}{|}}{RCH=COC_2H_5} \xrightarrow{Br_2} \underset{\underset{Br}{|}\;\underset{Br}{|}}{RCH-CR'-OC_2H_5} \xrightarrow{R''MgBr} \underset{\underset{Br}{|}\;\underset{R''}{|}}{RCH-CR'-OC_2H_5} \quad (31)$$

$$\underset{\underset{Br}{|}\;\underset{R''}{|}}{RCH-CR'-OC_2H_5} \xrightarrow{Zn\;dust} RCH=CR'R''$$

Of course, at the intermediate stage [Eq. (30)] the product may be converted to an olefin of the type RCH = CHR′.

$$\underset{\underset{Br}{|}\;\underset{R'}{|}}{RCH-CH-OC_2H_5} \xrightarrow[\substack{Propyl \\ alcohol}]{Zn\;dust} RCH=CHR' \quad (32)$$

The Boord synthesis combines Grignard coupling, dehydrogenation, and dehalodealkoxylation with zinc dust. The reaction can be modified to give either mono- [79], di- [80], or trisubstituted [81] olefins. The method has also been applied to the synthesis of 1,4-diolefins which are not easily obtained in pure isomeric form by ordinary dehydrohalogenation techniques.

2-7. Boord Method—Preparation of 1,4-Hexadiene [81]

(a) *Preparation of α-chloropropyl ethyl ether.*

$$CH_3CH_2CH=O + HCl + C_2H_5-OH \longrightarrow \underset{\underset{Cl}{|}}{CH_3CH_2-CH-OC_2H_5} \quad (33)$$

Equimolar amounts of propionaldehyde (58 gm) and absolute ethyl alcohol (46 gm) are placed in a short-stemmed separatory funnel which is immersed in a freezing mixture. Hydrogen chloride gas is carefully passed into the solution in such a manner that it does not pass through the aqueous layer which is slowly forming. This technique avoids violent agitation. The addition is stopped after there is a 5% excess of the theoretical gain in weight. The excess hydrogen chloride is removed at reduced pressure and the water white product is not distilled but used directly in the next step.

(b) *Bromination of α-chloropropyl ethyl ether.*

$$\underset{\underset{Cl}{|}}{CH_3-CH_2-CH-OC_2H_5} + Br_2 \longrightarrow \underset{\underset{Br}{|}\;\underset{Br}{|}}{CH_3CH-CH-OC_2H_5} + HCl \quad (34)$$

In a hood, the α-chloropropyl ether is cooled in an ice bath and 160 gm (1 mole) of bromine is added very slowly so that decolorization occurs before each

§ 2. Elimination Reactions

addition. The reaction is rapid but slows down at the end. The evolved gas is mainly hydrogen chloride. Distillation of the crude material under reduced pressure yields 90–97% of the α,β-dibromopropyl ethyl ether, bp 79°–82°C (20 mm), n_D^{20} 1.5000.

(c) Condensation with allylmagnesium bromide.

$$CH_3-\underset{Br}{CH}-\underset{Br}{CH}-OC_2H_5 + CH_2=CH-CH_2MgBr \longrightarrow CH_3-\underset{Br}{CH}-\underset{CH_2-CH=CH_2}{CH}-OC_2H_5$$

(35)

To a three-necked flask are added 75 gm of magnesium turnings and 200 ml of anhydrous ether. The stirred reaction mixture is cooled to 0°–5°C and 121 gm (1.0 mole) of allyl bromide in 570 ml of anhydrous ether is added dropwise over a period of 8–9 hr. In order to initiate the reaction a few pieces of magnesium are crushed under ether in a test tube and an ether solution of allyl bromide added. When the reaction starts, the test tube contents are added to the flask. When the addition of allyl bromide has been completed, the reaction mixture is allowed to remain at room temperature for ½ hr. The allylmagnesium bromide is filtered from the unused magnesium and placed in a 2-liter three-necked flask. While cooling with ice, α,β-dibromopropyl ethyl ether dissolved in an equal volume of ether is added slowly to a slight excess of the Grignard reagent. After the addition has been completed, the mixture is stirred for an additional 2 hr. The contents are hydrolyzed by pouring into a beaker of cracked ice containing dilute hydrochloric acid or ammonium chloride. The ether layer is separated, dried, distilled to remove ether, and the residual oil is subjected to steam distillation. The desired β-bromo ether separates in the distillate. Separation of the oil, drying over sodium hydroxide, and distillation under reduced pressure yields the product in 38–43% yield, bp 72°–75°C (15 mm), n_D^{20} 1.4592.

(d) 1,4-Hexadiene.

$$CH_3CH-\underset{Br}{CH}-\underset{CH_2-CH=CH_2}{CH}-OC_2H_5 + Zn \longrightarrow CH_3CH=CH-CH_2-CH=CH_2 + ZnBr(OC_2H_5)$$

(36)

To 100 gm (0.524 mole) of α-allyl-β-bromopropyl ethyl ether and 225 ml of *n*-propyl alcohol is added 100 gm (1.54 gm atom) of zinc dust. The mixture is heated to reflux for 15 hr. The contents are then distilled to yield 38 gm of a crude distillate. The distillate is washed five times with ice water (one-half of its volume of water each time), dried with calcium chloride, and distilled twice from metallic sodium to remove absorbed alcohol. The product bp 64.3°–64.6°C (745mm), n_D^{20} 1.4162 is obtained in 67% yield.

a. REACTIONS OF 1,1-DIHALOCYCLOPROPANES—INSERTION OF A CARBON ATOM BETWEEN THE ATOMS OF A DOUBLE BOND [82–87]

Reacting olefins in the presence of chloroform or bromoform with potassium *tert*-butoxide yields 1,1-dihalocyclopropanes. Heating the latter compounds with aqueous silver nitrate or silver acetate-acetic acid gives halo allyl alcohols or halo allyl acetates, respectively.

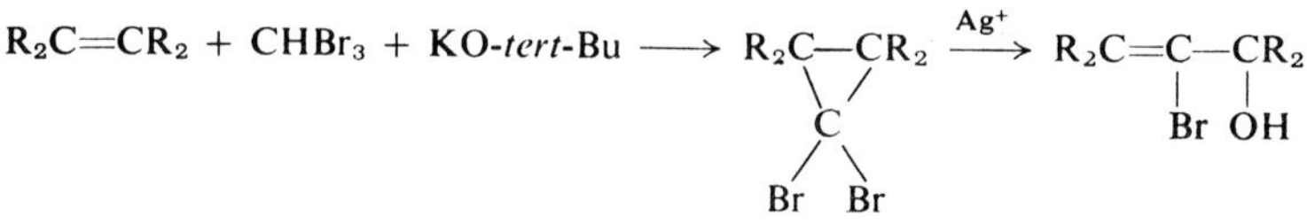

$$R_2C=CR_2 + CHBr_3 + KO\text{-}tert\text{-}Bu \longrightarrow R_2C\text{---}CR_2 \underset{Br\ Br}{\overset{C}{\diagdown\diagup}} \xrightarrow{Ag^+} R_2C=C\text{---}CR_2 \quad (37)$$
$$\underset{Br\ OH}{}$$

This reaction affords a general method for extending the carbon chain through insertion of a carbon atom between the olefinic double bond [82–87]. References to related reactions have been reported [84–87].

Cyclopentene yields 2-bromo- or 2-chloro-2-cyclohexene-1-ol when prepared from the 1,1-dibromo- or 1,1-dichlorobicyclo[3.1.0]hexane, respectively. Analogous results are obtained with the cyclohexene series giving 2-halo-2-cycloheptene-1-ol. When the R groups in the olefin are methyl groups, the reactivity is increased. In addition, as a result of steric strain, the bicyclohexane series is more reactive than the bicycloheptanes.

The pyrolysis and solvolysis of the *gem*-dihalocyclohexanes is dependent on the stereochemistry of these compounds. In the case of the bicyclo[3,1,0] adducts (I) and (II), the silver-ion catalyzed solvolysis promotes loss of the endo-oriented halogen and produces the ring-expanded halo allylic alcohol [84–87].

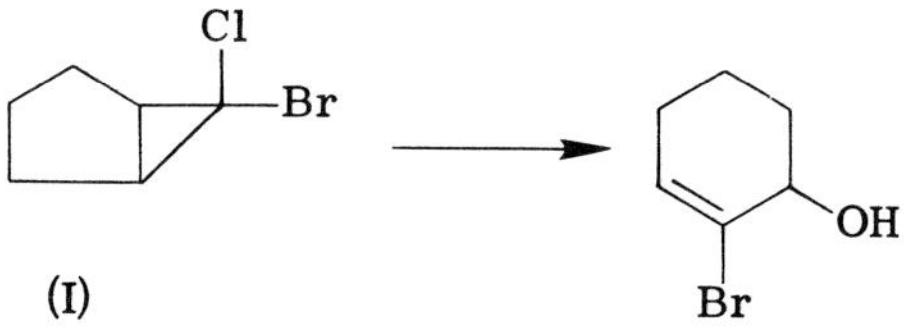

(I) → (cyclohexenol with Br and OH)   (38)

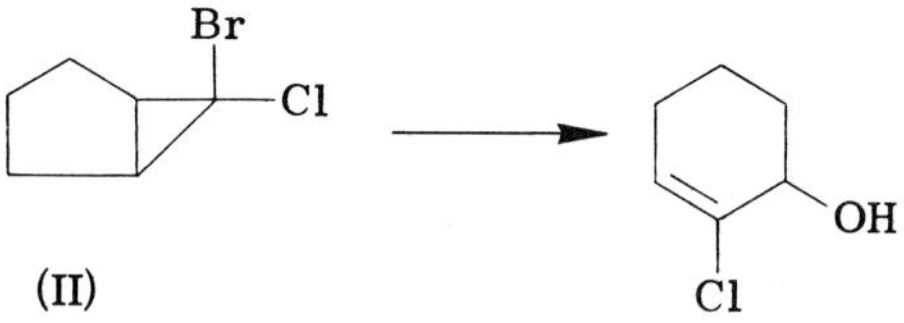

(II) → (cyclohexenol with Cl and OH) (39)

Pyrolysis of the 1,1-dihalocyclopropanes also gives halo allyl halides, which subsequently can react to give allyl alcohols, ethers, etc. [82–91].

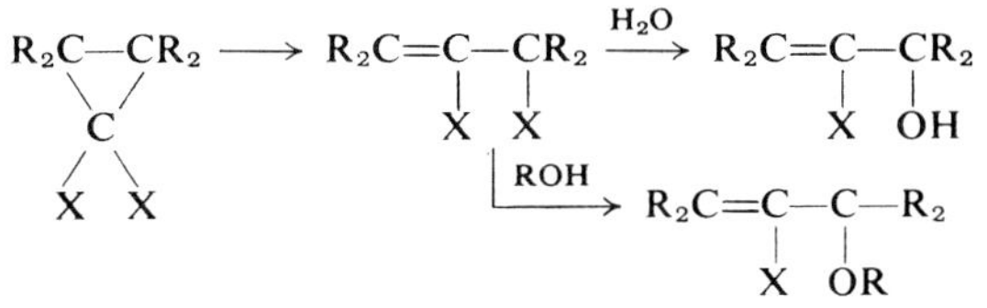

§ 2. Elimination Reactions

$$R_2C\!-\!CR_2 \longrightarrow R_2C\!=\!C\!-\!CR_2 \xrightarrow{H_2O} R_2C\!=\!C\!-\!CR_2$$
$$\underset{X\ \ X}{\overset{\diagdown\ \diagup}{C}} \qquad \underset{X\ \ X}{|\ \ |} \qquad \underset{X\ \ OH}{|\ \ |}$$
$$\xrightarrow{ROH} R_2C\!=\!C\!-\!C\!-\!R_2 \quad (40)$$
$$\underset{X\ \ OR}{|\ \ |}$$

2-8. *Preparation of 2-Chloro-3-hydroxycyclohexene from Cyclopentene* [92, 93]

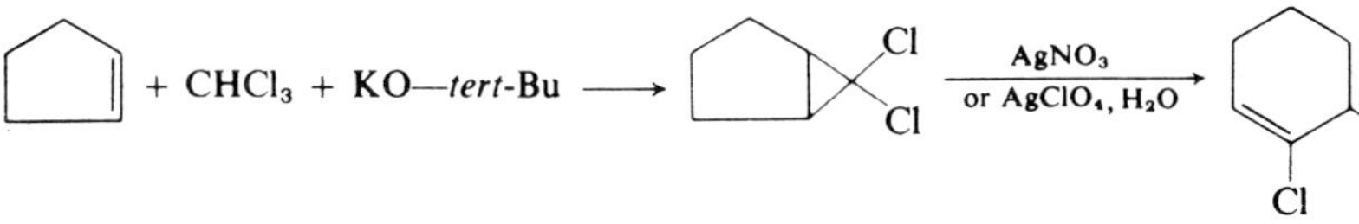

[cyclopentene] + CHCl$_3$ + KO—*tert*-Bu ⟶ [6,6-dichlorobicyclo[3.1.0]hexane] $\xrightarrow[\text{or AgClO}_4,\,\text{H}_2\text{O}]{\text{AgNO}_3}$ [2-chloro-3-hydroxycyclohexene] (41)

To a cooled (0°–5°C) flask containing 235.2 gm (2.1 moles) of freshly prepared, or a good commercial grade of, solid potassium *tert*-butoxide is added 204 gm (3.0 moles) of cyclopentene. Chloroform (298 gm, 2.5 moles) is added over a 2-hr period while the temperature is kept at 5°–10°C. The organic layer is then washed with water, dried, and distilled to yield 6,6-dichlorobicyclo-[3.1.0]hexane, bp 87°–90°C (61 mm), n_D^{27} 1.4907–1.4941, 39% based on cyclopentene.

To 34.9 gm (0.23 mole) of the above dichloride is added 100 ml of water containing 88 gm (0.425 mole) of dissolved silver perchlorate. The mixture is stirred for 5½ hr at 90°–100°C and filtered. The organic material is isolated by extracting with ether, drying, and distilling. The product, 2-chloro-3-hydroxycyclohexene is isolated in 25.6 gm (84%) yield, bp 80°C (11 mm), n_D^{23} 1.5093.

The dibromocyclopropanes are much more reactive and thus give better conversions to product. The dihalocyclopropanes described in Chapter 6 may all be rearranged to the allylic alcohol or acetate by a procedure analogous to that described above [92]. For additional references to this reaction, see Ref. [92].

2-9. *Preparation of 1,1-Diphenyl-2-bromo-3-acetoxy-1-propene* [85]

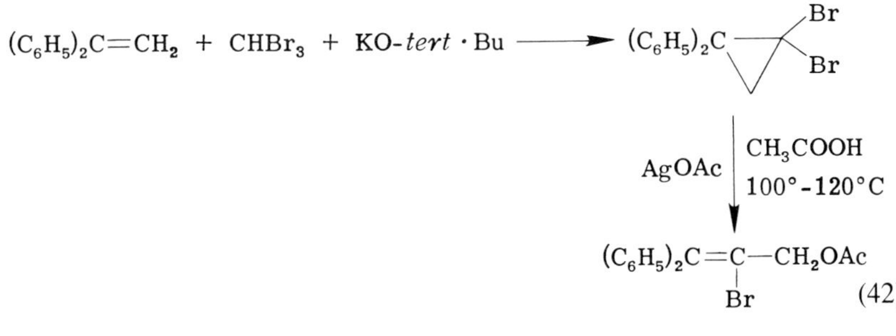

$$(C_6H_5)_2C\!=\!CH_2 + CHBr_3 + KO\text{-}tert\cdot Bu \longrightarrow (C_6H_5)_2C\!\underset{Br}{\overset{Br}{\diagup\!\diagdown}}$$

$$\Bigg\downarrow \text{AgOAc} \quad \begin{array}{l} CH_3COOH \\ 100°\text{-}120°C \end{array}$$

$$(C_6H_5)_2C\!=\!\underset{Br}{\overset{|}{C}}\!-\!CH_2OAc \quad (42)$$

To a cooled flask containing 25 gm (0.14 mole) of 1,1-diphenylethylene and 100 ml of dry pentane is added 28 gm (0.25 mole) of potassium *tert*-butoxide. The mixture is stirred at 0°C and then 66.0 gm (0.26 mole) of bromoform is added dropwise during 30–45 min. Stirring is continued for an additional 2-3 hr at room temperature and then 200 ml of water is carefully added. The yellow solid product is filtered, dried, and digested with 300 ml of refluxing isopropanol for ½ hr. On cooling, the product is filtered, washed with 100 ml of cool isopropanol, and dried to give 31–38 gm (63–78%) of a colorless, crystalline product, mp 151°–152°C.

To 17.6 gm (0.050 mole) of the above dibromide is added 12.5 gm (0.025 mole) of silver acetate and 50 ml of glacial acetic acid. The reaction mixture is warmed at the oil bath temperature of 100°–120°C for 24 hr and then cooled. Then 200 ml of ether is added and the mixture filtered. The ethereal filtrate is washed successively with two 100-ml portions of water, two 100-ml portions of aqueous saturated sodium carbonate, and finally with two 100-ml portions of water. The ether layer is dried over anhydrous sodium sulfate, concentrated on a rotary evaporator, and the residue distilled under reduced pressure to give 12.0 gm (72%), bp 142°–145°C (0.15 mm Hg), n_D^{22} 1.6020–1.6023.

D. DEHALOGENATION OF DIHALIDES

$$\underset{\underset{Br}{|}\underset{Br}{|}}{RCH-CH_2} + Zn\ dust \xrightarrow{Ethanol} RCH=CH_2 \qquad (43)$$

This reaction has very little preparative value since one usually prepares the dibromide from the olefin by the addition of bromine. Usually no isomerization of the carbon chain takes place in the regeneration of the olefin. Zinc dust and 95% ethanol are the most common dehalogenation reagents [94]. The reaction is usually carried out at the reflux temperature of the solvent.

In the case of 1,2,5-tribromopentane, 5-bromo-1-pentene is isolated in 71% yield [95].

$$\underset{\underset{Br}{|}}{Br-CH_2-CH_2-CH_2-CH-CH_2Br} \longrightarrow Br-CH_2-CH_2-CH_2-CH=CH_2 \qquad (44)$$

Thus, an isolated halogen is unaffected by zinc under the conditions of this reaction.

Other dehalogenating agents that have been reported are: zinc in methanol with a trace of zinc chloride [96–98], sodium iodide in acetone [99], trimethylphosphite [100], thiourea [101], and chromium(II) salts in DMF or complexed with ethylene diamine [102, 103]. Dibromides can also be dehalogenated by heat alone [104] or by use of tri-*n*-butyltin hydride [105].

§ 3. Condensation Reactions   59

Fluorocarbons have been dehalofluorinated by reaction with lithium [106] or methyl magnesium chloride [107].

1,1-Dihalocyclopropanes also react with sodium [90], magnesium [90], and methyl- or butyllithium [91] to give allenes.

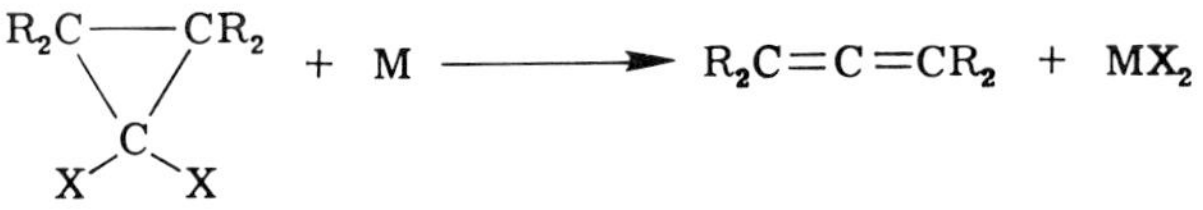

$$\begin{array}{c} R_2C \text{---} CR_2 \\ \diagdown \diagup \\ C \\ \diagup \diagdown \\ X X \end{array} + M \longrightarrow R_2C=C=CR_2 + MX_2 \qquad (45)$$

3. CONDENSATION REACTIONS

A. The Wittig Synthesis of Olefins [108–113]

In 1953 Wittig and Geissler discovered that methylene triphenylphosphorane reacted with benzophenone to give 1,1-diphenylethylene and triphenylphosphine oxide in almost quantitative yield. The phosphorane was prepared from triphenylmethylphosphonium bromide and phenyllithium.

$$[(C_6H_5)_3P\text{---}CH_3]Br \xrightarrow{C_6H_5\text{---}Li} (C_6H_5)_3P=CH_2 \xrightarrow{(C_6H_5)_2C=O} (C_6H_5)_2C=CH_2$$
$$+ (C_6H_5)_3PO \qquad (46)$$

The advantage of the Wittig method is that a carbonyl group is replaced specifically with a carbon carbon double bond. Furthermore, the reaction is carried out under mild alkaline conditions at low temperatures, which allow sensitive olefins to be easily prepared.

Other bases such as butyllithium, sodium amide, and alkali alkoxides can be substituted for phenyllithium. The solvent may be ether, tetrahydrofuran, or dimethylformamide. Polar solvents give better yields than do solvents such as benzene.

The reaction of yields of the type $R_3P = CHCOOC_2H_5$ has been used to prepare unsaturated esters [114, 115].

Carbenes can also add to triphenylphosphine. This leads to alkylidene phosphoranes [116–119] which can be used for the synthesis of 1,1-dihaloolefins.

Recently [120], vinyltriphenylphosphonium bromide has been reported to react with diethyl (3-oxobutyl)malonate in the presence of sodium hydride to give diethyl 4-methyl-3-cyclohexene dicarboxylate in 51% yield. Five- and six-membered cycloalkenes can be prepared using this procedure.

A modification of the Wittig reaction is the use of dimethyl sulfoxide as a solvent [121]. The yields of olefins that are obtained are superior to that obtained using the example of the Wittig reaction [122]. Sterically hindered ketones react with greater ease in dimethyl sulfoxide than in other solvents. In addition, this method gives excellent yields on the micro scale using a standard solution of methyl sulfinyl carbanion [123].

Several reviews on the Wittig reaction are worth consulting for more details [112, 125–126].

3-1. Wittig Method—Preparation of Methylenecyclohexane in Dimethyl Sulfoxide Solvent [121]

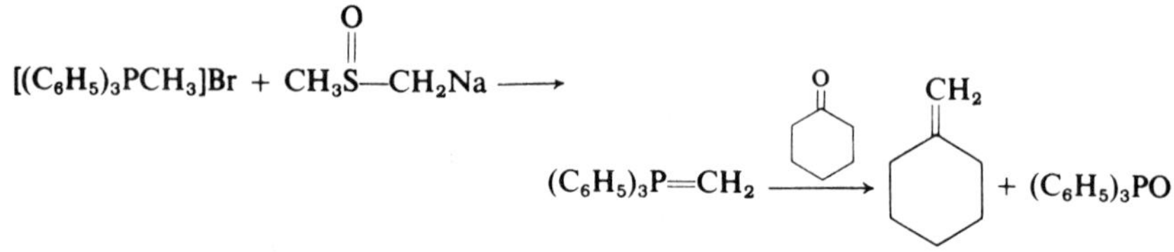

$$[(C_6H_5)_3PCH_3]Br + CH_3\overset{O}{\underset{\|}{S}}-CH_2Na \longrightarrow$$

$$(C_6H_5)_3P=CH_2 \longrightarrow \text{cyclohexanone} \longrightarrow \text{methylenecyclohexane} + (C_6H_5)_3PO$$

(47)

Sodium hydride (0.1 mole; 55% dispersion in oil) is washed with *n*-pentane to remove the traces of oil. The flask is flushed with nitrogen and 50 ml of dimethyl sulfoxide (DMSO) is added by means of a hypodermic syringe. The mixture is heated at 75°–80°C for approximately ¾ hr or until the hydrogen evolution ceases. To the resulting solution of methylsulfinyl carbanion which is cooled with ice is added 0.1 mole (35.7 gm) of methyltriphenylphosphonium bromide in 100 ml of warm DMSO. The solution turns red, and it is stirred for an additional 10 min. Freshly distilled cyclohexanone, 10.8 gm (0.11 mole), is added to the ylid and the reaction mixture is stirred at room temperature for 30 min. The immediate distillation of the mixture yields 8.10 gm (86.3%) of methylenecyclohexane, bp 42°C (105 mm), which is collected in a Dry Ice trap. The reported [122] boiling point is 99°–101°C (740 mm).

The preparation of 1,2-distyrylbenzene [127] in 84% yield is described as a representative example of the unmodified Wittig reaction.

3-2. Preparation of 1,2-Distyrylbenzene [127]

$$\text{o-}C_6H_4(CH_2Br)_2 + 2(C_6H_5)_3P \longrightarrow \left[\text{o-}C_6H_4(CH_2P(C_6H_5)_3)_2\right]^{2+} 2\,Br^- \xrightarrow{C_6H_5-CH=O}{2\,LiOC_2H_5}$$

$$\text{o-}C_6H_4(CH=CH-C_6H_5)_2 + 2\,LiBr + 2\,C_2H_5OH + 2\,(C_6H_5)_3PO$$

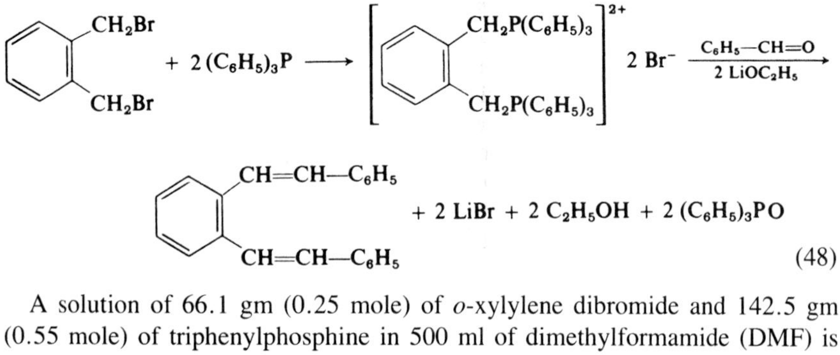

(48)

A solution of 66.1 gm (0.25 mole) of *o*-xylylene dibromide and 142.5 gm (0.55 mole) of triphenylphosphine in 500 ml of dimethylformamide (DMF) is heated under reflux. After the first 10–15 min, a colorless crystalline solid begins to separate and the refluxing is continued for 3 hr. The mixture is cooled, filtered, and the solid is washed with DMF and ether. After air-drying, 175.9 gm (89.4%) of pure o-xylylene bis(triphenylphosphonium) dibromide, mp > 340°C is obtained.

To a solution of 42.5 gm (0.054 mole) of o-xylylene bis(triphenylphosphonium)dibromide and 12.6 gm (0.119 mole) of benzaldehyde in 150 ml of absolute alcohol is added 500 ml of 1.4 M lithium ethoxide in ethanol. After standing at room temperature for 30 min. the solution is refluxed for 2 hr to yield a red-orange solution. Concentrating the mixture to 100 ml and adding 300 ml of water causes the precipitation of a yellow oil which is extracted with ether. Upon concentrating the ether solution, a mobile oil is isolated which is purified by column chromatography using alumina (Fisher A540, 2.5 × 55 cm). Elution with 250–300 ml of low-boiling petroleum ether gives an oil which solidifies on further evaporation of the residual solvent. The combined solids are recrystallized from ethanol to yield 12.7 gm (84%) of colorless crystals melting at 117°–119°C.

The use of phosphonate carbanions represents a further modification of the Wittig reaction which is useful in preparing sensitive olefins [128, 129]. In addition, the reagent is not affected by atmospheric oxygen.

$$\text{RCHO} + (C_2H_5O)_2P(O)CH_2R' \xrightarrow{\text{Base}} \text{RCH=CHR'} + \text{HOP(O)(OC}_2\text{H}_5)_2 \quad (49)$$

3-3. Preparation of trans-Stilbene with Phosphonate Carbanions [128, 130]

$$(C_2H_5O)_3P + C_6H_5CH_2Cl \longrightarrow (C_2H_5O)_2P(O)CH_2C_6H_5 + C_2H_5Cl$$

$$(C_2H_5O)_2P(O)CH_2C_6H_5 + \text{NaH} \longrightarrow [(C_3H_5O)_2\overset{-}{P}O\overset{+}{C}HC_6H_5]\text{Na} + H_2 \quad (50)$$

$$\xrightarrow{C_6H_5CH=O} C_6H_5CH=CHC_6H_5 + (C_2H_5O)_2\underset{\underset{OH}{|}}{P}-O$$

The phosphonates may be prepared by the Michaelis–Arbuzov reaction of the corresponding halides with triethyl or trimethyl phosphite [131, 132]. Forty grams of triethylphosphite is refluxed with an equimolar amount of benzyl chloride (30 gm) for 24 hr. Distillation of the mixture gives diethyl benzylphosphonate (50 gm), bp 170°–174°C (15 mm).

Diethyl benzylphosphonate (11.4 gm, 0.05 mole), 50% sodium hydride (2.4 gm, 0.05 mole), and 5.3 gm (0.05 mole) of benzaldehyde are added to 100 ml of dry 1,2-dimethoxyethane. The mixture is heated slowly. At 70°C a large evolution of gas occurs and a semisolid precipitate forms. The mixture is heated to 85°C and then refluxed for ½ hr. After the mixture has been cooled and water has been added, the mixture is filtered. The precipitate is recrystallized from ethyl alcohol to give 5.6 gm (62.6%) of white crystals, mp 124.5°C.

A more recent development involves the use of phosphonate carbanions in the preparation of acetylenes and α-chlorostilbenes in good yields by the process outlined in Eq. (51) [133].

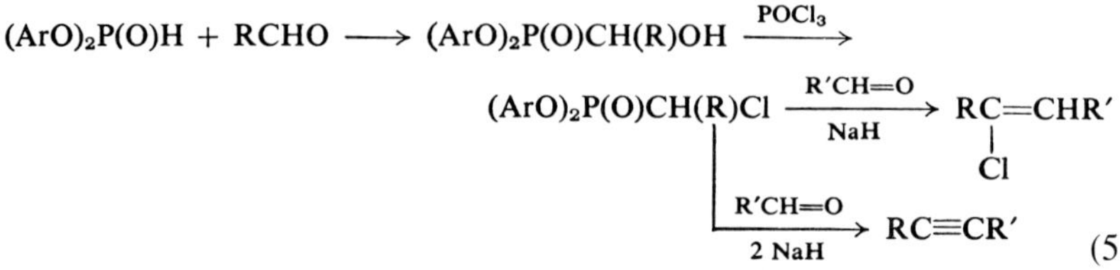

$$(ArO)_2P(O)H + RCHO \longrightarrow (ArO)_2P(O)CH(R)OH \xrightarrow{POCl_3}$$

$$(ArO)_2P(O)CH(R)Cl \xrightarrow[NaH]{R'CH=O} RC=CHR'$$
$$\qquad\qquad\qquad\qquad\qquad\qquad\qquad\qquad |$$
$$\qquad\qquad\qquad\qquad\qquad\qquad\qquad\qquad Cl$$

$$\xrightarrow[2\ NaH]{R'CH=O} RC\equiv CR' \qquad (51)$$

B. Condensations Involving Acetylenes (Vinylation Reactions)

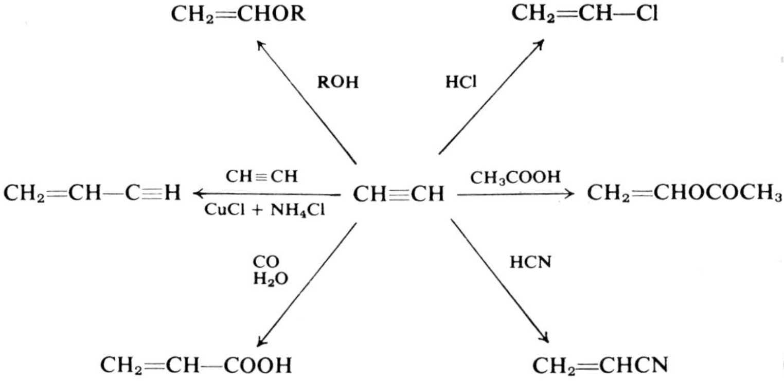

$$\text{CH}_2=\text{CHOR} \qquad\qquad \text{CH}_2=\text{CH}-\text{Cl}$$

$$\text{ROH} \qquad \text{HCl}$$

$$\text{CH}_2=\text{CH}-\text{C}\equiv\text{H} \xleftarrow[\text{CuCl} + \text{NH}_4\text{Cl}]{\text{CH}\equiv\text{CH}} \text{CH}\equiv\text{CH} \xrightarrow{\text{CH}_3\text{COOH}} \text{CH}_2=\text{CHOCOCH}_3$$

$$\text{CO} \qquad\qquad \text{HCN}$$
$$\text{H}_2\text{O}$$

$$\text{CH}_2=\text{CH}-\text{COOH} \qquad\qquad \text{CH}_2=\text{CHCN} \qquad (52)$$

Acetylene reacts with alcohols in the presence of basic catalysts under pressure to give vinyl ethers [134, 135]. Glycols, acids, and amines are condensed under similar conditions [134]. Mercury salts are effective catalysts for the condensation with hydrogen chloride and acids [136]. Very few simple laboratory procedures involving acetylene exist; however, one good example is the preparation of vinyl chloroacetate in 49% yield [137]. Yields up to 60% at atmospheric pressure and 86% under pressure have been reported when using 6% by weight of mercuric oxide catalyst [138]. The procedure given below was used by the authors to prepare vinyl chloroacetate in 32–42% yield [19]. Another method for the preparation of this material utilizes a method similar to that described for the preparation of vinyl caproate in Chapter 10, Esters.

Recently the preparation of vinyl nitriles has been reported to occur via the addition of an anion derived from a nitrile to an acetylene under mild conditions [139].

$$C_6H_5-\underset{CN}{\overset{\ }{CH}}-C_2H_5 + HC\equiv CH \longrightarrow C_6H_5-\underset{CN}{\overset{C_2H_5}{C}}-CH=CH_2 \qquad (53)$$

§ 3. Condensation Reactions

3-4 Preparation of Vinyl Chloroacetate [19, 140, 141]

$$HC \equiv CH + ClCH_2COOH \xrightarrow[H_2SO_4]{HgO} ClCH_2COOCH = CH_2 \quad (54)$$

To a 12-liter, three-necked flask equipped with a stirrer, thermometer, condenser, and gas inlet and outlet tubes are added 2500 gm (26 moles) of chloroacetic acid, 250 gm of yellow mercuric oxide, 100 gm of freshly prepared mercuric sulfate (from 100 gm of mercuric oxide and 26 ml of concentrated sulfuric acid), and 2.5 gm of hydroquinone. The flask is heated by a steam–hot water bath to approximately the melting point of chloroacetic acid (56°C). Acetylene is then bubbled in carefully (using a hood) until the addition of 26 moles (0.027 moles/min) has been completed (about 6 hr). After the first ½ hr the reaction mixture is cooled to 50°–56°C And the temperature is maintained for the remaining additional time. The crude black product is filtered; inhibited with 25 gm of hydroquinone, 2 gm of phenothiazine, and 2 gm of methylene blue; and distilled through a 1-ft column to give 1 kg of product (32%) bp 37°–38°C (16 mm), n_D^{20} 1.4436. The final product is inhibited with 0.25% hydroquinone before being stored. The omission of mercuric sulfate has also given good yields [140, 141].

Recently the facile trimerization of 2-butyne has been reported to give hexamethyl-Dewar benzene (hexamethylbicyclo[2.2.0]-2,5-hexadiene) when anhydrous aluminum chloride is used as the catalyst [142]. The use of cyclobutadieneiron tricarbonyl complex [143, 144] to give similar derivatives is summarized in Section 1, Introduction. The scope of the trimerization reaction has not yet been determined. As a result of the widespread interest in Dewar benzene derivatives, the preparation of hexamethyl–Dewar benzene is given below.

3-5. Preparation of Hexamethyl–Dewar Benzene (Hexamethylbicyclo[2.2.0]-2,5-hexadiene) from 2-Butyne [142]

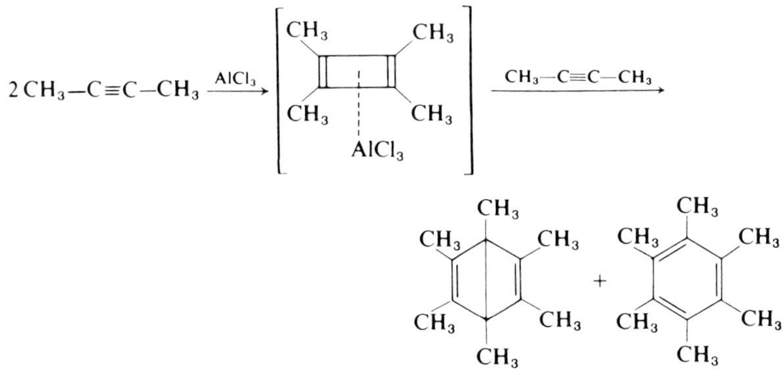

$$2\,CH_3-C \equiv C-CH_3 \xrightarrow{AlCl_3} \left[\begin{array}{c} CH_3 \quad\quad CH_3 \\ \square \\ CH_3 \quad\quad CH_3 \\ AlCl_3 \end{array} \right] \xrightarrow{CH_3-C \equiv C-CH_3}$$

hexamethyl-Dewar benzene + hexamethylbenzene (55)

To a 3-liter flask equipped with stirrer, dropping funnel, condenser, and drying tube are added 50 gm (0.375 mole) of freshly sublimed aluminum chloride and 1 liter of benzene (dried over sodium metal). Then 2-butyne is slowly added from a dropping funnel at temperatures no higher than 35°C until 1 kg (18.5 mole) has been added. The addition requires approximately 2.5 hr and the mixture is stirred for an additional 4 hr at 35°C. If the reaction is carried out at 20°C, then the reaction time is approximately 20 hr. The reaction mixture is cooled and 50 to 100 ml of water is added to decompose the catalyst. The addition of water is stopped when the initially brown mixture turns pale yellow. Approximately 200 gm (3.7 mole) of 2-butyne is removed by distillation at 200 mm pressure and is collected in two successive receivers cooled by a Dry Ice–methanol bath.

The 2-butyne-free benzene solution is washed with water until neutral, and is then fractionally distilled under reduced pressure to give 490 to 510 gm (61–64%) of hexamethylbicyclo[2.1.0]-2,5-hexadiene, bp 43°–45°C (15 mm), n_D^{20} 1.4479, mp 7°C. The distillation residue yields 150–165 gm (19–21%) of hexamethylbenzene.

Hexamethylbicyclo[2.1.0]-2,5-hexadiene (Hexamethyl–Dewar benzene) is stable for several months when kept refrigerated and for a long time at room temperature when kept protected from light.

The infrared spectrum of hexamethylbicyclo[2.1.0]-2,5-hexadiene shows a weak band at 1680 cm^{-1} (C=C in strained ring systems), and bands at 1370 cm^{-1} ($-CH_3$), 1060 cm^{-1} ($CH_3-C=C-$), 1280 1220, 735, and 660 cm^{-1}.

Trimerization in boiling benzene of 2-butyne yields 80–90% of hexamethylbenzene.

C. Condensation of Aldehydes and Ketones with Themselves or with Other Active Methylene Compounds

$$2\ RCH_2CHO \xrightarrow[OH^-]{Aldol} \left[RCH_2-\underset{}{\overset{OH}{CH}}-\underset{R}{\overset{}{CH}}CHO \right] \longrightarrow RCH_2-CH=\underset{R}{\overset{}{C}}-CHO \quad (56)$$

$$ArCHO + (RCH_2CO)_2 + RCOONa \xrightarrow{Perkin} ArCH=\underset{R}{\overset{}{C}}-COONa \quad (57)$$

$$ArCHO + CH_3COOC_2H_5 \xrightarrow[base]{Claisen} ArCH=CH-COOC_2H_5 \quad (58)$$

$$RCH_2CHO + CH_2(COOH)_2 \xrightarrow{Knoevenagel} RCH_2-CH=CH-COOH \quad (59)$$

$$RCHO + R'CH_2NO_2 \xrightarrow{OH^-} RCH=\underset{R'}{\overset{}{C}}-NO_2 \quad (60)$$

§ 3. Condensation Reactions

The aldol, Perkin, and Knoevenagel condensations of active methylene compounds with aldehydes give olefins which are probably derived from the intermediate alcohols, as is true in the aldol condensation shown above. The Perkin, Knoevenagel, and Claisen reactions are described in further detail in Chapter 9, Carboxylic Acids.

The aldol condensation [145] yields alcohols which in some cases dehydrate easily at room temperature upon acidification by acetic acid. For example, the condensation of benzaldehyde with butyraldehyde gives α-ethylcinnamaldehyde in 58% yield [146].

The condensation of methyl ketones with simultaneous dehydration to give olefinic ketones can be accomplished in 70–80% yields [147].

Alkyl ketones and aldehydes react under basic conditions with cyclopentadiene to produce fulvenes in good yield [148, 149]

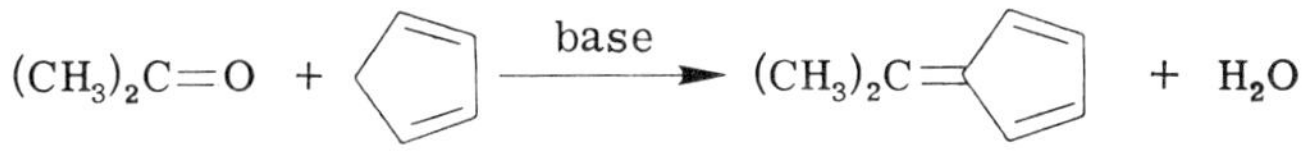

$$(CH_3)_2C=O + \text{(cyclopentadiene)} \xrightarrow{\text{base}} (CH_3)_2C=\text{(fulvene)} + H_2O \quad (61)$$

Aliphatic and aromatic aliphatic nitro compounds which contain an active methylene group can condense with carbonyl compounds. The nitro alcohols [150] can be subsequently dehydrated to the olefin compound, a reaction discussed further in Chapter 16, Nitro Compounds.

3-6. Aldol Condensation—Preparation of α-Ethylcinnamaldehyde [146]

$$C_6H_5CHO + C_2H_5CH_2CHO \longrightarrow C_6H_5CH=\underset{\underset{C_2H_5}{|}}{C}-CHO \quad (62)$$

To 15 gm of a 50% solution by weight of potassium hydroxide and 175 gm of ethanol is added 110 gm (1.1 mole) of benzaldehyde. The reaction mixture is stirred and cooled to 5°C while 50 gm (0.69 mole) of butyraldehyde is added over a period of 3 hr. Then the reaction mixture is allowed to remain over night at room temperature. Acidification, filtration, and then distillation yield 65 gm (58%) of α-ethylcinnamaldehyde, bp 111°–112°C (7 mm), n_D^{25} 1.5822.

D. Coupling and Grignard Reactions

Several olefins can be coupled with each other or with other materials by the elimination of hydrogen halide, sodium halide, magnesium halide, or carbon dioxide.

The reaction of cinnamic acid with aryl diazonium salts in the presence of sodium acetate and cuprous chloride leads to substituted stilbenes in low yields (Meerwein condensation) [151]. The aryl radical may have halo, nitro, alkyl, ether, or ester [151–154] substituents.

$$Ar'CH=CHCOOH + ArN_2{}^+X^- \xrightarrow[CaCl_2]{NaOAc} Ar'CH=CHAr + N_2 + CO_2 + HX \quad (63)$$

Grignard coupling is a good method for the preparation of 1-alkenes. The reaction is exothermic and it occurs readily at room temperature; however, short periods of heating may sometimes be required to complete the coupling process. For example, neopentylethylene is prepared in 85% yield by reacting allyl bromide with *tert*-butylmagnesium chloride in ether at temperatures below 5°C [94]. See also the Boord method for the coupling of allyl magnesium bromide, in Section 2-7 of this chapter.

Allylmagnesium halide can also be reacted with carbonyl compounds to give substituted allyl alcohols. The Barbier–Grignard procedure is recommended since it involves a one-step preparation of the Grignard reagent in the presence of the reactive carbonyl compound [155].

Recently, fluoro olefins have been prepared by the reaction of allylmagnesium bromide and fluoro olefins [156].

Vinylsodium may also find use in such reactions [157].

$$CH_2=CH-Na + RX \longrightarrow CH_2=CHR + NaX \qquad (64)$$

Wurtz-type coupling of allyl and methallyl chloride gives trienes in 30% yields. In this case, sodium amide in liquid ammonia can be used as the condensing agent [158–160].

$$2\ CH_2=\underset{\underset{CH_3}{|}}{C}-CH_2Cl + 2\ NaNH_2 \xrightarrow{\text{liq. NH}_3}$$

$$CH_2=\underset{\underset{CH_3}{|}}{C}-CH=CH-\underset{\underset{CH_3}{|}}{C}=CH_2 + 2\ NaCl + 2\ NH_3 \qquad (65)$$

Recently vinyl copper reagents have been reported to be useful in preparation of substituted vinyl compounds [161–165].

$$CH_3-CH=CH-Cl \xrightarrow[\text{2. CuI}]{\text{1. Li}} (CH_3-CH=CH)_2CuLi \xrightarrow{RX} RCH=CH-CH_3 \qquad (66)$$

$$(CH_2=CH)_2CuLi + RC\equiv C-COOR' \longrightarrow R-\underset{\underset{CH=CH_2}{|}}{C}=CH-COOR' \qquad (67)$$

$$R_1R_2C=C:Cu(MgBr_2) \xrightarrow{\Delta} R_1R_2C=CH-CH=CR_1R_2 \qquad (68)$$

The reductive coupling of carbonyl compounds by transition metal complexes has recently been shown to be a synthetically useful method for the preparation of olefins [166, 167].

$$2\ R_2C=O \longrightarrow \underset{R}{\overset{R}{\diagdown}}C=C\underset{R}{\overset{R}{\diagup}} \qquad (69)$$

Recently ketones have been reported to be converted to olefins in about 70% yields by the use of $CH_2(MgBr)_2$ prepared *in situ* [168].

§ 3. Condensation Reactions

$$Mg/Hg + CH_2Br_2 + R_2C=O \longrightarrow R_2C=CH_2 + (MgBr)^+(OMgBr)^- \quad (70)$$

The use of lithium alkyls to react with unsaturated carbonyl compounds is useful in giving unsaturated alcohols [169].

$$\underset{\underset{CH_3}{|}}{CH_2-C-COOCH_3} \xrightarrow[2.\ HCl/H_2O]{1.\ 2\ n\text{-BuLi}} CH_2=\underset{H_3C}{\overset{Bu}{\underset{|}{C}}}-\underset{Bu}{\overset{|}{C}}-OH \quad (71)$$

The reaction of allyl alcohol with acids or of allyl chloride with sodium carboxylates in dimethylformamide (DMF) gives good yields of allyl esters [170]. Omitting the use of DMF gives poor yields and in some cases no reaction.

3-7. Preparation of Diallyl Isophthalate [170]

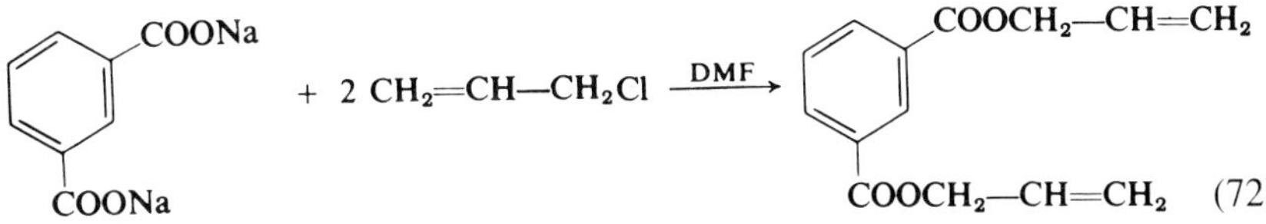

$$\text{(isophthalate-COONa)}_2 + 2\ CH_2=CH-CH_2Cl \xrightarrow{DMF} \text{(isophthalate-COOCH}_2-CH=CH_2)_2 \quad (72)$$

To a flask containing 1382 gm (6.6 moles) of disodium isophthalate are added 1515 gm (19.8 moles) of allyl chloride and about 5 liters of dimethylformamide. The mixture is refluxed for about 24 hr. Sodium chloride is filtered off and the filtrate is distilled to yield 1331 gm 82% (based on sodium isophthalate reacted) of diallyl isophthalate, bp 120°–135°C (1–4 mm), n_D^{20} 1.0221.

E. The Diels–Alder Reaction [171–174]

The Diels–Alder reaction involves the 1,4-*cis* addition of an olefinic compound to a conjugated diene to yield an olefin. The olefinic compound usually contains electron-withdrawing groups to activate its addition, e.g., carboxylic acid groups. For example, butadiene reacts with maleic anhydride to give tetrahydrophthalic anhydride in 90% yield [174]. The latter reaction is carried out in a vessel under pressure since butadiene is volatile. However, dienes such as 1-phenyl-1,3-butadiene or furan do not require pressure. In fact, the reaction of maleic anhydride with furan is exothermic and requires cooling [171].

3-8. Preparation of cis-Cyclohexene-1,2-dicarboxylic Anhydride [175] (cis-Tetrahydrophthalic Anhydride)

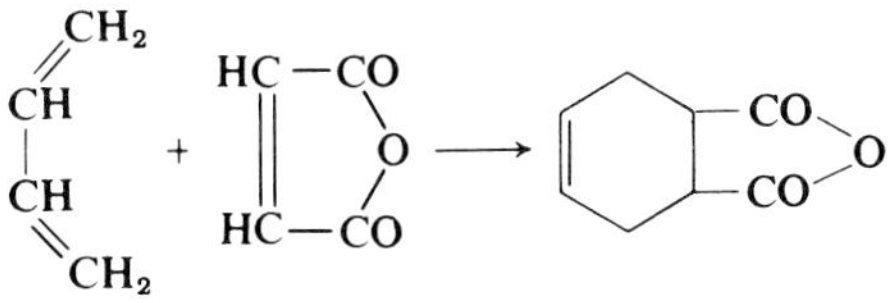

$$\text{butadiene} + \text{maleic anhydride} \longrightarrow \text{cis-tetrahydrophthalic anhydride} \quad (73)$$

To a carefully tared soda bottle is added 50 gm (0.51 mole) of maleic anhydride and 80 ml of benzene. The bottle is chilled to 0°C and 32 gm (0.59 mole) of butadiene is added. The bottle is capped and placed in an autoclave along with 100 ml of benzene in order to equalize the pressure on both sides of the bottle. The reaction mixture is allowed to remain at room temperature for 12 hr and is then heated for 5 hr at 100°C. The solid product is recrystallized from benzene–petroleum ether to yield 69.6 gm (90%), mp 101°–103°C.

Butadiene alone can be made to polymerize via a trans diradical process at room temperature. However, at elevated temperatures butadiene forms some cis isomer which reacts with itself by a Diels–Alder process to give 4-vinyl-1-cyclohexane [176]. Also prolonged storage of butadiene leads to the same product [177].

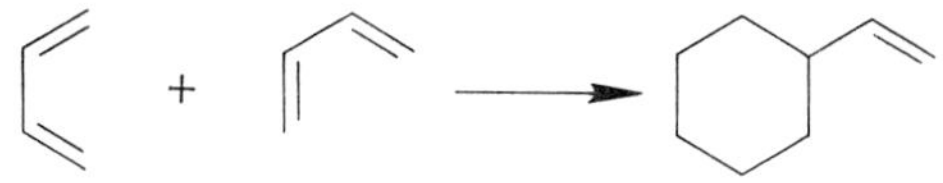

$$\text{butadiene} + \text{butadiene} \longrightarrow \text{4-vinyl-1-cyclohexene} \quad (74)$$

Some dienes like hexachlorocyclopentadiene behave as dienes and dienephiles in the Diels–Alder reaction [178, 179].

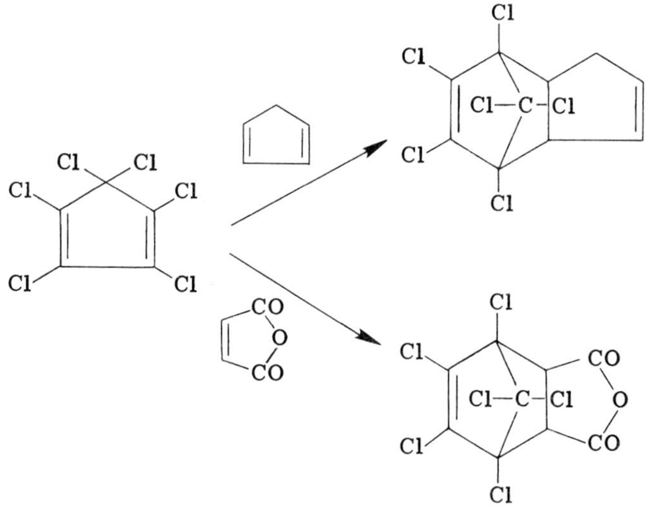

(75)

The Diels–Alder reaction is useful in giving various substituted bicyclic ring systems that are difficult to prepare by other techniques. For example, Grunewald and Davis recently reported that bridgehead substituted norbornenes can be prepared by this teachnique [180].

Diels–Alder reactions are reversible and when the products are unstable an equilibrium composition is observed consisting of diene, dieneophile, and adduct [181–183].

Several excellent reviews of the Diels–Alder reaction are worth consulting for additionl details [172, 174, 176, 181–183].

4. REDUCTION REACTIONS

$$R'C{\equiv}CR \xrightarrow{H_2} R'CH{=}CHR \qquad (76)$$

Acetylene derivatives can be catalytically reduced by several catalysts in the presence of hydrogen [184–186] or with sodium and liquid ammonia [184–186]. Pure *trans*-olefins in 90% yields result using the latter reagent with dialkyl acetylenes [184–186]. Acetylenic alcohols [187], ethers [188] and acids [189] are reduced to the corresponding olefinic compounds. *cis*-Olefins are obtained using palladium catalysts in hydrogenations [190, 191].

A more recent method [192–194] for the reduction of acetylenes is the use of a hydroborating agent of large steric requirements such as disiamylborane to convert internal as well as terminal acetylenes to vinylboron compounds in quantitative yields. The vinylboron compounds are protonated at room temperature by acetic acid to give *cis*-olefins in high purity from internal acetylenes and 1-olefins from terminal acetylenes.

$$RC{\equiv}CR + R_2'BH \longrightarrow RCH{=}\underset{R}{C}{-}BR_2' \xrightarrow{H^+} RCH{=}CHR \qquad (77)$$

4-1. Hydroboration of 1-Hexyne to 1-Hexene [192–194]

$$2\,CH_3{-}\underset{CH_3}{\overset{CH_3}{C}}{=}CH{-}CH_3 + NaBH + BF_3 \longrightarrow HB({-}\underset{CH_3}{\overset{H\;\;CH_3}{C{-}CH}}{-}CH_3)_2 \xrightarrow{CH_3(CH_2)_3{-}C{\equiv}CH}$$

$$CH_3(CH_2)_3CH{=}CHB({-}\underset{CH_3}{\overset{H\;\;CH_3}{C{-}CH}}{-}CH_3)_2 \xrightarrow{H^+} CH_3(CH_2)_3C{=}CH_2 \qquad (78)$$

To a dry three-necked flask are added 33.6 gm (0.48 mole) of 2-methyl-2-butene and 6.8 gm (0.18 mole) of sodium borohydride in 100 ml of diglyme. The reaction flask is immersed in an ice bath and to it is added dropwise with stirring 32.2 gm (0.24 mole) of boron trifluoride etherate. After the addition is completed the reaction mixture is stirred for 2 hr at 0°–5°C.

To the disiamylborane [bis-3-methyl-2-butylborane] prepared as above is added 16.4 gm (0.20 mole) of 1-hexyne as fast as possible while keeping the temperature below 10°C by use of an ice-water bath. After the addition, the flask is allowed to remain at 0°–5°C for ½ hr and then for 2 hr at room temperature. A small amount of ethylene glycol is added to decompose the residual hydride. While keeping the reaction mixture at 0°C, 100 ml of glacial acetic acid is added and then the mixture is allowed to remain at room temperature for 2 hr. The reaction mixture is poured into ice water, the upper layer is separated, washed with sodium hydroxide solution, washed with a saturated sodium chloride solution, dried, and distilled from a Claisen flask. The fraction boiling up to 80°C is collected and saturated with sodium chloride. The upper layer is decanted onto anhydrous potassium carbonate. Distillation through a Todd microcolumn yields 1-hexene, 12.0 gm (72%), bp 64°C (743 mm), n_D^{20} 1.3879.

Recently the stereospecific reductive alkylation of acetylenes by successive hydroalumination, and carbodemetalation has been reported [195–197].

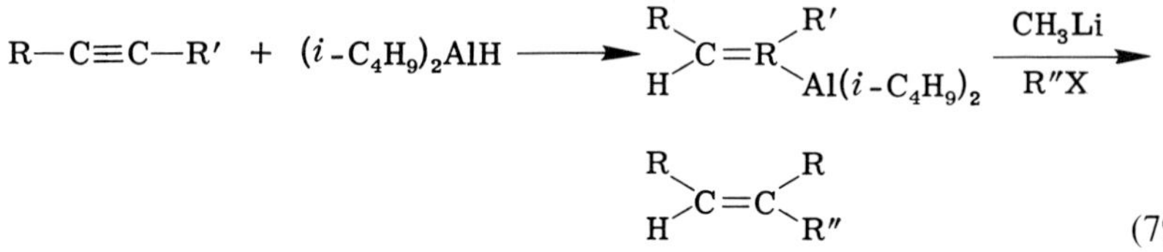

$$R-C{\equiv}C-R' + (i\text{-}C_4H_9)_2AlH \longrightarrow \underset{H}{\overset{R}{>}}C=R\underset{Al(i\text{-}C_4H_9)_2}{\overset{R'}{<}} \xrightarrow[R''X]{CH_3Li}$$

$$\underset{H}{\overset{R}{>}}C=C\underset{R''}{\overset{R}{<}} \qquad (79)$$

The Birch reduction of aromatic compounds yields 1,4-dihydroaromatics and cyclohexenes under more drastic conditions [198–200]. Benkeser [201] has improved the reaction by using lithium in low-molecular-weight amines. Iron has been found to catalyze the sodium-ammonia-alcohol reaction in the Birch reduction [202, 203].

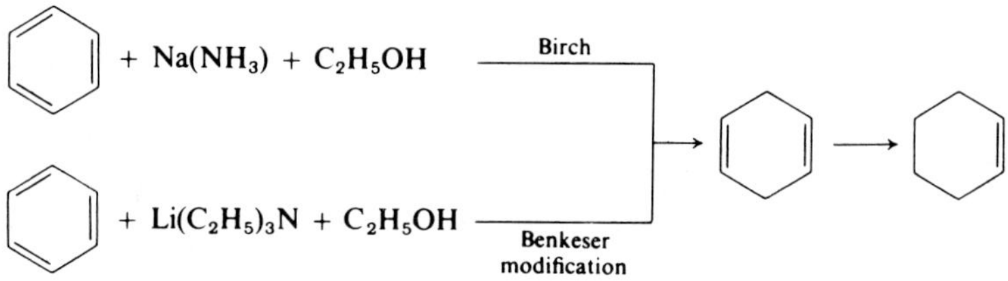

$$\text{benzene} + Na(NH_3) + C_2H_5OH \quad \xrightarrow{\text{Birch}}$$
$$\text{benzene} + Li(C_2H_5)_3N + C_2H_5OH \quad \xrightarrow{\text{Benkeser modification}} \longrightarrow \text{1,4-cyclohexadiene} \longrightarrow \text{cyclohexene} \qquad (80)$$

4-2. *Modified Birch Reduction with Lithium in an Amine–Alcohol System— Preparation of 2,5-Dihydroethylbenzene* [201]

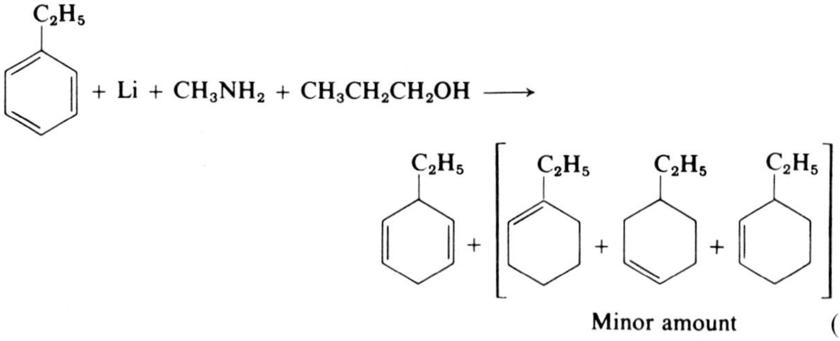

$$\text{ethylbenzene} + Li + CH_3NH_2 + CH_3CH_2CH_2OH \longrightarrow$$

$$\text{2,5-dihydroethylbenzene} + [\text{1,4-} + \text{2,3-} + \text{1,2-dihydro isomers}]$$

Minor amount (81)

In a three-necked flask fitted with a stirrer and a Dry Ice condenser are placed 21.2 gm (0.2 mole) of ethylbenzene, 30.0 gm (0.5 mole) of 1-propanol, and 300 ml of methylamine. Lithium wire, 3.15 gm (0.45 gm atom), is added in two

§ 4. Reduction Reactions

portions and it is entirely consumed within 30 min. Afterwards, the Dry Ice condenser is replaced with a water condenser and the methylamine is allowed to evaporate off in a hood. The reaction mixture is then hydrolyzed carefully by slowly adding water to the flask, extracting with ether, drying, stripping the solvent, and distilling the yield—16.6 gm (78%) of a crude mixture consisting of 89% 2,5-dihydroethylbenzene, 8% of 1-ethylcyclohexene, and 3% mixture of 3- and 4-ethylcyclohexene. Redistilling carefully in a Todd column (75:1 reflux ratio) affords 9.6 gm (46%) of purified 2,5-dihydroethylbenzene [203], bp 140°–142°C, n_D^{21} 1.4710.

Aromatic ethers have been reported to be electrochemically alkoxylated in an alcohol containing a suitable electrolyte. An example of this reaction has been recently reported in which 3,3,6,6-tetramethoxy-1,4-cyclohexadiene is prepared from *p*-dimethoxybenzene in 70% yield [204].

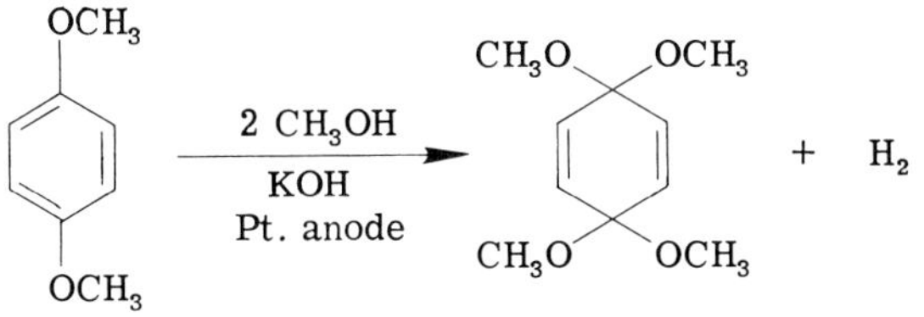

$$\text{p-dimethoxybenzene} \xrightarrow[\text{Pt. anode}]{\substack{2\ CH_3OH \\ KOH}} \text{3,3,6,6-tetramethoxy-1,4-cyclohexadiene} + H_2 \quad (82)$$

5. ISOMERIZATION REACTIONS

$$RCH=CH-CH_2R' \longrightarrow RCH_2CH=CHR' \quad (83)$$

Thermal isomerization over aluminum at 470°–480°C converts cyclohexene to alkyl cyclopentenes [205].

Other isomerizations such as the conversion of cis to trans can sometimes be carried out in the presence of a base, or acid by heating alone, by photochemical means or by the presence of pi-complexes of some metal salts [206].

The base-catalyzed isomerization of an alkene is illustrated, for example, by the conversion of 1,5-cyclooctadiene to 1,3-cyclooctadiene in almost quantitative yield [207].

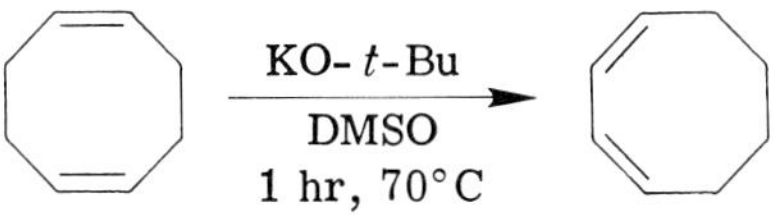

$$\text{1,5-cyclooctadiene} \xrightarrow[\substack{DMSO \\ 1\ hr,\ 70°C}]{KO\text{-}t\text{-}Bu} \text{1,3-cyclooctadiene} \quad (84)$$

Recently Gibson and Strassburger reported that *p*-toluenesulfinic acid catalyzes the cis–trans isomerization of olefins [208]. For example, treatment of a solution of methyl oleate in dioxane at reflux for 2 hr with 10 mol% of *p*-

toluenesulfinic acid gave an equilibrium mixture composed of 76% of the trans isomer.

Double bonds have been reported to be isomerized in unsaturated esters and enol ethers by means of base or by a photochemical process [209].

Hubert reported that α,ω diynes can be isomerized to the tetraene using KNH_2 or Al_2O_3 [210]. For example, 1,8-nonadiyne gives 1,3,5,7-nonatetraene.

Allylic alcohols may be prepared by a base $[LiN(C_2H_5)_2]$ rearrangement of the starting epoxides [211].

The Claisen rearrangement [212, 213] of allyl phenyl ethers to o-allylphenol in the presence of base is a reaction giving good yields. Ortho substituents on the allyl phenyl ethers do not effect the predominant formation of the ortho isomer in the rearranged product [214].

5-1. Claisen Rearrangement—Preparation of Allyl Phenyl Ether and Its Rearrangement to 2-Allylphenol [212, 213]

To a flask containing 188 gm (2.0 moles) of phenol are added 242 gm (2.0 moles) of allyl bromide, 280 gm (2.0 moles) of anhydrous potassium carbonate, and 300 gm of acetone. The mixture is refluxed on the steam bath for 8 hr, cooled, diluted with an equal volume of water, and extracted with ether. The ether extract is washed twice with 10% aqueous sodium hydroxide, dried, the solvent stripped off, and the residue distilled under reduced pressure to yield 230 gm (86%) of allyl phenyl ether, bp 85°C (19 mm).

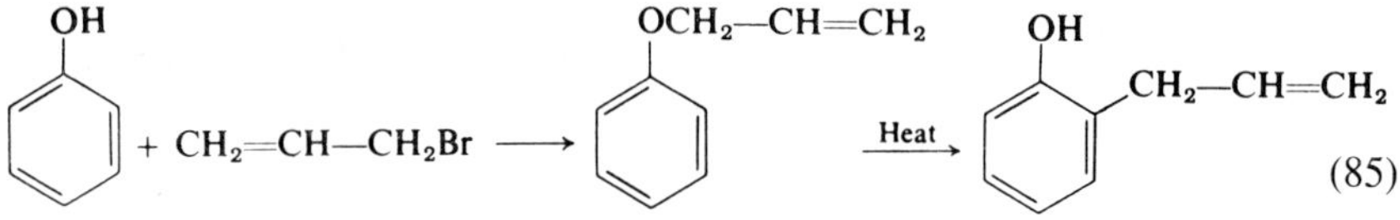

$$\text{PhOH} + CH_2{=}CH{-}CH_2Br \longrightarrow \text{PhOCH}_2{-}CH{=}CH_2 \xrightarrow{\text{Heat}} o\text{-}(CH_2{-}CH{=}CH_2)C_6H_4OH \quad (85)$$

The allyl phenyl ether is rearranged by boiling at 195°–200°C at atmospheric pressure under nitrogen until the refractive index of the liquid remains constant (5 to 6 hr to get n_D^{24} 1.55). The crude material is dissolved in 20% sodium hydroxide solution and extracted twice with 30°–60°C petroleum ether. The alkaline solution is acidified, extracted with ether, dried, the solvent stripped off, and the remaining liquid distilled under reduced pressure to yield 73% of 2-allylphenol, bp 103°–105.5°C (19 mm), n_D^{24} 1.5445.

6. MISCELLANEOUS METHODS

A. Elimination Reactions

(1) Pyrolysis of α-bromoacetates in the presence of trialkyl or triaryl phosphines [215].

(2) The Ramberg–Bäcklund reaction in which sulfones are converted to olefins [216, 217].
(3) Olefins via desulfurization of sulfones and α-halosulfones [218, 219].
(4) Desulfurization of thioacetals [220].
(5) Olefins by the base-catalyzed reaction of aliphatic sulfones and sulfoxides at 55°C [221].
(6) Pyrolysis of ethylene sulfones to stilbenes [222].
(7) Vinyl acetylenes by the base-catalyzed elimination of sulfonates [223].
(8) Desulfurization of thioketones to give olefins [224].
(9) Thermal decomposition of alkyllithium compounds [225].
(10) Thermal decarboxylation of diaryl fumarates to stilbene [226, 227].
(11) Pyrolysis of Δ^3-1,3,4-thiadiazolines [228, 229].
(12) The Leukert reaction and deamination to give olefins [230].
(13) Kishner eliminative reduction of haloketones [231].
(14) Dehydrogenation of saturated hydrocarbons with iodine [232].
(15) Dehydrogenation of ethylbenzene to styrenes [233].
(16) Pyrolysis of tetrahalomethanes to tetrahaloethylenes [234].
(17) Preparation of 1-methylcyclopropene from methallyl chloride and sodamide [235].
(18) Extrusion of the sulfur bridge [216].
(19) Preparation of methoxytrifluoroethylene. CAUTION: Explosive reaction [236].
(20) Preparation of fluoroethylenes [237].
(21) Preparation of nitroolefins via dehydration of nitroalcohols using methane sulfonyl chloride and triethylamine [238].

B. Condensation Reactions

(1) Preparation of alkyl vinyl ethers from ethylenes and alcohols [239].
(2) Thermal addition of acetylene to olefins to yield diolefins [240].
(3) 1,4-Hexadienes from ethylene and butadiene [241].
(4) Vinyl esters by the condensation of ethylene and acids [242].
(5) Coupling of thiocarbonyl compounds to diarylethylenes [243].
(6) Condensation of primary alkylmagnesium halides with carbon monoxide under pressure [244].
(7) Condensation of acetylene with carbon monoxide to give acrylic acids [245].
(8) Olefins from the condensation of aromatic aldehydes with dimethyl sulfone [246].
(9) Organometallic styrenes by the condensation of styrylmagnesium halide and halogenated organometallics [247].
(10) Ariens–Dorp Synthesis of α,β-unsaturated aldehydes [248].

(11) Trans vinylation of esters using vinyl acetate [249].
(12) Reaction of C_3 dicarbene with olefins to give allenes [250, 251].
(13) Reaction of vinylidene carbene with olefins to give allenes [252].
(14) Catalytic addition of ethylene to 1,3-dienes to give 1,4-dienes [253].
(15) Linear dimerization of butadiene with ferric chloride to trienes [254].
(16) Trimerization of isoprene to trimethylcyclododecatriene [255].
(17) Dimerization of isoprene [256].
(18) Photochemical conversion of cyclopentadiene to bicyclo-[2,1,0]pent-2-ene [257].
(19) Photosensitized cycloaddition of haloethylenes and 1,3-dienes [258].
(20) Conversion of diphenylacetylene to 2,3-diphenyl-1,3-butadiene via methyl sulfonyl carbanion [259].

C. Oxidation Reactions

(1) Oxidative condensation of 2-methoxy-4-nitrotoluene to stilbenes [260, 261].
(2) Preparation of acrylonitrile from propylene and ammonia [262].
(3) Oxidative coupling of 1,5-hexadiyne [263].

D. Reduction Reactions

(1) Reduction of acetylene halides to allenes [258].
(2) Reduction of acetylene compounds with chromous sulfate to *trans*-olefins [264].
(3) Synthesis of olefins from thionocarbonates by an alkylation-reduction sequence [265].
(4) Reductive cleavage of alcohols to olefins [266].
(5) Reductive elimination of epoxides to olefins with zinc–copper couple [267].
(6) Reduction of dialkylacetylenes to *cis*-olefins with lithium aluminum hydride and nickel chloride or titanium chloride [268, 269].

E. Isomerization and Rearrangement Reactions

(1) Trimerization of vinylacetylene [270].
(2) Thermal isomerization of 3,4-dimethyl-1,5-hexadiene to octadienes (Cope rearrangement) [271, 272].
(3) Thermal isomerization of cyclooctatetraene to styrene [233].
(4) Cope allyl-vinyl rearrangement [271].
(5) Rearrangements in allylic systems undergoing electrophilic and nucleophilic substitution [273].

(6) Thermal isomerization of alkynes to dienes [274].
(7) Thermal isomerization of small ring compounds to olefins [275].
(8) Photochemical isomerization of unsaturated esters [276].
(9) Isomerization via pi-complexes [277].

REFERENCES

1. E. J. Corey and A. K. Long, *J. Org. Chem.* **41**, 2208 (1978).
2. J. Rencroft and P. G. Sammes, *Q. Rev., Chem. Soc.* **25** (1), 135 (1971).
3. G. H. Coleman and H. F. Johnstone, *Org. Synth. Collect. Vol.* **1**, 183 (1941).
4. E. W. Garbisch, Jr., *J. Org. Chem.* **26**, 4165 (1961).
5. W. Karo, R. L. McLaughlin, and H. F. Hipsher, *J. Am. Chem. Soc.* **75**, 3233 (1953).
6. V. J. Traynelis, W. L. Hergenrother, J. R. Livingston, and J. A. Valicenti, *J. Org. Chem.* **27**, 2377 (1962).
7. R. B. Carlin and D. A. Constantine, *J. Am. Chem. Soc.* **69**, 50 (1947).
8. R. E. Miller and F. F. Nord, *J. Org. Chem.* **15**, 89 (1950).
9. E. Levas, *Ann. Chem. (Paris)* [12] **3**, 145 (1948).
10. G. B. Backman and L. L. Lewis, *J. Am. Chem. Soc.* **69**, 2022 (1947).
11. J. R. Dice, T. E. Watkins, and H. L. Schuman, *J. Am. Chem. Soc.* **72**, 1738 (1950).
12. N. Campbell and D. Kidd, *J. Chem. Soc.* p. 2154 (1954).
13. W. S. Allen and S. Bernstein, *J. Am. Chem. Soc.* **77**, 1028 (1955).
14. C. A. Walker, *Ind. Eng. Chem.* **41**, 2640 (1949).
15. H. H. Pines and W. O. Haig, *J. Am. Chem. Soc.* **83**, 2847 (1961).
16. G. P. Shulman, M. Trusty, and J. H. Vickers, *J. Org. Chem.* **28**, 907 (1963).
17. H. Pines, *J. Am. Chem. Soc.* **55**, 3892 (1933).
18. D. T. Mowry, M. Renoll, and W. F. Huber, *J. Am. Chem. Soc.* **68**, 1108 (1946).
19. Authors' laboratory.
20. J. J. Lamson, R. H. Hall, E. Stroiwas, and L. D. Yats, U.S. Patent 4,150,059 (1979).
21. A. J. Lundeen and R. Van Hoozer, *J. Am. Chem. Soc.* **85**, 2180 (1963).
22. M. A. Dolliver, T. L. Gresham, G. B. Kistiakowsky, and W. E. Vaughan, *J. Am. Chem. Soc.* **59**, 831 (1937).
23. J. Hine, E. L. Pollitzer, and H. Wagner, *J. Am. Chem. Soc.* **75**, 5607 (1953).
24. P. S. Skell and I. Starer, *J. Am. Chem. Soc.* **81**, 4117 (1959).
25. M. Julia, S. Julia, and S.-Y. Tchen, *Bull. Soc. Chim. Fr.* p. 1849 (1961).
26. M. Julia, S. Julia, and R. Guegan, *Bull. Soc. Chim. Fr.* p. 1072 (1960).
27. S. F. Brady, M. A. Ilton, and W. S. Johnson, *J. Am. Chem. Soc.* **90**, 2882 (1968).
28. E. C. Ashby, G. F. Willard, and A. B. Goel, *J. Org. Chem.* **44**, 1221 (1979).
29. J. P. Wibaut and A. J. Van Pelt, Jr., *Recl. Trav. Chim. Pays-Bas* **60**, 55 (1941).
30. L. T. Smith, C. H. Fisher, W. P. Ratchford, and M. L. Fein, *Ind. Eng. Chem.* **34**, 473 (1942).
31. W. J. Bailey and R. A. Bayloung, *J. Am. Chem. Soc.* **81**, 2126 (1959).
32. H. R. Nace, *Org. React.* **12**, 57 (1962).
33. G. L. O'Connor and H. R. Nace, *J. Am. Chem. Soc.* **77**, 1578 (1955).
34. W. Brandenberg and A. Galait, *J. Am. Chem. Soc.* **72**, 3275 (1950).
35. G. E. Illingworth and G. W. Lester, U.S. Patent 3,409,698 (1968).
36. K. L. Williamson, R. T. Keller, G. S. Fonken, J. Azmuskovicz, and W. S. Johnson, *J. Org. Chem.* **27**, 1612 (1962).
37. K. G. Rutherford, R. M. Ottenbrite, and B. K. Tang, *J. Chem. Soc. C* p. 582 (1971).
38. E. C. Ashby, G. F. Willard, and A. B. Goel, *J. Org. Chem.* **44**, 1221 (1979).
39. G. L. O'Connor and H. R. Nace, *J. Am. Chem. Soc.* **75**, 2118 (1953).
40. M. S. Newman and F. W. Hetzel, *J. Org. Chem.* **34**, 3604 (1969).

41. L. C. Roach and W. H. Daly, *Chem. Commun.* p. 606 (1970).
42. F. G. Bordwell and P. S. Landis, *J. Am. Chem. Soc.* **80,** 6382 (1958).
43. M. A. Dolliver, T. L. Gresham, G. B. Kistiakowsky, E. A. Smith, and W. E. Vaughan, *J. Am. Chem. Soc.* **60,** 440 (1938).
44. H. Scheibler and H. Banganz, *Justus Liebigs Ann. Chem.* **565,** 170 (1949).
45. P. G. Stevens and J. H. Richmond, *J. Am. Chem. Soc.* **63,** 3132 (1941).
46. I. N. Feit and W. H. Saunders, Jr., *Chem. Commun.* p. 610 (1967).
47. R. J. Baumgarten, *J. Chem. Educ.* **45,** 122 (1968).
48. L. D. Freedman, *J. Chem. Educ.* **43,** 662 (1966).
49. D. V. Banthorpe, E. D. Hughes, and C. Ingold, *J. Chem. Soc.* p. 4054 (1960).
50. J. H. Burckhalter and R. C. Fuson, *J. Am. Chem. Soc.* **70,** 4184 (1948).
51. F. C. Whitmore, A. H. Popkin, H. I. Bernstein, and J. P. Wilkins, *J. Am. Chem. Soc.* **63,** 124 (1941).
52. H. C. Brown and O. H. Wheeler, *J. Am. Chem. Soc.* **78,** 2199 (1956).
53. H. C. Brown and I. Moritani, *J. Am. Chem. Soc.* **78,** 2203 (1956).
54. R. A. Bartsch and J. F. Bunnett, *J. Am. Chem. Soc.* **90,** 408 (1968).
55. R. A. Bartsch, C. F. Kelly, and G. M. Pruss, *J. Org. Chem.* **34,** 662 (1971).
56. H. C. Brown and R. L. Klimisch, *J. Am. Chem. Soc.* **88,** 1425 (1966).
57. A. C. Cope and E. R. Trunbell, *Org. React.* **11,** 361 (1960).
58. D. J. Cram, M. R. V. Sahyun, and G. R. Knox, *J. Am. Chem. Soc.* **84,** 1734 (1962).
59. E. J. Corey and R. A. E. Winter, *J. Am. Chem. Soc.* **85,** 2677 (1963).
60. E. Block, "Reactions of Organosulfur Compounds." Academic Press, New York, 1978.
61. C. Walling and K. B. Wolfstirn, *J. Am. Chem. Soc.* **69,** 852 (1947).
62. T. W. Abbott and J. R. Johnson, *Org. Synth. Collect. Vol.* **1,** 440 (1941).
63. C. S. Marvel and D. W. Hein, *J. Am. Chem. Soc.* **70,** 1897 (1948).
64. R. H. Wiley and N. R. Smith, *J. Am. Chem. Soc.* **70,** 2296 (1948).
65. O. H. Wheeler and H. N. B. De Pabon, *J. Org. Chem.* **30,** 1473 (1965).
66. C. G. Overberger and R. E. Allen, *J. Am. Chem. Soc.* **68,** 722 (1946).
67. G. B. Bachman and H. M. Hellman, *J. Am. Chem. Soc.* **70,** 1772 (1948).
68. W. G. Young, R. T. Dillon, and H. J. Lucas, *J. Am. Chem. Soc.* **51,** 2528 (1929).
69. C. C. Price and J. M. Judge, *Org. Synth.* **45,** 22 (1965).
70. W. S. Emerson and T. M. Patrick, Jr., *Org. Synth. Collect. Vol.* **4,** 980 (1963).
71. N. H. Cromwell, *Org. Synth. Collect. Vol.* **3,** 125 (1955).
72. U. Steiner and H. Schinz, *Helv. Chim. Acta* **34,** 1126 (1951).
73. J. Cason, *J. Org. Chem.* **18,** 850 (1953).
74. D. Reisdorf and H. Normant, *C.R. Hebd. Seances Acad. Sci., Ser. C* **268,** 959 (1969).
75. F. Hunziker and F. X. Müllner, *Helv. Chim. Acta* **41,** 70 (1958).
76. T. Sasaki, S. Eguchi, and T. Ogawa, *J. Org. Chem.* **39,** 1927 (1974).
77. E. V. Dehmlow, *Angew. Chem., Int. Ed. Engl.* **13,** 170 (1924).
78. L. C. Swallen and C. E. Boord, *J. Am. Chem. Soc.* **52,** 654 (1930).
79. H. B. Dykstra, J. F. Lewis, and C. E. Boord, *J. Am. Chem. Soc.* **52,** 3396 (1930).
80. F. J. Soday and C. E. Boord, *J. Am. Chem. Soc.* **55,** 3293 (1933).
81. B. H. Shoemaker and C. E. Boord, *J. Am. Chem. Soc.* **53,** 1505 (1931).
82. P. S. Skell and S. R. Sandler, *J. Am. Chem. Soc.* **80,** 2024 (1958).
83. S. R. Sandler, *Diss. Abstr.* **21,** 61 (1960).
84. P. S. Skell, R. E. Glick, S. R. Sandler, and L. Gatlin, *4th Rep. Petrol. Res. Fund, 1959* p. 82 (1959).
85. S. R. Sandler, *Org. Synth.* **77,** 32 (1977).
86. S. R. Sandler, *J. Org. Chem.* **32,** 3876 (1967).
87. J. Reucroft and P. G. Sammes, *Q. Rev., Chem. Soc.* **25,** 135 (1971).

88. J. Sonnenberg and S. Winstein, *J. Org. Chem.* **27,** 748 (1962).
89. E. Bergman, *J. Org. Chem.* **28,** 2210 (1963).
90. W. von E. Doering and P. M. La Flame, *Tetrahedron Lett.* **2,** 75 (1958).
91. W. R. Moore and H. R. Ward, *J. Org. Chem.* **27,** 4179 (1962).
92. S. R. Sandler, *J. Org. Chem.* **32,** 3876 (1967).
93. S. R. Sandler, Ph.D. Thesis, Pennsylvania State University, University Park (1960).
94. F. C. Whitmore and A. H. Homeyer, *J. Am. Chem. Soc.* **55,** 4556 (1933).
95. C. L. Wilson, *J. Chem. Soc.* p. 50 (1945).
96. J. C. Sauer, *Org. Synth. Collect. Vol.* **4,** 268 (1963).
97. P. E. Spoerri and M. J. Rosen, *J. Am. Chem. Soc.* **72,** 4918 (1950).
98. R. N. Hazeldine, *J. Chem. Soc.* p. 2040 (1954).
99. L. F. Fieser, and M. Fieser, "Reagents in Organic Synthesis," Vol. 1, p. 1089. Wiley, New York, 1967.
100. S. Dershowitz and S. Proskauer, *J. Org. Chem.* **26,** 3595 (1961).
101. K. M. Ibne-Rasa, *Chem. Ind. (London)* p. 1418 (1966).
102. C. E. Castro and W. C. Kray, *J. Am. Chem. Soc.* **85,** 2768 (1963).
103. D. M. Singleton and J. K. Kochi, *J. Am. Chem. Soc.* **90,** 1582 (1968).
104. E. T. McBee, *J. Am. Chem. Soc.* **77,** 4942 (1955).
105. R. J. Strunk, P. M. DiGiacoma, K. Aso, and H. G. Kuivila, *J. Am. Chem. Soc.* **92,** 2849 (1970).
106. J. A. Beel, H. C. Clark, and D. Whyman, *J. Chem. Soc.* p. 4423 (1962).
107. E. S. Lo, *J. Org. Chem.* **36,** 364 (1971).
108. G. Wittig and G. Geisler, *Justus Liebigs Ann. Chem.* **580,** 44 (1953).
109. G. Wittig and U. Scholtkopf, *Chem. Ber.* **87,** 1318 (1954).
110. G. Wittig and W. Haag, *Chem. Ber.* **88,** 1654 (1955).
111. U. Schoellkopf, *Angew. Chem.* **71,** No. 8, 260 (1959).
112. A. Maercker, *Org. React.* **14,** 270 (1965).
113. M. Schlosser, G. Muller, and K. F. Christman, *Angew. Chem., Int. Ed. Engl.* **5,** 667 (1966).
114. H. O. House, V. K. Jones, and G. A. Frank, *J. Org. Chem.* **29,** 3327 (1962).
115. D. E. Bissing, *J. Org. Chem.* **30,** 1296(1965).
116. G. Wittig and M. Schlosser, *Angew. Chem.* **72,** 324 (1960).
117. A. J. Speziale, G. J. Marco, and K. W. Rats, *J. Am. Chem. Soc.* **82,** 1260 (1960).
118. D. Seyferth, S. O. Grim, and T. O. Read, *J. Am. Chem. Soc.* **82,** 1510 (1960).
119. S. A. Fuqua, W. G. Duncan, and R. M. Silverstein, *J. Org. Chem.* **30,** 1027 (1965).
120. E. E. Schweizer and G. J. O'Neill, *J. Org. Chem.* **30,** 2082 (1965).
121. R. Greenwald, M. Chaykovsky, and E. J. Corey, *J. Org. Chem.* **28,** 1128 (1963).
122. G. Wittig and U. Schoellkopf, *Org. Synth.* **40,** 66 (1960).
123. E. J. Corey and M. Chaykowsky, *J. Am. Chem. Soc.* **84,** 866 (1962).
124. E. J. Corey and H. Yamamoto, *J. Am. Chem. Soc.* **92,** 226 (1970).
125. S. Trippett, *Q. Rev., Chem. Soc.* **17,** 406 (1963).
126. J. Reucroft and P. G. Sammes, *Q. Rev., Chem. Soc.* **25,** 135 (1971).
127. C. E. Griffin, K. R. Martin, and B. E. Douglas, *J. Org. Chem.* **27,** 1627 (1962).
128. W. S. Wadsworth, Jr. and W. D. Emmons, *J. Am. Chem. Soc.* **83,** 1733 (1961).
129. D. H. Wadsworth, Jr., D. E. Schupp, III, E. J. Seus, and J. A. Ford, Jr., *J. Org. Chem.* **30,** 680 (1965).
129a. J. Boutagy and R. Thomas, *Chem. Rev.* **74,** 87 (1974).
129b. M. J. Jorgenson and T. Leung, *J. Am. Chem. Soc.* **90,** 3769 (1968).
129c. I. Agranat, M. Rabinovitz, and W.-C. Shaw, *J. Org. Chem.* **44,** 1936 (1979).
130. O. H. Wheeler and H. H. B. De Pabon, *J. Org. Chem.* **30,** 1473 (1965).
131. A. H. Ford-Moore and J. H. Williams, *J. Chem. Soc.* p. 1465 (1947).

132. G. M. Kosolapoff, "Organo Phosphorous Compounds," 1st ed., p. 121. Wiley, New York, 1950.
133. H. Zimmer, P. J. Bercz, O. J. Maltenieks, and M. W. Moore, *J. Am. Chem. Soc.* **87,** 2777 (1965).
134. W. E. Hanford and D. L. Fuller, *Ind. Eng. Chem.* **40,** 1171 (1948).
135. W. Reppe, *Experientia* **5,** 109 (1949).
136. H. E. Fierz-David and H. Zollinger, *Helv. Chim. Acta* **28,** 1130 (1945).
137. R. W. Wiley and G. M. Brauer, *J. Polym. Sci.* **3,** 708 (1948).
138. P. J. Pare, E. M. Smolin, and K. W. Saunders, *Ind. Eng. Chem.* **38,** 239 (1946).
139. M. Makosza, J. Czyzewski, and J. Jawdosiuk, *Org. Synth.* **55,** 99 (1976).
140. R. H. Wiley, *Org. Synth. Collect. Vol.* **3,** 853 (1955).
141. S. R. Sandler, *J. Chem. Eng. Data* **14,** 503 (1969).
142. W. Schäfer, *Angew. Chem., Int. Ed. Engl.* **5,** No. 7, 699 (1966).
143. R. Pettit, *Abstr., Natl. Org. Chem. Symp., 20th, 1967* pp. 21–32 (1967).
144. G. F. Emerson, L. Watts, and R. Pettit, *J. Am. Chem. Soc.* **87,** 131 (1965).
145. M. Baches, *Bull. Soc. Chim. Pr.* **1,** No. 5, 1101 (1934).
146. W. M. Kraft, *J. Am. Chem. Soc.* **70,** 3570 (1948).
147. W. Wayne and H. Adkins, *J. Am. Chem. Soc.* **62,** 3401 (1940).
148. G. H. McCain, *J. Org. Chem.* **23,** 632 (1958).
149. E. D. Bergmann, *Chem. Rev.* **68,** 41 (1968).
150. L. M. Long and H. D. Troutman, *J. Am. Chem. Soc.* **71,** 2470 (1949).
151. H. Meerwein, E. Buchner, and E. K. Van Emster, *J. Prakt. Chem.* [2] **152,** 237, 242, 256 (1939).
152. R. C. Fuson and H. G. Cooke, Jr., *J. Am. Chem. Soc.* **62,** 1180 (1940).
153. F. Bergmann and D. Schapiro, *J. Org. Chem.* **14,** 795 (1949).
154. F. Bergmann, J. Weizman and D. Schapiro, *J. Org. Chem.* **9,** 408 (1944).
155. M. P. Dreyfus, *J. Org. Chem.* **28,** 3269 (1963).
156. P. Tarrant and J. Heyes, *J. Org. Chem.* **30,** 1485 (1965).
157. D. J. Foster, German Patent 124,948 (1962).
158. M. S. Kharasch and E. Sternfeld, *J. Am. Chem. Soc.* **61,** 2318 (1939).
159. D. R. Howton, *J. Org. Chem.* **14,** 7 (1949).
160. M. S. Kharasch, W. Nudenberg, and E. Sternfeld, *J. Am. Chem. Soc.* **62,** 2034 (1940).
161. G. Linstrumelle, J. A. Krieger, and G. M. Whitesides, *Org. Synth.* **55,** 103 (1976).
162. E. J. Corey, C. V. Kim, R. H. K. Chen, and M. Takeda, *J. Am. Chem. Soc.* **94,** 4395 (1972).
163. E. J. Corey and R. L. Carney, *J. Am. Chem. Soc.* **93,** 7318 (1931).
164. G. M. Whitesides, C. P. Casey, and J. K. Krieger, *J. Am. Chem. Soc.* **93,** 1379 (1971).
165. A. Marfat, R. P. McGuirk, R. Kramer, and P. Helquist, *J. Am. Chem. Soc.* **99,** 253 (1977).
166. J. E. McMurry and L. R. Krepski, *J. Org. Chem.* **41,** 3929 (1976).
167. Y. Fujiwara, R. Ishikawa, F. Akiyama, and S. Teranishi, *J. Org. Chem.* **43,** 2477 (1978).
168. G. Cainelli, F. Bertini, P. Grasselli and G. Zubiani, Tetrahedron Lett. p. 5153 (1967).
169. P. J. Pearce, D. H. Richards, and N. F. Scilly, *Org. Synth.* **52,** 19 (1972).
170. K. C. Tsou, S. R. Sandler, and A. Astrup, U.S. Patent 3,069,459 (1962).
171. O. Diels, K. Alder, and P. Pries, *Chem. Ber.* **62,** 2081 (1929).
172. J. A. Norton, *Chem. Rev.* **31,** 319 (1942).
173. M. C. Kloetzel, *Org. React.* **4,** 1 (1948).
174. L. H. Flett and W. H. Gardner, "Maleic Anhydride Derivatives." Wiley, New York, 1952.
175. L. F. Fieser and F. C. Novello, *J. Am. Chem. Soc.* **64,** 802 (1942).
176. A. S. Onishchenko, "Diene Synthesis," Chapter 8, ref. 21. D. Davey Co., 257 Park Ave. S., New York, 1964.
177. R. E. Foster and R. S. Schreiber, *J. Am. Chem. Soc.* **70,** 2303 (1948).

§ References 79

178. H. Bluestone, R. E. Lidov, J. H. Knaus, and P. W. Howerton, U.S. Patent 2,576,666 (1951).
179. E. A. Prill, *J. Am. Chem. Soc.* **69,** 62 (1947).
180. G. L. Grunewald and D. P. Davis, *J. Org. Chem.* **43,** 3074 (1978).
181. H. Kwart and K. King, *Chem. Rev.* **68,** 415 (1968).
182. A. Wasserman, "Diels–Alder Reactions," Am. Elsevier, New York, 1965.
183. J. Sauer, *Angew. Chem., Int. Ed. Engl.* **6,** 16 (1967). L. L. Muller and J. Hamer, "1,2-Cycloaddition Reactions," Wiley (Interscience), New York, 1967.
184. A. L. Henne and K. W. Greenlee, *J. Am. Chem. Soc.* **65,** 2020 (1943).
185. A. T. Blomquist, L. H. Liu, and J. C. Bohrer, *J. Am. Chem. Soc.* **74,** 3643 (1952).
186. H. A. Dobson and R. A. Raphael, *J. Chem. Soc.* p. 3558 (1955).
187. H. S. Taylor and W. J. Shenk, *J. Am. Chem. Soc.* **63,** 2756 (1941).
188. R. Golse, *Ann. Chem. (Paris)* [12] **3,** 538 (1948).
189. K. Ahmad and F. M. Strong, *J. Am. Chem. Soc.* **70,** 1700 (1948).
190. E. F. Meyer and R. L. Burwell, Jr., *J. Am. Chem. Soc.* **85,** 2877, 2881 (1963).
191. D. J. Cram and N. L. Allinger, *J. Am. Chem. Soc.* **78,** 2518 (1956).
192. H. C. Brown and G. Zweifel, *J. Am. Chem. Soc.* **83,** 3834 (1961).
193. H. C. Brown and G. Zweifel, *J. Am. Chem. Soc.* **81,** 1512 (1959).
194. G. Zweifel, N. L. Polston, and C. C. Whiteney, *J. Am. Chem. Soc.* **90,** 6243 (1968).
195. J. J. Eisch and G. A. Damasevitz, *J. Org. Chem.* **41,** 2214 (1976).
196. J. Reucroft and P. G. Sammes, *Chem. Soc. Rev.* **25,** 135 (1971).
197. *Chem. Eng. News* **57,** April 16, p. 18 (1979).
198. A. J. Birch, *Q. Rev., Chem. Soc.* **4,** 69 (1950).
199. A. J. Birch, *Q. Rev., Chem. Soc.* **12,** 17 (1958).
200. E. Vogel, W. Klug, and A. Breuer, *Org. Synth.* **54,** 11 (1974).
201. R. A. Benkeser, M. L. Burrows, J. J. Hazdra, and E. M. Kaiser, *J. Org. Chem.* **28,** 1094 (1963).
202. H. L. Dryden, Jr., G. M. Webber, R. R. Burtner, and J. A. Cella, *J. Org. Chem.* **26,** 3237 (1961).
203. A. P. Krapcho and A. A. Bothner-By, *J. Am. Chem. Soc.* **81,** 3658 (1959).
204. P. Murgartha and P. Tissot, *Org. Synth.* **57,** 92 (1977).
205. H. Adkins and A. K. Roebuck, *J. Am. Chem. Soc.* **70,** 4041 (1948).
206. A. Maccoll, *in* "The Chemistry of Alkenes" (S. Patai, ed.), Chapter 3. Wiley (Interscience), New York, 1964.
207. P. D. Gardner, *J. Am. Chem. Soc.* **85,** 1553 (1963).
208. T. W. Gibson and P. Strassburger, *J. Org. Chem.* **41,** 791 (1976).
209. S. J. Rhoads, J. K. Chattopadhyay, and E. E. Waali, *J. Org. Chem.* **35,** 3352 (1970).
210. A. J. Hubert, *Chem. Ind. (London)* p. 975 (1968).
211. J. K. Crandall and L. C. Crawley, *Org. Synth.* **53,** 17 (1973).
212. D. S. Tarbell, *Org. React.* **2,** 8 (1944).
213. L. Claisen, *Justus Liebigs Ann. Chem.* **418,** 97 (1919).
214. E. N. Marvell, B. Richardson, R. Anderson, J. L. Stephenson, and T. Crandall, *J. Org. Chem.* **30,** 1032 (1965).
215. D. B. Denney, C. J. Rossi, and J. J. Vill, *J. Am. Chem. Soc.* **83,** 3336 (1961).
216. E. J. Corey and E. Black, *J. Org. Chem.* **34,** 1233 (1969).
217. C. Y. Meyers, A. M. Malte, and W. S. Matthews, *J. Am. Chem. Soc.* **91,** 7510 (1969).
218. N. P. Newreiter, *J. Org. Chem.* **30,** 1313 (1965).
219. L. A. L. Paquette, *J. Am. Chem. Soc.* **86,** 4383 (1964).
220. J. Fishman, M. Torigoe, and H. Gazik, *J. Org. Chem.* **28,** 1443 (1963).
221. J. E. Hofmann, T. J. Wallace, P. A. Argabright, and A. Schnesheim, *Chem. Ind. (London)* p. 1243 (1963).

222. L. Vargha and E. Kovacs, *Chem. Ber.* **75,** 794 (1942).
223. G. Eglinton and M. C. Whiting, *J. Chem. Soc.* p. 3650 (1950).
224. J. H. Wood, J. A. Bacon, A. W. Meibohm, W. H. Throckmorton, and G. P. Turner, *J. Am. Chem. Soc.* **63,** 1334 (1941).
225. W. H. Glaze, J. Lin, and E. G. Felton, *J. Org. Chem.* **30,** 1258 (1965).
226. R. Anschutz, *Ber. Dtsch. Chem. Ges.* **18,** 1945 (1855).
227. S. M. Spatz, *J. Org. Chem.* **26,** 4158 (1961).
228. D. H. R. Barton and B. J. Willis, *J. Chem. Soc., Perkin Trans. 1* p. 305 (1972).
229. D. H. R. Barton, F. S. Guziec, Jr., and I. Shahak, *J. Chem. Soc., Perkin Trans. 1* p. 1974 (1974).
230. R. Oda, *Rep. Inst. Chem. Res., Kyoto Univ.* **14,** 84 (1947).
231. N. J. Leonard and S. Gelfand, *J. Am. Chem. Soc.* **77,** 3277 (1955).
232. J. H. Raley, R. D. Millineux, L. H. Slaugh, and C. W. Bittner, *J. Am. Chem. Soc.* **85,** 3174, 3178, 3180 (1963).
233. J. N. Hornibrook, *Chem. Ind. (London)* p. 872 (1962).
234. M. Schmeisser, H. Schroter, H. Schilder, J. Massone, and F. Rosskopf, *Chem. Ber.* **95,** 1648 (1962).
235. F. Fisher and D. E. Applequist, *J. Org. Chem.* **30,** 2089 (1965).
236. A. W. Anderson, *Chem. Eng. News* **54,** April 12, p. 5 (1976).
237. J. C. Tatlow, *Chem. Ind. (London)* p. 522 (1978).
238. J. Melton and J. E. McMurry, *J. Org. Chem.* **40,** 2138 (1975).
239. *Chem. Eng. News* **43,** 42 (1965).
240. N. F. Cywinski, *J. Org. Chem.* **30,** 361 (1965).
241. *Chem. Eng. News* **43,** 41 (1965).
242. Imperial Chemical Industries, Australian Patent 257,018 (1965).
243. J. H. Wood, J. A. Bacon, A. W. Meibohm, W. H. Throckmorton, and G. P. Turner, *J. Am. Chem. Soc.* **63,** 1334 (1941).
244. F. G. Fischer and D. Stoffers, *Justus Liebigs Ann. Chem.* **500,** 253 (1933).
245. W. Reppe, *Justus Liebigs Ann. Chem.* **582,** 1 (1953).
246. G. A. Russell, H. Becker, and J. Schoeb, *J. Org. Chem.* **28,** 3584 (1963).
247. S. R. Sandler and K. C. Tsou, *J. Phys. Chem.* **68,** 300 (1964).
248. J. F. Arens and D. A. Dorp, *Nature (London)* **160,** 189 (1947).
249. W. O. Herrmann and W. Haehnel, U.S. Patent 2,245,131 (1941).
250. P. S. Skell and L. D. Westcott, Jr., *J. Am. Chem. Soc.* **85,** 1023 (1963).
251. P. S. Skell, L. D. Wescott, J. P. Goldstein, and R. R. Engel, *J. Am. Chem. Soc.* **87,** 2829 (1965).
252. H. D. Hartzler, *J. Am. Chem. Soc.* **83,** 4990, 4997 (1961).
253. G. Hata, *J. Am. Chem. Soc.* **86,** 3903 (1964).
254. H. Takahasi, S. Tai, and M. Yamaguchi, *J. Org. Chem.* **30,** 1661 (1965).
255. Mitsubishi Petrochikh, German Patent 2,833,367 (1979).
256. J. P. Neilan, R. M. Laine, N. Cortese, and R. F. Heck, *J. Org. Chem.* **41,** 3455 (1976).
257. A. H. Andrist, J. E. Baldwin, and R. K. Pinschmidt, Jr., *Org. Synth.* **55,** 15 (1976).
258. G. F. Hennion and J. J. Sheehan, *J. Org. Chem.* **71,** 1964 (1949).
259. I. Iwai and J. Ide, *Org. Synth.* **50,** 62 (1970).
260. G. R. Trenes, *J. Am. Chem. Soc.* **70,** 875 (1948).
261. A. G. Green and J. Baddiley, *J. Chem. Soc.* **93,** 1721 (1908).
262. *Chem. Eng. News* **42,** 46 (1964).
263. K. Stockel and F. Sondheimer, *Org. Synth.* **54,** 1 (1974).
264. C. E. Castro and R. D. Stevens, *J. Chem. Soc.* **86,** 4358 (1964).
265. E. Vedejs and E. S. C. Wu, *J. Org. Chem.* **39,** 3641 (1974).

References

266. I. Elphimoff-Felkin and P. Sarda, *Org. Synth.* **56,** 101 (1977).
267. S. M. Kupchan and M. Maruyama, *J. Org. Chem.* **36,** 1187 (1971).
268. P. W. Chum and S. E. Wilson, *Tetrahedron Lett.* p. 15 (1976).
269. E. C. Ashby and J. J. Lin, *Tetrahedron Lett.* p. 4481 (1977).
270. F. J. Hoover, O. W. Webster, and C. T. Hardy, *J. Org. Chem.* **26,** 2234 (1961).
271. A. C. Cope and E. M. Hardy, *J. Am. Chem. Soc.* **62,** 441 (1940).
272. G. S. Hammond and C. D. De Boer, *J. Am. Chem. Soc.* **86,** 899 (1964).
273. W. G. Young, *J. Chem. Educ.* **39,** 455 (1962).
274. W. D. Huntsman and H. J. Wristers, *J. Am. Chem. Soc.* **89,** 342 (1967).
275. R. Breslow, *in* P. DeMayo, "Molecular Rearrangements," (P. DeMayo, ed.). Part 1, Chapter 4, Wiley, New York, 1963; G. L. Closs and P. E. Pfeffer, *J. Am. Chem. Soc.* **90,** 2452 (1968).
276. R. R. Rando and W. von E. Doering, *J. Org. Chem.* **33,** 1671 (1968).
277. C. W. Bird, "Transition Metal Intermediates in Organic Synthesis," Chapter 3. Academic Press, New York, 1967.

CHAPTER 3 / ACETYLENES

1. Introduction	82
2. Elimination Reactions	83
2-1. Preparation of 1-Butyne from 1,2-Dibromobutane	84
2-2. Preparation of Propyne from 1-Bromo-1-propene	84
2-3. Preparation of p-Tolylacetylene from 1-p-Tolyl-1-chloroethylene	84
2-4. Preparation of Diphenylacetylene by Deoxygenation	86
3. Condensation Reactions	86
3-1. Preparation of 3-Nonyne	87
3-2. Preparation of 1-Ethynyl-1-cyclohexanol	87
3-3. Preparation of 1-Ethynyl-1-cyclohexanol from Lithium Acetylide Complexed with Ethylenediamine	87
3-4. Preparation of 1-Pentyne-3-ol	89
3-5. Preparation of p-Methoxydiphenylacetylene	90
3-6. Preparation of Di-p-nitrophenylacetylene	91
4. Oxidation Reactions	91
4-1. Preparation of 2,4-Hexadiyne-1,6-diol	91
4-2. Preparation of Cyclodecyne	92
5. Rearrangement Reactions	93
6. Miscellaneous Methods	94
References	94

1. INTRODUCTION

The two most important synthetic methods for introducing an acetylenic group into the molecule involve the elimination of hydrogen halides [Eq. (1)] or condensation with acetylenic derivatives.

$$\left. \begin{array}{l} RCX{=}CH_2 \\ RCX_2{-}CH_3 \end{array} \right\} \xrightarrow{\text{Base}} RC{\equiv}CH \tag{1}$$

Condensation reactions of alkyl halides and carbonyl compounds with organometallic derivatives of acetylene or with acetylene itself are quite useful in the laboratory and in industry [1].

$$R_2C{=}O + HC{\equiv}CH \longrightarrow R_2\underset{OH}{C}{-}C{\equiv}CH \tag{2}$$

$$RC{\equiv}C{-}M + R'X \longrightarrow RC{\equiv}CR' + MX \tag{3}$$

where M = metal.

Oxidation reactions involving coupling of acetylenic compounds are also important procedures for the preparation of diacetylenic compounds. In addition, the oxidation of dihydrazones from diketones is also a good procedure for preparing disubstituted acetylenes.

Terminal alkyl acetylenes undergo simultaneous polymerization, isomerization, and degradation at temperatures of 500°–600°C. In the presence of basic catalysts, isomerization to allenes occurs as shown in Eq. (4).

$$(CH_3)_2CH-C\equiv CH \xrightarrow[\substack{150°C \\ 6\ hr}]{\substack{C_2H_5OH \\ KOH,}} (CH_3)_2C=C=CH_2 \qquad (4)$$

Acetylene can be handled safely at atmospheric pressure but great care should be exercised in handling it. Special precautions are required for pressure reactions because of the explosion hazard.

Divinylacetylene is very hazardous because it reacts with oxygen in a rapid manner to form an explosive peroxidic polymer. This may also be true for other acetylenic compounds and great care should be exercised in handling them.

Several reviews [2–4b] are available which describe additional information on the synthesis and chemistry of acetylenes.

2. ELIMINATION REACTIONS

Hydrogen halides can be eliminated either from 1,1- , or 1,2-dihalogenated hydrocarbons or from 1-halo olefins to yield acetylenes in good yields.

$$\left.\begin{array}{c} RC=CH,\ RCH=CH-X \\ | \\ X \\ RCH-CH_2,\ RCH_2CHX_2 \\ |\ \ \ | \\ X\ \ \ X \end{array}\right\} \xrightarrow{Base} RC\equiv CH \qquad (5)$$

$$\begin{array}{ccc} RCH=C-R & RCX_2CH_2R,\ RCH-CHR \\ | & |\ \ \ | \\ X & X\ \ \ X \end{array} \xrightarrow{Base} RC\equiv CR \qquad (6)$$

The most frequently used bases in the above dehydrohalogenations are finely divided potassium hydroxide [5] and sodium amide [6]. Alcoholic potassium hydroxide [7] tends to cause the isomerization of 1-acetylenes to internal acetylenes (Favorskii rearrangement). Aromatic acetylenes are not effected. Sodium amide may cause the reverse rearrangement from internal acetylenes to 1-acetylenes [8]. Impure sodium amide may be an ineffective reagent and should not be used since dangerous (explosive) peroxides may be present [9].

Other methods used less frequently are the decomposition of quaternary ammonium salts [10], the diazo decomposition of vinylamines [11], and the thermal decomposition of tosylazoethylenes [12].

Recently Kocienski reported that dimethyl sulfoxide and potassium t-butroxide easily dehydrochlorinate pinacolone dichloride to t-butylacetylene [13].

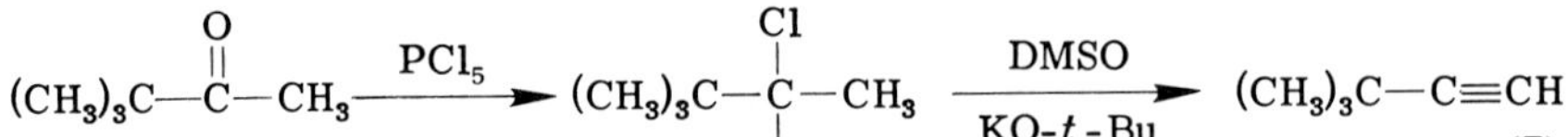

$$(CH_3)_3C-\overset{O}{\overset{\|}{C}}-CH_3 \xrightarrow{PCl_5} (CH_3)_3C-\overset{Cl}{\underset{Cl}{\overset{|}{C}}}-CH_3 \xrightarrow[KO-t-Bu]{DMSO} (CH_3)_3C-C\equiv CH \quad (7)$$

This is an improvement over the earlier procedure of Bartlett and Rosen [14] who used a sodium hydroxide melt for the dehydrochlorination reaction. The use of DMSO–KOt-Bu should also be useful in other dehydrochlorinations to give acetylenes.

2-1. Preparation of 1-Butyne from 1,2-Dibromobutane [15]

$$CH_3CH_2-\underset{Br}{\overset{|}{C}H}-\underset{Br}{\overset{|}{C}H_2} + KOH \xrightarrow{C_2H_5OH} CH_3CH_2-C\equiv CH \quad (8)$$

To a flask equipped with an addition funnel, mechanical stirrer, and a condenser whose exit is connected to a Dry Ice trap containing 145 gm (26 moles) of potassium hydroxide and 145 ml of 95% alcohol is added 100 gm (0.46 mole) of 1,2-dibromobutane, dropwise. The mixture is heated in an oil bath and the evolved ethylacetylene is passed through the reflux condenser into a cold trap ($-18°C$) to yield 17 gm (31%) of product, bp 18°C.

2-2. Preparation of Propyne from 1-Bromo-1-propene [16]

$$CH_3-CH=CH-Br + KOH \xrightarrow{C_4H_9OH} CH_3C\equiv CH + KBr + H_2O \quad (9)$$

In a flask equipped as in Procedure 2-1, a stirred refluxing solution of 450 gm (8.1 moles) of potassium hydroxide in 1 liter of n-butyl alcohol is treated dropwise over a period of 4 hr with 242 gm (2 moles) of 1-bromo-1-propene. Propyne, 68 gm (85%), bp 27°–31°C, is obtained by trapping the vapors from the condenser in the Dry Ice trap.

2-3. Preparation of p-Tolylacetylene from 1-p-Tolyl-1-chloroethylene [17]

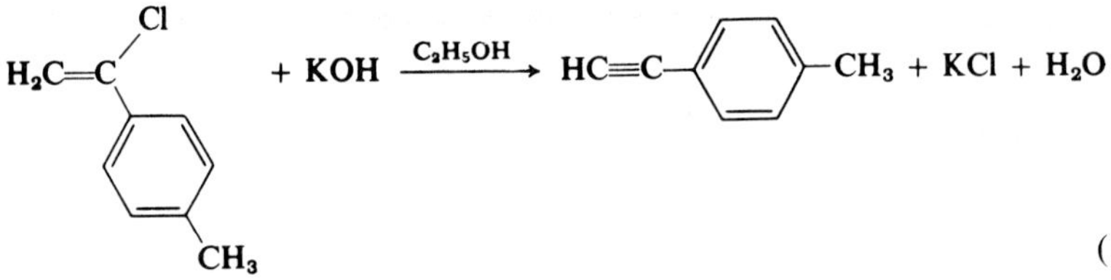

$$H_2C=C\overset{Cl}{\diagup}\text{-}C_6H_4\text{-}CH_3 + KOH \xrightarrow{C_2H_5OH} HC\equiv C\text{-}C_6H_4\text{-}CH_3 + KCl + H_2O \quad (10)$$

§ 2. Elimination Reactions

To a flask containing 50 gm (0.78 mole) of potassium hydroxide and 100 ml of dry ethanol is added 85 gm (0.56 mole) of 1-*p*-tolyl-1-chloroethylene. The mixture is refluxed for 24 hr and poured into 1 liter of ice water. The resulting oil layer is separated. The water layer is extracted with ether and the combined organic phases are dried over potassium hydroxide, concentrated, and distilled to yield 31 gm (48%) of *p*-tolylacetylene, bp 79°–82°C (31–33 mm).

Diphenylacetylenes (tolanes) can be obtained either by the dehydrohalogenation [7] of stilbene dibromide or by the rearrangement occurring in the dehydrohalogenation of unsymmetrical diaryl haloolefins [18].

$$\begin{array}{c} \text{ArCH—CHAr} \\ || \\ \text{Br}\text{Br} \end{array} \xrightarrow[\text{C}_2\text{H}_5\text{OH}]{\text{KOH}}$$

$$\longrightarrow \text{ArC}\equiv\text{CAr} \qquad (11)$$

$$(\text{Ar})_2\text{C}=\text{CHX} \xrightarrow{\text{KNH}_2}$$

Substituents such as halogen [19] or nitro [20] groups attached to the aromatic ring are stable during the dehydrohalogenation reaction.

While the alkaline hydroxides are the most commonly used dehydrohalogenation reagents, other bases are also effective in causing dehydrohalogenation to acetylenes. For example, sodium in alcohol has been used in the preparation of tolanes from benzal chloride [20].

$$2\ \text{C}_6\text{H}_5\text{CHCl}_2 \xrightarrow{\text{Na} + \text{C}_2\text{H}_5\text{OH}} \text{C}_6\text{H}_5\text{C}\equiv\text{CC}_6\text{H}_5 \qquad (12)$$

Solid potassium hydroxide has also been used for dehydrohalogenation. In this case, vinyl halide is dropped into KOH at 130°–200°C and the product is distilled from the reaction mixture at reduced pressure [21, 22]. Trans elimination is observed as with the other oxygen bases. Sodium amide, however, gives eliminations even from the cis position [23, 24]. Liquid ammonia is said to be a better solvent for cis elimination than aprotic solvents [25].

Organoalkali metal compounds can also be used to effect dehydrohalogenations. For example, two moles of phenyllithium react with β-chlorostyrene to yield phenylacetylene on hydrolysis [26]. The use of *n*-butyllithium in ether yields the same product at room temperature [27].

Phenyllithium also cleaves vinyl ethers at room temperature to yield acetylenes and lithium alkoxide at a rate slower than in dehydrohalogenations [28].

Zinc dust has also been used to effect the elimination of halogens. For example, zinc dust in alcohol converts 1,1,1,4,4,4-hexafluoro-2,3-dibromobutene to perfluoro-2-butyne in 90% yield [29, 30].

Some other preparative methods are the dehydrogenation–decarboxylation of cinnamic acid dibromide [31]. Arylpropiolic acids are also decarboxylated to aryl acetylenes when refluxed with water [32].

It has been reported [33] that the deoxygenation of α-diketones by triethyl phosphite yields diaryl- and alkylaryl-substituted acetylenes in yields ranging from 24 to 60%.

Dialkyl acetylenes could not be obtained by the above reaction.

2-4. Preparation of Diphenylacetylene by Deoxygenation [33]

$$2\ C_6H_5-\underset{\underset{O}{\|}}{C}-\underset{\underset{O}{\|}}{C}-C_6H_5 + 2\ (C_2H_5-O)_3P \xrightarrow[4\ hr]{215°C} C_6H_5-C\equiv C-C_6H_5 + 2\ (C_2H_5O)_3PO$$

(13)

Triethyl phosphite (1.69 gm, 0.01 mole) is added to 210 gm (0.01 mole) of benzil with stirring under nitrogen. The benzil dissolves with the evolution of heat. Then triethyl phosphite (8.30 gm, 0.05 mole) is added again, and the resulting mixture is heated in a sealed tube for 4 hr at 215°C under nitrogen. The reaction mixture is distilled under reduced pressure to yield 1.43 gm (81%) of diphenylacetylene, bp 113°–115°C (2 mm).

The preparation of small ring alkynes by elimination reactions has been reviewed [34]. The actual isolation of cyclic alkynes of heterocyclic rings and benzynes was not accomplished but their presence was inferred by capturing the intermediates as Diels–Alder adducts. It has been reported that cycloalkynes can be conveniently generated near room temperature by the dehydrochlorination of 1-chlorocycloalkanes with lithium piperidine [35]. Cyclooctyne is the smallest ring size cycloalkyne that is stable [36] (see also Preparation 4-2 in this chapter).

3. CONDENSATION REACTIONS

An important reaction for the preparation of substituted acetylenes involves the condensation of metallo acetylenes with alkyl halides or with carbonyl compounds.

Sodium acetylide [37, 38] (prepared from sodium amide) is useful for the condensations with primary alkyl halides. However, secondary, tertiary, and primary halides branched at the second carbon atom are dehydrohalogenated to olefins by the reagent [39, 40]. Iodides react at a faster rate than bromides and the latter faster than chlorides. Chlorides are rarely used. The bromides are more common for preparative reactions [41]. Sodium acetylide can also react with carbonyl compounds to yield acetylenic carbinols [42].

The synthesis of lithium acetylide–ethylenediamine complex has been reported; it is a white, free-flowing powder that is safe and stable up to about 45°C

§ 3. Condensation Reactions

[43–45]. This complex reacts with ketones to give excellent yields of ethynyl carbinols. The complex can either be prepared or obtained from a commercial source [46].

3-1. Preparation of 3-Nonyne [47]

$$C_5H_{11}-C\equiv CH + NaNH_2 \xrightarrow{NH_3} C_5H_{11}C\equiv C-Na \xrightarrow{C_2H_5Cl} C_5H_{11}C\equiv C-C_2H_5 \quad (14)$$

To a flask containing a cold well-stirred mixture of 1 liter of ammonia and 40 gm (1.0 mole) of sodamide is added 50 gm (0.52 mole) of 1-heptyne. The mixture is stirred for 1 hr and ammonia is added in order to maintain the volume. To the mixture is added 75 gm (1.2 mole) of ethyl chloride and stirring is continued for 3 hr. The reaction is worked up by adding water to separate the oil, washing the latter with water, drying, and distilling to yield 15 gm (23%) of 3-nonyne, bp 151°–154°C.

NOTE: The sodamide should be freshly prepared or obtained from a good commercial source. Long exposure to air and oxygen reduces the effectiveness of the material and may also cause the formation of explosive peroxides, oxides, etc.

3-2. Preparation of 1-Ethynyl-1-cyclohexanol [48]

$$HC\equiv C-Na + \underset{\text{cyclohexanone}}{\bigcirc}\!\!=\!\!O \xrightarrow[\text{2. } H_3O^+]{\text{1. } NH_3} \underset{\text{1-ethynyl-1-cyclohexanol}}{\bigcirc}\!\!\begin{matrix}OH\\ C\equiv CH\end{matrix} \quad (15)$$

To a flask containing a mixture of 5.1 moles of sodium acetylide, prepared during a period of 3 hr from 117 gm of sodium and acetylene in 3 liters of liquid ammonia at −50°C, is added dropwise 5.1 moles (500 gm) of cyclohexanone. The mixture is stirred overnight while a slow stream of dry acetylene is passed through the solution. The ammonia is then evaporated, the residue acidified with 200 gm of tartaric acid in 500 ml of water, and the mixture extracted with ether. The ether layer is dried and evaporated. The residue is fractionated to yield 518 gm (82%) of 1-ethynyl-1-cyclohexanol, bp 74°–77°C (15 mm), n_D^{20} 1.4823 and mp 31°–32°C.

3-3. Preparation of 1-Ethynyl-1-cyclohexanol from Lithium Acetylide Complexed with Ethylenediamine [44, 45]

Lithium acetylide ethylenediamine is available commercially [46] or can be made as described below.

To a flask equipped with a stirrer, dropping funnel, and condenser fitted with a T-tube for argon inlet and outlet is added 40.1 gm (0.40 mole) of lithium acetylide ethylenediamine, 200 ml of N,N-dimethylacetamide, and 200 ml of benzene. An argon atmosphere is maintained throughout the reaction. The mixture is warmed to 35°C and 39.2 gm (0.04 mole) of cyclohexanone is added dropwise over a period of 15 min with cooling to maintain 35°C. The mixture is then stirred at room temperature for 1¾ hr, hydrolyzed with 100 ml water, refluxed for 1 hr, cooled, and the organic layer separated to give upon distillation 41.7 gm (84%) of product, bp 178°C, mp 32°–33°C.

$$LiNHCH_2CH_2NH_2 + HC\equiv CH \longrightarrow$$

$$LiC\equiv CH \cdot H_2NCH_2CH_2NH_2 \xrightarrow[\text{2. H}_2\text{O}]{\text{1. Cyclohexanone}} \underset{}{\text{cyclohexyl}}\begin{matrix}OH\\C\equiv CH\end{matrix} + H_2NCH_2CH_2NH_2 \quad (16)$$

Lithium acetylide ethylenediamine [44]. To a flask equipped with a stirrer, dropping funnel, and a condenser fitted with a T-tube for argon inlet and outlet are added 13.9 gm (2 moles) of lithium metal powder and then 400 ml of dry benzene. To the stirred mixture under an argon atmosphere is added dropwise 120 gm (2 moles) of ethylenediamine over a period of 1 hr while maintaining gentle reflux. The mixture is refluxed for an additional 2 hr, cooled, filtered under argon pressure, and washed several times with hexane to give 66 gm (97%) of N-lithioethylenediamine, a white, free-flowing cyrstalline solid.

To a 500-ml two-necked stainless steel flask equipped with a gas inlet attachment, thermometer, and stirrer is added 92.4 gm (1.4 moles) of N-lithioethylenediamine followed by 350 ml of 1,4-dioxane. The mixture is stirred vigorously and 78 gm (3 moles) of acetylene is introduced over a period of 1 hr. The reaction is exothermic and external cooling is necessary to keep the temperature at 25 C. The addition is continued for ½ hr after the evolution of heat has ceased. The mixture is poured into 500 ml of hexane, filtered under an argon atmosphere, washed with pentane to remove ethylenediamine, and dried under argon pressure to give 122 gm (95%) of the off-white solid product.

Substituted acetylenes may be obtained by reacting acetylenic Grignard reagents with alkyl halides or alkyl sulfonates [49–51]. The Grignard reagent can be prepared readily by the reaction of ethylmagnesium bromide in ether with terminal acetylenes containing one unsubstituted position.

Propargyl halides can be used to couple with Grignard reagents to give acetylenic hydrocarbons in good yields [52].

Potassium hydroxide can also effect ethynylations of aldehydes and ketones to yield secondary and tertiary acetylenic carbinols or glycols [53].

$$R_1R_2C=O + CH\equiv CH \xrightarrow{KOH} \underset{OH \quad (A)}{R_1R_2C-C\equiv CH}$$

and/or

$$\underset{OH \qquad OH}{R_1R_2C-C\equiv C-C-R_1R_2}$$

(B) \qquad (17)

§ 3. Condensation Reactions

Potassium hydroxide ethynylations are usually performed in liquid ammonia under pressure (160–185 psig). The butyne-3-ols formed above (A) act as co-catalysts by forming a complex catalyst with potassium hydroxide and acetylene. Sodium or lithium hydroxides do not ordinarily form complex adducts which are able to act as catalysts and therefore fail to effect ethynylation. Under special conditions, sodium hydroxide adducts can be preformed with the butyne-3-ols and acetylene and will then give ethynylations. Rubidium hydroxide behaves as well as potassium hydroxide in the above reactions.

An atmospheric pressure ethynylation procedure for aldehydes has been reported [54] to be effected by finely dispersed potassium hydroxide in 1,2-dimethoxyethane at $-10°$ to $0°C$. Aldehydes that give good yields of acetylenic carbinols are propionaldehyde, butyraldehyde, isobutyraldehyde, 2-ethyl-2-hexanol, crotonaldehyde, 3-cyclohexenecarboxaldehyde, 3,4-dihydro-2,5-dimethyl-2-formyl-2H-pyran, and 2-methylbicyclo[2.2.1]-5-heptene-2-carboxaldehyde. However, methacrolein, phenylacetaldehyde, cinnamaldehyde, and 2,4-hexadienol gave no isolatable acetylenic carbinols. The above procedure is safe since it does not operate under pressure with acetylene. The use of a Waring Blendor enables one to obtain a finely dispersed potassium hydroxide for the ethynylations.

3-4. Preparation of 1-Pentyne-3-ol [54]

$$CH_3CH_2CH{=}O + CH{\equiv}CH \xrightarrow{KOH} CH_3CH_2\underset{\underset{OH}{|}}{CH}{-}C{\equiv}CH \qquad (18)$$

Potassium hydroxide is broken up to a fine powder in 1,2-dimethoxyethane at $-5°C$ during 0.5 hr. The suspension is transferred to a flask and diluted to obtain a concentration of 6.67 moles of potassium hydroxide in 1600 gm of 1,2-dimethoxyethane. After adding 11 gm of ethanol, the rapidly stirred mixture is saturated with acetylene at $-10°$ to $0°C$. While acetylene is continuously being added during the reaction (an excess was used), $3\frac{1}{3}$ moles of propionaldehyde containing 11 gm of ethanol is added during a 2-hr period. The mixture is allowed to stir an additional 0.5 hr at $-10°$ to $0°C$. Ice water (894 gm) is used to decompose the reaction mixture and the water layer is extracted with ether. The combined oil layer and ether are neutralized with carbon dioxide, filtered, and distilled to yield 226 gm (81%) of 1-pentyne-3-ol.

Diarylacetylenes (tolanes) have been prepared [55] by the oxidation of benzil dihydrazones with mercuric oxide [56], rearrangement of 1,1-diaryl-2-halo-ethylenes upon treatment with base [57, 58], dehydrohalogenation of stilbene dibromide [59], and by the reaction of base with 5,5-diaryl-3-nitroso-2-oxazolidones [60].

It has been found that tolanes can be prepared in good yields by coupling aryl iodides with cuprous acetylides in refluxing pyridine [61]. However, when the aryl iodide has an ortho nucleophilic substituent, cyclization to heterocycles such as isocoumarins, benzofurans, and indoles occurs under the reaction conditions.

The coupling reactions involving acetylenes are covered in Section 4, Oxidation Reactions, in this chapter. In addition, thionyl chloride has been reported to couple phenylethynylmagnesium bromide [62].

$$2\ p\text{-}CH_3\text{—}C_6H_4C\equiv C\text{—}MgBr\ +\ SOCl_2\ \xrightarrow{\text{THF } 0°C}$$

$$(p\text{-}CH_3C_6H_4C\equiv C\text{—})_2\ +\ 2\,MgBrCl\ +\ SO \quad (19)$$
$$\text{mp } 183°C$$

Phenylethynylmagnesium bromide has also been coupled with propargyl bromide in the presence of cuprous chloride using THF as a solvent [63].

$$C_6H_5\text{—}C\equiv C\text{—}MgBr\ +\ BrCH_2C\equiv CH\ \longrightarrow\ C_6H_5\text{—}C\equiv C\text{—}CH_2C\equiv CH \quad (20)$$

3-5. Preparation of p-Methoxydiphenylacetylene [61]

$$p\text{-}CH_3O\text{—}C_6H_4I\ +\ CuC\equiv C\text{—}C_6H_5\ \longrightarrow\ p\text{-}CH_3O\text{—}C_6H_4\text{—}C\equiv C\text{—}C_6H_5\ +\ CuI \quad (21)$$

Cuprous phenylacetylide is prepared by adding a mixture of an aqueous ammoniacal solution of 20.0 gm (0.105 mole) of cuprous iodide to 10.7 gm (0.105 mole) of phenylacetylene in 500 ml of ethanol. After standing for 15 min, the precipitate is filtered and washed five times each with water, ethanol, and ether. The yellow solid is dried in a rotary evaporator for 2 hr at 50°C (20 mm), yield 13.4 gm (77%).

Cuprous phenylacetylide, 5.0 gm (0.030 mole), is added portionwise to a flask containing 7.1 gm (0.030 mole) of p-iodoanisole in 100 ml of dry pyridine. Extremely dry conditions must be maintained throughout and nitrogen is used to flush the flask before and after each addition. The mixture is warmed to 120°C for 10 hr with stirring. The clear reddish amber liquid is then added to 300 ml of water, extracted with ether, and the combined ether extracts are washed successively three times each with dilute hydrochloric acid, 5% sodium bicarbonate, and water. The dried ether layer is concentrated under reduced pressure, and the residue is recrystallized from methanol (with a carbon treatment) to yield white platelets weighing 6.2 gm (99%) of p-methoxydiphenylacetylene, mp 58°–59°C.

The synthesis of tolanes under very mild conditions has been effected using appropriately substituted phosphonates [64]. Diphenyl phosphite is condensed with an aryl aldehyde to afford 1-hydroxy-1-arylmethane phosphonate. The latter is converted by means of $POCl_3$ to 1-chloro-1-arylmethane phosphonate which upon treatment with another equivalent of the aryl aldehyde and two equivalents of sodium hydride in DMSO or THF is converted to an acetylene. The scope and limitations of this reaction have not been completely investigated.

3-6. Preparation of Di-p-nitrophenylacetylene [64]

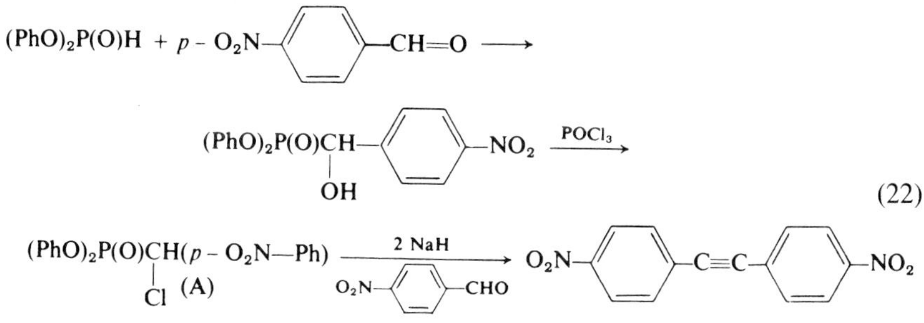

$(PhO)_2P(O)H + p\text{-}O_2N\text{-}C_6H_4\text{-}CH=O \longrightarrow$

$(PhO)_2P(O)CH(OH)\text{-}C_6H_4\text{-}NO_2 \xrightarrow{POCl_3}$

$(PhO)_2P(O)CH(Cl)(p\text{-}O_2N\text{-}Ph)$ (A) $\xrightarrow[O_2N\text{-}C_6H_4\text{-}CHO]{2\ NaH}$ $O_2N\text{-}C_6H_4\text{-}C{\equiv}C\text{-}C_6H_4\text{-}NO_2$ (22)

To a flask containing 20.2 gm (0.05 mole) of (A) and 7.55 gm (0.05 mole) of p-nitrobenzaldehyde are added 250 ml of dry DMSO and one equivalent of sodium hydride (1.2 gm). After the reaction has subsided, another equivalent of sodium hydride is added. After the reaction has been completed, the deep red mixture is diluted with water and twice extracted with 500-ml portions of ether. The combined extracts are dried, concentrated under reduced pressure, and the residue is dissolved in methanol. The solution is filtered and twice recrystallized from methanol–water to yield 8.8 gm (67%) of di-p-nitrophenylacetylene, mp 214°–215°C.

4. OXIDATION REACTIONS

Glaser [65] discovered that acetylenes are converted to diacetylenes by oxidative coupling of the copper(I) acetylides. Glaser employed a mixture of copper(I) chloride, ammonia, and ammonium chloride as the coupling reagent in the presence of oxygen.

Recently [66] it was found that propargyl alcohol or its esters can be coupled in good yields when amines are substituted for ammonia in the reaction mixture.

4-1. Preparation of 2,4-Hexadiyne-1,6-diol [66]

$$HC{\equiv}C\text{-}CH_2OH \xrightarrow{(O)} HOCH_2C{\equiv}C\text{-}C{\equiv}CCH_2OH \quad (23)$$

To a flask containing 25 gm (0.13 mole) of copper(I) chloride, 12.5 ml of concentrated ammonium hydroxide (0.18 mole based on NH_3), 40 gm (0.75 mole) of ammonium chloride, and 200 ml of water is added 11.2 gm (0.20 mole) of propargyl alcohol. The mixture is stirred under approximately 30 mm of oxygen pressure for 20 hr. The blue-green reaction mixture is acidified with dilute hydrochloric acid, diluted to 750 ml with water, and extracted with ether continuously for 24 hr. The ether solution is concentrated to yield 9.1 gm (83%) of 2,4-hexadiyne-1,6-diol, mp 111.5°–112°C.

Chodkiewicz and Cadiot reported that terminal acetylenes reacts with a 1-bromoacetylene in the presence of a cuprous salt catalyst and an amine to give diacetylenes [67].

$$RC\equiv CH + CR'\equiv CBr \xrightarrow[H_2NOH \cdot HCl]{Cu_2Cl_2, \ C_2H_5NH_2} RC\equiv C-C\equiv CR' + HBr \quad (24)$$

For example, where
R′ = CH=CH—COOH [68]
R = O$_2$N—C$_6$H$_4$ [68],
1-Hydroxycyclopentyl,
1-hydroxycyclohexyl [68]

Other oxidation procedures utilize potassium ferricyanide to oxidize cuprous acetylides to diacetylenes [69]. Dihydrazones of aryldiketones [70] or of aliphatic diketones [71] are oxidized by mercuric oxide to acetylenes. The use of lead tetraacetate in place of mercuric oxide has also been reported [72].

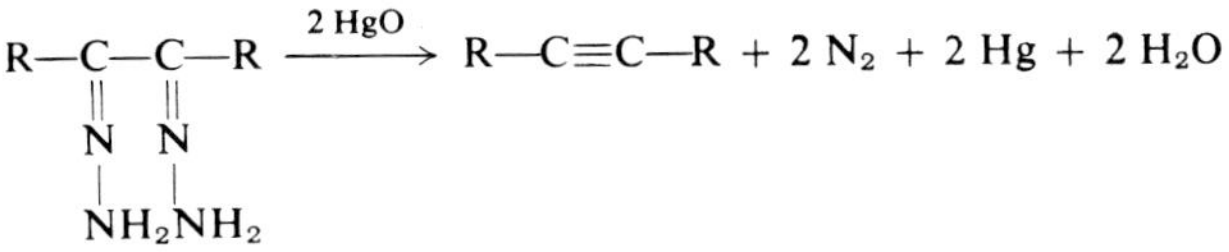

$$\underset{\underset{NH_2\,NH_2}{\underset{|\quad|}{N\quad N}}}{R-C-C-R} \xrightarrow{2\,HgO} R-C\equiv C-R + 2\,N_2 + 2\,Hg + 2\,H_2O \quad (25)$$

Furthermore, 2,7-dimethylocta-3,5-diyne-2,5-diol is prepared in good yield by the oxidative coupling of 3-methyl-1-butyn-3-ol in aqueous solution using a catalytic amount of cuprous chloride, solubilized by excess ammonium chloride [73]. Oxygen is more effective than air as the oxidant but the latter may be employed.

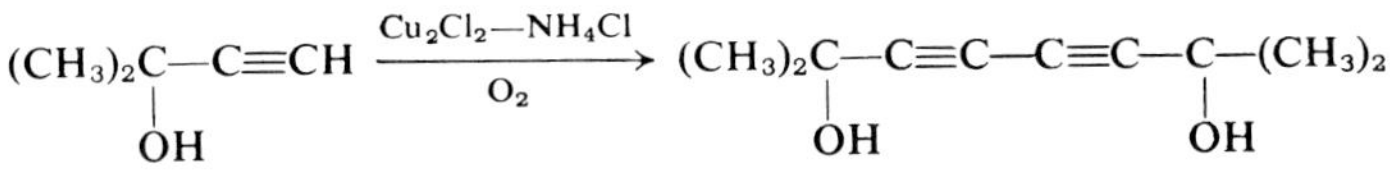

$$\underset{OH}{(CH_3)_2C}-C\equiv CH \xrightarrow[O_2]{Cu_2Cl_2-NH_4Cl} \underset{OH}{(CH_3)_2C}-C\equiv C-C\equiv C-\underset{OH}{C}-(CH_3)_2 \quad (26)$$

4-2. Preparation of Cyclodecyne [74]

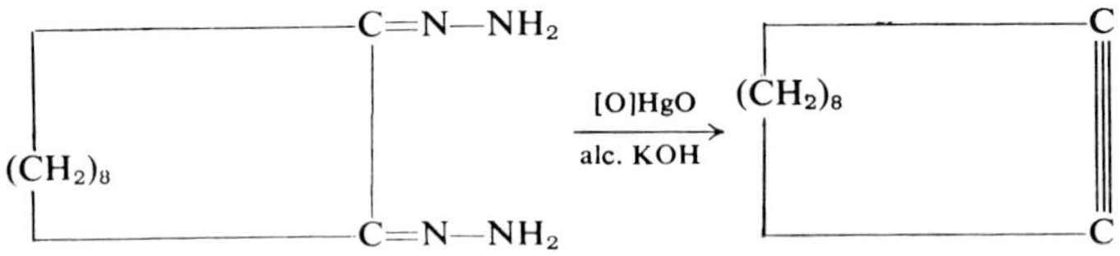

$$\begin{array}{c} \text{cyclic}(CH_2)_8\text{-C(=N-NH}_2\text{)-C(=N-NH}_2\text{)} \end{array} \xrightarrow[\text{alc. KOH}]{[O]HgO} (CH_2)_8\text{-C}\equiv\text{C (cyclic)} \quad (27)$$

To a flask containing a refluxing solution of 204 gm (1.04 mole) of sebacil dihydrazone in 1 liter of dry benzene is added 50 gm (0.23 mole) of yellow mercuric oxide. The reaction commences when 80 gm of anhydrous sodium sulfate and 2 ml of a saturated KOH solution in 95% ethanol are added. A vigorous evolution of nitrogen immediately occurs along with a simultaneous color change from orange to black. After the initial reaction subsides, further additions of mercuric oxide and alcoholic potassium hydroxide are made until 563 gm (2.6 moles) of mercuric oxide and 20 ml of KOH solution have been added. The mixture is stirred and refluxed for about 70 hr. The benzene solution is filtered, concentrated under reduced pressure, and the resulting residue distilled to yield 51.5 gm (36%) of cyclodecyne, bp 59°–60°C (5–6 mm), n_D^{20} 1.4903. (Sebacil is 1,2-cyclodecanedione.)

Oxidative coupling can be used to prepare polyacetylenic compounds [75].

$$HC\equiv C-R-C\equiv CH \longrightarrow [-C\equiv C-R-C\equiv C-]_n \qquad (28)$$

5. REARRANGEMENT REACTIONS

Acetylenic compounds undergo rearrangement either to allenes or internal acetylenes. This is illustrated by the preparation of 1-phenyl-1,3-pentadiene [76].

$$C_6H_5-C\equiv CMgBr + HC\equiv C-CH_2Br \xrightarrow[THF]{Cu_2Cl_2} C_6H_5-C\equiv C-CH_2-C\equiv CH$$

$$\downarrow base \qquad (29)$$

$$C_6H_5-C\equiv C-C\equiv C-CH_3 \xleftarrow{base} C_6H_5-C\equiv C-CH=C=CH_2$$

Evidence has been reported that the ethynylallenes may be isolated by quenching the reaction with acid. Furthermore, allenes can be prepared from acetylenes by the propargyl rearrangement reaction by prototropic, anionotropic displacement and

$$X-C-C\equiv C- + Y \longrightarrow \!\!\!>\!\!C=C=C-Y + X \qquad (30)$$

intermolecular rearrangement reactions [77]. On the other hand, allenes undergo base-catalyzed rearrangement to 2-alkynes [78].

Terminal alkynes also can be converted to 2-alkynes by the use of alcoholic potassium hydroxide at elevated temperature [79].

The dehydrochlorination of 1,1-diphenyl-2-chloroethylenes [80] and the thermal degradation [81] of 2,2-diphenyl-1-tosylazoethylenes rearrange to give 1,2-diphenylethylenes (tolanes).

Highly strained cycloalkynes have been reported to occur by the reaction of potassium *t*-butoxide on bromomethylenecycloalkanes [82].

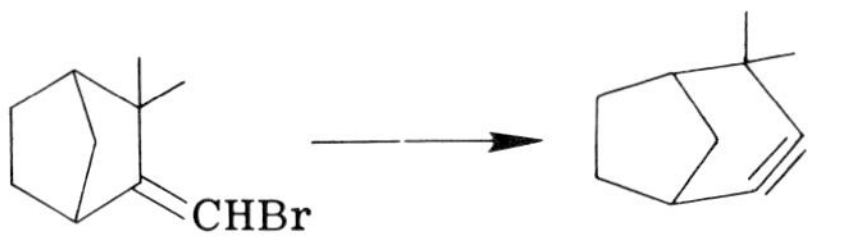

$$\text{(structure with CHBr)} \longrightarrow \text{(strained cycloalkyne structure)} \qquad (31)$$

94   3. Acetylenes

6. MISCELLANEOUS METHODS

(1) Isomerization of allenes [83].
(2) Oxidation of acetylene to vinylacetylene [84].
(3) Reppe synthesis of acetylenic amides [85].
(4) The synthesis of acetylenic amides [86].
(5) The Prevost reaction to give substituted acetylene [87].
(6) Exhaustive methylation to give acetylenes [88].
(7) Elimination of carbon monoxide to yield acetylenes [89].
(8) Pyrolysis of fluoromaleic anhydride to fluoroacetylene [90].
(9) Photolytic elimination of nitrogen and sulfonate of alkali salts of N-(p-toluenesulfonamide)triazoles to acetylenes [91].
(10) Pyrolysis of acylcarbomethoxymethylene triphenyl phosphoranes [92].
(11) Ketene acetals yield acetylenes [93].
(12) Pyrolysis of halogenated acrylic acid to acetylenes [94].
(13) Preparation of dialkynyl thioethers [95].
(14) Acetylenic sugars [96].
(15) Acetylenic derivatives of ruthenocene [97].
(16) Acetylenic sulfoxides [98].
(17) Preparation of 2-alkynoic esters [99].
(18) Preparation of lithium acetylides [100].
(19) Acetylenes by the reaction of iodine with lithium 1-alkynyltriorganoborates [101].
(20) Preparation of 1-pyridylacetylenes [102].
(21) Preparation of ethynyltriethylammonium bromide [103].
(22) Acetylenes by decomposition of acetylenic sulfones [104].
(23) Diacetylenes by the base-catalyzed decomposition of diynediols [105].
(24) Preparation of propiolonitrile [106].
(25) Condensation of allenes with ketones to give acetylenic alcohols [107].
(26) Preparation of allenes and acetylenes from ethynylalkanol acetates via organoboranes [108].

REFERENCES

1. J. J. Nedwick, *Ind. Eng. Chem., Process Des. Dev.* **1**, 137 (1962).
2. T. L. Jacobs, *Org. React.* **5**, 1 (1949).
3. R. A. Raphael, "Acetylenic Compounds in Organic Synthesis." Academic Press, New York, 1955.
4a. G. Kobrich, *Angew. Chem., Int. Ed. Engl.* **4**, 49 (1965).
4b. T. F. Rutledge, "Acetylenic Compounds," Van Nostrand-Reinhold, Princeton, New Jersey, 1968.
5. G. B. Bachman and A. J. Hill, *J. Am. Chem. Soc.* **56**, 2730 (1934).
6. R. Lespieau and M. Bourguel, *Org. Synth. Collect. Vol.* **1**, 191 (1941).
7. L. I. Smith and M. M. Falkof, *Org. Synth.* **22**, 50 (1942).
8. M. Bourguel, *Ann. Chim. (Paris)* [10] **3**, 191, 325 (1925).

§ References

9. T. H. Vaugh, *J. Am. Chem. Soc.* **56,** 2064 (1934).
10. C. R. Hauser, H. M. Taylor, and T. G. Ledford, *J. Am. Chem. Soc.* **82,** 1786 (1960).
11. D. V. Curtin, J. A. Kampmeier, and B. R. O'Connor, *J. Am. Chem. Soc.* **87,** 863 (1965).
12. G. Rosini and S. Cacchi, *J. Org. Chem.* **37,** 1856 (1972).
13. P. J. Kocienski, *J. Org. Chem.* **39,** 3285 (1974).
14. P. D. Bartlett and L. J. Rosen, *J. Am. Chem. Soc.* **64,** 543 (1942).
15. J. R. Johnson and W. L. McEwen, *J. Am. Chem. Soc.* **48,** 469 (1926).
16. G. B. Heisig and H. M. Davis, *J. Am. Chem. Soc.* **57,** 339 (1935).
17. L. I. Smith and H. H. Hoehn, *J. Am. Chem. Soc.* **63,** 1175 (1941).
18. G. H. Coleman and R. D. Maxwell, *J. Am. Chem. Soc.* **56,** 132 (1934).
19. C. Dufraisse and A. Dequesnes, *Bull. Soc. Chim. Fr.* [4] **49,** 1880 (1931).
20. P. Ruggli and F. Lang, *Helv. Chim. Acta* **21,** 138 (1938).
21. E. Ott, *Ber. Dtsch. Chem. Ges. B* **75,** 1517 (1942).
22. F. Krafft and L. Reuter, *Ber. Dtsch. Chem. Ges.* **25,** 2243 (1892).
23. E. Ott and G. Dittus, *Ber. Dtsch. Chem. Ges. B* **76,** 80 (1943).
24. H. G. Viehe, *Angew. Chem.* **75,** 638 (1963).
25. A. T. Bottini, B. J. King, and J. M. Lucas, *J. Org. Chem.* **27,** 3688 (1962).
26. G. Wittig and H. Witt, *Ber. Dtsch. Chem. Ges. B* **74,** 1474 (1941).
27. H. Gilman, W. Langham, and F. W. Moore, *J. Am. Chem. Soc.* **62,** 2327 (1940).
28. G. Wittig and G. Harborth, *Ber. Dtsch. Chem. Ges. B* **77,** 306 (1944).
29. A. L. Henne and W. G. Finnegan, *J. Am. Chem. Soc.* **71,** 298 (1949).
30. R. N. Haszeldine, *J. Chem. Soc.* p. 2504 (1942).
31. M. T. Bogert and D. Davidson, *J. Am. Chem. Soc.* **54,** 334 (1932).
32. M. M. Otto, *J. Am. Chem. Soc.* **56,** 1393 (1934).
33. T. Mukaiyama, H. Nambu, and T. Kumamoto, *J. Org. Chem.* **29,** 2243 (1964).
34. G. Wittig, *Angew. Chem.* **1,** 415 (1962).
35. L. K. Montgomery, A. D. Clouse, A. M. Crelier, and L. E. Applegate, *J. Am. Chem. Soc.* **89,** 3453 (1967).
36. G. Wittig and A. Krebs, *Chem. Ber.* **94,** 3260 (1961).
37. I. M. Heilbron, E. R. H. Jones, and B. C. L. Weedon, *J. Chem. Soc.* p. 81 (1945).
38. K. N. Campbell and B. K. Campbell, *Org. Synth. Collect. Vol.* **4,** 117 (1963).
39. T. H. Vaughn, G. F. Hennison, R. R. Vogt, and J. A. Nieuwland, *J. Org. Chem.* **2,** 1 (1938).
40. T. F. Rutledge, *J. Org. Chem.* **24,** 840 (1959).
41. A. L. Henne and K. W. Greenlee, *J. Am. Chem. Soc.* **67** 484 (1945).
42. A. W. Johnson, "The Chemistry of Acetylenic Compounds," Vol. 1, Edward Arnold & Co., London, 1946.
43. *Chem. Eng. News* **40,** 48 (1962).
44. O. F. Beumel, Jr. and R. F. Harris, *J. Org. Chem.* **29,** 1872 (1964).
45. O. F. Beumel, Jr. and R. F. Harris, *J. Org. Chem.* **28,** 2775 (1963).
46. Foote Mineral Co., Exton, Pennsylvania.
47. T. H. Vaughn, R. R. Vogt, and J. A. Nieuwland, *J. Am. Chem. Soc.* **56,** 2120 (1934).
48. N. A. Milas, N. S. MacDonald, and D. M. Black, *J. Am. Chem. Soc.* **70,** 1829 (1948).
49. J. R. Johnson, A. M. Schwartz, and T. L. Jacobs, *J. Am. Chem. Soc.* **60,** 1882 (1938).
50. S. D. Thorn, G. F. Hennion, and J. A. Nieuwland, *J. Am. Chem. Soc.* **58,** 796 (1936).
51. J. R. Johnson, T. L. Jacobs, and A. M. Schwartz, *J. Am. Chem. Soc.* **60,** 1887 (1938).
52. K. N. Campbell and L. T. Eby, *J. Am. Chem. Soc.* **62,** 1798 (1940).
53. R. J. Tedeschi, *J. Org. Chem.* **30,** 3045 (1965).
54. H. A. Stansbury, Jr. and W. R. Proops, *J. Org. Chem.* **27,** 279 (1962).
55. S. H. Harper, *in* "Chemistry of Carbon Compounds" (E. H. Rodd, ed.), Vol. III, pp. 1157–1158. Elsevier, Amsterdam, 1951.

56. A. C. Cope, D. S. Smith, and R. J. Cotter, *Org. Synth.* **34,** 42 (1954).
57. G. H. Coleman, W. H. Holst, and R. D. Maxwell, *J. Am. Chem. Soc.* **58,** 2310 (1936).
58. D. H. Curtin, E. W. Flynn, R. F. Wyström, and W. H. Richardson, *Chem. Ind. (London)* p. 1453 (1957).
59. L. I. Smith and M. M. Falkof, *Org. Synth. Collect. Vol.* **3,** 350 (1955).
60. M. S. Newman and A. Kutner, *J. Am. Chem. Soc.* **73,** 4199 (1951).
61. R. D. Stephens and C. E. Castro, *J. Org. Chem.* **28,** 3313 (1963).
62. A. Uchida, T. Nakazawa, I. Kondo, N. Iwata, and S. Matsuda, *J. Org. Chem.* **37,** 3749 (1972).
63. H. Taniguchi, I. M. Mathai, and S. I. Miller, *Org. Synth.* **50,** 97 (1970).
64. H. Zimmer, P. J. Bercz, O. J. Maltenieks, and M. W. Moore, *J. Am. Chem. Soc.* **87,** 2778 (1965).
65. C. Glaser, *Ber. Dtsch. Chem. Ges.* **2,** 422 (1869).
66. M. D. Cameron and G. E. Bennett, *J. Org. Chem.* **22,** 557 (1957).
67. W. Chodkiewicz, and P. Cadlot, *Ann. Chim. (Paris)* **2,** 819 (1957); *C. R. Hebd. Seances Acad. Sci.* **245,** 322 (1957).
68. J. Meier, *C. R. Hebd. Seances Acad. Sci.* **245,** 1634 (1957).
69. J. B. Conn, G. B. Kistiakowsky, and E. A. Smith, *J. Am. Chem. Soc.* **61,** 1868 (1939).
70. A. C. Cope, D. S. Smith, and R. J. Cotter, *Org. Synth.* **34,** 42 (1954).
71. A. T. Blomquist and L. H. Liu, *J. Am. Chem. Soc.* **75,** 2153 (1953).
72. A. Krebs, *Tetrahedron Lett.* No. 43, p. 4511 (1968).
73. R. J. Tedeschi and A. E. Brown, *J. Org. Chem.* **29,** 2051 (1964).
74. A. T. Blomquiest, R. E. Burge, Jr., and A. C. Sucsy, *J. Am. Chem. Soc.* **74,** 3636 (1952).
75. F. Sondheimer and R. Wolovsky, *J. Am. Chem. Soc.* **84,** 260 (1962). E. R. H. Jones, *Proc. Chem. Soc., London* p. 199 (1960).
76. I. M. Mathai, H. Tanigachi, and S. I. Miller, *J. Am. Chem. Soc.* **89,** 115 (1967).
77. D. R. Taylor, *Chem. Rev.* **67,** 317 (1967).
78. V. A. Engelhardt, *J. Am. Chem. Soc.* **78,** 107 (1956).
79. T. L. Jacobs, R. Akavie, and R. G. Cooper, *J. Am. Chem. Soc.* **73,** 1273 (1957).
80. G. H. Coleman and R. D. Maxwell, *J. Am. Chem. Soc.* **56,** 132 (1934); G. H. Coleman, W. H. Holst, and R. D. Maxwell, *ibid.* **58,** 2310 (1936).
81. G. Rosini and S. Cacchi, *J. Org. Chem.* **37,** 1856 (1972).
82. K. L. Erickson and J. Wolinsky, *J. Am. Chem. Soc* **87,** 1142 (1965).
83. M. Bouis, *Ann. Chim. (Paris)* [10] **9,** 459 (1928).
84. K. Sugino, Y. Aiya, and K. Ariga, *J. Soc. Chem. Ind., Jpn.* **46,** 573 (1943).
85. J. W. Copenhaver and M. H. Bigelow, "Acetylene and Carbon Monoxide Chemistry," Van Nostrand-Reinhold, Princeton, New Jersey, 1949.
86. N. R. Easton and R. D. Dillard, *J. Org. Chem.* **28,** 2465 (1963).
87. C. V. Wilson, *Org. React.* **9,** 332 (1957).
88. Y. M. Slobodin and N. A. Selezneva, *Zh. Obsheh, Khim.* **26,** 691 (1956).
89. R. Breslow, R. Haynie, and J. Mirra, *J. Am. Chem. Soc.* **81,** 247 (1959).
90. W. J. Middleton and W. H. Sharkey, *J. Am. Chem. Soc.* **81,** 803 (1959).
91. F. G. Wiley, *Angew. Chem.* **76,** 144 (1964); *Angew. Chem., Int. Ed. Engl.* **3,** 138 (1964).
92. G. Markl, *Chem. Ber.* **94,** 3005 (1961).
93. S. M. McElvain and P. L. Weyna, *J. Am. Chem. Soc.* **81,** 2579 (1959).
94. A. Michael, *Ber. Dtsch. Chem. Ges.* **34,** 4215 (1901).
95. L. Brandsma and J. F. Arens, *Recl. Trav. Chim. Pays-Bas* **82,** A1119 (1963).
96. R. B. Roy and W. S. Chilton, *J. Org. Chem.* **36,** 3242 (1971).
97. M. D. Rausch and A. Siegel, *J. Org. Chem.* **34,** 1974 (1969).
98. G. A. Russel and L. A. Ochrymowycz, *J. Org. Chem.* **35,** 2106 (1970).

§ References

99. E. C. Taylor, R. L. Robey, and A. McKillop, *Angew. Chem., Int. Ed. Engl.* **11,** 48 (1972).
100. K. Okuhara, *J. Org. Chem.* **41,** 1487 (1976).
101. H. C. Brown, J. A. Sinclair, and M. M. Midland, *J. Am. Chem. Soc.* **95,** 3080 (1973); M. M. Midland, J. A. Sinclair, and H. C. Brown, *J. Org. Chem.* **39,** 731 (1974).
102. G. R. Newkome and D. L. Koppersmith, *J. Org. Chem.* **38,** 4461 (1973).
103. R. Tanaka and S. I. Miller, *J. Org. Chem.* **36,** 3856 (1971).
104. W. E. Truce and G. C. Wolf, *J. Org. Chem.* **36,** 1727 (1971).
105. R. J. Tedeschi and A. E. Brown, *J. Org. Chem.* **29,** 2051 (1964).
106. L. J. Krebaum, *J. Org. Chem.* **31,** 4103 (1966).
107. D. Gaillerm-Dron, M. L. Capman, and W. Chodkiewicz, *Tetrahedron Lett.* No. 1, p. 37 (1972).
108. M. M. Midland, *J. Org. Chem.* **42,** 2650 (1977).

CHAPTER 4 / ALCOHOLS AND PHENOLS

1. Introduction	99
2. Hydrolysis Reactions	102
2-1. Preparation of p-Isopropylphenol by the Hydrolysis of a Diazonium Compound	103
2-2. Preparation of p-Isopropylphenol by the Potassium Hydroxide Fusion of Sodium Cumenesulfonate	104
2-3. Preparation of 2.3-Pentanediol by Hydrolysis of an Epoxide	104
2-4. Preparation of Dihydroxyoctadecanes by the Addition of Formic Acid to Oleyl Alcohol	105
3. Condensation Reactions	105
3-1. Synthesis of 2,2-Dimethyl-1-hexanol Using a Grignard Reaction	106
3-2. Synthesis of 1-Octanol Using a Grignard Reaction	107
3-3. Preparation of β-2-Hydroxyethyl Naphthyl Ether Using Ethylene Carbonate	107
3-4. Synthesis of 2-(2,4-Dimethylphenyl)ethanol by the Friedel–Crafts Reaction	108
3-5. Preparation of α,α-Dimethyl-β-hydroxypropionaldehyde by the Aldol Condensation Reaction	108
3-6. Synthesis by Anisoin by the Benzoin Condensation	109
4. Reduction Reactions	111
4-1. Synthesis of Dimethylphenylethanols Using Sodium Borohydride	112
4-2. Reduction of Ethyl Stearate to 1-Octadecanol	112
4-3. Preparation of 4-Chlorophenylmethylcarbinol Using the Meerwein Reduction Reaction	114
5. Oxidation Reactions	115
5-1. Oxidation of Phenylmagnesium Bromide to Phenol	115
5-2. General Procedure for the Markovnikov Hydration of Olefins by the Oxymercuration-Demercuration Procedure	116
5-3. Preparation of trans-2-Methylcyclopentanol by the Hydroboration Reaction	117
5-4. Preparation of Anisyl Alcohol by the Crossed Cannizzaro Reaction	119
5-5. Preparation of 1,1,1-Tris(hydroxymethyl)-2-methylpropane	119
6. Rearrangement Reactions	120
A. Rearrangement of Ethers	120
6-1. Claisen Rearrangement	120
6-2. Rearrangement of Aryl Alkyl Ethers to Alcohols	120
6-3. Rearrangement of n-Alkyl Phenyl Ethers	120
B. Rearrangement of Aryl Hydroxylamines	120
C. Rearrangement of Aryl Esters to Phenolic Ketones	121
D. Rearrangement of Epoxides to Allylic Alcohols	121
E. Sigmatropic Rearrangements of Olefinic Alcohols	121
7. Miscellaneous Methods	122
References	123

1. INTRODUCTION

The more common and synthetically useful methods for the preparation of alcohols depend on hydrolysis, condensation (Grignard reagents condensing with carbonyl compounds or alkylene oxides), and reduction reactions of carbonyl or alkylene oxide compounds. Oxidation reactions are also of importance and will be described below. The more specialized reactions will be treated briefly.

Hydrolysis reactions involving alkyl or activated aryl halides, and sulfonates, or diazonium compounds afford alcohols in good yields. The products in some cases may be contaminated by olefin by-products. The Bucherer reaction of an aromatic amine can be considered a hydrolysis reaction. The generalized hydrolysis-type reaction may be represented by Eq. (1).

$$R-Z + H_2O \longrightarrow ROH + ZOH \qquad (1)$$

The hydration of olefins can be considered a hydrolysis reaction since the olefin on reaction with sulfuric acid yields an alkyl sulfuric acid which on subsequent hydrolysis yields the alcohol. Sulfuric acid adds to olefins in accordance with the Markovnikov rule as illustrated in Eq. (2) for isobutylene.

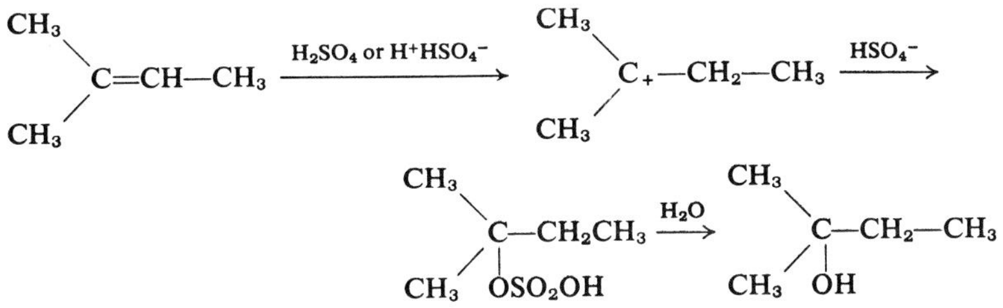

$$\begin{array}{c} CH_3 \\ \diagdown \\ C=CH-CH_3 \\ \diagup \\ CH_3 \end{array} \xrightarrow{H_2SO_4 \text{ or } H^+HSO_4^-} \begin{array}{c} CH_3 \\ \diagdown \\ C_+-CH_2-CH_3 \\ \diagup \\ CH_3 \end{array} \xrightarrow{HSO_4^-}$$

$$\begin{array}{c} CH_3 \\ \diagdown \\ C-CH_2CH_3 \\ \diagup | \\ CH_3 OSO_2OH \end{array} \xrightarrow{H_2O} \begin{array}{c} CH_3 \\ \diagdown \\ C-CH_2-CH_3 \\ \diagup | \\ CH_3 OH \end{array} \qquad (2)$$

Alkyl halides, aldehydes, ketones, and esters can be converted to alcohols of a higher carbon content by condensation reactions involving the Grignard reagent or a related organometallic (lithium compounds). The general scope of the reaction is illustrated below in Eq. (3).

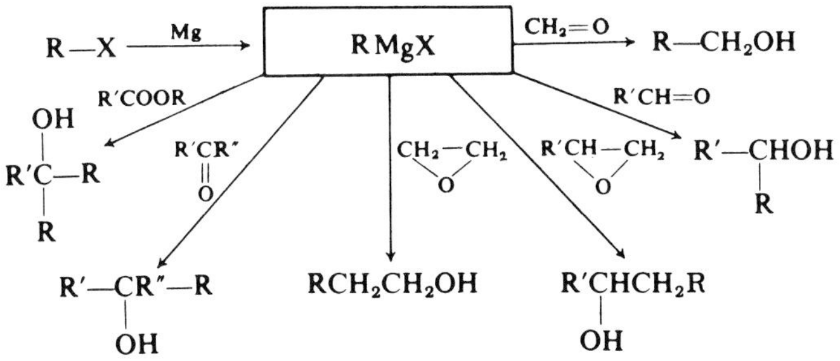

$$R-X \xrightarrow{Mg} \boxed{RMgX} \qquad (3)$$

The Friedel–Crafts condensation of ethylene oxide with aromatics gives aryl ethanols in good yields.

$$\text{ArH} + \underset{\underset{O}{\diagdown\diagup}}{CH_2 - CH_2} \xrightarrow{AlCl_3} \text{ArCH}_2\text{CH}_2\text{OH} \quad (4)$$

Some of the more specialized condensation reactions that are useful are the aldol condensation [Eqs. (5, 6)]. Reformatskii [Eq. (7)], acyloin [Eq. (8)], and benzoin condensations [Eq. (9)], alkylation of phenols by alcohols [Eq. (10)], Prins reaction [Eqs. (11, 12)], Tollens hydroxymethylation reaction [Eq. (13)], the condensation of ketones [Eq. (14)] and aldehydes [Eq. (15)] with acetylenes, and the Pinacol reaction [Eq. (16)].

$$2\, RCH_2C(R)=O \xrightarrow{OH^-} RCH_2-\underset{\underset{HO}{|}}{\overset{\overset{R}{|}}{C}}-\underset{\underset{R}{|}}{CH}-\overset{R}{C}=O \quad R=H,\ \text{alkyl or aryl} \quad (5)$$

$$RCH=O + R'CH_2NO_2 \xrightarrow{OH^-} RCH-\underset{\underset{OH}{|}}{\overset{\overset{R'}{|}}{CH}}-NO_2 \quad (6)$$

$$R_2C=O + RCH(Br)-COOC_2H_5 + Zn\ \text{or}\ Mg \longrightarrow R_2C-\underset{\underset{OH}{|}}{\overset{\overset{R}{|}}{CH}}-COOC_2H_5 \quad (7)$$

$$2\, RCOOC_2H_5 + Na \longrightarrow \underset{R-C-ONa}{\overset{R-C-ONa}{\|}} \longrightarrow \underset{R-C=O}{\overset{R-CH-OH}{|}} \quad (8)$$

$$2\, ArCH=O + NaCN \longrightarrow ArCH-CAr \quad (9)$$
$$\underset{OH}{|}\ \underset{O}{\|}$$

$$C_6H_5OH + ROH \xrightarrow{AlCl_3} R-C_6H_4OH \quad (10)$$

$$\underset{CH_3}{R_1-C}=\overset{R_2}{\underset{H}{C}} + CH_2=O + H_3O^+ \longrightarrow R_1-\underset{HO}{\overset{|}{C}}-\underset{CH_3\ R_2}{\overset{|}{CH}}-CH_2OH \quad (11)$$

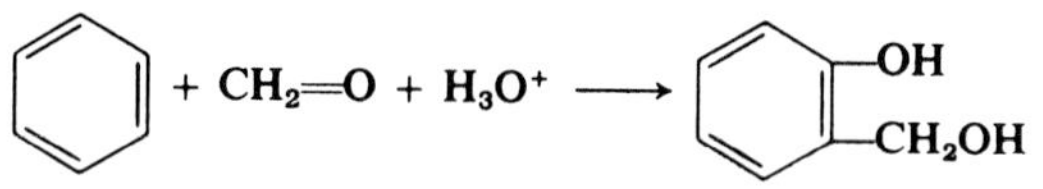

$$\underset{}{\bigcirc} + CH_2=O + H_3O^+ \longrightarrow \underset{}{\bigcirc}\begin{matrix}-OH\\-CH_2OH\end{matrix} \quad (12)$$

$$CH_3CH=O + 4\, CH_2=O + H_2O \xrightarrow[\text{Heat}]{Cu(OH)_2} C(CH_2OH)_4 + HCOOH \quad (13)$$

$$R_2C=O + HC\equiv CR' \xrightarrow{KOH} R_2C-\underset{OH}{\overset{|}{C}}\equiv CR' \quad (14)$$

§ 1. Introduction

$$\text{RCH=O} + \text{HC}\equiv\text{CR}' \xrightarrow{\text{KOH}} \underset{\underset{\text{OH}}{|}}{\text{RCH}}-\text{C}\equiv\text{CR}' \quad (15)$$

$$2\ \text{R}_2\text{C=O} \longrightarrow \underset{\underset{\text{HO}\ \ \text{OH}}{|\ \ \ |}}{\text{R}_2\text{C}-\text{CR}_2} \quad (16)$$

The reduction of carbonyl compounds or alkylene oxides by organometallic hydrides yields alcohols of known structure in good yields.

$$\underset{\underset{\text{O}}{\|}}{\text{R}-\text{C}-\text{R}'} + \text{NaBH}_4 \longrightarrow \underset{\underset{\text{OH}}{|}}{\text{R}-\text{CH}-\text{R}'} \quad (17)$$

$$\text{RCH=O} + \text{NaBH}_4 \longrightarrow \text{RCH}_2-\text{OH} \quad (18)$$

$$\underset{\text{O}}{\text{R}-\text{CH}-\text{CH}_2} + \text{LiAlH}_4 \longrightarrow \underset{\underset{\text{OH}}{|}}{\text{R}-\text{CH}-\text{CH}_3} \quad (19)$$

Lithium borohydride is selective in reducing aldehydes or ketones in the presence of esters or acids at 0°C. Refluxing for several hours reduces the esters and acids.

Sodium borohydride has the advantage over lithium borohydride and lithium aluminum hydride since it is soluble in water without decomposition and can effect reduction in either methanol or water of aldehydes and ketones. However, lithium aluminum hydride reacts violently with water and requires anhydrous conditions in order for it to be an effective reducing agent.

Other reducing agents are hydrogen in the presence of catalysts [Eq. (19)], zinc in the presence of alcoholic sodium hydroxide [Eq. (20)], and the Bouveault-Blanc method for esters using sodium metal in alcohol [Eq. (21)].

$$\text{RCHO} + \text{H}_2 + \text{Cat.} \longrightarrow \text{RCH}_2\text{OH} \quad (20)$$

$$\underset{\text{R}}{\overset{\text{R}}{>}}\text{C=O} \xrightarrow{\text{Zn alc. NaOH}} \text{R}_2\text{CHOH} \quad (21)$$

$$\text{RCOOR}' + \text{Na} + \text{ROH} \longrightarrow \text{RCH}_2\text{OH} + \text{ROH} \quad (22)$$

The Cannizzaro reaction can be considered a reduction-oxidation reaction of an aldehyde to the corresponding alcohol and acid.

$$\text{R}_3\text{C}-\text{CHO} + \text{NaOH} \longrightarrow \text{R}_3\text{CCH}_2\text{OH} + \text{R}_3\text{C}-\text{COOH} \quad (23)$$

The reaction occurs only if the aldehyde lacks an α-hydrogen atom.

The crossed Cannizzaro reaction is a modification using formaldehyde so that the desired aldehyde is reduced entirely to the corresponding alcohol whereas the formaldehyde is converted to sodium formate.

$$\text{RCH=O} + \text{CH}_2\text{=O} + \text{NaOH} \longrightarrow \text{RCH}_2\text{OH} + \text{HCOONa} \quad (24)$$

The Meerwein-Pondorf-Oppenaur-Verley reduction of aldehydes and ketones uses aluminum isopropoxide to reduce them to alcohols with acetone as the by-product.

$$R'CH{=}O + (CH_3)_2CHOH \xrightarrow{Al(iso\text{-}PrO)_3} R'CH_2OH + CH_3\overset{O}{\underset{\|}{C}}CH_3 \quad (25)$$

Formation of acetone is an indication of the extent of reaction. This reaction is not as popular as it was earlier since the borohydrides have been found more convenient to use in the laboratory as the reducing agent.

The oxidation of the Grignard reagents offers an alternate convenient method for the preparation of alcohols from halides.

$$RMgX \xrightarrow[\text{2. } H_2O]{\text{1. } O_2} ROH \quad (26)$$

In a related manner the oxidation of trialkylboranes prepared from olefins offers a convenient laboratory method for the conversion of an olefin into a saturated alcohol of the same carbon content.

$$RCH{=}CH_2 \xrightarrow[\text{2. } H_2O_2]{\text{1. } B_2H_6} RCH_2CH_2OH \quad (27)$$

The diborane can be made *in situ* using sodium borohydride and boron trifluoride. Isolation of the substituted boranes is not necessary, and oxidation of the crude material with hydrogen peroxide occurs with retention of configuration.

$$RCH{=}CH_2 \xrightarrow[\substack{BF_3 \\ \text{Ether}}]{NaBH_4} (RCH_2{-}CH_2)_3B \xrightarrow{H_2O_2} RCH_2CH_2OH \quad (28)$$

Olefins are also oxidized easily to glycols either by hydrogen peroxide or osmium tetroxide. Allylic alcohols result upon the selenium chloride oxidation of olefins. Substituted phenols are oxidized by persulfate ion or by hydrogen peroxide (Dakin reaction of phenolic aldehydes) to hydroxybenzenes.

The oxidation of olefins and acetylenes, photochemical, free radical, and enzyme reactions are more specialized methods for the preparation of substituted alcohols which have limited application in the laboratory.

Alcohols and phenols are also produced by specialized rearrangement reactions. Ethers have been rearranged to secondary and tertiary alcohols. Aromatic allylic ethers rearrange on heating to allyl phenols (Claisen rearrangement).

2. HYDROLYSIS REACTIONS

The ease of hydrolysis of aliphatic halides by dilute base or water is in the order of tertiary > secondary > primary halides. Iodides are more reactive than bromides and these in turn are more reactive than chlorides. Allyl halides [1] are hydrolyzed with ease whereas aromatic halides are unreactive unless activated by electron-attracting groups [2]. Benzyl halides show enhanced reactivity [3]. For example, refluxing for 4 hr a mixture of 500 gm of trimethylbenzyl chloride and 206 gm of calcium oxide in 1½ liters of water yielded 78% of trimethylbenzyl alcohol [4] [Eq. (28)].

In general, the activity toward hydrolysis depends on the stability of the intermediate carbonium ion.

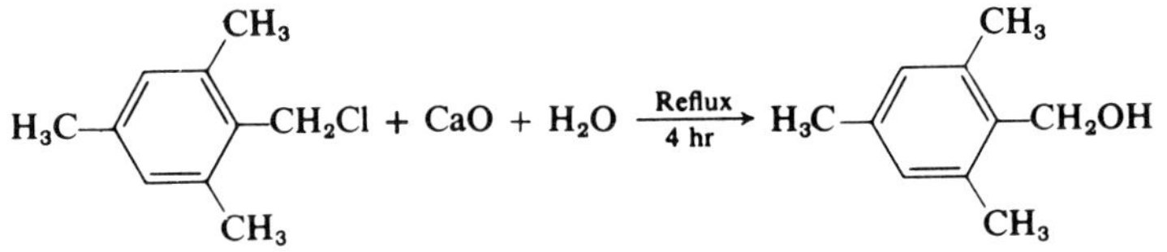

$$\text{(trimethylbenzyl-CH}_2\text{Cl)} + CaO + H_2O \xrightarrow[\text{4 hr}]{\text{Reflux}} \text{(trimethylbenzyl-CH}_2\text{OH)} \quad (29)$$

The hydrolysis of esters [5] yields alcohols but since the alcohols are usually precursors of the esters, this method has limited application. Arylsulfonates yield phenols on fusion with sodium hydroxide [6, 7]. Aryldiazonium compounds on hydrolysis with water yield phenols [8]. While this reaction is general, care must be taken to maintain all of the diazonium salt in solution since dried diazonium salts may explode on warming. (See Chapter 15 for further details on the preparation and handling of diazonium compounds.) Aliphatic diazonium compounds yield a mixture of isomerized alcohols and olefins on hydrolysis [9].

2-1. Preparation of p-Isopropylphenol by the Hydrolysis of a Diazonium Compound [8]

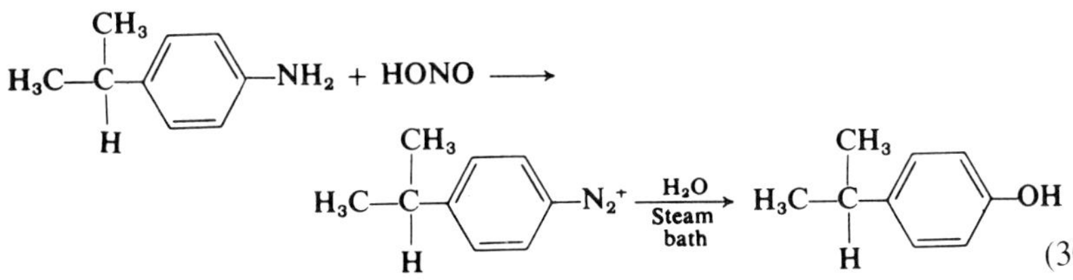

$$\text{p-isopropylaniline} + HONO \longrightarrow \text{diazonium} \xrightarrow[\text{Steam bath}]{H_2O} \text{p-isopropylphenol} \quad (30)$$

To a round-bottomed flask equipped with a mechanical stirrer, condenser, and dropping funnel is added 34 gm (0.25 mole) of p-isopropylaniline in 500 ml of water containing 62.5 gm of concentrated sulfuric acid. The compound is diazotized with 17.3 gm (0.25 mole) of sodium nitrite dissolved in 175 ml of water and slowly added while cooling with an ice bath. Approximately 2 gm of urea is added and the reaction mixture is allowed to come to room temperature. The mixture is then heated on a steam bath at 50°–60°C for several hours, cooled, and extracted with benzene. The benzene layer is extracted with 10% sodium hydroxide and the latter is acidified and again extracted with benzene. The final benzene extract is dried and evaporated to yield a crystalline product in the amount of 25 gm (73.5%), mp 60°C.

2-2. Preparation of p-Isopropylphenol by the Potassium Hydroxide Fusion of Sodium Cumenesulfonate [7]

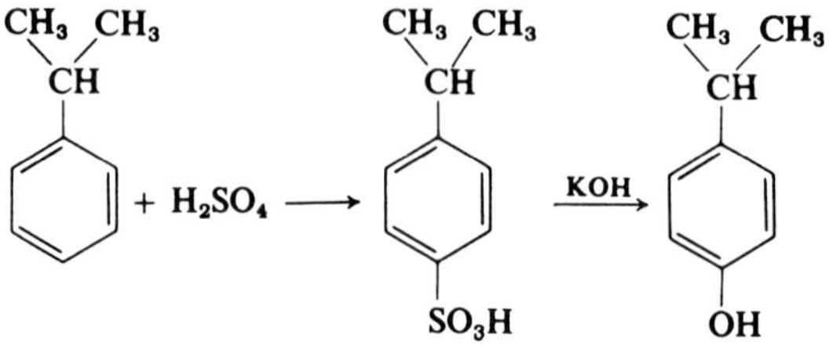

$$\text{cumene} + H_2SO_4 \longrightarrow \text{p-cumenesulfonic acid (SO}_3\text{H)} \xrightarrow{\text{KOH}} \text{p-isopropylphenol (OH)} \quad (31)$$

To a flask containing 360 gm (416 ml, 2.9 moles) of cumene is added 332 gm (3.4 moles) of 95% sulfuric acid. The contents are heated with stirring for 3 hr on a steam bath. The warm mixture is poured into 1 liter of water and 160 gm of sodium bicarbonate is added in small portions, followed by 420 gm (7.2 moles) of sodium chloride. On cooling, the sodium cumenesulfonate precipitate is filtered, washed with 160 ml of saturated sodium chloride solution, and dried for 2 days at 80°C to yield 436 gm (68%).

To a 3-liter iron kettle fitted with a mechanical stirrer and condenser are added 960 gm (17.1 moles) of potassium hydroxide and 40 ml of water. The mixture is heated to 250°C and, with stirring, over a ½-hr period is added 360 gm (1.62 moles) of the impure sodium p-cumenesulfonate. The temperature is raised to 325°C for 10 min and then the contents are poured into 3 liters of cracked ice. The resulting water solution is neutralized with sulfuric acid and is then steam-distilled. The distillate contains colorless crystals of p-isoproylphenol. The crystals are dried in a vacuum desiccator for 2 days. Yield 118 gm (53%), mp 58°–59°C.

Oxirane compounds are hydrolyzed by dilute aqueous acid to glycols with inversion of configuration [10]. Alkylene oxides may also be reduced to monoalcohols by hydride reducing agents, as will be discussed later.

2-3. Preparation of 2,3-Pentanediol by Hydrolysis of an Epoxide [10]

$$CH_3-\overset{O}{\overset{\diagup\diagdown}{CH-CH}}-C_2H_5 + H_2O + H_2SO_4 \longrightarrow CH_3-\underset{OH}{CH}-\underset{OH}{CH}-C_2H_5 \quad (32)$$

To a 2-liter flask equipped with a mechanical stirrer are added 240 gm (2.78 moles) of *trans*-2,3-epoxypentane, 1500 ml of water, and 0.2 ml of concentrated sulfuric acid. The mixture is stirred vigorously at room temperature for 8 hr. The initial two-phase system becomes homogeneous after 3 hr. The solution is then neutralized with sodium hydroxide and concentrated below 70°C using an aspirator. Distillation of the residue yields 236 gm (81%) of *erythro*-2,3-pentanediol, bp 89°C (10 mm), n_D^{20} 1.4431.

Epoxides have been converted to chlorohydrins by the reaction with equimolar amounts of ferric chloride at room temperature in the dark followed by reaction with water [11].

$$\underset{R_2}{\overset{R_1}{>}}\!\!C\!\!\underset{\diagdown}{\overset{O}{-}}\!\!C\!\!\underset{R_4}{\overset{R_3}{<}} + FeCl_3 \longrightarrow R_1\!-\!\underset{\underset{R_2}{|}}{\overset{\overset{Cl}{|}}{C}}\!-\!\underset{\underset{OFeCl_2}{|}}{\overset{\overset{R_3}{|}}{C}}\!-\!R_4 \xrightarrow{H_2O} R_1\!-\!\underset{\underset{R_2}{|}}{\overset{\overset{Cl}{|}}{C}}\!-\!\underset{\underset{OH}{|}}{\overset{\overset{R_3}{|}}{C}}\!-\!R_4 \quad (33)$$

Oxirane compounds have also been reported to react with equimolar amounts of hydrobromic acid to give bromo alcohols [12].

Aryl ethers are converted to phenols by reaction with hydrobromic acid or other strong acids such as trifluoroacetic acid [13].

Finally the hydration of olefins is an important laboratory and industrial process for the preparation of alcohols [14]. A typical preparation involves the reaction of the olefin with either acetic, formic, or trifluoroacetic acid followed by hydrolysis to the alcohol.

2-4. *Preparation of Dihydroxyoctadecanes by the Addition of Formic Acid to Oleyl Alcohol* [15]

To a flask equipped with a mechanical stirrer and condenser are added 15.0 gm (0.055 mole) of oleyl alcohol, 30 ml of anhydrous formic acid, and 0.15 ml of 70% perchloric acid. The reaction mixture is stirred and refluxed for 1 hr. The reaction mixture is poured into water and the reaction product is extracted into ether and washed until acid-free. The crude product of formoxyoctadecyl formates is isolated by concentration and weighs 15 gm (76% conversion). The latter is dissolved in 450 ml of 0.2 N alcoholic potassium hydroxide and the bulk of the alcohol is boiled off on the steam bath. Hot water is added to precipitate the crude dihydroxyoctadecane as an oil which solidifies on cooling. The solid product is remelted in fresh water and resolidifies to give 13.4 gm (84%). Recrystallization from 67 ml of petroleum ether gives 9.6 gm (60%), mp 62.5°–53.5°C.

3. CONDENSATION REACTIONS

The reaction of alkyl or aryl Grignard reagents with gaseous formaldehyde produces primary alcohols in good yields [16]. Benzylmagnesium halide is an exception since *o*-methylbenzyl alcohol is formed in 55% yields [17].

$$RMgX + CH_2\!=\!O \longrightarrow RCH_2OH \quad (34)$$

3-1. Synthesis of 2,2-Dimethyl-1-hexanol Using a Grignard Reaction [10]

$$C_4H_9-\underset{\underset{CH_3}{|}}{\overset{\overset{CH_3}{|}}{C}}-MgCl + CH_2=O \xrightarrow[NH_4Cl]{H_2O} C_4H_9-\underset{\underset{CH_3}{|}}{\overset{\overset{CH_3}{|}}{C}}-CH_2OH \quad (35)$$

The Grignard reagent is prepared from 3.2 moles of 2-chloro-2-methylhexane in 2.1 liters of ether in the usual manner (see Chapter 2) and then 3.3 moles of formaldehyde, obtained by heating 100 gm of paraformaldehyde at 165°–170°C, is forced over the surface of the rapidly stirred Grignard by means of a nitrogen stream. The reaction mixture is gently refluxed for 8 hr while stirring and then decomposed by pouring carefully into 2 liters of ice water containing 3.2 moles of ammonium chloride. The ether layer is separated and the water layer is further extracted with two portions of ether. The combined ether layers are dried with anhydrous sodium sulfate and potassium carbonate overnight, filtered, and distilled through a good fractionating column to yield 260 gm (62%) of 2,2-dimethyl-1-hexanol, bp 80°–82°C (14 mm Hg), n_D^{20} 1.4304.

Using higher aldehydes yields secondary alcohols [18].

$$RMgX + RCHO \longrightarrow \underset{\underset{OH}{|}}{RCHR} \quad (36)$$

Other active organometallics such as sodium [19] or lithium [20] acetylide give a similar reaction. The pure acetylides do not have to be isolated [21].

Use of ketones in the Grignard condensation reaction yields tertiary alcohols in good yields and this is a widely used reaction for this purpose [22]. Strong acids should be avoided in decomposing the Grignard complex since tertiary alcohols are susceptible to dehydration. Ammonium chloride is the preferred agent for hydrolysis [23]. Distillations should be carried out at low temperatures to avoid dehydration to the olefin.

Tertiary alcohols are also made by reacting Grignard reagents with esters [24], aryl halides [25], carbon dioxide [26], carbonates [27], or acids [28].

Primary alcohols result in reaction of the Grignard reagent and alkylene oxides. This method is often used to add two carbons from ethylene oxide to yield the primary alcohol in good yields [29] when ethylene oxide is present in 2:1 molar excess. Other alkylene oxides such as propylene oxide [30], cyclohexene oxide [31], styrene oxide [17], etc., yield secondary alcohols, in lower yields.

A one-step alternative to the Grignard reaction involves the reaction of organic halides with carbonyl compounds in the presence of lithium or sodium using tetrahydrofuran as a solvent [32].

3-2. Synthesis of 1-Octanol Using a Grignard Reaction [33]

$$C_5H_{11}CH_2MgBr + \underset{\underset{O}{\diagdown\diagup}}{CH_2\text{---}CH_2} \longrightarrow C_5H_{11}CH_2CH_2CH_2OH \quad (37)$$

To a solution of n-hexylmagnesium bromide in 700 ml of ether [prepared from 2 moles (330 gm) of n-hexylbromide, 2 moles (48 gm) of magnesium turnings, and 700 ml of anhydrous ether] is added 95 gm (2.2 moles) of liquid ethylene oxide over a period of 1 hr while the ether refluxes vigorously as a result of the heat of reaction. After the addition has been completed, the ether is distilled off until 275 ml have been collected. Then 330 ml of dry benzene is added and the distillation is continued until the temperature reaches 65°C. The mixture is then refluxed for 1 hr and then hydrolyzed with ice water containing 10% sulfuric acid. The benzene layer is separated, washed twice with 10% sodium hydroxide solution, and then distilled to remove the benzene solvent. The residue is distilled under reduced pressure to yield 185 gm (71%) of 1-octanol, bp 105°C (15 mm).

Organolithium reagents are useful for reacting with ketones to give alcohols [34]. These reagents also add to allyl alcohol using tetramethylethylenediamine as catalyst [35]. The use of lithium dimethylcopper with aldehydes to give alcohols is also a useful preparative procedure [36].

The condensation of ethylene carbonate or ethylene oxide with alcohols or phenols yields 2-hydroxymethyl ethers [33].

3-3. Preparation of β-2-Hydroxyethyl Naphthyl Ether Using Ethylene Carbonate [37]

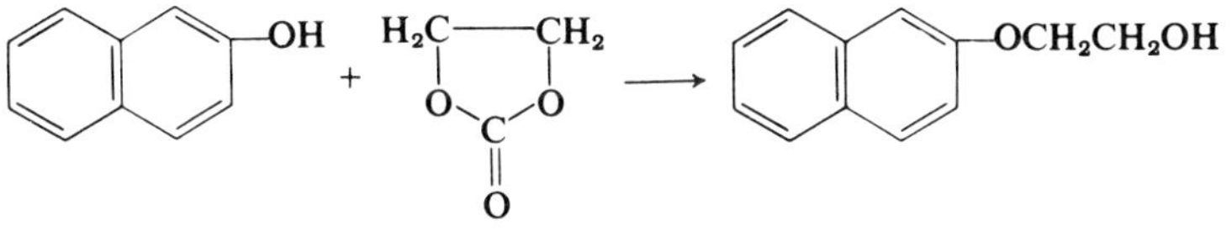

naphthyl-OH + ethylene carbonate ⟶ naphthyl-OCH$_2$CH$_2$OH   (38)

To a flask are added 35.2 gm (0.4 mole) of ethylene carbonate and 28.8 gm (0.2 mole) of β-naphthol. The mixture is heated in an oil bath at 195°C for 1½ hr. The reaction mixture is then cooled, diluted with 75 ml of ethanol, chilled, and poured into 300 ml of cold 3 N sodium hydroxide. The precipitate is filtered, washed with cold water, dried, and dissolved in benzene. The benzene is then partially distilled off to remove the remaining water. The benzene solution is filtered and then concentrated using an aspirator to yield 38.5 gm of 2-hydroxyethyl naphthyl ether, mp 72°–73°C. Despite the sharp melting point, this produce is not quite pure since the weight of product isolated would represent a 103% yield.

108   4. Alcohols and Phenols

The Friedel–Crafts reaction of aromatic hydrocarbons with ethylene oxide [38] has been used to afford xylylethanols [39, 40] in good yields.

3-4. Synthesis of 2-(2,4-Dimethylphenyl)ethanol by the Friedel–Crafts Reaction [39]

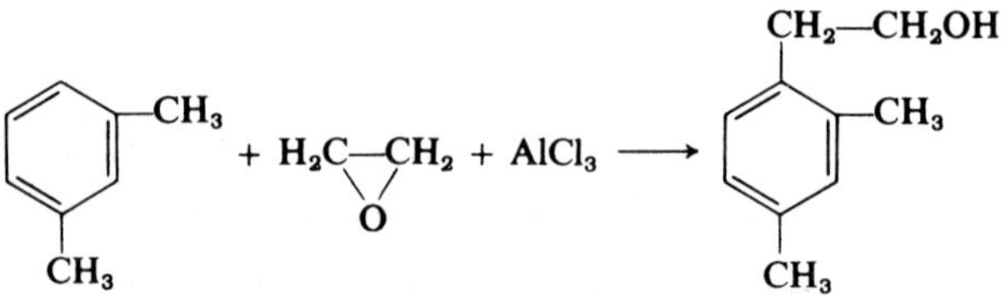

$$\text{m-xylene} + H_2C\underset{O}{-}CH_2 + AlCl_3 \longrightarrow \text{2-(2,4-dimethylphenyl)ethanol} \quad (39)$$

To a 3-liter, three-necked round-bottomed flask with a stirrer, dropping funnel, and gas outlet tube are added 5 moles of dry m-xylene (613 ml, 95%, Oronite Chemical Co.) and 120 gm (0.9 mole) of aluminum chloride. The flask is cooled in an ice–salt water bath while 2 moles (88 gm) of ethylene oxide dissolved in 5 moles (613 ml) of m-xylene is added over a period of 2 hr. The reaction mixture is allowed to stand at room temperature overnight and then 100 ml of concentrated hydrochloric acid in 400 ml of ice water is added. The xylene layer is separated, washed with aqueous sodium hydroxide solution, and then distilled to yield 125 gm (41% based on ethylene oxide) of 2-(2,4-dimethylphenyl)ethanol, bp 100°–104°C (1 mm Hg), n_D^{25} 1.5310.

The aldol condensation usually occurs with aldehydes or ketones with labile hydrogen atoms alpha to the carbonyl group or other activating group (nitro). The aldol (hydroxyaldehyde) or ketol (hydroxyketone) is formed in good yields under basic conditions. However, aldehydes lacking an α-hydrogen, such as benzaldehyde, are subject to simultaneous dehydration to the olefinic aldehyde or ketone. Mixed aldehydes or ketones give mixtures of products and thus purer products are obtained when one of the aldehydes does not contain an α-hydrogen atom.

3-5. Preparation of α,α-Dimethyl-β-hydroxypropionaldehyde by the Aldol Condensation Reaction [41]

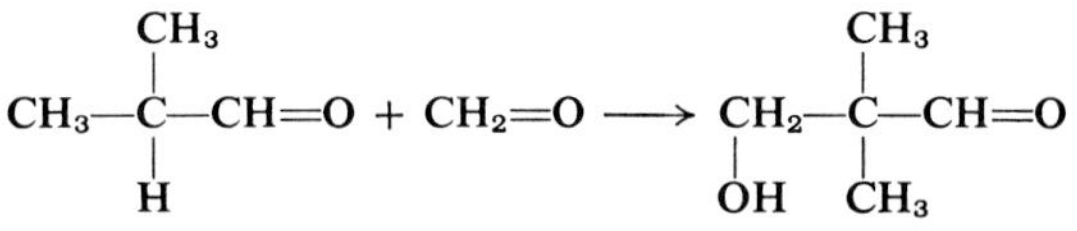

$$CH_3-\underset{\underset{H}{|}}{\overset{\overset{CH_3}{|}}{C}}-CH=O + CH_2=O \longrightarrow CH_2-\underset{\underset{CH_3}{|}}{\overset{\overset{CH_3}{|}}{C}}-CH=O \quad (40)$$
$$OH$$

To a cooled flask containing 200 gm (2.3 mole) of isobutyraldehyde and 224 gm of 40% formalin (3.0 moles) are added with stirring small portions of potassium carbonate at temperatures up to 20°C until 160 gm has been added. The mixture is stirred without cooling for 1 hr, at which time the temperature rises to 23°–25°C. The viscous liquid is extracted with ether, dried over sodium sulfate,

§ 3. Condensation Reactions

and then concentrated to yield a solid. Distillation of the aldol yields a product with bp 83°–86°C (15 mm).* Recrystallization from alcohol and vacuum drying at 60°C give a solid, mp 96°–97°C.

Methyl ketones condense to give ketols which usually dehydrate spontaneously. Diacetone alcohol can be made from acetone by refluxing in the presence of barium hydroxide.

Nitro groups activate methylene hydrogens and cause them to be capable of condensing with carbonyl compounds to give good yields of nitro alcohols from aliphatic or aromatic aldehydes [42].

$$RCH{=}O + H_2\overset{R}{\underset{|}{C}}{-}NO_2 \xrightarrow{\text{Base}} RCH{-}\underset{\underset{OH}{|}}{\overset{R}{\underset{|}{CH}}}{-}NO_2 \quad (41)$$

Alkaline alkoxides have been found to be good basic condensation catalysts [43, 44] in the above reaction [Eq. (40)].

The benzoin condensation [45] is the condensation of an aromatic aldehyde by means of alkali cyanides

$$2\ Ar\overset{H}{\underset{|}{C}}{=}O \xrightarrow{\text{NaCN}} Ar\overset{H}{\underset{\underset{HO}{|}}{C}}{-}\overset{}{\underset{\overset{\|}{O}}{C}}Ar \quad (42)$$

to give a hydroxy ketone of the type shown in Eq. (42).

It has been reported [46] that the benzoine condensation of anisaldehyde is reversible in the presence or absence of water, although the condensation of benzaldehyde [47] appears to go to completion.

3-6. *Synthesis of Anisoin by the Benzoin Condensation* [46]

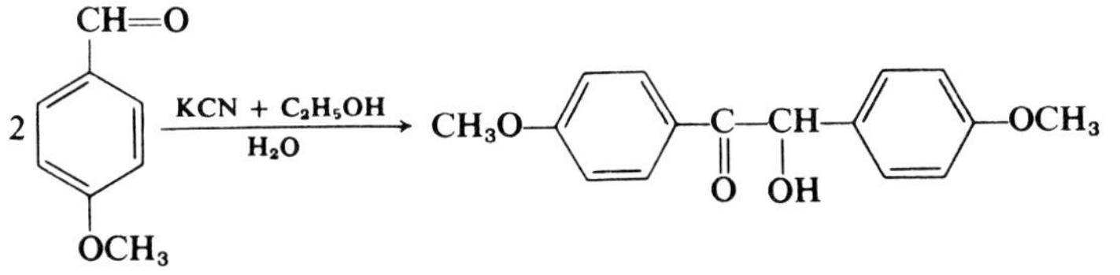

$$2\ \underset{OCH_3}{\underset{|}{\bigcirc}}\text{–CH=O} \xrightarrow[H_2O]{KCN + C_2H_5OH} CH_3O{-}\bigcirc{-}\underset{\underset{O}{\|}}{C}{-}\underset{\underset{OH}{|}}{CH}{-}\bigcirc{-}OCH_3 \quad (43)$$

CAUTION: Since this reaction involves potassium cyanide, all due precautions must be observed.

To a flask containing 22.4 gm (0.345 moles) of potassium cyanide, 90 ml of water, and 150 ml of 95% ethanol is added 112 gm (0.825 mole) of anisalde-

---

*The exact yield of this preparation is not given by Stiller *et al.* [41], but it is mentioned that *good* yields of product are obtained.

hyde. The mixture is refluxed for 5 hr, cooled, and then scratched to start crystallization. The mixture is stored in a refrigerator overnight and then filtered to yield 49.5 gm (44%), mp 105°–110°C. Literature mp 113°C [48].

Sodium metal in benzene or ether catalyzes the condensation of aliphatic [49–51] and aromatic esters [52] to give α-hydroxyketones (acyloins) in good yields. Dispersed sodium techniques [53] facilitate the ease of running this reaction in the laboratory.

$$\text{RCOOR}' \xrightarrow[\text{2. H}^+]{\text{1. Na}} \underset{\underset{\text{OH}}{|} \underset{\text{O}}{\|}}{\text{R—CH—C—R}} \tag{44}$$

The Friedel–Crafts condensation of phenols and aliphatic alcohols gives alkyl phenols in variable yields.

$$\text{ROH} + \text{C}_6\text{H}_5\text{OH} \xrightarrow[\text{BF}_3]{\text{AlCl}_3 \text{ or}} \text{R—C}_6\text{H}_5\text{OH} \tag{45}$$

In this reaction, primary alkyl groups isomerize to secondary alkyl groups. Tertiary alcohols [54] or olefins [55] may also be used for the alkylation. The ethers may be intermediates but then rearrange [56] in a manner similar to a Fries reaction [57] of phenolic esters. Straight-chain alkyl phenols can best be prepared by reduction of the acyl phenols [58, 59].

Phenols also condense with formaldehyde to give methylol and polymethylol phenols [60]. If the condensation is carried too far, methylol phenolic resins are obtained.

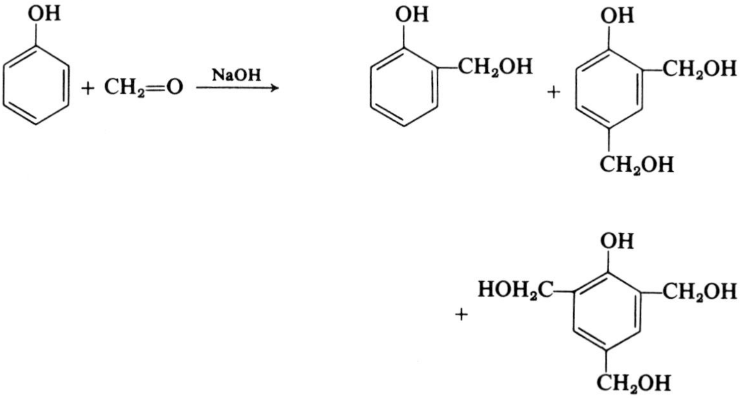

$$\underset{\text{OH}}{\text{C}_6\text{H}_5} + \text{CH}_2\text{=O} \xrightarrow{\text{NaOH}} \underset{\text{OH}}{\text{C}_6\text{H}_4}\text{—CH}_2\text{OH} + \underset{\underset{\text{CH}_2\text{OH}}{}}{\underset{\text{OH}}{\text{C}_6\text{H}_3}}\text{—CH}_2\text{OH}$$

$$+ \quad \text{HOH}_2\text{C}\underset{\underset{\text{CH}_2\text{OH}}{}}{\underset{\text{OH}}{-\text{C}_6\text{H}_2-}}\text{CH}_2\text{OH} \tag{46}$$

Phenols condense with paraformaldehyde in the presence of potassium *t*-butoxide in tetralin solvent at 80°C to give phenolic[1,1,1]metacyclophanes [61].

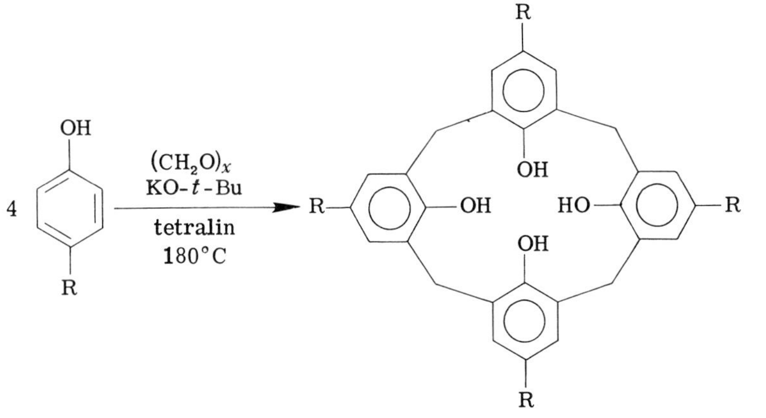

§ 4. Reduction Reactions

$$4 \text{ [4-R-phenol]} \xrightarrow[\text{tetralin} \\ 180°C]{(CH_2O)_x \\ KO\text{-}t\text{-}Bu} \text{[cyclic tetramer, R substituents]}$$

32-92% Yields   (47)

These compounds are not always cyclic tetramers [62].

4. REDUCTION REACTIONS

The reduction of carbonyl compounds to alcohols can either be effected using catalytic hydrogenation [63] or organometallic hydrides. The latter is preferred because of its simplicity and selectivity.

The metallic hydrides reduce aliphatic and aromatic carbonyl compounds in good yields to alcohols. The reaction conditions are convenient and the method is highly recommended for large-scale or small-scale preparation.

Some of the reduction methods are summarized in Table I [64].

Sodium borohydride should not be used in dimethylformamide since a violent explosion for this system has recently been reported [65].

The use of several reducing agents of the complex borohydride variety and of diboranes has been reviewed [64].

More recently, several newer reducing agents have been reported. They are: diisopinocamphenylborane [66], lanthanides/sodium borohydride [67]. $NaBH_3CN$ [68], diisobutylaluminum hydride [69], 9-borabicyclononane (9-BBN) [70], lithium cyanohydridoborate [71], tetrabutylammonium borohydride [72], and trimethoxyborohydride [73] to reduce carbonyl groups to alcohol groups.

2-Propanol on dried Woelm alumina at room temperature has been reported to be an effective reducing agent for diverse aldehydes without affecting other functional groups such as nitro, ester, amide, nitrile, primary and secondary

TABLE I

The Reduction of Carbonyl Compounds and Oxiranes by Hydrides or Boranes to Alcohol[a]

Compound functional group	Reducing agents							Product
	$NaBH_4$	$LiBH_4$	$NaBH_4$ + $AlCl_3$	$LiAlH_4$	$LiAlH_4$ [49] + $AlCl_3$	RNH_2 [50] + BH_3	B_2H_6	
RCH=O	+	+	+	+	+	+	+	RCH_2OH
RCOR	+	+	+	+	+	+	+	RCH_2OH
RCOOH	−	+	+	+	+	+	+	RCH_2OH
RCOOR	−	+	+	+	+	+	−	RCH_2OH
RCOCl	+	+	+	+	+	+	−	RCH_2OH
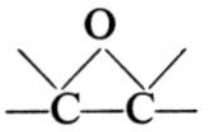 −C−C− (oxirane)	−	+	+	+	+		+	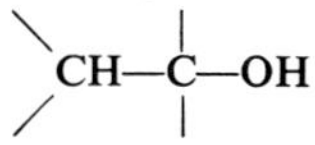 CH−C−OH

[a]The advantage of sodium borohydride over the other reducing agents shown in Table I is that it is relatively stable in aqueous or alcoholic solutions and especially in 25% molar amounts of sodium hydroxide.

iodide benzylic bromide, and olefinic sites. A comparison of this system to five different reducing agents also recently reported as shown in Table II.

4-1. Synthesis of Dimethylphenylethanols Using Sodium Borohydride [39]

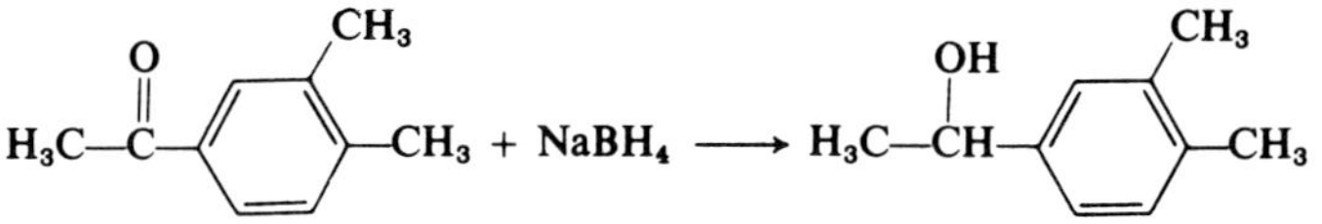

$$H_3C-\underset{O}{C}-C_6H_3(CH_3)-CH_3 + NaBH_4 \longrightarrow H_3C-\underset{OH}{CH}-C_6H_3(CH_3)-CH_3 \quad (48)$$

To a 1-liter flask containing 300 ml of isopropanol and 10 gm of sodium hydroxide is added 20 gm (0.54 mole) of sodium borohydride with cooling. To the solution is added dropwise a solution of 280 gm (1.9 moles) of 3,4-dimethylacetophenone in 150 ml of isopropanol at such a rate that the solution gently refluxes. After the addition, the solution is allowed to stand overnight. The solution is then poured into a 5-liter flask and with stirring 1 liter of 1 N sodium hydroxide is added. The oily layer is separated, dried, and vacuum-distilled. If a clear separation does not occur, then the solution is saturated with salt. The alcohol product is distilled through a 2-ft column to yield 177 gm (62%) of product, bp 101°–103°C (5 mm Hg), n_D^{20} 1.5284.

4-2. Reduction of Ethyl Stearate to 1-Octadecanol [40]

$$C_{17}H_{35}CO_2C_2H_5 + NaBH_4 + AlCl_3 \xrightarrow{\text{Diglyme}} C_{17}H_{35}CH_2OH \quad (49)$$

§ 4. Reduction Reactions

To a 1-liter flask equipped with a dropping funnel, mechanical stirrer, condenser, and drying tube are added 8.5 gm (0.25 mole) of sodium borohydride and 250 ml of diglyme. The contents are stirred to facilitate the solution of the borohydride and then 0.4 mole of ethyl stearate (bp 198°–199°C at 10 mm) is slowly added. The solution is vigorously stirred while 42 ml of a 2.0 M solution of aluminum chloride (0.084 mole) in diglyme is added at such a rate as to keep the temperature below 50°C. After all the aluminum chloride has been added, the reaction mixture is stirred for an hour at room temperature, followed by heating on a steam cone for 0.5–1.0 hr.

The reaction mixture is cooled to room temperature and poured into 500 gm of crushed ice containing 50 ml of concentrated hydrochloric acid. The precipitate is collected on a filter, washed with ice water, pressed, and dried under reduced pressure. The crude product is recrystallized from aqueous alcohol. The 1-octadecanol is obtained in 91% yield (98.2 gm), mp 58°–59°C

TABLE II

Amount of Carbonyl Reduction by Various "Selective" Reducing Agents[a]

	% reduction[b] of	
Reducing agent	RCH_2CHO	4-tert-butylcyclohexanone
Bu_4NBH_3CN[c]	90[c]	88[c]
$NaBH(OAc)_3$[d]	95[e]	65[e]
$LiAlH(OBu-t)_3$[f]	High[f]	High[f]
$LiBu_2$-BBN[g]	50–72[g]	63–77[g]
9-BBN	63[e]	64[e,h]
i-PrOH/Al_2O_3[i]	80–90	90
i-Pr_2CHOH/Al_2O_3[i]	84	13

[a] Reprinted from G. H. Posner, A. W. Runquist, and M. J. Chapdelaine, *J. Org. Chem.* **42**, 1202 (1977). Copyright 1977 by the American Chemical Society. Reprinted by permission of the copyright owner.

[b] Determined by GLC using calibrated internal standards; R = C_8H_{17}-n.

[c] R. O. Hutchins and D. Kandasamy, *J. Am. Chem. Soc.* **95**, 6131 (1973).

[d] G. W. Gribble and D. C. Ferguson, *J. Chem. Soc., Chem. Commun.* p. 535 (1973); N. Umino, T. Iwakuma, and N. Itoh, *Tetrahedron Lett.* p. 763 (1976).

[e] Determined in our laboratory.

[f] C. S. Sell, *Aust. J. Chem.* **28**, 1383 (1975).

[g] Y. Yamamoto, H. Toi, A. Sonoda, and S. I. Murohashi, *J. Am. Chem. Soc.* **98**, 1965 (1976).

[h] 40:60 *cis*–*trans*-4-*tert*-butylcyclohexanol.

[i] Woelm W-200 neutral dehydrated alumina.

The Meerwein–Pondorf–Verley reduction of carbonyl compounds uses a mixture of isopropanol and aluminum isopropoxide to reduce aldehydes and ketones to alcohols [74]. The extent of the reaction can be determined either by vapor phase chromatography or by distilling off the acetone as it is formed in the reaction.

$$R_2C=O + CH_3-\underset{\underset{OH}{|}}{CH}-CH_3 \underset{}{\overset{Al[OCH(CH_3)_2]_3}{\rightleftharpoons}} R_2CHOH + CH_3\overset{\overset{O}{\|}}{C}CH_3 \quad (50)$$

The higher boiling aldehydes or ketones remain behind and are later purified by distillation. Excellent yields are usually obtained in this reaction. For example, 4-chlorophenylmethylcarbinol is produced in 81% yield from the ketone [75].

4-3. Preparation of 4-Chlorophenylmethylcarbinol Using the Meerwein Reduction Reaction [75]

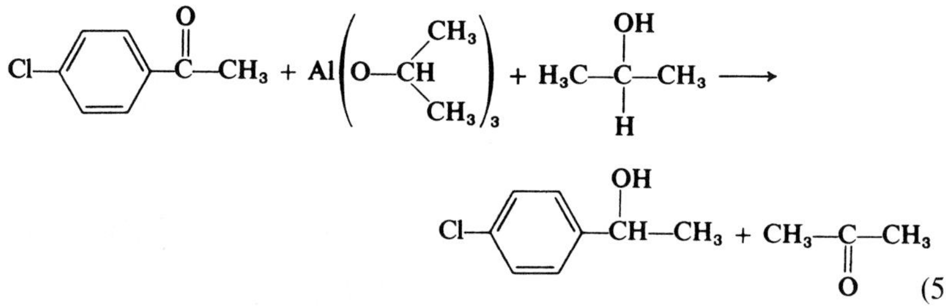

$$Cl-C_6H_4-\overset{\overset{O}{\|}}{C}-CH_3 + Al\left(O-\underset{CH_3}{\overset{CH_3}{\diagup}}CH\right)_3 + H_3C-\underset{\underset{H}{|}}{\overset{\overset{OH}{|}}{C}}-CH_3 \longrightarrow$$

$$Cl-C_6H_4-\underset{\underset{}{|}}{\overset{\overset{OH}{|}}{C}H}-CH_3 + CH_3-\underset{\underset{O}{\|}}{C}-CH_3 \quad (51)$$

To a flask containing aluminum isoproproxide (from 13.5 gm aluminum powder and 500 ml of dry, distilled isopropanol) is added a solution of 90 gm of *p*-chloroacetophenone in 620 ml of isopropanol. The mixture is heated for 6 hr and acetone is removed as it forms. The isopropanol solvent is then distilled off and the cooled residue is acidified with 400 ml of 10% hydrochloric acid. The organic layer is separated and the aqueous layer is extracted with two 150-ml portions of benzene. The combined organic phase, upon distillation, yields 74 gm (81%) of the carbinol, bp 81°–86°C (1 mm Hg), n_D^{20} 1.5420.

Some miscellaneous reducing agents are

(1) Lithium–ethylene for epoxides [76]
(2) Organotin hydrides for reduction of aldehyde and ketones [77]
(3) Iron chloride–sodium hydride for reducing carbonyl compounds [78]
(4) Silicon hydrides as reducing agents for carbonyl compounds [79]
(4) Organosilane–borontrifluoride reductions [80]
(5) Electrochemical reductions of ketones to alcohols [81]

5. OXIDATION REACTIONS

The oxidation of organometallics such as the Grignard reagent and organoboranes affords alcohols in good yields [82–84].

$$\text{RMgX} + \text{O}_2 \longrightarrow \text{ROMgX} \xrightarrow{\text{H}_3\text{O}^+} \text{ROH} \qquad (52)$$

$$\text{R}_3\text{Al} + \text{O}_2 \longrightarrow (\text{RO})_3\text{Al} \xrightarrow{\text{H}_3\text{O}^+} 3\,\text{ROH} \qquad (53)$$

$$\text{R}_3\text{B} + \text{H}_2\text{O}_2 \longrightarrow 3\,\text{ROH} \qquad (54)$$

The Grignard method is useful for the conversion of alkyl or aryl halides to the corresponding hydroxy compound. However, the borane method is used mainly to convert olefins to alcohols.

5-1. Oxidation of Phenylmagnesium Bromide to Phenol [82]

$$\text{C}_6\text{H}_5\text{MgX} + \text{O}_2 \longrightarrow \text{C}_6\text{H}_5\text{OOMgX} \xrightarrow{\text{RMgX}}$$

$$\text{C}_6\text{H}_5\text{OMgX} + \text{ROMgX} \xrightarrow[\text{H}_2\text{O}]{\text{H}_3\text{O}^+} \text{C}_6\text{H}_5\text{OH} + \text{ROH} \qquad (55)$$

The use of slightly more than an equimolar amount of an aliphatic Grignard reagent in the presence of the aryl Grignard reagent during oxidation gives improved yields of phenols with aliphatic alcohols as by-products.

To a flask containing 60 gm of magnesium (2.5 moles) and 1 liter of ether is added dropwise a mixture of 1 mole of bromobenzene and 1.5 moles of isopropyl bromide. Dry oxygen, carbon dioxide-free, is bubbled into an agitated solution of the Grignard reagent at a rate (50 ml/min) which causes gentle boiling of the ether. The exotherm ceases after approximately 10 hr. Oxygen gas is bubbled into the solution for an additional hour and then the mixture is allowed to stand overnight. The reaction mixture is decomposed by slightly less than 2.5 moles of aqueous sulfuric and at 0°C. The amount of water used is just enough to keep the magnesium salt in solution. The water layer is extracted three times with ether and the ether layers extracted with aqueous sodium hydroxide solution. The latter solution is further extracted with ether to remove isopropanol. Acidification of the sodium hydroxide solution yields 64% of phenol. Fractionation of the remaining ether yields isopropanol.

The reaction of alkyl halides and tosylates with an excess of potassium superoxide gives alcohols with inversion of configuration [85].

Aryl bromides can be converted to phenols by reaction of arylmagnesium bromide Grignard reagents with molybdenium peroxide-pyridine-hexamethylphosphoramide [86].

Olefins are conveniently hydroborated and then oxidized in good yields to alcohols of known configuration [84]. The diborane used for the hydroboration is generated from sodium borohydride and boron trifluoride etherate in diglyme.

$$3 \text{ RCH}=\text{CH}_2 + \text{NaBH}_4 + \text{BF}_3 \longrightarrow (\text{RCH}_2\text{—CH}_2)_3\text{B} \xrightarrow{\text{H}_2\text{O}} \text{RCH}_2\text{CH}_2\text{OH} \quad (56)$$

Cyclic olefins are hydrated in a cis manner [88].

The hydroboration–oxidation of olefins is a useful procedure for achieving anti-Markovnikov hydration of carbon–carbon double bonds [89].

$$\underset{\underset{\text{CH}_3}{|}}{\overset{\overset{\text{CH}_3}{|}}{\text{CH}_3\text{—C—CH}=\text{CH}_2}} \xrightarrow{\text{[HB]}} \xrightarrow{\text{[O]}} \underset{\underset{\text{CH}_3}{|}}{\overset{\overset{\text{CH}_3}{|}}{\text{CH}_3\text{—C—CH}_2\text{—CH}_2}}\text{—OH} \quad (57)$$

Recently it has been reported that oxymercuration of olefins, combined with reduction of the oxymercurial intermediate by sodium borohydride *in situ*, affords a mild and convenient method to achieve Markovnikov hydration of carbon–carbon bonds without observable rearrangement [87].

$$\underset{\underset{\text{CH}_3}{|}}{\overset{\overset{\text{CH}_3}{|}}{\text{CH}_3\text{—C—CH}=\text{CH}_2}} \xrightarrow{\text{Hg(OAc)}_2} \xrightarrow{\text{NaBH}_4} \underset{\underset{\text{CH}_3\text{OH}}{|}}{\overset{\overset{\text{CH}_3}{|}}{\text{CH}_3\text{—C—CH—CH}_3}} \quad (58)$$

A general procedure for this reaction follows, using the preparation of 2-hexanol as an example. Other representative examples are given in Table III.

5-2. *General Procedure for the Markovnikov Hydration of Olefins by the Oxymercuration–Demercuration Procedure* [87]

In a flask fitted with a stirrer is placed 31.9 gm (0.1 mole) of mercuric acetate. To the flask is added 100 ml of water followed by 100 ml of tetrahydrofuran. (See *Note* below.) Then 0.1 mole of 1-hexene is slowly added while a cooling bath maintains the temperature at 25°C. The reaction mixture is stirred for 10 min at 25°C to complete the oxymercuration. Then 100 ml of 3 *M* sodium hydroxide is added, followed carefully by a dropwise addition of 0.5 *M* sodium borohydride in 3.0 *M* sodium hydroxide until 100 ml has been added. (CAUTION: Since the reaction is exothermic, a cooling bath is required to maintain the temperature at 25°C throughout the reaction.)

The reduction of the oxymercurial intermediate is almost instantaneous. The mercury that is formed is allowed to settle and then sodium chloride is added to form a saturated solution. The upper layer consisting of tetrahydrofuran is separated and distilled to yield 96% of 2-hexanol.

Note. In the above procedure the mercuric acetate dissolves in the water to give a clear solution. When tetrahydrofuran is added, a yellow suspension forms which becomes ligher and finally colorless and clear. The disappearance of the yellow color provides an approximate indication of the time required to complete

TABLE III

THE MARKOVNIKOV HYDRATION OF REPRESENTATIVE OLEFINS BY THE OXYMERCURATION-DEMERCURATION PROCEDURE [87][a]

Olefin	Reaction time t_1[b]	t_2[c]	Product	Yield[d] (%)
1-Hexene	45 sec	10 min	2-Hexanol	96
1-Dodecene[e]	7 min	70 min	2-Dodecanol	91
cis-2-Pentene	45 sec	10 min	65% 2-, 35% 3-pentanol	98
trans-2-Pentene	65 sec	15 min	54% 2-, 46% 3-pentanol	95
2-Methyl-1-butene	10 sec	5 min	2-Methyl-2-butanol	90
2-Methyl-2-butene	20 sec	10 min	2-Methyl-2-butanol	95
3,3-Dimethyl-1-butene	2 min	20 min	3,3-Dimethyl-2-butanol	94
2,3-Dimethyl-2-butene	3.5 min	35 min	2,3-Dimethyl-2-butanol	86
2,4,4-Trimethyl-1-pentene	10 min	30 min	2,4,4-Trimethyl-2-pentanol	96
Cyclopentene	20 sec	1 hr	Cyclopentanol	91
Cyclohexene	55 sec	11 min	Cyclohexanol	99
Cyclooctene[e]	2 hr	3 hr	Cyclooctanol	88
1-Methylcyclopentene	20 sec	6 min	1-Methylcyclopentanol	93
1-Methylcyclohexene	20 sec	5 min	1-Methylcyclohexanol	100
Methylenecyclohexane	10 sec	5 min	1-Methylcyclohexanol	99
Styrene	25 sec	5 min	1-Phenylethanol	96
α-Methylstyrene	45 sec	10 min	2-Phenyl-2-propanol	95

[a] Reprinted from: H. C. Brown and P. Geoghegan, Jr., *J. Am. Chem. Soc.* **89**, 1522 (1967). Copyright 1967 by the American Chemical Society. Reprinted by permission of the copyright owner.

[b] Time for yellow color to disappear.

[c] Complete reaction time for the oxymercuration stage, before addition of the 3.0 M sodium hydroxide.

[d] Analysis by gas-liquid chromatography with an internal standard.

[e] Reaction heterogeneous, with olefin possessing only limited solubility in the reaction mixture. The longer reaction may be, in part, due to this factor.

the oxymercuration reaction. It is recommended that the reaction mixture be allowed to stir five to ten times the length of time required for the yellow color to vanish before initiating the reduction stage.

5-3. *Preparation of trans-2-Methylcyclopentanol by the Hydroboration Reaction* [84]

$$3 \text{ [methylcyclopentene]} + B_2H_6 \longrightarrow [\text{[methylcyclopentyl]}B]_3 \xrightarrow{H_2O_2} \text{[trans-2-methylcyclopentanol]} \quad (59)$$

To a flask containing 16.4 gm (0.2 mole) of 1-methylcyclopentene in 60 ml of tetrahydrofuran at 0°C is added gaseous diborane generated from 3.8 gm sodium borohydride in diglyme and boron trifluoride etherate over a period of 2 hr. After this has stood 1 hr at room temperature, some small chips of ice are added to hydrolyze the excess diborane. The flask is immersed in an ice bath, and 45 ml of 3 M sodium hydroxide is added followed by 25 ml of 30% hydrogen peroxide over a period of 1 hr. After 1 hr at room temperature, the organic layer is separated and the water layer extracted with ether. The combined organic layers are distilled to yield 17.0 gm (85%) of *trans*-2-methylcyclopentanol, bp 152°–153°C (745 mm), n_D^{20} 1.4488.

Several other examples are cited in a recent review [89] and the yields are generally good.

Recently Kabalka and Hedgecock reported that organoboranes are oxidized efficiently by trimethylamine N-oxide dihydrate [90].

$$R_3B + \overset{-}{O}-\overset{+}{N}R_3 \longrightarrow R_2B-OR + :NR_3 \quad (60)$$

Olefins can also be directly oxidized with hydrogen peroxide to give glycols in good to excellent yields. Performic acid yields the best results and it is produced *in situ* from 30% hydrogen peroxide and formic acid [91, 92].

Peracetic acid can be used to prepare mono hydroxy compounds by photochemical oxidation of activated tertiary hydrogens [93].

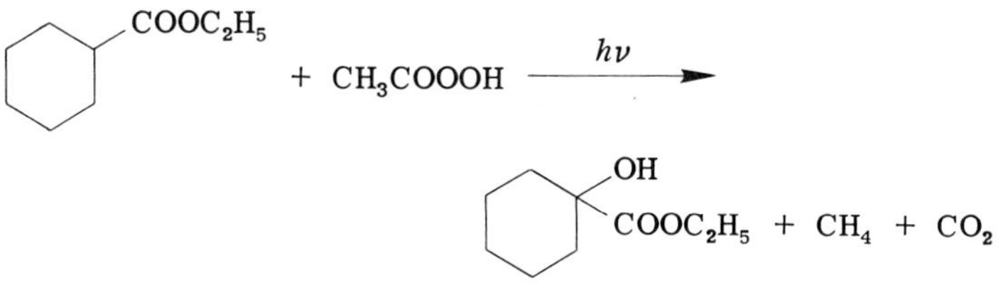

$$\text{C}_6\text{H}_{11}\text{COOC}_2\text{H}_5 + \text{CH}_3\text{COOOH} \xrightarrow{h\nu}$$

$$\text{(HO)C}_6\text{H}_{10}\text{COOC}_2\text{H}_5 + \text{CH}_4 + \text{CO}_2 \quad (61)$$

Allylic oxidation of olefins can be effected by catalytic acid stoichiometric selenium dioxide with *tert*-butyl hydroperoxide [94].

Aromatic compounds can be hydroxylated by use of hydrogen peroxide–aluminum chloride [95].

$$\text{ArH} + \text{H}_2\text{O}_2 + \text{AlCl}_3 \longrightarrow \text{ArOH} + \text{HCl} + \text{Al(OH)Cl}_2 \quad (62)$$

Peroxy trifluoroacetic acid–boron trifluoride has been reported as a source of positive hydroxyl and has been shown to be an excellent reagent for effecting electrophilic aromatic hydroxylations of alkylated benzene compounds. Mesitylene gives mesitol in 88% yield whereas benzene gives only traces of phenols [95].

The Cannizzaro reaction [96] of aldehydes lacking an α-hydrogen yields alcohols and acids by a bimolecular oxidation–reduction mechanism.

§ 5. Oxidation Reactions

$$2\ R_3C-CHO + NaOH \longrightarrow R_3C-CH_2OH + R_3C-COONa \quad (63)$$
$$50\% 50\%$$

In order to increase the yields of alcohol, formaldehyde may be added to the aldehyde (crossed Cannizzaro reaction). By this means, the higher aldehydes are reduced to alcohols while formaldehyde forms sodium formate as a by-product [97, 98].

$$R_3C-CHO + CH_2{=}O \xrightarrow{NaOH} R_3C-CH_2OH + HCOONa \quad (64)$$

5-4. Preparation of Anisyl Alcohol by the Crossed Cannizzaro Reaction [97]

$$CH_3O-\!\!\bigcirc\!\!-CH{=}O + CH_2{=}O + H_2O \xrightarrow{NaOH}$$

$$CH_3O-\!\!\bigcirc\!\!-CH_2OH + HCOOH \quad (65)$$

To a 2-liter flask are added 136 gm (1 mole) of anisaldehyde, 200 ml of methanol, and 100 ml (1.3 mole) of formalin solution. The mixture is heated to 65°C and then cooled while a solution of 120 gm (3 moles) of sodium hydroxide in 120 ml water is rapidly added, keeping the temperature between 65° and 75°C. The reaction mixture is heated to 70°C for ¾ hr and then refluxed for 20 min. The reaction is cooled and diluted with 300 ml of water. The organic layer is separated and the water phase is extracted with four portions of 150 ml of benzene. The combined organic phase is washed with water, dried, and distilled to yield 124 gm of anisylalcohol (90%), bp 134°–135°C (12 mm), $n_D^{2.5}$ 1.5420.

When the aldehyde contains an α-hydrogen in the crossed Cannizzaro reaction, one obtains hydroxymethyl groups in its place. Thus pentaerythritol is obtained from acetaldehyde and formaldehyde in good yield [99]. Di- and trimethylol compounds result when higher aldehydes are used [100–102].

5-5. Preparation of 1,1,1-Tris(hydroxymethyl)-2-methylpropane [100]

$$\underset{\underset{CH_3}{|}}{CH_3-CH}-CH_2-CH{=}O + 3\ CH_2{=}O \xrightarrow{Ca(OH)_2} \underset{\underset{CH_3}{|}}{CH_3-CH}-C(CH_2OH)_3 \quad (66)$$

To a flask containing 172 gm (2 moles) of isovaleraldehyde and 650 gm (8 moles of 37% formalin is slowly added while stirring 1 mole of solid calcium hydroxide at such a rate that the temperature remains approximately 55°–65°C. The reaction mixture is then heated to 32°C for 12 hr, to 50°–60°C for 24 hr, and to 85°C for 6 hr. The hot mixture is filtered and the filtrate concentrated to 600 ml. Approximately 1800 ml of 95% ethanol is then added and the sodium calcium formate is separated by filtration. The filtrate is distilled to remove the ethanol and upon distilling the residue there is obtained 156 gm (53%) of the triol, bp 170°–175°C (6 mm), mp 82°–82.1°C (after five recrystallizations from ether).

6. REARRANGEMENT REACTIONS

The rearrangement of aryl ethers, esters, or hydroxylamines gives phenols. In the case of alkyl aryl ethers, the products obtained are aryl alkyl alcohols. Epoxides can also be rearranged to allylic alcohols. Sigmatropic rearrangements of olefinic alcohols have also been reported. Some typical examples are described below.

A. Rearrangement of Ethers

6-1. Claisen Rearrangement [103–108]

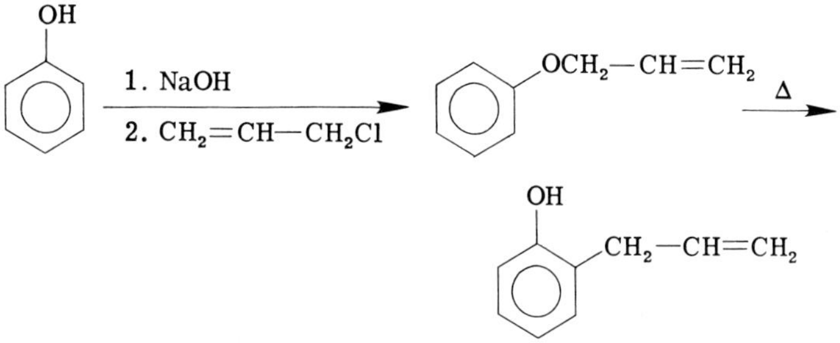

$$\text{C}_6\text{H}_5\text{OH} \xrightarrow[2.\ \text{CH}_2=\text{CH}-\text{CH}_2\text{Cl}]{1.\ \text{NaOH}} \text{C}_6\text{H}_5\text{OCH}_2-\text{CH}=\text{CH}_2 \xrightarrow{\Delta} \text{o-(CH}_2-\text{CH}=\text{CH}_2)\text{C}_6\text{H}_4\text{OH}$$

(67)

The Claisen rearrangement reaction can be directed to the para position by blocking the ortho positions. Blocking of only one ortho position still leads to some (approximately 10%) para substitution. A typical example of the Claisen rearrangement is described in Chapter 2, Olefins.

6-2. Rearrangement of Aryl Alkyl Ethers to Alcohols [109]

$$\text{C}_6\text{H}_5-\text{CH}_2\text{OR} \xrightarrow[\text{KNH}_2]{\text{LiR}} \text{C}_6\text{H}_5-\underset{R}{\text{C}}\text{HOH} \qquad (68)$$

6-3. Rearrangement of n-Alkyl Phenyl Ethers [110]

$$\text{RC}_6\text{H}_4\text{OR}' \xrightarrow[\substack{\text{heat,} \\ \text{pressure}}]{\text{BF}_3 \text{ in CCl}_4} \text{R}_1\text{R}'\text{C}_6\text{H}_3\text{OH} \qquad (69)$$

B. Rearrangement of Aryl Hydroxylamines [111]

The arylhydroxylamines are rearranged using strong acid to p-aminophenols in fair yields. The rearrangement is typical of those described by the type [112]:

$$C_6H_5\overset{R}{N}-X \xrightarrow{H^+} o \text{ and } p \text{ } X-C_6H_4NHR \qquad (70)$$

In the case of

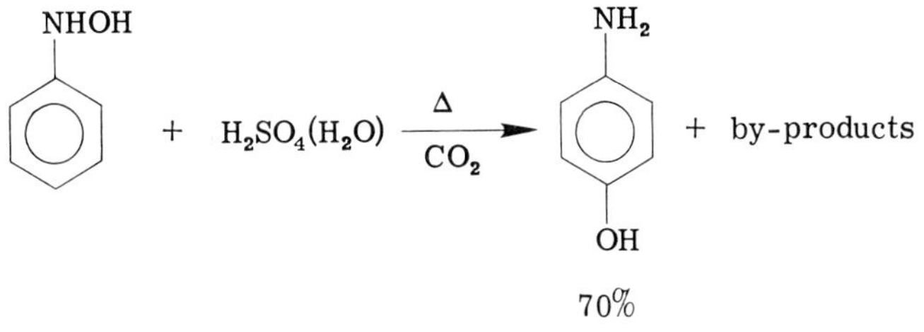

$$\text{C}_6\text{H}_5\text{NHOH} + H_2SO_4(H_2O) \xrightarrow[CO_2]{\Delta} p\text{-HOC}_6H_4NH_2 \text{ (70\%)} + \text{by-products} \qquad (71)$$

C. Rearrangement of Aryl Esters to Phenolic Ketones [113]

The Fries rearrangement is an acid-catalyzed, acyl-transfer reaction of phenolic esters to give a phenolic ketone.

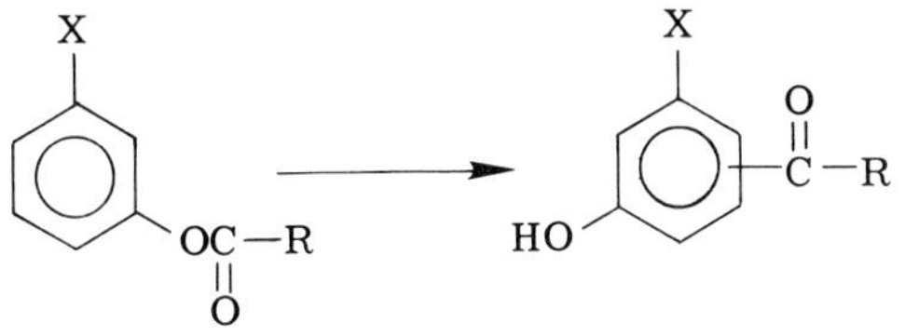

$$\text{X-C}_6H_4\text{-OC(O)R} \longrightarrow \text{X-C}_6H_3(\text{OH})\text{-C(O)R} \qquad (72)$$

D. Rearrangement of Epoxides to Allylic Alcohols [114]

Strong bases lead to the isomerization of epoxides to allylic alcohols. An example is the rearrangement of 1-methylcycloalkene oxide to the corresponding exocyclic methylene alcohols.

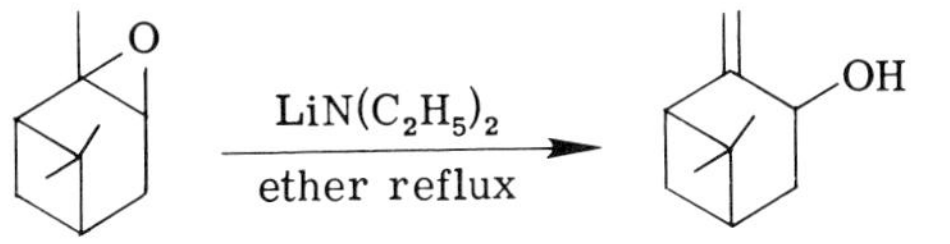

$$\text{2,3-Epoxy pinane} \xrightarrow[\text{ether reflux}]{LiN(C_2H_5)_2} \text{trans-Pinocarveol (90-95\%)} \qquad (73)$$

Recently 1-methylcyclohexene oxide has been rearranged to the allyl alcohol in a patented procedure [115].

E. Sigmatropic Rearrangements of Olefinic Alcohols

Allylic alcohols can be converted to allyl tributyl stannyl methyl ethers which on reaction with excess n-butyl lithium gives tin/lithium exchange and a smooth [2,3] sigmatropic rearrangement to the homoallylic alcohol [116].

7. MISCELLANEOUS METHODS

(1) Guerbert condensation of alcohols by sodium [117].
(2) The condensation of pyridines or quinolines with ketones [118].
(3) Hydrolysis of α-diazo ketones [119].
(4) Claisen rearrangement of allyl ethers [120].
(5) Cleavage of furans and pyrans [121].
(6) Cleavage of ethers [122].
(7) Dehydrogenation of cyclic ketones to phenols [123–125].
(8) Selenium dioxide oxidation of olefins or acetylenes [126] to yield unsaturated alcohols.
(9) Potassium persulfate oxidation of phenols to dihydroxybenzenes [127].
(10) The Friedel–Crafts oxygenation of toluene with diisopropyl peroxydicarbonate [128].
(11) Free radical reaction of primary and secondary alcohols with formaldehyde [129].
(12) Electrophilic hydroxylations with trifluoroperoxyacetic acid [130].
(13) Hydrazine reduction of α,β-epoxy ketones to allylic alcohols [131].
(14) The synthesis of cyclopropanols [132].
(15) Phenolic derivatives from perhaloacetones [133].
(16) Demjanov rearrangement [134].
(17) The condensation of olefins, acetylenes, or alcohols with carbon monoxide [135–137].
(18) The preparation of polyvinyl alcohol [138].
(19) A synthesis of homoallylic alcohols [139].
(20) The oxidation of epoxides by dimethyl sulfoxide. A simple synthesis of α-hydroxy ketones [140, 141].
(21) Use of preformed enolates in aldol condensations [142–146].

(22) Reaction of alkylmercuric halides with sodium borohydride in dimethylformamide saturated with molecular oxygen to produce alcohols [147].

(23) Degrading primary alcohols to their next lower homolog via formation of hydroperoxides and subsequent decomposition [148].

(24) Baeyer-Villiger oxidation of ketones to give alcohols [148].

(25) Baeyer-Villiger oxidation of methoxybenzaldehydes to methoxyphenols [149].

(26) Preparation of phenols by the acid decomposition of aromatic hydroperoxides [150].

(27) Preparation pf p,p'-biphenols by oxidative coupling and reduction [151].

(28) Conversion of trienes to angular substituted bicyclic alcohols [152].

(29) Preparation of phenols from diazonium compounds [153].

(30) Stereospecific route to allylic alcohols [154].

(31) Preparation of thioglycols [155].

(32) Preparation of acetylene alcohols via the condensation of carbonyl compounds and acetylenes [156].

(33) Stereospecific synthesis of homoallyl alcohols via Grignard type carbonyl addition of allyl halides using chromous salts [157].

(34) Process for preparing 2-hydroxybenzothiazoles [158].

(35) Preparation of perfluoroalkyl alcohols [159].

(36) Improved procedure for cleavage of ethers and ketals to alcohols [160].

(37) Demethylation of arylmethyl ethers using a volatile base [161].

(38) Copper-induced Pschorr cyclization to produce phenols [162].

(39) Directed aldol condensation to give threo-4-hydroxy-3-phenyl-2-heptanone [163].

(40) Ring opening dehydrohalogenation to produce (E)-4-hexen-1-ol [164].

(41) Direct lithiation of aromatic compounds and subsequent reaction with ketones to give aryl carbinols [165].

(42) Oxymercuration-reduction of olefins to give alcohols [166].

(43) Preparation of orcinol monomethyl ether [167].

(44) Synthesis of nitronyl alcohols [168].

(45) Preparation of perfluoroalkyl alcohols [169].

(46) Chlorine containing polyether polyol for preparation of polyurethane foams [170].

(47) Addition of carbon tetrachloride to allyl alcohol to give tetrachlorobutanol [171].

REFERENCES

1. M. W. Tamele, C. J. Ott, K. E. Marple, and G. W. Hearne, *Ind. Eng. Chem.* **33**, 115 (1941).
2. L. I. Smith, J. W. Opie, S. Wawzonek, and W. W. Pritchard, *J. Org. Chem.* **4**, 318 (1939).
3. J. N. Ashley, H. J. Barber, A. J. Ewins, G. Newbery, and A. D. Self, *J. Chem. Soc.* p. 103 (1942).

4. H. D. Hartough, *Ind. Eng. Chem.* **42,** No. 5, 903 (1950).
5. J. Cason, *Org. Synth.* **21,** 1 (1941).
6. W. W. Hartman, *Org. Synth. Collect. Vol.* **1,** 175 (1941).
7. R. L. Frank, R. E. Berry, and O. L. Shotwell, *J. Am. Chem. Soc.* **71,** 3891 (1949).
8. J. R. Stevens and R. H. Bentel, *J. Am. Chem. Soc.* **63,** 311 (1941).
9. F. C. Whitmore and R. S. Thorpe, *J. Am. Chem. Soc.* **63,** 1118 (1941).
10. H. J. Lucas, M. J. Schlatter, and R. C. Jones, *J. Am. Chem. Soc.* **63,** 25 (1941).
11. J. Kagan, B. E. Firth, N. Y. Shih, and C. G. Boyajian, *J. Org. Chem.* **42,** 343 (1977).
12. S. M. Linder, U.S. Patent 4,171,457 (1979).
13. J. P. March, Jr. and L. Goodman, *J. Org. Chem.* **30,** 2491 (1965); G. R. Pettit and D. M. Piatek, *ibid.* **25,** 721 (1960).
14. P. E. Peterson and E. V. P. Tao, *J. Org. Chem.* **29,** 2322 (1964); H. Schechter, F. Conrad, A. L. Daulton, and R. B. Kaplan, *J. Am. Chem. Soc.* **74,** 3052 (1952); M. Kanemara, T. Kimura, N. Ishii, and H. Kawashima, U.S. Patent 4,060,564 (1977); W. Webers, L. Sandhack, and U. Neier, Ger. Offen. 2,429,770 (1976); *Chem. Abstr.* **84,** 121114h (1976).
15. H. B. Knight, R. E. Koos, and D. Swern, *J. Am. Chem. Soc.* **75,** 6212 (1953).
16. F. C. Whitmore and J. M. Church, *J. Am. Chem. Soc.* **55,** 1120 (1933).
17. M. S. Newman, *J. Am. Chem. Soc.* **62,** 2295 (1940).
18. C. G. Overberger, J. H. Saunders, R. E. Allen, and R. Gander, *Org. Synth.* **28,** 28 (1948).
19. G. F. Hennion and J. J. Sheehan, *J. Am. Chem. Soc.* **71,** 1964 (1949).
20. G. Wittig and D. Waldi, *J. Prakt. Chem. [2]* **160,** 243 (1942).
21. L. P. Kirillova, A. V. Rechkina, and L. I. Vereshchagin, *Zh. Org. Khim.* **7** 2469 (1971); *Chem. Abstr.* **76,** 71934n (1972); A. F. Thompson, Jr., J. G. Burr, Jr., and E. N. Shaw, *J. Am. Chem. Soc.* **63,** 186 (1941).
22. C. S. Marvel, H. W. Johnston, J. W. Meier, T. W. Mastin, J. Whitson, and C. M. Himel, *J. Am. Chem. Soc.* **66,** 914 (1944).
23. R. V. Valkenburgh, K. W. Greenlee, J. M. Derer, and C. E. Boord, *J. Am. Chem. Soc.* **71,** 172 (1949).
24. J. M. Church, F. C. Whitmore, and R. V. McGrew, *J. Am. Chem. Soc.* **56,** 176 (1934).
25. F. C. Whitmore, J. S. Whitaker, W. A. Mosher, O. N. Breivik, N. R. Wheeler, C. S. Miner, Jr., L. H. Sutherland, R. B. Wagner, T. W. Clapper, C. E. Lewis, A. R. Lux, and A. H. Popkin, *J. Am. Chem. Soc.* **63,** 643 (1941).
26. G. Calingaert, H. Soroos, V. Hnizda, and H. Shapiro, *J. Am. Chem. Soc.* **66,** 1389 (1944).
27. A. A. Morton, J. R. Myles, and W. S. Emerson, *Org. Synth.* **23,** 95 (1943).
28. R. C. Huston and D. L. Bailey, *J. Am. Chem. Soc.* **68,** 1382 (1946).
29. E. E. Dreger, *Org. Synth. Collect. Vol.* **1,** 306 (1941).
30. R. C. Huston and C. O. Bostwilk, *J. Org. Chem.* **13,** 334 (1948).
31. P. D. Bartlett and C. M. Berry, *J. Am. Chem. Soc.* **56,** 2683 (1934).
32. P. J. Pearce, D. H. Richards, and N. F. Scilly, *J. Chem. Soc., Perkin Trans.* **1,** P. 1655 (1972).
33. T. H. Vaughn, R. J. Spahr, and J. A. Nieuwland, *J. Am. Chem. Soc.* **55,** 4208 (1933).
34. W. G. Dauben and D. M. Michus, *J. Org. Chem.* **42,** 682 (1977).
35. J. K. Crandall and A. C. Rojas, *Org. Synth.* **55,** 1 (1976).
36. E. Barreiro, J. L. Lache, J. Zweig, and P. Crabbé, *Tetrahedron Lett.* **28,** 2352 (1975).
37. W. W. Carlson and L. H. Cretcher, *J. Am. Chem. Soc.* **69,** 1952 (1947).
38. J. Colonge and P. Rochas, *Bull. Soc. Chim. Fr. [5]* **15,** 818, 822, 825, 827 (1948).
39. S. R. Sandler, Borden Chem. Co. Central Research Lab., Philadelphia, Pennsylvania, unpublished results (1960).
40. H. C. Brown and B. C. S. Rao, *J. Am. Chem. Soc.* **78,** 2587 (1956).

§ References

41. E. T. Stiller, S. A. Harris, J. Finkelstein, J. C. Keresztesy, and K. Folkers, *J. Am. Chem. Soc.* **62,** 1785 (1940).
42. L. M. Long and H. D. Troutman, *J. Am. Chem. Soc.* **71,** 2469 (1949).
43. J. Attenburrow, J. Elks, B. A. Hems, and K. N. Speyer, *J. Chem. Soc.* p. 510 (1949).
44. A. Lambert and A. Lowe, *J. Chem. Soc.* p. 1517 (1947).
45. W. S. Ide and J. S. Buck, *Org. React.* **4,** 269 (1948).
46. G. Sumrell, J. I. Stevens, and G. E. Goheen, *J. Org. Chem.* **22** 39 (1957).
47. R. Adams and C. S. Marvel, *Org. Synth. Collect. Vol.* **1,** 94 (1941).
48. I. M. Heilbron and H. M. Bunbury, "Dictionary of Organic Compounds" (rev. ed.), Vol. I. Oxford Univ. Press, London and New York, 1953.
49. J. M. Snell and S. M. McElvain, *Org. Synth. Collect. Vol.* **2,** 114 (1943).
50. J. C. Speck, JR. and R. W. Bost, *J. Org. Chem.* **11,** 788 (1946).
51. V. L. Hansley, *J. Am. Chem. Soc.* **57,** 2303 (1935).
52. M. S. Kharasch, E. Sternfield, and F. R. Mayo, *J. Org. Chem.* **5,** 362 (1940).
53. V. L. Hansley, *Ind. Eng. Chem.* **43,** 1759 (1951).
54. R. C. Huston and G. W. Hedrick, *J. Am. Chem. Soc.* **59,** 2001 (1937).
55. H. Hart, *J. Am. Chem. Soc.* **71,** 1966 (1949).
56. F. J. Sowa, H. D. Hinton, and J. A. Nieuwland, *J. Am. Chem. Soc.* **54,** 3694 (1932).
57. A. H. Blatt, *Org. React.* **1,** 342 (1942).
58. R. R. Read and J. Wood, Jr., *Org. Synth.* **20,** 57 (1940).
59. D. V. Nightengale and H. D. Radford, *J. Org. Chem.* **14,** 1089 (1949).
60. J. H. Freeman, *J. Am. Chem. Soc.* **74,** 6257 (1952).
61. T. B. Patrick and P. A. Egan, *J. Org. Chem.* **42,** 382 (1977).
62. C. D. Gutsche and E. Mathukrishnan, *J. Org. Chem.* **43,** 4905 (1978).
63. W. H. Carothers and R. Adams, *J. Am. Chem. Soc.* **46,** 1675 (1924).
64. E. Schenker, *Angew. Chem.* **73,** No. 3, 81 (1961).
65. D. A. Yeowell, R. L. Seaman, and J. Meutha, *Chem. Eng. News* Sept. 24, p. 4 (1979).
66. H. C. Brown and A. K. Mandal, *J. Org. Chem.* **42,** 2996 (1977).
67. J.-L. Luche, *J. Am. Chem. Soc.* **100,** 2226 (1978).
68. R. O. Hutchins and D. Kandasamy, *J. Org. Chem.* **40,** 2530 (1975).
69. E. Winterfeldt, *Synthesis* p. 617 (1975).
70. H. C. Brown, S. Krishnamurthy, and N. M. Yoon, *J. Org. Chem.* **44,** 1778 (1976); S. Krishnamurthy and H. C. Brown, *ibid.* **42,** 1197 (1977).
71. R. F. Borch and H. D. Durst, *J. Am. Chem. Soc.* **91,** 3996 (1969).
72. D. J. Raber and W. C. Guida, *J. Org. Chem.* **41,** 690 (1976).
73. R. A. Bell and M. B. Gravestock, *Can. J. Chem.* **47,** 2099 (1969).
000. E. L. Eliel and M. N. Rerick, *J. Am. Chem. Soc.* **82,** 1362 (1960).
000. H. Noth and H. Beyer, *Chem. Ber.* **93,** 1078 (1960).
74. A. L. Wilds, *Org. React.* **2,** 178 (1944).
75. C. S. Marvel and G. L. Schertz, *J. Am. Chem. Soc.* **65,** 2055 (1943).
76. E. M. Kaiser, C. G. Edmonds, S. D. Grabb, J. W. Smith, and D. Tramp, *J. Org. Chem.* **36,** 330 (1971).
77. N. M. Weinshenker, G. A. Crosby, and J. Y. Wong, *J. Org. Chem.* **40,** 1966 (1975).
78. T. Fujisawa, K. Sugimoto, and H. Ohta, *J. Org. Chem.* **41,** 1667 (1976).
79. M. P. Doyle and C. T. West, *J. Org. Chem.* **40,** 3821 (1975).
80. J. L. Fry, M. Orfanopoulos, M. G. Adlington, W. R. Dittman, Jr., and S. B. Silverman, *J. Org. Chem.* **43,** 374 (1978).
81. R. A. Benkeser and S. J. Mels, *J. Org. Chem.* **35,** 261 (1970); T. Shono, I. Nishiguchi, H. Ohmizer, and M. Mitani, *J. Am. Chem. Soc.* **100,** 545 (1978).

82. M. S. Kharasch and W. B. Reynolds, *J. Am. Chem. Soc.* **65**, 501 (1943).
83. A. M. Sladkov, V. A. Markevich, I. A. Yavich, L. K. Luneva, and V. N. Chernov, *Doki. Akad. Nauk. SSSR* **119**, 1159 (1958).
84. H. C. Brown and G. Zweifel, *J. Am. Chem. Soc.* **81**, 247 (1959).
85. J. S. Filippo, Jr., C. I. Chern, and J. S. Valentine, *J. Org. Chem.* **40**, 1678 (1975).
86. N. J. Lewis, S. Y. Gabhe, and M. R. DeLaMater, *J. Org. Chem.* **42**, 1479 (1977).
87. H. C. Brown and P. Geoghegan, Jr., *J. Am. Chem. Soc.* **89**, 1522 (1967).
88. H. Kono and J. Hooz, *Org. Synth.* **53**, 77 (1973).
89. G. Zweifel and H. C. Brown, *Org. React.* **13**, 1 (1963).
90. G. W. Kabalka and H. C. Hedgecock, Jr., *J. Org. Chem.* **40**, 1776 (1975).
91. D. Swern, G. N. Billen, and J. T. Scanlan, *J. Am. Chem. Soc.* **68**, 1504 (1946).
92. D. Swern, *Chem. Rev.* **45**, 25 (1949).
93. M. A. Umbreit and K. B. Sharpless, *J. Am. Chem. Soc.* **99**, 5526 (1977).
94. M. E. Kurz and G. J. Johnson, *J. Org. Chem.* **36**, 3184 (1971).
95. H. Hart and C. A. Buehler, *J. Org. Chem.* **29**, 2397 (1964).
96. T. A. Gelssman, *Org. React.* **2**, 94 (1944).
97. D. Davidson and M. T. Bogert, *J. Am. Chem. Soc.* **57**, 905 (1935).
98. D. Davidson and M. Weiss, *Org. Synth. Collect. Vol.* **2**, 590 (1943).
99. H. B. J. Schurink, *Org. Synth. Collect. Vol.* **1**, 425 (1941).
100. J. M. Derfer, K. W. Greenlee, and C. E. Boord, *J. Am. Chem. Soc.* **71**, 178 (1949).
101. R. W. Shortridge, *J. Am. Chem. Soc.* **70**, 984 (1948).
102. H. Wittcoff, *Org. Synth.* **31**, 101 (1951).
103. D. S. Tarbell, *Org. React.* **2**, 1 (1944).
104. S. J. Rhoads, in "Molecular Rearrangements" (P. de Mayo, ed.), Part 1, p. 660. Wiley (Interscience), New York, 1963.
105. E. N. Marvell, B. Richardson, R. Anderson, J. L. Stephenson, and T. Crandall, *J. Org. Chem.* **30**, 1032 (1965).
106. C. F. H. Allen and J. W. Gates, Jr., *Org. Synth. Collect. Vol.* **3**, 418 (1955).
107. D. F. Lohr, Jr. and L. B. Wakefield, U. S. Patent 4,060,563 (1977).
108. S. R. Sandler and W. Karo, "Organic Functional Group Preparations," Vol. 1, pp. 58–59. Academic Press, New York, 1968.
109. C. R. Hauser and S. W. Kantor, *J. Am. Chem. Soc.* **73**, 1437 (1951).
110. L. M. Kozlova, T. A. Chernyavskaya, I. Romadane, and V. G. Farafonova, *Latv. PSR Zinat. Akad. Vestis, Kim. Ser.* **3**, 319 (1973); *Chem. Abstr.* **79**, 78282a (1973); I. Romadane, L. M. Kozlova, T. A. Chernyavskaya, U.S.S.R. Patent 323,864 (1972); *Chem. Abstr.* **78**, 3957 (1973).
111. E. Bamberger, *Justus Liebigs Ann. Chem.* **390**, 131 (1912); L. F. Fieser and M. Fieser, "Reagents for Organic Synthesis," p. 1123. Wiley, New York, 1967; E. J. Behrman, *J. Am. Chem. Soc.* **89**, 2424 (1967).
112. J. Hine, "Physical Organic Chemistry," pp. 348–349. McGraw-Hill, New York, 1956.
113. A. W. Ralston, M. R. McCorke, and E. W. Segerbrett, *J. Org. Chem.* **6**, 750 (1941); A. F. Morey, F. G. Badder, and W. I. Award, *Nature (London)* **172**, 1186 (1953); A. Warshawsky, R. Kalir, and A. Patchornik, *J. Am. Chem. Soc.* **100**, 4544 (1978).
114. J. K. Crandall and L. C. Crawley, *Org. Synth.* **53**, 17 (1973).
115. K. Tanabe and K. Aruta, Japan Kokai 76/16,641 (1976); *Chem. Abstr.* **85**, 94545m (1976).
116. W. C. Still and A. Mitra, *J. Am. Chem. Soc.* **100**, 1927 (1978).
117. C. Weizman, E. D. Bergmann, and M. Sulzbacher, *J. Org. Chem.* **15**, 54 (1950).
118. B. Emmert and E. Asendorf, *Chem. Ber.* **72**, 1188 (1939).
119. F. Kipnis, H. Soloway, and J. Ornfelt, *J. Am. Chem. Soc.* **70**, 142 (1948).
120. D. S. Tarbell, *Org. React.* **2**, 1 (1944).

References

121. L. E. Schniepp, H. H. Geller, and R. W. Karff, *J. Am. Chem. Soc.* **69,** 672 (1947).
122. W. J. Close, B. D. Tiffany, and M. A. Spielman, *J. Am. Chem. Soc.* **71,** 1265 (1949).
123. E. C. Horning, M. G. Horning, and G. N. Walker, *J. Am. Chem. Soc.* **71,** 169 (1949).
124. H. Adkins, L. M. Richards, and J. W. Davis, *J. Am. Chem. Soc.* **63,** 1320 (1941).
125. L. Ruzicka, *Helv. Chim. Acta* **19,** 419 (1936).
126. N. Rabjohn, *Org. React.* **5,** 331 (1949).
127. S. M. Sethna, *Chem. Rev.* **49,** 91 (1951).
128. P. Kovacic and S. T. Morneweck, *J. Am. Chem. Soc.* **87,** 1566 (1965).
129. A. J. Davidson and R. O. C. Norman, *J. Chem. Soc.* p. 5404 (1964).
130. M. Oyama, *J. Org. Chem.* **30,** 2429 (1965).
131. P. S. Wharton and D. H. Bohlen, *J. Org. Chem.* **26,** 3615 (1961).
132. C. H. DePuy, G. M. Dappen, K. L. Eilers, and R. A. Klein, *J. Org. Chem.* **29,** 2813 (1964).
133. B. S. Farah, E. E. Gilbert, M. Litt, J. A. Otto, and J. P. Sibilia, *J. Org. Chem.* **30,** 1003 (1965).
134. R. Kotani, *J. Org. Chem.* **30,** 350 (1965).
135. H. Greenfield, J. H. Wotiz, and I. Wender, *J. Org. Chem.* **22,** 542 (1959).
136. N. Kutepow and H. Kindler, *Angew. Chem.* **72,** 802 (1960).
137. J. Falbe and F. Korte, *Chem. Ber.* **97,** 1104 (1964).
138. C. E. Schildknecht, "Vinyl and Related Polymers," pp. 341–343. Wiley, New York, 1952.
139. J. K. Crandall, D. B. Banks, R. A. Colyer, R. J. Watkins, and J. P. Arrington, *J. Org. Chem.* **33,** 423 (1968).
140. T. Cohen and T. Tsuji, *J. Org. Chem.* **26,** 1681 (1961).
141. T. M. Santosusso, R. D. Slockett, and D. Swern, *Abstr. 3rd Middle Atl. Am. Chem. Soc. Reg. Meet.* p. 59H (1968).
142. H. House, *J. Am. Chem. Soc.* **95,** 3310 (1973).
143. G. Stork, G. A. Kraus, and G. A. Garcia, *J. Org. Chem.* **39,** 3459 (1974).
144. T. Nakata, G. Schmid, B. Vranesic, M. Okigawa, T. Smith-Palmer, and Y. Kishi, *J. Am. Chem. Soc.* **100,** 2933 (1978).
145. D. A. Evans, C. E. Sacks, W. A. Kleschick, and T. R. Taber, *J. Am. Chem. Soc.* **101,** 6789 (1979).
146. C. H. Heathcock, C. T. Buse, W. A. Kleschick, M. C. Pirrung, J. E. Sohn, and J. Lampe, *J. Org. Chem.* **45,** 1066 (1980).
147. C. L. Hill and G. M. Whitesides, *J. Am. Chem. Soc.* **96,** 870 (1974).
148. N. C. Deno, W. E. Billups, K. E. Kramer, and R. T. Lastomirsky, *J. Org. Chem.* **35,** 3080 (1970).
149. I. M. Godfrey, M. V. Sargent, and J. A. Elix, *J. Chem. Soc., Perkin Trans. 1* No. 11, p. 1353 (1974).
150. V. Suzuki, T. Maki, and K. Mineta, Japan Kokai 75/50,311 (1975); *Chem. Abstr.* **83,** 96730a (1975).
151. A. S. Hay, *J. Org. Chem.* **34,** 1160 (1969).
152. H. C. Brown and E.-I. Negishi, *J. Am. Chem. Soc.* **91,** 1224 (1969).
153. T. Cohen, A. G. Dietz, Jr., and J. R. Miller, *J. Org. Chem.* **42,** 2053 (1977).
154. J. G. Duboudin, *J. Organomet. Chem.* **91,** (1), C1-C3 (1975).
155. K. H. Meyer and K. Nagel, Ger. Offen. 2,064,306 (1972); *Chem. Abstr.* **77,** 125958q (1972).
156. A. F. Thompson, Jr., J. G. Burr, Jr., and E. N. Shaw, *J. Am. Chem. Soc.* **63,** 186 (1941).
157. Y. Okude, S. Hirano, T. Hiyama, and H. Nozaki, *J. Am. Chem. Soc.* **99,** 3179 (1977).
158. J. J. D'Amico and R. W. Fuhrhop, U. S. Patent 4,150,027 (1979).
159. F. Millauer, Ger. Offen. 2,318,677 (1974); *Chem. Abstr.* **82,** 72523 (1975).
160. C. A. Smith and J. B. Brutzner, *J. Org. Chem.* **41,** 367 (1976).
161. I. M. Lockhart and N. E. Webb, *Chem. Ind. (London)* 1230 (1970).

162. A. H. Lewin and T. Cohen, *J. Org. Chem.* **32,** 3844 (1967).
163. R. A. Auerbach, D. S. Crumrine, D. L. Ellison, and H. O. House, *Org. Synth.* **54,** 49 (1974).
164. R. Paul, O. Riobe, and M. Maumy, *Org. Synth.* **55,** 62 (1976).
165. J. V. Hay and T. M. Harris, *Org. Synth.* **53,** 56 (1973).
166. J. M. Jerkunica and T. G. Traylor, *Org. Synth.* **53,** 94 (1973).
167. R. N. Mirrington and G. I. Feutrill, *Org. Synth.* **53,** 90 (1973).
168. E. G. Janzen and R. C. Zawalski, *J. Org. Chem.* **43,** 1900 (1978).
169. P. L. Coe, N. E. Milner, and J. A. Smith, *J. Chem. Soc., Perkin Trans. 1* No. 7, p. 654 (1975).
170. S. Fuzesi, M. Lapkin, and R. J. Polak, U.S. Patent 3,847,844 (1974).
171. N. E. Morganson, U.S. Patent 3,969,423 (1976).

CHAPTER 5 / **ETHERS AND OXIDES**

1. Introduction	129
2. Condensation Reactions	131
A. The Williamson Synthesis	132
2-1. Preparation of Triphenylmethyl Ethyl Ether	132
2-2. Preparation of Allyl Phenyl Ether	133
B. The Condensation of Alcohols with Aldehydes, Olefins, Acetylenes, Alkyl Sulfates, and Oxides	133
2-3. Preparation of Chloromethyl Methyl Ether	133
2-4. Preparation of Nitro-tert-butyl Methyl Ether	134
2-5. Preparation of Isobutyl Ethyl Ether	135
2-6. Preparation of 1-Ethoxy-2-propanol	135
C. The Reaction of Chloroethers with Olefins and Organometallic Reagents	136
D. The Condensation of Oxirane Compounds to Give Substituted Oxiranes (Epoxides)	136
2-7. Preparation of Glycidyl Benzoate	136
2-8. Preparation of the Diglycidyl Ether of 2,2',6,6'-Tetrabromobisphenol A [2,2-Di(3,5-dibromo-4-hydroxyphenyl)propane]	137
E. The Darzens Glycidic Ester Synthesis	138
2-9. Preparation of Phenylmethylglycidic Ester	139
3. Elimination Reactions	139
A. Ethers	139
B. Epoxides	140
3-1. Preparation of 2,3-Epoxy-1-propanol (Glycidol)	140
4. Oxidation Reactions	140
A. Peroxidation of Olefins to Give Oxiranes (Epoxides)	140
4-1. Preparation of 1-Hexene Oxide	142
4-2. Preparation of 2,3-Epoxy-trans-decalin	142
4-3. Preparation of Isophorone Oxide	142
5. Miscellaneous Methods	143
References	143

1. INTRODUCTION

Ethers and oxides differ in the type of chemical bonding of carbon to oxygen. Oxides are three-membered rings and are attached to adjacent carbon atoms in a given system whereas ethers, if they are cyclic, are not attached to adjacent carbon atoms as shown below.

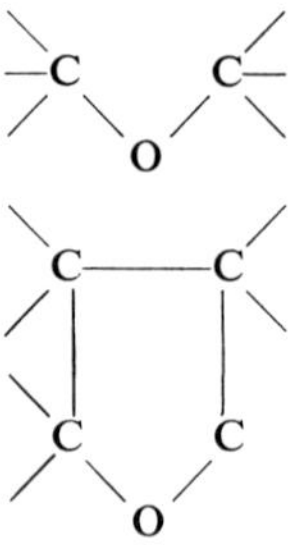

5. Ethers and Oxides

Ethers

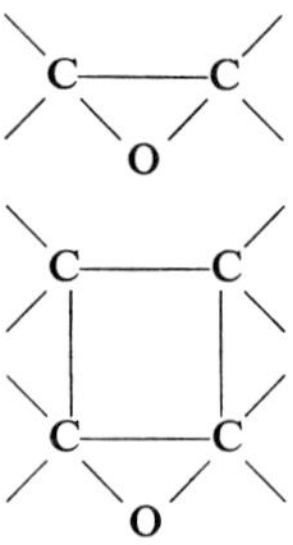

Oxides (Epoxides)

The common methods used to produce ethers in the laboratory are the Williamson synthesis and the dehydration of alcohols using acids or inorganic oxides such as alumina.

$$RONa + R'X \longrightarrow ROR' + NaX \qquad (1)$$

$$2 ROH \xrightarrow[\text{Heat}]{\text{H}^+ \text{ or Alumina}} ROR + H_2O \qquad (2)$$

Other methods involve the reaction of dialkyl sulfates with sodium alcoholates (or sodium phenolates) and the addition of olefins to alcohols or phenols using acid catalysts.

$$(RO)_2SO_2 + R'ONa \longrightarrow R'OR + ROSO_3Na \qquad (3)$$

$$ROH + R'CH=CH_2 \longrightarrow R'\underset{OR}{CH}-CH_3 \qquad (4)$$

The laboratory reactions with dialkyl sulfates is not preferred since it has been reported that these reagents may cause carcinogenic reactions.

Phenolic ethers can also be formed via the Claisen rearrangement as described earlier in Chapters 2 and 4 of this volume.

Ethers are also formed by the base- or acid-catalyzed condensation of oxides with themselves or with alcohols or phenols to give monomeric or polymeric systems.

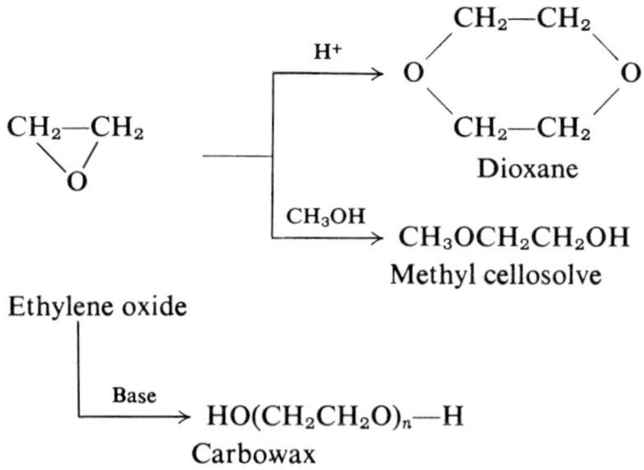

Ethylene oxide

$\xrightarrow{\text{H}^+}$ Dioxane

$\xrightarrow{\text{CH}_3\text{OH}}$ $CH_3OCH_2CH_2OH$
Methyl cellosolve

$\xrightarrow{\text{Base}}$ $HO(CH_2CH_2O)_n-H$
Carbowax

(5)

Oxides are usually formed in the laboratory by the peroxidation of olefins with H_2O_2, peracids, or by the dehydrohalogenation of halohydrins.

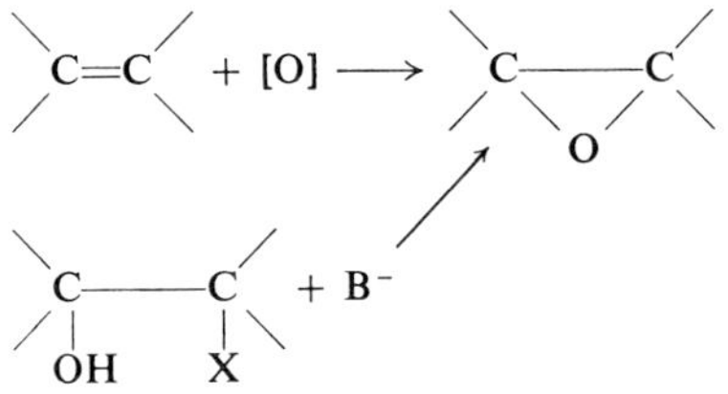

$$\begin{array}{c} \diagdown C=C \diagup + [O] \longrightarrow \diagdown C \text{---} C \diagup \\ \diagup \quad \diagdown \quad \quad \quad \diagup \diagdown_O \diagup \diagdown \\ \\ \diagdown C \text{---} C \diagup + B^- \\ \diagup | \quad | \diagdown \\ OH \quad X \end{array} \quad (6)$$

CAUTION: Ethers tend to absorb and react with oxygen from the air to form peroxides. These peroxides must be removed prior to distillation or concentration to prevent an explosive detonation. In some cases, heat, shock, or friction can also cause a violent decomposition.

Some epoxides have been reported to be carcinogenic, especially those containing electron-withdrawing groups.

Due to their volatility, the lower members of the ethers and oxides are extremely flammable and care should be taken in handling these materials.

2. CONDENSATION REACTIONS

The condensation reaction most widely used in the laboratory to give ethers is the Williamson synthesis.

$$RONa + R'X \longrightarrow ROR' + NaX \quad (7)$$

The reaction is applicable to the aromatic as well as the aliphatic series. In fact the reaction of epichlorohydrin with sodium phenolates can be considered a Williamson reaction and is the basis of the large-scale production of commercial epoxy resins. Crown ethers are also made via a Williamson synthesis.

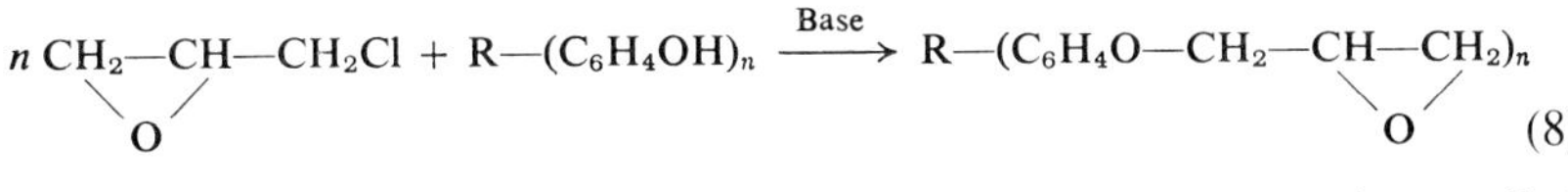

$$n\,CH_2\text{---}CH\text{---}CH_2Cl + R\text{---}(C_6H_4OH)_n \xrightarrow{\text{Base}} R\text{---}(C_6H_4O\text{---}CH_2\text{---}CH\text{---}CH_2)_n \quad (8)$$
$$\diagdown_O\diagup \qquad\qquad\qquad\qquad\qquad\qquad\qquad\qquad\qquad \diagdown_O\diagup$$

Alcohols or their salts also condense with aldehydes, olefins, acetylenes, alkyl sulfates, and oxides to give ethers.

Ethers or epoxides containing a reactive functional group can also be condensed with other functional groups as a means of introducing the ether functionality.

The Darzens glycidic ester synthesis involves the condensation (with the aid of bases) of chloroacetates, chloroketones, or other reactive halides with aldehydes or ketones to yield epoxy compounds.

The Ullmann reaction is used to condense phenoxides with aromatic halides by means of finely divided copper or copper salts to give diaryl ethers.

Dichlorocarbene can be inserted into C—H bonds alpha to an ether oxygen to give dichloromethyl ethers.

A. The Williamson Synthesis

$$RX + R'OH \longrightarrow ROR + HX \qquad (9)$$

The Williamson synthesis usually involves the use of the sodium salt of the alcohol and an alkyl halide [1]. Primary halides give the best yields since secondary and tertiary halides readily dehydrohalogenate to give olefins. However, triarylmethyl chlorides react with alcohols directly to give 97% yields of ethers [2]. Alkyl phenyl ethers are prepared from aqueous or alcoholic solutions of alkali phenolates and alkyl halides [3, 4]. Benzyl halides are easily replaced by alkoxy groups in high yields [5]. Polar solvents such as dimethylformamide favor the reaction [6]. Phase transfer catalysis has been reported to aid this reaction [7].

2-1. *Preparation of Triphenylmethyl Ethyl Ether* [2]

$$(C_6H_5)_3C—Cl + C_2H_5OH \longrightarrow (C_6H_5)_3C—OC_2H_5 + HCl \qquad (10)$$

To an Erlenmeyer flask containing 100 ml of absolute ethanol is added 27.9 gm (0.10 mole) of triphenylmethyl chloride. The flask is heated to get rid of hydrogen chloride. Upon cooling, 28.0 gm (97%) of the trityl ether separates, m.p. 83°C.

The preparation of 18-Crown-6 is reported to take place by the condensation of triethyleneglycol and its chloro analog [8].

The Ullmann condensation involves the formation of diaryl ethers by the reaction of alkali phenolates and aryl halides catalyzed by copper [9–11] or cuprous chloride [12, 13].

The condensation of chloromethyl methyl ether with alcohols in the presence of pyridine gives acetal derivatives of formaldehyde [14].

$$ROH + ClCH_2OCH_3 \xrightarrow{\text{Pyridine}} ROCH_2OCH_3 \qquad (11)$$

Refluxing a phenol with allyl bromide and anhydrous potassium carbonate in acetone for several hours yields allyl aryl ethers in good yields [15]. The reaction is unsatisfactory for phenolic aldehydes. The Williamson synthesis using sodium phenoxide and allyl bromide in methanol is more rapid than the above method and gives good results [16].

Epichlorohydrin (3-chloro-1,2-propylene oxide) has found widespread use for preparing [17, 18] glycidyl ethers (epoxides) of phenols and alkyl compounds. The diglycidyl ethers of phenols have been used to prepare polymers by reaction with amines [17]. Epichlorohydrin also condenses with alcohols to give chlorohydroxy ethers [19]. The use of phase transfer catalysts is also very effective in giving improved yields.

§ 2. Condensation Reactions

$$\text{ROH} + \text{Cl—CH}_2\text{—}\overset{\overset{\displaystyle O}{\diagup\diagdown}}{\text{CH}}\text{—CH}_2 \xrightarrow{\text{NaOH}} \text{RO—CH}_2\text{—}\overset{\overset{\displaystyle O}{\diagup\diagdown}}{\text{CH}}\text{—CH}_2 \quad (12)$$

2-2. Preparation of Allyl Phenyl Ether [20]

$$\text{C}_6\text{H}_5\text{OH} + \text{CH}_2\text{=CH—CH}_2\text{Br} \xrightarrow[\text{Acetone}]{\text{K}_2\text{CO}_3} \text{C}_6\text{H}_5\text{OCH}_2\text{—CH=CH}_2 \quad (13)$$

In a flask equipped with a reflux condenser are placed 18.8 gm (0.20 mole) of phenol, 24.2 gm (0.23 mole) of allyl bromide, 28.0 gm (0.20 mole) of potassium carbonate, and 200 ml of acetone. The contents are refluxed for 10 hr, cooled, treated with 200 ml of water, and extracted three times with 25-ml portions of ether. The combined ether extracts are washed with 10% aqueous sodium hydroxide and three times with 25-ml portions of a saturated NaCl solution, dried, and distilled to yield 22 gm (82%) of the product, bp 119.5%–120.5°C (30.2 mm), n_D^{25} 1.5210, ν_max 882 cm^{-1}.

B. The Condensation of Alcohols with Aldehydes, Olefins, Acetylenes, Alkyl Sulfates, and Oxides

CAUTION: Mono and bis(chloromethyl) ether have been reported to be very highly carcinogenic and may be present in all reactions involving the use of formaldehyde and hydrochloric acid. The handling of other chloromethyl ethers should also be done with great care and reactions should be carried out in a hood.

Chloromethyl methyl ether is prepared from a mixture of methanol, aqueous formaldehyde, and hydrogen chloride [21].

$$\text{CH}_3\text{OH} + \text{CH}_2\text{=O} + \text{HCl} \longrightarrow \text{CH}_3\text{OCH}_2\text{Cl} + \text{H}_2\text{O} \quad (14)$$

The reaction is applicable to higher aldehydes and primary or secondary alcohols [22].

Chloromethyl methyl ether adds to olefins in the presence of mercuric chloride [23] or zinc chloride [24] to give γ-chloroethers.

Alcohols also add to olefins and α-nitroolefins [25] in the presence of dilute sulfuric acid. For example, isobutylene and trimethylethylene give tertiary alkyl ethers. Primary alcohols are more reactive than secondary alcohols and tertiary alcohols are practically nonreactive. Ethyl-*tert*-amyl ether is prepared in 90% yields [26].

2-3. Preparation of Chloromethyl Methyl Ether [21, 27]

$$\text{CH}_3\text{OH} + \text{CH}_2\text{=O} + \text{HCl (gas)} \longrightarrow \text{CH}_3\text{OCH}_2\text{Cl} + \text{H}_2\text{O} \quad (15)$$

CAUTION: Chloromethyl methyl ether is considered highly carcinogenic.

In a 2-liter three-necked round-bottomed flask are placed 438 ml (350 gm 10.9 moles) of methanol and 682 ml (757 gm) of a 37% formaldehyde solution (9.34

moles). While cooling to 0°–5°C dry hydrogen chloride gas is bubbled through the solution for 2 hr. An oily layer separates and the hydrogen chloride gas is bubbled into the solution for another 2 hr longer. The chloromethyl methyl ether layer is separated, dried over calcium chloride, and fractionally distilled at atmospheric pressure to yield 252 gm (33.5%) of product, bp 55°–58°C. It should be noted that the product is a lachrymator which smells of hydrogen chloride. It should be protected from moisture.

2-4. Preparation of Nitro-tert-butyl Methyl Ether [25]

$$\underset{NO_2}{\underset{|}{CH_2}}-\underset{|}{\overset{CH_3}{\underset{|}{C}}}=CH_2 + CH_3OH \xrightarrow{CH_3ONa} \underset{NO_2}{\underset{|}{CH_2}}-\underset{OCH_3}{\overset{CH_3}{\underset{|}{C}}}-CH_3 \qquad (16)$$

To a flask containing 10.8 gm (0.2 mole) of sodium methoxide (from 4.6 gm of sodium) in 100 ml of methanol is added dropwise 20.2 gm (0.20 mole) 3-nitro-2-methyl-1-propene over a period of 30 min at room temperature. The reaction mixture is diluted with water, neutralized, and extracted with ether. The ether extract is then dried and distilled to yield 17.3 gm (65%) of nitro-tert-butyl methyl ether, bp 75°C (15 mm).

The reaction of acrylonitrile with aryl and aliphatic phenols produces cyanoethers (cyanoethylation) [28]. For example the cyanoethylation of resorcinol in the presence of Triton B gives a 40% yield of 1,3-bis(β-cyanoethoxy)benzene [29].

$$\underset{OH}{\underset{}{\text{[resorcinol]}}} + 2\,CH_2{=}CH{-}CN \longrightarrow \underset{OCH_2CH_2CN}{\underset{}{\text{[product]}}} \qquad (17)$$

In a related manner phenol and m-methoxyphenol give 67.5 and 76% yields of β-phenoxypropionitrile and m-methoxyphenylpropionitrile [29]. Other examples are found in Chapter 17.

Phenols give phenoxy derivatives at low temperatures when condensed with olefins in the presence of mineral acid [30] or of boron trifluoride [31]. Primary and secondary alcohols and phenols add to acrylic esters to give β-alkoxy and β-aryloxy propionates.

Alcohols react with acetylenes to give vinyl ethers [32, 33].

$$ROH + HC{\equiv}CH \longrightarrow ROCH{=}CH_2 \qquad (18)$$

Anisole is prepared by the alkylation of sodium phenoxide at 10°C with dimethyl sulfate (75%) [34]. Aliphatic ethers are made in a similar manner [35]. In addition acetylenic ethers are also prepared from acetylenic alcohols and dimethyl sulfate in the presence of sodium amide [36].

2-5. Preparation of Isobutyl Ethyl Ether [35]

$$\text{CH}_3\text{—CH(CH}_3\text{)—CH}_2\text{ONa} + (\text{C}_2\text{H}_5)_2\text{SO}_4 \longrightarrow$$

$$\text{CH}_3\text{—CH(CH}_3\text{)—CH}_2\text{OC}_2\text{H}_5 + \text{Na}(\text{C}_2\text{H}_5)\text{SO}_4 \quad (19)$$

To a flask is added 93 gm (1.25 moles) of dry isobutyl alcohol followed by 12.5 gm (0.54 gm atom) of sodium (small pieces). The exothermic reaction causes the mixture to reflux. After the reaction ceases it is heated by means of an oil bath at 120°–130°C for 2¾ hr. After this time, some of the sodium still remains unreacted. The mixture is cooled to 105°–115°C and 77.1 gm (0.5 mole) of diethyl sulfate is added dropwise over a 2-hr period. The reaction is exothermic and steady refluxing occurs while the addition proceeds. The mixture is refluxed for 2 hr after all the diethyl sulfate has been added. The reaction mixture is cooled to room temperature. Then to it are added an equal weight of crushed ice and a slight excess of dilute sulfuric acid. The ether is steam-distilled off, separated, washed three times with 30% sulfuric acid, washed twice with water, and then dried over potassium carbonate. The dried product is refluxed over sodium ribbon and then fractionally distilled to give 35.7 gm (70%) of isobutyl ethyl ether, bp 78°–80°C, n_D^{25} 1.3739.

Alcohols also react with epoxides to give hydroxy ethers by a trans opening of the ring.

$$\text{ROH} + \text{CH}_2\text{—CH—CH}_3 \xrightarrow[\text{or } H^+]{\text{NaOCH}_3} \text{ROCH}_2\text{—CH(OH)—CH}_3 \quad (20)$$

Cyclohexene oxide reacts with refluxing methanol in the presence of a catalytic amount of sulfuric acid to give *trans*-2-methoxycyclohexanol in 82% yield [37]. Unsymmetrical epoxides such as propylene oxide give a primary or secondary alcohol, depending on the reaction conditions. Base catalysis favors secondary alcohol formation whereas acid or noncatalytic conditions favor a mixture of the isomeric ethers [38–40]. Epichlorohydrin can be used in a similar manner to give chlorohydroxy ethers [19].

2-6. Preparation of 1-Ethoxy-2-propanol [40]

$$\text{CH}_3\text{—CH—CH}_2\text{(O)} + \text{C}_2\text{H}_5\text{OH} \xrightarrow{\text{NaOH}} \text{CH}_3\text{—CH(OH)—CH}_2\text{OC}_2\text{H}_5 \quad (21)$$

To a mixture of 2560 gm (55.5 moles) of absolute ethanol and 10 gm of sodium hydroxide at 76°–77°C is added 638 gm (11 moles) of propylene oxide over a period of 4 hr. The mixture is boiled for 2 additional hr until the temperature becomes steady at 80°C. Distillation of the neutralized liquid yields 770 gm (81.4%) of 1-ethoxy-2-propanol, bp 130°–130.5°C.

C. The Reaction of Chloroethers with Olefins and Organometallic Reagents

In a manner related to alcohols adding to olefins and acetylenes to give ethers, active chloroethers also can add to olefins to give substituted chloroethers [41–43]. For example, chloromethyl methyl ether adds to isobutylene under Friedel–Crafts catalysis to give 3-chloro-3-methylbutyl methyl ether in 60% yield [41]. (See page 133 for hazards on handling chloromethyl ethers.)

$$(CH_3)_2C=CH_2 + Cl-CH_2-OCH_3 \longrightarrow (CH_3)_2\underset{\underset{Cl}{|}}{C}-CH_2CH_2OCH_3 \quad (22)$$

Organometallics such as Grignard reagents couple with chloroethers to give substituted ethers [44].

$$RMgX + Cl-CH_2OCH_3 \longrightarrow RCH_2OCH_3 + MgXCl \quad (23)$$

One of the steps in the Boord synthesis involves the Grignard coupling with an α,β-dibromo ether to form a β-bromo ether [45, 46].

D. The Condensation of Oxirane Compounds to Give Substituted Oxiranes (Epoxides)

Oxirane compounds such as 3,4-epoxy-1-butene [47], 2,3-epoxy-1-propanol (glycidol) [48], 2,3-epoxy-1-chloropropane (epichlorohydrin) [49], and 2,3-epoxybutanoic acid [50] can be used to react with other functional groups while keeping the epoxy groups intact. Recently several pure diglycidyl esters have been prepared by the simultaneous addition of an acid chloride and triethylamine to glycidol at 0°–5°C in 25–94% yields [51].

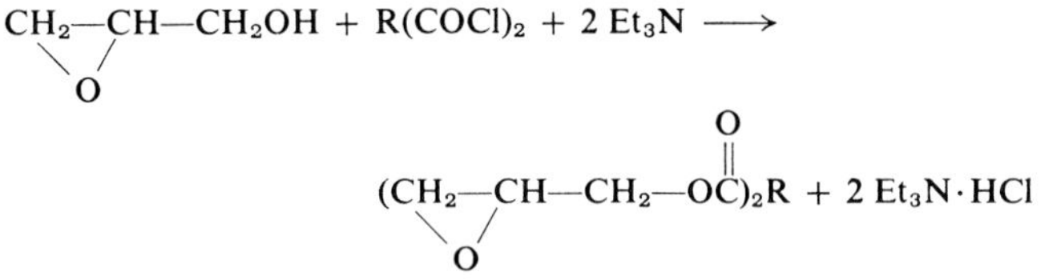

$$CH_2\overset{\diagdown}{\underset{O}{-}}CH\overset{\diagup}{-}CH_2OH + R(COCl)_2 + 2\ Et_3N \longrightarrow$$

$$(CH_2\overset{\diagdown}{\underset{O}{-}}CH\overset{\diagup}{-}CH_2-O\overset{O}{\overset{\|}{C}})_2R + 2\ Et_3N \cdot HCl \quad (24)$$

2-7. Preparation of Glycidyl Benzoate [27]

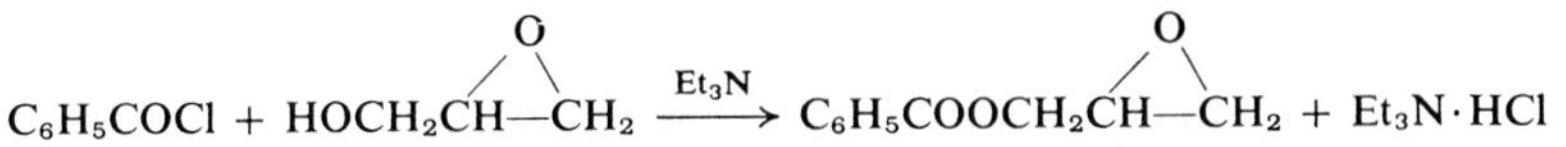

$$C_6H_5COCl + HOCH_2\overset{O}{\overset{\diagdown}{CH}\underset{}{-}}CH_2 \xrightarrow{Et_3N} C_6H_5COOCH_2\overset{O}{\overset{\diagdown}{CH}\underset{}{-}}CH_2 + Et_3N \cdot HCl \quad (25)$$

To a 2-liter flask are added 51.9 gm (0.7 mole) of glycidol and 400 ml of benzene. The flask is cooled to 0°C and a solution of 81.2 ml (0.7 mole) of benzoyl chloride in 60 ml of benzene is placed in one dropping funnel and in another is placed a solution of 97.0 ml (0.7 mole) of triethylamine in 40 ml of benzene. The solutions are added dropwise simultaneously over a 2-hr period and then stirred for an additional 2 hr. The solids are filtered and rinsed with two 50-ml portions of benzene. The filtrate is shaken twice with dilute hydrochloric acid, washed with water until neutral, and dried over sodium sulfate. The benzene is then removed using a water aspirator and the residue is vacuum-distilled to yield 90 gm (71%), bp 97.5°–98.5°C (0.8–0.9 mm). The reported [52] boiling point is 103°C (1 mm).

2-8. Preparation of the Diglycidyl Ether of 2,2′,6,6′-Tetrabromobisphenol A [2,2-Di(3,5-dibromo-4-hydroxyphenyl)propane] [17]

$$\text{HO}\underset{\text{Br}}{\overset{\text{Br}}{-}}\!\!\!\!\!\!\!\!\!\bigcirc\!\!\!-\underset{\text{CH}_3}{\overset{\text{CH}_3}{\text{C}}}-\underset{\text{Br}}{\overset{\text{Br}}{-}}\!\!\!\!\!\!\!\!\!\bigcirc\!\!\!-\text{OH} + 2\ \text{ClCH}_2\overset{\text{O}}{\overset{\diagup\diagdown}{\text{CH}-\text{CH}_2}} \xrightarrow{\text{NaOH}}$$

$$\overset{\text{O}}{\overset{\diagup\diagdown}{\text{H}_2\text{C}-\text{CH}}}-\text{CH}_2\text{O}-\underset{\text{Br}}{\overset{\text{Br}}{-}}\!\!\!\!\!\!\!\!\!\bigcirc\!\!\!-\underset{\text{CH}_3}{\overset{\text{CH}_3}{\text{C}}}-\underset{\text{Br}}{\overset{\text{Br}}{-}}\!\!\!\!\!\!\!\!\!\bigcirc\!\!\!-\text{O}-\text{CH}_2-\overset{\text{O}}{\overset{\diagup\diagdown}{\text{CH}-\text{CH}_2}} +$$

$$2\ \text{NaCl} + 2\ \text{H}_2\text{O} \quad (26)$$

To a 1-liter resin kettle equipped with a stirrer, condenser, and thermometer are added 136 gm (0.25 moles) of tetrabromobisphenol A (Michigan Chemical Corp.), 462.5 gm of epichlorohydrin (5.0 moles), and 2.5 ml water. The mixture is stirred until the solids dissolve and then 6 gm (0.15 moles) of sodium hydroxide pellets is added. The temperature is raised to 100°C and then lowered to 95°C while more sodium hydroxide (35 gm, 0.87 moles) is added portionwise in 6-gm batches. After the last addition, the reaction mixture became yellow and opaque. Stirring is continued until no further exotherm is observed. The excess epichlorohydrin is distilled off using a water aspirator while keeping the reaction mixture below 150°C. Benzene (25 ml) is added to precipitate sodium chloride and the mixture is filtered through a Büchner funnel. The salts are washed with another 25 ml of benzene and the washings together with the filtrate are concentrated under reduced pressure to yield 117 gm (71%) of a dark brown syrup having an epoxy equivalent weight of 331 (calculated 328).

The condensation reactions of 2,3-epoxy-1-chloropropane (epichlorohydrin) have assumed great commercial significance in the preparation of the epoxies for polymer applications in plastics, adhesives, and coatings. For example, the diepoxides of bisphenol A are prepared by condensing two moles of epichlorohydrin with a basic solution of the bisphenol A [53–55].

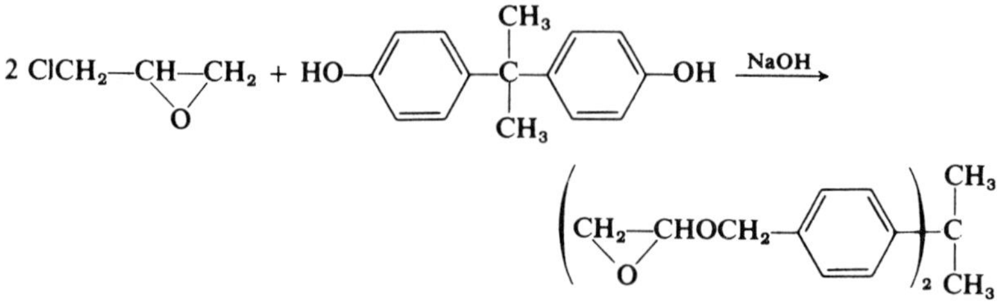

$$2\ ClCH_2\text{—}CH\underset{O}{\text{—}}CH_2 + HO\text{—}C_6H_4\text{—}C(CH_3)_2\text{—}C_6H_4\text{—}OH \xrightarrow{NaOH}$$

$$\left(CH_2\underset{O}{\text{—}}CHOCH_2\text{—}C_6H_4\text{—}\right)_2 C(CH_3)_2 \quad (27)$$

The synthesis of monoglycidyl aryl ethers has also been reported [56].

A review of the ether-forming reactions of epichlorohydrin (2,3-epoxy-1-chloropropane) has been published [57].

The literature on the synthesis of epoxy resins is extensive, and several good sources are available [18, 58a,b].

Anderson, Lindsay, and Reese reported that dichlorocarbenes insert into positions alpha to ethers to give dichloromethyl ethers [59]. Unsaturated ethers give dichlorocyclopropyl ethers. The use of sodium trichloroacetate or ethyl trichloroacetate is useful for generating the carbene.

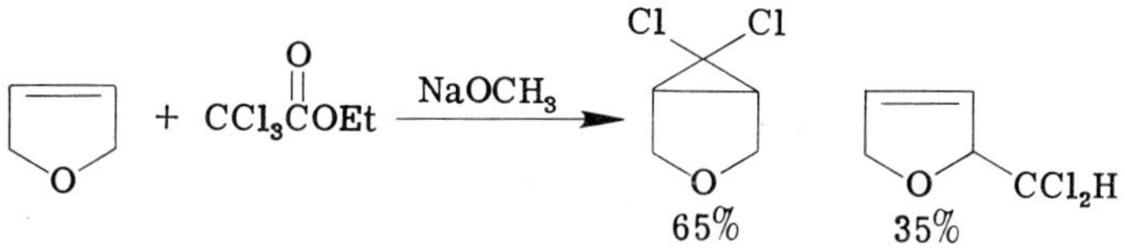

$$\text{furan} + CCl_3COEt \xrightarrow{NaOCH_3} \text{(dichlorocyclopropane-fused tetrahydrofuran)} \ 65\% + \text{(CCl_2H-substituted dihydrofuran)} \ 35\%$$

E. The Darzens Glycidic Ester Synthesis

The Darzens synthesis [60–62] involves the condensation of aldehydes or ketones with ethyl chloroacetate in the presence of sodium amide or ethoxide to give α,β-epoxyesters in one step. Aromatic ketones and aldehydes as well as aliphatic ketones give good yields. Yields from aliphatic aldehydes, however, are poor.

$$R_2C=O + Cl\text{—}CH_2\text{—}COOC_2H_5 \xrightarrow{Base} R\text{—}\underset{O}{C}\underset{|}{\overset{R}{|}}\text{—}CHCOOC_2H_5 \quad (28)$$

Ethyl dichloroacetate also condenses with aldehydes and ketones with the aid of magnesium amalgam to give α-chloro-β-hydroxyesters which, upon treatment with sodium ethoxide, give glycidic esters.

In place of chloroacetic esters other halogenated compounds have been found to give good yields of epoxides, e.g., α-chloroketones [63] with benzyl [64] or benzal halides [65].

$$RCHO + ClCH_2COR' \xrightarrow{C_2H_5ONa} RCH\underset{O}{\text{—}}CH\text{—}COR' \quad (29)$$

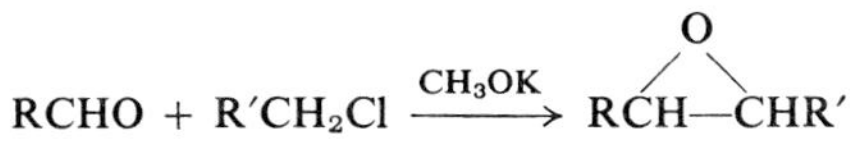

§ 3. Elimination Reactions

$$RCHO + R'CH_2Cl \xrightarrow{CH_3OK} R\overset{O}{\overset{|\!\!\diagdown}{CH}}\!\!-\!\!\overset{}{\overset{\diagup\!\!|}{CH}}R' \qquad (30)$$

2-9. Preparation of Phenylmethylglycidic Ester [66]

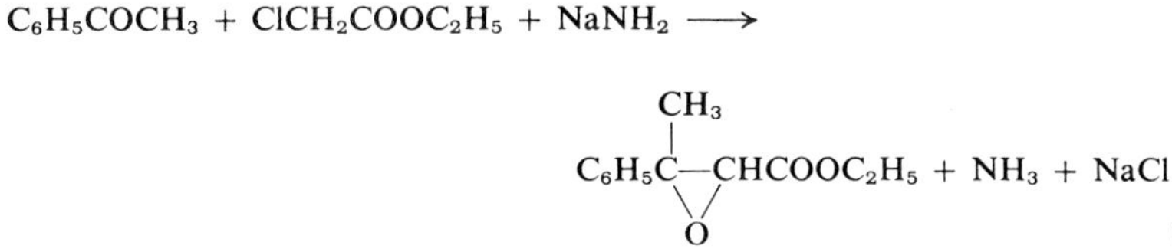

$$C_6H_5COCH_3 + ClCH_2COOC_2H_5 + NaNH_2 \longrightarrow$$

$$\overset{CH_3}{\underset{\overset{\diagdown\!|\!\diagup}{O}}{C_6H_5C\!\!-\!\!CHCOOC_2H_5}} + NH_3 + NaCl \qquad (31)$$

To a flask equipped with a stirrer, condenser, and thermometer are added 120 gm (1 mole) of acetophenone, 120 gm (1 mole) of ethyl chloroacetate, and 200 ml of dry benzene. (NOTE: The reaction should be carried out in a hood since a large volume of ammonia gas is given off.) During a period of 2 hr 47.2 gm (1.2 moles) of finely powdered sodium amide is added (strongly exothermic) while the temperature is kept at 15°–20°C. Following the addition, the reaction mixture is stirred for 2 hr at room temperature and then the red mixture is slowly poured into a beaker containing 700 gm of cracked ice while stirring by hand. The organic layer is separated and the aqueous layer is extracted with 200 ml of benzene. The combined organic layers are washed with three 300-ml portions of water, the last one containing 10 ml of acetic acid. The benzene solution is dried over sodium sulfate, filtered, concentrated, using an aspirator, to give a residue which upon fractionation under reduced pressure yields 128–132 gm (62–64%), bp 111°–113°C (3 mm).

3. ELIMINATION REACTIONS

A. Ethers

The elimination of water from two moles of alcohol yields an ether.

$$2\ ROH \xrightarrow{H^+} ROR + H_2O \qquad (32)$$

For example, adding *tert*-butyl alcohol to a mixture of ethanol and 15% aqueous sulfuric acid gives 95% *tert*-butyl ethyl ether [67].

Alcohols are also dehydrated over solid catalysts such as alumina and phosphoric acid [68–71].

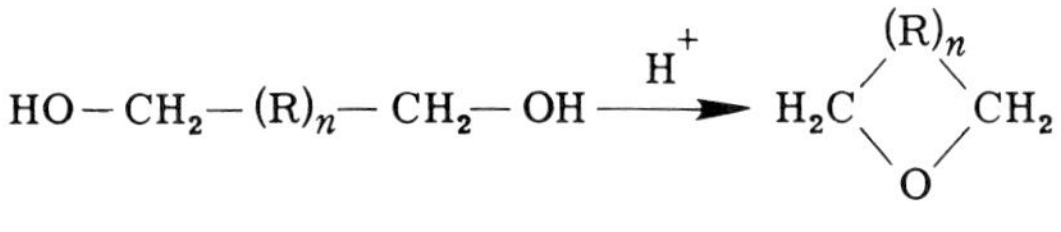

$$HO-CH_2-(R)_n-CH_2-OH \xrightarrow{H^+} H_2C\underset{O}{\overset{(R)_n}{\diagup\!\!\!\diagdown}}CH_2 \qquad (33)$$

B. Epoxides

Epoxides are produced by the dehydrohalogenation of halohydrins in strong alkaline solutions and distilling the products as they are formed. *cis*- and *trans*-2,3-epoxybutane are produced in 90% yield by the above procedure from 3-chloro-2-butanol [72]. Alkyl-substituted ethylene oxides are readily formed by a similar procedure. It is to be noted that alkyl groups enhance ring closure to the oxide, which occurs by a trans mechanism [73].

Aryl-substituted ethylene oxides, such as styrene oxide, are made by the reaction of iodine, water, and mercuric oxide with styrene (51% yield) [74].

$$C_6H_5-CH=CH_2 + I_2 + H_2O \longrightarrow C_6H_5-\underset{I}{CH}-\underset{OH}{CH_2} \xrightarrow{HgO} C_6H_5CH\underset{O}{-}CH_2 \quad (34)$$

Glycidol is prepared in good yields by the dehydrohalogenation of glycerol α-monochlorohydrin with potassium hydroxide [27, 48].

3-1. Preparation of 2,3-Epoxy-1-propanol (Glycidol) [27, 48]

$$\underset{Cl}{CH_2}-\underset{OH}{CH}-\underset{OH}{CH_2} + KOH \longrightarrow CH_2\underset{O}{-}CH-\underset{OH}{CH_2} + KCl + H_2O \quad (35)$$

To a 500-ml, three-necked flask, fitted with a stirrer and a condenser, is added 332 gm (3.0 moles) of 3-chloro-1,2-propanediol (glycerin α-monochlorohydrin).* The flask is cooled to 0°C, then potassium hydroxide powder is added in small portions over a 5-hr period at 0°C until 74 gm (1.32 moles) has been added. Stirring is continued for an additional 4 hr after the reaction flask has reached room temperature. The solids are filtered and the filtrate is distilled under vacuum to give 83.5 gm (97%), bp 44°–56°C (0.60–1.15 mm). The literature reports a 60% yield by a similar procedure with the following physical constants: bp 65°–66°C (2–2.5 mm).

An example of an industrial method of forming epoxides by dehydrochlorination is the reported process for preparing 4,4,4-trichloro-1,2-epoxybutane [75].

$$CCl_3-CH_2-\underset{Cl}{CH}-\underset{OH}{CH_2} \xrightarrow{NaOH} CCl_3-CH_2-CH\underset{O}{-}CH_2 \quad (36)$$

4. OXIDATION REACTIONS

A. Peroxidation of Olefins to Give Oxiranes (Epoxides) [76–79]

CAUTION: All organic peracid reactions should be conducted behind a safety shield because some reactions proceed with uncontrollable violence. Reactions

---

*Glycerin α-monochlorohydrin is available commercially. It may also be prepared from glycerin and concentrated hydrochloric acid as described by Rider and Hill [48].

§ 4. Oxidation Reactions

should first be run on a small scale, e.g., 0.1 mole or less, before scaling the preparation up. Efficient stirring and cooling should be provided.

Peracids and other peroxides can be destroyed by the addition of ferrous sulfate or sodium bisulfite [78].

Peracid-containing mixtures should not be distilled until the peracids have been eliminated.

The preparation of peracetic [80], performic [81], perbenzoic [82], and monoperphthalate [83] acids have been described. m-Chloroperbenzoic [84] acid is available; it has the advantage of being more stable than perbenzoic acid. Higher aliphatic peracids have been prepared in sulfuric acid as a solvent with 50% hydrogen peroxide [85].

The literature on epoxidation reactions with peracids is extensive. Several reviews are available which should be consulted [79, 86]. Peracids are used extensively for epoxidation of unsaturated compounds. The epoxidation reaction is stereospecific leading to a cis addition to the double bond.

The preformed–peracetic acid technique for epoxidations was first reported by Findley, Swern, and Scanlan for use in epoxidizing soybean oil [80]. Others have developed a similar preparation for synthesizing epoxy esters [87]. The *in situ* performic acid process for use in epoxidizing unsaturated oils was developed by Niederhauser [88]. The *in situ* process is reported to be safer to operate on a large scale but a reaction that involves peroxides is never free of hazards.

Emmons and Pagano claim that trifluoroperacetic acid is the only known peracid effective in the peroxidation of negatively substituted olefins such as methyl methacrylate [89]. A 1956 patent [90] claims that perpropionic acid is superior to perbenzoic acid, performic acid, or peracetic acid in the one-step conversion of cyclohexene to cyclohexene oxide (90%). The selective epoxidation of diolefins to epoxy vinyl monomers suitable for polymerization reactions has been described [91].

The synthesis of the isomeric 2-butene oxides has been reported earlier [92, 93] to involve the conversion of *cis*- and *trans*-2-butene to the halohydrins which on treatment with base give epoxides.

An improved method for preparing volatile [94] and nonvolatile [86] epoxides has been reported to involve the use of m-chloroperbenzoic acid in dioxane or chloroform at 0°C. Using this method, *cis*- and *trans*-2-butene oxides have been prepared in 52–60% yields [94]. The preparation of 1-hexene oxide using the latter reagent gives a 60% yield [94]. Diallyl esters of carboxylic acids also are epoxidized in good yields to produce glycidyl esters [51].

The use of alkyl hydroperoxides to oxidize olefins via metal catalysis is widespread. The large-scale industrial production of propylene oxide via the reaction of propylene with alkyl hydroperoxide is such a typical process where molybdenium salts are used as catalysts [95].

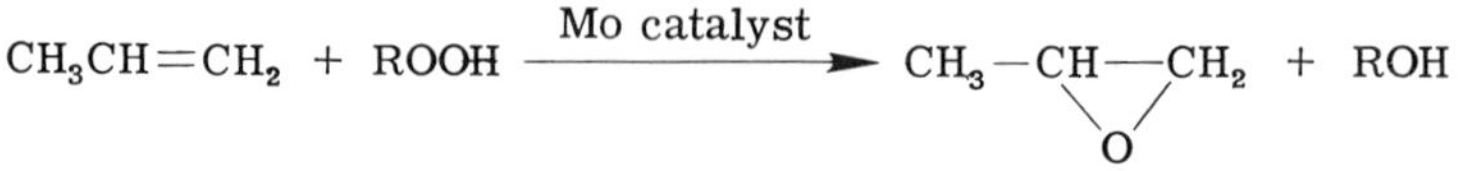

5. Ethers and Oxides

$$CH_3CH=CH_2 + ROOH \xrightarrow{\text{Mo catalyst}} CH_3-CH\underset{O}{-}CH_2 + ROH \quad (37)$$

Borate esters have also been reported to be effective catalysts [96].

A recent review on the industrial methods for epoxidation is worthwhile consulting [97].

4-1. Preparation of 1-Hexene Oxide [94]

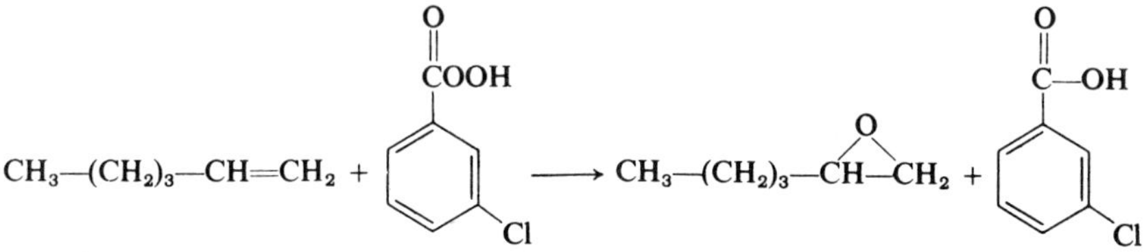

$$CH_3-(CH_2)_3-CH=CH_2 + \underset{Cl}{\underset{|}{C_6H_4}}-COOH \longrightarrow CH_3-(CH_2)_3-\underset{O}{CH-CH_2} + \underset{Cl}{\underset{|}{C_6H_4}}-COOH \quad (38)$$

To a round-bottomed flask are added 24.4 gm (0.119 mole) of m-chloroperbenzoic acid (85% pure) and 10.0 gm (0.119 mole) of 1-hexene in 300 ml of anhydrous diglyme. The flask is then placed in the refrigerator for 24 hr and afterward the reaction mixture is subjected to distillation through a 2-ft helices-packed column to give 7.05 gm (60%) of 1-hexene oxide, bp 116°–119°C, n_D^{20} 1.4051.

4-2. Preparation of 2,3-Epoxy-trans-decalin [98]

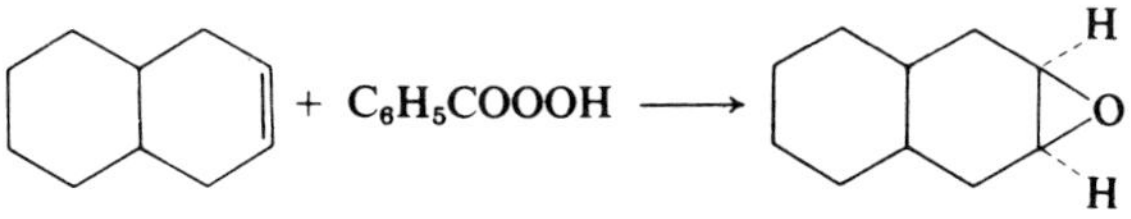

$$\text{trans-}\Delta^2\text{-decalin} + C_6H_5COOOH \longrightarrow \text{2,3-epoxy-trans-decalin} \quad (39)$$

To a flask containing 18 gm (0.13 mole) of perbenzoic acid in 320 ml of chloroform at 0°C is added 16.24 gm (0.119 mole) of trans-Δ^2-decalin. The flask is placed in the refrigerator (6°C) for 4 days and then the reaction mixture is extracted with 10% aqueous sodium hydroxide, washed with water, and dried over anhydrous sodium sulfate. The chloroform is removed and the residual oil is vacuum-distilled to yield 11.24 gm (62%), bp 88°–91°C (0.2 mm).

4-3. Preparation of Isophorone Oxide [99]

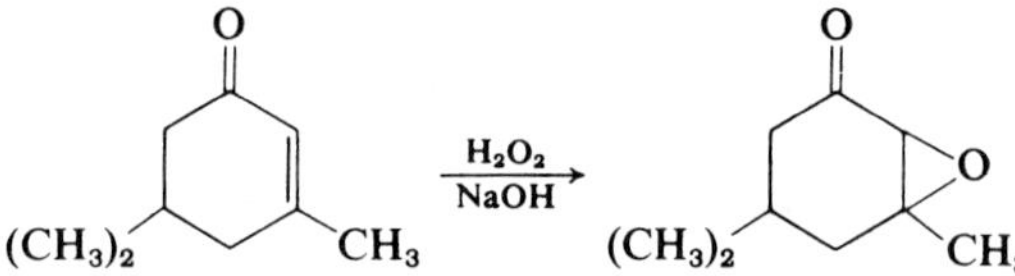

$$\text{isophorone} \xrightarrow[\text{NaOH}]{H_2O_2} \text{isophorone oxide} \quad (40)$$

To a flask containing a stirred mixture of 55.2 gm (0.4 mole) of isophorone and 115 ml (1.2 moles) of 30% aqueous hydrogen peroxide in 400 ml of methanol at 15°C is added 33 ml (0.2 mole) of 6 N aqueous sodium hydroxide over a period of 1 hr at 15°–20°C. The resulting mixture is stirred for 3 hr at 20°–25°C and then poured into 500 ml of water. The product is extracted with ether, dried over anhydrous magnesium sulfate, and distilled to yield 43.36 gm (70.4%), bp 70°–73°C (5 mm), n_D^{25} 1.4500–1.4510.

5. MISCELLANEOUS METHODS

(1) Vinyl transetherification using mercuric acetate, vinyl ether, and a given alcohol [100].

(2) Vinyl ethers from acetylene [101].

(3) The reaction of diazomethane and its derivatives with aldehydes and ketones [102].

(4) The reaction of acyl chlorides with acetals to give chloroethers [103–105].

(5) Propylene oxide via oxidation of propylene [106].

(6) The preparation of aryl t-butyl ethers using aryl halides and potassium t-butoxide in dimethyl sulfoxide [107, 108].

(7) Amine oxides via the reaction of organic hydroperoxides in the presence of Group VB and VIB transition metals [109].

(8) Molybdenum hexacarbonyl catalyzed hydroperoxide oxidation of olefins to epoxides [110].

(9) Triglycidyl isocyanurate synthesis using dispersed caustic [111].

(10) Solvent assisted Ullmann ether synthesis [112].

(11) Oxidative coupling of 2,6-xylenol with activated manganese dioxide [113] or polymerization with a copper-pyridine complex [114].

(12) The preparation of *trans*-stilbene oxide from *trans*-stilbene and peracetic acid [115].

(13) The preparation of cholesteryl oxides using monoperphthalic acids [116].

(14) Stereospecific epoxidation of dihydrophthalates [117].

(15) The use of peroxybenzimidic acid for epoxidation of olefins [118].

(16) The use of hydrogen peroxide–sodium orthovanadate to epoxidize maleic and fumaric acids [119].

(17) The use of N,N-dimethylformamide dimethyl acetal for dehydration of diols to epoxides [120].

(18) Stereospecific epoxidation catalyzed by enzymes [121].

(19) Ethylene oxide via the air oxidation of ethylene [122].

(20) Tritylation of alcohols with pyridinium trityl chloride [123].

(21) Preparation of acetals [124].

REFERENCES

1. A. I. Vogel, *J. Chem. Soc.* p. 616 (1948).
2. A. C. Nixon and G. E. K. Branch, *J. Am. Chem. Soc.* **58,** 492 (1936).

3. W. T. Olson, H. F. Hipscher, C. M. Buess, I. A. Goodman, I. Hart, J. H. Lamneck, Jr., and L. C. Gibbons, *J. Am. Chem. Soc.* **69**, 2451 (1947).
4. W. N. White and B. E. Norcross, *J. Am. Chem. Soc.* **83**, 3268 (1961).
5. W. J. Monacelli and G. F. Hennion, *J. Am. Chem. Soc.* **63**, 1772 (1941).
6. Du Pont Technical Bulletin, "A Review of Catalytic and Synthetic Applications for DMF/DMAC" (compiled by R. S. Kittila).
7. R. A. Dubois, *Diss. Abstr. Int. B.* **37**(1), 223 (1976); *Chem. Abstr.* **85**, 123273d (1976).
8. G. W. Gokel, D. J. Cram, C. L. Liotta, H. P. Harris, and F. L. Cook, *Org. Synth.* **57**, 30 (1977).
9. H. E. Ungnade and E. F. Orwoll, *Org. Synth.* **26**, 50 (1946).
10. F. O. Suter and F. D. Green, *J. Am. Chem. Soc.* **59**, 2578 (1937).
11. R. Q. Brewster and T. G. Groening, *Org. Synth. Collect. Vol.* **2**, 445 (1943).
12. H. Weingarten, *J. Org. Chem.* **29**, 977, 3624 (1964).
13. W. C. Hammana, *J. Chem. Eng. Data* **15**, 352 (1970).
14. W. Cocker, A. Lapworth, and A. Walton, *J. Chem. Soc.* p. 451 (1930).
15. D. S. Tarbell, *Org. React.* **2**, 22, 26 (1944).
16. Q. R. Bartz, R. F. Miller, and R. Adams, *J. Am. Chem. Soc.* **57**, 371 (1935).
17. S. R. Sandler and F. R. Berg, *J. Appl. Polym. Sci.* **9**, 3707 (1965).
18. A. M. Paquin, "Epoxydverbindungen und Epoxydharze," Springer-Verlag, Berlin, 1958.
19. Olin Matheson Chemical Corp., British Patent 1,005,657 (1965); *Chem. Abstr.* **64**, 2252 (1966); S. Fuzesi and F. A. Porica, U.S. Patent 3,255,126 (1966).
20. W. N. White and B. E. Norcross, *J. Am. Chem. Soc.* **83**, 3268 (1961).
21. C. S. Marvell and P. K. Porter, *Org. Synth. Collect. Vol.* **1**, 377 (1941).
22. B. H. Shoemaker and C. E. Board, *J. Am. Chem. Soc.* **53**, 1505 (1931).
23. F. Straus and W. Thiel, *Justus Liebigs Ann. Chem.* **525**, 152 (1936).
24. C. D. Nenitzescu and V. Przemetzki, *Chem. Ber.* **69**, 2706 (1936).
25. A. Lambert, C. W. Scaife, and A. E. Wilder-Smith, *J. Chem. Soc.* p. 1474 (1947).
26. T. W. Evans and K. R. Edlund, *Ind. Eng. Chem.* **28**, 1186 (1936).
27. S. R. Sandler, unpublished data (1964).
28. H. A. Bruson, *Org. React.* **5**, 79 (1952).
29. G. B. Buchman and H. A. Levine, *J. Am. Chem. Soc.* **70**, 599 (1948).
30. J. B. Niederl and S. Natelson, *J. Am. Chem. Soc.* **53**, 272, 1928 (1931).
31. F. J. Sowa, H. D. Hinton, and J. H. Niewland, *J. Am. Chem. Soc.* **54**, 2019, 3695 (1932).
32. W. Reppe, U.S. Patent 1,959,927 (1934).
33. J. Y. Johnson, British Patent 369,297 (1932).
34. G. S. Hiers and F. D. Hager, *Org. Synth. Collect. Vol.* **1**, 58 (1941).
35. E. M. Marks, D. Lipkin, and B. Bettman, *J. Am. Chem. Soc.* **59**, 946 (1937).
36. B. Gredy, *Bull. Soc. Chim. Fr.* [5]**3**, 1093 (1936).
37. S. Winstein and R. B. Henderson, *J. Am. Chem. Soc.* **65**, 2196 (1943).
38. S. Winstein and R. B. Henderson, in "Heterocyclic Compounds" (R. C. Elderfield, ed.), Vol. 1, pp. 22–42. Wiley, New York, 1950.
39. A. R. Sexton and E. C. Britton, *J. Am. Chem. Soc.* **70**, 3606 (1948).
40. H. C. Chitwood and B. T. Freure, *J. Am. Chem. Soc.* **68**, 680 (1946).
41. F. Straus and W. Thiel, *Justus Liebigs Ann. Chem.* **525**, 151 (1936).
42. C. D. Nenitzescu and V. Przemetzki, *Chem. Ber.* **69**, 2706 (1936).
43. C. D. Nenitzescu and V. Przemetzki, *Chem. Ber.* **74**, 676 (1941).
44. B. Gredy, *Bull. Soc. Chim. Fr.* [5]**3**, 1094 (1936).
45. L. C. Swollen and C. E. Boord, *J. Am. Chem. Soc.* **52**, 651 (1930).
46. B. H. Shoemaker and C. E. Boord, *J. Am. Chem. Soc.* **53**, 1505 (1931).
47. R. G. Kadesch, *J. Am. Chem. Soc.* **68**, 44 (1946).

48. T. H. Rider and A. J. Hill, *J. Am. Chem. Soc.* **52,** 1521 (1930).
49. G. Braun, *Org. Synth. Collect. Vol.* **2,** 256 (1943).
50. G. Braun, *J. Am. Chem. Soc.* **52,** 3185 (1930).
51. S. R. Sandler and F. R. Berg, *J. Chem. Eng. Data* **11,** 447 (1966).
52. A. C. Mueller and E. C. Shokal, U.S. Patent 2,772,296 (1956).
53. E. G. G. Weiner and E. Farenhorst, *Recl. Trav. Chim. Pays-Bas* **67,** 438 (1948).
54. E. G. G. Weiner and E. Farenhorst, U.S. Patent 2,467,171 (1949).
55. S. R. Sandler and F. R. Berg, *J. Appl. Polym. Sci.* **9,** 3708 (1965).
56. A. Fairbourne, G. P. Gibson, and D. W. Stephens, *J. Chem. Soc.* p. 1965 (1932).
57. "Epichlorohydrin," Tech. Booklet SC:49–35. Shell Chemical Corp, 1949.
58a. W. R. Sorenson and T. W. Campbell "Preparative Methods of Polymer Chemistry," pp. 307–313. Wiley (Interscience), New York, 1961.
58b. S. Sherman, J. Gannon, G. Buchi, and W. R. Howell, *Kirk-Othmer Encycl. Chem. Technol.*, 3rd Ed. Vol. 9, p. 267 (1980).
59. J. C. Anderson, D. G. Lindsay, and C. B. Reese, *J. Chem. Soc.* p. 4874 (1964).
60. G. Darzens, *C. R. Hebd. Seances Acad. Sci.* **203,** 1374 (1936).
61. M. S. Newman and B. J. Magerlein, *Org. React.* **5,** 443 (1952).
62. H. E. Zimmerman and L. Ahramjian, *J. Am. Chem. Soc.* **82,** 5459 (1960).
63. K. Freudenberg and W. Stoll, *Justus Liebigs Ann. Chem.* **440,** 41 (1924).
64. E. Bergmann and J. Hervey, *Ber. Dtsch. Chem. Ges. B* **62,** 902 (1929).
65. M. S. Newman and B. J. Magerlein, *Org. React.* **5,** 419 (1952).
66. C. F. H. Allen and J. Van Allan, *Org. Synth. Collect. Vol.* **3,** 727 (1955).
67. J. F. Norris and G. W. Rigby, *J. Am. Chem. Soc.* **54,** 2088 (1932).
68. R. H. Clarke, W. E. Graham, and A. G. Winter, *J. Am. Chem. Soc.* **47,** 2748 (1925).
69. V. N. Ipatieff and R. L. Burwell, Jr., *J. Am. Chem. Soc.* **63,** 969 (1941).
70. A. Frank and A. Kroupa, *Monatsch. Chem.* **69,** 1741 (1936).
71. R. C. Olberg, H. Pines, and V. N. Ipatieff, *J. Am. Chem. Soc.* **65,** 2260 (1943).
72. C. E. Wilson and H. J. Lucas, *J. Am. Chem. Soc.* **58,** 2396 (1936).
73. F. H. Norton and H. B. Haas, *J. Am. Chem. Soc.* **58,** 2147 (1936).
74. G. Golumbic and D. L. Cattle, *J. Am. Chem. Soc.* **61,** 996 (1939).
75. P. M. Pirawer (Olin Corp.), U.S. Patent 3,923,844 (1975).
76. D. Swern, *J. Am. Chem. Soc.* **69,** 1692 (1947).
77. D. Swern, *Chem. Rev.* **45,** 1 (1949).
78. D. Swern, *Org. React.* **7,** 378 (1953).
79. E. Searles, "Preparation, Properties, Reactions and Uses of Organic Peracids and Their Salts." F.M.C. Corporation, Inorg. Chem. Div., New York, 1964; B. Philips, "Peracetic Acid and Derivatives," 2nd ed. Union Carbide Chemicals Co., New York, 1957.
80. T. W. Findley, D. Swern, and J. T. Scanlan, *J. Am. Chem. Soc.* **67,** 412 (1945).
81. G. Toennies and R. P. Homiller, *J. Am. Chem. Soc.* **64,** 3054 (1942).
82. I. M. Kolthoff, T. S. Lee, and M. A. Mairs, *J. Polym. Sci.* **2,** 199 (1947).
83. H. Bohme, *Org. Synth.* **20,** 70 (1940); G. P. Payne, *ibid.* **42,** 77 (1962).
84. P. Brocklehurst and P. J. Pengiily, U.S. Patent 3,075,921 (1963) (available from F.M.C. Corporation, Inorg. Chem. Div., New York).
85. W. E. Parker, C. Ricciuti, C. L. Ogg, and D. Swern, *J. Am. Chem. Soc.* **77,** 4037 (1955).
86. J. G. Wallace, *Kirk-Othmer Encycl. Chem. Technol. 2nd Suppl. Vol.* 8, pp. 325–346 (1965).
87. D. E. Terry, U.S. Patent 2,458,484 (1949).
88. W. D. Niederhauser, U.S. Patent 2,458,160 (1949).
89. W. D. Emmons and A. S. Pagano, *J. Am. Chem. Soc.* **77,** 89 (1955).
90. F. P. Greenspan and R. J. Gall, U.S. Patent 2,745,848 (1956).
91. F. C. Frostick, Jr., B. Phillips, and P. S. Starcher, *J. Am. Chem. Soc.* **81,** 3350 (1959).

92. C. E. Wilson and H. J. Lucas, *J. Am. Chem. Soc.* **58,** 2397 (1936).
93. S. Winstein and H. J. Lucas, *J. Am. Chem. Soc.* **61,** 1576 (1939).
94. D. J. Pasto and C. C. Cumbo, *J. Org. Chem.* **30,** 1271 (1965).
95. A. O. Chong and K. B. Sharples, *J. Org. Chem.* **42,** 1587 (1977).
96. R. F. Wolf and R. K. Barnes, *J. Org. Chem.* **34,** 3441 (1969).
97. J. T. Lutz, Jr., *Kirk-Othmer Encycl. Chem. Technol., 3rd Ed.* Vol. 9, p. 251 (1980).
98. E. E. Smissman, T. L. Lemke, and O. Kristiansen, *J. Am. Chem. Soc.* **88,** 334 (1966).
99. H. O. House and R. L. Wasson, *J. Am. Chem. Soc.* **79,** 1488 (1957).
100. W.H. Watanabe and L. E. Conlon, U.S. Patent 2,760,990 (1956); *J. Am. Chem. Soc.* **79,** 2828 (1957).
101. S. A. Miller, "Acetylene," Vol. 2, Chapter 4. Academic Press, New York, 1966.
102. C. D. Gutsche, *Org. React.* **8,** 364 (1954).
103. H. W. Post, *J. Org. Chem.* **1,** 231 (1936).
104. F. Straus and H. Heinze, *Justus Liebigs Ann. Chem.* **493,** 203 (1932).
105. F. Straus and H. J. Weber, *Justus Liebigs Ann. Chem.* **493,** 120 (1932).
106. *Chem. Eng. News* **43,** No. 43, 40 (1968).
107. J. S. Bradshaw, N. B. Nielsen, and D. P. Rees, *J. Org. Chem.* **33,** 259 (1968).
108. M. R. V. Sahyun and D. J. Cram, *Org. Synth.* **45,** 89 (1965).
109. M. N. Sheng and J. W. Zajacek, *Abstr. 3rd Middle Atl. Reg. Am. Chem. Soc. Meet.* p. 61H (1968).
110. M. N. Sheng and J. W. Zajacek, *Abstr. 3rd Middle Atl. Reg. Am. Chem. Soc. Meet.* p. 60H (1968).
111. H. P. Price and G. E. Schroll, *J. Org. Chem.* **32,** 2005 (1967).
112. A. L. Williams, R. E. Kinney, and R. F. Bridger, *J. Org. Chem.* **32,** 2501 (1967).
113. E. McNelis, *J. Org. Chem.* **31,** 1255 (1966).
114. A. S. Hay, H. S. Blanchard, G. F. Endres, and J. W. Eustance, *J. Am. Chem. Soc.* **81,** 6335 (1959); A. S. Hay, *J. Polym. Sci.* **58,** 581 (1962).
115. D. J. Reif and H. O. House, *Org. Synth. Collect. Vol.* **4,** 860 (1963).
116. P. N. Chakravorty and R. H. Levin, *J. Am. Chem. Soc.* **64,** 2317 (1942).
117. S. A. Cerefice and E. K. Fields, *J. Org. Chem.* **41,** 355 (1976).
118. R. G. Carlson, N. S. Behn, and C. Cowles, *J. Org. Chem.* **36,** 3832 (1971).
119. M. A. Beg and I. Ahmad, *J. Org. Chem.* **42,** 1590 (1977).
120. C. Cortez and R. G. Harvey, *Org. Synth.* **58,** 12 (1978).
121. S. W. May and R. D. Schwartz, *J. Am. Chem. Soc.* **96,** 4031 (1974).
122. J. N. Cawse, J. P. Henry, M. W. Swartzlander, and P. H. Wadia, *Kirk-Othmer Encycl. Chem. Technol., 3rd Ed.* Vol. 9, pp. 432–471 (1980).
123. S. Hanessian and A. P. A. Staub, *Methods Carbohydr. Chem.* **7,** 63 (1976).
124. E. J. Corey, J. L. Gras, and P. Ulrich, *Tetrahedron Lett.* p. 809 (1976).

CHAPTER 6 / **HALIDES**

1. Introduction	148
2. Condensation Reactions	149
A. Conversion of Alcohols to Alkyl Halides	149
2-1. Preparation of n-Butyl Bromide	150
2-2. General Procedure for the Hydrochlorination of Alcohols	152
2-3. Hydrochlorination of 2-Methyl-endo-norborneol	154
B. The Chloromethylation Reaction	154
2-4. Preparation of Bis(chloromethyl)durene	155
C. Acid Chlorides	155
2-5. Preparation of Pyromellitoyl Chloride	156
2-6. Preparation of 2,5-Dichloroterephthaloyl Chloride	156
2-7. Preparation of Trimesoyl Chloride	157
2-8. Preparation of Trimellitic Anhydride Acid Chloride	157
2-9. Preparation of Trimellitoyl Chloride	158
D. Halogenation Reactions	158
2-10. Preparation of Bromobenzene	159
2-11. Bromination of p-Nitrophenol in Aqueous Solution with Bromine Chloride to Give 2,6-Dibromo-4-nitrophenol	160
E. Reaction of Olefins with Halogens and Halogen Derivatives	161
2-12. Bromination of Cyclohexene to 1,2-Dibromocyclohexane	162
2-13. Hydrochlorination of α-Methylstyrene to 2-Chloro-2-phenylpropane	163
2-14. Preparation of Tetrachlorobutanol by the Reaction of Carbon Tetrachloride with Allyl Alcohol	164
2-15. General Procedure for the Preparation of 1,1-Dihalocyclopropanes	165
2-16. Preparation of 1,1-Dibromo-2,2-diphenylcyclopropane	165
F. Halogenation of Aldehydes, Ketones, and Acids	167
2-17. Preparation of 1,3-Dibromoacetone	167
2-18. Hell–Volhard–Zelinsky Reaction—Preparation of α-Bromobutyric Acid	169
3. Elimination and Cleavage Reactions	169
3-1. Preparation of Allyl Iodide	170
A. The Sandmeyer and Schiemann Reactions	170
3-2. Sandmeyer Reaction—Preparation of p-Nitrochlorobenzene	171
3-3. Schiemann Reaction—Preparation of 3,5-Dimethyl-2-fluoro-1-bromobenzene	171
B. The Hunsdiecker Reaction	172
3-4. Preparation of p-Nitrobromobenzene	173
3-5. Modified Hunsdiecker Reaction—Preparation of 1-Bromohexane from Heptanoic Acid	173
4. Miscellaneous Methods	173
References	175

1. INTRODUCTION

Most halogenation reactions involve either condensation or elimination reactions. Practically any organic compound can be halogenated by these reactions.

$$X\text{---}R\text{---}Z \longleftarrow RZ + X_2 \longrightarrow RX + ZX \\ \longrightarrow R\text{---}Z\text{---}X \tag{1}$$

Olefins either add halogen to give 1,2-dihaloalkanes or react at the allylic hydrogen to give allyl halides.

The hydrogenator previously described (Chapter 1), which generates hydrogen automatically from sodium borohydride to achieve the quantitative hydrogenation of unsaturated compounds, has been adapted to the hydrochlorination of reactive alcohols and olefins. In this procedure hydrogen chloride is generated automatically as required to react with the alcohol or olefin, and the generation ceases when the reaction is complete. In this manner it is possible both to follow the rate of utilization of the hydrogen chloride and to convert the alcohol or olefin essentially quantitatively into product without excessive contact with the hydrochlorination reagent [1].

Selective chlorination reactions involving alcohols, ethers, and carboxylic acids have been reported to give ω-1 monochlorination products in good yields [2–4].

The reaction of halomethanes with olefins under basic conditions yields either mono or *gem*-dihalocyclopropanes in Eqs. (2) and (3).

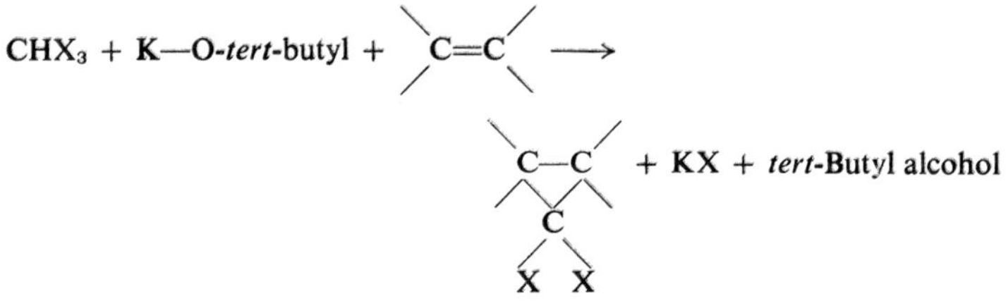

$$CHX_3 + K\text{---}O\text{-}tert\text{-butyl} + C{=}C \longrightarrow$$

$$ C\text{---}C + KX + tert\text{-Butyl alcohol} \tag{2}$$
$$ \underset{X\ \ X}{C}$$

$$CH_2X_2 + n\text{-Butyllithium} + C{=}C \xrightarrow{\text{Ether}}$$

$$ C\text{---}C + LiX + n\text{-Butyl halide} \tag{3}$$
$$ \underset{X\ \ H}{C}$$

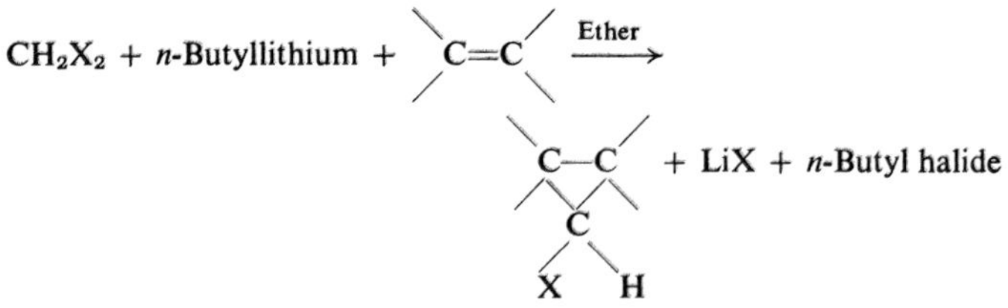

Haloalkanes also add free radically to olefins to give linear addition products and in some cases telomers.

$$CCl_4 + R\text{---}CH{=}CH_2 \longrightarrow R\text{---}\underset{Cl}{\underset{|}{CH}}\text{---}CH_2CCl_3 \tag{4}$$

Aromatic hydrocarbons undergo electrophilic substitution reactions where electron-donating substituents favor the reaction and influence the orientation. The halogen source may be Cl_2, Br_2, I_2, mixed halogens, PCl_5, PCl_3, P + halogen, $SOCl_2$, N-bromosuccinamide, and others.

The chloromethylation reaction is important in adding $Cl-CH_2-$ to aromatics and heterocycles.

The interest in fluorocarbons, with their high temperature stability and chemical resistance, has generated extensive research in the preparation of these compounds.

2. CONDENSATION REACTIONS

The condensation reactions for the introduction of a halogen atom involve the reaction of a halogen source HX, PX_3, PX_5, $SOCl_2$, RCOX, SF_4, X_2, HOX, or RX with alcohols, ethers, diazonium compounds, Grignard reagents, silver salts of acids, acids, amides, aromatic compounds, aldehydes, ketones, olefins, and amines. Many other organic compounds also undergo these reactions.

A. Conversion of Alcohols to Alkyl Halides

$$ROH + HX \longrightarrow RX + H_2O \qquad (5)$$

$$3\,ROH + PX_3 \longrightarrow 3\,RX + H_3PO_3 \qquad (6)$$

$$ROH + KI + H_3PO_4 \longrightarrow RI + KH_2PO_4 + H_2O \qquad (7)$$

$$ROH + P_2I_4 \longrightarrow RI \qquad (8)$$

$$ROH + CH_3I + (C_6H_5O)_3P \longrightarrow RI + C_6H_5-OH + (C_6H_5O)_2POCH_3 \qquad (9)$$

$$ROH + SOCl_2 \longrightarrow RCl + SO_2 + HCl \qquad (10)$$

$$ROH + RSO_2Cl \longrightarrow RCl \qquad (11)$$

$$3\,ROH + \text{cyanuric chloride} \longrightarrow 3\,RCl + \text{cyanuric acid} \qquad (12)$$

$$ROH + CHCl_3 \xrightarrow[H_2O]{NaOH} RCl \qquad (13)$$

Alkyl bromides are prepared from alcohols in good yield in the laboratory using aqueous hydrogen bromide [5], dry hydrogen bromide [6, 7], or by means of sodium bromide–sulfuric acid–water. The latter procedure is satisfactory for low-molecular-weight alcohols. Primary chlorides are synthesized using zinc chloride and hydrochloric acid [8].

Highly branched halides rearrange to tertiary halides. Primary halides

R$_2$CH—CH$_2$X can be obtained free of rearrangement products by using the parent alcohol in the presence of phosphorus tribromide or thionyl chloride in pyridine [9].

Tertiary halides are easily prepared by the reaction of hydrogen halide with the alcohol [1, 10].

The use of phosphorus halides (phosphorus and bromine or iodine) in the presence of pyridine can be used to prepare primary and secondary alkyl bromides [11], or iodides [12] free of rearrangement. Phosphorus pentachloride [13] or even better thionyl chloride [14] in pyridine is used to yield alkyl chlorides without rearrangement. Allylic alcohols, at low temperatures, give unrearranged allylic bromides [15].

Dibromides can also be produced using phosphorus tribromide [16].

Cyanuric chloride is useful for converting, primary, secondary, and tertiary alcohols to chlorides in good yield under mild conditions without rearrangement [17].

The reaction of dichlorocarbene (prepared from chloroform and aqueous sodium hydroxide and a phase transfer agent such as triethylbenzylammonium chloride) with alcohols leads to the corresponding chlorides in high yields. The reaction offers the preparative advantage in that it takes place under basic conditions and usually is run at room temperature [18].

2-1. Preparation of n-Butyl Bromide [15]

$$2\ CH_3CH_2CH_2CH_2OH + 2\ NaBr + H_2SO_4 \longrightarrow$$
$$2\ CH_3CH_2CH_2CH_2Br + Na_2SO_4 + 2\ H_2O \quad (14)$$

To a 2-liter round-bottomed flask equipped with a stirrer, dropping funnel, and reflux condenser and containing 270 ml of water is added with stirring 309 gm (5 moles) of sodium bromide powder. (The reverse addition causes caking.) To this solution is first added 178 gm (2.4 moles) of *n*-butyl alcohol, and then 218 ml of concentrated sulfuric acid is slowly added dropwise. The mixture is stirred vigorously or shaken to prevent the sulfuric acid from forming a layer. The mixture is refluxed for 2 hr and then distilled to yield the product. The water-insoluble layer is separated, washed with water, then washed with a sodium carbonate solution (5 gm/100 ml water), separated, dried over 5–10 gm of calcium chloride, and distilled to yield 298 gm (90%), bp 101°–104°C, n_D^{20} 1.4398.

Utilizing a commercial model of the H. C. Brown hydrogenator (see Chapter 1), it was found that tertiary alcohols and olefins could be hydrochlorinated at 0°C in approximately 100 sec. Some representative experimental data for various alcohols are given in Table I.

TABLE I

DATA FOR THE HYDROCHLORINATION OF REPRESENTATIVE TERTIARY ALCOHOLS[a]

Alcohol	Mmoles	Solvent	Temp (°C)	Time[b] (min)	Hydrogen chloride used (mmoles)	Alkyl chloride obtained (mmoles)	Product n_D^{20} or m p (°C)	Product Purity[c] by titration	Product Yield isolated (%)
tert-Butyl Alcohol	42.1	Neat	Room temp.	75	38.3	31	—	99	74
tert-Pentyl alcohol	39.1	Neat	Room temp.	15	37.3	38.2	1.4058	100	97
Triethylcarbinol	40.0	Neat	Room temp.	7.5	39.9	37.4	1.4330	100	94
Benzhydrol	10.0	CH_2Cl_2	0°	2.5	10.3	9.3	1.5957	95	93
Diphenylmethylcarbinol	10.0	CH_2Cl_2	0°	3.5	9.5	9.0	—	(60)[d]	(90)[d]
1-Methylcyclopentanol	18.0	n-C_5H_{12}	10°	3.5	17.5	17.5	1.4460	100	96
2-Methyl-endo-norborneol	20.1	n-C_5H_{12}	10°	26	20.3	17.6	20–25	98	88
1,2-Dimethyl-endo-norborneol	20.0	n-C_5H_{12}	0°	12	20.3	19.4	120–122	99	97
2-Phenyl-endo-norborneol	20.1	n-C_5H_{12}	0°	6.5	21.0	18.7	43–45	90	94
	10.0	n-C_5H_{12}	0°	1.8[b]	9.9	—	—	—	—
	10.0	n-C_5H_{12}	0°	2.2[b]	9.6	—	—	—	—
	10.0	CH_2Cl_2	0°	1.2[e]	10.1	—	—	—	—
	10.0	CH_2Cl_2	0°	1.0[d]	10.5	—	—	—	—

[a] Reprinted from H. C. Brown and N. H. Rei, *J. Org. Chem.* **31**, 1090 (1966). Copyright 1966 by the American Chemical Society. Reprinted by permission of the copyright owner.
[b] The reaction time is a function of the stirring rate. The values reported should be considered only as an indication of the very fast rates realized.
[c] No detectable vinyl proton and hydroxy group was detected in NMR and infrared spectra.
[d] The tertiary chloride is highly unstable and the product appears to lose hydrogen chloride during the aspiration procedure.
[e] The data indicate a faster rate in methylene chloride than in n-pentane.

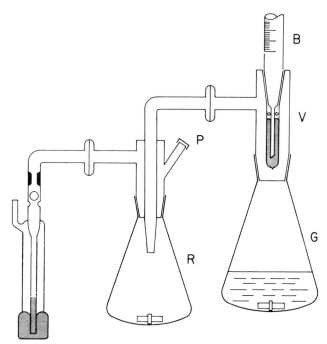

FIG. 1. Apparatus used for the automatic hydrochlorination of alcohols and olefins. [Reprinted from: H. C. Brown and N. H. Rei, *J. Org. Chem.* **31,** 1090 (1966)]. Copyright 1966 by the American Chemical Society. Reprinted by permission of the copyright owner.

2-2. General Procedure for the Hydrochlorination of Alcohols [1]*

It is necessary to modify the commercial model of the Brown hydrogenator (Delmar Scientific Laboratories, Maywood, Illinois 60154) to avoid possible corrosion of the hypodermic needle of the buret. A glass-tipped buret and the corresponding valve are substituted for these two items in the commercial unit. In this way the concentrated hydrochloric acid in the buret comes in contact only with glass and the mercury in the valve before it reacts with the concentrated sulfuric acid in the generator to produce anhydrous hydrogen chloride. The complete assembly is shown in Fig. 1.

The following procedure will convert from 10 to 100 mmoles of alcohol or olefin into the chloride. However, it can readily be modified to handle either smaller or larger quantities.

Approximately 200 ml of concentrated sulfuric acid is placed in a 500-ml modified Erlenmeyer flask, serving as the generator (G) (Fig. 1). A Teflon-

*Procedure 2-2 is reprinted from H. C. Brown and N. H. Rei, *J. Org. Chem.* **31,** 1090 (1966). Copyright 1966 by the American Chemical Society. Reprinted by permission of the copyright owner.

§ 2. Condensation Reactions

coated magnetic stirring bar is introduced. (It is important that the sulfuric acid be efficiently stirred to effect thorough mixing of the less-dense hydrochloric acid phase with the sulfuric acid.) The slight curvature of the flask bottom and the Teflon collar on the stirring bar are not essential for the present application, but they greatly facilitate smooth, efficient stirring of the contents.

The apparatus is assembled so that it is gas-tight. In the buret (B) is placed an adequate amount of concentrated hydrochloric acid. The height of the liquid column must be controlled so that it is supported by the column of mercury in the valve (V).

By means of a small rubber bulb attached to the top of the buret, pressure is supplied to force 3 to 4 ml of concentrated hydrochloric acid through the mercury seal to generate sufficient hydrogen chloride to flush the system. (The amount is selected to provide a volume of hydrogen chloride that is approximately twice the free volume of the system.)

The reaction flask (R), a 125-ml modified flask for the usual scale of the preparations, is commonly immersed in an ice bath to provide a convenient reaction temperature. However, in many cases it proved quite satisfactory to carry the reaction out at room temperature, and in some cases quite low temperatures are used.

The reaction is initiated by injecting, by means of a hypodermic syringe, through the introductory port (P), 10 to 1000 mmoles of the alcohol or olefin, neat, or as a concentrated solution in *n*-pentane, methylene chloride, or carbon tetrachloride. Reaction begins immediately. (It is important that the contents of the reaction flask be stirred vigorously by the usual Teflon-coated magnetic stirrer bar.) As soon as the pressure drops 10 to 20 mm below atmospheric, concentrated hydrochloric acid is drawn through the mercury seal in the valve (V) and it then reacts with the magnetically stirred sulfuric acid to generate hydrogen chloride to restore the pressure. The reaction proceeds automatically until absorption of hydrogen chloride ceases, as shown by the behavior of the mercury in the bubbler or the constancy of the quantity of hydrochloric acid remaining in the buret.

Experiments revealed that under the operating conditions each milliliter of the concentrated hydrochloric acid generated 11.0 ± 0.1 mmoles of hydrogen chloride. In the case of olefins, such as α-methylstyrene, 1 mole of hydrogen chloride was taken up for each mole of olefin. However, in the case of the alcohols, additional hydrogen chloride was utilized to saturate the water produced in the reaction. At 25°C each mole of alcohol utilized 1.29 moles of hydrogen chloride; at 0°C the corresponding figure was 1.35 moles.

In the case of liquid chlorides, prepared neat from the corresponding alcohols, it was adequate to remove with the aid of a capillary tube the small lower phase of water saturated with hydrogen chloride. After a brief aspiration to remove traces of dissolved hydrogen chloride, the organic phase appeared to be essentially pure tertiary chloride by all of the usual tests. In cases where solvents were employed, it was removed under vacuum to recover the chloride or the solution was cooled to low temperatures to precipitate the solid chloride [1].

2-3. Hydrochlorination of 2-Methyl-endo-norborneol [1]*

$$\text{norbornyl-CH}_3\text{-OH} + \text{HCl} \longrightarrow \text{norbornyl-CH}_3\text{-Cl} + \text{H}_2\text{O} \quad (15)$$

The apparatus is assembled and flushed with hydrogen chloride as is described in Procedure 2-2. The reaction flask is immersed in a bath at 10°C, and the reaction is initiated by injecting 20 mmoles (2.53 gm) of 2-methyl-endo-norborneol in 10 ml of n-pentane solution through the injection port. There is an immediate uptake of approximately 75% of the required amount of reagent. Then the reaction proceeds slowly to 100% utilization, being essentially complete in 26 min. The organic layer is withdrawn from the reaction flask and is subjected to gentle aspiration to remove dissolved hydrogen chloride. Fresh pentane is added to bring the volume to precisely 10.0 ml and a 1-ml aliquot is added to 80% aqueous ethanol at room temperature. Standard alkali is added to neutralize the hydrogen chloride produced, the amount corresponding to 97.4% of reactive chloride in solution. This quantity is listed as the purity (titration) in Table 1. The remaining solution is evaporated under vacuum to constant weight. There is obtained 2.54 gm, 17.5 mmoles, of 2-methyl-exo-norbornyl chloride, mp 20°–25°C. A weighed sample of the chloride is solvolyzed in 80% aqueous ethanol at 25°C. There is produced 98.6% of the calculated quantity of hydrogen chloride.

B. The Chloromethylation Reaction

Aromatic hydrocarbons react with formaldehyde and hydrogen chloride to yield benzyl chlorides [19, 20]. Alkyl and alkoxy groups facilitate the introduction of the chloromethyl group, whereas nitro, carboxyl, or halogen prevent or retard the reaction. Bis(chloromethyl) compounds are also formed as by-products and can become the major product in good yields by employing excess formaldehyde and hydrochloric acid [21]. Zinc chloride, phosphoric acid, and sulfuric acid can be used as catalysts [22].

Thiophene compounds also undergo chloromethylation, i.e., 2- and 3-chloromethyl thiophenes are thus produced [23].

*Procedure 2-3 is reprinted from H. C. Brown and N. H. Rei, *J. Org. Chem.* **31**, 1090 (1966). Copyright 1966 by the American Chemical Society. Reprinted by permission of the copyright owner.

§ 2. Condensation Reactions

Chloroethylation, chloropropylation [24], and bromoethylation [25, 26] have been reported.

CAUTION: Bis(chloromethyl) ether has been reported to be very highly carcinogenic [27] and may be present in all reactions involving the use of formaldehyde and hydrochloric acid. The handling of other chloromethyl ethers should also be done with great care and all reactions should be carried out in a hood.

2-4. Preparation of Bis(chloromethyl)durene [21]

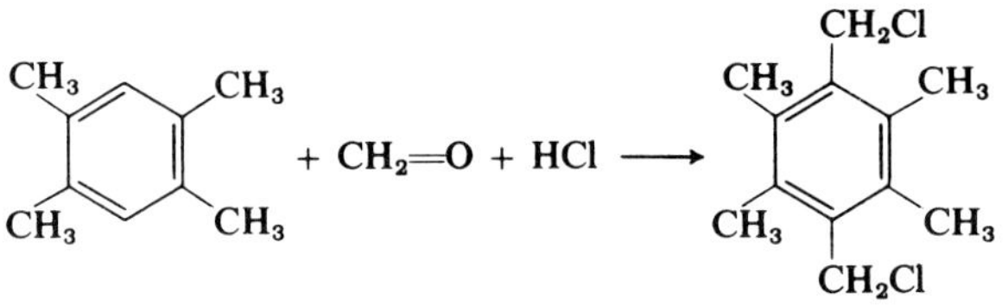

$$\text{durene} + CH_2=O + HCl \longrightarrow \text{bis(chloromethyl)durene} \quad (16)$$

To a flask containing 50 gm (0.37 mole) of durene (mp 79°–81°C) dissolved in 200 ml of a high-boiling petroleum fraction, bp 175°–190°C, are added 113 ml (1.5 moles) of 40% aqueous formaldehyde solution and 100 ml of concentrated hydrochloric acid. The mixture is heated and stirred on a steam bath while a slow stream of hydrogen chloride gas is bubbled through the mixture. After 6 hr the organic layer is separated and then set aside to slowly cool. The white needles of crude product are collected leaving monochloromethyldurene in solution. The mother liquor is treated again with CH_2—O + HCl. After a total of six such treatments, there is isolated a total of 69 gm (91%) of crude bis(chloromethyl)durene, mp 193°–194°C (from benzene). The yield of recrystallized material is 67%.

Aromatic alcohols also react with formaldehyde and hydrogen chloride to give chloromethyl ethers [28].

$$C_6H_5CH_2OH + HCl + CH_2=O \longrightarrow C_6H_5CH_2OCH_2Cl \quad (17)$$

C. Acid Chlorides

Thionyl chloride is generally used to prepare aliphatic and aromatic acid chlorides. Although acid chlorides are readily obtained by the reaction of thionyl chloride with trimesic acid and terephthalic acid, difficulty was encountered in preparing pyromellitoyl chloride and trimellitoyl chloride [29]. These have now been prepared by the use of phosphorus pentachloride [30], and their preparations are described below.

$$Ar(COOH)_n + n\ PCl_5 \xrightarrow[Cl_3C_6H_3]{500\ ml} Ar(COCl)_n + n\ HCl + n\ POCl_3 \quad (18)$$

Thionyl chloride has also been used to prepare aliphatic acid chlorides containing 11–19 carbon atoms [31].

2-5. Preparation of Pyromellitoyl Chloride [29, 30]

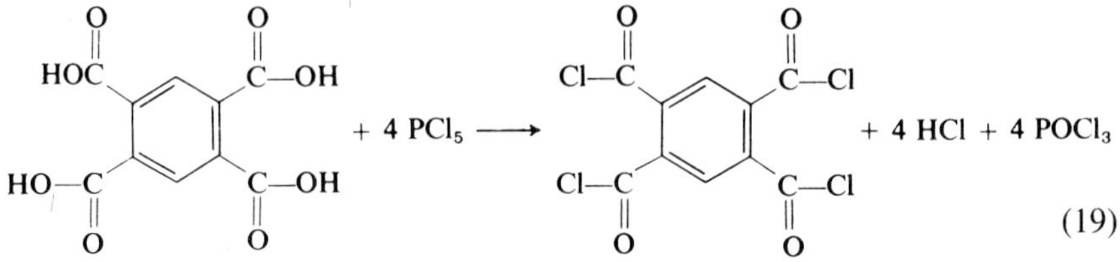

$$\text{Pyromellitic acid} + 4\,PCl_5 \longrightarrow \text{Pyromellitoyl chloride} + 4\,HCl + 4\,POCl_3 \quad (19)$$

To a 2-liter flask equipped with a stirrer, thermometer, and condenser with a sodium hydroxide trap are added 46 gm (0.181 mole) of pyromellitic acid, 151 gm (0.728 mole) of phosphorus pentachloride, and 333 ml of 1,2,4-trichlorobenzene. The mixture is stirred until the exotherm subsides and then another 46 gm (0.181 mole) of pyromellitic acid and 151 gm (0.728 mole) of phosphorus pentachloride are added. After stirring for a few minutes another 46 gm (0.181 mole) of pyromellitic acid, 151 gm (0.728 mole) of phosphorus pentachloride, and 167 ml of trichlorobenzene are added. The reaction mixture is stirred for 45 min and then the temperature is gradually raised to about 120°C (do not exceed 130°C). After about 6 hr at 120°C, the mixture turns into an amber clear liquid.

A distillation head is substituted for the condenser and the $POCl_3$ is removed at atmospheric pressure up to 130°C. Trichlorobenzene is removed under slight vacuum at 54°C (0.5 to 4 mm Hg) and then the temperature is raised in order to distill the product, bp 169°–173°C (0.5–1.25 mm Hg), 151 gm, (85%), mp 59°–62.5°C [32].

2-6. Preparation of 2,5-Dichloroterephthaloyl Chloride [29]

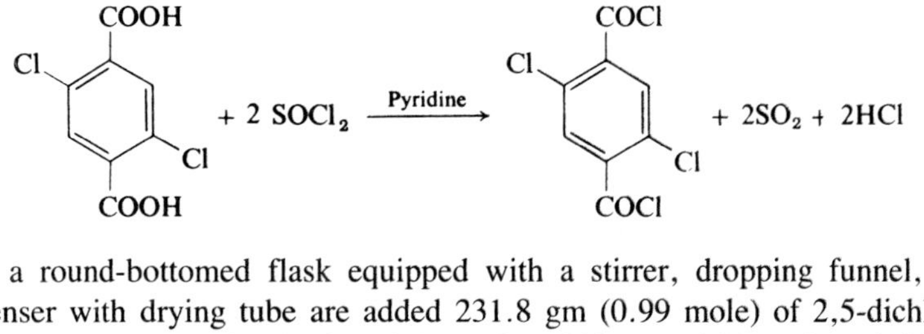

$$\text{2,5-dichloroterephthalic acid} + 2\,SOCl_2 \xrightarrow{\text{Pyridine}} \text{2,5-dichloroterephthaloyl chloride} + 2SO_2 + 2HCl \quad (20)$$

To a round-bottomed flask equipped with a stirrer, dropping funnel, and condenser with drying tube are added 231.8 gm (0.99 mole) of 2,5-dichloroterephthalic acid and 2 ml of pyridine. Then 800 gm (6.72 mole) of thionyl chloride is added dropwise. When all the material has been added, the mixture is refluxed for 24 hr while hydrochloric acid and sulfur dioxide are given off.
On standing overnight, crystals separate which are filtered and washed with n-hexane. The weight of the crystals is 78.8 gm, mp 68°C. Analysis of % Cl— calculated: 26.2 (hydrolyzable chlorine); found: 25.9. The remaining thionyl

§ 2. Condensation Reactions

chloride is distilled off to give a solid residue which, upon recrystallization from n-hexane, yields an additional 37.1 gm, mp 65°–67°C. The total yield was 115.9 gm (43.1%).

2-7. Preparation of Trimesoyl Chloride [29]

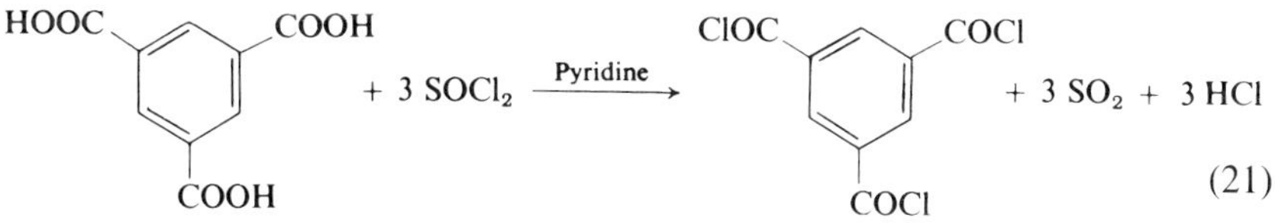

$$\text{HOOC-C}_6\text{H}_3(\text{COOH})_2 + 3\,\text{SOCl}_2 \xrightarrow{\text{Pyridine}} \text{ClOC-C}_6\text{H}_3(\text{COCl})_2 + 3\,\text{SO}_2 + 3\,\text{HCl} \quad (21)$$

To a flask as described above are added 420 gm (2 moles) of trimesic acid, 12 ml of pyridine, and 2856 gm (24 moles) of thionyl chloride. The mixture is refluxed for 20 hr, at which time the mixture becomes clear. The excess thionyl chloride is distilled off at atmospheric pressure and the residue is distilled to yield 459.2 gm of trimesoyl chloride (97%), bp 142°–147°C (0.7 mm).

Sodium carboxylate salts can react with PCl_3, PCl_5, $POCl_3$, or $SOCl_2$ to give acid chlorides. For example, sodium fluoroacetate and phosphorus pentachloride yield fluoroacetyl chloride in 63% yield [33].

Mono- and dianhydrides react with thionyl chloride to give acid chlorides. For example, phthalic anhydride [34] is converted to phthaloyl chloride in 86% yield [34]. Phosphorus pentachloride in 1,2,4-trichlorobenzene gives better yields of pyromellitoyl chloride from pyromellitic acid [29].

Benzoyl chloride [35] or phthaloyl chloride [36] react with lower boiling acids to give volatile acyl halides. For example, phthaloyl chloride converts maleic anhydride to fumaroyl chloride in 95% yield [36].

2-8. Preparation of Trimellitic Anhydride Acid Chloride [29]

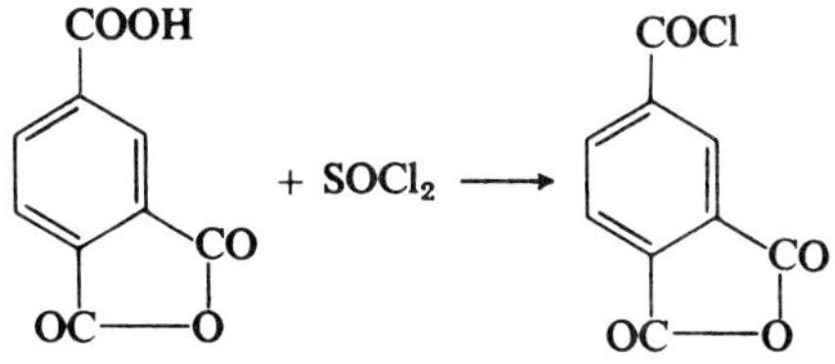

$$\text{COOH-C}_6\text{H}_3(\text{CO})_2\text{O} + \text{SOCl}_2 \longrightarrow \text{COCl-C}_6\text{H}_3(\text{CO})_2\text{O} \quad (22)$$

To a flask as previously described are added, with stirring, 115.2 gm of trimellitic anhydride (0.6 moles), 458 ml (750 gm, 6.3 moles) of thionyl chloride (added slowly with caution), and 3 ml of pyridine, and the temperature is slowly raised to reflux. After 10 hr the mixture becomes a clear yellow liquid. The thionyl chloride is distilled off at atmospheric pressure and the residue distilled under vacuum to yield 112 gm (71%), bp 128°–132°C (0.26–0.30 mm), mp 59°–62°. Analysis of % Cl—calculated: 16.86; found: 16.96.

2-9. Preparation of Trimellitoyl Chloride [29]

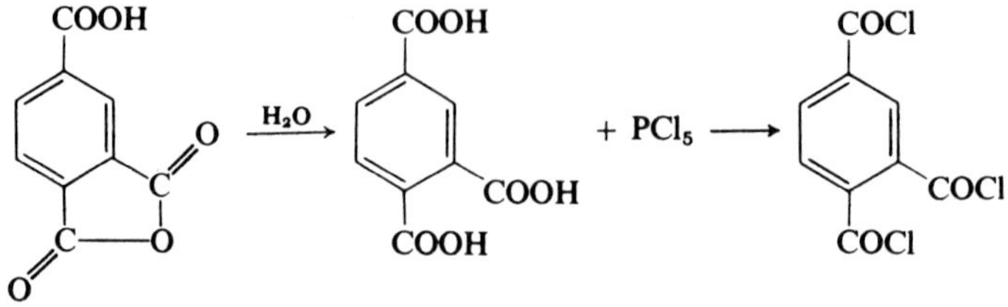

$$\text{trimellitic anhydride} \xrightarrow{H_2O} \text{trimellitic acid} + PCl_5 \longrightarrow \text{trimellitoyl chloride} \quad (23)$$

To a 3-liter flask equipped with a thermometer, condenser, and stirrer is added 500 ml of dry, distilled 1,2,4-trichlorobenzene. The reactants are added in three increments of 70 gm of trimellitic acid (0.33 mole) and 229 gm of phosphorus pentachloride (1.1 moles). The first increment shows no exotherm, and after 10 min the second increment is added. After 30 min the last increment is added, followed by 250 ml of 1,2,4-trichlorobenzene. The temperature is slowly raised to 130°C and held there for 30 min, at which time the reaction mixture becomes clear. The condenser is removed and a fractionating column and distillation head are attached. Distillation of the phosphorus oxychloride is begun at 55°–130°C. The trichlorobenzene is distilled off at 120°C and 1 mm pressure. Infrared lamps and a heating tape are used to facilitate the distillation. The product distills at 146°–148°C (1 mm), n_D^{20} 1.5960 and is obtained in 219.8 gm yield (70%).

Heating 0.2 M trimellitic anhydride (38 gm) with 100 ml of $SOCl_2$ and 3 drops of DMF for 120 hr yields 92.5% of trimellitoyl chloride of identical infrared spectrum with that of the product prepared from PCl_5 and trimellitic acid.

Oxalyl chloride reacts with acids to give acid chlorides while liberating hydrogen halide, carbon monoxide, and carbon dioxide [37]. High-molecular-weight olefinic acyl halides are prepared in good yields with the aid of oxalyl chloride [38].

The use of oxalyl chloride in the presence of aluminum chloride has been reported to be useful for preparing 4-alkyl and 4-halobenzoyl chlorides using methylene chloride as solvent at 20°–25°C [39].

The use of phthaloyl chloride and chlorosulfonic acid has been reported to be useful for converting esters to acyl chloride in good yields [40].

D. Halogenation Reactions

Halogenation of organic compounds is a common method of introducing a halogen functional group. Bromine and chlorine react in the liquid or gas phase. Fluorine, however, is too reactive and oxidation reactions occur. Sulfur tetrafluoride is an important reagent for the conversion of carboxyl group to CF_2 or CF_3 groups, as discussed later. Several newer methods of halogenation have been reviewed [41].

Hydrofluoric acid and lead dioxide or nickel dioxide react with toluenes bearing electronegative substituents (nitro, cyano, carboethoxy, etc.) to give the corresponding benzyl and/or benzal fluoride [42].

Aromatic compounds are conveniently chlorinated [43] or brominated [44, 45] in the laboratory. For example, durene is chlorinated at 0°C in chloroform to 57% monochlorodurene (mp 47°–48°C) [43]. In the absence of catalysts and in sunlight, alkylbenzenes are brominated [46, 47] or chlorinated [48] in the side chain [47, 48].

Trichlorocyanuric acid has been reported to be an effective laboratory chlorinating reagent for nuclear or side-chain halogenation of aromatic systems [49].

Naphthalene is brominated at room temperature in the absence of a catalyst to α-bromonaphthalene in 75% yield [50], whereas in the presence of an iron catalyst at 150°–165°C, β-bromonaphthalene is formed in 57% yield [51]. Bromine [50], bromine monochloride [52], iodine monobromide [53], and N-bromosuccinimide [54–56] have been employed as brominating agents.

Bromine monochloride reacts with p-nitrophenol to give 2,6-dibromo-4-nitrophenol, indicating that the more electrophilic bromine group reacts preferentially.

para-Bromination of aromatic amines can be achieved by 2,4,4,6-tetrabromo-2,5-cyclohexadien-1-one [57].

2-10. Preparation of Bromobenzene [45]

$$\text{C}_6\text{H}_6 + \text{Br}_2 \xrightarrow{\text{Fe}} \text{C}_6\text{H}_5\text{Br} + \text{HBr} + \text{some } p\text{-C}_6\text{H}_4\text{Br}_2 \tag{24}$$

To a 500-ml round-bottomed flask equipped with a reflux condenser, mechanical stirrer, and dropping funnel are added 33 gm (0.42 moles) of benzene and 3 gm of iron filings. From the dropping funnel containing 60 gm, 19 ml (0.38 mole), of bromine is added 2 ml of bromine, and the flask warmed until hydrogen bromide is evolved. The remaining bromide is added over a 1-hr period. After the addition, the flask is warmed to expel any red vapors. The product is washed several times with water and then it is steam-distilled. The first portion of the steam distillate is collected until crystals appear in the condenser. The receiver is changed and the second portion of distillate containing mainly p-dibromobenzene is collected. If a heavy crystalline deposit collects in the condenser and threatens to clog it, then the flow of water in the condenser is stopped until the crystals melt. The first portion of the steam distillate yields a heavy liquid which, after drying, is distilled at atmospheric pressure. The fraction boiling at 140°–170°C is collected and then redistilled to give 30 gm (50%), bp 150°–160°,

n_D 1.5604 (20°C). The total amount of p-dibromobenzene obtained from the second steam distillate fraction and the above distillation residue is decolorized and recrystallized from ethanol (4 ml per gram of p-dibromobenzene) to give 5–10 gm (11–22%), mp 87°–88°C.

2-11. Bromination of p-Nitrophenol in Aqueous Solution with Bromine Chloride to give 2,6-Dibromo-4-nitrophenol [52]

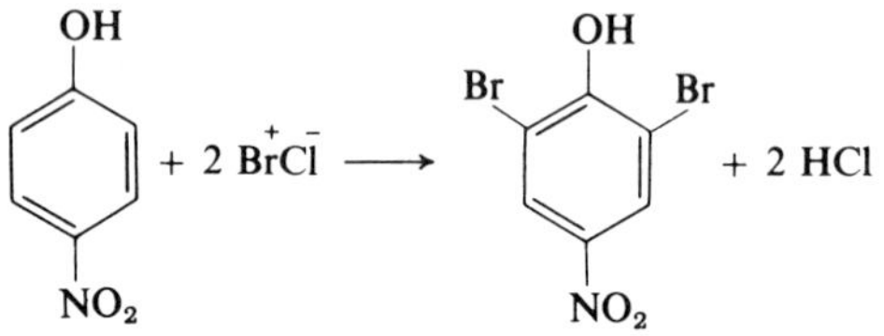

$$\text{p-O}_2\text{N-C}_6\text{H}_4\text{-OH} + 2\ \overset{+}{\text{Br}}\overset{-}{\text{Cl}} \longrightarrow \text{2,6-Br}_2\text{-4-O}_2\text{N-C}_6\text{H}_2\text{-OH} + 2\ \text{HCl} \qquad (25)$$

To a flask containing a vigorously stirred mixture of 13.9 gm (0.1 mole) of p-nitrophenol, 21.6 gm (0.21 mole) of sodium bromide, and 450 ml of water at 45°C (the heating mantle is removed) is slowly added a stream of chlorine gas. A white precipitate forms and the temperature rises to 50°–55°C. The chlorine gas is added just so long as no bromine color appears above the solution. An oil deposits on the side of the flask but it changes to a white solid as the reaction progresses. After 15–20 min the bromine color appears above the solution and the original white suspension turns orange. The chlorine gas is stopped and stirring is continued for 10–15 min at 50°C. If the bromine color fades, more chlorine gas is added until the color reappears. Stirring is then continued for an additional 15 min. When no further fading of the bromine color results, the suspension is cooled to 20°C and filtered. The light yellow precipitate is washed with cold water and dried for 16–20 hr in a vacuum oven at 60°C to give 28.6 gm (96%), mp 141°–142°C.

Iodobenzene is produced in 87% yield by the iodination in the presence of sodium persulfate in acetic acid [58]. HI as it is formed is oxidized back to free I_2, thus avoiding the reducing properties of HI. Iodine monochloride readily condenses phenols and amines [59]. m-Iodobenzoic acid is produced in 75% yields using iodine and silver sulfate in concentrated sulfuric acid [60].

Aliphatic compounds are halogenated preferentially at the tertiary hydrogens in the presence of light or free radical catalysts. Sulfuryl chloride is effective in the presence of peroxide [61]. The halogenation of alkenes at elevated temperatures leads to allyl-type monohalides [62]. In the laboratory n-butyl bromide is chlorinated with sulfuryl chloride (and benzoyl peroxide) to 1-bromo-4-chlorobutane in 35% yield [63–65].

Deno recently reported that R_2NCl is a selective chlorinating reagent and can be used to form 2-chloroalkanes from linear alkanes [2].

§ 2. Condensation Reactions

A novel method has been reported [66] for preparing organic iodides in high yield at ambient temperature by the direct interaction of an ether or alcohol with elementary iodine in the presence of a small quantity of diborane or boron hydride.

$$CH_3OH + 3 I_2 + B_2H_6 \longrightarrow 6 CH_3I + 2 H_3BO_3 + 3 H_2 \qquad (26)$$

$$6 C_6H_5OCH_3 + 3 I_2 + B_2H_6 \longrightarrow 6 CH_3I + 2 B(OC_6H_5)_3 + 3 H_2 \qquad (27)$$

Furthermore iodinated phenols are produced in good yield by dissolving the appropriate phenol in methyl alcohol or other solvent and adding 1 mole of morpholine and 3 moles of iodine. After 48 hr in the cold, the products were separated in about 90% yields [67].

Durene has been converted to iodoflurene by reaction with iodine and periodic acid dihydrate [68].

Aliphatic ethers are halogenated at low temperatures ($-20°C$) to monosubstituted products. Higher temperatures favor mixtures of polysubstituted products. For example, chlorination of ethyl ether at $-20°C$ with one equivalent of chlorine yields 42% of α-chloroethyl ethyl ether [69].

Aryl ethers in the presence of a solvent halogenate preferentially in the nucleus. Anisole reacts with phosphorus pentabromide to give 90% p-bromoanisole [70].

α-Chloro ethers are brominated readily in high yields to the dibromo ethers as is performed during the Boord synthesis. The products are unstable and have been used directly without purification in the Boord synthesis [71, 72].

E. Reaction of Olefins with Halogens and Halogen Derivatives

Bromine readily adds to a double bond at $-20°$ to $20°C$ to give dibromides in high yield, e.g., allyl bromide gives 98% 1,2,3-tribromopropane [73]. Chloroform, carbon tetrachloride, acetic acid, or ether are recommended solvents for the addition of halogen to olefins. Heat or sunlight favors dehydrohalogenation reactions.

Bromine adsorbed on a molecular sieve has been reported to be useful for the selective bromination of olefins. For example, when a cyclohexene-styrene mixture is treated with a 5A sieve that was previously saturated with bromine, only α,β-dibromostyrene (95% yield) is obtained and no trace of dibromocyclohexane is formed [74]. This selectivity may be due to the fact that 5A molecular sieves favor the free radical chain reaction which without molecular sieves give this same selectivity [75].

Chlorine adds trans to a double bond at low temperatures [76]. Elevated temperatures favor substitution reactions. Sulfuryl chloride [77] and phosphorus pentachloride [78] have been used as chlorination agents.

The chlorination of olefins using molybdenum chlorides [79] and cupric chloride [80] has also been reported. In addition, other transition metal chlorides and bromides show a similar reactivity to halogenate olefins [79].

Conjugated diolefins under 1,4-addition of halogen at low temperatures. Bromine gives mainly 1,4-addition [81], whereas chlorine in the liquid or vapor phase gives equal amounts of 1,2- and 1,4-addition products [82].

Mixtures of bromine and chlorine add to olefins such as cyclohexene, styrene, ethylene, stilbene, and cinnamic acid to give the bromochlorides. The products isolated are those expected for the electrophilic addition of a bromine atom and a nucleophilic addition of a chlorine atom [83].

Recently the feasibility of the addition of elemental fluorine to rather sensitive olefins has been demonstrated. The vicinal difluorides produced were predominantly *cis* [84].

2-12. Bromination of Cyclohexene to 1,2,Dibromocyclohexane [85]

$$\text{cyclohexene} + Br_2 \longrightarrow \text{1,2-dibromocyclohexane} \tag{28}$$

To a 2-liter flask equipped with a stirrer, separatory funnel, thermometer, and containing a solution of 123 gm (1.5 moles) of cyclohexene in 300 ml of carbon tetrachloride and 15 ml of absolute alcohol is added dropwise 210 gm (67 ml, 1.3 moles) of bromine in 145 ml of CCl_4 at such a rate (3 hr) that the temperature does not exceed $-1°C$. Higher temperatures cause substitution to occur. The carbon tetrachloride is distilled off using a water bath and the residue washed with 20 ml of 20% alcoholic potassium hydroxide, washed with water, dried, and immediately distilled under reduced pressure to yield 303 gm (85%) of product, bp 99°–103°C (16 mm), n_D^{25} 1.5495.

Olefins also brominate in the allylic position with N-bromosuccinimide (NBS) to give allyl bromides. The olefin is dissolved in anhydrous carbon tetrachloride and N-bromosuccinimide is added. Succinimide is an insoluble coproduct, whose isolation allows the determination of the extent of reaction [86]. Benzoyl peroxide and/or light allow one to use NBS to brominate conjugated dienes and terminal methyl groups [87]. Skell has reported that the succinimidyl radical is a chain carrier when brominations involving NBS are carried out in methylene chloride (0.25 M) and acetonitrile (0.8 M) in the presence of a bromine scavenger. Skell found that the NBS bromination of cyclohexene in CCl_4 gives only 3-bromocyclohexene but photobromination of cyclohexene with NBS in acetonitrile gives both the 3- and 4-bromocyclohexenes in 5:6 ratio [88].

Olefins have been converted to bromohydrins by the use of NBS and moist dimethyl sulfoxide [89].

§ 2. Condensation Reactions

TABLE II

Data on the Hydrochlorination of α-Methylstyrene at 0°C[a]

α-Methylstyrene (mmoles)	Solvent	Time (sec)	Hydrogen chloride used (mmoles)	Molar ratio HCl/olefin
40.1[b]	Neat	126	39.2	0.98
10.5	Neat	138	10.9	1.03
10.0	Neat	126	10.1	1.01
9.4	Neat	128	10.0	1.06
9.9	Neat	130	10.0	1.01
9.8	Neat	136	9.6	0.98
10.0	n-C_5H_{12}[c]	995	10.1	1.01
10.0	n-C_5H_{12}[c]	1100	10.3	1.03
10.0	CH_2Cl_2[c]	61	10.6	1.06
10.0	CH_2Cl_2[c]	123	10.5	1.05
10.0	CH_2Cl_2[c]	120	10.3	1.03

[a] Reprinted from H. C. Brown and N. H. Rei, *J. Org. Chem.* **31**, 1090 (1966). Copyright 1966 by the American Chemical Society. Reprinted by permission of the copyright owner.
[b] Isolated a 97.5% yield of *tert*-cumyl chloride, 99.5% pure by titration, n_D^{30} 1.5230.
[c] 2 M solution.

Methylene groups are brominated faster than methyl groups [90].

Hydrogen halides add according to Markovnikov's rule in the absence of peroxide to give halogen addition to the carbon with the fewer hydrogen atoms. In the presence of peroxides or oxygen the addition is reversed [91].

Recently the hydrochlorination of olefins has been described, using the H. C. Brown hydrogenator as discussed in Procedure 2-2. Using this apparatus, α-methylstyrene is converted to 2-chloro-2-phenylpropane in almost quantitative yield. Table II describes some representative experiments in the hydrochlorination of α-methylstyrene and the effect of solvent on the rate of reaction [1].

2-13. *Hydrochlorination of* α-*Methylstyrene to 2-Chloro-2-phenylpropane* [1]

$$C_6H_5-\underset{\underset{CH_3}{|}}{C}=CH_2 + HCl \longrightarrow C_6H_5-\underset{\underset{Cl}{|}}{\overset{\overset{CH_3}{|}}{C}}-CH_3 \qquad (29)$$

Using the apparatus described in Fig. 1 for Procedure 2-2, the flask is cooled to 0°C and the reaction is initiated by injecting 4.73 gm (0.04 mole) of α-methylstyrene, n_D^{20} 1.5381. After 10 min the excess dissolved hydrogen chloride is removed by applying gentle aspiration. The organic phase is extracted with 100 ml of pentane and concentrated. The crude product, 2-chloro-2-phenylpropane, is obtained in 98% yield (6.02 gm), n_D^{20} 1.5230.

Peroxides and aluminum chloride induce the addition of halogenated compounds to olefins. In peroxide-initiated reactions carbon tetrachloride, chloroform, bromotrichloromethane, etc., add to olefins containing a terminal double hood [92–93a]. For example, 1-octene reacts with CCl_4 to yield 85% 1,1,1,3-tetrachlorononene. This reaction is also applicable to substituted olefins (olefinic alcohols, ethers, acetates, etc.) (see preparation 2-14).

2-14. Preparation of Tetrachlorobutanol by the Reaction of Carbon Tetrachloride with Allyl Alcohol [93]

$$CCl_4 + CH_2=CH-CH_2OH \longrightarrow CCl_3-CH_2-\underset{Cl}{CH}-CH_2OH \quad (30)$$

To a three-necked flask equipped with a mechanical stirrer, condenser, and thermometer are added 230.8 gm (1.5 mole) of carbon tetrachloride, 29.0 gm (0.50 mole) of allyl alcohol, and 2.5 gm of 100-mesh iron powder. The mixture is stirred and refluxed for 7½ hr at 73°C. Gas chromatographic analysis indicates that it contains 20.5% 2,4,4,4-tetrachlorobutanol and 2% allyl alcohol. Omission of the iron gave no product after 7½ hr.

Aluminum chloride and other Friedel–Crafts catalysts aid in the condensation of alkyl halides and olefins. For example, *tert*-butyl chloride adds to ethylene to yield 75% 1-chloro-3,3-dimethylbutane [94]. In addition, by the Friedel–Crafts reactions, olefins and alcohols will react with aryl halides to give aryl halides in good yields [95].

Hypohalous acid also adds to olefins to give substituted chlorohydrins. The hydroxyl group attaches to the carbon atom with the smaller number of hydrogen atoms. In these reactions, hypohalous acid is generated from *tert*-butyl hypochlorite in dilute acetic acid [96, 97] or from calcium hypochlorite and mineral acids [96, 97]. An emulsifying agent gives improved yields [96, 97]. Styrene yields styrene chlorohydrin in 76% yield [96, 97].

Dihalocarbenes react with olefins to give 1,1-dihalocyclopropanes [98, 99].

$$:CHX \text{ or } :CX_2 + \overset{\diagdown}{\underset{\diagup}{C}}=\overset{\diagdown}{\underset{\diagup}{C}} \longrightarrow \underset{\underset{X\ \ H}{\diagup\diagdown}}{\underset{C}{\diagup\diagdown}}\overset{\diagdown}{\underset{\diagup}{C-C}}\overset{\diagdown}{\diagup} \text{ or } \underset{\underset{X\ \ X}{\diagup\diagdown}}{\underset{C}{\diagup\diagdown}}\overset{\diagdown}{\underset{\diagup}{C-C}}\overset{\diagdown}{\diagup} \quad (31)$$

Halocarbenes can be generated from haloforms and base, methylene halide and lithium alkyls, and by the decarboxylation of sodium trichloroacetate. Difluorocarbene [100] can also be generated by the photolysis or pyrolysis of difluorodiazirene. Mixed halocarbenes have also been reported [101].

2-15. General Procedure for the Preparation of 1,1-Dihalocyclopropanes [99]

The gem-dihalocyclopropanes are generally prepared by adding 1.0 mole of haloform to 1.0 mole of dry potassium tert-butoxide (M.S.A. Research Corp.) and 1.0 mole or more of the olefin in 200–300 ml of n-pentane, at 0°–10°C. After the addition has been completed, the temperature is raised to room temperature and the mixture stirred for several hours. Water is added and the organic layer separated, washed with water, dried, and concentrated at atmospheric pressure. The crude material is weighed, analyzed by gas-liquid chromatography, and distilled under reduced pressure to yield the pure product. Several examples are summarized in Table III. The detailed preparation of 1,1-dibromo-2,2-diphenylcyclopropane is given as an example (2-16).

2-16. Preparation of 1,1-Dibromo-2,2-diphenylcyclopropane [99, 102]

$$(C_6H_5)_2C=CH_2 + CHBr_3 + KO\text{-}tert\text{-}Bu \longrightarrow$$

$$(C_6H_5)_2C\underset{\underset{Br}{\diagup}\underset{Br}{\diagdown}}{\overset{\diagdown\diagup}{\underset{C}{|}}}CH_2 + KBr + tert\text{-Butyl alcohol} \quad (32)$$

To a flask containing 100 ml of dry pentane, 25 gm (0.14 mole) of 1,1-diphenylethylene, and 28 gm (0.25 mole) of potassium tert-butoxide at 0°C is added dropwise 66 gm (0.26 mole) of bromoform over a 1-hr period at 0°–10°C. During the reaction a yellow solid is formed which upon filtration at the end is obtained in 63% yield (30.8 gm), mp 140°–147°C (from isopropanol).

The 1,1-dihalocyclopropanes rearrange thermally or with the aid of electrophiles to give allyl halides [Eq. (33)] [101]. The preparations are described, in Chapter 2, Olefins. In the case of 1,1-dibromo-2,-diphenyl, cyclopropene reaction with silver acetate gives 1,1-diphenyl-2-bromo-3-acetoxy-1-propene in 72% yield [102].

$$\underset{X\ X}{\overset{\diagdown\diagup}{\underset{C}{\overset{|}{\diagup\diagdown}}}\overset{\diagup}{\underset{C}{\diagdown}}} \xleftarrow{\text{Heat}} \underset{X\ X}{\overset{\diagdown\diagup}{\underset{C}{\overset{C-C}{\diagup\diagdown}}}\overset{\diagup}{\diagdown}} + AgNO_3 + H_2O \longrightarrow \underset{OH\ X}{\overset{\diagdown\diagup}{\underset{C}{\overset{|}{\diagup\diagdown}}}\overset{\diagup}{\underset{C}{\diagdown}}} + AgX \quad (33)$$

The general nature of the chain extension of alkenes via gem-dihalocyclopropanes is illustrated with many examples of olefins [101–104].

The use of phase transfer catalysts allows the preparation of gem-dihalocyclopropanes from haloforms and aqueous sodium hydroxide systems [103].

TABLE III

METHOD OF PREPARATION AND PHYSICAL PROPERTIES OF SEVERAL 1,1-DIHALOCYCLOPROPANES [76a][a]

1,1-Dibromocyclopropane (except where noted)	Olefin (moles)	Solvent (ml) potassium-*tert*-butoxide, moles	Haloform (moles) C = CHCl₃ B = CHBr₃	Yield (%)	Bp[b] (mm) (°C)	n_D (temp., °C)
6,6-Dichlorobicyclo [3.1.0]-hexane	Cyclopentene (1.0)	None/0.75	0.75C	20	87°–90° (61)	1.4907–1.4941[c] (27.2°)
6,6-Dibromobicyclo[3.1.0]-hexane	Cyclopentene (1.0)	None/0.75	0.82B	42	63°–69° (2)	1.5560–1.5594 (18°)
7,7-Dichlorobicyclo[4.1.0]-heptane	Cyclohexene (1.0)	*n*-Pentane(400)/1.0	1.0C	18	67°–68° (6.0)	1.5038 (20°)
7,7-Dibromobicyclo[4.1.0]-heptane	Cyclohexene (1.0)	Cyclohexane (100)/0.88	1.18B	35	98°–100° (6–6.5)	1.5579[d] (23°)
2,2-Dimethyl	Isobutylene (1.0)	None/0.4	0.3B	28	47°–48° (11)	1.5136 (23°)
cis-2,3-Dimethyl	*cis*-2-Butene (1.0)	None/0.4	0.3B	90	55°–56° (11–12)	1.5188– 1.5206 (23°)
trans-2,3-Dimethyl	*trans*-2-Butene (1.0)	None/0.4	0.3B	90	55°–56° (11)	1.5110 (25°)
2,2,3-Trimethyl	2-Methyl-2-butene (1.0)	None/0.6	0.5B	50	48°–50° (3.8)	1.5167 (23°)
2,2,3,3-Tetramethyl	2,3-Dimethyl-2-butene (1.0)	*n*-Pentane(100)/1.0	0.9B	60	mp 75°–76°	—
2-Phenyl-	Styrene (1.0)	None/0.25	0.25B	55	118°–120° (5.7)	1.5996 (23°)
2,2-Diphenyl-	1,1-Diphenylethylene (0.14)	*n*-Pentane(100)/0.25	0.26B	63	mp 146°–147°[e]	—
2-*sec*-Butyl	4-Methyl-1-pentene (1.0)	None/0.25	0.25B	52	50° (1.0)	1.4992 (23°)[f]

[a] Reprinted in part from S. R. Sandler, *J. Org. Chem.* **32**, 3876 (1967). Copyright 1967 by the American Chemical Society. Reprinted by permission of the copyright owner.
[b] Boiling points and melting points are uncorrected.
[c] Analysis of C₆H₈Cl₂—calculated: C, 47.60; H, 5.30; found: C, 47.98; H, 5.18.
[d] Analysis of C₇H₁₀Br₂—calculated: C, 33.10; H, 4.00; found: C, 33.06; H, 4.15.
[e] Recrystallization from isopropanol.
[f] Analysis of C₇H₁₂Br₂—calculated: C, 32.80; H, 4.68; found: C, 33.22; H, 4.76.

§ 2. Condensation Reactions

Acetylenes can also be halogenated to give olefinic halides [104]. Dibromobutene diol is an example of a commercial product made by this procedure.

F. Halogenation of Aldehydes, Ketones, and Acids

Aldehydes and ketones are readily halogenated. Acetone is brominated to give bromoacetone [105], 1,3-dibromoacetone [106, 107], or 1,3,3-tribromoacetone [107].

Acetone reacts with chloroform to give 1,1,1-trichlo-2-methyl-2-propanol (chlorobutanol) [108]. Bromination of 3-methyl-2-butanone in methanol at 0° to 5°C gives 95% yields of 1-bromo-3-methyl-2-butanone [109].

Chloroketones are prepared using sulfuryl chloride [110] or by direct chlorination [111].

Aliphatic aryl ketones are halogenated exclusively in the aliphatic side chain. The bromination of acetophenone yields α-bromoacetophenone in 96% yield [112].

Copper(II) bromide in chloroform–ethyl acetate, brominates, hydroxyacetophenone almost quantitatively to ω-bromohydroxyacetophenone [113].

$$\text{HO—C}_6\text{H}_4\text{—}\overset{\overset{\text{O}}{\|}}{\text{C}}\text{—CH}_3 + 2\text{ CuBr} \xrightarrow[\text{EtOH}]{\text{CHCl}_3} \text{HO—C}_6\text{H}_4\text{—}\overset{\overset{\text{O}}{\|}}{\text{C}}\text{—CH}_2\text{Br}$$

$$+ \text{ HBr (100\%)} + 2 \text{ CuBr} \qquad (34)$$

Copper(II) chloride in dimethylformamide in the presence of lithium chloride chlorinates ketones in the α-position in good yields [114]; for example, propiophenone (89%) and methyl ethyl ketone (50–70%).

Aliphatic aldehydes or ketones are converted to *gem*-dihalides by means of phosphorus pentachloride [115].

$$\text{RCH}_2\overset{\overset{\text{O}}{\|}}{\text{C}}\text{R} + \text{PCl}_5 \longrightarrow \text{RCH}_2\text{CCl}_2\text{R} + \text{POCl}_3 \qquad (35)$$

Arylacetones are not converted to *gem*-dihalides but to chloroolefins and α-chloroketone [116].

Bromination of acetone in acetic acid–water solution at 60°C was found to give low yields of 1,3-dibromoacetone (3–21%). The preparation is given below not because it is an excellent preparative method but because it starts from inexpensive available raw materials. Improvements could be made, possibly through variation of the temperature or type of solvent.

2-17. *Preparation of 1,3-Dibromoacetone* [29, 117]

$$\text{CH}_3\overset{\overset{\text{O}}{\|}}{\text{—C—}}\text{CH}_3 + 2 \text{ Br}_2 \longrightarrow \text{BrCH}_2\overset{\overset{\text{O}}{\|}}{\text{C}}\text{—CH}_2\text{Br} + 2 \text{ HBr} \qquad (36)$$

To a 2-liter flask in a well-ventilated hood are added 99.0 gm (1.71 moles) of acetone, 98.2 gm (1.64 moles) of acetic acid, and 400 gm of water. To the stirred solution is added 475 gm (2.97 moles) of bromine from a pressure-equalized addition funnel at such a rate as to maintain the temperature at 60°C. The addition time is about 10 hr. The mixture is then cooled to room temperature. After 48 hr, the bottom layer is separated, dried over magnesium sulfate, and distilled under reduced pressure to yield 97.6 gm (30.4%), based on bromine used, of 1,3-dibromoacetone, bp 69°–71°C (1.5 mm), n_D^{22} 1.5423. CAUTION: The aqueous upper layer from the reaction mixture must be handled with extreme care since it contains large quantities of dissolved hydrogen bromide.

An alternate procedure involves the preparation of glycerol 1,3-dibromohydrin and oxidation (sulfuric acid–dichromate) of the latter to 1,3-dibromoacetone in 67% yield [29]. In this preparation, 1000 gm (10.8 moles) of glycerol, 25 gm (1.0 mole) of red phosphorus, and 1869 gm (23.4 moles) of bromine are reacted to produce glycerol 1,3-dibromohydrin in 59% yields. This procedure has the advantage that the possibility of isomer production is drastically reduced, a matter which is difficult to avoid in the direct bromination of acetone.

$$3\ HOCH_2\text{—}\underset{\underset{OH}{|}}{\overset{\overset{H}{|}}{C}}\text{—}CH_2OH + 2\ P + 3\ Br_2 \longrightarrow 2\ H_2PO_3 + 3\ Br\text{—}CH_2\text{—}\underset{\underset{OH}{|}}{\overset{\overset{H}{|}}{C}}\text{—}CH_2Br\quad (59\%)$$

$$\xrightarrow[H_2SO_4]{K_2Cr_2O_7} 3\ Br\text{—}CH_2\text{—}\underset{\underset{O}{\|}}{C}\text{—}CH_2Br\quad (67\%) \tag{37}$$

Aldehydes are halogenated on the aldehyde carbon as well as the α-carbon atom [118]. However, when there are no α-hydrogens, then halogenation occurs extensively on the aldehyde carbon atom, o-Chlorobenzaldehyde undergoes chlorination to give o-chlorobenzoyl chloride in 75–84% yield [119].

Acids and esters are halogenated conveniently at the α-position by means of red phosphorus and halogens or by phosphorus halides. The preparation of bromoacetic acid is carried out using acetic anhydride with pyridine [120]. The Hell–Volhard–Zelinsky reaction halogenates the acid in the α-position. Using two equivalents of halogen yields an α-halogenated acyl halide in one step [121]. Dicarboxylic acids give α,α'-dihalogenated acids [122]. The use of thionyl chloride and N-bromosuccinimide has been reported to be useful for preparing α-bromo acid chlorides. The latter procedure avoids the use of free bromine, the reaction time is considerably shorter (usually about 2 hr), and the work-up is simplified [123].

The α-chlorination of carboxylic acids has been reported to be mediated by chlorosulfonic acid. For example [124]:

§ 3. Elimination and Cleavage Reactions

$$C_6H_5-\overset{O}{\overset{\|}{C}}NH(CH_2)_5-\overset{O}{\overset{\|}{C}}OH + Cl_2SO_2 + ClSO_3H \xrightarrow[\text{1,2-dichloro-ethane}]{60°-70°C}$$

$$C_6H_5-\overset{O}{\overset{\|}{C}}NH(CH_2)_4\underset{Cl}{CH}-\overset{O}{\overset{\|}{C}}OH \qquad (38)$$

Deno reported that an increasing amount of terminal ω chlorination of fatty acids takes place by absorbing and aligning the reactants on alumina [3]. Deno reported selective chlorination at C-4 in butyric, hexanoic and octanoic acids by conducting the chlorination in 90% sulfuric acid [4].

2-18. Hell–Volhard–Zelinsky Reaction—Preparation of α-Bromobutyric Acid [125]

$$CH_3CH_2CH_2COOH + Br_2 \xrightarrow{P\ (red)} CH_3CH_2\underset{Br}{CH}-COOH \qquad (39)$$

To a heated (120°C) flask equipped with a condenser, mechanical stirrer, and pressure-equalizing dropping funnel, containing 15 gm (0.171 mole) of butyric acid and 0.4 gm (0.0129 mole) of red phosphorus is added dropwise 30 gm (0.188 mole) of bromine. A great amount of hydrogen bromide is evolved and the reaction is completed after 1½ hr. Distillation of the reaction mixture yields 23 gm (82%) of α-bromobutyric acid, bp 120°C (23 mm).

Sulfur tetrafluoride has the ability to replace carbonyl oxygen with fluorine in aldehydes, ketones, and carboxylic acids in 56–88% yields [126, 127]. However, the handling of sulfur tetrafluoride is somewhat problematical. The reaction is normally carried out in a stainless steel pressure vessel.

$$R_2C=O + SF_4 \longrightarrow R_2CF_2 + SOF_2 \qquad (40)$$
$$RCOOH + 2\,SF_4 \longrightarrow RCF_3 + HF + 2\,SOF_2 \qquad (41)$$
$$HOOC-R-COOH + 6\,SF_4 \longrightarrow CF_3-R-CF_3 \qquad (42)$$
$$HOOC-R-COOH + 3\,SF_4 \longrightarrow CF_3-R-COF \qquad (43)$$
$$H_2C=CH-COOH + SF_4 \longrightarrow H_2C=C-CF_3 \qquad (44)$$
$$HC\equiv C-COOH + SF_4 \longrightarrow HC-C-CF_3 \qquad (45)$$

3. ELIMINATION AND CLEAVAGE REACTIONS

The reactions involving elimination can be summarized as follows:

$$RX + X' \longrightarrow RX' + X \qquad (46)$$

6. Halides

$$R-Z + X \longrightarrow R-X + Z \quad (47)$$
$$R-Z-X \longrightarrow RX + Z \quad (48)$$

where X = Cl and X′ = I or F present in NaI and NaF, respectively; Z is $-N_2^+$ (diazonium) (Sandmeyer), N_2 (diazo), —COOAg or COOH (Hunsdiecker); and Z—X is $-N_2BF_4$ (Schiemann).

Ethers are cleaved to dihalides by halogen acids in the presence of Friedel–Crafts catalysts such as $ZnCl_2$.

The above reactions represent a few of the more common elimination–cleavage type reactions which are discussed below.

Conversion of alkyl and aryl chlorides to their iodide derivatives is effected in high yields. Allyl bromide or chloride is converted in good yields to allyl iodide by refluxing with a mixture of acetone and sodium iodide [29]. 2,4-Dinitrochlorobenzene in dimethylformamide is converted to 2,4-dinitroiodobenzene in 70% yields by refluxing with sodium iodide [128].

In a related manner, alkyl chlorides are converted to alkyl fluorides by refluxing with a mixture of potassium fluoride and N-methyl-2-pyrrolidone as a solvent. Alkyl compounds with three or more carbon atoms give high yields. Sodium fluoride is ineffective [129]. Furthermore, heating 3,6-dichlorophthalic anhydride with potassium fluoride at 260°–270°C for 1 hr gives 63% of 3,6-difluorophthalic anhydride [130].

3-1. Preparation of Allyl Iodide [29]

$$CH_2=CH-CH_2-Cl + NaI \xrightarrow{Acetone} CH_2=CH-CH_2I + NaCl \quad (49)$$

To a flask containing 4 liters of acetone is added 2270 gm (15.2 moles) of sodium iodide followed by the dropwise addition of allyl chloride until 925 gm (12.1 moles) has been added. The mixture is then refluxed for 23 hr, cooled, poured into 10 liters of deionized water, separated, washed with 2% sodium bisulfite, dried over Na_2SO_4, and distilled to yield 1407 gm (69%), bp 101°–103°C (20 mm), n_D^{20} 1.5545, d_4^{20} 1.844.

A. The Sandmeyer and Schiemann Reactions

The Sandmeyer reaction [131] involves the displacement of the diazonium group by a halogen to give an aryl halide. The diazonium salt is decomposed with a solution of the appropriate hydrogen halide acid and cuprous chloride or bromide. Copper powder and mineral acid can also be used effectively in the Gattermann modification [132]. For the preparation of chloro- and bromophenanthrenes, the ordinary conditions fail. They are successfully obtained by reacting the appropriate diazonium salt with a mixture of the halo mercuric and potassium salts [133]. In some cases copper salts are not necessary, as in the case of the preparation of aryl iodides using potassium iodide [134].

§ 3. Elimination and Cleavage Reactions

3-2. Sandmeyer Reaction—Preparation of p-Nitrochlorobenzene [135]

$$O_2N-C_6H_4NH_2 \xrightarrow{HCl+NaNO_2} O_2N-C_6H_4-N_2{}^+Cl^- \xrightarrow{CuCl} O_2N-C_6H_4Cl + N_2 \quad (50)$$

To a stirred flask containing 69 gm (0.5 mole) of *p*-nitroaniline, 108 ml of concentrated hydrochloric acid, and 200 gm of ice is slowly added 35 gm of sodium nitrite in 100 ml of water. The excess nitrite is destroyed with urea and the solution is filtered. The *p*-nitrobenzene diazonium chloride is diluted up to 500 ml and slowly added to 2 liters of acetone at 0°C. Argon free of oxygen is swept through the solution and then 8.0 gm (0.081 mole) of cuprous chloride and 4 gm (0.0945 mole) of lithium chloride in 200 ml of anhydrous acetone is added dropwise. Nitrogen is immediately evolved. The acetone is removed under reduced pressure and the residue is extracted with ether. Concentration of the ether solution yields 13.6 gm (68%) of *p*-nitrochlorobenzene, mp 82°–84°C. The mother liquor yields 21% nitrobenzene.

The Schiemann reaction involves the reaction of the diazonium chloride with fluoroboric acid to give a solid borofluoride which is isolated and then decomposed with heat to the aryl fluoride [136]. Heterocyclic fluorine compounds can also be prepared in an analogous fashion [137]. Fluorophenols and acid are best obtained by starting with the ethers and esters [136].

An improved fluorination of aromatic compounds involves the decomposition of the diazonium hexafluorophosphate [138].

3.3. Schiemann Reaction—Preparation of 3,5-Dimethyl-2-fluoro-1-bromobenzene [139]

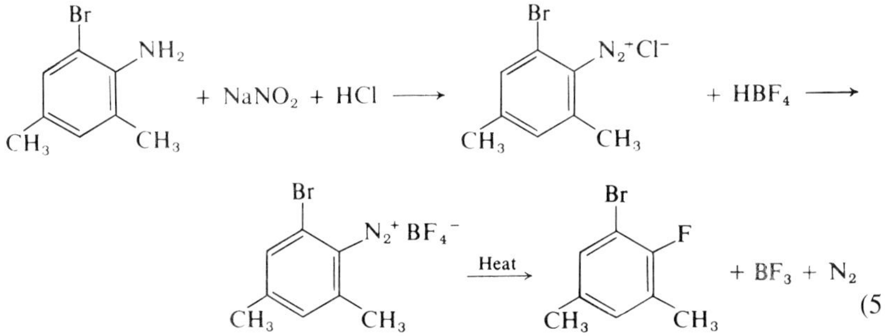

$$\text{(51)}$$

In a 300-ml beaker containing 20 ml of concentrated hydrochloric acid is suspended 5 gm (0.025 mole) of 3,5-dimethyl-2-amino-1-bromobenzene. The mixture is cooled to 0°C. The amine is diazotized by addition of a slight excess of a saturated nitrite solution, the end point being determined by potassium iodide starch test paper. The mixture is then filtered and to the filtrate are added 27.6

gm of a hydrofluoroboric acid solution prepared from 20 gm of 48% hydrofluoric acid and 7.6 gm of boric acid. The diazonium tetrafluoride precipitates as white needles in the amount 4.5 gm (60%). The latter salt is placed in a distilling flask and heated to 170°C, at which point decomposition starts. The liberated BF_3 is caught in a NaOH trap. After all the BF_3 is eliminated, water is added to the flask and the product is steam-distilled. Redistillation yields 3.0 gm (100%) of 3,5-dimethyl-2-fluoro-1-bromobenzene, bp 87°–89°C (11 mm), n_D^{20} 1.3100.

In a related manner, diazoketones can be decomposed in the presence of halogen acids to give halomethyl alkyl [140], aryl [141], or heterocyclic ketones [142].

The diazoketones are obtained from the acyl halide and diazomethane [143].

Ethers are cleaved by halogen acids to alcohols and alkyl halides. The Ziesel method is based on the hydrogen iodide cleavage of methoxy groups to give methyl iodide. Reacting Grignard reagents with chloromethyl methyl ether and cleaving with halogen halide are used to increase the carbon chain of an alkyl halide [144].

A convenient cleavage reagent employs orthophosphoric acid and potassium iodide [145].

Tetrahydrofuran [146] and tetrahydropyran [147] are cleaved to give 1,4- and 1,5-diahlo derivatives, respectively. With tetrahydrofuran and hydrochloric acid, the presence of zinc chloride catalyst drives the reaction to completion [146].

B. The Hunsdiecker Reaction

$$RCOOAg + X_2 \longrightarrow RX + CO_2 + AgX \quad (52)$$

Silver carboxylates react with chlorine or bromine to give carbon dioxide, silver halide, and an alkyl halide with one less carbon atom than the acid [148–151]. Aromatic silver carboxylates react in a similar fashion but nuclear halogenation is a competing side reaction [152]. The yields of alkyl halides obtained when employing this reaction are satisfactory. For example, methyl 5-bromovalerate is obtained in 65–68% yield [153].

$$CH_3OOC(CH_2)_4COOH + AgNO_3 \longrightarrow CH_3O_2C(CH_2)_4COOAg \xrightarrow{Br_2}$$
$$CH_3O_2C(CH_2)_3CH_2Br \quad (53)$$

A general method for the synthesis of perfluoroalkyl iodides and diiodides by the reaction of iodine on the corresponding silver perfluoro fatty acid salt has been described [154].

The yields of halides from substituted aromatic carboxylic acid increases as the electron-withdrawing capacity of the substituent increases. For example,

§ 4. Miscellaneous Methods

benzoic acid yields 18% bromobenzene whereas *o*-nitrobenzoic acid yields 95% *o*-nitrobromobenzene [155].

A modified Hunsdiecker reaction has been reported [156] where the free acid and red mercuric oxide were found to be more effective than the acid chloride and silver oxide. The method is also much simpler than the original Hunsdiecker method in that it does not require the preparation of the dry silver salt.

Another modification of the Hunsdiecker reaction involves the reaction of bromine with thalium carboxylates to give alkyl bromide in yields ranging from 84–98% [157, 157a].

3-4. Preparation of **p-***Nitrobromobenzene* **[155]**

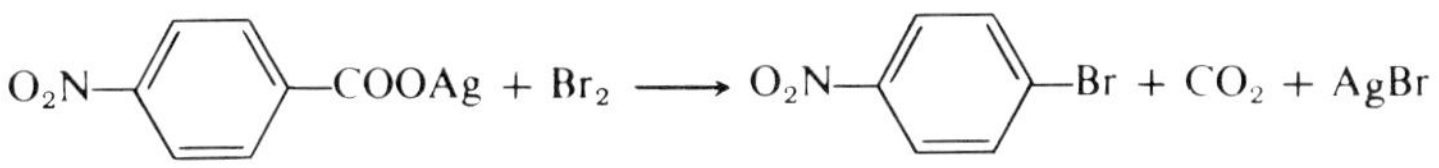

$$O_2N-C_6H_4-COOAg + Br_2 \longrightarrow O_2N-C_6H_4-Br + CO_2 + AgBr \quad (54)$$

To a flask containing 34 gm (0.125 mole) of silver *p*-nitrobenzoate suspended in 500 ml of carbon tetrachloride is added dropwise 20 gm (0.125 mole) of bromine at room temperature. There is no visible reaction at room temperature. The deep red solution is then refluxed for 3 hr, at which time the color gradually fades. The hot solution is filtered and the filtrate is washed with sodium bisulfite and sodium bicarbonate solutions. Evaporation of the carbon tetrachloride yields 20 gm (79%) of crystals, mp 126°–127°C. Acidification of the sodium bicarbonate extract yields 2 gm (10%) of recovered *p*-nitrobenzoic acid.

3-5. Modified Hunsdieker Reaction—Preparation of 1-Bromohexane from Heptanoic Acid **[156]**

$$CH_3(CH_2)_5COOH + HgO + Br_2 \ (CCl_4) \longrightarrow CH_3(CH_2)_4CH_2Br + CO_2 + HBr \quad (55)$$

To a flask containing a stirred solution of 13.0 gm (0.1 mole) of heptanoic acid and 22 gm (0.0945 mole) of red mercuric oxide in 150 ml of dry carbon tetrachloride is slowly added dropwise a solution of 16 gm (0.1 mole) of bromine in 50 ml of carbon tetrachloride. The mixture is refluxed for 1 hr and filtered. The filtrate is washed with 5% sodium hydroxide solution, washed with water, dried over magnesium sulfate, and fractionally distilled to yield 6.0 gm (36%) of 1-bromohexane, bp 150°–159°C, n_D^{28} 1.4470.

4. MISCELLANEOUS METHODS

(1) The chlorination of aryl aldehydes yielding aryl acid chlorides [158].
(2) *N*-Halogenation of amines [159].
(3) *N*-Halogenation of amides and imides [160].

(4) Preparation of halohydrins by the addition of hydrogen halide to epoxides [161].

(5) Reduction of polyhalides to lower halide derivatives [162].

(6) Chloroformylation of phenylacetylenes to β-chlorocinnamic aldehydes [163].

(7) Addition of fluorine to halogenated olefins with CoF_3 [164].

(8) Direct fluorination of 1,1-diphenylethylene with PbF_4 [165].

(9) The fluorination of bromofluoroethanes with bromine trifluoride in bromine solution [166].

(10) The halogenation of acenaphthene derivatives [167].

(11) The reaction of triphenylphosphine, carbon tetrachloride, and alcohols. A new synthesis of alkyl chlorides (under neutral conditions) [168].

(12) The preparation of aryl and alkyl halides using triphenylphosphine and a halogen [169, 170].

(13) Reaction of iodine isocyanate with dienes and acetylenes [171].

(14) Reaction of methylene chloride with cyclohexanone to give 1-(dichloromethyl)cyclohexanol [172].

(15) The addition of dichlorocarbene to *cis,cis*-1,5-cyclooctadiene [173].

(16) The addition of dihalocarbenes to steroidol olefins [174].

(17) Preparation of iodocyclopropanes from iodoform and olefins [175].

(18) The preparation of monofluorocyclopropanes [176].

(19) Preparation of *cis*-3,4-dichlorocyclobutene [177].

(20) The addition of dibromodifluoromethane to vinylidene fluoride [178].

(21) Reactions of olefins with carbon tetrachloride and carbon monoxide [179].

(22) Conversion of epoxides to halohydrins [180].

(23) Monohalogenation of primary nitroparaffins [181].

(24) Chlorination of phenols (solid state) [182].

(25) Halogen interchange in alkyl halides using molybdenum (V) chloride [183].

(26) 3-Chlorophthalic anhydride via chlorination of phthalic anhydride [184].

(27) Conversion of aromatic and α,β-unsaturated aldehydes to dichlorides [185].

(28) Hexachloroacetone/triphenylphosphine as a mild reagent for the production of allyl chlorides from the alcohols [186].

(29) *N*-Bromosuccinimide–DMF as a mild selective nuclear monobromination of reactive aromatic compounds [187].

(30) Preparation of vicinal chloroiodoalkanes by the reaction of olefins with copper(II) chloride and iodine or an iodine donor [188].

(31) Preparation of β-bromothiophenes [189].

(32) Chlorination of cyclopentadiene [190].

(33) Preparation of α-bromoacetals [191].

(34) Trifluoromethanesulfonyl chloride as a mild chlorinating agent [192].
(35) α-Bromination of aralkyl ketones [193].
(36) Controlled electrolytic reduction of polyhalogen compounds to give alkyl halides [194].
(37) *cis*-Dichloroalkanes from epoxides [195].
(38) Preparation of terminally halogenated olefins by the reaction of aldehydes, carbon tetrachloride, and triphenylphosphine [196].
(39) Preparation of halo substituted acyl halides by the reaction of olefins, carbon monoxide, with carbon tetrachloride in the presence of a free radical generating compound [197].
(40) Preparation of aromatic acid chlorides by the reaction of the corresponding carboxylic acid with carbon monoxide, chlorine, carbon tetrachloride, and tetrabutylphosphine oxide [198].

REFERENCES

1. H. C. Brown and N. H. Rei, *J. Org. Chem.* **31,** 1090 (1966).
2. N. C. Deno, E. J. Gladfelter, and D. G. Pohl, *J. Org. Chem.* **44,** 3728 (1979).
3. N. C. Deno, R. Fishbein, and C. Pierson, *J. Am. Chem. Soc.* **92,** 1451 (1970).
4. N. C. Deno, R. Fishbein, and J. C. Wyckoff, *J. Am. Chem. Soc.* **92,** 5274 (1970).
5. O. Kamm and C. S. Marvel, *Org. Synth. Collect. Vol.* **1,** 25 (1941).
6. E. E. Reid, J. R. Ruthoff, and R. E. Burnett, *Org. Synth. Collect. Vol.* **2,** 246 (1943).
7. F. C. Whitmore, R. W. Schiessler, C. S. Rowland, and J. N. Cosby, *J. Am. Chem. Soc.* **69,** 235 (1947).
8. A. M. Whaley and J. E. Coppenhaver, *J. Am. Chem. Soc.* **60,** 2497 (1938).
9. F. C. Whitmore and F. A. Karnatz, *J. Am. Chem. Soc.* **60,** 2533 (1938).
10. J. F. Norris and A. W. Olmsted, *Org. Synth. Collect. Vol.* **1,** 144 (1941).
11. H. A. Shonle, J. H. Waldo, A. K. Keltch, and H. W. Coles, *J. Am. Chem. Soc.* **58,** 585 (1936).
12. H. S. King, *Org. Synth. Collect. Vol.* **2,** 399 (1943).
13. G. A. Lutz, A. E. Bearse, J. E. Leonard, and F. C. Croxton, *J. Am. Chem. Soc.* **70,** 4135 (1948).
14. K. Ahmad, F. M. Bumpus, and F. M. Strong, *J. Am. Chem. Soc.* **70,** 3391 (1948).
15. H. L. Goering, S. J. Cristol, and K. Dittmer, *J. Am. Chem. Soc.* **70,** 3314 (1948).
16. J. D. Bartleson, R. E. Burk, and H. P. Lankelma, *J. Am. Chem. Soc.* **68,** 2513 (1946).
17. S. R. Sandler, *J. Org. Chem.* **35,** 3967 (1970).
18. I. Tabushi, Z.-I. Yushida, and N. Takahashi, *J. Am. Chem. Soc.* **93,** 1820 (1971).
19. R. C. Fuson and C. H. McKeaver, *Org. React.* **1,** 63 (1942).
20. A. Ginsburg, W. H. C. Rueggeberg, I. D. Tharp, H. A. Nottorf, M. R. Cannon, F. I. Carnahan, D. S. Cryder, G. H. Fleming, G. M. Goldberg, H. H. Haggard, C. H. Herr, T. B. Hoover, H. L. Lovell, R. G. Mraz, C. I. Noll, T. S. Oakwood, H. T. Patterson, R. E. Vanstrien, R. N. Walter, H. D. Zook, R. B. Wagner, C. A. Weisgerber, J. P. Wilkins, and F. C. Whitmore, *Ind. Eng. Chem.* **38,** 478 (1946).
21. M. J. Rhoad and P. J. Flory, *J. Am. Chem. Soc.* **72,** 2216 (1950).
22. G. M. Kosolapoff, *J. Am. Chem. Soc.* **68,** 1670 (1946).
23. J. M. Griffling and L. F. Salisbury, *J. Am. Chem. Soc.* **70,** 3416 (1948).
24. R. Quelet, *Bull. Soc. Chim. Fr.* [5C] **7,** 196, 205 (1940).
25. G. Kubiczek and L. Neugebauer, *Monatsh. Chem.* **81,** 917 (1950).

26. Y. Ogata and M. O. Kane, *J. Am. Chem. Soc.* **78,** 5423 (1956).
27. B. L. Van Duren, A. Sivak, B. M. Goldschmidt, C. Katz, and S. Melchionne, *JNCI, J. Natl. Cancer Inst.* **43,** 481 (1969).
28. D. S. Connor, G. W. Klein, and G. N. Taylor, *Org. Synth.* **52,** 16 (1972).
29. S. R. Sandler, unpublished results (1961).
30. Du Pont Information Bulletin, Philadelphia, Pennsylvania 19124. (1955).
31. A. W. Ralston, E. W. Segerbrecht, and S. T. Bauer, *J. Org. Chem.* **4,** 503 (1939).
32. E. O. Ott, A. Langenohl, and W. Zerweck, *Chem. Ber.* **70B,** 260 (1937). These authors report the following physical constants: m.p. 64, and b.p. 205 (12 mm Hg.). The acid chloride was prepared by the action of PCl_5 on the dithiodianhydride.
33. W. E. Truce, *J. Am. Chem. Soc.* **70,** 2828 (1948).
34. L. P. Kyrides, *J. Am. Chem. Soc.* **59,** 206 (1937).
35. W. H. Miller, A. M. Dessert, and G. W. Anderson, *J. Am. Chem. Soc.* **70,** 502 (1948).
36. L. P. Kyrides, *Org. Synth.* **20,** 51 (1940).
37. R. Adams and L. H. Ulich, *J. Am. Chem. Soc.* **42,** 599 (1920).
38. B. I. Daubert, H. H. Fricke, and H. E. Longenecker, *J. Am. Chem. Soc.* **65,** 2143 (1943).
39. M. E. Neubert and D. L. Fishel, *Org. Synth.* **55,** 2095 (1976).
40. W. J. Middleton, *J. Org. Chem.* **44,** 2291 (1979).
41. L. R. Belohlav and E. T. McBee, *Ind. Eng. Chem.* **52,** 1022 (1960).
42. A. E. Feiring, *J. Org. Chem.* **44,** 1252 (1979).
43. L. I. Smith and C. L. Moyle, *J. Am. Chem. Soc.* **58,** 1 (1936).
44. L. I. Smith and C. L. Moyle, *J. Am. Chem. Soc.* **55,** 1676 (1933).
45. R. Adams and J. R. Johnson, "Laboratory Experiments in Organic Chemistry," pp. 289–291. Macmillan, New York, 1949.
46. L. I. Smith, *Org. Synth. Collect. Vol.* **2,** 95 (1943).
47. J. R. Sampey, F. S. Fawcett, and B. A. Morehead, *J. Am. Chem. Soc.* **62,** 1839 (1940).
48. E. F. J. Atkinson and J. F. Thorpe, *J. Chem. Soc.* p. 1695 (1907).
49. E. C. Juonge, D. A. Beal, and W. P. Duncan, *J. Org. Chem.* **35,** 719 (1970).
50. H. T. Clarke and M. R. Brethen, *Org. Synth. Collect. Vol.* **1,** 121 (1941).
51. J. P. Wibaut, F. L. J. Sixma, and J. F. Suyver, *Recl. Trav. Chim. Pays-Bas* **68,** 525, 915 (1949).
52. C. O. Obenland, *J. Chem. Educ.* **41,** 566 (1964).
53. W. Militzer, *J. Am. Chem. Soc.* **60,** 256 (1938).
54. C. Djerassi, *Chem. Rev.* **43,** 271 (1948).
55. Buu-Hoï, *Justus Liebigs Ann. Chem.* **556,** 6 (1944).
56. L. Horner and E. H. Windelmann, *Angew. Chem.* **71,** 349 (1959).
57. G. J. Fox, G. Hallas, J. D. Hopworth, and K. N. Paskins, *Org. Synth.* **55,** 20 (1976).
58. F. B. Davis and R. Q. Brewster, *Org. Synth. Collect. Vol.* **1,** 323 (1941).
59. R. B. Sandin, W. V. Drake, and F. Leger, *Org. Synth. Collect. Vol.* **2,** 196 (1943).
60. D. H. Derbyshire and W. A. Waters, *J. Chem. Soc.* p. 3694 (1950).
61. M. S. Kharasch and H. C. Brown, *J. Am. Chem. Soc.* **61,** 2143 (1939).
62. H. P. A. Groll and G. W. Hearne, *Ind. Eng. Chem.* **31,** 1239, 1413, 1530 (1939).
63. P. G. Stevens, *J. Am. Chem. Soc.* **68,** 620 (1946).
64. H. B. Hass and H. C. Huffman, *J. Am. Chem. Soc.* **63,** 1233 (1941).
65. D. C. Sayles and E. F. Degering, *J. Am. Chem. Soc.* **71,** 3161 (1949).
66. G. F. Freeguard and L. H. Long, *Chem. Ind. (London)* p. 1582 (1964).
67. P. Chabrier, J. Seyden-Penne, and A. M. Fonace, *C.R. Hebd. Seances Acad. Sci.* **245,** 174 (1957).
68. H. Suzuki, *Org. Synth.* **51,** 94 (1971).
69. G. E. Hall and F. M. Ubertini, *J. Org. Chem.* **15,** 715 (1950).

70. W. Autenrieth and P. Muhlinghaus, *Chem. Ber.* **39,** 4098 (1906).
71. B. H. Shoemaker and C. E. Boord, *J. Am. Chem. Soc.* **53,** 1505 (1931).
72. L. C. Swallen and C. E. Boord, *J. Am. Chem. Soc.* **52,** 651 (1930).
73. J. R. Johnson and W. L. McEwen, *Org. Synth. Collect. Vol.* **1,** 521 (1941).
74. P. A. Rishood and D. M. Rathven, *J. Am. Chem. Soc.* **100,** 4919 (1978).
75. R. M. Dessau, *J. Am. Chem. Soc.* **101,** 1344 (1979).
76. H. J. Lucas and C. W. Gould, Jr., *J. Am. Chem. Soc.* **63,** 2541 (1941).
77. M. S. Kharasch and H. C. Brown, *J. Am. Chem. Soc.* **61,** 3432 (1939).
78. L. Spiegler and J. H. Tinker, *J. Am. Chem. Soc.* **61,** 940 (1939).
79. J. San Filippo, Jr., A. F. Sowinski, and L. J. Romano, *J. Am. Chem. Soc.* **97,** 1599 (1975).
80. P. P. Nicholas and R. T. Carroll, *J. Org. Chem.* **33,** 2345 (1968).
81. G. S. Skinner, G. Limperos, and R. H. Pettibone, *J. Am. Chem. Soc.* **72,** 1648 (1950).
82. R. F. Taylor and G. H. Morey, *Ind. Eng. Chem.* **40,** 432 (1948).
83. R. E. Buckles and J. L. Forrester, *J. Org. Chem.* **25,** 24 (1960).
84. R. F. Merritt and F. A. Johnson, *J. Org. Chem.* **31,** 1859 (1966).
85. H. R. Snyder and L. A. Brooks, *Org. Synth. Collect. Vol.* **1,** 171 (1943).
86. C. Djerassi, *Chem. Rev.* **43,** 271 (1948).
87. P. Karrer and W. Ringli, *Helv. Chim. Acta* **30,** 863, 1771 (1947).
88. J. C. Day, M. J. Lindstrom, and P. S. Skell, *J. Am. Chem. Soc.* **96,** 5616 (1974).
89. D. R. Dalton, V. P. Dutta, and D. C. Jones, *J. Am. Chem. Soc.* **90,** 5498 (1968).
90. K. Ziegler, A. Spaeth, E. Schoff, W. Schumann, and E. Winkelmann, *Justus Liebigs Ann. Chem.* **551,** 80 (1942).
91. F. R. Mayo and C. Walling, *Chem. Rev.* **27,** 351 (1940).
92. M. S. Kharasch and M. Sage, *J. Org. Chem.* **14,** 537 (1949).
92a. C. L. Osborn, T. V. Van Auken, and D. J. Trecker, *J. Am. Chem. Soc.* **90,** 5806 (1968).
92b. E. Tobler and D. J. Foster, *J. Org. Chem.* **29,** 2839 (1964).
92c. J. G. Traynham and T. M. Couvillon, *J. Am. Chem. Soc.* **89,** 3205 (1967).
92d. Y. Fujita, Y. Omura, F. Mori, K. Itoi, T. Nishida, Y. Tamai, S. Aihara, T. Hosogai, and F. Wada, U.S. Patent 4,053,380 (1977).
93. E. Smith, U.S. Patent 3,399,241 (1968).
93a. M. S. Kharasch, E. V. Jensen, and W. H. Urry, *J. Am. Chem. Soc.* **69,** 1100 (1947).
94. L. Schmerling, *J. Am. Chem. Soc.* **67,** 1152 (1945).
95. W. S. Emerson and V. E. Lucas, *J. Am. Chem. Soc.* **70,** 1180 (1948).
96. W. S. Emerson, *J. Am. Chem. Soc.* **67,** 516 (1945).
97. W. E. Hanby and H. N. Rydon, *J. Chem. Soc.* p. 1141 (1946).
98. P. S. Skell and A. Y. Garner, *J. Am. Chem. Soc.* **78,** 5430 (1956).
99. S. R. Sandler, *J. Org. Chem.* **32,** 3876 (1967).
100. R. A. Mitsch, *J. Am. Chem. Soc.* **87,** 758 (1965).
101. P. S. Skell and S. R. Sandler, *J. Am. Chem. Soc.* **80,** 2024 (1958).
101a. D. C. Duffey, R. C. Gueldner, B. R. Layton, and J. P. Minyard, Jr., *J. Org. Chem.* **42,** 1082 (1977).
102. S. R. Sandler, *Org. Synth.* **56,** 32 (1977).
103. J. Docky, *Synthesis* p. 441 (1973); E. V. Dehmlow, *Angew. Chem., Int. Ed. Engl.* **13,** 170 (1974); I. Crossland, *Org. Synth.* **59,** 2142 (1979); E. V. Dehmlow, *Chem. Ted.* April, p. 210 (1975).
104. J. A. Pincock and K. Yaks, *J. Am. Chem. Soc.* **90,** 5643 (1968).
104a. D. Seyferth, S. P. Hopper, and T. F. Jula, *J. Organomet. Chem.* **17,** 193 (1969).
105. P. A. Levine, *Org. Synth. Collect. Vol.* **2,** 88 (1943).
106. Authors' laboratory (1963).
107. F. Weygand and V. Schmid-Kowarzik, *Chem. Ber.* **82,** 333 (1949).

108. G. A. Harrington, U.S. Patent 2,462,389 (1946); G. A. Weizmann, E. Bergmann, and M. Sulzbacher, *J. Am. Chem. Soc.* **70,** 1189 (1948).
109. M. Gaudry and A. Marquet, *Org. Synth.* **55,** 24 (1976).
110. E. R. Buchman and H. Sargent, *J. Am. Chem. Soc.* **67,** 400 (1945).
111. H. Rabjohn and E. R. Rogers, *J. Org. Chem.* **11,** 781 (1946).
112. R. M. Cowper and L. H. Davidson, *Org. Synth. Collect. Vol.* **2,** 480 (1943).
113. L. C. King and G. K. Ostrum, *J. Org. Chem.* **29,** 3459 (1964).
114. E. M. Kosower, W. J. Cole, G. S. Wu, D. E. Lardy, and G. Meisters, *J. Org. Chem.* **28,** 630 (1963).
115. T. L. Jacobs, *Org. React.* **5,** 20 (1949).
116. R. Adams and C. W. Theobold, *J. Am. Chem. Soc.* **65,** 2208 (1943).
117. G. Braun, *Org. Synth.* **14,** 42 (1934).
118. H. Erlenmeyer and J. P. Jung, *Helv. Chim. Acta* **32,** 37 (1949).
119. H. T. Clarke and E. R. Taylor, *Org. Synth. Collect. Vol.* **1,** 155 (1943).
120. S. Natelson and S. P. Gottfried, *Org. Synth.* **23,** 37 (1943).
121. B. C. Saunders and G. J. Stacey, *J. Chem. Soc.* p. 1773 (1948).
122. J. M. Zanden, *Recl. Trav. Chim. Pays-Bas* **63,** 113 (1944).
123. D. N. Harpp and L. Q. Bao, *Org. Synth.* **55,** 27 (1976).
124. Y. Ogata, T. Sugimoto, and M. Inaishi, *Org. Synth.* **59,** 20 (1979).
125. C. F. Ward, *J. Chem. Soc.* p. 1164 (1922).
126. W. R. Hasek, W. C. Smith, and V. A. Engelhardt, *J. Am. Chem. Soc.* **82,** 543 (1960).
127. W. C. Smith, U.S. Patent 2,859,245 (1958) (to Du Pont).
128. J. F. Bunnett and R. M. Conner, *J. Org. Chem.* **23,** 305 (1958).
129. J. T. Maynard, *J. Org. Chem.* **28,** 112 (1963).
130. E. D. Bergmann, M. Bentov, and A. Levy, *J. Chem. Soc.* p. 1194 (1964).
131. H. H. Hodgson, *Chem. Rev.* **40,** 251 (1947).
132. C. S. Marvel and S. M. McElvain, *Org. Synth. Collect. Vol.* **1,** 170 (1941).
133. W. E. Bachmann and C. H. Boatner, *J. Am. Chem. Soc.* **58,** 2194 (1936).
134. H. J. Lucas and E. R. Kennedy, *Org. Synth. Collect. Vol.* **2,** 351 (1943).
135. J. K. Kochi, *J. Am. Chem. Soc.* **79,** 2946 (1957).
136. A. Roe, *Org. React.* **5,** 193 (1949).
137. A. Roe and G. F. Hawkins, *J. Am. Chem. Soc.* **71,** 1785 (1949).
138. M. S. Newman and R. H. B. Galt, *J. Org. Chem.* **25,** 214 (1960).
139. E. C. Kleiderer and R. Adams, *J. Am. Chem. Soc.* **53,** 1579 (1931).
140. J. R. Catch, D. F. Elliot, D. H. Hey, and E. R. H. Jones, *J. Chem. Soc.* p. 278 (1948).
141. W. D. McPhee and E. Klingsberg, *Org. Synth.* **26,** 13 (1946).
142. H. King and T. S. Work, *J. Chem. Soc.* p. 1307 (1940).
143. W. E. Bachmann and W. S. Struve, *Org. React.* **1,** 47 (1942).
144. B. Gredy, *Bull. Soc. Chim. Fr.* [5] **3,** 1094 (1936).
145. H. Stone and H. Schechter, *Org. Synth.* **31,** 66 (1951).
146. S. Fried and R. D. Kleene, *J. Am. Chem. Soc.* **63,** 2691 (1941).
147. J. B. Cloke and F. J. Pilgrim, *J. Am. Chem. Soc.* **61,** 2667 (1939).
148. C. Hunsdiecker, H. Hunsdiecker, and E. Vogt, U.S. Patent 2,176,181 (1939).
149. H. Hunsdiecker and C. Hunsdiecker, *Chem. Ber.* **75,** 291 (1942).
150. C. V. Wilson, *Org. React.* **9,** 332 (1957).
151. H. Kleinberg, *Chem. Rev.* **40,** 381 (1947).
152. W. G. Dauben and H. Tilles, *J. Am. Chem. Soc.* **72,** 3185 (1950).
153. C. F. H. Allen and C. V. Wilson, *Org. Synth.* **26,** 52 (1946).
154. M. Hauptschien and A. V. Grosse, *J. Am. Chem. Soc.* **73,** 2461 (1951).
155. R. A. Barnes and R. J. Prochaska, *J. Am. Chem. Soc.* **72,** 3188 (1950).

References

156. J. A. Davis, J. Herynk, S. Carroll, J. Bunds, and D. Johnson, *J. Org. Chem.* **30,** 415 (1965).
157. A. McKillop and D. Bromley, *J. Org. Chem.* **34,** 1172 (1969).
157a. G. M. Lampman and J. C. Aumiller, *Org. Synth.* **51,** 107 (1971).
158. H. T. Clarke and E. R. Taylor, *Org. Synth. Collect. Vol.* **1,** 155 (1941).
159. F. Flages, *Justus Liebigs Ann. Chem.* **547,** 25 (1941).
160. C. R. Hauser and W. B. Renfrow, *J. Am. Chem. Soc.* **59,** 122 (1937).
161. S. Winstein, E. Grunwald, R. E. Buckles, and C. Hanson, *J. Am. Chem. Soc.* **70,** 816 (1948).
162. H. W. Dougherty and G. T. Derge, *J. Am. Chem. Soc.* **53,** 1594 (1931).
163. W. Ziegenbein and W. Franke, *Angew. Chem.* **71,** 573 (1959).
164. D. A. Ransch, R. A. Davis, and D. W. Osborne, *J. Org. Chem.* **23,** 494 (1963).
165. R. F. Merritt, *J. Org. Chem.* **31,** 3871 (1966).
166. R. A. Davis and E. R. Larsen, *J. Org. Chem.* **32,** 3478 (1967).
167. B. M. Trost and D. R. Brittelli, *J. Org. Chem.* **32,** 2620 (1967).
168. R. G. Weiss and E. I. Snyder, *Abstr. 3rd Middle Atl. Reg. Am. Chem. Soc. Meet.* p. 78H (1968).
169. G. A. Wiley, R. L. Hershkowitz, B. M. Rein, and B. C. Chung, *J. Am. Chem. Soc.* **86,** 964 (1964).
170. J. P. Schaefer and D. S. Weinberg, *J. Org. Chem.* **30,** 2635 (1965).
171. B. E. Grimwood and D. Swern, *J. Org. Chem.* **32,** 3665 (1967).
172. H. Taguchi, H. Yamamoto, and H. Nozaki, *Org. Synth.* **59,** 2033 (1979).
173. L. F. Fieser and D. H. Sachs, *J. Org. Chem.* **29,** 1113 (1964).
174. R. A. Moss and D. J. Smudin, *J. Org. Chem.* **41,** 611 (1976).
175. T. A. Marolewski and N. C. Yang, *Org. Synth.* **52,** 132 (1972).
176. J. P. Oliver, U. V. Rao, and M. T. Emerson, *Tetrahedron Lett.* p. 3419 (1964).
177. R. Pettit and J. Henery, *Org. Synth.* **50,** 36 (1970).
178. F. A. Grindahl, W. X. Bajzer, and O. R. Pierce, *J. Org. Chem.* **32,** 603 (1967).
179. T. Susuki and J. Tsuji, *J. Org. Chem.* **35,** 2982 (1970).
180. J. Kagan, B. E. Firth, N. Y. Shih, and C. G. Boyajian, *J. Org. Chem.* **42,** 343 (1977).
181. A. S. Erickson and N. Kornblum, *J. Org. Chem.* **42,** 3764 (1977).
182. R. Lamartine and R. Perrin, *J. Org. Chem.* **39,** 1744 (1974).
183. J. San Filippo, Jr., A. F. Sowinski, and L. J. Romano, *J. Org. Chem.* **40,** 3295 (1975).
184. Z. Zweig and M. Epstein, *J. Org. Chem.* **43,** 3690 (1978).
185. M. S. Newman and P. K. Sujeethy, *J. Org. Chem.* **43,** 4367 (1978).
186. R. M. Magid, O. S. Fruchey, W. L. Johnson, and T. G. Allen, *J. Org. Chem.* **44,** 359 (1979).
187. R. H. Mitchell, Y.-H. Lai, and R. V. Williams, *J. Org. Chem.* **44,** 4733 (1979).
188. W. C. Baird, Jr., J. H. Sarridge, and M. Buza, *J. Org. Chem.* **36,** 2088 (1971).
189. M. G. Reinecke, H. W. Adickes, and C. Pyun, *J. Org. Chem.* **36,** 2690 (1970).
190. V. L. Heasley, P. D. Davis, D. M. Ingle, K. D. Rold, and O. E. Hensly, *J. Org. Chem.* **39,** 736 (1974).
191. T. Hamaoka and H. C. Brown, *J. Org. Chem.* **40,** 1189 (1975).
192. G. H. Hakimelahi and G. Just, *Tetrahedron Lett.* **38,** 3643 (1979).
193. J. Jacques and A. Margret, *Org. Synth* **53,** 111 (1973).
194. M. R. Rifi, *Org. Synth.* **52,** 22 (1972).
195. J. E. Oliver and P. E. Sonnet, *Org. Synth.* **58,** 64 (1978).
196. R. Rabinowitz, U.S. Patent 3,225,106 (1965).
197. L. Schmerling, U.S. Patent 3,484,482 (1969).
198. S. Matsunami, J. Kita, and S. Fujimura, Ger. Offen. 2,841,069 (1979); *Chem. Abstr.* **91,** 39154g (1979).

CHAPTER 7 / ALDEHYDES

1. Introduction 180
2. Oxidation Reactions 181
   A. Oxidation of Primary Alcohols 181
   2-1. Preparation of Propionaldehyde 182
   2-2. Preparation of Acrolein by Oxidation of Allyl Alcohol with Manganese Dioxide ... 182
   2-3. Preparation of 2-(N,N-Dimethylamino)-5-methylbenzaldehyde by a Modified Oppenauer Oxidation ... 183
   B. Oxidation of Glycols 184
   2-4. Preparation of Adipaldehyde Using Lead Tetraacetate 184
   C. Oxidiation of Olefins and Acetylenes 185
   2-5. Preparation of n-Octaldehyde from 1-Octyne 185
   D. Oxidation of Alkyl Groups 186
   2-6. Preparation of Heptaldehyde by the Dimethyl Sulfoxide Oxidation of Heptyl Tosylate ... 186
   2-7. Preparation of 3-Thiophenealdehyde by the Sommelet Reaction ... 187
3. Reduction Reactions 187
   A. Reduction of Nitriles 187
   3-1 Preparation of n-Octaldehyde by the Stephen Method 188
   3-2. Preparation of Pivalaldehyde by the Reduction of Trimethylacetonitrile with Lithium Triethoxyaluminum Hydride ... 189
   B. Reduction of Other Acid Derivatives 189
   3-3. Preparation of Cyclopropanecarboxaldehyde by the $LiAlH_4$ Reduction of 1-Acylaziridine ... 190
   C. Miscellaneous Reducing Agents 191
4. Condensation Reactions 191
   4-1. Preparation of Resorcylaldehyde by the Modified Gattermann Synthesis ... 192
   4-2. Preparation of Tolualdehyde by the Formylation of Toluene with Formyl Fluoride ... 193
   4-3. Preparation of 3-Formylindole by the Vilsmeier Method 194
   A. Miscellaneous Condensation Methods 195
5. Elimination Reactions 196
   5-1. Preparation of Hydrotropaldehyde by the Darzens Reaction 197
   A. Miscellaneous Elimination Reactions 198
   5-2. Preparation of Benzaldehyde by the McFadyen–Stevens Reaction ... 198
6. Miscellaneous Reactions 199
   References 201

1. INTRODUCTION

The laboratory preparations of aldehydes involve oxidation, reduction, condensation, and elimination reactions.

The oxidation of primary alcohols using chromic acid, chromic oxide–pyridine complex, manganese dioxide, etc., gives aldehydes in good yields. Allylic alcohols are readily oxidized at room temperature with active manganese dioxide.

Olefins, alkyl groups (Étard reaction), and alkyl halides (Sommelet reaction) may also be oxidized to aldehydes. Acetylenes can be converted to trivinylboranes which in turn are easily oxidized to aldehydes.

Several reduction methods are available to convert nitriles (Stephen reaction) or acyl chlorides (Rosenmund reduction) or amides to aldehydes. The use of lithium aluminum hydride at low temperatures or lithium triethoxyaluminum hydride appears to be a more effective means of reduction than the stannous chloride–HCl used in the Stephen method or palladium and hydrogen used in the Rosenmund reduction.

The Gattermann condensation raction and its modifications, the Vilsmeier reaction as well as the use of modified Friedel–Crafts and Grignard reactions, offer useful methods for the laboratory synthesis of aromatic and heterocyclic aldehydes.

The Darzens reaction and the McFadyen–Stevens reaction are also useful aldehyde syntheses involving elimination reactions.

Since aldehydes are usually highly reactive compounds, these preparations should be carried out under conditions that will reduce product losses by further reactions (such as oxidation to carboxylic acids).

Recently, the use of phase transfer catalysts has facilitated the oxidation of alcohols and primary alkyl halides to aldehydes.

Benzal halides are hydrolyzed to aldehydes via water and an iron catalyst. Aromatic aldehydes are also obtained by the partial oxidation of an aromatic side chain using chromic acid and acetic anhydride followed by hydrolysis of the diacetate.

Industrially, olefins are converted to aldehydes via the OXO process (olefin, carbon monoxide, hydrogen, and a catalyst). In addition, olefins can be oxidized to give the aldehyde (propylene→acrolein).

2. OXIDATION REACTIONS

A. Oxidation of Primary Alcohols

Primary alcohols are oxidized in a controlled manner with a mixture of sulfuric and chromic acids to give aldehydes. Unless the aldehyde is distilled from the reaction mixture, poor yields are obtained due to substantial oxidation to carboxylic acids. For low molecular weight alcohols [1a] or benzyl alcohols [1b] an aqueous medium is suitable. Aqueous acetic acid is used as a solvent for the oxidation of arylalkyl alcohols [2]. Olefinic aldehydes are also prepared by oxidation at 5°–20°C in good yields [3]. Active manganese dioxide is used to oxidize allylic alcohols to aldehydes in good yields at room temperature [4, 5]. Chromic oxide in sulfuric acid has been used to oxidize propargyl alcohol to propiolaldehyde in 35–41% yield [6a].

2-1. Preparation of Propionaldehyde [6b]

$$CH_3CH_2CH_2OH + K_2Cr_2O_7 + H_2SO_4 \longrightarrow CH_3CH_2CH=O \qquad (1)$$

To a well-stirred mixture of 100 gm (1.66 mole) of boiling isopropanol is added dropwise a solution of 164 gm (0.55 mole) of potassium dichromate and 120 ml of concentrated sulfuric acid in 1 liter of water. The addition takes 30 min and the aldehyde distills off as it is formed. The propionaldehyde is redistilled to yield 44–47 gm (45–49%), bp 48°–55°C, n_D^{20} 1.3636.

Ratcliffe [6c] has reported that 1-decanol is oxidized to 1-decanal in 63–66% yield, using dipyridine–chromium(VI) oxide prepared directly in methylene chloride at 20°C. This procedure has the advantage over the earlier reported procedure by Sarett and co-workers [6d] and Collins [7] in that pure dipyridine-chomium(VI) oxide does not have to be separately prepared. The latter is troublesome since it is hygroscopic and has a propensity to enflame during preparations.

2-2. Preparation of Acrolein by Oxidation of Allyl Alcohol with Manganese Dioxide [8]

$$CH_2=CH-CH_2OH + MnO_2 \longrightarrow CH_2=CH-CHO \qquad (2)$$

To a flask containing 2 gm (0.034 mole) of allyl alcohol is added a suspension of 20 gm (0.219 mole) of commercial manganese dioxide in 100 ml of petroleum ether (bp 40°–60°C). The mixture is stirred for ½ hr. The mixture is quickly filtered with suction and the precipitate is washed with fresh solvent. The combined filtrates are concentrated and distilled to afford a 60% yield of acrolein, isolated as the 2,4-dinitrophenylhydrazone, 4.7 gm, mp 165°C. Similarly cinnamyl alcohol is converted to cinnamaldehyde in 75% yield.

Active manganese dioxide gives better results. This reagent may be prepared from manganese sulfate. Manganese sulfate, 111.0 gm (0.482 mole, tetrahydrate) in 1.5 liters of water, and 1170 ml of a 40% sodium hydroxide solution are added simultaneously over a period of 1 hr to a hot stirred solution of 960 gm (5.91 mole) of potassium permanganate in 6 liters of water. After 1 hr the precipitate of brown manganese dioxide is filtered, washed with water until the wash water is colorless, dried at 100°–120°C, and ground into a fine powder to give 920 gm.

§ 2. Oxidation Reactions

The 2,4-dinitrophenylhydrazine (2,4-DNPH) solution (Brady's reagent) is prepared from 8 gm of 2,4-DNPH, 16 ml of concentrated H_2SO_4, and 200 ml of methanol.

Aldehydes and particularly unsaturated aldehydes are readily oxidized by air and therefore they should be stored in the presence of a protective nitrogen blanket in tightly sealed containers.

The Oppenauer reaction [9] is used for the conversion of aliphatic and aromatic alcohols to aldehydes. High-boiling aldehydes, aluminum alkoxide, and the alcohol of interest are mixed. The alcohol yields a volatile aldehyde which is removed. Examples of alcohols subjected to the reaction are benzyl alcohol and *n*-butyl alcohol, which give the aldehydes in good yields [10, 11]. A modified Oppenauer oxidation has been reported in which the aluminum alkoxide is replaced by the base potassium *tert*-butoxide, and the reaction is carried out in the presence of benzoquinone or benzophenone as hydrogen acceptors [12]. The modified procedure is especially useful for compounds containing basic nitrogen atoms, as in the alkaloid series.

Recently it has been reported that aryl carbinols are oxidized in good yield in dimethyl sulfoxide and air to the aldehydes. This reaction is also applicable to allylic and aliphatic alcohols [13a,b]. Carbodiimides in dimethyl sulfoxide–phosphoric acid have also been reported to be effective oxidants [14a,b,c].

Other oxidizing agents such as cupric acetate [15], selenium dioxide [16, 17], nitric acid [18], and nitrogen tetroxide [19a] have been used in more specialized preparations.

Corey and Kim reported that dimethyl sulfide and chlorine complexes or dimethyl sulfide and *N*-chlorosuccinimide complexes oxidize primary alcohols at low temperature to aldehydes in good yields [19b].

2-3. Preparation of 2-(N,N-Dimethylamino)-5-methylbenzaldehyde by a Modifed Oppenauer Oxidation [20]

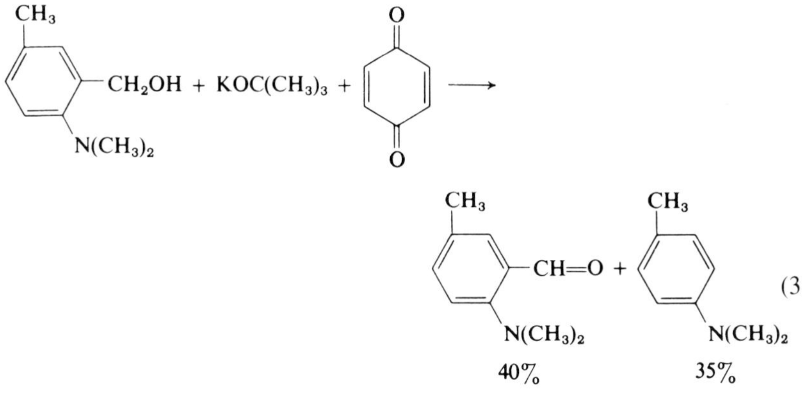

To a flask containing 400 ml of *tert*-butanol is added 20 gm (0.51 gm atoms) of potassium and, after completely reacting, the excess of *tert*-butanol is removed under reduced pressure. To the potassium *tert*-butoxide is then added 2 liters of dry benzene, 260 gm (2.4 mole) of benzophenone, and 33 gm (0.20 mole) of 2-(*N*,*N*-dimethylamino)-5-methylbenzyl alcohol. The mixture is refluxed for 23 hr under a nitrogen atmosphere, extracted with a solution of hydrochloric acid (150 ml concentrated acid and 500 ml of water), and the extract is made alkaline. The latter is extracted three times with 100 ml each of ether and the combined ether extracts are concentrated under reduced pressure to give a residue which, upon distillation, gives 9.5 gm (35%) of *N*,*N*-dimethyl-*p*-toluidine, bp 90°–97°C (12–15 mm) and 14.0 gm (40%) of the yellow aldehyde, bp 138°–142°C (16 mm).

B. Oxidation of Glycols

Peracetic acid [21] and lead tetraacetate [22] are useful where glycols are involved. Oxidative cleavage to two aldehydes occurs and the reaction is useful both as a preparative and analytical method. The method of Grundmann of reducing an acid chloride [23, 24a] involves as its final step the lead tetraacetate cleavage of a glycol to an aldehyde containing the same number of carbon atoms as in the starting material.

$$\text{RCOCl} \xrightarrow{\text{CH}_2\text{N}_2} \text{RCOCHN}_2 \xrightarrow{\text{CH}_3\text{COOH}} \text{RCOCH}_2\text{OCOCH}_3 \xrightarrow{\text{(H)}}$$

$$\underset{\underset{\text{OH}}{|}\underset{\text{OH}}{|}}{\text{RCH}-\text{CH}_2} \xrightarrow{\text{Pb(OAc)}_4} \text{RCH}=\text{O} + \text{CH}_2=\text{O} \qquad (4)$$

2-4. *Preparation of Adipaldehyde Using Lead Tetraacetate* [22]

$$\text{cyclohexane-1,2-diol} + \text{Pb(OAc)}_4 \longrightarrow \text{cyclohexane with two CH=O groups} \qquad (5)$$

To a flask containing 20 gm (0.17 mole) of 1,2-cyclohexanediol dissolved in 200 ml of dry benzene is added 50 gm of anhydrous potassium carbonate. While the flask is vigorously stirred, 76 gm (0.1 mole) of lead tetraacetate is added in 5 gm portions over a period of 1 hr. A nitrogen atmosphere is maintained throughout this preparation. The mixture is stirred an additional hour, filtered, and the salts extracted with benzene. The combined filtrates are dried, concentrated, and the residue is distilled under reduced pressure to yield 13.4 gm (68%) of colorless adipaldehyde, bp 68°–70°C (3 mm), n_D^{20} 1,4350.

The cleavage of di-*n*-butyl tartrate by lead tetraacetate to *n*-butyl glyoxalate in 77–87% yield has recently been described [24b].

§ 2. Oxidation Reactions

The condensation of chloral with certain hydrocarbons gives a trichloromethyl carbinol which can be oxidized to the aldehyde [25].

$$\text{ArCH}_3 + \text{Cl}_3\text{C}-\text{CH}=\text{O} \xrightarrow{\text{OH}^-} \text{ArCH}_2-\underset{\underset{\text{OH}}{|}}{\text{CHCCl}_3} \xrightarrow{\text{(O)}} \text{ArCH}_2-\text{CH}=\text{O} \quad (6)$$

C. Oxidation of Olefins and Acetylenes

Olefins are readily ozonized and, upon decomposition of the ozonides, aldehydes are obtained [26]. Cyclic olefins give dialdehydes when they are ozonized [27].

Ozonolysis of vinylpyridine and its derivatives gives pyridine aldehydes [28]. Oxidation of vinyl sulfides with oxygen gives aldehydes in fair yields [29].

Reaction of olefins or terminal acetylenes with sodium borohydride and boron trifluoride yields boranes, which, upon oxidation with alkaline hydrogen peroxide, yield aldehydes in good yields [30, 31a].

2-5. Preparation of n-Octaldehyde from 1-Octyne [30]

$$\text{CH}_3-\text{CH}=\underset{\underset{\text{CH}_3}{|}}{\text{C}}-\text{CH}_3 + \text{NaBH}_4 + \text{BF}_3 \longrightarrow (\text{CH}_3\text{CH}_2-\underset{\underset{\text{H}}{|}}{\overset{\overset{\text{CH}_3}{|}}{\text{C}}}-\text{CH}_3)_3\text{B} \quad (7)$$

$$(\text{CH}_3-\text{CH}_2-\underset{\underset{\text{H}}{|}}{\overset{\overset{\text{CH}_3}{|}}{\text{C}}}-\text{CH}_3)_3\text{B} + 3\,\text{C}_6\text{H}_{13}\text{C}\equiv\text{CH} \longrightarrow (\text{C}_6\text{H}_{13}\text{CH}=\text{CH}-)_3\text{B} \xrightarrow[\text{NaOH}]{\text{H}_2\text{O}_2}$$

$$[\text{C}_6\text{H}_{13}-\text{CH}=\text{CHOH}] \longrightarrow \text{C}_6\text{H}_{13}\text{CH}_2\text{CH}=\text{O} \quad (8)$$

A three-necked flask containing 33.6 gm (0.48 mole) of 2-methyl-2-butene in 180 ml of a 1.00 M solution of sodium borohydride in diglyme is placed in an ice bath while 34.1 gm (0.24 mole) of boron trifluoride etherate is added dropwise to the reaction mixture. The flask is allowed to remain at 0°C for 2 hr and then it is placed in an ice–salt water bath. To the disiamylborane in the flask is added at a rapid rate 22.0 gm (0.200 mole) of 1-octyne in 20 ml of diglyme while maintaining the reaction temperature below 10°C. The reaction mixture is warmed to room temperature and then oxidized at 0°C with 150 ml of a 15% solution of hydrogen peroxide while maintaining the pH at 7–8 with 3 N sodium hydroxide.

After the oxidation, the reaction mixture is neutralized and steam-distilled. The distillate is ether-extracted, dried, and distilled to yield 18 gm (70%) of n-octaldehyde, bp 83°–85°C (33 mm), n_D^{20} 1.4217.

Recently the use of chromyl chloride solutions in methylene chloride have been reported to oxidize terminal olefins to aldehydes in good yield [31b]. For example 2,4,4-trimethyl-1-pentene gives 2,4,4-trimethylpentanal in 70–78% yield (distilled).

D. Oxidation of Alkyl Groups

Aromatic alkyl groups are readily oxidized by chromium trioxide in acetic anhydride to crystalline diacetates which are hydrolyzed in fair yield to the corresponding aldehyde [32, 33]. The diacetates are stable toward further oxidation. Nitro, halo, and cyano groups do not interfere. o-Nitrotoluene is oxidized in 74% yield to o-nitrobenzaldehyde [33].

Manganese dioxide in 65% sulfuric acid is an effective oxidant [34].

The Étard reaction employing chromyl chloride in chloroform gives variable yields and the reagent should be handled carefully [35]. Fast removal of the aldehyde by solvent extraction is required in order to avoid further oxidation to the acid. CAUTION: Chromyl chloride is highly carcinogenic.

Selenium dioxide [36] or copper nitrate [37] have been used to oxidize benzyl halides directly. Selenium dioxide can also oxidize methyl ketones to glyoxals [38].

Oxidation of haloalkyl groups can be effectively carried out at room temperature in dimethyl sulfoxide to give aldehydes in good yields [39–41]. The reaction has recently been applied to the heterocyclic series [40].

The Sommelet reaction [42] of benzyl halides involves the reaction of hexamethylenetetramine at the boiling point of 60% ethanol or 50% acetic acid [43] to yield benzaldehyde in good yields. Steric hindrance of the chloromethyl group causes the reaction to fail. The reaction is also applicable to heterocycles such as 2-chloromethylthiophene to yield 2-thiophenealdehyde in 48–53% yield [44]. Recently 3-thiophenaldehyde has been prepared in 54–72% yield using the Sommelet reaction [45].

2-6. Preparation of Heptaldehyde by the Dimethyl Sulfoxide Oxidation of Heptyl Tosylate* [41]

$$CH_3(CH_2)_6I + AgOSO_2-\!\!\left\langle\!\!\bigcirc\!\!\right\rangle\!\!-CH_3 \longrightarrow$$

$$CH_3(CH_2)_6-OSO_2-\!\!\left\langle\!\!\bigcirc\!\!\right\rangle\!\!-CH_3 \xrightarrow[\substack{NaHCO_3 \\ 150°C,\ 3\ min}]{DMSO} CH_3(CH_2)_5-CH\!\!=\!\!O\ (70\%) \quad (9)$$

To a flask, protected from light, containing 11 gm (0.0394 mole) of silver tosylate in 100 ml of acetonitrile at 0°–5°C is added 7.0 gm (0.0309 mole) of 1-iodoheptane. The product is allowed to come to room temperature overnight, added to ice water, and then extracted with ether. The dried ether solution is

*Can also be prepared from heptyl alcohol and p-toluene sulfonyl chloride.

concentrated under reduced pressure to yield an oily residue which is added to a flask containing 150 ml of dimethyl sulfoxide and 20 gm of sodium bicarbonate at 150°C. Nitrogen is bubbled through the mixture. After 3 min at 150°C the reaction mixture is cooled rapidly to room temperature and the product, 6.9 gm (70%), is isolated as the 2,4-dinitrophenylhydrazine, mp 106°–107°C. For benzyl halides the same procedure is used except that the tosylate is heated for 5 min at 100°C in the dimethyl sulfoxide–sodium bicarbonate mixture. For example, *p*-methylbenzyl bromide gives a 65% yield of *p*-tolualdehyde.

2-7. Preparation of 3-Thiophenealdehyde by the Sommelet Reaction [42]

$$\underset{S}{\bigcirc}\!\!-\!CH_2Br \xrightarrow[\text{2. Steam distillation}]{1.\ C_6H_{12}N_4} \underset{S}{\bigcirc}\!\!-\!CH\!=\!O \qquad (10)$$

To a flask containing a solution of 114 gm (0.645 mole) of 3-bromomethylthiophene in 200 ml of chloroform is added 90 gm (0.642 mole) of hexamethylenetetramine. The mixture is refluxed for 1 hr, cooled, and the resulting salt is filtered off. The latter is washed with ether to yield 150 gm (73%), mp 120°–150°C (from ethanol). The salt (150 gm) is dissolved in ½ liter of hot water and rapidly steam-distilled until 1 liter of distillate is collected. The distillate is acidified, extracted with ether (three 100-ml portions), dried, concentrated, and distilled at atmospheric pressure to yield 35.8 gm (50%) of 3-thiophenealdehyde, bp 195°–199°C (744 mm), n_D^{20} 1.5860.

3. REDUCTION REACTIONS

Aldehydes can be prepared in good yield by reducing nitriles or derivatives of acids such as acid chlorides, amides, or esters. The latter derivatives provide an overall method for reducing carboxylic acids to aldehydes.

A. Reduction of Nitriles

Nitriles can be reduced to the intermediate imine which upon hydrolysis gives the aldehyde in good yield. Reducing agents that are commonly used are stannous chloride–hydrochloric acid (Stephen method) [46–48], lithium aluminum hydride [47], lithium triethoxyaluminum hydride [49, 50], Raney nickel and semicarbazide [51a], Raney nickel alloy and concentrated formic acid [51b], nickel and hydrogen [52, 53], diisobutylaluminum hydride [54, 55], and sodium amalgam and phenylhydrazine [56]. The latter reagent yields a phenylhydrazone, which gives the aldehyde upon hydrolysis.

Stannous chloride–HCl, LiAlH$_4$, and Li(OC$_2$H$_5$)$_3$AlH are easily accessible for laboratory synthesis problems. The latter two have recently been shown to be effective reducing agents for nitriles, giving high yields of aldehydes on hydrolysis of the aldimine intermediate.

The Stephen method involves the addition of the nitrile to an ether solution of stannous chloride saturated with hydrogen chloride. The nitrile is converted to an imino chloride, which is further reduced to an aldimine and then hydrolyzed to an aldehyde.

$$RCN + HCl \longrightarrow R\overset{Cl}{\underset{|}{C}}=NH \xrightarrow{SnCl_2 + HCl} RCH=NH \xrightarrow{H_2O} RCH=O \quad (11)$$

Aromatic nitriles [57] give excellent yields of aldehydes. However, the yields are quite variable from aliphatic nitriles. The reaction has not been successfully applied to many heterocyclic nitrile systems. For example, the reaction failed to give any aldehyde with 4-pyridinecarbonitrile [58].

3-1. Preparation of n-Octaldehyde by the Stephen Method [46]

$$C_6H_{13}CH_2CN + SnCl_2 + 3\ HCl \longrightarrow C_6H_{13}CH_2-CH=NH \cdot HCl + SnCl_4$$
$$\xrightarrow{H_2O} C_6H_{13}CH_2CH=O \quad (12)$$

To a flask are added 57 gm (0.30 mole) of stannous chloride and 200 ml of dry ether. The mixture is saturated with dry hydrogen chloride until it separates into two layers. The lower layer consists of stannous chloride dissolved in ethereal hydrogen chloride. Octanenitrile, 25 gm (0.18 mole), is added dropwise with vigorous stirring and after a few minutes the aldimine stannichloride begins to separate out of solution. The reduction is substantially completed after 2 hr. The aldehyde is formed by hydrolysis of the aldimine complex with warm water and isolated by steam distillation or extraction with ether. The aldehyde is redistilled to yield a colorless liquid, bp 65°C (11 mm).

Lithium triethoxyaluminum hydride [50] has the advantage over the Stephen method in that it reduces aromatic as well as aliphatic nitriles, the latter giving yields of approximately 70–90%. Some representative results of reductions of nitriles to aldehydes are cyclopropanecarbonitrile (69%), n-butyronitrile (69%), p-chlorobenzonitrile (92%), benzonitrile (96%), and capronitrile (69%).

Lithium aluminum hydride ($LiAlH_4$) also appears to have wide applicability in the reduction of nitriles to aldehydes. The ready availability of the reagent makes its use an attractive method [59, 60]. The method has been applied to the reduction of γ-dimethylamino-α,α-diphenylvaleronitrile to the aldehyde in 82% yield [61]. In addition m-hydroxybenzonitrile is reduced by $LiAlH_4$ at 0°C in an ether solution to the aldimine lithium aluminum salt. The latter is decomposed with excess dilute hydrochloric acid. The ether layer is separated, dried, evaporated, and the residue recrystallized from an alcohol–H_2O solution to yield 100% m-hydroxybenzaldehyde [60]. Amides are reduced in a similar fashion.

3-2. Preparation of Pivalaldehyde by the Reduction of Trimethylacetonitrile with Lithium Triethoxyaluminum Hydride [50]

$$(CH_3)_3C-CN + Li(C_2H_5O)_3AlH \longrightarrow [(CH_3)_3C-CH=N]Al(OC_2H_5)_3 \xrightarrow{H_3O^+}$$
$$(CH_3)_3C-CH=O \quad (13)$$

In a 1-liter flask equipped with a stirrer, condenser, dropping funnel, and thermometer is placed 10.2 gm (0.3 mole) of lithium aluminum hydride in 300 ml of ether. The flask contains a nitrogen atmosphere throughout the reaction period. To the stirred solution is added 39.6 gm (0.45 mole) of ethyl acetate over a period of 1¼ hr at 3°–7°C. The reaction mixture is stirred for an additional ½ hr and then 24.9 gm (0.30 mole) of trimethylacetonitrile is added over a period of 5 min. The temperature rises to 10°C, with the formation of a highly viscous solution. The reaction mixture is stirred for 1 hr at 0°C and then the solution is decomposed cautiously with 300 ml of 5 N H_2SO_4. The ether layer is separated and the water layer is extracted three times with 50 ml portions of ether. The ether extracts are washed with saturated sodium bicarbonate solution followed by eight washings with 30-ml portions of cold water in order to remove ethanol. The ether extracts are dried over sodium sulfate and distilled to yield 25.8 gm (74%) of pivalaldehyde, bp 70°–72.5°C (747 mm), n_D^{20} 1.3794.

B. Reduction of Other Acid Derivatives

$$R\overset{O}{\underset{\|}{C}}-Z \xrightarrow{(H)} RCH=O \quad (14)$$

The low-temperature reduction of acids, esters, lactones, lactams, and amides to aldehydes in fair yields has been reviewed [24a,b]. Carboxylic acids can be reduced by lithium in ethylamine to aldehydes [62]. Acid chlorides can be reduced to aldehydes by one of several methods.

The Rosenmund reduction [63a] of acid chlorides involves bubbling hydrogen through a warm xylene solution containing a 5% palladium suspended on barium sulfate catalyst. Careful control of the temperature is required in order to avoid reduction of the aldehyde. Acyl halides with nitro, halo, ester, or olefinic substituents do not interfere; however, hydroxyl groups need to be acylated before starting the reduction.

A modified Rosenmund reaction involves use of a pressure reactor and an amine for removal of the hydrogen chloride that is generated [63b].

A more convenient method of reducing aromatic acid chlorides to aldehydes in 70–90% yields has recently been demonstrated to involve the use of lithium tri-*tert*-butoxyaluminum hydride [64–66]. However, aliphatic acid chlorides gave yields of only 40–60%. Brown [67a] has been able to produce aliphatic aldehydes in up to 88% by starting with 1-acylaziridines and reducing them with lithium aluminum hydride [67a]. It was further shown that the 1-acylaziridine need not be isolated but could be prepared in solution by adding the acid chloride to an equimolar mixture of triethylamine and ethylenimine, separating the precipitated triethylammonium chloride, and then adding to the solution lithium aluminum hydride. Aliphatic acid chlorides have also been reduced to aldehydes via reduction of ester mesylates with sodium borohydride [67b].

3-3. Preparation of Cyclopropanecarboxyaldehyde by the LiAlH$_4$ Reduction of 1-Acylaziridine [67a]

$$\begin{array}{c}CH_2-CH_2 \\ \diagdown\diagup \\ CH \\ | \\ COCl\end{array} + \begin{array}{c}CH_2-CH_2 \\ \diagdown\diagup \\ N \\ | \\ H\end{array} \xrightarrow{Et_3N} \begin{array}{c}CH_2-CH_2 \\ \diagdown\diagup \\ CH \\ | \\ CO-N\end{array}\begin{array}{c}CH_2 \\ \diagup \\ \diagdown \\ CH_2\end{array} \xrightarrow{LiAlH_4} \begin{array}{c}CH_2-CH_2 \\ \diagdown\diagup \\ CH \\ | \\ CH=O\end{array} + \begin{array}{c}CH_2-CH_2 \\ \diagdown\diagup \\ N \\ | \\ H\end{array} \quad (15)$$

To a flask containing a stirred solution of 17.5 gm (0.40 mole) of ethylenimine and 40 gm (0.40 mole) of triethylamine in 200 ml of ether at 0°C (ice–salt bath) is added dropwise 42.2 gm (0.40 mole) of cyclopropanecarbonyl chloride over a period of 1 hr. CAUTION: Ethylenimine is highly carcinogenic. After another ½ hr the precipitated triethylaminehydrochloride is filtered off and washed with 100 ml of ether. The combined ether layer is cooled to 0°C and 80 ml of 1.25 M LiAlH$_4$ in ether is added dropwise over a period of ½ hr. After an hour a cold 5 N sulfuric acid is added cautiously in order to neutralize the reaction mixture. The ether is separated and the aqueous layer is extracted with ether. The combined ether layer is washed with water, sodium bicarbonate, water again, dried, and distilled to yield 16.8 gm (60%) of cyclopropanecarboxyaldehyde, bp 97°–100°C (740 mm), n_D^{20} 1.4302.

Acid chlorides have also been reduced to aldehydes by 10% palladium on charcoal using triethylsilane [67c].

The controlled reduction of amides and tertiary amides with lithium aluminum hydride followed by hydrolysis yields aldehydes. The following types of amides have been utilized in the latter reductions; N-acylcarbazoles [68], N-methylanilides [69], and carboxylpiperidines [70]. The latter two gave 60–90% and 80% yields of aldehydes, respectively.

The N-methylanilides are obtained by condensing methylphenylcarbamyl chloride with aromatic hydrocarbons in the presence of aluminum chloride [69].

$$ArH + Cl-\overset{O}{\underset{\underset{CH_3}{|}}{C}}-N-Ar \xrightarrow{AlCl_3} Ar\overset{O}{\underset{\underset{CH_3}{|}}{C}}-N-Ar \xrightarrow{LiAlH_4} ArCH=O + ArNH-CH_3 \quad (16)$$

The partial reduction of 1-acyl-3,5-dimethylpyrazoles also gives aldehydes [71]. Using ordinary conditions, i.e., with an excess of reducing agent present, tertiary amides are reduced to tertiary amines by lithium aluminum hydride [72].

C. Miscellaneous Reducing Agents

The use of sodium diisobutylaluminum dihydride has been reported to be effective in the reduction of esters, ethers, lactones, nitriles, and amides to aldehydes [73].

Sodium trimethoxyborohydride has also been used to reduce acid chlorides to aldehydes by carrying out the reaction at $-80°C$ in tetrahydrofuran [74].

The Sonn–Müller [24a,b, 75] method can be considered a reduction method since an acid anilide is first converted to an imido chloride which is then reduced and hydrolyzed to give an aldehyde.

$$RCONHC_6H_5 \xrightarrow{PCl_5} \underset{Cl}{RC{=}N{-}C_6H_5} \xrightarrow{SnCl_2 + HCl} \underset{H}{RC{=}NC_6H_5} \xrightarrow{H_3O^+} RCH{=}O \quad (17)$$

Aldehydes are obtained in good yields from aromatic anilides [76] which do not contain other substituents that can be effected by phosphorus pentachloride. The method is not applicable to simple aliphatic anilides but α,β-unsaturated anilides give fair yields of aldehydes. The method has not been applied successfully to heterocyclic systems [77] but further work in this area remains to be done.

4. CONDENSATION REACTIONS

The condensation reactions used to prepare precursors for aldehydes involve the Friedel–Crafts or Grignard reaction. Hydrolysis of these products yield the aldehydes.

The Friedel–Crafts reaction involves the reaction of a formyl derivative with aluminum chloride as shown in Eq. (18).

$$ArH + Cl\underset{}{-}\overset{X}{\underset{\|}{C}}\underset{}{-}R \xrightarrow[-HCl]{AlCl_3} Ar\overset{X}{\underset{\|}{C}}\underset{}{-}R \xrightarrow{H_3O^+} ArCH{=}O \quad (18)$$

where X = oxygen, NH, or Cl, and R = H, N-methylanilide, OCH_3, or SCH_3 groups. Thus, reagents for the Gattermann [78], Gattermann–Koch [79], Weygand [69], and Rieche [80, 81] reactions can be described as, respectively,

$$Cl\overset{O}{\underset{\|}{-}C}{-}H, \quad Cl\overset{NH}{\underset{\|}{-}C}{-}H, \quad Cl{-}C{-}\underset{\underset{O}{\|}}{N}(CH_3)C_6H_5, \quad Cl_2CHOCH_3, \text{ or } Cl_2CHCH_3$$

Olah and Kuhn [82] prepared formyl fluoride and found that it condensed with aromatic hydrocarbons in the presence of boron trifluoride to give aldehydes in 56–78% yields [Eq. (19)].

$$\text{ArH} + \text{F}-\overset{\overset{\text{O}}{\|}}{\text{C}}-\text{H} \xrightarrow{\text{BF}_3} \text{ArCH}=\text{O} + \text{HBF}_4 \tag{19}$$

The Vilsmeier–Haack [83] reaction involves a related condensation of $(\text{CH}_3)_2\text{N}$—CHO and derivatives with aromatic or heterocyclic compounds in the presence of POCl_3 to give good yields of aldehydes in the aromatic and heterocyclic series.

The Gattermann [78] synthesis employing zinc cyanide in one of its modifications is more suitable for laboratory synthetic problems since it avoids the use of liquid hydrogen cyanide or carbon monoxide (Gattermann–Koch reaction). The latter reaction is not suitable for formylating phenols or their ethers, as is true for the Gattermann reaction.

In the original Gattermann synthesis, one employs liquid hydrogen cyanide and hydrochloric acid catalyzed by zinc chloride to react with aromatic hydrocarbons to yield an aldimine hydrochloride. One obtains the aldehyde on hydrolysis.

Adams [84, 85] has modified the procedure by using zinc cyanide containing a trace of potassium chloride [86]. The synthesis of mesitaldehyde (2,4,6-trimethylbenzaldehyde) in 75–81% yields has been reported to be effected using zinc cyanide and mesitylene [87]. The use of NaCN gives poor yields [88].

$$\text{ArH} \xrightarrow[\text{or Zn(CN)}_2 + \text{HCl}]{\text{HCN} + \text{HCl} + \text{ZnCl}_2} \text{ArCH}=\text{O} \tag{20}$$

Since these syntheses utilize cyanides, careful attention to the toxicity hazards of cyanides [HCN or Zn(CN)_2] should be kept in mind. See Chapter 17, Nitriles (Cyanides). The reaction should be carried out in a well-ventilated hood and the experimentalist should be provided with rubber gloves and a gas mask.

4-1. Preparation of Resorcylaldehyde by the Modified Gattermann Synthesis [84]

$$\underset{\text{OH}}{\text{C}_6\text{H}_4}\text{-OH} + \text{Zn(CN)}_2 \xrightarrow[\text{2. H}_2\text{O}]{\text{1. HCl}} \underset{\substack{\text{OH} \\ \text{CH}=\text{O}}}{\text{C}_6\text{H}_3}\text{-OH} \tag{21}$$

To a flask equipped with a stirrer (mercury seal or ground glass), reflux condenser, gas inlet tube (with a large bore to prevent clogging), a safety trap for the hydrogen chloride, and an exit tube for the hydrogen chloride leading from the top of the condenser to a trap which is connected to an inverted funnel placed over sodium hydroxide solution is added 20 gm (0.182 mole) of resorcinol in 200 ml of dry ether. Dry zinc cyanide, 32.1 gm (0.273 mole), is added and, while stirring, the hydrogen chloride gas is bubbled in rapidly. The zinc cyanide dissolves and the solution becomes milky. The product begins to separate as an oil which hardens after 10–30 min. After 1½ hr the HCl gas is passed in more slowly for another ½ hr. The ether is then decanted and 100 ml of water is added to the imide hydrochloride. The mixture is slowly heated to the boiling point, filtered, and the filtrate cooled. The solid is filtered to give about a 50% yield. Upon allowing the second filtrate to stand for an additional 10–15 hr, another 45% yield of product is obtained. The total yield is 23.9 gm (95%) of resorcylaldehyde, mp 135°–136°C. The product is recrystallized from water using decolorizing charcoal to give a colorless product.

A similar method has recently been reported to be effective in giving good yields of alkylated resorcylaldehydes [89].

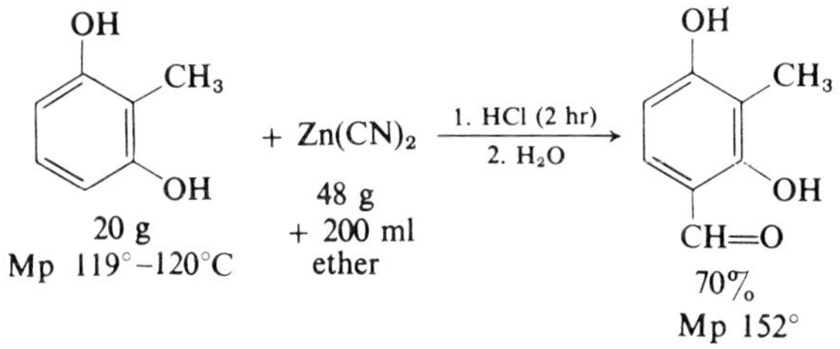

$$\text{(resorcinol with CH}_3\text{)} + \text{Zn(CN)}_2 \xrightarrow[\text{2. H}_2\text{O}]{\text{1. HCl (2 hr)}} \text{(resorcylaldehyde product)} \quad (22)$$

20 g, Mp 119°–120°C; 48 g + 200 ml ether; 70%, Mp 152°

4-2. Preparation of Tolualdehyde by the Formylation of Toluene with Formyl Fluoride [82]

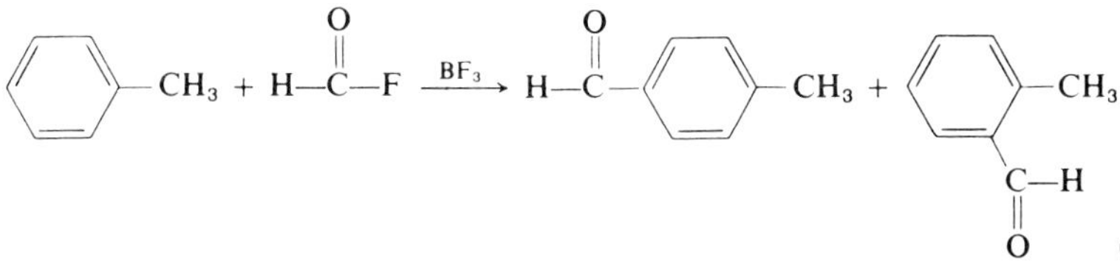

$$\text{C}_6\text{H}_5-\text{CH}_3 + \text{H}-\overset{\text{O}}{\underset{\|}{\text{C}}}-\text{F} \xrightarrow{\text{BF}_3} \text{H}-\overset{\text{O}}{\underset{\|}{\text{C}}}-\text{C}_6\text{H}_4-\text{CH}_3 + \text{C}_6\text{H}_4(\text{CH}_3)(\text{CHO}) \quad (23)$$

(a) Preparation of Formyl Fluoride from Formic Acid. To a flask containing 46 gm (1 mole) of anhydrous formic acid is added 60 gm (0.77 mole) of dry potassium hydrogen fluoride. (No exotherm or gas evolution is observed.) Benzoyl chloride, 116 ml (141 gm, 1 mole), is added dropwise to the stirred mixture, and the mixture is slowly warmed on a steam bath. During the heating period formyl fluoride distills off and passes through an ice–salt-cooled condenser, and it is collected in a Dry Ice–acetone trap. Formyl fluoride is redistilled to give 17 gm (35.4%), bp −29°C.

(b) Preparation of Tolualdehyde. To a flask cooled at −30° to −70°C and containing 92 gm (1 mole) of toluene is added 24 gm (0.5 mole) of formyl fluoride. Boron trifluoride is added to the cold solution to the saturation point

(0.5 mole). The complex is slowly warmed to room temperature while boron trifluoride evolves. The reaction mixture is washed with water, dried over calcium chloride, and fractionated to give 90 gm (75%) of ortho and para isomer mixture of tolualdehyde, bp 197°–205°C, n_D^{20} 1.547–1.549.

The Vilsmeier–Haack formylation reaction has been applied to alkoxy or N,N-dimethylamino derivatives of benzene [90]. naphthalene [91], stilbene [92], naphthols [93], and to activated heterocycles. Dimethylformamide [94] has been used as a convenient substitute of N-methylformanilide in the formylation of pyrrole [95] (78–79% yield of 2-pyrrolaldehyde), indole [96] (indole-3-aldehyde, yield 97%) [97], thiophene [98] (2-thiopheneladehyde, yield 71–74%) [99], anthracene [100], and stilbene [92].

$$(CH_3)_2NCH{=}O + POCl_3 \longrightarrow [(CH_3)_2NCH{=}O \cdot POCl_3] \xrightarrow{ArH}$$

$$[Ar{-}CH\begin{smallmatrix}OPOCl_2 \\ \cdot HCl \\ N(CH_3)_2\end{smallmatrix}] \xrightarrow{H_2O + CH_3COONa} Ar{-}CH{=}O \qquad (24)$$

4-3. Preparation of 3-Formylindole by the Vilsmeier Method [101]

$$\text{indole} + (CH_3)_2N\overset{O}{\overset{\|}{C}}H \xrightarrow[\text{2. NaOH, H}_2\text{O}]{\text{1. POCl}_3} \text{3-formylindole} \qquad (25)$$

To a flask containing 16 gm (0.22 mole) of dimethylformamide cooled to 10°–20°C and protected from moisture is added 5.0 ml (0.055 mole) of phosphorus oxychloride. Indole (5.85 gm, 0.059 mole) in 4 gm of dimethylformamide is then added slowly with stirring at such a rate as to keep the temperature at 20°–30°C. The mixture is kept at 35°C for 45 min and then it is poured into crushed ice. The clear solution is heated at 20°–30°C and 9.5 gm (0.24 mole) of sodium hydroxide in 50 ml water is added at such a rate that the solution remains acidic until approximately three-fourths of the alkali solution has been added. The last quarter is quickly added and the solution is boiled for 1 min. The resulting white crystals are filtered off, washed with five 25-ml portions of water, and dried to constant weight at 100°C at 10 mm to yield 6.93 gm (95.5%), mp 197°–199°C.

Another useful method in preparing aldehydes involves Grignard condensation reactions.

The Grignard reagent reacts with ethyl orthoformate to give acetals which upon hydrolysis in dilute acid give aldehydes in good yields. The reaction is run at room temperature and then refluxed before hydrolysis [102].

Ethyl formate [103], formic acid [104], and dialkylformamides [105] have been used to give aldehydes in fair yields. Aryllithium reagents [105] react with dialkylformamides and also with N-methylformanilide to give good yields of aldehydes.

$$\text{RMgX} + \text{HC(OC}_2\text{H}_5)_3 \longrightarrow \text{RCH(OC}_2\text{H}_5)_2 \xrightarrow{\text{H}_3\text{O}^+} \text{RCH}=\text{O}$$
$$\downarrow \xrightarrow{\text{R}_2\text{NCH}=\text{O}} \text{RCH}=\text{O}$$
$$\downarrow \xrightarrow{\text{HCOOR}} \text{RCH}=\text{O}$$
$$\downarrow \xrightarrow{\text{HCOOH}} \text{RCH}=\text{O} \qquad (26)$$

Ethoxymethyleneaniline used in place of the formates in the Grignard condensation reaction gives an imine which is also easily hydrolyzed to an aldehyde [106]. The reaction is an excellent method for the large-scale laboratory preparation of aldehydes in good yields. Ethoxymethyleneaniline is prepared from the dry silver salt of formanilide and ethyl iodide.

$$\text{C}_6\text{H}_5\text{MgX} + \text{C}_6\text{H}_5\text{N}=\text{CHOC}_2\text{H}_5 \longrightarrow \text{C}_6\text{H}_5\text{CH}=\text{NC}_6\text{H}_5 \xrightarrow{\text{H}_3\text{O}^+} \text{C}_6\text{H}_5\text{CH}=\text{O} \quad (27)$$

A. Miscellaneous Condensation Methods

The Reimer-Tiemann [107, 108] reaction involves the formylation of phenols when the phenol and chloroform are heated with an alkaline ethanolic solution for several hours. The mixture is then acidified and the product is isolated by recrystallization or steam distillation. The yields are usually below 50% and the para isomer predominates [109, 110].

$$\text{ArOH} + \text{CHCl}_3 \xrightarrow[\text{2. HCl}]{\text{1. NaOH}} \text{HO—ArCH}=\text{O} \qquad (28)$$

An example of this method is the reported preparation of 2-hydroxy-1-naphthaldehyde from β-naphthol, chloroform, and sodium hydroxide in 38–48% yields [111]. The Reimer-Tiemann reaction has also been applied to the preparation of indole-3-aldehyde [112] and 2-pyrroladehyde [113a].

The Claisen rearrangement has been reported to be useful in the preparation of certain aldehydes [113b].

The Wittig reaction has been utilized in the preparation of α,β-unsaturated aldehydes in good yields [113c].

Phase transfer catalysts have been reported to be useful for aiding the basic dimerization of crotonaldehyde [113d].

The Duff [114] reaction involves the heating of a phenolic compound with a mixture of glycerol, boric acid, and hexamine at 150°–160°C for about 30 min. The aldehyde is obtained on acidification and steam distillation. The reaction normally affords low yields. However, since only the ortho isomer is formed, it has found use in such specialty syntheses as the flavone series.

$$\text{ArOH} + (\text{CH}_2)_6\text{N}_4 \longrightarrow o\text{-HOArCH}=\text{O} \quad (29)$$

Olefins can be formylated by means of carbon monoxide, hydrogen, and a cobalt carbonyl catalyst under pressure [115]. The yields are good, but since pressure equipment and carbon monoxide are involved, it has found limited use in the laboratory. However, the reaction is the basis of several large-scale commercial processes.

$$\text{RCH}=\text{CH}_2 + \text{CO} + \text{H}_2 \xrightarrow[\text{Pressure}]{\text{Catalyst}} \text{RCH}_2\text{CH}_2\text{CH}=\text{O} \quad (30)$$

Ketones with an active methylene group can be formylated with ethyl formate [116–119]. Sodium metal is added to a mixture of the ketone and ethyl formate in ether to yield keto aldehydes in fair to good yields. Mixtures are obtained when an unsymmetrical ketone is used.

5. ELIMINATION REACTIONS

Some of the elimination or hydrolysis reactions which produce aldehyde are the cleavage of Schiff bases [120], the decarboxylation of α-keto acids [121], the decomposition of α-hydroxy acids [122], the hydrolysis of *gem*-dihalides [123–126], the hydrolysis of 2-alkoxy-3,4-dihydro-1,2-pyrans [127], the hydrolysis of oximes, semicarbazones, hydrazones, acetals, etc. [128, 129], the decomposition of glycol monoalkyl ethers [130], and the Darzens reaction [131].

Since Schiff bases are usually prepared from aldehydes, the most useful synthetic methods for preparing aldehydes by elimination reactions are the decomposition of α-hydroxy acids, the hydrolysis of *gem*-dihalides, and the decomposition of glycidic esters.

The decomposition of α-hydroxy acids is effected at the boiling point at reduced pressure. The reaction is particularly suitable for high-molecular-weight aliphatic aldehydes. Heptadecanal is obtained in almost quantitative yield [122]. The α-hydroxy acids are obtained by hydrolysis of the corresponding α-bromo acids.

The hydrolysis of *gem*-dihalides by calcium carbonate or sulfuric acid is easily effected for benzal halides. Aliphatic halides, require much higher temperatures. In both cases the aldehydes are obtained in good yield [123]. It has been reported that benzal halides upon reaction with four moles of morpholine yield a morpholine derivative which undergoes hydrolysis by aqueous hydrochloric acid to mono aldehydes in 82–91% yields [124].

Terephthalaldehyde is prepared in 81–84% yield by the bromination of *p*-xylene to α,α,α′,α′-tetrabromo-*p*-xylene followed by hydrolysis by aqueous sulfuric acid [125]. Similar *o*-phthalaldehyde is prepared in 71–80% yield from α,α,α′,α′-tetrabromo-*o*-xylene [126].

§ 5. Elimination Reactions

The Darzens glycidic ester synthesis involves the conversion of a ketone to an aldehyde with one more carbon atom than the initial ketone [131]. Ketones are first condensed with ethyl chloroacetate in the presence of sodium amide to give a glycidic ester. The glycidic ester is reacted with sodium ethoxide and then treated with aqueous acid using gentle reflux to afford the aldehyde in low yields [132, 133]. The intermediates need not necessarily be isolated [134].

A pyrolytic modification of the Darzens glycidic ester synthesis of aldehydes was reported [135] in which *tert*-butyl esters were used and yields of 45–63% were obtained. For example, α-phenylpropionaldehyde (hydrotropaldehyde) is prepared in 63% yield by pyrolysis of *tert*-butyl 2-methyl-2-phenylglycidate.

$$C_6H_5COCH_3 + ClCH_2COOC(CH_3)_5 \longrightarrow C_6H_5C(CH_3)\underset{O}{\diagdown\diagup}CHCOOC(CH_3)_3 \xrightarrow{Heat}$$

$$C_6H_5CH-CH=O + CO_2 + (CH_3)_2CH=CH_2$$
$$\quad\quad\quad | $$
$$\quad\quad CH_3 \quad\quad\quad\quad\quad\quad\quad\quad\quad\quad\quad (31)$$

5-1. Preparation of Hydrotropaldehyde by the Darzens Reaction [133]

$$C_6H_5\underset{\diagdown O \diagup}{\overset{CH_3}{\underset{|}{C}}}-CH-COOC_2H_5 \xrightarrow[\text{2. HCl}]{\text{1. NaOH}} C_6H_5\underset{\diagdown O \diagup}{\overset{CH_3}{\underset{|}{C}}}-CH-COOH \longrightarrow$$

$$\quad\quad\quad\quad\quad\quad\quad\quad\quad\quad\quad\quad\quad\quad CH_3$$
$$\quad\quad\quad\quad\quad\quad\quad\quad\quad\quad\quad\quad\quad\quad |$$
$$\quad\quad\quad\quad\quad\quad\quad\quad\quad\quad\quad\quad C_6H_5CH-CHO \quad (32)$$

To a flask containing a stirred solution of 274 gm (0.685 mole) of sodium hydroxide in 770 ml of water is added 708 gm of ethyl 2-methyl-2-phenyl glycidate [133].* The solution is stirred for 8 hr at 45°–50°C and then acidified to congo red. The glycidic acid is extracted with benzene and then steam-distilled using superheated steam at 180°C. Decarboxylation occurs and the aldehyde distills over during a period of about 5 hr. Redistillation of the aldehyde gives 268 gm (58%), bp 101°–102°C (21–22 mm).

The Krohnke [137] reaction of benzyl halides and other active halides involves the reaction with pyridine and then with *p*-nitrosodimethylaniline followed by acid hydrolysis of the nitrone to the aldehyde.

$$C_6H_5CH_2X \xrightarrow{\text{Pyridine}} [C_6H_5CH_2NC_6H_5]^+X^- \xrightarrow{(CH_3)_2NC_6H_4NO}$$
$$C_6H_5CH=NOC_6H_5N(CH_3)_2 \xrightarrow{H_3O^+} C_6H_5CH=O \quad (33)$$

By this means benzyl halides may be converted to aromatic aldehydes in good yields [138].

*Ethyl 2-methyl-2-phenylglycidate [133] can be prepared from acetophenone and ethyl chloroacetate by condensing with sodamide to give 62–64% yields, bp 111°–114°C (3 mm).

The thermal decarboxylation of mixtures of formic acid and other carboxylic acids over thorium oxide catalyst at 300°C gives [139, 140] aliphatic or aromatic [141] aldehydes in good to excellent yields. Titanium dioxide and magnesium oxide have also been used as catalysts. Aliphatic acids yield aldehydes when more than seven carbon atoms are present, and the yields improve with increasing molecular weight.

A. Miscellaneous Elimination Reactions

The McFadyen–Stevens reaction [24a,b, 136, 142] involves the conversion of carboxylic acid derivative to aldehydes by conversion of hydrazides to arylsulfonyl hydrazides and decomposition with sodium carbonate in ethylene glycol at 160°C.

Although the reaction can also be classified as a reduction reaction, it is included in this section since it involves the elimination of the hydrazide group.

Aromatic aldehydes are prepared in good yields, whereas the heterocyclic aldehydes are prepared in fair yields. Only recently has the method been successfully applied to the aliphatic series where a 16% yield of cyclopropane carboxyaldehyde [143] (as the 2,4-dinitrophenylhydrazone) was obtained. However, it has been shown that the McFadyen-Stevens reaction is applicable primarily to the preparation of aliphatic aldehydes in which there are no α-hydrogen atoms [144a]. The failure of the McFadyen-Stevens reduction in earlier investigations in the aliphatic series is purported to be due to the instability of the product under the reaction conditions rather than an inherent difference in the reactivity of the starting hydrazides. Short reaction times, of the order of 30 sec, are required. For example, N-p-toluenesulfonylapocamphane-1-carbohydrazide yields apocamphane-1-carboxaldehyde in 60% yield, isolated as the 2,4-dinitrophenylhydrazone, when subjected for 30 sec to the McFadyen-Stevens reaction conditions [144a].

5-2. *Preparation of Benzaldehyde by the McFadyen-Stevens Reaction* [136]

$$C_6H_5CONHNH_2 + C_6H_5SO_2Cl \xrightarrow{\text{Pyridine}}$$

$$C_6H_5CONHNHSO_2C_6H_5 \xrightarrow{Na_2CO_3} C_6H_5CHO \qquad (34)$$

A flask containing 25 ml of pyridine and 4 gm (0.0295 mole) of benzhydrazide is placed in an ice bath. The flask is stirred while 5.2 gm (0.0295 mole) of benzene sulfonyl chloride are slowly added. After 2 hr the solution is poured into a mixture of ice and hydrochloric acid. A pale yellow solid precipitates and is filtered off. The solid is washed with dilute hydrochloric acid and water and recrystallized from 80 ml of alcohol to yield 7 gm (86%) of colorless prismatic needles of 1-benzoyl-2-benzenesulfonylhydrazine, mp 192°–194°C dec. The 7 gm (0.0236 mole) of 1-benzoyl-2-benzenesulfonylhydrazine is dissolved in 140 gm of ethylene glycol and heated to 160°C while 6 gm (0.056 mole) of sodium carbonate is added in one portion. A brisk effervescence occurs and after 75 sec the reaction is stopped by the addition of hot water. CAUTION: The reaction vessel should be large enough to accommodate the foam produced. The cooled mixture is extracted with ether several times. The dried ethereal extract is distilled to yield 1.8 gm (73%) of benzaldehyde, bp 177°–178°C, n_D^{20} 1.5463.

The alkylation of metalated mercaptal compounds and their subsequent hydrolysis is a newer method of aldehyde synthesis [144b].

$$\underset{RS\quad SR}{\overset{Li}{\diagdown C \diagup}} \xrightarrow{R'X} \underset{RS\quad SR}{\overset{R'}{\diagdown C \diagup}} \xrightarrow[H_2O]{Hg^{II}} R'CH=O \qquad (35)$$

In a related manner, aldehydes are obtained from *sym* trithiane [144c].

$$\text{(trithiane)}-Li \xrightarrow[\substack{2.\ CH_3OH \\ HgCl_2/HgO}]{1.\ RX} (CH_3O)_2CHR \xrightarrow{H_3O^+} RCH=O \qquad (36)$$

6. MISCELLANEOUS REACTIONS

(1) The reaction of benzyl halides and sodium 2-propanenitronate to yield aldehydes [145, 146].
(2) Formylation of sodium acetylide with formates [147].
(3) Karrer synthesis of aldehydes [148].
(4) Isomerization of allyl alcohols [149].
(5) Hydrolysis of olefin 1,2-dibromides [150].
(6) Hydrolysis of amides and azides [151–153].
(7) Hydrolysis of aci-nitroparaffins [154].
(8) Grignard condensation of ethyl ethoxyacetate or ethyl phenoxyacetate followed by acid hydrolysis [155].
(9) Oxidation of benzhydrazides with potassium ferricyanide in ammonium hydroxide [156].
(10) Desulfurization with Raney nickel [157].
(11) Decomposition of Reissert compounds [158].
(12) Cleavage of *p*-dimethylaminophenylcarbinols with diazonium salts [159].
(13) The Wittig reaction [160].
(14) Reaction of pentavalent phosphorus esters with *gem*-dihalides [161].
(15) Condensation of lithium salts of Schiff bases with ketones [162].
(16) Reaction of diazonium salts with oximes [163].

(17) Hoffmann reaction with α-hydroxy amides [164].
(18) Arens-Dorp synthesis of α,β-unsaturated aldehydes [165].
(19) Borodine-Hunsdiecker reaction [166].
(20) Jones-Weedon synthesis of acetylenic-ethylenic aldehydes [167].
(21) Hydrolysis of acetylene [168].
(22) Zemplen degradation of sugars [169].
(23) Rearrangement of α,β-epoxy ketones [170].
(24) Acid azide degradation [171].
(25) Catalytic thermal aliphatic Claisen rearrangement by ammonium chloride [172].
(26) Pyrolysis of hydroxy esters [173].
(27) Preparation of aldehydes from ethers [174].
(28) Cleavage of unsaturated cyclic ethers; 5-hydroxypentanal from 2,3-dihydropyran [175].
(29) α-Bromoaldehydes from α-bromodimethyl acetals [176].
(30) Atmospheric pressure carbon monoxide conversion of olefins to aldehydes via hydroboration [177].
(31) Aldehydes from 2-benzyl-4,4,6-trimethyl-5,6-dihydro-1,3-(4H)-oxazine [178].
(32) Oxidation of olefins by supported chromium oxide [179].
(33) Oxidative clevage of olefins to give aldehydes by bis(triphenylalyl) chromate [180].
(34) Reaction of β-alkyl-9-borabicyclo[3.3.1]nonenes with carbon monoxide in the presence of lithium trimethoxyalaminohydride to give aldehydes [181].
(35) Aldehydes from formamides [182].
(36) Electrochemical reduction of aliphatic amides to aldehydes [183].
(37) Photochemical reduction of aromatic nitriles to aldehydes [184].
(38) Preparation of aldehydes from acid chlorides using copper tetrahydroborate complexes [185].
(39) Reaction of organoboranes with 2-bromoacrolein to prepare α-bromo aldehydes [186].
(40) Synthesis of 9,10-anthracene dicarboxaldehyde [187].
(41) Synthesis of γ,δ-unsaturated aldehydes and ketones by the Claisen rearrangement of allyl ethers catalyzed by ruthenium(II) [188].
(42) Synthesis of aldehydes via the Pummerer reaction [189].
(43) The synthesis of mono-, di-, and trialkylacetaldehydes via 2-thiazolines [190].
(44) The synthesis of aldehydes from dihydro-1,3-oxazines [191].
(45) 1,6-Hexanedial by the periodic acid oxidation of cyclohexane epoxide [192].
(46) Oxidation of primary alcohols to aldehydes via use of poly(p-methylmercaptostyrene) and chlorine [193].
(47) Rearrangement of cyclic epoxides to cyclic aldehydes [194].

References

1a. C. D. Hurd and N. Meinert, *Org. Synth. Collect. Vol.* **2**, 541 (1943).
1b. D. G. Lee and U. A. Spitzer, *J. Org. Chem.* **35**, 3589 (1970).
2. L. I. Smith, H. E. Ungnade, J. W. Opie, W. W. Pritchard, R. B. Carlin, and E. W. Kaiser, *J. Org. Chem.* **4**, 323 (1939).
3. C. J. Martin, A. I. Schepartz, and B. F. Daubert, *J. Am. Chem. Soc.* **70**, 2601 (1948).
4. H. Z. Barakat, M. E. Abdel-Wahab, and M. M. El-Sadi, *J. Chem. Soc.* p. 4685 (1956).
5. C. D. Robeson, *Eastman Org. Chem. Bull.* **32**, No. 1 (1960).
6a. J. C. Sauer, *Org. Synth. Collect. Vol.* **4**, 813 (1963).
6b. C. D. Hurd, R. N. Meinert, and L. V. Spence, *J. Am. Chem. Soc.* **52**, 138 (1930).
6c. R. W. Ratcliffe, *Org. Synth.* **55**, 84 (1976).
6d. G. I. Poos, G. E. Arth, R. E. Beyler, and L. H. Sarett, *J. Am. Chem. Soc.* **75**, 422 (1953).
7. J. C. Collins, W. W. Hess, and F. J. Frank, *Tetrahedron Lett.* p. 3363 (1968); J. C. Collins and W. W. Hess, *Org. Synth.* **52**, 5 (1972).
8. J. Attenburrow, A. F. B. Cameron, J. H. Chapman, R. M. Evans, B. A. Hems, A. B. A. Jansen, and T. Walker, *J. Chem. Soc.* p. 1094 (1952).
9. C. Djerassi, *Org. React.* **6**, 207 (1951).
10. R. R. Davies and H. H. Hodgson, *J. Soc. Chem. Ind., London* **62**, 109 (1943).
11. A. Lauchenauer and H. Schinz, *Helv. Chim. Acta* **32**, 1265 (1949).
12. R. B. Woodward, N. L. Wendler, and F. J. Brutschy, *J. Am. Chem. Soc.* **67**, 1425 (1965).
13a. V. J. Traynelis and W. L. Hergenrother, *J. Am. Chem. Soc.* **86**, 298 (1964).
13b. K. Omara, A. K. Sharma, and D. Swern, *J. Org. Chem.* **41**, 957 (1976).
14a. K. E. Pfitzner and J. G. Moffatt, *J. Am. Chem. Soc.* **85**, 3027 (1963).
14b. A. H. Fenselau and J. G. Moffatt, *J. Am. Chem. Soc.* **88**, 1762 (1966).
14c. J. D. Albright and L. Goldman, *J. Am. Chem. Soc.* **89**, 2416 (1967).
15. W. E. Evans, Jr., C. J. Carr, and J. C. Krantz, Jr., *J. Am. Chem. Soc.* **60**, 1628 (1938).
16. N. Rabjohn, *Org. React.* **5**, 331 (1949).
17. F. Weygand, K. G. Kinkel, and D. Tietjen, *Chem. Ber.* **83**, 394 (1950).
18. J. N. Ashley, H. J. Barber, A. J. Ewins, G. Newbery, and A. D. H. Self, *J. Chem. Soc.* p. 103 (1942).
19a. B. O. Field and J. Grundy, *J. Chem. Soc.* p. 1110 (1955).
19b. E. J. Corey and C. V. Kim, *J. Am. Chem. Soc.* **94**, 7586 (1972).
20. R. B. Woodward and E. C. Kornfeld, *J. Am. Chem. Soc.* **70**, 2508 (1948).
21. R. J. Speer and L. R. Mahler, *J. Am. Chem. Soc.* **71**, 1133 (1949).
22. J. English and G. W. Barber, *J. Am. Chem. Soc.* **71**, 3310 (1949).
23. C. Grundmann, *Justus Liebigs Ann. Chem.* **524**, 31 (1936).
24a. E. Mosettig, *Org. React.* **8**, 218 (1954).
24b. F. J. Wolf and J. Weylard, *Org. Synth. Collect. Vol.* **3**, 124 (1963).
25. L. N. Ferguson, *Chem. Rev.* **38**, 227 (1946).
26. C. R. Noller and R. A. Banneret, *J. Am. Chem. Soc.* **56**, 1563 (1934).
27. F. G. Fisher and K. Lowenberg, *Chem. Ber.* **66**, 6666 (1933).
28. R. H. Callighan and M. H. Wilt, *J. Org. Chem.* **26**, 4912 (1961).
29. W. E. Truce and R. J. Steltenkamp, *J. Org. Chem.* **27**, 2816 (1962).
30. H. C. Brown and G. Zweifel, *J. Am. Chem. Soc.* **83**, 3834 (1961).
31a. G. Zweifel and H. C. Brown, *Org. React.* **13**, 1 (1963).
31b. F. Freeman, R. H. DuBois, and T. G. McLaughlin, *Org. Synth.* **51**, 4 (1971).
32. A. F. Walton, R. S. Tipson, and L. H. Gretcher, *J. Am. Chem. Soc.* **67**, 1501 (1945).
33. S. M. Tsang, E. H. Wood, and J. R. Johnson, *Org. Synth. Collect. Vol.* **3**, 641 (1955).
34. C. S. Marvel, J. H. Saunders, and C. G. Overberger, *J. Am. Chem. Soc.* **68**, 1085 (1946).
35. H. D. Law and F. M. Perkin, *J. Chem. Soc.* p. 259 (1907).
36. C. H. Fisher, *J. Am. Chem. Soc.* **56**, 2056 (1934).

37. J. W. Baker, W. S. Nathan, and C. W. Shoppee, *J. Chem. Soc.* p. 1848 (1935).
38. H. A. Riley and A. R. Gray, *Org. Synth. Collect. Vol.* **2,** 509 (1943).
39. N. Kornblum, J. W. Powers, G. J. Anderson, W. J. Jones, H. O. Larson, O. Levand, and W. M. Weaver, *J. Am. Chem. Soc.* **79,** 6562 (1957).
40. E. J. Moriconi and A. J. Fritsch, *J. Org. Chem.* **30,** 1542 (1965).
41. N. Kornblum, W. J. Jones, and G. Anderson, *J. Am. Chem. Soc.* **81,** 4113 (1959).
42. E. Campaigne and W. A. Le Suer, *J. Am. Chem. Soc.* **70,** 1557 (1948).
43. S. J. Angyal, J. R. Tetaz, and J. W. Witson, *Org. Synth. Collect. Vol.* **4,** 690 (1963).
44. K. B. Wiberg, *Org. Synth. Collect. Vol.* **3,** 811 (1955).
45. E. Campaigne, R. C. Bourgeois, and W. C. McCarthy, *Org. Synth. Collect. Vol.* **4,** 918 (1963).
46. H. Stephen, *J. Chem. Soc.* **127,** 1874 (1925).
47. E. Mosettig, *Org. React.* **8,** 218 (1954).
48. E. N. Zilberman and P. S. Pyryalova, *Zh. Obshch. Khim.* **33,** No. 10, 3420 (1963).
49. H. C. Brown, C. J. Shoaf, and C. P. Garg, *Tetrahedron Lett.* No. 3, p. 9 (1959).
50. H. C. Brown and C. P. Garg, *J. Am. Chem. Soc.* **86,** 1085 (1964).
51a. H. Plieninger and G. Werst, *Angew. Chem.* **67,** 156 (1955).
51b. T. Van Es and B. Staskun, *Org. Synth.* **51,** 20 (1971).
52. G. Mignonae and P. Bourbon, U.S. Patent 2,945,862 (1960).
53. A. Galffe and R. Palland, *C. R. Hebd. Seances Acad. Sci.* **254,** 496 (1962).
54. L. I. Zakharkin and I. M. Khorlina, *Dokl. Akad. Nauk. SSSR* **116,** 422 (1957).
55. L. I. Zakharkin and I. M. Khorlina, *Zh. Obshch. Khim.* **34,** No. 3, 1029 (1964).
56. F. Henle, *Chem. Ber.* **38,** 1362 (1905).
57. J. W. Williams, *J. Am. Chem. Soc.* **61,** 2248 (1939).
58. H. H. Fox, *J. Org. Chem.* **17,** 555 (1952).
59. F. Weygand, G. Eberhardt, H. Linden, F. Schafer, and I. Eigen, *Angew. Chem.* **65,** 525 (1953).
60. L. Friedman, U.S. Patent 3,076,033 (1963); *Chem. Abstr.* **59,** 1534 (1963).
61. T. D. Perrine and E. L. May, *J. Org. Chem.* **19,** 775 (1954).
62. A. W. Burgstahler, L. R. Worder, and T. B. Lewis, *J. Org. Chem.* **28,** 2918 (1963).
63a. E. Mosetting and R. Mozingo, *Org. React.* **4,** 362 (1948).
63b. A. I. Rachlin, H. Garien, and D. P. Wagner, *Org. Synth.* **51,** 8 (1971).
64. H. C. Brown and R. F. McFarlin, *J. Am. Chem. Soc.* **80,** 5372 (1958).
65. H. C. Brown and B. C. Subba Rao, *J. Am. Chem. Soc.* **80,** 5377 (1958).
66. I. A. Pearl, *J. Org. Chem.* **24,** 736 (1959).
67a. H. C. Brown and A. Tsukamoto, *J. Am. Chem. Soc.* **83,** 4559 (1961).
67b. M. R. Johnson and B. Rickborn, *Org. Synth.* **51,** 11 (1971).
67c. J. E. Citron, *J. Org. Chem.* **34,** 1977 (1969).
68. G. Wittig and P. Hornberger, *Justus Liebigs Ann. Chem.* **577,** 11 (1952).
69. W. Weygand and E. Eberhardt, *Angew. Chem.* **64,** 458 (1952).
70. M. Mousseron, R. Jacquier, M. Mousseron-Conet, and R. Zagebaron, *Bull. Soc. Chim. Fr.* **19,** 1042 (1952).
71. W. Reid and F. J. Konjstein, *Angew. Chem.* **70,** 165 (1958).
72. R. F. Nystrom and W. G. Brown, *J. Am. Chem. Soc.* **70,** 3738 (1946).
73. L. I. Zakharkin and I. M. Khorlina, *Izv. Akad. Nauk SSSR, Ser. Khim.* No. 3, p. 465 (1964).
74. R. Greive and H. Buttner, *Chem. Ber.* **91,** 2452 (1958).
75. A. Sonn and E. Muller, *Chem. Ber.* **52,** 1927 (1919).
76. J. W. Cook, W. Graham, A. Cohen, R. W. Lapsley, and C. A. Lawrence, *J. Chem. Soc.* p. 322 (1944).
77. T. S. Work, *J. Chem. Soc.* p. 429 (1942).

References

78. W. E. Truce, *Org. React.* **9,** 37 (1957).
79. N. N. Crounse, *Org. React.* **5,** 290 (1949).
80. H. Gross, A. Rieche, and G. Matthey, *Chem. Ber.* **96,** 308 (1963).
81. H. Gross and G. Matthey, *Chem. Ber.* **97,** 2609 (1964).
82. G. Olah and I. Kuhn, *J. Am. Chem. Soc.* **82,** 2380 (1960).
83. A. Vilsmeier and A. Haack, *Chem. Ber.* **60,** 119 (1927).
84. R. Adams and I. Levine, *J. Am. Chem. Soc.* **45,** 2373 (1923).
85. R. Adams and E. Montgomery, *J. Am. Chem. Soc.* **46,** 1518 (1924).
86. R. T. Arnold and J. Sprung, *J. Am. Chem. Soc.* **60,** 1699 (1938).
87. R. C. Fuson, E. C. Horning, S. P. Roland, and M. L. Ward, *Org. Synth. Collect. Vol.* **3,** 549 (1955).
88. E. L. Niedzielski and F. F. Nord, *J. Org. Chem.* **8,** 147 (1943).
89. A. Bauer-Benedikt and H. Punzengruber, *Monatsh. Chem.* **81,** 772 (1950).
90. E. Campaigne and W. L. Archer, *Org. Synth. Collect. Vol.* **4,** 331 (1963).
91. L. F. Fieser, H. L. Hartwell, and J. E. Jones, *Org. Synth.* **20,** 11 (1940).
92. E. J. Seus, *J. Org. Chem.* **30,** 2818 (1965).
93. P. Ruggli and E. Burckhardt, *Helv. Chim. Acta* **23,** 447 (1940).
94. H. H. Bosshard and H. Zollinger, *Helv. Chim. Acta* **42,** 1619 (1959).
95. R. M. Silverstein and E. E. Ryskiewicz, *Org. Synth. Collect. Vol.* **4,** 831 (1963).
96. A. E. Shabria, E. E. Horne, J. B. Ziegler, and M. Tishler, *J. Am. Chem. Soc.* **68,** 1156 (1946).
97. P. N. James and H. R. Snyder, *Org. Synth. Collect. Vol.* **4,** 539 (1963).
98. A. W. Weston and R. J. Michaels, Jr., *J. Am. Chem. Soc.* **72,** 1422 (1950).
99. A. W. Weston and R. J. Michaels, Jr., *Org. Synth. Collect. Vol.* **4,** 539 (1963).
100. L. F. Fieser, J. L. Hartwell, J. E. Jones, J. H. Wood, and R. W. Bost, *Org. Synth. Collect. Vol.* **3,** 98 (1955).
101. G. F. Smith, *J. Chem. Soc.* p. 3842 (1954).
102. L. I. Smith and M. Bayliss, *J. Org. Chem.* **6,** 437 (1941).
103. L. Gattermann and F. Moffezzoli, *Ber. Dtsch. Chem. Ges.* **36,** 4152 (1903).
104. N. D. Zelinsky, *Chem.-Ztg.* **28,** 303 (1904).
105. E. A. Evans, *Chem. & Ind. (London)* p. 1596 (1957).
106. L. I. Smith and J. Nichols, *J. Org. Chem.* **6,** 489 (1941).
107. K. Reimer, *Ber. Dtsch. Chem. Ges.* **9,** 423 (1876).
108. J. Hine and J. H. Van der Veen, *J. Am. Chem. Soc.* **81,** 6446 (1959).
109. A. R. Russell, and L. B. Lockhart, *Org. Synth.* **22,** 63 (1962).
110. H. H. Hodgson and T. A. Jenkinson, *J. Chem. Soc.* p. 469 (1929).
111. A. R. Russell and L. B. Lockhart, *Org. Synth. Collect. Vol.* **3,** 463 (1955).
112. W. J. Boyd and W. Robson, *Biochem. J.* **29,** 555 (1935).
113a. H. Fischer, H. Beller, and A. Stern, *Ber. Dtsch. Chem. Ges. B* **61,** 1074 (1928).
113b. R. E. Ireland and D. J. Dawson, *Org. Synth.* **54,** 71 (1974).
113c. W. Nagata, T. Wakabayashi, and Y. Hayase, *Org. Synth.* **53,** 104 (1973).
113d. J. M. McIntosh, H. Khalil, and D. W. Pillon, *J. Org. Chem.* **45,** 3436 (1980).
114. J. C. Duff and V. I. Furness, *J. Chem. Soc.* p. 1512 (1951).
115. H. Adkins and G. Krsek, *J. Am. Chem. Soc.* **71,** 3051 (1949).
116. R. Levine, J. A. Conroy, J. T. Adams, and C. R. Hauser, *J. Am. Chem. Soc.* **67,** 1510 (1945).
117. R. S. Long, *J. Am. Chem. Soc.* **69,** 990 (1947).
118. B. P. Mariella, *J. Am. Chem. Soc.* **69,** 2670 (1947).
119. V. A. Petrov, *J. Chem. Soc.* p. 694 (1942).
120. R. Adams and G. H. Coleman, *Org. Synth. Collect. Vol.* **1,** 214 (1941).
121. R. P. Barnes, C. J. Pierce, and C. C. Cochrane, *J. Am. Chem. Soc.* **62,** 1084 (1940).
122. G. Darzens and G. Levy, *C. R. Hebd. Seances Acad. Sci.* **196,** 348 (1933).

123. W. L. McEwen, *Org. Synth. Collect. Vol.* **2**, 133 (1943).
124. M. Kerfanto, *C. R. Hebd. Seances Acad. Sci.* **254**, 4935 (1962).
125. J. M. Snell and A. Weissberger, *Org. Synth. Collect. Vol.* **3**, 788 (1955).
126. J. C. Bill and D. S. Tarbell, *Org. Synth. Collect. Vol.* **4**, 807 (1963).
127. R. I. Longley, Jr. and W. S. Emerson, *J. Am. Chem. Soc.* **72**, 3079 (1950).
128. L. C. Keagle and W. H. Hartung, *J. Am. Chem. Soc.* **68**, 1609 (1946).
129. C. F. H. Allen and C. O. Edens, Jr., *Org. Synth.* **25**, 92 (1945).
130. A. Behal, and M. Sommelet, *Bull. Soc. Chim. Fr.* [3] **31**, 300 (1904).
131. M. S. Newman and B. J. Magerlein, *Org. React.* **5**, 413 (1949).
132. M. S. Newman and R. D. Closson, *J. Am. Chem. Soc.* **66**, 1554 (1944).
133. C. F. H. Allen and J. Van Allan, *Org. Synth. Collect. Vol.* **3**, 733 (1955).
134. R. A. Barnes and W. M. Budde, *J. Am. Chem. Soc.* **68**, 2339 (1946).
135. E. P. Blanchard, Jr. and G. Buchi, *J. Am. Chem. Soc.* **85**, 955 (1963).
136. J. S. McFadyen and T. S. Stevens, *J. Chem. Soc.* p. 584 (1936).
137. F. Krohnke and E. D. Dorner, *Chem. Ber.* **69**, 2006 (1936).
138. P. Karrer and A. G. Epprecht, *Helv. Chim. Acta* **24**, 1039 (1941).
139. P. Sabatier and A. Mailhe, *C. R. Hebd. Seances Acad. Sci.* **158**, 986 (1914).
140. R. R. Davis and H. N. Hodgson, *J. Chem. Soc.* p. 84 (1943).
141. R. H. Herbst and R. H. Manske, *Org. Synth. Collect. Vol.* **2**, 389 (1943).
142. E. R. Buchman and E. M. Richardson, *J. Am. Chem. Soc.* **61**, 892 (1939).
143. J. D. Roberts, *J. Am. Chem. Soc.* **73**, 2959 (1951).
144a. M. Sprecher, M. Feldkimel, and M. Wilchek, *J. Org. Chem.* **26**, 3667 (1961).
144b. R. D. Balanson, V. M. Kobal, and R. R. Schumaker, *J. Org. Chem.* **42**, 393 (1977).
144c. D. Seebach and Λ. K. Beck, *Org. Synth.* **51**, 39 (1971).
145. H. B. Haas and M. L. Bender, *J. Am. Chem. Soc.* **71**, 1767 (1949).
146. H. B. Haas and M. L. Bender, *Org. Synth.* **30**, 99 (1950).
147. J. C. Lunt and F. Sondheimer, *J. Chem. Soc.* p. 3361 (1950).
148. P. Karrer, *Helv. Chim. Acta* **2**, 89 (1919).
149. G. W. Hearne, M. W. Tamele, and W. Converse, *Ind. Eng. Chem.* **33**, 805 (1941).
150. W. L. Evers, H. S. Rothrock, H. M. Woodburn, E. E. Stahly, and F. C. Whitmore, *J. Am. Chem. Soc.* **55**, 1136 (1933).
151. M. S. Newman, *J. Am. Chem. Soc.* **57**, 732 (1935).
152. T. Curtius, *J. Prakt. Chem.* **94**, 273 (1917).
153. I. J. Rinkes, *Rec. Trav. Chim. Pays-Bas* **48**, 960 (1929).
154. K. Johnson and E. F. Degering, *J. Org. Chem.* **8**, 10 (1943).
155. R. Stoermer, *Chem. Ber.* **39**, 2288 (1906).
156. L. Kalb and O. Gross, *Chem. Ber.* **59**, 727 (1926).
157. G. R. Pettit and E. E. Van Tamelen, *Org. React.* **12**, 356 (1962); G. Maerkl, *Tetrahedron Lett.* p. 1027 (1962).
158. E. Mosettig, *Org. React.* **8**, 218 (1954).
159. A. Sisti, J. Burgmaster, and M. Fudin, *J. Org. Chem.* **27**, 279 (1962).
160. G. Wittig and E. Knauss, *Angew. Chem.* **71**, 127 (1959).
161. H. J. Harwood and M. L. Becker, *Abstr., 145th Meet. Am. Chem. Soc.* p. 68g (1963).
162. G. Wittig, H. D. Frommeld, and P. Suchanek, *Angew. Chem.* **75**, 978 (1963).
163. W. F. Beech, *J. Chem. Soc.* p. 1297 (1954).
164. C. L. Arcus and D. G. Greenwood, *J. Chem. Soc.* p. 1937 (1953).
165. J. F. Arens and D. A. van Dorp, *Nature (London)* **160**, 189 (1947).
166. A. Borodine, *Ann. Chem.* **119**, 121 (1861).
167. E. R. H. Jones and B. C. L. Weedon, *J. Chem. Soc.* p. 937 (1946).
168. M. Kutscheroff, *Chem. Ber.* **14**, 1540 (1881).

169. E. Weitz and A. Scheffer, *Chem. Ber.* **54,** 2344 (1921).
170. E. Weitz and A. Scheffer, *Chem. Ber.* **64,** 2351 (1921).
171. M. S. Newman, *J. Am. Chem. Soc.* **57,** 723 (1935).
172. J. W. Ralls, R. E. Lunden, and G. F. Bailey, *J. Org. Chem.* **28,** 3521 (1963).
173. A. Horeau and A. Ormancey, *C. R. Hebd. Seances Acad. Sci.* **236,** 826 (1953).
174. S. O. Larvesson and C. Berglund, *Ark. Kemi* **17,** 465 (1961).
175. G. F. Woods, Jr., *Org. Synth. Collect. Vol.* **3,** 470 (1955).
176. P. Z. Bedonkian, *Org. Synth. Collect. Vol.* **3,** 127 (1955); see also *J. Am. Chem. Soc.* **66,** 1325 (1944).
177. H. C. Brown, R. A. Coleman, and M. W. Rathke, *J. Am. Chem. Soc.* **90,** 499 (1968).
178. I. R. Rolitzer and A. I. Meyers, *Org. Synth.* **51,** 24 (1971).
179. L. M. Baker and W. L. Carrick, *J. Org. Chem.* **33,** 616 (1968).
180. L. M. Baker and W. L. Carrick, *J. Org. Chem.* **91,** 774 (1969).
181. H. C. Brown, E. F. Knights, and R. A. Coleman, *J. Am. Chem. Soc.* **91,** 2144 (1969).
182. G. E. Niznik, W. H. Morrison, III, and H. M. Walborsky, *Org. Synth.* **51,** 31 (1971).
183. R. A. Bankeser, H. Watanabe, S. J. Mels, and M. A. Sabol, *J. Org. Chem.* **35,** 1210 (1970).
184. J. P. Ferris and F. R. Antonucci, *J. Am. Chem. Soc.* **94,** 8091 (1972).
185. T. N. Sorrell and P. S. Pearlman, *J. Org. Chem.* **45,** 3449 (1980).
186. H. C. Brown, G. W. Kabalka, M. W. Rathke, and M. Rogie, *J. Am. Chem. Soc.* **90,** 4165 (1968).
187. Y. Lin, S. N. Lang, Jr., C. M. Seifert, R. G. Child, G. O. Morton, and P. F. Fabio, *J. Org. Chem.* **44,** 470 (1979).
188. J. M. Reuter and R. G. Salomon, *J. Org. Chem.* **42,** 3360 (1977).
189. S. Friuchijina, T. Sakakibara, and G. Tsuchihashi, *Agric. Biol. Chem.* **40**(7), 1369 (1976); *Chem. Abstr.* **85,** 123501b (1976).
190. A. I. Meyers and J. L. Durandetta, *J. Org. Chem.* **40,** 2021 (1975).
191. A. I. Meyers, A. Nabeya, H. W. Adickers, I. R. Politzer, G. R. Malone, A. C. Korelesky, R. L. Nolen, and R. C. Portnoy, *J. Org. Chem.* **38,** 36 (1973).
192. J. P. Nagarkatti and K. R. Ashley, *Tetrahedron Lett.* p. 4599 (1973).
193. G. A. Crosby, *J. Am. Chem. Soc.* **97,** 2232 (1975).
194. T. Matsuda and M. Sugishita, *Bull. Chem. Soc. Jpn.* **40,** 174 (1967); B. Rickborn and R. M. Gerkin, *J. Am. Chem. Soc.* **93,** 1693 (1971); G. Magnusson and S. Thoren, *J. Org. Chem.* **38,** 1380 (1973).

CHAPTER 8 / KETONES

1. Introduction	206
2. Oxidation Reactions	207
2-1. Preparation of 4-Methylcyclohexanone	208
2-2. Preparation of 2-Phenycyclohexanone	208
2-3. Preparation of l-Menthone	209
2-4. Preparation of 2,3-Dimethylnaphthoquinone	210
2-5. Preparation of Methyl Neopentyl Ketone	211
2-6. Preparation of 2-Methylcyclohexanone	211
2-7. Preparation of 2-Ethylcyclohexanone by the Oppenauer Oxidation	212
2-8. Preparation of Δ^4-3-Cholestenone	213
3. Condensation Reactions	213
3-1. Preparation of 9-Acetylphenanthrene by the Grignard Method	214
3-2. Preparation of Acetophenone by the Grignard Method	215
3-3. Preparation of Dimethylacetophenones by the Friedel–Crafts Acylation Method	215
3-4. Preparation of Diphenylketene	217
3-5. Preparation of 3-Oxo-2,2-diphenylcyclobutyl Acetate by the Reaction of Diphenylketene with Vinyl Acetate	219
3-6. Preparation of 2-Heptanone from Acetylacetone	222
3-7. Preparation of 3-Isobutyl-2-heptanone from tert-Butylacetoacetate	222
4. Elimination Reactions	224
4-1. Preparation of Acetophenone by the Cleavage of β-Ketosulfoxides Derived from Esters	224
5. Rearrangement Reactions	226
5-1. Preparation of 3,3-Dimethyl-2-pentanone and 2,2-Dimethyl-3-pentanone	226
6. Miscellaneous Methods	227
References	229

1. INTRODUCTION

The oxidation of secondary alcohols by sulfuric–chromic acid at 20°–40°C gives good yields of ketones and is a widely used synthetic method. Several other oxidizing agents have been found to be quite useful. Mild oxidizing agents such as dimethyl sulfoxide and air have also been used.

$$R_2CHOH \longrightarrow R_2C=O \tag{1}$$

Triethylene compounds activated by halogen, carbonyl, double bond, aromatic rings, heterocyclic rings, etc., can be oxidized using several oxidants to ketones or quinones.

§ 1. Introduction

The hydroboration of olefins and their subsequent oxidation offers a novel method of specifically oxidizing olefins to ketones.

Ozonolysis of olefins and the use of other oxidants such as permanganate, permanganate–periodate, dichromate–sulfuric acid, and hydrogen peroxide–lead tetraacetate are also useful in preparing ketones by cleaving the olefin into two fragments.

The Oppenauer oxidation is a preferred method for oxidizing sensitive alcohols such as the sterols. Modifications have been reported which permit the reaction to be carried out at room temperature.

The condensation reactions involving the Grignard reagent and the Friedel–Crafts method are perhaps the most popular laboratory methods for introducing a ketone group into a molecule.

$$\text{RMgX} + \begin{cases} \text{R'COCl} \\ \text{R'CN} \\ (\text{R'CO})_2\text{O} \\ \text{R'CONH}_2 \end{cases} \longrightarrow \text{R}\overset{\overset{\text{O}}{\|}}{\text{C}}\text{R'} \quad (2)$$

$$\text{R}_2\text{Cd} + 2\,\text{RCOCl} \longrightarrow 2\,\text{RCOR} + \text{CdCl}_2 \quad (3)$$

$$\text{RCOCl} + \text{ArH} \xrightarrow{\text{AlCl}_3} \text{RCOAr} \quad (4)$$

Active methylene groups of esters, ketones, and other compounds can be alkylated, acylated, or self-condensed prior to hydrolysis or cleaved to give substituted mono- or diketones.

The cleavage of β-ketosulfoxides provides a novel method of converting an ester to a methyl ketone.

The thermal decarboxylation of acids is not a preferred laboratory method since high temperatures are involved and low yields of unsymmetrical ketones are obtained. However, new modifications have been described to extend this reaction to the preparation of unsymmetrical ketones.

The pinacol rearrangement is an effective method for rearranging completely alkylated 1,2-glycols. The reaction has been extended to 1,1'-dihydroxyalkanes to yield spiro ketones.

2. OXIDATION REACTIONS

The oxidation of secondary alcohols to ketones in good yields is effected by sulfuric–chromic acid mixtures. For water-soluble alcohols the reaction is carried out in aqueous solution at 20°–40°C [1, 2]. Insoluble aromatic alcohols are oxidized in an acetic acid solvent [3]. Some other oxidation reagents that have been used are nitric acid [4], copper sulfate in pyridine [5], cupric acetate in 70%

acetic acid [6], ferric chloride [6] in water, chromic anhydride in glacial acetic acid [7], chromic oxide in pyridine [8,9], chromic acid in aqueous ether solutions [10], dimethyl sulfoxide and air [11], and dinitrogen tetroxide in chloroform [12]. Oxidation reactions in organic chemistry have recently been reviewed [13a].

More recently methyl sulfide–N-chlorosuccinimide in toluene at 0°C gives good yields of ketones by the oxidation of primary and secondary alcohols, excluding allylic and dibenzylic alcohols which give halides (13b). The latter alcohols can be oxidized to ketones with a dimethyl sulfoxide–chlorine reagent [13c].

2-1. Preparation of 4-Methylcyclohexanone [14]

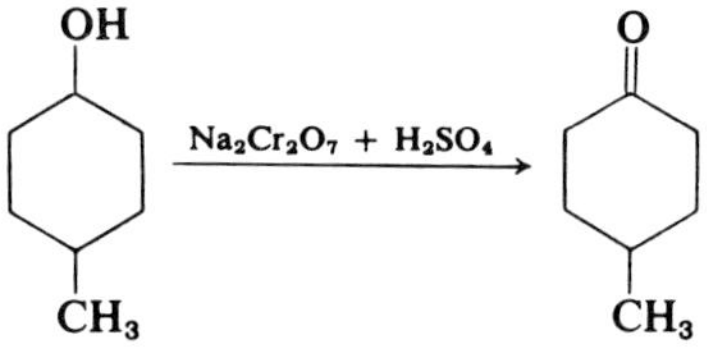

$$\underset{\text{CH}_3}{\text{4-methylcyclohexanol}} \xrightarrow{\text{Na}_2\text{Cr}_2\text{O}_7 + \text{H}_2\text{SO}_4} \underset{\text{CH}_3}{\text{4-methylcyclohexanone}} \quad (5)$$

To a flask containing a solution of 367 gm (1.4 moles) of sodium dichromate in 1¾ liters of water at 80°C is added 798 gm (7.0 moles) of 4-methylcyclohexanol. To the stirred mixture is added dropwise a solution of 367 gm (1.4 moles) of sodium dichromate and 1078 gm (11.0 moles) of sulfuric acid in 1¾ liters of water at such a rate to maintain the temperature at 80°C (approximately 12 hr). The ketone and water are distilled from the reaction mixture, the ketone is separated from the aqueous phase, and then distilled to give 549 gm (70%) of 4-methylcyclohexanone, bp 171.4°–171.5°C (760 mm), n_D^{20} 1.4462; d_{20}^{20} 0.917.

2-2. Preparation of 2-Phenylcyclohexanone [3]

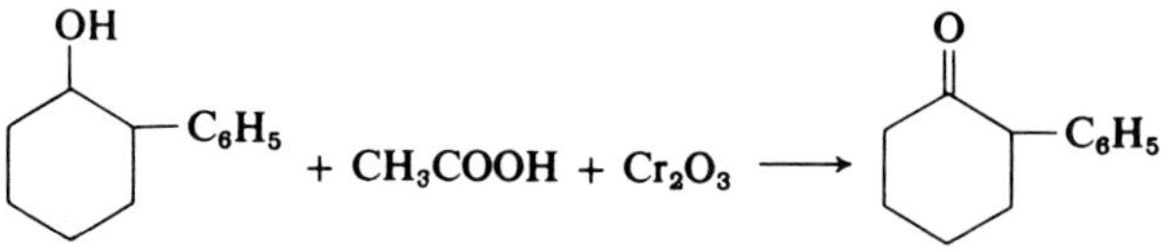

$$\underset{}{\text{2-phenylcyclohexanol}} + \text{CH}_3\text{COOH} + \text{Cr}_2\text{O}_3 \longrightarrow \underset{}{\text{2-phenylcyclohexanone}} \quad (6)$$

To a flask containing a stirred solution of 31 gm (0.176 mole) of cis-2-phenylcyclohexanol in 50 ml of glacial acetic acid is added dropwise a solution of 14.0 gm (0.092 mole) of chromic oxide in 50 ml of 80% acetic acid at such a rate as to maintain 50°C. The mixture is allowed to stand for 24 hr and then extracted with benzene [(1) 200-ml portion and (2) 100-ml portion]. The combined extracts are washed with aqueous sodium bicarbonate and then with water, dried, and then distilled under vacuum to yield 25.0 gm (80%), bp 155°–160°C (16 mm), mp 52.5°–54.5°C.

§ 2. Oxidation Reactions

The advantages of an immiscible ether layer during oxidation is that ketones capable of undergoing epimerization under the usual oxidation condition are extracted into the ether layer as soon as they are formed and are thus protected from further oxidation [15].

2-3. Preparation of l-Menthone [15]

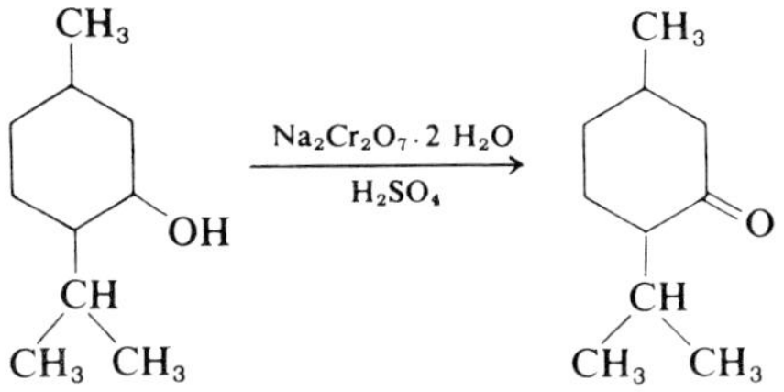

$$\text{menthol} \xrightarrow[\text{H}_2\text{SO}_4]{\text{Na}_2\text{Cr}_2\text{O}_7 \cdot 2\text{H}_2\text{O}} \text{menthone} \qquad (7)$$

To a flask containing 20 ml of ether and 7.80 gm (0.050 mole) of l-menthol is added dropwise over a period of 15 min a solution of chromic acid, prepared from 5.0 gm (0.0168 mole) of sodium dichromate dihydrate, and 3.75 ml (0.067 mole) of 96% sulfuric acid diluted to 25 ml, at such a rate as to maintain the temperature at 25°C. After 2 hr, the ether layer is separated and the aqueous phase is extracted with two 10-ml portions of ether. The combined extracts are washed with a saturated sodium bicarbonate solution and then with water. Distillation at reduced pressure yields 6.45 gm (84%) of l-menthone, bp 66°–67°C (4 mm), n_D^{20} 1.4500.

Dinitrogen tetroxide has been used to oxidize arylcarbinols to acetophenones in chloroform at 0°C in 88–98% yields [12]. Steam distillation or distillation at reduced pressure yields the ketone.

Recently it has been reported that active methylene bromides can be oxidized at room temperature in dimethyl sulfoxide to yield ketones or aldehydes [16a,b].

$$\text{>CHBr} \longrightarrow \text{RCHO} \quad \text{or} \quad \text{>C=O} \qquad (8)$$

Primary and secondary alkyl bromides have been reported to be oxidized by dimethyl sulfoxide–silver tetrafluoroborate at room temperature to give 55–95% aldehydes and ketones [16c]. Benzyl bromides are oxidized in fair yields using acetonitrile [16a].

Active methylene compound can be oxidized by manganese dioxide to ketones [17a].

Active methylene groups in ring systems such as adamantane has been reported to be oxidized to adamantanone by 98% sulfuric acid at 80°C [17b].

Ruthenium tetroxide [18] has been reported to be a powerful oxidizing agent that readily oxidizes aldehydes to acids, alcohols to aldehydes or ketones, and ethers to esters at low temperatures.

Careful oxidation of some aromatic hydrocarbons with chromic–sulfuric acid mixtures yields quinones in good yield [19]. Other oxidizing agents have also been found to be effective. Vanadium pentoxide and sodium chlorate mixtures have been used for anthracene [20]. Hydrogen peroxide (30%) in acetic acid (peracetic acid) has also been found to be effective [21].

Phenols, amino phenols, and aryl diamines yield quinones upon oxidation with chromic–sulfuric acid mixtures [22–23].

Quinones can also be prepared by ring closure of o-aroylbenzoic acid in the presence of concentrated sulfuric acid [20].

2-4. Preparation of 2,3-Dimethylnaphthoquinone [24]

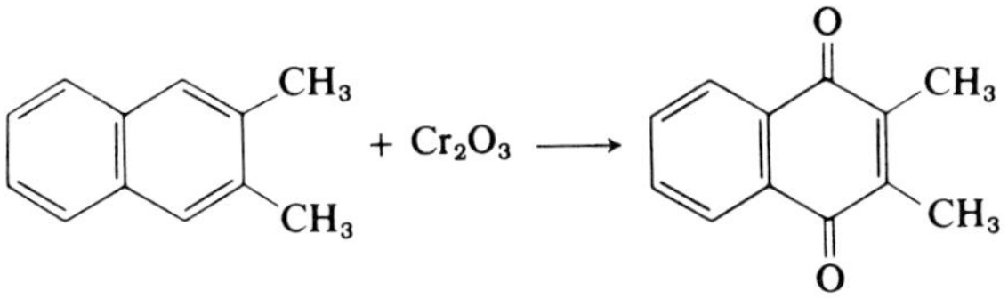

$$\text{2,3-dimethylnaphthalene} + Cr_2O_3 \longrightarrow \text{2,3-dimethylnaphthoquinone} \quad (9)$$

To a flask containing 10 gm (0.064 mole) of 2,3-dimethylnaphthalene in 350 ml of acetic acid at room temperature is added dropwise with stirring a solution prepared from 25.6 gm (0.169 mole) of chromic acid dissolved in 130 ml of acetic acid and 35 ml of water. The addition requires about 15 min and the temperature is maintained at 20°–30°C. After standing at room temperature for 3 days, the reaction mixture is diluted with 1.5 liters of water and then allowed to stand for several hours. The product is filtered, and then recrystallized from alcohol to yield 6–7.9 gm (60–80%) of material with a melting point of 126°–127°C.

Olefins are readily ozonized and the ozonides are decomposed to ketones [25, 26]. CAUTION: This method requires a controlled source of ozone and extreme caution since ozonides are potentially explosive. For the latter reason, the method has thus far found limited use in the laboratory. However, commercial ozonizers [27] have become available and this has increased the use of this oxidation procedure.

Potassium dichromate–sulfuric acid [28], potassium permanganate [28], and oxygen [29] have been used to oxidize olefins to ketones. In addition olefins can be oxidized to glycols with hydrogen peroxide or performic acid and then cleaved with lead tetraacetate [30] to ketones in good yields.

$$RCH_2CH{=}CHR \longrightarrow RCH_2\underset{\underset{O}{\|}}{C}{-}CH_2R \quad (10)$$

§ 2. Oxidation Reactions

$$R_2C=CR_2 + [O] \longrightarrow R_2\underset{OH}{C}-\underset{OH}{C}R_2 \longrightarrow 2\,R_2C=O \qquad (11)$$

2-5. Preparation of Methyl Neopentyl Ketone [31]

$$\underset{\underset{CH_3}{|}}{\overset{\overset{CH_3}{|}}{CH_3C}}-CH_2-\overset{\overset{CH_3}{|}}{C}=CH_2 + K_2Cr_2O_7 + H_2SO_4 \longrightarrow CH_3-\underset{\underset{CH_3}{|}}{\overset{\overset{CH_3}{|}}{C}}-CH_2-\overset{\overset{}{\|}}{\underset{O}{C}}-CH_3 \qquad (12)$$

To a flask containing 1176.9 gm (4.0 moles) of potassium dichromate in 800 ml of water is added 336.6 gm (3.0 moles) of technical diisobutylene, bp 101°–104°C, n_D^{25} 1.4060 (80% of 2,4,4-trimethyl-1-pentene). To this stirred mixture is added dropwise over a period of 5 days 1569 gm (16 moles) of concentrated sulfuric acid at such a rate as to maintain the temperature at 30°–35°C. The mixture is stirred for an additional day, and then steam-distilled to yield 248.6 gm of crude methyl neopentyl ketone. Distillation yields 105.8 gm of methyl neopentyl ketone, bp 124°–125°C (760 mm), n_D^{25} 1.4018. The total yield obtained is 155.2 gm (56.0%).

It has been reported [32a] that olefins can be hydroborated and then oxidized directly to ketones in good yields.

$$R_2C=CR_2 + NaBH_4 + ZnCl_2 \longrightarrow (R_2CHCR_2)_3B \xrightarrow[35°-37°C]{H_2CrO_4} R_2CH-\underset{\underset{O}{\|}}{C}R_2 \qquad (13)$$

More recently Brown and co-workers reported a general synthesis of ketones from alkenes via a stepwise hydroboration with hexylchloroborane and methyl sulfide [32b].

2-6. Preparation of 2-Methylcyclohexanone [32a]

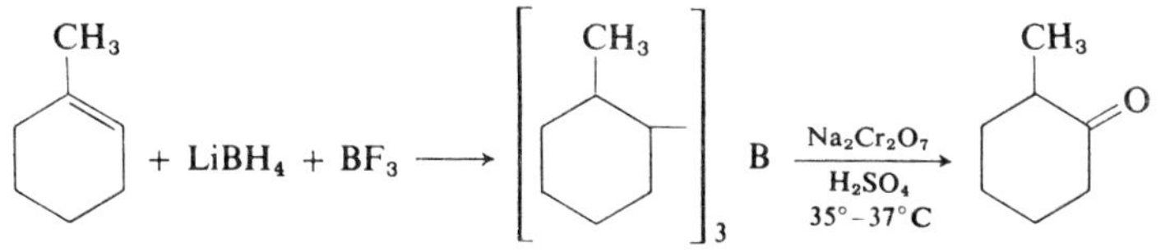

$$\underset{}{\text{(1-methylcyclohexene)}} + LiBH_4 + BF_3 \longrightarrow \left[\underset{}{\text{(methylcyclohexyl)}}\right]_3 B \xrightarrow[\substack{H_2SO_4 \\ 35°-37°C}]{Na_2Cr_2O_7} \underset{}{\text{(2-methylcyclohexanone)}} \qquad (14)$$

To a flask containing 4.8 gm (0.05 mole) of 1-methylcyclohexene and 0.38 gm (0.0225 mole) of lithium borohydride in 30 ml ether is added 0.95 ml (0.0075 mole) of boron trifluoride etherate in 4 ml of ether over a period of 15 min at 25°–30°C. After 2 hr, 5 ml of water is cautiously added in order to destroy the excess hydride. To the remaining reaction mixture is added a chromic acid solution [prepared from 11.0 gm (0.0369 mole) of sodium dichromate dihydrate and 8.25 ml (0.147 mole) of 96% sulfuric acid diluted to 45 ml with water] over a period of 15 min while the temperature is maintained at 25°–30°C. The reaction mixture is heated to reflux for 2 hr, the upper layer is separated, and the aqueous layer is extracted with two 10-ml portions of ether. After concentrating the ether, the residue is distilled to yield 4.36 gm (78%) of 2-methylcyclohexanone, bp 63°–64°C (24 mm), n_D^{20} 1.4487.

Trialkylboranes have been added to acetylacetylene in the presence of catalytic amounts of oxygen and then the hydrolysis of the allenic intermediate gave the α,β-unsaturated ketone in good yield [32c]. Other ketone syntheses via organo boron derivatives are also worth investigation [32c].

The Oppenauer oxidation [33, 34] employs mild conditions and, therefore, is the preferred method for sensitive compounds such as steroids. Aluminum isopropoxide, acetone, and the alcohol are refluxed for several hours to produce the ketone in excellent yields. Methyl ethyl ketone and cyclohexanone are the best oxidizing agents for high molecular weight alcohols such as the sterols [35]. Benzil is employed when the ketone can be distilled from the reaction mixture and p-benzoquinone or benzil are suitable oxidizing agents for ketones boiling at 100° to 200°C [35]. Aluminum or potassium *tert*-butoxides are also commonly used in place of aluminum isopropoxide.

2-7. *Preparation of 2-Ethylcyclohexanone by the Oppenauer Oxidation* [35]

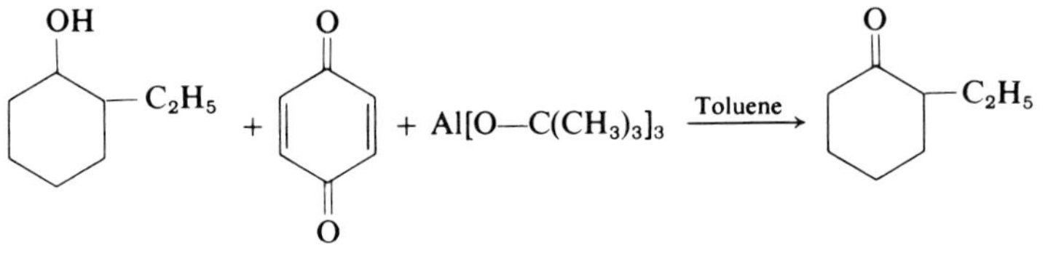

$$\text{(cyclohexanol-C}_2\text{H}_5\text{)} + \text{(p-benzoquinone)} + \text{Al[O—C(CH}_3)_3]_3 \xrightarrow{\text{Toluene}} \text{(2-ethylcyclohexanone)} \quad (15)$$

To a flask containing 65 gm (0.60 mole) of p-benzoquinone and 14.8 gm (0.06 mole) of aluminum *tert*-butoxide in 1 liter of toluene is added 15 gm (0.127 mole) of 2-ethylcyclohexanol. The flask is allowed to stand at 20°–25°C for 8 days. Then water is cautiously added and the resultant aluminum hydroxide is removed. The remaining toluene solution is washed with 5% sodium hydroxide solution and then with water. Distillation yields 11.4 gm (77%) of 2-ethylcyclohexanone, bp 178°–182°C (738 mm), n_D^{25} 1.4500.

A modified Oppenauer oxidation [36] employing fluorenone to oxidize quinine permits the reaction to be carried out in benzene solution at room temperature in ½ to 1 hr. Other alcohols such as cholesterol, deoxyajmaline, and yohimbine gave good yields of the ketone.

2-8. Preparation of Δ⁴-3-Cholestenone [36]

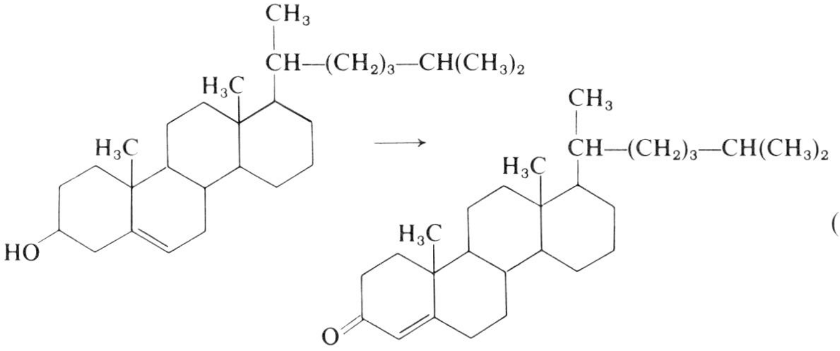

$$\text{(structure of cholesterol)} \longrightarrow \text{(structure of }\Delta^4\text{-3-cholestenone)} \tag{16}$$

To a flask containing potassium *tert*-butoxide prepared from 0.5 gm (0.013 mole) of potassium, 4.50 gm (0.025 mole) of fluorenone, and 40 ml of dry benzene is added 2.02 gm (0.00524 mole) of dry cholesterol. The reaction mixture is stirred under a nitrogen atmosphere for 1 hr at room temperature and then diluted with water and ether. The ether layer is separated, dried, and concentrated to give 6.35 gm of a yellow oil. Removal of 2.99 gm (0.0166 mole) of fluorenone is facilitated by recrystallization from cyclohexane. Chromatographing the filtrate on 100 gm of aluminum and elution with benzene gives Δ⁴-3-cholestenone which upon recrystallization from cyclohexane yields 0.90 gm (44%), mp 80°–82°C.

Other oxidation reagents such as dicyclohexylcarbodiimide and phosphoric acid in dimethyl sulfoxide [37] or acetic anhydride in dimethyl sulfoxide [38] yield ketones from secondary alcohols at room temperature in the 18–24 hr. The latter method appears to be particularly useful for the mild oxidation of sterically hindered hydroxyl groups such as in the indole alkaloids [39]. This reaction cannot be extended to phenols in order to obtain quinones [40, 41a].

More recently secondary alcohols have been reported to be oxidized to ketones by peroxo complexes of molybdenum and tungsten [41b]. The use of oxidative dehydrogenation of alcohols by cupric oxide [41c] or palladium salts [41d] has been reported to give good yields of ketones from secondary alcohols.

The oxidation of aryl alkyl ethers by stable nitronium or nitrosonium salts has been reported to give aryl ketones [41e].

3. CONDENSATION REACTIONS

The Grignard reaction and Friedel–Crafts reaction are the two condensation reactions used most frequently to prepare ketones in the laboratory.

The Grignard reagent reacts with nitriles to form ketimine salts which on hydrolysis yield ketones. Low molecular weight aliphatic nitriles give ketones contaminated with hydrocarbons derived from the acidic α-hydrogen of the nitrile [42]. The difficulty can be overcome to some extent by discarding the ethereal solution containing the hydrocarbon by-products before the hydrolysis of the ketimine salts [43]. Therefore, the Grignard procedure is successful with aromatic nitriles or high molecular weight aliphatic nitriles [44]. Low molecular weight aliphatic nitriles respond favorably with aromatic Grignard reagents [45, 46].

3-1. Preparation of 9-Acetylphenanthrene by the Grignard Method [47]

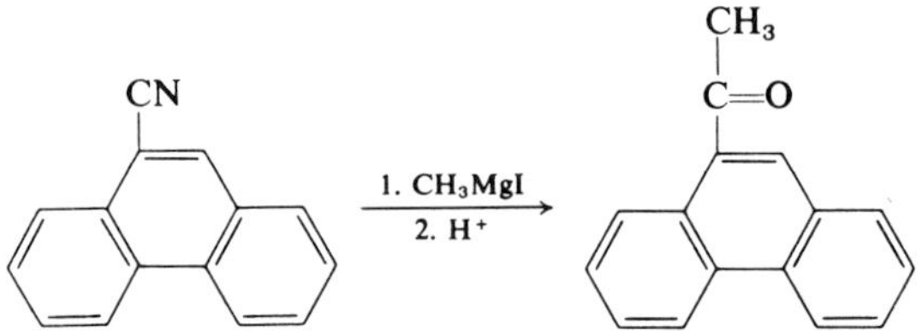

$$\text{9-cyanophenanthrene} \xrightarrow[\text{2. H}^+]{\text{1. CH}_3\text{MgI}} \text{9-acetylphenanthrene} \quad (17)$$

To a flask containing 0.11 mole of methylmagnesium iodide (prepared from 7 ml of methyl iodide in 30 ml of ether containing 2.7 gm of magnesium) is added 25 ml of benzene followed by 15 gm (0.074 mole) of 9-cyanophenanthrene. The reaction mixture is refluxed for 3 hr, cooled, and hydrolyzed with cold ammonium chloride solution. The benzene–ether layer is separated, shaken with cold dilute hydrochloric acid, and then refluxed for 1 hr in order to hydrolyze the ketimine hydrochloride to the ketone. The ketone, which precipitates as an oil, is separated, distilled under pressure, and then recrystallized from ethanol to yield 9.5 gm (58%), mp 73°–74°C.

Grignard reagents catalyzed by ferric chloride also give ketones upon reaction with acid chlorides [48], with anhydrides [49], esters [50], amides [51, 52], and with salts of carboxylic acids [50]. Low-temperature reaction ($-75°C$) gives good yields of ketones, when primary, secondary, and tertiary aliphatic or aromatic Grignard reagents are treated with acetic [49], propionic, or butyric anhydrides [53, 54].

Coupling the Grignard reagent with α-haloketones gives good yields of substituted ketones, as, for example, in the case of 2-chlorocyclohexanone [55, 56a].

Organolithium reagents with carboxylic acids represent another general method for synthesizing ketones [56b].

Cadmium [57], zinc alkyls [58a], or copper alkyls [58b] have been used to couple with acid halides to give good yields of ketones. Cadmium alkyls give good yields if the alkyl is not secondary or tertiary and if the formed ketone is not highly reactive [58a].

§ 3. Condensation Reactions

Tetra-organotin compounds have been used to react with acid chlorides to give ketones [58c]. Benzylchlorobis(triphenylphosphine)palladium(II) is used to catalyze the reaction [58c].

3-2. Preparation of Acetophenone by the Grignard Method [53]

$$C_6H_5MgBr + (CH_3CO)_2O \xrightarrow[\text{Ether}]{80°C} C_6H_5\overset{\overset{O}{\|}}{C}CH_3 \quad (18)$$

To a flask containing 40 gm (0.39 mole) of acetic anhydride in 100 ml of anhydrous ether cooled to $-80°C$ using Dry Ice–acetone is added 0.2 mole of an ether solution of phenylmagnesium bromide. The reaction mixture is stirred for 2–3 hr and then hydrolyzed with ammonium chloride solution. The ether layer is washed with water and alkali in order to remove the excess acetic anhydride and acetic acid. The remaining ether is dried and fractionally distilled to yield 16.8 gm (70%) of acetophenone, bp 202°C, n_D^{20} 1.5339.

The Friedel–Crafts [59] acylation [60] method is an excellent method for introducing a ketone group into an aromatic hydrocarbon molecule. Some of the acylating agents used vary in reactivity as follows [2,61]: $RCO^+BF_4^- > RCO_2ClO_3 > RCOOSO_3H > RCO$ halogen $> RCOOCOR' > RCOOR' > RCONR_2 > RCHO > RCOR'$. The acylation of polyalkylbenzenes with acetyl chloride, and aluminum chloride in carbon tetrachloride has been reported to give good yields [62]. Aluminum chloride is the most effective catalyst [63].

The acylation of heterocycles has been reported [64]. Diketones are prepared by the Friedel–Crafts method using adipyl chloride, benzene, and aluminum chloride [65].

3-3. Preparation of Dimethylacetophenones by the Friedel–Crafts Acylation Method [66]

$$\underset{\text{CH}_3}{\underset{|}{\text{C}_6\text{H}_3}}\text{—CH}_3 + AlCl_3 + CH_3COCl \xrightarrow{CCl_4} \underset{\underset{\text{COCH}_3}{|}}{\underset{\text{CH}_3}{\underset{|}{\text{C}_6\text{H}_2}}}\text{—CH}_3 + AlCl_3 + HCl \quad (19)$$

The method of Mowry [62] was used to react each of the xylene isomers with acetyl chloride to yield the corresponding acetophenones.

General Procedure: To a 5-liter three necked flask equipped with a stirrer, dropping funnel, condenser, and dry tube is added 1½ liter of dry, freshly distilled carbon tetrachloride. To this is added 454 gm (3.3 moles) or reagent grade granular aluminum chloride and the mixture is cooled to about 5°C by means of an ice-water bath. Acetyl chlorides (275 gm, 249 ml, 3.5 moles) are added dropwise to the cooled mixture over a 5–10 min interval. This addition is then followed with the dropwise addition of the appropriate xylene isomer [o-xylene, 300 ml (2.5 moles), m-xylene 300 ml (2.5 moles), or p-xylene 275 ml (2.3 moles)] at 10°–15°C, which takes about 1–2 hr. The mixture is stirred at room temperature for 2 hr and then allowed to stand overnight at room temperature.* The mixture is poured into a mixture of 5 kg of ice and 700 ml of concentrated hydrochloric acid. The lower carbon tetrachloride layer is separated, washed twice with 250 ml portions of water, once with 500 ml of 2% sodium hydroxide solution, and then several times with water until the washings are neutral. The carbon tetrachloride is distilled off at atmospheric pressure and the residue is fractionally distilled to give 280 gm (75%) of 3,4-dimethylacetophenone, bp 103°C (6 mm), n_D^{20} 1.6381; 140 gm (43%) of 2,5-dimethylacetophenone, bp 85°–86°C (3 mm), n_D^{20} 1.5291; and 225 gm (61%) of 2,4-dimethylacetophenone, bp 90°C (6 mm), n_D^{20} 1.5340.

The Fries reaction, which involves the rearrangement of phenolic esters to phenolic ketones, can be considered an intramolecular Friedel–Crafts reaction [67].

Acyl chlorides can be condensed with olefins [68] or acetylenes [69a] to give ketones.

$$R'COCl + RCH{=}CHR \longrightarrow RCO\underset{\underset{Cl}{|}}{\overset{\overset{H}{|}}{C}}{-}\underset{R}{\overset{|}{C}}HR \qquad (20)$$

$$R'COCl + CH{\equiv}CH \longrightarrow RCOCH{=}CHCl \qquad (21)$$

Organocopper such as methyl copper reacts with α,β-unsaturated ketones by adding to the double bond (1,4 conjugate adduct) [69b]. The same product is obtained using methyl magnesium bromide and cuprous iodide [69b].

Aldehydes also add free radically to olefins to give ketones [69c].

Keto acids are obtained by the condensation of succinic anhydride as in the case of benzene to give β-benzoylpropionic acid [70]. The reaction is applicable to other aromatic hydrocarbons and other aliphatic dibasic acid anhydrides [71].

Diketene condenses with benzene in the presence of aluminum chloride to yield benzoyl acetone [72]. The use of other diketenes may lead to other 1,3-diketones.

Ketenes such as butylethylketene condense readily with vinyl ethers [73], esters [73], and enamines [74a] to give substituted cyclobutanones.

$$C_2H_5{-}\underset{C_4H_9}{\overset{|}{C}}{=}C{=}O + CH_2{=}CH{-}O\overset{\overset{O}{\|}}{C}CH_3 \longrightarrow CH_3CO_2{-}\underset{\underset{CH_2}{|}}{\overset{\overset{C_4H_9\overset{C_2H_5}{\overset{|}{C}}{-}\overset{\overset{O}{\|}}{C}}{|}}{CH}} \qquad (22)$$

*The reaction may be worked up immediately but this is a convenient place to stop the reaction.

Some of the ketenes which undergo this reaction are $CH_2\!=\!C\!=\!O$ [74b], $(CH_3)_2C\!=\!C\!=\!O$, $(C_2H_5)(n\text{-}C_4H_9)C\!=\!C\!=\!O$, $(C_2H_5)(iso\text{-}C_4H_9)C\!=\!C\!=\!O$, and $(C_6H_5)_2C\!=\!C\!=\!O$ [74c]. Several examples of these reactions are summarized in Tables I and II.

Ketene may be conveniently generated by the pyrogenic decomposition of the reagent grade acetone in a generator described by Williams and Hurd [74d]. The dialkyl ketenes may be prepared by the pyrolysis of the corresponding anhydrides [74e]. Diphenylketene may be prepared by the dehydrogenation of diphenyl acetyl chloride or by the method described below [74f]. Haloketenes can also be prepared by dehydrohalogenation methods [74g].

Ketenes can also react with conjugated dienes such as cyclopentadiene to give the bicyclic ketone shown below [74b].

$$CH_2\!=\!C\!=\!O \ + \ \bigcirc \quad \xrightarrow[100°C]{\text{Toluene}} \quad \text{bicyclic ketone} \quad \text{or} \quad \text{bicyclic ketone} \qquad (23)$$

3-4. Preparation of Diphenylketene [74d]

$$\underset{\underset{NNH_2}{\|}}{C_6H_5C}\!-\!\overset{O}{\overset{\|}{C}}\!-\!C_6H_5 \ + \ HgO \ \longrightarrow \ \underset{\underset{N_2}{\|}}{C_6H_5\!-\!C}\!-\!COC_6H_5 \ + \ Hg \ + \ H_2O \qquad (24)$$

$$\longrightarrow N_2 \ + \ (C_6H_5)_2C\!=\!C\!=\!O$$

(a) Benzil monohydrazone. To a flask containing 158 gm (0.75 mole) of benzil is added 300 ml of alcohol. The contents are heated and then 45 gm (0.75 mole) of an 85% solution of hydrazine hydrate in water is slowly added with stirring. The product begins to separate from the hot solution after 33 gm has been added. After the addition is complete, the solution is refluxed for 5 min. The contents are cooled to 0°C, and the hydrazone is filtered off. The precipitate is washed with two 100-ml portions of cold alcohol and then dried to give an almost quantitative yield (168 gm), mp 149°–151°C with decomposition.

(b) Diphenylketene. To a mortar are added 56 gm (0.25 mole) of benzil monohydrazone, 81 gm (0.38 mole) of yellow mercuric oxide, and 35 gm of anhydrous calcium sulfate. The mixture is ground and blended together. The mixture is then added to a 1-liter three-necked flask fitted with a stirrer, condenser, and thermometer. To this flask, placed in a water bath, is now added 200 ml of dry thiophene-free benzene, and the suspension is stirred at 25°–35°C. The temperature is kept at 25°–35°C by adding ice to the water to control the initial highly exothermic reaction. The reaction mixture is stirred for 4 hr and then filtered through fine-grained filter paper. The precipitate is washed with dry benzene until the washings are colorless.

TABLE I

CYCLOADDUCTS OF DISUBSTITUTED KETENES[a]

$$RR'C{=}O + XCH{=}CHY \longrightarrow \underset{Y}{\overset{X}{\square}}\underset{}{\overset{R'}{\underset{O}{\biggr|}}}$$

R	R'	X	Y	Method[b]	Yield (%)	Bp (°C)	(mm)	n_D^{20}	% C Calcd.	% C Found	% H Calcd.	% H Found
CH₃	CH₃	OC₂H₅	H	A	80	82°–83°	38	1.4270	67.6	67.8	9.9	10.0
CH₃	CH₃	OC₂H₅	CH₃	A	64	90°–92°	34	—	69.2	68.8	10.3	10.3
CH₃	CH₃	OC₄H₉	H	A	65	72°–78°	6.5	1.4323	70.6	70.0	10.6	10.6
CH₃	CH₃	OCH₂CH(CH₃)₂	H	A	54	70°	6.9	1.4287	70.6	70.7	10.6	10.9
CH₃	CH₃	OCH(C₂H₅)(C₄H₉)	H	A	56	85°	0.7	1.4419	74.3	74.2	11.5	11.6
CH₃	CH₃	OCH₂CH₂Cl	H	A	70	91.5°–93.5°	4.9	1.4578	54.7	54.6	7.5	7.5[c]
CH₃	CH₃	OCH₂CH₂OC₆H₅	H	A	67	118°–119°	0.5	1.5104	71.8	71.7	7.7	7.8
CH₃	CH₃	OCH₂CH₂NHCOCH(CH₃)₂	H	A	80	62°–67°	0.003–0.007	1.4666	63.5	63.9	9.3	9.3[d]
CH₃	CH₃	OC₆H₄OCH₃(p)	H	A	46	134°–139°	3.5	1.5290	76.4	76.7	7.9	7.8
CH₃	CH₃	SCH₂CH₂OCOCH₃	H	A	41	119°–124°	1.5	—	55.6	55.5	7.4	7.4[e]
CH₃	CH₃	—OCH₂CH₂CH₂—		A	80	80°–90°	11	1.4632	70.1	70.4	9.2	9.2
CH₃	CH₃	—OCH(OC₂H₅)CH₂CH₂—		A	85	116°[f]	3	—	66.6	66.5	9.2	9.2
CH₃	CH₃	OC₂H₅	OC₂H₅	A	20	90°–92.5°	9.7	1.4355	64.5	63.9	9.7	9.7
C₂H₅	n-C₄H₉	OC₂H₅	H	B	81	80°	1.6	1.4443	72.8	72.8	11.1	11.3
C₂H₅	n-C₄H₉	OCOCH₃	H	C	30	97°	0.5	1.4554	67.9	67.8	9.5	9.4
C₂H₅	n-C₄H₉	CH₂OC₂H₅	H	C	31	76°	0.4	1.4497	73.5	73.0	11.4	11.4
C₂H₅	n-C₄H₉	CH₂OC₂H₅	H	C	15	133°–134°	0.4	—	78.4	78.3	9.3	9.4
C₂H₅	iso-C₄H₉	OCOCH₃	H	C	31	64°	0.6	1.4550	67.9	67.7	9.5	9.4
C₂H₅	iso-C₄H₉	—CH₂OCH₂—		C	30	—[g]	—	—	73.4	73.8	10.3	10.2
C₆H₅	C₆H₅	OCOCH₃	H	A	72	114°–115.5°[h]	—	—	77.1	77.1	5.7	5.8
C₆H₅	C₆H₅	SCH₂CH₂OCOCH₃	H	A	87	90°–91°[i]	—	—	70.6	70.7	5.9	6.0

[a] Reprinted from: R. H. Hasek, P. G. Gott, and J. C. Martin, *J. Org. Chem.* **29**, 1239 (1964). Copyright 1964 by the American Chemical Society. Reprinted by permission of the copyright owner.
[b] A, reaction in inert solvent at room temperature; B, reactants heated at 100°C (no solvent); C, reactants heated at 180°C (from ethyl alcohol).
[c] % Cl: calculated, 19.6; found, 19.5.
[d] % N: calculated, 6.2; found, 6.1.
[e] % S: calculated, 14.8; found, 15.3.
[f] Product solidified, mp 55°C (from ethyl alcohol).
[g] Purified by gas-liquid chromatography.
[h] Melting point (from benzene-hexane).
[i] Melting point (from ethyl alcohol).

The benzene solution of the diazo compound is poured into a separatory funnel. The separatory funnel is attached to a 125-ml Claisen distillation flask which is provided with a condenser for downward distillation. The flask is heated in an oil bath to 100°–110°C by means of a hot plate while the benzene solution is slowly dropped into the hot flask. Under these conditions the diazo compound is thermally transformed to diphenylketene while the benezene distills. The residue is fractionally distilled to yield 25 gm (59%) of diphenylketene, bp 119°–121°C (3.5 mm).

Diphenylketene is stored under a nitrogen atmosphere in the presence of a small crystal of hydroquinone in order to inhibit polymerization.

3-5. Preparation of 3-Oxo-2,2-diphenylcyclobutyl Acetate by the Reaction of Diphenylketene with Vinyl Acetate [73]

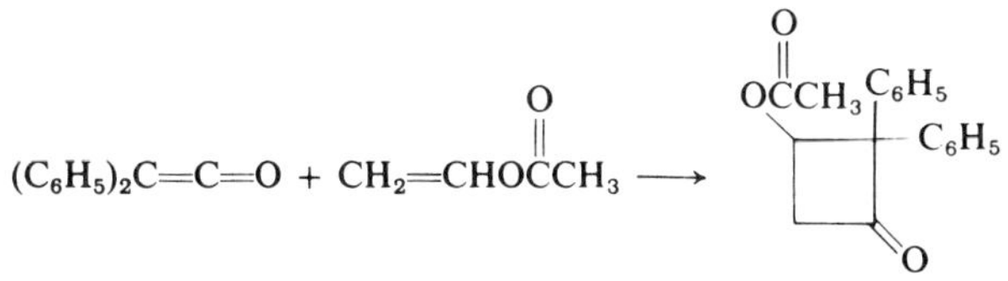

$$(C_6H_5)_2C=C=O + CH_2=CHOCCH_3 \longrightarrow \text{[cyclobutanone with OCOCH}_3, \text{ 2 } C_6H_5 \text{ substituents]} \quad (25)$$

To a flask containing 9.7 gm (0.05 mole) of diphenylketene is added 4.3 gm (0.05 mole) of vinyl acetate under a nitrogen atmosphere. The flask is sealed and after several days the mixture crystallizes. The solid is rinsed with cold hexane to give 11.4 gm (80%) of crude 3-oxo-2,2-diphenylcyclobutyl acetate. Recrystallization of the crude material from a mixture of benzene and hexane gives 10 gm (72%), mp 114°–115.5°C.

Existing carbonyl compounds can be alkylated and acylated to give new ketones [75a] or diketones [75b].

Simple ketones [75c] and diketones [76] are reacted with strong bases such as sodium amide or sodium alkoxides to form the enolate ion which is then coupled with a reactive alkyl halide.

$$RCH_2COCH_2R \xrightarrow[2.\ R'X]{1.\ Base} RCH_2COCHRR' \quad (26)$$

The diketone is first reacted with powdered sodium and then reacted with an alkyl iodide.

Acetylacetone can be alkylated [77] by means of an alkyl halide by use of bases and the resulting alkyl derivative can be cleaved by alcoholic base to give simple ketones.

Recently a one-stage procedure has been developed which is more convenient than the above two-stage process involving alkylation and then cleavage. The use of ethanolic potassium carbonate was used to prepare several ketones of the type CH_3COCH_2R where R is an alkyl or an arylalkyl group [78].

TABLE II

Cycloadducts of Ketenes with Enamines[a]

$$R^b R_2'^b C{=}O + R_2'''NCH{=}CR_2'' \longrightarrow \underset{H\quad R^b}{\overset{R_2''\quad O}{\underset{R_2'''N\quad R'^b}{\square}}}$$

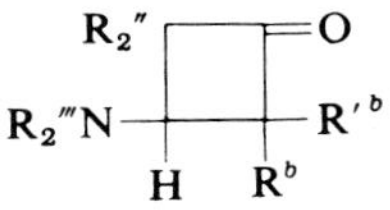

R^b	$R^{b'}$	R''	R''_2N	Bp °C (mm)	n_D^{20}	% C Calcd.	% C Found	% H Calcd.	% H Found	% N Calcd.	% N Found
CH$_3$	CH$_3$	CH$_3$	(CH$_3$)$_2$N	83°–85° (24)	1.4439	71.1	71.3	11.2	11.2	8.3	8.1
CH$_3$	CH$_3$	CH$_3$	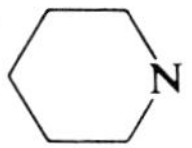 piperidino	95°–97° (4.2)	1.4705	74.5	74.3	11.0	11.2	6.7	6.5
CH$_3$	CH$_3$	CH$_3$	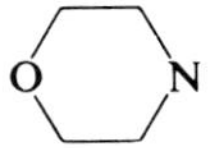 morpholino	58°–59.6°[c]	—	68.3	68.3	10.0	10.0	6.6	6.6
CH$_3$	CH$_3$	CH$_3$	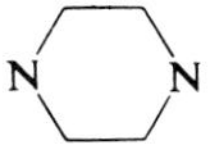 piperazino	256° dec.[d]	—	72.0	72.1	10.2	10.2	8.4	8.3
CH$_3$	CH$_2$	CH$_3$	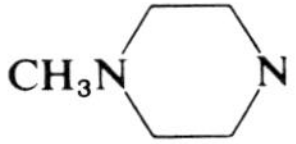 CH$_3$N-piperazino	105°–107° (3)[e]	—	69.7	70.2	10.7	10.8	12.5	12.4
CH$_3$	C$_2$H$_5$	CH$_3$	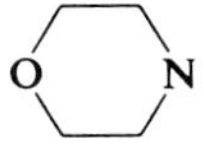 morpholino	101°–103° (1.3)	1.4736	69.4	69.6	10.2	10.4	6.2	6.3

C_2H_5	C_2H_5	CH_3	$(CH_3)_2N$	195°–198° (8)	1.4585	73.1	72.9	11.7	11.8	7.1	7.0
C_2H_5	C_2H_5	CH_3	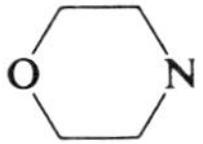 morpholine (N-O ring)	101°–104° (0.7)[e]	1.4794	70.3	69.9	10.5	10.5	5.9	5.9
C_2H_5	C_2H_5	C_2H_5	$(CH_3)_2N$	112° (5)	1.4662	74.7	74.1	12.0	11.6	6.2	5.9
C_2H_5	C_2H_5	C_2H_5	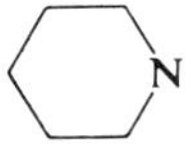 piperidine (N ring)	130°–132° (1.5)[e]	—	77.0	76.7	11.7	11.7	5.3	5.0
C_2H_5	C_4H_9	CH_3	$(CH_3)_2N$	194° (2)	1.4592	74.7	74.9	12.0	12.5	6.2	6.1
–$(CH_2)_5$–		CH_3	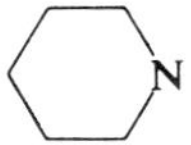 piperidine (N ring)	81°–82.5°[f]	—	77.1	77.1	10.8	11.0	5.6	5.4

[a] Reprinted from: R. H. Hasek and J. C. Martin, *J. Org. Chem.* **28**, 1468 (1963). Copyright (1963) by the American Chemical Society. Reprinted by permission of the copyright owner.
[b] Substituents on original dialkylketene.
[c] Melting point (from pentane).
[d] Bis adduct, melting point (from toluene).
[e] Solidified on standing.
[f] Melting point (from ethyl alcohol).

3-6. Preparation of 2-Heptanone from Acetylacetone [78]

$$CH_3COCH_2COCH_3 + n\text{-}C_4H_9I \xrightarrow{K_2CO_3,\ C_2H_5OH} CH_3COCH_2C_4H_9 \quad (27)$$

To a flask is added a mixture 11.0 gm (0.11 mole) of acetylacetone, 18.3 gm (0.10 mole) of *n*-butyl iodide, and 100 ml of anhydrous ethanol containing 13.8 gm (0.1 mole) of anhydrous potassium carbonate. The mixture is refluxed for 16 hr, shaken with 150 ml of water, and then extracted three times with 150 ml of ether. The ether extracts are combined, dried over anhydrous magnesium sulfate, concentrated, and the residue is distilled to yield 7.5 gm (60%), bp 150°–152°C (760 mm).

Acetoacetic ester can be alkylated and then cleaved by acid to give mono- and dialkylated methyl ketones in good yields.

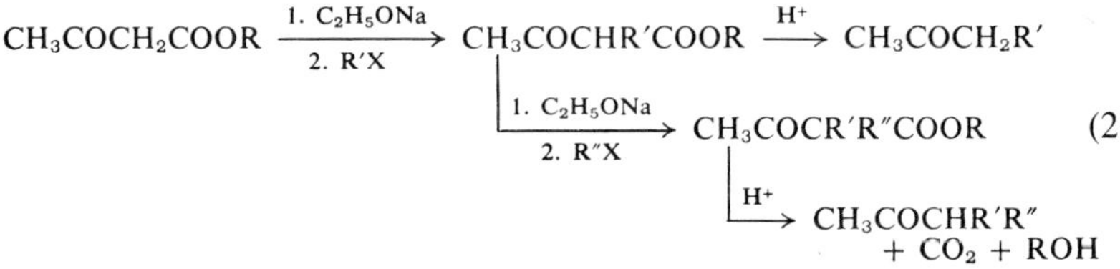

$$CH_3COCH_2COOR \xrightarrow[2.\ R'X]{1.\ C_2H_5ONa} CH_3COCHR'COOR \xrightarrow{H^+} CH_3COCH_2R'$$

$$\downarrow {1.\ C_2H_5ONa,\ 2.\ R''X}$$

$$CH_3COCR'R''COOR \quad (28)$$

$$\xrightarrow{H^+} CH_3COCHR'R'' + CO_2 + ROH$$

Primary alkyl bromides [79] give better yields of ketones than secondary alkyl bromides [80] when cleaved by acid [81].

The use of *tert*-butyl acetoacetate is similar to other acetoacetic esters except that *tert*-butyl esters decompose readily in the presence of acid to give isobutylene, carbon dioxide, and the methyl ketone [82, 83]. Products which are sensitive to hydrolysis may be prepared to advantage by this method.

3-7. Preparation of 3-Isobutyl-2-heptanone [82] from tert-Butylacetoacetate [84a]

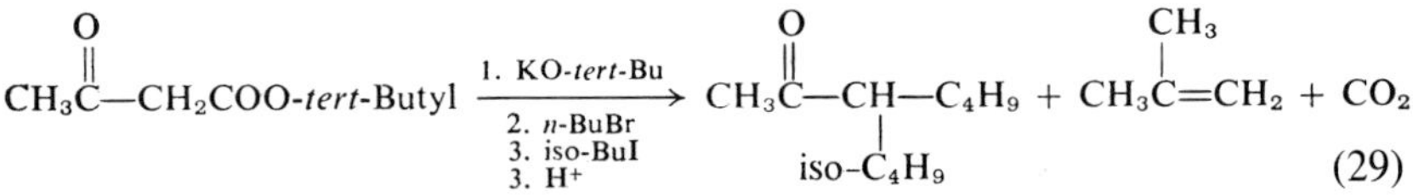

$$\underset{CH_3\overset{O}{\overset{\|}{C}}-CH_2COO\text{-}tert\text{-Butyl}}{} \xrightarrow[\substack{2.\ n\text{-BuBr}\\3.\ iso\text{-BuI}\\3.\ H^+}]{1.\ KO\text{-}tert\text{-Bu}} CH_3\overset{O}{\overset{\|}{C}}-\underset{iso\text{-}C_4H_9}{CH}-C_4H_9 + CH_3\overset{CH_3}{\overset{|}{C}}=CH_2 + CO_2 \quad (29)$$

To a flask equipped with a stirrer, condenser, and a thermometer containing 120 ml of *tert*-butyl alcohol is added 7.8 gm (0.2 gm atom) of potassium, and the mixture is stirred and refluxed until the metal reacts. The solution is cooled to about 50°C and 31.6 gm (0.20 mole) *tert*-butyl acetoacetate [84b] is added. The mixture is stirred for 2 min (all the potassium *tert*-butoxide dissolves) and 27.4 gm (0.20 mole) of *n*-butyl bromide is added. The mixture is refluxed for 1 hr*

*The reaction can be stopped here and worked up to give 2-heptanone.

§ 3. Condensation Reactions   223

and then 22.4 gm (0.2 mole) of fresh potassium *tert*-butoxide is added. After refluxing for ½ hr, 36.6 gm (0.20 mole) of isobutyl iodide is added and again the mixture is refluxed for 1 hr. Approximately 50–100 ml of *tert*-butyl alcohol is collected by distillation and the residue is warmed on a water bath with 5% of its weight of *p*-toluenesulfonic acid until gas evolution ceases (about 1 hr). The liquid is extracted with ether, washed with saturated sodium bicarbonate, dried, and distilled to yield 28.6 gm (84%), bp 80°C (10 mm), n_D^{25} 1.4237.

Esters having an α-hydrogen atom can also undergo self-condensation in the presence of a base to give substituted acetoacetic esters [85–87]. The reaction can also occur with mixed esters.

$$2\ RCH_2COOR \underset{}{\overset{NaOR}{\rightleftarrows}} RCH_2COCHRCOOR + ROH \quad (30)$$

β-Diketones are prepared [88] by acylation of ketones having a reactive methylene group using anhydrides [89] or esters [90]. The anhydrides are condensed with the aid of boron trifluoride and esters are condensed with the aid of a base to yield two types of β-diketones shown in Eq. (30).

$$RCH_2COCH_3 \begin{array}{c} \xrightarrow{BF_3 + (R'CO)_2O} RCH_2COCH_2COR' \\ \xrightarrow{Base + R'COOC_2H_5} \underset{\underset{O}{\parallel}}{R'C}-CHRCOCH_3 \end{array} \quad (31)$$

Using diethyl oxalate as the acylating agent, α,γ-diketo esters or a substituted glyoxalate are formed [91].

Enol esters of ketones can also be acylated with anhydrides in the presence of trifluoride to give diketones [92].

The acylation of sodium enolates of esters with acyl chloride gives α-substituted keto esters [93].

$$RCOCl + Na\overset{+}{\overset{-}{C}}R'_2COOC_2H_5 \longrightarrow RCOCR'_2COOC_2H_5 + NaCl \quad (32)$$

Methylene groups activated by nitrile groups can also be acylated with esters using sodium ethoxide to form β-keto nitriles in good yields [94].

β-Keto esters are also obtained by the Reformatskii reaction in good yields [95a].

$$R'COOR' + R_2CBrCOOC_2H_5 \xrightarrow{Zn} R'COCR_2COOC_2H_5 \quad (33)$$

The Dieckmann condensation is a useful reaction leading to large or small ring ketones [95b]. However, the yields vary from low to good [95c]. The Ziegler cyclization of the dinitrile has been reported to give much higher yields of the cyclic ketones [95d].

The hydration of acetylene can also be considered as a condensation involving water to give an intermediate alcohol which rearranges to the corresponding ketone [96a].

Cyclic ketones are obtained by the reaction of 1,3-dithiane with dihaloalkanes using n-butyl lithium followed by oxidation (H_2O and $HgCl_2$) of the dithiaspiroalkane [96b].

4. ELIMINATION REACTIONS

Aliphatic and aromatic ketones can be easily prepared in good yields by hydrolysis and decarboxylation of β-keto diesters obtained by the acylation of ethoxymagnesium diethylmalonate [97, 98].

$$RCOCl + C_2H_5OMgCH(COOC_2H_5)_2 \longrightarrow RCOCH(COOC_2H_5)_2$$
$$\xrightarrow{H^+} RCOCH_3 + 2 CO_2 + 2 C_2H_5OH \quad (34)$$

This method is useful for the preparation of nitro and chloroacetophenones which cannot be obtained by the Grignard, or Friedel–Crafts methods [98].

In a related manner the olefin elimination and decarboxylation of ethyl-*tert*-butyl acylmalonates is easily effected by p-toluenesulfonic acid to form β-keto esters [99]. This is related to the elimination with alkylated *tert*-butylacetoacetate discussed in Section 3-7.

$$RCOCl + C_2H_5OMgCH(COOC_2H_5)COO(\textit{tert}\text{-Butyl}) \longrightarrow$$
$$RCOCH(COOC_2H_5)(COO\text{-}\textit{tert}\text{-Butyl}) \xrightarrow{H^+}$$
$$RCOCH_2COOC_2H_5 + CH_2{=}C(CH_3)_2 + CO_2 \quad (35)$$

It has been reported [100] that esters can be converted in good yield to methyl ketones. The esters are reacted with methylsulfonyl carbanion to give β-keto sulfoxides, which are subsequently reacted with aluminum amalgam in 90% tetrahydrofuran–10% water mixture to give the ketones.

The reaction appears to offer a new and general route to ketones. See the discussion of the hazards involved with the use of the Corey–Chaykovsky reagent in Chapter 9, Carboxylic Acids.

The preparation of acetophenone given below, while not the most convenient method of preparing acetophenone, is representative of the technique employed.

4-1. *Preparation of Acetophenone by the Cleavage of* β-*Ketosulfoxides Derived from Esters* [100]

$$C_6H_5COOC_2H_5 + 2\ CH_3SOCH_3$$
$$\longrightarrow C_6H_5COCH_2SOCH_3 \xrightarrow{Al/Hg + H_2O} C_6H_5COCH_3 \quad (36)$$

§ 4. Elimination Reactions

(a) Preparation of ω-(methylsulfinyl)acetophenone [101]. To a flask containing 50 ml of refluxing *tert*-butyl alcohol is added 2 gm (0.05 gm atoms) of potassium metal. The flask is cooled to room temperature and 50 ml of dimethyl sulfoxide is added. The solution is promptly distilled at about 2 mm Hg while being heated at 65°–70°C until about 50 ml of dimethyl sulfoxide is distilled at about 43°C. Ethyl benzoate (7.5 gm, 0.6 mole) is added dropwise to the residue. The reaction mixture is agitated by means of a stream of oxygen-free nitrogen for 4 hr. The solvent is then removed under reduced pressure while heating to 75°C. Water (50 ml) and 100 ml of ether are added to the oily yellow residue. The aqueous layer is separated, acidified with hydrochloric acid to pH 5–6, and extracted with five 200-ml portions of chloroform. Evaporation of the chloroform yields a yellow oil. Removal of additional solvent at 2 mm leaves a solid residue which, after washing with 100 ml of ether, filtering, and drying gives 6.6 gm (72%) of colorless crystals, mp 85°C.

(b) Conversion of ω-(methylsulfinyl)acetophenone to acetophenone [100]. To a flask containing 2250 ml of a 2% mercuric chloride solution is added 8.9 gm (0.33 gm atom) of aluminum foil for a period of 10–15 sec. The amalgamated foil is rinsed with alcohol–ether and added immediately to another flask at 0°C containing 6.6 gm (0.033 mole) of ω-*(methylsulfinyl)acetophenone* and 400 ml of 10% aqueous tetrahydrofuran. The mixture is heated to the reflux temperature for 60–90 min while the reaction is vigorously stirred. The reaction mixture is extracted with ether, washed with water, dried, concentrated, and distilled to yield 4.0 gm (98%) of acetophenone, bp 202°C, n_D^{20} 1.5339.

The acylation of β-keto esters to form diacetyl esters which are then cleaved to a new β-keto ester [102] occurs in poor yield but an improved ammonolysis procedure has been reported [103].

Ketones are readily obtained by the hydrolysis of oximes [104]. Oximes are prepared by the reduction of nitroso or nitro compounds using zinc and acetic acids [105].

The thermal decarboxylation of acids to give ketones is not a preferred laboratory method since high temperatures are required, and often the method can be successfully applied only to the preparation of symmetrical ketones, otherwise mixtures are obtained [106, 107]. The pyrolysis of manganese salts of acids gives better yields of ketones than the corresponding calcium salts [108]. A recent study of the decomposition of ferrous salts of mixtures of aliphatic and aromatic carboxylic acids has been shown to lead to good yields of alkyl phenyl ketones [109a].

The synthesis of α-iodocarbonyl compounds can be effected by the reaction of enol silyl ethers with silver acetate–iodine [109b].

5. REARRANGEMENT REACTIONS

The pinacol–pinacolone rearrangement is the most popular rearrangement used to obtain ketones from α-diols. Dilute sulfuric acid suffices to cause the rearrangement to occur in good yield [110].

$$(CH_3)_2C(OH)-COH(CH_3)_2 \xrightarrow{H_3O^+} (CH_3)_3C-COCH_3 + H_2O \quad (37)$$

This reaction has been extended to aromatic α-diols such as benzopinacol [111]. Cyclic α-diols undergo ring expansions [112] and also lead in some cases to spiro ketones [113].

The treatment of 1-aminomethylcycloalkanols with nitrous acid gives ring-enlarged ketones. This reaction can be viewed as an extension of the Demjanov reaction to give a pinacol rearrangement [114].

5.1. Preparation of 3,3-Dimethyl-2-pentanone and 2,2-Dimethyl-3-pentanone [115a]

$$\begin{array}{c}
\text{CH}_3\ \text{CH}_3 \\
|\quad\ | \\
\text{CH}_3\text{CH}_2-\text{C}-\text{C}-\text{CH}_3 + \text{H}_3\text{O}^+ \longrightarrow \\
|\quad\ | \\
\text{OH}\ \text{OH}
\end{array}
\begin{array}{c}
\text{CH}_3 \\
| \\
\text{CH}_3\text{CH}_2-\text{C}-\text{C}-\text{CH}_3 \\
|\ \ \| \\
\text{CH}_3\text{O}
\end{array} \ (83\%)$$

$$+ \begin{array}{c}
\text{CH}_3 \\
| \\
\text{CH}_3\text{CH}_2-\text{C}-\text{C}-\text{CH}_3 \\
\|\ \ | \\
\text{O}\ \ \text{CH}_3
\end{array} \ (17\%) \quad (38)$$

To a flask containing 7.1 gm (0.053 mole) of 2,3-dimethyl-2,3-pentanediol is added 100 ml of 50% sulfuric acid which has been cooled to 5°C. The mixture is stirred at room temperature for 6 hr, added to about 100 gm of ice, and extracted with three 50-ml portions of ether. The combined ether layer is washed with 5% sodium carbonate, dried over sodium sulfate, evaporated to give a residue which, upon fractional distillation, gives 4.7 gm (78%), bp 125°–132°C. Vapor phase chromatography indicates a mixture of ketones consisting of 83% 3,3-dimethyl-2-pentanone and 17% of 2,2-dimethyl-3-pentanone.

The migratory aptitudes for various groups by the pinacol rearrangement is discussed by Dubois and Bauer [115b].

The stereochemical course in the pinacol rearrangement has been discussed by Mundy and Otzenberger [115c]. In particular it was noted that both *cis-* and *trans-*1,2 dimethylcyclohexane-1,2-diols give yields in excess of 90% of 1-

methyl-1-acetylcyclopentane. The pinacol rearrangement of bicyclic and tricyclic compounds have been reported [115d]. A synthetic approach to the 3-benzazepines through the pinacol-pinacolone rearrangement has been reported [115e].

A more specialized rearrangement reaction is the Serini reaction which involves a zinc-promoted rearrangement of 17-hydroxy-20-acetoxysterol derivatives to C-20 ketones [115f].

Other miscellaneous rearrangements are described in the next section.

6. MISCELLANEOUS METHODS

(1) Condensation of acetonitrile with polyhydroxyaromatic systems to yield phenolic ketones [116–118].

(2) Catalytic dehydrogenation of secondary alcohols to ketones [119].

(3) Oxidation of active methylene groups (activated by carbonyl, double bond, aromatic ring, or heterocyclic ring) [120–123].

(4) Hydration of acetylenic compounds to ketones [124].

(5) Condensation of diazomethane with aldehydes [125].

(6) Reduction of α,β-unsaturated ketones to ketones [126].

(7) Reduction of phenols to cyclohexanones [127].

(8) Decomposition of glycol monoalkyl ethers [128].

(9) Thermal decomposition of alkenyl esters of β-keto acids to yield γ,δ-olefinic ketones [129].

(10) Cyclopentenones from lactones [130].

(11) α-Ketoacids from acylaminopyruvic acid [131].

(12) Pyrolysis of glycidic acids [132].

(13) Hydrolysis of *gem*-dihalides [133].

(14) The conversion of acinitroparaffins to ketones [134].

(15) Isomerization of vinyl carbinols [135].

(16) Acylation of picoline [136].

(17) Oxidation of diarylacetylenes to α-diketones [137].

(18) Addition of aldehydes to olefins [138].

(19) Coupling of ketones to yield γ-diketones [139].

(20) Hydrolysis of β-Iminonitriles [140].

(21) Nuclear acylation of acylamines using phosphoric acid at 180°–200°C [141].

(22) Free radical addition of aldehydes to fluorolefins [142].

(23) Photoisomerization of cyclic ketones [143].

(24) Tetramethylcyclopropanone—a photochemical synthesis [144].

(25) Oxidation of allyl alcohol with dichlorodicyanoquinone [145].

(26) Homotropone [146].

(27) Methylene ketones [147].
(28) Opening of epoxides with DMSO-BF$_3$ [148].
(29) Photochemical addition of acyl radicals [149].
(30) The oxy-Cope rearrangement [150].
(31) Raney nickel desulfurization of monothioketals [151].
(32) Phosphine-catalyzed isomerization of epoxides [152].
(33) The reaction of phosphoranes with acid anhydrides to yield acetylenic ketones [153].
(34) N,N-dimethylacetamide as an acylating agent [154].
(35) Decomposition of tetrabutylphosphonium acylates [155].
(36) Synthesis of cyclopropenones [156].
(37) The reduction of α,β-olefinic ketones [157].
(38) The reduction of phenols to cyclohexanones [158].
(39) Hydrogenolysis of 1,3-diketone [158].
(40) Reduction of haloketones [159]
(41) Acylation of resorcinol [160].
(42) Acylation of aryl ethers [161].
(43) Hoesch synthesis [162].
(44) Wittig synthesis [163, 164].
(45) α,β-Epoxy ketones yield β-diketones [165].
(46) The Clemmensen reduction of 1,3-diketones [166].
(47) Organocadmium compounds react with dibasic acid chlorides to give diketones [167].
(48) Cyclization of δ-ethylenic acid chloride by stannic chloride [168].
(49) The preparation of hindered ketones by a photodecarboxylation process [169].
(50) The carbonylation of thexyldialkylboranes. A new general synthesis of ketones [170].
(51) Ring contraction via a Favorskii rearrangement to give cycloundecanone [171].
(52) Acid catalyzed isomerization of β,γ-unsaturated ketones [172].
(53) Preparation of tropolone [173].
(54) Rearrangement of bridgehead alcohols to polycyclic ketones by fragmentation-cyclization to give 4-protoadamantanone [174].
(55) A one carbon ring expansion of cycloalkanone to conjugated cycloalkenones such as 2-cyclohepten-1-one [175].
(56) One-pot syntheses of cyclopent-2-enones from furan derivatives [176].
(57) Preparation of propiophenone by the n-butyl lithium isomerization of α-vinylbenzyl alcohol [177].
(58) 1,4-Diketones via isoxazole intermediates [178].
(59) Oxidation of alcohols to ketones via nitroxyls [179].
(60) Rearrangements of ketones [180].

(61) Microbiological processes for production of ketones [181].
(62) Preparation of 1-bromo-3-methyl-2-butanone [182].
(63) α-Halogenation of certain ketones [183].
(64) Reaction of secondary alcohols with carbon tetrachloride in the presence of potassium hydroxide to give ketones [184].
(65) Use of dimethyl sulfoxide and acetic anhydride to oxidize secondary alcohols to ketones [185].
(66) Selective oxidation of secondary alcohols in the presence of primary alcohols to ketones [186].

REFERENCES

1. L. T. Sandborn, *Org. Synth. Collect. Vol.* **1,** 340 (1941).
2. W. A. Mosher and E. O. Langerak, *J. Am. Chem. Soc.* **71,** 286 (1949).
3. C. C. Price and J. V. Karabinos, *J. Am. Chem. Soc.* **62,** 1159 (1940)
4. P. D. Bartlett and A. Schneider, *J. Am. Chem. Soc.* **67,** 141 (1945).
5. H. T. Clarke and E. E. Dreger, *Org. Synth. Collect. Vol.* **1,** 87 (1941).
6. P. Ruggli and M. Herzog, *Helv. Chim. Acta* **29,** 111 (1946).
7. N. J. Leonard and R. M. Mader, *J. Am. Chem. Soc.* **72,** 5388 (1950).
8. G. I. Poos, G. E. Arth, R. E. Beyles, and L. H. Sarett, *J. Am. Chem. Soc.* **75,** 425 (1953).
9. J. R. Holum, *J. Org. Chem.* **26,** 4814 (1961).
10. H. C. Brown and C. P. Garg, *J. Am. Chem. Soc.* **83,** 2952 (1961).
11. V. J. Traynelis and W. L. Hergenrother, *J. Am. Chem. Soc.* **86,** 298 (1964).
12. J. Grundy, *J. Chem. Soc.* p. 5087 (1957).
13a. K. B. Wiberg, ed., "Oxidation in Organic Chemistry," Part A. Academic Press, New York, 1965.
13b. E. J. Corey and C. U. Kim, *J. Am. Chem. Soc.* **94** 7586 (1972); *J. Org. Chem.* **38,** 1233 (1973); E. J. Corey, C. U. Kim, and M. Takeda, *Tetrahedron Lett.* p. 4339 (1972).
13c. E. J. Corey and C. U. Kim, *Tetrahedron Lett.* p. 919 (1973).
14. F. K. Signaigo and P. L. Cramer, *J. Am. Chem. Soc.* **55,** 3326 (1933).
15. H. C. Brown and C. P. Garg, *J. Am. Chem. Soc.* **83,** 2952 (1961).
16a. N. Kornblum, J. W. Powers, G. J. Anderson, W. J. Jones, H. O. Larson, O. Levand, and W. H. Weaver, *J. Am. Chem. Soc.* **79,** 6562 (1957).
16b. W. W. Epstein and F. W. Sweat, *Chem. Rev.* **67,** 247 (1967).
16c. B. Ganem and R. K. Boeckman, Jr., *Tetrahedron Lett.* p. 917 (1974).
17a. E. F. Pratt and S. P. Suskind, *J. Org. Chem.* **28,** 638 (1963).
17b. H. W. Geluk and V. G. Keizer, *Org. Synth.* **53,** 8 (1973)
18. L. M. Berkowitz and P. N. Rylander, *J. Am. Chem. Soc.* **80,** 6682 (1958).
19. F. J. Moore and E. H. Huntress, *J. Am. Chem. Soc.* **49,** 1324 (1927).
20. H. W. Underwood, Jr. and W. L. Walsh, *J. Am. Chem. Soc.* **58,** 646 (1936).
21. F. R. Greenspan, *Ind. Eng. Chem.* **39,** 847 (1947).
22. L. F. Fieser, *J. Am. Chem. Soc.* **51,** 1896 (1929).
23. E. B. Vliet, *Org. Synth. Collect. Vol.* **1,** 482 (1941).
24. L. I. Smith and I. M. Webster, *J. Am. Chem. Soc.* **59,** 662 (1937).
25. P. S. Bailey, *Chem. Rev.* **58,** 925 (1958).
26. W. S. Knowles and Q. E. Thompson, *J. Org. Chem.* **25,** 1031 (1960).
27. Cole-Parmer Instrument Co., 7425 North Oak Park Avenue, Chicago, Illinois 60648.
28. G. W. Moersch and F. C. Whitmore, *J. Am. Chem. Soc.* **71,** 819 (1949).
29. W. H. Clement and C. M. Selwitz, *J. Org. Chem.* **29,** 241 (1964).

30. J. D. Roberts and C. W. Sauer, *J. Am. Chem. Soc.* **71**, 3925 (1949).
31. W. A. Mosher and J. Cox, Jr., *J. Am. Chem. Soc.* **72**, 3701 (1950).
32a. H. C. Brown and C. P. Garg, *J. Am. Chem. Soc.* **83**, 2951 (1961).
32b. S. V. Kulkarmi, H. D. Lee, and H. C. Brown, *J. Org. Chem.* **45**, 4542 (1980).
32c. A. Suzuki, S. Nozawa, M. Itoh, H. C. Brown, G. W. Kabalka, and G. W. Holland, *J. Am. Chem. Soc.* **92**, 3503 (1970).
32d. B. A. Calson and H. C. Brown, *J. Am. Chem. Soc.* **95**, 6876 (1973); *Org. Synth.* **58**, 24 (1978); H. C. Brown and H. W. Rathke, *ibid.* **89**, 2738 (1967).
33. R. Oppenauer, *Recl. Trav. Chim. Pays-Bas* **56**, 1397 (1937).
34. C. Djerassi, *Org. React.* **6**, 207 (1951).
35. H. Adkins and R. C. Franklin, *J. Am. Chem. Soc.* **63**, 2381 (1941).
36. E. W. Warnhoff and P. R. Warnhoff, *J. Org. Chem.* **28**, 1431 (1963).
37. K. E. Pfitzner and J. G. Moffatt, *J. Am. Chem. Soc.* **85**, 3072 (1963).
38. J. D. Albright and L. Goldman, *J. Am. Chem. Soc.* **87**, 4214 (1965).
39. J. D. Albright and L. Goldman, *J. Org. Chem.* **30**, 1107 (1965).
40. M. G. Burdon and J. G. Moffatt, *J. Am. Chem. Soc.* **87**, 4057 (1965).
41a. K. E. Pfitzner, J. P. Marino, and R. A. Olafson, *J. Am. Chem. Soc.* **87**, 4658 (1965).
41b. S. E. Jacobsen, D. A. Maccigrosso, and F. Mares, *J. Org. Chem.* **44**, 921 (1979).
41c. M. Y. Sheikh and G. Eadon, *Tetrahedron Lett.* No. 4, p. 257 (1972).
41d. R. J. Theissen, *J. Org. Chem.* **36**, 752 (1971).
41e. G. A. Olah, *Acc. Chem. Res.* **13**, 330 (1980).
42. C. R. Hauser and W. J. Humphlett, *J. Org. Chem.* **15**, 359 (1950).
43. F. F. Blicke, *J. Am. Chem. Soc.* **49**, 2847 (1927).
44. R. L. Shriner and T. A. Turner, *J. Am. Chem. Soc.* **52**, 1268 (1930).
45. C. R. Hauser, W. J. Humphlett, and M. J. Weiss, *J. Am. Chem. Soc.* **70**, 426 (1948).
46. W. J. Humphlett, M. J. Weiss, and C. R. Hauser, *J. Am. Chem. Soc.* **70**, 4030 (1948).
47. W. E. Bachman and C. H. Boatner, *J. Am. Chem. Soc.* **58**, 2098 (1936).
48. V. J. Cason and K. W. Kraus, *J. Org. Chem.* **26**, 1768 (1961).
49. M. S. Newman and A. S. Smith, *J. Org. Chem.* **13**, 592 (1948).
50. E. A. Braude and J. A. Coles, *J. Chem. Soc.* p. 2012 (1950).
51. F. C. Whitmore, C. I. Noll, and V. C. Meunier *J. Am. Chem. Soc.* **61**, 683 (1939).
52. E. A. Evans, *Chem. Ind. (London)* p. 1596 (1957).
53. M. S. Newman and W. T. Booth, *J. Am. Chem. Soc.* **67**, 154 (1945).
54. E. Campaigne and W. B. Reid, *J. Am. Chem. Soc.* **68**, 1663 (1946).
55. J. R. Dice, L. E. Loveless, and H. L. Gates, *J. Am. Chem. Soc.* **71**, 3546 (1949).
56a. G. P. Mueller and R. May, *J. Am. Chem. Soc.* **71**, 3313 (1949).
56b. M. J. Jorgenson, *Org. React.* **18**, 1 (1970).
57. J. Cason and K. W. Kraus, *J. Org. Chem.* **26**, 1772 (1961).
58a. G. A. Schmidt and D. A. Shirley, *J. Am. Chem. Soc.* **71**, 3804 (1949).
58b. G. H. Posner, C. E. Whitten, and P. McFarland, *J. Am. Chem. Soc.* **94**, 5106 (1972).
58c. D. Milstein and J. K. Stille, *J. Org. Chem.* **44**, 1613 (1979).
59. W. S. Johnson, *Org. React.* **2**, 114 (1944).
60. D. P. N. Satchell, *Q. Rev. Chem. Soc.* **17**, 160 (1963).
61. G. A. Olah, S. T. Kuhn, W. S. Talgyesi, and E. B. Baker, *J. Am. Chem. Soc.* **84**, 2733 (1962).
62. D. T. Mowry, M. Renoll, and W. F. Huber, *J. Am. Chem. Soc.* **68**, 1105 (1946).
63. O. C. Dermer, D. M. Wilson, F. M. Johnson, and V. H. Dermer, *J. Am. Chem. Soc.* **63**, 2881 (1941).
64. C. R. Noller and R. Adams, *J. Am. Chem. Soc.* **46**, 1889 (1924).
65. R. C. Fuson and J. T. Walker, *Org. Synth. Collect. Vol.* **2**, 169 (1943).

References

66. S. R. Sandler, unpublished results (1960).
67. A. H. Blatt, *Chem. Rev.* **27,** 413 (1940).
68. R. E. Christ and R. C. Fuson, *J. Am. Chem. Soc.* **59,** 893 (1937).
69a. C. C. Price and J. A. Pappalardo, *J. Am. Chem. Soc.* **72,** 2613 (1950).
69b. G. H Posner, *Org. React.* **19,** 1 (1972).
69c. M. S. Kharasch, W. H. Urry, and B. M. Kuderna, *J. Org. Chem.* **14,** 248 (1949); T. M. Patrick, Jr., *J. Org. Chem.* **17,** 1009 (1952); M. J. Onore, German Patent 2,122,750 (1971); B. M. Rein, J. E. Broderick, and B. Weinstein, German Patent 2,157,867 (1972).
70. L. F. Somerville and C. F. H. Allen, *Org. Synth. Collect. Vol.* **2,** 81 (1943).
71. E. Berliner, *Org. React.* **5,** 229 (1949).
72. A. B. Boese, Jr., *Ind. Eng. Chem.* **32,** 16 (1940).
73. R. H. Hasek, P. G. Gott, and J. C. Martin, *J. Org. Chem.* **29,** 1239 (1964).
74a. R. H. Hasek, J. C. Martin, *J. Org. Chem.* **28,** 1468 (1963).
74b. A. T. Blomquist and J. Kwiatck, *J. Am. Chem. Soc.* **73,** 2098 (1951).
74c. J. D. Roberts and C. M. Sharts, *Org. React.* **12,** 26, (1962).
74d. J. Williams and C. D. Hurd, *J. Org. Chem.* **5,** 122 (1940).
74e. R. H. Hasek and E. V. Elam, Canadian Patent 618,772 (1961).
74f. L. I. Smith and H. H. Hoehn, *Org. Synth.* **20,** 47 (1940).
74g. W. T. Brady, *J. Org. Chem.* **31,** 2676 (1966).
75a. B. O. Linn and C. R. Hauser, *J. Am. Chem. Soc.* **78,** 6066 (1956).
75b. C. -L. Mao and C. R. Hauser, *Org. Synth.* **51,** 90 (1971).
75c. H. Sobotka and J. D. Chanley, *J. Am. Chem. Soc.* **71,** 4136 (1949).
76. K. G. Hampton, R. J. Light, and C. R. Hauser, *J. Org. Chem.* **30,** 413 (1965).
77. K. G. Hampton, T. M. Harris, and C. R. Hauser, *J. Org. Chem.* **30,** 61 (1965).
78. J. Bornstein and F. Nunes, *J. Org. Chem.* **30,** 3324 (1965).
79. J. R. Johnson and F. D. Hager, *Org. Synth. Collect. Vol.* **1,** 351 (1941).
80. W. B. Renfrow, *J. Am. Chem. Soc.* **66,** 144 (1944).
81. W. M. Dehn and K. E. Jackson, *J. Am. Chem. Soc.* **55,** 4284 (1933).
82. W. B. Renfrow and G. B. Walker, *J. Am. Chem. Soc.* **70,** 3957 (1948).
83. W. B. Renfrow and A. Renfrow, *J. Am. Chem. Soc.* **68,** 1801 (1946).
84a. It can also be prepared according to the procedure described by Renfrow and Walker [82].
84b. This material is commercially available from Eastman Chemical Products, Kingsport, Tennessee.
85. C. R. Hauser and B. E. Hudson, Jr., *Org. React.* **1,** 266 (1942).
86. J. C. Shivers, M. L. Dillon, and C. R. Hauser, *J. Am. Chem. Soc.* **69,** 119 (1947).
87. F. C. Frostick, Jr., and C. R. Hauser, *J. Am. Chem. Soc.* **71,** 1350 (1949).
88. K. Kulka, *Am. Perfum.* **76,** No. 11, 23 (1961).
89. J. T. Adams and C. R. Hauser, *J. Am. Chem. Soc.* **67,** 284 (1945).
90. R. Levone, J. A. Conroy, J. T. Adams, and C. R. Hauser, *J. Am. Chem. Soc.* **67,** 1510 (1945).
91. E. E. Royals, *J. Am. Chem. Soc.* **67,** 1508 (1945).
92. F. G. Young, F. C. Frostick, Jr., J. J. Sanderson, and C. R. Hauser, *J. Am. Chem. Soc.* **72,** 3635 (1950).
93. C. R. Hauser and W. B. Renfrow, Jr., *Org. Synth. Collect. Vol.* **2,** 268 (1943).
94. J. B. Dorsch and S. M. McElvain, *J. Am. Chem. Soc.* **54,** 2960 (1932).
95a. M. S. Bloom and C. R. Hauser, *J. Am. Chem. Soc.* **66,** 152 (1944).
95b. N. J. Leonard and C. W. Schimelpfenig, Jr., *J. Org. Chem.* **23,** 1708 (1958).
95c. J. P. Schaefer and J. J. Bloomfield, *Org. React.* **15,** 1 (1967).
95d. K. Ziegler and R. Aurnhammer *Justus Liebigs Ann. Chem.* **513,** 43 (1934).
96a. R. J. Thomas, K. N. Campbell, and G. F. Hennion, *J. Am. Chem. Soc.* **60,** 719 (1938).
96b. D. Seebach and A. K. Beck, *Org. Synth.* **51,** 75 (1971).

8. Ketones

97. H. G. Walker and C. R. Hauser, *J. Am. Chem. Soc.* **68**, 1386 (1946).
98. G. A. Reynolds and C. R. Hauser, *Org. Synth.* **30**, 70 (1950).
99. D. S. Breslow, E. Baumgarten, and C. R. Hauser, *J. Am. Chem. Soc.* **66**, 1286 (1944).
100. E. J. Corey and M. Chaykovsky, *J. Am. Chem. Soc.* **86**, 1639 (1964).
101. H. D. Becker, G. J. Mikol, and G. A. Russell, *J. Am. Chem. Soc.* **85**, 3410 (1963).
102. R. L. Shriner and A. G. Schmidt, *J. Am. Chem. Soc.* **51**, 3636 (1929).
103. H. W. Anderson, I. F. Halverstadt, W. H. Miller, and R. O. Roblin, Jr., *J. Am. Chem. Soc.* **67**, 2197 (1945).
104. W. W. Hartman and L. J. Roll, *Org. Synth.* **23**, 1 (1943).
105. D. Nightengale and J. R. Janes, *J. Am. Chem. Soc.* **66**, 352 (1944).
106. P. Sabatier and A. Mailhe, *C. R. Hebd. Seances Acad. Sci.* **159**, 217 (1914).
107. R. H. Pickard and J. Kenyon, *J. Chem. Soc.* **99**, 57 (1911).
108. P. Mastagli, P. Lambert, and C. Hirigoyen, *C. R. Hebd. Seances Acad. Sci.* **248**, 1830 (1959).
109a. C. Granito and H. P. Schultz, *J. Org. Chem.* **28**, 879 (1963).
109b. G. M. Rubottom and R. C. Mott, *J. Org. Chem.* **44**, 1731 (1979).
109c. C. W. Jefford, J. Gunsher, D. T. Hill, P. Brun, J. LeGras, and B. Waegell, *Org. Synth.* **51**, 60 (1971).
110. G. A. Hill and E. W. Flosdorf, *Org. Synth. Collect. Vol.* **1**, 462 (1941).
111. W. E. Bachman, *Org. Synth. Collect. Vol.* **2**, 73 (1943).
112. A. Burger and W. B. Bennet, *J. Am. Chem. Soc.* **72**, 5414 (1950).
113. R. D. Sands and D. G. Botterton, *J. Org. Chem.* **28**, 2690 (1963).
114. M. Tiffeneau, P. Weill, and B. Tehoubar, *C. R. Hebd. Seances Acad. Sci.* **205**, 54 (1937).
115a. M. Stiles and R. P. Mayer, *J. Am. Chem. Soc.* **81**, 1495 (1959).
115b. J. E. Dubois and P. Bauer, *J. Am. Chem. Soc.* **98**, 6903, P. Bauer and J. E. Dubois, *J. Am. Chem. Soc.* **98**, 6999 (1976).
115c. B. P. Mundy and R. D. Otzenberger, *J. Chem. Educ.* **48**, 431 (1971).
115d. R. Bishop and W. Parker, *Tetrahedron Lett.* p. 2329 (1973).
115e. T. Kametami, H. Nemoto, K. Suzuki, and K. Fukumoto, *J. Org. Chem.* **41**, 2988 (1976).
115f. E. Ghera, *J. Org. Chem.* **35**, 660 (1970).
116. P. E. Spoerri and A. S. Dubois, *Org. React.* **5**, 387 (1949).
117. K. C. Gulati, S. R. Seth, and K. Venkuraman, *Org. Synth. Collect. Vol.* **2**, 522 (1943).
118. K. Hoesch, *Chem. Ber.* **60**, 389, 2537 (1927).
119. C. D. Hurd, H. Greengard, and A. S. Roe, *J. Am. Chem. Soc.* **61**, 3359 (1939).
120. H. H. Hart, A. Pilgrim, and W. J. Hurran, *J. Chem. Soc.* p. 93 (1936).
121. D. T. Mowry, *J. Am. Chem. Soc.* **67**, 1050 (1945).
122. F. C. Whitmore and G. W. Pedlow, Jr., *J. Am. Chem. Soc.* **63**, 758 (1941).
123. E. H. Huntress and H. C. Walter, *J. Am. Chem. Soc.* **70**, 3704 (1948).
124. R. J. Thomas, K. N. Campbell, and G. F. Hennion, *J. Am. Chem. Soc.* **60**, 7, 9 (1938).
125. D. W. Adamson and J. Kenner, *J. Chem. Soc.* p. 184 (1939).
126. L. W. Covert and H. Adkins, *J. Am. Chem. Soc.* **54**, 4116 (1932).
127. R. B. Thompson, *Org. Synth.* **27**, 21 (1947).
128. J. C. Barhan, *Bull. Soc. Chim. Fr.* [4] **49**, 1426 (1931).
129. W. Kimel and A. K. Cope, *J. Am. Chem. Soc.* **65**, 1992 (1943).
130. R. L. Frank, R. Armstrong, J. Kwiatek, and H. A. Price, *J. Am. Chem. Soc.* **70**, 1379 (1948).
131. R. H. Herbst and D. Shemin, *Org. Synth. Collect. Vol.* **2**, 1 and 519 (1943).
132. W. A. Yarnall and E. S. Wallis, *J. Org. Chem.* **4**, 270 (1939).
133. C. S. Marvel and W. M. Sperry, *Org. Synth. Collect. Vol.* **1**, 95 (1941).
134. K. Johnson and E. F. Degering, *J. Org. Chem.* **8**, 10 (1943).
135. R. Delaby, *Bull. Soc. Chim. Fr.* [4]**33**, 602 (1923).
136. M. J. Weiss and C. R. Hauser, *J. Am. Chem. Soc.* **71**, 2023 (1949).

References

137. P. Ruggli and H. Zaeslin, *Helv. Chim. Acta* **18,** 848 (1935).
138. M. S. Kharasch, W. H. Urry, and B. M. Kuderma, *J. Org. Chem.* **14,** 248 (1949).
139. M. S. Kharasch, H. C. McBay, and W. H. Urry, *J. Am. Chem. Soc.* **70,** 1269 (1948).
140. A. Dornow, I. Kuhlcke, and F. Baxmann, *Chem. Ber.* **82,** 254 (1949).
141. B. Staskun, *J. Org. Chem.* **29,** 2856 (1954).
142. H. Muramatsu and K. Inukai, *J. Org. Chem.* **27,** 1572 (1962).
143. S. Cremer and R. Srinivasan, *J. Am. Chem. Soc.* **56,** 4197 (1965).
144. N. J. Tureo, W. B. Hammond, and P. A. Leermakers, *J. Am. Chem. Soc.* **87,** 2774 (1965).
145. S. H. Burstein and H. J. Ringold, *J. Am. Chem. Soc.* **86,** 4952 (1964).
146. J. D. Holmes and R. Pettit, *J. Am. Chem. Soc.* **85,** 2531 (1963).
147. W. H. W. Lunn, *J. Org. Chem.* **30,** 2925 (1965).
148. T. Cohen and T. Tsuji, *J. Org. Chem.* **26,** 1681 (1961).
149. U. Schmidt, *Angew. Chem.* **77,** 216 (1965).
150. J. A. Berson and M. Jones, Jr., *J. Am. Chem. Soc.* **86,** 5019 (1964).
151. E. L. Eliel and S. Krishnamurthy, *J. Org. Chem.* **30,** 845 (1965).
152. D. E. Brissing and A. J. Speziale, *J. Am. Chem. Soc.* **87,** 1405 (1965).
153. P. A. Chopard, R. J. G. Searle, and F. H. Devitt, *J. Org. Chem.* **30,** 1015 (1965).
154. D. S. Deorha and P. Gupta, *Chem. Ber.* **97,** 616 (1964).
155. D. B. Denney and H. A. Kudergrab, *J. Org. Chem.* **28,** 1133 (1963).
156. R. Breslow, J. Posner, and A. Krebs, *J. Am. Chem. Soc.* **85,** 234 (1963).
157. L. W. Covert, R. Connor, and H. Adkins, *J. Am. Chem. Soc.* **54,** 1658 (1932).
158. J. H. Sprague and H. Adkins, *J. Am. Chem. Soc.* **56,** 2670 (1934).
159. L. Canonica and T. Bacchetti, *Atti Accad. Naz. Lincei, Cl. Sci. Fis. Mat. Nat.*, Rend. **10,** 479 (1951); *Chem. Abstr.*, **48** 6377 (1954).
160. M. Neneki and N. Sieber, *J. Prakt. Chem.* **23,** 147 (1881).
161. L. Gattermann, *Chem. Ber.* **22,** 1129 (1889).
162. T. Iwadare, Y. Yasunari, S. Tono-Oka, M. Gohde, and T. Irie, *J. Org. Chem.* **28,** 3206 (1963).
163. D. B. Denney and J. Song, *J. Org. Chem.* **29,** 495 (1964).
164. J. J. Bestmann, *Tetrahedron Lett.* No. 4 p. 7 (1960).
165. H. House and R. L. Wasson, *J. Am. Chem. Soc.* **78,** 4394 (1956).
166. N. J. Cusak and B. R. Davis, *J. Org. Chem.* **30,** 2062 (1965).
167. M. Renson and J. Bonhomme, *Bull. Soc. Chim. Belg.* **68,** 667 (1959).
168. O. Riove, *C. R. Hebd. Seances Acad. Sci.* **248,** 2774 (1959).
169. R. A. Finnegan and D. Knutson, *Chem. Ind. (London)* p. 1837 (1965).
170. H. C. Brown and E. Negishi, *J. Am. Chem. Soc.* **89,** 5285 (1967).
171. J. Wohllebe and E. W. Garbisch, Jr., *Org. Synth.* **56,** 107 (1977).
172. R. L. Cargill and J. W. Crawford, *J. Org. Chem.* **35,** 356 (1970); R. L. Cargill, D. M. Pond, and S. O. LeGrand, *ibid.* p. 359.
173. R. A. Minns, *Org. Synth.* **57,** 117 (1977).
174. Z. Majerski and Z. Hameršak, *Org Synth.* **59,** 147 (1980).
175. Y. Ito, S. Fuju, M. Nakatsuka, F. Kawamoto, and T. Saegusa, *Org. Synth.* **59,** 113 (1980).
176. R. D'Ascoli, M. D'Auria, C. Iavarone, G. Piancatelli, and A. Scettri, *J. Org. Chem.* **45,** 4502 (1980).
177. D. R. Dimmel, W. Y. Fu, and S. B. Gharpure, *J. Org. Chem.* **41,** 3092 (1976).
178. A. Barco, S. Benetti, G. P. Pollini, P. G. Baraldi, M. Guarneri, and C. B. Vicentini, *J. Org. Chem.* **44,** 105 (1979).
179. B. Ganem, *J. Org. Chem.* **40,** 1998 (1975).
180. A. Fry and M. Oka, *J. Am. Chem. Soc.* **101,** 6353 (1979).
181. L. E. Casida and D. A. Klein, U.S. Patent 3,625,824 (1971).

182. M. Gaudry and W. Marquet, *Org. Synth.* **55,** 24 (1976).
183. B. Modarai and E. Khosdel, *J. Org. Chem.* **42,** 3527 (1977).
184. C. Y. Meyers and W. S. Matthews, III, U.S. Patent 3,953,494 (1976).
185. K. Yamada, K. Kato, H. Nagase, and Y. Hirata, *Chem. Commun.* p. 65 (1976).
186. M. E. Jung and L. M. Speltz, *J. Am. Chem. Soc.* **98,** 7882 (1976).

CHAPTER 9 / CARBOXYLIC ACIDS

1. Introduction	236
2. Oxidation Reactions	238
A. Oxidation of Alkyl Side Chains	238
2-1. Preparation of o-Toluic Acid	238
B. Oxidations of Alcohols and Aldehydes	239
2-2. Preparation of n-Heptane-γ-carboxylic Acid	240
2-3. Preparation of 9-Anthroic Acid	241
2-4. Preparation of Furoic Acid	241
C. Catalytic Oxidation with Oxygen	241
2-5. Preparation of 6-Bromo-2-naphthyl-β-D-glucoronide	242
3. Oxidation of Olefins	242
3-1. Preparation of 2,3-Dimethylheptanoic Acid	243
3-2. Preparation of 5-Methylhexanoic Acid	244
3-3. Preparation of Homophthalic Acid	245
4. Oxidation of Ketones and Quinones	246
4-1. Preparation of Trimethylacetic Acid	246
4-2. Preparation of Diphenic Acid	247
4-3. Preparation of m-Chlorophenylacetic Acid	247
A. The Haloform Reactions	248
4-4. Preparation of tert-Butylacetic Acid	248
4-5. Preparation of Benzoic Acid	249
B. Willgerodt Reaction	249
4-6. Preparation of p-Methoxyphenylacetic Acid	250
4-7. Preparation of Isophthalic Acid	251
5. Bimolecular Oxidation–Reduction Reactions	251
5-1. Cannizzaro Reaction: Preparation of Furoic Acid	251
5-2. Benzilic Acid Rearrangement	252
6. Carbonation of Organometallic Reagents	253
6-1. Grignard Reagents—Preparation of α-Methylbutyric Acid	253
6-2 Lithium Reagents—Preparation of Fluorene-9-carboxylic Acid	254
6-3. Sodium Reagents—Preparation of Fluorene-9-carboxylic Acid	254
7. Carboxylation of the Aromatic Nucleus	255
7-1. The Kolbe–Schmitt Reaction—Preparation of β-Resorcylic Acid	256
A. The Friedel–Crafts Reaction	256
7-2. Preparation of β-Benzoylacrylic Acid	257
7-3. The Use of Oxalyl Chloride—Preparation of 9-Anthroic Acid	257
7-4. Preparation of β,β-Dianisylacrylic Acid Using Oxalyl Chloride	258
7-5. Preparation of β,β-Dianisylacrylic Acid Using Phosgene	259
8. Condensation Reactions	259
A. Perkin Reaction	259
8-1. Preparation of 2-Methyl-3-nitrocinnamic Acid	259
B. Knoevenagel Condensation	260

8-2. *Preparation of 2-Hexenoic Acid*	260
C. Other Condensation Reactions	261
8-3. *Claisen Condensation—Preparation of Ethyl Cinnamate and Cinnamic Acid*	261
8-4. *Malonic Ester Synthesis—Preparation of Ethyl n-Butylmalonate*	262
8-5. *Preparation of Pelargonic Acid*	263
D. Ethyl Acetoacetic Ester Synthesis	264
8-6. *Preparation of Caproic Acid*	264
E. Arndt–Eistert Rearrangement	264
8-7. *Preparation of Biphenyl-2-acetic Acid*	265
F. Strecker Amino Acid Synthesis	266
G. Condensation of Active Methylene Compounds with Chloral	266
8-8. *Preparation of 3-(2-Pyridyl)acrylic Acid*	267
H. Reformatskii Reaction	268
8-9. *Preparation of Ethyl 4-Ethyl-3-hydroxy-2-octanoate*	268
I. Diels–Alder Reaction	269
8-10. *Preparation of Tetrahydrophthalic Anhydride*	269
9. Hydrolysis of Acid Derivatives	270
A. Hydrolysis of Nitriles	270
9-1. *Preparation of 9-Phenanthroic Acid*	270
B. Hydrolysis of Amides	271
9-2. *Preparation of α-Phenylbutyric Acid*	271
C. Hydrolysis of Esters, Acyl Halides, Anhydrides, and Trihalides	271
9-3. *Preparation of m-Nitrobenzoic Acid*	272
9-4. *Preparation of Citraconic Acid*	273
D. Hydrolysis of 1,1,1-Trihalomethyl Derivatives	273
9-5. *Preparation of α-Methoxyisobutyric Acid*	273
10. Miscellaneous Methods	274
A. Oxidation Reactions	274
B. Hydrolysis and Elimination Reactions	275
C. Reactions Involving Organometallics	277
D. Reactions Using Carbon Monoxide	278
E. Condensation Reactions	280
F. Reduction Reactions	282
G. Enzyme Reactions	282
H. Substitution Reactions	283
References	283

1. INTRODUCTION

The common laboratory methods for the synthesis of carboxylic acids are oxidation, oxidation–reduction reactions, carbonation of organometallics, condensation reactions, and hydrolysis reactions.

Oxidation reactions are useful for the conversion of aliphatic side chains of aromatic compounds, primary alcohols, aldehydes, ketones, olefins, and a combination of one or more of the latter groups to a carboxylic acid.

§ 1. Introduction

$$\left.\begin{array}{l} ArCH_3 \\ RCH_2OH \\ RCH=O \\ R_2C=O \\ RCH=CH_2 \end{array}\right\} \xrightarrow{(O)} RCOOH \quad (1)$$

The haloform reaction is a good method for the oxidation of aliphatic methyl ketones or of methyl carbinols to carboxylic acids. The reaction is also applicable to aromatic methyl ketones and aryl methyl carbinols obtained by the Friedel–Crafts reaction and by the Grignard reaction, respectively.

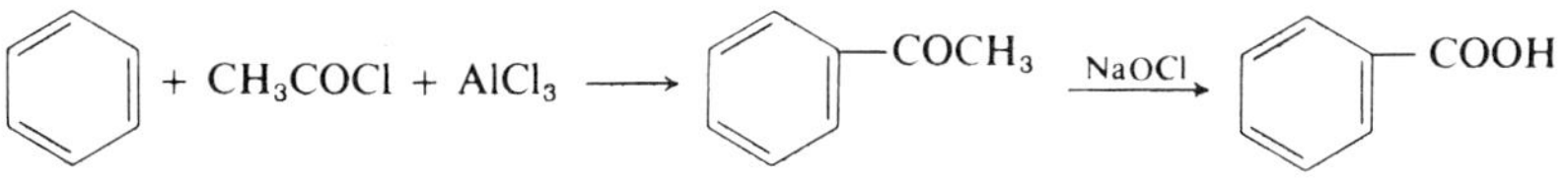

$$C_6H_6 + CH_3COCl + AlCl_3 \longrightarrow C_6H_5-COCH_3 \xrightarrow{NaOCl} C_6H_5-COOH \quad (2)$$

$$C_6H_5-CH=O + CH_3MgX \longrightarrow C_6H_5-CH(OH)CH_3 \xrightarrow{NaOCl} C_6H_5-COOH \quad (3)$$

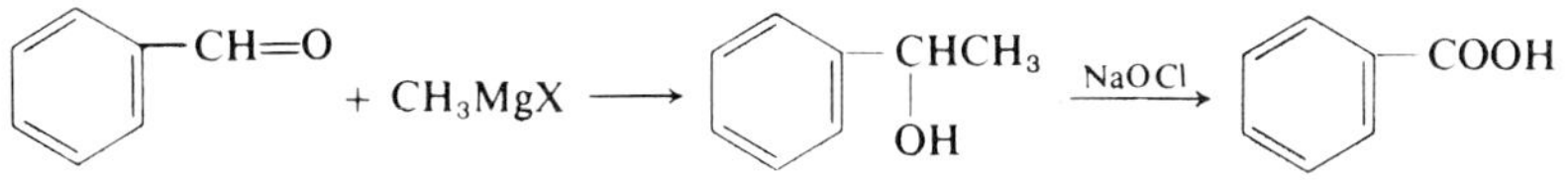

Oxidation–reduction reactions such as the Cannizzaro reactions are useful for the conversion of aldehydes lacking an α-hydrogen atom to a mixture of the acid and alcohol. The reaction is useful in those cases where an acid oxidation medium would degrade or rearrange the molecule.

Carbonation of the Grignard reagent and hydrolysis of the magnesium halide derivative is one of the most generally applicable laboratory methods for the preparation of carboxylic acids.

$$RX \xrightarrow[\text{ether}]{Mg} RMgX \xrightarrow{CO_2} RCOOMgX \xrightarrow{H_2O} RCOOH + MgXOH \quad (4)$$

where R = aliphatic or aromatic.

Carbonation of aryl or alkyl lithium compounds also affords carboxylic acids in good yields.

$$C_6H_5Li + CO_2 \longrightarrow C_6H_5COOLi \xrightarrow{H_2O} C_6H_5COOH + LiOH \quad (5)$$

Condensation reactions such as the Reformatskii reaction, Perkin reaction, malonic ester synthesis, and the Diels–Alder reaction afford acids or their esters in good yields.

The hydrolysis of nitriles in the aliphatic or aromatic series yields carboxylic acids. This method is useful for the conversion of primary aliphatic nitriles and aromatic nitriles to the carboxyl derivative. Since the nitriles are generally prepared from primary halides, the hydrolysis of nitriles represents another method of converting readily accessible organic raw materials to carboxylic acids.

$$RX + KCN \longrightarrow RCN \xrightarrow[H^+ \text{ or } OH^-]{H_2O} RCOOH \tag{6}$$

$$X-CH_2-R-CH_2-X + 2\,KCN \longrightarrow NC-CH_2-R-CH_2-CN$$

$$\xrightarrow[H_2O]{H^+ \text{ or } OH^-} HO\overset{O}{\overset{\|}{C}}-CH_2-R-CH_2\overset{O}{\overset{\|}{C}}OH \tag{7}$$

Other hydrolysis reactions are also used to prepare acids from esters, amides, acid chlorides, anhydrides, and haloketones.

More specialized reactions having limited synthetic value in the laboratory are listed briefly with references so that the reader may obtain more detailed information on the particular reaction. Some of these reactions are of industrial importance in the preparation of particular acids.

2. OXIDATION REACTIONS

A. Oxidation of Alkyl Side Chains

The liquid phase oxidation of aromatic alkyl side chains may be affected by aqueous sodium dichromate [1a, b] or with dilute nitric acid by refluxing the mixture for a prolonged period as in the case for the conversion of o-xylene to o-toluic acid. The procedure is illustrated in Eq. (8), a reaction which gives a 55% yield of o-toluic acid [1a, b]. With o-xylene, aqueous sodium dichromate yields phthalic acid.

2-1. *Preparation of o-Toluic Acid* [1b]

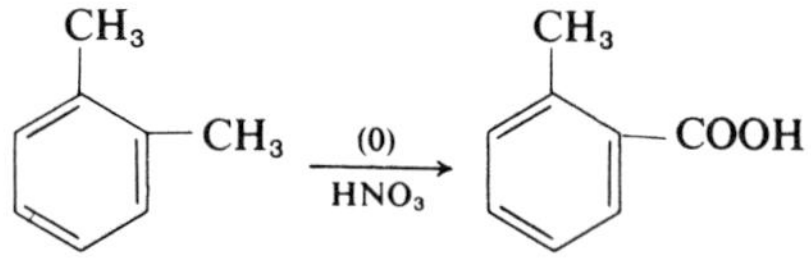

$$\underset{\text{CH}_3}{\text{o-xylene}} \xrightarrow[\text{HNO}_3]{(0)} \underset{\text{CH}_3}{\text{o-toluic acid}} \tag{8}$$

In a 5-liter, round-bottomed flask are placed 1.6 liters of water, 800 ml of concentrated nitric acid, and 400 ml (3 moles) of xylene (90%). A reflux condenser with an outlet to a gas absorption trap is attached to the flask. The mixture is refluxed by heating in an oil bath at 145°–155°C for 55 hr. At the end of this time, the organic layer has settled to the bottom of the flask. The hot reaction mixture is poured onto 1 kg of ice, and the precipitate is filtered off, washed with cold water, and filtered again. The wet product is dissolved by warming in 1 liter of 15% solution of sodium hydroxide. Residual o-xylene is separated by an ether extraction. The aqueous layer is then decolorized with Norite and acidified with concentrated hydrochloric acid. The crude material is recrystallized from aqueous ethanol to yield a tan product melting at 99°–101°C, in the amount of 218–225 gm (53–55%).

By control of reaction conditions, branched chains rather than linear alkyl side chains may be preferentially oxidized. Impurities found in the crude product will include intermediate oxidation stages, such as ketones, and the products obtained when all side chains are oxidized. A typical example is the preparation of p-toluic acid from p-cymene [2].

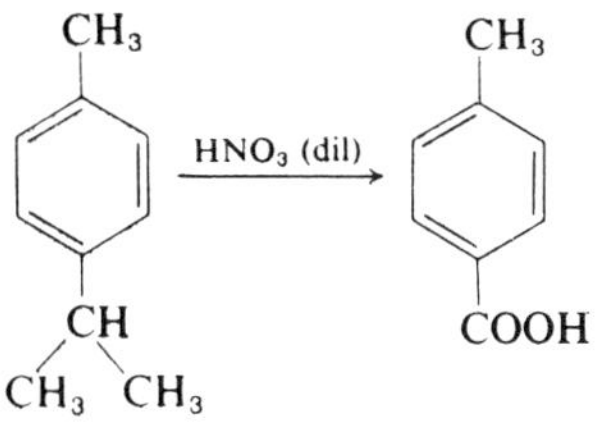

$$
\underset{\underset{CH_3\ CH_3}{CH}}{\overset{CH_3}{\text{C}_6H_4}} \xrightarrow{HNO_3\ (dil)} \underset{COOH}{\overset{CH_3}{\text{C}_6H_4}} \tag{9}
$$

Catalytic oxidations by oxygen [3] in the liquid phase are used to product aromatic acids in 25–68% from substituted or unsubstituted alkylbenzenes using the acetates of cobalt, lead, or manganese as catalysts [4]. Butyric acid serves as a solvent for the latter oxidations. Halogens and nitro groups attached to the aromatic ring are unaffected by the oxidation of the alkyl group [1a, b].

o- and p-nitro [5], o-chloro [6], and p-iodobenzoic acid [7] have been prepared by the oxidation of the substituted toluenes [1a, b].

The picolines (α and β) may be oxidized by permanganate to prepare picolinic acid and isonicotinic acid in 45–60% yields [8–10].

The alkyl group does not necessarily have to belong to an aromatic ring in order to be oxidized. Oxidation of a methyl group of trimethylacetic acid (pivalic acid) by heating for 7 hr with alkaline permanganate gives dimethylmalonic acid [11].

$$
\underset{\underset{CH_3}{|}}{\overset{\overset{CH_3}{|}}{CH_3-C-COOH}} \xrightarrow[MnO_4^-]{OH^-} C(CH_3)_2(COOH)_2 \tag{10}
$$

Recently tetrabutyl ammonium permanganate in pyridine has been used to give high yields of acids by the room temperature oxidation of alkyl aromatic compounds [12a].

B. Oxidations of Alcohols and Aldehydes

Alcohols and aldehydes can be oxidized directly to the carboxylic acid. In the case of primary alcohols, aldehydes are intermediates. The latter are oxidized in turn to acids.

$$\text{RCH}_2\text{OH} \xrightarrow{(O)} [\text{RCH}{=}\text{O}] \xrightarrow{(O)} \text{RCOOH} \qquad (11)$$

The oxidation of alcohols and aldehydes by potassium permanganate is illustrated below by the oxidation of β-ethyl-*n*-hexanol and α-ethyl-*n*-hexanal in basic permanganate to *n*-heptane-γ-carboxylic acid [12b].

Tetraalkylammonium permanganate has also been reported as an organic soluble oxidizing agent for alcohols and aldehydes [12a]. The oxidation reaction is carried out rapidly in pyridine at room temperature and some typical yields of acids from the starting carbonyl or alcohol compounds are *m*-nitrobenzaldehyde (95%) and *p*-chlorobenzaldehyde (99%).

Aqueous potassium permanganate has the disadvantage of requiring a large excess of the reagent for oxidation and organic substrates are usually poorly soluble in it.

2-2. Preparation of n-Heptane-γ-carboxylic Acid [12b]

$$\begin{array}{c}
\text{CH}_3\text{CH}_2\text{CH}_2\text{CH}_2\text{—CH—CH}_2\text{OH} \xrightarrow{(O)} \\
\quad\quad\quad\quad\quad\quad\quad | \\
\quad\quad\quad\quad\quad\quad \text{C}_2\text{H}_5 \\
\text{CH}_3\text{CH}_2\text{CH}_2\text{CH}_2\text{—CH—CH}{=}\text{O} \xrightarrow{(O)} \\
\quad\quad\quad\quad\quad\quad | \\
\quad\quad\quad\quad\quad \text{C}_2\text{H}_5
\end{array} \longrightarrow \begin{array}{c} \text{CH}_3\text{CH}_2\text{CH}_2\text{CH}_2\text{—CH—COOH} \\ | \\ \text{CH}_2 \\ | \\ \text{CH}_3 \end{array} \qquad (12)$$

(a) By oxidation of β-ethyl-n-hexanol. Potassium permanganate (340 gm) in water (3000 ml) is added to a well-stirred mixture of β-ethyl-*n*-hexanol (130 gm) and sodium hydroxide (30 gm) dissolved in water (250 ml). After 12 hr of stirring, the mixture is acidified with sulfuric acid and sulfur dioxide. The sulfur dioxide is passed through the reaction mixture until the MnO_2 has completely dissolved. The solution is then extracted with ether. Upon evaporation of the ether solution, the residue consisted of *n*-heptane-γ-carboxylic acid (107 gm, 74%); bp 119°–121°C (14 mm), n_D^{18} 1.4255.

(b) By oxidation of α-ethyl-n-hexanal. Potassium permanganate (150 gm) in water (2.5 liters) is added to a well-stirred mixture of α-ethyl-*n*-hexanal (128 gm), sodium carbonate (23 gm), and water (300 ml). After 7 hr of stirring, *n*-heptane-γ-carboxylic acid (113 gm, 78%) is obtained by a work-up procedure similar to the one given for the oxidation of the corresponding alcohol.

Silver oxide is perhaps the best reagent for the preparation of pure acids from aldehydes. Its advantage is that it does not attack other easily oxidizable groups in the molecule. 9-Anthroic acid is prepared from the corresponding aldehyde in 72% yield using this method [13a]. An alternative method for preparing this specific compound is also given below (Procedure 7-3). The silver oxide procedure is applicable to oxidizing a wide variety of aliphatic aldehydes to carboxylic acids in good yields [13b].

2-3. Preparation of 9-Anthroic Acid [13a]

(9-anthraldehyde) + Ag$_2$O $\xrightarrow{C_2H_5OH}$ (9-anthroic acid)   (13)

A mixture of 18 gm of silver nitrate in 300 ml of 50% ethanol to which has been added 8 gm of sodium hydroxide is refluxed with 10.3 gm (0.05 mole) of anthraldehyde for 4 hr. After dilution with two volumes of hot water and filtration, the hot filtrate is acidified to yield 8 gm (72%) of the acid, mp 204°–206°C.

Sulfuric acid–potassium dichromate can also be used to oxidize aldehydes to acids, as is illustrated by the oxidation of furfural to furoic acid [14a, b].

2-4. Preparation of Furoic Acid [14a]

furfural $\longrightarrow$ furoic acid   (14)

In a round-bottomed flask equipped with a mechanical stirrer, dropping funnel, and reflux condenser are placed 100 gm of furfural, 100 gm of potassium dichromate, and 10 gm of water. On a steam bath the flask is heated to 100°C while a mixture of 200 gm of sulfuric acid and 100 gm of water is added during 30–45 min. The heat of reaction is such that the steam bath is removed after a short time. When reaction is complete, the reaction mixture is cooled and nearly neutralized with sodium hydroxide. Then it is completely neutralized with sodium carbonate. The chromium hydroxide that is filtered off weighs 56 gm after drying. The filtrate is made acid with sulfuric acid and the dark brown precipitate of furoic acid is filtered. More furoic acid is obtained by concentrating the filtrate. One hundred and five grams of crude material is collected. The product is recrystallized from water to yield 87 gm of white crystals of furoic acid (75%), mp 131.5°C.

An alternative procedure for the preparation of furoic acid by oxidation of furfural with hydrogen peroxide in pyridine has been reported [14b].

C. Catalytic Oxidation with Oxygen

The direct, catalytic oxidation of an alcohol with oxygen has much to recommend it in terms of simplicity of purification of crude products. The excess oxidizing agents or reduced products from the oxidizing agent usually present when "chemical" oxidizing agents are used involve the removal of catalyst and separation of water (and possibly some hydrogen peroxide) by the simplest techniques. An example showing the potential scope of this technique is the oxidation of a glucopyranoside (a primary alcohol) to a glucoronide (a carboxylic acid) [15].

2-5. Preparation of 6-Bromo-2-naphthyl-β-D-glucoronide [15]

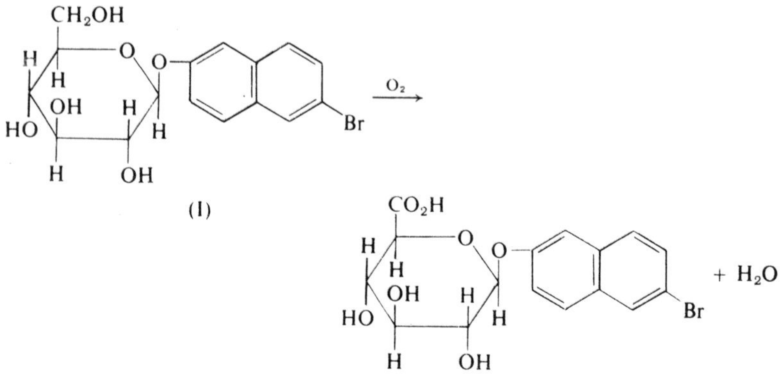

$$\text{(I)} \xrightarrow{O_2} \text{(II)} + H_2O \quad (15)$$

6-Bromo-2-naphthyl-β-D-glucopyranoside(I) is finely ground. Ten grams of compound(I) and 2.0 gm of platinum black are suspended in 3 liters of distilled water with a drop of Dow–Corning Antifoam A. The reaction flask is provided with an efficient stirrer and four interconnected gas inlet tubes—one in each quadrant of the apparatus. The mixture is stirred vigorously and maintained at 90°–100°C on a steam bath while oxygen is bubbled in. The reaction mixture is maintained at a pH of 7 to 7.5 by addition of small amounts of sodium carbonate at 15-min intervals. By adjusting the gas flow and stirring rate, frothing can be controlled. After approximately 3½ hr, the solution is filtered to remove the catalyst and unreacted glucopyranoside. The filtrate is cooled and evaporated to a volume of 60 cc at reduced pressure at 50°–60°C. The residue is decolorized with 2 gm of charcoal, cooled to 0°C, and acidified with 6 N hydrochloric acid to a pH of 2. The precipitated 6-bromo-2-naphthyl-β-D-glucoronide (II) is collected and washed with ice cold water and 5 ml of ether. The product is dried in a vacuum oven at 70°C for 1 hr. Yield 6 gm (54%), mp 163°–165°C (dec.). Upon recrystallization from ethanol–water, lustrous pink flakes, mp 168°–169°C (dec.), are obtained.

3. OXIDATION OF OLEFINS

The oxidation of the double bond provides a useful method for the preparation of several acids. Alkaline permanganate oxidation is frequently employed. For example, 3,7-dimethyl-1-octene yields 2,6-dimethylheptanoic acid in 45% yield [16].

§ 3. Oxidation of Olefins

3-1. Preparation of 2,3-Dimethylheptanoic Acid [16]

$$CH_3-\underset{\underset{CH_3}{|}}{CH}-CH_2-CH_2-CH_2-\underset{\underset{CH_3}{|}}{CH}-CH=CH_2 + KMnO_4 + NaHCO_3 \xrightarrow[7°C]{Acetone}$$

$$CH_3-\underset{\underset{CH_3}{|}}{CH}-CH_2-CH_2-CH_2-\underset{\underset{CH_3}{|}}{CH}-COOH \quad (16)$$

To a solution of 3,7-dimethyl-1-octene (165.5 gm) in acetone (1.3 liters) containing 37 gm powdered sodium bicarbonate at 7°C is added powdered potassium permanganate in small portions until 700 gm has been added. The latter addition takes about 4 hr and the solution is vigorously stirred. After acetone has been removed by distillation, 2 liters of water is added, followed by dilute sulfuric acid and solid sodium bisulfite (400 gm). The resulting clear solution is extracted with 500 ml ether. After the ether has been evaporated, the residue is fractionated to afford a 45% yield of 2,6-dimethylheptanoic acid, bp 115°C (3 mm).

Tetrabutylammonium permanganate in pyridine has also been reported as a room-temperature oxidizing agent for olefins [12a].

The Barbier–Wieland degradation is a classical method for the removal of one carbon atom from the chain. Using this method, pentadecanoic acid has been prepared in 58% yield from palmitic acid [17].

$$RCH_2COOH \xrightarrow{C_2H_5OH} RCH_2COOC_2H_5 \xrightarrow{2C_6H_5MgBr} RCH_2-\underset{\underset{OH}{|}}{C}(C_6H_5)_2 \xrightarrow{Heat}$$

$$RCH=C(C_6H_5)_2 \xrightarrow{CrO_3} RCOOH \quad (17)$$

Djerassi [18] has reviewed a modification of the Barbier–Wieland degradation in which three carbon atoms are removed. The olefin is brominated in the allylic position with N-bromosuccinimide and the product is then dehydrohalogenated to the diene. Upon oxidation, an acid with three fewer carbon atoms is obtained [18].

$$RCH_2-CH_2-CH=C(C_6H_5)_2 \xrightarrow{NBS} RCH_2CHBr-CH=C(C_6H_5)_2 \longrightarrow$$

$$RCH=CH-CH=C(C_6H_5)_2 \xrightarrow{CrO_3} RCOOH \quad (18)$$

The ozonization of the olefinic double bond is sometimes employed to produce acids, but it is not as popular in the laboratory as the other oxidative methods. In this procedure, alkaline silver oxide reacts with ozonides at 95°C to give acids in good yields. 1-Tridecene produces lauric acid in 94% yield [19].

Hydrogen peroxide solutions of 30% strength are also used to decompose ozonides. For example, the ozonide of cyclohexene, decomposed with hydrogen peroxide, affords adipic acid in 60% yield [20]. This procedure should become more popular in the laboratory inasmuch as commercial ozonization equipment has become available. The ozonization reaction has been reviewed [21].

The preparation of 5-methylhexanoic acid by the ozonization of 6-methyl-1-heptene and decomposition of the ozonide by acidic hydrogen peroxide gives yields which average 67% [20].

3-2. *Preparation of 5-Methylhexanoic Acid* [20]

$$\underset{\substack{|\\CH_3}}{CH_3-CH-CH_2-CH_2-CH_2CH=CH_2} \xrightarrow{O_3}$$

$$\underset{\substack{|\\CH_3}}{CH_3-CH-CH_2-CH_2-CH_2-CH-CH_2} \underset{O_3}{\diagdown\diagup} \xrightarrow[H_2SO_4]{H_2O_2} \underset{\substack{|\\CH_3}}{CH_3CH-CH_2CH_2-CH_2-COOH}$$

(19)

A solution of 0.5 mol of 6-methyl-1-heptene in 200 ml of methylene chloride is cooled to $-73°C$ and subjected to a stream of 6% ozonized oxygen at 20 liters/hr for 12 hr [22]. CAUTION: Ozonides can explode. The solution is added over a ½-hr period to a suspension of 39.5 gm of zinc dust in 300 ml of 50% acetic acid, contained in a 1-liter flask. A great deal of heat is evolved at this stage. The methylene chloride is allowed to distill through the condenser and is caught in a cold receiver. The mixture is refluxed for 1 hr, stirred until cold, then extracted twice with 200 ml of ether. The ether extract is washed with a solution of potassium iodide until all peroxides have disappeared. This step is said to be an essential step to avoid explosions later. The ozonide is dissolved in acetic acid and the solution is slowly dropped into a mixture of 114 gm of 30% hydrogen peroxide, 5 ml of concentrated sulfuric acid, and 200 ml of water. Behind a shield, cautious heating is applied progressively. This phase needs careful attention because the reaction becomes vigorous and requires intermittent cooling. Refluxing is continued for 2 hr. After cooling to ice temperature, extraction with ether is performed. The ether layer is then extracted with a solution of sodium hydroxide. The latter is acidified, extracted with ether, dried, and distilled. The acid is collected at $200°–210°C$. Redistillation yields a fraction bp $204°–207°C$ (752 mm), n_D^{20} 1.4220, d_4^{20} 0.9162; amide mp $99.5°–100°C$; *p*-bromophenacyl ester mp $72.5°–73°C$. The yields average 67%.

Cycloolefins react with ozone in an aqueous emulsion in the presence of alkaline hydrogen peroxide to give α,ω-alkanedicarboxylic acid in a one-step process in good yields. The method is illustrated by the preparation of homophthalic acid (90%) from indene [23].

3-3. Preparation of Homophthalic Acid [23]

[indene] + H_2O_2 + NaOH + Emulsifier + O_3 ⟶

[benzene ring]–C(=O)–ONa / –CH$_2$COONa $\xrightarrow{H_3O^+}$ [benzene ring]–COOH / –CH$_2$COOH (20)

The ozone is generated by an electric discharge in an oxygen stream, using a Welsbach T23 ozonator. Its concentration in oxygen, which serves also as a carrier gas, averages 2.8–3.0 wt%. The ozone output of the generator is determined by percolation of the ozone–oxygen mixture through a 2% potassium iodide solution within a timed period; the amount of liberated iodine is titrated with 0.1 N thiosulfate solution. The generator is calibrated for a fixed gas flow rate, pressure, and discharge voltage for the number of grams of ozone per hour.

An emulsion of 80 gm of indene in 600 ml of distilled water is established by the addition of the hydrocarbon and 1 gm of the Brij 30 emulsifier to the well-stirred aqueous phase in an indented, three-necked, 1500-ml reaction flask. The reactor provided with a high-speed stirrer, a gas inlet tube, gas vent, and thermometer are placed in an ice bath to maintain the reaction temperature at 10°C. A saturated aqueous solution of 56 gm of sodium hydroxide (2 mol equiv.) and 142 gm of 30% hydrogen peroxide (1.5 mole equiv., 0.5 mole excess) is added and ozone is introduced into the vigorously stirred emulsion.

The reaction is interrupted after the absorption of 23 gm of ozone, leaving 25 gm of indene in excess as a safety margin for the prevention of overozonization. Addition of 15 gm of sodium chloride, after 1 additional hour of stirring, accelerates the deemulsification, allowing some peroxy polymers and excess indene to separate from the aqueous phase. Acidification of the clear aqueous solution with hydrochloric acid precipitates the homophthalic acid, which is filtered, washed with water, and dried at 70°C and 60 mm. The weight of the crude homophthalic acid is 80 gm (90%), mp 173°C. Steam distillation of the peroxy polymer–indene mixture yields 16 gm of indene. Acidification of the residual solution yields another 6 gm of homophthalic acid and 5 gm of homophthalide.

In the absence of hydrogen peroxide under similar conditions 38 gm of crude homophthalide and 42 gm of crude homophthalic acid are isolated.

In the case of the formation of water-soluble carboxylic acids, the acidified aqueous reaction phase is evaporated to dryness, and the residue—a mixture of sodium chloride and carboxylic acid—is extracted with anhydrous ethanol. Evaporation of the solvent leaves the crude acid, which is purified either by washing with ether, by recrystallization, or by esterification and distillation.

The choice of emulsifying agent is limited by the strong oxidizing conditions. Polyoxyethylated lauryl alcohol, Brij 30, is recommended.

4. OXIDATION OF KETONES AND QUINONES

In most cases the oxidation of ketones yields the acid derived by cleaving the compound into two acids. Lower acid by-products are neglected and the acid of higher molecular weight is usually of interest. For example, trimethylacetic acid (pivalic acid) is prepared in 75% yield by the oxidation of pinacolone using chromic anhydride in aqueous acetic acid [24].

4-1. Preparation of Trimethylacetic Acid [24]

$$(CH_3)_3C-\underset{\underset{O}{\|}}{C}-CH_3 \xrightarrow{Cr_2O_3 + CH_3COOH + H_2O} (CH_3)_3C-COOH \tag{21}$$

Pinacolone (2.9 moles) is dissolved in 150 ml of glacial acetic acid and added to a flask equipped with a stirrer, dropping funnel, and thermometer. Two moles of chromic anhydride (200 gm) is dissolved in 100 ml of water and 250 ml of glacial acetic acid is added when solution in the water has been completed. The chromic anhydride solution is added dropwise to the ketone over a period of 5 hr at a temperature of 100°C. Oxidation did not proceed at 50°C or 80°C. The reaction mixture is diluted with 2 liters of water and neutralized. The mixture is then steam-distilled and the isolated material fractionated to yield 75% trimethylacetic acid, bp 164°C (760 mm); anilide mp and mixed mp 129°C; and 18% unreacted ketone. No *tert*-butyl alcohol is detected.

Adipic acid is obtained in 60% yield from the oxidation of cyclohexanone using nitric acid and a vanadium pentoxide catalyst [25].

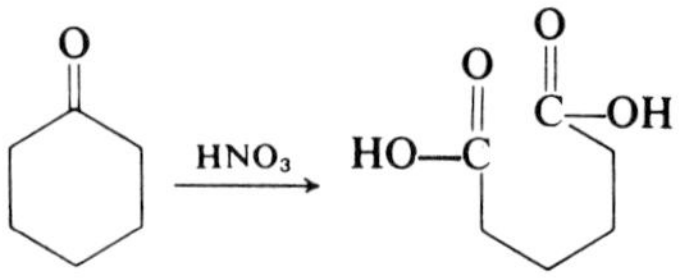

$$\text{cyclohexanone} \xrightarrow{HNO_3} HO-\underset{\underset{O}{\|}}{C}-\cdots-\underset{\underset{O}{\|}}{C}-OH \tag{22}$$

The oxidation of phenanthraquinone by chromic acid yields diphenic acid in 70–85% yields [26, 27].

§ 4. Oxidation of Ketones and Quinones

Diphenic acid has been prepared by Bischoff and Adkins [27] using the Schmitz method [28].

4-2. Preparation of Diphenic Acid [27]

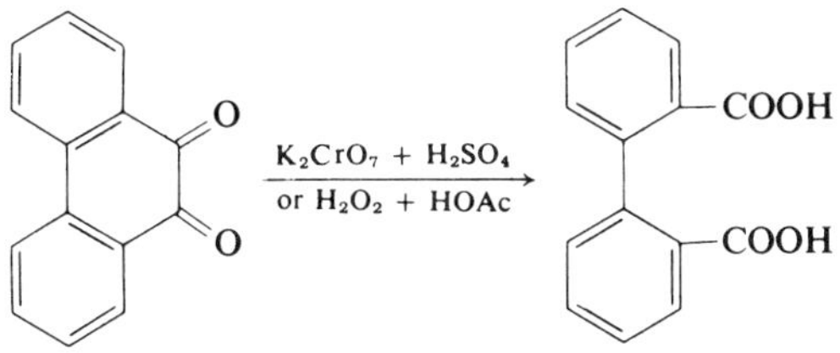

$$\text{phenanthraquinone} \xrightarrow[\text{or } H_2O_2 + HOAc]{K_2CrO_7 + H_2SO_4} \text{2,2'-biphenyldicarboxylic acid (COOH, COOH)} \quad (23)$$

To a cold solution of 200 gm of potassium dichromate, 500 gm of water and 300 gm of concentrated sulfuric acid is added 50 gm of phenanthraquinone. The mixture, in a round-bottomed flask fitted with an air condenser, is cautiously heated for 1 hr to avoid oxidation to carbon dioxide and water. The reaction mixture is then maintained between 105° and 110°C for approximately 20 hr with occasional agitation. The reaction mixture is then cooled, poured into cold water, and the precipitated product filtered off. By repeated washes with cold 5% sulfuric acid, the product may be substantially freed of excess oxidizing agent and reduced side products. The product is then washed and recrystallized from water or glacial acetic acid. Yield, 85%, mp 228°C (uncorrected).

NOTE: Roberts and Johnson [29] recommend that commercial phenanthraquinone be free from chromium compounds and that it be allowed to stand for several hours in concentrated sulfuric acid prior to oxidation. They also claim that sodium bichromate affords a better yield of product than does the potassium salt.

α-Keto acids are oxidized in a basic solution of 30% hydrogen peroxide to carboxylic acids with one less carbon atom. For example, m-chlorophenylpyruvic acid is oxidized to m-chlorophenylacetic acid in 57% yields [30].

4-3. Preparation of m-Chlorophenylacetic Acid [30]

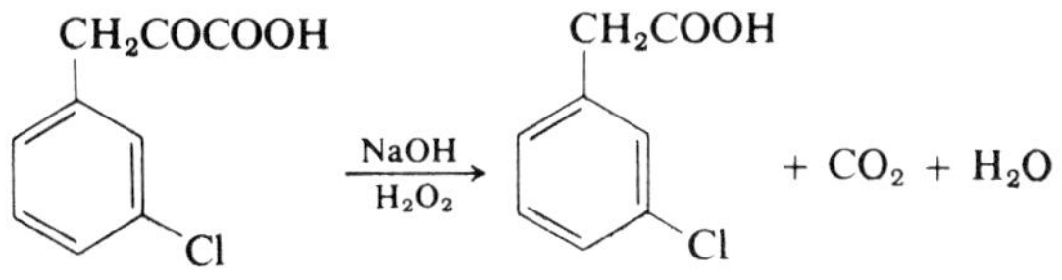

$$\text{Ar-CH}_2\text{COCOOH} \xrightarrow[H_2O_2]{NaOH} \text{Ar-CH}_2\text{COOH} + CO_2 + H_2O \quad (24)$$

(Ar = m-chlorophenyl)

Six grams of m-chlorophenylpyruvic acid is dissolved in a solution of 8.0 gm of sodium hydroxide in 25 ml of water. Ten grams of ice is added and then the slow addition of hydrogen peroxide solution is begun (7.5 gm 30% hydrogen peroxide in 15 ml of water) with cooling and shaking. After standing 5 hr, the solution is cautiously acidified with hydrochloric acid, and, while still warm, extracted with benzene. The extract is dried and the benzene is evaporated. Upon recrystallizing from aqueous ethanol, there is obtained 3.0 gm of large pearly leaflets, melting at 74°C (57%).

A. The Haloform Reactions

Acetyl groups and methylcarbinols are converted to carboxyl groups by substitution of the three hydrogens of the methyl groups by halogen which is subsequently hydrolyzed.

$$RCOCH_3 + 3\ NaOH + 3\ X_2 \longrightarrow [RCOCX_3] + 3\ H_2O + 3\ NaX$$
$$\downarrow NaOH$$
$$RCOONa + CHX_3 \quad (25)$$

In order to ensure good yields, it is desirable that no similarly replaceable hydrogens be present in the R group. However, methylene groups are not easily affected by the reagents since β-phenylisovaleric acid is obtained in 84% yield from 4-methyl-4-phenyl-2-pentanone [31].

The reagents that are used for the haloform reaction are chlorine in sodium hydroxide at 55°–80°C [32–34], aqueous sodium or potassium hypochlorite [35], commercial bleaching agents [36], iodine in sodium hydroxide, and bromine in sodium hydroxide at 0°C, which is illustrated below for the conversion of 4,4-dimethylpentan-2-one to *tert*-butylacetic acid [37].

4-4. *Preparation of* **tert-***Butylacetic Acid [37]*

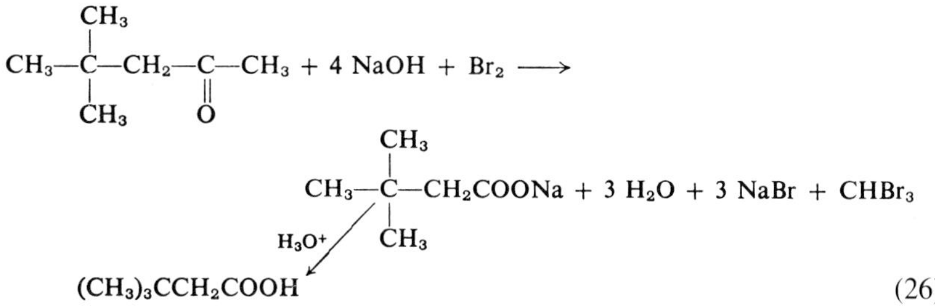

$$\underset{\underset{CH_3}{|}}{\overset{\overset{CH_3}{|}}{CH_3-C-CH_2-\underset{\underset{O}{\|}}{C}-CH_3}} + 4\ NaOH + Br_2 \longrightarrow$$

$$\underset{\underset{CH_3}{|}}{\overset{\overset{CH_3}{|}}{CH_3-C-CH_2COONa}} + 3\ H_2O + 3\ NaBr + CHBr_3$$

$$\xrightarrow{H_3O^+} (CH_3)_3CCH_2COOH \qquad\qquad (26)$$

To a flask containing 1 kg of ice and a solution of 525 gm of sodium hydroxide in 2 liters of ice water is added 240 ml of bromine (4.75 moles) from a dropping funnel during 1 hr. Following the preparation of sodium hypobromite, 171 gm of 4,4-dimethylpentanone-2 is added during about 10 min. The solution is stirred for 14 hr and then it is steam-distilled. Stirring is continued during the distillation to prevent bumping. The distillate comprises about 600 ml of water and 175 gm of a mixture of bromoform and carbon tetrabromide. The residue from the steam distillation is acidified with 600 ml of concentrated sulfuric acid. Steam distillation yields 151 gm of an oil. Ether extraction of the distillate yields 17 gm of an additional portion of the oil. Fractional distillation of the oil gives 155 gm of *tert*-butylacetic acid (89%), bp 96°C (26 mm); mp 6°–7°C; n_D^{20} 1.4096.

Another example of the haloform reaction is the oxidation of propiophenone to benzoic acid in 96% yield [38].

4-5. Preparation of Benzoic Acid [38]

$$C_6H_5-\overset{O}{\underset{\|}{C}}-CH_2-CH_3 + NaOBr \longrightarrow C_6H_5COOH + CH_3COOH + NaBr \quad (27)$$

To 300 ml of a rapidly stirred sodium hypobromite solution (0.512 mole) at 22°C is added 20.1 gm (0.15 mole) of propiophenone over a 5-min period. Stirring is rapid so that the solution exists as an emulsion throughout the reaction. Best yields are obtained under these conditions. Stirring is continued for 2½ hr and the mixture kept at 24°–25°C by immersing the flask in an ice bath when necessary. The unreacted hypobromite is destroyed with sodium bisulfite and the basic solution is extracted with ether to remove the unreacted ketone. Acidification of the aqueous phase with concentrated hydrochloric acid yields 17.6 gm (96%) of benzoic acid, mp 121.5°–122°C alone and when mixed with an authentic sample.

B. Willgerodt Reaction

The Willgerodt reaction is useful in the preparation of arylacetic acids and amides from substituted methyl aryl ketones or vinyl aromatic compounds. The aliphatics and acetylenes give lower yields. The conversion is effected by heating aromatic compounds at 160°–200°C in an aqueous solution under pressure using ammonium polysulfide [39, 40]. In the Kindler [41] modification, the ketone or styrene is refluxed with a mixture of sulfur and an amine, usually morpholine, to give a thioamide, $ArCH_2CSNR_2$.

$$\left. \begin{array}{l} RCOCH_3 \\ RCH=CH_2 \\ RC{\equiv}CH \end{array} \right\} \longrightarrow RCH_2CSNH_2 \longrightarrow RCH_2COOH \quad (28)$$

Schwenk and Block [42] also suggested the use of morpholine as an amine, which permits operations in ordinary laboratory equipment. The reaction appears to be quite general for aromatic monomethyl ketones. Substitutions such as nitro, amino, hydroxy, or second acetoxy groups interfere with the standard reaction, probably because these functional groups are capable of reacting with sulfur, polysulfides, or other components of the reaction mixture.

A typical example of the Schwenk and Block procedure for the preparation of p-methoxyphenylacetic acid is given by Solmssen and Wenis [43].

4-6. Preparation of p-Methoxyphenylacetic Acid [43]

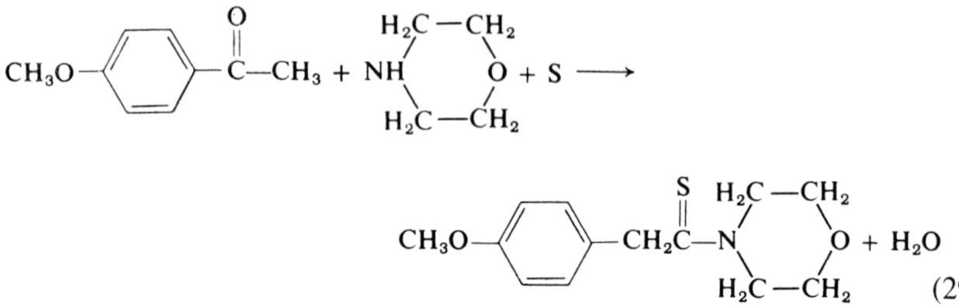

$$CH_3O\text{-}C_6H_4\text{-}\underset{O}{\underset{\|}{C}}\text{-}CH_3 + NH\begin{pmatrix}H_2C\text{-}CH_2\\H_2C\text{-}CH_2\end{pmatrix}O + S \longrightarrow$$

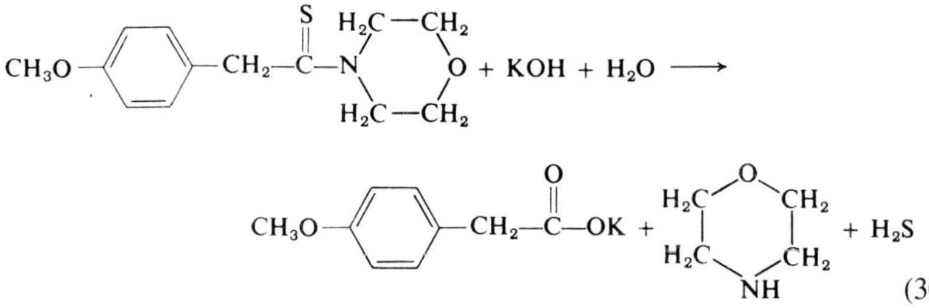

$$CH_3O\text{-}C_6H_4\text{-}CH_2\underset{S}{\underset{\|}{C}}\text{-}N\begin{pmatrix}H_2C\text{-}CH_2\\H_2C\text{-}CH_2\end{pmatrix}O + H_2O \quad (29)$$

$$CH_3O\text{-}C_6H_4\text{-}CH_2\underset{S}{\underset{\|}{C}}\text{-}N\begin{pmatrix}H_2C\text{-}CH_2\\H_2C\text{-}CH_2\end{pmatrix}O + KOH + H_2O \longrightarrow$$

$$CH_3O\text{-}C_6H_4\text{-}CH_2\underset{O}{\underset{\|}{C}}\text{-}OK + \begin{pmatrix}H_2C\\H_2C\end{pmatrix}\underset{NH}{O}\begin{pmatrix}CH_2\\CH_2\end{pmatrix} + H_2S \quad (30)$$

$$CH_3O\text{-}C_6H_4\text{-}CH_2\underset{O}{\underset{\|}{C}}\text{-}OK + HCl \longrightarrow$$

$$CH_3O\text{-}C_6H_4\text{-}CH_2CO_2H + KCl + H_2O \quad (31)$$

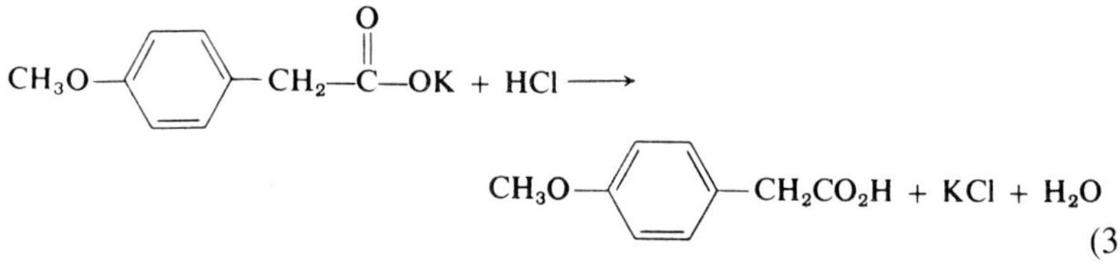

In a hood, a mixture of 879 gm (5.86 moles) of p-methoxyacetophenone, 281.5 gm (8.79 moles) of morpholine, and 281.5 gm (8.79) of flowers of sulfur as refluxed overnight with efficient stirring.

The warm reaction mixture is poured into 1 liter of warm absolute ethanol containing 2% benzene. A crop of the thiomorpholide separates on cooling. Evaporation of the mother liquor affords additional quantities of material. Total yield 1173 gm (80%).

All of the thiomorpholide is refluxed overnight with 13.2 liters of a 10% solution of potassium hydroxide in water (hood). The reaction mixture is the cooled and acidified to a pH of 3 to 5. Approximately 570 gm of crystalline product separates.

The product is dissolved in hot ethanol containing 2% benzene, treated with activated charcoal, filtered, and precipitated by cautious addition of distilled water to the boiling alcohol solution until a slight cloudiness appears. Upon cooling, 474 gm of product is obtained. Yield 49%.

§ 5. Bimolecular Oxidation–Reduction Reactions

A more recent modification of the Willgerodt reaction describes the preparation of acids from aromatic hydrocarbons using aqueous base and sulfur [44]. An example is the preparation of isophthalic acid from *m*-xylene using aqueous ammonia, sulfur, and water.

4-7. Preparation of Isophthalic Acid [44]

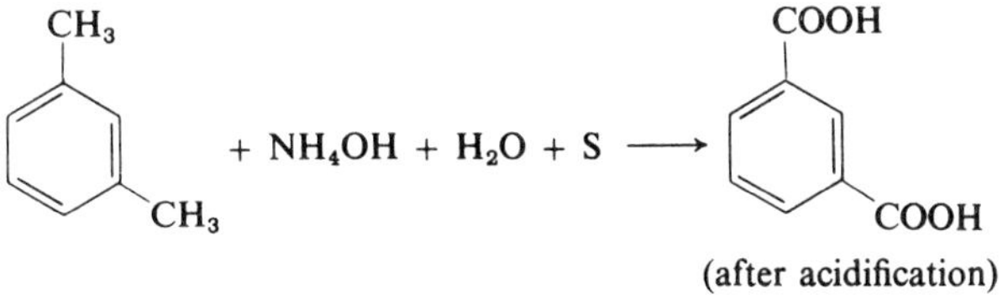

$$\text{m-xylene} + NH_4OH + H_2O + S \longrightarrow \text{isophthalic acid} \quad (32)$$

(after acidification)

A 2.5-liter autoclave is charged with 42.5 gm (0.4 mole) of *m*-xylene (98%), 243 gm of 28% aqueous ammonia (4 moles of NH_3), 250 g of water (23.6 moles H_2O), and 96 gm (3.0 gm atoms) of sulfur, and heated to 316°C before shaking is begun. After 30 min, the maximum pressure of 176 atm is obtained, indicating completion of the reaction. Shaking is continued an additional 5 min. Steam distillation of the reaction mixture removes hydrogen sulfide and ammonia and reduces the pH to 7. No xylene is recovered. Sulfur from the polysulfide decomposition is filtered off and 2 moles of sodium hydroxide is added to saponify amides. The solution is steam-distilled until the vapors test neutral to moist pH paper, and the remaining solution of nonvolatiles is adjusted to pH 7 with dilute hydrochloric acid. After thorough washing and drying further acidification gives 57.0 gm (86%) of isophthalic acid, neutral equivalent 82.9.

5. BIMOLECULAR OXIDATION–REDUCTION REACTIONS

Aldehydes that lack an α-hydrogen and therefore cannot undergo an aldol condensation undergo the Cannizzaro reaction in the presence of a strong base, giving the alcohol and the corresponding carboxylic acid. Furfural in the presence of sodium hydroxide yields 72–76% furfuryl alcohol and 73–76% furoic acid upon acidifying [45–47].

5-1. Cannizzaro Reaction: Preparation of Furoic Acid [45]

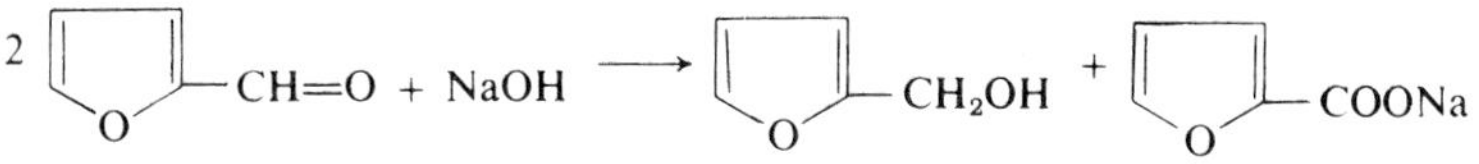

$$2\ \text{furfural–CH=O} + NaOH \longrightarrow \text{furfuryl–CH}_2\text{OH} + \text{furoate–COONa} \quad (33)$$

One kilogram (10.4 moles) of redistilled furfural is placed in a flask which is surrounded by an ice bath and cooled to 5°–8°C. While stirring, 625 gm of 33.3% of sodium hydroxide (5.2 moles) is added dropwise at such a rate to keep the reaction temperature below 20°C. After the addition, the mixture is stirred for an additional hour. The mixture is cooled to 0°C and pressed as dry as possible on a suction filter. The sodium 2-furancarboxylate is transferred to a beaker, triturated therein with a 200–250-ml portion of cold water, cooled to −5°C, and again filtered. The trituration of the solid with water is repeated.

The combined filtrates are distilled at 25 mm nearly to dryness using a heating bath, the temperature of which is kept below 145°C. The water is distilled away under vacuum and the residue of furfuryl alcohol is shaken with sodium bisulfite solution to remove any remaining furfural. Fractionation yields 367–390 gm (72–76%), bp 83°C (24 mm), n_D^{21} 1.4869.

All the solid residues (sodium 2-furancarboxylate) are dissolved in warm water and filtered from a small amount of dark insoluble material. The filtrate is acidified with concentrated hydrochloric acid, cooled to 0°C, and filtered. The solid is washed twice with a little ice water and dried. A yield of 420–440 gm (73–76%) of white furoic acid is obtained, mp 132°C (recrystallized from carbon tetrachloride).

The reaction of α-diketones with strong bases yields the rearranged α-hydroxy carboxylic acids. This reaction is known as the benzilic acid rearrangement and is illustrated below using benzoin [48].

5-2. Benzilic Acid Rearrangement [48]

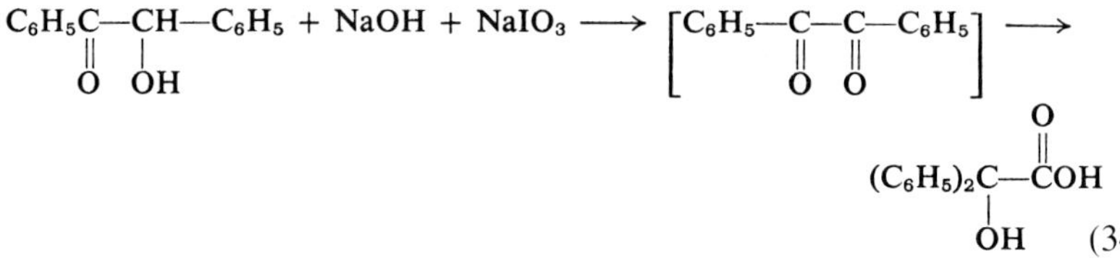

$$C_6H_5\underset{\underset{O}{\|}}{C}-\underset{\underset{OH}{|}}{CH}-C_6H_5 + NaOH + NaIO_3 \longrightarrow \left[C_6H_5-\underset{\underset{O}{\|}}{C}-\underset{\underset{O}{\|}}{C}-C_6H_5 \right] \longrightarrow$$

$$(C_6H_5)_2\underset{\underset{OH}{|}}{C}-\overset{\overset{O}{\|}}{C}OH \quad (34)$$

To a solution of 20 gm of sodium hydroxide and 7 gm of sodium iodate in hot water is added 20 gm of benzoin. The mixture is stirred and a purple color becomes evident. The mixture is heated while being stirred until the purple color fades and then sufficient water is added to dissolve the solids and form a clear solution. Concentrated hydrochloric acid is added, and the iodine and traces of benzoic acid are expelled by boiling. On cooling, the solution is filtered, and 20 gm of dry crude benzilic acid is obtained. Recrystallization from benzene yields 18 gm (90%) of benzilic acid.

In the above example, benzoin is oxidized directly to benzilic acid via the benzil intermediate. A similar rearrangement occurs when α-epoxy ketones are refluxed with 30% aqueous sodium hydroxide [49].

6. CARBONATION OF ORGANOMETALLIC REAGENTS

Carbonation of the Grignard reagent and organometallic compounds is a useful laboratory method for the conversion of most halides to acids containing one additional carbon atom. One technique involves pouring the ether solution of the organometallic into excess crushed Dry Ice [50]. Carbon dioxide is sometimes required for tertiary Grignard reagents [51]. A low temperature and rapid stirring produce high yields of acids [52]. The main by-products are symmetrical ketones and tertiary alcohols formed by the reaction of the organometallic compound with the carboxylic acid salt. Jetwise addition of the organometallic compound to excess powdered Dry Ice greatly reduces the amount of these products [53, 54]. Yields range from 50% to 85% of the carboxylic acids. Of the other organometallic reagents the lithium and sodium derivatives have been the most popular and have also given good yields of carboxylic acids under suitable conditions.

The carbonation of the Grignard reagent has advantages over the procedure involving hydrolysis of nitriles since the latter is applicable only to primary halides if reasonable yields are to be obtained. The Grignard method has limitations since tertiary acids above dimethylethylacetic acid cannot be prepared by carbonation of the appropriate Grignard compounds because the latter reagents prepared from higher halides react abnormally, yielding mixtures of alkanes and alkenes. The Wurtz type of Grignard reagent coupling becomes increasingly more important as higher halides are involved [55].

Recently the use of highly reactive magnesium has been reported to be useful for the preparation of Grignard reagents from a variety of aromatic and aliphatic halides, including fluorides. These have been carbonated to convert them to carboxylic acids in fair-to-good yields [56]. For example, fluorobenzene is converted to benzoic acid in 69% yield and 1-chloronorboronene is converted to norboronene 1-carboxylic acid in 63% yield.

6-1. Grignard Reagents—Preparation of α-Methylbutyric Acid [54]

$$C_2H_5CHClCH_3 + Mg \longrightarrow C_2H_5-CH(MgCl)CH_3 \xrightarrow[H_2O]{CO_2} C_2H_5-CH(COOH)CH_3$$

(35)

sec-Butylmagnesium chloride is prepared in 400 ml of ether from 13.4 gm (0.55 gm atom) of magnesium shavings and 46 gm (0.5 mole) of sec-butyl chloride. A stream of carbon dioxide is passed through the solution at $-5°$ to $-12°C$. After 1½ hr the temperature drops from $-5°$ to $-12°C$ and does not rise on increasing the flow rate of carbon dioxide. The drop in temperature is taken as the end point for the carbonation. The reaction mixture is hydrolyzed with 500 ml of 25% sulfuric acid while cooling with ice. The water layer is extracted with ether, the ether washed with 150 ml of 25% sodium hydroxide solution, and the aqueous layer is acidified to yield an oil which is separated. The acid is distilled to yield the product at 173°–174°C, 39–44 gm (76–86% based on *sec*-butyl chloride used).

6-2. Lithium Reagents—Preparation of Fluorene-9-carboxylic Acid [57a]

$$n\text{-BuLi} + \text{[fluorene]} \longrightarrow$$

$$\text{[9-lithiofluorene]} \xrightarrow{CO_2} \text{[fluorene-9-COOLi]} \quad (36)$$

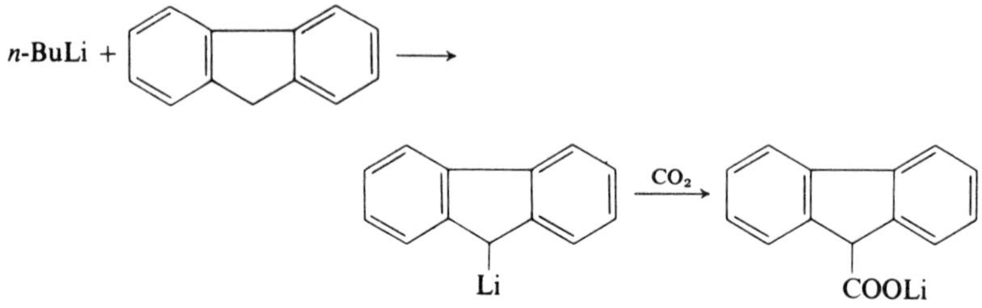

A solution of 1 mole of *n*-butyllithium in 500 ml of ether is treated portionwise with 0.75 mole of fluorene. The solution turns an orange color accompanied by vigorous evolution of butane. The mixture is refluxed for 1 hr and then poured jetwise onto crushed Dry Ice. As soon as the mixture warms up to room temperature, the unreacted lithium is skimmed off and 2 liters of water is added cautiously. The insoluble residue is filtered off and the organic layer is extracted three times with 300-ml portions of luke-warm 2% sodium hydroxide. Acidification of the combined aqueous solutions precipitates the desired acid. The yield is 118 gm (75%), mp 228°–230°C, based upon fluorene.

6-3. Sodium Reagents—Preparation of Fluorene-9-Carboxylic Acid [57a]

$$\text{PhCl} + \text{Na} \longrightarrow \text{PhNa} \xrightarrow{\text{fluorene}}$$

$$\text{[9-sodiofluorene]} \xrightarrow{CO_2} \text{[fluorene-9-COONa]} \quad (37)$$

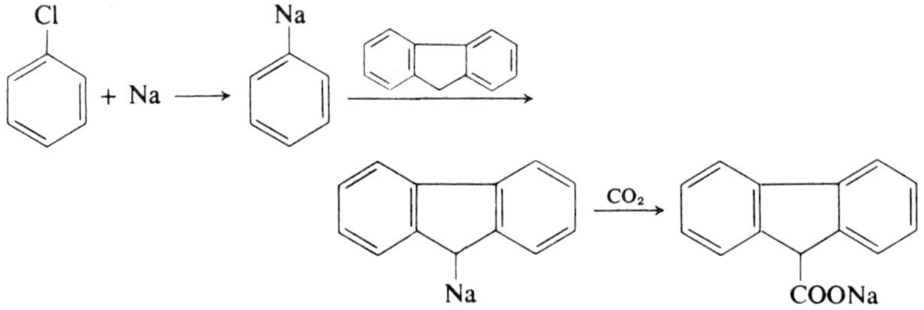

Phenylsodium is prepared by heating a mixture of 11.5 gm (0.5 atom) of powdered sodium with 22.5 gm (0.2 mole) of chlorobenzene and 200 ml of thiophene-free benzene under a nitrogen atmosphere at 60°–65°C until the reaction starts. The reaction is only mildly exothermic, as evidence by a small rise in temperature after removing the heating bath. The induction period generally requires 30–50 min. The stirred mixture is held at 50°–55°C for 2 hr. At once 28.9 gm (0.174 mole) of fluorene is added and the mixture is refluxed for 2 hr. The reaction mixture is then cooled to room temperature and 200 ml of ether is added. Then the mixture is carbonated by pouring it jetwise over Dry Ice. The excess sodium is destroyed by the addition of 100 ml of 50% alcohol. The solution is diluted with 300 ml of water and undissolved material is filtered off. The organic layer is separated and extracted three times with 300-ml portions of lukewarm 2% sodium hydroxide. Upon acidification of the combined aqueous solution, 21.5 gm of the acid is isolated, mp 227°C. A similar procedure has been described in ref [57b].

It has recently been reported that the Corey-Chayakowski reagent (sodium hydride in dimethyl sulfoxide, also called dimsyl sodium) metalates certain reactive hydrogen compounds. Thus, treatment of acetophenone or phenylacetylene with this reagent, followed by treatment with Dry Ice, yielded benzoylacetic acid and phenylpropiolic acid, respectively [58]. The reagent also provides as a rapid saponification system for esters. Further exploration of its applicability to various synthetic operations would be of great interest. In this connection, the reader's attention is directed to a report of a violent explosion when a scale-up of a C-methylation of isoquinoline was attempted with this reagent [59]. It has been reported [60] that when a solution of the DMSO anion was held at elevated temperatures for a long time the anion decomposed at 70°–80°C, producing the methyl sulfinate ion, sulfide ion, and other products. This decomposition was exothermic and was accompanied by formation of a precipitate. The accumulation of the precipitate produced a soft gel and at this time the temperature abruptly rose to 40°–85° above the bath temperature and then dropped again, showing no further exotherm. Therefore, it is suggested, when preparing DMSO anion, that the reaction be vigorously stirred in a heating bath so that the cooling effect of the bath is available if the reaction is held long enough to encounter the abrupt temperature rise.

7. CARBOXYLATION OF THE AROMATIC NUCLEUS

Heating the alkali salts of resorcinol and α-naphthol with carbon dioxide gives excellent yields of the carboxylic acids (Kolbe–Schmitt reaction) [61–63].

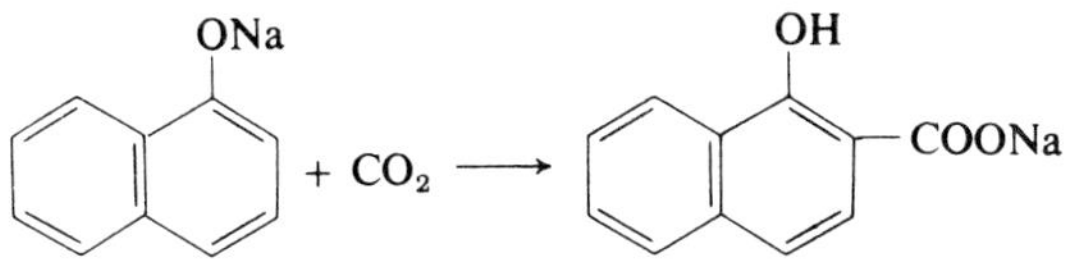

$$\text{naphthyl-ONa} + CO_2 \longrightarrow \text{naphthyl(OH)-COONa} \tag{38}$$

When salicylic acid is heated to 240°C with potassium carbonate, an 80% yield of *p*-hydroxybenzoic acid results because of a carboxyl group migration [63, 64].

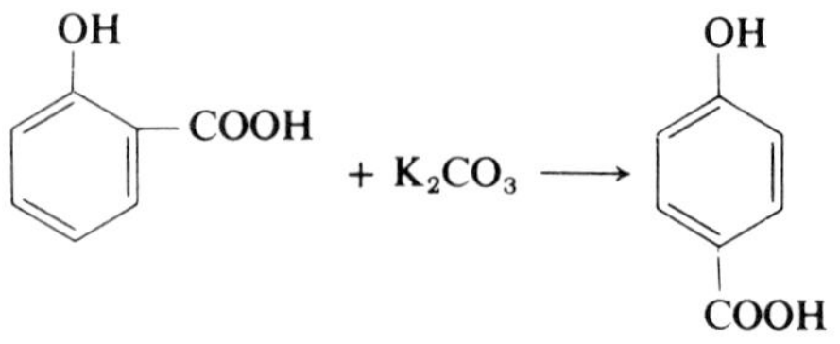

$$\underset{\text{OH}}{\text{C}_6\text{H}_4}\text{COOH} + \text{K}_2\text{CO}_3 \longrightarrow \underset{\text{COOH}}{\underset{\text{OH}}{\text{C}_6\text{H}_4}} \qquad (39)$$

It has been reported that the reaction of aromatic compounds with sodium palladium(II) malonate in a mixed solvent of acetic acids and acetic anhydride or carbon tetrachloride gives aromatic acids in good yields, together with lower yields of aromatic dimers [65].

7-1. Kolbe–Schmitt Reaction—Preparation of β-Resorcylic Acid [61]

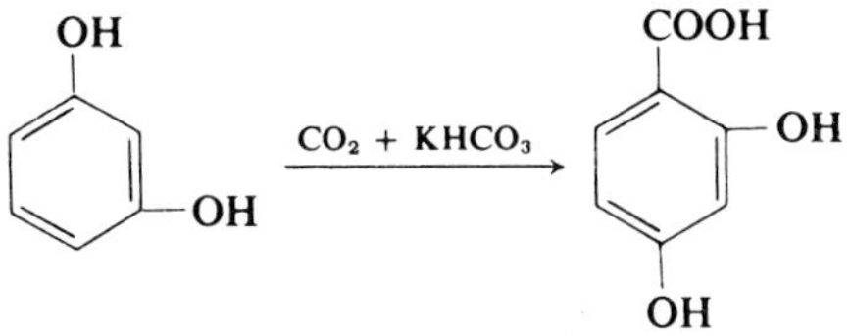

$$\underset{\text{OH}}{\underset{\text{OH}}{\text{C}_6\text{H}_4}} \xrightarrow{\text{CO}_2 + \text{KHCO}_3} \underset{\text{OH}}{\underset{\text{OH}}{\overset{\text{COOH}}{\text{C}_6\text{H}_3}}} \qquad (40)$$

To a 5-liter flask equipped with a reflux condenser is added a solution containing 200 gm (1.8 moles) of resorcinol, 1 kg (9.9 moles) of potassium bicarbonate (or sodium carbonate in equimolar amount), and 2 liters of water. The mixture is heated slowly at 80°–100°C in an oil bath for 4 hr. Then the temperature of the bath is raised so that the contents reflux vigorously for 30 min while a rapid stream of carbon dioxide is passed through.

The hot solution is acidified by adding 900 ml of concentrated hydrochloric acid from the dropping funnel, which is connected to the gas inlet tube reaching the bottom of the flask.

After being cooled to room temperature, the flask is chilled in an ice bath. The resorcylic acid crystallizes in colorless prisms, which, on exposure to air, may turn pink due to free resorcinol. The crude yield is 225 gm. Extraction of the filtrate with ether yields an additional 35 gm of crude resorcylic acid. The crude resorcylic acid (260–270 gm) is dissolved in 1 liter of boiling water, boiled with 25 gm of Norite, filtered, and cooled in an ice–salt water bath. The solution is stirred vigorously and the crystalline product is obtained. Yield 160–170 gm (57–60%), mp 216°–217°C.

A. Friedel–Crafts Reaction

The Friedel–Crafts reaction with maleic anhydride and benzene proceeds quite readily. Careful attention to details in the preparation are required to avoid prolonged contact of the product with aqueous hydrochloric acid which gives rise to β-benzoyllactic acid, while prolonged treatment with aqueous alkali affords

§ 7. Carboxylation of the Aromatic Nucleus   257

acetophenone. Inhalation of β-benzoylacrylic acid dust must be avoided because of its sternutatory action.

7-2. Preparation of β-Benzoylacrylic Acid [66]

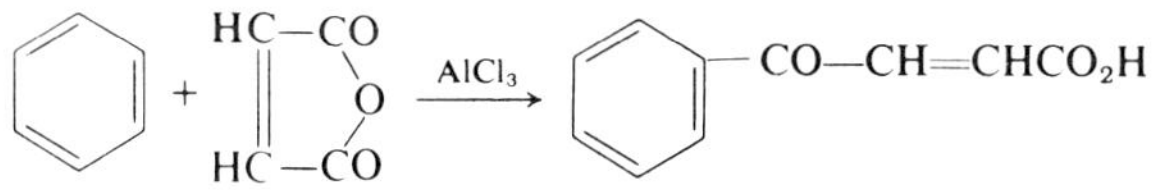

$$\text{C}_6\text{H}_6 + \underset{\text{HC—CO}}{\underset{\|}{\text{HC—CO}}}\!\!>\!\!\text{O} \xrightarrow{\text{AlCl}_3} \text{C}_6\text{H}_5\text{—CO—CH=CHCO}_2\text{H} \quad (41)$$

To a mixture of 200 ml of anhydrous, thiophene-free benzene and 49 gm (0.5 moles) of maleic anhydride at room temperature, 132 gm (1 mole) of aluminum chloride (J. T. Baker, *granular,* anhydrous C.P. grade) is added in small portions. The reaction temperature rises to 40°–45°C and is maintained at this level during the addition. Then the reaction mixture is heated on a steam bath for 2 to 3 hr, cooled rapidly, and added to an excess of ice and 1:1 hydrochloric acid. The benzene layer is separated and freed of benzene by steam distillation. On cooling, the supernatant liquid is decanted from the semisolid residue. The crude product is dissolved in 5% sodium carbonate solution, filtered, and acidified with efficient cooling. The precipitating solid is washed with cold water and dried. Yields as high as 91% have been reported. The product, recrystallized from water, forms a colorless hydrate, mp 60°–62°C. Recrystallized from benzene (with minimum heating perod), the anhydrous product, mp 94°–96°C, is obtained

Direct introduction of the carboxyl group into the aromatic ring has also been accomplished with phosgene and oxalyl chloride [67–69]. Thus, for example, 9-anthroic acid is prepared in 67% yield from anthracene by heating to 240°C with oxalyl chloride in nitrobenzene [68].

Certain activated carbon atoms may be conveniently acylated with oxalyl chloride to afford an intermediate which is readily decarboxylated to afford a carboxylic acid.

7-3. The Use of Oxalyl Chloride—Preparation of 9-Anthroic Acid [68]

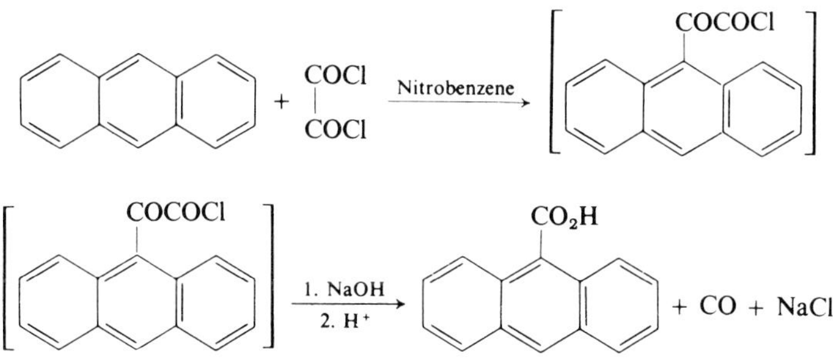

$$\text{anthracene} + \underset{\text{COCl}}{\overset{\text{COCl}}{|}} \xrightarrow{\text{Nitrobenzene}} [\text{9-(COCOCl)-anthracene}]$$

$$[\text{9-(COCOCl)-anthracene}] \xrightarrow[\text{2. H}^+]{\text{1. NaOH}} \text{9-(CO}_2\text{H)-anthracene} + \text{CO} + \text{NaCl} \quad (42)$$

A mixture of 50 gm of anthracene, 30 ml of oxalyl chloride, and 150 ml of dry nitrobenzene is heated at such a rate that only a gentle reflux is observed at all times. Over a period of 5 to 6 hr, the temperature is raised from 120° to 240°C. After steam distillation to remove the nitrobenzene, 100 ml of 1 N sodium hydroxide solution and enough water to make 700 ml of reaction mixture are added. The mixture is refluxed for ½ hr. Insoluble materials (from which 11 gm of anthracene can be isolated) are removed by filtration. The filtrate is extracted with ligroin (bp 30°–60°C), treated with activated charcoal, and filtered hot. The charcoal is washed with 2 N sodium carbonate solution. Acidification of the combined filtrate and carbonate washings gives 41.6 gm (67%) of 9-anthroic acid, mp 208°–212°C.

Carboxylation of benzene is effected in 15–58% yields by treating with liquid phosgene and aluminum chloride [70].

$$C_6H_6 \xrightarrow[AlCl_3]{COCl_2} C_6H_5COCl \cdot AlCl_3 \xrightarrow{H_2O} C_6H_5COOH \qquad (43)$$

Dimethaniline reacts directly with phosgene to yield p-dimethylaminobenzoic acid in 50% yield [71]. β,β-Diarylacrylic acids can similarly be prepared using either phosgene or oxalyl chloride on 1,1-diarylethylenes [72].

7-4. Preparation of β,β-Dianisylacrylic Acid Using Oxalyl Chloride [72]

$$\begin{array}{c} CH_3O-C_6H_4 \\ \diagdown \\ C=CH_2 \xrightarrow{(COCl)_2} \\ \diagup \\ CH_3O-C_6H_4 \end{array}$$

$$\begin{array}{c} CH_3O-C_6H_4 \\ \diagdown \\ C=CH-COCl \xrightarrow[\text{2. Acid}]{\text{1. Na}_2\text{CO}_3} \\ \diagup \\ CH_3O-C_6H_4 \end{array} \begin{array}{c} CH_3O-C_6H_4 \\ \diagdown \\ C=C-COOH \\ | \\ H \\ CH_3O-C_6H_4 \end{array} \qquad (44)$$

One mole of the 1,1-diarylethylene and 8 to 9 moles of oxalyl chloride are refluxed until the evolution of hydrogen chloride ceases. In the case of 1,1-di(p-anisyl)ethylene, the reaction is complete in ½ hr at room temperature.

Excess oxalyl chloride is removed under reduced pressure and the syrupy residue is stirred into an ice-cold sodium carbonate solution. The acid chloride requires 1 to 2 hr for hydrolysis.

The mixture is then boiled with a large amount of water (about 1 liter per 50 gm of substituted ethylene) to dissolve the sodium salt and to separate it from tarry by-products. Charcoal is added and the solution is filtered. Part of the sodium salt crystallizes from the cool filtrate and, upon acidification, yields a pure sample of the desired acid. The remainder of the acid is recovered from the filtrate by acidification and purified by recrystallization. Dianisylacrylic acid is obtained in 75% yield, mp 142°C.

The ethylenes containing halogenated phenyl groups were partially converted to tarry material by the long reflux time necessary to cause them to react. Lower yields are obtained in these cases.

7-5. Preparation of β,β-Dianisylacrylic Acid Using Phosgene [72]

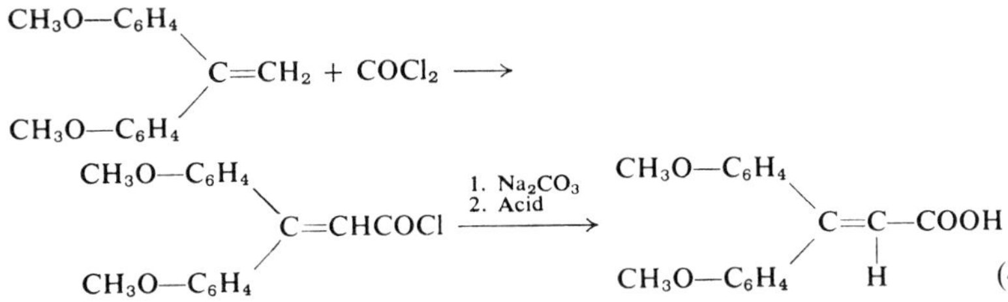

$$\begin{array}{c} CH_3O-C_6H_4 \\ \diagdown \\ C=CH_2 + COCl_2 \longrightarrow \\ \diagup \\ CH_3O-C_6H_4 \end{array}$$

$$\begin{array}{c} CH_3O-C_6H_4 \\ \diagdown \\ C=CHCOCl \xrightarrow[\text{2. Acid}]{\text{1. Na}_2\text{CO}_3} \\ \diagup \\ CH_3O-C_6H_4 \end{array} \begin{array}{c} CH_3O-C_6H_4 \\ \diagdown \\ C=C-COOH \\ \diagup| \\ CH_3O-C_6H_4 H \end{array} \quad (45)$$

In a hood, with suitable precautions, a slow stream of phosgene is bubbled during 10 hr through a solution of 10 gm of 1,1-di(p-anisyl)ethylene in 50 ml of benzene. The solvent is distilled off under reduced pressure and the residue is dissolved in cold sodium carbonate solution. The sodium salt of the acid is dissolved by heating and filtered from the insoluble, neutral material. Acidification of the filtrate precipitates 3.5 gm (30%) of β,β-dianisylacrylic acid, mp 139°–140°C.

8. CONDENSATION REACTIONS

A. Perkin Reaction

The Perkin reaction is the base-catalyzed reaction of an active methylene group of an anhydride and the aldehyde group. Basic catalysts such as the sodium salt of the acid corresponding to the anhydride, potassium carbonate, or tertiary amines may be used satisfactorily [73]. This reaction is very useful for the preparation of substituted cinnamic acids such as those containing halo, methyl, and nitro groups [74, 75].

The preparation of 2-methyl-3-nitrocinnamic acid is representative of the reaction and its conditions.

8-1. Preparation of 2-Methyl-3-nitrocinnamic Acid [74]

A mixture of 75 gm of m-nitrobenzaldehyde, 98 gm of propionic anhydride, and 48 gm of sodium propionate is heated at 170°C in an oil bath for 5 hr. The reaction mixture is poured into water and saturated sodium carbonate solution added to strong alkalinity. The tarry liquid is boiled with decolorizing charcoal for 10 min and then filtered. The alkaline mixture is poured into dilute hydro-

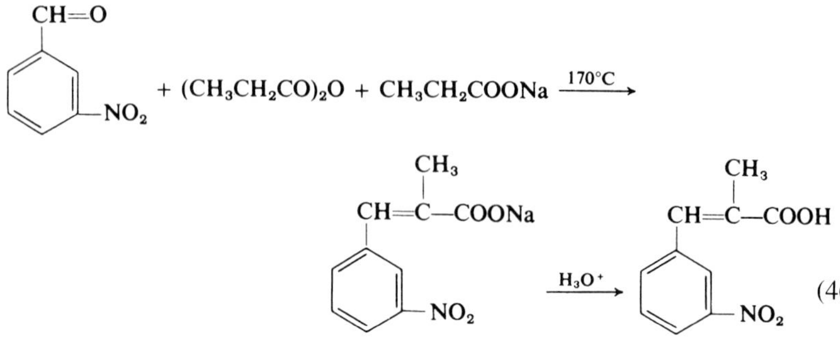

$$\underset{NO_2}{\underset{|}{C_6H_4}}-CH=O + (CH_3CH_2CO)_2O + CH_3CH_2COONa \xrightarrow{170°C}$$

$$\underset{NO_2}{\underset{|}{C_6H_4}}-CH=\underset{CH_3}{\underset{|}{C}}-COONa \xrightarrow{H_3O^+} \underset{NO_2}{\underset{|}{C_6H_4}}-CH=\underset{CH_3}{\underset{|}{C}}-COOH \quad (46)$$

chloric acid. A white curdy precipitate results. This is filtered by suction and allowed to dry overnight. The crude acid is recrystallized from 85% ethanol to give 70 gm of white needles, mp 199.5°–200.5°C (corr.).

B. Knoevenagel Condensation

The condensation of aldehydes with the active methylene group of malonic acid to give α,β- and β,γ-olefinic acids is called the Knoevenagel condensation. Decarboxylation occurs at room temperature or heating to 100°C to give the unsaturated acids [76]. The reaction is catalyzed by pyridine, and an example is the preparation of 2-hexenoic acid in 64% yield.

Triethanolamine is the best catalyst for the preparation of β,γ-olefinic acids such as 3-hexenoic acid [77].

$$RCH_2CH=O + CH_2(COOH)_2 \begin{array}{c} \xrightarrow{\text{Pyridine}} RCH_2CH=CH-COOH \\ \xrightarrow{\text{Triethanolamine}} RCH=CH-CH_2COOH \end{array} \quad (47)$$

Substituted benzaldehydes and malonic acids give cinnamic acids in excellent yields [78].

8-2. Preparation of 2-Hexenoic Acid [79]

$$CH_3CH_2CH_2CH=O + CH_2(COOH)_2 + \text{Pyridine} \longrightarrow$$
$$CH_3CH_2CH_2CH=CH-COOH \quad (48)$$

To 200 gm (1.92 moles) of dry malonic acid in 200 ml of anhydrous pyridine is added 158 ml (1.75 moles) of freshly distilled n-butyraldehyde. The reaction is allowed to proceed at 25°C for 24 hr, at 40°–45°C for an additional 24 hr, and finally at 60°C for 3 hr. The reaction mixture is chilled, acidified with 6 N sulfuric acid, and the nonaqueous phase collected and the aqueous phase extracted with three 100 ml portions of ether. The organic layer and the ether extracts are dried over calcium chloride, filtered, the solvent removed, and the residue is allowed to crystallize at 0°C. The crystals are collected to give 130 gm (64%) of crude 2-hexenoic acid, mp 30°–32°C.

C. Other Condensation Reactions

The Knoevenagel reaction is related to the general class of reactions in which the condensation of aldehydes with active methylene compounds is catalyzed by base. The Claisen condensation is also an example and, for one typical case, may involve the condensation of ethyl acetate and aromatic aldehydes catalyzed by sodium sand. Benzaldehyde yields ethyl cinnamate in 74% yield [80].

$$\text{ArCH=O} \begin{array}{l} \xrightarrow{\text{Acetic anhydride / Sodium acetate}} \text{(Perkin)} \\ \xrightarrow{\text{Malonic acid / Pyridine}} \text{(Knoevenagel–Doebner)} \quad \text{ArCH=CH—COOH} \\ \xrightarrow{\text{Ethyl acetate / Sodium sand}} \text{(Claisen)} \end{array} \qquad (49)$$

The Stobbe condensation is used to condense ketones with diethyl succinate by a variety of basic reagents to give isopropylidene succinates [81].

Acetacetic ester [82] and pyruvic acid [83] are some other compounds containing active methylene groups that undergo base-catalyzed condensations with aldehydes to give olefinic β-keto esters and α-keto acids, respectively.

A one-carbon homologation of aldehydes and ketones to carboxylic acids has been reported to involve the Horner–Emmons modification of the Wittig reaction using diethyl-*tert*-butoxy(cyano)methylphosphonate $(EtO)_2$POCH(CN)O*t*-Bu to produce ethyl-*tert*-butoxyacrylonitriles. The *tert*-butyl ether group is cleaved by zinc chloride in refluxing acetic anhydride, and the α-acetoxyacrylonitrile is converted to the acid by solvolysis [84].

8-3. *Claisen Condensation: Preparation of Ethyl Cinnamate and Cinnamic Acid* [80]

$$C_6H_5CHO + CH_3COOC_2H_5 \xrightarrow{C_2H_5ONa} C_6H_5CH=CHCOOC_2H_5 + H_2O \qquad (50)$$

In a two-necked flask equipped with a reflux condenser and mechanical stirrer are placed 400 ml of dry xylene and 29 gm of clean sodium. The sodium is melted by means of an oil bath and the stirrer is used to powder the sodium under the xylene. Care should be taken not to splash any of the sodium onto the walls of the flask above the solvent. The oil bath is removed and the stirring is continued until the sodium powder has been formed completely and no more liquid sodium remains. The xylene is decanted and to the sodium is added 460 ml (4.7 moles) of absolute ethyl acetate containing 3–4 ml of absolute alcohol. The flask is quickly cooled to 0°C and 106 gm (1 mole) of pure benzaldehyde is slowly (1 to

1½ hr) added by means of a dropping funnel. The temperature is kept between 0° to 5°C, being careful not to exceed 5°C for the best yields. The reaction commences as soon as the benzaldehyde is added, as evidenced by the production of a reddish color on the sodium particles. The stirring is continued about 1 hr after the addition to complete the reaction of all the sodium particles. Glacial acetic (90–95 ml) acid is now added slowly and water is cautiously added to the mixture. Care should be exercised in hydrolyzing free sodium that has caked on top of the flask. The ester layer is separated, and the water layer is extracted with about 25–50 ml of ethyl acetate. The combined ester portions are washed with 300 ml of 6 N hydrochloric acid and then dried with sodium sulfate. The ethyl acetate is distilled off and the remaining liquid is distilled under reduced pressure to yield ethyl cinnamate, bp 128°–133°C (6 mm), 120–130 gm (68–74%). During the distillation, a reddish brown semisolid mass sometimes appears in the flask. This mass melts down if the oil bath is heated to 220°–230°C and the distillation continued smoothly.

Acid hydrolysis of ethyl cinnamate yields cinnamic acid.

Monoalkylation of malonic ester occurs with primary and some secondary halides in 75–90% yield [85–87]. An example of the reaction is given for the preparation of ethyl n-butylmalonate [85].

8-4. Malonic Ester Synthesis—Preparation of Ethyl-n-Butylmalonate [85]

$$CH_2(COOC_2H_5)_2 + NaOC_2H_5 \longrightarrow NaCH(COOC_2H_5)_2 \xrightarrow{n\text{-BuBr}} n\text{-BuCH}(COOC_2H_5)_2 \quad (51)$$

A sodium ethoxide solution freshly prepared from 25 liters of anhydrous ethanol and 115 gm (5 gm atom) of sodium is warmed to 50°C and stirred while 825 gm of diethyl malonate is added. To the clear solution is added slowly 685 gm of n-butyl bromide. The reaction commences almost immediately and considerable heat is generated. The addition rate is adjusted so that the reaction does not become violent. Cooling may be necessary. After the addition, the reaction mixture is refluxed until neutral to moist litmus (about 2 hr). Then a distillation column is attached to the flask and approximately 2 liters of alcohol are distilled off in 6 hr, using a water bath. The residue is treated with about 2 liters of water and shaken thoroughly. The upper layer of n-butylmalonic ester is separated and distilled under reduced pressure. First, a low-boiling portion is collected, consisting of alcohol, water, and n-butyl bromide; then a small intermediate fraction of unchanged malonic ester comes over; and finally n-butylmalonic ester boiling at 130°–135°C (20 mm). The first fraction amounts to less than 100 ml, while the main fractions weigh 860–970 gm (80–90%). All the reagents used should be highly purified in order to achieve the maximum yield.

The second hydrogen atom may be replaced by an alkyl group in 60–85% yield [88]. This technique has been used to prepare polymethylene dibasic acids

§ 8. Condensation Reactions

by the alkylation of malonic esters with polymethylene dibromides followed by hydrolysis and decarboxylation [89]. Dialkylated esters may be separated from monoalkylated compounds by refluxing for 2 hr with 50% potassium hydroxide solution. Under these conditions, the monoalkyl malonates are saponified, whereas the dialkylated compounds are unaffected [90]. The substituted malonic esters are saponified and the free acids decarboxylated in excellent yields by refluxing with the concentrated hydrochloric acid [91] or by heating to 170°–190°C until the evolution of carbon dioxide ceases [86]. Monoalkylmalonic acids require lower temperatures to decarboxylate (98°–123°C) [90]. The preparation of perlargonic acid is an example of the synthesis of a monoalkylmalonic acid [92].

8-5. Preparation of Pelargonic Acid [92]

$$CH_2(COOC_2H_5)_2 + C_7H_{15}Br + C_4H_9ONa \longrightarrow$$
$$C_7H_{15}CH(COOC_2H_5)_2 + C_4H_9OH + NaBr \quad (52)$$

$$C_7H_{15}CH(COOC_2H_5)_2 + 2\ H_2O \xrightarrow{KOH} C_7H_{15}CH(CO_2K)_2 + 2\ C_2H_5OH \quad (53)$$

$$C_7H_{15}CH(CO_2K)_2 + 2\ HCl \longrightarrow C_7H_{15}CH_2CO_2H + CO_2 + 2\ KCl \quad (54)$$

In a flask are placed 2.5 liters of anhydrous butyl alcohol and 115 gm of small pieces of cleanly cut sodium, added one at a time. Stirring may be necessary to facilitate solution. After the sodium has dissolved completely, the solution is cooled to 70°–80°C, and then 800 gm (5 moles) of redistilled ethyl malonate (bp 135°–136°C/100 mm) is added rapidly with stirring. After heating the reaction solution to 80°–90°C, 913 gm (5.1 moles) of pure heptyl bromide (bp 179°–180°C) is added. The bromide is added slowly until the sodium bromide begins to precipitate and then it is added at such a rate that the butyl alcohol refluxes gently. Usually about 1 hr is required for the addition of the heptyl bromide. The mixture is refluxed until it is neutral to litmus (about 1 hr). The entire mixture is transferred to a 12-liter flask and a solution of 775 gm (12.5 moles) of 90% potassium hydroxide in an equal weight of water is added slowly with shaking. The mixture is heated carefully, with occasional shaking, until refluxing starts. The refluxing is continued until saponification is complete (about 4–5 hr). The butyl alcohol is steam-distilled off. To the residue is carefully added 1350 ml (15.5 moles) of concentrated hydrochloric acid, with shaking, and the mixture is refluxed for about 1 hr. After cooling the water layer is separated and discarded. The oil layer is transferred to a 3-liter flask equipped with an air-cooled condenser and heated to 180°C by means of an oil bath. In about 2 hr the evolution of carbon dioxide ceases, the oil is decanted from a small amount of solid material, and is then distilled to yield 525–590 gm (66–75%), bp 140°–142°C (12 mm), mp 12°–12.5°C. If the small amount of solid material is also treated with 200–300 ml of concentrated hydrochloric acid, an additional small amount of oil is obtained which is added to the oil to be distilled.

D. Ethyl Acetoacetic Ester Synthesis

The alkylation of ethyl acetoacetate with a variety of alkyl halides affords intermediates that may be converted to α-alkyl-substituted acetic acids. The alkylation itself is widely discussed [93]. A typical preparation is that of ethyl n-butylacetoacetate by Marvel and Hager [93].

Depending on hydrolytic conditions, the alkylation products may be converted to methyl ketones or acids according to the following reaction schemes:

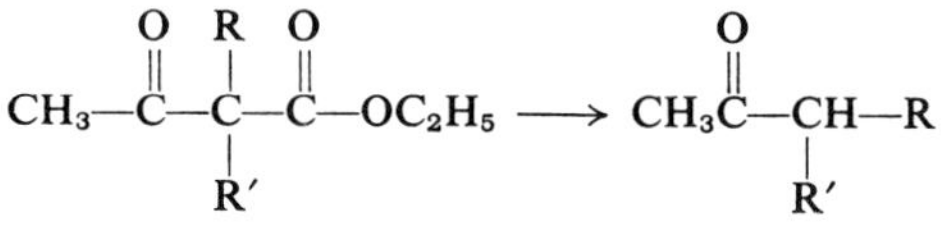

$$CH_3-\underset{}{\overset{O}{\overset{\|}{C}}}-\underset{\underset{R'}{|}}{\overset{R}{\overset{|}{C}}}-\overset{O}{\overset{\|}{C}}-OC_2H_5 \longrightarrow CH_3\overset{O}{\overset{\|}{C}}-\underset{\underset{R'}{|}}{CH}-R \qquad (55)$$

or

$$CH_3-\underset{}{\overset{O}{\overset{\|}{C}}}-\underset{\underset{R'}{|}}{\overset{R}{\overset{|}{C}}}-\overset{O}{\overset{\|}{C}}-OC_2H_5 \longrightarrow R-\underset{\underset{R'}{|}}{CH}-\overset{O}{\overset{\|}{C}}-OH \qquad (56)$$

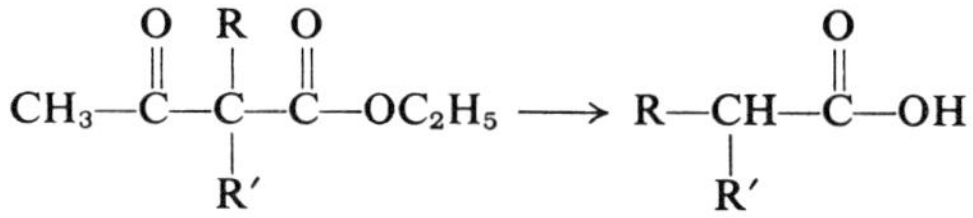

These two reactions are competitive in nature, and reaction conditions must be selected such that the desired product predominates in yield. Generally speaking, the concentration of alkali used in hydrolysis appears to have the more profound effect on the course of the reaction. Higher concentrations of alkali tend to favor acid formation.

While the temperature of the reaction has an effect on relative yields, this effect appears to be slight [94]. Additional research, using the statistical design of experiments to study simultaneous variation of reaction variable, would be of considerable interest.

8-6. Preparation of Caproic Acid [94]

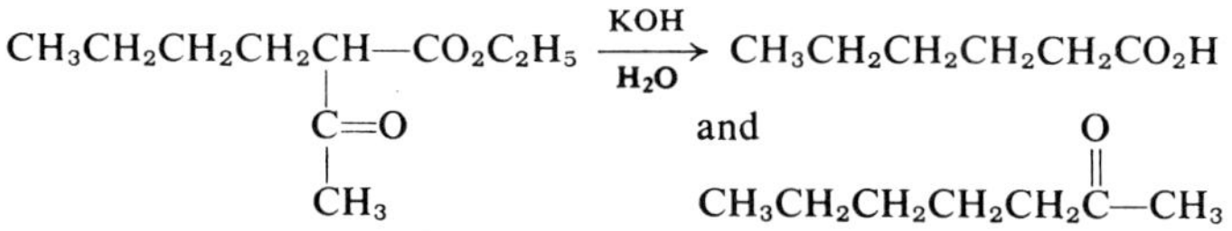

$$CH_3CH_2CH_2CH_2\underset{\underset{\underset{CH_3}{|}}{\underset{C=O}{|}}}{CH}-CO_2C_2H_5 \xrightarrow[H_2O]{KOH} CH_3CH_2CH_2CH_2CH_2CO_2H$$

and

$$CH_3CH_2CH_2CH_2CH_2\overset{O}{\overset{\|}{C}}-CH_3 \qquad (57)$$

To a solution of 59.5 gm of potassium hydroxide in 100 gm of an aqueous solution maintained at 75°C is added over a 1-hr period 12.3 gm of ethyl n-buylacetoacetate. Heating is continued for 5 hr with efficient stirring.

The alkaline solution is then diluted with 250 cc of water and the alcohol and ketone by-product are distilled off until no more water-insoluble product distills over.

The alkaline residue is then cooled and acidified by careful addition of 50% sulfuric acid, care being taken that the mixture does not become unduly hot during acidification.

The caproic acid is then extracted with three 100-cc portions of ether, the combined extracts are dried with sodium sulfate, and the ether is distilled off on a steam bath. The sodium sulfate may also be extracted a few times with anhydrous ether. The crude caporic acid remaining is distilled at reduced pressure, the fraction boiling from 105° to 110°C at 16 mm being collected. A yield of 64.6% of the theoretical amount of caproic acid is isolated.

E. Arndt–Eistert, Rearrangement

Acid chlorides are reacted with diazomethane to yield the diazoketone which upon reaction with silver oxide rearranges to the next higher homolog of the acid. Biphenyl-2-acetic acid is produced in 86% by the method described below [95].

8-7. Preparation of Biphenyl-2-acetic Acid [95]

$$\text{C}_6\text{H}_5\text{-C}_6\text{H}_4\text{-COOH} \xrightarrow{(\text{COCl})_2} \text{C}_6\text{H}_5\text{-C}_6\text{H}_4\text{-COCl} \xrightarrow{\text{CH}_2\text{N}_2} \text{C}_6\text{H}_5\text{-C}_6\text{H}_4\text{-COCHN}_2 \xrightarrow[\text{Na}_2\text{S}_2\text{O}_3]{\text{Ag}_2\text{O}} \text{C}_6\text{H}_5\text{-C}_6\text{H}_4\text{-CH}_2\text{COOH}$$

(58)

Biphenyl-2-carboxylic acid (1 mole) dissolved in dry benzene is treated with 1.6 mole of oxalyl chloride and kept at 30°C for 1 hr or until no further evolution of gas. The solvent and excess of reagent are removed under reduced pressure at 35°C. Ether is added twice and evaporated under reduced pressure to ensure the removal of the reagent and hydrogen chloride. The acid chloride is added dropwise to an etheral solution of diazomethane (see Chapter 15, Diazo and Diazonium Compounds) of equal molarity and then cooled. After keeping the reaction mixture overnight at room temperature, the ether is removed under reduced pressure. A 61% yield of yellow crystals of ω-diazo-o-phenylacetophenone is obtained, mp 106°C (from alcohol).

To 0.6 mole of silver oxide in 0.84 mole of sodium thiosulfate in 1 liter of warm water is added a 1 M dioxane solution of ω-diazo-o-phenylacetophenone. The mixture is stirred for 3 hr at room temperature while an additional quantity of freshly precipitated silver oxide (equal in amount to that used initially) is added in portions at intervals and the temperature is kept at 50°C for 1 hr. The solution is cooled and filtered and the residue washed with a 1% sodium hydroxide solution. This slowly deposits a flocculent precipitate upon acidifying. The residue is again treated with about one-half the amount of silver oxide used above and worked up as before to yield a further quantity of product. The total yield of biphenyl-2-acetic acid is 86% mp 116°C (from benzene).

Wilds and coworkers [96] have discussed the influence of highly hindered acyl chlorides on the Arndt–Eistert synthesis. The diazomethanes derived from such acyl chlorides fail to rearrange normally with any of the three conventional catalysts, silver oxide–methanol, silver benzoate–triethylamine–methanol, or tertiary amines–high-boiling solvents. Under special reaction conditions, abnormal reaction products were isolated.

F. Strecker Amino Acid Synthesis

The conversion of a carbonyl compound to an α-amino acid with one additional carbon atom makes use of sodium cyanide and ammonium chloride. Since sodium cyanide and hydrogen cyanide are involved in the reaction, great care in handling is required. Use of hood, gloves, rubber aprons, and gas masks is strongly recommended. After the reaction, all pieces of equipment used should be carefully cleaned with alkaline potassium permanganate solution and plenty of water.

A typical example of the Strecker synthesis is given in Eqs. (59)–(63) [97].

$$C_2H_5COC_2H_5 + NaCN + NH_4Cl \longrightarrow (C_2H_5)_2C(OH)CN + NaCl + NH_3 \quad (59)$$

$$(C_2H_5)_2C(OH)CN + NH_3 \longrightarrow (C_2H_5)_2C(NH_2)CN + H_2O \quad (60)$$

$$(C_2H_5)_2C(NH_2)CN + H_2O + HCl \xrightarrow{\text{Cold}} (C_2H_5)_2C(NH_2\cdot HCl)CONH_2 \quad (61)$$

$$(C_2H_5)_2C(NH_2\cdot HCl)CONH_2 + H_2O + HCl \longrightarrow$$
$$(C_2H_5)_2C(NH_2\cdot HCl)CO_2H + NH_4Cl \quad (62)$$

$$(C_2H_5)_2C(NH_2\cdot HCl)CO_2H + \tfrac{1}{2}Pb(OH)_2 \longrightarrow$$
$$(C_2H_5)_2C(NH_2)CO_2H + \tfrac{1}{2}PbCl_2 + \tfrac{1}{2}H_2O \quad (63)$$

G. Condensation of Active Methylene Compounds with Chloral

A procedure for producing carboxylic acids with an increase of the carbon skeleton by two carbon atoms that has been somewhat neglected in recent years involves an aldol condensation of an active methylene compound with chloral, with ultimate hydroysis of the terminal trichloromethyl group to a carboxylic acid or, by variation of the hydrolysis medium, to an ester. One problem which must be worked out for each specific application of this reaction is an evaluation of the exact nature of the acid formed. As indicated in the equation, either a substituted acrylic acid, a substituted lactic acid, or a mixture of both may form. For example, an alternate method for the preparation of β-benzoylacrylic acid, described earlier under the Friedel–Crafts reaction, is the condensation of

§ 8. Condensation Reactions

acetophenone with chloral to give 1,1,1-trichloro-2-hydroxy-3-benzoylpropane, followed by hydrolysis to the corresponding acid and dehydration [98].

$$C_6H_5\overset{O}{\underset{\|}{C}}-CH_3 + H\overset{O}{\underset{\|}{C}}CCl_3 \longrightarrow C_6H_5\overset{O}{\underset{\|}{C}}-CH_2-\underset{OH}{\overset{}{C}H}-CCl_3 \longrightarrow$$

$$C_6H_5\overset{O}{\underset{\|}{C}}-CH_2-\underset{OH}{\overset{}{C}H}-COOH \longrightarrow C_6H_5\overset{O}{\underset{\|}{C}}-CH=CH-COOH \quad (64)$$

An example originally due to Einhorn [99] and modified by Tullock and McElvain [100] is represented in Eq. (65).

8-8. Preparation of 3-(2-Pyridyl)acrylic Acid [100]

(Pyridyl)-CH$_3$ + H-$\overset{O}{\underset{\|}{C}}$-CCl$_3$ ⟶ (Pyridyl)-CH$_2$-$\underset{OH}{\overset{}{C}H}$-CCl$_3$ $\xrightarrow{(H_2O)}$

(Pyridyl)-CH$_2$-$\underset{OH}{\overset{}{C}H}$-CO$_2$H $\xrightarrow{H^+}$ (Pyridyl)-CH=CH-CO$_2$H (65)

In a hood, a mixture of 528 gm of chloral and 1050 ml of α-picoline is heated at 112°–113°C for 36 to 40 hr. The reaction mixture is then cooled to below 95°C and the excess α-picoline is separated by distillation at 10–20 mm and at a temperature no higher than 95°C. The black, viscous residue is poured into a large beaker while still warm and extracted twice with 400-ml portions of Skellysolve (bp 100°–140°C).

The black residue is extracted three times with 800-ml portions of hot water containing enough hydrochloric acid to maintain a pH between 3 and 5 at the end of each extraction. The acidic solutions are filtered and neutralized with sodium carbonate. Additional trichloro base precipitates from this aqueous solution as an oil which soon solidifies. The solid is taken up in hot Skellysolve which is combined with the previous Skellysolve extracts. The carbinol may be recrystallized from this solvent and also treated with activated charcoal. The yield amounts to 67%. After repeated crystallizations, a white product is obtained, mp 85°–86°C, which is identified as 1,1,1-trichloro-2-hydroxy-3-(2-pyridyl)propane. Hydrolysis of the latter trichloromethyl derivative in refluxing alcoholic potassium hydroxide, followed by acidification, affords primarily 3-(2-pyridyl) acrylic acid which may be recrystallized from water, mp 202°–203°C, while hydrolysis in acid yields 3-(2-pyridyl)lactic acid, mp 124°–125°C (from absolute alcohol). The 3-(2-pyridyl)lactic acid may be dehydrated to 3-(2-pyridyl)acrylic acid by heating at 130°–140°C under reduced pressure [99].

H. Reformatskii Reaction

The Reformatskii reaction involves the reaction of the product of an α-halo ester and activated zinc in the presence of an anhydrous organic solvent, with a carbonyl compound, followed by hydrolysis. The reaction is very similar in nature to the Grignard reaction except the carbonyl reagent is added at the start. It has been suggested that Grignard reactions might be conducted in a similar manner [101].

Magnesium has been used in some reactions in place of zinc but poor yields resulted since the more reactive organomagnesium reagents attack the ester group. With zinc, this latter reaction is not appreciable and the organozinc reagent attacks the carbonyl group of aldehydes and ketones to give the β-hydroxy ester.

α-Bromo esters react satisfactorily, but β- and γ-derivatives of saturated esters give poor yields unless activated by an unsaturated group in such a manner as to yield allylic bromides [102].

The reaction of α-lithiated acid salts with carbonyl compounds has been reported to offer an improved route to the β-hydroxy acids usually obtained by the hydrolysis of the β-hydroxy ester products of Reformatskii reactions [103].

8-9. Preparation of Ethyl 4-Ethyl-3-hydroxy-2-octanoate [104]

$$n\text{-}C_4H_9-\underset{\underset{C_2H_5}{|}}{CH}-CHO + Br\underset{\underset{CH_3}{|}}{CH}COOC_2H_5 + Zn \longrightarrow$$

$$C_4H_9\underset{\underset{C_2H_5}{|}}{CH}-\overset{\overset{OZnBr}{|}}{CH}-\underset{\underset{CH_3}{|}}{CH}-COOC_2H_5 \quad (66)$$

$$2\ C_4H_9\underset{\underset{C_2H_5}{|}}{CH}-\overset{\overset{OZnBr}{|}}{CH}-\underset{\underset{CH_3}{|}}{CH}-COOC_2H_5 + H_2SO_4 \longrightarrow$$

$$2\ C_4H_9\underset{\underset{C_2H_5}{|}}{CH}-\overset{\overset{OH}{|}}{CH}-\underset{\underset{CH_3}{|}}{CH}-COOC_2H_5 + ZnBr_2 + ZnSO_4 \quad (67)$$

To a flask containing a nitrogen atmosphere are added freshly sandpapered zinc foil strips, and 750 ml of thiophene-free benzene (dried). To further ensure that the flask and contents are dry, 175–200 ml of benzene is distilled off. Distillation is interrupted and the benzene is refluxed while a solution of 64.1 gm (0.5 mole) of 2-ethylhexanal and 271.5 gm (1.5 mole) of ethyl α-bromoporpionate in 500 ml of dried benzene is placed in the dropping funnel. The first 50 ml of the solution is added to the flask at once. In most cases the reaction starts immediately, as evidenced by the darkening of the zinc surface and the formation

§ 8. Condensation Reactions

of a cloudy solution. However, approximately 15 min may elapse before the reaction starts. When the reaction has started, the remainder of the solution is added in about 1 hr, and the solution then refluxed for 2 hr with continuous stirring. When the solution has cooled to room temperature, 750 ml of 12 N sulfuric acid is added and the solution is stirred vigorously for 1 hr. The benzene layer is separated and the aqueous layer extracted several times with benzene. The combined benzene layers are washed with 500 ml of water, saturated sodium bicarbonate solution, and then with water. The benzene layer is dried over anhydrous sodium sulfate and the benzene distilled off under aspirator pressure. The product at this point is ethyl 4-ethyl 3-hydroxy-2-octanoate and is obtained from the distillation in 87% yield (100 gm), bp 122°–124°C (4.9 mm), n_D^{25} 1.4415.

I. Diels–Alder Reaction

The Diels–Alder reaction [105, 106] is a 1,4-addition of an olefinic compound to a conjugated diene. The diene system may be part of an aliphatic, aromatic, or heterocyclic nucleus such as furan. The olefinic compound usually contains one or more groups that activate the double bond [107], although this is not always necessary. For example, ethylene is condensed with butadiene at 200°C to give cyclohexene [108]. Triple bonds may replace double bonds in both the diene and the dienophile. Cis addition of the dienophile to the diene occurs and several reactions of the above type have been shown to be reversible [109].

Maleic anhydride condenses with butadiene to give Δ^4-tetrahydrophthalic anhydride. The latter can be hydrolyzed to the diacid [110].

8-10. Preparation of Tetrahydrophthalic Anhydride **[110]**

$$\begin{array}{c}CH_2\\ \|\\ HC\\ |\\ HC\\ \|\\ CH_2\end{array} + \begin{array}{c}O\\ \|\\ HC-C\\ \diagdown\\ O\\ \diagup\\ HC-C\\ \|\\ O\end{array} \longrightarrow \begin{array}{c}CH_2\\ \diagup \diagdown\\ HCHCC=O\\ \|\diagdown\diagup\\ O\\ HCHC\diagup\\ \diagdown\diagupC=O\\ CH_2\end{array} \longrightarrow \begin{array}{c}CH_2O\\ \diagup\diagdown\|\\ HCHC-C-OH\\ \||\\ HCHC-C-OH\\ \diagdown\diagup\|\\ CH_2O\end{array} \quad (68)$$

A flask containing 500 ml of dry benzene and 196 gm (2 moles) of maleic anhydride is heated with a pan of hot water while butadiene is introduced rapidly (0.6–0.8 liter/min) from a commercial cylinder. The flask is stirred rapidly and the heating is stopped after 3–5 min when the temperature of the solution reaches 50°C. In 15–25 min, the reaction causes the temperature of the solution to reach 70°–75°C. The absorption of the rapid stream of butadiene is nearly complete in 30–40 min. The addition of butadiene is continued at a slower rate for a total of 2½ hr. The solution is poured into a 1-liter beaker which is covered and kept at

0°–5°C overnight. The product is collected on a larger Buchner funnel and washed with 250 ml of 35°–60°C bp petroleum ether. A second crop is obtained by diluting the filtrate with an additional 250 ml of petroleum ether. Both crops are dried to constant weight in an oven at 70°–80°C to yield 281.5–294.5 gm (96–97%, mp 99°–102°C). Recrystallization from ligroin or ether raises the melting point to 103°–104°C.

9. HYDROLYSIS OF ACID DERIVATIVES

A. Hydrolysis of Nitriles

Nitriles can be hydrolyzed to carboxylic acids by refluxing in concentrated solutions of sulfuric acid or sodium hydroxide [111]. For example, a solution of potassium hydroxide and ethylene glycol monoethyl ether is used to prepare 9-phenanthroic acid in 98% yield from 9-cyanophenanthrene [112].

9-1. Preparation of 9-Phenanthroic Acid **[112]**

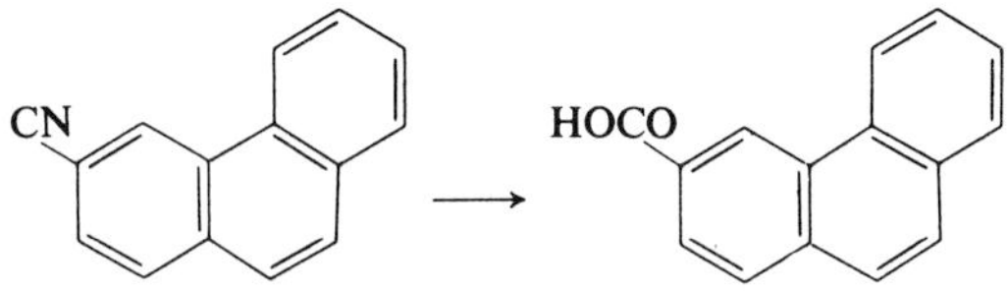

$$CN\text{-phenanthrene} \longrightarrow HOCO\text{-phenanthrene} \tag{69}$$

To a hot solution of 350 gm of 9-cyanophenanthrene in 1400 ml of ethylene glycol monoethyl ether is added a solution of 350 gm of potassium hydroxide in 160 ml of water. The solution is refluxed while a slow stream of carbon dioxide-free air is bubbled through it until 1 ml of 0.1 N hydrochloric acid is not neutralized within 5 min by the exit air carrying the liberated ammonia. Approximately 6 hr is required to attain this condition. The solution is cooled and poured into a stirred solution of 610 ml of concentrated hydrochloric acid in 5250 ml of water. After standing overnight, the precipitated 9-phenanthroic acid is filtered off and washed thoroughly with water; yield 374 gm (98%), mp 246°–248°C. After one recrystallization from glacial acetic acid, a sample melted at 252°–253°C.

Acid hydrolysis has been used to prepare phenylacetic acid in 78% yield [113]. Acids are conveniently prepared from halides by their conversion to nitriles and subsequent hydrolysis. The nitrile does not need to be isolated [114, 115]. Phosphoric acid (100%) is a good solvent for the difficult hydrolysis of nitriles [116].

B. Hydrolysis of Amides

Amides may be hydrolyzed by an acid or basic medium. For example α-phenylbutyramide is hydrolyzed by aqueous sulfuric acid to give α-phenylbutyric acid in 90% yield [117].

9-2. Preparation of α-Phenylbutyric Acid [118]

$$\underset{\text{CH}_3\text{CH}_2-\overset{|}{\text{CHCONH}_2}}{\overset{\text{C}_6\text{H}_5}{|}} \xrightarrow[\text{H}_2\text{O}]{\text{H}_2\text{SO}_4} \underset{\text{CH}_3\text{CH}_2-\overset{|}{\text{CH}}-\text{COOH}}{\overset{\text{C}_6\text{H}_5}{|}} \quad (70)$$

A mixture of 600 gm of α-phenylbutyramide, 1 liter of water, and 400 ml of concentrated sulfuric acid is vigorously stirred and boiled with refluxing for 2 hr. Another liter of water is added and the mixture is cooled. The oily layer is dissolved in 12% sodium hydroxide and the acid precipitated with 30% sulfuric acid, separated, and distilled under reduced pressure. The yield is 530–554 gm (88–90%), bp 136°–138°C (3 mm), mp 42°C.

Sodium hydroxide can also be used to hydrolyze amides. For example, 2-and 4-dibenzofurylacetic acid is obtained in 87% yield by this method [119].

Amides that are hydrolyzed with difficulty can be hydrolyzed with some success with 100% phosphoric acid [120].

Nitrous acid can also be used to hydrolyze amides. Trialkylacetic acids have been made by this method [121].

Recently amides have been reported to be hydrolyzed in high yields by use of aqueous sodium peroxide (in less than 2 hr at 50°–80°C.) [122].

C. Hydrolysis of Esters, Acyl Halides, Anhydrides, and Trihalides

Hydrolysis of esters is effected by refluxing with aqueous or alcoholic alkali hydroxides. Acid-catalyzed hydrolysis is an equilibrium reaction favoring ester formation. Sterically hindered esters are hydrolyzed with difficulty.

The Corey–Chayakowski reagent ("dimsyl sodium"—a reagent prepared in the absence of air by the reaction of sodium hydride with dimethyl sulfoxide) is said to hydrolyze esters extremely rapidly [58, 123]. This reaction deserves further exploration. As mentioned previously, an explosion hazard may exist in the handling of this reagent [see 59, 123].

Partial saponification of malonic ester occurs with cold potassium hydroxide to give an 82% yield of potassium ethyl malonate [124].

Hindered esters are hydrolyzed at room temperature using 1 equivalent of ester, 2 equivalents of water, and 8 equivalents of potassium tertiary butoxide [125].

Potassium superoxide in benzene has been reported to react with carboxylic esters and yield, after aqueous work-up, carboxylic acids. The value of this reagent is that amides and nitriles are not affected by it [126].

Olefinic esters require mild conditions in order to avoid isomerization. Aqueous alcoholic sodium hydroxide is used for the preparation of 3-ethyl-3-pentenoic acid in 56% yield [127]. Ice-cold sodium carbonate solution has been found to be an effective medium for the hydrolysis of acid chlorides to acids [128]. However, the hydrolysis of acid halides is not a common method for the production of acids.

Heating anhydrides with the theoretical amount or an excess of water yields the free acid. The preparation of citraconic acid is given below as an example of this method [129].

Alkaline hydrolysis must be used with caution with certain esters because of interfering side reactions. For example, α-halo esters may also be hydrolyzed further under alkaline conditions to α-hydroxy esters; olefinic esters may be rearranged; keto esters may be cleaved (cf. ethyl acetoacetate-based syntheses).

Particularly in the aliphatic series, occasional difficulty is experienced with foam formation. Sometimes spreading a thin layer of silicone stopcock grease just above the stirred liquid in the reaction flask aids in controlling the foam. Dow–Corning's Antifoam emulsion in low concentrations, other commercial "anti-foams," or addition of a very low concentration of a cationic surfactant is frequently effective in reducing the foam problem. In the simple aromatic acid series, this problem is usually not significant.

9-3. Preparation of m-Nitrobenzoic Acid [130]

$$\text{m-O}_2\text{N-C}_6\text{H}_4\text{-CO}_2\text{CH}_3 + \text{NaOH} \longrightarrow \text{m-O}_2\text{N-C}_6\text{H}_4\text{-CO}_2\text{Na} + \text{CH}_3\text{OH} \quad (71)$$

$$\text{m-O}_2\text{N-C}_6\text{H}_4\text{-CO}_2\text{Na} + \text{HCl} \longrightarrow \text{m-O}_2\text{N-C}_6\text{H}_4\text{-CO}_2\text{H} + \text{NaCl} \quad (72)$$

A mixture of 80 gm (2 moles) of sodium hydroxide in 320 ml of water and 181 gm (1 mole) of methyl m-nitrobenzoate is refluxed until the ester has disappeared (5 to 10 min). The reaction mixture is then diluted with an equal volume of cold water, cooled, and poured slowly with stirring into 250 ml of concentrated hydrochloric acid. After the solution has cooled to room temperature, the m-nitrobenzoic acid is filtered off and dried. The product should be completely ether-soluble and melt at 140°C. Yield 150–160 gm (90–96% of the theoretical amount).

§ 9. Hydrolysis of Acid Derivatives

To purify the product further, it may be recrystallized by dissolving in approximately 15 times its weight of 1% aqueous hydrochloric acid. The final product is light cream-colored; loss on recrystallization is approximately 5%.

The acyl chlorides prepared from the action of oxalyl chloride on diarylethylenes are hydrolyzed to β,β-diarylacrylic acids by stirring with sodium carbonate solution [72].

$$(C_6H_5)_2C=CH-COCl \xrightarrow{Na_2CO_3} (C_6H_5)_2C=CHCOONa \xrightarrow{H_3O^+}$$
$$(C_6H_5)_2C=CH-COOH \quad (73)$$

9-4. Preparation of Citraconic Acid [108]

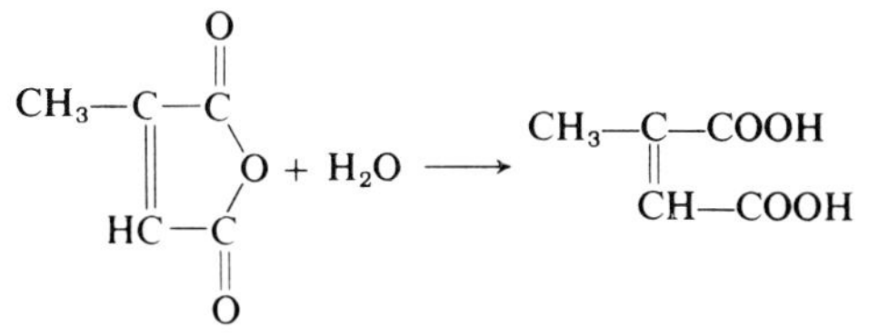

$$\begin{matrix} CH_3-C-\overset{O}{\overset{\|}{C}} \\ \overset{\|}{}O \\ HC-\underset{O}{\underset{\|}{C}} \end{matrix} + H_2O \longrightarrow \begin{matrix} CH_3-\overset{\|}{\underset{\|}{C}}-COOH \\ CH-COOH \end{matrix} \quad (74)$$

To 22.4 gm (0.2 mole) of citraconic anhydride is added 4.00 ml of water, and the mixture is heated until a homogeneous solution is formed. The mixture is allowed to stand for 48 hr at room temperature. The solid mass is ground to a powder, washed with 50 ml of benzene, dried in air, and then dried for 24 hr in a vacuum desiccator over phosphorus pentoxide. The yield was 24.4 gm (94%) of citraconic acid which melts at 92°–93°C.

D. Hydrolysis of 1,1,1-Trihalomethyl Derivatives

The hydrolysis of trihalides yields acids; however, the method is not very practical since the trihalides are usually not readily available. The addition of carbon tetrachloride to olefins is a good method of obtaining trihalides (see also Chapter 6, Halides). Recently it has been reported that chloroform can be added to aldehydes via sodium hydride in THF solution at 0°–5°C. The resulting alkyl(trichloromethyl)methanol containing solution on addition to hot methanolic potassium hydroxide yielded fair yields of α-methoxy carboxylic acids [131]. Acetone condenses with chloroform to give acetone–chloroform. The hydrolysis of acetone–chloroform by methanolic potassium hydroxide is illustrated below and is reported to give 70.8% yield of α-methoxyisobutyric acid [132]. (See also pp. 266–267 of this chapter for additional methods.)

9-5. Preparation of α-Methoxyisobutyric Acid [132]

$$\underset{\underset{(CH_3)_2C-CCl_3}{|}}{OH} + 4\ KOH + CH_3OH \longrightarrow \underset{\underset{(CH_3)_2C-COOH}{|}}{OCH_3} \quad (75)$$

To a vigorously stirred cold solution of 448 gm of potassium hydroxide in 250 ml of water and 1 liter of methanol is slowly added a solution of 355 gm of acetone–chloroform in 700 ml of methanol. A violent reaction occurs which requires constant cooling. The mixture is stirred for 1 hr at room temperature and for 2 hr at the reflux temperature. The potassium chloride is filtered off and washed with methanol. The filtrate is distilled under reduced pressure to remove the methanol and water, and the residue is treated with sulfuric acid (Congo red). The inorganic precipitate is filtered, washed with ether, and the aqueous solution extracted with ether. The ether layers are combined and the ether removed to give a residue which upon distillation at reduced pressure gives 167 gm (70.8%), bp 98°–99°C (20 mm). The same result is obtained using methanol–sodium methoxide.

Meyers [133] has reported the reaction of lithiooxazolines with alkyl halides and hydrolysis of the products to give homologated acetic acids.

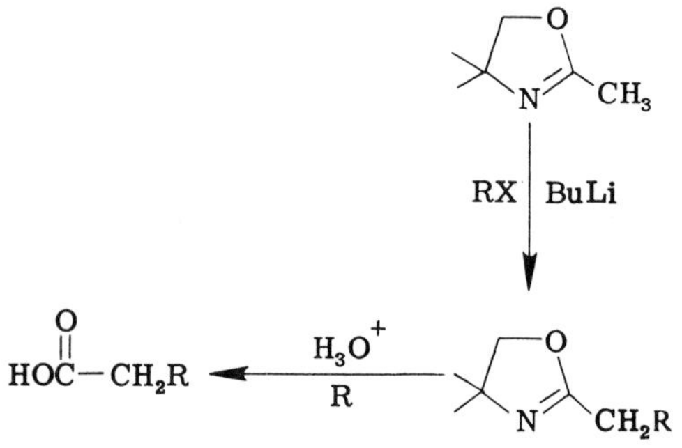

$$\text{oxazoline} \xrightarrow[\text{BuLi}]{\text{RX}} \text{alkylated oxazoline} \xrightarrow{\text{H}_3\text{O}^+ / R} \text{HOC(O)}-\text{CH}_2\text{R} \tag{76}$$

10. MISCELLANEOUS METHODS

A. Oxidation Reactions

(1) Oxidation of ketones using potassium hydroxide [134].

$$CH_3C_6H_4COC_6H_5 + KOH \longrightarrow CH_3C_6H_4COOH\ (87\%) + C_6H_5COOH + C_6H_6 \tag{77}$$

(2) Oxidation of aromatic ring [135].

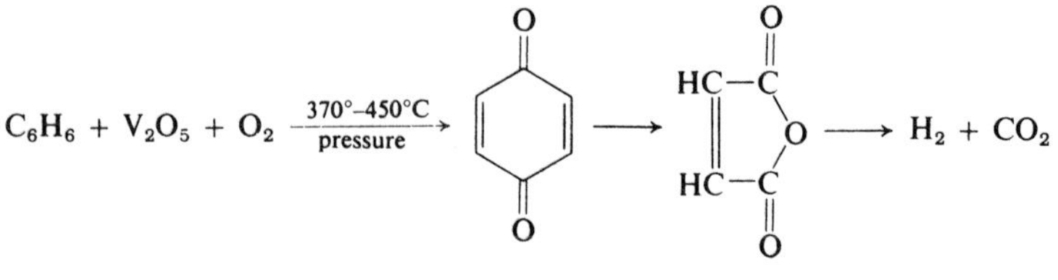

$$C_6H_6 + V_2O_5 + O_2 \xrightarrow[\text{pressure}]{370°-450°C} \text{benzoquinone} \longrightarrow \text{maleic anhydride} \longrightarrow H_2 + CO_2 \tag{78}$$

§ 10. Miscellaneous Methods

(3) Oxidation of tertiary alcohols [136].

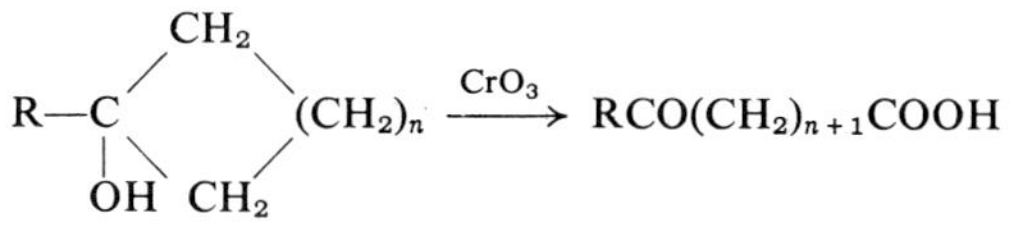

$$R-\underset{\underset{OH}{|}}{C}\underset{CH_2}{\overset{CH_2}{<}}(CH_2)_n \xrightarrow{CrO_3} RCO(CH_2)_{n+1}COOH \quad (79)$$

(4) Oxidation of aromatic acetylenes [137].

$$C_6H_5-C\equiv CH \xrightarrow[\substack{CH_2Cl_2 \\ Na_2HPO_4}]{CF_3COOOH} C_6H_5-CH_2COOH + C_6H_5COOH\ (25\%) \quad (80)$$

(5) Oxidation of alkyl side chains using nitrogen dioxide and selenium dioxide [138].

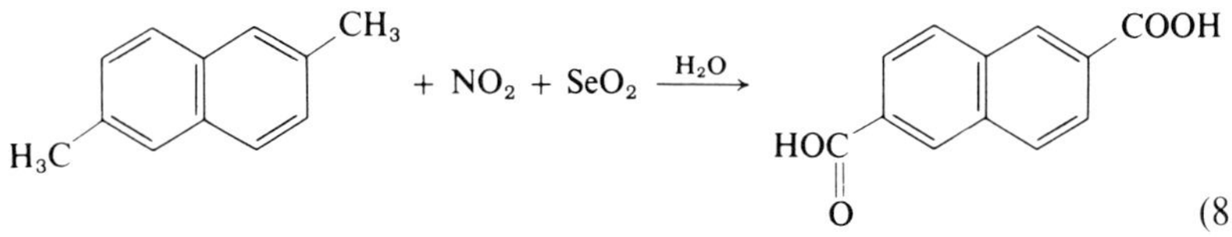

2,6-dimethylnaphthalene + NO_2 + SeO_2 $\xrightarrow{H_2O}$ 6-carboxy-2-naphthoic acid (81)

(6) Cyanozonolysis of olefins to yield hydroxy acids [139].

Cyclooctene + HCN emulsion in H_2O $\xrightarrow{O_3}$ 2,9-Dihydroxy sebacic acid (82)

(7) Base-catalyzed autoxidation of hydrocarbons in diphenyl sulfoxide [140].

o-Xylene + O_2 $\xrightarrow[21\ hr]{(C_6H_5)_2SO}$ 54.1% o-Toluic acid + 37.5% Phthalic acid (83)

(8) Oxidation of aryloxyethanol with oxygen gas in an aqueous alkaline medium in the presence of platinum or palladium metal containing catalyst to give acyloxyacetic acid [141].

(9) Preparation of m-chloroperbenzoic acid [142].

B. Hydrolysis and Elimination Reactions

(1) Diazoacetic ester synthesis [143].

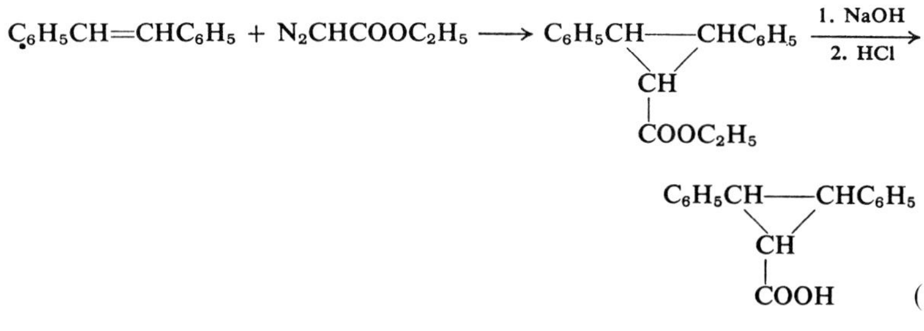

$$C_6H_5CH=CHC_6H_5 + N_2CHCOOC_2H_5 \longrightarrow \underset{\underset{COOC_2H_5}{|}}{\underset{CH}{C_6H_5CH\text{——}CHC_6H_5}} \xrightarrow[2.\ HCl]{1.\ NaOH}$$

$$\underset{\underset{COOH}{|}}{\underset{CH}{C_6H_5CH\text{——}CHC_6H_5}} \quad (84)$$

(2) Hydrolysis of primary nitro compounds [144].

$$RCH_2NO_2 \xrightarrow{H_3O^+} RCONH-OH \xrightarrow{H_3O^+} RCOOH \qquad (85)$$

(3) Hydrolysis of α-keto dihalides [145].

$$C_6H_5-COCHCl_2 \xrightarrow{NaOH} C_6H_5\underset{OH}{CH}COONa \xrightarrow{H^+} C_6H_5\underset{OH}{CH}-COOH \qquad (86)$$

(4) Hydrolysis of hydantoins [146].

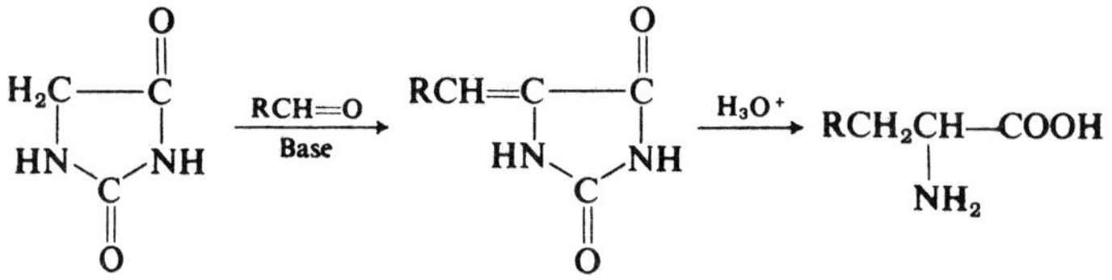

$$\underset{\text{hydantoin}}{\begin{array}{c}H_2C-C=O \\ HN\diagup C\diagdown NH \\ \| \\ O\end{array}} \xrightarrow[\text{Base}]{RCH=O} \underset{}{\begin{array}{c}RCH=C-C=O \\ HN\diagup C\diagdown NH \\ \| \\ O\end{array}} \xrightarrow{H_3O^+} RCH_2\underset{NH_2}{CH}-COOH \qquad (87)$$

(5) Hydrolysis of the reaction products of oxazolines and aldehydes [147].

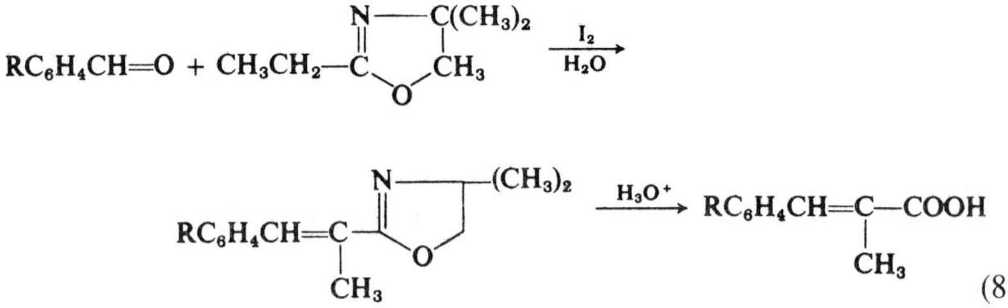

$$RC_6H_4CH=O + CH_3CH_2-\underset{O}{\overset{N-C(CH_3)_2}{C}}\diagdown CH_3 \xrightarrow[H_2O]{I_2}$$

$$RC_6H_4CH=\underset{CH_3}{C}-\underset{O}{\overset{N-(CH_3)_2}{C}} \xrightarrow{H_3O^+} RC_6H_4CH=\underset{CH_3}{C}-COOH \qquad (88)$$

(6) Von Richter reaction [148].

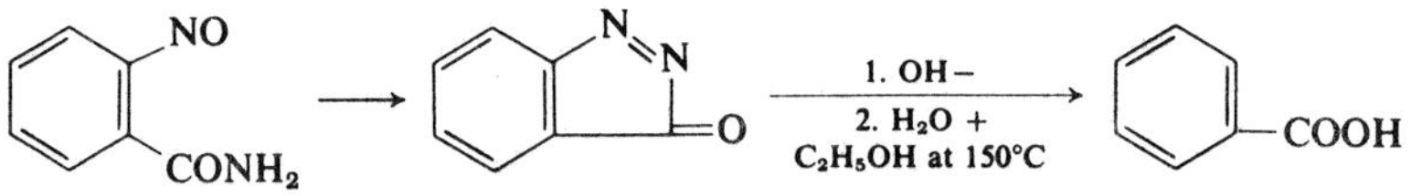

[o-nitrobenzamide] → [benzotriazine-oxide] $\xrightarrow[\text{2. } H_2O + C_2H_5OH \text{ at } 150°C]{1.\ OH^-}$ [C_6H_5-COOH] (89)

(7) Reaction of perhalogenated cyclopropanes and ketones with alkali [149].

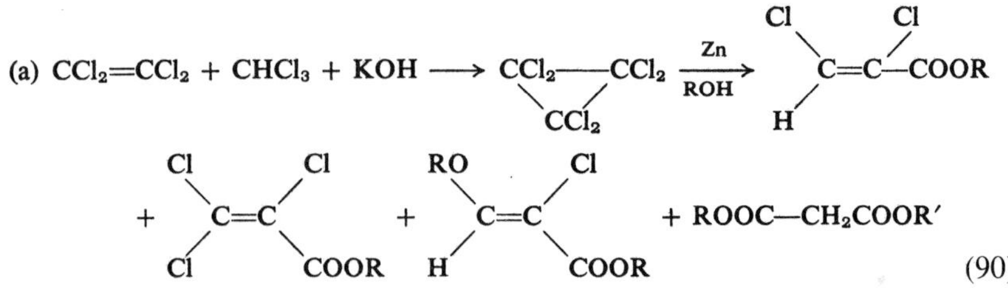

(a) $CCl_2=CCl_2 + CHCl_3 + KOH \longrightarrow \underset{CCl_2}{CCl_2-CCl_2} \xrightarrow[ROH]{Zn} \underset{H}{\overset{Cl}{\diagdown}}C=\underset{}{\overset{Cl}{C}}-COOR$

$+ \underset{Cl}{\overset{Cl}{\diagdown}}C=\underset{COOR}{\overset{Cl}{C}} + \underset{H}{\overset{RO}{\diagdown}}C=\underset{COOR}{\overset{Cl}{C}} + ROOC-CH_2COOR' \qquad (90)$

(b) $CClF_2C-CClF_2 + 4\ NaOH \longrightarrow$
$\qquad\ \ \|$
$\qquad\ \ O$
$\qquad\qquad\qquad CClF_2COONa + 2\ NaF + NaCl + CO + 2\ H_2O$ (91)

$CClF_2C-CCl_2F + NH_4OH$
$\quad\ \ \|$
$\quad\ \ O$ $\qquad\qquad\qquad\longrightarrow CClF_2COONa(NH_4) + CHCl_2F$ (92)

$CClF_2C-CCl_2F + NaOH$
$\quad\ \ \|$
$\quad\ \ O$

(8) Decarboxylation of dicarboxylic acids [150]; for example, 3-chlorocyclobutane dicarboxylic acid → 3-chlorocyclobutane carboxylic acid.
(9) Ester cleavages with S_N2-type dealkylation [151]. An example is the use of lithium iodide in pyridine to cleave methyl phenylacetate to phenylacetic acid in 93% yield.
(10) Hindered ester cleavage with boron tribromide [152].

C. Reactions Involving Organometallics

(1) Reaction of alkyl aluminum and carbon dioxide [153].

$$R_3Al + 3\ CO_2 \longrightarrow 3\ RCOOH \qquad (93)$$

(2) Reactions of organosodium compounds with carbon dioxide [154].

$$RNa + CO_2 \longrightarrow RCOONa \xrightarrow{H_3O^+} RCOOH \qquad (94)$$

$$RC_6H_4X + Na + CO_2 \text{ in } C_6H_6 \xrightarrow[\text{pressure}]{30\ \text{lbs}} R-C_6H_4COOH\ (76\%) \qquad (95)$$

$\qquad\ \ CH_3 \qquad\qquad\qquad\qquad\quad CH_3$
$\qquad\ \ \ |\qquad\qquad\qquad\qquad\qquad\ \ |$
$CH_3-C=CH_2 + \text{Amylsodium} \longrightarrow CH_2=C-CH_2Na \xrightarrow[\text{2. Aq.KOH}]{1.\ CO_2}$

$\qquad\qquad\qquad\qquad\qquad\qquad\qquad\qquad\qquad\ \ CH_3$
$\qquad\qquad\qquad\qquad\qquad\qquad\qquad\qquad\qquad\ \ \ |$
$\qquad\qquad\qquad\qquad\qquad\qquad\qquad\qquad CH_3-CH-CH_2COOH$ (96)

$C_6H_5CH=CH_2 + Na \longrightarrow [C_6H_5CH-CH_2] \longrightarrow \text{Dimer} \xrightarrow{CO_2}$

$$\left(\begin{matrix} C_6H_5-CH-CH_2- \\ | \\ COOH \end{matrix}\right)_2 \qquad (97)$$

(3) Reactions of organometallics with fluoroaromatic compounds [155].

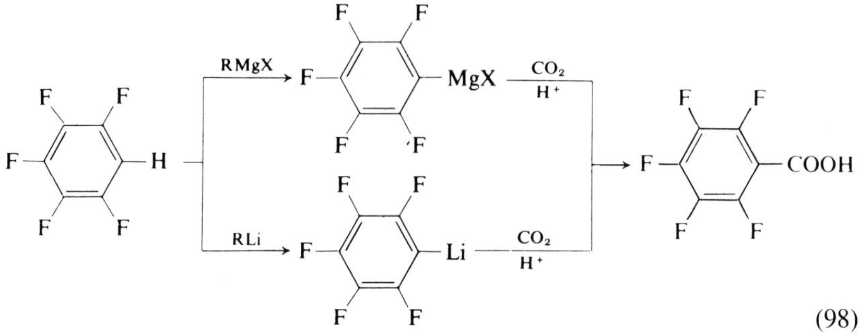

$$\underset{\text{F}}{\underset{\text{F}}{\bigodot}}\text{—H} \xrightarrow[\text{RLi}]{\text{RMgX}} \begin{array}{c} \text{F}\!-\!\underset{\text{F}}{\underset{\text{F}}{\bigodot}}\!-\!\text{MgX} \xrightarrow{\text{CO}_2 / \text{H}^+} \\ \text{F}\!-\!\underset{\text{F}}{\underset{\text{F}}{\bigodot}}\!-\!\text{Li} \xrightarrow{\text{CO}_2 / \text{H}^+} \end{array} \text{F}\!-\!\underset{\text{F}}{\underset{\text{F}}{\bigodot}}\!-\!\text{COOH}$$

(98)

$$\underset{\text{F}}{\underset{\text{F}}{\bigodot}}\text{H} + n\text{-BuLi} \xrightarrow[-70°C]{\text{THF}} \text{Li}\!-\!\underset{\text{F}}{\underset{\text{F}}{\bigodot}}\!-\!\text{Li} \xrightarrow{\text{CO}_2 / \text{H}^+} \text{HOC}\!-\!\underset{\text{F}}{\underset{\text{F}}{\bigodot}}\!-\!\text{COH} \quad (84\%)$$

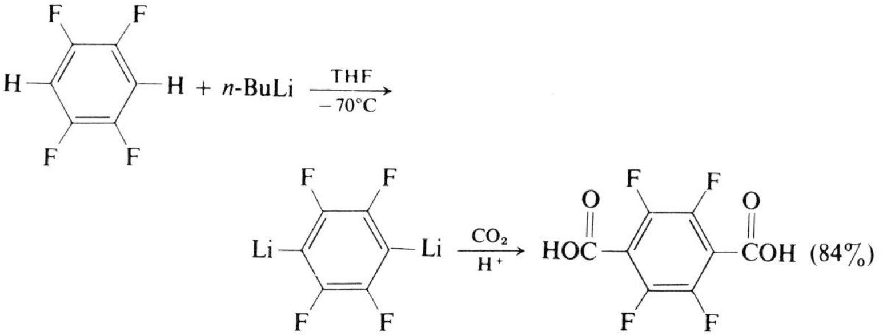

(99)

(4) Reaction of halobenzoic acids with activated ketones [156].

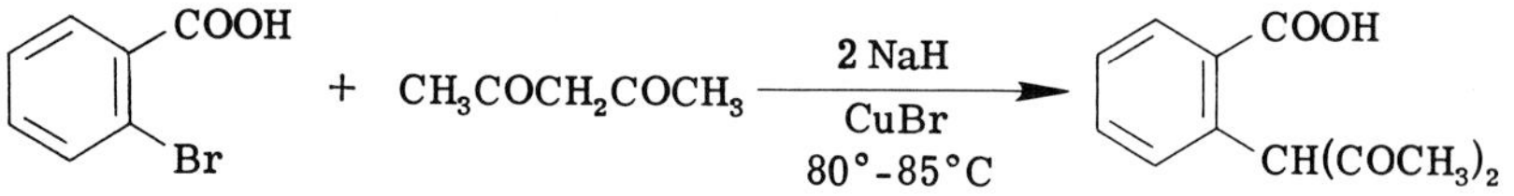

$$\underset{\text{Br}}{\bigodot}\!\!\!\text{—COOH} + \text{CH}_3\text{COCH}_2\text{COCH}_3 \xrightarrow[\text{CuBr} \atop 80°-85°C]{\text{2 NaH}} \underset{\text{CH(COCH}_3)_2}{\bigodot}\!\!\!\text{—COOH}$$

(100)

D. Reactions Using Carbon Monoxide

(1) Reaction of carbon monoxide with bases [157].

$$\text{NaOH} + \text{CO} \longrightarrow \text{HC}\overset{\overset{\text{O}}{\|}}{\text{O}}\text{Na} \xrightarrow{\text{H}^+} \text{HCOOH} \tag{101}$$

(2) Carbonylation of olefins [158]

$$\text{RCH}\!=\!\text{CH}_2 + \text{CO} + \text{H}_2 \xrightarrow[\text{Pressure}]{\text{Cat.}} \underset{\text{CH}_3}{\text{RCH}\!-\!\text{COOH}} \tag{102}$$

(3) Reaction of olefin-palladium chloride complex with carbon monoxide [159].

§ 10. Miscellaneous Methods

$$CH_3\text{—}CH\text{=}CH_2 + PdCl_2 + CO \xrightarrow{H_2O} CH_3\text{—}\underset{\underset{Cl}{|}}{CH}\text{—}CH_2COOH \quad (103)$$

(4) Reaction of carbon monoxide with alcohols [160].

$$ROH + CO + Steam \longrightarrow RCOOH \quad (104)$$

(5) Reaction of aryl halides and carbon monoxide [161].

$$RC_6H_4X + CO + Ni(NO_3)_2 \longrightarrow RC_6H_4COOH \quad (105)$$

(6) Reaction of carbon monoxide with acetylenes [162].

$$RC\equiv CH + CO + H_2O \xrightarrow[150°C]{Ni \text{ or } NiI_2} RCH\text{=}CH\text{—}COOH \quad (106)$$

(7) Reaction of carbon monoxide with ketones [163].

$$(CH_3)_2C\text{=}O + H_2O + CO \xrightarrow{87\% H_3PO_4} CH_3COOH + (CH_3)_2CHCOOH \quad (107)$$

(8) Reaction of carbon monoxide with ethers [164].

$$\underset{O}{\bigcirc} + Ni(CO)_4 + NiI_2 + CO \longrightarrow 74\% \; (CH_2)_4(COOH)_2 \quad (108)$$

(9) Reaction of carbon monoxide with alkanes [165].

(a) $\quad RCH_3 + H_2O_2 + FeSO_4 + CO \longrightarrow RCOOH \quad (109)$

(b)  $\quad \triangle + H_2SO_4 + CO \longrightarrow CH_3\text{—}CH\underset{\diagdown COOH}{\diagup ^{CH_3}} \quad (110)$

(10) Reaction of polyhalomethanes with carbon monoxide [166].

$$CCl_4 + AlCl_3 + CO \longrightarrow CCl_3COOH \quad (111)$$

(11) Disproportionation using carbon monoxide [167].

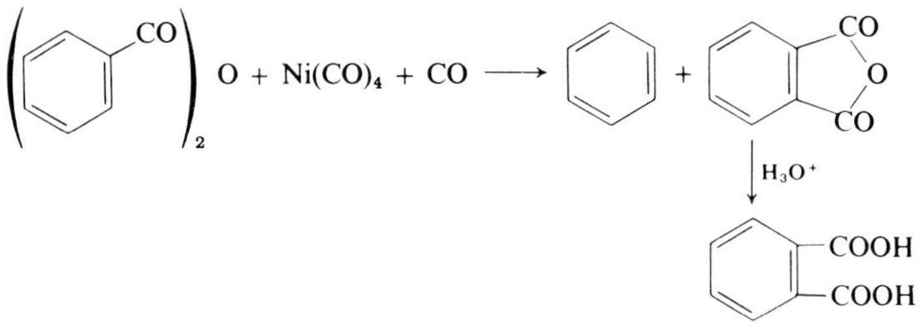

$$\left(\underset{}{\bigcirc}\text{—CO} \right)_2 O + Ni(CO)_4 + CO \longrightarrow \bigcirc + \underset{}{\bigcirc}\underset{CO}{\overset{CO}{\diagup O}} \xrightarrow{H_3O^+} \underset{}{\bigcirc}\underset{COOH}{\overset{COOH}{}} \quad (112)$$

(12) Reaction of carbon monoxide with linear and cyclic olefins in the presence of HF [168].

$$\text{RCH=CH}_2 + \text{CO} + \text{HF aq.} \xrightarrow[20°\text{-}50°C]{\text{CO} \\ 1000\text{-}2500 \text{ psig}} \text{RCHCH}_3 \quad (113)$$
(linear and cyclic) $\quad\quad\quad\quad\quad\quad\quad\quad\quad\quad\quad\quad\quad$ COOH

(13) Carbonylation of vinylmercurials using aqueous THF and 10% Pd/C to give carboxylic acids in high yield [169].

(14) Carbonylation of alcohols, olefins, and halides over homogeneous catalystd of RH or In [170].

E. Condensation Reactions

(1) Favorskii rearrangement [171].

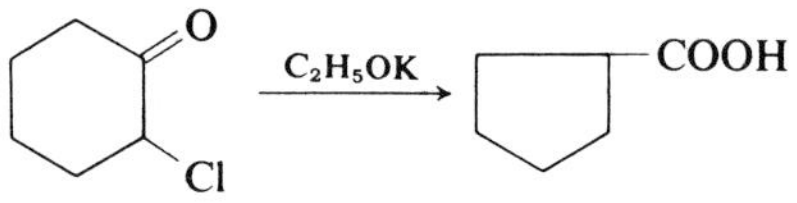

$$\text{[cyclohexanone with Cl]} \xrightarrow{C_2H_5OK} \text{[cyclopentane-COOH]} \quad (114)$$

(2) Carboxylation of amines [172].

$$\text{RNH}_2 + 2\text{ NaCN} + 2\text{ CH}_2\text{=O} + 2\text{ H}_2\text{O} \longrightarrow \text{RN(CH}_2\text{COONa)}_2 \xrightarrow{H_2O} \\ \text{RN(CH}_2\text{COOH)}_2 \quad (115)$$

(3) Pschorr synthesis [173].

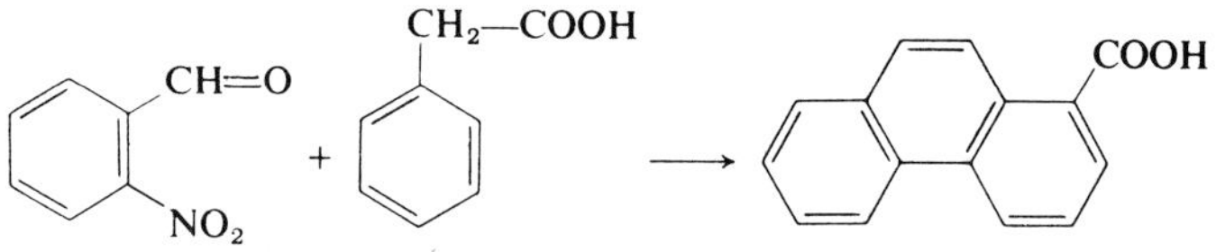

$$\text{[o-NO}_2\text{-C}_6\text{H}_4\text{-CH=O]} + \text{[C}_6\text{H}_5\text{-CH}_2\text{-COOH]} \longrightarrow \text{[phenanthrene-COOH]} \quad (116)$$

(4) Ullmann reaction [174].

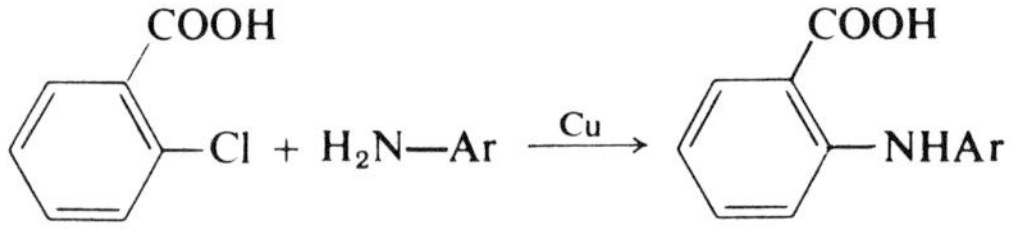

$$\text{[o-COOH-C}_6\text{H}_4\text{-Cl]} + \text{H}_2\text{N-Ar} \xrightarrow{Cu} \text{[o-COOH-C}_6\text{H}_4\text{-NHAr]} \quad (117)$$

(5) Williamson synthesis [175].

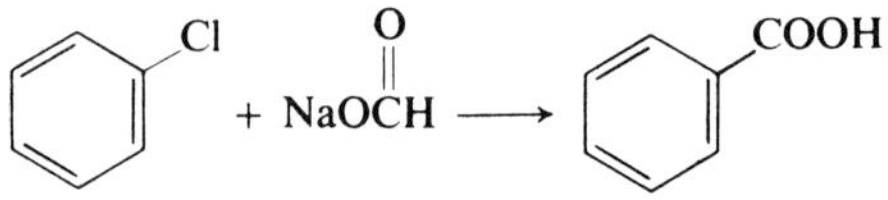

$$\text{[C}_6\text{H}_5\text{-Cl]} + \text{NaOCH=O} \longrightarrow \text{[C}_6\text{H}_5\text{-COOH]} \quad (118)$$

§ 10. Miscellaneous Methods

(6) Reaction of oxalyl bromide with olefins and tertiary alkanes [176].

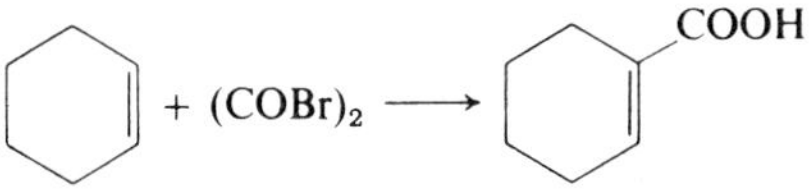

$$\text{C}_6\text{H}_{10} + (\text{COBr})_2 \longrightarrow \text{C}_6\text{H}_9\text{-COOH} \quad (119)$$

(7) Wittig synthesis [177, 178].

$$(\text{C}_6\text{H}_5)_3\text{P}=\underset{R}{\text{C}}-\text{COOR} + \text{R}'\text{CH}=\text{O} \longrightarrow \text{R}'\text{CH}=\underset{R}{\text{C}}-\text{COOR} \xrightarrow{\text{H}_2\text{O}}$$

$$\text{R}'\text{CH}=\underset{R}{\text{C}}\;\text{COOH} \quad (120)$$

(8) Reaction of 2-pyrones with cyanide ion [179].

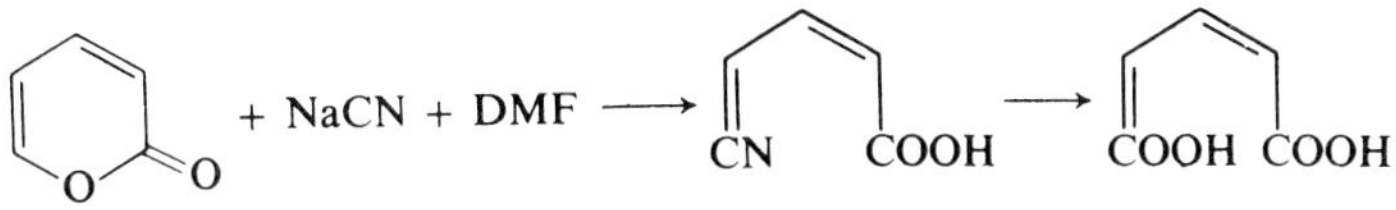

$$\text{2-pyrone} + \text{NaCN} + \text{DMF} \longrightarrow \underset{\text{CN} \quad \text{COOH}}{\text{diene}} \longrightarrow \underset{\text{COOH} \quad \text{COOH}}{\text{diene}} \quad (121)$$

(9) Electrolytic hydrodimerization of derivatives of α,β-unsaturated acids [180].

$$\text{R}-\text{CH}=\text{CHCN} \longrightarrow \begin{array}{c} \text{R}-\text{CH}-\text{CH}_2\text{CN} \\ | \\ \text{R}-\text{CH} \\ | \\ \text{CH}_2\text{CN} \end{array} \xrightarrow[\text{OH}^-]{\text{H}_2\text{O}} \begin{array}{c} \text{R}-\text{CH}-\text{CH}_2\text{COOH} \\ | \\ \text{R}-\text{CH} \\ | \\ \text{CH}_2\text{COOH} \end{array} \quad (122)$$

(10) Wislicenus synthesis of aliphatic dibasic acids [181].

$$2\;\begin{array}{c} \text{CH}_2-\text{CH}_2\text{COOH} \\ | \\ \text{I} \end{array} \xrightarrow{2\,\text{Ag}} \begin{array}{c} \text{CH}_2\text{CH}_2\text{COOH} \\ | \\ \text{CH}_2\text{CH}_2\text{COOH} \end{array} + 2\;\text{AgI} \quad (123)$$

(11) Berthelot-Goldschmidt process for oxalates [182].

$$2\;\text{HCOONa} \xrightarrow{\text{Heat}} (\text{COONa})_2 + \text{H}_2 \quad (124)$$

(12) Hayashi rearrangement of substituted o-benzoylbenzoic acids [183].

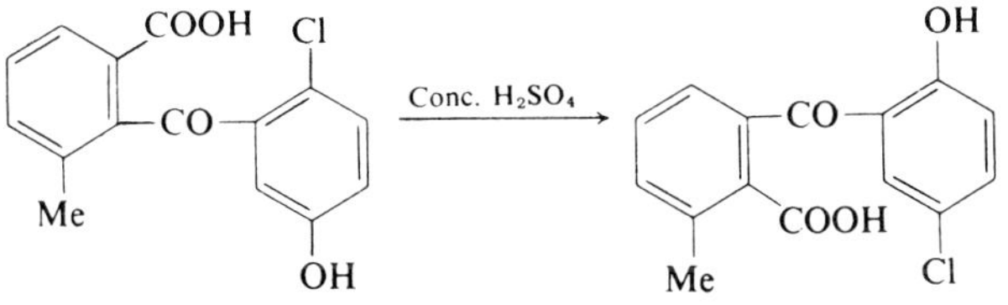

$$\text{(aryl-COOH, Me, CO, Cl, OH)} \xrightarrow{\text{Conc. H}_2\text{SO}_4} \text{(aryl-OH, CO, COOH, Me, Cl)} \quad (125)$$

(13) Ivanov reaction [184].

$$C_6H_5CH_2COOMgCl + 1.5\ C_2H_5MgBr \xrightarrow[0°C]{Ether}$$

$$\underset{\underset{MgBr}{|}}{C_6H_5CH}-COOMgCl + C_2H_6 \xrightarrow{R_2C=O} R_2C\underset{\underset{OH}{|}}{-}CH\underset{COOH}{\overset{C_6H_5}{\diagup}} \quad (126)$$

(14) Reaction of acrylic acid with diazonium salts [185].

$$CH_2=CHCOOH + ArN_2^+Br^- \xrightarrow[HBr]{CuBr} ArCH_2\underset{\underset{Br}{||}}{CH}-COOH \quad (127)$$

(15) Photochemical Wolff rearrangement [186].

F. Reduction Reactions

(1) Reduction of unsaturated acids to give saturated acids [187].

$$RCH=CH-COOH \xrightarrow{H_2} RCH_2CH_2COOH \quad (128)$$

(2) Kindler synthesis of substituted phenylacetic acids [188].

$$\underset{\underset{OH}{|}}{Ar-CHCOOH} \xrightarrow[AcOH + H_2SO_4]{Pd/H_2} ArCH_2COOH \quad (129)$$

(3) Reduction of phthalic acid to *trans*-1,2-dihydrophthalic acid [189]: application of the Meerwein reaction.

G. Enzyme Reactions [58, 59, 190, 191]

(1) Preparation of itaconic acid [190].

$$\text{Starches, Sugars + carbohydrates} \xrightarrow[\substack{+O_2\ (air),\ room \\ temp. + pressure}]{Aspergillus\ terreus} \underset{\underset{CH_2COOH}{|}}{CH_2=CCOOH}$$
$$+$$
$$\text{Minor amounts of other acids} \quad (130)$$

(2) Preparation of L-threonine by a fermentation process utilizing genetic engineering technology [192].

(3) Microbiological preparation of (S)-(+)-2,3-dihydroxy-3-methylbutanoic acid [193].

H. Substitution Reactions

(1) α-Chlorination of carboxylic acids: Preparation of ε-benzoylamino-2-chlorocaproic acid [194].

REFERENCES

1a. H. E. Zaugg and R. T. Rapala, *Org. Synth.* **27,** 84 (1947).
1b. H. E. Zaugg and R. T. Rapala, *Org. Synth. Collect. Vol.* **3,** 820 (1955).
2. W. F. Tuley and C. S. Marvel, *Org. Synth. Collect. Vol.* **3,** 822 (1955).
3. C. E. Senseman and J. J. Stubbs, *Ind. Eng. Chem.* **23,** 1129 (1931).
4. P. Hill and W. F. Short, *J. Chem., Soc.* p. 1125 (1935).
5. O. Kamm and A. O. Matthews, *Org. Synth. Collect. Vol.* **1,** 392 (1941).
6. H. T. Clarke and E. R. Taylor, *Org. Synth. Collect. Vol.* **2,** 135 (1943).
7. P. P. J. Sah and C.-L. Hsu, *Recl. Trav. Chim. Pays-Bas* **59,** 351 (1940).
8. A. W. Singer and S. M. McElvain, *Org. Synth.* **20,** 79 (1940).
9. D. G. Leis and B. C. Curran, *J. Am. Chem. Soc.* **67,** 79 (1945).
10. C. F. Koelsch, *J. Am. Chem. Soc.* **65,** 2464 (1943).
11. P. D. Bartlett, G. L. Fraser, and R. B. Woodward, *J. Am. Chem. Soc.* **63,** 594 (1941).
12a. *J. Chem. Soc., Chem. Commun* p. 253 (1978).
12b. J. Kenyon and B. V. Platt, *J. Chem. Soc.* p. 636 (1939).
13a. R. R. Burtner and J. W. Cusic, *J. Am. Chem. Soc.* **65,** 265 (1943).
13b. E. J. Corey, N. W. Gilman, and B. E. Ganem, *J. Am. Chem. Soc.* **90** 5616 (1968); S. C. Thomason and D. G. Kubler, *J. Chem. Educ.* **45,** 546 (1968).
14a. C. D. Hurd, J. W. Garrett, and E. N. Osborne, *J. Am. Chem. Soc.* **55,** 1084 (1933).
14b. H. Baba, *Kagaku Kenkyusho Hokoku* **33,** 168 (1957).
15. A. M. Seligman, K. C. Tsou, S. H. Rutenberg, and R. B. Cohen, *J. Histochem. Cytochem.* **2,** 211 (1954).
16. L. I. Smith and G. F. Rouault, *J. Am. Chem. Soc.* **65,** 745 (1943).
17. H. E. Fierz-David and W. Kuster, *Helv. Chim. Acta* **22,** 87 (1939).
18. C. Djerassi, *Chem. Rev.* **38,** 526 (1946).
19. F. Ansinger, *Chem. Ber.* **75,** 656 (1942).
20. A. L. Henne and P. Hill, *J. Am. Chem. Soc.* **65,** 753 (1943).
21. L. Long, Jr., *Chem. Rev.* **27,** 437 (1940).
22. A. L. Henne, *J. Am. Chem. Soc.* **51,** 2676 (1929).
23. M. I. Fremery and E. K. Fields, *J. Org. Chem.* **28,** 2537 (1963).
24. W. A. Mosher and F. C. Whitmore, *J. Am. Chem. Soc.* **70,** 2545 (1948).
25. B. A. Ellis, *Org. Synth. Collect. Vol.* **1,** 18 (1941).
26. H. W. Underwood, Jr. and E. L. Kochmann, *J. Am. Chem. Soc.* **46,** 2071 (1924).
27. F. Bischoff and H. Adkins, *J. Am. Chem. Soc.* **45,** 1031 (1923).
28. H. J. Schmitz, *Justus Liebigs Ann. Chem.* **193,** 160 (1878).
29. R. C. Roberts and T. B. Johnson, *J. Am. Chem. Soc.* **47,** 1399 (1925).
30. J. S. Buck and W. S. Ide, *J. Am. Chem. Soc.* **54,** 3308 (1932).
31. J. Colonge and L. Pichat, *Bull. Soc. Chim. Fr* [5] **16,** 180 (1949).
32. A. M. Van Arendonk and M. E. Cupery, *J. Am. Chem. Soc.* **53,** 3184 (1931).
33. D. Pressmen and H. J. Lucas, *J. Am. Chem. Soc.* **62,** 2069 (1940).
34. M. S. Newman and H. L. Holmes, *Org. Synth. Collect. Vol.* **2,** 428 (1943).
35. L. I. Smith, W. W. Pichard, and L. J. Spillane, *Org. Synth.* **23,** 27 (1943).
36. L. F. Fieser and J. Cason, *J. Am. Chem. Soc.* **61,** 1742 (1939).

37. A. H. Homeyer, F. C. Whitmore, and V. H. Wallingford, *J. Am. Chem. Soc.* **55,** 4211 (1933).
38. R. Levine and J. R. Stephens, *J. Am. Chem. Soc.* **72,** 1642 (1950).
39. M. Carmack and M. A. Spielman, *Org. React.* **3,** 95 (1946).
40. A. C. Ott, L. A. Mattano, and G. H. Coleman, *J. Am. Chem. Soc.* **68,** 2633 (1946).
41. K. Kindler and T. Li, *Chem. Ber.* **74,** 321 (1941).
42. E. Schwenk and E. Block, *J. Am. Chem. Soc.* **64,** 3051 (1942).
43. U. V. Solmssen and E. Wenis, *J. Am. Chem. Soc.* **70,** 4200 (1948).
44. W. G. Toland, Jr., D. L. Hagmann, J. B. Wilkes, and F. J. Brutschy, *J. Am. Chem. Soc.* **80,** 5423 (1958).
45. C. D. Hurd, J. W. Garrett, and E. N. Osborne, *J. Am. Chem. Soc.* **55,** 1083 (1933).
46. V. Franzen, *Chem. Ber.* **90,** 2036 (1957).
47. St. Pancescu, *Stud. Cercet. Chim.* **8,** 623 (1960).
48. T. W. Evans and W. M. Dehn, *J. Am. Chem. Soc.* **52,** 3645 (1930).
49. E. Rohrmann, R. G. Jones, and H. A. Shonle, *J. Am. Chem. Soc.* **66** 1856 (1944).
50. T. L. Jacobs, S. Winstein, R. B. Henderson, J. Bond, J. W. Rolls, D. Seymour, and W. H. Florsheim, *J. Org. Chem.* **11,** 229 (1946).
51. C. T. Lester and J. R. Proffitt, Jr., *J. Am. Chem. Soc.* **71,** 1878 (1949).
52. H. Gilman and H. H. Parker, *J. Am. Chem. Soc.* **46,** 2816 (1924).
53. H. Gilman and P. R. Van Ess, *J. Am. Chem. Soc.* **55,** 1258 (1933).
54. H. Gilman and C. H. Kirby, *Org. Synth. Collect Vol.* **1,** 361 (1941).
55. C. Schuerch, Jr. and E. H. Huntress, *J. Am. Chem. Soc.* **70,** 2824 (1948).
56. R. D. Rieke, S. E. Bales, P. M. Hudnall, and G. S. Poindexter, *Org. Synth.* **59,** 85 (1979).
57a. R. R. Burtner and J. W. Cusic, *J. Am. Chem. Soc.* **65,** 264 (1943).
57b. R. A. Heacock, R. L. Wain, and F. Wightman, *Ann. Appl. Biol.* **46,** 352 (1958).
58. G. G. Price and M. C. Whiting, *Chem. Ind. (London)* p. 775 (1963).
59. F. A. French, *Chem. Eng. News* **44,** 48 (1966).
60. W. MacGregor, private communication, Crown Zellerbach Corp. Camas, Washington 98607 (1967).
61. M. Nierenstein and D. A. Clibbens, *Org. Synth Collect. Vol.* **2,** 557 (1943).
62. A. S. Lindsay and H. Jeskey, *Chem. Rev.* **57,** 583 (1957).
63. C. A. Buehler and W. E. Cate, *Org. Synth. Collect. Vol.* **2,** 341 (1943).
64. A. J. Rostron and A. M. Spivey, *J. Chem. Soc.* p. 3092 (1964).
65. T. Sakakibara and Y. Odaira, *J. Org. Chem.* **41,** 2049 (1970).
66. D. Papa, E. Schwenk, F. Villani, and E. Klingsberg, *J. Am. Chem. Soc.* **70,** 3359 (1948).
67. N. O. Calloway, *Chem. Rev.* **17,** 356 (1935).
68. H. G. Latham, Jr., E. L. May, and E. Mosettig, *J. Am. Chem. Soc.* **70,** 1079 (1948).
69. W. Treibs and H. Orttmann, *Chem. Ber.* **93,** 545 (1960).
70. W. H. C. Rueggeberg, R. K. Frantz, and A. Ginsburg, *Ind. Eng. Chem.* **38,** 624 (1946).
71. D. S. Breslow, *J. Am. Chem. Soc.* **72,** 4245 (1950).
72. F. Bergmann, M. Weizmann, E. Dimant, J. Patai, and J. Szmurkowicz, *J. Am. Chem. Soc.* **70,** 1612 (1948).
73. J. R. Johnson, *Org. React.* **1,** 210, 248–254 (1942).
74. R. W. Maxwell and R. Adams, *J. Am. Chem. Soc.* **52,** 2967 (1930).
75. F. Bock, G. Lock, and K. Schmidt, *Monatsh. Chem.* **64,** 399 (1934).
76. S. E. Boxer and R. P. Linstead, *J. Chem. Soc.* p. 740 (1931).
77. R. P. Linstead, E. G. Noble, and E. J. Boorman, *J. Chem. Soc.* p. 559 (1933).
78. W. E. Bachmann, *J. Org. Chem.* **3,** 444 (1938).
79. C. Niemann and C. T. Redemann, *J. Am. Chem. Soc.* **68,** 1933 (1946).
80. C. S. Marvel and W. O. King, *Org. Synth. Collect. Vol.* **1,** 252 (1941).
81. W. S. Johnson and G. H. Daub, *Org. React.* **6,** 1 (1951).

82. A. C. Cope and C. M. Hofmann, *J. Am. Chem. Soc.* **63,** 3456 (1941).
83. M. Reimer, *J. Am. Chem. Soc.* **46,** 785 (1924).
84. S. E. Dinizo, R. W. Freerksen, W. E. Pabst, and D. S. Watt, *J. Am. Chem. Soc.* **99,** 182 (1977).
85. R. Adams, and R. H. Kamm, *Org. Synth. Collect. Vol.* **1,** 250 (1941).
86. E. B. Vliet, C. S. Marvel, and C. M. Hsueh, *Org. Synth. Collect. Vol.* **2,** 416 (1943).
87. E. M. Reid and J. R. Ruhoff, *Org. Synth. Collect. Vol.* **2,** 474 (1943).
88. F. F. Blicke and R. E. Feldkamp, *J. Am. Chem. Soc.* **66,** 1087 (1944).
89. K. R. Tsai and M. S. Newman, *J. Org. Chem.* **45,** 4785 (1980).
90. J. F. Norris and H. F. Tucker, *J. Am. Chem. Soc.* **55,** 4697 (1933).
91. R.Adams and C. S. Marvel, *J. Am. Chem. Soc.* **42,** 310 (1920).
92. E. M. Reid and J. R. Ruhoff, *Org. Synth Collect. Vol.* **2,** 474 (1943).
93. C. S. Marvel and F. D. Hager, *Org. Synth. Collect. Vol.* **1,** 248 (1941).
94. N. L. Drake and R. W. Riemenschneider, *J. Am. Chem. Soc.* **52,** 5005 (1930).
95. A. Schönberg and F. L. Warren, *J. Chem. Soc.* p. 1840 (1939).
96. A. L. Wilds, J. Van der Berghe, C. H. Winestock, R. L. von Trebra, and N. F. Woolsey, *J. Am. Chem. Soc.* **84,** 1503 (1962).
97. R. E. Steiger, *Org. Synth. Collect. Vol.* 3 66 (1955).
98. W. Koenigs and E. Wagstaffe, *Ber. Dtsch. Chem. Ges.* **26,** 558 (1893).
99. A. Einhorn, *Justus Liebigs Ann. Chem.* **265,** 221 (1891); A. Einhorn and A. Liebrecht, *Ber. Dtsch. Chem. Ges.* **20,** 1592 (1887).
100. C. W. Tullock and S. M. McElvain, *J. Am. Chem. Soc.* **61,** 961 (1939).
101. M. P. Dreyfuss, *J. Org. Chem.* **28,** 3269 (1963).
102. S. Akiyoshi, K. Okuno, and S. Nagahama, *J. Am. Chem. Soc.* **76,** 902 (1954).
103. G. W. Moersch and A. R. Burkett, *J. Org. Chem.* **36,** 1149 (1971).
104. K. L. Rinehart, Jr. and E. G. Perkins, *Org. Synth.* **37,** 37 (1957).
105. M. C. Kloetzel, *Org. React.* **4,** 1 (1948).
106. S. Seltzer, *J. Am. Chem. Soc.* **85,** 1360 (1963).
107. K. Alder, *Justus Liebigs Ann. Chem.* **564,** 79, 96,. 109, 120 (1949).
108. L. M. Joshel and L. W. Butz, *J. Am. Chem. Soc.* **63,** 3350 (1941).
109. M. C. Kloetzel and H. L. Herzog, *J. Am. Chem. Soc.* **72,** 1991 (1950).
110. A. C. Cope and E. C. Herrick, *J. Am. Chem. Soc.* **72,** 984 (1950).
111. J. P. Fleury and J. Lichtenberger, *Bull. Soc. Chim. Fr.* p. 565 (1963).
112. M. A. Goldberg, E. P. Odas, and G. Carsch, *J. Am. Chem. Soc.* **69,** 261 (1947).
113. R. Adams and A. F. Thal, *Org. Synth. Collect. Vol.* **1,** 436 (1941).
114. R. Adams and C. S. Marvel, *J. Am. Chem. Soc.* **42,** 310 (1920).
115. A. A. Noyes, *J. Am. Chem. Soc.* **23,** 393 (1901).
116. G. Berger and S. C. J. Olivier, *Recl. Trav. Chim. Pays-Bas* **46,** 600 (1927).
117. T. T. Chu and C. S. Marvel, *J. Am. Chem. Soc.* **55,** 2842 (1933).
118. R. L. Shriner, S. G. Ford, and L. J. Roll, *Org. Synth. Collect. Vol.* **2,** 140 (1943).
119. L. Gilman and S. Avakian, *J. Am. Chem. Soc.* **68** 2104 (1946).
120. G. Berger and S. C. J. Olivier, *Recl. Trav. Chim. Pays-Bas* **46,** 600 (1927).
121. C. L. Carter and S. N. Slater, *J. Chem. Soc.* p. 131 (1946).
122. H. L. Vaughn and M. D. Robbins, *J. Org. Chem.* **40,** 1187 (1975).
123. W. Roberts and M. C. Whiting, *J. Chem. Soc.* p. 1290 (1965).
124. D. S. Breslow, E. Baumgarten, and C. R. Hauser, *J. Am. Chem. Soc.* **66,** 1287 (1944).
125. P. G. Grossman and W. N. Schenk, *J. Org. Chem.* **42,** 918 (1977).
126. J. S. Filippo, Jr., L. J. Romano, C.-I. Chern, and J. S. Valentine, *J. Org. Chem.* **41,** 5861 (1976).
127. J. Colonge and D. Joly, *Ann. Chim. (Paris)* [11] **18,** 312 (1943).

128. F. Bergmann, M. Weizmann, E. Dimant, J. Patai, and J. Szmurkowicz, *J. Am. Chem. Soc.* **70,** 1612 (1948).
129. R. L. Shriner, S. G. Ford, and L. J. Roll, *Org. Synth. Collect. Vol.* **1,** 391 (1932).
130. O. Kamm and J. B. Segus, *Org. Synth. Collect. Vol.* **1,** 391 (1932).
131. E. L. Compere, Jr. and A. Shockravi, *J. Org. Chem.* **43,** 2702 (1978).
132. C. Weizmann, M. Sulzbacher, and E. Bergmann, *J. Am. Chem. Soc.* **70,** 1154 (1948).
133. A. I. Meyers, G. Knaus, and K. Kaman, *J. Am. Chem. Soc.* **96,** 268 (1974); A. I. Meyers and G. Knaus, *ibid.* p. 6508; A. I. Meyers, D. L. Temple, R. L. Nolen, and E. D. Mihalich, *J. Org. Chem.* **39,** 2278 (1974); G. R. Malone and A. I. Meyer, *ibid.* p. 618.
134. N. S. Kozlov, P. N. Fedoseev, and V. S. Lazarev, *J. Gen. Chem. (USSR)* **6,** 485 (1936).
135. V. V. Pigulevskii and L. I. Gulyaeva, *Trans. Exp. Res. Lab. "Khemgas"* Materials on Cracking and Chem. Treatment of Cracking Products (USSR) **3,** 185 (1936); *Chem. Abstr.,* **31,** 5315 (1937).
136. H. Adkins and A. K. Roebuck, *J. Am. Chem. Soc.* **70,** 4044 (1948); L. R. Fieser and J. Szmuszkovicz, *ibid.* p. 3352.
137. R. N. McDonald and P. A. Schwab, *J. Am. Chem. Soc.* **86,** 4866 (1964); J. K. Stille and D. D. Whitehurst, *ibid.* p. 4871.
138. J. J. Melchiore and H. R. Moyer *Abstr., 148th Meet. Am. Chem. Soc.* p. 25a (1964).
139. E. K. Fields, *Abstr., 148th Meet. Am. Chem. Soc.* p. 26u (1964).
140. T. J. Wallance, A. Schriesheim, and N. Jacobson, *J. Org. Chem.* **29,** 2907 (1964).
141. H. Flege and K. Wedemeyer, U.S. Patent 4,238,625 (1980).
142. R. N. McDonald, R. N. Steppel, and J. E. Dorsey, *Org. Synth.* **50,** 15 (1970).
143. J. K. Blatchford and M. Orchin, *J. Org. Chem.* **29,** 839 (1964).
144. S. B. Lippincott and H. B. Haas, *Ind. Eng. Chem.* **31,** 118 (1939).
145. R. C. Fuson, H. Gray, and J. J. Gonza, *J. Am. Chem. Soc.* **61,** 1937 (1939).
146. R. A. Jacobson, *J. Am. Chem. Soc.* **68,** 2628 (1946).
147. H. L. Wehrmeister, *J. Org. Chem.* **27,** 4418 (1962).
148. K. M. Ibne-Rash and E. Koubek, *J. Org. Chem.* **28,** 3240 (1963).
149. S. W. Tobey and R. West, *J. Am. Chem. Soc.* **86,** 56 (1964); C. B. Miller and C. Woolf, U.S. Patents 2,827,485, 2,827,486 (1958).
150. G. M. Lampman and J. C. Aumiller, *Org. Synth.* **51,** 73 (1971).
151. J. McMurry, *Org. Synth.* **24,** 187 (1976).
152. A. M. Felix, *J. Org. Chem.* **39,** 1427 (1974).
153. L. I. Zakharkin and V. V. Gavrilenko, *Dokl. Akad. Nauk SSSR* **118,** 713 (1958).
154. A. A. Morton and J. R. Stevens, *J. Am. Chem. Soc.* **53,** 4028 (1931); A. A. Morton, M. L. Brown, M. E. T. Holden, R. L. Letsinger, and E. E. Magat, *ibid.,* **67,** 2224 (1945); C. E. Frank, J. R. Leebrick, L. F. Moormeier, J. A. Scheben, and O. Homberg, *J. Org. Chem.* **26,** 307 (1961).
155. R. D. Harper, Jr., E. J. Soloski, and C. Tamborski, *J. Org. Chem.* **29,** 2385 (1964).
156. H. Bruggiak, S. J. Ray, and A. McKillop, *Org. Synth.* **58,** 52 (1978).
157. F. Adickes and G. Schäfer, *Ber Dtsch. Chem. Ges.* **65,** 950 (1932).
158. H. Adkins and R. W. Rosenthal, *J. Am. Chem. Soc.* **72,** 4550 (1950).
159. J. Tsaji, M. Morikawa, and J. Kiji, *J. Am. Chem. Soc.* **86,** 4851 (1964).
160. D. H. Newitt and S. A. Momer, *J. Chem. Soc.* p. 2945 (1949).
161. W. W. Prichard and G. E. Tabet, U.S. Patent, 2,565,462 (1951); *Chem. Abstr.* **46,** 2578 (1952).
162. W. Reppe, *Mod. Plast.* **23,** No. 3, 162 (1945).
163. C. W. Bird, *Chem. Rev.* **62,** 283 (1962).
164. W. Reppe, H. Kröper, H. J. Pistor, and O. Weissbarth, *Justus Liebigs Ann. Chem.* **582,** 87 (1953).

165. D. D. Coffmann, R. Cramer, and W. E. Mochel, *J. Am. Chem. Soc.* **80,** 2882 (1958); J. Falbe, R. Paatz, and F. Korte, *Chem. Ber.* **97,** 3088 (1964).
166. C. W. Theobald, U.S. Patent 2,378,048 (1945).
167. W. W. Prichard, U.S. Patent 2,680,750 (1954).
168. J. R. Norell, *J. Org. Chem.* **37,** 1971 (1972).
169. R. C. Larock, *J. Org. Chem.* **40,** 3237 (1975).
170. Monsanto Co., Japan Kokai, Tokkyo Koko 80/92,339 (1980). *Chem. Abstr.*, *94,* 14603b (1981).
171. C. Friedel and J. M. Crafts, *C. R. Hebd. Seances Acad. Sci.*, **84,** 1392, 1450 (1887).
172. A. E. Faworskii, *J. Chem. Soc.* pp. 64i, 391 (1893) (abstr.).
173. A. E. Martell and F. C. Bersworth, *J. Org. Chem.* **15,** 46, 255 (1950).
174. R. Pschorr, *Chem. Ber.* **29,** 496 (1896).
175. F. Ullmann, *Chem. Ber.* **36,** 2382 (1903).
176. A. W. Williamson, *J. Chem. Soc.* **4,** 229 (1852).
177. W. Treibs and H. Orttmann, *Naturwissenschaften* **45,** 85 (1958).
178. H. J. Bestmann and H. Schulz, *Chem. Ber.* **95,** 2921 (1962).
179. G. Vogel, *Chem. Ind. (London)* p. 1829 (1962).
180. M. M. Baizer and J. D. Anderson, *J. Electrochem. Soc.* **111,** 215 (1964).
181. J. Wislicenus, *Justus Liebigs Ann. Chem.* **149,** 215 (1896).
182. R. E. Kirk and D. F. Othmer, eds., "Encyclopedia of Chemical Technology," Vol. 9, p. 661. Wiley (Interscience), New York, 1952.
183. M. Hayashi, *J. Chem. Soc.* p. 2516 (1927).
184. E. F. Pratt and E. Werble, *J. Am. Chem. Soc.* **72,** 4638 (1950).
185. G. H. Cleland, *Org. Synth.* **51,** 1 (1971).
186. T. H. Wheeler and J. Meinwald, *Org. Synth.* **52,** 53 (1972).
187. H. A. Smith, D. M. Alderman, and F. W. Nadig, *J. Am. Chem. Soc.* **67,** 272 (1945).
188. K. Kindler and D. Kwok, *Justus Liebigs Ann. Chem.* **554,** 9 (1943).
189. R. N. McDonald and C. E. Reineke, *Org. Synth.* **50,** 50 (1970).
190. Charles Pfizer & Co., Inc., British Patent 603,866 (1948).
191. I. Hirao, T. Matsuura, and A. K. Ota, *Polym. Rep.* **83,** 33 (1965).
192. *Jpn. Chem. Rep.* Issue No. 1, December, p. 6, Ref. No. 386-12-21 (1980) (Ajinomoto Co. Development).
193. D. J. Aberhart, *J. Org. Chem.* **45,** 5218 (1980).
194. Y. Ogata, T. Sugimoto, and M. Inaishi, *Org. Synth.* **59,** 20 (1979).

CHAPTER 10 / ESTERS

1. Introduction	289
2. Condensation Reactions	290
A. The Reaction of Alcohols with Carboxylic Acids	291
2-1. Preparation of Methyl Acetate	292
2-2. Preparation of Amyl Acetate	292
B. The Reaction of Acyl Halides with Hydroxy Compounds	293
2-3. Preparation of tert-Butyl Acetate	294
2-4. Preparation of Diglycidyl Isophthalate	294
C. The Reaction of Anhydrides with Hydroxy Compounds	295
2-5. Preparation of Methyl Hydrogen Phthalate	295
D. The Reaction of Halides with Salts of Carboxylic Acids	296
2-6. Preparation of p-Ethylbenzyl Acetate	297
E. Ester Interchange	297
2-7. Preparation of Cellosolve Acrylate	297
2-8. Preparation of Vinyl Caproate	298
F. The Reaction of Carboxylic Acids with Olefins	299
2-9. Preparation of tert-Butyl Acetate	299
2-10. Preparation of tert-Butyl Acrylate	300
G. Alkylation Reactions	300
2-11. Preparation of Diethyl sec-Butylmalonate	301
H. The Preparation of Lactones	302
2-12. Preparation of α-Ethylbutyrolactone	302
I. The Reaction of Lactones with Alcohols	303
2-13. Preparation of Ethyl α-Ethyl-γ-hydroxybutyrate	303
J. Miscellaneous Condensation Reactions	303
a. The Acid-Catalyzed Reaction of Nitriles with Alcohols	303
b. The Reaction of Diazomethane with Carboxylic Acids	304
c. Esterification of Carboxylic Acids with Trialkyloxonium Salts	304
d. The Reformatskii Reaction	304
e. The Diels–Alder Reaction	304
3. Oxidation Reactions	304
A. Oxidation of Primary Alcohols	304
3-1. Preparation of n-Butyl n-Butyrate (Dichromate Oxidation)	305
B. The Tischtschenko–Cannizzaro Reaction of Aldehydes	305
3-2. Preparation of n-Butyl n-Butyrate	305
C. Direct Oxidation of Aldehydes and Ketones	306
3-3. Preparation of Ethyl 6-Hydroxyhexoate (Caro's Acid Oxidation)	306
D. Oxidation of Ethers	306
4. Reduction Reactions	307
5. Rearrangements	307
A. The Arndt–Eistert Rearrangement	307
B. The Favorskii Rearrangement	307

5-1. Preparation of Methyl 1-Methylcyclohexanecarboxylate   308
6. Miscellaneous Methods   308
   References   309

1. INTRODUCTION

The most common laboratory method for the preparation of esters utilizes the condensation between a carboxylic acid and an alcohol catalyzed by acids such as HCl, H_2SO_4, BF_3, or p-toluenesulfonic acid. The problem of catalysis is receiving continued attention and several new catalyst systems are briefly mentioned.

The esterification reaction is an equilibrium reaction and it can be displaced toward the product side by removal of water, or by the use of an excess of one of the reactants. The use of acetone dimethylacetal, which reacts with the water formed to produce methanol and acetone, allows the preparation of methyl esters in high yield. Primary and secondary alcohols are esterified in good yield but tertiary alcohols give very low yields. Neopentyl alcohol reacts normally and is esterified with either acids or acid chlorides.

The preferred method for the preparation of tertiary esters is based on the interaction of the acid halides and the tertiary alcohol or an olefin and a carboxylic acid. For example, *tert*-butyl acrylate is made by the condensation of isobutylene and acrylic acid.

Some other common condensation methods may be summarized by Eqs. (1), (2), and (3).

$$RCO(Z) + R'OH \longrightarrow RCOOR' + ZOH$$
$$\text{where } Z = OH, Cl, OR, (OCOR), \text{ and lactones} \quad (1)$$

$$R\underline{C}HCOOR + RX \longrightarrow RCH-COOR \quad (2)$$
$$\qquad\qquad\qquad\qquad\qquad\quad |$$
$$\qquad\qquad\qquad\qquad\qquad\quad R$$

$$RCOONa \text{ (or Ag)} + R'X \longrightarrow RCOOR' \quad (3)$$

The condensation of carboxylic acids with diazomethane leads to methyl esters.

$$RCOOH + CH_2N_2 \longrightarrow RCOOCH_3 + N_2 \quad (4)$$

Little use has been made of this technique on a preparative scale since diazomethane is a yellow toxic gas which may explode violently if undiluted or on contact with rough glass surfaces.

Dimethyl sulfate can be used in place of diazomethane to form methyl esters of carboxylic acids through the sodium salt.

$$2\ RCOONa + (CH_3)_2SO_4 \longrightarrow 2\ RCOOCH_3 + Na_2SO_4 \quad (5)$$

However, dimethyl sulfate is acidic in nature and may not be as satisfactory as diazomethane for methylating sensitive acids. Dimethyl sulfate is also very toxic and is a suspected carcinogen.

Transesterification reactions are valuable methods for the preparation of vinyl esters. In these reactions the alcohol portion of the ester is exchanged [Eq. (6)]. In another type of transesterification, acid moieties are exchanged [Eq. (7)].

$$\text{RCOOR}' + \text{R}''\text{OH} \rightleftharpoons \text{RCOOR}'' + \text{R}'\text{OH} \qquad (6)$$

$$\text{RCOOR}' + \text{R}''\text{COOH} \rightleftharpoons \text{R}''\text{COOR}' + \text{RCOOH} \qquad (7)$$

The other methods of oxidation and reduction are more specialized reactions and are treated briefly.

The Arndt–Eistert, Favorskii, and Darzens syntheses are described in more detail elsewhere in this text.

2. CONDENSATION REACTIONS

Several of the more common condensation methods for the preparation of esters involve either the condensation of alcohols or alkyl halides with carboxylic acids or their derivatives as shown below.

$$\left.\begin{array}{l}\text{RCOOH}\\(\text{RCO})_2\text{O}\\\text{RCOCl}\\\text{RCOOR}\\(\text{and Lactones})\end{array}\right\} + \text{R}'\text{OH} \longrightarrow \text{RCOOR}' \qquad (8)$$

$$\text{CH}_2=\text{C}=\text{O} + \text{R}'\text{OH} \longrightarrow \text{CH}_3\text{COOR}' \qquad (9)$$

$$\text{RCOONa} + \text{R}'\text{X} \longrightarrow \text{RCOOR}' \qquad (10)$$

$$\text{R}\bar{\text{C}}\text{H}-\text{COOR} + \text{R}'\text{X} \longrightarrow \text{RCH}-\text{COOR} \qquad (11)$$
$$\qquad\qquad\qquad\qquad\qquad\qquad\qquad |$$
$$\qquad\qquad\qquad\qquad\qquad\qquad\qquad \text{R}'$$

$$^-\text{CH}-(\text{COOR})_2 + \text{R}'\text{X} \xrightarrow{\text{Malonic Ester synthesis}} \text{R}'\text{CH}(\text{COOR})_2 \qquad (12)$$

In addition, olefins react with carboxylic acids in the presence of catalytic amounts of sulfuric acid or boron trifluoride. This method allows one to prepare *tert*-butyl esters [Eq. (13)].

$$(\text{CH}_3)_2\text{C}=\text{CH}_2 + \text{RCOOH} \longrightarrow \text{RCOOC}(\text{CH}_3)_3 \qquad (13)$$

Acetylenes also condense with acids to give unsaturated esters.

$$\text{CH}\equiv\text{CH} + \text{RCOOH} \longrightarrow \text{RCOOCH}=\text{CH}_2 \qquad (14)$$

§ 2. Condensation Reactions

The latter is a widely used commercial method for the preparation of vinyl esters such as vinyl acetate, vinyl benzoate, and vinyl formate.

A. The Reaction of Alcohols with Carboxylic Acids

$$\text{R'OH} + \text{RCOOH} \xrightleftharpoons{\text{Acid catalyst}} \text{RCOOR'} + \text{H}_2\text{O} \quad (15)$$

Primary alcohols give better yields of esters than secondary alcohols and tertiary alcohols, and phenols react only to a very small extent [1, 2]. Acid catalysts are used in small amounts. The mixture is refluxed for several hours and the equilibrium is shifted to the right by the use of a large excess of either the alcohol or acid and the removal of water. Azeotropic distillation of water, the use of a Dean and Stark trap, or a suitable drying agent helps to increase the rate of reaction. No acid catalysts are required for the preparation of esters of formic acid [1, 2] or of benzyl alcohol [3].

Polyesters are prepared by heating diols and dicarboxylic acids with an acid catalyst.

Some useful acid catalysts are sulfuric acid, hydrogen chloride, p-toluenesulfonic acid [4, 5], and methane sulfonic acid (Pennwalt).

In addition, trifluoracetic anhydride has been a useful acid catalyst for the esterification of phenols [6, 7]. The use of heavy metal salts as effective esterification catalysts has been widely reported. For example, $BuSnO_2H$ and Bu_2SnO are equal or superior to $p\text{-MeC}_6\text{H}_4\text{SO}_3\text{H}$ as esterification catalysts and do not effect appreciable dehydration of secondary alcohols [8]. It has been reported that catalytic activities generally descend in the order Sn^{+4}, Co^{+3}, Fe^{+3}, Al, Bi, Cr, Sn^{+2}, Cu, Co^{+2}, Pb, Fe^{+2}, Zn, Ni, Mn, Cd, Mg, Ba, K, and ClO_4, SO_4, $PhSO_3$, Cl, Zr, I, NO_3 [9]. The use of 1 wt% of $Fe_2(SO_4)_3$ is recommended as a noncorrosive catalyst in steel [5].

The use of stannous salts of carboxylic acids [$Sn(OOCR)_2$, where RCOO = 2-ethylhexoate, n-octoate, laurate, palmitate, stearate, and oleate] as catalysts for the preparation of polyesters has been reported to give colorless products with low acid numbers. Stannous oxide gives the same results but there is a short induction period before the catalyst becomes effective. The catalysts are present in 5×10^{-4} to 1×10^{-2} mole of catalyst per 100 gm of polyester and the temperature of the reaction is up to 200°C [10].

Metal oxides of the Group V metals such as tetraalkyltitanate esters, sodium alkoxy titanates, and alkaline earth salts of weak acids are among some other catalysts used to give polyesters without dehydration to the ether or olefin [11–14].

Trifluoromethanesulfonic anhydride [15] is an effective esterification catalyst, as is trifluoroacetic anhydride. However, in an organic medium, trifluoromethanesulfonic acid is a much stronger acid than trifluoroacetic acid or perchloric acid. When the anhydride is used, the esterification is exothermic, and heating for only a short period of time may be required.

Dihydric alcohols readily yield cyclic ethers under esterification conditions and the usual catalysts are not effective. The use of boric acid as a catalyst allows one to prepare esters of mono- or dibasic acids [16]. For example, esters of 1,4-butanediol and 2,5-hexanediol can be prepared by this method.

The use of orthophosphoric acid [17] allows one to prepare less-colored esters from "oxo synthesis" alcohols, which may contain sulfur impurities.

Ortho substituents in the aromatic acids retard esterification by the conventional method, but they can be esterified by dissolving in 100% sulfuric acid and pouring the solution into the desired alcohol [18]. This method is applicable to other unreactive systems and was successively applied to heterocyclic acids [19], polybasic acids [20], long-chain aliphatic acids [21], and several other unreactive substituted aromatic acids [22].

A simple esterification employing azeotropic removal of water by means of a Dean and Stark trap can be used in an introductory organic laboratory course. This technique is widely used in industry for the preparation of polyesters. The preparation of γ-chloropropyl acetate in 93–95% yield [23] and of n-amyl acetate in 71% [24a] yield have been described.

2-1. Preparation of Methyl Acetate [1]

$$CH_3OH + CH_3COOH \underset{}{\overset{H^+}{\rightleftharpoons}} CH_3\overset{O}{\underset{\|}{C}}OCH_3 + H_2O \qquad (16)$$

To a flask are added 48 gm (1.5 mole) of absolute methanol, 270 gm (4.5 moles) of glacial acetic acid, and 3.0 gm of concentrated sulfuric acid. The mixture is refluxed for 5 hr and then fractionated to give 112 gm of crude ester, bp 55°–56°C. The crude ester is washed successively with a saturated salt solution, a sodium bicarbonate solution until the effervescence ceases, and a saturated salt solution, dried, and distilled to yield 92 gm (83%), bp 56°C (754 mm), n_D^{25} 1.3594.

2-2. Preparation of Amyl Acetate [24a]

$$CH_3COOH + CH_3CH_2CH_2CH_2CH_2OH \overset{H^+}{\rightleftharpoons}$$
$$CH_3CH_2CH_2CH_2CH_2O\overset{O}{\underset{\|}{C}}CH_3 + H_2O \qquad (17)$$

To the round-bottomed flask are added 15 gm (0.25 mole) of glacial acetic acid, 17.6 gm (0.20 mole) of n-amyl alcohol, 30 ml of benzene, and 0.15 gm of p-toluenesulfonic acid catalyst. A Dean and Stark trap is filled with benzene and the contents of the flask are refluxed for 1 hr. The water that is produced remains in the trap as a bottom layer. The reaction mixture is extracted with sodium bicarbonate solution in order to remove the excess acid, washed with water, and then with a saturated sodium chloride solution. The organic layer is fractionated in order to remove the benzene–water azeotrope and the residue is transferred to another flask containing a small loose plug of steel wool in the neck. The product is isolated by simple distillation in 71% yield (20.7 gm), bp 141°–146°C, n_D^{26} 1.4012.

Although a "simple" procedure for the preparation of methyl esters has been reported using dimethyl sulfate [24b], the 1980 NIOSH *Registry of Toxic Effects of Chemical Substances* lists this chemical on p. 669 as a suspected carcinogen. (It also has a oral LD50 of 250 mg/kg in rats.)

B. The Reaction of Acyl Halides with Hydroxy Compounds

$$R'COCl + ROH \longrightarrow R'COOR + HCl \quad (18)$$

Alcohols and phenols react with acid chlorides. The reaction is facilitated by the use of a tertiary amine [25], pyridine [26], aluminium alcoholate [27a], or lithium alcoholate [27b] to react with the liberated acid. Tertiary alcohols and phenols give good yields of esters [28]. Acid halides of aromatic polycarboxylic acids [29], olefinic acyl halides, [30a], and others give good yields to esters [30b]. The alcohol portion of the ester may have an epoxy group (glycidyl) or a cyano group, as in glyconitrile formed from formaldehyde and sodium cyanide [29].

It has been reported [1] that vinyl esters are obtained by the reaction of aliphatic and aromatic acid halides with acetaldehyde in the presence of pyridine. It was postulated that acetaldehyde enolizes to vinyl alcohol in the presence of pyridine, which subsequently reacts with the acid halide [31].

$$CH_3CH{=}O \xrightarrow{\text{Pyridine}} [CH_2{=}CHOH] \xrightarrow[\text{pyridine}]{RCOCl} CH_2{=}CH{-}O\overset{O}{\overset{\|}{C}}R \quad (19)$$
$$+ \text{Pyridine hydrochloride}$$

This reaction has also been referred to in *Organic Syntheses,* "Methods of Preparation" section [32].

The present authors have independently tried to repeat these experiments with acetyl chloride, benzoyl chloride, and other acid chlorides as described by Sladkov and Petrov [31] but did not obtain any vinyl esters as determined by careful analysis via gas chromatography. Therefore, this procedure is not recommended as a preparative method but as an area for further research.

2-3. Preparation of tert-Butyl Acetate [33]

$$(CH_3)C-OH + CH_3\overset{O}{\underset{\|}{C}}-Cl \xrightarrow{N,N\text{-Dimethylaniline}}$$

$$(CH_3)_3C-O\overset{O}{\underset{\|}{C}}CH_3 + N,N\text{-Dimethylaniline hydrochloride} \quad (20)$$

To a flask equipped with a reflux condenser, ground glass stirrer (or mercury-sealed stirrer), and a dropping funnel are added 74 gm (1 mole) of *tert*-butyl alcohol, 120 gm (1.0 mole) of dry N,N-dimethylaniline, and 200 ml of dry ether. To the stirred mixture is slowly added dropwise 78.5 gm (1 mole) of acetyl chloride at such a rate that the ether vigorously refluxes. (Cooling may be necessary.) After the addition, the mixture is warmed on a water bath for 2 hr and then allowed to stand for several hours. The ether layer is separated from the N,N-dimethylaniline hydrochloride solid precipitate. The ether layer is extracted with portions of 10% sulfuric acid until the extract does not cloud when made alkaline. The ether layer is dried over anhydrous sodium sulfate and then distilled to yield 63–76% (73–88 gm) *tert*-butyl acetate, bp 93°–98.5°C, n_D^{25} 1.3820.

Under special reaction conditions, the reaction of an acyl halide may be used to prepare esters of glycidol. When the procedure outlined above is used, the tertiary amines catalyze the well-known polymerization of the epoxy group. The reaction is quite sudden and highly exothermic. Therefore, both the acyl halide and the tertiary amine are gradually added to glycidol when ester formation is desired. Care must be taken that no excess of the tertiary amine is present in the reaction flask at any time. In this manner, well-crystallized diglycidyl isophthalate was prepared.

2-4. Preparation of Diglycidyl Isophthalate [25]

$$\underset{\text{COCl}}{\underset{|}{\text{C}_6\text{H}_4}}\text{-COCl} + 2\ \text{HOCH}_2\text{CH}\overset{O}{-}\text{CH}_2 \xrightarrow{(C_2H_5)_3N} \underset{\text{COOCH}_2\text{CH}\overset{O}{-}\text{CH}_2}{\underset{|}{\text{C}_6\text{H}_4}}\text{-COOCH}_2\text{CH}\overset{O}{-}\text{CH}_2 \quad (21)$$

A 1-liter resin kettle is fitted with stirrer, thermometer, condenser with drying tube, and two 300-ml dropping funnels. All the equipment is carefully dried and flushed with nitrogen for 10 min. Glycidol (47.1 gm, 1.0 mole) and 200 ml of benzene are placed in the flask and cooled, with stirring, to 0°C in an ice water–methanol bath. A solution of isophthaloyl chloride (101.5 gm, 0.5 mole) in 150 ml of benzene is placed in one dropping funnel and a solution of triethylamine (101 gm, 1.0 mole) in 150 ml of benzene is placed in the second dropping funnel. The dropwise addition of the acid chloride is begun first, then the dropwise addition of the triethylamine solution is started. The rates are controlled so that the flask temperature does not exceed 5°C and so that the acid chloride addition is slightly faster than that of the triethylamine. Complete addition requires 3 hr. Stirring is continued for 3 hr longer while the flask is allowed to warm to room temperature. The solids (triethylamine hydrochloride) are filtered, rinsed with 50 ml of benzene, and dried. The weight of triethylamine hydrochloride is 123 gm (theoretical weight, 137.5 gm).

The filtrate and benzene rinsings are washed in a separatory funnel with 200 ml of saturated sodium chloride, twice with 200-ml portions of distilled water, and dried over anhydrous calcium chloride. The salt is removed by gravity filtration and the benzene is stripped from the filtrate under reduced pressure using a warm water (40°–45°C) bath. The residue, a white solid, is mixed with petroleum ether and filtered. The crude product weighs 111 gm (theoretical yield, 139 gm) and melts at 46°–53°C. The product is dissolved in 700 ml of a 1:1 petroleum ether–benzene solution, stirred with about 5 gm of activated charcoal, filtered, and cooled. About 25 ml of additional petroleum ether is added. The recrystallized material is filtered and dried in a vacuum oven at room temperature. Final yield: 36 gm (25.9%), mp 60°–63°C.

C. The Reaction of Anhydrides with Hydroxy Compounds

$$(R'CO)_2O + ROH \longrightarrow R'COOR + R'COOH \tag{22}$$

Alcohols and phenols react with anhydrides to give esters, especially when catalyzed by acids such as sulfuric [34] or p-toluenesulfonic acid [35]. Acetic anhydride reacts with *tert*-butyl alcohol to give *tert*-butyl acetate in 60% yield [36]. Phenols are also acetylated in an aqueous alkaline solution in yields above 90% [37, 38]. Cyclic anhydrides of dibasic acids give monoesters upon reaction with alcohols [39–41a].

2-5. Preparation of Methyl Hydrogen Phthalate [40]

$$C_6H_4(CO)_2O + CH_3OH \longrightarrow C_6H_4\begin{matrix}COOH \\ COOCH_3\end{matrix} \tag{23}$$

To a flask are added 74 gm (0.5 mole) of phthalic anhydride and 50 ml (1.25 mole) of absolute methanol. The mixture is refluxed for 2 hr and the excess methanol is removed by distillation. To the residue is added 25 ml of dry benzene, and the distillation is continued to remove the last traces of methanol. The residue is filtered through a cotton plug and then diluted to 300 ml with benzene. Petroleum ether (bp 30°–60°C) is added to give a total volume of 600 ml, and crystallization can now be seen to start. The flask is put in the refrigerator overnight and the next day the product is filtered, washed twice with 50-ml portions of fresh petroleum ether, and air-dried to give 75 gm (83%) of methyl hydrogen phthalate, mp 82°–82.5°C.

D. The Reaction of Halides with Salts of Carboxylic Acids

$$R'X + NaOCR \longrightarrow R'OCR + NaX \qquad (24)$$

(where each OCR group has a C=O)

The reaction of activated halides with sodium acetate or silver acetate leads to good yields of esters. The reaction is not of preparative value for ordinary halides since dehydrohalogenation is a competing reaction, especially with secondary and tertiary halides.

An exception to this has been reported when esters are prepared in good yield by the reaction of aliphatic or aromatic potassium carboxylate salts with alkyl iodides in acetone or water/acetone mixtures (16 hr in refluxing solvent at 56°C) [41b]. Although alkyl iodides only were used, the latter authors suggest that the bromides and chlorides should also be effective.

Benzyl halides [42, 43], 2-(chloromethyl)thiophene [44], and 1,1-dihalocyclopropanes (especially bicyclic 1,1-dibromo compounds) [45] react to give esters.

It has been reported that α-bromoparaffins react with metal salts of fatty acids in the presence of fatty acids at 170°–190°C to give esters [46].

Aromatic allyl esters are prepared in good yield from sodium aryl carboxylates and allyl chloride using dimethylformamide as a solvent to accelerate the rate of reaction [47a–c]. The use of triethylamine in organic solvents has also been reported to be a useful reaction medium [47b].

In a related manner, the Simonini reaction [Eq. (25)] and Prevost reaction [Eqs. (26), (27)] yield esters when the silver carboxylates are treated with iodine [48]. Equimolar amounts of iodine and silver carboxylate yield the Hunsdiecker reaction to give alkyl halides (Eq. 28) [48].

$$2\,RCOOAg + X_2 \longrightarrow RCOOR + CO_2 + 2\,AgX \qquad (25)$$

$$RCOOAg + X_2 + R'CH\!=\!CHR'' \longrightarrow \underset{\underset{OCOR}{|}}{R'CH}\!-\!CHXR'' + AgX \qquad (26)$$

$$2\,RCOOAg + X_2 + R'CH\!=\!CHR'' \longrightarrow \underset{\underset{OCOR}{|}}{R'CH}\!-\!\underset{\underset{OCOR}{|}}{CH}\!-\!R'' + 2\,AgX \qquad (27)$$

$$RCOOAg + X_2 \longrightarrow RX + CO_2 + AgX \qquad (28)$$

For example, silver phenylacetate reacts with iodine in ether exothermically. Removal of the solvent followed by heating the residue at 80°C for 1 hr affords a yield of 68% of benzyl phenylacetate and 10% of phenylacetic acid [49].

Small ring compounds give low yields of mixed products [50].

2-6. Preparation of p-Ethylbenzyl Acetate [43]

$$p\text{-}C_2H_5\text{—}C_6H_4\text{—}CH_2Cl + CH_3COONa \longrightarrow p\text{-}C_2H_5\text{—}C_6H_4\text{—}CH_2O\overset{\overset{O}{\|}}{C}CH_3 \qquad (29)$$

To a flask are added 110 gm (1.34 moles) of fused sodium acetate, 800 ml of glacial acetic acid, and 223 gm (1.44 moles) of p-ethylbenzyl chloride. The mixture is refluxed for ½ hr, cooled, filtered, the salts are washed with 200 ml of glacial acetic acid, and 25 gm of fused sodium acetate is added to the combined filtrates. The filtrates are refluxed for an additional 2½ hr, diluted with water, extracted with benzene, and distilled to yield 238 gm (93%) of p-ethylbenzyl acetate, bp 117°–127°C (14 mm), n_D^{25} 1.5018. Redistillation increased the boiling point to 130°–132°C (15 mm), n_D^{25} 1.5042.

E. Ester Interchange

$$\text{RCOOR}' + \text{R}''\text{OH} \underset{}{\overset{H^+ \text{ or } OH^-}{\rightleftarrows}} \text{RCOOR}'' + \text{R}'\text{OH} \qquad (30)$$

The exchange of alcohol fragments is catalyzed by either acid [51] or base [52]. This reaction involves a reversible equilibrium which can be shifted to the right either by employing a large excess of alcohol R″OH or by removing the lower boiling alcohol R′OH. The low-boiling alcohol can also be removed as an azeotrope, as in the preparation of cellosolve acrylate.

2-7. Preparation of Cellosolve Acrylate [51]

$$CH_2\!=\!CH\text{—}COOCH_3 + HOCH_2CH_2\text{—}OC_2H_5 \longrightarrow$$
$$CH_2\!=\!CH\text{—}COOCH_2CH_2OC_2H_5 + CH_3OH \qquad (31)$$

To a flask equipped with a stirrer, Vigreux column, and distillation head are added 45 gm (0.50 mole) of cellosolve (2-ethoxyethanol), 86 gm (1.0 mole) of methyl acrylate, 2.0 gm of hydroquinone, and 1.0 gm of p-toluenesulfonic acid. The mixture is heated to reflux for about 8 hr. At first the reflux temperature is 81°C and then drops to 65°C. The methanol–methyl acrylate azeotrope is then removed as it is formed at 64°–65°C but not higher. After 9–10 hr, the crude reaction mixture is analyzed by gas chromatography (3 meter Apiezon L 0.3/1.0 on firebrick) which, in this case, indicated the presence of 43% cellosolve acrylate and 15% unreacted cellosolve. Distillation of this material yields 40 gm (56%), bp 174°C, specific gravity 20°/20° 0.9834.

In a similar manner methyl acrylate has been used to prepare esters of a host of higher carbon chain alcohols in good yields using an acid catalyst [52, 53].

In a typical experiment using a basic catalyst, as in the preparation of ethyl benzoate from methyl benzoate, one shakes for 1 hr at room temperature 20 gm (0.147 mole) of methyl benzoate and 100 ml of ethanol containing 0.3 gm (0.05

mole) of potassium ethylate. The mixture is neutralized with dilute sulfuric acid and extracted with ether. The ether layer is washed with a dilute sodium carbonate solution and then twice with water. The ether layer is dried, concentrated, and the residue distilled to yield 96% of ethyl benzoate [54].

The relative exchange tendency of a series of primary and secondary alcohols has been determined [55]. Ethyl esters of terephthalic [56], succinic, malonic, and oxalic acids are prepared from their corresponding methyl esters. However, dimethyl phthalate has been found not to react.

Some specific catalysts useful for the ester interchange reaction have been found to be the alkyl orthotitanates [57], sodium hydroxide [58], and alkali metal alcoholates [59].

The orthotitanates catalyze the ester interchange reaction more rapidly and at lower temperatures than do the titanium ester acrylates or the condensed esters of titanium such as propyl polytitanate [59].

Exchange reactions can also involve an ester and an acid [Eq. (32)].

$$\text{RCOOR'} + \text{R''COOH} \rightleftharpoons \text{R''COOR'} + \text{RCOOH} \tag{32}$$

For example, methyl acrylate reacts with formic acid under acid catalysis (sulfuric acid) to give methyl formate and acrylic acid [60]. The advantage of this method is that it permits the preparation of acids free from water. This general method has found applicability in preparing vinyl esters from vinyl acetate and carboxylic acid in the presence of catalytic amounts of sulfuric acid and mercuric acetate [61].

$$\text{CH}_3\text{COOCH}=\text{CH}_2 + \text{RCOOH} \rightleftharpoons \text{CH}_3\text{COOH} + \text{RCOOCH}=\text{CH}_2 \tag{33}$$

Mercuric oxide has also been described as a catalyst [62].

A major problem of this method of preparing a vinyl ester arises from the fact that in all cases, except in the preparation of vinyl formate and vinyl trifluoroacetate, the desired product is higher boiling than the starting ester. Consequently, the essential starting reagent (vinyl acetate) is removed from the reacting system during the product isolation step. This tends to drive the equilibrium in the direction of the starting materials.

Isopropenyl acetate behaves in a manner similar to vinyl acetate on reacting with carboxylic acids in the presence of mercury salts and sulfuric acid catalysts [63, 64a].

The use of alcohols with other functional groups such as alkyl thio groups has also been reported to be effective in transesterifications [64b].

2-8. Preparation of Vinyl Caproate [65a]

$$\text{CH}_2=\text{CH}-\text{O}\overset{\text{O}}{\underset{\|}{\text{C}}}\text{CH}_3 + \text{CH}_3(\text{CH}_2)_4\text{COOH} \rightleftharpoons$$

$$\text{CH}_2=\text{CHO}\overset{\text{O}}{\underset{\|}{\text{C}}}(\text{CH}_2)_4\text{CH}_3 + \text{CH}_3\text{COOH} \tag{34}$$

To a flask equipped with a stirrer, thermometer, reflux condenser, and a nitrogen gas inlet tube are added 215 gm (2.5 moles) of vinyl acetate and 51 gm (0.5 mole) of caproic acid. The flask is warmed to dissolve the caproic acid, and then 2.0 gm of mercuric acetate (about 4% based on the carboxylic acid reactant) is added. The mixture is stirred or agitated manually for ½ hr and 0.2 ml of 100% sulfuric acid (prepared by carefully mixing 7.3 gm of fuming sulfuric acid which contains 30% SO_3 and 10 gm of 95% H_2SO_4) is added. The solution is refluxed for 3½ hr and then neutralized with sodium acetate trihydrate. The excess vinyl acetate and acetic acid are distilled off at atmospheric pressure at 70°–80°C until the distillation flask temperature reaches 125°C. The distillation is continued under reduced pressure at 100 mm to give 25.6 gm (40%), bp 98°–99°C, n_D^{30} 1.4159.

The transesterification process is also useful for preparing linear polyesters [65b].

F. The Reaction of Carboxylic Acids with Olefins

$$RCOOH + CH_2=C(CH_3)_2 \longrightarrow RCOOC(CH_3)_3 \quad (35)$$

The preparation of esters of tertiary alcohols is accomplished by the reaction of an olefin with a carboxylic acid. For example, isobutylene condenses with malonic acid to give 58–60% of di-*tert*-butyl malonate [66], with monoethyl maleate to give 53–58% of ethyl *tert*-butyl malonate [67], and with acrylic acid to give 47% *tert*-butyl acrylate [51, 68].

Acids also have been esterified with propene [3] and trimethylethylene [51]. These reactions need to be carried out under strictly anhydrous conditions and they are acid-catalyzed (H_2SO_4, or BF_3).

Carboxylic acids also add to acetylenes to give alkenyl esters. The commercial production of vinyl esters involves this synthesis.

$$R'COOH + RC\equiv CH \longrightarrow R'COOCR=CH_2 \quad (36)$$

One of the catalysts is BF_3—HgO. See Chapter 2, Olefins, for a preparative example involving this method.

2-9. Preparation of **tert-**Butyl Acetate [69]

$$CH_3COOH + CH_2=C(CH_3)_2 \longrightarrow CH_3COOC(CH_3)_3 \quad (37)$$

To a 500-ml Pyrex pressure bottle containing 26 gm (0.45 mole) of glacial acetic acid and 2 ml of concentrated sulfuric acid is added 50 gm (0.89 mole) of liquid isobutylene (liquified by the passage of the gas through a Dry Ice trap). The bottle is stoppered and allowed to remain at room temperature over night. Then it is chilled in an ice–salt water bath, opened, and poured into a cold solution of 40 gm (1.0 mole) of sodium hydroxide in 500 ml of ice water. The organic layer is separated, washed with dilute alkali, dried over potassium carbonate, and distilled through a 6-inch Vigreux column to give 26.5 gm (53%) of tert-butyl acetate, bp 94°–97°C (738 mm), n_D^{25} 1.3820.

2-10. Preparation of **tert-Butyl Acrylate** [51]

$$CH_2{=}CH{-}COOH + CH_2{=}C(CH_3)_2 \xrightarrow{H^+} CH_2{=}CHCOOC{-}(CH_3)_3 \quad (38)$$

Method A. To a 1-liter Pyrex pressure bottle or a Hoke steel pressure cylinder are added 62.4 gm (0.867 mole) of glacial acrylic acid and 4 ml of concentrated sulfuric acid. The bottle or cylinder is cooled in an isopropanol–Dry Ice bath while 100 gm (1.96 mole) of liquid isobutylene (liquified by passage through a Dry Ice trap) is added. The bottle or cylinder is sealed, allowed to stand at room temperature for 24 hr, vented carefully in a hood, opened, and the contents poured into a separatory funnel containing 300 ml of a saturated sodium carbonate solution. The crude organic layer (96 gm. 96%) is separated and dried over sodium sulfate. (A vapor chromatograph of the crude material on a 1-meter silicone no. 200 firebrick at 125°C column indicated it to be 99% pure *tert*-butyl acrylate.) Distillation through a 6-inch Vigreux column gives 57 gm (57%), bp 30°–31°C (25 mm), n_D^{25} 1.4092; reported [70] bp 117°–120°C (759 mm).

Method B. To a 500-ml round bottomed flask equipped with a condenser and a gas inlet tube are added 62.4 gm (0.867 mole) of glacial acrylic acid and 4 ml of concentrated sulfuric acid. A slow stream of isobutylene is bubbled into the solution at room temperature for abour 4½ hr until 56 gm (1.0 mole) has been added. The reaction mixture is then worked up as in Method A. The crude product weighs 93.6 gm and after being distilled yields 58 gm (58%), bp 33°–34°C (30 mm), n_D^{25} 1.4092.

Both samples were identical to a sample prepared from the acid chloride and *tert*-butyl alcohol.

tert-Butyl acrylate has also been prepared from *tert*-butyl alcohol and acrylic acid [70], from acetylene and *tert*-butyl alcohol [71–74], and from isobutylene and acrylic acid [75a].

Recently the process for preparing long chain esters of carboxylic acids has been reported to be effected by condensing olefins and carboxylic acids in the presence of a manganic carboxylic acid salt or oxide, a zirconyl carboxylic acid salt, or zirconium oxide [75b].

G. Alkylation Reactions

$$RCH_2{-}COOR \xrightarrow[\text{Base}]{R'X} RCHR'{-}COOR \quad (39)$$

Esters with a α-hydrogen undergo alkylation with primary aliphatic bromides [76], diethyl sulfate [77], and ethyl *p*-toluenesulfonate [78]. Malonic ester [79]

§ 2. Condensation Reactions

can be alkylated with ethylene bromide or trimethylene dibromide to give ring closure to give diesters of 1,1-cyclopropane and 1,1-cyclobutanedicarboxylic acids [80, 81]. In addition, five- and six-membered rings have also been formed in this manner [82a].

The condensation of methyl bromoacetate with methyl nitroacetate is effected by methanolic sodium methoxide in dimethylacetamide as solvent to give dimethyl nitrosuccinate in 23–27% yields [82b].

Dimethylformamide has been recommended as a useful solvent for the alkylation of enolate anions since is shows an accelerating effect on the rate of alkylation [83, 84].

The base-catalyzed alkylation of olefins with an active hydrogen compound is known as the Michael reaction. Compounds which furnish an active hydrogen atom are malonic, cyanoacetic, and acetoacetic esters, nitroparaffins [85], benzyl cyanide [86], malononitrile [87], cyanoacetamide [88], sulfones [89], ketones [90], and methylpyridines [91]. The olefinic compound taking part in this reaction may be one in which the double bond is in an α-position to an ester [92], aldehyde [93], ketone [94], cyananide [95], nitro compound [96], or sulfone [97]. α- and γ-vinylpyridine also undergo this reaction [98a].

$$CH_3NO_2 + \underset{\underset{C(COOC_2H_5)_2}{\parallel}}{CH-CH_3} \xrightarrow{C_6H_5CH_2\overset{+}{N}(CH_3)_3OH} \underset{NO_2}{CH_2}-\underset{CH(COOC_2H_5)_2}{CH}-CH_3 \quad (40)$$

Grignard reagents also add to activated double bonds such as ethyl acrylate to give conjugate addition products [98b].

2-11. Preparation of Diethyl sec-Butylmalonate [79]

$$CH_2(COOC_2H_5)_2 + NaOC_2H_5 \longrightarrow$$

$$\underset{Na^+}{^-CH-(COOC_2H_5)_2} \xrightarrow{\overset{CH_3}{\underset{C_2H_5}{\diagdown}}CH-Br} \overset{CH_3}{\underset{C_2H_5}{\diagdown}}CHCH(COOC_2H_5)_2 \quad (41)$$

To a 2-liter three-necked flask equipped with a wide bore condenser, drying tube, dropping funnel, and a mercury-sealed stirrer is added 700 ml of absolute alcohol followed by the gradual addition of 35 gm (1.52 gm atom) of sodium metal. When the sodium has reacted, the stirred flask is warmed with a hot water bath or a steam cone while 250 gm (1.56 moles) of diethyl malonate is added in a steady stream. After the ester addition, 210 gm (1.53 moles) of sec-butyl bromide is added at such a rate as to maintain the mixture refluxing. The mixture is stirred and refluxed for an additional 48 hr, distilled to remove alcohol, treated with 200 ml of water, shaken, and the ester layer is separated. Distillation through a Vigreux column affords 274–278 gm (83–84%), bp 110°–120°C (18–20 mm).

Diethyl *tert*-butylmalonate has been prepared by the condensation of diethylamalonate with acetone and subsequent reaction with methyl magnesium iodide [98c].

H. The Preparation of Lactones

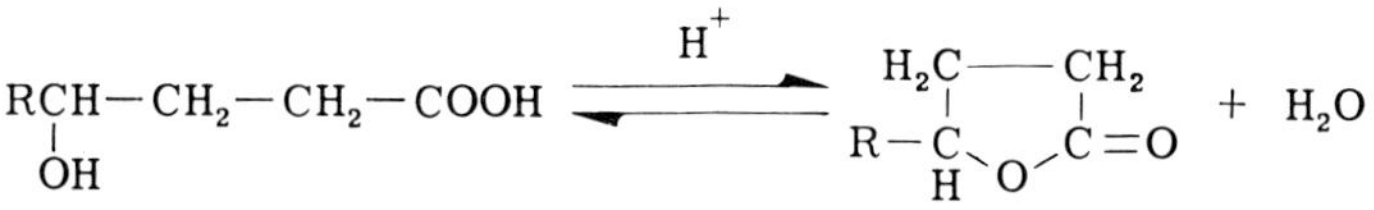

$$RCH(OH)-CH_2-CH_2-COOH \xrightleftharpoons{H^+} \underset{R-CH}{\overset{H_2C-CH_2}{\underset{O}{|\quad\quad|}}}C=O + H_2O \quad (42)$$

Intramolecular esterification of hydroxy acids yields lactones. The reaction is catalyzed by hydrogen ion. Removal of the water formed favors the reaction, especially from γ- and δ-hydroxy acids [99]. However, β-hydroxy acids do not form lactones directly by this method. ε-Hydroxy acids cyclize only with difficulty.

Another method of preparation of lactones is by the conversion of β,γ-olefinic acids to γ-lactones using boiling 50% sulfuric acid [100].

$$R-CH=CH-CH_2COOH \xrightarrow[H_2SO_4]{H_2O} RCH-CH_2-CH_2-C=O \quad (43)$$
$$\underset{\quad\quad\quad\quad\quad O \quad\quad\quad\quad\quad}{\underbrace{\quad\quad\quad\quad\quad\quad\quad\quad\quad\quad\quad}}$$

The reaction of halo acids with base also yields γ-lactones [101]. In addition silver salts of halo acids yield lactones [102a].

$$X-CH_2CH_2-CH_2COOH \xrightarrow{Base} CH_2-CH_2-CH_2-C=O \quad (44)$$

2-12. Preparation of α-Ethylbutyrolactone [99]

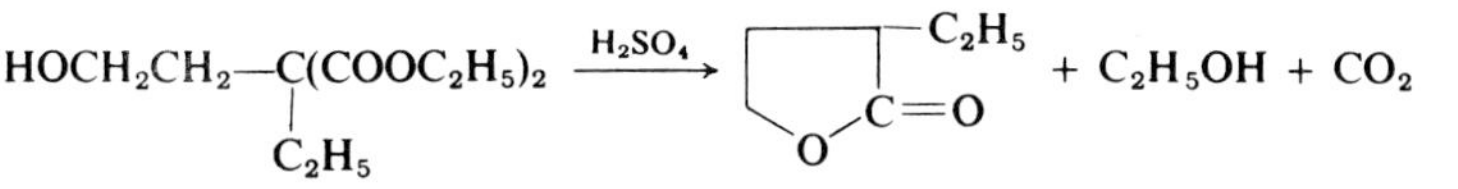

$$HOCH_2CH_2-C(COOC_2H_5)_2(C_2H_5) \xrightarrow{H_2SO_4} \text{(lactone with } C_2H_5, C=O) + C_2H_5OH + CO_2 \quad (45)$$

To a heated 2-liter flask containing a boiling solution of 100 gm (2.5 moles) of sodium hydroxide in 200 ml of water is added from a dropping funnel 364 gm (1.57 moles) of ethyl β-hydroxyethyldiethylmalonate at such a rate that two layers do not form. After the addition has been completed, a mixture of 110 ml of concentrated sulfuric acid in 150 ml water is slowly added. The reaction mixture is refluxed for 5 hr, and the lactone which separates is removed. The water layer is extracted with 75 ml of benzene and then concentrated to 250 ml by distillation. To concentrated water solution is extracted again with two 75-ml portions of benzene. The combined product layer and benzene extracts are dried over sodium sulfate and distilled to yield 156 gm (88%) of α-ethylbutyrolactone, bp 213°–216°C (740 mm).

Recently the preparation of γ-butyrolactones has been reported by the photochemical addition of alcohols to unsaturated esters [102b].

I. The Reaction of Lactones with Alcohols

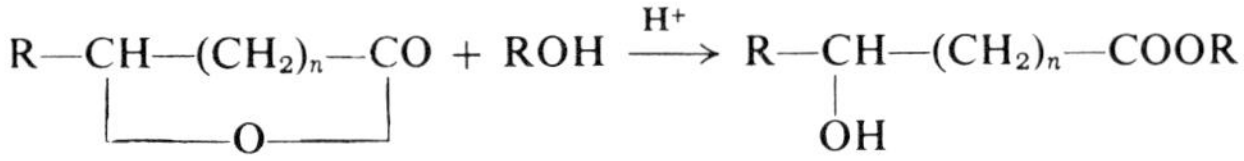

$$\underset{\underset{O}{\underbrace{\qquad\qquad}}}{R-CH-(CH_2)_n-CO} + ROH \xrightarrow{H^+} R-CH-(CH_2)_n-COOR \underset{OH}{|} \qquad (46)$$

The reaction of lactones with alcohols yields hydroxy esters. For example, α-ethyl-γ-butyrolactone reacts with ethanol to give ethyl α-ethyl-γ-hydroxybutyrate in 48% yield [99]. By use of hydrogen halides and alcohols the appropriate halo esters are prepared [103].

2-13. *Preparation of Ethyl α-Ethyl-γ-hydroxybutyrate* [99]

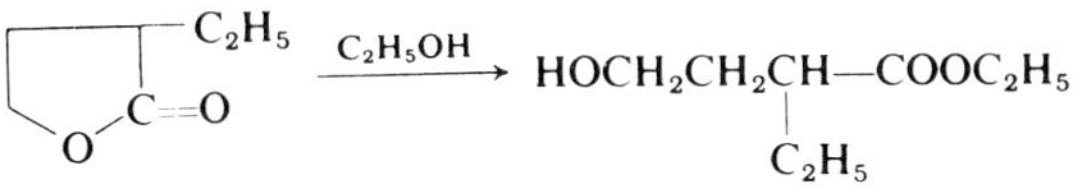

$$\xrightarrow{C_2H_5OH} HOCH_2CH_2CH-COOC_2H_5 \underset{C_2H_5}{|} \qquad (47)$$

To a flask containing 500 ml of absolute ethanol is added 150 gm (1.32 moles) of α-ethylbutyrolactone. The mixture is saturated with dry hydrogen chloride and then allowed to stand for 3 days. The alcohol is removed under reduced pressure and the residue is poured into 500 ml of ice water. The organic layer is separated and the water layer is extracted with five 50-ml portions of ether. The combined organic layer and ether layers are dried over sodium sulfate and distilled to yield 175 gm (84%) of ethyl α-ethyl-γ-hydroxybutyrate, bp 78–80°C (8 mm).

J. Miscellaneous Condensation Reactions

a. THE ACID-CATALYZED REACTION OF NITRILES WITH ALCOHOLS

$$RCH_2CN + R'OH \xrightarrow{H^+} RCH_2COOR' \qquad (48)$$

Nitriles are converted to esters by heating with an alcohol containing an acid catalyst (H_2SO_4, etc.). The reaction is quite general for the aliphatic, aromatic, and heterocyclic cyano compounds [104].

Alkyl acrylates [105] are made in this manner by the reaction of ethylene cyanohydrin with alcohols. The latter dehydrates under the experimental conditions to give the acrylate bond.

$$\text{HO-CH}_2\text{CH}_2\text{CN} + \text{R OH} \xrightarrow{\text{H}^+} \text{CH}_2\!\!=\!\!\text{CHCOOR}' \qquad (49)$$

b. The Reaction of Diazomethane with Carboxylic Acids

$$\text{RCOOH} + \text{CH}_2\text{N}_2 \longrightarrow \text{RCOOCH}_3 + \text{N}_2 \qquad (50)$$

This is an excellent method for the preparation of small amounts of methyl esters in high yields from difficult-to-prepare acids. The reaction usually is conducted in an ethereal solution at room temperature and the completion of the reaction is evidenced by the cessation of the evolution of nitrogen. The yellow color of diazomethane also disappears at the end of the reaction. The disadvantage of the method is that it requires diazomethane, which must be handled with great care in dilute solutions [106a].

c. Esterification of Carboxylic Acids with Trialkyloxonium Salts

$$\text{R}'\text{COOH} + \text{R}_3\overset{+}{\text{O}}\text{BF}_4^- \longrightarrow \text{R}'\text{COOR} \qquad (51)$$

This is a mild and convenient method for the preparation of esters from sterically hindered carboxylic acids using equivalent amounts of trialkyloxonium tetrafluoroborate in dichlormethane. The reaction proceeds in almost quantitative yield [106b].

Other possibly useful methylating agents for carboxylic acids are trimethylsulfonium hydroxide [106c] and pentamethoxyphosphorane [106d].

d. The Reformatskii Reaction

The Reformatskii reaction is useful for the formation of α, β-unsaturated esters by the condensation of α-haloesters with an aldehyde or ketone in the presence of zinc metal [106e].

e. The Diels–Alder Reaction

The condensation of dienes with dieneophiles such as unsaturated esters of maleic or fumaric acids leads to cyclic unsaturated diesters. In some cases dienes are not necessary, as in the case of sulfolene [106f].

3. OXIDATION REACTIONS

A. Oxidation of Primary Alcohols

The oxidation of primary alcohols in acid media is in many cases accompanied by esterification. For example, the chromic acid oxidation of n-butyl alcohol yields n-butyl n-butyrate [107].

§ 3. Oxidation Reactions

3-1. Preparation of n-Butyl n-Butyrate (Dichromate Oxidation) [107, 108]

$$6\ CH_3(CH_2)_2CH_2OH + 2\ Na_2Cr_2O_7 + 8\ H_2SO_4 \longrightarrow$$
$$3\ CH_3CH_2CH_2COOCH_2(CH_2)_2CH_3 + 2\ Cr_2(SO_4)_3 + 2\ Na_2SO_4 + 14\ H_2O \quad (52)$$

In a four-necked flask equipped with a stirrer, thermometer, condenser, and a separatory funnel is placed 243 ml (4.3 moles) of concentrated sulfuric in 240 ml of water. To this mixture is added 240 gm (3.25 moles) of technical grade n-butyl alcohol.

With cooling is added a solution of 320 gm (1.07 moles) of crystalline sodium dichromate in 200 ml of water at such a rate as to maintain the temperature at 20°C.

The resulting solution is diluted with an equal volume of water, whereupon an oil separates. The oil is washed three times with water, separated, and then treated with some Na_2SO_4 in order to remove the last traces of water. The crude oil is distilled to yield 170–175 ml of the desired product, boiling at 150°–170°C (5 mm).

The crude ester is washed with five 15-ml portions of 60% sulfuric acid, then with sodium hydroxide, dried, and redistilled to yield 96–110 gm (41–47%). bp 162°–166°C (5 mm). This material contains a small amount of n-butyl alcohol. If a higher purity is desired, the fractionation should be repeated several times.

B. The Tischtschenko–Cannizzaro Reaction of Aldehydes

The Tischtschenko reaction, like the Cannizzaro reaction, involves an intramolecular oxidation–reduction reaction which is applicable to most aldehydes lacking an α-hydrogen atom [109]. In some cases, as for example n-butyraldehyde, the aldehyde has an α-hydrogen atom and still undergoes the Tischtschenko reaction as described below.

$$2\ RCH{=}O \longrightarrow RCOOCH_2R \quad (53)$$

The most suitable catalyst is aluminum alkoxide and only a few mole percent is required. More basic catalysts yield an aldol condensation reaction. Mildly basic catalysts such as $Mg(OC_2H_5)_2$ or $Ca(OC_2H_5)_2$ lead to the formation of a trimeric glycol ester $RCH_2CH(OH)CHR{-}CH_2OCOCH_2R$ in preference to the simple esters. Examples of the Tischtschenko reaction are the preparation of benzyl benzoate [110] and furfuryl furoate [111] from benzaldehyde and furfural, respectively.

3-2. Preparation of n-Butyl n-Butyrate [109]

$$2\ CH_3CH_2CH_2CH{=}O \xrightarrow{Al(OC_2H_5)_3} CH_3CH_2CH_2\overset{\overset{\displaystyle O}{\|}}{C}{-}OCH_2CH_2CH_2CH_3 \quad (54)$$

To an Erlenmeyer flask are added 144 gm (2.0 moles) of *n*-butyraldehyde and 7.2 gm of aluminum ethoxide. The flask is stoppered and quickly placed in an ice water bath for several hours. Then it is allowed to stand at room temperature for 48 hr. The reaction mixture is fractionally distilled under reduced pressure without further treatment to yield 117.5 gm (81.6%) of *n*-butyl *n*-butyrate, bp 150°–170°C (5 mm). The crude material is washed with 10% sodium carbonate solution, extracted with ether, dried over sodium sulfate, and fractionally distilled to yield a purified material, bp 162°–166°C (5 mm).

C. Direct Oxidation of Aldehydes and Ketones

The direct oxidation of aldehydes and ketones with peracids yields esters by the addition of an oxygen atom between the C—C bonds [112, 113a].

$$R-\overset{O}{\underset{\|}{C}}-R' \xrightarrow{RCOOOH} R\overset{O}{\underset{\|}{C}}-O-R' \tag{55}$$

3-3. Preparation of Ethyl 6-Hydroxyhexoate (Caro's Acid Oxidation) [113a]

$$\text{cyclohexanone} \xrightarrow[\text{2. } C_2H_5OH + H_2SO_4]{\text{1. } K_2S_2O_8 + H_2SO_4} C_2H_5O\overset{O}{\underset{\|}{C}}-(CH_2)_4-CH_2OH \tag{56}$$

To a well-stirred mixture of 71 ml concentrated sulfuric acid in 24 ml of water at 10°–12°C is added 50 gm (0.185 mole) of potassium persulfate. Then 100 ml of ethanol is added at 10°–12°C. While keeping the temperature at 10°C, 9.5 gm of (0.097 mole) cyclohexanone in 15 ml of ethanol is added over a 1-hr period. After another 15 min, the cooling bath is lowered and the mixture is stirred for 1 hr. The mixture is diluted to 1 liter with water, filtered, saturated with ammonium sulfate, and extracted with ether. After removing the ether, the residue is mixed with 25 ml of ethanol and 2 ml of sulfuric acid and refluxed for 6 hr. The ester is extracted with ether and upon distillation yields 7.4 gm (50%), bp 134°C (15 mm).

The oxidation of allylic alcohols by active manganese dioxide leads to conjugated aldehydes. In the presence of HCN and CN⁻ the aldehyde is converted to the cyanohydrin and is further oxidized by manganese dioxide to an acyl cyanide leading in an alcoholic medium to an ester [113b].

D. Oxidation of Ethers

Ruthenium tetroxide oxidizes ethers at 10°–15°C to esters in quantitative yield. *n*-Butyl ether, when treated this way, is reported to give a quantitative yield of *n*-butyl *n*-butyrate and tetrahydrofuran gives γ-butyrolactone [114].

4. REDUCTION REACTIONS

The reduction of aromatic esters to cycloaliphatic esters with hydrogen using platinum occurs at low temperatures and pressures [115]. The use of Raney nickel catalysts allows one to reduce phenolic esters in alcohol solution when catalyzed by sodium ethoxide [116].

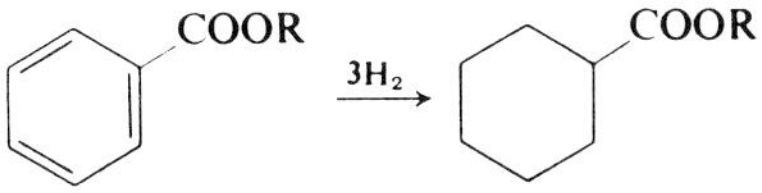

$$\text{C}_6\text{H}_5\text{COOR} \xrightarrow{3H_2} \text{C}_6\text{H}_{11}\text{COOR} \tag{57}$$

Olefinic esters may also be hydrogenated over platinum [117], palladium [118], or nickel [119] catalysts in preference to the ester function.

$$\text{RCH=CHCOOR} \xrightarrow[\text{Cat.}]{H_2} \text{RCH}_2\text{CH}_2\text{COOR} \tag{58}$$

5. REARRANGEMENTS

A. The Arndt–Eistert Rearrangement

The preparation of diazoketones and their solvolysis with anhydrous alcohols in the presence of silver oxide leads to esters with one carbon atom more than the starting acid halide.

$$\text{RCOCl} + \text{CH}_2\text{N}_2 \longrightarrow \text{RCOCHN}_2 \xrightarrow[\text{Ag}_2\text{O}]{\text{R'OH}} \text{RCH}_2\text{COOR'} + \text{N}_2 \tag{59}$$

The progress of the reaction can be followed by the amount of nitrogen evolved [120, 121].

Benzyl esters are prepared by heating the diazoketone with benzyl alcohols in the presence of a tertiary amine [122].

An example of a preparation using the Arndt–Eistert reaction is given in Chapter 9, Carboxylic Acids.

B. The Favorskii Rearrangement

α-Halo ketones react with sodium alkoxides and rearrange in anhydrous ether to esters [123, 124].

Dibromoketones, under similar conditions, give α,β-olefinic esters in good yields [125a, b].

5-1. Preparation of Methyl 1-Methylcyclohexanecarboxylate [123]

$$\text{CH}_3-\underset{\underset{O}{\|}}{C}-\overset{Br}{\underset{}{\bigcirc}} \xrightarrow[\text{Ether}]{\text{CH}_3\text{ONa}} \text{CH}_3\text{O}-\underset{\underset{O}{\|}}{C}-\overset{\text{CH}_3}{\underset{}{\bigcirc}} \quad (60)$$

To a round-bottomed flask fitted with a mercury-sealed stirrer and containing 11.3 gm (0.21 mole) of sodium methoxide (from a freshly opened bottle) suspended in ether is added 42 gm (0.21 mole) of methyl α-bromocyclohexyl ketone over a period of 1 to 3 hours at 0°C. The reaction mixture is stirred for an additional 2–8 hr under anhydrous conditions. Water is added and the ether layer separated, dried over sodium sulfate, and the ether distilled through a 3-ft helix-packed column. The residue is fractionated to yield 25.9 gm (79%) bp 35°C (3 mm), 1.4456.

The chemistry of other rearrangements involving esters has recently been covered by Kwart and King [125c].

6. MISCELLANEOUS METHODS

(1) The reaction of ketenes with hydroxy compounds [126].
(2) The reaction of phosgene with hydroxy compounds [127, 128].
(3) The reaction of alkyl chlorosulfites or alkyl sulfates with carboxylic acid salts [129, 130].
(4) The cleavage of α-keto esters [131].
(5) The cleavage of β-keto esters [132].
(6) Decarboxylation of alkyl hydrogen malonates [133].
(7) The reaction of diazoketones with carboxylic acids [134].
(8) Pyrolysis of tetramethylammonium salts of carboxylic acids [135].
(9) The reaction of aldehydes with anhydrides or acyl halides [136, 137].
(10) Acid halide cleavage of ethers [138].
(11) The reaction of diazoacetic esters with olefins [139, 140].
(12) The reaction of diazonium salts with carboxylic acids [141].
(13) Reduction of α- and β-keto esters [142].
(14) Reduction of acyl chloromalonates [143].
(15) Alcoholyses of benzotrihalides [144].
(16) The Darzens reaction [145].
(17) The preparation of vinyl acetate by the oxyacetylation of ethylene [146].
(18) The preparation of aromatic esters by the Friedel–Crafts reaction of aromatics and diaroyl peroxides [147, 148].
(19) The preparation of lactones via the hydroformylation of unsaturated acids [149].

(20) The preparation of lactones via the hydroformylation of unsaturated alcohols [150].

(21) Oxidative cleavage of cyclopropene by lead tetraacetate and thallium triacetate [151].

(22) Addition of acetyl hypobromite to styrene and its derivatives [152].

(23) Metal halide-catalyzed orthoester formation starting with chloroform or carbon tetrachloride and alcohols [153].

(24) Esterification of carboxylic acids (including hindered acids) with triethyl orthoformate [154].

(25) A method of esterification of hindered acids using trifluoracetic anhydride [155].

(26) The reaction of diethyl malonate with styrene oxide [156].

(27) The transesterification of lower vinyl esters with carboxylic acids of high molecular weight using salts of metals of the platinum family as catalysts in anhydrous medium [157].

(28) Carboxylation of allene [158].

(29) The synthesis of coumarins [159].

(30) Irreversibility of benzilic ester rearrangement [160].

(31) The preparation and hydrolytic stability of trialkylacetic acid esters [161].

(32) The Prins reaction [162].

(33) Reaction of olefins with silver salts and iodine (the Prévost reaction) [163].

(34) The reaction of ethyl bromoacetate with organoboranes. A convenient procedure for the conversion of olefins to esters [164].

(35) Carboxylation of alcohols to give carboxylic acids and esters [165].

(36) Claisen rearrangement via enolate formation [166].

(37) Carboxylation of olefins to give carboxylic acid esters [167].

(38) Carboxylation of ethers to give carboxylic acid esters [168].

(39) Preparation of di-*t*-butyl dicarbonate [169].

(40) Preparation of ethyl pyrrole-2-carboxylate [170].

(41) Acylation of chloral hemiacetals [171].

(42) Preparation of ethyl thiazole-4-carboxylate [172].

(43) γ-Ketoesters from aldehydes via diethyl acylsuccinates [173].

(44) Photoinduced alcoholysis of the trichloroacetyl group [174].

(45) Partial cleavage of diesters by lithium in pyridine [175].

REFERENCES

1. A. I. Vogel, *J. Chem. Soc.* pp. 624, 644, 654 (1948).
2. G. H. Jeffery and A. I. Vogel, *J. Chem. Soc.* pp. 658, 674 (1948).
3. T. J. Thompson and G. J. Leuck, *J. Am. Chem. Soc.* **44**, 2894 (1922).
4. A. Weissberger and C. J. Kibler, *Org. Synth. Collect. Vol.* **4**, 610 (1955).
5. M. L. Peterson and J. W. Way, *Ind. Eng. Chem.* **50**, No. 9, Part II, 1335 (1958).
6. M. L. Peterson and J. W. Way, *Ind. Eng. Chem.* **51**, No. 9, Part II, 1081 (1959).

7. E. J. Bourne, M. Stacey, J. C. Tatlow, and J. M. Tedder, *J. Chem. Soc.* p. 2976 (1949).
8. B. F. Goodrich Co., British Patent 810,381 (1959); *Chem. Abstr.* **53**, 15982 (1959).
9. B. V. Emel'yabov and L. M. Polyaninova, *Tr. Khim. Khim. Tekhnol.* **1**, 670 (1958); *Chem. Abstr.* **54**, 7003 (1960).
10. B. A. Dombrow and G. Fasman, U.S. Patent 3,162,616 (1964).
11. J. C. Shivers, Canadian Patent 563,070 (1958).
12. J. R. Whinfield, *Nature (London)* **158**, 930 (1946).
13. J. R. Whinfield and J. T. Dickson, British Patent 578,079 (1946).
14. A. Kibler, C. J. A. Bell, and J. G. Smith, U.S. Patent 2,901,466 (1959).
15. T. Granstad and R. N. Haszeldine, *J. Chem. Soc.* p. 4069 (1957).
16. N. V. Kutepow and W. Himmele, German Patent 1,014,979 (1957).
17. Esso Research and Engineering Co., German Patent 1,007,757 (1957).
18. M. S. Newman, *J. Am. Chem. Soc.* **63**, 2431 (1941).
19. K. Hofmann, *J. Am. Chem. Soc.* **67**, 421 (1945).
20. R. O. Clinton and S. C. Laskowski, *J. Am. Chem. Soc.* **70**, 3135 (1948).
21. J. R. Ruhoff, *Org. Synth. Collect. Vol.* **2**, 292 (1943).
22. W. C. Ault, J. K. Weil, G. C. Nutting, and J. C. Cowan, *J. Am. Chem. Soc.* **69**, 2003 (1947).
23. C. F. H. Allen and F. W. Spangler, *Org. Synth. Collect. Vol.* **3**, 203 (1955).
24a. W. H. Puterbaugh, C. M. Vanselow, K. Nelson, and E. J. Shrawder, *J. Chem. Educ.* **40**, No. 7, 349 (1963).
24b. A. V. R. Rao, M. N. Deshmukh, and L. Sivadasan, *Chem. Ind. (London)* p. 164 (1981).
25. S. R. Sandler and F. Berg, *J. Chem. Eng. Data* **11**, 447 (1966).
26. C. R. Hauser, *Org. Synth.* **24**, 19 (1944).
27a. N. O. V. Sonntag, U.S. Patent 2,801,256 (1959).
27b. G. P. Crowther, E. M. Kaiser, R. A. Woodruff, and C. R. Hauser, *Org. Synth.* **51**, 96 (1971).
28. S. G. Cohen and A. Schneider, *J. Am. Chem. Soc.* **63**, 3386 (1941).
29. D. T. Mowry, *J. Am. Chem. Soc.* **66**, 371 (1944).
30a. E. B. Womack and J. McWhirter, *Org. Synth.* **20**, 77 (1940).
30b. G. R. Jurch, Jr., M. D. Johnston, Jr., J. W. Perry, and T. E. Detty, *J. Chem. Educ.* **57**, 743 (1980).
31. A. M. Sladkov and G. S. Petrov, *Zh. Obshch. Khim.* **24**, 450 (1954); *Chem. Abstr.* **49**, 6093 (1955).
32. D. Swern and E. F. Jordan, Jr., *Org. Synth. Collect. Vol.* **4**, 977 (1963).
33. B. Abramovitch, J. C. Schrivers, B. E. Hudson, and C. R. Hauser, *J. Am. Chem. Soc.* **65**, 986 (1943).
34. W. W. Prichard, *Org. Synth.* **28**, 68 (1948).
35. A. C. Cope and E. C. Herrick, *Org. Synth. Collect. Vol.* **4**, 304 (1963).
36. R. H. Baker and F. G. Bordwell, *Org. Synth.* **24**, 18 (1944).
37. F. D. Chattaway, *J. Chem. Soc.* p. 2495 (1931).
38. E. R. Marshall, J. A. Kuck, and R. C. Elderfield, *J. Org. Chem.* **7**, 450 (1942).
39. J. Cason, *Org. Synth.* **25**, 19 (1945).
40. E. L. Eliel and A. W. Burgstahler, *J. Am. Chem. Soc.* **71**, 2252 (1949).
41a. R. Adams and J. M. Wilkinson Jr., *J. Am. Chem. Soc.* **65**, 2207 (1943).
41b. G. G. Moore, T. A. Foglia, and T. J. McGahan, *J. Org. Chem.* **44**, 2425 (1979).
42. I. D. Tharp, H. Nottorf, C. H. Herr. T. B. Hoover, R. B. Wagner, C. A. Weisgerber, J. P. Wilkins, and F. C. Whitmore, *Ind. Eng. Chem.* **39**, 1300 (1947).
43. W. S. Emerson, J. W. Heyd, V. E. Lucas, W. I. Lyness, G. R. Owens, and R. W. Shortridge *J. Am. Chem. Soc.* **69**, 1906 (1947).
44. W. S. Emerson and T. M. Patrick, *J. Org. Chem.* **14**, 792 (1949).

45. P. S. Skell and S. R. Sandler, *J. Am. Chem. Soc.* **80** 2024 (1958); S. R. Sandler, *J. Org. Chem.* **32,** 3876 (1967).
46. G. Hogele, German Patent 1,035,640 (1958).
47a. K. C. Tsou, S. R. Sandler, and A. Astrup, U.S. Patent 3,069,459 (1962).
47b. A. Suzai, Japan Kokai 76/08,212 (1976); *Chem. Abstr.* **84,** 180853 (1976).
47c. N. H. Morlyan, S. M. Gabrielyan, M. K. Mardoyan, L. O. Rostomyan, and L. O. Esayan, U.S.S.R. Patent 389,076 (1973); *Chem. Abstr.* **79,** 125884d (1973).
48. C. V. Wilson, *Org. React.* **9,** 332 (1957).
49. H. Wieland and F. G. Fischer, *Justus Liebigs Ann. Chem.* **446,** 49 (1925–1926).
50. J. D. Roberts and H. E. Simmons, Jr., *J. Am. Chem. Soc.* **73,** 5487 (1951).
51. S. R. Sandler, Author's laboratory, unpublished results (1966).
52. C. E. Rehberg, *Org. Synth. Collect. Vol.* **3,** 146 (1955).
53. C. E. Rehberg and C. H. Fisher, *J. Am. Chem. Soc.* **66,** 1203 (1944).
54. M. Reimer and H. R. Downes, *J. Am. Chem. Soc.* **43,** 945 (1921).
55. G. B. Hatch and H. Adkins *J. Am. Chem. Soc.* **59,** 1694 (1937).
56. D. H. Meyer, U.S. Patent 2,873,292 (1959).
57. J. H. Haslam, U.S. Patent 2,822,348 (1958).
58. Hedley & Co., Ltd., British Patent 779,456 (1957).
59. Unilever, Ltd., British Patent 796,808 (1958).
60. C. E. Rehberg, *Org. Synth. Collect. Vol.* **3,** 33 (1955).
61. W. J. Toussaint and L. G. MacDowell, U.S. Patent 2,299,862 (1942); R. L. Adelman. *J. Org. Chem.* **14,** 1057 (1949).
62. W. O. Hermann, E. Baum, and W. Haehnel, German Patent 654,282 (1937).
63. H. J. Hagemeyer, Jr. and D. C. Hall, *Ind. Eng. Chem.* **41,** 2920 (1949).
64a. E. S. Rothman, S. Serota, J. Perlstein, and D. Swern, *J. Org. Chem.* **27,** 3123 (1962).
64b. M. A. Korshunov, R. L. Kuzovlova, and I. V. Zuraeva, U.S.S.R. Patent 309,007 (1971); *Chem. Abstr.* **75,** 14025e (1971).
65a. For a related procedure describing vinyl esters, see D. Swern and E. F. Jordan, Jr., *Org. Synth. Collect. Vol.* **4,** 977 (1963) and W. J. Toussaint and L. G. MacDowell, Jr., U.S. Patent 2,299,862 (1942).
65b. G. Salee and J. C. Rosenfeld, U.S. Patent 4,255,555 (1981).
66. A. L. McCloskey, G. S. Fonken, R. Wikluiber, and W. S. Johnson, *Org. Synth. Collect. Vol.* **4,** 261 (1963).
67. R. E. Strube, *Org. Synth. Collect. Vol.* **4,** 417 (1963).
68. British Patent 814,360 (1959) (to Badische Anilin & Soda Fabrik A.-G.); *Chem. Abstr.* **54,** 16387 (1960).
69. W. S. Johnson, A. L. McCloskey, and D. A. Dunnigan, *J. Am. Chem. Soc.* **72,** 516 (1950).
70. French Patent 818,740 (1937) (to Rohm & Haas Co.).
71. H. T. Neher, E. H. Spreecht, and A. Neuman, U.S. Patent 2,582,911 (1952).
72. A. Neuman, H. T. Neher, and E. H. Sprecht, U.S. Patent 2,778,848 (1957).
73. A. Neuman, H. T. Neher, and E. H. Sprecht. British Patent 769,563 (1957).
74. H. Lautenschlager and H. H. Friederich. German Patent 1,046,030 (1958).
75a. British Patent 814,360 (1959) (to Badische Anilin & Soda Fabrik A.-G.).
75b. E. J. Broderick and B. M. Rein, U.S. Patent 3,641,120 (1972).
76. N. Weiner, *Org. Synth. Collect. Vol.* **2,** 279 (1943).
77. C. D. Hurd, R. N. Jones, and F. H. Blunck, *J. Am. Chem. Soc.* **57,** 2034 (1935).
78. D. H. Peacock and P. Tha, *J. Chem. Soc.* p. 2304 (1928).
79. C. S. Marvel, *Org. Synth.* **21,** 60 (1941).
80. G. B. Heisig and F. H. Slodola, *Org. Synth.* **23,** 16 (1943).
81. J. Cason and C. F. H. Allen, *J. Org. Chem.* **14,** 1036 (1949).

82a. T. L. Jacobs and W. H. Horsheim, *J. Am. Chem. Soc.* **72,** 258 (1950).
82b. S. Zen and E. Kaji, *Org. Synth.* **57,** 60 (1977).
83. H. E. Zaugg, D. A. Dunnigan, R. J. Michaels, L. R. Swett, T. S. Wang, A. H. Sommers, and R. W. DeNet, *J. Org. Chem.* **26,** 644 (1961).
84. H. E. Zaugg, *J. Am. Chem. Soc.* **82,** 2903 (1960).
85. M. C. Kloetzel, *J. Am. Chem. Soc.* **70,** 3571 (1948).
86. C. F. Koelsch, *J. Am. Chem. Soc.* **65,** 437 (1943).
87. E. P. Kohler, A. Graustein, and D. R. Merrill, *J. Am. Chem. Soc.* **44,** 2536 (1922).
88. E. P. Kohler and B. L. Souther, *J. Am. Chem. Soc.* **44,** 2903 (1922).
89. R. Connor, C. L. Fleming, and T. Clayton, *J. Am. Chem. Soc.* **58,** 1386 (1936).
90. R. Connor and D. B. Andrews, *J. Am. Chem. Soc.* **56,** 2713 (1934).
91. M. J. Weiss and C. R. Hauser, *J. Am. Chem. Soc.* **71,** 2026 (1949).
92. N. J. Leonard and D. L. Felley, *J. Am. Chem. Soc.* **72,** 1542 (1950).
93. D. T.Warner and O. A. Moe, *J. Am. Chem. Soc.* **71,** 2586 (1949).
94. M. C. Kloetzel, *J. Am. Chem. Soc.* **69,** 2272 (1947).
95. H. A. Bruson, *Org. React.* **5,** 79 (1949).
96. C. T. Bahner and H. T. Kite, *J. Am. Chem. Soc.* **71,** 3597 (1949).
97. G. D. Buckley, J. L. Charlish, and J. D. Rose, *J. Chem. Soc.* p. 1514 (1947).
98a. V. Boekelheide and S. Rothchild, *J. Am. Chem. Soc.* **71,** 882 (1949).
98b. S-H. Liu, *J. Org. Chem.* **42,** 3209 (1977).
98c. E. L. Eliel, R. O. Hutchins, and M. Knoeber, Sr., *Org. Synth.* **50,** 38 (1970).
99. E. R. Meineke and S. M. McElvain, *J. Am. Chem. Soc.* **57,** 1444 (1935).
100. L. F. Fieser and A. M. Seligman, *J. Am. Chem. Soc.* **58,** 2484 (1936).
101. W. E. Hanford and R. Adams, *J. Am. Chem. Soc.* **57,** 922 (1935).
102a. R. B. Wagner, *J. Am. Chem. Soc.* **71,** 3217 (1949).
102b. S. Majeti, *J. Org. Chem.* **37,** 2914 (1972).
103. R. P. Linstead and E. M. Meade, *J. Chem. Soc.* p. 935 (1934).
104. F. F. Blicke and R. F. Feldkamp, *J. Am. Chem. Soc.* **66,** 1087 (1944).
105. W. Bauer, U.S. Patent 1,829,208 (1931).
106a. T. J. De Boer and H. J. Backer, *Org. Synth. Collect. Vol.* **4,** 250 (1963).
106b. D. J. Raber, P. Gariano, Jr., A. V. Brod., A. Gariano, W. C. Guida, and M. D. Herbst, *J. Org. Chem.* **44,** 1149 (1979).
106c. K. Yanauchi, T. Tanabe, and M. Kinoshita, *J. Org. Chem.* **44,** 638 (1979).
106d. D. B. Denney, R. Melis, and A. D. Pendse, *J. Org. Chem.* **43,** 4672 (1978).
106e. M. W. Rathke, *Org. React.* **22,** 423 (1975).
106f. T. E. Sample, Jr. and L. F. Hatch, *Org. Synth.* **50,** 43 (1970); F. Näf and R. Decorzant, U.S. Patent 4,218,347 (1980).
107. G. R. Robertson, *Org. Synth. Collect. Vol.* **1,** 138 (1941).
108. J. Reilly and W. J. Hickinbottom, *Sci. Proc. R. Dublin Soc.* [N.S.] **16,** 246 (1921).
109. E. J. Villani and F. F. Nord, *J. Am. Chem. Soc.* **69,** 2605 (1947).
110. O. Kamm and W. F. Kamm, *Org. Synth. Collect. Vol.* **1,** 104 (1941).
111. E. R. Nielsen, *J. Am. Chem. Soc.* **66,** 1230 (1944).
112. S. L. Friess, *J. Am. Chem. Soc.* **71,** 14 (1949).
113a. R. Robinson and L. H. Smith, *J. Chem. Soc.* p. 373 (1937).
113b. E. J. Corey, N. W. Gilman, and B. E. Gunem, *J. Am. Chem. Soc.* **90,** 5616 (1968).
114. L. M. Berkowitz and P. N. Rylander, *J. Am. Chem. Soc.* **80,** 6682 (1958).
115. L. W. Covert, R. Connor, and H. Adkins, *J. Am. Chem. Soc.* **54,** 1651 (1932).
116. H. E. Ungnade and F. J. Morriss, *J. Am. Chem. Soc.* **70,** 1898 (1948).
117. J. Cason, C. E. Adams, L. L. Bennett, and U. D. Register, *J. Am. Chem. Soc.* **66,** 1764 (1944).

118. R. Robinson and J. Walker, *J. Chem. Soc.* p. 193 (1936).
119. C. S. Marvel, R. L. Myers, and J. H. Saunders, *J. Am. Chem. Soc.* **70,** 1695 (1948).
120. W. E. Bachmann and W. S. Struve, *Org. React.* **1,** 38, 52 (1942).
121. F. F. Blicke and M. F. Zienty, *J. Am. Chem. Soc.* **63,** 2945 (1941).
122. A. L. Wilds and A. L. Meader, Jr., *J. Org. Chem.* **13,** 763 (1948).
123. R. B. Wagner and J. A. Moore, *J. Am. Chem. Soc.* **72,** 2887 (1950).
124. J. G. Aston and R. B. Greenburg, *J. Am.Chem. Soc.* **62,** 2590 (1940).
125a. R. B. Wagner and J. A. Moore, *J. Am. Chem. Soc.* **72,** 974 (1950).
125b. A. Abad, M. Arnó, J. R. Pedio, and E. Seoane, *Chem. Ind. (London)* pp. 157–158 (1981).
125c. H. Kwart and K. King, in "The Chemistry of Carboxylic Acids and Esters" (S. Patai, ed.), Chapter 8, pp. 341–373. Wiley (Interscience), New York, 1969.
126. H. V. Claborn and L. I. Smith, *J. Am. Chem. Soc.* **61,** 2727 (1939).
127. A. C. Farthing, *J. Chem. Soc.* p. 3213 (1950).
128. M. Gomberg and H. R. Snow, *J. Am. Chem. Soc.* **47,** 201 1925).
129. M. S. Newman and W. S. Fones, *J. Am. Chem. Soc.* **69,** 1046 (1947).
130. C. M. Suter and P. B. Evans, *J. Am. Chem. Soc.* **60,** 537 (1938).
131. P. A. Levene and G. M. Meyers, *Org. Synth. Collect. Vol.* **2,** 288 (1943).
132. W. B. Renfrow and G. B. Walker, *J. Am. Chem. Soc.* **70,** 3957 (1948).
133. J. C. Shivers, B. E. Hudson, and C. R. Hauser, *J. Am. Chem. Soc.* **66,** 309 (1944).
134. R. G. Linville and R. C. Elderfield, *J. Org. Chem.* **6,** 271 (1941).
135. R. C. Fuson, J. C. Corse, and E. C. Horning, *J. Am. Chem. Soc.* **61,** 1290 (1939).
136. C. D. Hurd and F. O. Green, *J. Am. Chem. Soc.* **63,** 2201 (1941).
137. L. H. Ulich and R. Adams, *J. Am. Chem. Soc.* **43,** 660 (1930).
138. G. F. Hennion, H. D. Hinton, and J. A. Nieuwland, *J. Am. Chem. Soc.* **55,** 2857 (1933).
139. P. S. Skell and R. M. Etter, *Chem. Ind. (London)* p. 624 (1958).
140. A. Burger and W. L. Yost, *J. Am. Chem. Soc.* **70,** 2198 (1948).
141. H. Gilman, W. G. Bywater, and P. T. Parker, *J. Am. Chem. Soc.* **57,** 885 (1935).
142. M. S. Newman and H. M. Walborsky, *J. Am. Chem. Soc.* **72,** 4296 (1950).
143. A. C. Cope and G. Field, *J. Org. Chem.* **14,** 856 (1949).
144. G. M. La Fave and P. G. Scheurer, *J. Am. Chem. Soc.* **72,** 2464 (1950).
145. R. H. Hunt, L. J. Chinn, and W. S. Johnson, *Org. Synth. Collect. Vol.* **4,** 459 (1963).
146. R. Robinson, U.S. Patent 3,190,912 (1965); German Patent 1,196,644 (1966) to Farbenfab, Bayer A.G.).
147. J. T. Edward, H. S. Chang, and S. A. Samad, *Can. J. Chem.* **40,** 804 (1962).
148. D. Z. Denney, T. M. Valega, and D. B. Denney, *J. Am. Chem. Soc.* **86,** 46 (1964).
149. J. Falbe, N. Huppes, and F. Korte, *Chem. Ber.* **97,** 863 (1964).
150. J. Falbe, H. J. Steiner, and F. Korte, *Chem. Ber.* **98,** 886 (1965).
151. R. J. Ouellette, A. South, Jr., and D. L. Shaw, *J. Am. Chem. Soc.* **87,** 2602 (1965).
152. H. Haubenstock and C. Vander Werf, *J. Org. Chem.* **29,** 2993 (1964).
153. M. E. Hill, D. T. Carty, D. Tegg, J. C. Butler, and A. F. Stang, *J. Org. Chem.* **30,** 411 (1965).
154. H. Cohen and J. D. Mier, *Chem. Ind. (London)* p. 349 (1965).
155. R. C. Parish and L. M. Stock, *J. Org. Chem.* **30,** 927 (1965).
156. C. H. De Puy, F. W. Breitbeil, and K. L. Eilers, *J. Org. Chem.* **29,** 2810 (1964).
157. French Patent 1,304,569 (1963). (to Consortium for Electrochemische Industrie, G.m.b.H.).
158. T. J. Kealy and R. E. Benson, *J. Org. Chem.* **26,** 3026 (1961).
159. D. G. Crosby and R. V. Berthold, *J. Org. Chem.* **27,** 3083 (1962).
160. J. E. Eastham and J. Selman, *J. Org. Chem.* **26,** 293 (1961).
161. M. Coopersmith, A. J. Rutkowski, and S. J. Fusco, *Ind. Eng. Chem., Prod. Res. Dev.* **5,** 46 (1966).

162. L. J. Dolby and M. J. Schwartz, *J. Org. Chem.* **30,** 3581 (1965); P. R. Stapp, U.S. Patent 3,954,842 (1976).
163. P. S. Ellington, D. G. Hey, and G. D. Meakins, *J. Chem. Soc. C* p. 1327 (1966).
164. H. C. Brown, H. M. Rogić, M. W. Rathke, and G. W. Kabalka, *J. Am. Chem. Soc.* **90,** 818 (1968).
165. J. D. Holmes, U.S. Patent 4,218,340 (1980).
166. R. E. Ireland, R. H. Mueller, and A. K. Willard, *J. Am. Chem. Soc.* **98,** 2868 (1976).
167. R. J. Fanning, U.S. Patent 3,976,670 (1976); P. Hofmann, K. Kosswig, and W. Schaefer, *Ind. Eng. Chem., Prod. Res. Dev.* **19,** 330 (1980); J. F. Kniffon, *J. Org. Chem.* **41,** 2885 (1976).
168. B. Stouthamer and A. Kwantes, U.S. Patent 3,607,914 (1971).
169. B. M. Pope, Y. Yamamoto, and D. S. Tarbell, *Org. Synth.* **57,** 45 (1977).
170. D. M. Bailey, R. E. Johnson, and N. F. Albertson, *Org. Synth.* **51,** 100 (1971).
171. B. Moll, B. Hess, and M. Augustin, *J. Prakt. Chem.* **316,** 315 (1974); *Chem. Abstr.* **81,** 77424 (1974).
172. G. D. Hartman and L. M. Weinstock, *Org. Synth.* **59,** 183 (1979).
173. P. A. Wehrli and V. Chu, *Org. Synth.* **58,** 79 (1978).
174. Y. Izawa, H. Tomioka, M. Natsume, S. Beppu, and H. Tsujii, *J. Org. Chem.* **45,** 4835 (1980).
175. J. McMurry, *Org. React.* **24,** 187 (1976).

CHAPTER 11 / AMIDES

1. Introduction	316
2. Dehydration of Ammonium Salts	318
2-1. Preparation of Butyramide	318
2-2. Preparation of Olemorpholide	320
2-3. Preparation of N-Benzyloctamide	321
2-4. Preparation of N,N-Bis(2-chloroethyl)-2-(dichloroacetamido)acetamide	322
3. Condensation Reactions	323
A. Condensations with Acyl Halides	323
3-1. Preparation of N-Methylacrylamide	326
3-2. Preparation of N-(1,1-Dihydroperfluorobutyl)acrylamide	327
3-3. Preparation of 4,4',4''-Trichlorotribenzamide	328
3-4. Preparation of N-tert-Butyl-N-cyclohexyl-2-chloroacetamide	329
3-5. Preparation of Poly(hexamethylenesebacamide) (Nylon 6/10)	329
3-6. Preparation of N,N-Dimethylbenzamide	330
B. Condensations with Acid Anhydrides	330
3-7. Preparation of N-Acetyl-L-cysteine	331
3-8. Preparation of N-Allylacetamide	332
3-9. Preparation of 2-Phenyl-3-(2'-furyl)propionamide	333
3-10. Preparation of N,N'-Di(trifluoroacetyl)-meso-2,6-diaminopimelic Acid	334
3-11. Preparation of α-Ethylglutarimide	335
C. Condensations with Esters	336
3-12. Preparation of N-Ethylperfluorobutyramide	336
3-13. Preparation of N-(2-Methoxy-4-nitrophenyl)benzoylacetamide (Use of Continuous Reactor)	338
3-14. Preparation of Monoacetylethylenediamine	339
3-15. Preparation of N-(2-Hydroxyethyl)acetamide	340
D. Transamidation Reactions	340
a. Reactions between Amines and Amides	340
3-16. Preparation of 2-Iodoformanilide	340
3-17. Preparation of N-(α-Naphthyl)acetamide	341
b. Reactions between Acids and Amides	341
3-18. Preparation of Oleamide	341
E. N- and C-Alkylation and N-Acylation of Amides	342
3-19. Preparation of N-Hydroxymethylphenylacetamide	342
3-20. Preparation of 2-(4-Methoxyphenyl)-N-methylacetamide	342
3-21. Preparation of N-(9-Bromononyl)phthalimide	343
3-22. Preparation of N-(o-Tolyl)-2',6'-acetoxylidide (Goldberg Reaction)	345
3-23. Preparation of N-Stearoylsuccinimide	345
3-24. Preparation of N,N-Diethyl-1-hydroxy-4-methylcyclohexaneacetamide	346
F. Grignard Condensation Reactions	347
3-25. Preparation of N-(2-Furyl)benzamide	347
4. Addition Reactions	347
4-1. Preparation of 4-Nitro-4-methylpimeldiamide	347

5. Hydration Reactions with Nitriles	348
5-1. Preparation of cis-β-Chloroacrylamide	348
5-2. Preparation of 6-Methoxy-2-nitrobenzamide	349
5-3. Preparation of 2-exo-Hydroxy-4-azahomoadamantan-5-one	349
5-4. Preparation of Cyclohexylcarboxamide	350
5-5. Preparation of N,N'-Diisopropylfumaramide	351
6. Oxidation of Alkylene Amines	351
7. Reduction Reactions	351
8. Rearrangement Reactions	352
A. Beckmann Rearrangement	352
8-1. Preparation of Phenanthridone	352
B. The Schmidt and Curtius Rearrangements	352
C. The Wolff Rearrangement	353
8-2. Preparation of 11-Bromoundecanoamide	353
9. Miscellaneous	354
References	355

1. INTRODUCTION

A convenient and, in many cases, most economical method of preparing amides involves the thermal dehydration of ammonium or amine salts of carboxylic acids. Many variations of this reaction have been reported. For example, since the preparation of the amine salt of the gaseous amines is troublesome, mixtures of the carboxylic acids and ureas may be subjected to a combined thermal dehydration and decarboxylation.

By use of entraining agents, amine salts may be dehydrated under modest reaction conditions. In anhydrous THF, many acids may be converted to amides by reaction with appropriate amines in the presence of o-nitro-phenylthiocyanate and tri-n-butylphosphine.

The use of carbodiimides with amines and carboxylic acids is of particular interest in the preparation of polypeptides as well as in the synthesis of other amides from starting materials bearing other functional groups which may be capable of undergoing competing reactions with one of the reagents (e.g., acyl halides are usually not suitable for the preparation of amides of aminocarbinols, while aminocarbinols with carboxylic acids and a carbodiimide yield only amides).

A very general method for the preparation of amides involves the reaction of ammonia or amines with acyl halides or anhydrides. In this reaction, an excess of the amine may serve to "scavenge" the hydrogen halide (or carboxylic acid in the case of anhydrides) generated. Strong organic bases may also serve to react with the hydrogen halide. The reaction of diacid chlorides with diamines is also of importance in the preparation of polyamides such as Nylon. In the Schotten–Baumann reaction, aqueous inorganic bases such as sodium hydroxide or potassium carbonate are used as hydrogen halide scavengers.

§ 1. Introduction

Mixed anhydrides prepared from carboxylic acids and alkyl chlorocarbonates are of particular value for the preparation of amides of sensitive acids such as N-acylated amino acids. The reaction of cyclic dianhydrides with amines yields imides. If diamines are used with dianhydrides, polyimide resins are produced.

The aminolysis of esters is another useful procedure for the preparation of amides, particularly in the presence of glycols as reaction promoters.

Transamidations of amides with amines, as well as of carboxylic acids with amides, have been reported.

Amides may be converted to N-alkyl and N-acyl derivatives by appropriate condensation reagents. C-Alkylations of amides can also be carried out to modify the carbon skeleton of a compound. N-arylamides may be prepared by the Goldberg reaction of an amide with an aryl halide in the presence of potassium carbonate and cuprous iodide.

Since α-haloesters undergo the Reformatskii reaction, α-haloamides may also be subjected to this reaction to give amides.

Another condensation reaction which has been used in the preparation of amides is the Grignard reaction with isocyanates.

Active hydrogen compounds, such as aliphatic nitro-compounds may be added to the double bond of acrylamide in a Michael-type addition to produce saturated amides.

Amides have been prepared by the hydration of nitriles. With secondary or tertiary alcohols or with certain olefins, nitriles react to produce N-substituted amides (Ritter reaction).

A novel oxidation of Schiff bases of arylmethylamines to aromatic amides has been described.

The reduction of aromatic nitro compounds in the presence of molybdenum hexacarbonyl and a carboxylic acid leads to the formation of N-arylcarboxamides.

The Beckmann, Schmidt, and Wolff rearrangements have also been used to form amides.

In the preparation of amides the reported procedures do not always take several factors adequately into account, which may result in difficulties in the isolation and purification of the product. Even the syntheses discussed below do not always take these into consideration.

The first of these is that, contrary to the impression gained from the preparation of amides for identification purposes, not all of them are solids. Thus, for example, such commercially available compounds as formamide, N,N-dimethylformamide, N,N-dimethylacetamide, N-*tert*-butylformamide, and N,N-dimethyllauramide are liquids at room temperature. N-Acetylethylene-1,2-diamine, several N-alkyl acrylamides, and a number of N-alkyl amides of perfluorinated acids are also liquids.

The second factor which frequently leads to difficulties in isolation and purification arises from the exceptional solvent properties of amides—both liquid and solid. Many ionic compounds such as salts, water, and a large variety of covalent compounds, including aromatic hydrocarbons, have an appreciable solubility in many amides. The amides, in turn, may exhibit an appreciable solubility in very diversified solvents. Clearly, this situation may bedevil a synthesis with extremely complex solubility distribution coefficient problems. Vapor phase chromatography has been used in our laboratories to advantage in determining whether the amide has been adequately separated from co-products and whether a layer from a phase separation should be retained because it still contains product or whether it should be discarded.

Fortunately, many amides are quite thermally stable. Consequently, fractional distillation can frequently be used to purify those amides which do not melt at elevated temperatures that they seriously clog the condensers.

2. DEHYDRATION OF AMMONIUM SALTS

The reaction of carboxylic acids with ammonia normally produces its ammonium salt. The dehydration of ammonium salts may be used to prepare amides.

The classical dehydration of ammonium acetate to acetamide is accelerated by acetic acid, consequently the usual procedure for the preparation of acetamide involves careful distillation of an ammonium acetate–acetic acid mixture. The distillate is a solution of acetic acid and water. Finally excess acetic acid is distilled out followed by acetamide (yield 87–90% of theory, mp 81°C). The product may be recrystallized from a benzene–ethyl acetate mixture [1].

In the preparation of the required ammonium carboxylates, ammonium carbonate or ammonium bicarbonate is often a more convenient base than aqueous or anhydrous ammonia [1].

Although other methods may often be more convenient, the dehydration of amine salts has wide application, as is illustrated below.

2-1. Preparation of Butyramide [2]

$$CH_3CH_2CH_2\overset{O}{\underset{\|}{C}}-OH + NH_3 \longrightarrow [CH_3CH_2CH_2\overset{O}{\underset{\|}{C}}-O]^-NH_4^+ \quad (1)$$

$$[CH_3CH_2CH_2\overset{O}{\underset{\|}{C}}-O]^-NH_4^+ \longrightarrow CH_3CH_2CH_2\overset{O}{\underset{\|}{C}}-NH_2 + H_2O \quad (2)$$

In a hood, to a 250-ml three-necked flask fitted with an empty, electrically heated distillation column maintained between 85° and 90°C and topped with a

§ 2. Dehydration of Ammonium Salts

vacuum distillation head, a gas inlet tube, and a thermometer is placed 88 gm (1 mole) of butyric acid. The butyric acid is heated to 185°C while a steady stream of ammonia is passed directly from a cylinder through the gas inlet tube into the butyric acid for 7 hr. At these temperatures the water formed from the reaction is swept through the apparatus into the distillation head where it may be partially condensed and collected. Then the ammonia flow is stopped, and the contents are distilled under reduced pressure at 130°–145°C (22 mm). Yield of crude product, 86.5 gm (84%). The product may be crystallized from benzene or ether. To separate the mother liquor, centrifugation is recommended. The product is dried in a vacuum desiccator over sulfuric acid, mp of purified product, 114.8°C.

By a similar procedure, a variety of N,N-dimethylamides has been prepared. Table I gives preparative details for this procedure.

Free carboxylic acids may also be converted to amides by heating the acid with urea as the source of the amino function [3]. The reactions must be carried out with considerable care since large volumes of gas are evolved during the reaction and since urea tends to sublime into the reflux condenser during the preparation. The preparation of heptamide according to Eq. (3) is described in detail in the literature [4].

TABLE I

PREPARATION OF ALIPHATIC AMIDES AND DIMETHYLAMIDES BY DEHYDRATION [2]

Acid	Flask temp. (°C)	Condenser temp. (°C)	Amine addition time (hr)	Yield (%)	Product b p [°C (mm)]	M P (°C)
Amide of						
Acetic	170°–190°	80°	3–5	96	210°–220° (atm)	81.5°
Propionic	185°	80°	5.5	93	200°–220° (atm)	81.3°
Butyric	185°	85°–90°	7	84	130°–145° (22)	114.8°
Valeric	180°	90°	15	82	100°–130° (6)	105.8°
Caproic	160°	90°	7	75	135°–150° (10)	101.5°
Heptoic	160°–190°	90°	4–7	75	130°–150° (7)	96.5°
Caprylic	180°	90°	11	80	135°–155° (4)	106.0°
Dimethylamide of						
Formic (95%)	95°	60°	3	73	130°–165° (atm)	—
Acetic	150°	80°	3	84	165°–175° (atm)	—
Propionic	155°	80°	3	78	165°–178° (atm)	—
Butyric	155°	85°	2.5	84	180°–194° (atm)	—
Valeric	165°	85°	3	87	205°–215° (atm)	—
Caproic	155°	85°	3	88	220°–230° (atm)	—
Heptoic	160°	85°	3.5	84	165°–175° (95)	—

$$2\ CH_3(CH_2)_5CO_2H + H_2NCONH_2 \longrightarrow 2\ CH_3(CH_2)_5CONH_2 + CO_2 + H_2O \quad (3)$$

The dehydration of some amine salts of carboxylic acids may also be carried out by an azeotropic water removal procedure similar to the one commonly used in ester preparations. In Preparation 2-2, to the total volume of the reagents about 10 vol% of benzene is added as the water entraining agent. If higher reflux temperatures are required, toluene or xylene may be favored. CAUTION: Benzene is considered to be carcinogenic.

2-2. Preparation of Olemorpholide [5]

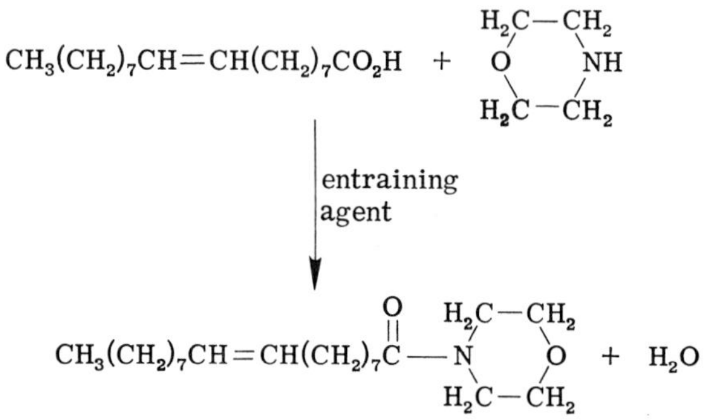

$$CH_3(CH_2)_7CH=CH(CH_2)_7CO_2H\ +\ \underset{\underset{H_2C-CH_2}{\diagup\ \diagdown}}{\overset{H_2C-CH_2}{\diagdown\ \diagup}}O\quad NH \quad (4)$$

$$\downarrow \text{entraining agent}$$

$$CH_3(CH_2)_7CH=CH(CH_2)_7\overset{O}{\overset{\|}{C}}-N\underset{H_2C-CH_2}{\overset{H_2C-CH_2}{\diagup\ \ \diagdown}}O\ +\ H_2O$$

In a reflux apparatus equipped with a Dean–Stark trap between the reaction flask and the condenser are placed 667 gm (2.4 mole) of oleic acid, 225 gm (2.6 moles) of morpholine, and 100 ml of benzene. The mixture is heated to reflux and maintained at the reflux temperature until no further water is collected in the Dean–Stark trap. Then 23 gm of additional morpholine is added to the reaction mixture and heating is continued until no further water is formed.

The benzene is then distilled off. The flask residue, which is said to contain about 4% unreacted oleic acid, is cooled to room temperature and dissolved in a suitable quantity of carbon tetrachloride. This solution is treated with a 50-ml portion of 4% alcoholic potassium hydroxide. Upon addition of water to this solution, the carbon tetrachloride solution of the product separates. The product solution is extracted by a similar procedure five times. The carbon tetrachloride solution is again separated. The solvent is evaporated. The residual product contains only 0.15% free acid. The product may be distilled between 223°–226°C at 1.3 mm.

Amides also have been formed by passing acrylic or methacrylic acids with primary or secondary amines in a molar ratio of 1:1.1 over an activated alumina catalyst at 250°C for a few seconds. The reaction products may be inhibited against uncontrolled polymerization by the addition of small quantities of hydroquinone. Table II gives the physical properties of several N-alkyl-substituted amides prepared by this method.

§ 2. Dehydration of Ammonium Salts

TABLE II
Alkyl-Substituted Acrylamides and Methacrylamides—Catalytic Dehydration of Amines in the Presence of Free Acids [6]

N-Alkylacrylamide	Boiling range at 1 mm (°C)
Isopropyl	85–88
Cyclohexyl	109–112
Phenyl	112–115
N-Alkylmethacrylamide	
Methyl	70–73
Ethyl	79–82
Dimethyl	50–55

In anhydrous tetrahydrofuran, the reaction of a large variety of carboxylic acids with amines in the presence of o-nitrophenylthiocyanate and tri-n-butylphosphine led to high yields of amides.

The reaction is thought to involve the formation of a reactive intermediate from the interaction of the thiophosphonic cyanide and the carboxylic acid:

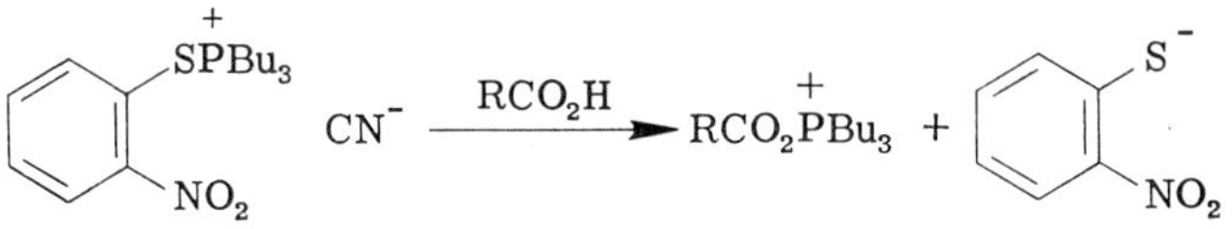

$$\text{o-O}_2\text{N-C}_6\text{H}_4\text{-SPBu}_3^+ \; CN^- \xrightarrow{RCO_2H} RCO_2PBu_3^+ + \text{o-O}_2\text{N-C}_6\text{H}_4\text{-S}^- + HCN \quad (5)$$

$$RCO_2PBu_3^+ + R'NH_2 \longrightarrow RCONHR' + [Bu_3POH]^+ \quad (6)$$

Yields of a variety of amides usually are greater than 90%. The formation of several lactams from amino acids by this method affords yields in the range of 73 to 97%. However, acrylic acid with benzylamine afforded only a 17% yield under comparable conditions [7]. While this method is very attractive, present procedures have been carried out only on a millimolar scale as indicated in Preparation 2-3 and the suitability of this approach to synthesize substantial quantities remains to be demonstrated.

2-3. Preparation of N-Benzyloctamide [7]

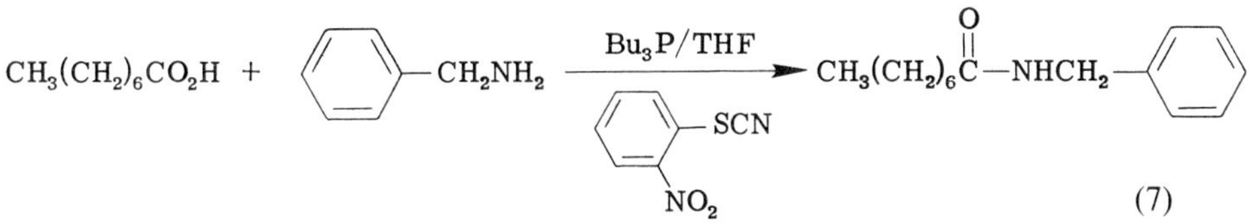

$$CH_3(CH_2)_6CO_2H + C_6H_5\text{-}CH_2NH_2 \xrightarrow[\text{o-O}_2\text{N-C}_6\text{H}_4\text{-SCN}]{Bu_3P/THF} CH_3(CH_2)_6\overset{O}{\underset{\|}{C}}\text{-}NHCH_2\text{-}C_6H_5 \quad (7)$$

The following reaction may be carried out conveniently in a small Erlenmeyer flask closed with a rubber serum cap. The various reagents may then be injected as solutions, if necessary, through the serum cap with syringes of suitable size. During the injection of the larger volumes of reagents, the flask may be vented conveniently by puncturing the cap with a syringe needle attached to a drying tube which is left in place during the addition stage.

To a solution of 0.202 gm (1.4 mmole) of octanoic acid in 10 ml of freshly distilled tetrahydrofuran (from lithium aluminum hydride) is added 0.379 gm (2.1 mmole) of *o*-nitrophenylthiocyanate. Then 230 μl (2.1 mmole) of benzylamine and 524 μl (2.1 mmole) of freshly distilled tri-*n*-butylphosphine are added. The reaction mixture is maintained at room temperature for 7 hours. Then the solvent is removed under reduced pressure. The residue is placed on a chromatographic column containing 80 gm of silica gel. Elution with 230 ml of benzene followed by hexane–ether (1:1) permitted the isolation of 0.325 gm of crystallized *N*-benzyloctamide, mp 65.0°–65.5°C (99% yield).

While mechanistically, probably quite unrelated to the reactions discussed above, the preparation of amides in the presence of carbodiimides may be looked upon as a reaction of an acid and an amine. While this reaction is of particular value for polypeptide syntheses, it is believed to be quite generally applicable. Of particular interest is the fact that the reaction appears to be specific for amino groups. Hydroxy-containing starting materials do not usually experience *O*-acylation [8, 9]. The reaction is characterized by high yields. The course of the reaction can be followed readily by the precipitation of highly insoluble substituted ureas. While the original work generally refers to the use of dicyclohexylcarbodiimide, other carbodiimides have become commercially available and their applicability should be explored further.

CAUTION: Carbodiimides must be handled with extreme care since free amino groups in the tissue of the operator may readily be acylated. This may cause physiological effects which are potentially hazardous.

An example of the reaction is given here mainly to illustrate the technique involved. Usually much better yields may be anticipated [10].

2-4. Preparation of N,N-*Bis(2-chloroethyl)-2-(dichloroacetamido)acetamide* [10]

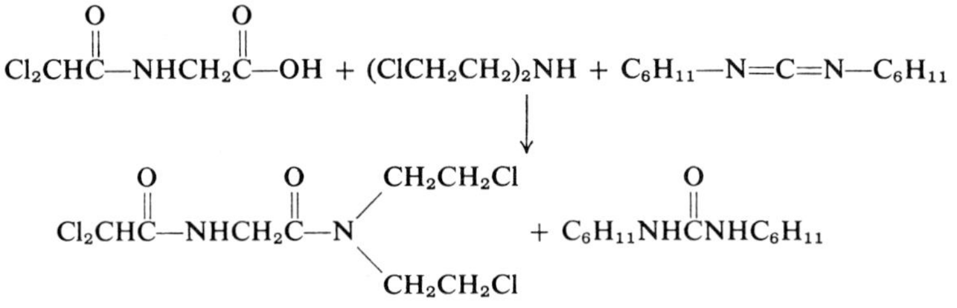

$$Cl_2CHC(O)-NHCH_2C(O)-OH + (ClCH_2CH_2)_2NH + C_6H_{11}-N=C=N-C_6H_{11}$$

$$\downarrow$$

$$Cl_2CHC(O)-NHCH_2C(O)-N(CH_2CH_2Cl)_2 + C_6H_{11}NHC(O)NHC_6H_{11} \quad (8)$$

CAUTION: The carbodiimide must be handled so that neither liquid nor vapor comes in contact with personnel by spillage or inhalation.

To a solution of 7.4 gm (0.051 moles) of bis(2-chloroethyl)amine in 60 ml of tetrahydrofuran is added 9.7 gm (0.051 moles) of N-dichloroacetylglycine. To this solution is added dropwise, with stirring, over a 30-min period a solution of 10.76 gm (0.051 moles) of N,N'-dicyclohexylcarbodiimide in 60 ml of tetrahydrofuran. After an additional stirring period of 30 min, the precipitated N,N'-dicyclohexylurea is removed by filtration and washed with tetrahydrofuran.

The combined filtrates are evaporated to dryness and the residue is crystallized from a mixture of ethyl acetate and petroleum ether. Yield 3.0 gm (18.6%), mp 62.5°–63.5°C. Other preparations of this compound gave yields up to 38.5%. In other amide or polypeptide preparations such as those reported by Sheehan and Hess [8, 9], much higher yields have been obtained.

3. CONDENSATION REACTIONS

A. Condensations With Acyl Halides

The reaction of acyl chlorides with ammonia, primary, or secondary amines is probably the most generally applicable method of preparing amides in the laboratory. A large variety of acid chlorides are available commercially, others are readily prepared from the acids. Not only halides of carboxylic acids but also those of sulfonic and phosphonic acids and of picric acid may be converted to amides. Both aliphatic and aromatic amines may be subjected to acylation with acid halides. The reactions of aliphatic acid chlorides have been extensively reviewed [11].

The reactions are usually quite rapid, which is desirable in the preparation of derivatives for the identification of amines or acids (acid chlorides). The reaction conditions used, however, do not always produce very high yields. For example, the acylation of 33 amines with 2-propylpentanoyl chloride (dipropylacetyl chloride) afforded yields ranging from 11% to 98%, of which one-half of the amides were recovered in yields of 40% to 69% and one-third in yields of 80% to 98% (Table III).

Since the reaction of an acid chloride with amines is highly exothermic, often quite violent, reactions must be carefully controlled from the standpoint of safety. It is customary to carry this reaction out at ice temperatures. Even so, local overheating may take place as an acid chloride is being added to the amine. This may cause loss of amine content, particularly if large-scale laboratory reactions are attempted. By cooling, the reaction rate is also materially reduced. There-

TABLE III

Yields and Properties of N-Substituted Dipropylacetamides[a] [12]

$$\begin{array}{c} CH_3CH_2CH_2 \\ \diagdown \\ CH\text{—}CONHR \\ \diagup \\ CH_3CH_2CH_2 \end{array}$$

RNH$_2$	mp (°C)	bp (°C 760 mm)	n_D^{20}	Yield (%)
Propylamine	77.3°	—	—	64
Isopropylamine	133°	—	—	53
Butylamine	69.5°	—	—	58
tert-Butylamine	114°	—	—	80
Isopentylamine	38°	—	—	16
Dodecylamine	65.5°	—	—	48
Hexadecylamine	79°	—	—	44
Amphetamine	104°	—	—	59
N-Aminopropylmorpholine	78°	—	—	52
Dimethylaminoethylamine	131°[b]	—	—	62
Diethylaminoethylamine	76°[b]	—	—	62
Diisopropylaminopropylamine	127°[b]	—	—	81
Dimethylaminopropylamine	109°[b]	—	—	11
Aniline	108°	—	—	44
2,3-Xylidine	161°	—	—	69
2,4-Xylidine	140°	—	—	98
2,5-Xylidine	155°	—	—	53
2,6-Xylidine	245°	—	—	57
3,4-Xylidine	109°	—	—	49
m-Chloroaniline	93°	—	—	29
o-Chloroaniline	92°	—	—	90
p-Chloroaniline	174.5°	—	—	89
p-Anisidine	164.5°	—	—	11
p-Phenetidine	153°	—	—	92
Procaine	91°	—	—	90
Cyclohexylamine	175°	—	—	48
Ethanolamine	64°	143°	1.445	40
Iminodipropionitrile	73°	—	—	51
Ethylenimine	—	143°	1.453	38
Piperidine	—	161°	1.470	96
Morpholine	—	152°	1.469	93
Piperazine	70.5°	—	—	88

[a] Reprinted by permission of *Bull. Soc. Chim. Fr.*, the copyright owner.
[b] Oxalate salt.

§ 3. Condensation Reactions

fore, adequate time for completion of the reaction must be allowed for isolation of an optimum yield.

The reaction itself is usually carried out in the presence of a base. This base serves to neutralize the hydrogen chloride formed as a coproduct. When no base is present, hydrochlorides of the amides may form which may be discarded during the product work-up.

Since the base often forms salts in a very finely divided state, considerable product may be occluded, and, in our experience, it is well to attempt to dissolve the salts in water, even though it appears quite dry after its initial separation from the reaction mixtures, and to extract the aqueous solution with solvents. Among the extraction solvents to be considered are aromatic hydrocarbons. Ether should be used only with caution since some simple amine hydrochlorides appear to have significant solubility in this solvent. Obviously, once this solvent is stripped off, a product residue containing both an amide and an amine hydrochloride will be troublesome to separate.

The typical procedures cited below were selected as representative of the variety of reaction conditions which may be used. In some cases, an excess of the amine is used as the base to neutralize hydrogen chloride. In others, particularly when the amine is available only in limited quantities, either strong organic bases such as trimethylamine, triethylamine, quinoline, or pyridine, or aqueous alkali is used.

In our own laboratories, we have frequently used triethylamine as the base to take up the evolving hydrogen chloride. From this experience, we should like to call to the reader's attention the fact that triethylamine is quite water soluble below 18°C and is surprisingly limited in water solubility at 25°C. Since the product work-up frequently involves water washes, if the water is at room temperature, excess triethylamine may separate as a distinct layer when the expectation is that it remain in the aqueous layer. Therefore, careful attention to the temperature of the extraction solvents and the water is recommended when this particular base is used.

Aqueous ammonia has been used to prepare an unsubstituted amide such as diallylacetamide from diallylacetyl chloride [13]. By a similar procedure the acid chlorides of substituted succinomononitriles have been converted to amides [14].

In the presence of strong bases, amine hydrochlorides may be acylated [15].

When one of the volatile amines has to be used, a method of measuring the amount of amine utilized is required. We have never found bubbling a gaseous amine directly into an acyl chloride solution to be satisfactory. Aside from the potential violence of reaction, addition rates are difficult to control, amine hydrochlorides may clog the delivery tube, and the reaction mixture may "suck back," even into the gas cylinder.

A technique which is frequently satisfactory consists of bubbling the amine, in an efficient hood, from the gas cylinder, through a suitable gas trap into a tared flask containing the requisite weight of reaction solvent (usually benzene). From time to time the gas flow is shut off, the delivery tube is removed from the flask, and the flask is weighed. This procedure is continued until the desired weight of amine is dissolved in the solvent. As the concentration of amine increases, the solution may be cooled in an ice–salt bath. The amine solution may be stored for short periods of time in an ice–salt bath until ready to use. Naturally this operation, as well as subsequent ones, must be carried out entirely in a hood and the operator should be equipped with a suitable gas mask, gloves, and rubber apron.

If desired, the amine solution may be added gradually to the reaction vessel containing acyl chloride using a pressure-equilizing addition funnel. Usually, however, it is simply transferred to the reaction flask and the acyl chloride is added gradually. This reduces the possibility of forming secondary and tertiary amides considerably. The preparation of N-methylacrylamide is an example of this procedure.

3-1. Preparation of N-Methylacrylamide [16]

$$2\ CH_3NH_2 + CH_2=CHC\underset{Cl}{\overset{O}{\diagdown}} \longrightarrow CH_2=CH-C\underset{NHCH_3}{\overset{O}{\diagdown}} + CH_3NH_2 \cdot HCl \quad (9)$$

To a well-chilled solution of 122.5 gm (4 moles) of methylamine in 500 ml of dry benzene is added a solution of 176 gm (1.94 moles) of freshly distilled acrylyl chloride in 100 ml of benzene over a 2.75-hr period while a reaction temperature below 5°C is maintained. The solid which forms is filtered off. The filtrate is preserved. The solid is dissolved in water. The aqueous solution is extracted with two portions of benzene. The benzene extracts are combined with the filtrate. The benzene, excess amine, and water are removed from the product solution by distillation. The high-boiling residue is then fractionally distilled. The fraction boiling between 92° and 97°C (4–5 mm) is collected and redistilled at 79°C (0.7 mm). The yield is 107 gm (66% of theory), n_D^{20} 1.4730. The product may be stored after inhibition with 0.1% of p-methoxyphenol. By-products are believed to be the addition products of hydrogen chloride or of methylamine to the double bond.

Frequently amides have to be prepared from an amine which is available in limited amounts. Under these circumstances, the use of an excess of the amine merely to neutralize hydrogen chloride is uneconomical. For this reason, a strong base, such as triethylamine, is added to the reaction system to neutralize the generated hydrogen chloride while the weaker amine undergoes acylation. In our own experience, for example, a series of N-alkyl-N-(1,1-dihydroperfluorobutyl)acrylamides was prepared by this method [16, 17].

§ 3. Condensation Reactions

The preparation of N-(1,1-dihydroperfluorobutyl)acrylamide is illustrative of the general procedure using liquid amines, not only for the preparation of fluorinated amides but for amides in general. The general applicability of this procedure for substituted methacrylamides is indicated in the papers by Sokolova and co-workers [18].

The use of barium chloride in the procedure given here is optional; phenothiazine is used as a polymerization inhibitor and benzene or toluene may be substituted for ether as a reaction solvent to some advantage. Recrystallization solvents may be varied with different amides (e.g., aqueous ethanol is frequently used).

3-2. Preparation of N-(1,1-Dihydroperfluorobutyl)acrylamide [16]

$$CF_3CF_2CF_2CH_2NH_2 + CH_2{=}CH{-}C\overset{O}{\underset{Cl}{\diagdown}} + (C_2H_5)_3N$$

$$\downarrow$$

$$CF_3CF_2CF_2CH_2NH{-}\overset{O}{\underset{\|}{C}}{-}CH{=}CH_2 + (C_2H_5)_3N \cdot HCl \quad (10)$$

To a solution of 46 gm (0.23 moles) of 1,1-dihydroperfluorobutylamine, 30 gm (0.3 moles) of triethylamine, 0.1 gm of phenothiazine in 100 ml of anhydrous, peroxide-free, ether is added 1 gm of powdered dry barium chloride. The mixture is vigorously stirred in an ice–salt bath while a solution of 23 gm (0.255 moles) of freshly distilled acrylyl chloride in 50 ml of anhydrous, peroxide-free ether is added slowly.

The solid is filtered off and the filtrate is freed of ether and other volatiles by distillation. The residue is recrystallized from textile spirits. The yield is 49 gm (89% of theory), mp 57.4°–57.6°C. The product may also be purified by distillation at 35°C (2 mm).

The general procedure of treating amines with acyl chlorides in a solvent with triethylamine as a scavenger of hydrogen chloride appears to be reasonably gentle. For example, *trans*-2,3-dimethylaziridine was acetylated with *p*-nitrobenzoyl chloride to produce the acetylated product with retention of the E-form in a 98% yield [19]. More recently, chiral amides were produced by the N-acylation of (*S*)-(−)-prolinol in the presence of triethylamine [20].

The preparation of tertiary amides (i.e., triacylamines) is frequently troublesome. In some cases, amides may be acylated with acid chlorides in an excess of pyridine [21]. This preparation evidently is not satisfactory when strong electron-withdrawing groups (such as NO_2^-) are present in the acyl portions of the amide. A method using lithium nitride with acyl halides may be generally applicable for the preparation of tertiary amides in which all three acyl groups are identical, although conditions for obtaining higher yields remain to be worked out [22].

3-3. Preparation of 4,4',4''-Trichlorotribenzamide [22]

$$Cl-C_6H_4-\overset{O}{\underset{\|}{C}}-Cl + Li_3N \longrightarrow Cl-C_6H_4-\overset{O}{\underset{\|}{C}}-N(-\overset{O}{\underset{\|}{C}}-C_6H_4-Cl)_2 \quad (11)$$
(with the N also bearing a $C(=O)-C_6H_4-Cl$ group)

In a dry 200-ml four-necked flask fitted with condenser, stirrer, thermometer, and an additional funnel is placed a suspension of 1.2 gm (0.034 mole) of lithium nitride in 40 ml of dry diglyme. To this is added a solution of 25 gm (0.142 mole) of p-chlorobenzoyl chloride over a 2-hr period while maintaining the temperature between 35° and 40°C. Stirring is then continued for 6 hr. The precipitating product is washed in turn with 100 ml of a 5% sodium carbonate solution and with water. Then the product is recrystallized first from tetrahydrofuran, then from chloroform. Yield 6.84 gm (48.7%). mp 234°–235°C.

By similar procedures were prepared: tribenzamide (36% yield), mp 206–207°C; 4,4',4''-trinitrobenzamide (25% yield), mp 231°–232°C; 4,4',4''-trimethoxybenzamide (yield not reported) mp 159°–160°C (crop I), mp 196°–198°C (crop II), ir and elementary analysis indicate polymorphic forms.

The preparation of amides by reaction of an amine and an acyl chloride in the presence of aqueous alkali is the well-known "Schotten–Baumann" reaction [23]. Since the rate of reaction of the acyl chloride with amines is greater than the rate of hydrolysis of the acyl chloride, amide formation is favored. In some cases, the stability of acyl chlorides to aqueous caustic solution is surprising. For example, benzoyl chloride may be kept in contact with sodium hydroxide solutions for long periods of time. We presume that a thin layer of sodium benzoate forms rapidly and that this acts as a protective coating unless the benzoyl chloride is stirred or shaken to disturb the protective layer. As a matter of fact, the Schotten–Baumann reaction (and the analogous Hinsberg reaction with aromatic sulfonyl chlorides) appears to be most satisfactory when acyl chlorides are used which are relatively insoluble in water. Many years ago, we noted that the addition of a small percentage of a surfactant (e.g., sodium lauryl sulfate as well as cationic surfactants) assisted in the dispersion of the acyl chloride with an increase in reaction rate. These may have been early examples of phase-transfer reactions.

The Schotten–Baumann reaction is widely used, particularly for the prepara-

§ 3. Condensation Reactions 329

tion of derivatives for the identification of small amounts of amines or acyl chlorides by shaking the reagents together in a glass-stoppered bottle. Sonntag [11] extensively reviewed this reaction.

The preparation of *N-tert*-butyl-*N*-cyclohexyl-2-chloroacetamide is given here as an example of the Schotten–Baumann procedure to indicate the complexity of the amine which may be subjected to the reaction. The example makes use of potassium carbonate instead of the more usual sodium hydroxide as the base.

3-4. Preparation of N-tert-Butyl-N-cyclohexyl-2-chloroacetamide [24]

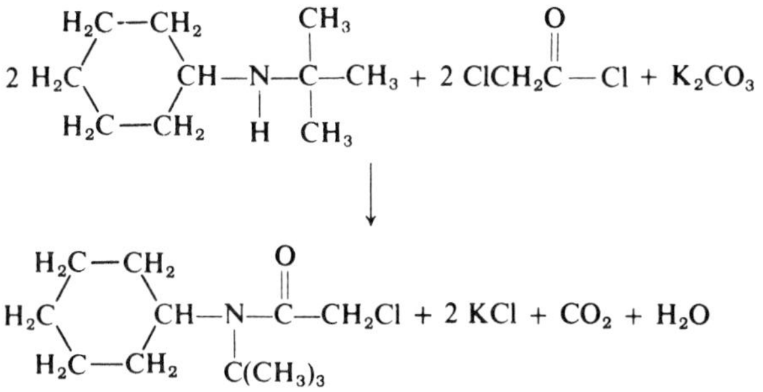

$$2\ \text{cyclohexyl-NH-C(CH}_3)_3 + 2\ \text{ClCH}_2\text{C(O)Cl} + \text{K}_2\text{CO}_3$$

$$\downarrow$$

$$\text{cyclohexyl-N(C(CH}_3)_3)\text{-C(O)CH}_2\text{Cl} + 2\ \text{KCl} + \text{CO}_2 + \text{H}_2\text{O} \quad (12)$$

To a dispersion of 46.5 gm (0.3 moles) of *N-tert*-butylcyclohexylamine, 42 gm (0.3 moles) of potassium carbonate, 100 ml of water, 400 gm of ice, 300 ml of benzene, and 100 ml of ether is added with vigorous agitation, over a 30-min period, 39.5 gm (0.35 moles) of 2-chloroacetyl chloride while maintaining the reaction temperature between $-4°$ and $+2°C$. From the organic phase, 45 gm of *N-tert*-butyl-*N*-cyclohexyl-2-chloroacetamide is isolated (yield 65% of theory), bp 123°–125°C (0.5 mm).

An interesting variation of the Schotten–Baumann procedure is involved in the so-called "interfacial polycondensation" procedure for the preparation of polyamides. Sorenson and Campbell [25] give a number of examples of this procedure. Their description of the Morgan and Kwolek [26] "Nylon Rope Trick" makes such an interesting demonstration that it is described below. The general procedure may be used to prepare various nylons as well as simple amides (with suitable modification of isolation procedures).

3-5. Preparation of Poly(hexamethylenesebacamide) (Nylon 6/10) [26]

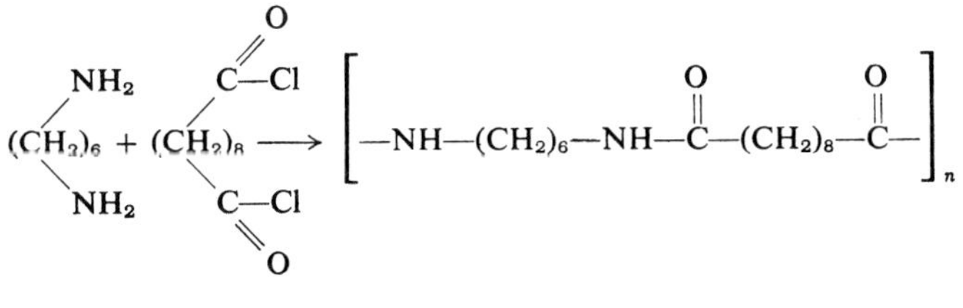

$$\text{H}_2\text{N-(CH}_2)_6\text{-NH}_2 + \text{ClC(O)-(CH}_2)_8\text{-C(O)Cl} \longrightarrow \left[-\text{NH-(CH}_2)_6\text{-NH-C(O)-(CH}_2)_8\text{-C(O)-} \right]_n \quad (13)$$

In a tall-form beaker is placed a solution of 3.0 ml of freshly distilled sebacoyl chloride in 100 ml of distilled tetrachloroethylene. Over this solution, a solution of 4.4 gm of hexamethylenediamine in 50 ml of water is carefully placed without disturbing the surface of the acid chloride solution. The polymeric film at the interface may be grasped with tweezers or a glass rod and pulled out of the beaker as a "rope" which forms continuously. The process stops when one of the reagents is exhausted. The polymer may be washed with 50% acetone in water and dried under reduced pressure at 60°C.

A novel method for the preparation of N,N-dimethylamides uses dimethylformamide as the source of the dimethylamino group [27]. The procedure appears to be applicable to both acyl halides and anhydrides (with a trace of concentrated sulfuric acid as a catalyst).

3-6. *Preparation of* N,N-*Dimethylbenzamide* [27]

$$\text{C}_6\text{H}_5\text{-C(=O)-Cl} + \text{(CH}_3\text{)}_2\text{N-C(=O)H} \longrightarrow \text{C}_6\text{H}_5\text{-C(=O)-N(CH}_3\text{)}_2 + \text{CO} + \text{HCl} \quad (14)$$

A charge of 14.3 gm (0.102 moles) of benzoyl chloride and 15 gm (0.206 moles) of dimethylformamide is heated at 150°C for 4 hr.

The product is then isolated by fractional distillation at 157°–158°C (35 mm). The yield of N,N-dimethylbenzamide is 14.6 gm (98% of theory), mp 40°–41°C.

B. Condensations with Acid Anhydrides

The acylation of amines by acid anhydrides is very similar to the acylation with acyl halides. The reactivity of the anhydrides may be somewhat lower than that of the corresponding acyl halides, consequently a longer reaction period is usually desirable.

The general availability of anhydrides is limited. Their preparation is often more difficult than the preparation of the corresponding acyl halide. Furthermore, if an unusual acid has to be converted, in effect, two moles of acid are required to prepare one mole of amide. As will be indicated below, by use of mixed anhydrides, this difficulty may be overcome at least in some cases. Occasionally use of an anhydride may be indicated, if the product or intermediate reagents are sensitive to hydrogen chloride or if the course of the reaction must be modified.

A case in point is the acylation of L-cysteine. When acyl halides are reacted with L-cysteine, acylation of the S atom predominates [28, 29]. With acetic anhydride, on the other hand, N-acylation is possible. This example is typical of the technique used in acylations with anhydrides.

In reactions using anhydrides, large excesses of the reagent should be avoided, since appreciable quantities of the diacetyl derivative may form [30].

3-7. Preparation of N-Acetyl-L-cysteine [31]

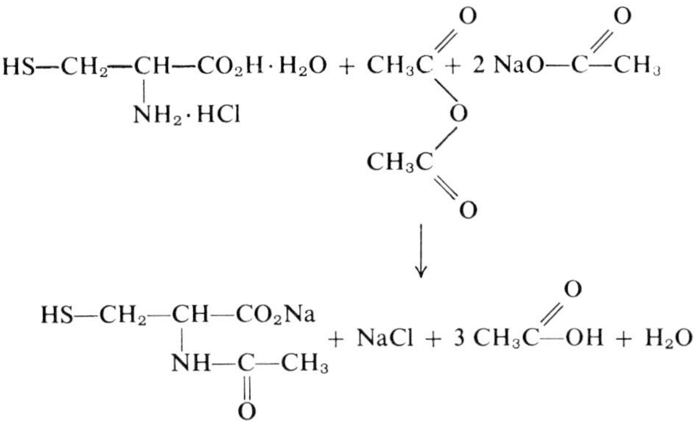

$$HS-CH_2-CH(NH_2 \cdot HCl)-CO_2H \cdot H_2O + CH_3C(=O)-O-C(=O)CH_3 + 2\,NaO-C(=O)-CH_3 \longrightarrow$$

$$HS-CH_2-CH(NH-C(=O)-CH_3)-CO_2Na + NaCl + 3\,CH_3C(=O)-OH + H_2O \quad (15)$$

$$HS-CH_2-CH(NH-C(=O)-CH_3)-CO_2Na + HCl \longrightarrow HS-CH_2-CH(NH-C(=O)-CH_3)-CO_2H + NaCl \quad (16)$$

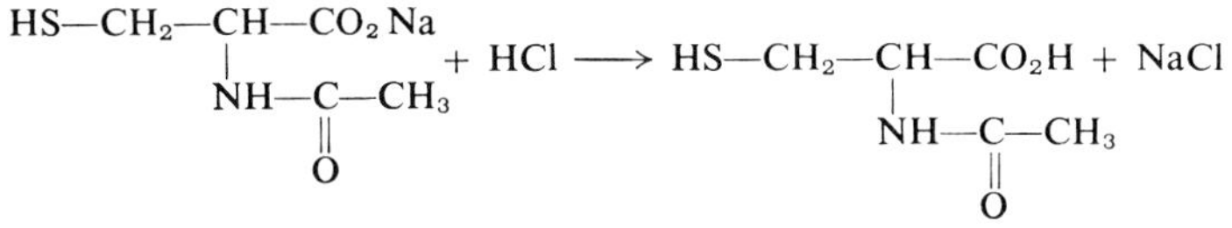

To a stirred suspension of 35.7 gm (0.2 mole) of L-cysteine hydrochloride monohydrate in 87 ml of 91% aqueous tetrahydrofuran, under nitrogen, is added 54.4 gm (0.4 mole) of sodium acetate trihydrate. While this treatment reduces the temperature to 9°C after a 20-min reaction time, the curdy mixture is cooled further to 3°–6°C and 20.8 gm (0.21 mole) of acetic anhydride is added dropwise. The suspension is stirred at room temperature for approximately 22 hr, then refluxed for 4 hr. To liberate the free acid, the mixture is cooled to 5°–10°C and anhydrous hydrogen chloride (8 gm) is bubbled into the mixture. To facilitate handling, additional THF is added, the sodium chloride is separated by filtration, and the product is isolated by cautious concentration of the filtrate under reduced pressure at 40°–50°C. The residual oil is crystallized from 35 ml of warm water (45°–50°C). Yield 26.3 gm (80.5%), in two crops, mp 109°–110°C.

Many acetylations with acetic anhydride are carried out at relatively low reaction temperatures. Then, for good measure, the reaction mixture is refluxed (for example, cf. Preparation 3-8). With modern gas–liquid chromatography and/or thin-layer chromatography procedures, it should be possible to evaluate the need for the heating step. The fact that as complex an amide as *N*-acetyl-3-*endo*-(aminomethyl)bicyclo[3.3.1]non-6-ene is prepared at room temperature within 30 min [32] indicates to us that many acetylations with acetic anhydride, and, for that matter, with acetyl chloride, may be carried out quite rapidly at modest temperatures. Even in preparation of *N*-allylacetamide (Preparation 3-8), the separate heating step may possibly be eliminated since the product is isolated by distillation.

3-8. Preparation of N-Allylacetamide [33]

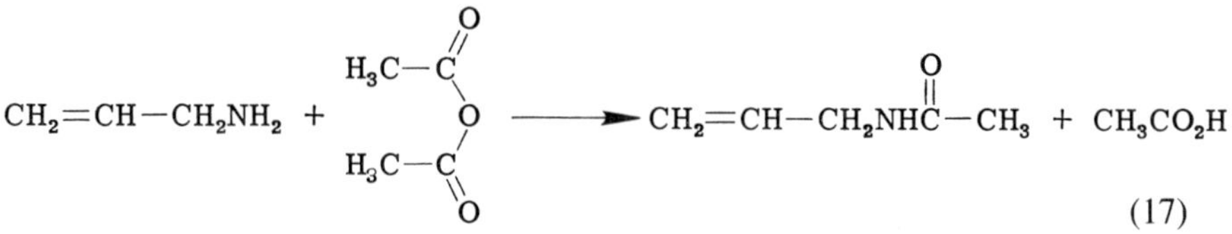

$$CH_2=CH-CH_2NH_2 + (CH_3CO)_2O \longrightarrow CH_2=CH-CH_2NHCOCH_3 + CH_3CO_2H \quad (17)$$

To 75 gm (0.735 mole) of acetic anhydride, cooled to 0°C, is added dropwise with stirring, over a 1-hr period, 25 gm (0.43 mole) of allylamine. The stirred reaction mixture is then heated at approximately 100°C for 15 hr.

Upon fractional distillation, the product is isolated at 87°C (5-mm Hg) as a colorless oil. Yield: 40.7 gm (94%). IR (CHCl$_3$): 3460 (NH free), 3340 (NH bonded), 1680 (C=O) cm^{-1} NMR (CDCl$_3$): δ 1.9(S,CH$_3$CO), 3.72 (t, J=6H$_z$, CH$_2$—N), 5.75 (m; CH=), 4.88 (m, =CH), 5.1 (dq, J=6.2 H$_z$, =CH), 6.2 (br s, NH).

Above, we stated that, in general, the use of anhydrides for the preparation of amides "wastes" one of the acyl moieties. However, this statement does not necessarily apply to cases involving cyclic anhydrides. In many cases these reactants are used to prepare cyclic imides—a matter to be discussed in greater detail below. One other aspect is of interest.

From either (+)- or (−)-tartaric acids, the corresponding (+) or (−)-diacetoxysuccinic anhydride (I) may be synthesized:

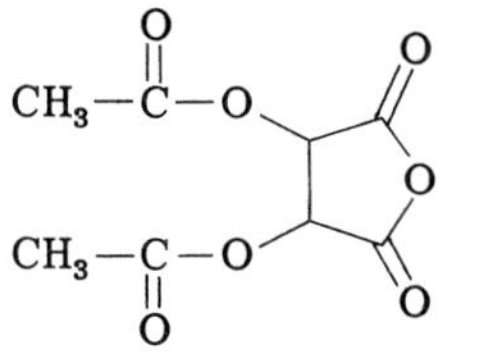

(I)

Upon reaction with substituted anilines, followed by saponification, the optically active tartaric acids (II) are produced:

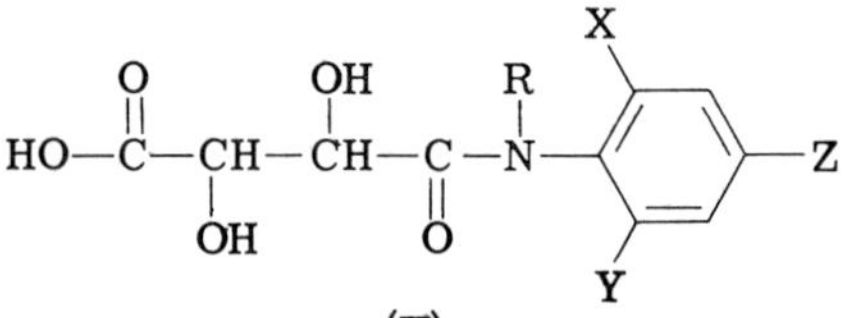

(II)

These acids are said to be exceptionally valuable in resolving several bases that are difficult to separate into their optical isomers such as ethyl-6,7-dimethoxy-1,2,3,4-tetrahydro-1-isoquinoline acetate [34].

§ 3. Condensation Reactions   333

Some acids, such as amino acids, are not conveniently converted to acyl halides for amide formation through their carboxylate function. Normal "dimeric" anhydrides also are difficult to prepare and purify even after the amino group has been suitably blocked. In the authors' laboratories, conversion of the carboxylate function of an acetylated amino acid to a mixed anhydride with isopropyl chlorocarbonate has been quite successful in the preparation of amides according to the reactions scheme shown in Eqs. (18) and (19). This technique has been applied to the preparation of amides from mixed anhydrides in general.

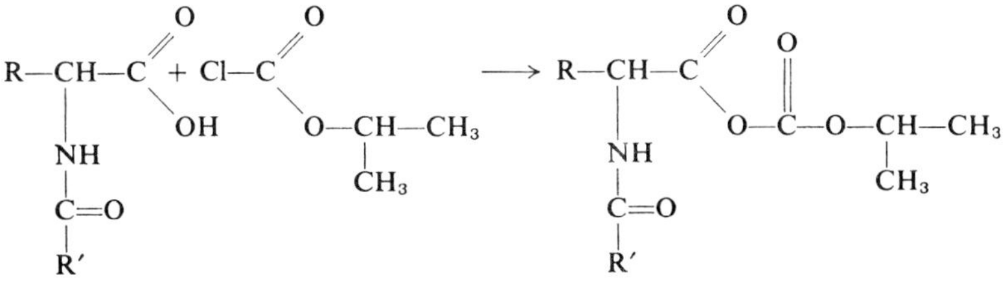

$$R-\underset{\underset{\underset{R'}{C=O}}{NH}}{CH}-\overset{O}{\overset{\|}{C}}-OH + Cl-\overset{O}{\overset{\|}{C}}-O-\underset{CH_3}{CH}-CH_3 \longrightarrow R-\underset{\underset{\underset{R'}{C=O}}{NH}}{CH}-\overset{O}{\overset{\|}{C}}-O-\overset{O}{\overset{\|}{C}}-O-\underset{CH_3}{CH}-CH_3 \qquad (18)$$

$$R-\underset{\underset{\underset{R'}{C=O}}{NH}}{CH}-\overset{O}{\overset{\|}{C}}-O-\overset{O}{\overset{\|}{C}}-O-\underset{CH_3}{CH}-CH_3 + R'-NH_2 \longrightarrow R-\underset{\underset{\underset{R'}{C=O}}{NH}}{CH}-\overset{O}{\overset{\|}{C}}-NHR' \qquad (19)$$

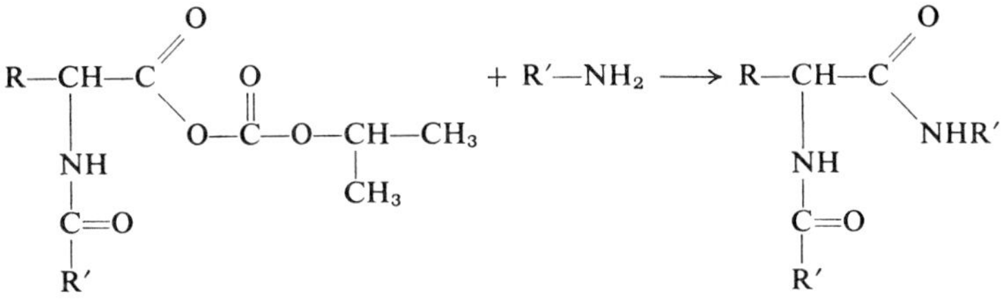

In this preparation, the exact fate of the isopropyl carbonate moiety depends somewhat on work-up conditions. In most cases, it may be presumed that carbon dioxide and isopropanol are formed.

A related example of the use of mixed anhydrides, based on ethyl chlorocarbonate, is the preparation of the rather complex amide 2-phenyl-3-(2'-furyl)propionamide [35]. As indicated above, not only can unsubstituted amides be prepared by this means, but also various substituted amides.

3-9. Preparation of 2-Phenyl-3-(2'-furyl)propionamide [35]

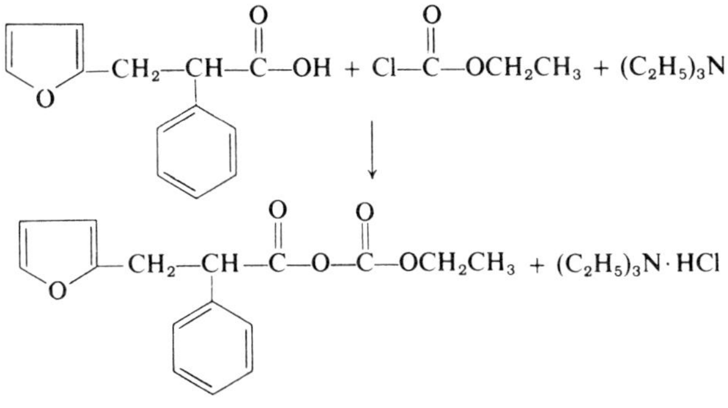

$$\text{Furyl}-CH_2-\underset{C_6H_5}{CH}-\overset{O}{\overset{\|}{C}}-OH + Cl-\overset{O}{\overset{\|}{C}}-OCH_2CH_3 + (C_2H_5)_3N$$

$$\downarrow$$

$$\text{Furyl}-CH_2-\underset{C_6H_5}{CH}-\overset{O}{\overset{\|}{C}}-O-\overset{O}{\overset{\|}{C}}-OCH_2CH_3 + (C_2H_5)_3N \cdot HCl \qquad (20)$$

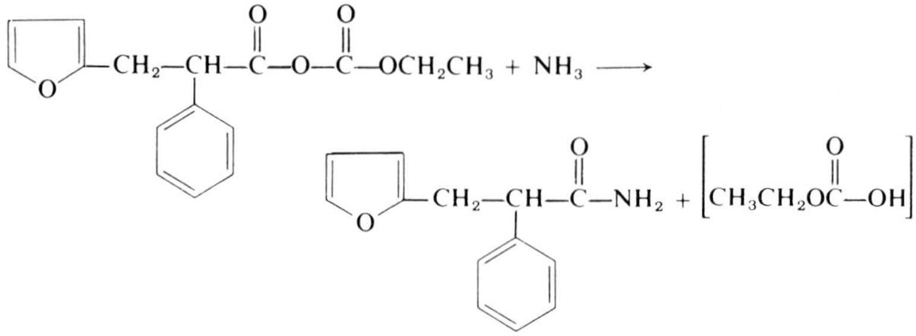

$$\text{furyl-CH}_2\text{-CH(Ph)-C(=O)-O-C(=O)-OCH}_2\text{CH}_3 + \text{NH}_3 \longrightarrow$$

$$\text{furyl-CH}_2\text{-CH(Ph)-C(=O)-NH}_2 + [\text{CH}_3\text{CH}_2\text{OC(=O)-OH}] \quad (21)$$

In a reaction setup, freed and protected from humidity, 50 gm (0.27 moles) of 2-phenyl-3-(2'-furyl)propionic acid and 30 gm (0.3 moles) of triethylamine in 140 ml of anhydrous chloroform are cooled with stirring in a Dry Ice–acetone mixture. To this mixture, a solution of 30 gm (0.28 moles) of ethyl chlorocarbonate in 75 ml of anhydrous chloroform is added dropwise. After 1 hr, while maintaining the temperature below 0°C, a stream of anhydrous ammonia is passed through the stirred reaction mixture (hood, suitable gas traps). Within approximately 30 min, the solution is saturated with respect to the ammonia. After the mixture has warmed to room temperature, 500 ml of water are added, and the chloroform layer is separated. The chloroform solution of the product is treated in turn with 4 N hydrochloric acid and water and is then dried with magnesium sulfate. The solvent is then evaporated. The residue is recrystallized from a solution of 80 ml of benzene and 35 ml of petroleum ether. Yield 40 gm (80% of theory), mp 100°–101°C.

Aqueous ammonia has been substituted successfully for the anhydrous gas in another preparation [36].

Related to the use of mixed anhydrides is the use of mixtures of acids and anhydrides. For example, by use of a mixture of formic acid and acetic anhydride, formamides have been prepared, and by the use of trifluoroacetic acid and trifluoroacetic anhydride, trifluoroacetamides have been prepared [37].

3-10. *Preparation of* **N,N'-Di(trifluoroacetyl)-meso-2,6-diaminopimelic Acid** [37]

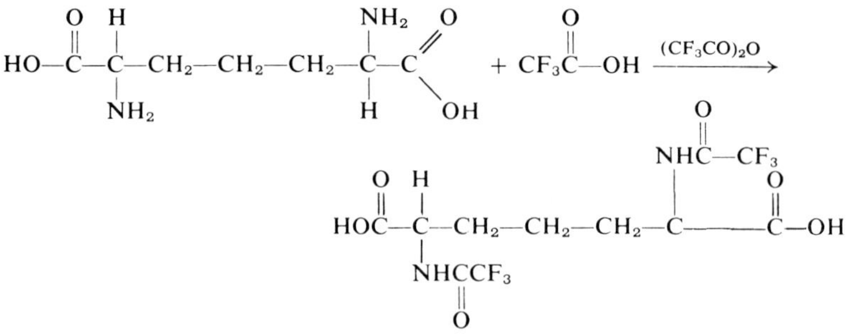

$$\text{HO-C(=O)-CH(NH}_2\text{)-CH}_2\text{-CH}_2\text{-CH}_2\text{-CH(NH}_2\text{)-C(=O)-OH} + \text{CF}_3\text{C(=O)-OH} \xrightarrow{(\text{CF}_3\text{CO})_2\text{O}}$$

$$\text{HOC(=O)-CH(NHC(=O)CF}_3\text{)-CH}_2\text{-CH}_2\text{-CH}_2\text{-CH(NHC(=O)CF}_3\text{)-C(=O)-OH} \quad (22)$$

§ 3. Condensation Reactions

To 190 mg (1 mmole) of *meso*-2,6-diaminopimelic acid dissolved in 2.24 ml of trifluoroacetic acid, stirred in an ice bath, is added dropwise 0.75 ml (5.4 mmole) of trifluoroacetic anhydride. The reaction temperature is allowed to rise gradually to room temperature. After a 16-hr period, the reaction mixture is again cooled to 0°C and another 0.75 ml of trifluoroacetic anhydride is added. Stirring is continued for 2 hr at 20°C. The reaction mixture is evaporated to dryness (hood). The residue is dissolved in anhydrous diethyl ether. The product is then precipitated from solution with petroleum ether. Yield 98% of theory. After recrystallization from ether–petroleum ether, the melting point of the product is 200°–201°C.

The acylation of amines with cyclic anhydrides, such as *o*-phthalic anhydride, pyromellitic dianhydride, or maleic anhydride usually leads to imide formation. We have observed in our own laboratories that the reaction goes in two steps and somewhat drastic conditions are usually required for ring closure to the imide stage. The usual reaction conditions are similar to those used with straight-chain anhydrides, possibly using acetic anhydride as a solvent. However, the molten dianhydride may also be used at times.

3-11. Preparation of α-Ethylglutarimide [38]

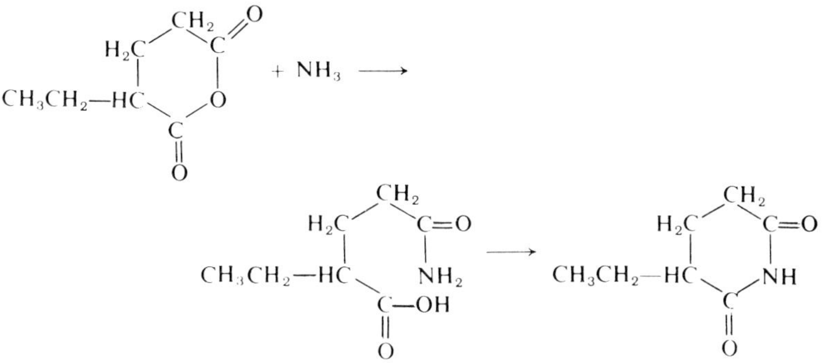

$$\text{α-ethylglutaric anhydride} + NH_3 \longrightarrow \text{amide-acid intermediate} \longrightarrow \text{α-ethylglutarimide} \qquad (23)$$

Twenty grams (0.14 mole) of α-ethylglutaric anhydride is heated to 130°–140°C. Through a suitable set of gas traps, anhydrous ammonia is rapidly passed through the anhydride (hood) for 1½ hr. The temperature of the reaction mixture is then raised slowly to 250°C. After a heating period of 25 min, the imide distills out of the reaction setup between 265° and 280°C. The product may be recrystallized from an acetone–ether mixture. Yield 85% of theory, mp 107°–108°C.

Dianhydrides, such as pyromellitic dianhydride, react with diamines to give complex polymers known as "polyimide resins" which are finding industrial applications.

C. Condensations With Esters

The aminolysis of esters has been used widely for the preparation of amides. Water and certain solvents, such as glycols, promote the reaction [39], although the presence of glycols may interfere with the isolation of amides at times. In the case of olefinic esters, addition of amine to the double bond also takes place during aminolysis, an example of the Michael condensation discussed in Chapter 13, Amines.

The aminolysis of esters may be carried out with aqueous ammonia [40], gaseous ammonia, and a variety of amines [41–43]. In the case of malonic esters, the methyl esters appear to react more readily, although ethyl esters have been used. Sodium methoxide is generally used as catalyst in either case [44].

With gaseous amines, the addition techniques mentioned under acyl halide reactions may be applied. With liquid amines, the preparation of N-ethylperfluorobutyramide may serve as a model synthesis. As to final purification procedures, solid amides may be recrystallized (alcohol, alcohol–water, petroleum ether, etc. are typical solvents), or, in many cases, distilled under reduced pressure. Liquid amides are best purified by distillation.

3-12. Preparation of N-Ethylperfluorobutyramide [41]

$$CF_3CF_2CF_2\overset{O}{\overset{\|}{C}}-OCH_3 + C_2H_5NH_2 \longrightarrow CF_3CF_2CF_2\overset{O}{\overset{\|}{C}}-NHC_2H_5 + CH_3OH \quad (24)$$

To an ice-cooled solution of 2736 gm (12 moles) of methyl perfluorobutyrate in 2 liters of ether (peroxide-free) is added dropwise 750 gm (16.7 moles) of cold ethylamine. After completion of the addition, the reaction mixture is allowed to warm to room temperature and the solvent and excess amine are separated by distillation under reduced pressure and with moderate warming. The residue is fractionally distilled; the fraction boiling quite sharply at 168°C at atmospheric pressure (759 mm) is collected. Yield 2665 gm (92.5%).

The aminolysis of esters may be carried out conveniently in the continuous reactor designed by Allen, Humphlett, and co-workers [45–51]. Figure 1 shows the apparatus assembled from standard laboratory equipment [52]. In this setup, columns A_1 and A_2 are packed with 1/8-inch glass helices and heated individually with 275-watt flexible heating tapes, 6 ft long and 1/2 inch wide, controlled by variable transformers; B is a distillation head; C, a receiving flask; D, an addition funnel; E, a thermometer; F, a condenser; G, a vented receiving flask for the more volatile coproduct; and H, an adapter. Parts B and F can be joined to adapter H with two clampable standard ground spherical joints rather than standard tapered joints. This facilitates proper assembly of the equipment.

Preparation 3-13 illustrates the use of this apparatus for the preparation of an amide. In general, the procedure consists of adding the reactants to the top of the column. As the material passes down the heated column, the product forms to some extent and passes down the column with excess reagents. Meanwhile, the volatile coproducts (ethanol in the example given) passes up the column and out of the reaction zone to condense finally in flask G.

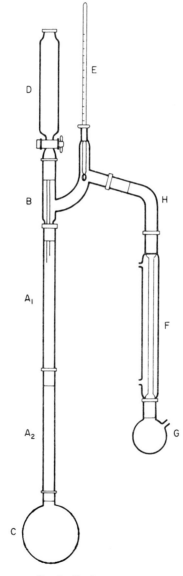

FIG. 1. Continuous reactor.

The desired product is separated from the mixture formed in the receiver, C. The unreacted materials from receiver C are recycled to funnel D with additions of the reactant which has been used up. Then the whole cycle is repeated. One of the virtues of this process is that the product is subjected to the reaction temperature for a relatively short time. Unreacted materials are recycled. By contrast, in the usual reflux procedure, reactants and products are heated together for prolonged periods of time. This frequently leads to by-product formation and thermal degradation of the product. The general applicability of this procedure, not only for the formation of amides but for many other types of functional groups, should be explored further—particularly since this approach will lend itself to scale-up procedures.

3-13. Preparation of N-(2-Methoxy-4-nitrophenyl)benzoylacetamide (Use of Continuous Reactor) [45, 48, 51]

$$\text{C}_6\text{H}_5\text{–CO–CH}_2\text{–CO–OC}_2\text{H}_5 + \text{H}_2\text{N–C}_6\text{H}_3(\text{OCH}_3)(\text{NO}_2) \longrightarrow$$

$$\text{C}_6\text{H}_5\text{–CO–CH}_2\text{–CO–NH–C}_6\text{H}_3(\text{OCH}_3)(\text{NO}_2) + \text{C}_2\text{H}_5\text{OH} \quad (25)$$

Using the apparatus shown in Figure 1, a small amount of xylene is admitted to the column while its temperature is adjusted to approximately 135°C. Then a solution of 2-methoxy-4-nitroaniline in ethyl benzoylacetate in a ratio of 8 gm of amine to 50 ml of ester is passed from the addition funnel down the heated column at a rate of approximately 10 ml/min. Ethanol distills up the column and is collected in receiver G, while the product and excess ester drops into receiver C. In 30 min about 46 gm of the amine is processed. Then the column is rinsed with 50 ml of fresh ester. The product precipitates in the receiver on cooling and is separated by filtration. The filtrate may be recycled after the addition of further quantities of the amine. The filtered product is washed with a solution of 50% petroleum ether in xylene; the yield is 89% of a yellow product, mp 178.5°–180°C. One recrystallization with acetic acid does not improve the melting point.

The commercial introduction of aprotic solvents such as various liquid amides and dimethyl sulfoxide has facilitated a variety of syntheses which were previously difficult to accomplish. The synthesis of potassium terephthalamate is a case in point.

In formamide, the potassium methyl terephthalate is efficiently ammonolyzed to the terephthalamate salt. Other amides such as methylformamide or dimethylformamide to not exhibit this accelerating effect [53].

Recently pentafluorophenyl acetate has been suggested as an acetylating agent of amines. Under mild conditions, in dimethylformamide, this reagent is said to be selective in its ability to act on amino groups in preference to hydroxyl groups. With this reagent, ethanolamine was converted to N-(2-hydroxyethyl)acetamide in a 78% yield at room temperature. Only under more vigorous conditions and in the presence of a strong base such as triethylamine the N,O-acetylated product N-(2-acetoxyethyl)acetamide was formed (84% yield) [54]. This reagent may have merit in peptide chemistry and in the case of some specialized amines, but we question whether such an expensive phenolic ester is justified for the acetylation of most amines. Surely simpler acetate esters may accomplish the same thing less expensively and equally selectively (after all, a transesterification is generally considered to be more sluggish than an ammonolysis of an ester).

Monoacetylated diamines are prepared by using an excess of the diamine over the ester as described below [55].

3-14. Preparation of Monoacetylethylenediamine [55]

$$NH_2CH_2CH_2NH_2 + C_2H_5O-\overset{O}{\underset{\|}{C}}-CH_3 \longrightarrow$$

$$NH_2CH_2CH_2NH-\overset{O}{\underset{\|}{C}}-CH_3 + C_2H_5OH \quad (26)$$

A mixture of 500 gm (6 moles) of ethyl acetate and 1550 gm (18 moles) of commercial 70% aqueous ethylenediamine is prepared and allowed to stand several days after the mixture has become homogeneous. The solution is then distilled, collecting the fraction boiling between 115°–130°C at 5 mm. Upon redistillation, 365 gm (60%) of monoacetylethylenediamine is collected between 125° and 130°C.

Phosphoramides have been suggested as reagents for the conversion of malonic esters to ester amides and sym-secondary amides [56]. An example of this reaction is given in Eq. (27) for the preparation of alkyl N,N-dimethylmalonamate.

$$CH_3CH_2-O\overset{O}{\underset{\|}{C}}-CH_2-\overset{O}{\underset{\|}{C}}-OCH_2CH_3 + [(CH_3)_2N]_3P \longrightarrow$$

$$(CH_3)_2N-\overset{O}{\underset{\|}{C}}-CH_2\overset{O}{\underset{\|}{C}}-OCH_2CH_3 + C_2H_5OP[N(CH_3)_2]_2 \quad (27)$$

Although, the yield for this reaction is reported to be 75.5%, the critical reagent, hexamethylphosphoramide, is considered to be highly carcinogenic. Therefore this reaction should not be used.

The acetylation of amines with isopropenyl acetate appears to be a transition between the highly exothermic reactions of acyl halides and anhydrides with amines on the one hand and the reaction of amines with more conventional esters on the other. While this reagent is of particular value in the preparation of enol acetate, it has been used for the preparation of amides. One interesting aspect of its use is that acetone forms as a coproduct which may distill off as the reaction proceeds. Isopropenyl acetate and other isopropenyl esters may also be used to *N*-acylate amides and imides. By the judicious selection of starting amides and isopropenyl esters, tertiary amides with three different acyl groups may be synthesized. This may very well be one of very few reaction systems which permits the synthesis of this rare group of tertiary amides. Preparation 3-23, below, is an example of the use of higher isopropenyl esters in the acylation of cyclic imides.

3-15. Preparation of N-(2-Hydroxyethyl)acetamide [58]

$$HO-CH_2CH_2-NH_2 + CH_2=C(CH_3)-O-C(=O)-CH_3 \longrightarrow$$

$$HO-CH_2CH_2-NHC(=O)-CH_3 + CH_3-C(=O)-CH_3 \quad (28)$$

To 100 gm (1.0 moles) of isopropenyl acetate in a small still is gradually added 63 gm (1.03 moles) of ethanolamine. A vigorous reaction takes place and acetone is distilled off continuously. The residue is distilled at 135°–140°C (1 mm). Yield 85.5 gm (81% of theory).

D. Transamidation Reactions

a. REACTIONS BETWEEN AMINES AND AMIDES

Formanilides have been prepared by a base-catalyzed transamidation with dimethylformamide [59]. In this example, DMF serves as the source of the carboxylate-function in the presence of amines.

3-16. Preparation of 2-Iodoformanilide [59]

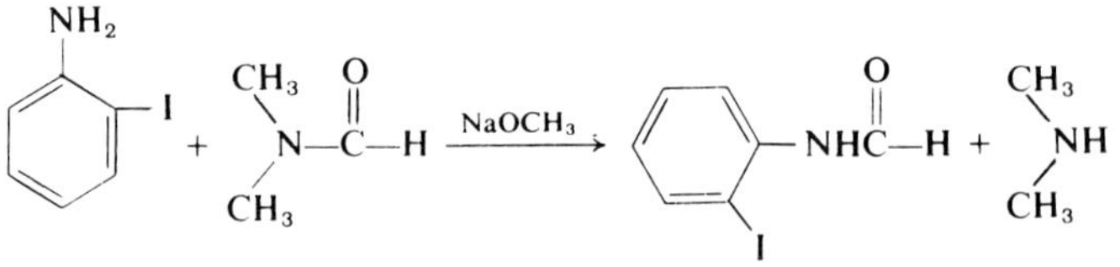

$$\text{2-I-C}_6\text{H}_4\text{-NH}_2 + (CH_3)_2N-C(=O)-H \xrightarrow{\text{NaOCH}_3} \text{2-I-C}_6\text{H}_4\text{-NHC(=O)-H} + HN(CH_3)_2 \quad (29)$$

§ 3. Condensation Reactions

A mixture of 16.2 gm (0.3 mole) of sodium methoxide and 32.9 gm (0.15 mole) of 2-iodoaniline in 150 ml of dimethylformamide is refluxed for 30 min. The product is then isolated by diluting the reaction mixture with water. Yield 25.2 gm (68%), mp 113°–113.5°C.

A more general method of acylating amides consists of the rapid heating of an amide with the hydrochloride of the desired aliphatic or aromatic amine [60].

Preparation 3-17 is given here as an illustration only of the method used in reacting an amine hydrochloride with an amide. CAUTION: The salt α-naphthylamine hydrochloride and its free base α-naphthylamine are considered highly carcinogenic and are no longer available.

3-17. Preparation of N-(α-Naphthyl)acetamide [60]

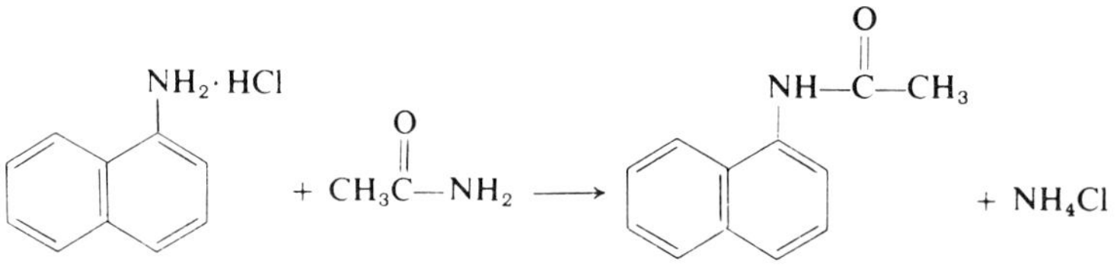

$$\text{Naphthyl-NH}_2 \cdot \text{HCl} + \text{CH}_3\overset{\overset{\text{O}}{\|}}{\text{C}}-\text{NH}_2 \longrightarrow \text{Naphthyl-NH}-\overset{\overset{\text{O}}{\|}}{\text{C}}-\text{CH}_3 + \text{NH}_4\text{Cl} \quad (30)$$

A mixture of 17.9 gm (0.1 mole) of α-naphthylamine hydrochloride and 10 gm (0.17 mole) of acetamide is heated until the precipitation of ammonium chloride is complete (a few minutes). The ammonium chloride is separated by repeated washing with warm water. The product is isolated by filtration, followed by recrystallization from ethanol. Yield 14.8 gm (80%), mp 132°C.

b. REACTIONS BETWEEN ACIDS AND AMIDES

At elevated temperature acidolysis of amides (or transamidation) with carboxylic acids may be carried out [61]. The general applicability of this reaction requires further study.

3-18. Preparation of Oleamide [61]

$$\text{CH}_3-(\text{CH}_2)_7-\text{CH}=\text{CH}-(\text{CH}_2)_7-\overset{\overset{\text{O}}{\|}}{\text{C}}-\text{OH} + \text{NH}_2\overset{\overset{\text{O}}{\|}}{\text{C}}-\text{H} \longrightarrow$$
$$\text{CH}_3(\text{CH}_2)_7-\text{CH}=\text{CH}-(\text{CH}_2)_7-\overset{\overset{\text{O}}{\|}}{\text{C}}-\text{NH}_2 + \text{HO}\overset{\overset{\text{O}}{\|}}{\text{C}}\text{H} \quad (31)$$

A mixture of 28.3 gm (0.1 mole) of oleic acid and 5.4 gm (0.12 mole) of formamide is heated in a reflux setup fitted with gas inlet tube, thermometer, and a steam-heated condenser for about ½ hr to 185°C. The temperature is then raised to 230°C (1 hr) and held at that temperature for ½ hr. The reaction mixture is then cooled and dissolved in 65 ml of acetone, treated with activated charcoal, and filtered. Enough acetone is then added to raise the solvent-to-solute ratio to 3 ml/gm and the solution is cooled to 0°C. The product precipitates. Yield 16.7 gm (59%). Upon recrystallization from acetone, a 50% overall yield (14.0 gm) is obtained, mp 75.0°–75.5°C.

E. N- and C-Alkylation and N-Acylation of Amides

N-Hydroxymethylated amides are readily prepared from an amide and formalin [62]. Other aldehydes react similarly.

3-19. Preparation of N-Hydroxymethylphenylacetamide [62]

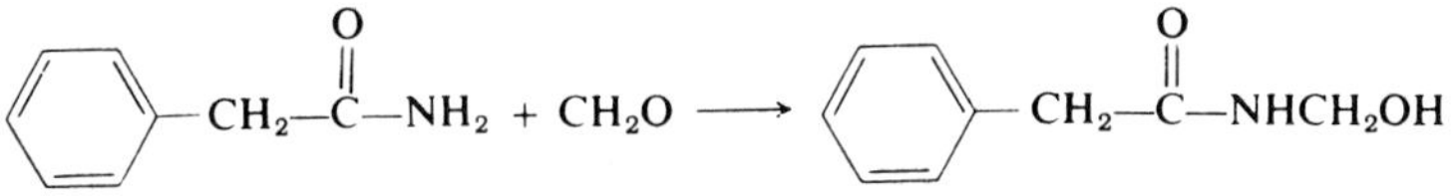

$$\text{C}_6\text{H}_5-\text{CH}_2-\overset{\text{O}}{\underset{\|}{\text{C}}}-\text{NH}_2 + \text{CH}_2\text{O} \longrightarrow \text{C}_6\text{H}_5-\text{CH}_2-\overset{\text{O}}{\underset{\|}{\text{C}}}-\text{NHCH}_2\text{OH} \quad (32)$$

In a water bath, a dispersion of 50 gm (0.37 mole) of phenylacetamide, 50 ml of a 4% aqueous solution of potassium carbonate, and 40 ml of 40% formalin are warmed until completely dissolved. After cooling to room temperature and allowing the mixture to stand for 12 hr, the product separates and is collected by filtration.

The crude product is washed in turn with dilute sodium hydroxide and with water. After the crystalline mass has been dried, it may be recrystallized from toluene. Yield 51 gm (83%), mp 78°C.

With sufficiently strong bases, the unsubstituted amino group of an amide exhibits its acidic character, a fact which may be utilized to prepare N-alkyl-substituted amides [63]. This reaction is, of course, of particular value in the case of phthalimides which may be reacted with alkyl halides to give N-alkyl-phthalimides in the Gabriel amine synthesis.

3-20. Preparation of 2-(4-Methoxyphenyl)-N-methylacetamide [63]

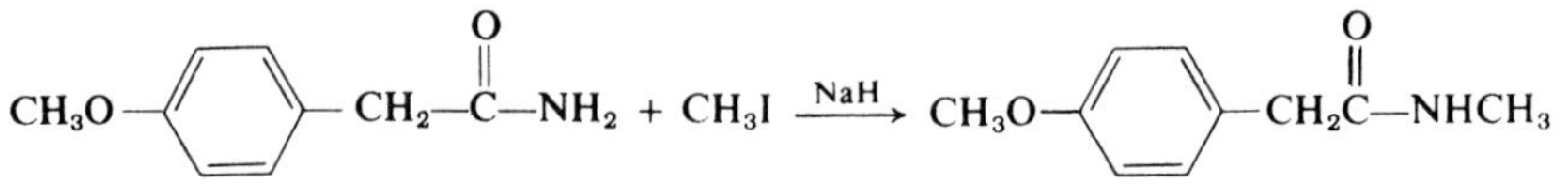

$$\text{CH}_3\text{O}-\text{C}_6\text{H}_4-\text{CH}_2-\overset{\text{O}}{\underset{\|}{\text{C}}}-\text{NH}_2 + \text{CH}_3\text{I} \xrightarrow{\text{NaH}} \text{CH}_3\text{O}-\text{C}_6\text{H}_4-\text{CH}_2\overset{\text{O}}{\underset{\|}{\text{C}}}-\text{NHCH}_3 \quad (33)$$

In a dry nitrogen atmosphere, to 1.4 gm (0.058 mole) of sodium hydride in 50 ml of anhydrous xylene is added 8.3 gm (0.05 mole) of 2-(4-methoxyphenyl)acetamide in 200 ml of anhydrous xylene. The mixture is stirred and refluxed under nitrogen for 24 hr. The reaction mixture is cooled, 20 gm (0.14 mole) of methyl iodide is added, and refluxing is continued for 8 hr. The mixture is then filtered hot. The filtrate is evaporated to dryness. The residue is distilled under reduced pressure. Yield 7.2 gm (80%), bp 137°–140°C (3.5 mm), mp 57.9°C.

The classical alkylation of phthalimide has been reported to be catalyzed by dimethylformamide [64]

3-21. *Preparation of* N-*(9-Bromononyl)phthalimide* [64]

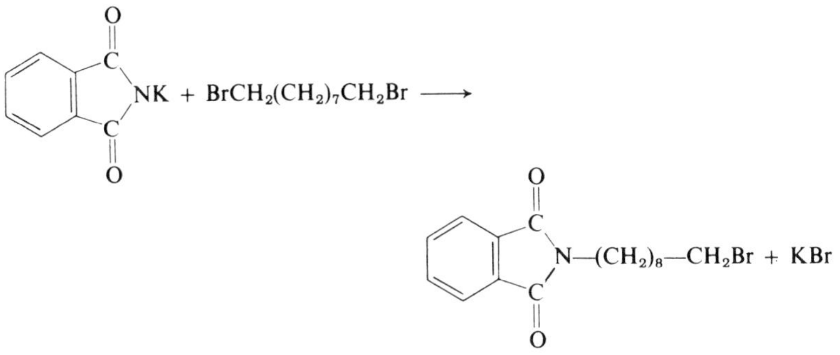

$$\text{Phthalimide-NK} + \text{BrCH}_2(\text{CH}_2)_7\text{CH}_2\text{Br} \longrightarrow \text{Phthalimide-N—(CH}_2)_8\text{—CH}_2\text{Br} + \text{KBr} \quad (34)$$

A dispersion of 8.45 gm (0.05 mole) of potassium phthalimide, 57.2 gm (0.2 mole) of 1,9-dibromononane, and 3.3 gm of dimethylformamide is heated for 1½ hr at 160°C. The precipitated potassium bromide is separated by filtration, and the excess dibromononane and dimethylformamide are removed by distillation under reduced pressure. The residue is fractionated under reduced pressure. The solid portion of the distillate is recrystallized from ethanol. Yields vary from 40 to 78%, depending on minor variations in procedure; mp 37.5°C.

The arylation of acetamides with aryl bromide in the presence of potassium carbonate and cuprous iodide has been called the Goldberg reaction. This process has recently been revived since it appears to be more general in applicability than the Ullman condensation and does not require imidoyl chlorides as intermediates as does the Chapman rearrangement—two more reactions which may also be considered for the formation of *N*-arylamides [65]. One limitation of the Goldberg reaction which was recently discovered was that not only aryl bromides but also aryl chlorides arylate acetamides. This fact was uncovered when Freeman, Butler, and Freedman [65] reacted 4[chloroacetanilide (I) with 1-bromo-4-chlorobenzene (II), saponified the reaction mixture, and isolated three amines as indicated in Eqs. (35–39).

In the product mixtures, product VI was present to the extent of 28.6%; product VII, 62.5%; and product VIII, 8.5%. As indicated, some of the arylations involved reactions of chlorosubstituents, others the bromo substituents.

§ 3. Condensation Reactions

The preparation of N-(o-tolyl)-2',6'-acetoxylidide is an illustration of the Goldberg reaction using reagents with ortho substituents that might be expected to interfere with the reaction but, in fact, do not.

3-22. Preparation of N-(o-Tolyl)-2',6'-acetoxylidide (Goldberg Reaction) [65]

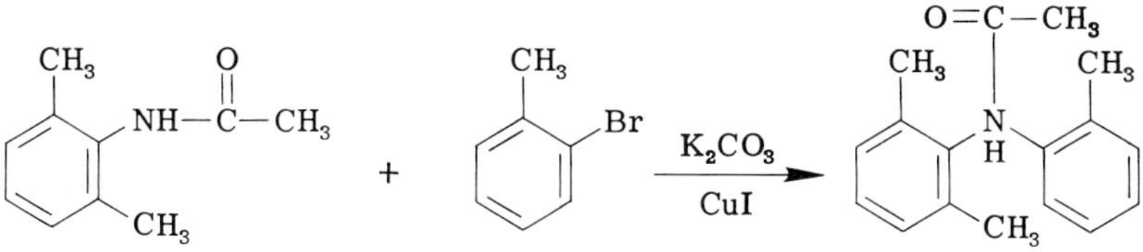

<chemical scheme (40)>

A mixture of 20.0 gm (0.123 mole) of 2',6'-acetoxylidide, 31.5 gm (0.184 mole) of o-bromotoluene, and 2.5 gm of cuprous iodide are heated at approximately 130°C until the mixture becomes stirrable. Then 17.0 gm of anhydrous potassium carbonate is added. The reaction mixture is then cooled and the crude product is extracted with 600 ml of benzene.

The solvent is removed by cautious distillation on a rotary evaporator. The residual unreacted o-bromotoluene is distilled off by distillation of the residue under reduced pressure. The residue is recrystallized twice from ether. Yield 26.2 gm (84%, mp 147°–150°C. HNMR (CDCl$_3$) δ1.93 (s, 3, COCH$_3$), 2.25 (5,6 ArCH$_3$) 2.35 (s, 3, ArCH$_3$), 6.60–7.40 ArH), IR(Nujol) 167 cm^{-1}(C=O).

Amides and imides may be acylated with higher isopropenyl esters to give tertiary amides [57]. This procedure permits the preparation of tertiary amides with three different acyl groups, although usually better yields are obtained when two similar acyl groups are present. This procedure has been applied to cyclic imides (e.g., succinimide, maleimide, phthalimide, barbituric acids), cyclic imide-amides (e.g., spirohydantoins), N-alkyl amides, and N-aryl amides with enol esters of long-chain esters.

3-23. Preparation of N-Stearoylsuccinimide [57]

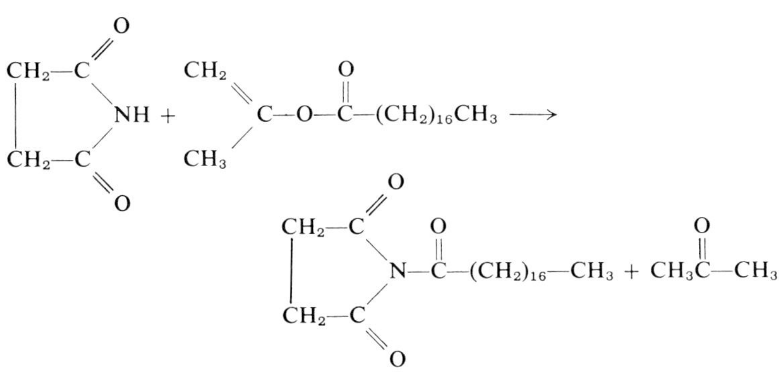

<chemical scheme (41)>

In a flask submerged in a heating bath, 11.9 gm (0.12 mole) of succinimide, 38.9 gm (0.12 mole) of isopropenyl stearate, and 100 mg of 4-toluenesulfonic acid monohydrate are heated for 20 min at 190°C, using a magnetic stirrer as soon as the mixture becomes fluid. The reaction is considered complete when the evolution of acetone ceases. The reaction mixture is cooled, and the residue is recrystallized from pentane (charcoal). Yield 38 gm (87%), mp 95°–96°C.

Many amides may be classified as active methylene compounds. Consequently C-alkylation of the active methylene group with alkyl halides has been carried out. Typically, sodium amide in liquid ammonia is used for the alkylation of N,N-disubstituted amides and N-alkylated-2-pyrrolidone [66].

In the Reformatskii reaction, N,N-dimethyl-α-haloamides may be substituted for the usual α-halo esters to give N,N-disubstituted β-hydroxyamides [67].

3-24. Preparation of N,N-Diethyl-1-hydroxy-4-methylcyclohexaneacetamide [67]

$$\begin{array}{c}CH_3CH_2\\ \diagdown\\ N-\overset{\overset{\displaystyle O}{\|}}{C}-CH_2Br\\ \diagup\\ CH_3CH_2\end{array} + \begin{array}{c}O\\ \|\\ C\\ \diagup\diagdown\\ CH_2CH_2\\ ||\\ CH_2CH_2\\ \diagdown\diagup\\ CH\\ |\\ CH_3\end{array} \xrightarrow{Zn}$$

$$CH_3-CH\begin{array}{c}CH_2-CH_2\\ \diagup\diagdown\\ \\ \diagdown\diagup\\ CH_2-CH_2\end{array}\begin{array}{c}OH\\ |\\ C\\ \end{array}CH_2-\overset{\overset{\displaystyle O}{\|}}{C}-N\begin{array}{c}CH_2CH_3\\ \diagup\\ \\ \diagdown\\ CH_2CH_3\end{array} \quad (42)$$

In thoroughly dried equipment, a calculated quantity of 95% zinc–5% copper alloy turnings, a crystal of iodine, and 15 ml of a solution prepared by dissolving 38.8 gm (0.2 mole) of N,N-diethyl-α-bromoacetamide (CAUTION: lachrymatory and sternutatory) and 22.4 gm (0.2 mole) of 4-methylcyclohexanone in 150 ml of toluene are cautiously heated to reflux with vigorous stirring until reaction starts. After the reaction has subsided, the remainder of the solution is added at such a rate that gentle refluxing is maintained. Then the mixture is heated for an additional hour, cooled, and cautiously poured into 200 ml of cold 10% sulfuric acid. The copper and zinc-containing co-products are removed by filtration. The organic layer is washed in turn with a 5% solution of sulfuric acid, potassium carbonate solution, and a saturated salt solution until neutral. The solution is dried, the solvent is evaporated under reduced pressure, and the product is distilled at 104°–105°C (0.3 mm). Yield 25 gm (59%).

F. Grignard Condensation Reactions

The reaction of isocyanates with Grignard reagents often affords a good yield of amides [68].

3-25. Preparation of N-(2-Furyl)benzamide [68]

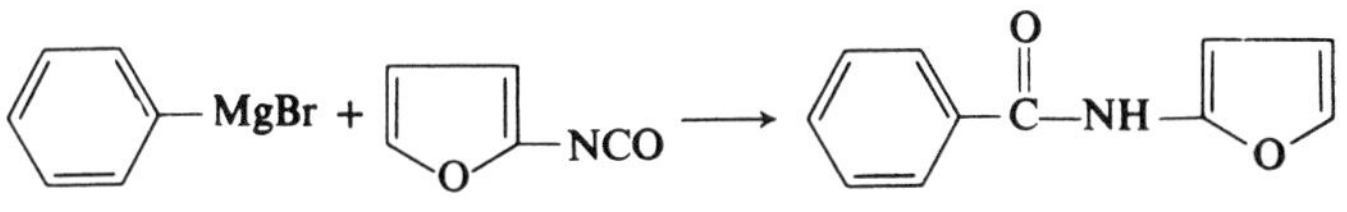

$$\text{C}_6\text{H}_5-\text{MgBr} + \text{(2-furyl)}-\text{NCO} \longrightarrow \text{C}_6\text{H}_5-\overset{\text{O}}{\underset{\|}{\text{C}}}-\text{NH}-\text{(2-furyl)} \quad (43)$$

In thoroughly dried equipment, 18 gm (0.1 mole) of phenylmagnesium bromide is prepared in 75 ml of anhydrous ether. The solution is cooled to 0°C and 8.9 gm (0.082 mole) of 2-furyl isocyanate is added dropwise with stirring. The resultant product is hydrolyzed with an excess of a 15% solution of ammonium chloride in water and then extracted with ether. The combined ethereal solutions are dried over Drierite. The volume of the solution is reduced to 50 ml by distillation; crystals of N-(2-furyl)benzamide separate and are isolated by filtration. A single recrystallization from benzene affords a nearly white product, yield 12.3 gm (80%), mp 124.5°C.

4. ADDITION REACTIONS

Acrylonitrile has been used to cyanoethylate certain active hydrogen compounds, which on hydrolysis gives the amide. Acrylamide may occasionally be used as an intermediate in a Michael-type condensation. This process was originally patented by Bruson in 1945 [69].

With 40% potassium hydroxide, the addition of nitroethane to two moles of acrylamide afforded 4-nitro-4-methylheptanediamide. This reaction is satisfactory when carried out with nitroethane, 1-nitropropane, and 1-nitrobutane. Phenylnitromethane gave only 4-phenyl-4-nitrobutyramide, which is the result of the addition of one mole of phenylnitromethane to acrylamide [70].

4-1. Preparation of 4-Nitro-4-methylpimeldiamide [70]

$$2\ \text{CH}_2=\text{CH}-\overset{\text{O}}{\underset{\|}{\text{C}}}-\text{NH}_2 \xrightarrow[\text{CH}_3\text{CH}_2\text{NO}_2]{40\%\ \text{KOH}} \text{NH}_2-\overset{\text{O}}{\underset{\|}{\text{C}}}-\text{CH}_2\text{CH}_2-\underset{\underset{\text{CH}_3}{|}}{\overset{\overset{\text{NO}_2}{|}}{\text{C}}}-\text{CH}_2\text{CH}_2-\overset{\text{O}}{\underset{\|}{\text{C}}}-\text{NH}_2 \quad (44)$$

In 250 ml of dioxane, 80 gm (1.12 mol) of acrylamide and 30 gm (0.40 mol) of nitroethane are stirred. Then 40% aqueous potassium hydroxide is added cautiously until the pH of the mixture is approximately 8.5. After the initial heat of reaction has subsided, the green reaction mixture is warmed with stirring at 50°C for about 10 hr, until the mixture is distinctly orange in color. Then the reaction mixture is cautiously neutralized with dilute hydrochloric acid. The solvents are evaporated to dryness. The product is recrystallized from isopropanol. Yield 34.2 gm, 39%, mp 134.5°–135°C.

5. HYDRATION REACTIONS WITH NITRILES

The acidic hydration of nitriles is carried out in the presence of a limited amount of water to minimize the hydrolysis of the resultant amide to the free acid. One method of controlling the water concentration involves the use of a solution of water in sulfuric acid corresponding approximately to sulfuric acid monohydrate (84.5% H_2SO_4) in composition. This acid tends to form an amide acid sulfate salt from which the free amide may be isolated by neutralization (with calcium oxide in commercial practice [71], or with ammonia as in the example below [72]).

This method may be used with sensitive nitriles such as acrylonitrile [71]. It even permits retention of steric configurations, as in the hydration of cis-β-chloroacrylonitrile to cis-β-chloroacrylamide, cited here as a typical example of the procedure [72], although the reported yield is not as high as it might be.

While, in principle, hydrochloric acid may be used in the hydration of nitriles, in the case of acrylonitrile [73], some of the hydrogen chloride may add to the double bond.

5-1. *Preparation of* cis-β-*Chloroacrylamide* [72]

$$\underset{Cl}{\overset{H}{\diagdown}}C=C\underset{CN}{\overset{H}{\diagup}} + H_2SO_4 + H_2O \longrightarrow \underset{Cl}{\overset{H}{\diagdown}}C=C\underset{CONH_2H_2SO_4}{\overset{H}{\diagup}}$$

$$\underset{Cl}{\overset{H}{\diagdown}}C=C\underset{CONH_2\cdot H_2SO_4}{\overset{H}{\diagup}} + NH_3 \longrightarrow \underset{Cl}{\overset{H}{\diagdown}}C=C\underset{CONH_2}{\overset{H}{\diagup}} + NH_4HSO_4 \quad (45)$$

To a solution of 113.2 gm of 95.4% sulfuric acid and 14.6 gm of water (equivalent to 1.1 moles of H_2O in an 85.3% H_2SO_4 solution) is gradually added 87.5 gm (1 mole) of cis-β-chloroacrylonitrile while maintaining the reaction temperature between 85°C and 90°C, first by cooling with an ice bath, later, as the reaction subsides, by heating. Warming is continued for 90 min after completion of the addition. The reaction is then cooled to 40°C and poured, with vigorous stirring, into a mixture of 400 gm of ice and 145 ml of concentrated

§ 5. Hydration Reactions with Nitriles

ammonia solution while maintaining the reaction temperature below 35°C with external cooling.

The crude product is filtered off. The solid is then extracted with several portions of hot acetone. The acetone solution of the product is then evaporated to dryness. The residue is recrystallized from the ethyl acetate. Yield 51.5 gm (48.9% of theory), mp 111°–112°C.

The alkaline hydration of nitriles to the amide stage in the presence of hydrogen peroxide has been described [74–76]. While this procedure has been applied widely with excellent results, careful control of reaction conditions must be exercised and precautions against explosions must be taken since large volumes of oxygen may evolve suddenly.

5-2. Preparation of 6-Methoxy-2-nitrobenzamide [77].

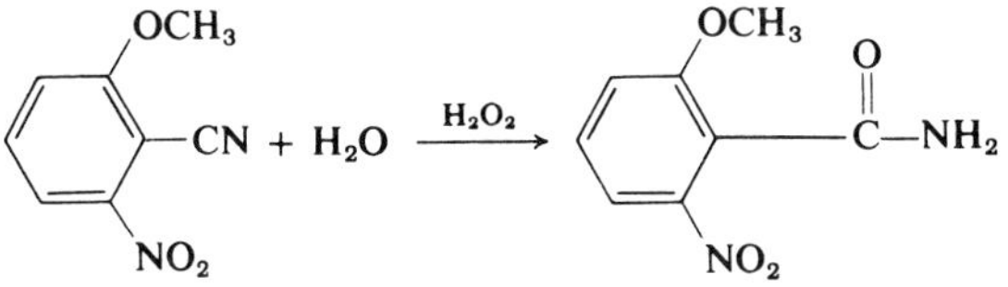

$$\text{2-OCH}_3\text{-6-NO}_2\text{-C}_6\text{H}_3\text{-CN} + \text{H}_2\text{O} \xrightarrow{\text{H}_2\text{O}_2} \text{2-OCH}_3\text{-6-NO}_2\text{-C}_6\text{H}_3\text{-C(O)NH}_2 \quad (46)$$

To 1 gm (0.0056 mole) of 6-methoxy-2-nitrobenzonitrile in 50 ml of ethanol is added 0.8 ml of 6 N sodium hydroxide and 50 ml of 10% hydrogen peroxide. The mixture is heated cautiously for 1 hr at 40°–50°C. Then the reaction system is evaporated to dryness. The residue is extracted with ether; the ether solution is then cautiously evaporated to dryness. The new residue is recrystallized from ethanol to yield 1 gm (91%), mp 195°C.

In a recent application of this reaction, a suitably substituted epoxynitrile was converted to a hydroxy-lactam in good yield (Preparation 5-3).

5-3. Preparation of 2-exo-Hydroxy-4-azahomoadamantan-5-one [32]

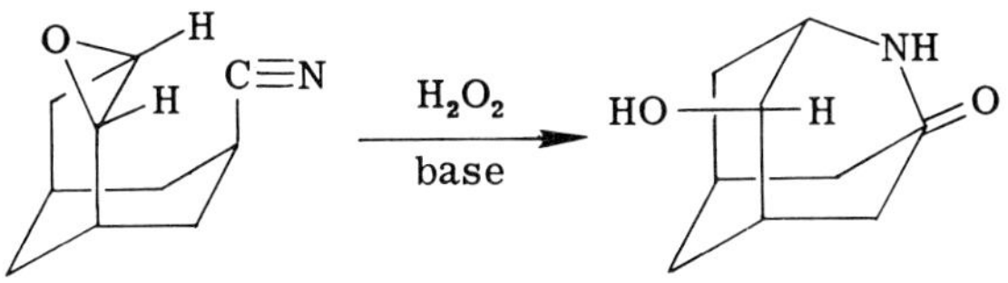

$$\text{epoxybicyclo[3.3.1]nonane-carbonitrile} \xrightarrow[\text{base}]{\text{H}_2\text{O}_2} \text{2-exo-hydroxy-4-azahomoadamantan-5-one} \quad (47)$$

A mixture of 1.63 gm (0.010 mole) of 6,7-*exo*-epoxybicyclo[3.3.1]nonane-3-*endo*-carbonitrile (for synthesis, see Ref. [32] also) in 20 ml of ethanol is cooled in an ice-water bath. To this are added 4 gm (0.035 mole) of 30% aqueous hydrogen peroxide (CAUTION: oxidizing agent, contact with the skin can lead to severe burns) and 0.6 mole of 6 N-sodium hydroxide. The reaction system is stirred at the ice bath temperature for 0.5 hr. The mixture is then allowed to remain at room temperature for 25 hr. Then the solvent is evaporated on a rotary evaporator. To the residue is added 30 ml of water. The solution is saturated with sodium chloride and extracted with 4 × 100 ml portions of chloroform. The chloroform extract is dried with sodium sulfate, anhydrous. The solution is filtered. The solvent is evaporated off and the residue is sublimed at 190°C (0.05 mm). The yield is 1.41 gm (78%), mp 302°–307°C (sealed tube). IR (CHCl$_3$) 3595, 3410, 3275, 2970, 2850, 1647, 1465, 1445, 1355, 1300, 1120, 1070, 1035, 990, 980, 955, 915, 900, 800 cm^{-1}.

A new variation on the base-induced hydration of nitriles to amides makes use of sodium superoxide in dimethylsulfoxide. This particular reaction is catalyzed by nitrobenzene. The reaction mechanism involved is not as yet well understood [78].

5-4. Preparation of Cyclohexylcarboxamide [78]

$$\text{Cyclohexyl-CH-CN} \xrightarrow[\text{DMSO}]{\text{NaO}_2} \text{Cyclohexyl-CH-C(=O)-NH}_2 \quad (48)$$

A three-necked flask, which is fitted with two pressure equalizing addition tubes, a two-way stopcock, and a magnetic stirrer, is swept with dry air while being assembled and maintained under a slight positive pressure of dry air throughout the process.

By use of a glove bag, under dry nitrogen, 15 ml of dimethylsulfoxide (which had been dried over a 4A-type molecular sieve) is introduced into the flask, 1.78 gm (0.032 mole) of 90+% pure sodium superoxide is introduced into one of the addition tubes, and 1.0 gm (0.0092 mole) of cyclohexyl cyanide is introduced into the other addition tube. The apparatus is then removed from the glove bag and mounted on a magnetic stirrer plate. With stirring, the nitrile is added to the dimethylsulfoxide. After it has completely dissolved, the sodium superoxide is introduced into the flask. The system is stirred at room temperature for 7 hr. After this, the mixture is poured into ice water and extracted with ethyl acetate.

The ethyl acetate extract is washed with cold water and then dried over anhydrous sodium sulfate. The solvent is then evaporated off at room temperature. The residue weighed 0.98 gm, mp 178°–182°C. Upon crystallization from water, 0.93 gm is isolated (85% yield), mp 185°–186°C. IR (KBr) 3340, 3180, 1630 cm^{-1}.

In the Ritter reaction, nitriles are reacted with secondary or tertiary alcohols or olefins in strongly acidic systems to produce N-alkylated amides. A large variety of olefins (e.g., 2-methyl-2-butene, mesityl oxide, chalcone) and of nitriles and

§ 7. Reduction Reactions 351

dinitriles have been subjected to the reaction which formally resembles an alcoholysis of nitriles [79–84]. Among the electrophilic reagents used in the reaction are not only concentrated sulfuric acid but also nitronium or nitryl fluoroborate (NO_2BF_4) [85, 86] and nitrosonium hexafluorophosphate ($NOPF_6$) [87]. With the latter reagent, adamantane reacted readily with many different nitriles to afford the corresponding N-(1-adamantyl)amides in high yields directly [87].

The example of the Ritter reaction given here makes use of more conventional reagents.

5-5. Preparation of N,N-Diisopropylfumaramide [85]

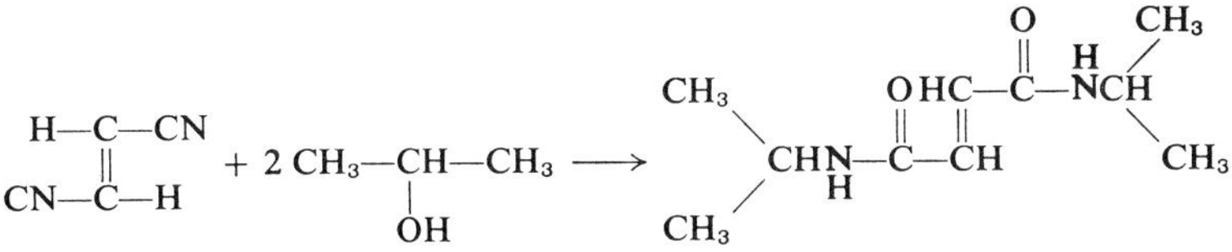

$$\begin{array}{c} H-C-CN \\ \parallel \\ CN-C-H \end{array} + 2\,CH_3-CH(OH)-CH_3 \longrightarrow \begin{array}{c} CH_3 \\ \diagdown \\ CHN-C-CH \\ \diagup \quad \parallel \quad \parallel \\ CH_3 \quad H \quad O \end{array} \begin{array}{c} O \quad CH_3 \\ \parallel \quad H \diagup \\ OHC-C-NCH \\ \parallel \quad \diagdown \\ CH \quad CH_3 \end{array} \quad (49)$$

A solution of 3.9 gm (0.05 mole) of fumaronitrile (CAUTION: lachrymator and vesicant) in 20 ml of concentrated sulfuric acid is prepared at room temperature. While maintaining this solution below 45°C with an ice bath, 6.0 gm (0.1 mole) of isopropanol is added over a 20-min period. The reaction mixture is then cautiously poured into cold water. The crude product which precipitates is filtered off, slurried with 50 ml of dilute sodium carbonate solution, washed with water, and then recrystallized from water. Yield 80%, mp 320°C after darkening at 225°C. A further crop may be obtained by neutralizing the acid filtrate.

6. OXIDATION OF ALKYLENE AMINES

Recently a method was devised for the oxidation of the methylene group of arylmethylamines to amides [Eq. (52)].

$$ArCH_2NH_2 \longrightarrow Ar\overset{\overset{\displaystyle O}{\parallel}}{C}-NH_2 \quad (50).$$

To accomplish this, the arylmethylamine is converted to a Schiff base by reaction with 2,6-di-t-butyl-p-benzoquinone. Oxygenation of dispersions of the Schiff bases in tert-butanol–potassium tert-butoxide or in ethanol–potassium hydroxide with gaseous oxygen yielded amides that could readily be isolated from the reaction mixture [88].

7. REDUCTION REACTIONS

Nitroarenes have been reduced under mild conditions in the presence of molybdenum hexacarbonyl in a carboxylic acid to form anilides directly (Eq. 53).

$$\text{ArNO}_2 \xrightarrow[\text{RCO}_2\text{H}]{\text{Mo(CO)}_6} \text{ArNHCR} \quad \overset{\text{O}}{\underset{\|}{}} \tag{51}$$

It has been postulated that arylamines form during the reductions. These amines are then promptly acetylated. While azo compounds are also convertible to anilides under the reaction conditions [89], no azoarenes have been detected during the molybdenum hexacarbonyl/carboxylic acid reaction.

8. REARRANGEMENT REACTIONS

A. Beckmann Rearrangement

Oximes of aldehydes and ketones, oxime ethers, and esters are susceptible to rearrangement by a variety of strong acidic reagents such as phosphorus pentachloride, benzenesulfonyl chloride, sulfuric acid, acetyl chloride, phosphorus oxychloride, chloral, hydrogen chloride, and polyphosphoric acid [90]. The products of the Beckmann rearrangement of oximes are amides.

8-1. Preparation of Phenanthridone [90]

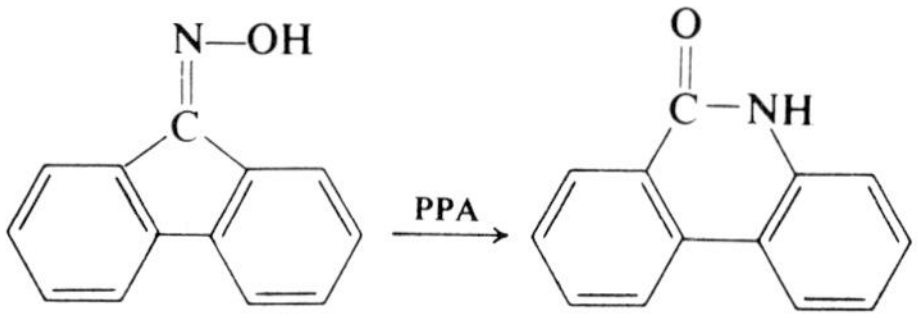

$$\text{fluorenone oxime} \xrightarrow{\text{PPA}} \text{phenanthridone} \tag{52}$$

A mixture of 2.0 gm (0.01 mole) of fluorenone oxime and 60 gm of polyphosphoric acid is heated to 175°–180°C with manual stirring. The resultant solution is cooled and diluted with 300 ml of water. The precipitated product is filtered off, washed with water, and dried. Yield 1.85 gm (43%) of phenanthridone, mp 286°–289°C.

A review of the Beckmann rearrangement appeared in *Organic Reactions* [91].

The rearrangement of amidoximes with benzenesulfonyl chloride yields substituted ureas. This variation on the Beckmann rearrangement has been termed Tiemann rearrangement [92] (Eq. 53).

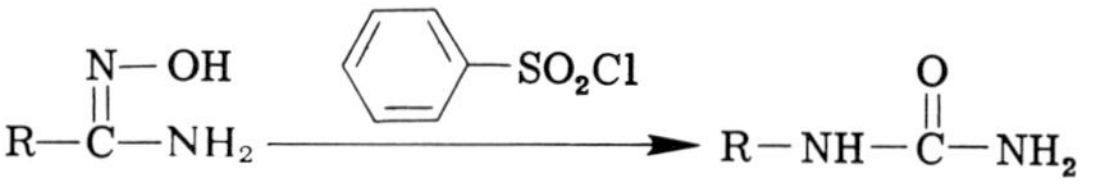

$$\text{R} - \underset{\underset{\text{NH}_2}{\|}}{\overset{\text{N}-\text{OH}}{\text{C}}} \xrightarrow{\text{C}_6\text{H}_5\text{SO}_2\text{Cl}} \text{R} - \text{NH} - \overset{\text{O}}{\underset{\|}{\text{C}}} - \text{NH}_2 \tag{53}$$

B. The Schmidt and Curtius Rearrangements

Under certain conditions, both the Schmidt and the Curtius rearrangements can afford amides. Since these reactions involve derivatives of hydrazoic acids, very serious explosion hazards exist; therefore, we cite only a few references to these reactions and caution the reader that very serious accidents have occurred in the preparation and handling of azides [93–97].

C. The Wolff Rearrangement

The Wolff rearrangement [98] of diazoketones in the presence of ammonia gives rise to amides. By going through this modification of the reaction, the Arndt-Eistert synthesis [99] may be used to convert an acid to an amide of the acid with one additional methylene group in the chain. Since both the formation and the handling of diazomethane, used in the reaction, is extremely hazardous, because of both explosions and toxicity, only small-scale reactions should be attempted, with extreme precautions, even though much of the literature gives little indication that an element of danger exists (see Chapter 15, Diazo Compounds).

8-2. Preparation of 11-Bromoundecanoamide [100]

$$BrCH_2(CH_2)_8C(=O)Cl + 2\,CH_2N_2 \longrightarrow BrCH_2(CH_2)_8C(=O)CHN_2 + CH_3Cl + N_2 \qquad (54)$$

$$BrCH_2(CH_2)_8C(=O)CHN_2 + NH_3 \xrightarrow{Ag^+} BrCH_2(CH_2)_9C(=O)NH_2 + N_2 \qquad (55)$$

CAUTION: Diazomethane is extremely toxic and a potential explosion hazard.

To 8.4 gm (0.2 moles) of diazomethane (see Chapter 15, Diazo Compounds) in 350 ml of anhydrous ether is added, with agitation, and while maintaining the temperature between 0° and 5°C, 9.1 gm (0.033 moles) of 10-bromodecanoyl chloride. The reaction mixture is then stored for 16 hr under refrigeration. The diazoketone is isolated by evaporation of the solvent under reduced pressure at a temperature maintained below 30°C. A yellow solid, mp approximately 30°C, is isolated. Yield 12.5 gm (98%).

To 12.5 gm (0.043 moles) of the diazoketone, dissolved in 100 ml of freshly distilled, warm dioxane in a large flask are added 30 gm (0.35 moles) of 20% aqueous ammonia and 6 ml of 10% aqueous silver nitrate. The reaction mixture is warmed in a boiling water bath. A vigorous evolution of gases takes place at once. After the solution has turned from yellow to an opaque brown, heating is continued for 25 min, followed by filtration of the solution.

Upon cooling, the product separates as a microcrystalline solid which is recrystallized with charcoal treatment from aqueous alcohol. Yield 5.7 gm (49%), mp 105°C.

9. MISCELLANEOUS

1. Synthesis of amidoximes [101].

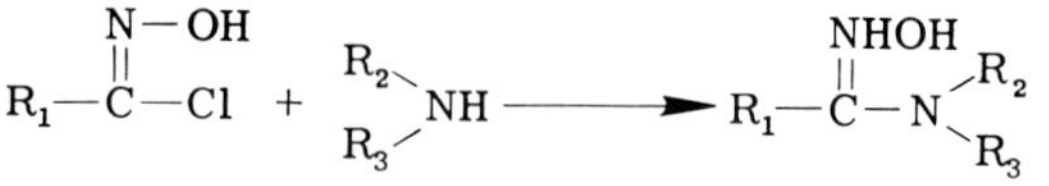

$$R_1-\underset{\overset{\|}{N-OH}}{C}-Cl + \underset{R_3}{\overset{R_2}{\diagdown}}NH \longrightarrow R_1-\underset{\overset{\|}{NHOH}}{C}-N\underset{R_3}{\overset{R_2}{\diagup}} \qquad (56)$$

2. Synthesis of poly(acrylamidoxime) [102].

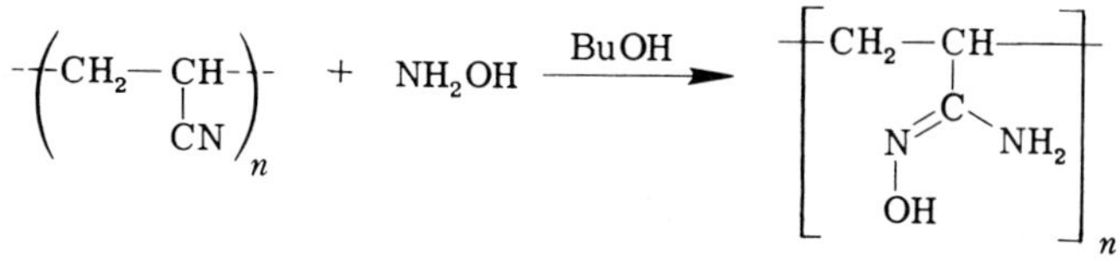

$$-(CH_2-CH-)_n\text{ (CN)} + NH_2OH \xrightarrow{BuOH} [CH_2-CH-C(=N-OH)-NH_2]_n \qquad (57)$$

3. Photoamidation of nonterminal olefins [103].

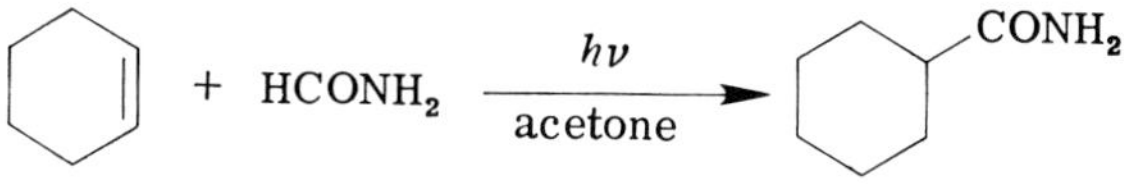

$$\text{cyclohexene} + HCONH_2 \xrightarrow[\text{acetone}]{h\nu} \text{cyclohexyl-CONH}_2 \qquad (58)$$

4. Synthesis of β-ketoamides [104].

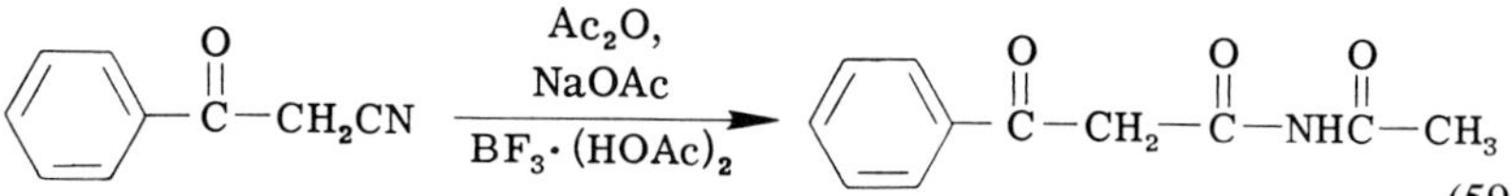

$$Ph-CO-CH_2CN \xrightarrow[BF_3\cdot(HOAc)_2]{Ac_2O, NaOAc} Ph-CO-CH_2-CO-NHC-CH_3 \qquad (59)$$

5. Bucherer hydantoin synthesis [105].

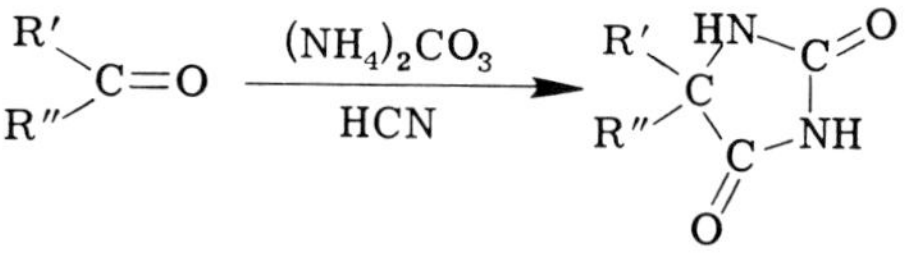

$$\underset{R''}{\overset{R'}{\diagdown}}C=O \xrightarrow[HCN]{(NH_4)_2CO_3} \text{hydantoin} \qquad (60)$$

6. Passerini reaction [106].

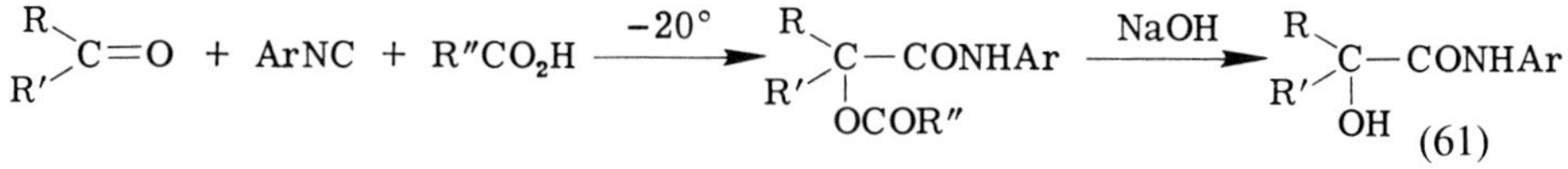

$$\underset{R'}{\overset{R}{\diagdown}}C=O + ArNC + R''CO_2H \xrightarrow{-20°} \underset{R'}{\overset{R}{\diagdown}}\underset{OCOR''}{\overset{|}{C}}-CONHAr \xrightarrow{NaOH} \underset{R'}{\overset{R}{\diagdown}}\underset{OH}{\overset{|}{C}}-CONHAr \qquad (61)$$

7. Bodroux synthesis [107].

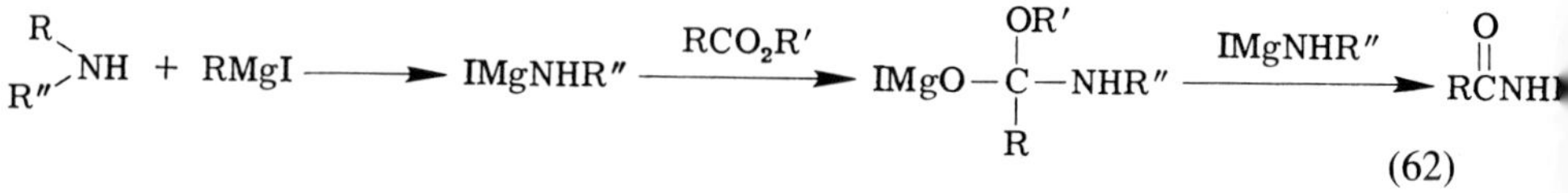

$$\underset{R''}{\overset{R}{\diagdown}}NH + RMgI \longrightarrow IMgNHR'' \xrightarrow{RCO_2R'} IMgO-\underset{R}{\overset{OR'}{\underset{|}{C}}}-NHR'' \xrightarrow{IMgNHR''} RCNHR'' \qquad (62)$$

§ 9. Miscellaneous

8. Phthalimidation of aromatic nuclei [108].

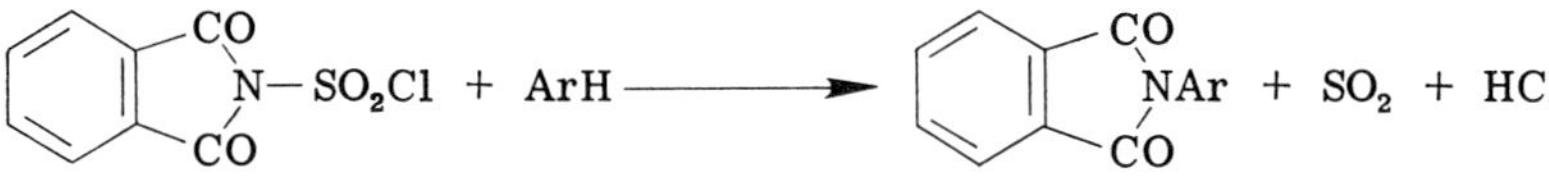

$$\text{Phth-N-SO}_2\text{Cl} + \text{ArH} \longrightarrow \text{Phth-NAr} + \text{SO}_2 + \text{HCl} \quad (63)$$

9. Chapmann rearrangement [109].

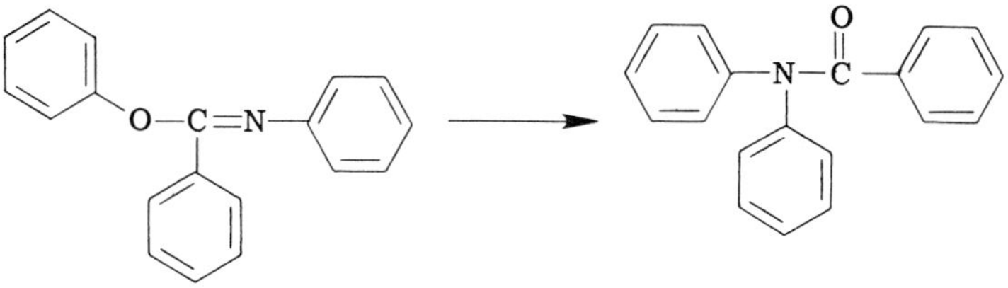

(64)

10. Newman-Kwart rearrangement [109].

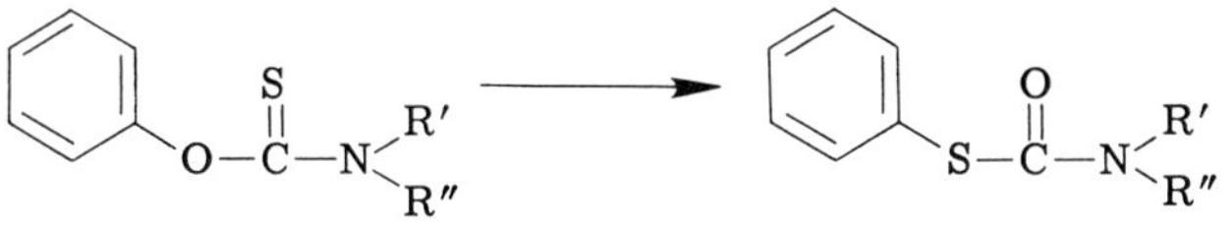

(65)

11. Leuckart N-acrylamide synthesis [110].

$$\text{ArH} + \text{Ar—NCO} \longrightarrow \text{Ar}\overset{O}{\underset{}{C}}\text{—NH—Ar} \quad (66)$$

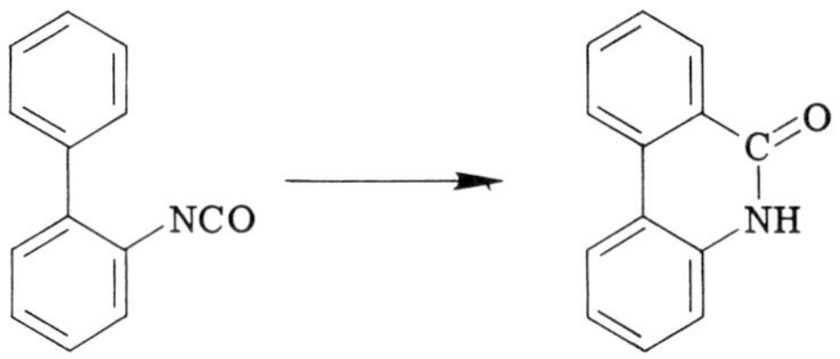

(67)

12. Enamide formation from N-allylamides [33].

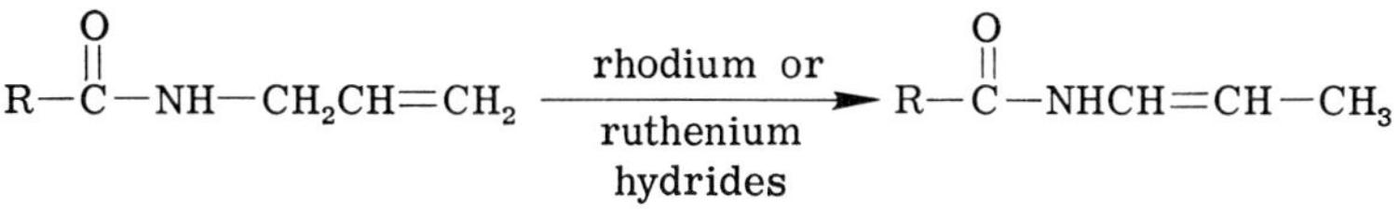

$$\underset{O}{\overset{\|}{R-C}}-NH-CH_2CH=CH_2 \xrightarrow[\text{ruthenium hydrides}]{\text{rhodium or}} \underset{O}{\overset{\|}{R-C}}-NHCH=CH-CH_3 \quad (68)$$

References

1. G. H. Coleman and A. M. Alvarado, *Org. Synth. Collect. Vol.* **1**, 3 (1948).
2. J. A. Mitchell and E. E. Reid, *J. Am. Chem. Soc.* **53**, 1879 (1931).
3. E. Cherbulier and F. Landolt, *Helv. Chim. Acta* **29**, 1438 (1946).
4. J. L. Guthrie and N. Rabjohn, *Org. Synth. Collect. Vol.* **4**, 513 (1963).

5. F. C. Magne, R. R. Mod, and E. L. Skau, *Ind. Eng. Chem.* **50,** 617 (1958).
6. H. W. Coover and N. H. Shearer, Jr., U. S. Patent 2,719,177 (1955).
7. P. A. P. A. Grieco, D. S. Clark, and G. P. Withers, *J. Org. Chem.* **44,** 2945 (1979).
8. J. C. Sheehan and G. P. Hess, *J. Am Chem. Soc.* **77,** 1067 (1955).
9. J. C. Sheehan, M. Goodman, and G. P. Hess, *J. Am. Chem. Soc.* **78,** 1367 (1956).
10. I. Levi, H. Blondal, J. W. R. Weed, A. C. Frosst, H. C. Reilly, F. A. Schmid, K. Sugiura, G. S. Tarnowski, and C. C. Stock., *J. Med. Chem.* **8,** 715 (1965).
11. N. O. V. Sonntag, *Chem. Rev.* **52,** 237 (1953).
12. J. L. Benoit-Guyod, A. Boucherle, and G. Carraz, *Bull. Soc. Chim. Fr.* No. 6, p. 1660 (1965).
13. R. Wojtowski, *Rocz. Chem.* **38,** No. No. 2, 319 (1964).
14. R. Robin, *Bull. Soc. Chim. Fr.* 8, 2286 (1965).
15. G. Vasvari, *Int. J. Appl. Radiat. Isot.* **16,** 327 (1965).
16. B. D. Halpern and W. Karo, *U.S. Air Force Syst. Command, Res. Technol. Div., Air Force Mater. Lab., Tech. Rep., WADD* 55–206 (1956).
17. B. D. Halpern, W. Karo, and P. Levine, U.S. Patent 2,957,914 (1960).
18. T. A. Sokolova, *Zh. Obshch. Khim.* **27,** 2205 (1957); M. M. Koton, T. A. Sokolova, M. N. Savitskaya, and T. M. Kiseleva, *ibid.* p. 2239.
19. H. W. Heine, D. C. King, and L. A. Portland, *J. Org. Chem.* **31,** 2662 (1966).
20. P. E. Sonnet and R. R. Heath, *J. Org. Chem.* **45,** 3137 (1980).
21. Q. E. Thompson, *J. Am. Chem. Soc.* **73,** 5841 (1951).
22. F. P. Baldwin, E. J. Blanchard, and P. E. Koenig, *J. Org. Chem.* **30,** 671 (1965).
23. H. Staudinger and N. Kon, *Justus Liebigs Ann. Chem.* **384,** 114 (1911).
24. J. F. Olin, French Patent 1,353,476 (1964).
25. W. R. Sorenson and T. W. Campbell, "Preparative Methods of Polymer Chemistry," p. 82. Wiley (Interscience), New York, 1961.
26. P. W. Morgan and S. L. Kwolek, *J. Chem. Educ.* **36,** 182 (1959).
27. G. M. Coppinger, *J. Am. Chem. Soc.* **76,** 1372 (1954).
28. A. Berger, J. Noguchi, and E. Katchalski, *J. Am. Chem. Soc.* **78,** 4483 (1956).
29. L. Zervas, J. Photaki, and N. Ghelis, *J. Am. Chem. Soc.* **85,** 1337 (1963).
30. R. A. B. Bannard, *Can. J. Chem.* **42,** No. 4, 744 (1964).
31. T. A. Martin, J. R. Corrigan, and C. W. Waller, *J. Org. Chem.* **30,** 2839 (1965).
32. A. Hassner, T. K. Morgan, Jr., and A. R. McLaughlin, *J. Org. Chem.* **44,** 1999 (1979).
33. J. K. Stille and Y. Becker, *J. Org. Chem.* **45,** 2139 (1980).
34. T. A. Montzka, T. L. Pindell, and J. D. Matiskella, *J. Org. Chem.* **33,** 3993 (1968).
35. M. Pesson, S. Dupin, M. Antoine, D. Humberto, and M. Joannic, *Bull. Soc. Chim. Fr.* No. 8, p. 2266 (1965).
36. M. S. Atwal, L. Bauer, S. N. Dixit, J. E. Gearien, and R. W. Morris, *J. Med. Chem.* **8,** 566 (1965).
37. E. Bricas, P. Dezélléé, C. Gansser, P. Lefrancier, C. Nicot, and J. van Heijenport, *Bull. Soc. Chim. Fr.* No. 6, p. 1813 (1965).
38. E. C. Kornfeld, R. G. Jones, and T. V. Parke, *J. Am. Chem. Soc.* **71,** 158 (1949).
39. M. Gordon, J. G. Miller, and A. R. Day, *J. Am. Chem. Soc.* **70,** 1245 (1949).
40. B. B. Corson, R. W. Scott, and C. E. Vose, *Org. Synth. Collect. Vol.* **1,** 179 (1948).
41. B. D. Halpern, W. Karo, L. Laskin, P. Levine, and J. Zomlefer, *U.S. Air Force Syst. Command, Res. Technol. Div., Air Force Mater. Lab., Tech. Rep. WADD* **WADD-TR-54–264** (1954).
42. H. Gilman and R. G. Jones, *J. Am. Chem. Soc.* **65,** 1458 (1943).
43. D. R. Husted and A. N. Ahlbrecht, *116th Am. Chem. Soc. Meet.* Paper 17, Org. Chem. Div. (1949).
44. P. B. Russell, *J. Am. Chem. Soc.* **72,** 1853 (1954).

45. W. J. Humphlett, U.S. Patent 2,784,225 (1957).
46. C. F. H. Allen, J. R. Byers, Jr., W. J. Humphlett, and D. D. Reynolds, *J. Chem. Educ.* **32**, 394 (1955).
47. C. F. H. Allen, J. R. Byers, Jr., W. J. Humphlett, and D. D. Reynolds, *Org. Chem. Bull.* **30**, No. 1 (1958).
48. C. F. H. Allen and W. J. Humphlett, U.S. Patent 3,133,044 (1964).
49. W. J. Humphlett and C. V. Wilson, Jr., *J. Org. Chem.* **26**, 2507 (1961).
50. C. F. H. Allen and W. J. Humphlett, *Bull. Soc. Chim. Fr.* p. 961 (1961).
51. C. F. H. Allen and W. J. Humphlett, Org. Sym. Coll. Vol. **4**, 80 (1963).
52. Drawing based on suggestions furnished by Prof. Allen, whose assistance is gratefully acknowledged.
53. C. S. Rondestvedt, Jr., *J. Org. Chem.* **42**, 3118 (1977).
54. L. Kisfaludy, T. Mohacsi, M. Low, and F. Dexler, *J. Org. Chem.* **44**, 654 (1979).
55. S. R. Aspinall, *J. Am. Chem. Soc.* **63**, 852 (1941).
56. R. Burgada, *C. R. Hebd. Seances Acad. Sci.* **258**, 1532 (1964).
57. E. S. Rothman, S. Serota, and D. Swern, *J. Org. Chem.* **29**, 646 (1964).
58. H. J. Hagemeyer, Jr. and D. C. Hull, *Ind. Eng. Chem.* **41**, 2920 (1949).
59. G. R. Petit and E. G. Thomas, *J. Org. Chem.* **24**, 895 (1939).
60. A. Galat and G. Elion, *J. Am. Chem. Soc.* **65**, 1566 (1943).
61. E. T. Roe, J. T. Scanlan, and D. Swern, *J. Am. Chem. Soc.* **71**, 2215 (1949).
62. R. S. Haworth, R. McGillivray, and D. H. Peacock, *J. Chem. Soc.* p. 1496 (1950).
63. W. S. Fones, *J. Org. Chem.* **14**, 1099 (1949).
64. H. B. Donahoe, R. J. Seiwald, M. M. C. Neuman, and K. K. Kimura, *J. Org. Chem.* **22**, 68 (1957).
65. H. S. Freeman, J. R. Butler, and L. D. Freedman, *J. Org. Chem.* **43**, 4975 (1978).
66. P. G. Gassman and B. L. Fox, *J. Org. Chem.* **31**, 982 (1966).
67. N. L. Drake, C. M. Eaker, and W. Shenk, *J. Am. Chem. Soc.* **70**, 677 (1948).
68. H. M. Singleton and W. R. Edwards, Jr., *J. Am. Chem. Soc.* **60**, 540 (1938).
69. H. A. Bruson, U.S. Patent 2,370,142 (1945).
70. J. M. Patterson, M. W. Barnes, and R. L. Johnson, *J. Org. Chem.* **31**, 3103 (1966).
71. American Cyanamid Co., British Patent 631,592 (1949).
72. F. Scotti and E. J. Frazza, *J. Org. Chem.* **29**, 1800 (1964).
73. R. A. Barnes, E. R. Kraft, and L. Gordon, *J. Am. Chem. Soc.* **71**, 3523 (1949).
74. R. L. West, *J. Am. Chem. Soc.* **42**, 1662 (1920).
75. J. V. Murray and J. B. Cloke, *J. Am. Chem. Soc.* **56**, 2749 (1934).
76. C. R. Noller, *Org. Synth. Collect. Vol.* **2**, 586 (1943).
77. T. Takahashi and Y. Hamada, *J. Pharm. Soc. Jpn.* **75**, 755 (1955).
78. N. Kornblum and S. Singaram, *J. Org. Chem.* **44**, 4727 (1979).
79. J. J. Ritter and P. P. Minieri, *J. Am. Chem. Soc.* **70**, 4045 (1948).
80. J. J. Ritter and J. Kalish, *J. Am. Chem. Soc.* **70**, 4048 (1948).
81. H. Plaut and J. J. Ritter, *J. Am. Chem. Soc.* **73**, 4076 (1951).
82a. J. J. Ritter and F. X. Murphy, *J. Am. Chem. Soc.* **74**, 763 (1952).
82b. T. Clarke, J. Devine, and D. W. Dicker, *J. Am. Oil Chem. Soc.* **41**, No. 1, 78 (1964).
83. P. J. Scheuer, H. C. Botelho, and C. Pauling, *J. Org. Chem.* **22**, 674 (1957).
84. T. R. Benson and J. J. Ritter, *J. Am. Chem. Soc.* **71**, 4128 (1949).
85. M. L. Scheinbaum and M. Dines, *J. Org. Chem.* **36**, 3641 (1971).
86. R. D. Bach, J. W. Holubka, and T. A. Taaffee, *J. Org. Chem.* **44**, 1739 (1979).
87. G. A. Olah and B. G. B. Gupta, *J. Org. Chem.* **45**, 3532 (1980).
88. A. Nishinaga, T. Shimizu, and T. Matsuura, *J. Chem. Soc., Chem. Commun.* p. 970 (1979).
89. T.-L. Ho, *J. Org. Chem.* **42**, 3755 (1977).

90. E. C. Horning, V. L. Stromberg, and H. A. Lloyd, *J. Am. Chem. Soc.* **74,** 5153 (1952).
91. L. G. Donaruma and W. Z. Heldt, *Org. React.* **11,** 1 (1960).
92. R. E. Plapinger and O. O. Owens, *J. Org. Chem.* **21,** 1186 (1956).
93. H. Wolff, *Org. React.* **3,** 307 (1946).
94. P. A. S. Smith, *Org. React.* **3,** 337 (1946).
95. W. E. McEwen, W. E. Conrad, and C. A. Vander Werf, *J. Am. Chem. Soc.* **74,** 1168 (1952).
96. S. C. Dickerman and A. J. Besozzi, *J. Org. Chem.* **19,** 1855 (1954).
97. R. T. Conley, *J. Org. Chem.* **23,** 1330 (1958).
98. L. Wolff, *Justus Liebigs Ann. Chem.* **394,** 25 (1912).
99. W. E. Bachmann and W. S. Struve, *Org. React.* **1,** 38 (1942).
100. F. J. Buckle, F. L. M. Pattison, and B. C. Saunders, *J. Chem. Soc.* p. 1478 (1949).
101. D. F. Bushey and F. C. Hoover, *J. Org. Chem.* **45,** 4198 (1980).
102. M. B. Colella, S. Siggia, and R. M. Barnes, *Anal. Chem.* **52,** 967 (1980).
103. D. Elad and J. Rokach, *J. Org. Chem.* **30,** 3361 (1965).
104. J. F. Wolfe and C.-L. Mao, *J. Org. Chem.* **31,** 3069 (1966).
105. H. A. Henze and W. C. Craig, *J. Org. Chem.* **10,** 2 (1945); H. A. Henze and W. B. Leslie, *ibid.* **15,** 901 (1950).
106. R. H. Baker and D. Stanonis, *J. Am. Chem. Soc.* **73,** 699 (1951).
107. H. L. L. Bussett and C. R. Thomas, *J. Chem. Soc.* p. 1188 (1954); C. G. Stuckwisch and J. S. Hilliard, *J. Med. Chem.* **8,** 734 (1965).
108. R. A. Lidgett, E. R. Lynch, and E. B. McCall, *J. Chem. Soc.* p. 3754 (1965).
109. H. M. Relles and G. Pizzolato, *J. Org. Chem.* **33,** 2249 (1968); H. L. Wehrmeister, *J. Org. Chem.* **30,** 664 (1965).
110. R. Leuckart, *Ber. Dtsch. Chem. Ges.* *18,* 873 (1885); J. M. Butler, *J. Am. Chem. Soc.* **71,** 2578 (1949).

CHAPTER 12 / CYANATES, ISOCYANATES, THIOCYANATES, AND ISOTHIOCYANATES

1. Introduction	359
A. Nomenclature	361
2. Cyanates	362
2-1. Preparation of Phenyl Cyanate	363
3. Isocyanates	364
A. Condensation Reactions	364
3-1. Preparation of p-*Nitrophenyl Isocyanate*	364
a. The Use of Inorganic Cyanates	365
3-2. Preparation of Allyl Isocyanate	366
B. Decomposition Reactions	366
3-3. Preparation of Cyclohexyl Isocyanate	366
C. Exchange Reactions	367
3-4. Preparation of n-*Hexyl Isocyanate*	367
D. Rearrangement Reactions	368
E. Oxidation Reactions	368
F. Pyrolysis Reactions	368
G. Miscellaneous Reactions	368
4. Thiocyanates	369
A. Condensation Reactions	369
4-1. Preparation of Undecyl Thiocyanate	369
B. Addition Reactions	369
4-2. Preparation of cis- and trans-3-Thiocyanoacrylamide	370
C. Miscellaneous Reactions	370
5. Isothiocyanates	371
A. Condensation Reactions	371
5-1. Preparation of p-*Chlorophenyl Isothiocyanate*	372
5-2. Preparation of p-*Nitrophenyl Isothiocyanate*	373
5-3. Preparation of n-*Butyl Isothiocyanate*	374
B. Decomposition Reactions	374
C. Miscellaneous Reactions	374
References	375

1. INTRODUCTION

Since the cyanates were only first isolated in 1960, satisfactory methods of synthesis are quite limited.

The best available method appears to be the reaction of cyanogen chloride and phenol under such conditions that the level of an added base is carefully limited at all times.

$$\text{ArOH} + \text{ClCN} \xrightarrow{(C_2H_5)_3N} \text{ArOCN} + [(C_2H_5)_3\overset{+}{\text{NH}}]\overset{-}{\text{Cl}} \tag{1}$$

Isocyanates are prepared by the reaction of amines with phosgene.

$$\text{RNH}_2 + \text{COCl}_2 \longrightarrow \text{RNCO} + 2\,\text{HCl} \tag{2}$$

Convenient laboratory methods involve the reaction of alkyl halides or dialkyl sulfate with inorganic cyanates such as silver cyanate.

$$\text{R}_2\text{SO}_4 + 2\,\text{KOCN} \longrightarrow 2\,\text{RNCO} + \text{K}_2\text{SO}_4 \tag{3}$$

$$\text{RX} + \text{AgOCN} \longrightarrow \text{RNCO} + \text{AgX} \tag{4}$$

The decomposition of urethanes either thermally or in the presence of phosphorus pentachloride has been reported. An interesting new reaction involves the decomposition of phosphoramidate anions.

Quite recently a more direct preparation has been developed in which aliphatic or aromatic amines are reacted with carbon monoxide in the presence of $PdCl_2$ [1]. The reaction is slow but it has the potential of offering a convenient method of preparing isocyanates in the laboratory.

$$\text{RNH}_2 + \text{CO} + \text{PdCl}_2 \longrightarrow \text{RNCO} + \text{Pd} + 2\,\text{HCl} \tag{5}$$

In some cases the carbonylation of amines gives urethanes instead [2].

More recently a process has been patented for preparing isocyanates by the reaction of nitro compounds with carbon monoxide in the presence of a rhodium oxide catalyst [3].

$$\text{ArNO}_2 + \text{CO} \xrightarrow[\substack{\text{Rh}_2\text{O}_3 \\ \text{CH}_3\text{CN}}]{\text{RhO}_2 \text{ or}} \text{ArNCO} \tag{6}$$

Recently a process has been reported for preparing aromatic isocyanates by the reaction of aromatic nitro compounds with carbon monoxide in alcohol solution to first form the urethanes. Thermal decomposition of the urethane gives the isocyanate [4].

Furthermore, the reaction of carbon monoxide with a metal salt of carbamic acid has been used to give isocyanuric acid and derivatives or isocyanates [5].

A recent review on isocyanates is worth consulting for additional details, especially as to commercial uses and manufacturing processes [4].

Since isothiocyanates are generally higher boiling than the corresponding isocyanates, exchange reactions of the type illustrated in Eq. (7) are possible.

§ 1. Introduction

$$\text{RNCS} + \text{ArNCO} \longrightarrow \text{RNCO} + \text{ArNCS} \tag{7}$$

The classical Curtius, Lossen, Hofmann, and Schmidt rearrangements may be used to prepare isocyanates, although normally isocyanates are considered only intermediates in these reactions.

The thiocyanates are usually prepared by the condensation of alkyl halides (or sulfates) with potassium or ammonium thiocyanate.

$$\text{RBr} + \text{KSCN} \longrightarrow \text{RSCN} + \text{KBr} \tag{8}$$

Aromatic thiocyanates have also been prepared by the action of potassium thiocyanate or cuprous thiocyanate on diazonium salts.

The isothiocyanates have been prepared by the condensation of amines with carbon disulfide in the presence of a base to yield the dithiocarbamate which, in turn, is decomposed by reagents such as lead nitrate to the isothiocyanate [see Eqs. (26) and (27)].

Just as amines react with phosgene to give isocyanates, thiophosgene may be used to prepare isothiocyanates [see Eq. (31)].

CAUTION: All compounds considered in this chapter should be handled with great care. Aside from the fact that many have strong, unpleasant odors, many exhibit strong physiological reactions. Many are lachrymators and/or vesicants.

A. Nomenclature

Esters of cyanic acid are referred to as cyanates and have been assigned structure (I). The related and better known isocyanates have structure (II).

$$\begin{array}{cc} \text{R—O—C} \equiv \text{N} & \text{R—N} = \text{C} = \text{O} \\ \text{(I)} & \text{(II)} \\ \text{Cyanates} & \text{Isocyanates} \end{array}$$

Similarly among the sulfur analogs, thiocyanates have structure (III) while the isothiocyanates have structure (IV).

$$\begin{array}{cc} \text{R—S—C} \equiv \text{N} & \text{R—N} = \text{C} = \text{S} \\ \text{(III)} & \text{(IV)} \\ \text{Thiocyanates} & \text{Isothiocyanates} \end{array}$$

Other isomers, whose preparation is beyond the scope of this chapter, are the fulminates (V) and the nitrile oxides (VI). Of these, probably because of the extreme explosion hazards, the electronic structure of the fulminates has not been settled completely.

$$\begin{array}{cc} \text{R—ONC} & \text{R—C} \equiv \text{N} \rightarrow \text{O} \\ \text{(V)} & \text{(VI)} \\ \text{Fulminates} & \text{Nitrile oxides} \end{array}$$

2. CYANATES

Up until 1960, the true cyanates, ROCN, had not been isolated. Thus, the reaction of sodium phenolates with cyanogen halides was shown to follow the course given in Eqs. (9) and (10) by both Nef [6], and Hantzsch and Mai [7].

$$\text{Ar—ONa} + \text{ClCN} \longrightarrow \text{ArO—}\underset{\|}{\overset{\text{N Na}}{\text{C}}}\text{—Cl} \xrightarrow[\text{2. H}^+]{\text{1. RO}^-} \text{ArO—}\underset{\|}{\overset{\text{NH}}{\text{C}}}\text{—OR} + \text{NaCl} \quad (9)$$

$$3 \text{ ArO—}\underset{\|}{\overset{\text{NH}}{\text{C}}}\text{—OR} \longrightarrow \underset{\text{ArO—C}\diagdown_{\text{N}}\diagup\text{C—OAr}}{\overset{\overset{\text{OAr}}{|}}{\underset{\|}{\text{N}\diagup\text{C}\diagdown\text{N}}}} + 3 \text{ ROH} \quad (10)$$

In 1960, however, a few sterically hindered cyanates were produced by essentially the same process [8]. All these procedures involved the addition of cyanogen halides to a reaction medium in which an excess of alcoholates or phenolates was present. Therefore, imido diesters form rapidly in the basic medium followed by trimerization unless steric factors prevent this reaction.

In 1964, Grigat and Pütter began the publication of an extensive series of papers on the chemistry of cyanates in *Chemische Berichte*. In their first paper of the series [9a], Grigat and Pütter point out that, if reaction conditions are such that the base is never present in excess, true cyanates can be prepared from phenolic compounds unless several electron-withdrawing groups are present in the aromatic nucleus. Even some aliphatic cyanates could be prepared by this procedure, provided strongly acidic alcohols such as the trihaloethanols or enols were used as starting materials.

These investigators found that purified cyanates are stable for several weeks. They can be recrystallized and in some cases can even be distilled.

The reactions may be carried out in acetone or other solvents such as ether, carbon tetrachloride, benzene, acetonitrile, or ethyl acetate. The cyanogen halides used may be either cyanogen chloride or, if a higher boiling point or higher reaction temperature is desired, cyanogen bromide.

CAUTION: These compounds are extremely toxic.

The usual base used in the reaction is triethylamine. With inorganic bases, such a sodium hydroxide, aqueous media have been used, although yields usually are lower. The range of phenols used in the Grigat and Pütter procedure is quite extensive, including phenol, 2-methylphenol, 3-methylphenol, 4-methylphenol, various other mono- and dialkylphenols, naphthols, chlorophenols, nitrophenols, methoxyphenols, trihydroxyquinolines, trihaloethanols, hydroquinones, hydroxybiphenyls, etc.

Depending on the substituents present in the aromatic nucleus, the cyanates show more or less tendency to trimerize in the presence of mineral acids, Lewis acids, bases, as well as phenolic impurities. The reaction proceeds quite smoothly and, therefore, the cyanates become a very useful starting point for the preparation of a large variety of triaryl esters of cyanuric acid.

2-1. Preparation of Phenyl Cyanate [9a]

$$C_6H_5-OH + ClCN \xrightarrow{(C_2H_5)_3N} C_6H_5-OCN + (C_2H_5)_3\overset{+}{N}H\overset{-}{Cl} \quad (11)$$

In a suitable hood and with precautions for handling a toxic material such as cyanogen chloride, a solution of 94.1 gm (1.0 mole) of phenol in 250 ml of acetone is cooled to 0°C. To the cold solution, 65.1 gm (approximately 1.05 moles) of liquid cyanogen chloride is added. While cooling is continued in an ice bath, 101.2 gm (1.0 moles) of triethylamine is added dropwise with vigorous stirring at such a rate that the temperature never exceeds 10°C. After the addition has been completed, stirring is continued for an additional 10 min, and the triethylamine hydrochloride is separated by filtration and extracted three times with 100 ml portions of acetone. The acetone solutions are combined and evaporated under reduced pressure. (CAUTION: The evaporating solvent may contain cyanogen chloride and due precautions must be taken that the exhaust from the aspirator or pump be properly treated.) The residue is then subject to distillation under reduced pressure to yield 112 gm (94%), bp 55°C (0.4 mm).

In the case of cyanates which are solids, such as the dicyanate derived from hydroquinone, the acetone solution of the reaction mixture is poured into an excess of ice water with vigorous stirring, whereupon the triethylamine hydrochloride goes into solution, allowing the dicyanate to precipitate.

Another recent method for the preparation of phenyl cyanates makes use of sodium azides according to the reaction scheme of Eq. (12) [10].

$$RO-C\overset{S}{\underset{Cl}{\diagdown}} + NaN_3 \longrightarrow \underset{N\diagdown N\diagup N}{RO-C-S} \xrightarrow[-S]{-N_2} ROCN \quad (12)$$

Since we consider reactions involving sodium azide potentially extremely hazardous, mention of this process is made here for reference only.

The chemistry of cyanates has recently been reviewed [11] and the preparation of various alkyl cyanates has been reported [12].

3. ISOCYANATES

A. Condensation Reactions

In view of the fact that the diisocyanates of aromatic diamines are of great industrial importance today, undoubtedly the most common method of preparation of diisocyanates, as well as of monoisocyanates, is based on the reaction of amines with phosgene. CAUTION: Phosgene is highly toxic and should be handled with great care in well-ventilated hoods [13a]. From the standpoint of laboratory procedures, suitable models for the preparation of many isocyanates are given in the literature [13b]. In connection with these preparations, precaution must be taken, since the amines, isocyanates, and phosgene are very toxic. Also, it is important in the laboratory preparations to keep moist air, water, and protic solvents out of contact with the isocyanates.

The reaction of aromatic amines with phosgene to produce isocyanates has wide applicability. The example of the preparation of *p*-nitrophenyl isocyanate illustrates the general procedure. Even complex aromatic amines such as fluorescein amine may be subjected to the reaction to produce fluorescein isocyanate, which has found some application in biochemical research [14].

In the preparation of aliphatic isocyanates, the volatility of the amines may lead to poor yields when phosgene is bubbled into the reactant. For this reason, thoroughly dried hydrochloride salts of the amines, suspended in high-boiling inert solvents, are substituted for the free amine [15]. Vapor phase preparation of isocyanates has also been reported [16].

To minimize such side reactions as the formation of substituted ureas, the reaction is generally carried out in an excess of phosgene. Symmetrical ureas are converted to aliphatic isocyanates in high yield by the reaction of phosgene with the urea at 150°C. and higher [17].

CAUTION: Phosgene is a war gas and, being extremely toxic, must be handled with extreme care. All work must be carried out in a well-ventilated hood. The operator must wear a gas mask suitable for use with phosgene, rubber gloves, and a rubber apron. It is also recommended that in the general work area warning filter papers which change color on contact with phosgene be prepared [13a].

3-1. Preparation of **p-*Nitrophenyl Isocyanate*** **[13b]**

$$O_2N-C_6H_4-NH_2 + Cl-\underset{\underset{O}{\|}}{C}-Cl \longrightarrow O_2N-C_6H_4-NCO + 2\,HCl \quad (13)$$

CAUTION: This reaction must be carried out in a well-ventilated hood, behind a shield, and, as a minimum, observing the precautions outlined above.

The reaction system is assembled as follows: The phosgene cylinder is connected through a mercury pressure regulator in turn to a gas wash bottle filled with cottonseed oil to remove chlorine, a wash bottle containing concentrated sulfuric acid, and an empty flask large enough to hold the contents of the reaction flask. To this is connected a gas delivery tube in a 5-liter three-necked flask fitted also with an addition funnel and a gooseneck leading to a condenser. The condenser is connected to a filter flask which serves as a distillate receiver. The vent of this receiver is connected to a gas wash bottle containing 20% sodium hydroxide solution and a safety trap. Finally, this trap is connected to an aspirator which permits drawing the phosgene into the reaction flask by slightly reducing the pressure inside the system.

In the three-necked flask is placed 500 ml of dry, ethanol-free ethyl acetate. This solvent is then saturated with phosgene. While a slow stream of phosgene is passed into the system throughout the reaction, a solution of 150 gm (1.09 moles) of *p*-nitroaniline in 1500 ml of ethyl acetate is added slowly through the addition funnel over a 3-hr period.

The *p*-nitroaniline is added in such a manner that the *p*-nitroaniline hydrochloride which forms initially is allowed to dissolve in the reaction medium. If necessary, the flask is gently warmed.

While the addition is being completed, the solution is gently boiled to assist in breaking up the *p*-nitroaniline hydrochloride lumps. After completion of the addition, the stream of phosgene is continued for 5 min. The phosgene stream is then turned off. The addition train is removed, boiling chips are added to the flask, the stopper which had carried the gas delivery tube is replaced by a solid stopper, and the ethyl acetate solvent is distilled off by the careful application of heat. To the brown residue is added 800 ml of hot carbon tetrachloride and the insoluble disubstituted urea by-product is separated by filtration.

Then about two-thirds of the solvent is distilled from the filtrate. The solution is then chilled and the crystals of *p*-nitrophenyl isocyanate are filtered off and stored in a tightly closed container. A further crop of product may be obtained by work-up of the mother liquor. The product may be recrystallized from dry carbon tetrachloride. Yield 160 gm ± 10 gm (83–95%), mp 56°–57°C, bp 160°–162°C (18 mm), light yellow needles.

a. THE USE OF INORGANIC CYANATES

The reaction of an alkyl halide or a dialkyl sulfate with inorganic cyanates often leads to alkyl isocyanates. The yields from these preparations are highly variable, although the reaction is quite generally applicable. We believe that part of the problem of the preparation involves the low order of solubility of the inorganic cyanate in the reaction medium. In the light of recent work [18], the use of such dipolar aprotic solvents as dimethyl sulfoxides, dimethylformamide, *N*-methylpyrrolidone, hexamethylphosphoramide, and acetonitrile in conjunction with the cyanates should be explored further.

In our own laboratory we have favored the use of silver cyanate with alkyl halides [19]. This reagent seems quite generally applicable and has even been used in the preparation of α-keto isocyanates from acid chlorides [20]. In our own example below, the yields reported are not necessarily representative of the best reaction conditions. When silver cyanate is used in conjunction with an alkyl halide, silver halide is formed which tends to coat silver cyanate and, thereby, reduces the extent of the reaction. It is possible to recover the contaminated silver cyanate by grinding it to separate silver halide from it to some extent and then reusing the compound in another preparation.

3-2. Preparation of Allyl Isocyanate [21]

$$CH_2=CHCH_2I + AgOCN \longrightarrow CH_2=CH-CH_2-N=C=O + AgI \quad (14)$$

To 2000 gm (12 moles) of allyl iodide (readily prepared by the addition of allyl bromide to a solution of sodium iodide in acetone followed by the filtration of sodium bromide and evaporation of the solvent) in 2 liters of xylene at 10°–15°C, 1950 g (13 moles) finely ground silver cyanate is added portionwise over a 3-hr period, during which the temperature is maintained at 10°–15°C. After the addition has been completed, the reaction mixture is heated for 3 hr at 75°C. After cooling, the mixed silver iodide–silver cyanate is separated by filtration and the filtrate is subjected to distillation. The fraction distilling between 80°–95°C is recovered and subjected to redistillation to yield 419 gm (42%) of allyl isocyanate, boiling at 84°C.

B. Decomposition Reactions

The decomposition of urethanes to give isocyanates has been known for some time [22] and, with the potential availability of a large variety of urethanes, particularly by a patented method [23], the use of this procedure should expand [24]. Both treatments with phosphorus pentachloride and thermal decompositions have been used for the preparation of isocyanates [25, 26].

An interesting reaction is based on the decomposition of the phosphoramidate anion [27].

Alkyl aryl ureas can also be pyrolyzed to alkylisocyanates [28].

3-3. Preparation of Cyclohexyl Isocyanate [27]

$$[(C_2H_5O)_2P(O)NC_6H_{11}]^- + CO_2 \longrightarrow \left[(C_2H_5O)_2P(O)N\overset{\overset{O}{\|}}{\underset{C_6H_{11}}{C}}-O \right]^- \longrightarrow$$

$$C_6H_{11}N=C=O + [(C_2H_5O)_2P(O)O]^- \quad (15)$$

To a slurry of 4.8 gm (0.1 mole) of 50% sodium hydride in 100 ml of dry 1,2-

§ 3. Isocyanates 367

dimethoxyethane, maintained at a temperature below 30°C, is added 23.5 gm (0.1 mole) of diethyl N-cyclohexylphosphoramidate (prepared from diethyl phosphorochloridate and cyclohexylamine.) The mixture is stirred at room temperature until it becomes homogeneous and the gas evolution has ceased. Then carbon dioxide gas is slowly passed through the homegeneous solution at approximately 0°–5°C while cooling is maintained with an ice bath. After the absorption of carbon dioxide has ceased, the mixture is heated at 80°C for half an hour until the formation of gummy precipitate has ceased. Generally, the optimum decomposition temperature of the reaction can be judged by the formation of this precipitate. The mixture is then cooled and the product solution is separated by decantation, and preserved under anhydrous conditions. The precipitate is washed with additional dry 1,2-dimethoxyethane. The solvent layers are combined, and the solvent is eliminated by distillation. The residue is then fractionally distilled, giving 8.7 gm of cyclohexyl isocyanate, bp 170°C, yield 70%.

C. Exchange Reactions

By use of exchange reactions the alkyl group of one isocyanate can be converted to another by the use of an appropriate N,N'-dialkyl ureas [29] according to Eq. (16).

$$\text{R-NH-}\overset{\overset{\text{O}}{\|}}{\text{C}}\text{-NHR} + 2\,\text{R'NCO} \longrightarrow \text{R'NH-}\overset{\overset{\text{O}}{\|}}{\text{C}}\text{-NHR'} + 2\,\text{RNCO} \quad (16)$$

Since the selection of substituted ureas may be limited, this reaction is of limited application. However, the exchange reaction of isocyanate with an isothiocyanate offers some possibilities for wider applicability [30]. This reaction is based on the observation that isothiocyanates normally have higher boiling points than isocyanate. The preparation of n-hexyl isocyanate is an example of this reaction.

3-4. *Preparation of* n-*Hexyl Isocyanate* [30]

$$\text{C}_6\text{H}_{13}\text{N=C=S} + \underset{\text{Cl}}{\text{C}_6\text{H}_4}\text{-N=C=O} \longrightarrow \text{C}_6\text{H}_{13}\text{N=C=O} + \underset{\text{Cl}}{\text{C}_6\text{H}_4}\text{-N=C=S} \quad (17)$$

A mixture of 135 gm (0.94 mole) of n-hexyl isothiocyanate and 460.5 gm (3 moles) of 3-chlorophenyl isocyanate is heated in a flask fitted with a short glass-packed fractionation column. While the distillation is maintained at total reflux, heating is continued until the head temperature remains constant at 161°C while the still pot temperature is maintained in the range of 210°–230°C. Then the product is slowly distilled, over a 10-hr period, at a head temperature of 163°–164°C to yield 77.5 gm of *n*-hexyl isocyanate (61%), bp 164°C.

D. Rearrangement Reactions

Other preparations of isocyanates include the use of the Curtius rearrangement [31–33]. We do not recommend the use of this reaction because of potential explosion hazards in connection with azides. Mention is made of this method only for reference.

The Lossen rearrangement of hydroxamic acids has been used for the preparation of isocyanates but is believed to be of only limited applicability [34].

Other rearrangement reactions such as the Hofmann degradation of amides and the Schmidt degradation are in the classical literature but are of only limited applicability either because of low yields or because of the hazards involved in the use of hydrazoic acid [35, 36].

Cyanates can be rearranged to the corresponding isocyanate [12].

E. Oxidation Reactions

The oxidation of isocyanides either with ozone [37], air [38], or with halogens in the presence of dimethyl sulfoxide [39] is currently of only limited applicability since the isocyanides are available only in limited quantities and because of the difficulties in the preparation and handling of isocyanides.

Formamides can also be oxidized (dehydrogenated) with the aid of various metal catalysts to isocyanates [40].

F. Pyrolysis Reactions

Isocyanates may also be obtained by the pyrolysis of acetoacetyl hydroxamates at 350°–400°C [41].

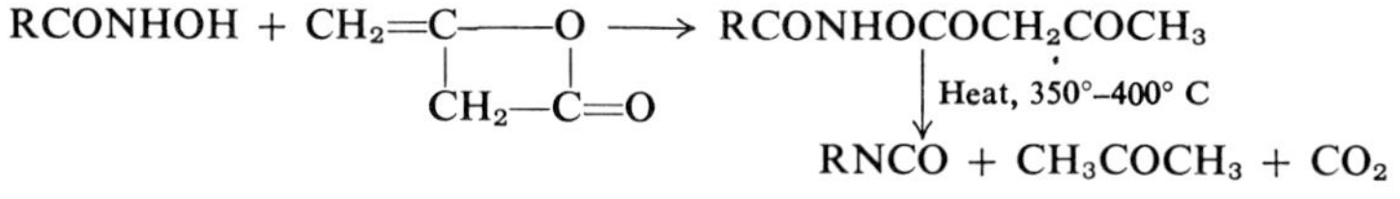

$$\text{RCONHOH} + \underset{\underset{\text{CH}_2-\text{C}=\text{O}}{|}}{\text{CH}_2=\text{C}-\text{O}} \longrightarrow \text{RCONHOCOCH}_2\text{COCH}_3 \xrightarrow{\text{Heat, 350°–400° C}} \text{RNCO} + \text{CH}_3\text{COCH}_3 + \text{CO}_2 \quad (18)$$

The pyrolysis of urethanes at 160°–250°C. also gives monoisocyanates [42].

G. Miscellaneous Reactions

1. Preparation of isocyanates by the reaction of *N*-haloamides with tertiary amines at 20°–180°C [43].

2. Isocyanates by the thermal decomposition of dialiphatic substituted furoxan [44].

3. Isocyanates and isothiocyanates by the reaction of substituted azomethines with monofunctional isocyanates or isothiocyanates [45].

4. Preparation of organic isocyanates by the reaction of primary amines, and carbon dioxide with hexamethyldisilazane in the presence of an acidic catalyst [46].

4. THIOCYANATES

The sulfur analogs of both cyanates and isocyanates are thermally and hydrolytically more stable than the oxygen analogs. Both series of sulfur analogs have been known for some time from natural and synthetic sources.

A. Condensation Reactions

The preparation of alkyl thiocyanates by reaction of cyanogen chloride with a mercaptan according to Eq. (19) has been used [9b].

$$R-SH + ClCN \longrightarrow RSCN + HCl \qquad (19)$$

It is believed that this reaction should be reinvestigated in the light of the preparative procedures for cyanates [9a].

Perhaps the most widely used preparation of thiocyanates involves the reaction of an alkyl halide with either potassium thiocyanate or ammonium thiocyanate [47]. The reaction is not confined to simple alkyl halides but has also been used for the preparations with dihalides [48], chlorohydrins [49], secondary alkyl halides [50], and acyl halides (in the preparation of acyl thiocyanates) [51]. The preparation of undecyl thiocyanate is a typical example of the procedure. The preparation of thiocyanates has recently been reviewed [11, 52].

4-1. Preparation of Undecyl Thiocyanate [47]

$$CH_3(CH_2)_9CH_2Br + KSCN \longrightarrow CH_3(CH_2)_9CH_2SCN + KBr \qquad (20)$$

To 145.5 gm (1.5 moles) of potassium thiocyanate dispersed in 340 ml of ethanol heated to reflux, 235 gm (1 mole) or undecyl bromide are added gradually. After addition has been completed, refluxing is continued for 2 hr. The reaction mixture is then cooled to room temperature, diluted with water, and the product extracted with diethyl ether. After the ether solution has been dried with calcium chloride, the ether is evaporated off and the residue fractionally distilled under reduced pressure to yield 184 gm (86.5%), bp 160°–161°C (10 mm).

B. Addition Reactions

The addition of thiocyanic acid to multiple bonds has recently been described [53]. This reaction evidently must be handled with considerable care since the intermediates and final products may be dangerous vesicants and since hydrogen cyanide may accidently be evolved.

The preparation of both cis- and trans-3-thiocyanoacrylamide indicates the method of separating the geometric isomers.

4-2. Preparation of cis- and trans-3-Thiocyanoacrylamide [53]

$$HC\equiv C-\overset{O}{\underset{\|}{C}}-NH_2 + NH_4SCN + H_2SO_4 \longrightarrow$$

$$NCS-CH=CH-\overset{O}{\underset{\|}{C}}-NH_2 + NH_4HSO_4 \quad (21)$$

CAUTION: The intermediates and products may be vesicants, and HCN may evolve during the reaction, processing, and during melting point determinations.

In a suitable hood at 0°C, a solution of 1.5 gm (0.02 mole) of ammonium thiocyanate in 10 ml of 2 M sulfuric acid (0.02 moles) is prepared. While maintaining a temperature of 0°C, 0.69 gm (0.01 mole) of propiolamide is added. After 1 hr at 0°C, the crude product is filtered off and washed with a few milliliters of ice water. The precipitate is then moistened, under nitrogen, with 15 ml of 2 N sodium hydroxide solution at 0°C and filtered rapidly. The insoluble fraction is the trans isomer; yield 0.21 gm (17%), mp (after recrystallization from water) 193°–194°C dec. (red melt).

The cis isomer is isolated from the filtrate by neutralizing with 6 ml of chilled 5 N hydrochloric acid. The precipitating cis isomer is filtered off and washed. Yield 0.86 gm (71%), mp (after crystallization from cold dimethyl sulfoxide on dilution with water) 153°–154°C dec. (HCN evolves).

C. Miscellaneous Reactions

The reaction of sulfenyl halide (RSX) with formamide, in the presence of thionyl chloride, has recently been patented as a method of preparing thiocyanates [54].

Another method of preparing thiocyanates involves the photosulfenchlorination of cycloaliphatic hydrocarbons according to Eqs. (22) and (23) [55, 56].

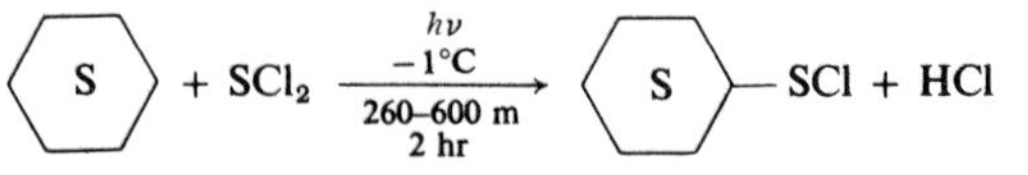

$$\langle S \rangle + SCl_2 \xrightarrow[260-600 \text{ m} \atop 2 \text{ hr}]{h\nu \atop -1°C} \langle S \rangle-SCl + HCl \quad (22)$$

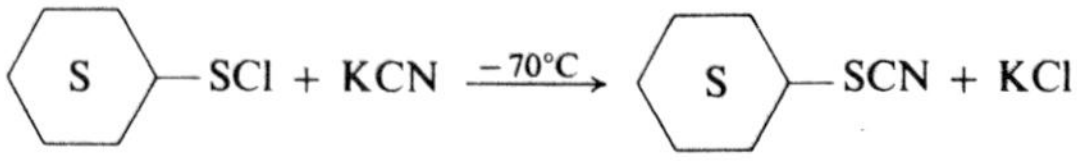

$$\langle S \rangle-SCl + KCN \xrightarrow{-70°C} \langle S \rangle-SCN + KCl \quad (23)$$

Aromatic thiocyanates have been prepared by the reaction of potassium thiocyanate or cuprous thiocyanate with diazonium salts [57].

In the absence of water and in the presence of easily substituted aromatic compounds, such as phenols or aniline derivatives, cupric thiocyanate may be used to substitute the thiocyanate group directly on the aromatic nucleus group according to Eq. (24) [57].

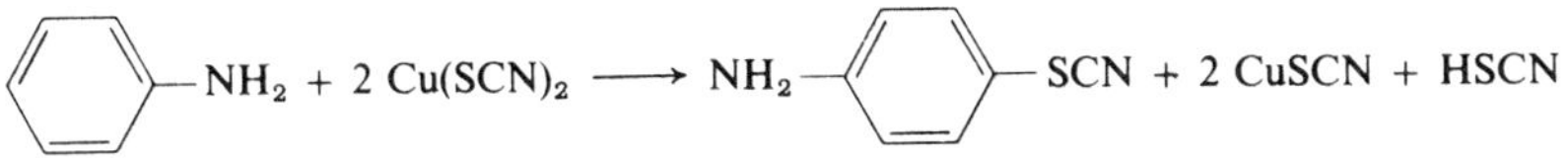

$$\text{C}_6\text{H}_5-\text{NH}_2 + 2\ \text{Cu(SCN)}_2 \longrightarrow \text{NH}_2-\text{C}_6\text{H}_4-\text{SCN} + 2\ \text{CuSCN} + \text{HSCN} \tag{24}$$

Under certain circumstances aromatic thiocyanate may be prepared by the reaction scheme (25) for which detailed experimental procedures are given in *Organic Syntheses* [58].

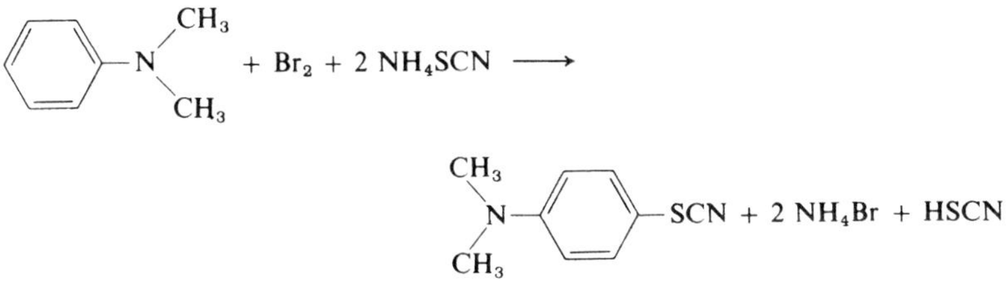

$$\text{C}_6\text{H}_5-\text{N}(\text{CH}_3)_2 + \text{Br}_2 + 2\ \text{NH}_4\text{SCN} \longrightarrow$$

$$(\text{CH}_3)_2\text{N}-\text{C}_6\text{H}_4-\text{SCN} + 2\ \text{NH}_4\text{Br} + \text{HSCN} \tag{25}$$

Thiocyanogen $(\text{SCN})_2$ has been used to add to unsaturated compounds to give thiocyanates [59].

5. ISOTHIOCYANATES

A. Condensation Reactions

Classically, perhaps the most widely used method for the preparation of isothiocyanate involves the reaction of amines with carbon disulfide in the presence of a base such as ethanolic aqueous ammonia or sodium hydroxide to form the appropriate salt of a dithiocarbamate. The conversion of a dithiocarbamate to an isothiocyanate may be carried out by a variety of reagents such as copper sulfate, ferrous sulfate, zinc sulfate, or lead nitrate. A procedure involving the use of lead nitrate for the general preparation of isothiocyanates has been described [60]. This reaction involves the reactions shown in Eqs. (26) and (27).

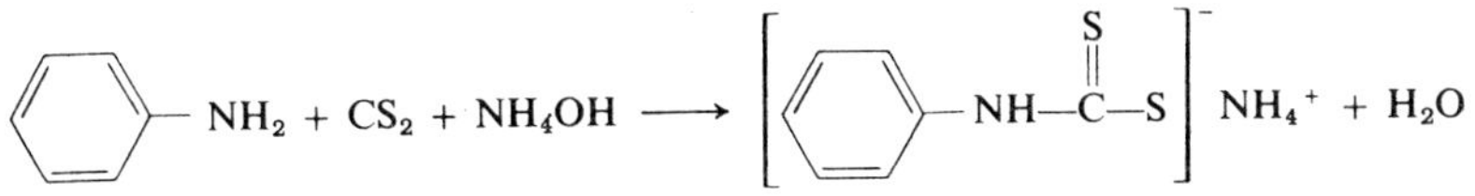

$$\text{C}_6\text{H}_5-\text{NH}_2 + \text{CS}_2 + \text{NH}_4\text{OH} \longrightarrow \left[\text{C}_6\text{H}_5-\text{NH}-\overset{\underset{\parallel}{\text{S}}}{\text{C}}-\text{S}\right]^- \text{NH}_4^+ + \text{H}_2\text{O} \tag{26}$$

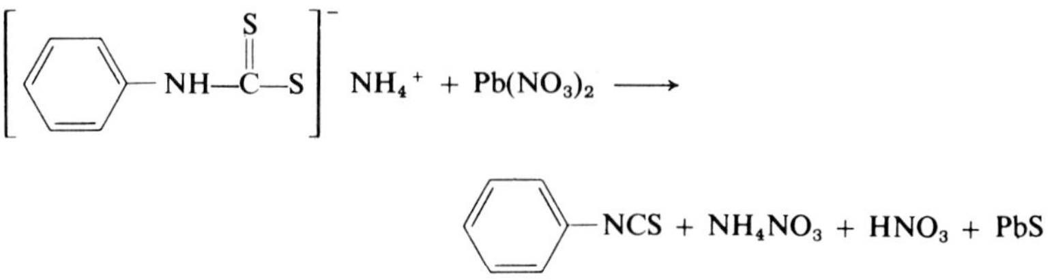

$$\left[\underset{}{\underset{}{\bigcirc}}-\text{NH}-\overset{\overset{S}{\|}}{C}-S\right]^{-} \text{NH}_4{}^+ + \text{Pb}(\text{NO}_3)_2 \longrightarrow$$

$$\underset{}{\underset{}{\bigcirc}}-\text{NCS} + \text{NH}_4\text{NO}_3 + \text{HNO}_3 + \text{PbS} \quad (27)$$

Instead of ammonia, aqueous sodium hydroxide [61] or strong organic bases have been used [62]. Alkyl isothiocyanates are prepared by the oxidation of dithiocarbamates with hydrogen peroxide [63] or oxygen [64].

The dithiocarbamate may be decomposed by the formation of a carboethoxy derivative in the so-called Kaluza reaction [62], as illustrated below. Generally, the reaction has been carried out for both aliphatic [61] and aromatic isothiocyanates [62], although evidently it cannot be used with aromatic amines containing strong electron-withdrawing groups such as *p*-cyano or *p*-nitro groups. In this reaction, sufficient time has to be allowed for the formation of dithiocarbamate (e.g., 15 min for the reaction of *N,N*-dimethylaniline to as long as 7 days for the reaction of β-naphthylamine).

5-1. Preparation of p-Chlorophenyl Isothiocyanate [62]

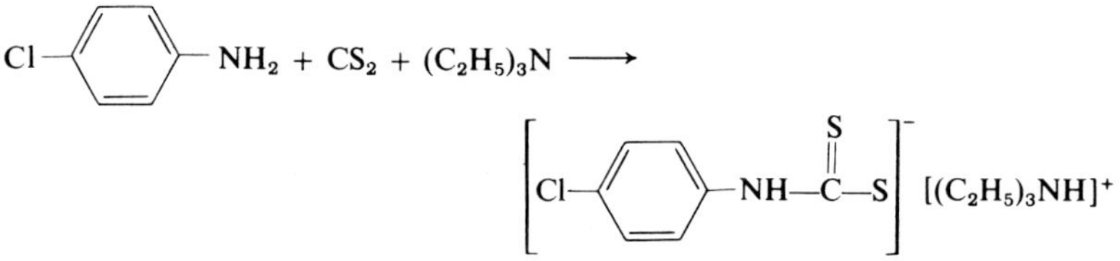

$$\text{Cl}-\underset{}{\bigcirc}-\text{NH}_2 + \text{CS}_2 + (\text{C}_2\text{H}_5)_3\text{N} \longrightarrow$$

$$\left[\text{Cl}-\underset{}{\bigcirc}-\text{NH}-\overset{\overset{S}{\|}}{C}-S\right]^{-} [(\text{C}_2\text{H}_5)_3\text{NH}]^+ \quad (28)$$

$$\left[\text{Cl}-\underset{}{\bigcirc}-\text{NH}-\overset{\overset{S}{\|}}{\underset{\underset{S}{|}}{C}}\right]^{-} [(\text{C}_2\text{H}_5)_3\text{NH}]^+ + \text{Cl}-\overset{\overset{O}{\|}}{C}-\text{OC}_2\text{H}_5 \longrightarrow$$

$$\text{Cl}-\underset{}{\bigcirc}-\text{NH}-\overset{\overset{S}{\|}}{C}-S-\overset{\overset{O}{\|}}{C}-\text{OC}_2\text{H}_5 + [(\text{C}_2\text{H}_5)_3\text{NH}]^+\text{Cl}^- \quad (29)$$

$$\text{Cl}-\underset{}{\bigcirc}-\text{NH}-\overset{\overset{S}{\|}}{C}-S-\overset{\overset{O}{\|}}{C}-\text{OC}_2\text{H}_5 \xrightarrow{\text{base}}$$

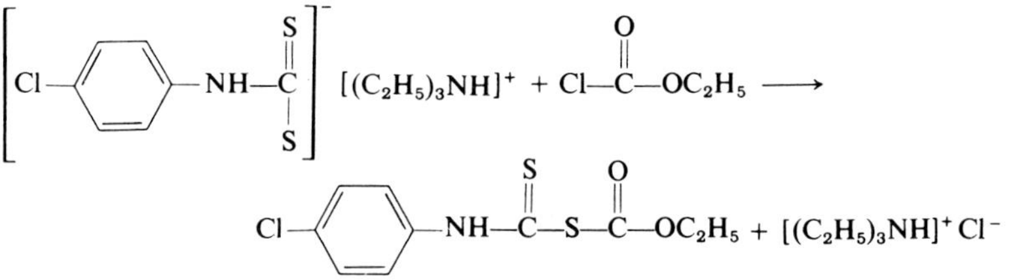

$$\text{Cl}-\underset{}{\bigcirc}-\text{NCS} + \text{COS} + \text{C}_2\text{H}_5\text{OH} \quad (30)$$

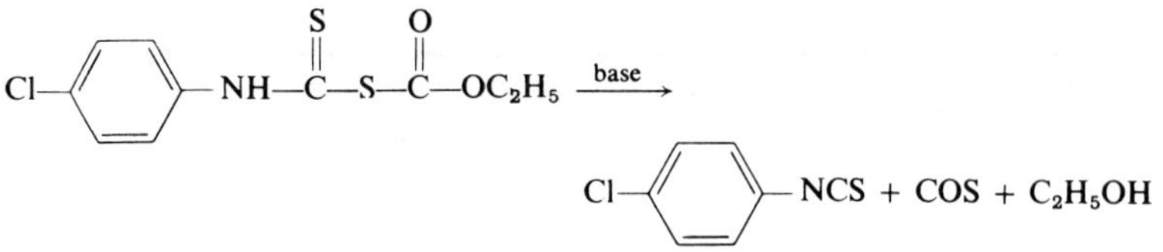

In a well ventilated hood, 12.8 gm (0.1 mole) of *p*-chloraniline is dissolved in a minimum amount of benzene and treated with 6.6 ml (0.1 mole) of carbon disulfide and 14 ml (0.1 mole) of triethylamine. The solution is then cooled to

0°C and the low temperature is maintained for 72 hr until the formation of the triethylammonium dithiocarbamate salt has been completed. The solution is then filtered and the solid is washed with anhydrous ether and airdried. Yield 83%.

The salt is then dissolved in approximately 75 ml of chloroform, treated with 14 ml (0.1 mole) of triethylamine, and cooled again to 0°C. To this solution is then added dropwise 10.2 ml (0.1 mole) of ethyl chlorocarbonate over a 15-min period with hand stirring. The resulting solution is stirred at 0°C for 10 min and is then allowed to warm to room temperature during a 1-hr period. The solution is then washed with 3 M hydrochloric acid solution, twice with water, and is then dried with sodium sulfate.

The chloroform is then evaporated under reduced pressure and the residual p-chlorophenyl isothiocyanate is recrystallized from ethanol. Yield of this step is 70% of theory, mp 46.5°C.

A quite general reaction for the preparation of isothiocyanates involves the use of thiophosgene. Since this material is a liquid, its handling is somewhat simpler than that of phosgene used in the synthesis of isocyanates. Its toxicity is believed to be as great as that of phosgene, if not greater. Consequently, due precautions must be taken in handling thiophosgene. This reaction is quite general in the aromatic series and has been used for the preparation of fluorescein isothiocyanate. It appears to fail in the naphthalene series. A typical procedure is the one for the preparation of p-nitrophenyl isothiocyanate [65].

5-2. Preparation of p-Nitrophenyl Isothiocyanate [65]

$$3\ NO_2-C_6H_4-NH_2 + Cl-\underset{\underset{S}{\|}}{C}-Cl \longrightarrow NO_2-C_6H_4-NCS + 2\left[NO_2-C_6H_4-NH_3\right]^+ Cl^- \quad (31)$$

CAUTION: Thiophosgene should be handled with extreme care, care also being taken for the disposal of residues. The highly toxic nature of thiophosgene must always be kept in mind.

A mixture of 20 gm (0.174 mole) of thiophosgene and 72 gm (0.522 mole) of p-nitroaniline suspended in 800 ml of dry benzene is refluxed for 1 hr with vigorous agitation. The mixture is then cooled and the precipitate of p-nitroaniline hydrochloride is filtered off. The benzene solution is then concentrated under reduced pressure. When the volume has been reduced to approximately 100 ml, a precipitate of 1 gm of p-nitroaniline is formed which is filtered off. Upon further reduction of the volume, 20 gm (71%) of p-nitrophenyl isothiocyanate is separated, mp 112°–115°C.

Toward the end of the precipitation from the concentrated mother liquor, contamination by di-*p*-nitrophenylthiourea takes place. This thiourea is relatively insoluble in cold benzene and therefore *p*-nitrophenylisothiocyanate can be separated by extraction with benzene, if necessary.

Isothiocyanates may also be formed from the phosphoramidate anion [27].

5-3. Preparation of n-Butyl Isothiocyanate [27]

$$[(C_2H_5O)_2P(O)NCH_2CH_2CH_2CH_3]^- \xrightarrow{CS_2}$$

$$[(C_2H_5O)_2P(O)-\underset{\underset{S-C=S}{|}}{N}CH_2CH_2CH_2CH_3]^-$$

$$\downarrow$$

$$[(C_2H_5O)_2P(O)S]^- + CH_3CH_2CH_2CH_2NCS \qquad (32)$$

To a slurry of 4.8 gm (0.1 mole) of 50% sodium hydride in 100 ml of 1,2-dimethoxyethane, 20.9 gm (0.1 mole) of diethyl *N-n*-butyl phosphoramidate is added. The mixture is stirred at room temperature until gas evolution has ceased. Then 7.6 gm (0.1 mole) of carbon disulfide is added and the solution is refluxed gently for 0.5 hr. The mixture is then stripped of solvent and the residue is distilled, giving a liquid distillate, identified as *n*-butyl isothiocyanate. Yield 8.7 gm (75%), bp 167°–170°C.

B. Decomposition Reactions

Aryl isothiocyanates may also be prepared by the thermal decomposition of monoarylthiourea according to Eq. (33) [66].

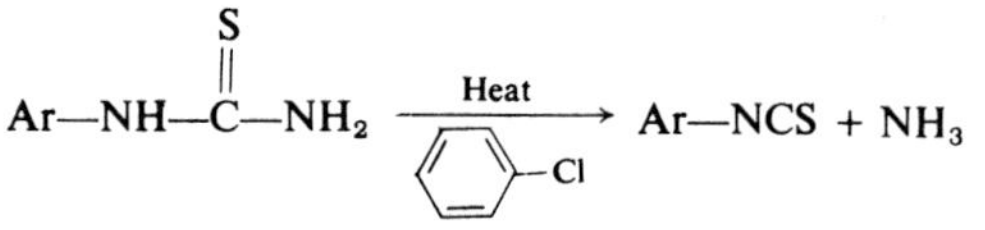

$$Ar-NH-\overset{\overset{S}{\|}}{C}-NH_2 \xrightarrow[\text{Cl-C}_6H_5]{\text{Heat}} Ar-NCS + NH_3 \qquad (33)$$

C. Miscellaneous Reactions

1. Polyisobutenyl isothiocyanates have been reported and used for reaction with amines to give the corresponding thiocarbamates [67].
2. The synthesis and infrared and ultraviolet spectra of phenyl isothiocyanates having a heterocyclic substituent have been reported [68].
3. Alkyl halides are reacted with alkali metal or ammonium isothiocyanate in acetone to give alkyl thiocyanates [69].
4. Preparation of methylisothiocyanate by the reaction of methyl chlorocarbonate with alkali metal or ammonium salt of *N*-methyldithiocarbamic acid [70].

5. Reaction of metal dithiocarbamate with cyanogen halide to give alkyl isothiocyanate, metal halide, and thiocyanic acid [71].

6. A. method of preparing aliphatic isothiocyanates by reacting aliphatic nitro compounds with a sulfur-containing carbon compound in the presence of metal carbonyl catalyst [72].

REFERENCES

1. E. W. Stern and M. C. Spector, *J. Org. Chem.* **31**, 396 (1966); U.S. Patent 3,405,156 (1968); P. M. Henry, U.S. Patent 3,641,092 (1972).
2. R. Becker, J. Grolig, and C. Rasp, U.S. Patent 4,297,501 (1981).
3. G. C. Licke, U.S. Patent 4,070,391 (1978).
4. D. H. Chadwick and T. H. Cleveland, *Kirk-Othmer Encycl. Chem. Technol., 3rd Ed.* **13**, 789–818 (1981).
5. R. K. Jordan, U.S. Patent 4,255,453 (1981).
6. J. U. Nef, *Justus Liebigs Ann. Chem.* **287**, 310 (1895).
7. A. Hantzsch and L. Mai, *Ber. Dtsch. Chem. Ges.* **28**, 2466 (1895).
8. R. Stroh and H. Gerber, *Angew. Chem.* **72**, 1000 (1960).
9a. E. Grigat and R. Pütter, *Chem. Ber.* **97**, 3012 (1964); **98**, 1359, 2619 (1965).
9b. C. Kosel, German Patent 1,270,553 (1968); *Chem. Abstr.* **69**, 96212u (1968); K. Kottke, F. Fredrich, and R. Pohloudek-Fabini, *Arch. Pharm. (Weinheim, Ger.)* **300** (7) 583 (1967); *Chem. Abstr.* **68**, 21884k (1968).
10. D. Martin, *Angew. Chem.* **76**, 303 (1964).
11. G. Entenmann, *Method. Chim.* **6**, 789 (1975); *Chem. Abstr.* **85**, 32108n (1976).
12. E. E. Flagg, U.S. Patent 4,216,316 (1980).
13a. The Merck Index, 9th Ed. p. 7149 (1976) published by Merck & Co. Rahway, N.J.
13b. R. L. Shriner and R. F. B. Cox, *J. Am. Chem. Soc.* **53**, 1601 (1931); R. L. Shriner and W. H. Horne, *ibid.* p. 3186; R. L. Shriner, W. H. Horne, and R. F. B. Cox, *Org. Synth. Collect. Vol.* **2**, 453 (1944).
14. A. H. Coons and M. H. Kaplan, *J. Exp. Med.* **91**, 1 (1954).
15. M. W. Farlow, *Org. Synth. Collect. Vol.* **4**, 521 (1963).
16. F. Mashio and T. Nomachi, *J. Chem. Soc. Jpn., Ind. Chem. Sect.* **56**, 289 (1953); *Chem. Abstr.* **48**, 10634 (1954).
17. M. Crochemore, U.S. Patent 4,151,193 (1979).
18. P. A. Argabright, H. D. Rider, and R. Sieck, *J. Org. Chem.* **30**, 3317 (1965).
19. J. J. Donleavy and J. English, Jr., *J. Am. Chem. Soc.* **62**, 218 (1940).
20. A. J. Hill and W. M. Degnan, *J. Am. Chem. Soc.* **62**, 1595 (1940).
21. W. Karo, Author's Laboratory, unpublished results (1963).
22. H. Wenker, *J. Am. Chem. Soc.* **58**, 2608 (1936).
23. S. Beinfest, P. Adams, and J. Halpern, U.S. Patents 2,837,560, 2,837,561 (1958).
24. W. Reichman, K. Konig, and J. Koster, U.S. Patent 4,195,031 (1980).
25. R. G. Arnold, J. A. Nelson, and J. J. Verbanc, *Chem. Rev.* **57**, 47 (1957).
26. C. V. Wilson, *Org. Chem. Bull.* **35**, Nos. 2–3 (1963).
27. W. S. Wadsworth, Jr. and W. D. Emmons, *J. Org. Chem.* **29**, 2816 (1964).
28. S. Schwartzman and D. A. Lima, U.S. Patent 4,141,913 (1979).
29. W. Bunge, *Angew. Chem.* **72**, 1002 (1960).
30. W. E. Erner, *J. Org. Chem.* **29**, 2091 (1964).
31. H. M. Singleton and W. R. Edwards, Jr., *J. Am. Chem. Soc.* **60**, 540 (1938).
32. P. A. S. Smith, *Org. React.* **3**, 337 (1946).
33. W. Lwowski and G. T. Tisue, *J. Am. Chem. Soc.* **87**, 4022 (1965).

34. T. Mukaiyama and H. Nohira, *J. Org. Chem.* **26**, 782 (1961).
35. E. S. Wallis and J. F. Lane, *Org. React.* **3**, 267 (1946).
36. H. Wolff, *Org. React.* **3**, 307 (1946).
37. H. Feuer, H. Rubinstein, and A. T. Nielsen, *J. Org. Chem.* **23**, 1107 (1958).
38. M. Tatsuno, Japanese Patent 70/19,884 (1970).
39. H. W. Johnson, Jr. and P. H. G. Daughhottee, Jr., *J. Org. Chem.* **29**, 246 (1964).
40. Asahi Chem. Industry Co., Ltd. Japan Kokai-Tokkyo Koho 81/100,752 (1981); *Chem. Abstr.* **95**, 203302j (1981); J. E. Lyons, U.S. Patent 3, 960, 914 (1976).
41. T. Mukaiyama and H. Nohira, *J. Org. Chem.* **26**, 782 (1961).
42. W. Reichmann, K. König, and J. Koster, U.S. Patent 4,195,031 (1980).
43. H. Zengel, M. Bergfeld, R. Zielke, and E. Klimesch, U.S. Patent 4,238,404 (1980).
44. J. Crosby, R. M. Paton, and R. A. C. Rennie, U.S. Patent 3,925,435 (1975).
45. W. Merz, U.S. Patent 3,444,231 (1969).
46. V. D. Sheludykov, A. D. Kirilin, V. F. Mironor, S. N. Glushakov, and Y. S. Karpman, U.S. Patent 4,192,815 (1980).
47. P. Allen, Jr., *J. Am. Chem. Soc.* **57**, 198 (1935).
48. M. A. Youtz and P. P. Perkins, *J. Am. Chem. Soc.* **51**, 3510 (1929).
49. H. D. Vogelsang, T. Wagner-Jauregg, and R. Rebling *Justus Liebigs Ann. Chem.* **568**, 192 (1950).
50. A. P. Terent'ev and A. I. Gershenovich, *Zh. Obshch. Khim.* **23**, 204 (1953).
51. J. C. Ambelang and T. B. Johnson, *J. Am. Chem. Soc.* **61**, 632 (1939).
52. S. Harusawa and T. Shiorin, *Yuki Gosei Kagaku Kyokaishi* **39**(8), 741 (1981); *Chem. Abstr.* **95**, 167873k (1981).
53. W. D. Crow and N. J. Leonard, *J. Org. Chem.* **30**, 2660 (1965).
54. E. Kühle, German Patent 1,157,603 (1963); *Chem. Abstr.* **60**, 6787e (1964).
55. E. Müller and E. W. Schmidt, *Chem. Ber.* **96**, 3050 (1963).
56. E. Müller and E. W. Schmidt, *Chem. Ber.* **97**, 2614 (1964).
57. H. P. Kaufman and K. Kühler, *Chem. Ber.* **67B**, 944 (1934).
58. R. Q. Brewster and W. Schroeder, *Org. Synth. Collect. Vol.* **2**, 574 (1948).
59. V. R. Kartashov, E. V. Skorobogatova, and I. V. Bodrikov, *Zh. Org. Khim.* **9**(1), 214 (1973); *Chem. Abstr.* **78**, 110264g (1973).
60. F. B. Dains, R. Q. Brewster, and C. P. Olander, *Org. Synth. Collect. Vol.* **1**, 447 (1948).
61. M. L. Moore and F. S. Crossley, *Org. Synth. Collect. Vol.* **3**, 599 (1955).
62. J. F. Hodgkins and W. Preston Reeves, *J. Org. Chem.* **29**, 3098 (1964).
63. G. Giesselmann, W. Schwarze, and W. Weigert, U.S. Patent 3,787,472 (1974).
64. A. G. Zeiler and H. Babad, U.S. Patent 3,923,852 (1975); A. G. Zeiler, U.S. Patents 3,923,851, 3,923,853 (1975).
65. E. Dyer and T. B. Johnson, *J. Am. Chem. Soc.* **54**, 781 (1932).
66. J. Cymerman-Craig, M. Moyle, and R. A. White, *Org. Synth. Collect. Vol.* **4**, 700 (1963).
67. W. R. Song, U.S. Patent 4,303,539 (1981).
68. A. Martron, I. Skacani, and I. Kanalova, *Chem. Zvesti* **27** (6), 808 (1973).
69. J. Rehor, Czech Patent 181,053 (1980). *Chem. Abstr.* **93**, 132072e (1980).
70. H. Werres, U.S. Patent 3,406,191 (1968).
71. G. Giesselmann, G. Schreyer, and R. Vanheertum, U.S. Patent 4,089,887 (1978).
72. P. H. Scott and E. H. Kober, U.S. Patent 3,953,488 (1976).

CHAPTER 13 / **AMINES**

1. Introduction	378
2. Condensation Reactions	380
A. Hofmann Alkylation of Ammonia and Amines	380
a. Treatment of Amines with Halides	380
2-1. Preparation of n-Butylamine	382
b. Reaction of Tosylates with Amines	382
2-2. Preparation of N-Methyldihydropyran-2-methylamine	383
c. Secondary Amines	383
2-3. Preparation of N-tert-Butylbenzylamine	384
d. Tertiary Amines	384
2-4. Preparation of Allyldiethylamine	385
e. N,N-Dimethylalkylamines	385
f. N,N-Dimethylalkylamines by the Reduction of Quaternary Ammonium Salts	385
2-5. Preparation of N,N-Dimethyl-(+)-neomenthylamine	386
g. Cyclic Tertiary Amines	387
h. Monoalkyl Derivatives of Diamines	387
2-6. Preparation of Monomethylethylenediamine (Method 1)	388
2-7. Preparation of Monoethylethylenediamine (Method 2)	389
B. Miscellaneous Condensation Reactions	390
a. Hydroboration	390
2-8. Preparation of trans-2-Methylcyclohexylamine	391
b. Delépine Reaction	391
2-9. Preparation of β-Alanine	392
c. Gabriel Condensation and Ing–Manske Modification	392
2-10. Preparation of α,δ-Diaminoadipic Acid	393
2-11. Preparation of 1,8-Diamino-3,6-dioxaoctane	394
d. Ritter Reaction	396
2-12. Preparation of 1-Methylcyclobutylamine	396
e. Hydrolysis of Amides	397
f. Addition of Amines to Double Bonds	398
2-13. Preparation of Ethyl β-Methylaminopropionate	398
g. The Mannich Reaction	400
2-14. Preparation of 6-Benzyloxy-5-methoxygramine	401
3. Reduction Reactions	401
A. Reduction of Amides	402
3-1. Preparation of N-Ethyl-1,1-dihydroheptafluorobutylamine	402
B. Reduction of Nitriles	403
3-2. Preparation of Cyclopropylmethylamine	404
3-3. Preparation of [2-(2-Fluoro-3,4-dimethoxyphenyl)ethyl] amine	404
C. Reduction of Nitro Compounds	405
a. Reduction of Activated Iron and Water	405
3-4. Preparation of 2-Chloroaniline	406
b. Reduction with Iron and Ferrous Sulfate	406

3-5.	Preparation of 2-Amino-2-methyl-1-propanol	407
c.	Catalyzed Reduction with Hydrazine Hydrate	407
3-6.	Preparation of Di(2-amino-4-chlorophenyl) Sulfone	407
d.	Reduction with Tin and Hydrochloric Acid	408
3-7.	Preparation of Aniline	408
e.	Reduction with Sodium Hydrosulfite	409
3-8.	Preparation of 2,4-Diamino-1-naphthol Dihydrochloride	409
f.	Reduction with Sodium Sulfide and Related Compounds	410
3-9.	Preparation of 2-Ethylaniline(2-Aminoethylbenzene)	410
g.	Reduction with Lithium Aluminum Hydride	411
h.	Reduction with Tervalent Phosphorus Reagents	411
i.	Selective Catalytic Hydrogenation of Aromatic Nitro Compounds	411
D.	The Leuckart Reaction	412
a.	Classical Leuckart Reaction	412
3-10.	Preparation of 1-(Methyl-3-phenylpropyl)piperidine	412
b.	Clarke–Eschweiler Modification	413
3-11.	Preparation of 1-Methyl-2-(p-tolyl)piperidine	413
4.	Rearrangement and Related Reactions	414
A.	Benzidine Rearrangement	414
4-1.	Preparation of 3,3'-Dibromobenzidine	415
B.	Hofmann Rearrangement	415
4-2.	Preparation of Nonylamine	416
C.	Schmidt Reaction	418
4-3.	Preparation of 3,4-Dimethylaniline	418
D.	Curtius Reaction	418
4-4.	Preparation of (±)-trans-2-Cyclohexyloxycyclopropylamine	419
5.	Miscellaneous Preparations	420
	References	429

1. INTRODUCTION

Syntheses of amines have perhaps received more attention than the preparation of any other functional group in organic chemistry. Thus, for example, Houben-Weyl [1] devotes one volume of over 1000 pages to procedures for the preparation of amino compounds. Clearly, a comprehensive survey of all of these methods would be beyond the scope of the present work. Emphasis is therefore placed on selecting a relatively small number of processes which have fairly general applicability. Methods used in the preparation of heterocyclic nitrogen compounds are, as a rule, not presented.

It is a common practice to discuss syntheses of primary, secondary, and tertiary amines as well as quaternary ammonium salts separately. While there is much merit to such an arrangement, particularly from the standpoint of teaching organic chemistry, this approach leads to considerable repetition. Therefore, the arrangement in this chapter is by reaction type with reference to the class (or classes) of amine which may be prepared by a particular method where appropriate.

§ 1. Introduction

The syntheses are presented here under the headings condensation reactions, reductive methods, and rearrangements. The best known condensation reaction for the preparation of amines is the alkylation of ammonia and various amines with alkyl halides and dialkyl sulfates (Hofmann alkylation). By this means, primary, secondary, and tertiary amines, as well as quaternary ammonium salts can be prepared. Special reactions used for the preparation of primary amines free of secondary and tertiary amines are the Delépine reaction (alkylation of hexamethylenetetramine followed by hydrolysis), the alkylation of phthalimides followed by hydrolysis with aqueous alkali (Gabriel condensation) or with hydrazine (Ing–Manske modification), and the reaction of organoboranes with either chloramine or hydroxylamine-O-sulfonic acid.

The alcoholysis of nitriles with secondary alcohols or related branched olefins readily produces amides which, upon hydrolysis, yield primary amines (Ritter reaction). A variety of amines may also be produced by an application of the Michaels condensation in which amines are added to olefinic bonds conjugated with carbonyl groups. The condensation of secondary amines with formaldehyde and an active hydrogen compound is an application of the Mannich reaction to the preparation of tertiary amines.

The reduction of many nitrogen compounds produces amines. The reduction of appropriately substituted amides may be used to prepare primary, secondary, or tertiary amines. Nitriles and nitro compounds are reduced to primary amines. Particularly the nitro compounds are a useful source for the production of both aliphatic and aromatic primary amines by reduction. A large variety of reducing agents have been used in these preparations.

The reductive alkylation of carbonyl compounds with formic acid and either amine or ammonium salts or amides of formic acid has been used to prepare primary, secondary, or tertiary amines under rather drastic conditions (Leuckart reaction). For the preparation of methylated tertiary amines, mild conditions have been used in the Eschweiler–Clarke reaction. In this modification of the Leuckart reaction, amines are reductively alkylated with formalin and an excess of formic acid.

Among the rearrangement reactions for the preparation of amines are the benzidine rearrangement of 1,2-diarylhydrazines to 4,4'-diamines. The Hofmann rearrangement of amides with hypohalides and the related Schmidt and Curtius reactions have also been used to prepare amines.

Despite the fact that amines are very commonly handled both in the laboratory and in industry, considerable care should be exercised. Many amines, particularly the aliphatic amines, have pungent odors. Many of them are believed to be toxic. Some compounds, such as aniline, are believed to be absorbed through the skin, which may give rise to physiological reactions.

Several compounds which were considered quite ordinary laboratory reagents for many years and which were important industrial intermediates, particularly in the dye industry, have recently been labeled carcinogenic. Among these are 1- and 2-naphthylamine, their salts, and their simple derivatives (which may be readily hydrolyzed to the naphthylamines) and 4,4'-benzidine. Handling of these compounds is prohibited. In fact, to the best of our knowledge, benzidine is no longer manufactured in the United States.

2. CONDENSATION REACTIONS

A. Hofmann Alkylation of Ammonia and Amines

a. TREATMENT OF AMINES WITH HALIDES

The treatment of ammonia and amines with alkyl halides, dialkyl sulfate, or alkyl *p*-toluenesulfonates we classify as "Hofmann alkylations" to distinguish this fundamental method of preparing amines from other reactions which were discovered by that great organic chemist and which also have his name associated with them.

The treatment of ammonia with alkyl halide normally gives rise to a mixture of primary, secondary, and tertiary amines, and the quaternary ammonium halide salt. The reaction sequence is usually given by Eqs. (1-4).

$$RX + NH_3 \longrightarrow RNH_2 + HX \quad (1)$$

$$RX + RNH_2 \longrightarrow R_2NH + HX \quad (2)$$

$$RX + R_2NH \longrightarrow R_3N + HX \quad (3)$$

$$RX + R_3N \longrightarrow R_4N^+ X^- \quad (4)$$

Fortunately, from the preparative standpoint, when R is ethyl or a higher alkyl group, the boiling points of the various amines are sufficiently far apart to permit their separation by fractional distillation. By adjustment of the mole ratio, reaction temperatures, times, and other reaction conditions, it is frequently possible to control the reaction so that any one of these classes of amines becomes the predominant product.

By the judicious selection of amines and alkyl halides (or dialkyl sulfates), amines with a variety of different alkyl groups may be prepared. Under ordinary laboratory conditions, aromatic halides do not undergo the reaction unless the halogen is sufficiently activated by other substituents. Whereas the preparation of aniline from chlorobenzene with ammonia is ordinarily not a useful laboratory process, activation of the chlorine by one or more ortho or para nitro groups permits the reaction to proceed. Thus, for example, 2,4-dichloro-3-nitropyridine is readily monoaminated to 4-amino-2-chloro-3-nitropyridine and a small amount of 2-amino-4-chloro-3-nitropyridine [2].

Under somewhat more forcing conditions, aromatic ortho-nitro-fluoro com-

§ 2. Condensation Reactions

pounds may be aminated, as indicated in the preparation of 3-amino-4,6-difluoro-2-nitroacetanilide and its related 4,6-difluoro-2-nitro-1,3-phenylenediamine [3].

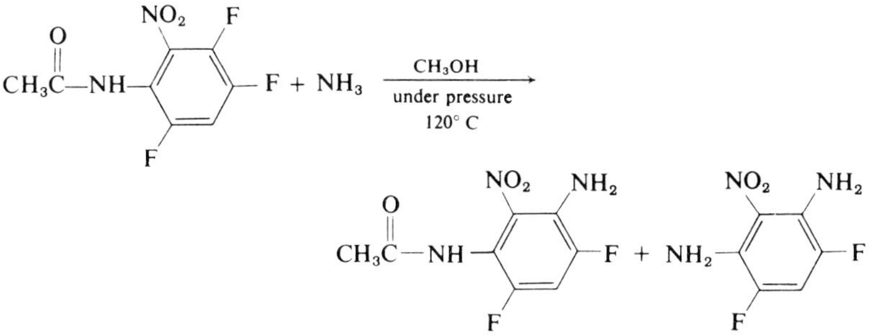

$$\text{CH}_3\text{C(O)NH-C}_6\text{H}(\text{NO}_2)(\text{F})_2\text{-F} + \text{NH}_3 \xrightarrow[120°\text{C}]{\text{CH}_3\text{OH, under pressure}} \text{CH}_3\text{C(O)NH-C}_6\text{H}(\text{NO}_2)(\text{NH}_2)(\text{F})\text{-F} + \text{NH}_2\text{-C}_6\text{H}(\text{NO}_2)(\text{NH}_2)(\text{F})\text{-F} \quad (5)$$

While the normal activity of halides is in the usual order I > Br > Cl, the chloro compounds are generally inert. However, in the aromatic series, nitro groups ortho or para to a chlorine substituent increase the reactivity of the chlorine sufficiently to afford an 85% yield in a reaction according to the following equation [4]:

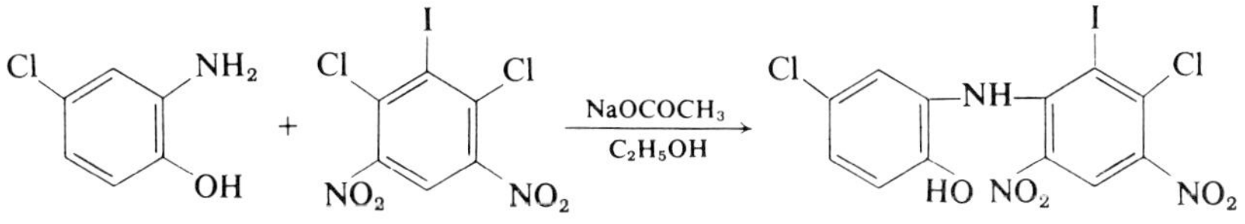

$$\text{Cl-C}_6\text{H}_3(\text{NH}_2)(\text{OH}) + \text{Cl-C}_6\text{H}(\text{I})(\text{Cl})(\text{NO}_2)_2 \xrightarrow[\text{C}_2\text{H}_5\text{OH}]{\text{NaOCOCH}_3} \text{Cl-C}_6\text{H}_2(\text{OH})\text{-NH-C}_6\text{H}(\text{I})(\text{Cl})(\text{NO}_2)_2 \quad (6)$$

Because of lower volatility and cost, N-methylation is often carried out with dimethyl sulfate rather than with methyl iodide, despite the hazardous nature of the reagent.

In general, Hofmann alkylations are carried out with appropriate alkyl halides or dialkyl sulfates. These reagents may have to be prepared from the related alcohols, often by rather troublesome methods. A more convenient conversion of alcohols to amines involves the alkylation of amines with toluenesulfonate esters of alcohols. These "tosylates" are generally prepared quite easily and may then be used as alkylating agents [5–18]. If the alcohol which is to be converted to an amine has the proper structural features, the Ritter reaction (see below) is another useful approach to the preparation of primary amines (by hydrolysis of the amide formed in the reaction) and of some secondary amines (by reduction of the amides).

As far as the nitrogen-bearing reaction component is concerned, liquid ammonia, aqueous ammonia, some aqueous amines, and aliphatic and aromatic amines both neat or in one of a host of solvents usually undergo Hofmann alkylations. Since the halides are usually water-insoluble, reactions in aqueous media are sometimes difficult although the addition of alcohols or anionic surfactants and vigorous agitation is helpful. Phase-transfer catalysts in the case of alkylation reactions may be problematical. Many phase-transfer catalysts are cationic surfactants such as quaternary ammonium salts. Consequently there may be a scrambling of alkyl groups in the course of the reaction.

In connection with product work-up, the fact that many amine hydrochlorides are soluble in chloroform while ammonium chloride is insoluble is sometimes useful.

To prepare a primary amine by alkylation of ammonia, the level of ammonia is kept high to reduce the formation of secondary and tertiary amines. The preparation of n-butylamine is a typical example of the procedure.

2-1. Preparation of n-Butylamine [11]

$$CH_3CH_2CH_2CH_2Br + NH_3 \longrightarrow CH_3CH_2CH_2CH_2NH_2 + HBr \qquad (7)$$

In a 12-liter flask fitted with a stirrer, addition funnel, gas inlet tube, and reflux condenser is placed 8 liters of 90% ethanol. The reaction system is placed in a hood and ammonia is run in with constant stirring until the flask has gained about 300 gm. Then 68.5 gm (0.5 mole) of purified n-butyl bromide is added rapidly. Then, while a slow stream of ammonia is passed through the flask, an additional 1438.5 gm (10.5 moles) of n-butyl bromide is added continuously at a rate of approximately 17 gm/hr. After the addition has been completed, the flask is stirred for an additional 2 days. Then the reaction mixture is distilled to remove ethanol, and, after approximately 4 liters of ethanol have been distilled off, the flask is cooled and the precipitating ammonium bromide is separated. Another 4 liters of ethanol are then distilled off and more ammonium bromide is filtered off. About 1 liter of solution remains in the flask. To this is added 1 liter of water and the last traces of ethanol are removed by distillation. If necessary, this step is repeated until all of the ethanol has been removed.

To the cooled residue is added a cold solution of 240 gm (6 moles) of sodium hydroxide in 1 liter of water and the mixture is distilled until the lower boiling fraction has come over. This distillate is dried over fused potassium hydroxide.

The pot residue is cooled and the remaining amine layer is separated and combined with the distillate.

The dried amine is then fractionally distilled through a glass-helix-packed distillation column. The fraction boiling at 76.5°C (742 mm) is collected. Yield 191.7 gm (47% of theory), n_D^{20} 1.4008.

b. REACTION OF TOSYLATES WITH AMINES

An example of the preparation of a secondary amine using a tosylate as the alkylating agent is the preparation of N-methyldihydropyran-2-methylamine.

2-2. Preparation of N-Methyldihydropyran-2-methylamine [8]

$$\text{(dihydropyranyl)}-CH_2OH + H_3C-\text{(C}_6H_4\text{)}-SO_2Cl \longrightarrow \text{(dihydropyranyl)}-CH_2O-SO_2-\text{(C}_6H_4\text{)}-CH_3 \quad (8)$$

$$\text{(dihydropyranyl)}-CH_2O-SO_2-\text{(C}_6H_4\text{)}-CH_3 + CH_3NH_2 \longrightarrow \text{(dihydropyranyl)}-CH_2NHCH_3 \quad (9)$$

To 34.5 gm (0.30 mole) of dihydropyran-2-methanol dissolved in 200 ml of pyridine, is added 75 gm (0.39 mole) of *p*-toluenesulfonyl chloride. The reaction mixture is warmed to 50°C for ½ hr. After this time, the mixture is cooled to room temperature to produce a white precipitate of pyridinium hydrochloride. This co-product is removed by filtration and the excess pyridine is separated by evaporation under reduced pressure. The residual product is recrystallized from ethanol. Yield of dihydropyran-2-methyl tosylate is 48 gm (59%), mp 47°–48°C.

In a cooled steel pressure bomb are placed 27 gm (0.107) mole) of dihydropyran-2-methyl tosylate and 14 gm (0.45 mole) of methylamine dissolved in 200 ml of absolute methanol. The bomb is sealed and heated with shaking to 125°C for 1 hr. After cooling, the sealed bomb is carefully opened in a hood and the reaction mixture is concentrated in a vacuum evaporator. After removal of the solvent, the semisolid is made basic with 20% solution of sodium hydroxide in water and continuously extracted with ether for 48 hr. The ether layer is dried with potassium carbonate and reduced in volume to yield a crude product. The crude product is purified by distillation under reduced pressure to afford 6.3–8.4 gm (30–40% of theory) of *N*-methyldihydropyran-2-methylamine, bp 60°C (17 mm). During the distillation, the pot temperature should not exceed 150°C to minimize spontaneous decomposition of the crude product.

Sulfonate esters may be used to prepare a variety of amines. For example, polyvinyl alcohol has been converted to a partially benzene-sulfonated polymeric ester. Upon treatment with one of a variety of amines, such as piperidine, morpholine, ethylamine, etc., N-substituted vinylamine polymers were isolated which also contained intramolecular cyclic ether units [6].

c. Secondary Amines

The Hofmann alkylation of primary amines to give secondary amines has been widely used. By adjusting the molar ratio to be 1 mole of alkyl halide to at least 2 moles of amine, the principal product is a secondary amine.

Steric factors often enter into the specificity of the reaction. Thus, for example, the reaction of *tert*-butylamine with alkyl halides affords the secondary amines in good yield. These secondary amines are, presumably, of low reactivity toward further alkylation.

Water may exercise a catalytic effect in some of these alkylation reactions. The more reactive alkyl halides such as benzyl chloride and the *n*-alkyl halides gave rise to a higher yield of amines than less reactive alkyl halides. A typical example of secondary amine formation is the preparation of *N-tert*-butylbenzylamine [12].

2-3. Preparation of N-tert-Butylbenzylamine [12]

$$H_3C-\underset{\underset{CH_3}{|}}{\overset{\overset{CH_3}{|}}{C}}-NH_2 + \underset{}{\bigcirc}-CH_2Cl \longrightarrow H_3C-\underset{\underset{CH_3}{|}}{\overset{\overset{CH_3}{|}}{C}}-NH-CH_2-\bigcirc \quad (10)$$

To a reaction flask containing 438 gm (6 moles) of *tert*-butylamine at reflux temperature is added rapidly 380 gm (3 moles) of benzyl chloride. During approximately 1 hr of heating, the solid formed during this period is redissolved. After cooling, the resulting reaction mixture is treated with 750 ml of 4 N aqueous sodium hydroxide solution. The oily amine layer is separated, dried over fused potassium hydroxide, and purified by fractional distillation under reduced pressure to yield 137 gm (84%) of *N-tert*-butylbenzylamine, bp 91°C (12 mm), along with a 4.5% yield of the *N-tert*-butyldibenzylamine, bp 142°–145°C (3 mm).

Highly hindered secondary and tertiary amines have been prepared by reaction of alkyl halides in the presence of copper powder [13], as illustrated in Eq. (11).

$$CH\equiv C-\underset{\underset{CH_3}{|}}{\overset{\overset{CH_3}{|}}{C}}-NH_2 + Cl-\underset{\underset{CH_3}{|}}{\overset{\overset{CH_3}{|}}{C}}-C\equiv CH \xrightarrow[\substack{KOH \\ 25°-30°C \\ 8 \text{ days}}]{\substack{Cu \\ 40\% \text{ aq.}}} CH\equiv C-\underset{\underset{CH_3}{|}}{\overset{\overset{CH_3}{|}}{C}}-NH-\underset{\underset{CH_3}{|}}{\overset{\overset{CH_3}{|}}{C}}-C\equiv CH \quad (11)$$

This procedure appears to fail in the case of even more highly hindered compounds. In this case, an excess of the primary amine, dimethylformamide as a solvent, along with copper–cuprous chloride, brings about secondary amine formation at 4°C within 3 days [14].

d. Tertiary Amines

A typical example of the preparation of a tertiary amine by alkylation of a secondary amine is the preparation of allyldiethylamine given below. This preparation is carried out in anhydrous benzene solution, which has much to recommend it. The same reference also gives an example of the preparation of a tertiary amine by alkylation of an aqueous solution of dimethylamine. However, the yields obtained from this reaction are small.

2-4. Preparation of Allyldiethylamine [15]

$$CH_2=CH-CH_2Br + CH_3CH_2NHCH_2CH_3 \longrightarrow$$

$$\underset{CH_3CH_2}{\overset{CH_3CH_2}{N}}-CH_2CH=CH_2 + HBr \quad (12)$$

In a 1-liter three-necked flask fitted with a mechanical stirrer, addition funnel, reflux condenser, and a thermometer, a solution of 200 gm (2.74 moles) of diethylamine in 240 ml of anhydrous benzene is treated dropwise with 165 gm (1.37 moles) of allyl bromide. The reaction flask is cooled intermittently during the addition to maintain a temperature between 45° and 50°C. After the addition has been completed, the reaction mixture is heated to reflux in a water bath at 80°C for 2 hr. After cooling, 150 ml of concentrated hydrochloric acid and 100 ml of water are added. The aqueous layer is separated, the benzene layer is extracted twice with 50 ml portions of 10% hydrochloric acid, and the acid extract and aqueous phases are combined. The acidic phase is then extracted with 50 ml of benzene. All benzene extracts are then discarded. The aqueous solution is cooled and made alkaline by the slow addition of 500 ml of 40% sodium hydroxide. The cooled aqueous phase is separated from the product layer and extracted two times with 50-ml portions of ether. The ether extract is combined with the amine layer. After drying the combined amine and ether extract with solid potassium hydroxide, the organic mixture is fractionally distilled through a glass-helix-packed column to yield 90.4 gm (80% of theory), bp 111°C (760 mm).

e. N,N-DIMETHYLALKYLAMINES

Since the preparation of N,N-dimethylalkylamines is of some importance, a number of modifications of the usual alkylation procedures are available. This is of particular importance, since dimethylamine is a gas and, therefore, somewhat troublesome to handle. One approach is to substitute dimethylformamide for dimethylamine. However, this procedure has only limited value, since, for example, an attempt to prepare N,N-dimethylbenzylamine from benzyl chloride and dimethyl formamide afforded a 36% yield of N,N-dimethylbenzylamine and a 34% yield of N-methyldibenzylamine [16].

f. N,N-DIMETHYLALKYLAMINES BY THE REDUCTION OF QUATERNARY AMMONIUM SALTS

While many N,N-dimethylalkylamines are prepared via the so-called Eschweiler–Clarke reaction (see below), Cope and co-workers [17] have found that good yields are also obtained by exhaustively methylating an amine to the quaternary ammonium salts and then reducing the resulting product with lithium aluminum hydride as described below.

2-5. Preparation of N,N-Dimethyl-(+)-neomenthylamine [17]

$$\text{Cyclohexane-NH}_2 + 3\ CH_3I \xrightarrow[CH_3OH]{Na_2CO_3} \text{Cyclohexane-}\overset{+}{N}(CH_3)_3 I^- \xrightarrow[THF]{LiAlH_4} \text{Cyclohexane-}N(CH_3)_2 + CH_4 \quad (13)$$

(a) Preparation of N,N-Dimethyl-(+)-neomenthylamine methiodide. A mixture of 14.3 gm (0.1 mole) of (+)-neomenthylamine, 25.2 gm (0.3 mole) of sodium carbonate, 42.6 gm (0.3 mole) of methyl iodide, and 150 ml of methanol is heated under reflux with vigorous stirring for 75 hr. After 24 and 48 hr, additional portions of methyl iodide are added until a total of approximately 65 gm (0.45 mole) of methyl iodide has been used. The reaction mixture is then evaporated to dryness under reduced pressure using a rotary evaporator. The residual solid is extracted three times with 150-ml portions of boiling chloroform. The combined extracts are cooled, filtered, and evaporated to dryness. The residual crude methiodide is recrystallized once from a mixture of acetone and pentane. Yield 28.2 gm (90% of theory).

(b) Reduction of quaternary ammonium salts. In a three-necked flask fitted with an efficient reflux condenser, topped by a drying tube, magnetic stirrer, and electric heating mantle, is placed 175 ml of freshly distilled, anhydrous tetrahydrofuran. To the solvent is added, with caution, 7.8 gm (0.2 mole) of powdered lithium aluminum hydride. The mixture is heated under reflux for 1 hr. Then 11.7 gm (0.04 mole) of N,N-dimethyl-(+)-neomenthylamine methiodide in finely ground form is introduced in a single portion and the mixture is heated under reflux with stirring until the evolution of methane has ceased. The reaction mixture is then cooled and cautiously hydrolyzed by gradual addition of 20 ml of water. To isolate the product, 100 ml of ether is added and the mixture is refluxed for 2 hr. After cooling, the solid is separated by filtration and the filter cake is washed repeatedly with ether. The filtrate and the washings are concentrated by distillation. The residue is washed with five 50-ml portions of water to remove the tetrahydrofuran and the remaining organic material is dissolved in pentane. The pentane solution is extracted with two 45-ml portions of 2 N hydrochloric acid and two 25-ml portions of water. The aqueous acidic washings are combined and heated with a solution of 10 gm of sodium hydroxide in 50 ml of water. The amine is separated by layering pentane over the water and the

§ 2. Condensation Reactions

aqueous layer is repeatedly extracted with pentane. The combined extracts are dried over magnesium sulfate. After removal of the solvent by distillation, the residue is distilled through a semimicro vacuum distillation column. Yield 4.9 gm of N,N-dimethyl-(+)-neomenthylamine (75% of theory), bp 75°–76°C (4.5 mm).

The treatment of amines with an excess of an alkyl halide, as shown in the first step of Eq. (13), is assumed to lead to the formation of the corresponding quaternary salt. In general, this appears to be the case. However, this matter requires further examination, in the light of the work of Melhado [18], when new amines are to be quaternized. It was found that when gramines were treated with methyl iodide, both mono- and bis-indolyl products were formed. The ratio of mono to bis products varied with substituents and with the reaction solvent systems [Eq. (14)] [18].

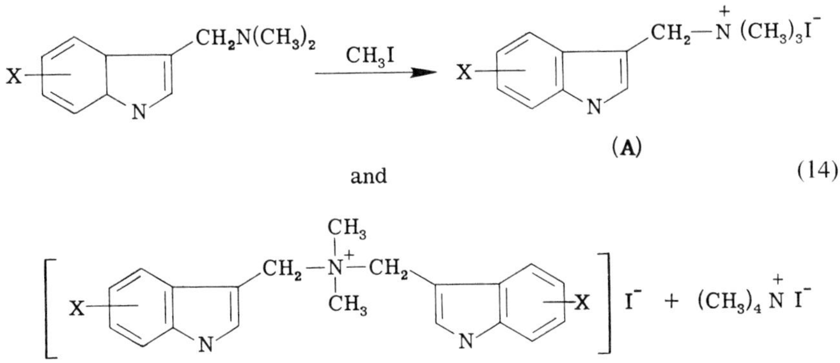

$$\text{X}-\text{indole}-\text{CH}_2\text{N}(\text{CH}_3)_2 \xrightarrow{\text{CH}_3\text{I}} \text{X}-\text{indole}-\text{CH}_2-\overset{+}{\text{N}}(\text{CH}_3)_3 \text{I}^-$$

(A)

and (14)

$$\left[\text{X}-\text{indole}-\text{CH}_2-\overset{\overset{\text{CH}_3}{|}}{\underset{\underset{\text{CH}_3}{|}}{\overset{+}{\text{N}}}}-\text{CH}_2-\text{indole}-\text{X}\right] \text{I}^- + (\text{CH}_3)_4 \overset{+}{\text{N}} \text{I}^-$$

(B)

When $X = H$, and reaction is carried out in absolute ethanol, the mole ratio of A to B is 2 to 1. In neat methyl iodide, A to B is 5 to 1.

When $X = 4\text{-NO}_2$, in absolute ethanol A to B is 3 to 1. In neat methyl iodide, A to B is 1 to 5.

g. Cyclic Tertiary Amines

The reaction of 1,5-dihaloalkanes or related compounds with primary amines yields cyclic tertiary amines [19]. If the distance separating the two halogens is much smaller, the two halogen substituents may react separately with two molecules of amine to give α,ω-diamines. This reaction is particularly facile when secondary amines rather than primary amines are used [20].

h. Monoalkyl Derivatives of Diamines

The preparation of monoalkyl derivatives of diamines represents something of a synthetic problem. Two methods are discussed here. Method 1 [21] (Procedure 2-6) involves acetylation of ethylenediamines under mild conditions to the monoacetyl ethylenediamine, protection of the second amino group by a Schotten–Baumann reaction with benzenesulfonyl chloride, followed by N-alkylation of the acetamido group, and recovery of the N-alkylethylenediamine by acid hydrolysis. Method 2 [22] involves the Hofmann alkylation of a primary amine with 2-bromoethylamine hydrobromide (Procedure 2–7).

2-6. Preparation of Monomethylethylenediamine (Method 1) [21]

$$NH_2-CH_2CH_2-NH_2 + CH_3\overset{O}{\underset{\|}{C}}-OC_2H_5 \longrightarrow$$

$$NH_2-CH_2CH_2NH\overset{O}{\underset{\|}{C}}CH_3 + C_2H_5OH \quad (15)$$

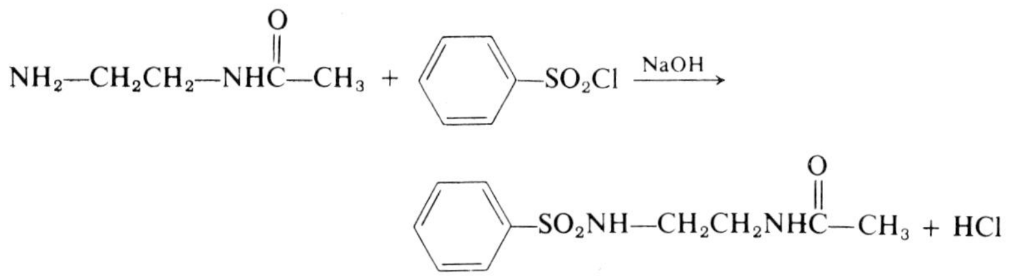

$$NH_2-CH_2CH_2-NH\overset{O}{\underset{\|}{C}}-CH_3 + \text{C}_6\text{H}_5-SO_2Cl \xrightarrow{NaOH}$$

$$\text{C}_6\text{H}_5-SO_2NH-CH_2CH_2NH\overset{O}{\underset{\|}{C}}-CH_3 + HCl \quad (16)$$

$$\text{C}_6\text{H}_5-SO_2-NH-CH_2CH_2NH-\overset{O}{\underset{\|}{C}}-CH_3 + CH_3I \xrightarrow{KOH}$$

$$\text{C}_6\text{H}_5-SO_2NH-CH_2CH_2\underset{\underset{CH_3}{|}}{N}\overset{O}{\underset{\|}{C}}-CH_3 + KI + H_2O \quad (17)$$

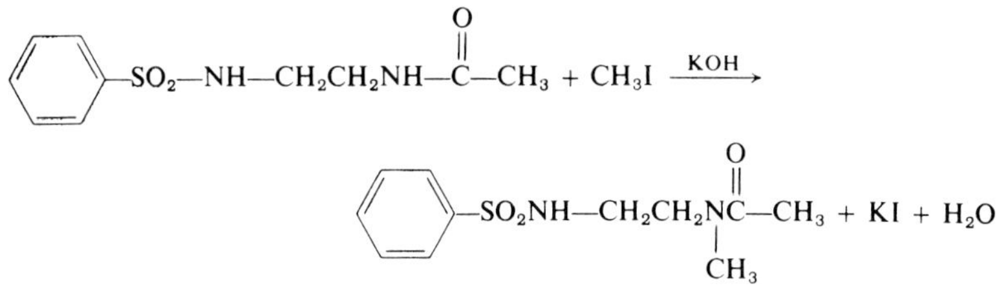

$$\text{C}_6\text{H}_5-SO_2NH-CH_2CH_2-\underset{\underset{CH_3}{|}}{N}-\overset{O}{\underset{\|}{C}}-CH_3 + 2\,H_2O \xrightarrow{2\,HCl}$$

$$HCl\cdot NH_2-CH_2CH_2-NHCH_3\cdot HCl + \text{C}_6\text{H}_5-SO_3H + CH_3CO_2H \quad (18)$$

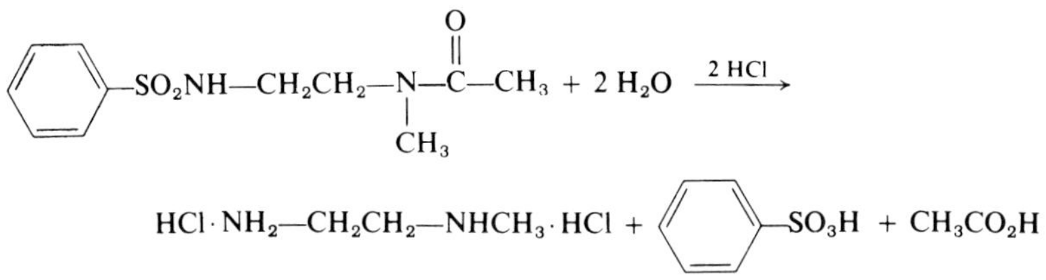

$$HCl\cdot NH_2CH_2CH_2NHCH_3\cdot HCl + 2\,NaOH \longrightarrow$$
$$NH_2CH_2CH_2NHCH_3 + 2\,NaCl + 2\,H_2O \quad (19)$$

To 306 gm (3 moles) of monoacetylethylenediamine (for preparation see Chapter 11, Amides) dissolved in 306 gm of water are added slowly and simultaneously 530 gm (3 moles) of benzenesulfonyl chloride and 1200 gm (3 moles) of a 10% aqueous solution of sodium hydroxide. After standing for several hours, this solution is faintly acidified with mineral acid and the product is separated by filtration. The N-benzenesulfonyl-N'-acetylethylenediamine is recrystallized from dilute ethanol, enough of the ethanol being used to prevent oiling out of the amide. A trace of the dibenzenesulfonylethylenediamine arising from traces of ethylenediamine present in the starting material is separated by filtration of the hot recrystallizing mixture. The pure N-benzenesulfonyl-N'-acetylethylenediamine separates on cooling and is collected by filtration. Yield, 500 gm (69%), mp 103°C.

To a boiling solution of 35 gm (0.53 mole) of 85% potassium hydroxide dissolved in 200 ml of absolute alcohol is added 121 gm (0.5 mole) of N-benzenesulfonyl-N'-acetylethylenediamine. To this solution is added 142 gm (1 mole) of methyl iodide dropwise over a 15 min period. The mixture is refluxed for 2 hr after which the precipitated potassium iodide is separated by filtration of the cooled reaction mixture. The filtrate is steam-distilled until the excess methyl iodide and the ethanol are completely removed.

The remaining alkylated sulfonamide is refluxed for 12 hr with 500 ml of concentrated hydrochloric acid which is replenished from time to time with additional amounts of concentrated acid. The hydrolysate is distilled nearly to dryness under reduced pressure and, after addition of an excess of sodium hydroxide, a concentrated water solution of the amine is distilled over. Dry sodium hydroxide is added to the distillate until the amine appears as a separate phase, which is then drawn off and dried over fresh sodium hydroxide. Finally the product may be dried by refluxing over metallic sodium. After cooling and filtration, the amine is obtained as a water-white liquid by fractional distillation of the filtrate under reduced pressure from a fresh piece of sodium; bp 115°–116°C (157 mm). The yield of monomethylethylenediamine is 28 gm (80% of theory).

2-7. *Preparation of Monoethylethylenediamine (Method 2)* [22]

$$C_2H_5NH_2 + Br-CH_2CH_2NH_2 \cdot HBr \longrightarrow HBr \cdot C_2H_5NHCH_2CH_2NH_2 \cdot HBr \quad (20)$$

$$HBr \cdot C_2H_5NHCH_2CH_2NH_2 \cdot HBr + 2\ NaOH \longrightarrow$$
$$C_2H_5-NHCH_2CH_2NH_2 + 2\ NaBr + 2\ H_2O \quad (21)$$

To a solution of 112.5 gm (2.5 moles) of ethylamine in 340 gm of water is added a solution of 102.5 gm (0.5 mole) of 2-bromoethylamine hydrobromide in 100 ml of water. The resulting mixture is refluxed gently for 12 hr. The mixture is then cooled and treated with solid sodium hydroxide until the base no longer dissolves. During this step, two layers form. The upper layer is separated and the lower aqueous layer is extracted several times with ether. Ether extract is combined with the upper amine layer. The combined organic materials are dried over anhydrous potassium hydroxide and fractionally distilled through a short fractionating column packed with glass helices. After the ether has been separated, monoethylethylenediamine is separated by distillation at 125°–127°C (743 mm). Yield 17.6 gm (40% of theory).

In some cases, alkylation of secondary amines may be carried out in liquid ammonia in the presence of sodium amide with ferric nitrate monohydrate as a catalyst [23]. The following equation, for the preparation of 1-methylindole, is representative of the process:

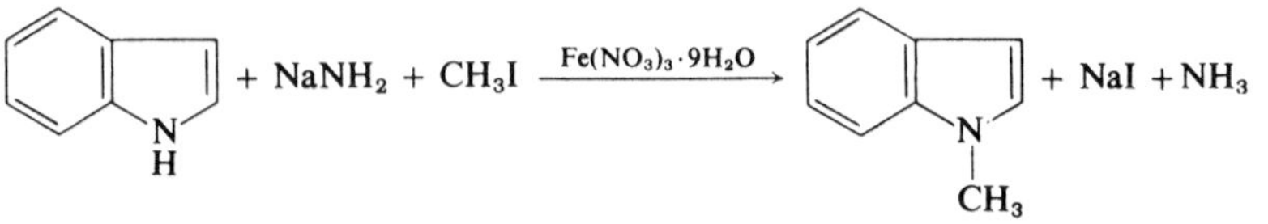

$$\text{indole} + NaNH_2 + CH_3I \xrightarrow{Fe(NO_3)_3 \cdot 9H_2O} \text{1-methylindole} + NaI + NH_3 \quad (22)$$

By way of contrast, heterocyclic amines may be aminated in liquid ammonia with sodium or potassium amide to form primary amines (Equation 23). This process is known as the Chichibabin reaction [24–26].

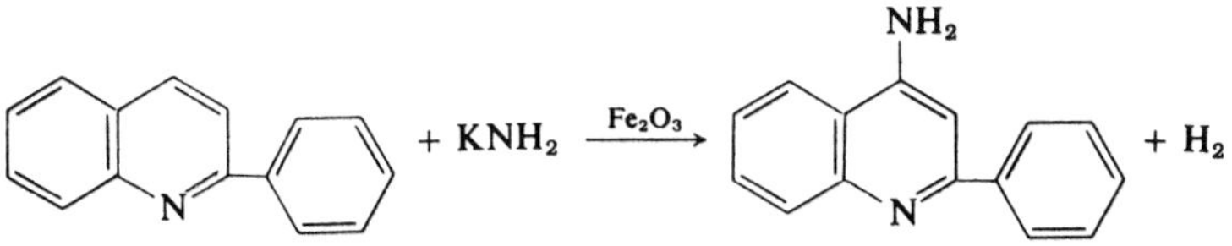

$$\text{2-phenylquinoline} + KNH_2 \xrightarrow{Fe_2O_3} \text{4-amino-2-phenylquinoline} + H_2 \quad (23)$$

[Ref. 24]

B. Miscellaneous Condensation Reactions

Methods for the preparation of primary amines free of secondary and tertiary amines have occupied the efforts of organic chemists since the days of Hofmann. Several of these have at least a formal resemblance to the Hofmann alkylation and therefore are given here along with other nonreductive condensation reactions.

a. Hydroboration

Organoboranes derived from terminal olefins or relatively unhindered olefins are readily converted to the corresponding amine by treatment with chloroamine or hydroxylamine-*O*-sulfonic acid [27].

Since this early work had been carried out in tetrahydrofuran, in which hydroxylamine-*O*-sulfonic acid is insoluble, the reaction could not be carried out with relatively hindered olefins. Consequently, it was not possible to take advantage of the highly stereospecific nature of this application of the hydroboration reaction for all types of olefins. More recent work, however, making use of diglyme as the solvent, has extended the reaction [28]. A typical example of the reaction is the preparation of *trans*-2-methylcyclohexylamine. In Eq. (24), the symbol HB represents the hydroboration reagent. The intermediate hydroborated product is only sketched in for clarity. While the directions of Preparation 2-8 follow closely those of Ref. 28, we suggest that consideration be given to modifying the apparatus. We believe that by the use of serum caps over a standard taper 29/42 joint, nitrogen may be introduced into the flask through a syringe needle. The addition of various reagents may also be done by use of a syringe fitted with an appropriate needle through a serum cap.

2-8. *Preparation of* trans-2-*Methylcyclohexylamine* [28]

$$\underset{}{\overset{CH_3}{\bigcirc}} \xrightarrow{HB} \underset{}{\overset{CH_3}{\underset{B-}{\bigcirc}}} \xrightarrow{H_2NOSO_3H} \underset{}{\overset{CH_3}{\underset{NH_2}{\bigcirc}}} \quad (24)$$

A dry 250-ml flask equipped with a dropping funnel, condenser, and magnetic stirrer is flushed with nitrogen. A solution of 0.78 gm (20.6 mmoles) of sodium borohydride and 20 ml of diglyme is introduced, followed by 4.8 gm (50 mmoles) of 1-methylcyclohexene. The flask is immersed in an ice-water bath and hydroboration is carried out by the dropwise addition of 3.90 gm (27.5 mmoles) of boron trifluoride etherate. The solution is then stirred at room temperature for 3 hr. Then a solution of 6.22 gm (55 mmoles) of hydroxylamine-O-sulfonic acid in 25 ml of diglyme is added, and the solution is heated to 100°C for 3 hr. The reaction mixture is cooled, cautiously treated with 20 ml of concentrated hydrochloric acid, and the poured into 200 ml of water. The acidic phase is extracted with ether to remove diglyme and the residual boronic acid. The aqueous solution is then made strongly alkaline with sodium hydroxide and the amine is extracted with ether. The ether extract is dried over potassium hydroxide and the dried product solution is fractionally distilled. After removal of the ether, 5 gm (45% of theory) of *trans*-2-methylcyclohexylamine is isolated, bp 148°C (750 mm).

[Reprinted from: M. W. Rathke, N. Inoue, K. R. Varma, and H. C. Brown, *J. Am. Chem. Soc.* **88,** 2870 (1966). Copyright 1966 by the American Chemical Society. Reprinted by permission of the copyright owners.]

b. DELÉPINE REACTION

The original procedure of Delépine [29] for the preparation of primary amines involves the reaction of hexamethylenetetramine in chloroform with an alkyl halide to give a quaternary ammonium complex which was hydrolyzed under acidic conditions to give the salt of a primary amine. Under these conditions, alkyl iodides react reasonably readily while the chlorides and bromides react much more slowly and conversion to the iodide was recommended. A modification involves the solution of hexamethylenetetramine in a large volume of 95% alcohol to which sodium iodide is added. To this solution alkyl chloride or bromide could be added. In effect, the alkyl iodide was thus formed *in situ* [30]. In another modification of the reaction, the complex of hexamethylenetetramine and the alkyl halide is decomposed by refluxing in an aqueous ammonia solution in the presence of formalin to form a methyleneimine derivative of the ultimate amine as an intermediate. This imine is then hydrolyzed under acidic conditions to form the desired amine hydrochloride [31]. A modification of the more classical Delépine reaction is given here [32]. This procedure illustrates the wide applicability of the reaction.

2-9. Preparation of β-Alanine [32]

$$(CH_2)_6N_4 + BrCH_2CH_2CO_2H \xrightarrow{HCO_3^-}$$

$$[(CH_2)_6N_4]^+[CH_2CH_2CO_2]^- \xrightarrow[C_2H_5OH]{H^+} H_2NCH_2CH_2CO_2H \quad (25)$$

A solution of 5 gm (0.032 mole) of β-bromopropionic acid in 15 ml of water and 10 ml of ethanol is neutralized with 2.74 gm (0.34 mole) of sodium bicarbonate. Then a solution of 4.57 gm (0.033 mole) of hexamethylenetetramine in 10 ml of water is added and the solution is maintained at room temperature for 15 hr. After 60 ml of ethanol is added to the point of faint turbidity, followed by scratching the glass, the betaine complex crystallizes from the solution. The crystals are chilled in an ice bath for 2 hr and filtered. Yield 9 gm. A second crop of 0.5 gm may be obtained from the mother liquor by partial evaporation of the solvent.

The betaine complex is refluxed with 120 ml of ethanol and 15 ml of concentrated hydrochloric acid for 15 hr. The mixture is evaporated to dryness under reduced pressure at 50°C, and the residue is extracted several times with ethanol. The filtered extract is evaporated to dryness. To the residue, 75 ml of water is added, and the aqueous mixture is refluxed for ½ hr. To isolate the product, the cooled aqueous solution is first treated with an excess of silver oxide to remove chloride ions. After filtration, the filtrate is saturated with hydrogen sulfide gas (hood). The precipitated silver sulfide is removed by centrifugation. The colorless solution is concentrated under reduced pressure to a few milliliters and diluted with ethanol to the point of crystallization. After chilling and filtering 2.5 gm of β-alanine (85%) is separated, mp 199°–200°C with decomposition.

c. GABRIEL CONDENSATION AND ING–MANSKE MODIFICATION

Another method for the preparation of primary amines involves the alkylation of potassium phthalimide according to procedures of Gabriel to give *N*-alkylphthalimides, which, on hydrolysis, afford the primary amine and phthalic acid. Since the hydrolysis is sometimes difficult, Ing and Manske [33] developed a modification in which the decomposition of the *N*-alkylphthalimide is carried out in the presence of hydrazine. While the alkylation step has been carried out either without solvent or in the presence of nonpolar high-boiling solvents, the modification [34] suggests the use of dimethylformamide as the solvent. In this solvent the reaction can often be carried out at relatively low temperatures and exhibits a mildly exothermic character. For less reactive alkyl halides, the reaction temperature may be varied.

The use of a phase-transfer catalyst such as hexadecyltributylphosphonium bromide permits the alkylation of potassium phthalimide in toluene in a liquid–solid system [35, 36].

For the sake of applicability, the preparation of α,δ-diaminoadipic acid using the Ing–Manske modification is given here, although the same reference indicates that, at least in this case, the Gabriel condensation and hydrolysis affords a higher yield.

2-10. Preparation of α,δ-Diaminoadipic Acid [34]

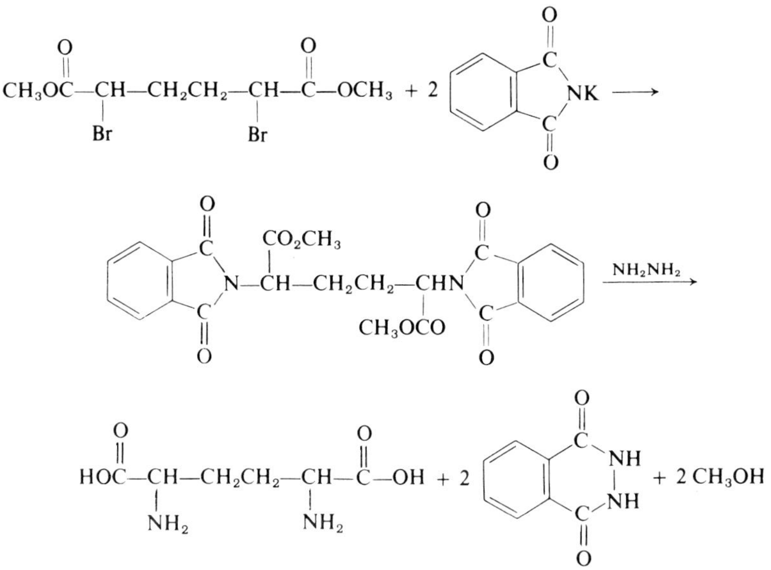

$$\text{CH}_3\text{OC}-\underset{\text{Br}}{\text{CH}}-\text{CH}_2\text{CH}_2-\underset{\text{Br}}{\text{CH}}-\text{C}-\text{OCH}_3 + 2\ \text{(phthalimide-NK)} \longrightarrow$$

$$\text{(Phth)}-\text{N}-\underset{\text{CH}_3\text{OCO}}{\overset{\text{CO}_2\text{CH}_3}{\text{CH}}}-\text{CH}_2\text{CH}_2-\underset{}{\text{CHN}}-\text{(Phth)} \xrightarrow{\text{NH}_2\text{NH}_2}$$

$$\text{HOC}-\underset{\text{NH}_2}{\text{CH}}-\text{CH}_2\text{CH}_2-\underset{\text{NH}_2}{\text{CH}}-\text{C}-\text{OH} + 2\ \text{(phthalhydrazide)} + 2\ \text{CH}_3\text{OH} \quad (26)$$

A mixture of 69 gm (0.21 mole) of dimethyl α,δ-dibromoadipate, 87 gm (0.47 mole) of potassium phthalimide, and 260 ml of dimethylformamide is gently heated. A mildly exothermic reaction starts at 50°C; however, sufficient heat is applied to maintain the reaction mixture for 40 min at 90°C. Then the reaction mixture is cooled, diluted with 300 ml of chloroform, and poured into 1200 ml of water. The chloroform layer is separated and the aqueous phase is extracted twice with 100-ml portions of chloroform. The combined chloroform extract is washed with 200 ml of 0.1 N sodium hydroxide, and 200 ml of water. Then the solution is dried over sodium sulfate. The chloroform is removed by concentration under reduced pressure to the point of incipient crystallization. The immediate addition of 300 ml of ether induces a rapid crystallization. The product is collected on a filter and washed with ether. Yield 87 gm (90.2%), mp over the range 160°–185°C. After three crystallizations from ethyl acetate and one from benzene, an apparently pure stereoisomer is obtainable which melts at 210.7°–211.4°C.

In a hood, a mixture of 4.64 gm (0.01 mole) of dimethyl α,δ-diphthalamidoadipate (mp 160°–185°C), 50 ml of methanol, and 1.2 ml (0.02 mole) of an 85% aqueous hydrazine solution (CAUTION: Toxic reagent) is heated under reflux for 1 hr. After cooling, 25 ml of water is added and the methanol is removed by concentration under reduced pressure. After 25 ml of concentrated hydrochloric acid has been added to the residual aqueous suspension, the mixture is heated under reflux for 1 hr. After cooling to 0°C, crystalline phthalhydrazide is removed by filtration. The filtrate is concentrated under reduced pressure to remove hydrochloric acid and the moist residue is dissolved in 50 ml of water. A small amount of insoluble matter is removed by filtration and the clear filtrate is neutralized with 2 N sodium hydroxide. After cooling at 0°C for 12 hr, 1.4 gm (79.5%) of α,δ-diaminoadipic acid is obtained.

[Reprinted from J. C. Sheehan and W. A. Bolhofer, *J. Am. Chem. Soc.* **72**, 2786 (1950). Copyright 1950 by the American Chemical Society. Reprinted by permission of the copyright owner.]

Using a phase-transfer catalyst, long chain diamines such as 1,20-diamino-3,6,9,12,15,18-hexaoxaeicosane ($H_2NCH_2(CH_2OCH_2)_6CH_2NH_2$) have been prepared. The method of preparation is analogous to that of 1,8-diamino-3,6-dioxaoctane given in Preparation 2-11 [36].

2-11. *Preparation of 1,8-Diamino-3,6-dioxaoctane* [36].

$$2 \; \underset{CO}{\overset{CO}{\diagup}}\!\!\text{NK} + ClCH_2(CH_2OCH_2)_2CH_2Cl \xrightarrow[\text{toluene}]{\text{P-T catalyst}}$$

$$\underset{CO}{\overset{CO}{\diagup}}\!\!N-CH_2(CH_2OCH_2)_2CH_2-N\!\!\overset{OC}{\underset{OC}{\diagdown}} \xrightarrow{NH_2NH_2 \cdot H_2O}$$

$$NH_2-CH_2(CH_2OCH_2)_2CH-NH_2 \qquad (27)$$

In a 500-ml Erlenmeyer flask containing a magnetic stirrer and fitted with a reflux condenser, to a dispersion of 92 gm (0.5 mole) of potassium phthalimide and 10 gm (0.02 mole) of hexadecyltri-*n*-butylphosphonium bromide in 200 ml of toluene is added 31.4 ml (0.2 mole) of 1,8-dichloro-3,6-dioxaoctane. The mixture is heated for 10 hr at 100°C with rapid stirring. The solid which forms is filtered off and extracted once with a small quantity of toluene. The toluene extract is combined with the toluene solution formed during the reaction. The solid filtrate is discarded. The mother liquor is evaporated on a rotary evaporator. The crude product from this evaporation is washed with small volumes of de-

§ 2. Condensation Reactions

ionized water and then recrystallized from a large volume of ethanol. The product, 1,8-diphthalamido-3,6-dioxaoctane, is air dried at 100°C at atmospheric pressure. Yield: 60.2 gm (74%), mp 183°–185°C.

In a 500-ml Erlenmeyer flask fitted with a reflux condenser and containing a magnetic stirrer, a suspension of 45 gm (0.11 mole) of 1,8-diphthalamido-3,6,dioxaoctane in 200 ml of ethanol is warmed to reflux. Then 12.8 ml (0.25 mole) of 85% hydrazine monohydrate (CAUTION: Toxic reagent) is added. Refluxing is continued for 1 hr. The reaction mixture is then cooled and 6 N hydrochloric acid is added until the mixture is slightly acidic. The mixture is again heated to reflux for 0.5 hr, cooled, and filtered. The filtrate is preserved. The solid on the filter is washed with ethanol. The ethanol wash is combined with the filtrate. The filtrate is freed of solvent by gentle warming in a rotary evaporator.

To the residue is added a solution of 14 gm (0.25 mole) of potassium hydroxide in ethanol. The insoluble potassium chloride which forms is filtered off and washed with ethanol. The ethanolic filtrate is again evaporated in the rotary evaporator. The oily residue is treated with chloroform. At this stage more potassium chloride and phthalhydrazide precipitates. The solids are separated by filtration and washed with a small quantity of chloroform. The chloroform solutions are combined and evaporated to dryness. The residue is distilled under reduced pressure. Yield: 14.4 gm (68.8%), bp 95°C (1 mm); n_D^{18} 1.4623.

The Schweizer reaction is a new method which is related to the Ing–Manske modification since it leads to olefinic phthalimides which may be converted to allylic amines. In this process, vinyltrialkylphosphonium salts, aldehydes, and sodiophthalimide are reacted in tetrahydrofuran to give allylic phthalimides [Eqs. (28, 29)]. When the phosphonium salt is vinyl tri-n-butylphosphonium salt, the product is the E-stereoisomer. When vinyltriphenylphosphonium salts are used, the Z-isomer predominates [37].

(28)

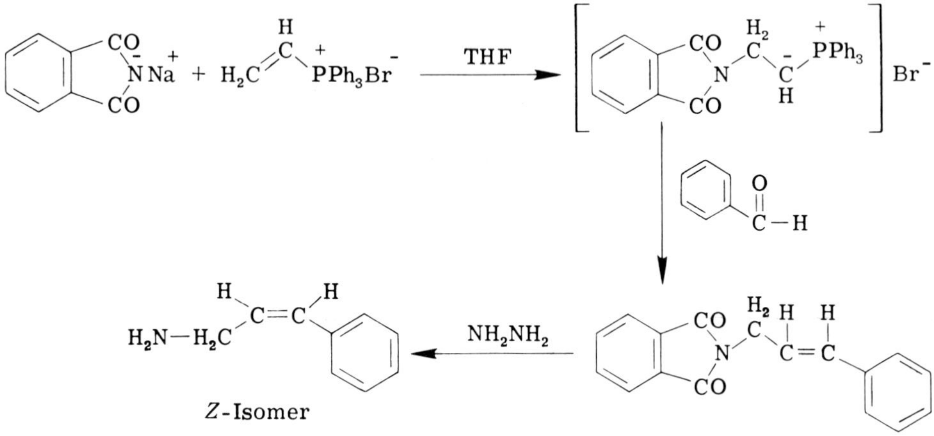

13. Amines

$$\text{(29)}$$

d. Ritter Reaction

As is discussed in the chapter on amides, nitriles may be reacted with certain types of alcohols or olefins in strongly acidic media to afford an alkylated amide. Since amides may be hydrolyzed to yield a free amine (as, for example, in Ferris et al. [38]), a facile method for converting olefins and alcohols to amines is available. The major drawback to the Ritter reaction is the fact that only those alcohols are suitable for the reaction which have branched chains on the carbon adjacent to the hydroxyl group. Thus, 2-propanol is the lowest aliphatic alcohol which undergoes the Ritter reaction. Olefins which are to be subjected to the Ritter reaction also require branching. The example here cited indicates the simplicity of the Ritter reaction as a means of preparing amines.

2-12. Preparation of 1-Methylcyclobutylamine [39]

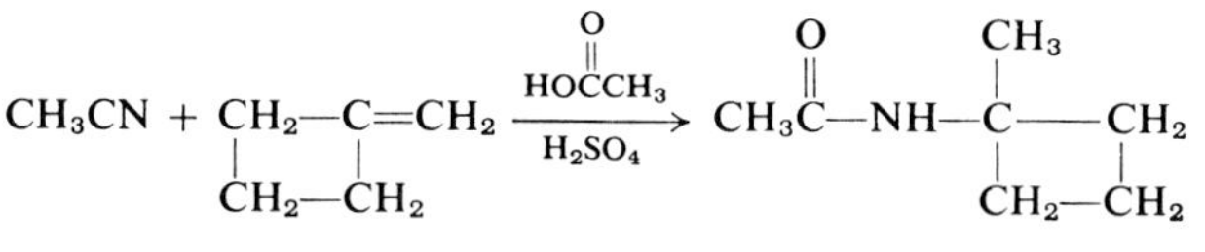

$$\text{(30)}$$

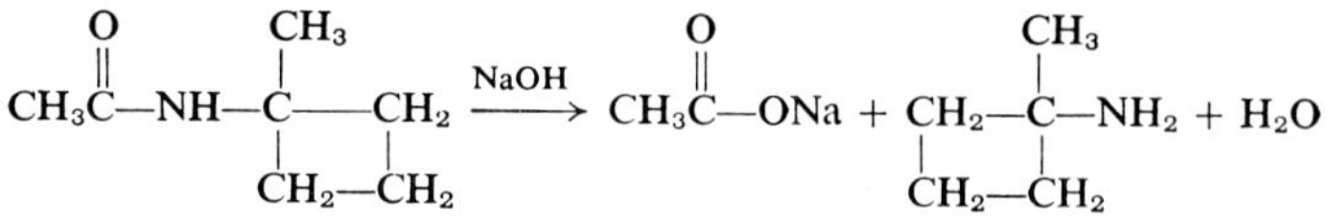

$$\text{(31)}$$

In a 500-ml three-necked flask equipped with a mechanical stirrer, dropping funnel, and reflux condenser, immersed in an ice-salt bath, are placed 9.0 gm (0.22 mole) of acetonitrile, 100 ml of glacial acetic acid, and 20 gm of concentrated sulfuric acid. To the cooled solution, 13.6 gm (0.20 mole) of methylenecyclobutane is slowly added with stirring. After the addition has been completed, stirring is continued for 1 hr at 20°C. The solution is then cooled, diluted with 300 ml of water, and sufficient sodium carbonate is added to render the solution basic. The aqueous phase is then extracted with five 50-ml portions of ether. The etheral solution is dried, filtered, and evaporated to give 17.7 gm (70% of theory) of white crystalline N-(1-methylcyclobutyl)acetamide.

To 400 ml of a $4N$ solution of potassium hydroxide in ethylene glycol is added 10.0 gm (0.076 mole) of N-(1-methylcyclobutanol)acetamide. The reaction mixture is heated to reflux for 48 hr. After this reaction period, the materials boiling below 180°C at atmospheric pressure are distilled off. This distillate is extracted continuously with ether. The ether extract is dried over potassium hydroxide. After removal of the ether, the residue is distilled to afford 3.1 gm (46% of theory) of 1-methylcyclobutylamine, bp 85.5°–86.0°C (464 mm).

e. Hydrolysis of Amides

Ordinarily, the hydrolysis of amides is not considered as a method of synthesis of amines. In most cases, amides are prepared from the interaction of amines and various reagents and, consequently, the reversal of this preparation is primarily of interest from the analytical rather than the synthetic standpoint. And yet, we have already described at least two procedures in which the hydrolysis of amides leads to amines which had not been used to prepare the original amide—the Gabriel–Ing–Manske reaction and the Ritter reaction.

There are other situations in which it may be appropriate to use an amide as a starting point for the substitution of a hydrogen on the nitrogen. For example, a synthesis of N-3,5-dichlorophenylaniline makes use of the reaction sequence shown in Eq. (32) [40].

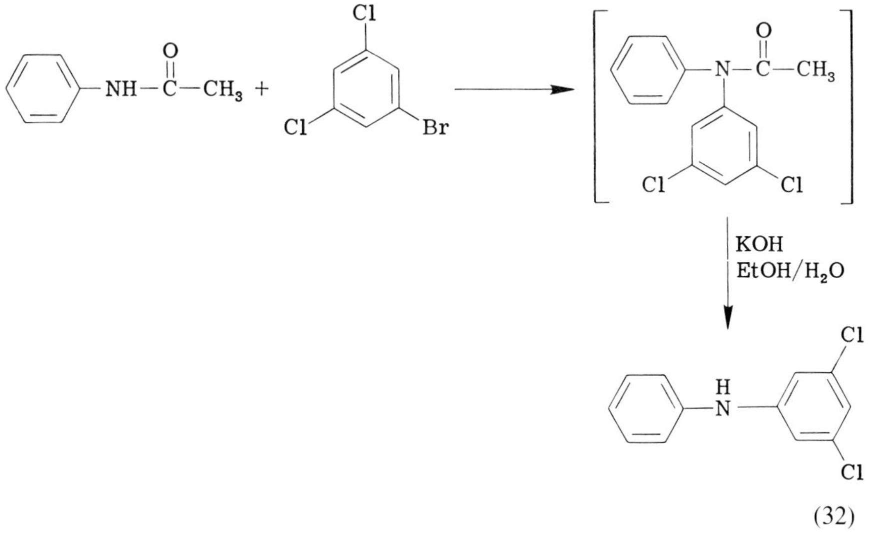

(32)

More recently, acetanilide derivatives such as 2'-methylacetanilide or 3',5'-dimethylacetanilide have been treated with bromobenzene in the presence of potassium carbonate and cuprous iodide to prepare N-arylated acetanilide which was not isolated but appeared to be converted directly to 2-methyldiphenylamine of 3,5-dimethyldiphenylamine, respectively, in more than 70% yield [41].

f. ADDITION OF AMINES TO DOUBLE BONDS

Activated olefinic bonds such as the double bonds in acrylic acid derivatives readily react with amines to yield saturated products. In our own experience, for example, acrylyl chloride treated with amines affords substantial amounts of alkylaminopropionic acid derivatives, formed by the addition of the amine to the double bond. The addition of amines to the double bond of mesityl oxide is also used in one of the diazoalkane syntheses.

The reaction itself is believed to be an example of the Michael condensation. A typical example is the preparation of ethyl β-methylaminopropionate [42].

2-13. Preparation of Ethyl β-Methylaminopropionate [42]

$$CH_2=CH-\overset{O}{\underset{\|}{C}}-OC_2H_5 + CH_3NH_2 \longrightarrow CH_3NHCH_2CH_2\overset{O}{\underset{\|}{C}}-OC_2H_5 \quad (33)$$

In a hood, to a cooled solution of 34.1 gm (0.1 mole) of methylamine in approximately 185 ml of absolute ethanol is added with stirring and cooling 100 gm (1.0 mole) of ethyl acrylate. The solution is allowed to stand at room temperature for 2 days. The product is then isolated by fractional distillation. After the ethanol has been removed from the solution, the crude product is collected at 61°–72°C (17 mm), yield 63.5 gm (48.5%). By repeated fractional distillations, a product is obtained with bp 68°–68.4°C (18 mm).

By a similar procedure, amines may even be added to acetylenic carboxylic acids or their derivatives. Thus, for example, addition of dimethylamine to dimethyl acetylenedicarboxylates give rise to high yields of dimethyl dimethylaminomaleate [43].

In the presence of a palladium chloride catalyst and benzoquinone, aromatic secondary amines aminate electron-deficient olefins such as methyl acrylate, acrylonitrile, or methyl vinyl ketones to produce vinologous arylamine esters, nitriles, or ketones. Primary amines such as aniline produce precipitates—possibly by forming bisamino palladium dichloride complexes [44]. Equations (34–36) indicate the types of products which are said to be produced by this reaction. We are not giving detailed procedures for these preparations since the literature deals only with reactions on a millimolar scale which may be difficult to scale up.

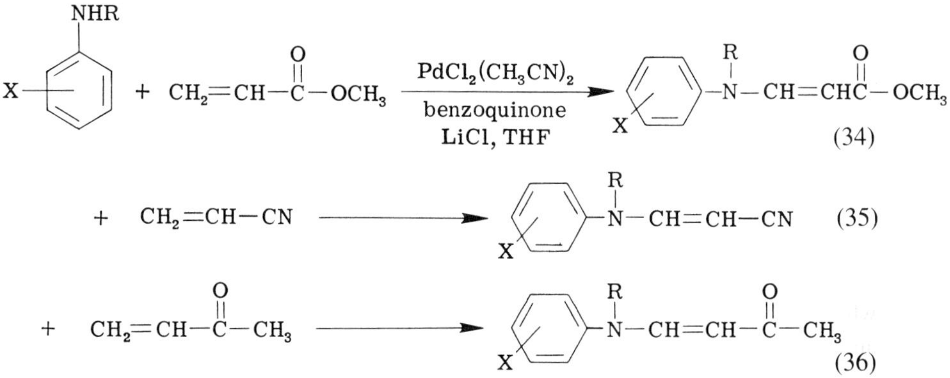

§ 2. Condensation Reactions

$$\text{X-C}_6\text{H}_4\text{-NHR} + CH_2=CH-\overset{O}{\overset{\|}{C}}-OCH_3 \xrightarrow[\text{LiCl, THF}]{\text{PdCl}_2(CH_3CN)_2, \text{ benzoquinone}} \text{X-C}_6\text{H}_4-\overset{R}{\overset{|}{N}}-CH=CH\overset{O}{\overset{\|}{C}}-OCH_3 \quad (34)$$

$$+ CH_2=CH-CN \longrightarrow \text{X-C}_6\text{H}_4-\overset{R}{\overset{|}{N}}-CH=CH-CN \quad (35)$$

$$+ CH_2=CH-\overset{O}{\overset{\|}{C}}-CH_3 \longrightarrow \text{X-C}_6\text{H}_4-\overset{R}{\overset{|}{N}}-CH=CH-\overset{O}{\overset{\|}{C}}-CH_3 \quad (36)$$

A novel route to the preparation of secondary and tertiary amines is a recent extension of Reppe chemistry. In this procedure olefins, carbon monoxide, water, and a nitrogen source are reacted in the presence of rhodium-based catalysts under pressure. Equation (37) is representative of this reaction [45].

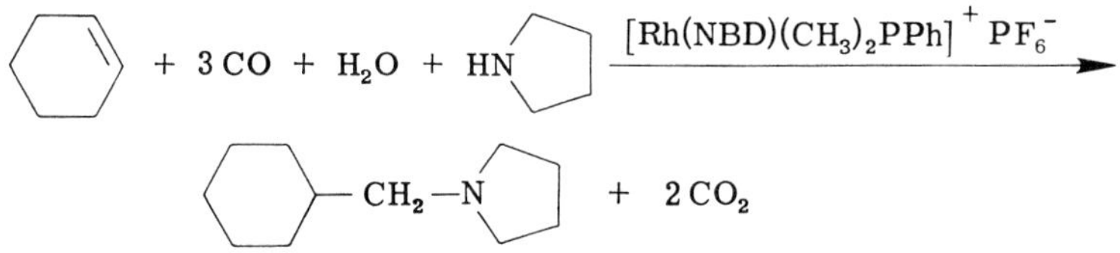

$$\text{cyclohexene} + 3\,CO + H_2O + HN\text{(pyrrolidine)} \xrightarrow{[\text{Rh(NBD)(CH}_3)_2\text{PPh}]^+ \text{PF}_6^-}$$

$$\text{cyclohexyl-}CH_2-N\text{(pyrrolidine)} + 2\,CO_2 \quad (37)$$

Other reactions involving the reaction of olefins to produce amines are shown in Eqs. (38–39).

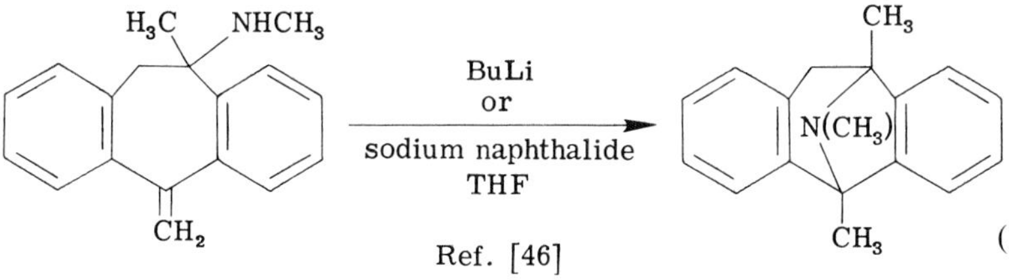

$$\text{(dibenzocycloheptane with } H_3C, NHCH_3 \text{ and } CH_2) \xrightarrow[\text{THF}]{\text{BuLi or sodium naphthalide}} \text{(rearranged bridged structure with } CH_3, N(CH_3), CH_3) \quad (38)$$

Ref. [46]

$$\overset{H_3C}{\underset{H_3C}{>}}N-CH_2-CH_2-N\overset{CH_3}{\underset{CH_3}{<}} + CH_2=CH-CH_3 \xrightarrow[\substack{150°C \\ \text{pressure}}]{\text{EtLi ether}} \overset{H_3C}{\underset{H_3C}{>}}N-\overset{H}{\underset{CH_3}{\overset{|}{C}}}\overset{CH_3}{\underset{}{}}$$

67.5%

and

$$\overset{H_3C}{\underset{H_3C}{>}}N-CH=CH_2$$

Ref. [47] 20% (39)

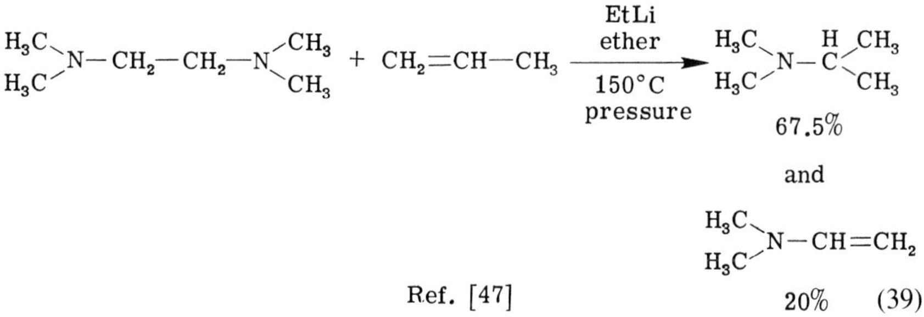

g. The Mannich Reaction

The replacement of a reactive hydrogen with a methylamino or a substituted methylamino group is referred to as a Mannich reaction. The usual reaction conditions involve condensation of the reactive hydrogen compound with formaldehyde and with a nitrogen compound such as ammonia, a primary, or a secondary amine (usually as the hydrochloride salt) [48]. The overall reaction may be represented by the following equation:

$$R-\underset{R}{N}H + H-\overset{O}{\underset{\|}{C}}-H + R'H \longrightarrow R-\underset{R}{N}-CH_2-R' + H_2O \quad (40)$$

where R'H represents the active hydrogen compound, usually a ketone, an acid, an ester, a phenol, or other reactive methylene compound such as nitroalkanes [49] or monosubstituted acetylenic compounds [50]. Even orthophosphorous acid may be used as the active hydrogen component [51]. In this case nitrogen-containing phosphonic acids are produced.

The amine component appears to be virtually unlimited in the aliphatic amine series although secondary amines are probably used most frequently. Instead of free ammonia, however, salts are used, particularly ammonium acetate. Equation (41) is an illustration of the reaction using ammonium acetate and a nitro compound [52].

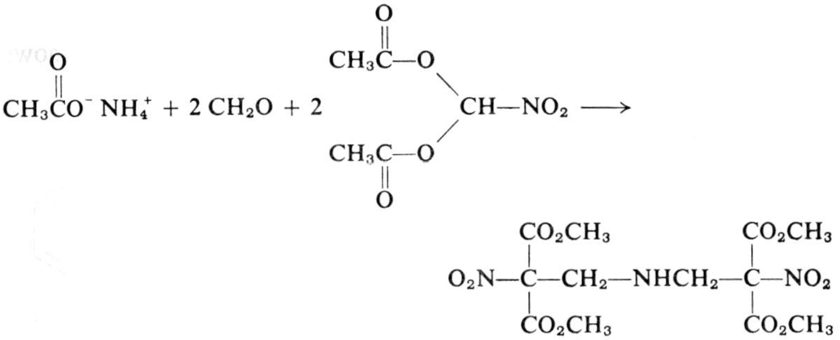

$$CH_3\overset{O}{\underset{\|}{C}}O^- NH_4^+ + 2\,CH_2O + 2\begin{matrix}CH_3C-O\\ \| \\ O\end{matrix}\!\!\diagdown_{\!CH-NO_2}\!\!\diagup^{\!CH_3C-O}_{\!\|}_{\!O} \longrightarrow$$

$$O_2N-\underset{\underset{CO_2CH_3}{|}}{\overset{\overset{CO_2CH_3}{|}}{C}}-CH_2-NHCH_2-\underset{\underset{CO_2CH_3}{|}}{\overset{\overset{CO_2CH_3}{|}}{C}}-NO_2 \quad (41)$$

Aromatic amines evidently do not successfully undergo the Mannich reactions. However, many tertiary Mannich bases prepared from simpler aliphatic compounds undergo exchange reactions with both primary and secondary aromatic amines [53]. These exchange reactions are evidently widely applicable and thus extend the range of products which can be prepared. A general reaction scheme representing this reaction is given in Eqs. (42) and (43).

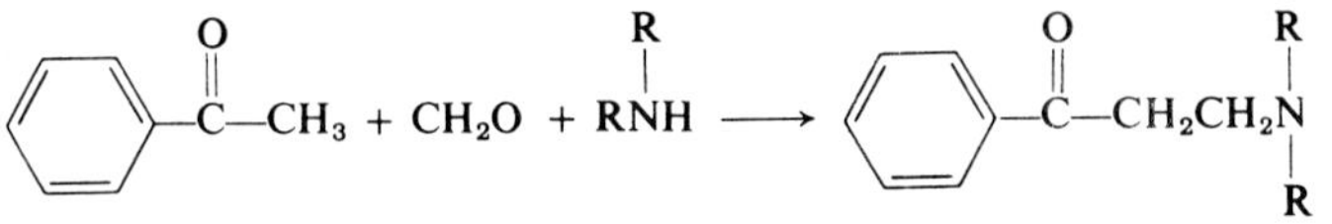

$$\text{Ph}-\overset{O}{\underset{\|}{C}}-CH_3 + CH_2O + R\underset{}{N}H \longrightarrow \text{Ph}-\overset{O}{\underset{\|}{C}}-CH_2CH_2\underset{R}{\overset{R}{N}} \quad (42)$$

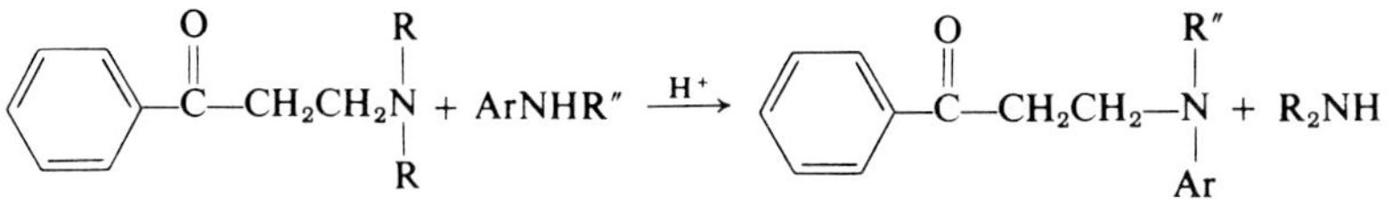

$$\underset{R}{\underset{|}{\text{Ph-C(=O)-CH}_2\text{CH}_2\text{N}}}\text{-R} + \text{ArNHR}'' \xrightarrow{H^+} \underset{Ar}{\underset{|}{\text{Ph-C(=O)-CH}_2\text{CH}_2\text{-N}}}\text{-R}'' + R_2\text{NH} \quad (43)$$

The preparation of 6-benzyloxy-5-methoxygramine is an example of the Mannich reaction [Eq. (44)].

2-14. Preparation of 6-Benzyloxy-5-methoxygramine [54]

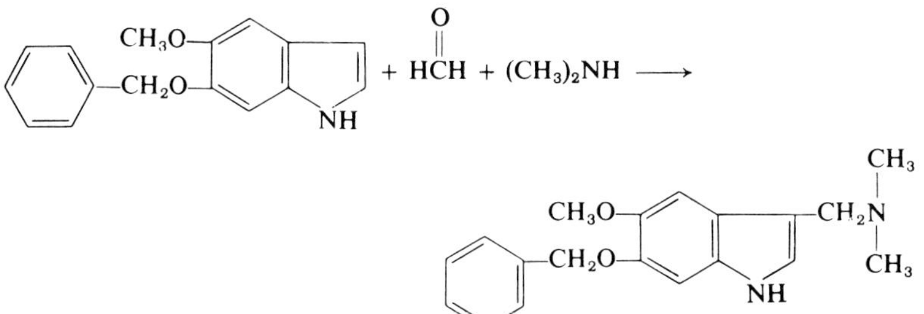

[6-benzyloxy-5-methoxyindole] + HCHO + (CH₃)₂NH ⟶ [6-benzyloxy-5-methoxygramine with -CH₂N(CH₃)₂ group] (44)

To a stirred mixture of 40 ml of dioxane, 40 ml of glacial acetic acid, 30 ml of 36% aqueous formaldehyde (350 mmoles), and 7.0 ml of 25% aqueous dimethylamine (390 mmoles) cooled to 5°C in a large flask is added dropwise a solution of 8.1 gm (32 mmoles) of 6-benzyloxy-5-methoxyindole in 70 ml of dioxane, with vigorous stirring. After the addition has been completed, the reaction solution is kept at 5°C for 2 hr and allowed to warm to room temperature. After standing overnight at room temperature in the dark, the mixture is diluted with 400 ml of water, treated with charcoal, filtered, and made alkaline with a 20% aqueous solution of sodium hydroxide. The reaction mixture is again allowed to stand overnight, whereupon 8.5 gm (79% of theory) of a needlelike crystalline product is obtained, mp 131°–134°C. Upon recrystallization from toluene and hexane, the melting point is raised to 135°–136°C.

3. REDUCTION REACTIONS

Amines have been prepared by the reduction of a variety of nitrogen compounds with many reducing agents. Among the compounds reduced are amides; nitriles; quaternary ammonium salts (cf. page 387; azides [56]; azines, hydrazo compounds, and hydrazones [57, 58]; azo compounds [59]; azomethines or Schiff's bases [60]; azomethinium salts (König's salt) [61]; enamines [62, 63]; nitro and nitroso compounds [64]; and oximes [63, 65], oxime ethers, and oxime esters [66, 67]. Hydrazides also have been reduced with lithium aluminum hydride using ethylmorpholine as a solvent [68].

Nitrogen-containing heterocyclic compounds such as azanaphthalenes have been reduced with sodium borohydride in trifluoroacetic acid [69]. Thioamides, which are readily prepared from amides by treatment with P_2S_5, are reduced to amines with a borohydride or a cyanoborohydride [70]. Tetrabutylammonium trihydrocyanoborates are said to be useful and selective reagents for the reductive amination of aldehydes and ketones [71]. The use of sodium cyanoborohydride for the reductive amination of aldehydes and ketones and the reductive alkylation of amines and hydrazones has been reviewed [63].

Below are discussed the reductions of some of the more common nitrogen compounds.

A. Reduction of Amides

Primary, secondary, and tertiary amines are readily prepared by the reduction of the corresponding amides. Obviously, this method will always result in a product in which one carbon attached to the nitrogen cannot be branched since it originated as a carbonyl carbon. The method is quite generally applicable. Thus, for example, methylamines have been prepared by the reduction of substituted formamides [72], and other amines can be prepared by the judicious selection of acyl and amino residues of the amides.

The basic method of reduction of amides described here is a modification of the reduction of acids and their derivatives using lithium aluminum hydride [73]. It was found in the laboratory at Polysciences, Inc. that while the reduction of fluorine-containing amides with lithium aluminum hydride is quite exothermic, the reaction, in actual fact, was far from complete when the reaction mixture was worked up shortly after completion of the exothermic phase of the reaction. On the other hand, when the reaction mixture was maintained at room temperature for several days, yields improved considerably. The authors have not checked this phenomenon for non-fluorine-containing amides, however, we suggest that a prolonged room temperature reaction period be tried in the preparation of other amines where yields are normally low.

3-1. Preparation of N-Ethyl-1,1-dihydroheptafluorobutylamine [74]

$$CF_3CF_2CF_2\overset{\overset{O}{\|}}{C}-NH-C_2H_5 + [H] \xrightarrow{LiAlH_4} CF_3CF_2CF_2CH_2NHC_2H_5 + H_2O \quad (45)$$

CAUTION: During one of several preparations of this compound, a minor explosion occurred while the reaction mixture was being stirred. A more serious explosion took place when a related compound was reduced with lithium aluminum hydride [75]. Although these were the only two explosions encountered in many related reductions, the reaction should be carried out behind an adequate barricade. In the directions given, the ether content is higher than previously published to assist in dissolution of the reducing agent.

§ 3. Reduction Reactions   403

In a hood, behind a heavy safety shield, in a 5-liter three-necked flask equipped with additional funnel, gas inlet tube, explosion-proof mechanical stirrer, and a long reflux condenser connected to a drying tube, is placed 76 gm (2.0 moles) of finely powdered lithium aluminum hydride dissolved in 2 liters of anhydrous ether under a dry nitrogen atmosphere. The hydride solution is cooled in an ice-water bath and a solution of 241 gm (1.0 mole) of N-ethyl-2,2,3,-3,4,4,4-heptafluorobutyramide in 500 ml of anhydrous ether is added gradually with stirring, at such a rate that the ether refluxes gently. After the addition has been completed, the reaction mixture is allowed to warm to room temperature with stirring. After it has reached room temperature, the reaction mixture is allowed to remain at room temperature for 5 days without additional stirring. At the end of this time, 200 ml of anhydrous ethyl acetate are cautiously added to the reaction mixture to decompose the excess lithium aluminum hydride. Then 1 liter of water is added. The ether layer is separated, the aqueous layer and solid sludges are extracted several times with ether, and the ether extracts are combined and subjected to distillation. After removal of the ether, the fraction boiling between 96° and 99°C is collected. Yield 204.3 gm (93% of theory)

Upon refractionating, N-ethyl-1,1-dihydroheptafluorobutylamine was found to have a boiling point of 98.5°C (751 mm).

Unreduced amide starting material could, at times, be isolated from the distillation residues.

It has been reported that the decomposition of lithium aluminum hydride with ethyl acetate, as recommended above, may lead to some N-alkylation of products [76]. The extent of this N-alkylation will have to be evaluated in each preparation.

Cyclic amides have been reduced with lithium aluminum hydride in tetrahydrofuran [77].

The addition of ynamines to carbonyl compounds forms derivatives of acrylamides which may be reduced with lithium aluminum hydride to allylamines which, in turn, may be further hydrogenated to amines [78].

B. Reduction of Nitriles

Nitriles have been reduced with hydrogen and various catalysts [79] and also by chemical means. Among the chemical reducing agents, sodium and alcohol has been used [39]. Deuterated amines have been produced from nitriles by reduction with lithium aluminum deuteride [80]. The preparation of cyclopropylmethylamine with lithium aluminum hydride is a related example of this reduction [81].

3-2. Preparation of Cyclopropylmethylamine [81]

$$\triangleright\!\!-CN + [H] \xrightarrow{LiAlH_4} \triangleright\!\!-CH_2NH_2 \qquad (46)$$

Using precautions indicated for the reduction of amides with lithium aluminum hydride, to a slurry of 52.5 gm (1.4 moles) of lithium aluminum hydride in 1 liter of tetrahydrofuran is added 93.3 gm (1.4 moles) of cyclopropyl cyanide at such a rate as to cause moderate refluxing. The mixture is stirred vigorously for 5½ hr and left overnight. Then the reaction mixture is worked up by the addition of an excess of an aqueous sodium hydroxide solution, the mixture is filtered, and the filter cake is washed well with tetrahydrofuran. Distillation of the filtrates and washings at atmospheric pressure afforded 78.4 gm (79%) of cyclopropylmethylamine, bp 84°–86°C (760 mm).

More recently, diborane–tetrahydrofuran has been suggested as a convenient reducing agent of nitriles. The preparation of [2-(2-fluoro-3,4-dimethoxyphenyl)ethyl] amine is representative of the procedure [82].

3-3. Preparation of [2-(2-Fluoro-3,4-dimethoxyphenyl)ethyl] amine [82]

$$\underset{\underset{OCH_3}{CH_3O}}{\overset{CH_2CN}{\underset{F}{\bigcirc}}} \xrightarrow{BH_3/THF} \underset{\underset{OCH_3}{CH_3O}}{\overset{CH_2CH_2NH_2}{\underset{F}{\bigcirc}}} \qquad (47)$$

With suitable safety precautions, to a 2-liter three-necked flask fitted with a reflux condenser, addition funnel, and stirrer, containing 259 ml of a solution containing BH_3 in tetrahydrofuran, is added slowly, with stirring and with external cooling with an ice bath, a solution of 25.9 gm (0.133 mol) of 2-fluoro-3,4-dimethoxyphenylacetonitrile in 259 ml of tetrahydrofuran. After the addition has been completed, the reaction mixture is heated at reflux for 1.5 hr.

The reaction system is again cooled in an ice bath and 250 ml of methanol is slowly added. Then the reaction mixture is again refluxed for 0.5 hr, transferred to a rotary evaporator, and concentrated. The residual oil is twice dissolved in methanol and evaporated to dryness. Then 50 ml of 10% hydrochloric acid is slowly added. The mixture is refluxed for 0.5 hr and concentrated. The residue is taken up in a mixture of 2.5 liters of water and 50 ml of 10% hydrochloric acid. The solution is extracted twice with ether. The ether extracts are discarded.

The aqueous solution is cooled and made basic by the cautious addition of 40% aqueous sodium hydroxide. The product is extracted from the aqueous system by first partially saturating the system with sodium chloride and then extracting the product with three portions of ether. The ether extracts are combined, dried with anhydrous magnesium sulfate, and concentrated to yield 19.9 gm (76% of theory) of a colorless oil. ^{1}H NMR (CDCl$_3$) δ 1.23 (br s, 2, NH$_2$), 2.77 (m, 4, CH$_2$CH$_2$), 3,82 (s, 3, OCH$_3$), 3.88 (s, 3, OCH$_3$), 6.63 (m, 2, ArH).

The catalytic hydrogenation of nitriles is complex. For example, hydrogenation of benzonitrile may produce a product mixture containing benzylamine, dibenzylamine, tribenzylamine, *N*-benzylidenebenzylamine, benzamide, benzaldehyde, benzyl alcohol and toluene.

Using a platinum-catalyzed hydrogenation of benzonitrile without a solvent, in the presence of water to prevent catalyst poisoning, a commercially useful process for the synthesis of dibenzylamine has been developed [83].

C. Reduction of Nitro Compounds

Particularly in the aromatic series, many amines have been prepared by the reduction of the corresponding nitro compounds. With the advances in the preparation of aliphatic nitro compounds, the techniques for reducing aliphatic nitro compounds have also been developed to afford aliphatic amines. In the aliphatic series, the techniques permit the preparation of a very large variety of amines since aliphatic nitro compounds are active methylene compounds which undergo typical reactions of active methylene groups, such as the various condensation reactions, giving rise to a variety of branched compounds involving the carbon adjacent to the nitro group.

A large number of reducing agents has been used for the reduction of nitro groups. For example, stannous chloride has been used to reduce nitrosulfones [84] as well as aromatic nitro compounds [85]. Both nitroso and nitro compounds have been reduced with sodium borohydride in the presence of palladium on carbon [86]. Many examples of the reduction of nitro groups with hydrogen using palladium on carbon as catalyst exist. In this reduction, it has been shown that a nitro group reacts in preference to an acetamido group [87].

Nitroso compounds may also be reduced to amines by various reagents [64].

The classical Bechamp method of reducing nitro compounds involves reaction with iron and acetic acid [88].

a. REDUCTION WITH ACTIVATED IRON AND WATER

Both from the laboratory and the industrial standpoints, reduction of aromatic nitro compounds with iron or iron compounds is of considerable importance. By use of a previously "activated" iron, many nitro aromatic compounds have been successfully reduced under very mild conditions (from the point of view of possible hydrolysis of substituents which may be present on the ring).

The preparation of 2-chloroaniline illustrates the method of activating iron for this reaction [89]. In a personal communication, Professor C. F. H. Allen suggested that a mixture of iron powder and rusty iron filings frequently works just as well as specially activated iron [90].

3-4. *Preparation of 2-Chloroaniline* [89]

$$\underset{}{\text{2-O}_2\text{N-C}_6\text{H}_4\text{-Cl}} \xrightarrow{\text{Fe} / \text{H}_2\text{O}} \underset{}{\text{2-H}_2\text{N-C}_6\text{H}_4\text{-Cl}} \tag{48}$$

(a) Activation of iron. In a hood in an open beaker cooled by an ice-water bath, 56 gm (1.0 mole) of granulated iron (40 mesh, i.e., approximate diameter of 0.64 mm) is rapidly stirred. To the metal, 10 ml of concentrated hydrochloric acid is added very slowly at such a rate that excessive heat is not generated in the course of the ensuing reaction. Stirring is at such a rate that the formation of lumps is prevented. After addition has been completed, the iron powder is allowed to dry thoroughly.

(b) Reduction. In a three-necked flask provided with a reflux condenser and an efficient stirrer, 5 gm (0.032 mole) of 2-chloronitrobenzene is dissolved in 200 ml of benzene. The third neck of the flask is closed and the benzene solution is heated almost to boiling on a steam bath. Then the iron produced as described in step *(a)* is introduced while maintaining vigorous stirring. In the subsequent steps, vigorous stirring and refluxing are maintained. After ½ hr, 1 ml of water is added to the reaction mixture. Thereafter, small quantities of water are introduced from time to time at such a rate that at the end of 7 hr 20 ml of water have been added. Refluxing is continued for an additional hour. The hot solution is then filtered, the residual iron is extracted three times with hot benzene, and the benzene extracts are combined with the filtered product solution. Crude amines may be obtained by distilling the benzene from the product, followed by vacuum distillation or crystallization of the amine.

Alternatively the benzene solution is cooled and gaseous hydrogen chloride is passed into the solution to precipitate the amine as the insoluble hydrochloride salt. The salt may then be collected on a filter and recrystallized from alcohol. The hydrochloride salt may then be placed in a suitable distillation apparatus. Concentrated sodium hydroxide solution is added and the product is isolated by steam distillation. In this particular preparation, the product was isolated as the hydrochloride salt, yield 3.7 to 4.8 gm (71–92%).

b. REDUCTION WITH IRON AND FERROUS SULFATE

While aliphatic nitro compounds have been reduced with iron and acid, sometimes with indifferent results, a more convenient method of reduction makes use of iron powder and ferrous sulfate as the reducing agent.

3-5. Preparation of 2-Amino-2-methyl-1-propanol [91]

$$\underset{\underset{NO_2}{|}}{\overset{\overset{CH_3}{|}}{HO-CH_2-C-CH_3}} \xrightarrow{Fe/FeSO_4} \underset{\underset{NH_2}{|}}{\overset{\overset{CH_3}{|}}{HO-CH_2-C-CH_3}} \quad (49)$$

In a three-necked 3-liter flask fitted with a condenser, a sealed stirrer, a dropping funnel, and a thermometer reaching into the liquid, is placed a mixture of 280 gm (1 mole) of ferrous sulfate heptahydrate, 200 gm of iron powder, 500 ml of water, and 10 gm of concentrated sulfuric acid. The mixture is heated to reflux and a solution of 238 gm (2 moles) of 2-methyl-2-nitro-1-propanol in 240 gm of water is added with vigorous agitation at a rate of 120 gm of solution per hour. Heating and stirring is continued for 1 hr after the nitrohydroxy compound has been added. After the reaction mixture has cooled to room temperature, 100 gm of calcium hydroxide is added and the mixture is stirred for 1 hr. The mixture is filtered and the solid is washed with four 50-ml portions of water. Five grams of barium hydroxide is added to the combined filtrate and the mixture is stirred for 30 min. This treatment converts dissolved calcium sulfate into insoluble barium sulfate. The mixture is filtered and sufficient ammonium carbonate is added to precipitate the residual calcium and barium compounds completely. The mixture is again filtered. The filtrate is then distilled through a glass-helix-packed distillation column at atmospheric pressure until all the water has been removed. The residue is then distilled under reduced pressure to afford 80 gm (90%) of product, bp 45°C (10 mm), mp 30°–31°C.

c. Catalyzed Reaction with Hydrazine Hydrate

Alcohol solutions of nitro compounds are readily reduced by hydrazine hydrate in the presence of a hydrogenation catalyst such as Raney nickel, palladium on charcoal, platinum, iron, or copper powder. The latter two metals are less satisfactory as catalysts [92]. In this reaction, the concentration of hydrazine appears to be critical. If the reaction is carried out in a large volume of alcohol, in effect, the concentration of hydrazine hydrate is low. Under these conditions, the sole product is an amine. If, however, the hydrazine hydrate concentration is increased, good yields of intermediate products such as hydrazo or azo compounds may be isolated. Furthermore, nitrosobenzene can be reduced to aniline in alcoholic solution with hydrazine hydrate without the necessity of any catalyst. While ethanol has been used in this reaction [92], other alcohols, such as ethylene glycol, may be used as indicated in the preparation of di(2-amino-4-chlorophenyl) sulfone [93].

3-6. Preparation of Di(2-amino-4-chlorophenyl) Sulfone [93]

$$\underset{Cl\ \ O_2N\ \ NO_2\ \ \ \ \ \ Cl}{\text{Ar–SO}_2\text{–Ar}} \xrightarrow[\text{HOC}_2\text{H}_4\text{OH}]{\text{NH}_2\text{NH}_2\cdot\text{H}_2\text{O},\ \text{Ni(R)}} \underset{Cl\ \ H_2N\ \ NH_2\ \ \ \ \ \ Cl}{\text{Ar–SO}_2\text{–Ar}} \quad (50)$$

(CAUTION: Hydrazine hydrate is a toxic reagent.)

In a hood, a solution of 3.8 gm (10 mmoles) of di(2-nitro-4-chlorophenyl) sulfone in 150 ml of ethylene glycol is stirred on a water bath which maintains the reaction mixture below 35°C. To this solution, 1.5 ml of 95% hydrazine and a small quantity of Raney nickel is added. The reaction mixture is stirred for 24 hr while maintaining the temperature below 35°C. The reaction mixture is then poured into water. The precipitated product and Raney nickel are removed by filtration and recrystallized from 95% ethanol. Yield of di(2-amino-4-chlorophenyl) sulfone is 3.07 gm (97%), mp 165°C.

d. REDUCTION WITH TIN AND HYDROCHLORIC ACID

The preparation of aniline from nitrobenzene by the tin and hydrochloric reduction is a standard laboratory exercise. A typical procedure is that of Adams and Johnson [94], adapted to the more convenient equipment of the research laboratory.

3-7. *Preparation of Aniline* [94]

$$\text{C}_6\text{H}_5\text{NO}_2 \xrightarrow{\text{Sn/HCl}} \text{C}_6\text{H}_5\text{NH}_2 \cdot \text{HCl} + \text{SnCl}_2 \quad (51)$$

$$\text{C}_6\text{H}_5\text{NH}_2 \cdot \text{HCl} + \text{NaOH} \longrightarrow \text{C}_6\text{H}_5\text{NH}_2 + \text{NaCl} + \text{H}_2\text{O} \quad (52)$$

CAUTION: Since aniline is reputed to be absorbed through the skin, care in handling of the product as well as the intermediate reaction mixtures must be exercised. The use of plastic gloves, goggles, rubber aprons, etc., is strongly recommended.

In a 1-liter three-necked flask fitted with a sealed, explosion-proof mechanical stirrer having a Teflon blade, an efficient reflux condenser, and an addition funnel, are placed 25 gm (0.2 mole) of nitrobenzene and 45 gm (0.38 mole) of granulated tin. The reaction mixture is stirred vigorously while, in small portions (not exceeding 10 ml each) at the beginning, 120 gm (100 ml) of concentrated hydrochloric acid is added. The reaction mixture becomes warm and the reaction is controlled by alternately raising a steam bath or a water bath around the flask to maintain gentle evolution of gases. When the initial reaction moderates, further hydrochloric acid may be added in 15-ml portions. The addition is continued with careful control of the evolution of gas. If cooling is excessive, the reaction may be slowed down too much and sudden violent reactions may take place when additional hydrochloric acid is added. After all of the acid has been added, the reaction mixture is warmed on a steam bath for 30 min. During the course of the reaction, the double salt of aniline and stannous or stannic chloride occasionally precipitates, particularly during the cooling phases.

§ 3. Reduction Reactions

At the end of the heating period, no odor of nitrobenzene should be detectable and a few drops of the reaction mixture should form a perfectly clear solution when poured into water. If some residual nitrobenzene is found, additional tin and hydrochloric acid may be added.

The reaction mixture is cooled to room temperature, the reflux condenser is set down for distillation, the flask is cooled in an ice bath, and a cold solution of 75 gm of sodium hydroxide in 100 ml of water is slowly added. [Since 40–50% aqueous sodium hydroxide is commercially available, an equivalent volume of such a solution may be used. (CAUTION: Protective clothing, gloves, and goggles must be used when handling highly concentrated caustic solutions.)]. When the reaction mixture is strongly alkaline, the aniline separates as an oil. The addition funnel is then replaced by a steam inlet tube, and the reaction mixture is steam-distilled. The distillate is saturated with sodium chloride and the aniline layer is separated. If desired, the aqueous layer may be extracted several times with ether and the ether extracts combined with the aniline. The solution of ether and aniline is dried over anhydrous potassium hydroxide for at least 12 hr. The product solution is then decanted from the drying agent, ether is distilled off on a steam bath, and the residue is finally distilled at atmospheric pressure, preferably under a nitrogen atmosphere. Yield 16–18 gm (85–95%), bp 180°–185°C.

e. REDUCTION WITH SODIUM HYDROSULFITE

Aromatic nitro compounds, which, because of the presence of other functional groups, are water-soluble, may be reduced conveniently with sodium hydrosulfite. Use of this reagent frequently leads to clean-cut preparations of amino compounds without the complications arising from the formation of the double salts often associated with acid and metal reductions.

In the directions as given here, exact yield data for the conversion of Martius Yellow to 2,4-diamino-1-naphthol are not furnished since use is made of moist intermediate products. The dry free base is readily oxidized; it is therefore converted to its salt as rapidly as possible and used as its salt solution in subsequent reactions.

3-8. Preparation of 2,4-Diamino-1-naphthol Dihydrochloride [95]

Martius Yellow ammonium salt (1-O$^-$NH$_4^+$, 2-NO$_2$, 4-NO$_2$-naphthalene) + 6 Na$_2$S$_2$O$_4$ + 8 H$_2$O ⟶ 2,4-diamino-1-naphthol (1-OH, 2-NH$_2$, 4-NH$_2$-naphthalene) + 11 NaHSO$_3$ + Na(NH$_4$)SO$_3$ (53)

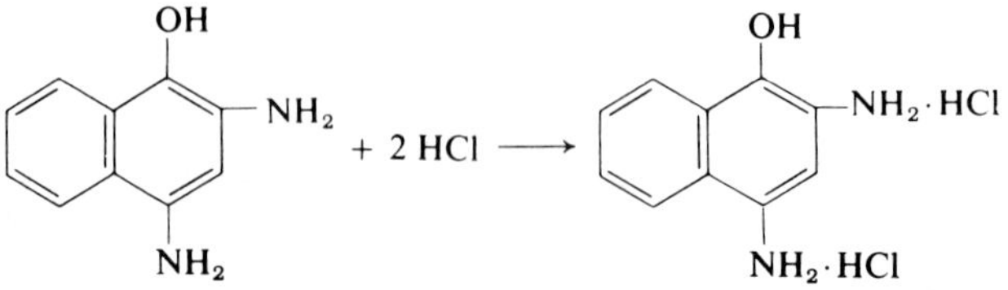

$$\text{[2-amino-1-naphthol with NH}_2\text{ at 4-position]} + 2\text{HCl} \longrightarrow \text{[dihydrochloride salt]} \quad (54)$$

Approximately 7.4 gm of the moist ammonium salt of 2,4-dinitro-1-naphthol [95] is dissolved in 200 ml of water. To this solution is added 40 gm of sodium hydrosulfite. The solution is stirred until the original orange color disappears and a crystalline tan precipitate is formed (5–10 min). The reaction mixture is then cooled in ice. While the reaction mixture cools, a solution of 2 gm of sodium hydrosulfite in 100 ml of water is prepared in one beaker. In another beaker a solution of 6 ml of concentrated hydrochloric acid in 25 ml of water is prepared. The precipitate is collected by suction filtration using the sodium hydrosulfite solution for rinsing and washing, avoiding even briefly sucking air through the filter cake after the reducing agent has been drained away. The solid is then rapidly washed into the beaker containing the dilute hydrochloric acid and stirred to convert all of the diamine to the dihydrochloride. The acid solution is clarified by filtration through a moist filter bed of decolorizing charcoal. The resulting aqueous solution may then be used to carry out further reactions in the Martius Yellow experiment.

f. Reduction with Sodium Sulfide and Related Compounds

To produce amino compounds by reduction of nitro compounds, acid conditions are normally used. Under alkaline conditions only a few reductions have been carried out.

Polyaminostyrenes have been produced by the reduction of polynitrostyrene with an aqueous solution of sodium disulfide in an autoclave at 150°C [96].

Aminofluorescein has been produced by reduction of nitrofluorescein with sodium sulfides [97].

A mixture of sodium hydroxide and sulfur in acetone–methanol solution may be used to reduce the aromatic nitro compounds [98]. It is stated that alcoholic or aqueous media will not bring about reduction, but that mixtures of acetone and methanol are essential. A typical example is the preparation of 2-ethylaniline.

3-9. Preparation of 2-Ethylaniline (2-Aminoethylbenzene) [95]

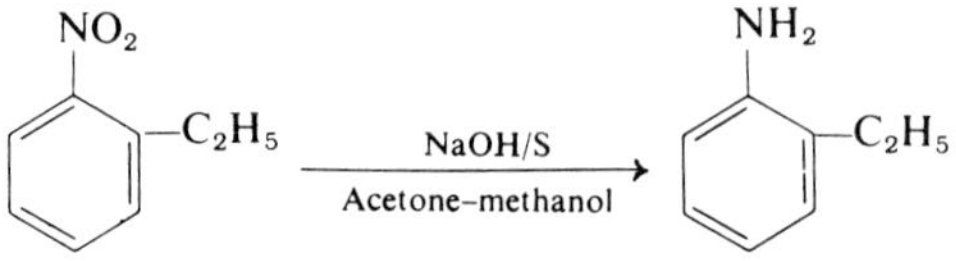

$$\text{2-nitroethylbenzene} \xrightarrow[\text{Acetone-methanol}]{\text{NaOH/S}} \text{2-ethylaniline} \quad (55)$$

§ 3. Reduction Reactions

In a round-bottomed flask, a mixture of 3 gm of sodium hydroxide, 1.5 gm of sulfur, 10 ml of acetone, 10 ml of methanol, and 3.75 gm (24.8 mmoles) of 2-ethylnitrobenzene is heated on a water bath for 1½ hr. After that period, the solvent is distilled off and the amine is separated by steam distillation. The steam distillate is saturated with salt, the amine layer is separated, and the aqueous solution is extracted with benzene. The benzene solution and the amine are combined, dried over potassium hydroxide, and then subjected to fractional distillation in a small still. The yield of 2-ethylaniline is 2.0 gm (69%), bp 215°–216°C.

g. Reduction with Lithium Aluminum Hydride

Nitro compounds such as the β-nitrostyrenes are readily reduced with lithium aluminum hydride to form aliphatic amines as illustrated in Eq. (56) [99].

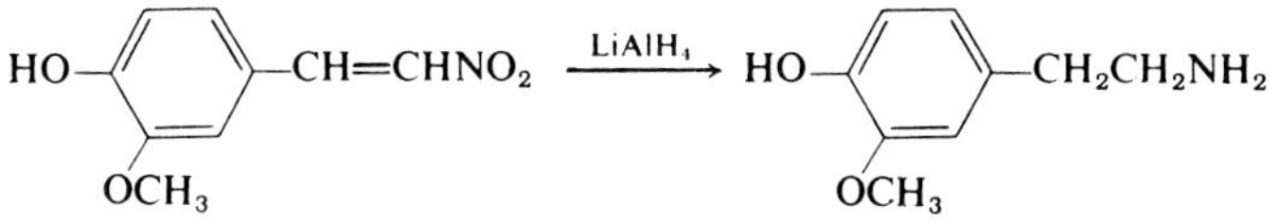

$$\text{HO-}\underset{\text{OCH}_3}{\text{C}_6\text{H}_3}\text{-CH=CHNO}_2 \xrightarrow{\text{LiAlH}_4} \text{HO-}\underset{\text{OCH}_3}{\text{C}_6\text{H}_3}\text{-CH}_2\text{CH}_2\text{NH}_2 \quad (56)$$

Simple aromatic nitro compounds are reduced bimolecularly to azobenzenes [Eq. (57)] [100].

$$\text{C}_6\text{H}_5\text{-NO}_2 \xrightarrow{\text{LiAlH}_4} \text{C}_6\text{H}_5\text{-N=N-C}_6\text{H}_5 \quad 73.5\% \quad (57)$$

With increased steric hindrance, amine formation predominates as shown in Eq. (58).

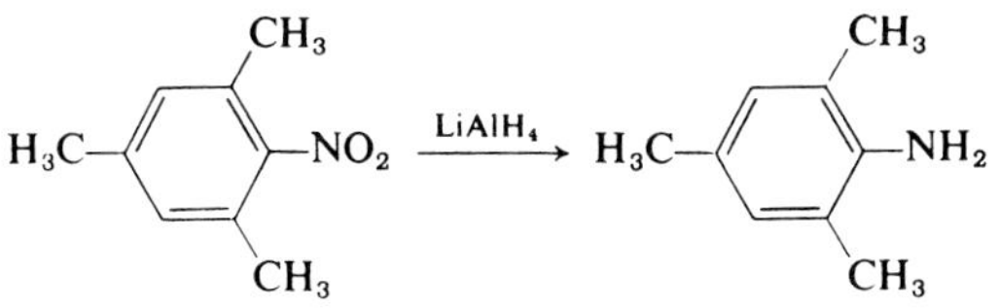

$$\text{H}_3\text{C-}\underset{\text{CH}_3}{\overset{\text{CH}_3}{\text{C}_6\text{H}_2}}\text{-NO}_2 \xrightarrow{\text{LiAlH}_4} \text{H}_3\text{C-}\underset{\text{CH}_3}{\overset{\text{CH}_3}{\text{C}_6\text{H}_2}}\text{-NH}_2 \quad (58)$$

h. Reduction with Tervalent Phosphorus Reagents

The reduction of aromatic nitro compounds with tervalent phosphorus reagents such as triethyl phosphide leads to complex products. The nature of these products are thought to indicate the intermediacy of a triplet nitrene [101].

i. Selective Catalytic Hydrogenation of Aromatic Nitro Compounds

The hydrogenation of aromatic nitro compounds which also bear acetylenic side chains, such as 3-nitrophenylacetylene in the presence of a ruthenium catalyst, selectively reduces the nitro groups without significantly reducing the acetylenic group [102].

D. The Leuckart Reaction

a. CLASSICAL LEUCKART REACTION

The reaction of formic acid or a variety of formic acid derivatives, such as formate salts and formamides, with ammonia or a variety of amines, as well as various amine derivatives and salts such as ammonium formate salts, and carbonyl compounds, results in the reductive alkylation of the amine in which the entering alkyl group is derived from the carbonyl compound. This reaction is known as the Leuckart reaction [103]. By proper selection of reagents, primary, secondary, and tertiary amines may be prepared. In general this reaction is carried out at elevated temperatures without further solvents. More recent work indicates that magnesium chloride and ammonium sulfate are particularly useful catalysts in the preparation of tertiary amines by the Leuckart reaction [104].

This method appears to be suitable for the preparation of a wide variety of tertiary amines including those derived from α,β-unsaturated carbonyl compounds [104].

3-10. Preparation of 1-(Methyl-3-phenylpropyl)piperidine [104]

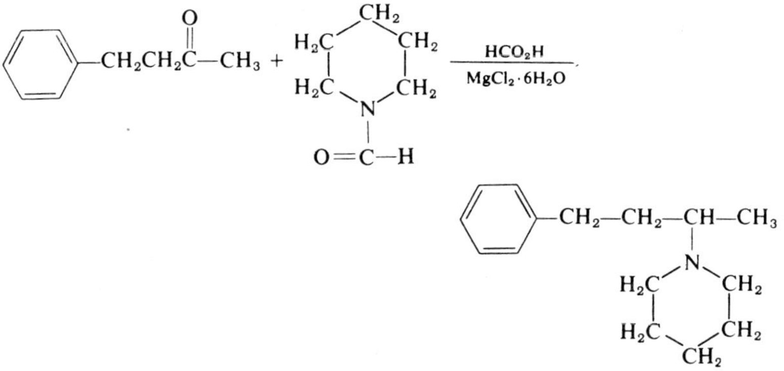

$$\text{Ph-CH}_2\text{CH}_2\text{C(O)-CH}_3 + \text{formpiperidide} \xrightarrow[\text{MgCl}_2 \cdot 6\text{H}_2\text{O}]{\text{HCO}_2\text{H}} \text{Ph-CH}_2\text{-CH}_2\text{-CH(CH}_3\text{)-piperidine} \quad (59)$$

In a three-necked flask equipped with a mechanical stirrer, thermometer, and a condenser set downward for distillation, are mixed 19.25 gm (0.13 mole) of 4-phenyl-2-butanone, 58.8 gm (0.52 mole) of formpiperidide, 4.47 gm (0.022 mole) of magnesium chloride hexahydrate, and 7.03 gm (0.13 mole) of 85% formic acid. The mixture is heated gradually and volatile constituents are removed until the pot temperature approximates the boiling point of the formpiperidide. A reflux condenser is then installed in place of the distillation condenser and the mixture is refluxed with stirring for 8 hr. Then the reaction mixture is poured into dilute hydrochloric acid. Unreacted ketone is removed by steam distillation. The ketone-free aqueous residue is then made strongly basic and the amine is distilled out with steam. The steam distillate is saturated with sodium chloride and extracted with ether. The ether extract, after drying over solid potassium hydroxide, is distilled to remove the solvent. The residual amine is then fractionally distilled at 176°–177°C at 25 mm. Yield 15.3 gm (57%).

b. ESCHWEILER–CLARKE MODIFICATION

A modification of the Leuckart reaction for the preparation of methylated amines which requires less drastic reaction conditions, known as the Eschweiler–Clarke reaction, involves the use of aqueous formaldehyde and an excess of formic acid as reducing agent, along with the appropriate amine. Usually the reaction conditions are such that only the methylated tertiary amine can be isolated [17]. An example of this reaction is the preparation of 1-methyl-2-(p-tolyl)piperidine [105].

3-11. Preparation of 1-Methyl-2-(p-tolyl)piperidine [106].

$$\text{H}_3\text{C}-\text{C}_6\text{H}_4-\text{HC}\underset{\text{NH}}{\overset{\text{CH}_2\text{CH}_2}{\diagup\diagdown}}\text{CH}_2 + \text{CH}_2\text{O} \xrightarrow{\text{HCO}_2\text{H}} \text{H}_3\text{C}-\text{C}_6\text{H}_4-\text{HC}\underset{\underset{\text{CH}_3}{|}}{\overset{\text{CH}_2\text{CH}_2}{\diagup\diagdown}}\text{CH}_2 \quad (60)$$

A mixture of 17.5 gm (0.1 mole) of 2-(p-tolyl)piperidine, 30 gm (0.59 mole) of 90% formic acid, and 25 gm (0.24 mole) of 35% formaldehyde is heated under reflux for 12 hr on a steam bath. The reaction mixture is cooled and 15 ml of concentrated hydrochloric acid is added. Heating under reflux is then continued for another 5 hr. Then the reaction mixture is cooled, made strongly basic with sodium hydroxide solution, and extracted with benzene. The benzene extract is dried over anhydrous sodium or potassium hydroxide. The benzene solution is then distilled. After removal of the solvent, the product distills at 117°–118°C (6 mm). Yield 17.8 gm (94%).

As stated before, the normal product from an Eschweiler–Clarke reaction is the tertiary amine. It has been observed that *tert*-butylamine subjected to the reaction will afford significant amounts of secondary amine, *tert*-butylmethylamine. By application of methods of experimental design, optimum conditions were found for the preparation of this secondary amine [106]. Table I shows the mole ratios found to be optimum for the preparation of *tert*-butyldimethylamine and *tert*-butylmonomethylamine. We particularly recommend the paper of Meiners *et al.* [106] to the reader, inasmuch as it is one of the few reports describing a statistical design of experiments to a problem in organic synthesis, a procedure which is now finding more extensive application in the laboratory as well as in industry.

With sterically hindered secondary amines, the Eschweiler–Clarke methylation is said either to fail or to lead to decomposition products [13].

TABLE I

REACTION CONDITIONS FOR THE ESCHWEILER–CLARKE REACTION TO OPTIMIZE SECONDARY OR TERTIARY AMINE PRODUCTION[a]

Major product	Mole ratio formic acid to starting amine	Mole ratio formaldehyde to starting amine	Reaction temperature (°C)	Reaction time (hr)	Yield (%)
tert-Butyldimethylamine	4:1	2.5:1	90–100	2	95
tert-Butylmethylamine	3:1	1.25:1	50	6	60[b]

[a] Meiners et al. [106].
[b] Based on tert-butylamine consumed. Conversion was only 46%.

4. REARRANGEMENT AND RELATED REACTIONS

A. Benzidine Rearrangement

The rearrangement of aromatic hydrazo compounds (1,2-diarylhydrazines) with no substitution in the 4,4'-positions to the benzidines (i.e., 4,4'-diamines) is readily accomplished. *In this connection it must be pointed out again that benzidine itself is considered to be highly carcinogenic* and, to the best of our knowledge, is no longer being produced in the United States. Whether other benzidines have similar properties is not known. We suggest, however, that extreme caution be exercised in the handling of the benzidine derivatives themselves as well as the intermediate reaction mixtures. Since the reaction is frequently carried out by shaking a stoppered flask containing an ether solution by hand, the hazard of breaking the flask or of the product solutions oozing out around the stopper is great.

The procedure for the benzidine rearrangement is given here for reference only. The methods for disposing of residues, reaction solvents, etc. need careful consideration before the reaction is attempted, even if the benzidine is thought to be "safe." Since the hazards are thought to be associated even at levels of parts-per-million or less of benzidine, ordinary analytical procedures and ordinary disposal methods are probably quite inadequate.

If one or both of the 4,4'-positions of the hydrazo compound is substituted, the normal benzidine rearrangement does not take place. The isolated reaction products are amino derivatives of diphenylamine called semidines [107].

The formation of semidine is illustrated in Eq. (61).

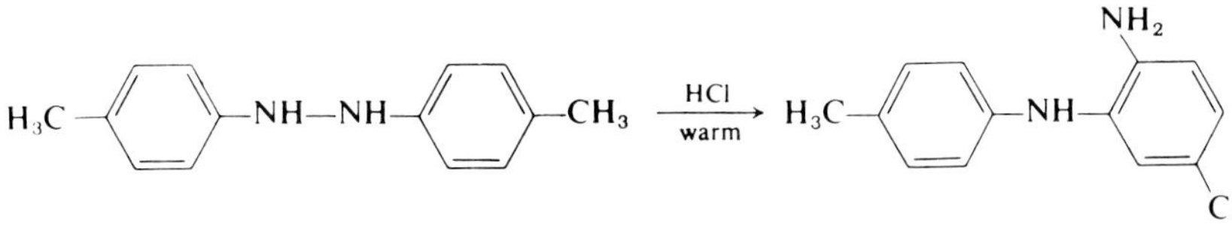

§ 4. Rearrangement and Related Reactions

$$H_3C\text{-}C_6H_4\text{-}NH\text{-}NH\text{-}C_6H_4\text{-}CH_3 \xrightarrow[\text{warm}]{\text{HCl}} H_3C\text{-}C_6H_4\text{-}NH\text{-}C_6H_3(NH_2)(CH_3) \quad (61)$$

The preparation of 3,3'-dibromobenzidine illustrates the normal procedure of the reaction [108]. As noted above, the procedure is only given here for reference.

4-1. Preparation of 3,3'-Dibromobenzidine [108]

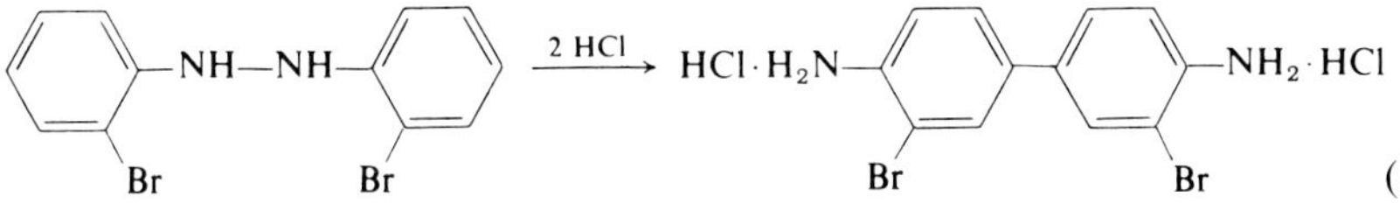

$$(2\text{-Br-C}_6H_4)\text{-NH-NH-}(2\text{-Br-C}_6H_4) \xrightarrow{2 \text{ HCl}} HCl\cdot H_2N\text{-}(3\text{-Br-C}_6H_3)\text{-}(3\text{-Br-C}_6H_3)\text{-}NH_2\cdot HCl \quad (62)$$

$$H_2N\text{-}(3\text{-Br-C}_6H_3)\text{-}(3\text{-Br-C}_6H_3)\text{-}NH_2\cdot HCl \xrightarrow{\text{NaOH}}$$

$$H_2N\text{-}(3\text{-Br-C}_6H_3)\text{-}(3\text{-Br-C}_6H_3)\text{-}NH_2 + 2\text{ NaCl} + 2\text{ H}_2O \quad (63)$$

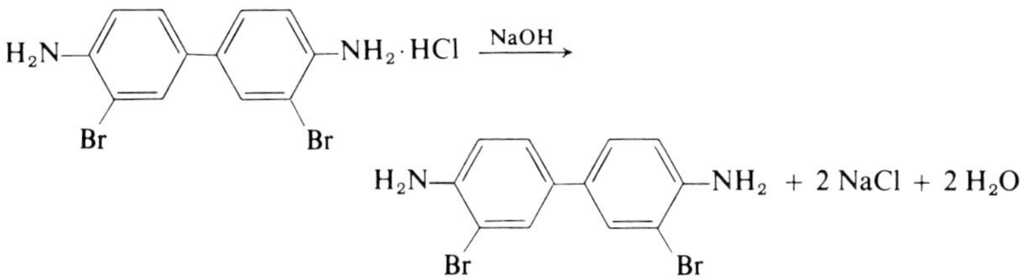

With suitable safety precautions, in a flask fitted with a suitable stopper is placed 25 ml of ice-cold concentrated hydrochloric acid. To this acid is added in very small increments a solution of 4 gm (0.011 mole) of 2,2'-dibromohydrazobenzene in 50 ml of ether. The addition is carried out slowly and, after each small portion of solution added, the reaction mixture is vigorously shaken. Shaking is continued for 1 hr after addition has been completed. The reaction mixture is cooled and the precipitating dihydrochloride is separated by filtration. The precipitate is washed repeatedly with ether.

The suspension of the white product is heated with an excess of 10% aqueous sodium hydroxide solution on a steam bath. The reaction mixture is then cooled and the free base is extracted from the mixture with ether. The extract is dried over calcium sulfate. After the separation of the drying agent, the ether is evaporated off and the residual 3,3'-dibromobenzidine is recrystallized from ethanol to give 3 gm (25%) of nearly white solid, mp 127°–129°C.

B. Hofmann Rearrangement

The Hofmann rearrangement with hypohalites, the Schmidt reaction, and the Curtius reaction are three closely related methods of preparing amines. While all of them have been used extensively, since they afford a means of converting carboxylic acid derivatives to amines with one less carbon atom, they all are considered somewhat hazardous reactions. Of the three, probably the Hofmann rearrangement is the least hazardous. Both the Schmidt and the Curtius reactions may, under certain circumstances, be extremely hazardous. The Hofmann rearrangement has been reviewed [109, 110]. The overall reaction is represented in Eqs. (64) and (65).

$$RC(=O)-NH_2 + NaOX \longrightarrow [RC(=O)-N\diagup\diagdown] \longrightarrow R-N=C=O \quad (64)$$

$$R-N=C=O + R'OH \longrightarrow R-NH-C(=O)OR' \xrightarrow{H_2O} RNH_2 + R'C(=O)OH \quad (65)$$

While many preparations involve the use of sodium hypobromite [111], the use of sodium hypochlorite has much to recommend it.

In the case of higher molecular weight fatty acid derivatives, ureas form when some of the free amine product reacts with the isocyanate intermediate of the reaction. Therefore, by the use of inert solvents such as dioxane, the mutual solubilities of isocyanates and the product amine are reduced and the amine may be isolated, as will be shown in the example given below [110].

4-2. Preparation of Nonylamine [110]

$$CH_3(CH_2)_8C(=O)-NH_2 + NaOCl \longrightarrow CH_3(CH_2)_7CH_2NH_2 \quad (66)$$

(a) Preparation of standard hypochlorite solution. In a distilling flask equipped with a dropping funnel is placed 6.7 gm of potassium permanganate. The side arm of the flask is joined by a glass-to-glass connection to a tube dipping below the surface of a solution containing 16 gm (0.4 mole) of sodium hydroxide dissolved in 100 ml of water and cracked ice contained in a graduated cylinder. The cylinder is mounted with an ice bath.

Fifty milliliters of concentrated hydrochloric acid is admitted slowly through the dropping funnel so as to produce a slow stream of chlorine. When all the acid has been added, the dropping funnel is closed and the content of the flask is heated gently with a small flame until the reflux point is a little below the juncture of the side arm. A Pyrex–wool plug below the side arm serves to prevent entrained acid from being carried over. The hypochlorite solution is then made up to 160 ml. The solution contains slightly over 0.1 mole of sodium hypochlorite and 0.2 mole of excess sodium hydroxide.

(b) Hofmann rearrangement. To a round-bottomed flask, equipped with reflux condenser, magnetic stirrer, thermometer, and addition funnel, containing 17.1

§ 4. Rearrangement and Related Reactions

gm (0.1 mole) of capramide in 80 ml of purified dioxane is added 160 ml of the hypochlorite solution prepared as above. The mixture is stirred vigorously and heated to 45°C. The temperature then rises spontaneously to 65°C in 2 min. Stirring is continued without external heating for 2 hr, at which time the temperature has dropped back to 42°C. On cooling, an oily product layer separates. This is removed and the aqueous layer is extracted with benzene. The oily amine product layer and the benzene extract are combined. Then the base is extracted into 100 ml of 1N hydrochloric acid. The resulting acid layer is separated and to it is added a concentrated sodium hydroxide solution. The liberated base is again taken up in benzene, and the benzene extract is dried over potassium hydroxide pellets and filtered. The benzene solution is then saturated with hydrogen chloride gas. The mixture is then concentrated to a volume of 50 ml with an air stream and 50 ml of dry ether is added. The amine hydrochloride is collected on a filter. Yield 11.9 gm (66.4%), mp 185°–186°C.

[Reprinted from: E. Magnien and R. Baltzly, *J. Org. Chem.* **23**, 2029 (1958). Copyright 1958 by the American Chemical Society. Reprinted by permission of the copyright owner.].

A modification by Jeffreys [112] involves the isolation of the intermediate urethane. The problem of this procedure is a general one of hydrolyzing urethanes to free amines. This appears to be associated with the low solubility of urethanes in the reaction medium.

Prolonged heating with an ethanolic solution of aqueous sodium hydroxide (10 gm of NaOH in 100 ml of 90% ethanol per 5 gm of urethane) may constitute an optimum hydrolysis condition. Even less-soluble urethane may be hydrolyzed in glycol media [110].

In a more recent example of the Hofmann rearrangement, a sodium hypochloride solution was prepared by passing chlorine from a suitably trapped cylinder into a solution of sodium hydroxide in water containing enough ice to maintain the formation temperature below 5°C. Addition to this solution of potassium *p*-phthalamate rapidly led to a Hofmann rearrangement. The reaction had to be initiated by warming. Then an exothermic reaction took place within a few minutes [113].

Among new reagents which are said to bring about Hofmann rearrangements under mild conditions are iodine pentafluoride [114], and 1,1-bis(trifluoroacetoxy)iodobenzene [115]. The latter reagent, used in an acetonitrile–water solution at room temperature, is said to convert amides to amines under extremely mild conditions without the necessity of isolating the isocyanate. In affinity chromatography, one method of "activating" a polyacrylamide packing involves conversion to an azide and reaction on the substrate by a variation of the Curtius reaction (see below). We suggest that the method of Loudon and co-workers [115] also be considered for this technique.

C. Schmidt Reaction

The reaction of carbonyl compounds with hydrazoic acid in concentrated sulfuric acid resembles the Curtius reaction discussed below. This reaction is known as the Schmidt reaction. The choice of carbonyl compounds which may be used in this reaction is wide. Carboxylic acids yield primary amines, aldehydes yield nitriles of formamides of amines, ketones form amides [116].

As mentioned elsewhere, we consider the use of hydrazoic acid hazardous and make mention of this reaction only for reference purposes. A typical example of the reaction using a methyl ketone is given below [Eq. (67)].

4-3. Preparation of 3,4-Dimethylaniline [117]

$$\underset{\underset{CH_3}{CH_3}}{\overset{\overset{O}{\overset{\|}{C-CH_3}}}{\bigcirc}} + HN_3 \xrightarrow{H_2SO_4} \left[H_3C-\underset{CH_3}{\bigcirc}-NH\overset{\overset{O}{\|}}{C}-CH_3 \right] \longrightarrow H_3C-\underset{CH_3}{\bigcirc}-NH_2 \quad (67)$$

In a hood, behind an adequate shield, a solution of 21.4 gm (0.145 mole) of 3,4-dimethylacetophenone in 100 ml of dry benzene is mixed with 30 ml of concentrated sulfuric acid. With stirring, 191 ml of a 4.1% solution of hydrazoic acid in benzene is added over a 50-min period while maintaining the temperature between 38° and 41°C. After all the solution has been added, the mixture is allowed to stir for an additional 5 min; then it is cooled and poured into a separatory funnel. The sulfuric acid layer is cautiously poured into 400 ml of water and made alkaline with ammonium hydroxide (approximately 120 ml required). The yellow oil which separates out solidifies on cooling. This precipitate is filtered off, washed once with water, and refluxed for 2 hr with 75 ml of concentrated hydrochloric acid. The resulting solution is poured into 250 ml of water and extracted with ether. The aqueous solution is made alkaline with 6N sodium hydroxide solution. The alkaline solution is extracted with ether; the ether solution is dried over potassium hydroxide and distilled to yield 7.0 gm of crude 3,4-dimethylaniline, bp 138°–143°C (55 mm), mp 47°C.

On recrystallization from low-boiling petroleum ether, 5.7 gm of pure 3,4-dimethylaniline is obtained, mp 50°–51°C. Yield 32.4%.

D. Curtius Reaction

Closely related to the Hofmann rearrangement is the Curtius reaction. In this reaction, acid azides are decomposed to isocyanates. The isocyanates, in turn, may be converted to urethanes, ureas, amides, or amines. Upon hydrolysis of the

§ 4. Rearrangement and Related Reactions 419

Curtius rearrangement products, in effect, a carboxyl group may be converted to an amino group [118].

As indicated before, we consider the Curtius reaction a hazardous reaction and discuss it here primarily for purposes of reference. Considerable detail on the conditions for carrying out the reaction as well as a detailed comparison of the effectiveness of the Curtius, Hofmann, and Schmidt rearrangements has been given [118]. Acid azides may be prepared by the reaction of acyl halides or acyl anhydrides with sodium azides (a reaction in which explosions have been reported) or by treatment of hydrazides with sodium nitrite. Since we believe that the acyl azides themselves are explosive in some cases, extreme care must be exercised in these reactions. The preparation of $(\pm)$-*trans*-2-cyclohexyloxycyclopropylamine is an example of the reaction involving sodium azide [119].

4-4. Preparation of ($\pm$)-trans-2-Cyclohexyloxycyclopropylamine [119]

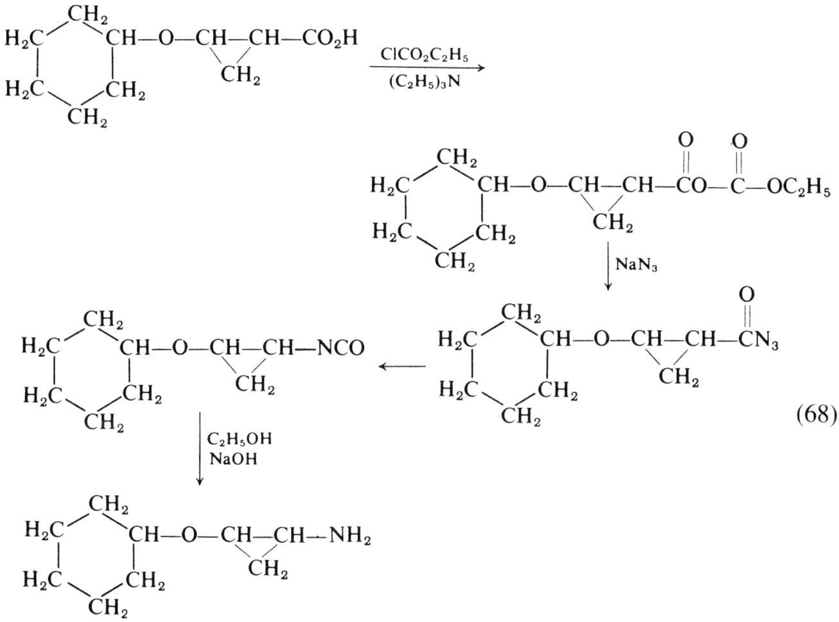

<chemical scheme>

(68)

To a stirred solution of 60 gm (0.33 mole) of $(\pm)$-*trans*-2-cyclohexyloxycyclopropane carboxylic acid in 150 ml of acetone and 75 ml of water at $-5°C$, a solution of 40.5 gm (0.4 mole) of triethylamine in 300 ml of acetone is added. After stirring for a short time, a solution of 43.5 gm (0.4 mole) of ethyl chloroformate in 100 ml of acetone is added and stirring at $-5°C$ is continued for 30 min.

The following steps are carried out behind adequate barricades. To the above reaction mixture, a solution of 32.5 gm of sodium azide in 200 ml of water is added while maintaining the temperature between −5°C and 0°C. Stirring is continued for an additional 2 hr between −5° and 0°C. The mixture is then poured into an ice-cold saturated sodium chloride solution and extracted several times with ether. The combined extracts are dried over magnesium sulfate and filtered. To the filtrate is added 1 liter of absolute ethanol. The solution is heated gently on a steam bath to distill the ether off slowly. The resulting solution is then refluxed for 6 hr. The ethanol is then removed by distillation under reduced pressure on a water bath to yield 65 gm of crude urethane. This product is treated with 300 ml of a 40% aqueous sodium hydroxide solution and refluxed for 36 hr. The solution is cooled and exhaustively extracted with ether. The extract is washed with water, dried, filtered, and concentrated. The residual oil is distilled under reduced pressure and the amine is collected at 50°–60°C (1 mm). Yield 22 gm (47% overall from the acid).

[Reprinted from: J. Finkelstein *et al.*, *J. Med. Chem.* **9,** 319 (1966). Copyright 1966 by the American Chemical Society. Reprinted by permission of the copyright owner.]

An example of the formation of an acid azide by the direct reaction of cyclobutanecarboxylic acid with sodium azide prior to a Curtius rearrangement is given in Ref. 120. The formation of an acid diazide from an acid halide derived from 4-nitro-4-methylheptanedioic acid prior to rearrangement is given in Ref. 121.

Examples of reactions involving the formation of azides from hydrazides are to be found in Ref. 122 and 123.

The treatment of aromatic carboxylic acids with nitromethane in polyphosphoric acid leads to aromatic amines. This reaction is said to be an example of the less-well-known Lossen rearrangement [109, 118, 124].

5. MISCELLANEOUS PREPARATIONS

1. Formation of a quaternary ammonium salt [125]. References illustrate alkylation or arylation of tertiary amines with alkyl halides and esters of strong mineral acids, with mixtures of halogens and olefins or ketones, with onium-compounds, alkylene oxides, β-lactones, alkynes and alkenes, with esters of amino-acids and preparation of polymeric quaternary ammonium compounds.

2. N-Alkylation of amines and organic copper reagents [126].

$$CH_3(CH_2)_6-NH-CH_2CH_2CH_2CH_3 \xrightarrow[(O_2)]{n\text{-}Bu_2CuLi} CH_3(CH_2)_3-\underset{\underset{CH_3}{\overset{(CH_2)_6}{|}}}{N}-(CH_2)_3CH_3 \qquad (69)$$

3. Synthesis of tri- and hexamine macrocycles [127a]

$$H_2N-CH_2CH_2-NH_2 \xrightarrow{TsCl} Ts-NH-CH_2CH_2-NH_2 \quad (70)$$

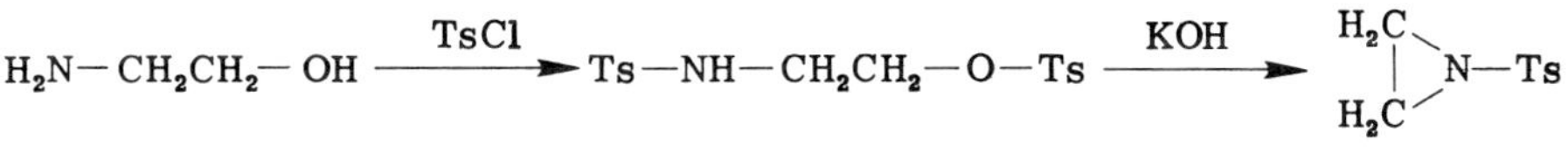

$$H_2N-CH_2CH_2-OH \xrightarrow{TsCl} Ts-NH-CH_2CH_2-O-Ts \xrightarrow{KOH} \begin{array}{c} H_2C \\ | \\ H_2C \end{array}\!\!\!\!\diagdown\!\!\!N-Ts \quad (71)$$

$$Ts-NH-CH_2CH_2-NH_2 + \begin{array}{c} H_2C \\ | \\ H_2C \end{array}\!\!\!\!\diagdown\!\!\!N-Ts \longrightarrow Ts-NH(CH_2)_2NH-(CH_2)_2-NH-Ts$$

$$\downarrow \; C_6H_5-C(=O)-Cl$$

$$Bz-N[(CH_2)_2-NH-Ts]_2 \quad (72)$$

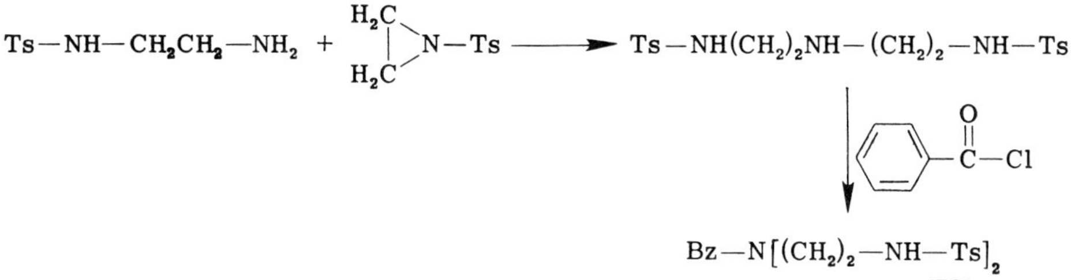

$$Bz-N[(CH_2)_2-NH-Ts]_2 \xrightarrow{2\,NaOEt} Bz-N[(CH_2)_2N^--Ts]_2 + 2\,Na^+ + 2\,EtOH \quad (73)$$

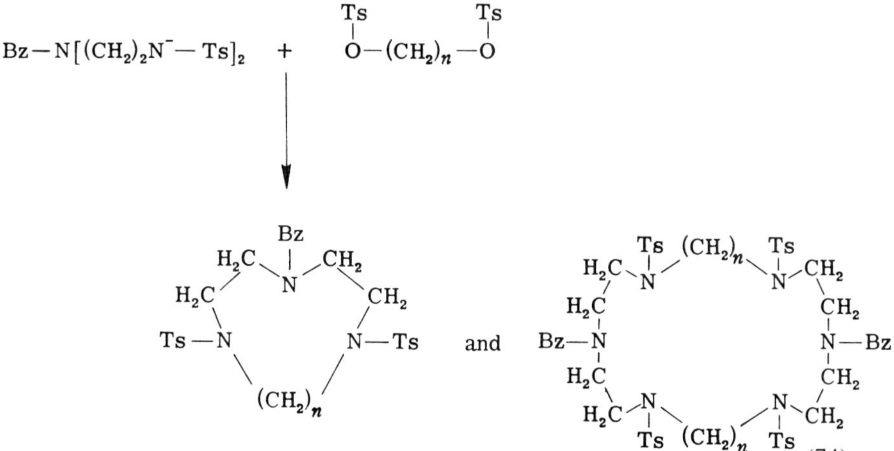

$$Bz-N[(CH_2)_2N^--Ts]_2 \;+\; \overset{Ts}{\underset{}{O}}-(CH_2)_n-\overset{Ts}{\underset{}{O}}$$

↓

(triamine macrocycle with Bz, Ts–N, Ts–N, (CH$_2$)$_n$) and (hexamine macrocycle with Ts, Bz, Ts groups and (CH$_2$)$_n$ bridges) (74)

Hydrolysis of the benzoylamide–tosylamide with acid leads to the salts of the free amino-macrocycles. 2,3-Dihydropyran has also been used as a protective agent of glycols which may be used to couple with intermediate compounds analogous to those formed in Eq. (73) [cf. 127b].

4. Cyclization of α,ω-aliphatic diamines with a ruthenium catalyst [128].

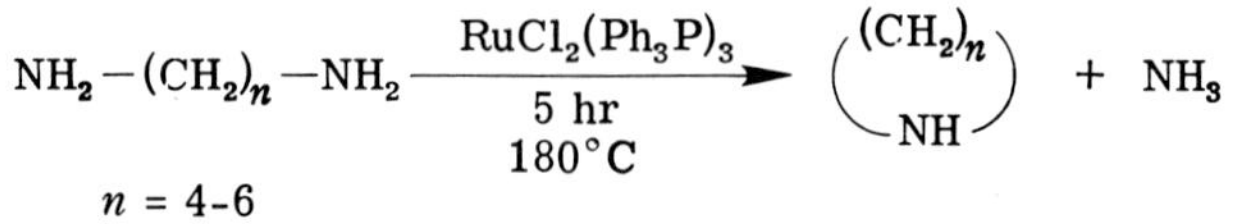

$$NH_2-(CH_2)_n-NH_2 \xrightarrow[180°C]{RuCl_2(Ph_3P)_3 \atop 5\ hr} \underset{NH}{(CH_2)_n} + NH_3 \qquad (75)$$

$$n = 4-6$$

5. Decarboxylation of amino acids [129].

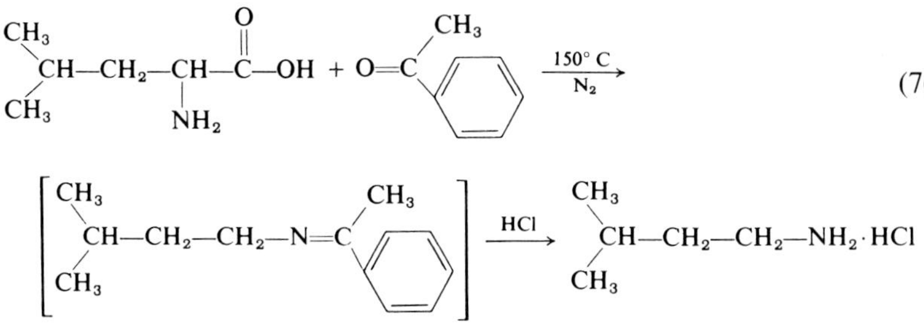

$$\underset{CH_3}{\overset{CH_3}{\diagdown}}CH-CH_2-\underset{NH_2}{CH}-\overset{O}{\overset{\|}{C}}-OH + O=C\underset{}{\diagup}\hspace{-2pt}\bigcirc \xrightarrow[N_2]{150°C} \qquad (76)$$

$$\left[\underset{CH_3}{\overset{CH_3}{\diagdown}}CH-CH_2-CH_2-N=C\underset{}{\overset{CH_3}{\diagup}}\hspace{-2pt}\bigcirc\right] \xrightarrow{HCl} \underset{CH_3}{\overset{CH_3}{\diagdown}}CH-CH_2-CH_2-NH_2\cdot HCl$$

6. Decarboxylation of aromatic carboxylic acids in the presence of ammonia and Cu(II) ions [130].

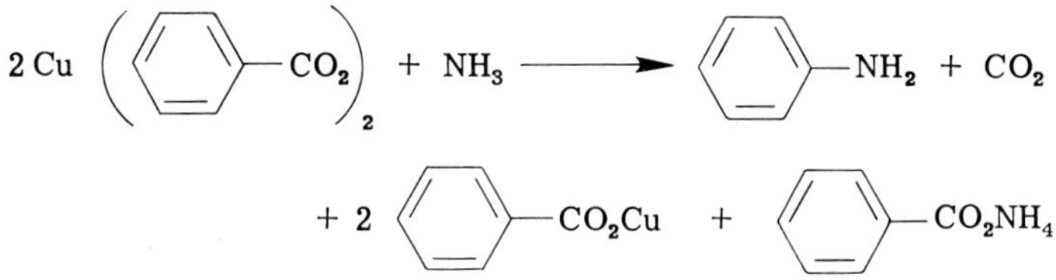

$$2\ Cu\left(\bigcirc\!\!-CO_2\right)_2 + NH_3 \longrightarrow \bigcirc\!\!-NH_2 + CO_2$$

$$+\ 2\ \bigcirc\!\!-CO_2Cu\ +\ \bigcirc\!\!-CO_2NH_4 \qquad (77)$$

$$2\ \bigcirc\!\!-CO_2Cu\ +\ 2\ \bigcirc\!\!-CO_2NH_4 + \tfrac{1}{2}O_2 \longrightarrow$$

$$2\ Cu\left(\bigcirc\!\!-CO_2\right)_2 + NH_3 + H_2O \qquad (78)$$

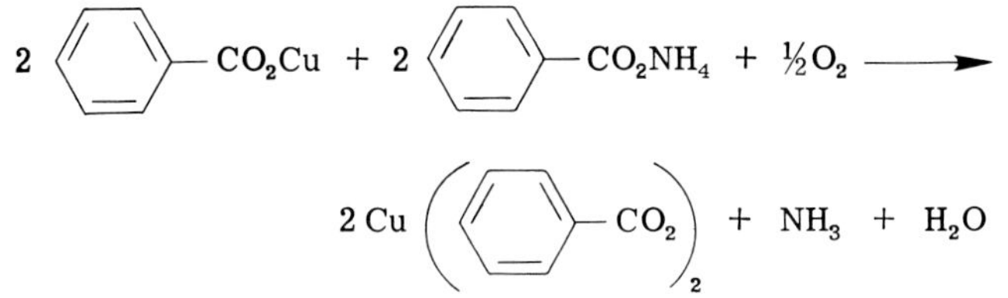

7. Methanolysis of trimethylsilyl derivatives for preparation of unsubstituted dihydropyridines [131].

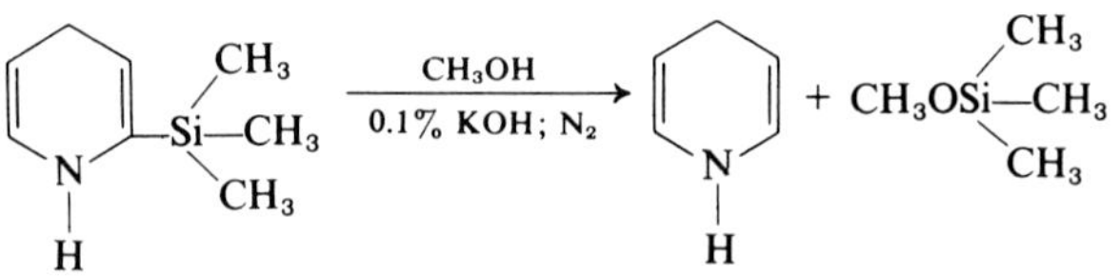

$$\underset{H}{\overset{}{\bigcirc_N}}\!\!-\underset{CH_3}{\overset{CH_3}{\diagdown}}\hspace{-2pt}Si-CH_3 \xrightarrow[0.1\%\ KOH;\ N_2]{CH_3OH} \underset{H}{\overset{}{\bigcirc_N}} + CH_3O\underset{CH_3}{\overset{CH_3}{\diagdown}}\hspace{-2pt}Si-CH_3 \qquad (79)$$

8. Direct amination of aromatic hydrocarbons with trichloramine [132].

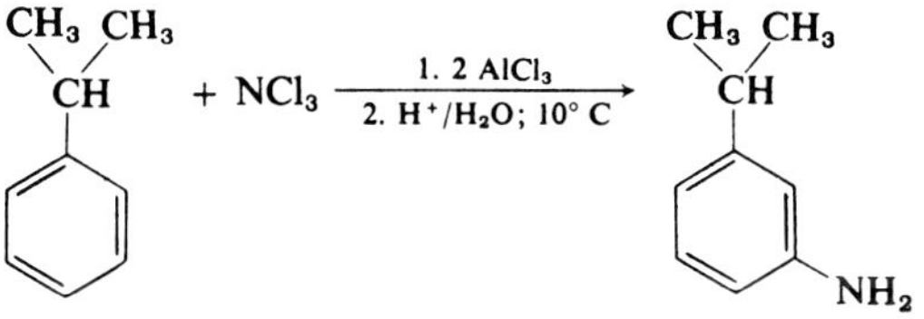

§ 5. Miscellaneous Preparations

[Reaction (80): isopropylbenzene + NCl₃, 1. 2 AlCl₃; 2. H⁺/H₂O; 10°C → 3-isopropylaniline] (80)

NOTE: Precautions must be exercised in working with haloamines

9. Addition of dialkyl-N-chloramines to olefins [133].

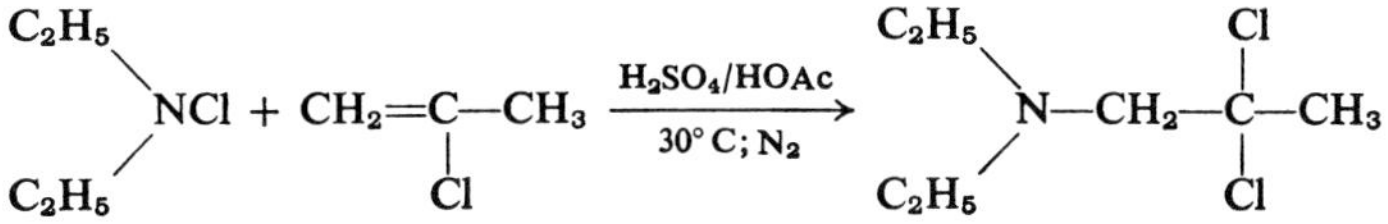

$$(C_2H_5)_2NCl + CH_2=C(Cl)-CH_3 \xrightarrow[30°C; N_2]{H_2SO_4/HOAc} (C_2H_5)_2N-CH_2-C(Cl)(Cl)-CH_3 \quad (81)$$

10. N-Vinylation with vinyl acetate [134].

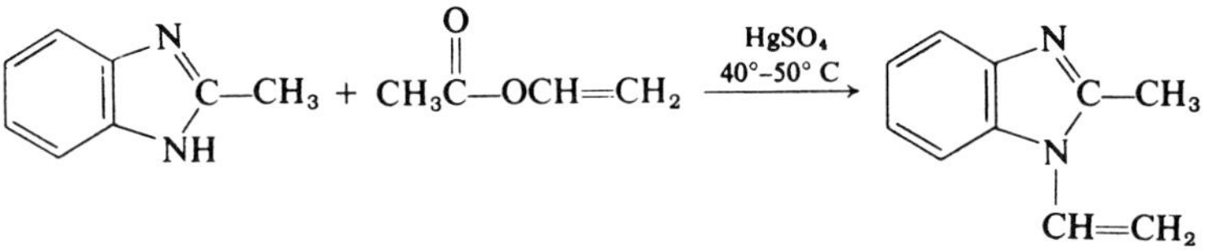

[Reaction (82): 2-methylbenzimidazole + CH₃C(O)OCH=CH₂, HgSO₄, 40°–50°C → 1-vinyl-2-methylbenzimidazole] (82)

11. Preparation of secondary alkylamines free of primary and tertiary amines (from calcium cyanamide) [135].

$$CaNCN + 2\ NaOH \longrightarrow Na_2NCN + Ca(OH)_2 \quad (83)$$

$$Na_2NCN + 2\ RX \longrightarrow R_2NCN + 2\ NaX \quad (84)$$

$$R_2NCN + H_2O \xrightarrow{H^+ \text{ or } OH^-} R_2NH + CO_2 + NH_3 \quad (85)$$

12. The Bucherer reaction [136]. The Bucherer reaction is a reversible replacement of the hydroxyl group of naphthol derivatives by amino groups in the presence of an aqueous sulfite or bisulfite. Under more vigorous conditions, N-alkyl and N,N-dialkylaminonaphthalene derivatives may also be prepared, e.g.,

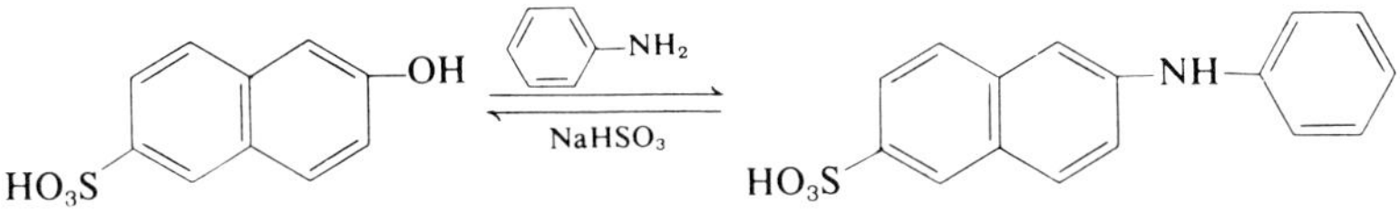

[Reaction (86): 6-hydroxy-2-naphthalenesulfonic acid + aniline, NaHSO₃ → 6-(phenylamino)-2-naphthalenesulfonic acid] (86)

In a nonaqueous system using fused zinc chloride and ammonium chloride, 2-aminonaphthalene and 2-naphthol were reacted to produce 1,2′-dinaphthylamine [137]:

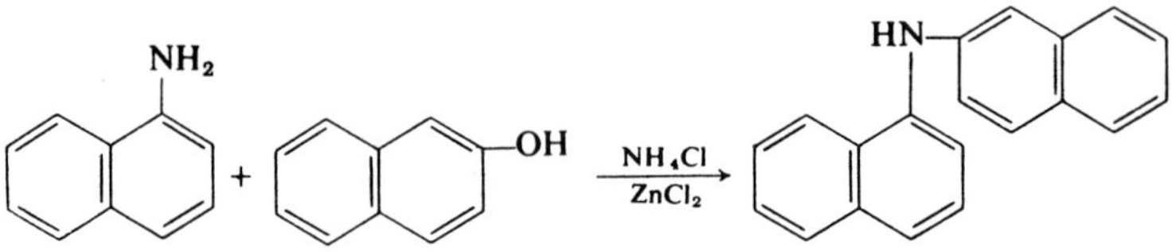

NOTE: 2-aminonaphthalene is a carcinogen. It is no longer available in the U.S. Equation (87) is cited for reference only.

13. Alkylation of ammonia and amines with thioethers [138, 139].

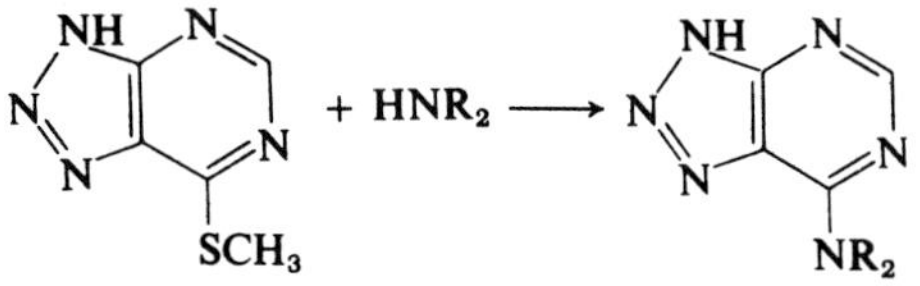

(88)

14. Aminoalkylation of phenols with 3,4-dihydro(2H)1,3-benzoxazines [140].

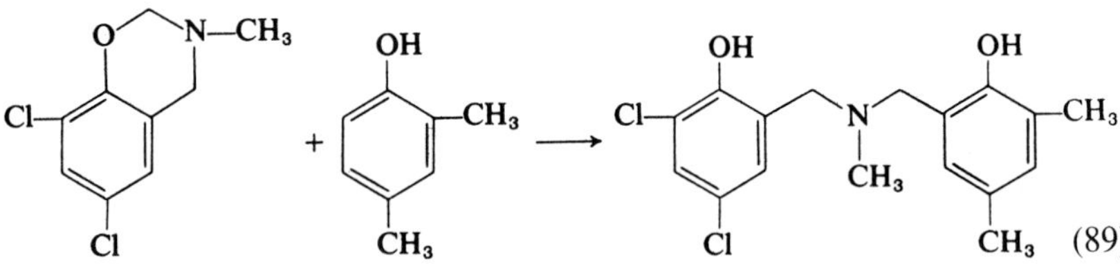

(89)

15. Exchange of amino groups of α-aminotetrahydropyrans [141].

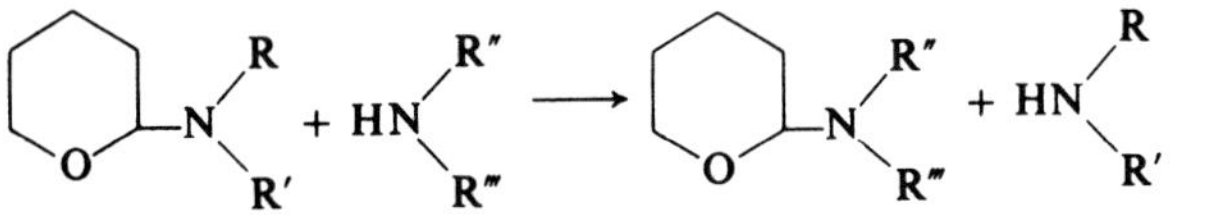

(90)

16. Preparation of tetrakis(dimethylamino)methane [142].

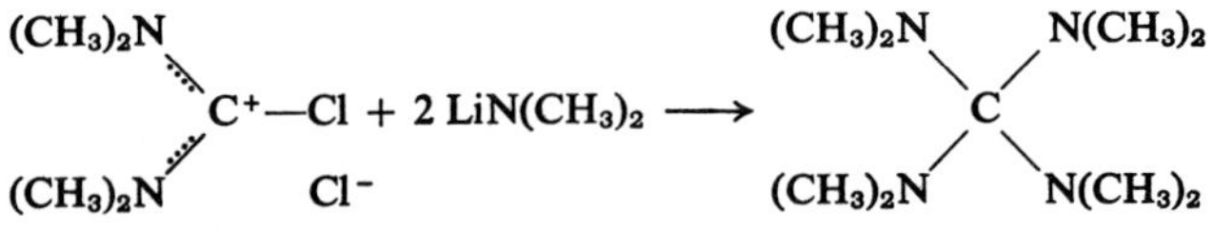

(91)

17. Preparation of β-alkylaminocrotonic esters [143].

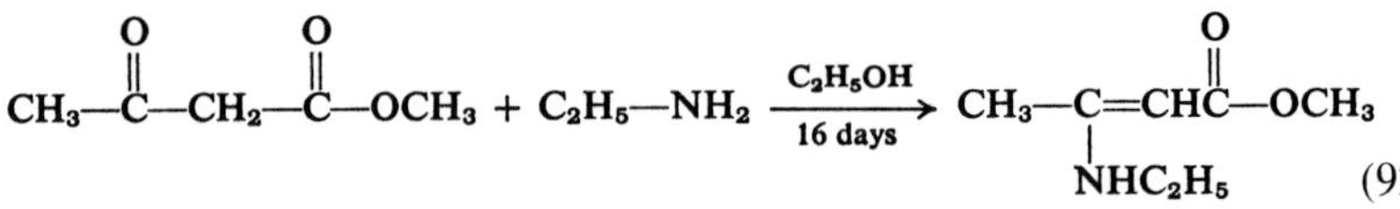

(92)

See also no. 29 below.

18. Reaction of dialkylmagnesium compounds with monochloramine [144].

§ 5. Miscellaneous Preparations

$$R_2Mg + NH_2Cl \longrightarrow RNHMgCl + RH \tag{93}$$

$$RNHMgCl + H_2O \longrightarrow RNH_2 + MgCl(OH) \tag{94}$$

NOTE: Precautions must be exercised when working with monochloramine.

19. Reaction of dimethylaminoethyl chloride with Grignard reagents [145].

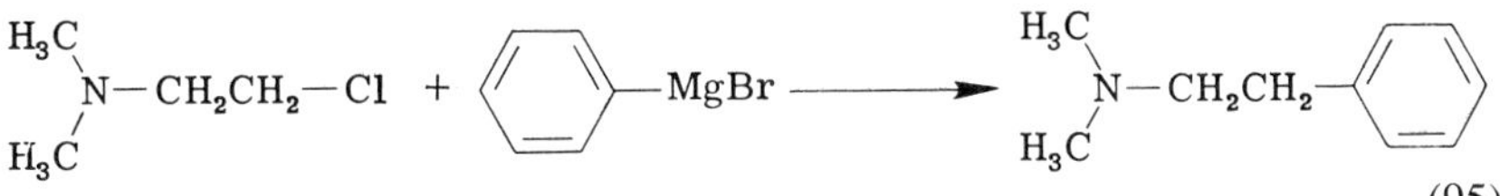

$$\underset{H_3C}{\overset{H_3C}{>}}N-CH_2CH_2-Cl + \underset{}{\bigcirc}-MgBr \longrightarrow \underset{H_3C}{\overset{H_3C}{>}}N-CH_2CH_2-\underset{}{\bigcirc} \tag{95}$$

20. Reaction of allylic Grignard reagents with unsaturated amines [146].

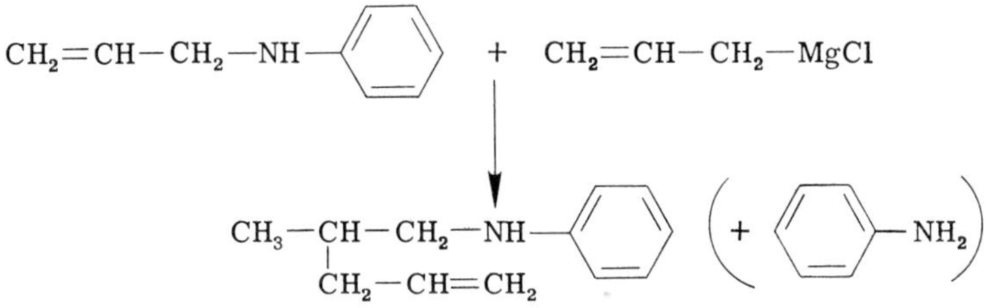

$$CH_2=CH-CH_2-NH-\bigcirc + CH_2=CH-CH_2-MgCl$$

$$\downarrow$$

$$\underset{\underset{CH_2-CH=CH_2}{|}}{CH_3-CH-CH_2-NH-\bigcirc} \quad \left(+ \bigcirc-NH_2 \right) \tag{96}$$

21. Reductive alkylation of ammonia and amines [147]. The reaction of ammonia, primary, and secondary amines with carbonyl compounds in the presence of a reducing agent, usually hydrogen and a hydrogenation catalyst, is an important method for the preparation of a variety of substituted amines. Since the preparations usually require high pressure equipment, details of the procedure are not given here. The reaction may be represented by the following set of equations:

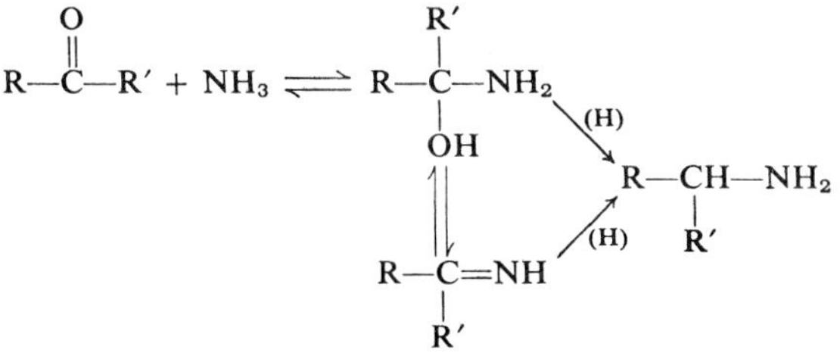

$$R-\overset{O}{\underset{}{\overset{\|}{C}}}-R' + NH_3 \rightleftharpoons R-\underset{\underset{OH}{|}}{\overset{\overset{R'}{|}}{C}}-NH_2 \xrightarrow{(H)} R-CH-NH_2 \atop R' \tag{97}$$

$$R-\overset{}{\underset{R'}{\overset{}{C}}}=NH \xrightarrow{(H)}$$

The scope and limitations of the reaction have been reviewed [147]. Alcohols may also be used and probably go through the carbonyl intermediate. An example using sodium borohydride as the reducing agent is [148].

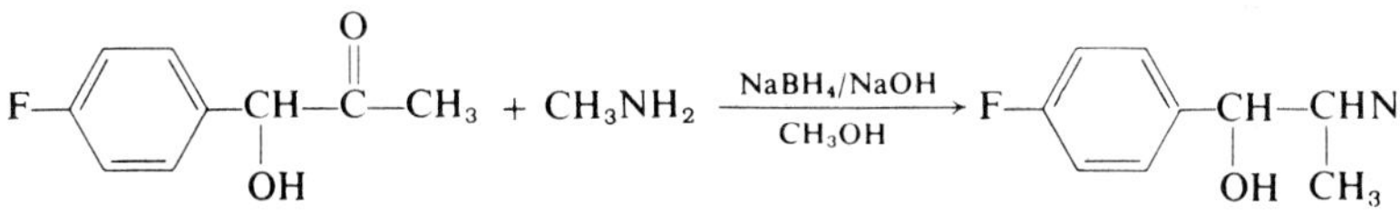

$$F-\bigcirc-\underset{\underset{OH}{|}}{CH}-\overset{O}{\overset{\|}{C}}-CH_3 + CH_3NH_2 \xrightarrow[CH_3OH]{NaBH_4/NaOH} F-\bigcirc-\underset{\underset{OH}{|}}{CH}-\underset{\underset{CH_3}{|}}{CH}NHCH_3 \tag{98}$$

22. Reductive alkylation of amines with esters [149].

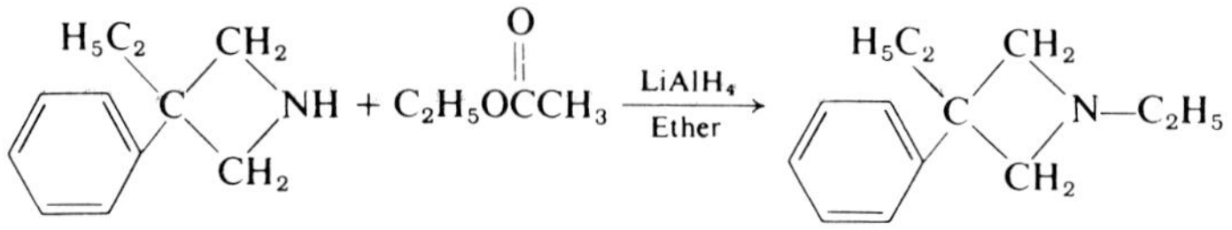

$$\text{3-Ethyl-3-phenylazetidine} + C_2H_5OCCH_3 \xrightarrow[\text{Ether}]{\text{LiAlH}_4} \text{(N-ethyl product)} \quad (99)$$

22. Reductive ammonation of ozonolysis products [150].

$$C_8H_{17}CH=CH_2 \xrightarrow[\substack{2.\ NH_3\ (\text{liq.})\\ \text{Raney nickel}\\ H_2\ \text{at 2900 psi}\\ 150°\ C}]{1.\ O_3/O_2\ \text{at}\ -15°\ C} C_8H_{17}CH_2NH_2 \quad (100)$$

23. Reductive amination of polynuclear aromatic hydrocarbons (under conditions of the Birch reduction) [151].

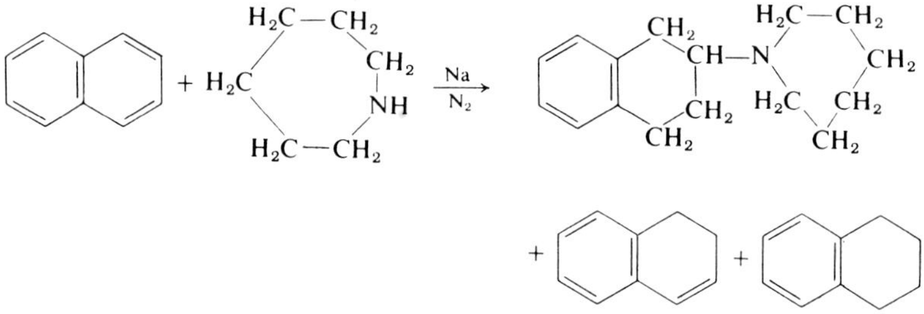

(101)

24. Reductive alkylation of amines with rhodium sulfide catalyst [152].

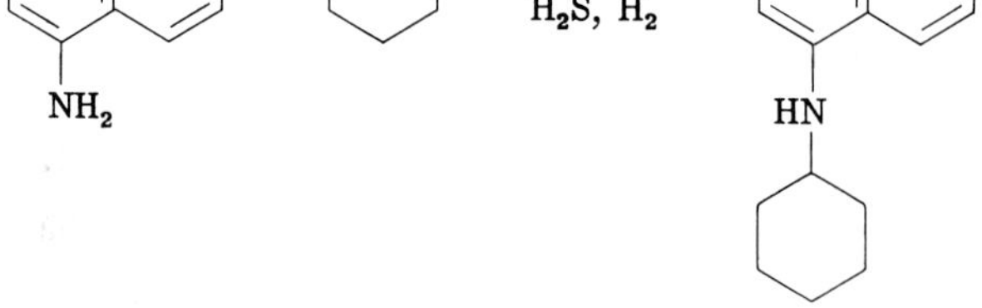

(1,5-diaminonaphthalene) + cyclohexanone $\xrightarrow[\text{H}_2\text{S, H}_2]{\text{rhodamine on C}}$ (N,N'-dicyclohexyl product) (102)

25. Rearrangement of alkylanilines [153].

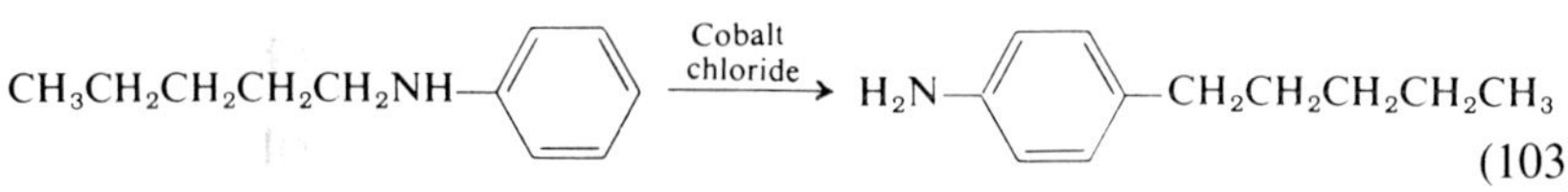

$$CH_3CH_2CH_2CH_2CH_2NH\text{—C}_6H_5 \xrightarrow{\text{Cobalt chloride}} H_2N\text{—C}_6H_4\text{—}CH_2CH_2CH_2CH_2CH_3 \quad (103)$$

§ 5. Miscellaneous Preparations   427

26. α-Substitution of *N,N*-dialkylanilines and triethylamine [154].

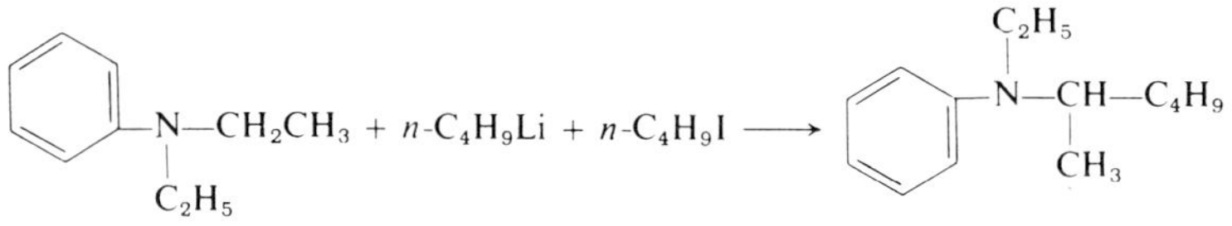

$$\text{Ph-N(C}_2\text{H}_5\text{)-CH}_2\text{CH}_3 + n\text{-C}_4\text{H}_9\text{Li} + n\text{-C}_4\text{H}_9\text{I} \longrightarrow \text{Ph-N(C}_2\text{H}_5\text{)-CH(CH}_3\text{)-C}_4\text{H}_9 \quad (104)$$

$$(\text{C}_2\text{H}_5)_2\text{NCH}_2\text{CH}_3 + n\text{-C}_4\text{H}_9\text{I} + n\text{-C}_4\text{H}_9\text{Li} \longrightarrow (\text{C}_2\text{H}_5)_2\text{N-CH(CH}_3\text{)-(CH}_2\text{)}_3\text{CH}_3 + n\text{-C}_8\text{H}_{18} \quad (105)$$

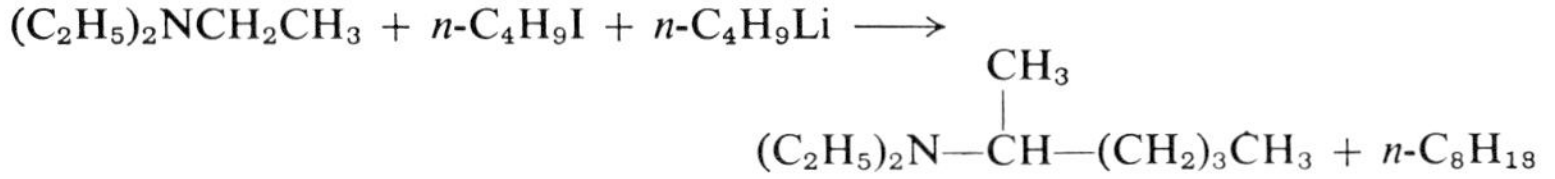

27. Hofmann–Löffler reaction [155].

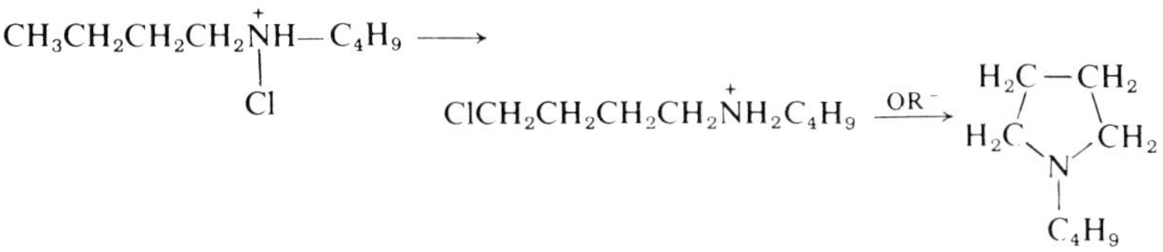

$$\text{CH}_3\text{CH}_2\text{CH}_2\text{CH}_2\overset{+}{\text{N}}\text{H(Cl)}-\text{C}_4\text{H}_9 \longrightarrow \text{ClCH}_2\text{CH}_2\text{CH}_2\text{CH}_2\overset{+}{\text{N}}\text{H}_2\text{C}_4\text{H}_9 \xrightarrow{\text{OR}^-} \text{pyrrolidine-N-C}_4\text{H}_9 \quad (106)$$

28. N-Hydroxyalkylation reaction (amino alcohol preparation) [156].

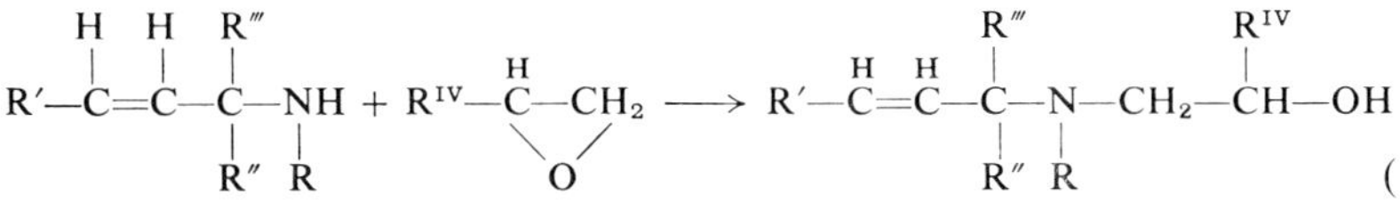

$$\text{R}'\text{-CH=CH-CR}''\text{R}'''\text{-NH} + \text{R}^{\text{IV}}\text{-CH(O)CH}_2 \longrightarrow \text{R}'\text{-CH=CH-CR}''\text{R}'''\text{-N(R}^{\text{IV}}\text{)-CH}_2\text{-CH(OH)} \quad (107)$$

$$\text{Ph-CH(O)CH}_2 + \text{RNH}_2 \longrightarrow \text{Ph-CH(OH)-CH}_2\text{NHR} \longrightarrow \text{Ph-CH-NR-CH}_2 \quad (108)$$

$$\text{Ph-CH(O)CH}_2 + \text{H}_2\text{N-CH}_2\text{CH}_2\text{-NH}_2 \longrightarrow$$
$$\text{H}_2\text{N-CH}_2\text{-CH}_2\text{-NH-CH}_2\text{CH(OH)-Ph} +$$
$$\text{Ph-CH(OH)-CH}_2\text{NH-CH}_2\text{CH}_2\text{-NH-CH}_2\text{-CH(OH)-Ph} \quad (109)$$

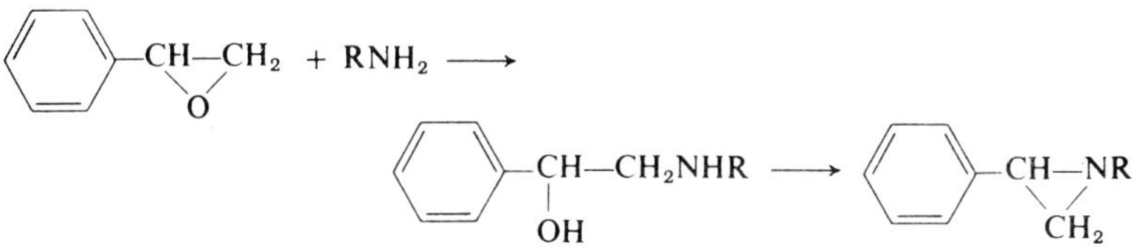

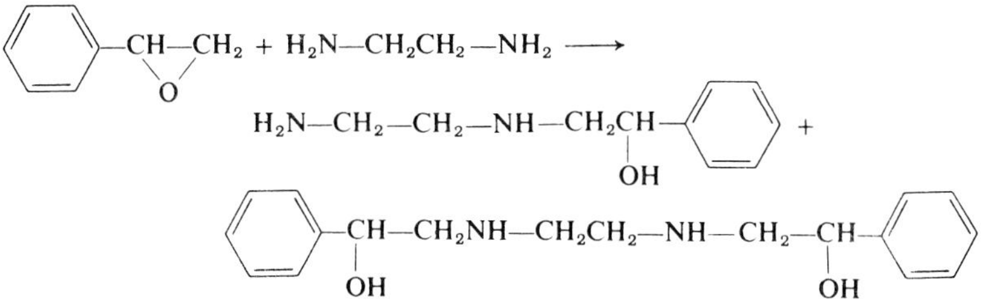

29. Enamine preparation [157] (see also no. 17 above).

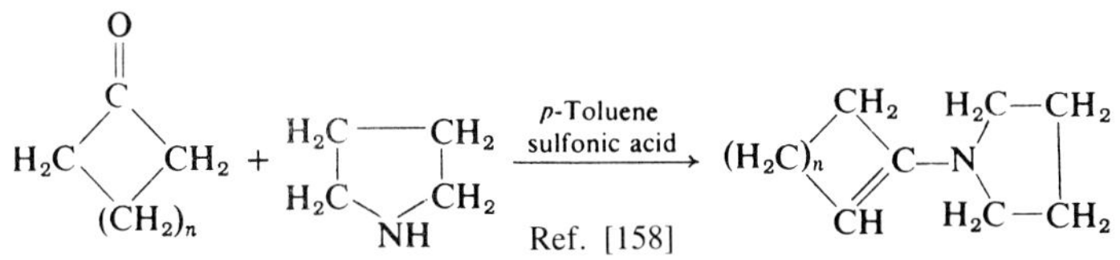

$$\text{cyclic ketone (CH}_2\text{)}_n + \text{pyrrolidine} \xrightarrow[\text{Ref. [158]}]{p\text{-Toluene sulfonic acid}} \text{enamine} \quad (110)$$

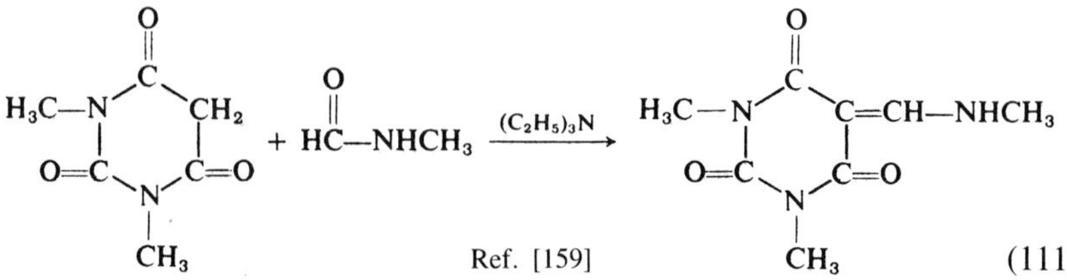

Some recent references to the synthesis of enamino derivatives are given in Ref. 160.

30. Ynamine preparation. Ref. 161 is a review of the synthesis and reactions of alkynylamines.

31. Tertiary amine preparation by electrolytic reduction of quaternary ammonium salts [162].

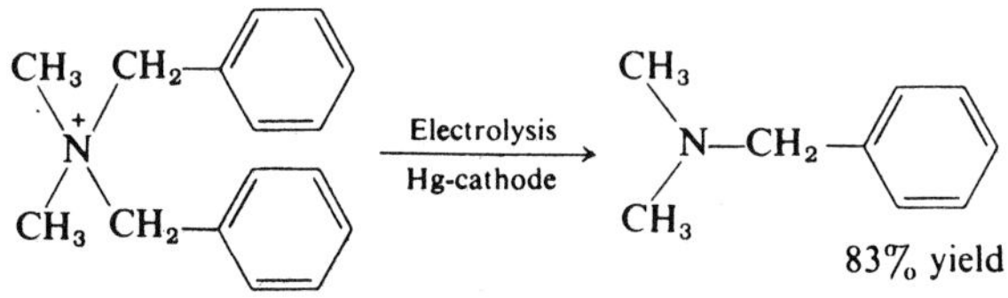

32. Preparation of hindered tertiary amines in the presence of alkyl lithium [163].

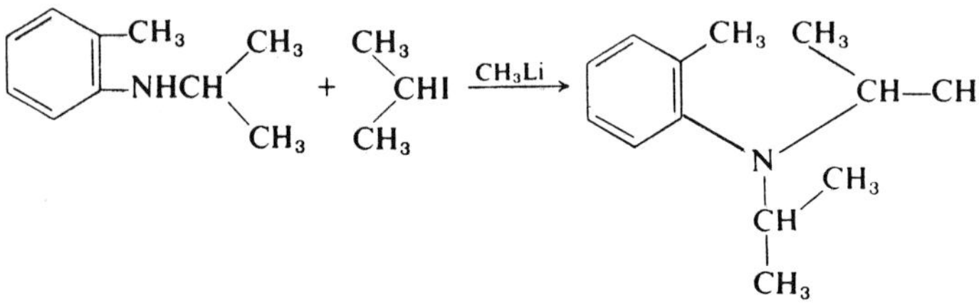

33. Neber rearrangement of amidoxime sulfonates [164].

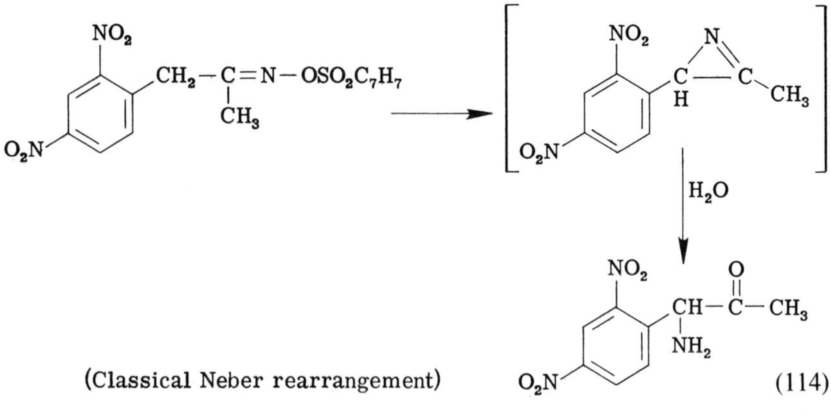

(Classical Neber rearrangement)

Under certain conditions, 2-amino-1-azirines may be isolated:

$$R-CH_2-\underset{\underset{NH_2}{|}}{\overset{\overset{NOSO_2-C_6H_4-OCH_3}{||}}{C}} \xrightarrow[CH_3OH]{NaOCH_3} \underset{R}{\overset{H}{>}}\!\!\!\underset{C-NH_2}{\overset{N}{<}} \quad (115)$$

REFERENCES

1. "Houben-Weyl, Methoden der organischen Chemie" (E. Mueller, ed.), Vol. XI, Part 1. Thieme, Stuttgart, 1957.
2. J. A. Montgomery and K. Hewson, *J. Med. Chem.* **9**, 105, 354 (1966).
3. J. A. Montgomery and K. Hewson, *J. Med. Chem.* **8**, 737 (1965).
4. A. B. Sen and A. K. Roy, *J. Indian Chem. Soc.* **37**, 647 (1960).
5. V. C. Sekera and C. S. Marvel, *J. Am. Chem. Soc.* **55**, 345 (1933).
6. D. D. Reynolds and W. O. Kenyon, *J. Am. Chem. Soc.* **72**, 1591 (1950).
7. A. Burger, S. E. Zimmerman, and E. J. Ariens, *J. Med. Chem.* **9**, 469 (1966).
8. H. Franke and R. Partch, *J. Med. Chem.* **9**, 643 (1966); R. Partch, private communication (1966).
9. K. Tamaki, S. Yada, and S. Kudo, *Yuki Gosei Kagaku Kyokaishi* **30**(2), 175 (1972); *Chem. Abstr.* **76**, 126065 (1972).
10. I. Ganea and R. Taranu, *Stud. Univ. Babes-Bolyai, Ser. Chem.* **16** (2), 89 (1971); *Chem. Abstr.* **76**, 112 868 (1972).
11. F. C. Whitmore and D. P. Langlois, *J. Am. Chem. Soc.* **54**, 3441 (1932).
12. N. Bortnick, L. S. Luskin, M. D. Hurwitz, W. E. Craig, L. J. Exner, and J. Mirza, *J. Am. Chem. Soc.* **78**, 4039 (1956).
13. G. F. Hennion and C. V. DiGiovanna, *J. Org. Chem.* **30**, 2645 (1965).
14. I. E. Kopka, Z. A. Fataftah, and M. W. Rathke, *J. Org. Chem.* **45**, 4616 (1980).
15. A. C. Cope and P. H. Towle, *J. Am. Chem. Soc.* **71**, 3423 (1949).
16. G. M. Coppinger, *J. Am. Chem. Soc.* **76**, 1372 (1954).
17. A. C. Cope, E. Ciganek, L. J. Fleckenstein, and M. A. P. Meisinger, *J. Am. Chem. Soc.* **82**, 4651 (1960).
18. L. L. Melhado, *J. Org. Chem.* **46**, 1920 (1981).
19. A. D. Lourie and A. R. Day, *J. Med. Chem.* **9**, 311 (1966).
20. M. Kerfanto, *C. R. Hebd. Seances Acad. Sci.* **254**, 493 (1962).
21. S. R. Aspinall, *J. Am. Chem. Soc.* **63**, 852 (1941).
22. R. E. O'Gee and H. M. Woodburn, *J. Am. Chem. Soc.* **73**, 1370 (1951).
23. K. T. Potts and J. E. Saxton, *Org. Synth.* **40**, 68 (1960).
24. Review by M. T. Leffler, *Org. React.* **1**, 91 (1942); F. W. Bergstrom, *J. Org. Chem.* **3**, 424 (1938).
25. H. Beschke, *Aldrichimica Acta* **14**,(1), 14 (1981).
26. H. J. W. van der Haak, H. C. van der Plas, and B. van Veldhuizen, *J. Org. Chem.* **46**, 2134 (1981).
27. H. C. Brown, W. R. Heydkamp, E. Breuer, and W. S. Murphy, *J. Am. Chem. Soc.* **86**, 3565 (1964).
28. M. W. Rathke, N. Inoue, K. R. Varma, and H. C. Brown, *J. Am. Chem. Soc.* **88**, 2870 (1966); related reviews are: G. Zweifel and H. C. Brown, *Org. React.* **13**, 1 (1963); H. C. Brown *Tetrahedron* **12**, 117 (1961); "Hydroboration." Benjamin, New York, 1962. L. Verbit and P. J. Heffron, *J. Org. Chem.* **32**, 3199 (1967).
29. M. Delépine, *C. R. Hebd. Seances Acad. Sci.* **120**, 501 (1895); **124**, 292 (1897); *Bull. Soc. Chim. Fr.* **17**, No. 3, 290 (1897); **31**, No. 4, 106 (1922).

30. A. Galat and G. Elion, *J. Am. Chem. Soc.* **61,** 3585 (1939).
31. J. Graymore, *J. Chem. Soc.* p. 1116 (1947).
32. N. L. Wendler, *J. Am. Chem. Soc.* **71,** 375 (1949).
33. H. R. Ing and R. H. F. Manske, *J. Chem. Soc.* p. 2348 (1926).
34. J. C. Sheehan and W. A. Bolhofer, *J. Am. Chem. Soc.* **72,** 2785 (1950).
35. D. Landini and F. Rolla, *Synthesis* p. 389 (1976).
36. E. Ciuffarin, M. Isola, and P. Leoni, *J. Org. Chem.* **46,** 3064 (1981).
37. A. I. Meyers, J. P. Lawson, and D. R. Carver, *J. Org. Chem.* **46,** 3119 (1981).
38. A. F. Ferris, O. L. Salerni, and B. A. Schultz, *J. Med. Chem.* **9,** 391 (1966).
39. E. F. Cox, M. C. Caserio, M. S. Silver, and J. D. Roberts, *J. Am. Chem. Soc.* **83,** 2719 (1961).
40. J. Hebký, O. Řádek, and J. Kejha, *Collect. Czech. Chem. Commun.* **24,** 3988 (1959).
41. H. S. Freeman and L. D. Freedman, *J. Org. Chem.* **46,** 5373 (1981).
42. R. W. Holley and A. D. Holley, *J. Am. Chem. Soc.* **71,** 2124 (1949).
43. A. N. Kurtz, W. E. Billups, R. B. Greenlee, H. F. Hamil, and W. T. Pace, *J. Org. Chem.* **30,** 3141 (1965).
44. J. J. Bozell and L. S. Hegedus, *J. Org. Chem.* **46,** 2561 (1981).
45. F. Jachimowicz and J. W. Raksis, *J. Org. Chem.* **47,** 445 (1982).
46. B. E. Evans and P. S. Anderson, *J. Org. Chem.* **46,** 140 (1981).
47. H. Lehmkuhl and D. Reinehr, *J. Organomet. Chem.* **55,** 215 (1973).
48. See review by F. F. Blicke, *Org. Reac.* **1,** 303 (1942); B. Reichert, "Die Mannich Reaktion." Springer-Verlag, Berlin and New York, 1960.
49. T. F. Cummings and J. R. Shelton, *J. Org. Chem.* **25,** 419 (1960).
50. E. R. H. Jones, I. Marszak, and H. Bader, *J. Chem. Soc.* p. 1578 (1947).
51. K. Moedritzer and R. R. Irani, *J. Org. Chem.* **31,** 1603 (1966).
52. T. Okuda, *Bull. Chem. Soc. Jpn.* **32,** 1165 (1959).
53. J. C. Craig, M. Moyle, and L. F. Johnson *J. Org. Chem.* **29,** 410 (1964).
54. R. G. Taborsky, P. Delvigs, and I. H. Page, *J. Med. Chem.* **8,** 855 (1965).
55. A. K. Bose, J. F. Kistner, and L. Farber, *J. Org. Chem.* **27,** 2925 (1962).
56. M. Imazawa and F. Eckstein, *J. Org. Chem.* **44,** 2039 (1978).
57. G. Losse and J. Müller, *J. Prakt. Chem.* **[4] 12,** 285 (1961).
58. H. E. Smith, B. Y. Padilla, J. R. Neergaard, and F. M. Chen, *J. Org. Chem.* **44,** 1690 (1979).
59. Y. F. Shealy, *J. Org. Chem.* **26,** 4433 (1961).
60. G. N. Walker and M. A. Moore, *J. Org. Chem.* **26,** 432 (1961); J. H. Billman and J. W. McDowell, *ibid.* p. 1437; R. G. Hiskey and R. C. Northrop, *J. Am. Chem. Soc.* **83,** 4798 (1961); J. C. Sheehan and R. E. Chandler, *ibid.* p. 4795; L. G. Humber, *J. Med. Chem.* **9,** 441 (1966).
61. R. Grewe and W. J. Bonin, *Chem. Ber.* **94,** 234 (1961).
62. C. R. Hauser, H. M. Taylor, and T. G. Ledford, *J. Am. Chem. Soc.* **82,** 1786 (1960).
63. C. F. Lane, *Synthesis* p. 135 (1975).
64. F. Troxler, F. Seemann, and A. Hofmann, *Helv. Chim. Acta* **42,** 2073 (1959); G. Levin, G. Kalmus, and F. Bergman, *J. Org. Chem.* **25,** 1752 (1960); J. G. Murphy, *ibid.* **26,** 3104 (1961); R. A. Odum and M. Brenner, *J. Am. Chem. Soc.* **88,** 2074 (1966).
65. M. Freifelder, M. D. Smart, and G. R. Stone, *J. Org. Chem.* **27,** 2209 (1962); L. M. Rice, E. C. Dobbs, and C. H. Grogan, *J. Med. Chem.* **8,** 825 (1965).
66. K. D. Berlin, N. K. Roy, R. T. Claunch, and D. Bude, *J. Am. Chem. Soc.* **90,** 4494 (1968).
67. H. Fauer and D. M. Braunstein, *J. Org. Chem.* **34,** 1817 (1969).
68. B. J. R. Nicolaus, L. Mariani, G. Gallo, and E. Testa, *J. Org. Chem.* **26,** 2253 (1961).
69. R. C. Bugle and R. A. Osteryoung, *J. Org. Chem.* **44,** 1719 (1979).
70. R. J. Sundberg, C. P. Walters, and J. D. Bloom, *J. Org. Chem.* **46,** 3730 (1981).

71. R. O. Hutchins and M. Markowitz, *J. Org. Chem.* **46,** 3571 (1981).
72. D. F. Heath and A. R. Mattocks, *J. Chem. Soc.* p. 4226 (1961).
73. R. F. Nystrom and W. G. Brown, *J. Am. Chem. Soc.* **69,** 2548 (1947).
74. B. D. Halpern and W. Karo, U.S. Patent 3,032,587 (1962).
75. W. Karo, *Chem. Eng. News* **33,** 1368 (1955).
76. W. B. Wright, Jr., *J. Org. Chem.* **25,** 1033 (1960).
77. M. Winn and H. E. Zaugg, *J. Org. Chem.* **33,** 3779 (1968).
78. R. Fuks and H. G. Viehe, *Chem. Ber.* **103,** 564 (1970).
79. M. Freifelder and R. B. Hasbroack, *J. Am. Chem. Soc.* **82,** 696 (1960); F. E. Gould, G. S. Johnson, and A. F. Ferris, *J. Org. Chem.* **25,** 1659 (1960).
80. E. A. Halevi, M. Nussim, and A. Ron, *J. Chem. Soc.* p. 866 (1963).
81. P. M. Carabateas and L. S. Harris, *J. Med. Chem.* **9,** 6 (1966).
82. D. L. Ladd and J. Weinstock. *J. Org. Chem.* **46,** 203 (1981).
83. H. Greenfield, *Ind. Eng. Chem., Prod. Res. Dev.* **15** (2), 156 (1976).
84. W. F. Hart, M. E. McGreal, and P. E. Thurston, *J. Org. Chem.* **27,** 338 (1962).
85. H. Oelschläger and O. Schreiber, *Justus Liebigs Ann. Chem.* **641,** 81 (1961).
86. T. Neilson, H. C. S. Wood, and A. G. Wylie, *J. Chem. Soc.* p. 371 (1962).
87. G. N. Walker and M. A. Klett, *J. Med. Chem.* **9,** 624 (1966).
88. H. Koopman, *Recl. Trav. Chim. Pays-Bas* **80,** 1075 (1961).
89. S. E. Hazlet and C. A. Dornfeld, *J. Am. Chem. Soc.* **66,** 1781 (1944).
90. C. F. H. Allen, private communication.
91. M. Senkus, *Ind. Eng. Chem.* **40,** 506 (1948).
92. D. Balcom and A. Furst, *J. Am. Chem. Soc.* **75,** 4334 (1953); A. Furst and R. E. Moore, *ibid.* **79,** 5492 (1957).
93. H. H. Szmant and R. Infante, *J. Org. Chem.* **26,** 4173 (1961).
94. R. Adams and J. R. Johnson, "Laboratory Experiments in Organic Chemistry," 4th ed., p. 319. Macmillan, New York, 1949.
95. L. F. Fieser, "Experiments in Organic Chemistry," 3rd ed. p. 234. Heath, Boston, Massachusetts, 1955.
96. A. Skogseid, U.S. Patent 2,592,349 (1952).
97. R. M. McKinney, J. T. Spillane, and G. W. Pearce, *J. Org. Chem.* **27,** 3986 (1962).
98. L. Legradi, *Chem. Ind. (London)* p. 1496 (1965).
99. F. A. Ramirez and A. Burger, *J. Am. Chem. Soc.* **72,** 2781 (1950).
100. F. A. L. Anet and J. M. Muchowski, *Can. J. Chem.* **38,** 2526 (1960).
101. J. I. G. Cadogan and M. T. Todd, *J. Chem. Soc. C* p. 2808, and references therein (1969).
102. A. Onopchenko, E. T. Subourin, and C. M. Selwitz, *J. Org. Chem.* **44,** 1233 (1979).
103. M. L. Moore, *Org. React.* **5,** 301 (1959); see also Meiners *et al.* [106] for references to reviews on the mechanism of the Leuckart reaction.
104. J. F. Bunnett and J. L. Marks, *J. Am. Chem. Soc.* **71,** 1587 (1949); J. F. Bunnett, J. L. Marks, and H. Moe, *ibid.* **75,** 985 (1953).
105. L. D. Quin and F. A. Shelburne, *J. Org. Chem.* **30,** 3137 (1965).
106. A. F. Meiners, C. Bolze, H. L. Scherer, and F. V. Morriss, *J. Org. Chem.* **23,** 1122 (1958).
107. R. Adams and J. R. Johnson, "Laboratory Experiments in Organic Chemistry," 4th ed., p. 359. Macmillan, New York, 1949.
108. H. R. Snyder, C. Weaver, and C. D. Marshall, *J. Am. Chem. Soc.* **71,** 289 (1949).
109. E. S. Wallis and J. F. Lane, *Org. React.* **3,** 267 (1946).
110. E. Magnien and R. Baltzly, *J. Org. Chem.* **23,** 2029 (1958).
111. A. C. Cope, T. T. Foster, and P. H. Towle, *J. Am. Chem. Soc.* **71,** 3929 (1949).
112. E. Jefferys, *Ber. Dtsch. Chem. Ges.* **30,** 898 (1897); *Am. Chem. J.* **22,** 14 (1899).
113. C. S. Rodestvedt, Jr., *J. Org. Chem.* **42,** 3168 (1977).

114. T. E. Stevens, *J. Org. Chem.* **31**, 2025 (1966).
115. A. S. Radhakrishna, M. E. Parham, R. M. Riggs, and G. M. Loudon, *J. Org. Chem.* **44**, 1746 (1979).
116. H. Wolff, *Org. React.* **3**, 307 (1946).
117. F. R. Benson, L. W. Hartzel, and W. L. Savell, *J. Am. Chem. Soc.* **71**, 1111 (1949).
118. P. A. S. Smith, *Org. React.* **3**, 337 (1946).
119. J. Finkelstein, E. Chiang, F. M. Vane, and J. Lee, *J. Med. Chem.* **9**, 319 (1966).
120. J. Casanova, Jr., N. D. Werner, and R. E. Schuster, *J. Org. Chem.* **31**, 3473 (1966).
121. J. M. Patterson, M. W. Barnes, and R. L. Johnson, *J. Org. Chem.* **31**, 3103 (1966).
122. J. Finkelstein, E. Chiang, and J. Lee, *J. Med. Chem.* **9**, 440 (1966).
123. H. A. Staab and F. Vögtle, *Chem. Ber.* **98**, 1691 (1965).
124. G. B. Bachman and J. E. Goldmacher, *J Org. Chem.* **29**, 2576 (1964).
125. J. Goerdeler in "Houben-Weyl, Methoden der organischen Chemie" (E. Mueller, ed.), Vol. XI, Part 2, pp. 591ff. Thieme, Stuttgart, 1958; W. W. Evans and J. G. Umberger, Belgian Patent 618,228 (1962); *Chem. Abstr.* **58**, 13341 (1963); C. Kimura and K. Murai, *Kogyo Kagaku Zasshi* **68**, 504 (1965); *Chem. Abstr.* **65**, 3716 (1966); C. H. Huang, S. Shimizu and H. Adachi, German Offen. 2,008,643 (1970); *Chem. Abstr.* **131**, 823 (1970); A. T. Bottini, *Sel. Org. Transform.* **1**, 89 (1970); *Chem. Abstr.* **74**, 63601 (1971); H. Z. Sommer, H. I. Lipp, and L. L. Jackson, *J. Org. Chem.* **36**, 824 (1971); T. Ohgaki and T. Yamamoto, *Kogyo Kagaku Zasshi* **74**, 416 (1971); K. J. Valan, German Offen. 2,103,899 (1971); *Chem. Abstr.* **75**, 132,919 (1971); E. S. Barabas and M. M. Fein, German Offen. 2,103,898 (1971); *Chem. Abstr.* **75**, 152537 (1971); A. L. Barney, U.S. Patent 2,677,699 (1954), patent reissued: RE 24,164 (1956); M. Hayek, U.S. Patent 2,723,256 (1955); H. E. Winberg, U.S. Patent 2,744,130 (1956); R. S. Shelton, M. G. Van Caupen, C. H. Tilford, H. C. Lang, L. Nisonger, F. J. Bandelin, and H. L. Ruben-Koenig, *J. Am. Chem. Soc.* **68**, 753 (1946).
126. H. Yamamoto and K. Maruoka, *J. Org. Chem.* **45**, 2739 (1980).
127a. A. E. Martin, T. M. Ford, and J. E. Bulkowski, *J. Org. Chem.* **47**, 412 (1982).
127b. A. E. Martin and J. E. Bulkowski, *J. Org. Chem.* **47**, 415 (1982).
128. Bui-The-Khai, C. Concilio, and G. Porzi, *J. Org. Chem.* **46**, 1759 (1981).
129. G. Chatelus, *C. R. Hebd. Seances Acad. Sci.* **248**, 1588 (1961); **254**, 136 (1962).
130. G. G. Arzoumanidis and F. C. Rauch, *J. Org. Chem.* **46**, 3930 (1981).
131. N. C. Cook and J. E. Lyons, *J. Am. Chem. Soc.* **87**, 3283 (1965).
132. P. Kovacic, J. A. Levisky, and C. T. Goralski, *J. Am. Chem. Soc.* **88**, 100 (1966); P. Kovacic and J. T. Gormish, *ibid.* p. 3819; V. L. Heasley, P. Kovacic, and R. M. Lange, *J. Org. Chem.* **31**, 3050 (1966).
133. R. S. Neale, *Tetrahedron Lett.* p. 483 (1966); R. S. Neale and M. R. Walsh, *J. Am. Chem. Soc.* **87**, 1255 (1965); R. S. Neale, M. R. Walsh, and N. L. Marcus, *J. Org. Chem.* **30**, 3683 (1965).
134. H. Hoff, U. Wyss, and H. Lüssi, *Helv. Chim. Acta* **43**, 135 (1960).
135. W. Traube and A. Engelhardt, *Chem. Ber.* **44**, 3149 (1911); E. B. Vliet, *J. Am. Chem. Soc.* **46**, 1307 (1924).
136. For review, see N. L. Drake, *Org. React.* **1**, 105 (1942); W. A. Cowdrey, *J. Chem. Soc.* p. 1046 (1946); H. Seeboth, *Angew. Chem., Int. Ed. Engl.* **6**, 307 (1967).
137. E. Lieber and S. Somasekhara, *J. Org. Chem.* **24**, 1775 (1959).
138. R. Weiss, R. K. Robins, and C. W. Noell, *J. Org. Chem.* **25**, 765 (1960).
139. H. Bredereck, O. Smerz, and R. Gompper, *Chem. Ber.* **94**, 1883 (1961).
140. W. J. Burke, J. L. Bishop, E. C. Mortensen Glennie, and W. N. Bauer, Jr., *J. Org. Chem.* **30**, 3423 (1965).
141. C. Glacet and D. Veron, *Bull. Soc. Chim. Fr.* No. 6, p. 1789 (1965).
142. H. Weingarten and W. A. White, *J. Am. Chem. Soc.* **88**, 2885 (1966).

143. P. C. Anderson and B. Staskun, *J. Org. Chem.* **30,** 3033 (1965).
144. G. H. Coleman and R. F. Blomquist, *J. Am. Chem. Soc.* **63,** 1692 (1941).
145. P. M. G. Bavin, C. R. Ganellin, J. M. Loynes, P. D. Miles, and H. F. Ridley, *J. Med. Chem.* **9,** 790 (1966).
146. H. G. Rickey, Jr., L. M. Moses, M. S. Domalski, W. F. Erickson, and A. S. Heyn, *J. Org. Chem.* **46,** 3773 (1981).
147. W. S. Emerson, *Org. React.* **4,** 174 (1948).
148. J. Weichet, J. Hodrová, and L. Bláha, *Collect. Czech. Chem. Commun.* **26,** 2040 (1961).
149. E. Testa; *et al., Justus Liebigs Ann. Chem.* **633,** 56 (1960); A. Segre and R. Viterbo, *Experientia* **14,** 54 (1958).
150. K. A. Pollart and R. E. Miller, *J. Org. Chem.* **27,** 2392 (1962).
151. R. C. Bansal, E. J. Eisenbraun, and P. W. Flanagan, *J. Am. Chem. Soc.* **88,** 1837 (1966).
152. H. Greenfield and F. S. Dovell, *J. Org. Chem.* **31,** 3053 (1966).
153. W. J. Hickinbottom, *J. Chem. Soc.* p. 1119 (1937).
154. A. R. Lepley and A. G. Giumanini, *J. Org. Chem.* **31,** 2055 (1966); A. R. Lepley and W. A. Khan, *ibid.* p. 2061; see also A. R. Lepley, W. A. Khan, A. B. Giumanini, and A. G. Giumanini, *ibid.* pp. 2047, 2051.
155. For review, see M. E. Wolff, *Chem. Rev.* **63,** 55 (1963); R. S. Neale, M. R. Walsh, and N. L. Marcus, *J. Org. Chem.* **30,** 3683 (1964).
156. R. D. Dillard and N. R. Easton, *J. Org. Chem.* **31,** 122 (1966); J. N. Wells, A. V. Shirodkar, and A. M. Knevel, *J. Med. Chem.* **9,** 195 (1966); W. R. Roderick, H. J. Platte, and C. B. Pollard, *ibid.* p. 181.
157. For review, see J. Szmuszkovicz, *Adv. Org. Chem.* **4,** 1 (1963).
158. M. E. Kuehne, *J. Am. Chem. Soc.* **81,** 5400 (1959).
159. H. Bredereck, F. Gompper, F. Ellenberger, K. H. Popp, and S. Simchen, *Chem. Ber.* **94,** 1241 (1961).
160. Y. L. Chen, P. S. Mariano, G. M. Little, D. O'Brien, and P. L. Huesmann, *J. Org. Chem.* **46,** 4643 (1981); R. W. Thompson and J. Swistock, *ibid.* p. 4907; R. F. Parcell and J. P. Sanchez, *ibid.* p. 5055; C. Paradisi, M. Prato, U. Quintily, and G. Scorrano *ibid.* p. 5156.
161. H. G. Viehe, *Angew. Chem., Int. Ed. Engl.* **6** (9), 767 (1967).
162. L. Horner and A. Mentrup, *Justus Liebigs Ann. Chem.* **646,** 49, 65 (1961).
163. W. G. Young, F. F. Caserio, Jr., and D. D. Brandon, Jr., *J. Am. Chem. Soc.* **82,** 6163 (1960).
164. J. A. Hyatt, *J. Org. Chem.* **46,** 3953 (1981).

CHAPTER 14 / HYDRAZINE DERIVATIVES, HYDRAZONES, AND HYDRAZIDES

1.	Introduction	434
2.	Hydrazines	438
	A. Alkylation of Hydrazine Derivatives	439
2-1.	*Preparation of Ethylhydrazine*	440
2-2.	*Preparation of 2,4-Dinitrophenylhydrazine*	441
2-3.	*Preparation of 4,4'-Dihydrazinooctafluorobiphenyl*	441
2-4.	*Preparation of 1-Methyl-2-isopropylhydrazine*	444
	B. Syntheses Involving Formation of Nitrogen–Nitrogen Bonds	445
	a. Modified Raschig Synthesis	445
2-5.	*Preparation of Cyclohexylhydrazine Hydrogen Sulfate*	445
2-6.	*Preparation of 2-Pentylhydrazine*	446
	C. Reductive Methods	448
	a. Reduction of N-Nitrosoamines	448
2-7.	*Preparation of 1,1-Diethylhydrazine*	448
2-8.	*Preparation of* N-tert-*Butyl*-N-*phenylhydrazine*	449
2-9.	*Preparation of 1,1-Diisobutylhydrazine*	450
	b. Reduction of Hydrazones	451
2-10.	*Preparation of 1-Ethyl-2-isopropylhydrazine*	451
	c. Reduction of Diazonium Salts	452
	d. Reduction of Azo Compounds	453
2-11.	*Preparation of 1,2-Diphenylhydrazine by Diazine Reduction*	453
	e. Bimolecular Reduction of Nitro Compounds	454
2-12.	*Preparation of 1,2-Di(2-bromophenyl)hydrazine*	454
	f. Oxidation Processes	454
2-13.	*Preparation of 1,1,2,2-Tetra(4-fluorophyenyl)hydrazine*	455
3.	Hydrazones	455
3-1.	*Preparation of Benzaldehyde* p-*Nitrophenylhydrazone*	455
4.	Hydrazides	457
4-1.	*Preparation of Terephthalic Dihydrazide*	457
4-2.	*Preparation of Polyacrylic Hydrazide(I)*	458
4-3.	*Preparation of Polyacrylic Hydrazide(II)*	459
5.	Miscellaneous Preparations	460
	References	463

1. INTRODUCTION

The syntheses of the following six types of hydrazine derivatives are described in this chapter:

§ 1. Introduction

(1) Substituted hydrazines with a free amino group, e.g., the simple monoaliphatic and monoaromatic hydrazines and the 1,1-disubstituted hydrazines, sometimes referred to as N,N-disubstituted hydrazines or "unsymmetrically disubstituted hydrazines."

(2) 1,2-Disubstituted hydrazines, frequently referred to as "hydrazo compounds" or "symmetrically disubstituted hydrazines."

(3) Trisubstituted hydrazines.

(4) Tetrasubstituted hydrazines.

(5) Hydrazones.

(6) Hydrazides.

Hydrazine and its derivatives are well known as reagents used for the identification of carbonyl compounds. Of these, phenylhydrazine, 2,4-dinitrophenyl hydrazine, and Girard's reagent $[(CH_3)_3N^+CH_2CONHNH_2]Cl^-$ are of particular importance. Many heterocyclic compounds with two adjacent nitrogens, such as pyrazoles and pyrazolines, may be considered hydrazine derivatives, and methods of preparation of hydrazine derivatives may be applicable to their synthesis. The detailed treatment of synthetic methods for such heterocyclic compounds is beyond the scope of this work, although a few examples of the synthesis of some heterocyclic compounds will be indicated.

References 1–12 are a selection of reviews on the chemistry of hydrazine derivatives.

An overview of the methods of preparing various classes of substituted hydrazines and hydrazine derivatives is given in Table 1. In this connection it must be pointed out that monomethyl- and monoethylhydrazine are usually prepared by special methods such as ones involving quaternization of benzalazines [13,14].

Hydrazine itself may be subjected to a variety of facile substitution reactions because of its great nucleophilic character. It may react with alkyl halides and activated aryl halides to yield substituted hydrazines. With acyl halides, esters, and amides it may form hydrazides. With carbonyl compounds, hydrazones form. Since hydrazine is bifunctional in nature, both amino groups may undergo reaction. In the case of the reaction of hydrazine with two moles of a carbonyl compound, the products are termed "azines."

Many of the characteristic reactions of hydrazine may also be carried out with substituted hydrazines to afford a wide variety of products. In the case of reactions of substituted hydrazine derivatives, azine formation is ordinarily not possible.

In the case of the reaction of substituted hydrazines, such as phenylhydrazine with glucose or certain other carbohydrates, one molecule of phenylhydrazine reacts with the terminal aldehyde group to form a phenylhydrazone. A second molecule of phenylhydrazine oxidizes the penultimate carbinol grouping to a carbonyl and a third molecule of phenylhydrazine converts this second carbonyl

TABLE I
Synthesis of Substituted Hydrazines and Hydrazine Derivatives

Product type	Reaction type	Reactant	Reagent	Sample procedures	References
Monosubstituted hydrazine	Condensation	Hydrazine (excess)	Dialkyl sulfate	2–1	5,15
	Condensation	Hydrazine (excess)	Higher alkyl halide		15–17
	Condensation	Hydrazine	Activated aryl halide	2–2,2–3	18–20
	Condensation	Primary amine	Chloramine	2–5	25–29
	Condensation	Primary amine	Hydroxyl-amine-O-sulfonic acid	2–6	30–31
	Reduction	N-Alkyl-N-nitrosoamine	—		34–42
	Reduction	Hydrazone of aldehyde	—		23,45,47
	Reduction	Hydrazone of ketone	—		16,20,48
	Reduction	Aromatic diazonium salts	—		50–53
1,1-Disubstituted hydrazine					
	Condensation	Hydrazine	Alkyl halide (2 moles)		17
	Condensation	Secondary amine	Chloramine		26–28
	Reduction	N,N-Dialkyl-N-nitrosoamine	—	2–7,2–8,2–9	34–43
1,2-Disubstituted hydrazine					
	Condensation	Hydrazine	Activated aryl halide (2 moles)		5
	Condensation	Dihydrazide	Alkyl halide(s)	2–4	22–24
	Reduction	Monosubstituted hydrazone of aldehyde			23,45,47
	Reduction	Azo compounds	—	2–11	54–55
	Reduction (bimolecular)	Aromatic nitro compounds	—	2–12	56
1,1,2-Trisubstituted hydrazine	Condensation	Hydrazine	Alkyl halide (3 moles)		17

§ 1. Introduction

TABLE I (*Continued*)

Product type	Reaction type	Reactant	Reagent	Sample procedures	References
	Reduction	Disubstituted hydrazone of aldehydes	—		45
	Reduction	Monosubstituted hydrazone of ketones	—		[c]
	Reduction	Acyl hydrazone of ketone	—	2–10	46
Tetrasubstituted hydrazine	Condensation	Hydrazine	Alkylhalide (4 moles)[a]		17
	Reduction	Disubstituted hydrazone of ketone	—		[c]
	Oxidation	Secondary amine	—	2–13	59
Hydrazone	Condensation	Hydrazine	Aldehyde or ketone	3–1	15,61,62
	Condensation	Hydrazide	Aldehyde or ketone		46
	Condensation	Substituted hydrazine	Aldehyde or ketone		61,62
	Exchange reaction	Substituted hydrazone	Hydrazine		69
Hydrazide	Condensation	Hydrazine	Ester	4–1,4–2	61,73,75,79
	Condensation	Hydrazine	Acyl halide[b]		76
	Condensation	Hydrazine	Amide	4–3	77,85
	Condensation	Substituted hydrazine	Esters		[c]
	Condensation	Substituted hydrazine	Acyl halide		[c]

[a]Excess of alkyl halide may lead to monoquaternized compounds [17].
[b]Frequently leads to 1,2-diacyl hydrazines [22,23].
[c]Suggested reaction procedures.

to a phenylhydrazone. This class of di-phenylhydrazones is called an "osazone," a series of compounds not discussed in the present chapter.

Substituted hydrazines can be prepared by the reaction of amines with chloramine and with hydroxylamine-*O*-sulfonic acid. Of these two reagents, hydroxylamine-*O*-sulfonic acid has been introduced more recently and the scope of its reaction is still being explored. The present authors believe that this is a more convenient reagent than chloramine. By its use, not only mono- but also 1,1-disubstituted hydrazine should be preparable. The synthesis and utilization of N-substituted chloramines and hydroxylamine-O-sulfonic acids require further exploration.

Caution: In working with hydrazine and many substituted hydrazines, one must bear in mind that many of these compounds are highly toxic. They may also be carcinogenic. Hydrazine has also been suggested as a component in rocket fuel. Consequently, extreme safety precautions must be taken.

Several classes of hydrazines have been prepared by a variety of reductive procedures of N-nitrosoamines and hydrazones of various carbonyl compounds. The latter procedure is of particular value, since by the judicious selection of the carbonyl compound and of a substituted hydrazine, hydrazones may be prepared which can lead to new hydrazines with up to three different substituents.

Other functional groups which have been reduced to substituted hydrazines are azides, carbazates (acyl hydrazones), azo compounds, and ketazines.

Since aromatic diazonium salts may be reduced to mono-aryl-substituted hydrazines, a simple route to a vast variety of N-arylhydrazines is available.

The use of hydrazine and substituted hydrazines in the preparation of hydrazones by reaction with carbonyl compounds is well known.

Of some interest is the fact that hydrazides also react with carbonyl compounds in an analogous fashion. Since poly-hydrazides can be prepared from polymeric esters, resins can be prepared which are capable of separating carbonyl compounds from organic mixtures. Such resins might be used in a "carbonyl-exchange" column.

Polyhydrazides can be converted to poly-azides. These polymers find application in affinity chromatography and other biochemical procedures.

The best method of preparing hydrazides is by the reaction of hydrazines with esters. Acyl halides and amides may also be reacted with hydrazines. In the case of the Ing–Manske reaction, N-substituted phythalimides are reacted with hydrazine to generate a primary amine and the cyclic phthalhydrazide.

2. HYDRAZINES

While some of the synthetic procedures given below may apply specifically to one or the other classes of hydrazine derivatives under discussion, many of the reactions, with suitable modifications, may be applied to the preparation of several types of hydrazine derivatives.

CAUTION: All reactions involving hydrazine, its hydrates, or its salts are extremely hazardous. Hydrazine, its salts, and its hydrates are toxic and considered cancer suspect agents. Extreme care should be exercised in working with these and related compounds and their derivatives.

A. Alkylation of Hydrazine Derivatives

The alkylation of hydrazine by an aliphatic halide or a dialkyl sulfate produces in the first step a monoalkyl hydrazine, according to Eq. (1). The alkyl-substituted nitrogen is substantially more basic than the unsubstituted nitrogen. Consequently this nitrogen is more readily alkylated a second time than that of the adjacent free amino group. Therefore, simply alkylation reactions tend to produce 1,1-dialkyl hydrazines [Eq. (2)]. Even so, this reaction appears to be of preparative value if the alkylating agent contains four or more carbon atoms [5,15,16,17].

$$NH_2-NH_2 + RX \longrightarrow R-NH-NH_2 + HX \tag{1}$$

$$R-NH-NH_2 + RX \longrightarrow \underset{R}{\overset{R}{\diagdown}}N-NH_2 + HX \tag{2}$$

In the aromatic series, simple halides such as chlorobenzene do not normally react with hydrazine. The more reactive halides, however, such as the nitrochlorobenzenes [18] do react with hydrazine to yield monosubstituted aromatic hydrazine derivatives in good yield. Since the mono-aryl-substituted amino group of the hydrazine is less basic than the unsubstituted amino group, treatment of hydrazine with an excess of a reactive aryl leads to 1,2-diaryl-hydrazines. Also, reaction of monoarylhydrazines with alkyl halides results in the formation of 1-aryl-2-alkylhydrazines.

The lower alkylhydrazines such as methylhydrazine and ethylhydrazine are usually prepared by indirect methods. Probably the most convenient method of preparing methylhydrazine involves quaternization of one of the nitrogen atoms of benzalazine followed by hydrolysis, according to the reaction scheme of Eqs. (3)–(5).

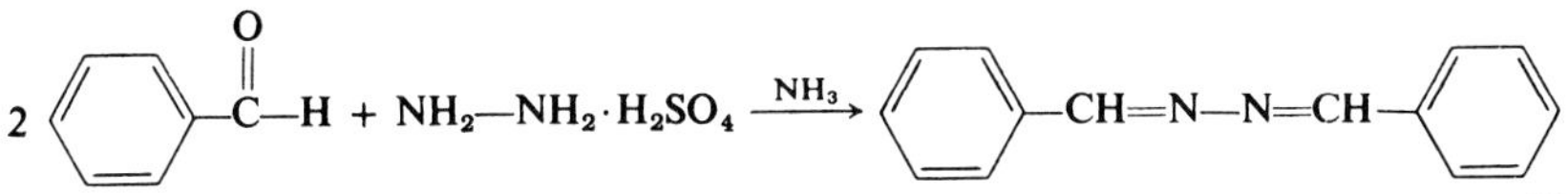

$$2\ Ph-\underset{\overset{\|}{O}}{C}-H + NH_2-NH_2 \cdot H_2SO_4 \xrightarrow{NH_3} Ph-CH=N-N=CH-Ph \tag{3}$$

$$Ph-CH=N-N=CH-Ph + (CH_3)_2SO_4 \longrightarrow \left[Ph-CH=\overset{+}{\underset{\underset{CH_3}{|}}{N}}-N=CH-Ph\right]\left[CH_3SO_4\right]^{-} \tag{4}$$

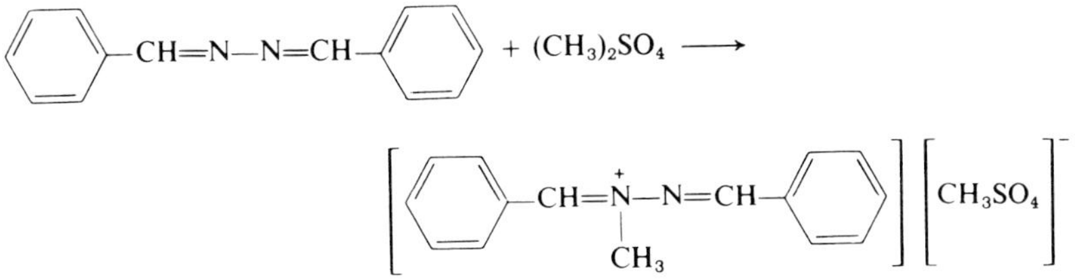

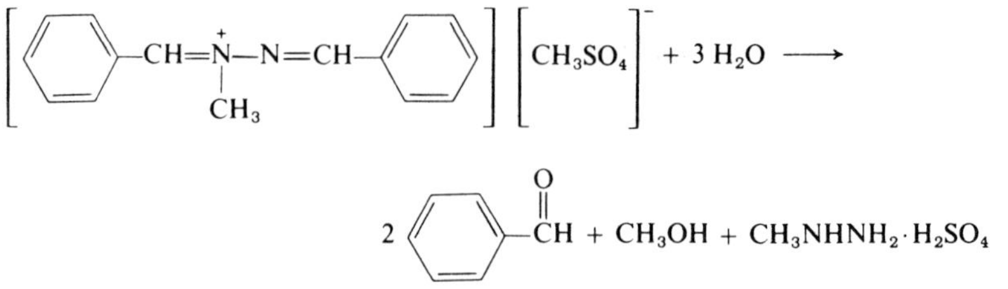

$$\left[\underset{CH_3}{\underset{|}{\bigcirc\!\!\!-CH=\overset{+}{N}-N=CH-\bigcirc}}\right]\left[CH_3SO_4\right]^- + 3\,H_2O \longrightarrow$$

$$2\,\bigcirc\!\!\!-\overset{\overset{O}{\|}}{CH} + CH_3OH + CH_3NHNH_2\cdot H_2SO_4 \quad (5)$$

While this reaction was originated by Thiele [13], detailed directions for the preparation of methylhydrazine have been described more recently by Hatt [14].

The monoethylation of hydrazines given here was found possible when an excess of hydrazine in ethanol was used [15].

2-1. *Preparation of Ethylhydrazine* [15]

$$NH_2-NH_2 + (CH_3CH_2)_2SO_4 \xrightarrow[C_2H_5OH]{KOH} CH_3CH_2NHNH_2 + C_2H_5OH + KHSO_4 \quad (6)$$

CAUTION: Both hydrazine and diethyl sulfate are extremely toxic and cancer suspect agents.

In a hood, in a three-necked flask fitted with a mechanical stirrer, an empty distillation column, topped with a total reflux–partial take-off distillation head, an addition funnel, and means to maintain a nitrogen atmosphere are placed 35 gm (0.63 mole) of potassium hydroxide, 30 ml (0.93 mole) of anhydrous hydrazine (95% active), and 60 ml of absolute ethanol. This mixture is cooled in an ice bath while being stirred mechanically.

From the addition funnel, 33 ml (0.25 mole) of acid-free diethyl sulfate is slowly added while maintaining a low temperature. After completion of the addition, the mixture is heated in a bath to 165°C to separate a mixture of hydrazine and ethylhydrazine by distillation. The distillate is cooled and made strongly acid by cautious addition of concentrated hydrochloric acid. The precipitating hydrazine hydrochloride is separated by filtration from the hot solution. The filtrate is then concentrated to half volume. A small amount of hydrochloric acid is added and the solution is allowed to cool. The precipitate that forms is washed in turn with small portions of concentrated hydrochloric acid, alcohol, and ether. The product is then dried in a vacuum desiccator over calcium chloride. A further crop of product may be obtained by concentrating the mother liquors to yield a total of 21 gm (87%) of ethylhydrazine hydrochloride.

The free hydrazine (see warnings above) may be prepared in 80% yields by treatment of the hydrochloride salt with base. The anhydrous compound is obtained by distillation from an excess of barium oxide.

2-2. Preparation of 2,4-Dinitrophenylhydrazine [19]

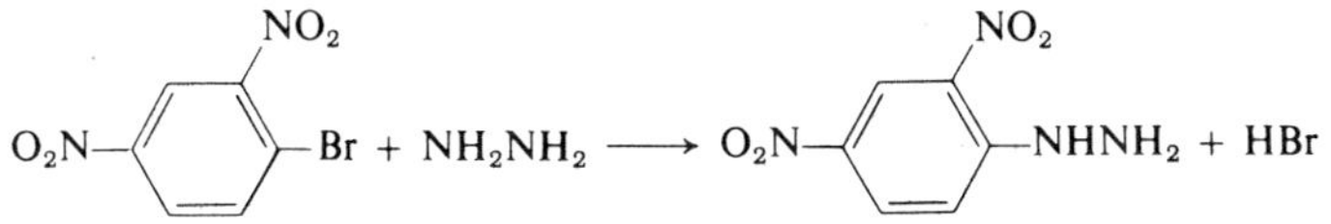

$$O_2N\text{-}C_6H_3(NO_2)\text{-}Br + NH_2NH_2 \longrightarrow O_2N\text{-}C_6H_3(NO_2)\text{-}NHNH_2 + HBr \quad (7)$$

CAUTION: See warnings given above concerning hydrazine and its related compounds.

A solution of 0.5 gm (0.002 mole) of 2,4-dinitrobromobenzene in 7.5 ml of 95% ethanol is heated almost to boiling. To this is added a solution of 0.5 ml of "100% hydrazine hydrate" [which is 64% in hydrazine] (0.010 mole of hydrazine) in 2.5 ml of 95% ethanol. The resultant solution rapidly turns deep red-purple as it is allowed to cool without being disturbed for approximately ½ hr. The crystals which separate are collected on a Büchner funnel and washed with a small quantity of 95% ethanol to yield 0.365 gm (91%), mp 200°–202°C. The product may be recrystallized from ethyl acetate (orange platelets).

Perfluorinated aromatic systems evidently may be subjected to nucleophilic attack by hydrazine to yield substitution products [20]. In the preparation and handling of these compounds, strong alkalies should be avoided to prevent complex internal oxidation–reduction reactions.

2-3. Preparation of 4,4'-Dihydrazinooctaflourobiphenyl [20]

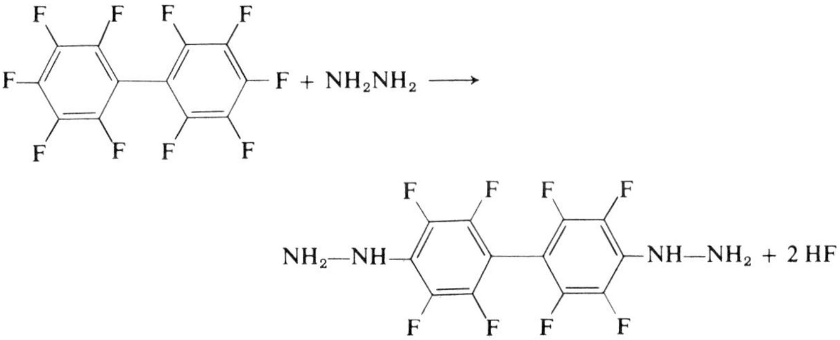

$$C_{12}F_{10} + NH_2NH_2 \longrightarrow NH_2\text{-}NH\text{-}C_{12}F_8\text{-}NH\text{-}NH_2 + 2\,HF \quad (8)$$

CAUTION: See warnings given above concerning hydrazine and its related compounds.

In a 1-liter two-necked flask fitted with a mechanical stirrer and reflux condenser, a mixture of 100.2 gm (0.3 mole) of decafluorobiphenyl, 40.3 gm (1.2 moles) of anhydrous hydrazine (95+% pure), and 450 ml of absolute ethanol is stirred and heated at reflux temperature for 21 hr.

The mixture is then distilled under reduced pressure to approximately half its original volume and decanted into 700 ml of water. The precipitate is filtered off and heated with stirring in 800 ml of a 50% solution of 95% ethanol in benzene.

Upon cooling, 87.8 gm (77%) of 4,4′-dihydrazinooctafluorobiphenyl is isolated, mp 208°–211°C.

In his review of the chemistry of hydrazine, Schiessl [12] mentions that reaction of hydrazine with epoxides forms hydroxyalkylhydrazines [Eq. (9)].

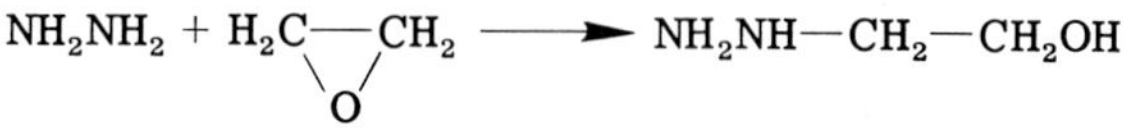

$$NH_2NH_2 + H_2C\underset{O}{-}CH_2 \longrightarrow NH_2NH-CH_2-CH_2OH \quad (9)$$

With aziridines like ethylenimine a β-aminoalkylhydrazine is said to form [Eq. (10)].

$$NH_2NH_2 + H_2C\underset{\underset{H}{N}}{-}CH_2 \longrightarrow NH_2NH-CH_2CH_2NH_2 \quad (10)$$

With sulfones, a sulfoalkylhydrazine is produced [Eq. (11)].

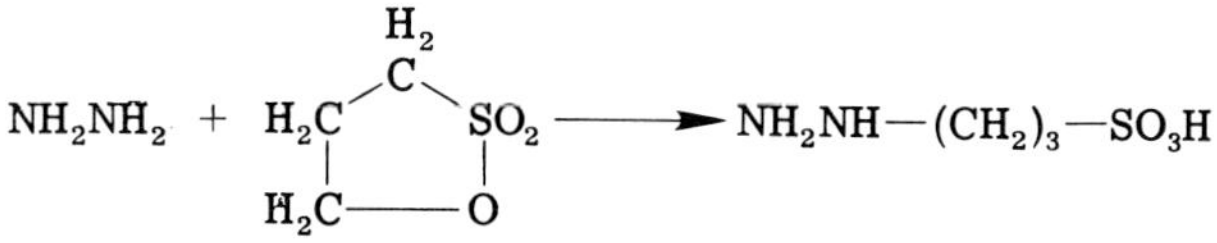

$$NH_2NH_2 + \text{(cyclic sulfone)} \longrightarrow NH_2NH-(CH_2)_3-SO_3H \quad (11)$$

Hydrazine may also be cyanoethylated as indicated in Eq. (12).

$$NH_2NH_2 + CH_2=CHCN \longrightarrow NH_2NH-CH_2CH_2CN \quad (12)$$

Recently, 3-alkyl (or aryl)-1,2,4,5-tetrazine, when treated with hydrazine hydrate, was found to undergo a Chichibabin reaction. Whereas, in the usual Chichibabin reaction, a metal amide reacts with a heterocyclic compound to produce an amine, in this case, a hydrazine derivative is formed according to Eq. (13) near room temperature. Yields are in the range of 9–15% [21].

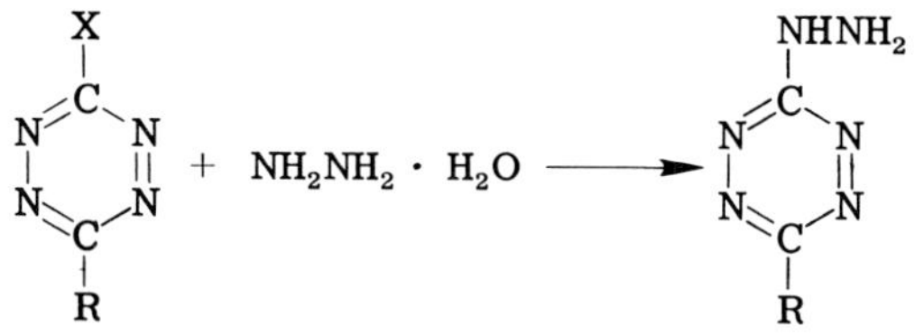

$$\text{(X-substituted tetrazine)} + NH_2NH_2 \cdot H_2O \longrightarrow \text{(NHNH}_2\text{-substituted tetrazine)} \quad (13)$$

where R may be methyl, ethyl, *t*-butyl, or phenyl, and X may be H, NH_2, Cl, or Br.

As stated above, the alkylation of hydrazine with alkyl halides tends to favor 1,1-dialkylation. By variations in the reaction conditions, either monoalkyl or dialkyl derivatives may be formed. With a large excess (5 molar) of hydrazine, comparatively low reaction temperatures (100°–120°C), and use of long chain

§ 2. Hydrazines

alkyl halides, monoalkylation is favored. Intermediate reaction conditions tend to favor the formation of mixtures. For example, treatment of 13.5 moles of hydrazine hydrate with 4.0 moles of chlorotriethylamine hydrochloride in an aqueous potassium carbonate solution resulted in a 59% yield of the monoalkylated hydrazine and 6% of the dialkylated hydrazine, as shown in Eq. (14) [16] (see also Procedure 2-1)

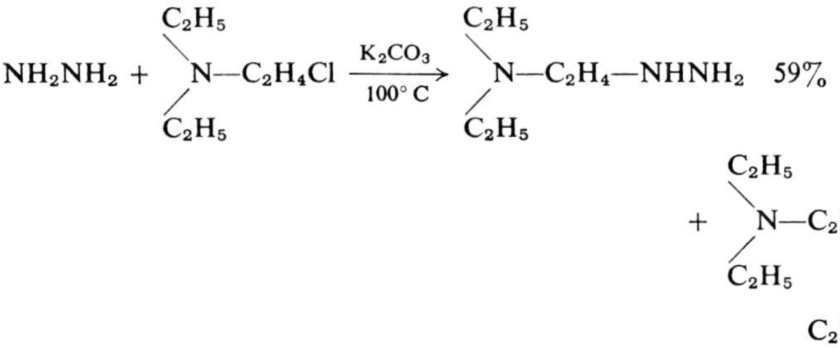

$$NH_2NH_2 + (C_2H_5)_2N-C_2H_4Cl \xrightarrow[100°C]{K_2CO_3} (C_2H_5)_2N-C_2H_4-NHNH_2 \quad 59\%$$

$$+ (C_2H_5)_2N-C_2H_4-N(C_2H_4-N(C_2H_5)_2)-NH_2 \quad 6\% \quad (14)$$

The lower 1,2-dialkylhydrazines have been prepared by alkylation of dihydrazides such as dibenzoylhydrazine [22]. A typical example is the preparation of 1,2-dimethylhydrazine dihydrochloride according to Eqs. (14–16) [22].

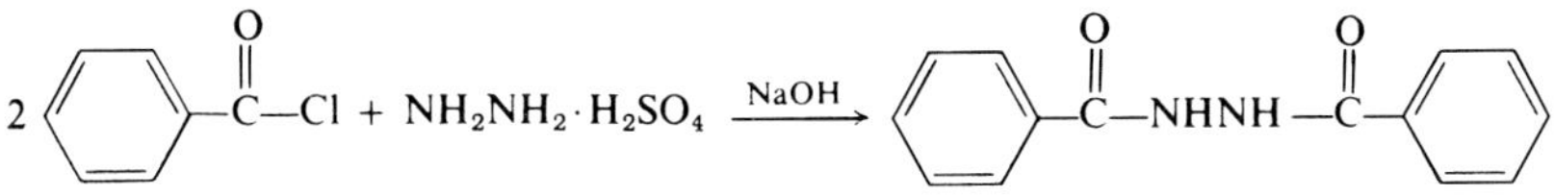

$$2\ Ph-CO-Cl + NH_2NH_2 \cdot H_2SO_4 \xrightarrow{NaOH} Ph-CO-NHNH-CO-Ph \quad (15)$$

$$Ph-CO-NHNH-CO-Ph + 2\ (CH_3)_2SO_4 + 2\ NaOH \longrightarrow$$

$$Ph-CO-N(CH_3)-N(CH_3)-CO-Ph + 2\ NaHSO_4 + 2\ CH_3OH \quad (16)$$

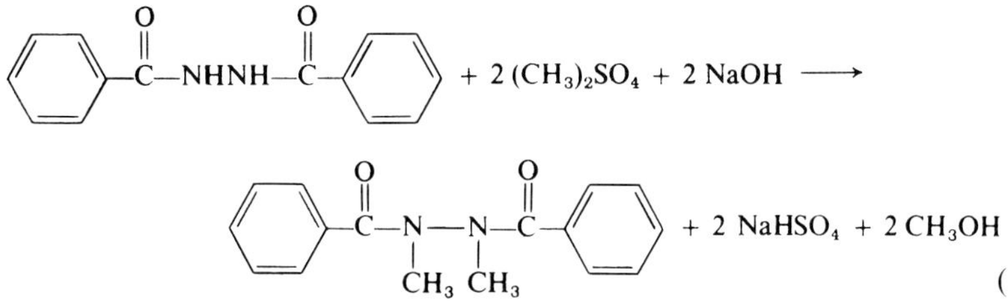

$$Ph-CO-N(CH_3)-N(CH_3)-CO-Ph + 2\ HCl + 2\ H_2O \longrightarrow$$

$$2\ Ph-CO_2H + CH_3NHNHCH_3 \cdot 2\ HCl \quad (17)$$

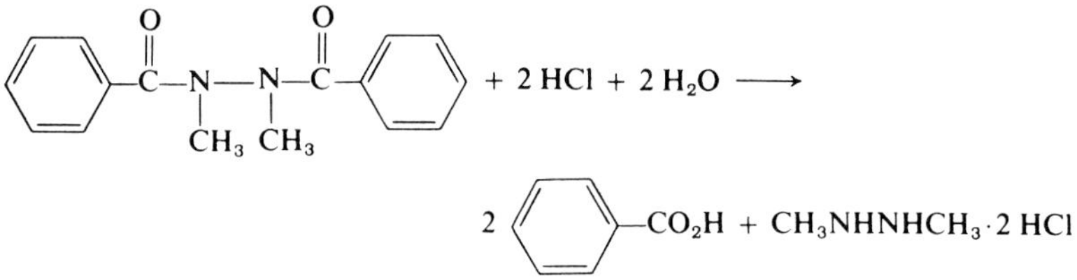

This procedure may be used to prepare a wide variety of 1,2-dialkylhydrazines in which the two alkyl groups are different. Since it has been shown that monoalkylhydrazines upon benzoylation with benzoyl chloride form dibenzoyl derivatives, the monoalkylated dibenzoylhydrazine may then be alkylated further with a new and different alkyl halide. Upon hydrolysis of this dihydrazide, a 1-alkyl-2-alkylhydrazine may be isolated [23,24].

2-4. Preparation of 1-Methyl-2-isopropylhydrazine [24]

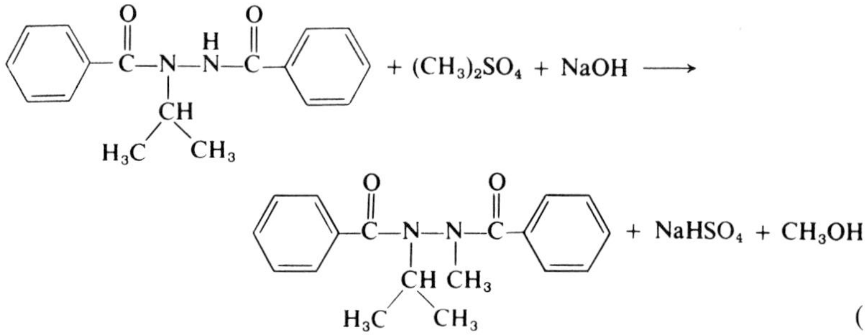

$$\text{PhC(O)-N(CH(CH}_3)_2)-NH-C(O)Ph} + (CH_3)_2SO_4 + NaOH \longrightarrow$$

$$\text{PhC(O)-N(CH(CH}_3)_2)-N(CH_3)-C(O)Ph} + NaHSO_4 + CH_3OH \quad (18)$$

$$\text{PhC(O)-N(CH(CH}_3)_2)-N(CH_3)-C(O)Ph} + 2 H_2O \xrightarrow{HCl}$$

$$(CH_3)_2CH-NHNH-CH_3 + 2\; C_6H_5-CO_2H \quad (19)$$

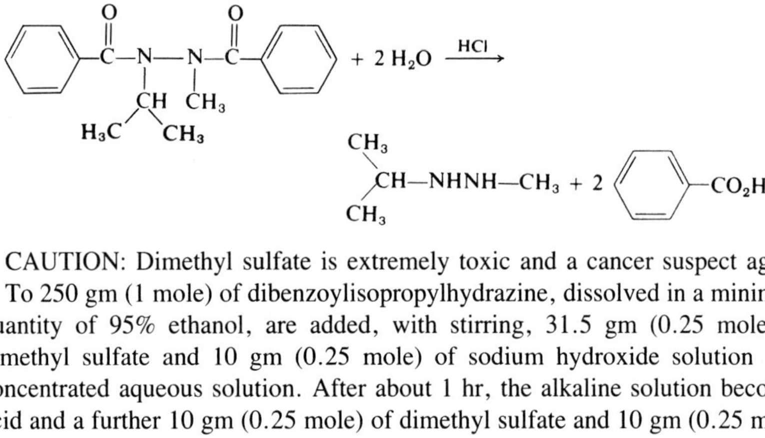

CAUTION: Dimethyl sulfate is extremely toxic and a cancer suspect agent.

To 250 gm (1 mole) of dibenzoylisopropylhydrazine, dissolved in a minimum quantity of 95% ethanol, are added, with stirring, 31.5 gm (0.25 mole) of dimethyl sulfate and 10 gm (0.25 mole) of sodium hydroxide solution as a concentrated aqueous solution. After about 1 hr, the alkaline solution becomes acid and a further 10 gm (0.25 mole) of dimethyl sulfate and 10 gm (0.25 mole) of sodium hydroxide solution are added. This procedure is repeated until approximately a 10% to 20% excess of dimethyl sulfate and sodium hydroxide is present. The product is precipitated by adding four volumes of cold water with constant stirring to the reaction mixture. The yield of crude product is 210 gm (80%), mp 63°-68°C. The crude 1,2-dibenzoyl-1-methyl-2-isopropylhydrazine product is purified by dissolving it in alcohol and reprecipitating it with water. The yield is reduced to about 70% of theory. By repeated recrystallization, this substituted dihydrazide may be purified to a melting point of 76.25°–76.75°C.

The purified product is now hydrolyzed by adding five times its weight of concentrated hydrochloric acid and heating the mixture at 90°–95°C. The resulting benzoic acid is removed by filtration and the remaining filtrate is concentrated to a small volume at reduced pressure. Since the hydrochloride salt of the hydrazine is quite hygroscopic, it is not possible to isolate the product from the aqueous syrup.

To prepare the free hydrazine, the syrupy residue is treated with an equal volume of concentrated sodium hydroxide solution under oxygen-free nitrogen in a vacuum distillation apparatus. The free base is distilled under reduced pressure, in an oxygen-free nitrogen atmosphere into a receiver containing solid sodium hydroxide. This product still contains water. The product from this first distillation is stored in a sealed container over sodium hydroxide for 24 hr. The upper liquid layer is then separated carefully and again distilled into a receiver containing aluminum amalgam as a drying agent. This procedure is repeated several times finally to form a product in 50% yield (44 gm). The boiling point of the purified product is 79.5°–79.9°C (371 mm).

Upon reaction of hydrazine with low-molecular-weight alkyl halides, alkylation can proceed until one nitrogen is quaternized. When tri- or tetrasubstituted propyl or butylhydrazines are to be prepared, forcing conditions such as elevated temperature are usually required [17].

B. Syntheses Involving Formation of Nitrogen–Nitrogen Bonds

a. MODIFIED RASCHIG SYNTHESIS

The preparation of hydrazine by the addition of chloramine to ammonia in an alkaline aqueous solution has been termed the "Raschig hydrazine synthesis" [25].

The substitution of various amines for the ammonia in the Raschig synthesis permits the preparation of a variety of N-substituted hydrazines [26,27]. Factors influencing the yield of such substituted hydrazines are the mole ratio of amine to chloramine; the presence of gelatin to deactivate metals which may be present in the reaction system; the presence of a nonvolatile base; and the reaction temperature [26]. The preparation of cyclohexylhydrazine sulfate is representative of the modified Raschig process.

2-5. Preparation of Cyclohexylhydrazine Hydrogen Sulfate [26,27]

$$NaOCl + NH_3 \longrightarrow NaOH + NH_2Cl$$

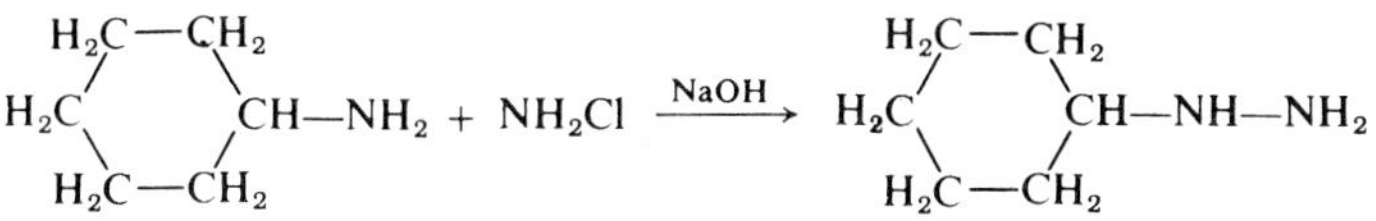

$$\text{cyclohexyl-NH}_2 + NH_2Cl \xrightarrow{NaOH} \text{cyclohexyl-NH-NH}_2 \qquad (20)$$

CAUTION: Since chloramine may be a particularly hazardous reagent, all due safety precautions must be taken. The use of a good hood and safety shields should be minimum precautions.

(a) Preparation of a chloramine solution. To 8.57 gm (0.115 mole) of sodium hypochlorite in 100 ml of water is added a solution of 0.25 gm of gelatin dissolved in 50 ml of water. The solution is placed in a three-necked flask equipped with a mechanical stirrer, a thermometer, dropping funnel, and an outlet vent. The apparatus is placed in an ice bath and cooled to 0°C. Then, with cautious shaking, 100 ml of cold 1.5 N aqueous ammonia is added dropwise. After each addition of the ammonia has been completed, the flask is cautiously shaken, care being taken to avoid excessive evolution of gas.

The solution prepared in this manner is found, by analysis, to be 0.16 M with respect to its chloramine content.

(b) Preparation of the substituted hydrazine salt. While maintaining ice temperatures, to this chloramine solution is added cautiously 30.2 gm (0.3 mole) of cyclohexylamine. The reaction mixture is stirred mechanically while it is allowed to warm to room temperature over a 1-hr period. Then it is heated on a steam bath for an additional 30 min. The reaction mixture is chilled in an ice bath, whereupon the excess amine and cyclohexylhydrazine form a separate liquid phase. The product mixture is separated and subjected to fractional distillation to separate the excess cyclohexylamine. To the aqueous cyclohexylhydrazine fraction, an excess of dilute sulfuric acid is added. Cyclohexylhydrazine hydrogen sulfate is isolated by evaporation of this aqueous solution. The product is recrystallized by dissolving the crude in methanol followed by precipitation upon addition of ether. Yield 5.8 gm (60%), mp 117°C.

Similar reactions have been described for the preparation of 1,1,1-trisubstituted hydrazinium hydrochlorides [28,29]. Therefore, except for the possible hazard of handling chloramine, this method appears to be of wide applicability for the preparation of 1-alkyl, 1,1-dialkyl and 1,1,1-trialkylhydrazinium salts.

A closely related method of establishing a hydrazine functionality from an amine group involves the use of hydroxylamine-*O*-sulfonic acid. This is perhaps the most convenient method of preparing alkylhydrazines from ethylamine and higher amines. In a typical example, 2-aminopentane was converted to 2-pentylhydrazine [30]. This procedure, it seems to us, differs only in minor detail from the general procedure of Gever and Hayes [31].

2-6. *Preparation of 2-Pentylhydrazine* [30]

$$CH_3CH_2-CH_2-\underset{\underset{CH_3}{|}}{CH}-NH_2 + NH_2OSO_3H \xrightarrow{KOH}$$

$$CH_3CH_2CH_2-\underset{\underset{CH_3}{|}}{CH}-NHNH_2 + KHSO_4 \quad (21)$$

§ 2. Hydrazines

To a mixture of 46 gm (0.53 mole) of 2-pentylamine and 9.2 gm (0.164 mole) of potassium hydroxide in 150 ml of water, heated to reflux, is added dropwise with stirring over a ½ hr period, a solution of 9.5 gm (0.084 mole) of hydroxylamine-O-sulfonic acid in 50 ml of water. The reaction mixture is concentrated to half volume under reduced pressure. The solution is then transferred to a separatory funnel, layered with ether, and 10 ml of 10 N sodium hydroxide solution is cautiously added. The ether layer is seperated, and the aqueous system is repeatedly extracted with ether. The combined ether extracts are dried with potassium hydroxide. After evaporation of the ether, the residual oil is fractionally distilled under reduced pressure. The product boils between 56°–60°C at 11 mm. The yield is not reported in the patent. However, yields normally run between 30% and 60% by this general procedure [31].

In the procedure of Gever and Hayes [31], the isolation of the reaction product involves conversion to the substituted hydrazone of benzaldehyde, separation of the excess benzaldehyde by steam distillation, hydrolysis of the hydrazone, and the precipitation of the hydrazine as the oxalate salt.

The closely related preparation of 1-aminopyridinium iodide has been described recently [32]. This reference indicates the wide applicability of this technique to the preparation of substituted hydrazines (see also [31]).

A novel method for the preparation of 1,1-disubstituted hydrazines uses an extension of the Curtius rearrangement to establish the required nitrogen-to-nitrogen bond. In this process, diaryl or dialkyl carbamyl azides are formed from the corresponding carbamyl chloride. The azide is heated by reflux in t-butanol for several days followed by hydrolysis of the intermediate hydrazide [Eqs. (22–24)]. For the preparation of very pure 1,1-diphenylhydrazine by this method, the overall yield was reported to be on the order of 67% [33].

$$\begin{array}{c} Ar \\ Ar \end{array}\!\!N-\overset{\overset{O}{\|}}{C}-Cl + NaN_3 \longrightarrow \begin{array}{c} Ar \\ Ar \end{array}\!\!N-\overset{\overset{O}{\|}}{C}-N_3 \quad (22)$$

$$\begin{array}{c} Ar \\ Ar \end{array}\!\!N-\overset{\overset{O}{\|}}{C}-N_3 \xrightarrow{t\text{-BuOH}} \begin{array}{c} Ar \\ Ar \end{array}\!\!N-NH-\overset{\overset{O}{\|}}{C}-O\overset{\overset{CH_3}{|}}{\underset{\underset{CH_3}{|}}{C}}-CH_3 \quad (23)$$

$$\begin{array}{c} Ar \\ Ar \end{array}\!\!N-NH-\overset{\overset{O}{\|}}{C}-O-\overset{\overset{CH_3}{|}}{\underset{\underset{CH_3}{|}}{C}}-CH_3 \xrightarrow[\text{2. NaOH}]{\text{1. HCl}} \begin{array}{c} Ar \\ Ar \end{array}\!\!N-NH_2 \quad (24)$$

Since the reactions with sodium azide are particularly hazardous and the reaction indicated in Eq. (23) requires refluxing for 5 days, this method is not given in greater detail here. We believe that the concept of rearrangement reactions for nitrogen-to-nitrogen bonds merits further research. To be noted is the use of t-butanol in Eq. (23). Evidently the hydrolysis of the analogous primary alkyl esters is quite difficult under either acidic or basic conditions while the t-butyl ester is readily saponified completely [33].

C. Reductive Methods

a. REDUCTION OF N-NITROSOAMINES

In the reduction of N-nitrosoamines to 1,1-disubstituted hydrazines, the substituents may be either aliphatic, aromatic, or aliphatic and aromatic groups. The earliest examples of such reductions used zinc and acetic acids as reducing agents. Typical examples of this procedure are detailed in *Organic Syntheses* [34,35].

High pressure hydrogenations using platinum on carbon or palladium on carbon catalysts in the presence of a variety of salts to prevent complete reduction of the nitroso compound to the amine have been described [36]. The liquid phase high-pressure hydrogenation in the presence of Raney nickel catalyst has also been described [37].

N-Nitrosoamines have been reduced with lithium aluminum hydride. Among these compounds are the N-nitroso derivatives of relatively simple primary aromatic and aliphatic amines, N-nitrosopiperidine derivatives, N-nitrosopyrrolidine derivatives and N-nitrosopyrroles [38]. N-nitroso derivatives of secondary amines [39,40], and N-nitrosodiethylhydroxylamine [41].

It has been noted that the reduction with lithium aluminum hydride of nitrosoamines is characterized by a long induction period followed by a rather violent initial reaction [38,42]. By the careful control of the addition of reagents and the use of oversized equipment, the reaction can be kept under control.

It should be pointed out, when considering syntheses involving N-nitrosoamines, that many of these compounds are quite toxic and possibly carcinogenic. Due precautions must be taken when handling such compounds.

2-7. Preparation of 1,1-Diethylhydrazine [43]

$$\underset{C_2H_5}{\overset{C_2H_5}{\diagdown}}N-NO \xrightarrow{LiAlH_4} \underset{C_2H_5}{\overset{C_2H_5}{\diagdown}}N-NH_2 \qquad (25)$$

CAUTION: N-nitrosoamines are cancer suspect agents. The reduction itself may also be violent. Appropriate precautions are urged.

In a hood behind an explosion shield, in a 3-liter three-necked flask fitted with a condenser protected with a drying tube, an addition funnel, and a ground glass stirrer, are placed 1 lb. of fresh dry ether and 27 gm (0.71 mole) of lithium aluminum hydride. The mixture is cautiously heated to reflux (possibly with an infrared lamp) for 1½ hr and then cooled in an ice bath. With very vigorous stirring and with ice cooling, a solution of 51 gm (0.5 mole) of N-nitrosodiethylamine in 50 ml of anhydrous ether is then added in the course of 2½ hr. [While Lemal, Menger, and Coats recommend the addition in this manner, the present authors suggest that a small quantity of the N-nitrosodiethylamine solution be added first and then, only after evidence of reaction has been noted, should the rest of the material be added with similar precautions.] In any event, a vigorous reaction is noted, which is characterized by considerable evolution of heat and frothing. Vigorous stirring is maintained at all times. When the reaction finally has subsided, an additional 150 ml of ether is added and the mixture is stirred for 45 min at 0°C, followed by 45 min at room temperature.

The flask is then cooled in ice and, dropwise, 60 ml of water is added with great caution. [The present authors suggest that an excess of ethyl acetate be used rather than water in decomposing the excess lithium aluminum hydride.] Then 300 ml of 20% sodium hydroxide solution is added and the ether layer is decanted from the suspension. After extracting the aqueous suspension with several proportions of ether, the ether solutions are combined and dried over potassium carbonate. The solvent is evaporated. The residue is distilled fractionally from potassium hydroxide through a jacketed Vigreux column. The product boils at 98°–99°C. Yield 25.5 gm (58%). The product is stored in a tight container under refrigeration.

Lemal, Menger, and Coats [43] found that a hindered starting material such as N-nitrosodiisopropylamine is not satisfactorily reduced by lithium aluminum hydrate in ether. However, substitution of tetrahydrofuran (THF) for ether as a solvent led to a reasonable yield of 1,1-diisopropylhydrazine. A recent example of the use of THF as the reaction solvent is the preparation of N-tert-butyl-N-phenylhydrazine. To be noted in this preparation of a hydrazine with a hindering substituent is the use of prolonged heating of the reaction mixture. Our own experience with lithium aluminum hydride reduction of amides has been that reaction times are frequently underestimated. Consideration of much longer reaction periods are suggested.

2-8. Preparation of N-tert-Butyl-N-phenylhydrazine [44]

$$CH_3-\underset{\underset{CH_3}{|}}{\overset{\overset{CH_3}{|}}{C}}-\underset{\underset{C_6H_5}{}}{N}-NO \xrightarrow{\underset{THF}{LiAlH_4}} CH_3-\underset{\underset{CH_3}{|}}{\overset{\overset{CH_3}{|}}{C}}-\underset{\underset{C_6H_5}{}}{N}-NH_2 \quad (26)$$

With the usual safety precautions, in a 1-liter three-necked flask fitted as in Preparation 2-6, to 200 ml of anhydrous tetrahydrofuran is added 1.7 gm (0.045 mole) of lithium aluminum hydride. After the reagent has dissolved, the mixture is heated to a gentle reflux and a solution of 4.0 gm (0.023 mole) of N-nitroso-N-tert-butylaniline in 100 ml of anhydrous tetrahydrofuran is slowly added. The reaction mixture is warmed at reflux for 16 hr. After this period, the condenser is set down for distillation and the solvent is partially removed by evaporative distillation. The residue is cooled in an ice bath. The excess lithium aluminum hydride is destroyed by the careful addition of water.

The reaction mixture is filtered. The solid on the filter paper is repeatedly washed with benzene. The benzene is combined with the filtrate. The organic layer is separated from the filtrated and placed in a distillation apparatus. The benzene is removed by evaporative distillation. The residue is distilled at 91°–93°C/5.0 mm. Yield: 2.1 gm (62%). NMR δ 1.12 (s, 9), 3.34 (s, 2), 7.10 (s, 5); IR (neat) 3325, 3050, 2960, 770, 700 cm^{-1}.

Instead of reduction of N-nitrosoamines with lithium aluminum hydride, with its tendency to long induction periods and sudden evolution of heat, the reduction of N-nitrosoamines with sodium and alcohol or with sodium in liquid ammonia in the presence of alcohol has been reported [42].

A typical example is the preparation of 1,1-diisobutylhydrazine [42].

2-9. Preparation of 1,1-Diisobutylhydrazine [42]

$$\begin{array}{c}
\text{CH}_3-\text{CH}-\text{CH}_2-\text{N}-\text{NO} \\
|| \\
\text{CH}_3\text{CH}_2 \\
| \\
\text{CH}-\text{CH}_3 \\
| \\
\text{CH}_3
\end{array} \xrightarrow{\text{Na/C}_2\text{H}_5\text{OH}} \begin{array}{c}
\text{CH}_3-\text{CH}-\text{CH}_2-\text{N}-\text{NH}_2 \\
|| \\
\text{CH}_3\text{CH}_2 \\
| \\
\text{CH}-\text{CH}_3 \\
| \\
\text{CH}_3
\end{array} \quad (27)$$

To a three-necked flask fitted with a reflux condenser protected against atmospheric moisture, and a mechanical stirrer, a solution of 33 gm (0.25 mole) of diisobutylnitrosoamine in 250 ml of absolute ethanol is cooled to 0°C. Over a period of 4 hr, a total of 23 gm (1 gm atom) of sodium is added. Since the reduction is slow toward the end of the reaction, the solution is allowed to warm to room temperature. Upon completion of the sodium addition, low boiling petroleum ether is added to the solution and then water is slowly added until phase separation has occurred. The reaction mixture is then extracted repeatedly with a total of 1.5 liters of low-boiling petroleum ether (30°–40°C). The extract is dried overnight over anhydrous sodium sulfate. The solvent is then removed at atmospheric pressure and the residue is fractionated under reduced pressure to give 13.2 gm (36.6%) of 1,1-diisobutylhydrazine, bp 58.5°–59.5°C (15 mm).

b. REDUCTION OF HYDRAZONES

Hydrazine derivatives generally react readily with ketones and aldehydes to form hydrazones, and since the readily prepared hydrazides also react with carbonyl compounds to produce the corresponding acylhydrazones (1-acyl-2-alkylidenehydrazines), reductive procedures permit the preparation of a wide variety of 1,2-dialkylhydrazines [16]. For example, by such procedures the bis-dimethylhydrazone of glyoxal has been converted to a bis-hydrazine according to the following equations [45]:

$$2 \begin{array}{c} CH_3 \\ \diagdown \\ N-NH_2 \\ \diagup \\ CH_3 \end{array} + \begin{array}{c} O \quad O \\ \parallel \quad \parallel \\ HC-C-H \end{array} \longrightarrow \begin{array}{c} CH_3 \\ \diagdown \\ N-N=CH-CH=N-N \\ \diagup \\ CH_3 \end{array} \begin{array}{c} CH_3 \\ \diagdown \\ \diagup \\ CH_3 \end{array} \quad (28)$$

$$\begin{array}{c} CH_3 \\ \diagdown \\ N-N=CH-CH=N-N \\ \diagup \\ CH_3 \end{array} \begin{array}{c} CH_3 \\ \diagup \\ \diagdown \\ CH_3 \end{array} \xrightarrow{[H]} \begin{array}{c} CH_3 \\ \diagdown \\ N-NH-CH_2-CH_2-NH-N \\ \diagup \\ CH_3 \end{array} \begin{array}{c} CH_3 \\ \diagdown \\ \diagup \\ CH_3 \end{array} \quad (29)$$

A typical preparation making use of an acylhydrazone as the starting material, illustrative of this general procedure, is the preparation of 1-ethyl-2-isopropylhydrazine.

2-10. Preparation of 1-Ethyl-2-isopropylhydrazine [46].

$$\begin{array}{c} O \\ \parallel \\ CH_3CNHNH_2 \end{array} + \begin{array}{c} CH_3 \\ \diagdown \\ C=O \\ \diagup \\ CH_3 \end{array} \longrightarrow$$

$$\begin{array}{c} O \\ \parallel \\ CH_3C-NH-N=C \end{array} \begin{array}{c} CH_3 \\ \diagup \\ \diagdown \\ CH_3 \end{array} \xrightarrow{[H]} CH_3CH_2NHNHCH \begin{array}{c} CH_3 \\ \diagdown \\ \diagup \\ CH_3 \end{array} \quad (30)$$

To a flask fitted with a stirrer, a reflux condenser, and addition funnel and containing 56 gm (0.76 mole) of acethydrazide is added 50 gm (0.86 mole) of acetone. The mixture warms appreciably then a homogeneous solution results. This solution is cooled and the resulting solid mass is broken up and triturated with an additional 10 ml of acetone to ensure completeness of reaction.

After filtration, the product is recrystallized from isopropanol and dried in a vacuum over at 50°C to yield 71 gm (82%) of 1-acetyl-2-isopropylidenehydrazine, mp 134°–135°C.

In the thimble of a Soxhlet extractor, 71 gm (0.62 mole) of 1-acetyl-2-isopropylidenehydrazine is loaded. The extractor and its condenser are attached to a 5-liter three-necked flask also equipped with a magnetic stirrer and drying tube and containing 52.4 gm (1.38 moles) of lithium aluminum hydride suspended in 2 lb of anhydrous ether. The reaction setup is placed behind an adequate shield and the reaction is carried out by refluxing the ether through the Soxhlet extractor. Reflux is continued until all of the acylhydrazone has been extracted from the thimble and no further reaction is noted. The lithium aluminum hydride complex is then carefully decomposed by the cautious addition of water or ethyl acetate, the precipitate is filtered off and repeatedly washed with anhydrous ether. The ether washings are then combined and concentrated at reduced pressure. The residual hydrazine is isolated by fractional distillation in a nitrogen atmosphere to yield 38.6 gm (61%), bp 113°–115°C (740 mm).

In the preparation of hydrazino-analogs of amino acids, a pyruvic acid derivative has been converted to a hydrazone. The hydrazone was then reduced with sodium amalgam and water [47].

The preparation of monoalkylhydrazines in high yield and good purity using diborane in tetrahydrofuran as the reducing agent of *tert*-butyl alkylidinecarbazole has recently been described [48]. Equations (31–33) illustrate the reaction scheme starting with cyclohexanone. To be noted again is the use of *tert*-butyl ester which facilitates the hydrolytic last step of the reaction sequence.

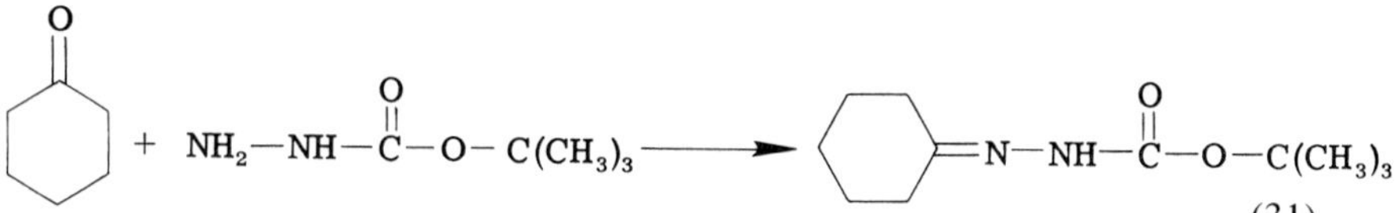

$$\text{cyclohexanone} + NH_2-NH-\overset{O}{\underset{\|}{C}}-O-C(CH_3)_3 \longrightarrow \text{cyclohexylidene}=N-NH-\overset{O}{\underset{\|}{C}}-O-C(CH_3)_3 \quad (31)$$

$$\text{cyclohexyl}-NH\overset{O}{\underset{\|}{C}}OC(CH_3)_3 \xrightarrow{BH_3/THF} \text{cyclohexyl}-NHNH\overset{O}{\underset{\|}{C}}OC(CH_3)_3 \quad (32)$$

$$\text{cyclohexyl}-NHNH\overset{O}{\underset{\|}{C}}OC(CH_3)_3 \xrightarrow{6N-HCl} \text{cyclohexyl}-NHNH_2 \cdot HCl \quad (33)$$

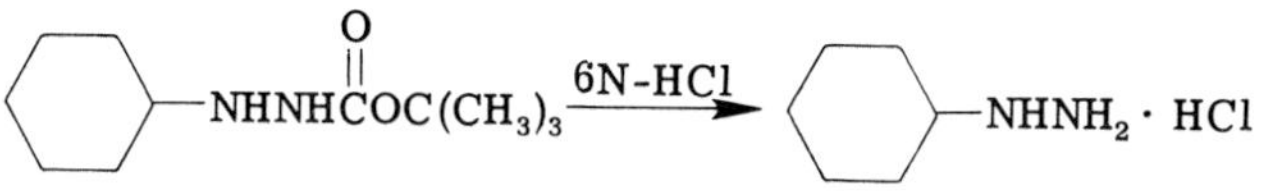

Both 1-alkyl- and 1,2-dialkylhydrazines have been prepared from the corresponding hydrazones or azines by hydrogenation over platinum [23,49].

c. REDUCTION OF DIAZONIUM SALTS

In the aromatic series, hydrazine derivatives are conveniently prepared by the reduction of diazonium salts. The reduction is usually carried out with freshly prepared sodium bisulfite [50,51]. For example, 2,4,6-tri-*tert*-butylphenylhydrazine has been prepared from the corresponding amine by this procedure according to Eqs. (34) and (35) [50].

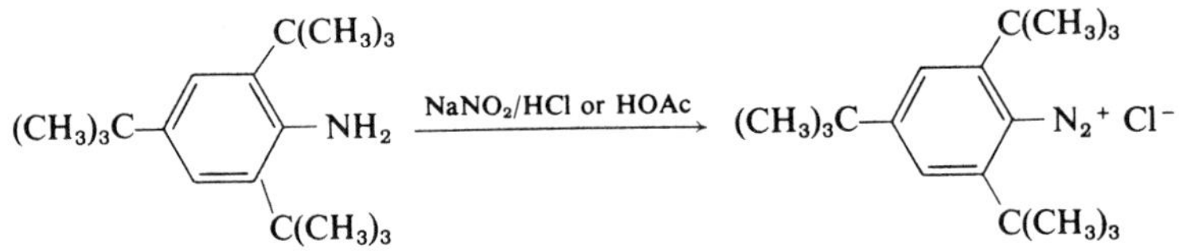

§ 2. Hydrazines

$$(CH_3)_3C\text{-}C_6H_2(C(CH_3)_3)_2\text{-}NH_2 \xrightarrow{NaNO_2/HCl \text{ or } HOAc} (CH_3)_3C\text{-}C_6H_2(C(CH_3)_3)_2\text{-}N_2^+\ Cl^- \quad (34)$$

$$(CH_3)_3C\text{-}C_6H_2(C(CH_3)_3)_2\text{-}N_2^+\ Cl^- \xrightarrow[\text{2. HCl/MeOH}]{\text{1. NaHSO}_3} (CH_3)_3C\text{-}C_6H_2(C(CH_3)_3)_2\text{-}NHNH_2\cdot HCl \quad (35)$$

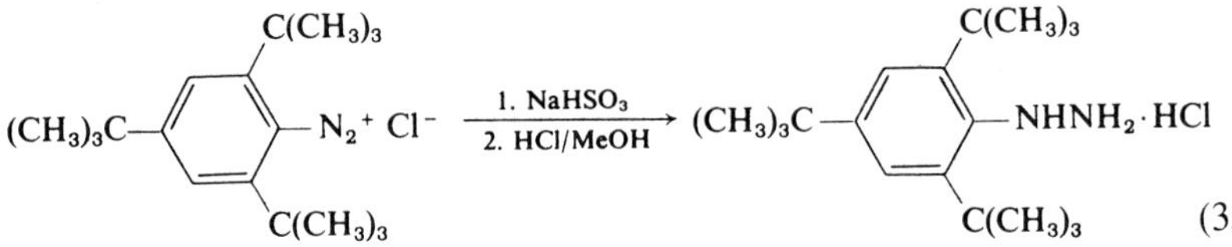

In this reduction, the preparation of sodium bisulfite from aqueous sodium hydroxide and gaseous sulfur dioxide to a phenolphthalein end-point appears to be critical. Coleman [52], in his description of a similar preparation in *Organic Syntheses*, indicates that an excessively alkaline reducing agent causes tar formation during the synthesis with considerable loss in yield.

A variation of the reduction of diazonium salts using stannous chloride has also been described [53].

d. REDUCTION OF AZO COMPOUNDS

Aromatic azo, azoxy, as well as nitroso compounds have been reduced to 1,2-diarylhydrazines (hydrazoarenes) with iodine and magnesium iodide [54].

A novel reduction of azobenzene to 1,2-diphenylhydrazine (hydrazobenzene) involves the *in situ* formation of a diazene(diimide) when azobenzene and hydrazine hydrate is treated with benzeneseleninic anhydride [55]. This method should be explored further as a reductive procedure.

2-11. *Preparation of 1,2-Diphenylhydrazine by Diazine Reduction* [55]

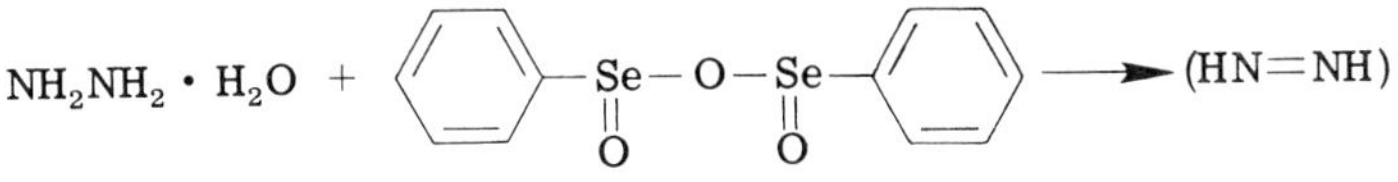

$$NH_2NH_2 \cdot H_2O + Ph\text{-}Se(O)\text{-}O\text{-}Se(O)\text{-}Ph \longrightarrow (HN=NH) \quad (36)$$

$$Ph\text{-}N=N\text{-}Ph + (HN=NH) \xrightarrow{DMF} Ph\text{-}NH\text{-}NH\text{-}Ph + N_2 \quad (37)$$

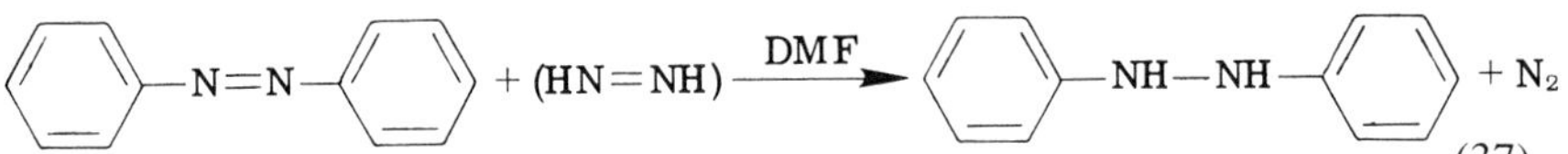

CAUTION: Benzeneseleninic anhydride is toxic and should be handled with care. In the absence of solvents, hydrazine hydrate reacts vigorously with the seleninic anhydride. Seleninic acid and benzhydrazide, as a solid mixture, may decompose violently after a short induction period. Consequently all work with seleninic compounds must be carried out with all due safety precautions. Hydrazine hydrate is extremely toxic and a cancer suspect agent.

To a solution of 0.91 gm (0.50 mmole) of azobenzene in 5 ml of dimethylformamide and 0.232 gm (4.66 mmole) of hydrazine hydrate is added over a 15-min period a solution of 0.456 gm (1.27 mmole) of benzeneseleninic anhydride in 3 ml of dimethylformamide. After 5 min, the solvent is removed under reduced pressure. The residue is taken up in carbon tetrachloride (caution). The product is isolated by preparative thin-layer chromatography. Yield: 0.94 gm (102%).

e. BIMOLECULAR REDUCTION OF NITRO COMPOUNDS

The bimolecular reduction of aromatic nitro compounds to yield 1,2-disubstituted hydrazines is a classical reaction of considerable applicability. The preparation of 1,2-di(2-bromophenyl)hydrazine is a typical example of this reaction.

2-12. *Preparation of 1,2-Di(2-bromophenyl)hydrazine* [56]

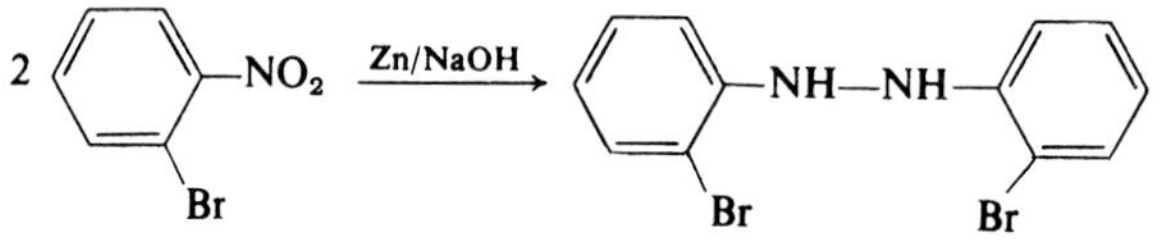

$$2 \;\; \text{C}_6\text{H}_4(\text{Br})\text{-NO}_2 \xrightarrow{\text{Zn/NaOH}} \text{(Br)C}_6\text{H}_4\text{-NH-NH-C}_6\text{H}_4(\text{Br}) \quad (38)$$

In a three-necked flask fitted with an efficient mechanical stirrer and a reflux condenser are placed 20 gm (0.1 mole) of *o*-nitrobromobenzene and 5 ml of a 50% aqueous sodium hydroxide solution. The mixture is stirred at 60°C. Then 15 gm (0.23 gm-atom) of zinc dust is added in very small portions at such a rate that the temperature is maintained between 70° and 80°C. Upon completion of the addition the sludge is diluted with 50 ml of water and 30 ml of a 20% aqueous sodium hydroxide solution. Then 20 gm (0.31 gm-atom) of zinc is added rapidly and the mixture is stirred at 70°–80°C until it is nearly colorless. The cooled mixture is poured slowly with vigorous stirring into 200 ml of a 25% solution of sulfuric acid maintained at 10°C with a large ice bath. The mixture is filtered. The dry residue is extracted with three 30-ml portions of ether. After drying the combined ether solutions over sodium sulfate, the ether extract is evaporated and the residue is recrystallized from petroleum ether (bp 40°–60°C), diluted with a little benzene. Yield 10.8 gm (57%), mp 97°–98°C.

Other bimolecular reductions of nitro-aromatic compounds have included an electrolytic procedure in the presence of water-insoluble organic solvents to prevent coating of the cathode by the product [57] and hydrogenation on a nickel or platinum catalyst in the presence of pyridine [58].

f. OXIDATION PROCESSES

One method of preparing tetraarylhydrazines is the bimolecular oxidation of amines according to the following reaction scheme. The preparation of 1,1,2,2-tetra(4-fluorophenyl)hydrazine is an example of this process.

2-13. Preparation of 1,1,2,2-Tetra(4-fluorophenyl)hydrazine [59]

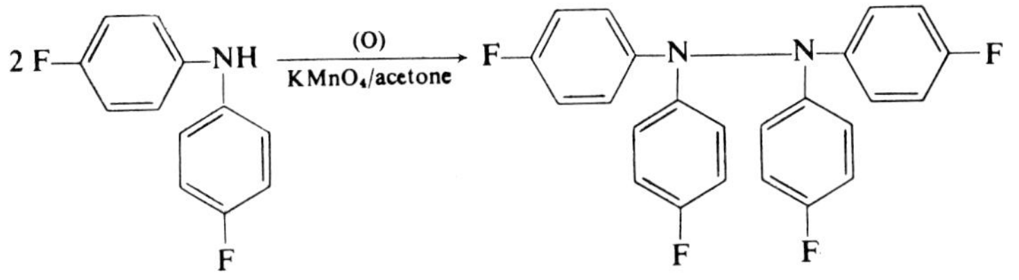

$$2\ F{-}C_6H_4{-}NH{-}C_6H_4{-}F \xrightarrow[\text{KMnO}_4/\text{acetone}]{(O)} (4\text{-}F\text{-}C_6H_4)_2N{-}N(C_6H_4\text{-}4\text{-}F)_2 \quad (39)$$

In a flask fitted with a thermometer, mechanical stirrer, addition funnel, and reflux condenser and containing 6.15 gm (0.03 mole) of di(4-fluorophenyl)amine, 10 gm of anhydrous sodium sulfate, and 50 ml of purified acetone at 0°C is added to a solution of 1.58 gm (0.01 mole) of potassium permanganate in sufficient purified acetone, dropwise, while maintaining the reaction temperature at 0°C. After the addition has been completed, stirring is continued for approximately 3 hr, until the permanganate color has disappeared. The reaction mixture is then filtered, the solid on the funnel is washed repeatedly with acetone, and the solution is combined with the filtrate. The filtrate is concentrated at room temperature under reduced pressure. The nearly colorless crystalline residue is then dissolved in cold benzene and precipitated with petroleum ether. Yield 3.4 gm (55%), mp 129°–130°C (dec).

By a similar procedure, the three isomeric 1,1'-, 2,2'-, and 1,2'-dinaphthylamines have been oxidized to the corresponding symmetrical tetranaphthylhydrazines by a neutral permanganate oxidation [60].

3. HYDRAZONES

The preparation of hydrazones by the reaction of hydrazine derivatives with carbonyl compounds is well known and extensively described in most laboratory manuals. A typical example, used in our laboratory, is the preparation of benzaldehyde *p*-nitrophenylhydrazine [61] given below.

3-1. Preparation of Benzaldehyde p-Nitrophenylhydrazone [61]

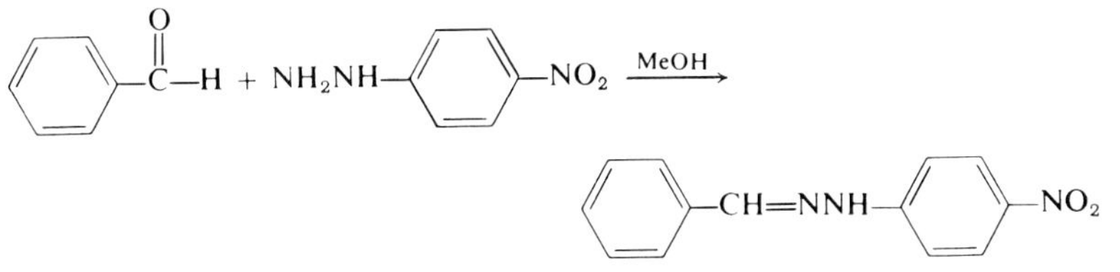

$$C_6H_5{-}CHO + NH_2NH{-}C_6H_4{-}NO_2 \xrightarrow{\text{MeOH}} C_6H_5{-}CH{=}NNH{-}C_6H_4{-}NO_2 \quad (40)$$

In a 22-liter flask, cooled in an ice bath, is placed 1000 gm (6.45 moles) of *p*-nitrophenylhydrazine and 10 liters of methanol. The solution is stirred while

800 gm (7.55 moles) of benzaldehyde is added dropwise. During this addition, the temperature within the reaction flask is kept below 0°C by use of an ice–salt bath. After the addition of benzaldehyde is completed, stirring is continued for 16 hr while the reaction flask is allowed to warm up to room temperature. The product is then filtered off and freed of excess benzaldehyde by repeated washing with cold methanol. After drying, the product may be recrystallized from ethanol. Yield 1484 gm (94%), mp 195°C.

A procedure using essentially the same approach but using a small amount of glacial acetic acid as a catalyst for the reaction has also been reported [62–64].

Recent studies on the mechanism of phenylhydrazone formation from a ketone and phenylhydrazine bears out the empirical observation that slightly acidic conditions are useful. It was found that the rate-determining attack of the nucleophilic reagent takes place under acidic conditions with the formation of an intermediate carbinolhydrazine. The dehydration of the carbinolhydrazine to a hydrazone is rate determining under neutral or basic conditions. Equation (41) outlines the steps of hydrazone formation:

$$RNHNH_2 + \;\;\;C=O \xrightleftharpoons{acid} RNH-NH-\overset{|}{\underset{|}{C}}-OH \xrightarrow{base} RNH-N=C \;\;\; + H_2O \quad (41)$$

Some hindered ketones, such as camphor, are difficult to convert to the 2,4,6-triisopropylbenzenesulfonylhydrazone. In acetonitrile as a solvent, and with an excess of hydrochloric acid, hydrazones of hindered ketones have been prepared [67]. Tosylhydrazones of hindered carbonyl compounds have been prepared by refluxing the carbonyl compound with a 10–20% molar excess of *p*-toluenesulfonylhydrazine in a minimal quantity of absolute ethanol (app. 200 ml of solvent per mole of hydrazine) for up to 24 hr without any acid present [68].

Aldehyde and ketones usually react readily with *N*-alkyl- or *N*,*N*-dialkylhydrazines to give the corresponding N-alkylated hydrazones whereas pure *N*-unsubstituted hydrazones are difficult to prepare. However, it has been found that, in many cases, an exchange reaction between *N*,*N*-dimethylhydrazones and hydrazine (anhydrous) in an anhydrous medium such as anhydrous ethanol will lead to hydrazones of the type [69]

$$\underset{R}{\overset{R}{\diagdown}}C = N - NH_2$$

Hydrazones may also be formed by the addition of diazonium salts to active methylene compounds (Japp–Klingemann reaction) [Eq. (42)] [70–72].

$$ArN \equiv N^{\oplus} \; Cl^{\ominus} + \underset{R}{\underset{|}{\text{Py-N}}} \;\; CH-\overset{O}{\overset{||}{C}}-OC_2H_5 \longrightarrow Ar-NHN=C \underset{R}{\underset{|}{\text{Py-N}}} \quad (42)$$

Other functional groups, related to the hydrazones, should also be mentioned. A procedure for the synthesis of an acylhydrazone from a hydrazone has been given above (procedure 2-10) [46]. When an excess of a carbonyl compound is used to treat hydrazine, both nitrogens may react to form azines. With an excess of hydrazine, the simple hydrazones are more apt to predominate in the product mixture as intermediates which may spontaneously form tars or azines, particularly in the presence of water or acids [23,69].

4. HYDRAZIDES

In general, hydrazides may be prepared by many of the methods analogous to those used in the preparation of amides. For example, hydrazine salts of carboxylic acids and reactions of hydrazine with esters, acyl halides, acyl anhydrides, and amides may be used to produce hydrazides. A reaction analogous to the Hofmann degradation is the formation of hydrazides from ureides (acylureas) [73] (Eq. 43).

$$R-\overset{O}{\underset{\|}{C}}-NH\overset{O}{\underset{\|}{C}}-NH_2 \xrightarrow{NaOX} R-\overset{O}{\underset{\|}{C}}-NHNH_2 \qquad (43)$$

The preparation of hydrazides by interaction of esters and hydrazine hydrate is quite straightforward and proceeds in good yields. In many cases, simple addition of hydrazine hydrate to the liquid esters suffices to cause the precipitation of the hydrazide [74]. If the esters are insoluble or solids, a more prolonged treatment is required usually in the presence of an alcohol, as in the preparation of terephthalic hydrazide given below:

4-1. Preparation of Terephthalic Dihydrazide [61,75]

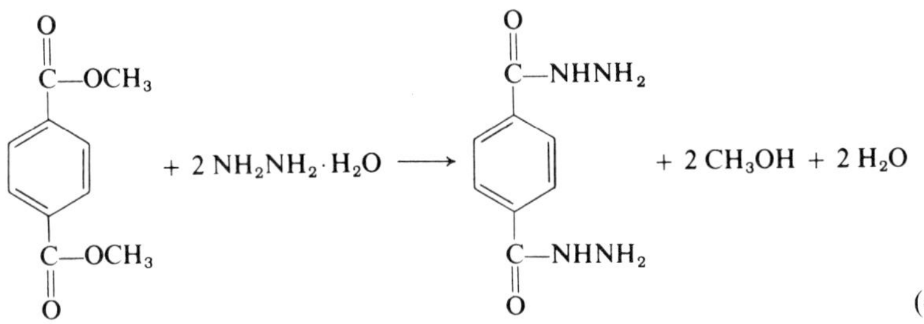

$$\underset{\substack{\| \\ O}}{\overset{O}{\underset{\|}{C}-OCH_3}}\!$$

p-C₆H₄(COOCH₃)₂ + 2 NH₂NH₂·H₂O ⟶ p-C₆H₄(CONHNH₂)₂ + 2 CH₃OH + 2 H₂O (44)

CAUTION: Hydrazine hydrate is extremely toxic and a cancer suspect agent.

For convenient handling, a quantity of dimethyl terephthalate is pulverized in a blender. In a flask, 232.8 gm (1.2 moles) of dimethyl terephthalate is slurried with 2760 ml of methanol. To this reaction mixture is added a solution of 504 gm (8.4 moles) hydrazine hydrate (85%) in 240 ml of methanol. The composition is stirred for 16 hr at room temperature. The solid product is then separated by filtration. The solid is repeatedly washed with cold methanol and finally dried in a vacuum oven to yield 222 gm (95%), mp over 330°C.

Other acid derivatives such as acid chlorides [76] or amides [77] may also be used for the preparation of hydrazides. With acid chlorides the diacyl hydrazine is frequently formed [22,23].

Thiophosphoric dihydrazides have been prepared by the reaction of such acid chlorides as ethyl phosphorodichloridothionate or phenyl phosphonothioic dichloride with hydrazine, as for example in Eq. (45) [78].

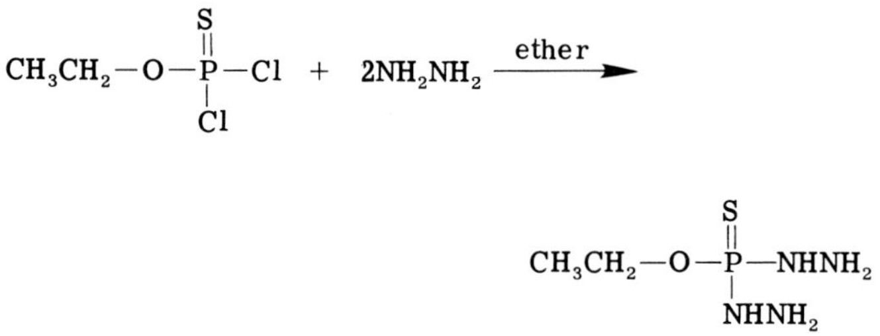

$$CH_3CH_2-O-\underset{Cl}{\overset{\underset{\|}{S}}{P}}-Cl \ + \ 2NH_2NH_2 \ \xrightarrow{\text{ether}}$$

$$CH_3CH_2-O-\underset{NHNH_2}{\overset{\underset{\|}{S}}{P}}-NHNH_2 \quad (45)$$

The reaction of esters and hydrazines may be extended to the preparation of polyacrylic hydrazide using polymethyl acrylate as a starting material.

Since methyl acrylate is difficult to polymerize without some cross-linking, the usual products isolated are somewhat cross-linked. Even the hydrazides prepared from a modest molecular weight polymer of methyl acrylate, which is not cross-linked, tend to cross-link on standing. The preparation below is an example of the preparation of a polymeric hydrazide.

4-2. Preparation of Polyacrylic Hydrazide(I) [79]

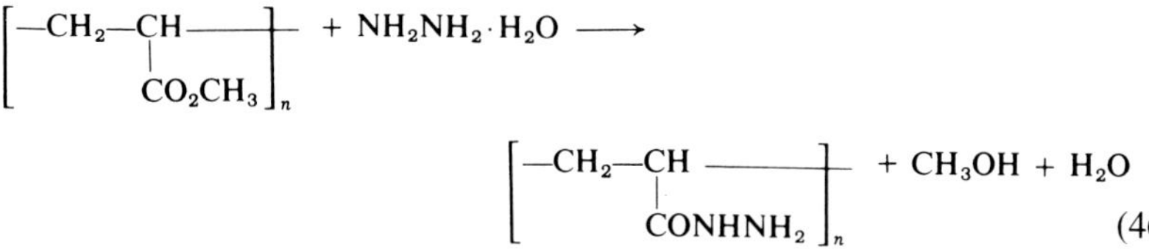

$$\left[-CH_2-\underset{CO_2CH_3}{\overset{|}{CH}}- \right]_n + NH_2NH_2 \cdot H_2O \longrightarrow$$

$$\left[-CH_2-\underset{CONHNH_2}{\overset{|}{CH}}- \right]_n + CH_3OH + H_2O \quad (46)$$

CAUTION: Hydrazine hydrate in extremely toxic and a cancer suspect agent.

In an Erlenmeyer flask, 5 gm of polymethyl acrylate (molecular weight approx. 80,000), which has been pulverized, and 50 gm of hydrazine hydrate are warmed on a water bath until a homogeneous solution is formed. To this solution is added 500 ml of methanol containing 1 ml of glacial acetic acid. The product thereupon precipitates.

To purify the product, the polyacrylic hydrazide is dissolved in 50 ml of water and again precipitated with methanol. This procedure is repeated several times. The product is finally dried in a vacuum desiccator over sulfuric acid at room temperature. The product is stored in the cold in a vacuum desiccator with the exclusion of light. The stability of this polymer, as a non-cross-linked raw material, is poor. In general, this preparation must be carried out quite rapidly, with a minimum amount of exposure to heat during the preparation and with a minimum exposure to methanol.

In a field of enzymology, polyacrylic hydrazide has become of considerable importance in "enzyme immobilization." Here polyacrylic hydrazide [usually prepared from poly(acrylamide)] is converted to poly(acrylic azide) by treatment with nitrous acid. The polymeric azide forms insoluble covalent derivatives with reactive functional groups of proteins such as enzymes. In several cases, the resulting polymer retains some of biochemical activities of the enzyme which may then be studied conveniently. Refs. 80–85 are a few references to this topic.

The preparation of the hydrazide of a cross-linked copolymer of acrylamide and methyl acrylate is an example of the formation of a resin which has been used to couple with trypsin or with acylase [85]. The fine structures of the polymers have not been determined. The assumption is that since an excess of hydrazine was used in the preparation, both the methyl acrylate and the acrylamide units are converted to the hydrazide.

4-3. *Preparation of Polyacrylic Hydrazide(II)* [85]

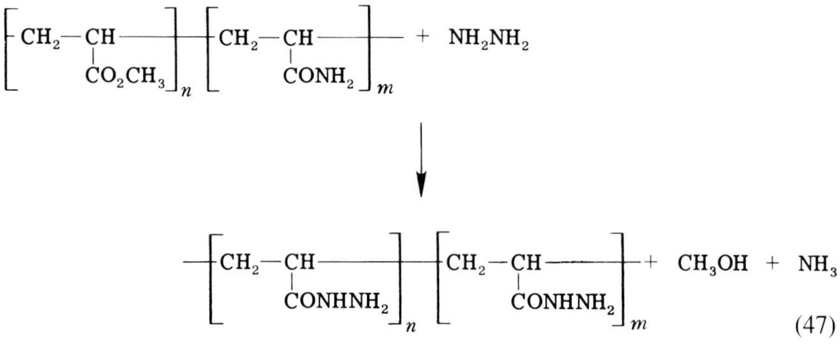

$$\left[\begin{array}{c} CH_2-CH \\ | \\ CO_2CH_3 \end{array} \right]_n \left[\begin{array}{c} CH_2-CH \\ | \\ CONH_2 \end{array} \right]_m + NH_2NH_2$$

$$\downarrow$$

$$\left[\begin{array}{c} CH_2-CH \\ | \\ CONHNH_2 \end{array} \right]_n \left[\begin{array}{c} CH_2-CH \\ | \\ CONHNH_2 \end{array} \right]_m + CH_3OH + NH_3 \quad (47)$$

CAUTION: 95% Hydrazine is extremely toxic and a cancer suspect agent.

In a hood, in a 200-ml Erlenmeyer flask fitted with a reflux condenser and a magnetic stirrer-heater, 7 gm of a cross-linked copolymer of acrylamide and methyl acrylate (prepared by the polymerization of 4.8 gm of acrylamide, 4.8 gm of methyl acrylate, and 0.4 gm of N,N'methylenebisacrylamide) are stirred with 50 ml of 95% hydrazine. The mixture is heated at 70°C for 9 hr. The product is filtered off and washed three times with absolute ethanol. The yield is reported to be 6 gm.

The general procedure of hydrazide preparation from esters has been extended to the amino acid series, where it is used as a means of protecting carboxylic acids in peptide syntheses [86].

5. MISCELLANEOUS PREPARATIONS

(1) Rearrangement of 2,3-dihydro-1,2-diazepin-4-ol to furfurylhydrazine [87]

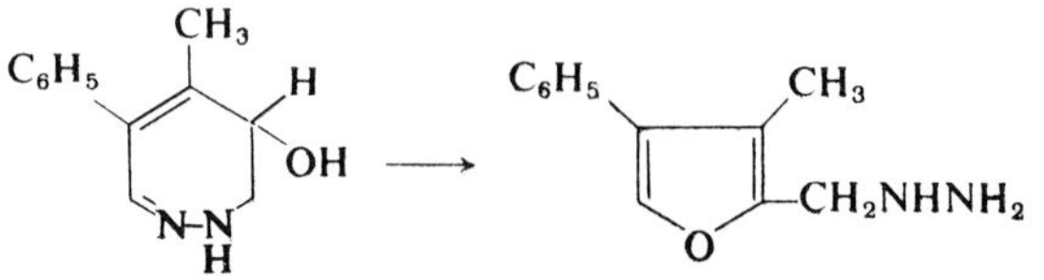

$$\text{(48)}$$

(2) Hydrazines by hydrolysis of hydrazides or substituted carbazates [88]

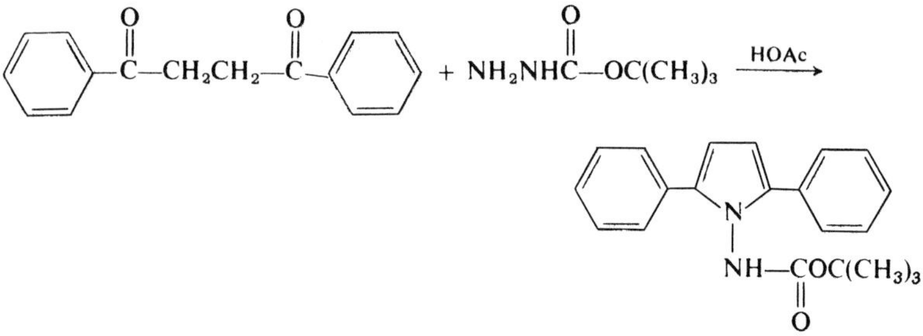

$$\text{(49)}$$

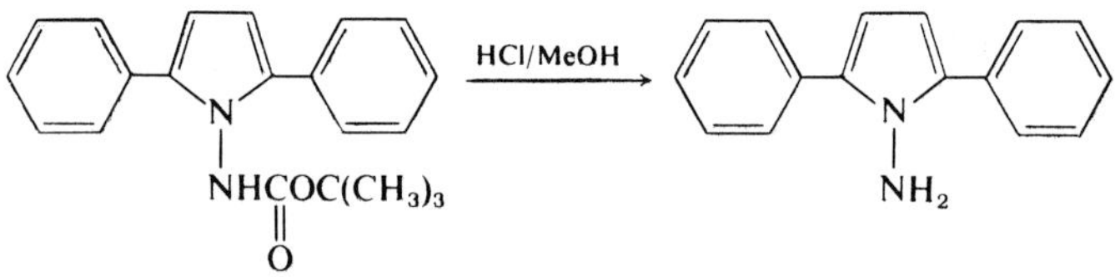

$$\text{(50)}$$

(3) Hydrazines by hydrolysis of diaziridines obtained from azomethines (Schiff bases) [89]

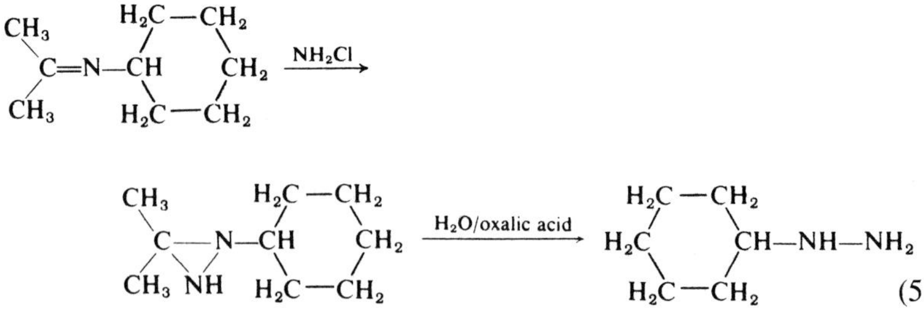

$$\text{(51)}$$

(4) Preparation of 1,2-Diarylbishydrazine by reduction of azoxy-*o*-anisole with sodium hydroxide and Raney nickel and formalin in methanol solution [90].

(5) Preparation of tetrasubstituted hydrazines from symmetrical tetrazines [91].

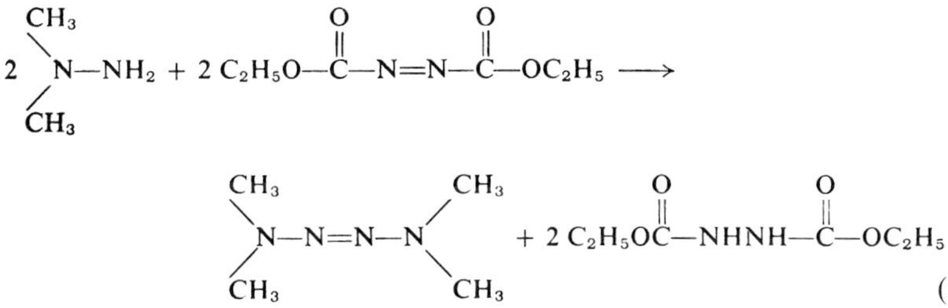

§ 5. Miscellaneous Preparations

$$2 \begin{array}{c} CH_3 \\ \diagdown \\ N-NH_2 \\ \diagup \\ CH_3 \end{array} + 2\ C_2H_5O-\overset{O}{\underset{\|}{C}}-N=N-\overset{O}{\underset{\|}{C}}-OC_2H_5 \longrightarrow$$

$$\begin{array}{c} CH_3 \qquad CH_3 \\ \diagdown \qquad \diagup \\ N-N=N-N \\ \diagup \qquad \diagdown \\ CH_3 \qquad CH_3 \end{array} + 2\ C_2H_5O\overset{O}{\underset{\|}{C}}-NHNH-\overset{O}{\underset{\|}{C}}-OC_2H_5 \quad (52)$$

$$\begin{array}{c} CH_3 \qquad CH_3 \\ \diagdown \qquad \diagup \\ N-N=N-N \\ \diagup \qquad \diagdown \\ CH_3 \qquad CH_3 \end{array} \xrightarrow{120°-140°\ C} \begin{array}{c} CH_3 \quad CH_3 \\ \diagdown \quad \diagup \\ N-N \\ \diagup \quad \diagdown \\ CH_3 \quad CH_3 \end{array} + N_2 \quad (53)$$

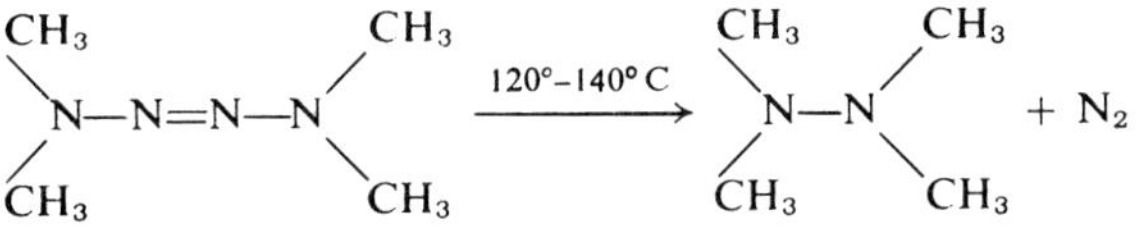

(6) Hydrazone formation by exchange of carbonyl functions [92].

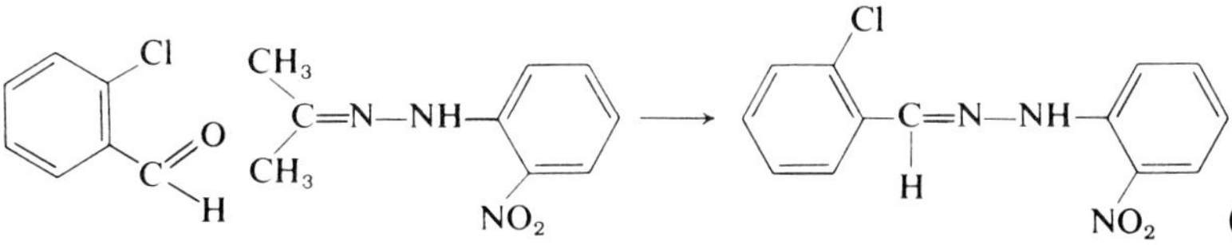

(54)

(7) Preparation of 1-substituted hydrazides by addition of sodamide to N-chloroacetanilides [93].

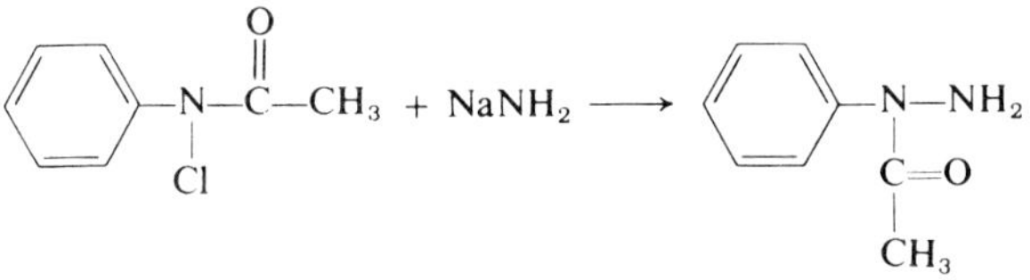

(55)

(8) Preparation of azines and dihydrazides from malonic ester derivatives [94].

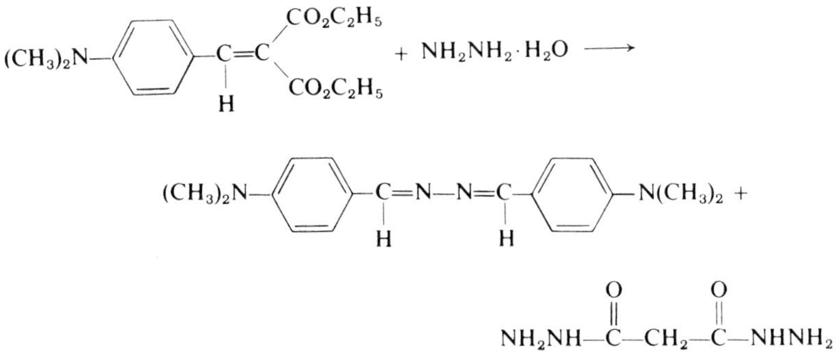

$$NH_2NH-\overset{O}{\underset{\|}{C}}-CH_2-\overset{O}{\underset{\|}{C}}-NHNH_2 \quad (56)$$

(9) Preparation of cyanoethylated hydrazines [16].

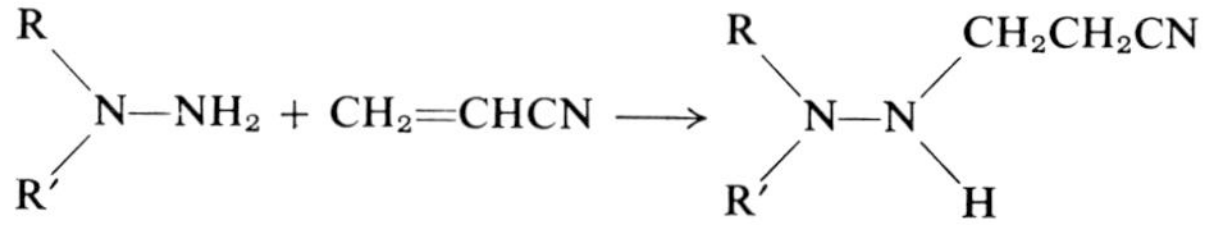

$$\begin{matrix} R \\ R' \end{matrix}\!\!N\!-\!NH_2 + CH_2\!\!=\!\!CHCN \longrightarrow \begin{matrix} R \\ R' \end{matrix}\!\!N\!-\!N\!\!\begin{matrix} CH_2CH_2CN \\ H \end{matrix}$$

(57)

(10) Preparation of isomeric bis(organosilyl)hydrazines [95].

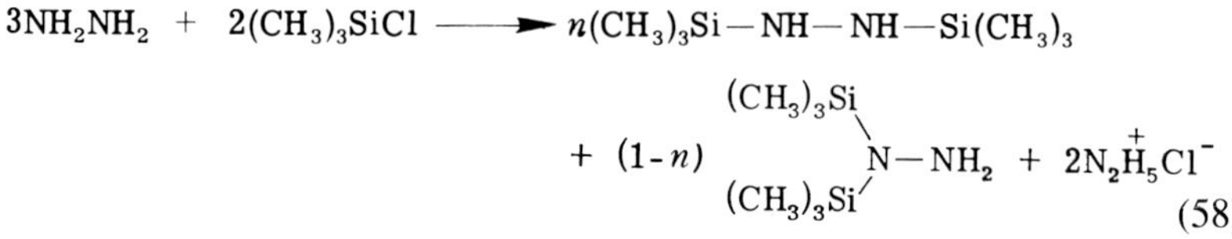

$$3NH_2NH_2 + 2(CH_3)_3SiCl \longrightarrow n(CH_3)_3Si\!-\!NH\!-\!NH\!-\!Si(CH_3)_3$$

$$+ (1-n) \begin{matrix} (CH_3)_3Si \\ (CH_3)_3Si' \end{matrix}\!\!N\!-\!NH_2 + 2N_2\overset{+}{H}_5Cl^-$$

(58)

(11) Preparation of ester p-Tolyhydrazones by reaction of orthoformates or orthocarbonates with p-tosylhydrazine [96].

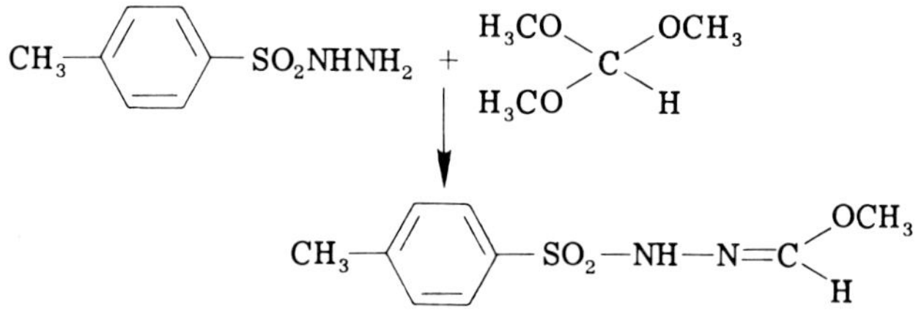

$$CH_3\!-\!\!\!\left\langle\!\!\!\bigcirc\!\!\!\right\rangle\!\!-\!SO_2NHNH_2 + \begin{matrix} H_3CO \\ H_3CO \end{matrix}\!\!C\!\!\begin{matrix} OCH_3 \\ H \end{matrix}$$

$$\downarrow$$

$$CH_3\!-\!\!\!\left\langle\!\!\!\bigcirc\!\!\!\right\rangle\!\!-\!SO_2\!-\!NH\!-\!N\!\!=\!\!C\!\!\begin{matrix} OCH_3 \\ H \end{matrix}$$

(59)

(12) Preparation of high-molecular-weight aromatic polyhydrazides by reaction of p-aromatic dihydrazides with aromatic acyl dichlorides [97].

(13) Preparation of N-Thionylphenylhydrazine [98].

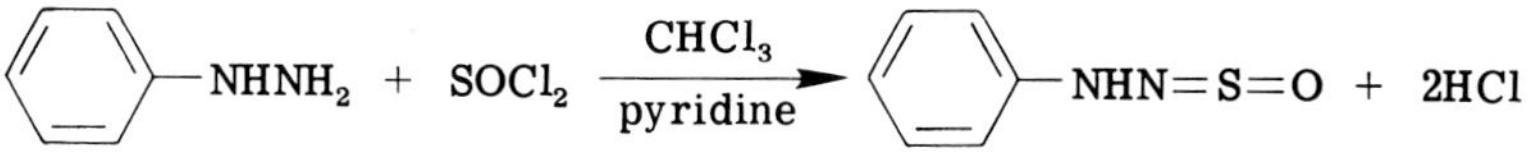

$$\left\langle\!\!\!\bigcirc\!\!\!\right\rangle\!\!-\!NHNH_2 + SOCl_2 \xrightarrow[\text{pyridine}]{CHCl_3} \left\langle\!\!\!\bigcirc\!\!\!\right\rangle\!\!-\!NHN\!\!=\!\!S\!\!=\!\!O + 2HCl$$

(60)

(14) Chemistry of "extremely congested hydrazines" [99].

(15) The Neber rearrangement of quaternized hydrazines [100].

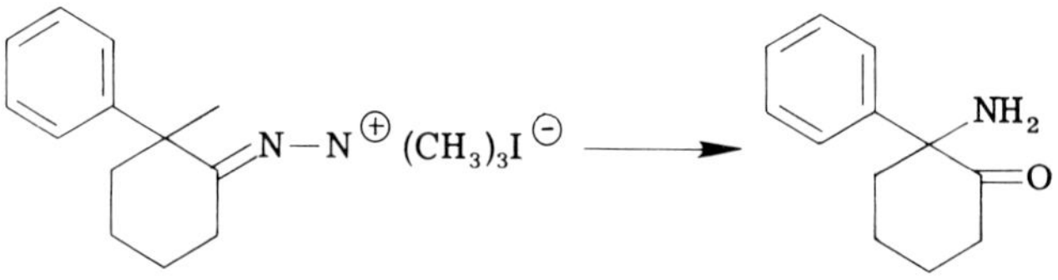

(61)

(16) Preparation of hydrazinated poly(vinylpyridine) [101] (cf. Ref. [21] above for a related Chichibabin reaction).

$$\left[\begin{array}{c}\text{CH}_2-\text{CH}-\\ \begin{array}{c}\diagdown\\ \text{N}\end{array}\end{array}\right]_n + \text{NH}_2\text{NH}_2 \longrightarrow \left[\begin{array}{c}\text{CH}_2-\text{CH}-\\ \begin{array}{c}\diagdown\\ \text{N}\end{array}\text{NHNH}_2\end{array}\right]_n \quad (62)$$

References

1. H. O. Wieland, "Die Hydrazine" *in* "Chemie in Einzeldarstellungen" (J. Schmidt, ed.), Enke, Stuttgart, 1913.
2. L. F. Audrieth and B. A. Ogg, "The Chemistry of Hydrazine." Wiley, New York, 1951.
3. C. C. Clark, "Hydrazine." Mathieson Chem. Corp., Baltimore, 1953.
4. R. A. Reed, *R. Inst. Chem., Rep.* No. 5 (1957).
5. N. V. Sidgwick, *in* "The Organic Chemistry of Nitrogen" (T. W. J. Taylor and W. Baker, eds.), pp. 378ff. Oxford Univ. Press, London and New York, 1942. "N. V. Sidgwick's Organic Chemistry of Nitrogen." 3 ed., Chapter 15. Oxford Univ. Press, London and New York, 1966, and Japan Hydrazine Co. in U.S. Patent 4,310,696 (1/12/82).
6. L. I. Smith, *Chem. Rev.* **23**, 193 (1938).
7. U. V. Solmssen, *Chem. Rev.* **37**, 490 (1945).
8. H. H. Sisler, G. M. Omietanski, and B. Rudner, *Chem. Rev.* **57**, 1021 (1957).
9. E. Enders, *in* "Houben-Weyl's Methoden der organischen Chemie," 4th ed., Vol. X/2, pp. 169, 750. Thieme Verlag, Stuttgart, 1967.
10. E. Müller, *in* "Houben-Weyl's Methoden der organischen Chemie," 4th ed., Vol. X/2. p. 121. Thieme Verlag, Stuttgart, 1967.
11. S. Patai, ed., "Chemistry of Hydrazo-, Azo-, and Azoxy Groups. Wiley, New York, 1975.
12. H. W. Schiessl, *Aldrichimica Acta* **13**(2), 33 (1980).
13. J. Thiele, *Justus Liebigs Ann. Chem.* **376**, 244 (1910).
14. H. H. Hatt, *Org. Synth. Collect. Vol.* **2**, 395 (1943).
15. R. D. Brown and R. A. Kearley, *J. AM. Chem. Soc.* **72**, 2762 (1950).
16. E. F. Elslager, E. A. Weinstein, and D. F. Worth, *J. Med. Chem.* **7**, 493 (1964).
17. C. Westphal, *Ber. Dtsch. Chem. Ges. B.* **74B**, 739, 1365 (1941).
18. C. F. H. Allen, *Org. Synth. Collect. Vol.* **2**, 228 (1943).
19. A. Ault, *J. Chem. Educ.* **42**, 267 (1965).
20. D. G. Holland, G. J. Moore, and C. Tamborski, *J. Org. Chem.*, **29**, 1562, 3042 (1964); D. G. Holland and C. Tamborski, *ibid.*, **31**, 280 (1966).
21. A. Counotte-Potman, H. C. van der Plas, B. van Veldhuizen, and C. A. Landheer, *J. Org. Chem.* **46**, 5102 (1981).
22. H. H. Hatt, *Org. Synth. Collect. Vol.* **2**, 208 (1943).
23. H. L. Lochte, W. A. Noyes, and J. R. Bailey, *J. Am. Chem. Soc.* **44**, 2556 (1922).
24. H. C. Ramsperger, *J. Am. Chem. Soc.* **51**, 918 (1929).
25. F. Raschig, *Ber. Dtsch. Chem. Ges.* **40**, 4587 (1907).
26. L. F. Audrieth and L. H. Diamond, *J. Am. Chem. Soc.* **76**, 4869 (1954).
27. L. H. Diamond and L. F. Audreith, *J. Am. Chem. Soc.* **77**, 3131 (1955).
28. G. M. Omietanski and H. H. Sisler, *J. Am. Chem. Soc.* **78**, 1211 (1956).
29. P. R. Steyermark and J. L. McClanahan, *J. Org. Chem.* **30**, 935 (1965).

30. J. Druey, P. Schmidt, K. Eichenberger, and M. Wilhelm, Swiss Patent 372,685 (1963); *Chem. Abstr.* **61,** 5517e (1964).
31. G. Gever and K. Hayes, *J. Org. Chem.* **14,** 813 (1949).
32. R. Gosi and A. Meuwsen, *Org. Synth.* **43,** 1 (1963).
33. N. Koga and J.-P. Anselme, *J. Org. Chem.* **33,** 3963 (1968).
34. W. W. Hartman and L. J. Roll, *Org. Synth. Collect. Vol.* **2,** 418 (1943).
35. H. H. Hatt, *Org. Synth. Collect. Vol.* **2,** 211 (1943).
36. G. W. Smith and D. N. Thatcher, *Ind. Eng. Chem., Prod. Res. Dev.* **1,** 117 (1962).
37. P. Besson, A. Nallet, and G. Luiset, French Patent 1,364,573 (1964); *Chem. Abstr.* **61,** 11892b (1964).
38. C. G. Overberger, L. C. Palmer, B. S. Marks, and N. R. Byrd, *J. Am. Chem. Soc.* **77,** 4100 (1955).
39. G. Neurath, B. Pirmann, and M. Dünger, *Chem. Ber.* **97,** 1631 (1964).
40. J. Neurath and M. Dünger, *Chem. Ber.* **97,** 2713 (1964).
41. R. J. Hedrich and R. T. Major, *J. Org. Chem.* **29,** 2486 (1964).
42. H. Zimmer, L. F. Audrieth, M. Zimmer, and R. A. Rowe, *J. Am. Chem. Soc.* **77,** 790 (1955).
43. D. M. Lemal, F. Menger, and E. Coats, *J. Am. Chem. Soc.* **86,** 2395 (1964).
44. M. DeRosa and P. Haberfield, *J. Org. Chem.* **46,** 2639 (1981).
45. H. W. Stewart, Belgian Patent 630,723 (1963); *Chem. Abstr.* **61,** 14876c (1964).
46. L. Spialter, D. H. O'Brien, G. L. Untereiner, and W. A. Rush, *J. Org. Chem.* **30,** 3278 (1965).
47. J. D. Benigni and D. E. Dickson, *J. Med. Chem.* **9,** 439 (1966).
48. N. I. Ghali, D. L. Venton, S. C. Hung, and G. C. LeBreton, *J. Org. Chem.* **46,** 5413 (1981).
49. H. L. Lochte, J. Bailey, and W. A. Noyes, *J. Am. Chem. Soc.* **43,** 2597 (1921).
50. F. E. Condon and G. L. Mayers, *J. Org. Chem.* **30,** 3946 (1965).
51. N. Kornblum, *Org. React.* **2,** 287 (1944).
52. G. H. Coleman, *Org. Synth. Collect. Vol.* **1,** 442 (1948).
53. D. S. Tarbell, C. W. Todd, M. C. Paulson, E. G. Lindstrom, and V. P. Wystrach, *J. Am. Chem. Soc.* **70,** 1381 (1948).
54. W. E. Bachman, *J. Am. Chem. Soc.* **53,** 1524 (1931).
55. T. G. Back, S. Collins, and R. G. Kerr, *J. Org. Chem.* **46,** 1564 (1981).
56. H. R. Snyder, C. Weaver, and C. D. Marshall, *J. Am. Chem. Soc.* **71,** 289 (1949).
57. T. Sekine, *Denki Kagaku* **28,** No. 8(1960); *Chem. Abstr.* **61,** 14187g (1964).
58. V. P. Shmonina and D. V. Sokolskii, U.S.S.R. Patent 165,174 (1964); *Chem. Abstr.* **62,** 2737d (1965).
59. F. A. Neugebauer and P. H. H. Fischer, *Chem. Ber.* **98,** 844 (1965).
60. E. Lieber and S. Somasekhara, *J. Org. Chem.* **24,** 1775 (1959).
61. Authors' Laboratory.
62. H. C. Yoa and P. Resnick, *J. Org. Chem.* **30,** 2832 (1965).
63. R. Fusco and F. Sannicolo, *J. Org. Chem.* **46,** 90 (1981).
64. T. Iida and F. C. Chang, *J. Org. Chem.* **46,** 2786 (1981).
65. L. do Amaral and M. P. Bastos, *J. Org. Chem.* **36,** 3412 (1971).
66. M. P. Bastos, L. D. Alves, G. de Oliveira Neto, and L. do Amaral, *J. Org. Chem.* **46,** 3342 (1981).
67. A. R. Chamberlin, J. E. Stemke, and F. T. Bond, *J. Org. Chem.* **43,** 147 (1978).
68. R. O. Hutchins, C. A. Milewski, and B. E. Maryanoff, *J. Am. Chem. Soc.* **95,** 3662 (1973).
69. G. R. Newkome and D. L. Fishel, *J. Org. Chem.* **31,** 677 (1966).
70. R. L. Frank and R. R. Phillips, *J. Am. Chem. Soc.* **71,** 2804 (1949).
71. R. R. Phillips, *Org. React.* **10,** 143 (1959).
72. S. M. Parmeter, *Org. React.* **10,** 1 (1959).

73. P. Schestakov, *Ber. Dtsch. Chem. Ges.* **45,** 32 73 (1912).
74. P. A. S. Smith, *Org. React.* **3,** 337 (1946).
75. W. Sweeney, private communication to W. R. Sorenson and T. W. Campbell; see "Preparative Methods of Polymer Chemistry," p. 103. Wiley (Interscience), New York, 1961.
76. F. K. Velichko, B. I. Keda, and S. D. Polikarpova, *Zh. Obshch. Khim.* **34,** 2356 (1964); *Chem. Abstr.* **61,** 9400b (1964).
77. T. A. Geissman, M. J. Schlatter, I. D. Webb, and J. D. Roberts, *J. Org. Chem.* **11,** 741 (1946).
78. L. A. Cates, Y. M. Cho, L. K. Smith, L. Williams, and T. L. Lemke, *J. Med. Chem.* **19,** 1133 (1976).
79. W. Kern, T. Hucke, R. Holländer, and R. Schneider, *Makromol. Chem.* **22,** 31 (1957).
80. H. Weetall, ed., "Enzymology," Vol. 1, pp. 428ff. Dekker, New York, 1975.
81. J. K. Inman and H. M. Dintzis, *Biochemistry* **8,** 4074 (1969).
82. P. Cuatrecasas, *J. Biol. Chem.* **245,** 3059 (1970).
83. S. A. Baker, P. S. Somers, R. Epton, and J. V. McLaren, *Carbohydr. Res.* **14,** 287 (1970).
84. R. Epton, J. V. McLaren, and T. H. Thomas, *Biochim. Biophys. Acta* **328,** 418 (1973).
85. Y. Ohno and H. A. Stahmann, *Macromolecules* **4,** 350 (1971).
86. H. J. Cheung and E. R. Blout, *J. Org. Chem.* **30,** 315 (1965).
87. J. A. Moore, R. W. Medeiros, and R. L. Williams, *J. Org. Chem.* **31,** 52 (1966).
88. L. Carpino, *J. Org. Chem.* **30,** 736 (1965).
89. E. Schmitz, *Angew. Chem.* **73,** 23 (1961); E. Schmitz and D. Habisch, *Chem. Ber.* **95,** 680 (1962).
90. H. Hiyama and O. Manabe, Japanese Patent 20,844 (1964); *Chem. Abstr.* **62,** 11736b (1965).
91. O. Diels, *Ber. Dtsch. Chem. Ges.* **56,** 1932 (1923).
92. F. Sparatore and F. Pagani, *Gazz. Chim. Ital.* **91,** 1294 (1961).
93. W. F. Short, *J. Chem. Soc.* **119,** 1446 (1921).
94. C. N. O'Callaghan and D. Twomey, *Proc. R. Ir. Acad.* **63,** No. 12, 217 (1964).
95. R. West, M. Ishikawa, and R. E. Bailey, *J. Am. Chem. Soc.* **88,** 4648 (1966).
96. R. M. McDonald and R. A. Krueger, *J. Org. Chem.* **31,** 488 (1966).
97. A. H. Frazer, U.S. Patent 3,130,182 (1964).
98. L. B. Pearce, M. H. Feingold, K. F. Cerny, and J.-P. Anselme, *J. Org. Chem.* **44,** 1881 (1979).
99. S. F. Nelsen and W. P. Parmelee, *J. Am. Chem. Soc.* **102,** 2732 (1980); *J. Org. Chem.* **46,** 3453 (1981).
100. R. F. Parcell and J. P. Sanchez, *J. Org. Chem.* **46,** 5229 (1981).
101. A. Rembaum, U.S. Patent 4,267,234 (1981).

CHAPTER 15 / **DIAZO AND DIAZONIUM COMPOUNDS**

1. Introduction	467
2. Aliphatic Diazo Compounds	467
A. Decomposition of N-Nitroso Compounds with Alkalies	469
a. Decomposition of N-Alkyl-N-nitroso-p-toluenesulfonamides	469
2-1. Preparation of Diazomethane from N-Methyl-N-*nitroso*-p-*toluenesulfonamide*	469
2-2. Preparation of Phenyldiazomethane	470
b. Decomposition of Bis(N-methyl-N-nitroso)terephthalamide	471
2-3. Preparation of Diazomethane	472
c. Decomposition of N-Alkyl-N-nitrosoacyl Amides	472
2-4. Preparation of Diazoethane	473
d. Decomposition of N-Nitroso-N-alkylurethanes	473
2-5. Preparation of Diazoethane	474
e. Decomposition of N-Alkyl-N-nitrosoureas	474
2-6. Preparation of Diazopropane	475
f. Decomposition of 1-Alkyl-1-nitroso-3-nitroguanidine	475
2-7. Preparation of Diazo-n-pentane	476
g. Decomposition of N-Nitroso-β-alkylaminoisobutyl Ketones	476
B. Diazotization of Primary Aliphatic Amines with Activating Substituents in the α-Position	477
2-8. Preparation of 1,1,1-Trifluoro-2-diazopropane	477
C. Reaction of "Dichlorocarbene" with Hydrazine	478
2-9. Preparation of Diazomethane from Hydrazine	479
D. Oxidation of Hydrazones	479
2-10. Preparation of 2-Diazopropane	480
2-11. Preparation of 1-Phenyldiazopropane	483
E. Diazo Transfer Reactions	483
2-12. Preparation of Methyl α-Diazo-α [(methylthio)carbonyl]acetate	485
F. Condensation Reactions	486
2-13. Preparation of 3-β-Acetoxy-16-β-diazoacetylisopregn-5-en-20-one	486
2-14. Preparation of Ethyl 2-Diazo-3-hydroxybutyrate	487
3. Aromatic Diazonium Salts	488
a. Preparation of Diazonium Salt Solutions in Aqueous Media	488
3-1. Preparation of Benzenediazonium Chloride	489
b. Stabilized Diazonium Salts	489
3-2. Preparation of Aryldiazonium Fluoroborate	490
c. Other Stabilized Diazonium Salts	490
3-3. Preparation of 1-Anthraquinonediazonium Chloride	490
3-4. Preparation of Poly(styrenediazonium Chloride)	491
d. Diazotization in Nonaqueous Media	492
3-5. Preparation of N,N-Dicyclohexylbenzamide-o-diazonium Fluoroborate	492

| 4. Miscellaneous Preparations of Diazo Compounds | 492 |
| 5. References | 493 |

1. INTRODUCTION

Although aliphatic diazo compounds and aromatic diazonium compounds do not resemble each other significantly in chemical behavior, electronic structure, or end uses, these two classes are treated together here because of space limitations. A number of reviews on the aliphatic diazo compounds have appeared from time to time (for representative references, see [1–13]).* The chemistry, properties, and uses of aromatic diazonium compounds have been reviewed extensively (Refs. [14–22] are representative of these).

The preparation of diazo hydrocarbons is generally carried out by the alkaline decomposition of N-nitroso-N-alkyl derivatives such as the N-nitroso derivatives of sulfonamides, amides, and phthalamides, urethanes, ureas, nitroguanidines and β-alkylaminoisobutyl ketones. Of these, the decompositions of N-alkyl-N-nitroso-p-toluenesulfonamide and of bis (N-methyl-N-nitroso)terephthalamide are perhaps the most convenient laboratory methods for the preparation of diazomethane.

Also of interest in the preparation of diazoalkanes are the diazotization of primary amines with activating substituents in the α-position, reaction of hydrazine or hydrazides with dichlorocarbene, diazo-group transfer reactions, the oxidation of hydrazones, and condensation reactions of active methylene compounds.

Aromatic diazonium salts are generally prepared by diazotization of aromatic amines in aqueous systems with nitrous acid. In nonaqueous systems, diazotizations have been carried out with isoamyl nitrite.

2. ALIPHATIC DIAZO COMPOUNDS

The aliphatic diazo compounds find application as intermediates in a variety of organic reactions such as the Arndt–Eistert synthesis. Because several explosions have been reported in the preparation of such materials as diazomethane, great care must be taken during the preparation of all diazo compounds, and the isolation of the pure compounds should normally be avoided. The scale of reaction should be kept small and extreme precautions against explosion hazards must be taken. Furthermore, diazomethane and presumably many other diazo compounds as well as many of the intermediates used in the preparation of diazo

---

*Added in proof: T. H. Black, *Aldrichimica Acta*. **16** (1), 3 (1983).

compounds are toxic. Some of the compounds used as intermediates may initially cause sensitization so that upon further contact severe physiological reactions may take place. Intermediates may also be carcinogenic. For example, nitrosomethylurea is considered a potent carcinogen [23].

In preparations involving the use of aliphatic diazo compounds, due caution must be exercised in the transfer of the diazoalkane solution to subsequent reaction systems and in the isolation of the final product. Particular attention must also be paid to the disposal of by-products, to the handling of residues in the reaction flasks, and to traces of diazo compounds in the apparatus. Extensive notes on safety and health considerations for handling of diazomethane are given by Gutsche [6], Moore and Reed [24], and De Boer and Backer [25]. The recommendations made for diazomethane should be applied to all the diazoalkanes and related compounds. To be kept in mind are the following points:

(1) The starting materials, particularly the N-nitroso derivatives used as starting materials, may be toxic, may cause skin irritations, and other serious allergic reactions on contact. They may also be carcinogenic. Therefore protection against contact, inhalation, and spillage must be provided.

(2) The diazo compounds may be both toxic and explosive; therefore, hoods and other provision for protection against explosive hazards must be provided, such as heavy shields, heavy gloves, protective goggles, and helmets.

(3) The explosions due to diazo compounds may be initiated by a variety of factors such as exposure to sunlight or other strong light, contact with sharp edges, corners, ground surfaces, chipped glass surfaces, sticks of potassium hydroxide, or crystalline side products. For this reason it is generally recommended that ground glass equipment not be used. Only clean, new glassware should be used. The flame-polishing of ground glass joints has also been suggested [26].

In one report of a serious diazomethane explosion, an investigation into the accident concluded that a static charge had built up on the chemist's polyester-and-cotton lab coat when the chemist had walked about in the laboratory just prior to pouring a 3% diazomethane solution. When the solution was poured, a spark may have discharged and initiated the explosion [27]. The use of all-cotton lab coats was recommended at that time. However, today polyester and other synthetics are so generally worn that other measures to prevent static discharge will have to be found.

The early directions for the preparation of diazo compounds used to call for shaking the intermediates with alkalies by hand. In view of the explosion hazards involved, Moore and Reed [24] recommend that Teflon-coated magnetic stirrers be used instead.

Recently, new designs for equipment have been suggested for the codistillation of diazomethane with ether using a special Dry Ice condenser to assure reasonably complete condensation of the diazomethane–ether azeotrope [26].

This equipment is commercially available [28]. An apparatus which allows the generation of diazomethane and its isolation without codistillation with ether using a hot bath has also been described. This equipment is probably most useful for the preparation of derivatives on a millimolar scale [29].

A. Decomposition of *N*-Nitroso Compounds with Alkalies

The base-catalyzed decompositions of a large variety of *N*-nitroso compounds to aliphatic diazo compounds are well known. Primary emphasis has been on methods for the generation of diazomethane. A number of these will be mentioned below.

a. Decomposition of *N*-Alkyl-*N*-nitroso-*p*-toluenesulfonamides

The method of DeBoer and Backer makes use of *N*-methyl-*N*-nitroso-*p*-toluenesulfonamide as the source of diazomethane [25,30–32]. The starting material is available from suppliers of specialty organic chemicals, it is reasonably stable at room temperature, and it seems to give fewer allergic reactions than some of the other nitroso compounds used in the preparation of diazomethane. It is believed to have a lower degree of carcinogenicity than many of the other proposed intermediates [33].

De Boer and Backer described their procedure in *Organic Syntheses* [25]. Despite the fact that these authors demonstrated that the maximum yield of diazomethane is obtained when the molar ratio of potassium hydroxide to *N*-methyl-*N*-nitroso-*p*-toluenesulfonamide is 0.8 to 1 [31]. The usual practice appears to be to use equimolar quantities.

Preparation 2-1 is a recent adaptation of the De Boer–Backer method using the Hudlicky apparatus with its special Dry Ice condenser [26].

2-1. *Preparation of Diazomethane from N-Methyl-N-nitroso-p-toluenesulfonamide* [26]

$$H_3C\text{-}C_6H_4\text{-}SO_2N(NO)\text{-}CH_3 + ROH \xrightarrow{KOH} CH_2N_2 + H_3C\text{-}C_6H_4\text{-}SO_2OR + H_2O \quad (1)$$

In a hood, behind a safety shield, in an apparatus (illustrated in Refs. [26] and [28]) consisting of a 500-ml three-necked round bottom flask immersed in a water bath and equipped with a Teflon-coated magnetic stirring bar, a thermometer, a 500-ml addition funnel with pressure equilizer, and a Dry Ice reflux condenser, arranged like a distillation head with a Teflon take-off stopcock and a liquid overflow trap connected by a ground glass joint to a receiver for the product and also fitted with a vent which is connected to a Dry Ice trap, are placed 10 gm (0.32 mole) of potassium hydroxide dissolved in 30 ml of water, 105 ml of Carbitol (diethylene glycol monoethyl ether), and 30 ml of ether. After filling the condenser with Dry Ice and acetone, to the addition funnel is charged a solution of 64.2 gm (0.3 mole) N-methyl-N-nitroso-p-toluenesulfonamide in 375 ml of ether. After attaching a 500-ml flask as a receiver and cooling it in an ice–water bath, the magnetic stirrer is started and the water bath is heated to 60°C. Then the addition of the nitroso compound is begun at such a rate that all of the yellow vapor formed is completely condensed when the take-off stopcock is closed. Upon opening the stopcock, the first portion of the diazomethane–ether distillate is allowed to flow into the receiver at such a rate that no vapor is permitted to pass the overflow trap. The bath temperature is gradually raised to approximately 80°C until the addition has been completed. Distillation is continued until only colorless ether collects in the overflow trap. The process is complete within approximately 100 min. Dry-Ice consumption in this time is on the order of 2 kg. The yield is said to be approx. 0.27 mole (90% of theory).

On a microgram scale, carboxylic acids have been esterified with diazomethane in a column procedure using N-methyl-N-nitroso-p-toluenesulfonamide. This procedure is particularly attractive for handling small amounts of materials which are to be subject to gas chromatography [34].

Other diazoalkanes may be prepared from nitrosated N-alkyl-p-toluenesulfonamides. For example, phenyldiazomethane may be prepared by this method [32]. In this reaction, N-benzyl-p-toluenesulfonamide is prepared by a conventional Schotten–Baumann type reaction followed by nitrosation and decomposition with sodium methylate.

2-2. Preparation of Phenyldiazomethane [32]

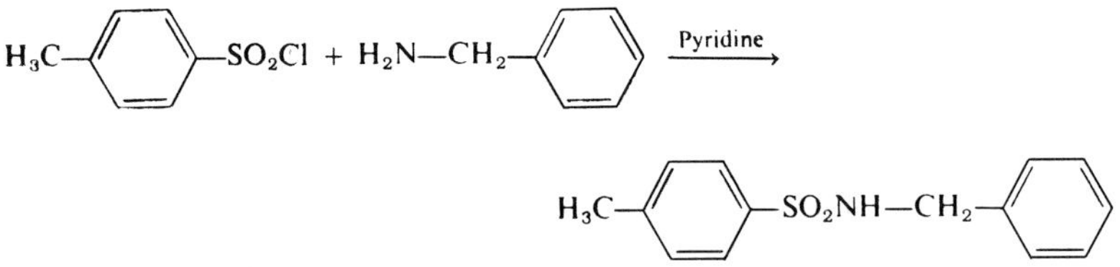

$$H_3C-C_6H_4-SO_2Cl + H_2N-CH_2-C_6H_5 \xrightarrow{\text{Pyridine}}$$

$$H_3C-C_6H_4-SO_2NH-CH_2-C_6H_5 \quad (2)$$

(Preparation not described)

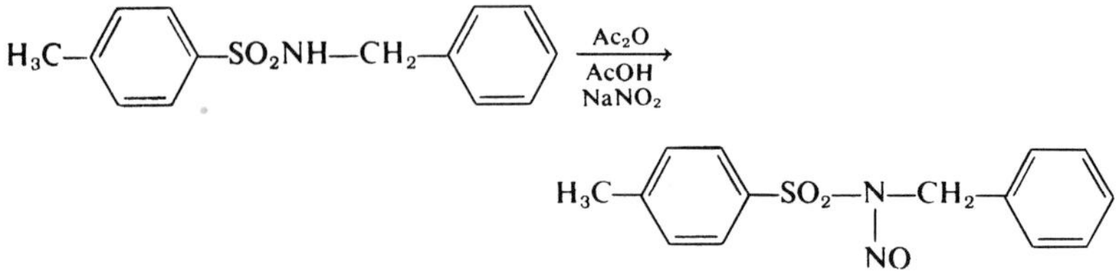

$$H_3C-C_6H_4-SO_2NH-CH_2-C_6H_5 \xrightarrow[\text{AcOH} \atop \text{NaNO}_2]{Ac_2O}$$

$$H_3C-C_6H_4-SO_2-\underset{NO}{N}-CH_2-C_6H_5 \quad (3)$$

§ 2. Aliphatic Diazo Compounds

$$H_3C-\langle\bigcirc\rangle-SO_2-\underset{\underset{NO}{|}}{N}-CH_2-\langle\bigcirc\rangle \xrightarrow[\text{Ether}]{CH_3ONa} \langle\bigcirc\rangle-CHN_2 \quad (4)$$

(a) Nitrosation of N-benzyl-p-toluenesulfonamide. In a hood, in a three-necked flask equipped with stirrer, thermometer, and an addition funnel, a solution of 10.5 gm (0.04 mole) of *N*-benzyl-*p*-toluenesulfonamide in 50 ml of glacial acetic acid and 200 ml of acetic anhydride is cooled to 5°C and 60 gm (0.85 mole) of granulated sodium nitrite is added portionwise while maintaining a temperature below 10°C at all times (approx. 6 hr). The green reaction mixture is then stirred overnight. It is then poured into a large quantity of ice water with rapid stirring. The reaction mixture is cooled in an ice bath for 1 hr. The resulting pale yellow precipitate is filtered, washed thoroughly with water, and dried under reduced pressure.

The yield after recrystallization from ethanol is 9.4 gm (81%) of *N*-nitroso-*N*-benzyl-*p*-toluenesulfonamide, mp 90°–92°C.

(b) Preparation of phenyldiazomethane. With due precautions against explosion hazards, in an apparatus equipped with a Teflon-coated magnetic stirrer, a reflux condenser topped with a drying tube, a mixture of 1.35 gm (0.025 mole) of sodium methylate, 5 ml of methanol, and 30 ml of ether is rapidly stirred. To this is added gradually 7.25 gm (0.025 mole) of *N*-nitroso-*N*-benzyl-*p*-toluenesulfonamide over a 1-hr period. The mixture is then stirred under cautious reflux for 15 to 20 min. After cooling the mixture, 50 ml of water is added to dissolve the salts and the aqueous layer then is separated, due precautions being taken to decontaminate the water layer prior to discarding it. The ether solution of phenyldiazomethane is washed three times with additional portions of water and dried over anhydrous sodium sulfate. The solution is of sufficient purity for most synthetic purposes. The yield of phenyldiazomethane may be estimated by conventional methods for analyzing aliphatic diazo compounds such as titration with organic solutions of benzoic acid. Yield is approximately 60%.

b. DECOMPOSITION OF BIS(*N*-METHYL-*N*-NITROSO)TEREPHTHALAMIDE

One of the methods of preparing diazomethane in good yield uses bis(*N*-methyl-*N*-nitroso)terephthalamide, according to Eq. (5). This preparation has the additional advantage that the starting material is commercially available [24,35].

While Gutsche and Kinoshita [35] make mention of the preparation of diazomethane by this procedure, detailed directions are given by Moore and Reed [24]. In connection with this procedure, recent directions indicate that the decomposition should be carried out with ether containing 3.5% ethanol, such as U.S.P. ether, rather than anhydrous ether, in order to reduce frothing and foaming [36]. Since the compound may be explosive, it is usually supplied in mineral oil.

2-3. Preparation of Diazomethane [24]

$$CH_3-N(NO)-\overset{O}{\underset{\|}{C}}-C_6H_4-\overset{O}{\underset{\|}{C}}-N(NO)-CH_3 + 2\ NaOH \longrightarrow$$

$$2\ CH_2N_2 + NaO_2C-C_6H_4-CO_2Na + 2\ H_2O \tag{5}$$

With safety precautions previously indicated, in a hood behind a safety shield, a 5-liter round- bottomed flask is fitted with a rubber stopper and gooseneck to a long condenser. The condenser is connected through a two-hold rubber stopper to a 5-liter round-bottomed receiving flask by means of an adapter to which a length of 9-mm tubing has been sealed to permit collection of distillate under the surface of approximately 200 ml of anhydrous ether placed in the receiver. The vent hole of the receiver is protected with a drying tube. The receiving flask is cooled in a salt–ice mixture.

In the reaction flask is placed a mixture of 450 ml of Carbitol, 3 liters of ether (U.S.P. grade), and 600 ml of a 30% aqueous solution of sodium hydroxide.

This mixture is thoroughly cooled in an ice–salt bath to 0°C or lower. Then 180 gm (0.5 mole) of a 70% solution of bis(N-methyl-N-nitroso)terephthalamide in mineral oil is added all at once. The condensing and receiving system is connected to the flask immediately, the outside of the flask is carefully dried and then surrounded with a heating mantle, while the receiving flask continues to be cooled. The yellow color of diazomethane is observed almost immediately. Distillation of ether and diazomethane proceeds at a rate of approximately 1 liter per hour. After approximately 2 hr, the distillate is virtually colorless. The distillate contains between 15 and 18 gm of diazomethane (76–86%).

c. Decomposition of N-Alkyl-N-nitrosoacyl Amides

The molecular weights of both the bis(N-methyl-N-nitroso)terephthalamide and the N-methyl-N-nitroso-p-toluenesulfonamides are quite high, considering the low molecular weight of diazomethane derived from them. Therefore, considerable interest exists in preparing diazomethane from low molecular weight starting materials. The alkaline decomposition of N-methyl-N-nitrosoacetamide has been described [37,38] and may be of some value for generation of diazomethane *in situ*. However, this compound, as well as its formamide analog, has a very strong irritating effect on the mucous membranes.

It has been shown that the N-N'-dialkyl-N,N'-dinitrosooxamides are suitable from the standpoint of molecular weight and may be decomposed to diazoalkanes. These N-nitroso compounds are said to be relatively stable and readily prepared by nitrosation of the corresponding N,N'-dialkyloxamide with nitrogen dioxide in carbon tetrachloride solution. The preparation of diazoethane is an example of this reaction [39].

2-4. Preparation of Diazoethane [39]

$$\underset{\underset{NO\ O\ O\ NO}{|\ ||\ ||\ |}}{C_2H_5-N-C-C-N-C_2H_5} \xrightarrow{2OR^-} 2\ CH_3CHN_2 + 2\ \underset{\underset{O\ O}{||\ ||}}{ROC-COR} + 2\ OH^- \quad (6)$$

Using safety precautions similar to those described above, in an apparatus equipped with a thermometer, addition funnel, and distillation connections typical of diazomethane preparation, to 100 ml of butanol, rapidly stirred with a Teflon-coated magnetic stirrer, is added 3 gm (0.13 gm/atom) of sodium. The reaction mixture is heated to 60°C to effect solution. At this temperature, 10.2 gm (0.05 mole) of N,N'-diethyl-N,N'-dinitrosooxamide in 250 ml of ether is added over a period of approximately 20 min. The rate of addition is controlled by the rate at which the mixture of ether and diazoethane distills from the reaction mixture. After addition has been completed, a small amount of additional ether is added to the reaction flask and distillation is continued until the distillate is colorless. The ether–diazoethane distillate is collected in a flask cooled to $-30°C$. Yield 2.9 gm of diazoethane in ether solution (51% of theory). This reaction has been carried out for the preparation of diazoethane, diazomethane, diazopropane, and diazobutane. If desired, conditions may be modified somewhat to distill only the diazoalkanes. This may be accomplished by use of higher boiling reaction solvents such as Cellosolve, if the explosion hazard is not too high [39].

Trimethylsilyldiazomethane has been prepared by the decomposition of the corresponding N-nitrosoacetamide using 2-phenylethylamines as the base [40]. This indicates that the use of N-nitrosoamides is a reasonably general reaction. However, as usual, the hazards associated with intermediates must be carefully evaluated prior to their synthesis and use.

d. DECOMPOSITION OF N-NITROSO-N-ALKYLURETHANES

A method originally developed by H. von Pechmann for the preparation of diazoalkanes makes use of the decomposition of N-nitroso-N-alkylurethane in an alkaline medium. The method may be considered an extension of the decomposition of N-nitrosoamides since urethanes are a special type of amide. Unfortunately some of the starting materials decompose spontaneously and also may exhibit strong physiological effects on the skin and mucous membranes. As a matter of fact, since N-nitrosomethylurethane is a potent carcinogen, all homologs should be considered suspect and this method of preparation should be replaced whenever possible by reactions involving safer reagents [33]. A typical example for the preparation of diazomethane from N-nitrosomethylurethane according to Eq. (7) is given by McPhee and Klingsberg [41].

$$\underset{\underset{NO}{|}}{CH_3-N-COOC_2H_5} + 2\ KOH \longrightarrow CH_2N_2 + K_2CO_3 + C_2H_5OH + H_2O \quad (7)$$

This method may be used for the preparation of higher diazohydrocarbons such as diazoethane, the diazopropanes, diazobutane [42], and 1,2-diphenylcyclopropenyldiazomethane [43]. A procedure representative of the method is given here.

2-5. Preparation of Diazoethane [42]

$$C_2H_5N\text{—}CO_2C_2H_5 + 2\ KOH \longrightarrow C_2H_5N_2 + K_2CO_3 + C_2H_5OH + H_2O \quad (8)$$
$$|$$
$$NO$$

In a hood, with suitable safety precautions, a 1-liter three-necked flask is equipped with an addition funnel, mechanical stirrer, and a gooseneck leading to a condenser set downward for distillation. The receiver is arranged much as described for the preparation of diazomethane above except that a 500-ml filter flask containing some ether is considered adequate for this preparation. The receiver is chilled in a salt–ice bath.

In the three-necked flask are placed 100 ml of anhydrous ether and a solution of 25 gm (0.45 mole) of potassium hydroxide in 100 ml of *n*-propanol. The solution is warmed on a water bath until the ether begins to distill. Then, while the waterbath is held at 50°C, from the addition funnel a solution of 25 gm (0.17 mole) of *N*-ethyl-*N*-nitrosourethane in 75 ml of anhydrous ether is added at such a rate that no serious frothing takes place (about 5 min). After the addition has been completed, anhydrous ether is added and distilled until the distillate is colorless. A total of approximately 400 ml of a deep orange solution is obtained. Yield approximately 75% of the theoretical amount of diazoethane.

The preparation of the newly discovered α-diazosulfones of the type

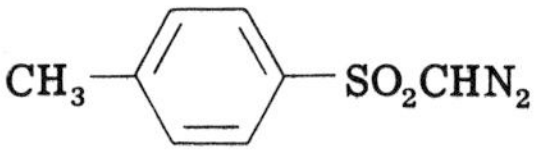

$$CH_3\text{—}\langle\text{C}_6H_4\rangle\text{—}SO_2CHN_2$$

is also based on the decomposition of the corresponding *N*-nitrosourethane, using alumina as the basic reagent [44].

e. Decomposition of *N*-Alkyl-*N*-nitrosoureas

Another method for the preparation of diazoalkanes involves the decomposition of *N*-alkyl-*N*-nitrosoureas [Eq. (9)] [45].

$$\begin{array}{c}\text{O}\\\|\\CH_3N\text{—}C\text{—}NH_2 + KOH \longrightarrow CH_2N_2 + KCNO + 2\ H_2O\\|\\NO\end{array} \quad (9)$$

This synthesis is of particular interest for *in situ* evolution of diazoalkanes. It has also been used for the preparation of 1-diazopropane [46].

§ 2. Aliphatic Diazo Compounds

More complex urea derivatives have also been nitrosated and decomposed. For example, the decomposition of N-nitroso-N-(2,2-diphenylcyclopropyl)urea at 0°C with lithium ethoxide in ether at 0°C afforded 1,1-diphenylallene, the decomposition product of 2,2-diphenyldiazocyclopropane, lithium 2,2-diphenylcyclopropyldiazotate, and lithium cyanate [47]. These products have led to the proposal of a new mechanism of this decomposition.

Preparation 2-6 is an application of the process to the formation of diazopropane.

2-6. Preparation of Diazopropane [46]

$$CH_3CH_2CH_2N(NO)-C(=O)-NH_2 + KOH \longrightarrow CH_3CH_2CHN_2 + KCNO + 2\,H_2O \quad (10)$$

Using safety precautions as indicated at the beginning of this chapter, 1.0 gm (0.0076 mole) of N-nitroso-N-1-propylurea is added during 2 min to a mixture of 10 ml of anhydrous ether and 3 ml of 40% (0.021 mole) potassium hydroxide at 0°C. After standing at 0°C for 30 min, the ether layer is decanted onto potassium hydroxide pellets. After drying for 2 hr at 0°C, the solution is filtered and used. By titration with benzoic acid the yield is estimated at 52% of 1-diazopropane.

In a recent report on the preparation of diazoketones, diazoalkanes solutions in ether were prepared by essentially the same procedure as given in Preparation 2-6. The ethereal solution was decanted off, cooled to 0°C and then treated dropwise with appropriate acyl halides. By this procedure, 1-diazo-5,7-octadiene-2-one was prepared in 100% yield. IR (neat) 2092 and 1633 cm^{-1} [48].

f. DECOMPOSITION OF 1-ALKYL-1-NITROSO-3-NITROGUANIDINE

Decomposition of 1-alkyl-1-nitroso-3-nitroguanidine to the diazoalkanes according to Eq. (11) has been reported [49].

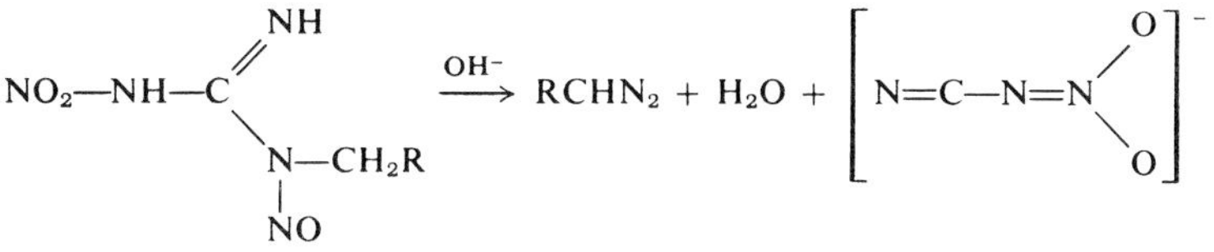

$$NO_2-NH-C(=NH)(N(NO)-CH_2R) \xrightarrow{OH^-} RCHN_2 + H_2O + \left[N=C(-N=N(O)(O))\right]^- \quad (11)$$

While this procedure lends itself to the usual technique of isolating the diazoalkanes up to diazo-*n*-butane by codistillation with ether, the higher diazohydrocarbons may also be isolated by a somewhat simpler technique, as illustrated below.

2-7. Preparation of Diazo-n-pentane [49]

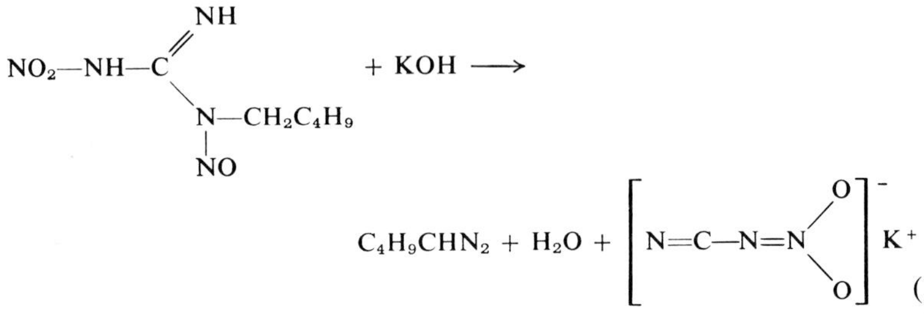

$$NO_2-NH-C\begin{pmatrix} NH \\ N-CH_2C_4H_9 \\ | \\ NO \end{pmatrix} + KOH \longrightarrow$$

$$C_4H_9CHN_2 + H_2O + \left[N=C-N=N\begin{pmatrix} O \\ O \end{pmatrix} \right]^- K^+ \quad (12)$$

All of the glassware in the following procedure is cooled to 0° to −4°C prior to use, and reactions are carried out with the safety precautions as outlined in previous sections of this chapter.

To a solution of 22.4 gm (0.4 mole) of potassium hydroxide in 22.4 ml of water and layered with 250 ml of freshly distilled ether, cooled to −4°C, is added portionwise over a 5–7 min period 20.3 gm (0.1 mole) of 1-nitroso-1-pentyl-3-nitroguanidine, with vigorous stirring by a Teflon-coated stirrer. The solid coproduct is filtered off and the ether layer is separated in a separatory funnel.

The ether layer is washed twice with 200-ml portions of water and dried over potassium hydroxide pellets. The yields of product, determined on the moist ether solution, were between 56 and 58%.

The solid coproduct may explode when treated above its melting point and should, therefore, be disposed of promptly and safely.

g. Decomposition of N-Nitroso-β-alkylaminoisobutyl Ketones

The nitrosation of secondary amines derived from the reaction of mesityl oxide with a primary amine is considered one of the more general methods for the preparation of diazoalkanes [50].

A typical reaction using the procedure is outlined in Eqs. (13)–(15).

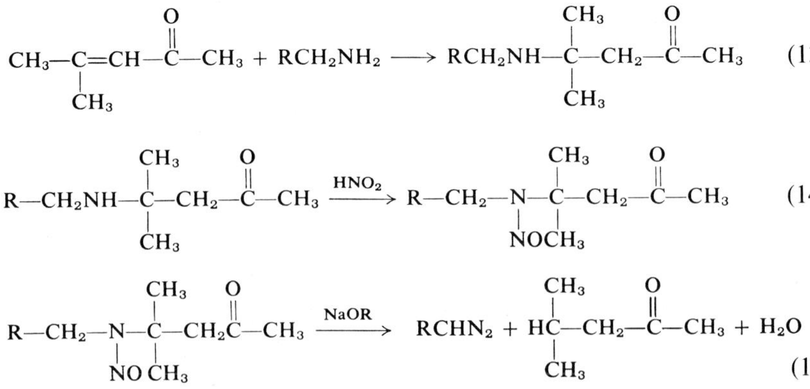

$$CH_3-\underset{\underset{CH_3}{|}}{C}=CH-\overset{O}{\overset{\|}{C}}-CH_3 + RCH_2NH_2 \longrightarrow RCH_2NH-\underset{\underset{CH_3}{|}}{\overset{CH_3}{\overset{|}{C}}}-CH_2-\overset{O}{\overset{\|}{C}}-CH_3 \quad (13)$$

$$R-CH_2NH-\underset{\underset{CH_3}{|}}{\overset{CH_3}{\overset{|}{C}}}-CH_2-\overset{O}{\overset{\|}{C}}-CH_3 \xrightarrow{HNO_2} R-CH_2-\underset{\underset{NOCH_3}{|}}{\overset{CH_3}{\overset{|}{N-C}}}-CH_2-\overset{O}{\overset{\|}{C}}-CH_3 \quad (14)$$

$$R-CH_2-\underset{\underset{NO\ CH_3}{|}}{\overset{CH_3}{\overset{|}{N-C}}}-CH_2\overset{O}{\overset{\|}{C}}-CH_3 \xrightarrow{NaOR} RCHN_2 + H\underset{\underset{CH_3}{|}}{\overset{CH_3}{\overset{|}{C}}}-CH_2-\overset{O}{\overset{\|}{C}}-CH_3 + H_2O \quad (15)$$

This reaction does not afford high yields in the case of heptyl and octylamine derivatives and fails for cyclobutylamine and cyclobutylmethylamine. However, it is considered a fairly general reaction procedure [51].

B. Diazotization of Primary Aliphatic Amines with Activating Substituents in the α-Position

A method, attributed to Curtius, for the preparation of diazo compounds from primary amines with keto, cyano, sulfonic acid, trifluoromethyl, and ethyl carboxy groups in the α-position involves direct diazotization. Since aliphatic diazo compounds are generally unstable in the presence of strong mineral acids, procedures are usually carried out in a way so as to separate the diazo compounds from aqueous acidic media as they are formed. Traditionally the reactions are carried out in an ether dispersion [52], although a more recent preparation of ethyl α-diazoacetate [53] recommends the use of methylene chloride as the reaction solvent which protects diazoacetic ester from decomposition by mineral acids. Equation (16) outlines the general reaction scheme:

$$\text{HCl} \cdot \text{NH}_2\text{CH}_2\overset{\overset{\text{O}}{\|}}{\text{C}}\text{—OC}_2\text{H}_5 + \text{NaNO}_2 \xrightarrow{\substack{\text{Ether} \\ \text{or CH}_2\text{Cl}_2}} \text{N}_2\text{CH}\overset{\overset{\text{O}}{\|}}{\text{C}}\text{—OC}_2\text{H}_5 + \text{NaCl} + 2\,\text{H}_2\text{O} \quad (16)$$

As always in dealing with diazo compounds, the potential hazards of explosions and toxicity are to be kept in mind. The reaction product should be maintained at ice or Dry Ice temperatures. The methods of *Organic Syntheses* [52,53] may be applied to a variety of esters of the α-amino acids. Particular care must be used in handling the methyl esters since some evidently detonate with extreme violence on heating.

Preparation 2-8 is an example of the diazotization reaction.

2-8. Preparation of 1,1,1-trifluoro-2-diazopropane [54]

$$\underset{\underset{\text{CH}_3}{|}}{\text{CF}_3\text{—CHNH}_2\cdot\text{HCl}} + \text{NaNO}_2 \longrightarrow \underset{\underset{\text{CH}_3}{|}}{\text{CF}_3\text{—CN}_2} + \text{NaCl} + 2\,\text{H}_2\text{O} \quad (17)$$

To a suitable apparatus (as indicated for previous preparations), containing 14.9 gm (0.10 mole) of 1-methyl-2,2,2-trifluoroethylamine hydrochloride, cooled to 0°C, is added a solution of 6.9 gm (0.10 mole) of sodium nitrite in 100 ml of water and 50 ml of ether. The mixture is vigorously stirred with a magnetic stirrer for 15 min at 0°C. The yellow organic layer which forms is separated by siphoning. The aqueous layer is repeatedly extracted with ether until it is colorless. The combined ether extracts are washed with cold 5% sodium carbonate solution in water and then dried with anhydrous sodium sulfate. The product solution should be stored at 0°C to reduce detonation hazards.

In a study of the preparation of 1-phenyl-2,2,2-trifluorodiazoethane, three approaches were considered: (1) diazotization of the corresponding amine; (2) basic decomposition of the *p*-toluenesulfonylhydrazone of 2,2,2-trifluoroacetophenone, and (3) oxidation of 2,2,2-trifluoroacetophenone hydrazone. In this particular case, the yield was poorest (17%) when the diazotization of the amine was attempted. The best yield (84%) resulted from the oxidation of the hydrazone [55].

In general, the use of the diazotization reaction of primary aliphatic amines appears to be sensitive to both the exact structure of the amines involved and the nature of the reaction medium [56]. It is also interesting to note that the diazotization of aminotetracyanocyclopentadienide

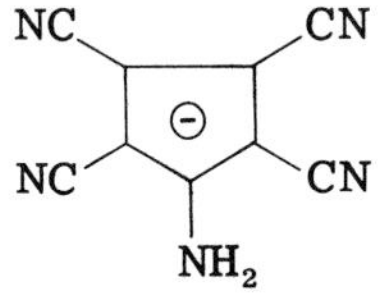

leads to a product which may be considered a resonance hybrid of a diazonium and a diazo compound [57].

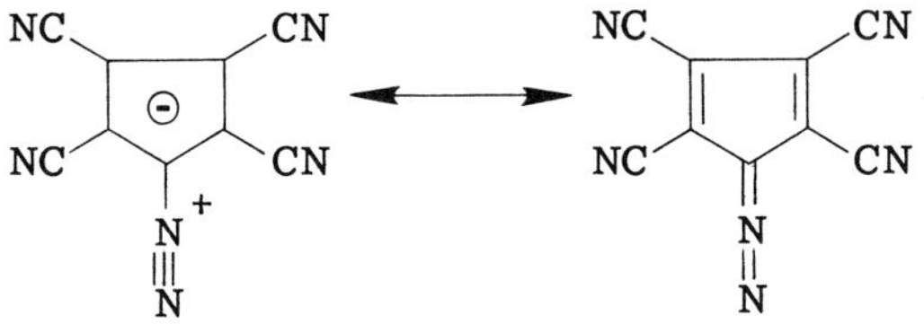

$$\tag{18}$$

C. Reaction of "Dichlorocarbene" with Hydrazine

Staudinger and Küpfer [58] reported the formation of diazomethane when an ethanolic solution of hydrazine and chloroform was treated with a base. The yield was only on the order of 25%. Since hydrazine is relatively inexpensive and of low molecular weight when compared to the *N*-nitroso derivatives discussed above, reexamination of the Staudinger–Küpfer process was of interest. It was found that in the presence of 18-crown-6-ether as a phase transfer catalyst and specifically with potassium hydroxide as a base, a 48% yield of diazomethane could be obtained. Use of sodium hydroxide reduced the yield as did substitution of quarternary ammonium salts for the crown ether. The product was generally contaminated with hydrazine since this reactant evidently forms an azeotropic composition with diazomethane. The hydrazine may be removed from di-

§ 2. Aliphatic Diazo Compounds

azomethane by extraction with water with a concomitant reduction of yield to 38% [59].

2-9. Preparation of Diazomethane from Hydrazine [59]

$$NH_2-NH_2 \cdot H_2O + CHCl_3 \xrightarrow[\text{18-crown-6-ether}]{\text{KOH}} CH_2N_2 + 3HCl + H_2O \quad (19)$$

CAUTION: Hydrazine hydrate is extremely toxic and is considered a cancer suspect agent.

In a hood with all precautions for the handling of hydrazine hydrate and diazomethane is placed the reaction vessel. This reactor consists of a 500-ml two-necked flask to the vertical neck of which a 3 × 28 cm column has been sealed. The lower 18 cm of this column has been "Vigreauxed." A side arm leading to a condenser has been sealed above the Vigreaux column. An adapter has been attached to the condenser, which extends almost to the bottom of a 500-ml ice-cooled Erlenmeyer receiving flask which initially contains 50 ml of ether. To the top of the column a 250-ml addition funnel is attached. The second neck of the reactor flask is used to insert a magnetic stirring bar and to add reagents.

To the reaction flask is added in turn 80 gm (1.4 mol) of potassium hydroxide pellets and 20 ml of water. After the mixture has been cooled, 48 gm (0.4 mole) of chloroform, 200 ml of ether, 0.2 gm (0.8 millimole) of 18-crown-6-ether, and 11.76 gm (0.2 mol) of 85% hydrazine hydrate are added. The addition neck is closed with a rubber stopper. While stirring vigorously, the reaction mixture is heated. Within about 15 minutes the reaction mixture turns yellow. As the reaction continues, diazomethane–ether codistills with foaming. After approximately 100 ml of the distillate has been collected, ether is added through the addition funnel at the top of the column at a rate comparable to the rate of distillation, so that approximately 100 ml of ether is maintained in the reaction flask throughout. After 400–600 ml of distillate has been collected and the vapors in the column are colorless, the reaction is considered complete. The yield is estimated by the reaction of an aliquot of the distillate with benzoic acid to form methyl benzoate. The yield is reported to be 48% ± 4.

Methylene chloride may be substituted for ether in this reaction. A methylene chloride solution of the reaction product (250 ml) may be extracted with 25 ml of water to separate an aqueous solution of the excess hydrazine. In this procedure some product is also lost. The overall yield, based on hydrazine consumed, is then 38%.

D. Oxidation of Hydrazones

The oxidation of hydrazones to diazoalkanes was suggested by Curtius in 1891 and was applied to the preparation of diazopropane by Staudinger and Gaule

[60]. Since then, some controversy has arisen as to the efficiency of the original procedure; consequently, a variety of reaction conditions, oxidizing agents, etc. have been suggested. Reference (61) indicates that the oxidation proceeds smoothly at low temperatures using the traditional yellow mercuric oxide, provided that basic reaction conditions are maintained and that the reaction products are separated as rapidly as possible from mercury-containing coproducts and unused yellow mercuric oxide. Yields are frequently quite high [55].

Preparation 2-10 of 2-diazopropane is an example of the procedure. Some procedures incorporate anhydrous sodium sulfate in the reaction mixture to maintain anhydrous conditions during this oxidation [54,62]. It would appear that the basic catalyst contributes more to efficient reaction than a drying agent.

The use of petroleum ether (boiling range 30°–60°C) instead of ether is said to facilitate separation of by-products such as ketazines [63].

The addition of catalytic amounts of alcoholic potassium hydroxide may also be useful [64].

The use of silver oxide in tetrahydrofuran [65] or of lead tetraacetate [66,67] as oxidizing agents has been described. The latter reference refers to the preparation of perfluoroalkyldiazomethanes. Manganese dioxide has also been suggested [68].

2-10. *Preparation of 2-Diazopropane* [61]

$$\begin{array}{c} CH_3 \\ \diagdown \\ \diagup \\ CH_3 \end{array} C{=}NNH_2 + HgO \longrightarrow \begin{array}{c} CH_3 \\ \diagdown \\ \diagup \\ CH_3 \end{array} CN_2 + H_2O + Hg \qquad (20)$$

In a suitable reaction flask containing an efficient Teflon-coated magnetic stirrer are placed 120 gm (1.33 moles) of yellow mercuric oxide, 350 ml of anhydrous ether, and 9 ml of a 3 N ethanolic potassium hydroxide solution. The mixture is maintained at 0°C by external cooling while a solution of 30 gm (0.42 mole) of freshly distilled acetone hydrazone in 100 ml of dry ether cooled to 0°C is added as rapidly as the vigor of the reaction will permit (approximately 5 min). Stirring is continued at the ice temperature for another 5 min, during which time the concentration of the product reaches a maximum.

Throughout the following operations, the temperature is maintained at 0°C. The reaction mixture is filtered rapidly through a cotton plug into a receiver containing 200 gm of anhydrous potassium hydroxide pellets. The residue on top of the filter is washed repeatedly with ether and the ether washes are combined with the filtrate solution. After 10 min the solution is refiltered into another receiver containing potassium hydroxide to give a usable solution of 2-diazopropane.

If desired, the product may be flash-distilled at $-20°C$ at a pressure of 10 mm

§ 2. Aliphatic Diazo Compounds

of mercury using a Dry Ice condenser. The original ether solution does contain soluble acetone azine and mercury compounds which gradually percipitate as the 2-diazopropane solution is distilled.

The oxidation of hydrazones is not limited to the preparation of diazoalkanes. Diazoketones, for example, may be prepared by the reaction involving a monohydrazone. The preparation of benzoyldiazomethane is a case in point. Equations (21)–(24) show the formation of 2-phenylglyoxal monohydrazone by the reaction of phenacyl bromide with hydrazine at 60°C. The product is then oxidized at 20°C with manganese dioxide [69].

$$R-\overset{O}{\overset{\|}{C}}-CH_2Br + 2N_2H_4 \longrightarrow \left[R-\overset{O}{\overset{\|}{C}}-CH_2NHNH_2 \right] + N_2H_4 \cdot HBr \quad (21)$$

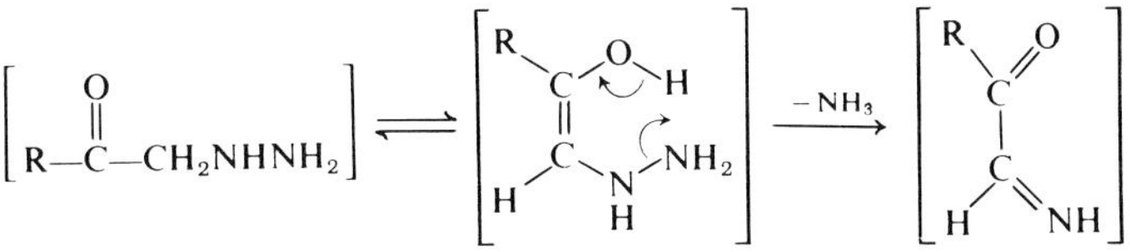

$$\left[R-\overset{O}{\overset{\|}{C}}-CH_2NHNH_2 \right] \rightleftharpoons \left[\begin{array}{c} R \diagdown \overset{O}{\diagup} H \\ C \\ \| \\ C-N \diagdown NH_2 \\ H \quad H \end{array} \right] \xrightarrow{-NH_3} \left[\begin{array}{c} R \diagdown \overset{O}{\diagup} \\ C \\ \| \\ C \\ H \diagup \diagdown NH \end{array} \right] \quad (22)$$

$$\begin{array}{c} R \diagdown \overset{O}{\diagup} \\ C \\ \| \\ C \\ H \diagup \diagdown NH \end{array} + N_2H_4 \longrightarrow R-\overset{O}{\overset{\|}{C}}-\underset{H}{\overset{}{C}}=N-NH_2 + NH_3 \quad (23)$$

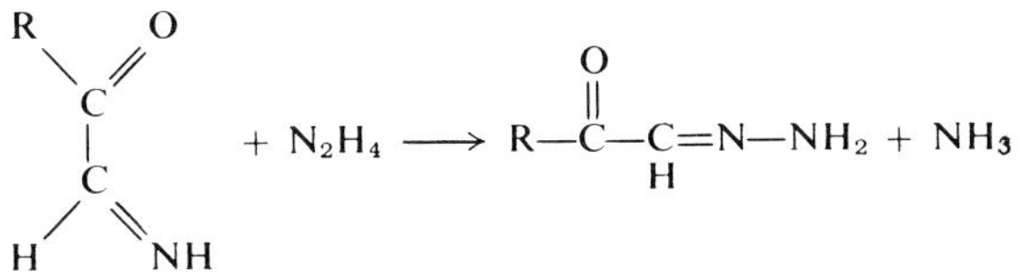

$$R\overset{O}{\overset{\|}{C}}-\underset{H}{\overset{}{C}}=N-NH_2 \xrightarrow{MnO_2} R-\overset{O}{\overset{\|}{C}}-CHN_2 \quad (24)$$

By this procedure a large variety of diazo ketones of the type

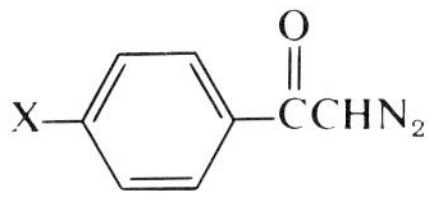

$$X-\underset{}{\bigcirc}-\overset{O}{\overset{\|}{C}}CHN_2$$

where X is *p*-ethyl, *p*-isopropyl, *p-tert*-butyl, *p*-cyclohexyl, *p*-fluoro, *p*-cyano, etc., have been prepared.

Diazoacetate esters, such as methyl phenyldiazoacetate, have also been prepared by the oxidation of the hydrazone of a keto-ester with lead tetraacetate [70]. Equations (25) to (26) illustrate the reaction sequence. To be noted in this process is that either the E or the Z isomers of the hydrazone may be used for the oxidation.

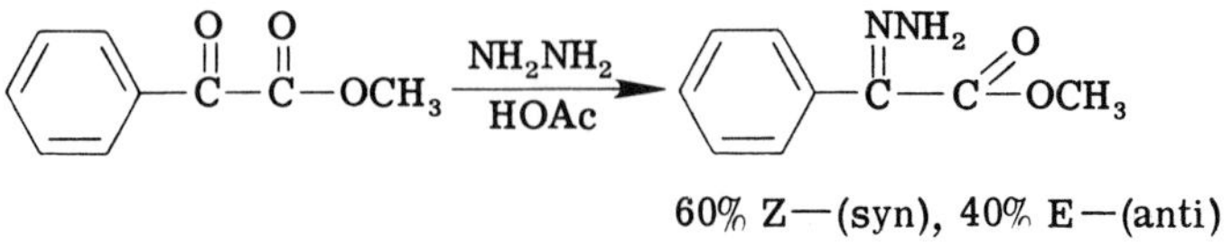

$$\text{Ph-}\underset{\text{O}}{\overset{\text{O}}{\text{C}}}\text{-}\underset{\text{O}}{\overset{\text{O}}{\text{C}}}\text{-OCH}_3 \xrightarrow{\text{NH}_2\text{NH}_2 / \text{HOAc}} \text{Ph-}\underset{\text{NNH}_2}{\overset{}{\text{C}}}\text{-}\overset{\text{O}}{\text{C}}\text{-OCH}_3$$

$$60\% \text{ Z—(syn), } 40\% \text{ E—(anti)} \quad (25)$$

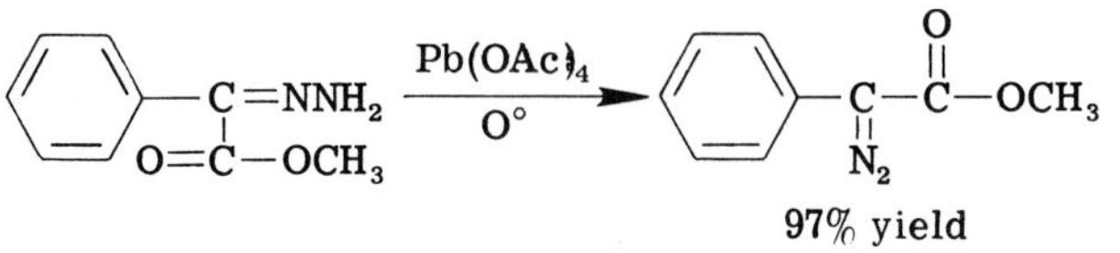

$$\text{Ph-}\underset{\text{O=C-OCH}_3}{\overset{\text{C=NNH}_2}{}} \xrightarrow[0°]{\text{Pb(OAc)}_4} \text{Ph-}\underset{\text{N}_2}{\overset{}{\text{C}}}\text{-}\overset{\text{O}}{\text{C}}\text{-OCH}_3$$

$$97\% \text{ yield} \quad (26)$$

Methyl cyanodiazoacetate was prepared by the silver oxide oxidation of methyl cyanoglyoxylate hydrazone [70].

The toluene-*p*-sulfonylhydrazones ("tosylhydrazones") of aldehydes and ketones, in the presence of base generate diazo compounds. This reaction has been termed the Bamford–Stevens reaction [71]. The tosylhydrazone of diketones produce diazoketones [72,73]. The reaction may be considered an internal oxidation–reduction as shown in Eq. (27) [72].

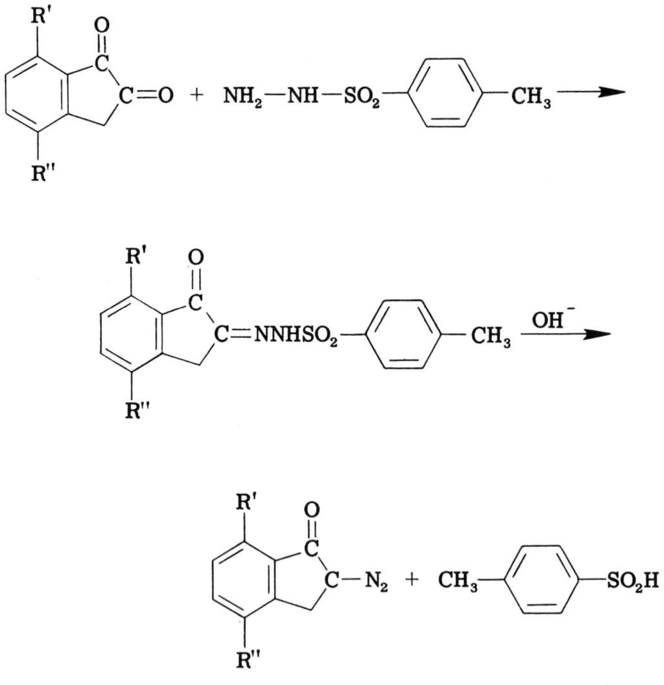

$$\text{(indanone with R', R'' substituents)} \text{ C=O} + \text{NH}_2\text{-NH-SO}_2\text{-C}_6\text{H}_4\text{-CH}_3 \longrightarrow$$

$$\text{(indanone with R', R'')} \text{ C=NNHSO}_2\text{-C}_6\text{H}_4\text{-CH}_3 \xrightarrow{\text{OH}^-}$$

$$\text{(indanone with R', R'')} \text{ C-N}_2 + \text{CH}_3\text{-C}_6\text{H}_4\text{-SO}_2\text{H} \quad (27)$$

The preparation of 1-phenyldiazopropane is an adaptation of the Bamford–Stevens procedure by Farnum which is said to give good yields rapidly [74].

2-11. Preparation of 1-Phenyldiazopropane [74]

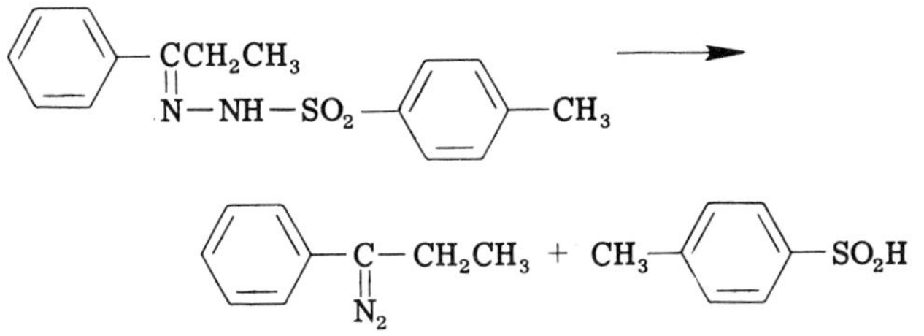

$$\text{Ph}-\underset{\underset{\text{N}-\text{NH}-\text{SO}_2-\text{C}_6\text{H}_4-\text{CH}_3}{\|}}{\text{C}\text{CH}_2\text{CH}_3} \longrightarrow$$

$$\text{Ph}-\underset{\underset{\text{N}_2}{\|}}{\text{C}}-\text{CH}_2\text{CH}_3 + \text{CH}_3-\text{C}_6\text{H}_4-\text{SO}_2\text{H} \qquad (28)$$

In a hood, behind a safety shield, and with other safety precautions, in a 100-ml flask protected by a Drierite drying tube, 1 gm (3.3 mmole) of propiophenone toluene-*p*-sulfonylhydrazone and 0.180 gm (3.3 mmole) of sodium methylate are stirred with a magnetic stirrer with 10 ml of freshly distilled pyridine. The mixture is stirred and heated at 60°–65°C for 1 hr. During this period the supernatent solution gradually turns red.

The reaction mixture is poured into 50 ml of an ice–water mixture. The product is extracted with two 20-ml portions of pentane. The combined pentane extracts are washed with four 25-ml portions of cold water and once with saturated aqueous sodium chloride. The organic solution is dried at 5°C. over anhydrous sodium sulfate, filtered, and evaporated under reduced pressure at room temperature. The residual oil is 90% pure. The yield is 0.29 gm (55% of theory).

E. Diazo Transfer Reactions

The syntheses of diazo compounds which have been discussed so far may be looked upon as reactions of a precurser in which two adjacent nitrogen atoms are bonded to the carbon atom which ultimately forms the $C-N_2$ moiety such as illustrated in Eqs. (29–31).

$$-\underset{|}{\text{CH}}-\text{NH}-\overset{\overset{\text{O}}{\|}}{\text{C}}-\text{R} \xrightarrow{\text{HNO}_2} -\underset{|}{\text{CH}}-\underset{\underset{\text{NO}}{|}}{\text{N}}-\overset{\overset{\text{O}}{\|}}{\text{C}}-\text{R} \longrightarrow -\underset{|}{\text{C}}=\text{N}_2 \qquad (29)$$

$$-\underset{|}{\text{CH}}=\text{NH}_2 \xrightarrow{\text{HNO}_2} \left[-\underset{|}{\text{CH}}-\text{NH}-\text{NO}\right] \longrightarrow -\underset{|}{\text{C}}=\text{N}_2 \qquad (30)$$

$$-\underset{|}{\text{CH}}=\text{N}-\text{NH}-\text{SO}_2-\text{Ar} \longrightarrow -\underset{|}{\text{C}}=\text{N}_2 \qquad (31)$$

to cite only three examples.

It has been found that a preformed N_2 group can be transferred from one reagent to another with an exchange of two hydrogens, hence the term "diazo transfer reaction" [7,8].

A convenient source of N_2-donor is a sulfonyl azide, particularly the readily prepared p-toluene-sulfonyl azide ("tosyl azide"). The acceptor molecule usually features an active methylene group as the source of the hydrogens which must be exchanged.

An example of the reaction is the preparation of diazocyclopentadiene [Eqs. (32–35)] [75].

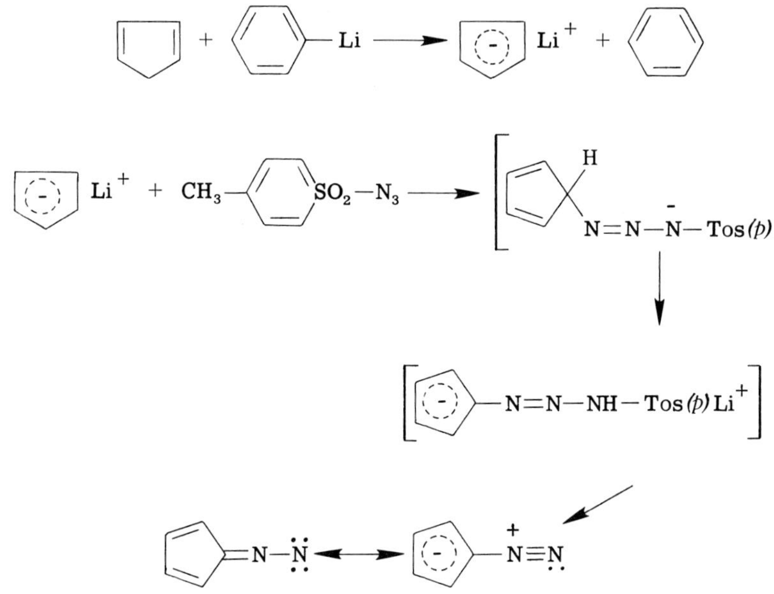

$$\text{C}_5\text{H}_6 + \text{C}_6\text{H}_5-\text{Li} \longrightarrow [\text{C}_5\text{H}_5]^- \text{Li}^+ + \text{C}_6\text{H}_6 \quad (32)$$

$$[\text{C}_5\text{H}_5]^- \text{Li}^+ + \text{CH}_3-\text{C}_6\text{H}_4-\text{SO}_2-\text{N}_3 \longrightarrow \left[\text{C}_5\text{H}_5\!\!\begin{array}{c}\text{H}\\\diagdown\\ \text{N}=\text{N}-\overset{-}{\text{N}}-\text{Tos}(p)\end{array}\!\!\text{Li}^+ \right] \quad (33)$$

$$\downarrow$$

$$\left[[\text{C}_5\text{H}_4]^- - \text{N}=\text{N}-\text{NH}-\text{Tos}(p)\text{Li}^+ \right] \quad (34)$$

$$\text{C}_5\text{H}_4\!=\!\text{N}-\ddot{\text{N}}: \longleftrightarrow [\text{C}_5\text{H}_4]^- - \overset{+}{\text{N}}\!\equiv\!\text{N}: \quad (35)$$

Pauson and Williams [76] also used phenyllithium to initiate the electrophilic attack by the azide. However, amines such as diethylamine, ethanolamine [77, 78], and piperidine [79] have been used. In the preparation of α-diazo-β-disulfones, sodium hydroxide in ethanol has been used [80]. Instead of p-tosyl azide, methane sulfonyl azide has been suggested for the preparation of diazoanthrone [81].

In his extensive review on the subject, Regitz [8] lists the following classes of compounds which have been prepared by a diazo transfer reaction: diazocyclopentadienes; diazocyclohexadienes; 2-diazo-1,3-diketones; diazomalonic esters and α-diazo-β-oxocarboxylic esters; diazomethylene disulfones; α-diazo-β-oxosulfonyl compounds; α-diazocarbonyl compounds from arylmethyl ketones, arylacetate esters, α-en-β-amino ketones, 1,3-diketones, or β-oxo-aldehydes; α-diazoimines; and α-diazoimonium salts.

The use of a diazo transfer reaction for the synthesis of t-butyl diazoacetate is described in *Organic Syntheses* [82].

The diazo transfer reaction has been used to prepare quite complex diazo

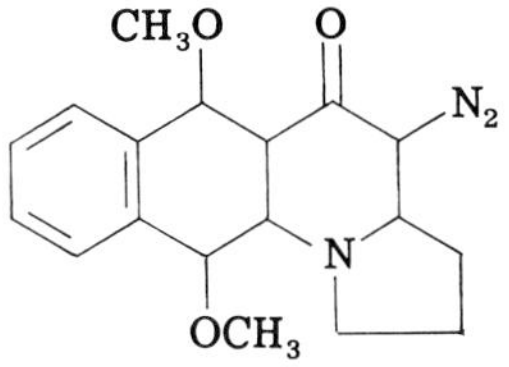

§ 2. Aliphatic Diazo Compounds

compounds, such as 4-diazo-6-11-dimethoxy-2,7,3a,4-tetrahydrobenzo[g]pyrrolo[1,2-a]quinolin-5(1H)-one [83].

A recent example of the reaction which uses triethylamine as the base is given in Preparation 2-12 [84].

2-12. Preparation of Methyl α-Diazo-α-[(methylthio)carbonyl]acetate [84]

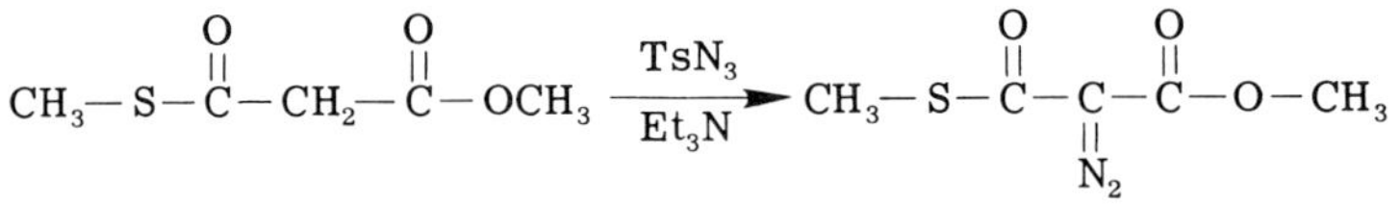

$$CH_3-S-\overset{O}{\underset{\|}{C}}-CH_2-\overset{O}{\underset{\|}{C}}-OCH_3 \xrightarrow[Et_3N]{TsN_3} CH_3-S-\overset{O}{\underset{\|}{C}}-\underset{\underset{N_2}{\|}}{C}-\overset{O}{\underset{\|}{C}}-O-CH_3 \quad (36)$$

In a hood, with suitable safety precautions, in a 500-ml flask containing a magnetic stirrer and fitted with a reflux condenser topped with a drying tube and a pressure equalizing addition funnel is placed a solution of 10 gm (0.07 mole) of methyl [(methylthio)carbonyl]acetate and 13.2 gm (0.07 mole) of tosyl azide in 250 ml of anhydrous ether. The reaction mixture is cooled to 0°C with stirring. While maintaining the temperature at 0°C, 6.8 gm (0.07 mole) of triethylamine is added dropwise. Stirring is continued overnight.

The reaction mixture is cooled to 0°C. Then 100 ml of pentane is added. The precipitated tosylamide is filtered off and discarded. The filtrate is again treated with 100 ml of pentane and refiltered to separate further quantities of tosylamide. The total tosylamide isolated is 17.1 gm (91% of theory). The filtrate is concentrated under reduced pressure. The crude yellow product is recrystallized from cyclohexane. Yield: 9.4 gm (77% of theory) mp 53–55°C. NMR (CDCl$_3$) 3.83 (s, 3H, OCH$_3$), 2.35 (s, 3H, SCH$_3$), IR (CHCl$_3$) 2900, 2140 (diazo), 1710 (br, C=O), 1620 (C=N), 1450, 1330, 1130, 955.

By analogous procedures compounds of the type

$$R-S-\overset{O}{\underset{\|}{C}}-C(N_2)-\overset{O}{\underset{\|}{C}}-OCH_3$$

were prepared where R = phenyl, yield: 70%, mp 94°–95°C;
= CH$_2$=CH—CH$_2$—, yield: 63%, bp 79°–81°C (0.005 mm);
= (CH$_3$)$_3$C—, yield: 96%, NMR (CDCl$_3$) 3.78 (s, 3H, OCH$_3$), 2.50 (s, 9H, *tert*-butyl), IR (neat) 2960, 2125 (diazo), 1718 (C=O), 1610 (C=N) 1440, 1320, 960.

F. Condensation Reactions

The interaction of diazoalkanes with acid halides to give diazo ketones is well known as the first phase in the Arndt–Eistert synthesis [3,42]. In this procedure an acyl halide is added to an excess of diazomethane so that the hydrogen halide formed in the course of the reaction can react with the excess of diazomethane to yield halomethane, which does not interfere with the subsequent course of the reaction. If the order is reversed, the hydrogen halide reacts with the diazoketones to give halogenated ketones, which is undesirable. Tertiary amines are sometimes used as scavengers for hydrogen halide. A variety of acyl halides may be used in the preparation of diazo ketones, including acid chlorides of olefinic acids [85].

Whereas most examples of diazo ketone syntheses use the higher acid chlorides, interestingly enough formyl fluoride has been used to prepare 2-diazoacetaldehydes [86] according to the following reaction:

$$\underset{\text{HCF}}{\overset{\text{O}}{\|}} + 2\,CH_2N_2 \xrightarrow{-79\,C^\circ} N_2CH\overset{\text{O}}{\underset{\|}{C}}-H + CH_3F + N_2 \qquad (37)$$

An example drawn from steroid chemistry is the preparation of 3-β-acetoxy-16-β-diazoacetylisopregn-3-en-20-one [87].

2-13. Preparation of 3-β-Acetoxy-16-β-diazoacetylisopregn-5-en-20-one [87]

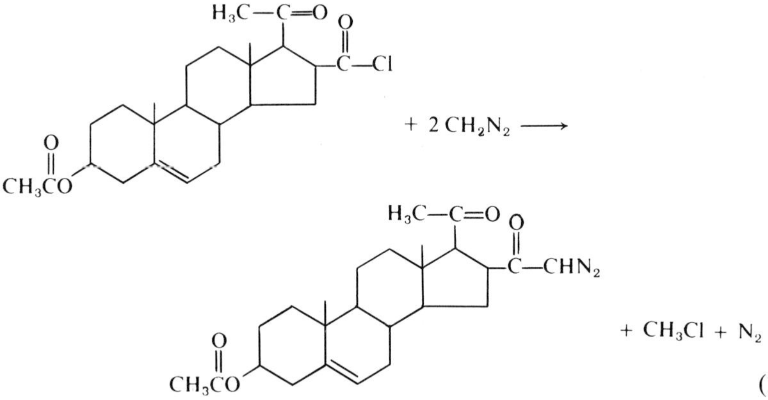

$$\text{Steroid-C(=O)-Cl} + 2\,CH_2N_2 \longrightarrow \text{Steroid-C(=O)-CHN}_2 + CH_3Cl + N_2 \qquad (38)$$

To a solution of 10 gm (0.024 mole) of 3-β-acetoxyisopregn-5-en-20-one-16-β-carboxylic acid in 360 ml of anhydrous benzene is added 9.28 gm (0.078 mole) of freshly distilled thionyl chloride dropwise at 0°C with agitation. The mixture is refluxed for 2 hr and the excess of thionyl chloride is removed by distillation under reduced pressure. Then 100 ml of benzene is added to the residue and the reaction mixture is evaporated to dryness. This operation is repeated three times.

The residue is finally dissolved in 100 ml of anhydrous benzene and to it is

§ 2. Aliphatic Diazo Compounds

slowly added an ether solution of 3.3 gm (0.079 mole) of diazomethane. After the solution has been kept at room temperature overnight, the excess diazomethane is removed by evaporation (hood).

The residue is crystallized from a mixture of acetone and hexane. Yield 9.4 gm, mp 151°–160°C. By repeated crystallization the melting point may be raised to 157°–158°C.

The preparation of diazo ketones by the oxidation of monohydrazones of diketones has been mentioned above [69].

Under basic conditions, the carbon of the diazo grouping which bears the two nitrogens may participate in aldol-condensations with aldehydes. For example, ethyl diazoacetate will react with an aldehyde to form a α-diazo-β-hydroxy ester. Diazo ketones also undergo this aldol-type condensation. Aldehyde and ketone enamines condense with diazo-acetic ester in alcohol solution or in aprotic solvents in the presence of metallic ions to form diazo-β-amino esters [88]. In the presence of lithium, trimethylsilyldiazomethane condenses with carbonyl compounds to produce 1-diazotrimethylsilyl-2-alkanols [40].

In the presence of triethyloxonium fluoroborate, a carbalkoxymethylene group from an alkyl diazoacetate ester may be inserted into a carbonylalkyl or a carbonylaryl bond to produce homologs of the aliphatic or aromatic ketones:

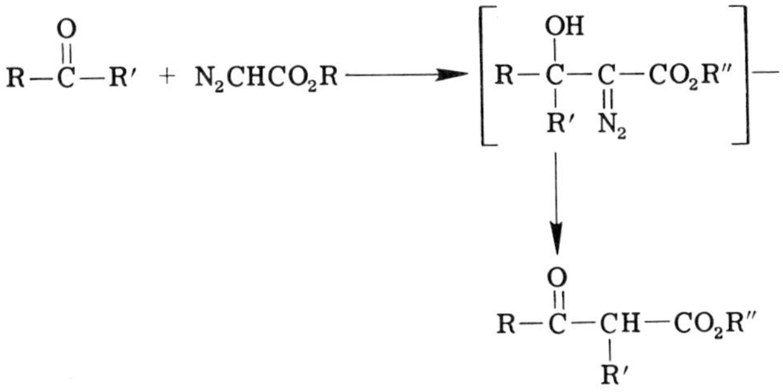

$$R-\underset{\underset{O}{\|}}{C}-R' + N_2CHCO_2R'' \longrightarrow \left[R-\underset{R'}{\overset{OH}{\underset{|}{C}}}-\underset{N_2}{\overset{\|}{C}}-CO_2R'' \right] \longrightarrow R-\underset{\underset{O}{\|}}{C}-\underset{R'}{\overset{|}{C}H}-CO_2R'' \quad (39)$$

In the case of unsymmetrically substituted ketones, the insertion is predominantly into the least highly substituted —C—C bond.
$$\underset{O}{\|}$$

Diazoacetonitrile, 2,2,2-trifluorodiazoethane, dimethyl diazomethylphosphonate, and *tert*-butyl diazoacetate have been used in this condensation reaction. In some cases, the use of antimony pentachloride as a catalyst at −78°C is said to be advantageous [89].

2-14. *Preparation of Ethyl 2-Diazo-3-hydroxybutyrate* [88]

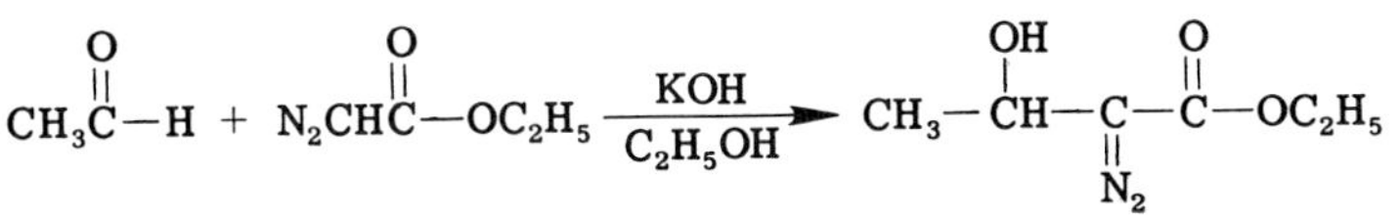

$$CH_3\underset{\underset{O}{\|}}{C}-H + N_2CH\underset{\underset{O}{\|}}{C}-OC_2H_5 \xrightarrow[C_2H_5OH]{KOH} CH_3-\underset{\underset{}{}}{\overset{OH}{\underset{|}{C}H}}-\underset{N_2}{\overset{\|}{C}}-\underset{\underset{O}{\|}}{C}-OC_2H_5 \quad (40)$$

In a hood, with appropriate safety precautions, to an ice-cooled 100-ml flask fitted with a magnetic stirrer and an addition funnel, with protection against atmospheric moisture, is charged 5.4 gm (0.047 mole) of ethyl diazoacetate, 10 ml of ethanol, and 1 ml of a 3% ethanolic potassium hydroxide solution. To this stirred, ice-cooled mixture is added dropwise 3.0 gm (0.068 mole) of acetaldehyde. After 15 min another 2 ml of 3% ethanolic potassium hydroxide is added, followed by the dropwise addition of another 3.0 gm (0.068 mole) of acetaldehyde. Upon completion of these additions, the mixture is stirred for another 30 min.

Then water is added. The reaction mixture is extracted with methylene chloride. The methylene chloride extract is evaporated under reduced pressure. The residue of ethyl α-diazo-β-hydroxybutyrate weighed 6.7 gm (90%). IR (neat) 3460 (m, OH); 2110 (s, C=N_2); 1684 (br s, C=O); pmr δ 1.29 (t,3, J= 7.0 Hz, ethyl, Me); 1.35 (d, 3, J=6.5 Hz, Me), 4.21 (q, 2, J=7.0 Hz, CH_2), 4.80 (q, 1, J=6.5 Hz, CH).

3. AROMATIC DIAZONIUM SALTS

a. Preparation of Diazonium Salt Solutions in Aqueous Media

In normal laboratory practice, diazonium salts are used as intermediates in the preparation of a variety of aromatic compounds. Since many diazonium salts may detonate when warmed or when dry, they are usually used in solution without isolation. Obviously, since diazonium compounds are used widely in the dye industry, dry diazonium salts of considerable stability can be prepared. Particularly those diazonium salts which contain electron-withdrawing groups may be converted to relatively stable salts. Even in those cases, it would be well if materials were handled with considerable care. In particular, the compounds should be stored in a cool dark place. Furthermore, since many diazonium salts are quite sensitive to ultraviolet light, preparation, handling, and storage in a cool, shaded area are recommended.

In general, the most common method of preparing diazonium salts involves the treatment of a soluble aromatic amine salt in aqueous mineral acid at low temperature with sodium nitrite. In this connection it should be kept in mind that many aromatic amines require considerable purification prior to use since they are subject to air oxidation. Frequently, purification of the amine in the presence of traces of sodium hydrosulfite in a recrystallizing solvent is helpful in overcoming discolorations due to oxidation. Also to be kept in mind in preparing diazonium salts is the fact that the hydrochloride salts of many aromatic amines are more soluble at low temperatures than at high temperatures.

The purity even of reagent grades of sodium nitrite is sometimes questionable, making it difficult to weigh out an equivalent amount of the reagent for a given preparation. It is recommended, therefore, that the diazotization be carried out with an excess of sodium nitrite solution on hand from the beginning of the reaction. The reaction is carried out in an excess of mineral acid with an excess of sodium nitrite. Starch–iodide paper is used to control the addition of the nitrite. When the paper turns blue, addition is considered completed.

Since the usual reactions of diazonium salts do not proceed satisfactorily with an excess of nitrous acid present, the addition of either crystalline sulfamic acid or urea assists in destroying the excess nitrous acid.

While it is useful to surround the reaction flask with an ice bath during diazotization and storage of the diazonium salt for subsequent reactions, the use of ice cubes or cracked ice *in* the reaction flask is strongly to be recommended to minimize problems of local overheating.

The basic method of preparing diazonium salts such as benzenediazonium chloride is illustrated below.

3-1. *Preparation of Benzenediazonium Chloride* [90]

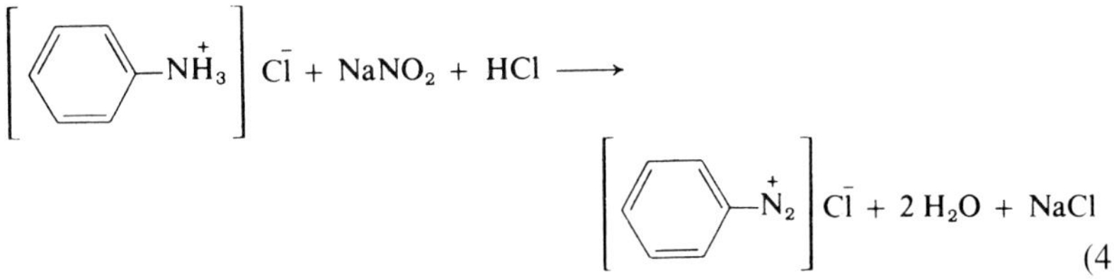

$$[C_6H_5-\overset{+}{N}H_3]\,Cl^- + NaNO_2 + HCl \longrightarrow$$

$$[C_6H_5-\overset{+}{N}_2]\,Cl^- + 2\,H_2O + NaCl \quad (41)$$

In a hood, in an open 5-liter flask fitted with a mechanical stirrer and a dropping funnel are placed 1000 gm of cracked ice, 1.5 liters of water, 279 gm (3.0 moles) of aniline, and 916 gm (9.0 moles) of concentrated hydrochloric acid. The reaction flask is placed in a large ice–salt bath. After the flask content has cooled to 0-5°C, while stirring vigorously, a solution of 218 gm (3.0 moles) of 95% sodium nitrite in 215 ml of water is added while maintaining the reaction temperature about 0°C. If necessary, pieces of ice may be added to the reaction mixture from time to time to maintain the temperature. Toward the end of the addition the reaction mixture is checked with starch–iodide paper for excess nitrous acid. The instant appearance of a blue color indicates an excess of nitrous acid.

If necessary, the excess nitrous acid may be destroyed by adding small quantities of urea until no further positive starch–iodide test is observable.

The resulting solution should be maintained at ice temperature until used in a subsequent reaction. Prolonged storage must be avoided.

b. STABILIZED DIAZONIUM SALTS

Many fluoroborate salts of aromatic diazonium compounds have a high degree of stability. A generalized preparation is cited here to indicate how some storable salts are prepared.

3-2. Preparation of Aryldiazonium Fluoroborate [91]

$$\text{ArNH}_2 + \text{HNO}_2 \xrightarrow{\text{HX}} \text{ArN}_2{}^+\text{X}^- + 2\,\text{H}_2\text{O} \qquad (42)$$

$$\text{ArN}_2{}^+\text{X}^- + \text{HBF}_4 \longrightarrow \text{ArN}_2{}^+\text{BF}_4{}^- + \text{HX} \qquad (43)$$

To 0.2 mole of the aromatic amine dissolved in or slurried with 10 ml of concentrated hydrochloric acid and 25 ml of water, cooled to 0°C in an ice-salt bath, 4 ml of a 5 N solution of sodium nitrite is added dropwise with stirring at such a rate as to maintain the temperature between 0° and 5°C. The resulting diazonium chloride solution is freed of excess nitrous acid by the addition of small quantities of urea. Then the solution is rapidly filtered with suction and 5 ml of a 48% solution of fluoroboric acid is added. The diazonium fluoroborate salt precipitates after several minutes and is separated by filtration.

c. Other Stabilized Diazonium Salts

The stabilization of some diazonium salts may also be carried out by the addition of naphthalene-1,5-disulfonic acid or 2-naphthol-1-sulfonic acid to the hydrochloric salt solution produced after diazotization [92]. The usual procedure simply involves addition of a slurry of acids to the diazonium salt solutions and, if necessary, assisting the precipitation of the product by the addition of sodium chloride. Some substituted amines may be diazotized in concentrated sulfuric acid or in glacial acetic acid at relatively high temperatures [93].

In our own laboratory we have even prepared 1-anthraquinone-diazonium chloride at 45°–50°C and found the product to be reasonably stable.

3-3. Preparation of 1-Anthraquinonediazonium chloride [94]

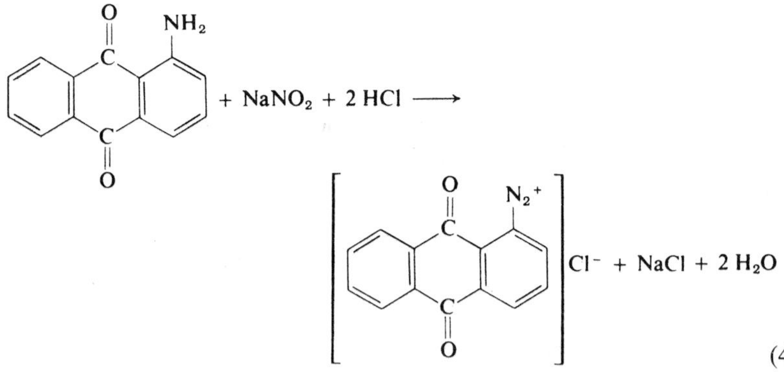

$$\text{[1-aminoanthraquinone]} + \text{NaNO}_2 + 2\,\text{HCl} \longrightarrow \left[\text{[1-anthraquinone-N}_2{}^+\text{]}\right]\text{Cl}^- + \text{NaCl} + 2\,\text{H}_2\text{O} \qquad (44)$$

In a 12-liter flask, 273 gm (1.25 moles) of 1-aminoanthraquinone and 685 ml of concentrated hydrochloric acid are mixed at 30°C. Then a solution of 86.5 gm (1.25 moles) of sodium nitrite in 450 gm of distilled water is added slowly and the temperature of the reaction product is raised to 50°C. After diazotization has been completed, 8 liters of warm distilled water is added and the solution is filtered while hot. The insoluble material is washed with 500 ml of warm water and discarded. To the warm filtrate is added approximately 2400 gm of sodium chloride. The solution is stirred until all of the sodium chloride has gone into solution. Upon cooling the 1-anthraquinone diazonium chloride precipitates. The product is collected by filtration and washed with saturated sodium chloride solution. The product may then be air-dried. Yield 285 gm (85.5% of theory), melting point range 117°–118°C with violent explosion.

In the field of immunochemistry, the fractionation of proteins is of importance. One technique involves the conjugation of an antigen protein to a polymer. This conjugated resin is then used as a column packing which may be used to fractionate antibodies. Many systems have been described. Among these has been the use of poly(styrenediazonium chloride). The preparation of this material is given here.

3-4. Preparation of Poly(styrenediazonium chloride) [95]

$$\left[\begin{array}{c}\text{CH}-\text{CH}_2\\ \big|\\ \text{C}_6\text{H}_4\\ \big|\\ \text{NH}_2\end{array}\right]_x + x\,\text{NaNO}_2 + 2x\,\text{HCl} \longrightarrow \left[\begin{array}{c}\text{CH}-\text{CH}_2\\ \big|\\ \text{C}_6\text{H}_4\\ \big|\\ ^+\text{N}_2 + \text{Cl}^-\end{array}\right]_x + x\,\text{NaCl} + 2x\,\text{H}_2\text{O} \quad (45)$$

A 2% solution (0.002 mole) of poly(aminostyrene) in 10 ml of 2N hydrochloric acid is cooled to between 0° and −5°C. The solution is then diazotized by the slow addition of 1.5 ml of 14% (0.0038 mole) solution of sodium nitrite. The excess of nitrous acid is destroyed by the addition of 2 ml of a 6% solution (0.002 mole) of urea previously cooled to 0°C. The reaction mixture is maintained between −5° and 0°C with stirring for about 1 hr until a test with starch–iodide paper no longer gives a positive test for nitrous acid. This solution may then be used to carry out reactions typical of diazonium salts. It is our belief that if a cross-linked poly(aminostyrene) were used in this preparation, diazotization would proceed on the slurry, resulting in a diazotized resin which may be particularly useful as a column packing in inmunochemistry.

Recent kinetic studies have shown that the thermal decomposition of certain diazonium salts is retarded by crown ethers. In these studies, cryptates were also considered but the cryptate used presumably had too small a "cavity" to permit the penetration of the diazonium group [96,97]. The stabilization of diazonium salts by complexation with crown ethers, at present is of considerable academic interest. The high current cost of crown ethers probably precludes their use in commercial applications at this time.

d. DIAZOTIZATION IN NONAQUEOUS MEDIA

Isoamyl nitrite has also been used to diazotize aromatic amines [98,99]. By the use of this reagent, solutions of aromatic amines in organic solvents such as acetic acid, propionic acid, or higher alcohols may be diazotized [100]. Again, care must be taken in the isolation of diazonium salts from this reaction mixture since they may present explosion hazards.

3-5. Preparation of N,N-Dicyclohexylbenzamide-O-diazonium Fluoroborate [98]

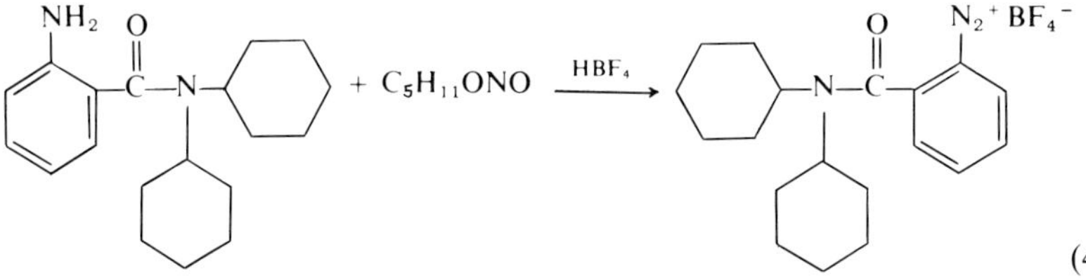

$$\text{o-amino-}N,N\text{-dicyclohexylbenzamide} + C_5H_{11}ONO \xrightarrow{HBF_4} \text{product with } N_2^+ BF_4^-$$

(46)

A solution of 3.0 gm (0.01 mole) of o-amino-N,N-dicyclohexylbenzamide and 5.4 gm (0.03 mole) of 48–50% fluoroboric acid in 50 ml of ethanol is cooled below 0°C. The solution is treated with 1.30 gm (0.011 mole) of freshly distilled isoamyl nitrite. After 30 min, 200 ml of cold ether is added to the solution, which is allowed to remain at 0°C for an additional 30 min. The precipitating product is filtered and washed with cold ether. Yield 3.2 gm (80% of theory), mp 100°–102°C with decomposition. The solid gradually changes color on standing at room temperature but may be stored for approximately 1 week under ether at 0°C without significant discoloration.

Anhydrous diazonium salts have also been prepared by the treatment of Schiff bases such as benzylidene aniline with nitrogen tetroxide in ether solution [101]. Since the precipitated product is a nitrate salt, extreme caution must be exercised in using this reaction. The products usually are explosive and sensitive to shock in the dry state.

Aromatic nitroso compounds upon treatment with nitrosyl chloride, in air, also are said to form diazonium nitrates [102].

4. MISCELLANEOUS PREPARATIONS OF DIAZO COMPOUNDS

(1) Forster reaction of α-oximinoketones with chloramine [72,103,104].

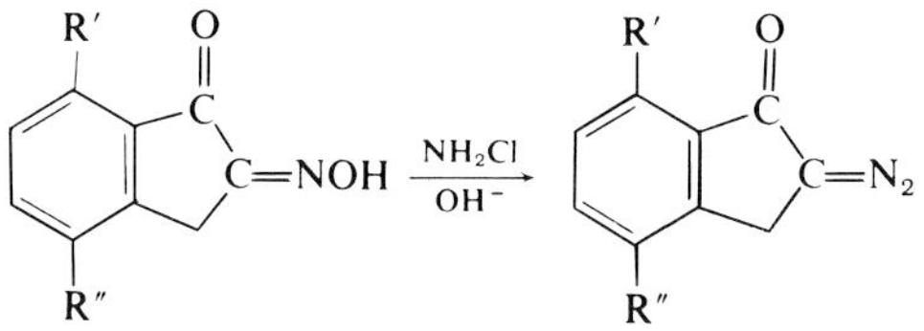

$$\underset{R''}{\underset{|}{\text{R'}}}\text{(indanone)}\text{C=NOH} \xrightarrow[\text{OH}^-]{\text{NH}_2\text{Cl}} \underset{R''}{\underset{|}{\text{R'}}}\text{(indanone)}\text{C=N}_2 \tag{47}$$

or

$$\underset{H}{\underset{|}{R-\overset{O}{\overset{\|}{C}}-C=NOH}} \xrightarrow[\text{OH}^-]{\text{NH}_2\text{OSO}_3\text{H}} \underset{H}{\underset{|}{R-\overset{O}{\overset{\|}{C}}-CN_2}}$$

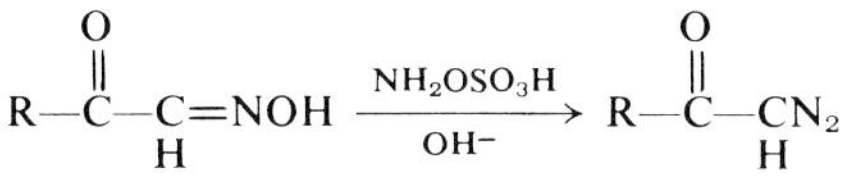

(2) Alkaline cleavage of azibenzil (i.e., benzoylphenyldiazomethane) [105].

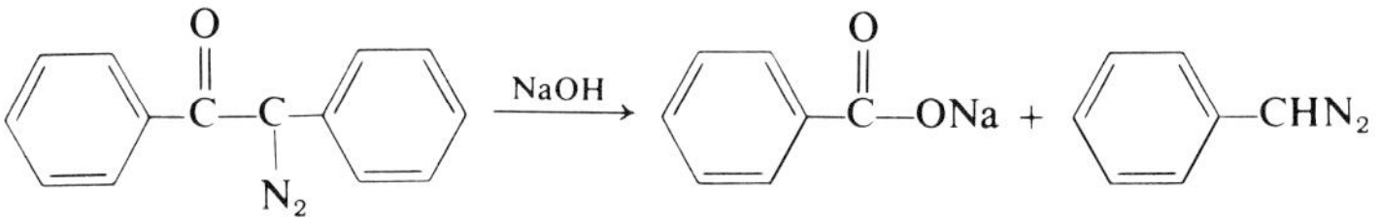

$$\text{Ph-CO-C(N}_2\text{)-Ph} \xrightarrow{\text{NaOH}} \text{Ph-CO-ONa} + \text{Ph-CHN}_2 \tag{48}$$

(3) Decomposition of hydroxytriazoles [106].

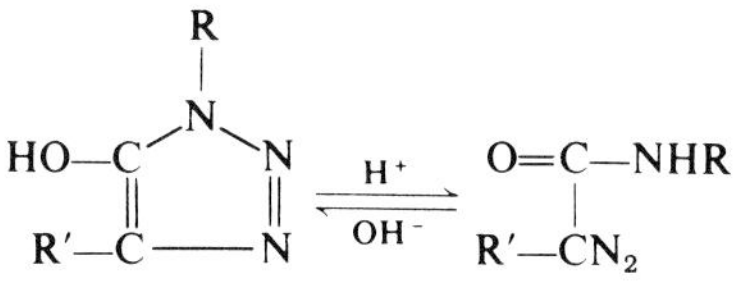

$$\underset{R'-C\overset{}{\underset{}{\longrightarrow}} N}{\overset{HO-C\overset{R}{\underset{}{\overset{|}{N}}}\searrow N}{\|}} \underset{\text{OH}^-}{\overset{\text{H}^+}{\rightleftharpoons}} \underset{R'-CN_2}{O=C-NHR} \tag{49}$$

(4) Preparation of isodiazomethane [107].

$$\text{CH}_2\text{N}_2 \xrightarrow[\substack{\text{2. Hydrolysis} \\ \text{at } -80°\text{C}}]{\text{1. CH}_3\text{Li}} \text{CHN}_2\text{H} \tag{50}$$

(5) Preparation of cyclodiazomethanes (diazines) [108].

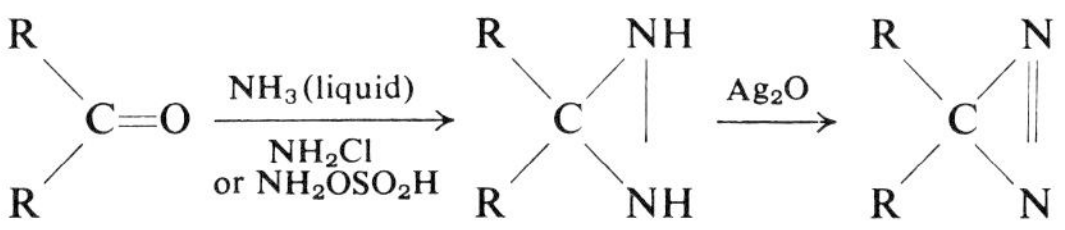

$$\underset{R}{\overset{R}{\diagdown}}\text{C=O} \xrightarrow[\text{NH}_2\text{Cl or NH}_2\text{OSO}_2\text{H}]{\text{NH}_3 \text{(liquid)}} \underset{R}{\overset{R}{\diagdown}}\text{C}\overset{\text{NH}}{\underset{\text{NH}}{\diagup}} \xrightarrow{\text{Ag}_2\text{O}} \underset{R}{\overset{R}{\diagdown}}\text{C}\overset{\text{N}}{\underset{\text{N}}{\diagup}} \tag{51}$$

This preparation could not be duplicated in a recent attempt. [109]

(6) Complexation and activation of diazenes and diazo compounds by transition metals [110].

REFERENCES

1. L. I. Smith, *Chem. Rev.* **23**, 193 (1938).
2. B. Eistert, *Z. Angew. Chem.* **54**, 99 (1941).
3. W. E. Bachman and W. S. Struve, *Org. React.* **1**, 38 (1942).
4. B. Eistert, *in* "Newer Methods of Preparative Organic Chemistry," p. 513. Wiley, New York,

(translation) (see B. Eistert, *in* "Neuere Methoden der Preparativen organischen Chemie," p. 361. Verlag Chemie, Weinheim, 1949).
5. A. F. McKay et al., *Chem. Rev.* **51**, 331 (1952).
6. C. D. Gutsche, *Org. React.* **8**, 391 (1954).
7. M. Regitz, *Synthesis* **7**, 351 (1972).
8. M. Regitz, *Angew. Chem., Int. Ed. Engl.* **6**, 733 (1967).
9. M. Regitz, I. K. Korobitsyna, and L. C. Redina, *Method. Chim.* **6**, 205 (1975).
10. A. L. Fridman, F. A. Gabitov, O. B. Kremeleva, and V. S. Zalesov, *Nauh. Tr. Perm. Formatsevot.* **8**, 3 (1975); *Chem. Abstr.* **85**, 5524 (1976).
11. R. F. Muraca, "Diazo Group" in *Treatise Anal. Chem.* Vol. 15, p. 347 (1976); I. M. Kolthoff and P. J. Elving, ed., Wiley, New York, 1971.
12. I. D. R. Stevens, *Org. React. Mech.* p. 327 (1976); *Chem. Abstr.* **85**, 62264 (1976).
13. R. Fields, M. S. Gibson, and G. Holt, *in* "Rodd's Chemistry of Carbon Compounds" (M. F. Ansell, ed.), 2nd ed., Vol. 1, Parts A–B, Suppl. 125. Elsevier, Amsterdam, 1975.
14. P. H. Groggins, "Unit Processes in Organic Synthesis," p. 129. McGraw-Hill, New York, 1947.
15. K. H. Saunders, "The Aromatic Diazo-Compounds and Their Technical Applications," p. 1. Longmans, Green, New York, 1947.
16. W. E. Bachman and R. A. Hoffman, *Org. React.* **2**, 224 (1944).
17. N.Kornblum, *Org. React.* **2**, 262 (1944).
18. H. Zollinger, "Azo and Diazo Chemistry, Aliphatic and Aromatic Compounds," Wiley (Interscience), New York, 1961.
19. J. H. Ridd, *Q. Rev., Chem. Soc.* **15**, 418 (1961).
20. K. Schank, *Method. Chim.* **6**, 159 (1975).
21. M. Yamamoto, *Senvyo To Yakuhin* **21**(2), 30; (3), 58 (1976); *Chem. Abstr.* **85**, 178948 (1976).
22. S. Patai, ed., "The Chemistry of Diazonium and Diazo Compounds." Wiley, New York, 1978.
23. A. Graffi and F. Hoffman, *Acta Biol. Med. Ger.* **16**, K–1 (1966); *Org. Synth. Collect. Vol.* **2**, 165 (1943).
24. J. A. Moore and D. E. Reed, *Org. Synth. Collect. Vol.* **5**, 351 (1973).
25. T. J. D Joer and H. J. Backer, *Org. Synth. Collect. Vol.* **4**, 250 (1963).
26. M. P Jlicky, *J. Org. Chem.* **45**, 5377 (1980).
27. Anonymous, *Catalyst* **53**, 202 (1968).
28. Aldrich Chemical Co., *Aldrichimica Acta* **14**(3), 59 (1981).
29. H. M. Fales and T. M. Jaouni, *Anal. Chem.* **45**, 2302 (1973).
30. T. J. De Boer and H. J. Backer, *Recl. Trav. Chim. Pays-Bas* **73**, 229 (1954).
31. T. J. De Boer and H. J. Backer, *Recl. Trav. Chim. Pays-Bas* **73**, 582 (1954).
32. C. G. Overberger and J.-P. Anselme, *J. Org. Chem.* **28**, 592 (1963).
33. H. Druckrey and D. Preusmann, *Nature (London)* **125**, 1111 (1962); see also insert in *Org. Synth. Collect. Vol.* **3**, 119 (1955).
34. D. P. Schwartz and R. S. Bright, *Anal. Biochem.* **61**, 271 (1974).
35. C. D. Gutsche and K. Kinoshita, *J. Org. Chem.* **28**, 1762 (1963).
36. Prod. Bull. "EXR-01". Du Pont Explosives Dept., E. I. du Pont de Nemours & Co., Inc., Gibbstown, New Jersey.
37. C. F. d'Alelio and E. E. Reid, *J. Am. Chem. Soc.* **59**, 109 (1937).
38. K. Heyns and O. Woyrsch, *Chem. Ber.* **86**, 76 (1953); K. Heyns and W. V. Bebenburg, *ibid.*, p. 82.
39. H. Reimlinger, *Chem. Ber.* **94**, 2547 (1961).
40. U. Schöllkopf and H. U. Scholz, *Synthesis* No. 4, p. 271 (1976).
41. W. D. McPhee and E. Klingsberg, *Org. Synth. Collect. Vol.* **3**, 119 (1955).

42. A. L. Wilds and A. L. Meader, Jr., *J. Org. Chem.* **13**, 763 (1948).
43. E. H. White, G. E. Maier, R. Graete, U. Zirngibl, and E. W. Friend, *J. Am. Chem. Soc.* **88**, 611 (1966).
44. A. M. van Leusen and J. Strating, *Org. Synth.* **57**, 94 (1977).
45. F. Arndt, *Org. Synth. Collect. Vol.* **2**, 165 (1943).
46. J. R. Dyer, R. B. Randall, Jr., and H. M. Deutsch, *J. Org. Chem.* **29**, 3423 (1964).
47. W. M. Jones, D. L. Muck, and T. K. Tandy, Jr., *J. Am. Chem. Soc.* **88**, 68 (1966).
48. T. Hudlicky, F. J. Koszyk, T. M. Kutchan, and J. P. Sheth, *J. Org. Chem.* **45**, 5020 (1980).
49. A. F. McKay, W. L. Ott, G. W. Taylor, M. N. Buchanan, and J. F. Crooker, *Can. J. Res., Sect. B.* **28**, 683 (1950).
50. C. E. Redemann, F. O. Rice, R. Roberts, and H. P. Ward, *Org. Synth. Collect. Vol* **3**, 244 (1955).
51. E. C. S. Jones and J. Kenner, *J. Chem. Soc.* p. 363 (1933); J. Kenner *et al.*, *ibid.* p. 286 (1936); p. 1551 (1937); p. 181 (1939).
52. E. Womack and A. B. Nelson, *Org. Synth. Collect. Vol.* **3**, p. 392 (1955).
53. N. E. Searle, *Org. Synth. Collect. Vol.* **4**, 424 (1963).
54. R. A. Shepard and P. L. Sciaraffa, *J. Org. Chem.* **31**, 964 (1966).
55. R. A. Shepard and S. E. Wentworth, *J. Org. Chem.* **32**, 3197 (1967).
56. J. H. Bayless and L. Friedman, *J. Am. Chem. Soc.* **89**, 147 (1967); A. J. Jurewicz and L. Friedman, *ibid.* p. 149.
57. O. W. Webster, *J. Am. Chem. Soc.* **88**, 4055 (1966).
58. H. Staudinger and O. Küpfer, *Ber. Dtsch. Chem. Ges.* **45**, 501 (1912).
59. D. T. Sepp, K. V. Scherer, and W. P. Weber, *Tetrahedron Lett.* **34**, 2983 (1974).
60. H.Staudinger and A. Gaule, *Ber. Dtsch. Chem. Ges.* **49**, 1897 (1916).
61. A. C. Day, P. Raymond, R. M. Southam, and M. C. Whiting, *J. Chem. Soc. C* p. 467 (1966).
62. R. W. Murray and A. M. Trozzolo, *J. Org. Chem.* **29**, 1268 (1964).
63. L. I. Smith and K. L. Howard, *Org. Synth. Collect. Vol.* **3**, 351 (1955).
64. C. D. Nenitzescu and E. Solomonica, *Org. Synth. Collect. Vol.* **2**, 496 (1943).
65. E. Ciganek, *J. Org. Chem.* **30**, 4366 (1965).
66. E. Ciganek, *J. Org. Chem.* **30**, 4198 (1965): *J. Am. Chem. Soc.* **87**, 652 (1965).
67. D. M. Gale, W. J. Middleton, and C. G. Krespan, *J. Am. Chem. Soc.* **88**, 3617 (1966).
68. J. Diekmann, *J. Org. Chem.* **30**, 2272 (1965).
69. S. Hauptmann, M. Kluge, K. D. Seidig, and H. Wilde, *Angew. Chem.* **77**, 678 (1965).
70. E. Ciganek, *J. Org. Chem.* **35**, 862 (1970).
71. W. R. Bamford and T. S. Stevens, *J. Chem. Soc.* p. 4735 (1952).
72. M. P. Cava, R. L. Litle, and D. K. Napier, *J. Am. Chem. Soc.* **80**, 2257 (1958).
73. P. Yates and R. J. Crawford, *J. Am. Chem. Soc.* **88**, 1561, 1562 (1966).
74. D. G. Farnum, *J. Org. Chem.* **28**, 870 (1963).
75. W. von E. Doering and C. H. De Puy, *J. Am. Chem. Soc.* **75**, 5955 (1953).
76. P. L. Pauson and B. J. Williams, *J. Chem. Soc.* p. 4153 (1961); also P. L. Pauson *et al.*, *ibid.* pp. 4158, 4162.
77. T. Weil and M. Cais, *J. Org. Chem.* **28**, 2472 (1963).
78. D. Lloyd and F. I. Wasson, *J. Chem. Soc. C* p. 408 (1966).
79. M. Regitz, *Chem. Ber.* **97**, 2742 (1964).
80. F. Klages and K. Bott, *Chem. Ber.* **97**, 735 (1964).
81. G. Cauquis, M. Rastoldo, and G. Reverdy, *Bull. Soc. chim. Fr.* p. 1263 (1965).
82. M. Regitz, J. Hocker, and A. Liedhegener, *Org. Synth. Collect. Vol.* **5**, 179 (1973).
83. S. N. Falling and H. Rapoport, *J. Org. Chem.* **45**, 1260 (1980).
84. V. Georgian, S. K. Boyer, and B. Edwards, *J. Org. Chem.* **45**, 1686 (1980).
85. M. M. Fawzi and C. D. Gutsche, *J. Org. Chem.* **31**, 1390 (1966).

86. F. Kaplan and G. K. Meloy, *J. Am. Chem. Soc.* **88,** 950 (1966).
87. J. L. Mateos, A. Dosal, and C. Carbajal, *J. Org. Chem.* **30,** 3578 (1965).
88. E. Wenkert and C. B. McPherson, *J. Am. Chem. Soc.* **94,** 8084 (1972).
89. W. C. Mock and M. E. Hartman, *J. Org. Chem.* **42,** 459 (1977).
90. W. W. Hartman and J. B. Dickey, *Org. Synth. Collect. Vol.* **2,** 163 (1943).
91. R. J. Cox and J. Kumamoto, *J. Org. Chem.* **30,** 4254 (1965).
92. H. H. Hodgson and E. Marsden, *J. Chem. Soc.* p. 207 (1940).
93. R. Howe, *J. Chem. Soc. C* p. 478 (1966).
94. W. Karo, unpublished procedure.
95. L. H. Kent and J. H. R. Slade, *Biochem. J.* **77,** 12 (1960).
96. R. A. Bartsch and P. N. Juri, *J. Org. Chem.* **45,** 1011 (1980).
97. R. A. Bartsch and P. Čársky, *J. Org. Chem.* **45,** 4782 (1980).
98. J. Lipowitz and T. Cohen, *J. Org. Chem.* **30,** 3891 (1961).
99. F. M. Logullo, A. H. Seitz, and L. Friedman, *Org. Synth. Collect. Vol.* **5,** 54 (1973).
100. J. S. Hanker, L. Katzoff, L. D. Aronson, M. L. Seligman, H. R. Rosen, and A. M. Seligman, *J. Org. Chem.* **30,** 1779 (1965).
101. R. M. Scribner, *J. Org. Chem.* **29,** 3429 (1964).
102. S. Torimitsu and M. Ohno (Toyo Rayon Co. Ltd.), Japanese Patent 66/10,017 (1966); *Chem. Abstr.* **65,** 15557 (1966).
103. W. Rundel, *Angew Chem.* **74,** 469 (1962).
104. J. Meinwald, P. G. Gassman, and E. G. Miller, *J. Am. Chem. Soc.* **81,** 4751 (1959).
105. P. Yates and B. L. Shapiro, *J. Org. Chem.* **23,** 759 (1958).
106. O. Dimroth, *Justus Liebigs Ann. Chem.* **335,** 1 (1904); **338,** 143 (1905); **364,** 197 (1909); **373,** 336 (1910); **377,** 127 (1910); **399,** 91 (1913).
107. E. Müller, P. Kästner, and W. Runkel, *Chem. Rev.* **98,** 711 (1965).
108. E. Schmitz and R. Ohme, *Chem. Ber.* **95,** 795 (1962).
109. C. G. Overberger and J.-P. Anselme, *J. Org. Chem.* **29,** 1188 (1964); R. A. Shepard and S. W. Wentworth, *ibid.* **32,** 3197 (1967).
110. A. Albini and H. Kirsch, *Top. Curr. Chem.,* **65** (Theor. Inorg. Chem. 2), 105 (1976).

CHAPTER 16 / NITRO COMPOUNDS

1. Introduction		498
2. Aliphatic Nitro Compounds		500
	A. Direct Nitration	500
	a. Nitration of Olefins with Dinitrogen Tetroxide	501
2-1.	Preparation of 1-Nitrocyclooctene	502
	b. Nitration of Olefins with Acetyl Nitrate	503
2-2.	Preparation of 3-Nitro-2-butyl Acetate	505
	c. Nitration of Olefins with Nitric Acid–Sodium Nitrite	505
2-3.	Preparation of 6-Nitrocholesteryl Acetate	506
	d. Nitration of Active Methylene Compounds with Nitric Acid	506
2-4.	Preparation of Diethyl Nitromalonate	506
	e. Nitration of Acetic Anhydride with Nitric Acid	507
2-5.	Preparation of Tetranitromethane	508
	f. Chloronitration of Olefins with Nitryl Chloride	509
2-6.	Preparation of Methyl 2-Chloro-3-nitropropionate	509
	g. Nitration of Active Methylene Compounds with Alkyl Nitrates in Alkaline Media	510
2-7.	Preparation of 2-Nitrocyclooctanone and Amyl 8-Nitrooctanoates	511
	h. Nitration of Cyclic Enol Acetates with Nitric Acid/Acetic Anhydride and of Cyclic Enolates with Alkyl Nitrates	512
2-8.	Preparation of 2-Nitro-4-methylcyclohexanone	512
2-9.	Preparation of 2-Nitro-3(and 4)-methylcyclopentanone	513
	B. Indirect Nitration	514
	a. Silver Nitrite Condensations	514
2-10.	Preparation of 1-Nitrooctane	514
	b. Sodium Nitrite Condensations	516
2-11.	Preparation of 2-Nitrooctane (in DMF)	517
2-12.	Preparation of 2-Nitrooctane (in DMSO)	518
2-13.	Preparation of Ethyl α-Nitrovalerate (in DMSO)	518
2-14.	Preparation of 1-Nitrocyclohexene	519
	C. Oxidation of Oximes and Amines	520
2-15.	Preparation of 4-Nitroheptane	520
2-16.	Preparation of 2-Nitro-2,4,4-trimethylpentane	522
	D. Condensation of Active Methylene Compounds	523
	a. Alkylations and Acylations	523
2-17.	Preparation of Ethyl α-Nitro-α-carbethoxy-β-(3-indole)-propionate	523
	b. Aldol Condensations	524
2-18.	Preparation of 2,2,2-Trinitroethanol	525
2-19.	Preparation of 2,4-Dibenzyloxy-5-methoxy-β-nitrostyrene	526
	c. Michael Condensations	526
2-20.	Preparation of 4-Nitro-4-methylheptadiamide	527
	d. Mannich Reactions	527
2-21.	Preparation of N-(2-Nitrobutyl)diethylamine	528
	e. (4 + 2) Reactions	529

2-22. Preparation of 2-endo-*Nitronorbornene*	529
3. Aromatic Nitro Compounds	529
A. Direct Nitration	529
a. Nitration with Mixed Acid or Nitric Acid	529
3-1. *Preparation of* p-*Nitrobromobenzene*	530
3-2. *Preparation of 2,4-Dinitrobromobenzene*	531
3-3. Preparation of 93.5% m,m'-Dinitrobenzophenone and 6.5% m,p'-Dinitrobenzophenone	534
b. Nitration with Acyl Nitrates	535
3-4. Preparation of Mixed Chloronitrobenzenes	537
c. Nitration with Nitronium Tetrafluoroborate and Related Systems	537
3-5. *Preparation of 2-Methyl-5-nitrobenzonitrile*	538
d. Nitration with Oxides of Nitrogen	538
B. Indirect Nitration	539
a. Displacement of Sulfonic Acid Groups	539
3-6. *Preparation of 2,4-Dinitro-1-naphthol*	539
b. Replacement of the Diazonium Groups by Nitro Groups (Sandmeyer Reaction)	540
c. Nitration with Tetranitromethane	540
C. Oxidation Methods	541
a. Oxidation of Aromatic Amines	541
3-7. *Preparation of 3,5-Dichloro-4-nitrobenzonitrile*	541
b. Oxidation of Nitroso Compounds	542
4. Miscellaneous Methods	542
References	544

1. INTRODUCTION

The nitration of organic compounds, particularly of aromatic compounds, is probably the most widely studied organic reaction, both from the theoretical and the technological standpoint. The literature abounds with review articles, monographs, and research papers dealing with various aspects of the preparation of nitro compounds. Therefore, only a somewhat random selection of recent papers reviewing the chemistry of nitro compounds is suggested [1–34]. The reader is directed particularly to the review of Hass and Riley [1], the chapter by Kornblum in *Organic Reactions* [14], and the book by Sir Christopher K. Ingold which reviews the extensive researches of Hughes, Ingold, and co-workers in this field [17]. The present chapter is divided into two major sections, the first section deals with aliphatic nitro compounds and the second section deals with aromatic nitro compounds. Within each section direct nitrations are discussed first, followed by indirect methods of preparation.

While the direct nitration of aliphatic hydrocarbons enjoys considerable importance in the industrial sphere, in the laboratory it has only limited value since complex mixtures of products are usually formed. Treatment of olefins with dinitrogen tetroxide may lead to dinitroparaffins and mixtures of nitro nitrites.

§ 1. Introduction

The latter may be oxidized to nitro nitrates or hydrolyzed to nitro alcohols or thermally degraded to olefinic nitro compounds under the reaction conditions. With solutions of acetyl nitrate, olefins may be converted to acetate esters of β-nitro alcohols produced from the olefins.

Active methylene compounds have been nitrated directly with nitric acid. In this synthesis one or more labile hydrogens are replaced by a nitro group.

Olefins may be chloronitrated with nitryl chloride. In this reaction, the nitro group appears to add to the olefinic carbon atom bearing the larger number of hydrogens.

Under alkaline conditions, active methylene compounds have been nitrated with alkyl nitrates. Under the conditions of the reaction, considerable cleavage of the reaction product takes place.

Among the indirect methods of preparing aliphatic nitro compounds are the reaction of various alkyl halides with silver nitrite or sodium nitrite, the oxidation of oximes and amines with peroxytrifluoroacetic acid, and the potassium permanganate oxidation of tertiary amines (see Table 1, page 515, which details the classes of nitroalkanes prepared by these methods).

Since aliphatic nitroalkanes are active methylene compounds, these may be used as starting materials for the preparation of more complex products by typical reactions of methylene compounds such as alkylations, aldol condensations, Michael condensations, and Mannich reactions.

Aromatic compounds are readily nitrated with nitric acid–sulfuric acid mixtures, acyl nitrates, nitronium tetrafluoroborate, and the oxides of nitrogen.

Among the indirect methods of preparing aromatic nitro compounds are the displacement of sulfonic acid groups with nitro groups and the replacement of diazonium groups.

By oxidation, aromatic amines and nitroso compounds may be converted to nitro compounds.

CAUTION: It is generally known that trinitrotoluene is a high explosive and that the treatment of glycerol, cellulose, and other carbohydrates with nitric acid/sulfuric acid leads to other explosives. The nitration of many other compounds may form products (or intermediates) which may be explosives, and precautions against detonations must be taken. Alkali salts of nitromethane or nitromethane derivatives are sensitive to impact when dry [35]. In fact, the addition of bases or acids to nitromethane is said to render the compound susceptible to detonation according to a recent publication [36]. An explosion during the nitrolysis of an acetyl compound may have been caused by the formation of acetyl nitrate in the reaction mixture. In general, mixtures of concentrated nitric acid and acetic anhydride or trifluoroacetic anhydride may form potentially explosive compositions in the reaction such as, for example, acetyl nitrate, as well as traces of tetranitromethane [37]. These safety notes [37] indicate that explosion hazards are quite prevalent in reaction systems designed to produce nitro compounds and due precautions must be taken quite generally.

Animal studies have implicated 2-nitropropane as a possible human carcinogen. Therefore, the use of approved respirators and full-body protective clothing is indicated for handling this compound and perhaps for related compounds [38a].

In our discussion below, the hazards may not always be pointed up, but it should be understood that safety shields, even barricades, respirators, and full-body protective clothing, should always be used when working with nitro compounds.

2. ALIPHATIC NITRO COMPOUNDS

With the introduction of vapor phase nitration of paraffins, the chemistry of aliphatic nitro compounds has become of considerable commercial importance. However, in the laboratory, this procedure has only limited applicability.

The procedures discussed here are only those which produce well-defined products rather than the complex mixtures normally obtained in vapor phase reactions.

A. Direct Nitration

We use the term "direct nitration" for reactions of aliphatic hydrocarbons, olefins, and related materials with nitric acid, oxides of nitrogen, and related reagents. In these reactions a proton is normally removed from the starting material and a nitro group is introduced in its place.

From the standpoint of laboratory preparations, the direct nitration of saturated hydrocarbons with nitric acid or oxides of nitrogen is not very satisfactory. Both in liquid phase and in vapor phase nitrations, complex mixtures of products are normally formed. The products found in the reaction mixture may include mononitro compounds, polynitro compounds, and a large variety of oxidation products. On an industrial scale the separation of the products of nitration by distillation or other techniques can be carried out to afford useful products.

The direct nitration of cycloparaffins may be of some use from a laboratory standpoint. Thus, for example, the vapor phase nitration of cyclohexane with nitrogen dioxide is said to afford a 69.3% yield of crude nitrocyclohexane [38b]. In the liquid phase, a heterogeneous reaction can be carried out at about 10 atm of pressure with cyclohexane and a 35% solution of nitric acid to produce a 40% yield of nitrocyclohexane along with substantial quantities of such by-products as cyclohexyl nitrite and cyclohexyl nitrate, a dinitrocyclohexane, nitrocyclohex-

§ 2. Aliphatic Nitro Compounds   501

ane, and other nitrated products. Also found in the product mixture are quantities of dicarboxylic acids such as adipic, glutaric, and succinic acids [38c].

a. NITRATION OF OLEFINS WITH DINITROGEN TETROXIDE

The direct nitration of olefins with dinitrogen tetroxides has been investigated in some detail [39]. The first products of reaction appear to be dinitroparaffins and a mixture of nitro nitrites. The latter products may be oxidized to the nitro nitrates in the course of the reaction, or upon subsequent hydrolysis these may be converted to nitro alcohols. Evidently, difficulties in obtaining reproducible results in these reactions are overcome by the use of purified dinitrogen tetroxide. The reaction is carried out in the liquid phase. Solvents are said to have a profound influence on the details of the reaction. Thus, solvents such as diethyl ether, dioxane, ethyl acetate, and other esters appear to moderate the oxidizing action of dinitrogen tetroxide and permit the formation of the dinitro and nitro-nitrite products normally expected. Since the possibility exists of some dinitrogen tetroxide being reduced to the trioxide in the course of the reaction, which leads to more complex reaction mixtures, oxygen is often introduced into the reaction system to afford control over the nitrating agent. Naturally, this may occasionally lead to some difficulties. For example, in the nitration of iso-butylene, the introduction of additional oxygen leads to the formation of considerable amounts of nitro-nitrate compounds, which makes the separation of 1,2-nitroisobutane by crystallization difficult.

Purification of dinitrogen tetroxide is carried out by cautious fractional distillation of crude liquid dinitrogen tetroxide in an oxygen stream at 30°C. During this distillation the lower oxides are separated. Nitric acid spray and moisture are removed from the vapor by passing the distillate through silica gel and through phosphorus pentoxide. The pure substance is condensed in a receiver to a clear white solid at −70°C. The receivers, as well as the rest of the distillation equipment, should have their outlets protected with phosphorus pentoxide and calcium chloride drying tubes to prevent contamination by atmospheric moisture. Liquid dinitrogen tetroxide may be stored at room temperature in closed stainless steel pressure bottles.

The organic nitro compounds may be explosive and suitable precautions must be taken. Such products as dinitroethane, for example, are said to have the explosive power of 91.1% blasting gelatin [39]. However, this particular product is said to be insensitive to friction, impact, and explosive initiation. Even so, it is our opinion that no effort should be spared in taking protective measures against explosions in handling such nitro compounds.

The products of a typical nitration of isobutylene are given in Eq. (1). In this reaction scheme, the hydrolysis product indicated results from the fact that the crude product mixture is added to water after stripping off solvents and excess dinitrogen tetroxide [39].

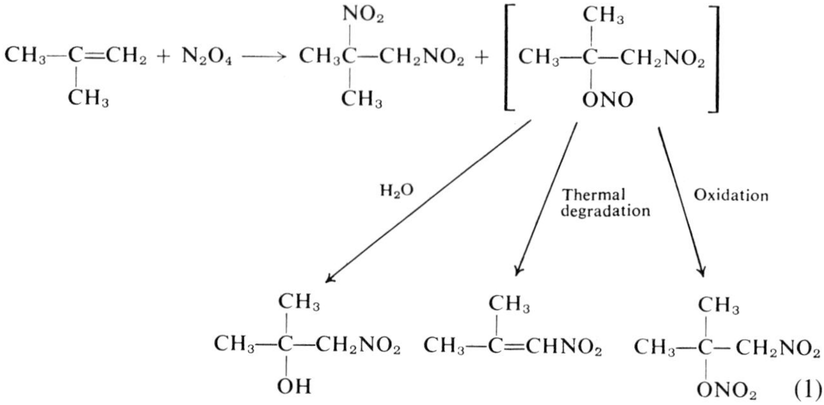

$$CH_3-\underset{\underset{CH_3}{|}}{C}=CH_2 + N_2O_4 \longrightarrow CH_3-\underset{\underset{CH_3}{|}}{\overset{\overset{NO_2}{|}}{C}}-CH_2NO_2 + \left[CH_3-\underset{\underset{ONO}{|}}{\overset{\overset{CH_3}{|}}{C}}-CH_2NO_2 \right]$$

$$\xrightarrow{H_2O} \quad \xrightarrow{\text{Thermal degradation}} \quad \xrightarrow{\text{Oxidation}}$$

$$CH_3-\underset{\underset{OH}{|}}{\overset{\overset{CH_3}{|}}{C}}-CH_2NO_2 \quad CH_3-\overset{\overset{CH_3}{|}}{C}=CHNO_2 \quad CH_3-\underset{\underset{ONO_2}{|}}{\overset{\overset{CH_3}{|}}{C}}-CH_2NO_2 \quad (1)$$

It has been found that dinitro paraffins may be treated with a base to obtain nitro olefins and that nitro nitrites similarly may be converted to nitro olefins. Thus, the nitration of olefins may be used as a starting point for the preparation of mononitroolefins in surprisingly good yields. The preparation of 1-nitrocyclooctene given below is an example of this reaction. As stated before, safeguards against explosion hazards should be used.

The author of Preparation 2-1 evidently used purified dinitrogen tetroxide from laboratory supply sources without difficulty. However, if necessary, the purification procedure outlined above should be used.

2-1. Preparation of 1-Nitrocyclooctene [40]

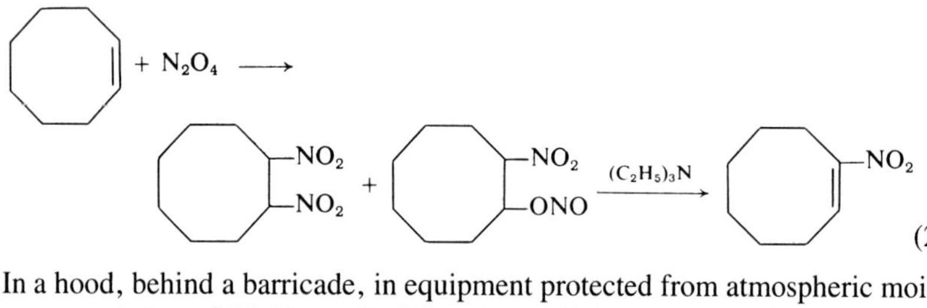

cyclooctene + $N_2O_4 \longrightarrow$ [1,2-dinitrocyclooctane] + [2-nitrocyclooctyl nitrite] $\xrightarrow{(C_2H_5)_3N}$ 1-nitrocyclooctene (2)

In a hood, behind a barricade, in equipment protected from atmospheric moisture, to a solution of 39.28 gm (0.427 mole) of dinitrogen tetroxide in 150 ml of dry ether at 10° ± 1°C, while bubbling 13 millimoles of oxygen through the solution, is added 44.4 gm (0.40 mole) of cyclooctene (which contained 4.6% cyclooctane) over a 24-min period. Then 24 ml of ether is added and, while the solution is being stirred for a ½ hr at 10°C, 121 gm (1.2 moles) of triethylamine is added with external cooling of the reaction flask. The reaction mixture is then maintained at 24°C for ½ hr, cooled to 3°C, diluted with 150 ml of ether, quenched by the cautious addition of a solution of 1.2 moles of acetic acid in 200 ml of water. The etheral solution is separated, washed in turn with water, sodium bicarbonate solution, and again with water. The solution is then cautiously evaporated under reduced pressure. The cyclooctane which had been added with the cyclooctene is removed during the evaporation under reduced pressure. The residue is a yellow liquid weighing 61.0 gm and represents a 96% yield (based on infrared analysis).

Pure 1-nitrocyclooctene is obtained by chromatography on silica gel and subsequent distillation, bp 60°C (0.2 mm), n_D^{20} 1.5116. After the solution has stood at room temperature for a period of several weeks, a slow decomposition appears to take place with the simultaneous precipitation of a solid.

According to a German patent application [41], the treatment of olefins with an excess of oxygen leads to the intermediate formation of a nitro-peroxynitrate which ultimately is converted to a nitro-nitratoalkane.

Nitric oxide (NO_2) has also been studied as a possible nitrating agent of olefinic compounds [42]. The reaction products obtained represent a very complex mixture. Thus, for example, the reaction of 2-methyl-2-butene affords at least 17 products, of which the three major products are 2-methyl-3-nitro-1-butene (24.6%); 2-methyl-1-nitro-2-butene (22.7%); and 2-methyl-3-nitro-2-butene (21.7%). Among the minor components of the reaction mixture are 2-methyl-3-nitro-2-butanol (5.9%); 2-methyl-2-nitro-2-butyl nitrate (1.0%); 2-methyl-3-nitro-2-nitrosobutane (0.3%); and 2,3-dinitro-2-methylbutane (1.5%), distillation residue (17.3%). Other products identified in the reaction mixture are acetaldehyde, acetone, methyl isopropyl ketone, acetic acid, nitromethane, water, and nitrogen.

By analogy to other work, it is expected that the nitro group would appear at the less-substituted carbon atom of the original olefinic bond. In addition, abnormal products were formed in this reaction, for example, 2-methyl-1-nitro-2-butene evidently formed from a NO_2-catalyzed allylic rearrangement of the normal product. The nitro nitroso compound is also believed to undergo an intramolecular oxygen shift ("nitrosite rearrangement") to account for observed ketoxime formation.

In the presence of amides (other than dimethylformamide), the reaction of cycloalkenes with nitric oxide–oxygen mixtures (one mole of nitric oxide to 20 moles of oxygen) at 0°C in aromatic hydrocarbon solutions produces good yields of 2-nitrocycloalkanones [43].

b. Nitration of Olefins with Acetyl Nitrate

Acetyl nitrate is readily prepared from 70% nitric acid and an excess of acetic anhydride. Nitration of certain olefins with this reagent represents a convenient procedure for the preparation of β-nitroacetates, β-nitro alcohols and β- and/or α-nitroalkenes. Alternative procedures for some of these products would be the condensation of aldehydes with nitroalkanes discussed below. Analogous reactions involving the condensation of ketones with nitroalkanes are rather difficult. By use of acetyl nitrate, however, many of the products expected from such interactions are readily accessible [44].

In the introduction to this chapter, the hazards of synthesizing and utilizing acetyl nitrate were mentioned [37]. The preparations given here must be carried out with extreme precautions in view of the explosive hazards.

While many olefins are readily nitrated by acetyl nitrate, by concerted reactions involving a protonated form of acetyl nitrate, a small amount of sulfuric acid has a strong accelerating effect and is particularly useful in the case of an olefin which is difficult to nitrate.

The reaction of isobutylene in Eq. (3) serves to illustrate some of the major products isolated from reactions with acetyl nitrate. This equation should be compared with Eq. (1) for the comparable process with dinitrogen tetroxide.

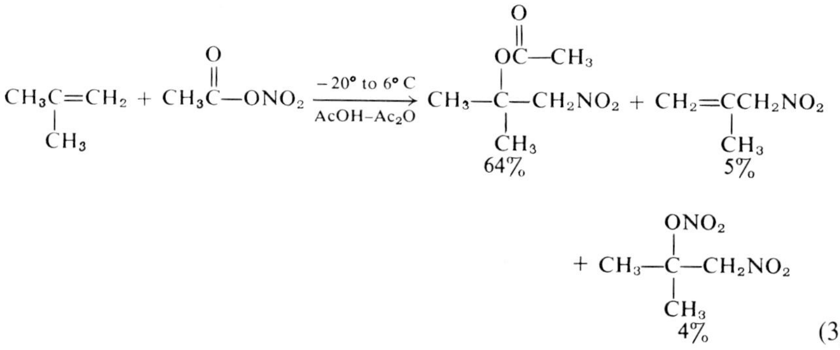

$$CH_3C(CH_3)=CH_2 + CH_3C(O)-ONO_2 \xrightarrow[AcOH-Ac_2O]{-20° \text{ to } 6°C} CH_3-C(OC(O)CH_3)(CH_3)-CH_2NO_2 + CH_2=C(CH_3)CH_2NO_2$$
$$64\% \qquad 5\%$$
$$+ CH_3-C(ONO_2)(CH_3)-CH_2NO_2$$
$$4\% \qquad (3)$$

In Preparation 2-2, the scale of the reaction should be maintained as indicated. If larger scale operations are desired, the olefin should be added in several stages so as not to permit the reaction temperature to rise above approximately 15°C. If the reaction mixture is allowed to rise above this range, difficulty in the isolation of reaction products is often experienced. In no instance should the maximum reaction temperature be permitted to approach 60°C since the vigorous decomposition of the acyl nitrate occurs at this temperature. Generally, however, the reaction should be carried out as rapidly as possible.

It is to be noted, that the initial reaction of acetic anhydride and nitric acid to form acetyl nitrate must be carried out at at about room temperature, since at lower temperatures, this reagent does not form. (CAUTION: Pure acetyl nitrate should not be isolated since explosions have been reported in the isolation of the pure material.) Prior to the addition of olefin, the solution of acyl nitrate in acetic anhydride is cooled to $-20°C$.

The preparation of 3-nitro-2-butyl acetate and related products is an illustration of the reaction technique [44].

2-2. Preparation of 3-Nitro-2-butyl Acetate [44]

$$\begin{array}{c} \text{H} \quad \text{CH}_3 \\ | \quad | \\ \text{C}=\text{C} \\ | \quad | \\ \text{CH}_3 \, \text{H} \end{array} + \text{CH}_3\overset{\overset{\text{O}}{\|}}{\text{C}}-\text{ONO}_2 \longrightarrow \begin{array}{c} \text{NO}_2 \quad \text{CH}_3 \quad \text{O} \\ \backslash \quad | \quad \| \\ \text{CH}-\text{CH}-\text{OCCH}_3 \\ / \\ \text{CH}_3 \end{array} + \begin{array}{c} \text{NO}_2 \\ \backslash \\ \text{CH}-\text{CH}=\text{CH}_2 \\ / \\ \text{CH}_3 \end{array} \quad (4)$$

CAUTION: Acetyl nitrate is considered a potential explosive.

Behind a barricade, to 500 ml of acetic anhydride at 15°C is added slowly, with rapid stirring and cooling, 67.5 gm (0.75 mole) of nitric acid (70.4% in water). As soon as the temperature in the flask approaches 20°C, the temperature of the cooling bath surrounding the reaction flask is lowered further to −25°C ± 5°C and the remainder of the nitric acid is added at such a rate that the temperature in the reactor is maintained between 20° and 25°C. The addition of nitric acid requires between 3 and 5 min. Then the solution is cooled quickly with rapid stirring to −20°C, and 16.8 gm (0.3 mole) of liquid *trans*-2-butene is added all at once. As soon as the temperature of the reaction mixture begins to rise, the temperature of the cooling bath is quickly lowered to −60°C to −70°C by use of a Dry Ice–acetone bath. The temperature of the reaction mixture rises rapidly to 11°C and then gradually falls back to −20°C within less than 5 min. The pale yellow reaction mixture is then poured into 1.5 liters of cold water. The mixture is stirred periodically until hydrolysis of the excess acetic anhydride has been completed. The reaction product is then stirred vigorously with 800 ml of a saturated sodium chloride solution and 800 ml of ether. The ether layer is washed four times with 800 ml of water and dried over anhydrous magnesium sulfate for 20 min. The drying agent is removed by filtration, the ether is removed from the filtrate under reduced pressure, and the reaction products are distilled under reduced pressure with a vacuum-jacketed distillation column to give the following fractions: (1) bp 39°–44°C (20 mm), 8.3 gm of 3-nitro-1-butene (27%). (2) bp 57°–58°C (0.6 mm), 18.0 gm (38%) of 3-nitro-2-butyl acetate. In the i.r. spectrum, 3-nitro-2-butyl acetate has a medium peak about 1037 cm^{-1}. [The 3-nitro-1-butene, upon redistillation, exhibited a boiling point of 43°–43.5°C (20 mm).]

c. Nitration of Olefins with Nitric Acid–Sodium Nitrite

A procedure for the nitration of cholesteryl acetate utilizes nitric acid along with sodium nitrite as the nitrating agent. This procedure is said to be "quite capricious with yields from 20 to 50% even under identical conditions" [45]. The variation was attributed to variations in the quality of the sodium nitrite and of the nitric acid by Haire and Boswell. We believe that this procedure merits further study since this reagent combination strikes us as being substantially less hazardous than work with acetyl nitrate. In the original procedure (preparation 2-3) solid sodium nitrite is added proportionwise to nitric acid and cholesteryl acetate with the evolution of brown fumes. This procedure undoubtedly leads to a reaction system that could be difficult to reproduce since the simple interaction of nitrite ions with acids and the reduction of nitrate ions by a solid nitrite are among the reactions that may be taking place. A more reproducible reaction system might involve the dropwise addition of nitric acid to a dispersion or solvent solution of sodium nitrite and cholesteryl acetate or other olefinic reagent.

2-3. Preparation of 6-Nitrochlolesteryl Acetate [45]

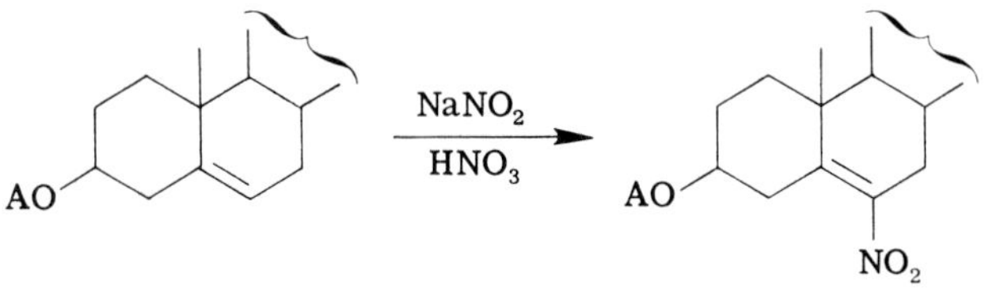

$$\text{cholesteryl acetate} \xrightarrow[\text{HNO}_3]{\text{NaNO}_2} \text{6-nitrocholesteryl acetate} \quad (5)$$

Use a hood, with suitable safety precautions. To a 3-liter flask cooled in an ice bath containing a vigorously stirred dispersion of 100 gm (0.23 mole) of cholesteryl acetate in 1700 ml of concentrated nitric acid is added, over a 30-min interval, 100 gm of sodium nitrite in small increments. The pink mixture gradually becomes yellow and brown toxic fumes evolve. The reaction mixture is stirred overnight at room temperature. The insoluble product is filtered off on a sintered-glass funnel and washed repeatedly with cold water. The yellow solid is recrystallized from methanol/ether. The yield of 6-nitrocholesteryl acetate varies between 20 and 50%. For example, in one case 52.5 gm were isolated (47% yield), mp 102°–104°C. IR (CHCl$_3$) 2959 (s), 2874, 1724 (s), 1517 (s), 1466, 1437, 1370, 1252, 1040 cm^{-1}.

d. NITRATION OF ACTIVE METHYLENE COMPOUNDS WITH NITRIC ACID

Active methylene compounds have been nitrated directly with nitric acid. Since the nitro group thus introduced also activates the carbon adjacent to it, the product resulting from the direct nitration of an active methylene compound is frequently highly reactive and consequently may be quite unstable—in fact, explosive. Therefore, as with all aliphatic nitro compounds, precaution must be taken to reduce the hazards of explosion to personnel and equipment. A typical example of the reaction is given in the following preparation.

2-4. Preparation of Diethyl Nitromalonate [46]

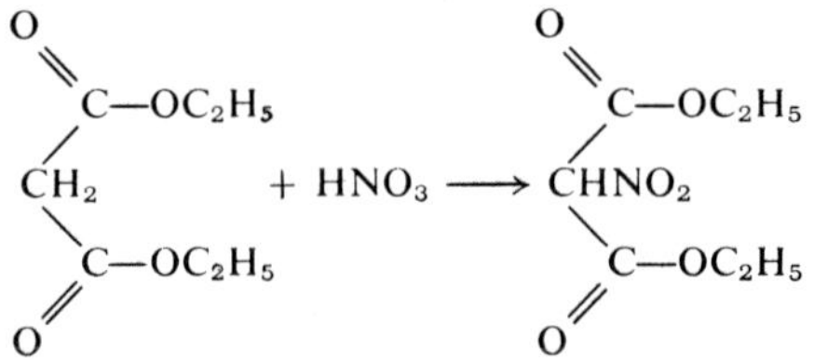

$$\begin{array}{c} \text{C(O)OC}_2\text{H}_5 \\ | \\ \text{CH}_2 \\ | \\ \text{C(O)OC}_2\text{H}_5 \end{array} + \text{HNO}_3 \longrightarrow \begin{array}{c} \text{C(O)OC}_2\text{H}_5 \\ | \\ \text{CHNO}_2 \\ | \\ \text{C(O)OC}_2\text{H}_5 \end{array} \quad (6)$$

§ 2. Aliphatic Nitro Compounds

In a 500-ml three-necked flask fitted with a dropping funnel, stirrer, thermometer, and an outlet protected by a drying tube is placed 80.0 gm (0.5 mole) of diethyl malonate. The flask is cooled by tap water at 12°C and 184 ml of fuming nitric acid (density: 1.5 gm/ml) is added at a rate sufficient to maintain the temperature between 15° and 20°C. The addition requires approximately 1 hr. After this period, the mixture is allowed to stir for 3½ hr at 15°C. The solution is then poured into 1 liter of ice and water and the ester is extracted with two portions of toluene, the first being 200 ml, the second 100 ml. The combined toluene extracts are washed twice with water and then with 200 ml portions of 5% aqueous urea solution until a starch–potassium iodide test for the oxide of nitrogen is negative.

To isolate the product from its toluene solution, the toluene solution is extracted with several portions of a 10% solution of sodium carbonate in water until the acidification of a test portion of the aqueous extract shows that no more nitro ester is being extracted by the sodium carbonate solution.

The sodium carbonate extracts are combined and washed once with 200 ml of toluene. The aqueous solution is then carefully acidified to Congo Red paper with concentrated hydrochloric acid while cooling by the occasional addition of ice. The ester is collected by extracting in turn with 500-, 200-, and 100-ml portions of toluene. The toluene solutions are washed twice with 200 ml portions of water and then again with a 5% aqueous urea solution, checking again with starch–potassium iodide test papers for complete absence of oxides of nitrogen. The toluene solution is then dried over magnesium sulfate. For many purposes this solution may be used as is. To assay the solution, an aliquot is added to an equal volume of ethanol and titrated to a phenolphthalein end point with 1 N sodium hydroxide.

In this case, the assay depends on the acidic nature of the nitro esters. The assay indicates a yield of 94.1 gm (91.7%) of product. To isolate the pure ester, the toluene is evaporated under reduced pressure, and the residue is distilled at 81°–83°C (0.3 mm).

[Reprinted from: D. I. Weisblat and D. A. Lyttle, *J. Am. Chem. Soc.* **71**, 3079 (1949). Copyright 1949 by the American Chemical Society. Reprinted by permission of the copyright owner.]

e. Nitration of Acetic Anhydride with Nitric Acid

The formation of acetyl nitrate by reaction of nitric acid in an excess of acetic anhydride has been discussed above. When the mole ratio of nitric acid to acetic anhydride becomes 1:1, and reaction is allowed to proceed at ordinary temperatures for several days, good yields of tetranitromethane are obtained. Since tetranitromethane has received attention as a nonacidic nitrating agent with high specificity, directions for its preparation are given here although we must caution the reader that the product is considered to be explosive [37] and toxic.

2-5. Preparation of Tetranitromethane [47,48]

$$4 \text{ CH}_3\overset{\overset{O}{\|}}{C}-O-\overset{\overset{O}{\|}}{C}-CH_3 + 4 \text{ HNO}_3 \longrightarrow C(NO_2)_4 + CH_3\overset{\overset{O}{\|}}{C}-OH + CO_2 \qquad (7)$$

CAUTION: The product is explosive and toxic. Reactions must be carried out in a hood, behind an adequate baricade.

In a 250-ml Erlenmeyer flask surrounded by an ice–water bath, resting on a magnetic stirring apparatus, and containing a Teflon-coated magnetic stirrer, and a thermometer reaching nearly to the bottom of the flask, is placed 31.5 gm (0.5 mole) of anhydrous nitric acid. (This anhydrous nitric acid is readily obtained by carefully distilling laboratory grades of concentrated nitric acid with a specific gravity of 1.4 or higher from an equal volume of concentrated sulfuric acid twice.) The flask is cooled, and, with stirring, 51 gm (0.5 mole) of acetic anhydride is added from a buret in 2-ml portions. An exothermic reaction ensues: however, the reaction temperature must never be allowed to rise above 10°C. As more and more acetic anhydride is added, the vigor of the reaction decreases and somewhat larger portions of acetic anhydride may be added at any one time. It is advisable not to add acetic anhydride in portions larger than 5 ml at a time. After all the anhydride has been added, stirring and cooling are continued for a short while. The thermometer is removed from the flask, and a small beaker is inverted over the neck of the flask. The flask is kept in the ice–water bath while the ice melts and is then allowed to stand at room temperature for 7 days. During this time, carbon dioxide is said to evolve gradually and continuously. The reaction mixture gradually changes from colorless to brown. After this time, the reaction mixture is poured into 200 ml of water. Most of the tetranitromethane forms a heavy oily layer which may be removed by means of a separatory funnel. A small quantity of product is retained in the dilute acetic acid mixture and may be separated by means of steam distillation.

Water or steam treatment of the product should be kept to a minimum since small quantities of nitroform may be produced by reaction of water. While the preparation in *Organic Syntheses* [48] indicates purification by steam distillation, for many applications it would seem to be sufficient to purify the product obtained directly from the reaction by rapidly washing it with dilute sodium carbonate solution followed by a water wash and drying with sodium sulfate. Under no circumstances should the product be distilled because of explosive hazards. Tetranitromethane must be kept out of contact with aromatic compounds except under carefully controlled reaction conditions since explosions are said to occur at times.

Chattaway [47] reports yields of 18–20 gm of pure dried tetranitromethane (approximately 80%), while Liang reports only 14–16 gm (57–65%).

§ 2. Aliphatic Nitro Compounds

f. Chloronitration of Olefins with Nitryl Chloride

Nitryl chloride has been added to a variety of olefinic compounds to yield nitro compounds. For example, the addition of nitryl chloride to vinyl bromide affords 2-bromo-2-chloronitroethane [Eq. (8)] [49].

Nitro ketones and aldehydes have also been prepared by the reaction of enol acetates and ethers with nitryl chloride according to Eqs. (9) and (10) [50].

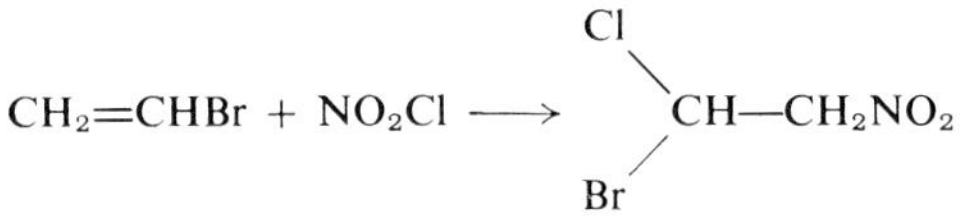

$$CH_2=CHBr + NO_2Cl \longrightarrow \underset{Br}{\overset{Cl}{CH}}-CH_2NO_2 \quad (8)$$

$$R-\underset{OCCH_3}{\underset{\|}{\overset{}{C}}}=CHR' + NO_2Cl \longrightarrow R-\underset{\underset{\|}{OC-CH_3}}{\overset{Cl}{\underset{|}{C}}}-\overset{R'}{\underset{|}{CH}}-NO_2 \quad (9)$$

$$R-\underset{\underset{\|}{OCCH_3}}{\overset{Cl}{\underset{|}{C}}}-\overset{R'}{\underset{|}{CH}}-NO_2 + H_2O \longrightarrow R-\overset{O}{\underset{\|}{C}}-\underset{R'}{\underset{|}{CHNO_2}} + CH_3CO_2H + HCl \quad (10)$$

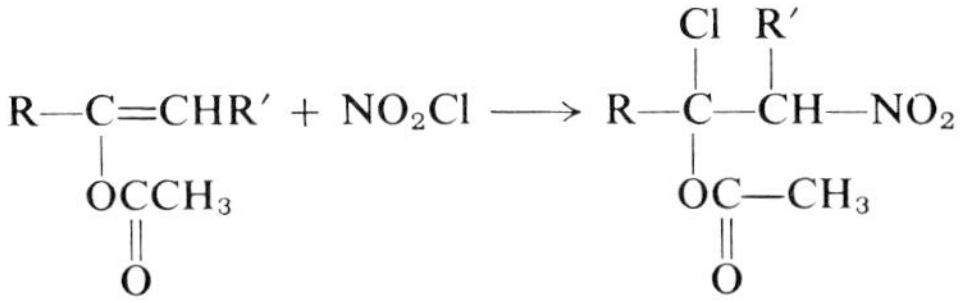

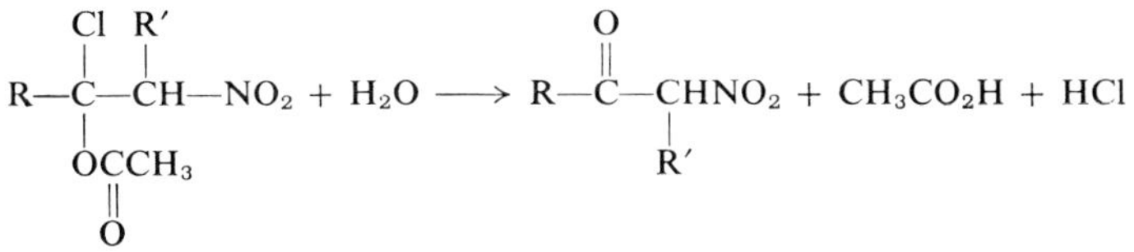

The preparation of methyl 2-chloro-3-nitropropionate is an example of the reaction involving an olefinic ester [51]. In connection with this description, the preparation of nitryl chloride is also given.

2-6. *Preparation of Methyl 2-Chloro-3-nitropropionate* [51]

$$HNO_3 + ClSO_3H \longrightarrow NO_2Cl + H_2SO_4 \quad (11)$$

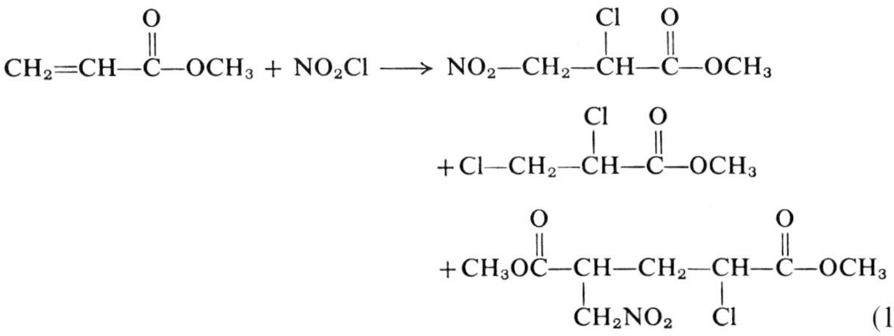

$$CH_2=CH-\overset{O}{\underset{\|}{C}}-OCH_3 + NO_2Cl \longrightarrow NO_2-CH_2-\overset{Cl}{\underset{|}{CH}}-\overset{O}{\underset{\|}{C}}-OCH_3$$

$$+ Cl-CH_2-\overset{Cl}{\underset{|}{CH}}-\overset{O}{\underset{\|}{C}}-OCH_3$$

$$+ CH_3O\overset{O}{\underset{\|}{C}}-\underset{CH_2NO_2}{\underset{|}{CH}}-CH_2-\underset{Cl}{\underset{|}{CH}}-\overset{O}{\underset{\|}{C}}-OCH_3 \quad (12)$$

CAUTION: The following reaction should be carried out in a hood, behind shields.

(a) Preparation of nitryl chloride. In a 500-ml three-necked flask equipped with a pressure equalizing dropping funnel, an efficient motor-driven glass stirrer sealed with sulfuric acid, and a condenser cooled with Dry Ice–acetone, to which a receiver of 150 ml capacity is connected which in turn is also cooled with Dry Ice–acetone, is placed 100 gm (1.4 moles) of fuming nitric acid (specific gravity 1.50, 91.6% acid content). The acid is cooled to 0°C and then 123 gm of fuming sulfuric acid (containing 30% sulfur trioxide) is added dropwise. The mixed acids are stirred vigorously at 0°C while 170.0 gm (97.5% ml, 1.47 moles) of chlorosulfonic acid is added slowly over a ¾-hr period, at such a rate that no brown nitrous fumes appear above the reaction mixture. Nitryl chloride, bp $-17°$ to $-50°C$, 95–108 gm (80–90%), is collected in the receiver.

(b) Nitration reaction. A three-necked flask equipped with a liquid-sealed stirrer and a series of drying tubes as well as a gas inlet tube is cooled to 0°C. The gas inlet tube is connected to a vessel containing nitryl chloride. In the flask is placed 86 gm (1.0 mole) of anhydrous methyl acrylate. With vigorous stirring, while maintaining the temperature at 0°C, 97.8 gm (1.2 moles) of nitryl chloride is distilled into the reaction flask over a 2-hr period. The mixture is then stirred for an additional hour at room temperature. The excess nitryl chloride now containing chlorine and nitrogen dioxide is removed by distillation under reduced pressure. During this distillation, the reaction mixture changes color from orange-red to light yellow-orange. The product is distilled under nitrogen at reduced pressure to yield three fractions. Yield of the fraction boiling range 43°–80°C (18-22 mm) is 22.2 gm (7.1%), the second fraction with bp range 68°–110°C (2-4 mm) is 251 gm (75.1%), and a pot residue of 43 gm (12.1%).

The first two fractions are combined and redistilled under nitrogen at reduced pressure through a 24 × 2 cm helix-packed column. The first fraction isolated from this distillation has been identified after rerectification as methyl 2,3-dichloropropionate, bp 72.5°C (21 mm). Yield 9.3 gm (5.9%). The second fraction is methyl 2-chloro-3-nitropropionate, bp 88°C at 4 mm. Yield 103.6 gm (62.1%). The high-boiling residue has been identified as dimethyl 2-chloro-4-nitromethylpentanedioate, bp 131°C (0.8 mm). Yield 5–10%.

g. Nitration of Active Methylene Compounds with Alkyl Nitrates in Alkaline Media

About the turn of the century, Wislicenus and co-workers [52] originated a method of preparing nitro compounds under alkaline conditions without drastic modification of the carbon skeleton. In the original procedure, the condensation of an active methylene compound in the presence of sodium hydroxide with ethyl nitrate yielded the *aci* form of the nitro compound. Thus, for example, phenylacetonitrile was converted to a nitro derivative which, upon decarboxylation, afforded phenylnitromethane according to Eqs. (13) and (14).

§ 2. Aliphatic Nitro Compounds

$$\text{Ph-CH}_2\text{CN} + \text{C}_2\text{H}_5\text{ONO}_2 \xrightarrow{\text{NaOC}_2\text{H}_5}$$

$$\text{Ph-}\underset{\underset{\text{CN}}{|}}{\text{C}}\text{=NO}_2^-\text{Na}^+ \xrightarrow{\text{NaOH}} \text{Ph-}\underset{\underset{^-\text{CO}_2\text{Na}^+}{|}}{\text{C}}\text{=NO}_2^-\text{Na}^+ \qquad (13)$$

$$\text{Ph-}\underset{\underset{^-\text{CO}_2\text{Na}^+}{|}}{\text{C}}\text{=NO}_2^-\text{Na}^+ \xrightarrow{2\,\text{HX}} \text{Ph-CH}_2\text{NO}_2 + \text{CO}_2 + 2\,\text{NaX} \qquad (14)$$

More recent investigations have shown that this reaction has wide applicability. The course of the reaction is somewhat more complex and the nature of the products isolated from the mononitration of ketones is complicated by cleavage reactions of the intermediates [53]. The preparation of 2-nitrocyclooctanone illustrates the modern modification of the procedure along with some of the problems associated with the cleavage of the cyclooctanone ring [53].

This procedure uses potassium *t*-butoxide in tetrahydrofuran (THF) in excess, along with equivalent amounts of a nitrate ester and a ketone. A modification which is said to be a greatly simplified procedure has also been reported [54]. This method is a nitration of ketones with amyl nitrate in liquid ammonia at about $-33°\text{C}$. Procedure 2-7 is the reaction in THF. We are describing this method here in more detail since the alternative, reaction in liquid ammonia, requires more elaborate safety precautions.

2-7. Preparation of 2-Nitrocyclooctanone and Amyl 8-Nitrooctanoates [53]

$$\text{cyclooctanone} + \text{RONO}_2 \xrightarrow[2.\ \text{HOCOCH}_3]{1.\ \text{KOC(CH}_3)_3}$$

$$\text{2-nitrocyclooctanone} + \text{NO}_2(\text{CH}_2)_7\text{C(=O)}-\text{OR} + \text{KOCOCH}_3 \qquad (15)$$

To a solution of 18.0 gm (0.16 mole) of sublimed potassium *tert*-butoxide in 90 ml of tetrahydrofuran at $-50°\text{C}$ is added 12.6 gm (0.10 mole) of cyclooctanone in 60 ml of tetrahydrofuran with stirring. To the heterogeneous mixture is added dropwise 14.4 gm (0.11 mole) of mixed amyl nitrates dissolved in 30 ml of tetrahydrofuran over a 20-min period while maintaining the temperature between $-45°$ and $-50°\text{C}$. At the end of this period, a solution of 30.0 gm (0.5 mole) of acetic acid in 100 ml of absolute ether is added rapidly. The reaction mixture is stirred at $0°\text{C}$ for 12 hr. The resulting potassium acetate is filtered off and the filtrate is evaporated in a rotating evaporator under reduced pressure.

Distillation of the residue affords 6.0 gm (35%) of 2-nitrocyclooctanone at 73°–74°C (0.2 mm), and 9.6 gm (37%) of mixed amyl 8-nitrooctanoates at bp 140°–145°C (0.2 mm). Unreacted cyclooctanone was found in the Dry Ice trap by gas–liquid chromatography of the content in the trap.

h. NITRATION OF CYCLIC ENOL ACETATES WITH NITRIC ACID/ACETIC ANHYDRIDE AND OF CYCLIC ENOLATES WITH ALKYL NITRATES

Enol acetates of cyclic ketones have been nitrated in acetic anhydride solution with nitric acid. The products that are isolated are the 2-nitroketones. In the case of cyclohexanone derivatives, yields are very good. The stereochemistry of the reaction is controlled by the geometry of the transition state of the enol acetate.

The 2-nitrocyclopentanones, on the other hand, are sensitive to the acidic reaction media. Consequently, alkaline reaction conditions are preferred. The enolates of the cyclopentanones are the starting materials of choice rather than the related enol acetates of the cyclopentanones. However, the nitration of ketone enolates does not lead to a regioselective synthesis of 2-nitroketones since the intermediate potassium enolate does not retain the desired geometry [55,56].

2-8. Preparation of 2-Nitro-4-methylcyclohexanone [55]

$$\text{1-acetoxy-4-methylcyclohexene} + HNO_3 \xrightarrow{Ac_2O} \text{2-nitro-4-methylcyclohexanone} + CH_3C(=O)-OH \qquad (16)$$

CAUTION: Interaction between nitric acid and acetic anhydride may lead to explosive intermediates; cf. Ref. 37.

In a hood, in a 50-ml round-bottomed flask equipped with a magnetic stirrer, an addition funnel, a thermometer, and an opening for venting, a mixture of 3.08 gm (0.02 moles) of 1-acetoxy-4-methylcyclohexene and 7.0 gm (0.068 mole) of acetic anhydride maintained at 15°–20°C with a cooling bath is treated dropwise with 1.4 ml (0.022 mole) of concentrated nitric acid with vigorous stirring. Stirring is continued for 1 hour beyond the completion of addition.

The residue of acetic acid, acetyl nitrate, and acetic anhydride is separated from the reaction mixture by cautious vacuum distillation behind a barricade at a pot temperature which never exceeds 40°C. The crude product is 40% of the cis-isomer and 60% of the trans-isomer. Upon vacuum distillation of the crude product, 2.62 gm (83%) of a clear yellow liquid, bp 85°–88°C/0.2 mm of mercury is isolated. The product is a mixture of the cis- and trans-isomers.

2-9. Preparation of 2-Nitro-3(and 4)-methylcyclopentanone [56]

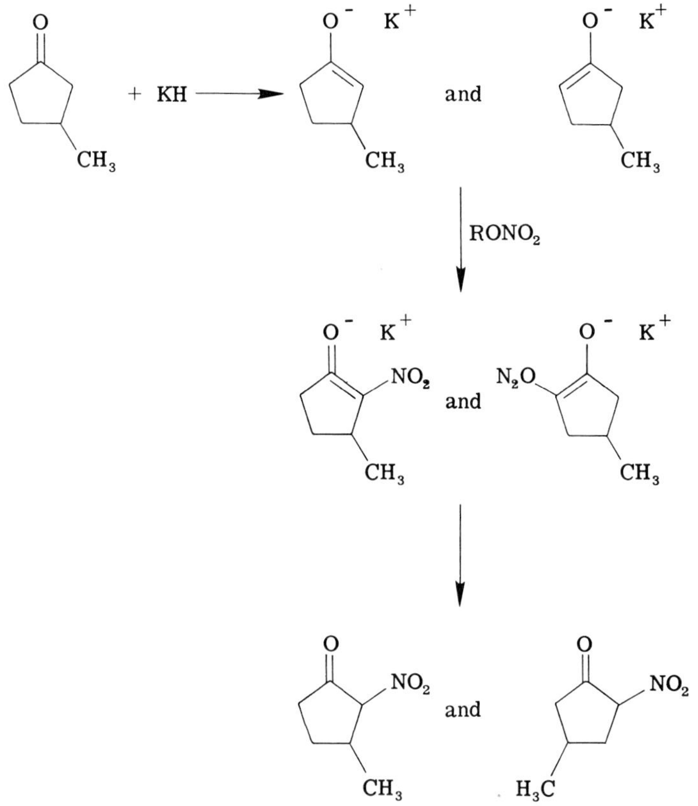

(17)

To 30 ml of dimethoxyethane which had been dried over calcium hydride, in a 100-ml three-necked, round-bottomed flask, which had been dried in an oven and cooled with protection against the incursion of atmospheric moisture, is added 2.24 gm (0.028 mole) of potassium hydride in a 50% mineral oil dispersion. (A "glove bag" is useful in manipulating this and some subsequent steps.)

The flask is then equipped with a dried pressure-equalizing addition funnel, magnetic stirrer, a thermometer, and gas inlet and outlet tubes which are protected against atmospheric moisture. The suspension of potassium hydride in dimethoxyethane is stirred at room temperature for 10–15 min. Then the flask is cooled in an ice–water bath and flushed with nitrogen. While the temperature is maintained between 0° and 20°C, a solution of 2.45 gm (0.025 mole) of 3-methylcyclopentanone in 15 ml of dry dimethoxyethane is added dropwise. The reaction mixture is stirred for an additional 10 min. Then the reaction mixture is cooled to −70°C in a Dry Ice–acetone bath and a solution of 36 gm (0.0275 mole) of amyl nitrate in 10 ml of dry dimethoxyethane is added dropwise over a 15-min period. The reaction mixture is then allowed to come to room temperature. Stirring at room temperature is continued for 2 hr.

The reaction mixture is cooled in an ice–water bath to 0°C and 2.5% aqueous hydrochloric acid is cautiously added until the solution is barely acid to litmus. The organic layer is separated and preserved. The aqueous layer is extracted several times with ether. The organic layer and the ether extracts are combined, washed with water, and dried over anhydrous magnesium sulfate. The solvents are then removed on a rotary evaporator. The residue is distilled in a molecular still at a pot temperature between 45° and 50°C (0.5-0.6 mm of mercury). The yield is 2.0 gm (56%) of a clear yellow liquid which is approximately equimolar 2-nitro-3-methylcyclopentanone and 2-nitro-4-methylcyclopentanone.

B. Indirect Nitration

The reaction of alkyl halides or dialkyl sulfates with both silver nitrite and sodium nitrite has been extensively studied by Kornblum and co-workers (see review by Kornblum [4]). These indirect methods of nitration offer procedures for the preparation of primary, secondary, and tertiary nitro compounds. Table I indicates the types of products to be obtained by various procedures.

a. SILVER NITRITE CONDENSATIONS

The reaction of alkyl halides with silver nitrite (Victor Meyer reaction) is of value in the preparation of primary nitroalkanes. In the case of secondary nitroalkanes, reactions are slow and yields are low. The reaction is of little value for the preparation of tertiary nitroalkanes.

The reaction may be carried out at relatively low temperatures, and, since it is exothermic in nature, diethyl ether is used as a diluent. As may be true with all reactions involving silver salts, it is believed to be preferable to carry out the reaction in a darkroom. Since these silver salts are not sensitized for photographic purposes, their sensitivity is primarily to the blue and ultraviolet end of the spectrum. Bright yellowish green or yellowish orange "safe-lights" used in making photographic prints are therefore suitable for illumination in a dark preparative laboratory.

In the preparation of 1-nitrooctane given below, details for the preparation of silver nitrite are also given.

2-10. Preparation of 1-Nitrooctane [57]

$$AgNO_3 + NaNO_2 \longrightarrow AgNO_2 + NaNO_3 \tag{18}$$

$$CH_3(CH_2)_6CH_2I + AgNO_2 \longrightarrow$$
$$CH_3(CH_2)_6CH_2NO_2 \text{ (83\%)} + AgI + CH_3(CH_2)_6CH_2ONO \text{ (11\%)} \tag{19}$$

(a) Preparation of silver nitrite. In a darkroom, well protected from "actinic

§ 2. Aliphatic Nitro Compounds

TABLE I
Nitro Compound Types Formed by Indirect Procedures

Starting material	Reagent	Product type
RCH_2X	$AgNO_2$	Primary nitroalkanes (particularly suitable where R contains electron-withdrawing groups)
$ICH(R)—CO_2R'$	$AgNO_2$	α-Nitro esters (limited to iodo compounds) (method of choice for ethyl α-nitroacetate)
RCH_2X	$NaNO_2$/DMF or DMSO	Primary nitroalkanes (less useful when R contains electron-withdrawing groups)
$RR'CHX$	$NaNO_2$/DMF or DMSO	Secondary nitroalkanes (reagent of choice) (fails for cyclohexyl halides)
$Br—CH(R)—CO_2R'$	$NaNO_2$/DMF or DMSO	α-Nitro esters (generally useful process) (not suitable for ethyl α-nitroacetate)
$R—CH{=}NOH$	$CF_3C(O)—OOH$	Primary nitroalkanes
$R(R')C{=}NOH$	$CF_3C(O)—OOH$	Secondary nitroalkanes
$R(R')C{=}NOH$	1. N-Bromosuccinimide 2. $NaBH_4$	Secondary nitroalkanes
$R—CH(R')—NH_2$	$CH_3C(O)—OOH$	Secondary nitroalkanes
$R'(R)(R'')C—NH_2$	$CH_3C(O)—OOH$	Tertiary nitroalkanes
$R'(R)(R'')C—NH_2$	$KMnO_4$	Tertiary nitroalkanes

radiation," to a solution of 76 gm (1.1 moles) of sodium nitrite in 250 ml of distilled water is added gradually a solution of 169.9 gm (1 mole) of silver nitrate in 500 ml of distilled water with vigorous shaking. The mixture is allowed to stand for 1 hr in the dark. The yellow precipitate is filtered with suction, stirred well with 250 ml of distilled water, and filtered. This washing operation is repeated two more times. The silver nitrite is then dried in a vacuum desiccator over potassium hydroxide pellets to yield 134 gm (86%).

(a) Preparation of 1-nitrobutane. In a darkroom, well protected from "actinic radiation," in a three-necked flask fitted with a stirrer, dropping funnel, and reflux condenser, protected by a drying tube, are placed 100 gm (0.65 mole) of silver nitrite and 150 ml of anhydrous ether. The mixture is cooled to 0°C with an ice–water bath and then, with vigorous stirring, 120 gm (0.5 mole) of freshly distilled 1-iodooctane is added over a 2-hr period. Stirring at ice temperatures is continued for 24 hr. Then the cooling bath is removed and stirring is continued for 36 hr more. Periodically, the disappearance of 1-iodooctane should be checked by vapor phase chromatography to reduce the time requirements.

At the end of the reaction period, the ether solution should give a negative Beilstein test for halogens.

The silver salts are then removed by filtration and washed thoroughly with more ether. The combined ether solutions are freed of solvent by distillation at atmospheric pressure (in daylight) through a glass-helix-packed distillation column.

The residue is distilled under reduced pressure. The first fractions, at 51°C (5 mm), are the pale yellow 1-octyl nitrite; yield 8.4 gm (11%). After a small intermediate fraction passes over, the main fraction of 1-nitrooctane is obtained at 71.5° to 72°C (3 mm), yield 64.9 gm (83%). This product is colorless.

b. SODIUM NITRITE CONDENSATIONS

Whereas the older literature indicated that the reaction of alkyl halides with sodium nitrite afforded primarily nitrite esters rather than nitro compounds, more recent investigations have shown that, by use of dimethylformamide [58] or dimethyl sulfoxide [59], good yields of either primary or secondary nitro compounds may be obtained. In these solvents, sodium nitrite is soluble, at least to some extent, which promotes the desired course of the reaction. Since more concentrated solutions are possible in dimethyl sulfoxides, the latter solvent offers some advantage in terms of shorter reaction times. The solubility of sodium nitrite in dimethylformamide is enhanced by the addition of urea. With open-chain secondary bromides, somewhat higher yields are obtained in dimethylformamide than in dimethyl sulfoxide. By the silver nitrite methods, yields are somewhat higher; however, for commercial operations the cost of the reagent outweighs this advantage. Table 1 indicates the situations in which silver nitrite is the reagent choice.

§ 2. Aliphatic Nitro Compounds

Phloroglucinol, which is indispensable in the preparation of α-nitro esters, as a scavenger of nitrite esters, has a strong retarding influence on the reaction in dimethyl sulfoxide. However, the reaction in this solvent is intrinsically so rapid that the reaction still proceeds at a reasonable rate even in the presence of phloroglucinol [59]. With secondary bromides [58] and also cyclopentyl and cycloheptyl iodides, urea in dimethylformamide is used along with a scavenger for nitrite esters. Suitable scavengers are phloroglucinol, catechol, and resorcinol. Of these, phloroglucinol appears to be the most satisfactory. The use of such scavengers is mandatory in the preparation of α-nitro esters. In preparations involving the use of dimethyl sulfoxide, the use of urea is omitted.

Phloroglucinol is added in the preparation of α-nitro esters in dimethyl sulfoxide [59,60]. We cite examples of preparations both in dimethyl sulfoxide and dimethylformamide.

While alkyl bromides or alkyl iodides are generally used in this preparation, tosylate derivatives of alcohol may also be used [58,61]. However, in a recent publication, 1-ethenyl-5,5-(ethylenedioxy)-1-methyl-2-[[(p-toluenesulfonyl)-oxy]-methyl]cyclohexane was deliberately converted to the corresponding 2-(iodomethyl) derivative prior to conversion to the 2-nitromethyl compound using the sodium nitrite/DMF procedure [62].

2-11. Preparation of 2-Nitrooctane (in DMF) [58]

$$CH_3(CH_2)_5-\underset{Br}{CH}-CH_3 + NaNO_2 \xrightarrow[2.\,H_2O]{1.\,DMF} CH_3(CH_2)_5-\underset{NO_2}{CH}-CH_3 + NaBr$$

$$+ CH_3(CH_2)_5-\underset{OH}{CH}-CH_3 \quad (20)$$

In a hood, behind a shield, to a stirred mixture of 600 ml of dimethylformamide, 36 gm (0.52 mole) of sodium nitrite, 40 gm (0.67 mole) of urea, and 40 gm (0.3 mole) of anhydrous phloroglucinol cooled in a water bath is added with vigorous stirring 58 gm (0.30 mole) of 2-bromooctane while maintaining a temperature no higher than 25°C in the reaction flask. The stirring at room temperature is continued for 45 hr. Then the reaction mixture is poured into 1.5 liters of ice water, layered over with 100 ml of petroleum ether (bp 35°–37°C). The aqueous phase is extracted four times with 100-ml portions of petroleum ether, after which the extracts are washed with four 75-ml portions of water and dried over anhydrous magnesium sulfate. The petroleum ether is then removed by distillation under reduced pressure through a column, heat being applied to the distillation flask with a bath which is gradually raised to approximately 65°C. Distillation of the residue through a 60 × 1 cm externally heated distillation column packed with ⅛ inch glass helices and equipped with a total reflux-variable take-off head yielded 15.0 gm (35%) of 2-octanol at 45°C (1 mm) and 27.6 gm (58%) of 2-nitrooctane at 67°C (3 mm).

When 2-bromooctane was converted to 2-nitrooctane using dimethyl sulfoxide as the solvent, the yields dropped to 46% from the 58% in dimethylformamide. However, the reaction time was 2 hr in dimethyl sulfoxide against 45 hr in dimethylformamide. When 2-iodooctane was used in dimethyl sulfoxide instead of 2-bromooctane, yields comparable to those obtained in dimethylformamide were observed. The example cited involves the use of 2-iodooctane.

2-12. Preparation of 2-Nitrooctane (in DMSO) [59]

$$CH_3(CH_2)_5-\underset{I}{CH}-CH_3 + NaNO_2 \xrightarrow{DMSO} CH_3(CH_2)_5\underset{NO_2}{CH}-CH_3 + NaI$$

$$+ CH_3(CH_2)_5-\underset{ONO}{CH}-CH_3 \quad (21)$$

NOTE: In handling dimethyl sulfoxide, extreme precautions must be taken. The compound is readily absorbed through the skin, and, once absorbed, is said to act as a pain killer. In addition, materials dissolved in dimethyl sulfoxide may also be carried into the body to produce their own physiological actions, in addition to those of dimethyl sulfoxide itself. Soap residues, for example, carried into the body by this solvent are said to have caused severe physiological effects.

In a hood and behind a shield, to a 500-ml flask immersed in a water bath and held at room temperature and containing 225 ml of dimethyl sulfoxide and 36 gm (0.52 mole) of sodium nitrite is added with stirring 71.2 gm (0.30 mole) of 2-iodooctane while maintaining the temperature in the flask at room temperature. Stirring is continued for 4 hr. Then the reaction mixture is poured into 600 ml of ice water and layered over with 100 ml of petroleum ether (bp 35°–37°C). After separation, the aqueous phase is further extracted with four 100-ml portions of petroleum ether. The combined extracts are then washed with four 100-ml portions of water and dried over anhydrous magnesium sulfate. The drying agent is removed by filtration; the petroleum ether solution is separated by distillation through a small column. The residue is subjected to distillation through a fractionation column under reduced pressure. At 32°C (2 mm), a 14.0-gm (30% yield) fraction of 2-octyl nitrites is collected, followed by a small intermediate fraction boiling between 53° and 56°C at 1 mm. Finally, 2-nitrooctane is isolated at 61°C (1 mm). Yield 27 gm (58%).

2-13. Preparation of Ethyl α-Nitrovalerate (in DMSO) [60]

$$CH_3CH_2CH_2\underset{Br}{CH}-CO_2C_2H_5 + NaNO_2 \xrightarrow{DMSO}$$

$$CH_3CH_2CH_2\underset{NO_2}{CH}-CO_2C_2H_5 + NaI \quad (22)$$

To a solution of 250 ml of dimethyl sulfoxide, 36 gm (0.52 mole) of sodium nitrite, and 52 gm (0.32 mole) of phloroglucinol dihydrate in a 500-ml three-necked flask equipped with a sealed stirrer is added 60.9 gm (0.3 mole) of ethyl α-bromovalerate. The flask is stoppered and the contents are maintained at room temperature and stirred for 1½ hr. A slurry forms after about ½ hr. If necessary, a small amount of dimethyl sulfoxide is added to facilitate stirring. The reaction is then poured into 600 ml of ice water layered over with 200 ml of diethyl ether. After separation of the upper layer, the aqueous phase is extracted four times with 75-ml portions of ether. The combined extracts are washed with four 100-ml portions of water and then dried over anhydrous magnesium sulfate. After filtration, the ether is removed by distillation at atmospheric pressure through a small distillation column. The residual liquid is distilled through a short column at reduced pressure. At 62°C (1 mm), a 34.2 gm (87%) yield of colorless ethyl α-nitrovalerate is collected.

Sodium chloroacetate, treated with sodium nitrite, leads to the formation of sodium nitroacetate in modest yield. Upon decarboxylation, nitromethane forms in 35–38% yield [63]. The reaction may be represented by Eq. (23).

$$ClCH_2-CO_2Na + NaNO_2 \longrightarrow O_2N-CH_2CO_2Na \longrightarrow$$
$$CH_3NO_2 + NaHCO_3 \quad (23)$$

At one time this reaction was considered the best method of preparing nitromethane. Today, of course, this material is commercially available from the vapor phase nitration of methane.

A general, mild process for the synthesis of cyclic nitro olefins was considered by Corey and Estreicher to be the key to a whole range of synthetic methodologies [64]. Indeed, in their brief paper on the synthesis of 1-nitrocyclohexene, an interesting range of transformations of this reagent is indicated. Their method of synthesizing 1-nitrocyclohexene from cyclohexene is widely applicable to the preparation of both cyclic and acyclic nitro olefins. We are classifying the process as an indirect one since it involves a nitromercuration step. Preparation 2-14 is given in general terms much like the original reference.

2-14. Preparation of 1-Nitrocyclohexene **[64]**

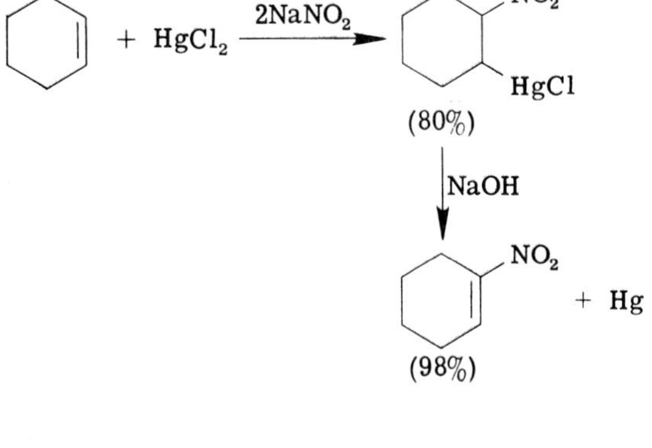

(24)

A mixture of 8.2 gm (0.1 mole) of cyclohexene, 27.2 gm (0.1 mole) of mercuric chloride, and 13.8 gm (0.2 mole) of sodium nitrite in water is stirred at 25°C for 30 hr. The resultant nitromercurial precipitate is filtered off (yield 29.2 gm, 80%).

The precipitate is dispersed in methylene chloride and treated, with stirring for 5 min, with 40 ml of 2.5 N aqueous sodium hydroxide at 25°C. The reaction mixture is then acidified with 1 N hydrochloric acid (approx. 100 ml). Upon filtration through Celite, the quantitative recovery of metallic mercury is possible. The product is extracted from the filtrate and isolated by distillation. The yield for the novel, base-catalyzed demercuration step is said to be better than 98% (approx. 10 gm).

C. Oxidation of Oximes and Amines

Aldoximes and ketoximes have been successfully oxidized with peroxytrifluoroacetic acid [55]. The reaction appears to be quite general except for derivatives of aldehydes and ketones which are highly hindered sterically, such as those of pinacolone and of trimethylacetaldehydes. It is believed that the oxidation proceeds through an *aci*-nitroparaffin stage which then rearranges to the nitroparaffin. In order to accomplish this tautomeric change, the solvent, acetonitrile, functions as a base. The reaction also has to be buffered. Sodium bicarbonate is most satisfactory for oxidation of aliphatic oximes while dibasic sodium phosphate gives high yields in the aromatic and heterocyclic oxime series. The course of the reaction has been found to be seriously affected by the rates of addition of peracids as well as the degree of heating. Generally speaking, addition rates of peracids should be relatively low and overheating must be avoided.

The procedure is particularly attractive for the preparation of secondary nitroalkanes from ketoximes. The example cited is representative of the technique. Aldoximes will afford primary nitroalkanes. It should be pointed out that with benzaldehyde oxime, phenylnitromethane is produced, which we classify as an aliphatic nitro compound since the nitro group is on a nonaromatic carbon atom. From this point of view, it follows that the oxidation of any oximes can lead only to aliphatic nitro compounds, if the reaction takes place at all. Since the directions call for the use of 90% hydrogen peroxide, extreme care must be exercised. The material will give severe burns upon coming in contact with the skin. It may initiate fires when in contact with some organic materials, hence spillage must be avoided. It may also decompose violently under uncontrollable conditions on finely divided solids.

2-15. *Preparation of 4-Nitroheptane* [65]

$$CH_3CH_2CH_2-\underset{\underset{\displaystyle CH_2CH_2CH_3}{\|}}{C}{\overset{\displaystyle N-OH}{}} \xrightarrow{CF_3CO_3H} CH_3CH_2CH_2-\underset{\underset{\displaystyle }{|}}{\overset{\displaystyle NO_2}{C}H}-CH_2CH_2CH_3 \quad (25)$$

§ 2. Aliphatic Nitro Compounds

CAUTION: The handling of 90% hydrogen peroxide should be done with great care. Spillage must be avoided and proper shielding and hoods should be used. The handling of peroxytrifluoroacetic acid should also be carried out with great care.

(a) Preparation of peroxytrichloroacetic acid solution. To 50 ml of acetonitrile maintained in a water bath at about room temperature are added 34.0 ml (0.24 mole) of trifluoroacetic anhydride and 5.5 ml (0.2 mole) of 90% hydrogen peroxide. The materials are gently stirred until ready for use.

(b) Oxidation of dipropyl ketone oxime. To a well-stirred suspension of 47 gm (0.55 mole) of sodium bicarbonate in a solution of 2 gm (0.033 mole) of urea, 12.9 gm (0.1 mole) of dipropylketoxime, and 200 ml of acetonitrile, the solution of peroxytrifluoroacetic acid prepared as above is added over an 80-min period. Throughout the addition and for 1 hr after, the solution is heated to a gentle reflux on a water bath. The solution is then cooled to room temperature and poured into 600 ml of cold water. The resulting solution is extracted with four 100-ml portions of methylene chloride. The combined extracts are washed with three 100-ml portions of 10% sodium bicarbonate solution and dried over magnesium sulfate. The solvent is evaporated under reduced pressure and the residue is fractionally distilled through a packed column at reduced pressure to yield 9.3 gm (64%), 4-nitroheptane, bp 58°–60°C (3 mm).

An interesting conversion of oximes to nitro compounds involves reaction of an oxime with N-bromosuccinimide, oxidation of the resultant nitroso compounds, and a reductive dehalogenation according to the scheme of Eqs. (26–28) [66]. While the process is somewhat complex, there may be synthetic situations in which this procedure is particularly useful.

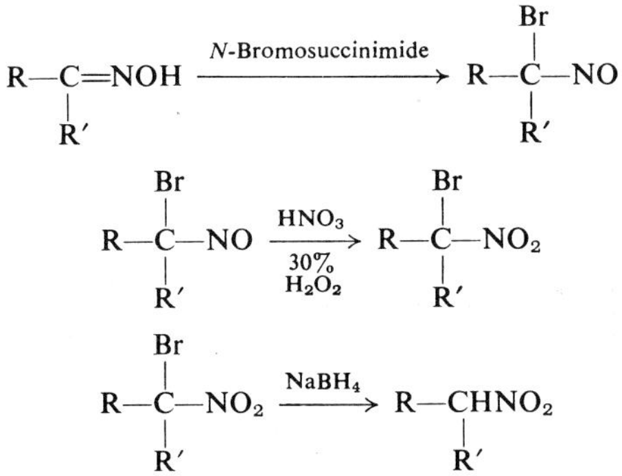

$$\underset{R'}{R-C=NOH} \xrightarrow{N\text{-Bromosuccinimide}} \underset{R'}{\overset{Br}{R-C-NO}} \qquad (26)$$

$$\underset{R'}{\overset{Br}{R-C-NO}} \xrightarrow[30\% \ H_2O_2]{HNO_3} \underset{R'}{\overset{Br}{R-C-NO_2}} \qquad (27)$$

$$\underset{R'}{\overset{Br}{R-C-NO_2}} \xrightarrow{NaBH_4} \underset{R'}{R-CHNO_2} \qquad (28)$$

A variation of this method of synthesizing secondary nitroparaffins involves chlorination of a ketoxime in a step analogous to Eq. (26), oxidation with ozone of the chloronitroso derivative to a chloronitro compound, and dehalogenation by catalytic hydrogenation in the presence of sodium hydroxide. The method is said to be quite general in applicability. Nitroparaffins such as 6-nitroundecane have been made by this method [67].

Anhydrous peracetic acid, at low reaction temperatures, has been used as a fairly satisfactory agent for the oxidation of amines to nitro compounds [68].

In the aliphatic series, yields appear to be best when amines with extensive branching on the number 1 carbon atom are used. Therefore, this method is usable for the preparation of tertiary nitro compounds as well as secondary and cyclic nitro compounds. Primary nitro compounds are not obtained in large yield. It is believed that the reaction proceeds stepwise from an amine through the corresponding hydroxylamine to the nitrosoalkane and finally to the nitroalkane [68].

The preparation of tertiary nitroparaffin by oxidation with potassium permanganate has been found possible. In this connection, it was found important that an appropriate solvent such as 80% acetone–20% water be used since many of the amines in question have limited solubility in an aqueous medium. The pH of the reaction is controlled by the addition of magnesium sulfate (dry powder).

A typical example of this procedure is given in the preparation of 2-nitro-2,4,4-trimethylpentane.

2-16. Preparation of 2-Nitro-2,4,4-trimethylpentane [69]

$$CH_3-\underset{\underset{CH_3}{|}}{\overset{\overset{CH_3}{|}}{C}}-CH_2-\underset{\underset{NH_2}{|}}{\overset{\overset{CH_3}{|}}{C}}-CH_3 \xrightarrow{KMnO_4} CH_3-\underset{\underset{CH_3}{|}}{\overset{\overset{CH_3}{|}}{C}}-CH_2-\underset{\underset{NO_2}{|}}{\overset{\overset{CH_3}{|}}{C}}-CH_3 \quad (29)$$

To a solution of 25.8 gm (0.2 mole) of 2-amino-2,4,4-trimethylpentane ("*tert*-octylamine") in 500 ml of acetone is added 125 ml of water. To the well-stirred solution, 30 gm of magnesium sulfate (purified, dry powder) is added in one portion and then 190 gm (1.2 moles) of potassium permanganate is added in portions over the course of 1 hr. During the addition care must be taken that the permanganate does not cake up on the bottom of the flask. After addition of the oxidizing agent has been completed, stirring is continued at 25°–30°C for 48 hr. The stirred mixture is then subjected to reduced pressure using a water aspirator while maintaining 25°–30°C in the reaction flask by external warming. When stirring becomes difficult, stripping is discontinued. Then about 100–200 ml of water is added and the mixture is steam-distilled. The pale blue distillate is taken up in petroleum ether (bp 35°–37°C) and dried over magnesium sulfate. The organic solution is distilled at reduced pressure through a small glass-helix-packed fractionation column to afford 24.3 gm (77% yield) of the colorless nitro compound, bp 54°C (3 mm).

§ 2. Aliphatic Nitro Compounds

It is interesting to note that the permanganate oxidation of 1,3,5,7-tetraminoadamantane produced 1,3,5,7-tetranitroadamantane in 45% yield. This is believed to be the first case of the oxidation process to produce a compound with more than two nitro groups [70].

D. Condensation of Active Methylene Compounds

a. ALKYLATIONS AND ACYLATIONS

Aliphatic nitro compounds which have a hydrogen on the carbon bearing the nitro function are active methylene compounds. Typical condensations associated with active methylene compounds may be carried out. By use of appropriate aliphatic compounds, a very wide variety of new nitro compounds may be prepared. Since nitro compounds may be considered precursors of amines, these reactions open up a variety of routes to unusual amines. One example of alkylation reactions involves reaction of a variety of nitroalkanes or nitro acetate esters with gramine as a means of preparing derivatives of tryptamine. A typical example of the reaction is given in Eq. (30) [71].

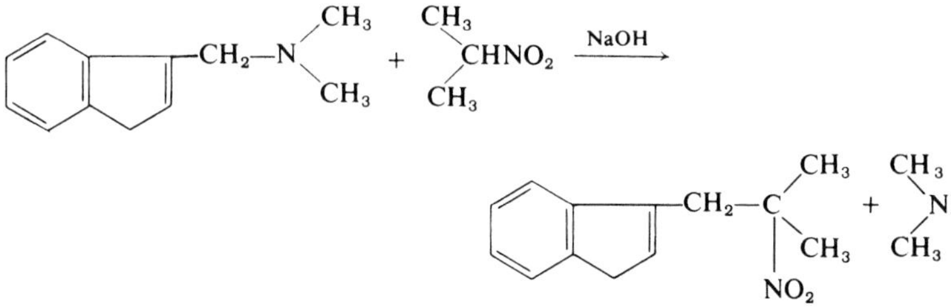

$$\text{(indene)}-CH_2-N(CH_3)_2 + CHNO_2(CH_3)_2 \xrightarrow{NaOH} \text{(indene)}-CH_2-C(CH_3)(NO_2)(CH_3) + HN(CH_3)_2 \quad (30)$$

Particularly active as a reactive methylene compound is diethyl nitromalonate. In this case the reaction site is activated not only by two carboxylate groups but also by a nitro group. A typical reaction is the preparation of ethyl α-nitro-α-carbethoxy-β-(3-indole)propionate [46].

2-17. Preparation of Ethyl α-Nitro-α-carbethoxy-β-(3-indole)propionate [46]

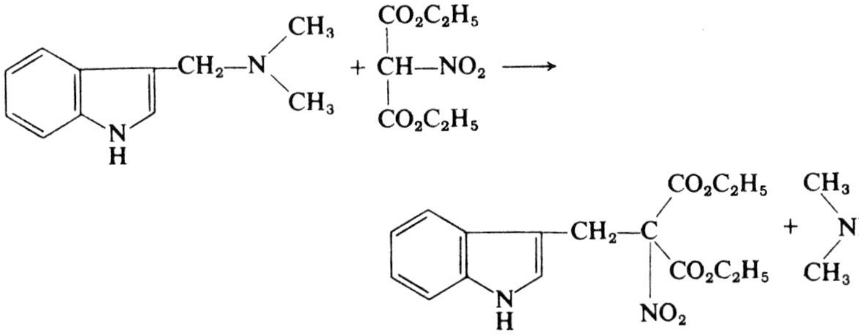

$$\text{(indole)}-CH_2-N(CH_3)_2 + \underset{CO_2C_2H_5}{\overset{CO_2C_2H_5}{CH-NO_2}} \longrightarrow \text{(indole)}-CH_2-\underset{NO_2}{\overset{CO_2C_2H_5}{C(CO_2C_2H_5)}} + HN(CH_3)_2 \quad (31)$$

To a 500-ml three-necked flask fitted with a stirrer, nitrogen inlet, a thermometer reaching into the reaction mixture, and an efficient reflux condenser, are added 43.3 gm (0.25 mole) of freshly distilled ethyl nitromalonate, 250 ml of toluene, and 51.3 gm (0.25 mole) of gramine. With a vigorous stream of nitrogen passing through the well-stirred mixture to ensure rapid and complete removal of the evolving dimethylamine, the mixture is heated rapidly to vigorous reflux. Evolution of dimethylamine begins at about 90°–95°C and is very rapid at the boiling point. Refluxing, the flow of nitrogen, and stirring are continued until the evolution of dimethylamine ceases (usually after about 3 hr). The solution is then cooled, extracted twice with 50-ml portions of 10% hydrochloric acid, washed with 50 ml of water, extracted with two 50-ml portions of 5% sodium hydroxide solution, and washed twice with water. The toluene solution is dried over magnesium sulfate and concentrated at reduced pressure. The last traces of solvent are removed by heating under nitrogen at 80°C (0.5) mm with stirring. The product is a light red color thick syrup and weighs 80.5 gm (96.5%).

(Reprinted from: D. I. Weisblat and D. A. Lyttle, *J. Am. Chem. Soc.* **71**, 3079 (1949). Copyright 1949 by the American Chemical Society. Reprinted by permission of the copyright owner.]

C-acylations of primary nitroparaffins to α-nitroketones have also been accomplished under basic conditions using acyl cyanides as acylating reagents of choice [72]. Suitable solvents for the reaction appear to be *tert*-butyl alcohol, ether, and tetrahydrofuran when lithium nitronates rather than sodium nitronates are being acylated.

Formally, the nitroalkylation of aromatic compounds resembles an arylation of a nitroparaffin. However, the reaction is promoted by unusual oxidizing agents such as manganese(III) acetate [73] or cerium (IV) ions [74]. The reaction is thought to be free radical in nature [Eqs. (32) and (33)].

$$R-CH_2NO_2 + 2M^nX_n \longrightarrow R-\overset{\cdot}{C}HNO_2 + 2M^{n-1}X_{n-1} + HX \quad (32)$$

$$R-\overset{\cdot}{C}HNO_2 + ArH \longrightarrow Ar-\underset{R}{\overset{|}{C}H}-NO_2 + H^+ \quad (33)$$

where M^nX_n may be $Mn(III)OAc_3$, $Ce(NH_4)_2(NO_3)_6$, $Ce(OAc)_4$, etc.

The reaction is evidently limited by the difficulty in preparing Mn(III) salts. Ceric ammonium nitrate, while it promotes the nitromethylation of aromatic hydrocarbons extensively, also promotes the formation of many side products. Cerium(IV) acetate, which does not promote by-product formation, is somewhat difficult to generate and leads to lowered yields [74].

b. Aldol Condensations

Aldehydes and cyclic ketones react under a variety of conditions with nitroalkanes to give typical aldol condensation products. While such catalysts as zinc chloride have been used for the reaction, usually alkaline conditions are the most suitable. Under gentle reaction conditions, the intermediate nitro alcohols may be isolated [75]. Under more vigorous conditions, dehydration of the nitro alcohol takes place to produce olefinic nitro compounds. For example, 5-nitro-4-octanol has been prepared by the process of Eq. (34) [75].

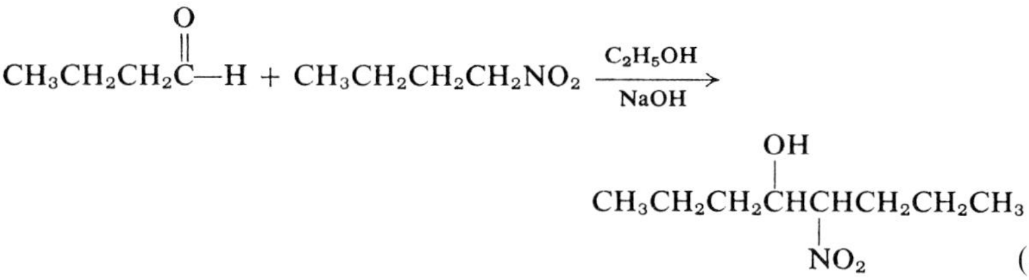

$$CH_3CH_2CH_2\overset{O}{\overset{\|}{C}}-H + CH_3CH_2CH_2CH_2NO_2 \xrightarrow[\text{NaOH}]{C_2H_5OH}$$

$$CH_3CH_2CH_2\underset{\underset{NO_2}{|}}{\overset{\overset{OH}{|}}{C}H}CHCH_2CH_2CH_3 \quad (34)$$

The reaction of benzaldehyde and nitromethane in alkaline medium represents a standard procedure for the preparation of β-nitrostyrene [76].

Aldol condensations involving nitromethane and nitroethane may give rise to explosive intermediates or products. Therefore, such reactions must be carried out with very small quantities only behind adequate shielding and with extreme care.

The preparation of 2,2,2-trinitroethanol is given here for two reasons: (1) it gives a method of forming potassium nitroform from tetranitromethane, and (2) it shows the aldol condensation of potassium nitroform with formalin [77].

2-18. Preparation of 2,2,2-Trinitroethanol [77]

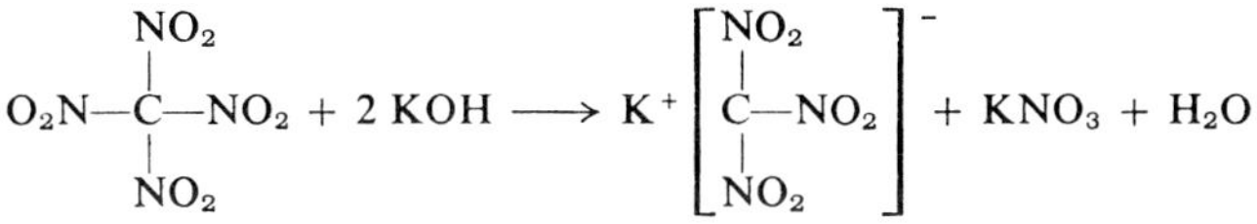

$$O_2N-\underset{\underset{NO_2}{|}}{\overset{\overset{NO_2}{|}}{C}}-NO_2 + 2\,KOH \longrightarrow K^+\left[\underset{\underset{NO_2}{|}}{\overset{\overset{NO_2}{|}}{C}}-NO_2\right]^- + KNO_3 + H_2O \quad (35)$$

$$K^+\left[\underset{\underset{NO_2}{|}}{\overset{\overset{NO_2}{|}}{C}}-NO_2\right]^- + CH_2O \xrightarrow{H_2O} NO_2-\underset{NO_2}{\overset{NO_2}{\diagdown}}C-CH_2OH + KOH \quad (36)$$

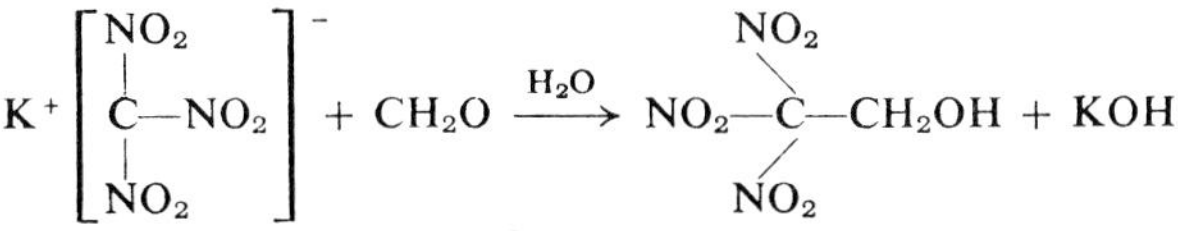

Behind a shield, in a hood, to a solution of 64.4 gm (1.0 mole) of potassium hydroxide in 60 ml of water and 140 ml of ethanol, is added dropwise 98 gm (0.50 mole) of tetranitromethane with stirring, maintaining the temperature between 15° and 20°C. The yellow potassium nitroform is collected, washed with 50 ml of ice water, 50 ml of ethanol, and finally with another 50 ml of ice water. (CAUTION: Potassium nitroform cannot be stored and must be handled with great care since it may be explosive.) The precipitate is immediately added to 150 ml of water and 45 ml of 38% formalin. Then 50 ml of concentrated hydrochloric acid is added in one portion. The mixture is stirred for 3 hr at room temperature and extracted with methylene chloride. The methylene chloride extracts are dried with anhydrous magnesium sulfate and filtered. The filtrate is distilled azeotropically to remove all possible water. Finally concentration is accomplished by evaporation under reduced pressure with a rotary evaporator. Crystallization of the residue from carbon tetrachloride gives 78.8 gm (86%) of 2,2,2-trinitroethanol as colorless needles, mp 70°–71°C.

The aldol condensation involving nitroalkanes is carried out in buffered media such as methylamine hydrochloride–sodium carbonate [78] or glacial acetic acid–ammonium acetate [79].

In the preparation of 2,4-dibenzyloxy-5-methoxy-β-nitrostyrene, ammonium acetate alone has been used [80].

2-19. Preparation of 2,4-Dibenzyloxy-5-methoxy-β-nitrostyrene [80]

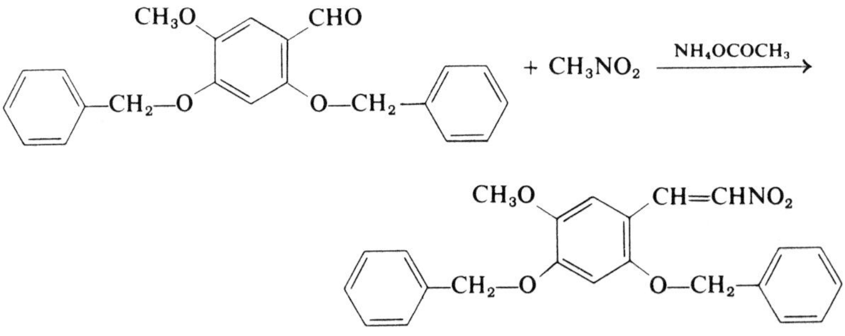

$$\text{(Ar)CH}_2\text{O-Ar-OCH}_2\text{(Ar)} \text{ with CHO and CH}_3\text{O substituents} + CH_3NO_2 \xrightarrow{NH_4OCOCH_3}$$

$$\text{product with CH=CHNO}_2 \tag{37}$$

To a solution of 2 gm (0.0057 mole) of 2,4-dibenzyloxy-5-methoxybenzaldehyde in 20 ml of nitromethane is added 170 mg of ammonium acetate. The stirred mixture is heated for 5 hr on a steam bath. The mixture is then cooled in a Dry Ice–acetone bath until crystallization is complete. The yellow solid is collected and air-dried overnight. Yield 1.9 gm (84.8%), mp 132.5°–134°C. The product may be recrystallized from methylcyclohexane to give 1.7 gm (76%) of 2,4-dibenzyloxy-5-methoxy-β-nitrostyrene as orange feathery plates (mp 133°–135°C).

c. MICHAEL CONDENSATIONS

The addition of active methylene groups of aliphatic nitro compounds to conjugated systems is an example of the Michael condensation.

An example of the reaction is the condensation of nitroalkenes with an α,β-unsaturated aldehyde such as acrolein [81]. Nitroform adds to benzalacetophenone to afford 4,4,4-trinitro-3-phenylbutyrophenone, presumably by a 1,4-addition mechanism [82]. The reaction with nitroform is somewhat complex since some reactions also involve the loss of a nitro-group. Thus, for example, upon reaction with naphthaquinone, 2-dinitromethyl-1,4-naphthaquinone is isolated [82].

The reaction of a nitroparaffin and a more complex unsaturated ketone, 1,5-diphenyl-2,4-pentadiene-1-one, was found to take place in the 1,4-manner rather than the expected 1,6-addition as follows [83]:

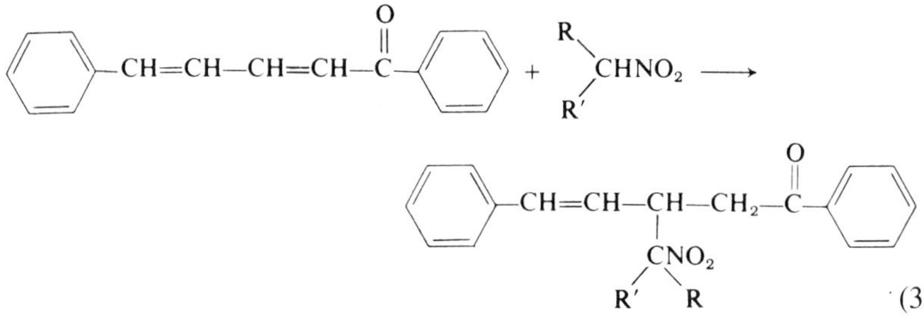

$$\text{Ph–CH=CH–CH=CH–C(=O)–Ph} + \text{R}\backslash\text{R'}/\text{CHNO}_2 \longrightarrow$$

$$\text{Ph–CH=CH–CH(CR R' NO}_2\text{)–CH}_2\text{–C(=O)–Ph} \quad (38)$$

The Michael condensation of ethyl nitroacetate with ethyl acrylate using aqueous benzyltrimethylammonium hydroxide as a catalyst afforded diethyl α-nitroglutarate in 66% yield [84]. Examples involving 1,4-additions to acrylonitrile and acrylamide are also available [85].

Since primary nitroalkanes have two labile hydrogens, the possibility exists that the product derived from a Michael condensation is still capable of reacting with additional quantities of the α,β-unsaturated reagent. This is the rationale in the preparation of 4-nitro-4-methylheptadiamide [85].

2-20. Preparation of 4-Nitro-4-methylheptadiamide [85]

$$\text{CH}_2\text{=CHC(=O)–NH}_2 + \text{CH}_3\text{CH}_2\text{NO}_2 + \text{CH}_2\text{=CHC(=O)–NH}_2 \longrightarrow$$

$$\text{NH}_2\text{–C(=O)–CH}_2\text{CH}_2\text{–C(NO}_2\text{)(CH}_3\text{)–CH}_2\text{CH}_2\text{–C(=O)–NH}_2 \quad (39)$$

In a hood, behind a shield, to a stirred mixture of 80 gm (1.12 moles) of acrylamide, 30 gm (0.40 mole) of nitroethane, and 250 ml of dioxane is added aqueous 40% potassium hydroxide until the pH reaches approximately 8.5. After the initial exothermic reaction has subsided, the reaction mixture is warmed at 50°C until the green color becomes orange (~ 10 hr). The reaction mixture is neutralized with diluted hydrochloric acid and evaporated to dryness. The residue is recrystallized from isopropanol. Yield 33.9 gm (39%), mp 134.5°–135°C.

Michael-type condensations of nitroparaffins with acrylonitrile, acrylic esters, or amides have also been carried out in liquid ammonia [86].

d. Mannich Reactions

As a potential method for the preparation of 2-nitro-1-alkenes, a series of nitroamines were prepared by the Mannich reaction of a secondary amine, formaldehyde, and a nitroalkane. The preparation of the nitroamine appears to be quite general except that, in the case of a reaction involving nitroethane and diethylamine, a violent explosion occurred when the crude product was redistilled under reduced pressure. When, however, piperidine was substituted for diethylamine, the Mannich reaction with nitroethane proceeded without difficulty. In general, it is wise to carry out these reactions as well as the purification of the reaction product with suitable precautions against violent reactions and violent explosions. It was further found that reactions involving nitromethane under a variety of conditions with formaldehyde and piperidine afforded disubstituted products. The thermal decomposition of some of the nitroamine hydrochlorides produced by this method afforded 2-nitro-1-alkenes [87].

2-21. Preparation of N-(2-Nitrobutyl)diethylamine [87]

$$(C_2H_5)_2NH + CH_2O \longrightarrow (C_2H_5)_2NCH_2OH \tag{40}$$

$$(C_2H_5)_2N-CH_2OH + CH_3CH_2CH_2NO_2 \longrightarrow CH_3CH_2CH(NO_2)-CH_2-N(C_2H_5)_2 \tag{41}$$

In a hood, behind an adequate shield, in a 1-liter flask, fitted with stirrer, thermometer, dropping funnel, and condenser, a solution of 73 gm (1 mole) of freshly redistilled diethylamine in an equal volume of water is treated dropwise while rapidly stirring with 84 ml (1.0 mole) of a 36% solution of formalin, over a period of ½ hr. The temperature is maintained at 18°C during the addition. After the addition of formalin has been completed, the reaction mixture is then stirred at room temperature for an additional ½ hr. To this solution of N-hydroxymethyldiethylamine is added all at once, with rapid stirring, 89 gm (1 mole) of 1-nitropropane. This addition is accompanied by an 8°–10° rise in temperature. Rapid stirring of the mixture is continued for 4 hr. The two-layer mixture is then extracted with 100 ml of ether. To the aqueous layer is then added 10 gm of sodium chloride followed by extraction with 20 ml of ether. The ether solutions are combined, dried with anhydrous magnesium sulfate, and distilled under reduced pressure. After removal of the ether, the nitroamine is obtained by distillation under nitrogen to yield 138 gm (79%) of a pale yellow liquid boiling at 79° (2 mm) which was identified as N-(2-nitrobutyl)diethylamine.

A study of the intermediates and of the kinetics of the Mannich reaction of nitroalkanes was carried out by Fernandez [88].

The use of a Mannich reaction, particularly with secondary nitroalkanes, formaldehyde, and p-toluidine has been suggested as a method of preparing solid derivatives of secondary nitroalkanes for identification purposes [89].

e. (4 + 2) Reactions

Examples of Diels–Alder reactions of 1-nitrocyclohexene with linear dienes were cited by Corey and Estreicher [64]. Recently, nitroethylene was also found to undergo (4 + 2) additions with cyclopentadiene and related compounds. The reagent is also capable of undergoing Michael additions and ene reactions [90]. The impetus to this work arose from the realization that standard solutions of nitroethylene in benzene may be preserved without decomposition at approximately 10°C for at least 6 months [90].

The preparation of 2-*endo*-nitronorbornene is an example of a (4 + 2) reaction with nitroethylene.

2-22. Preparation of 2-endo-Nitronorbornene [90]

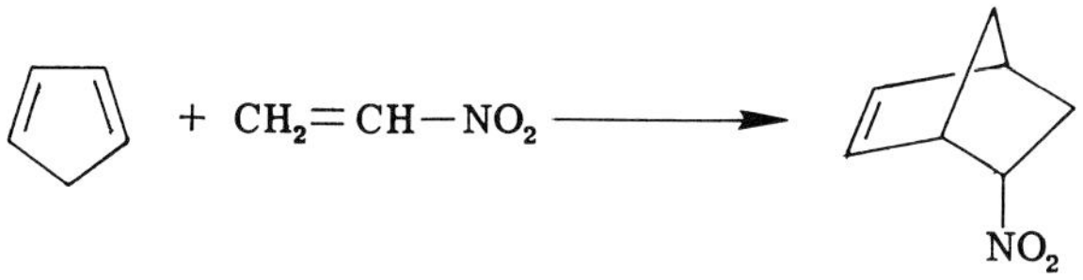

$$\text{cyclopentadiene} + CH_2=CH-NO_2 \longrightarrow \text{2-endo-nitronorbornene} \quad (42)$$

To 60 gm (0.909 mole) of freshly cracked cyclopentadiene is added, with stirring and cooling to −15°C (ice–salt bath), a dry ether solution of 20 gm (0.273 mole) of nitroethylene dropwise. The reaction mixture is stored overnight. The solvents are removed by evaporation under reduced pressure. The residue is distilled under reduced pressure. The product is isolated at 60°–65°C (1.8 mm of mercury). Yield: 39 gm (99.6%).

3. AROMATIC NITRO COMPOUNDS

A. Direct Nitration

a. Nitration with Mixed Acid or Nitric Acid

The nitration of aromatic compounds has been studied extensively for many years [8,10–31,34].

The literature describes a variety of reaction conditions and reagents ranging from the use of dilute nitric acid in the case of highly reactive aromatic compounds such as phenols, through concentrated nitric acids, the fuming nitric acids, nitric acid in a variety of solvents, and for particularly difficult to nitrate materials, potassium nitrate with concentrated sulfuric acid [91]. This latter method evidently has some application in the nitration of benzonitrile and benzoic acid, although the method is quite hazardous. Recently, "super acid" has also been used in nitrations of aromatic compounds.

In view of the hazardous reagents frequently used in these nitration reactions and the fact that many aromatic nitro compounds are explosive in nature, great care must be exercised in handling these reactions. There also was a report that 4-nitrodiphenyl was suspected of being carcinogenic [92]. It is prudent, therefore, to consider all nitro compounds circumspectly.

By far the most generally useful and most commonly used nitrating procedure involves the use of solutions of nitric acid in sulfuric acid, commonly referred to as "mixed acid" [93,94].

Extensive studies of the nitration in mixed acid have indicated that the nitrating agent is the nitronium ion NO_2^+ formed according to Eq. (43) [22].

$$HNO_3 + 2H_2SO_4 \rightleftharpoons NO_2^+ + 2\,HSO_4^- + H_3O^+ \qquad (43)$$

Theoretically and experimentally, the most rapid nitrations take place in systems which are above 90% in sulfuric acid.

Preparative procedures usually use larger concentrations of nitric acid than required by the theoretical requirements in an attempt to compromise the rate of reaction with the isolation of practical quantities of product. The use of sulfuric acid has the following advantages. (1) It will combine with water generated in the reaction and thus prevent the dilution of nitric acid. Since dehydration may also be accomplished with phosphorus pentoxide, the fact that phosphorus pentoxide in the presence of nitric acid does not increase the rate indicates that chemical dehydration alone is not a particularly important factor as far as obtaining the maximum rate of reaction is concerned. Procedures involving the azeotropic removal of water during aromatic nitrations have recently been proposed [95,96]. (2) Many organic compounds are soluble in concentrated sulfuric acid. (3) It also diminishes the oxidizing action of nitric acid. (4) In some cases, it is believed that the intermediate formation of a sulfonic acid takes place which subsequently is readily replaced by a nitro group.

Many examples of nitration of aromatic compounds with mixed acid are found in standard laboratory manuals, e.g., the preparation of nitrobenzene from benzene, p-bromonitrobenzene from bromobenzene, and of m-dinitrobenzene from nitrobenzene [93]. In addition, the preparations of 4-nitrocumene from cumene, 3-nitro-1,2-dimethylbenzene from o-xylene and o- and p-nitroethylbenzene are given by Shirley [91].

3-1. Preparation of p-Nitrobromobenzene [93]

$$\text{PhBr} + HNO_3/H_2SO_4 \longrightarrow \text{p-Br-C}_6H_4\text{-NO}_2 + \text{o-Br-C}_6H_4\text{-NO}_2 \qquad (44)$$

In a 200-ml flask, cooled with an ice–water bath, to 28 gm of concentrated nitric acid is cautiously added 37 gm of concentrated sulfuric acid. While maintaining the flask at room temperature, there is added, in 2–3 ml portions, 16 gm of bromobenzene with vigorous shaking of the flask (rubber gloves, rubber apron, and face shield must be worn throughout the preparation). During the addition, the reaction temperature is maintained between 50° and 60°C by cooling the flask in running water as necessary.

After all the bromobenzene has been added and the temperature no longer tends to rise, the flask is warmed on a steam bath for ½ hr. The reaction mixture is then cooled to room temperature and poured into approximately 100 ml of cold water. The crude product (which contains some ortho isomer) is filtered off, washed with water, and pressed dry.

To purify the product, the crude material is recrystallized from 100–125 ml of hot ethanol. The ortho isomer, being more soluble, remains in solution while the para isomer crystallizes out on cooling. Yield 10–14 gm (50–70%) mp 126°–127°C.

Under more vigorous reaction conditions, dinitration takes place with mixed acids. The semimicro preparation of 2,4-dinitrobromobenzene illustrates this procedure.

3-2. Preparation of 2,4-Dinitrobromobenzene [97]

$$\text{C}_6\text{H}_5\text{Br} + 2\,\text{HNO}_3 \xrightarrow{\text{H}_2\text{SO}_4} \text{C}_6\text{H}_3\text{Br}(\text{NO}_2)_2 + 2\,\text{H}_2\text{O} \quad (45)$$

In a hood, the mixed acid is prepared as follows. In a 50-ml Erlenmyer flask cooled in an ice–water bath, to 5 ml of concentrated nitric acid is added cautiously 15 ml of concentrated sulfuric acid. The flask is then heated to 85°–90°C, and 2.0 ml (3.0 gm) (0.19 mole) of bromobenzene is added in three or four portions during 1 min. The reaction mixture is swirled well after each addition. The temperature rises to 130°–135°C. The reaction is allowed to stand with occasional swirling for 5 min and is then cooled to nearly room temperature. Then the reaction mixture is poured over 100 gm of ice. The resulting mixture is stirred until the product solidifies. The lumps are crushed and the crude product is collected by filtration. The crude product is washed in turn with cold water, sodium bicarbonate solution, and again with cold water. To recrystallize the product, it is dissolved in 50 ml of hot 95% ethanol. The solution is then allowed to cool. Since the product tends to separate as an oil, crystallization is promoted by swirling the reaction mixture vigorously as soon as the oil appears. After crystals have started to form, the solution is cooled in cold water and finally in an ice bath. Crystallization is completed in about 5 min. The product is collected by suction filtration. The final product is then washed with a small quantity of cold ethanol. Yield 3.3 gm (70%), mp 69°–71°C.

In the preparation of poly(nitrostyrene) by nitration of poly(styrene), vigorous reaction conditions are desirable. Since such polymers may subsequently be reduced to poly(aminostyrene) and diazotized for immunochemical applications, the product must not contain sulfonic acid groups. This precludes the use of "mixed acids" for nitration. As an alternative, Kent and Slade [98] heated poly(styrene) in fuming nitric acid until dissolved on a water bath, followed by heating at 100°C on a steam bath until the nitrogen content of the polymer is 9–10%. The product is isolated by precipitation with water.

Nitration of biphenyl-2-carboxylic acid with fuming nitric acid at 0°C afforded 4,4'-dinitrobiphenyl-2-carboxylic acid while the mother liquor contained traces of 2',4-dinitrobiphenyl-2-carboxylic acid. The same starting material, when nitrated with concentrated nitric acid, afforded 4'-nitrobiphenyl-2-carboxylic acid as the major product while the mother liquor contained 2'-nitrobiphenyl-2-carboxylic acid [99].

Nitration of 4-hydroxyquinolines with nitric acid resulted in 4-hydroxy-3-nitroquinolines [100]: Various 2-alkylindoles as well as 2-phenylindoles have also been nitrated with concentrated nitric acid [101,102]. 1-Alkylpyrroles have been nitrated with mixed acids to form polynitro-1-alkylpyrroles.

As is well known, substituents on an aromatic ring have a profound effect both on the rate of formation and on the nature of the isomers formed. The meta-orientating substituents such as the nitro, carboxyl, and sulfonic acid groups deactivate the ring by withdrawing electrons from it by induction and further by reducing the electron density in the ortho and para position by a resonance effect. Para-directing groups such as methyl, ethyl, fluoromethyl, acetamino, amino, methoxy, and phenol activate the ring and increase the electron density in the ortho and para positions. The halogens, on the other hand, tend to deactivate the aromatic ring by withdrawing electrons from the ring as a whole by the inductive effect, but, because of their high electron density, the halogens, in effect, deactivate the ortho and para positions less than the meta positions. Consequently they lead to ortho–para-substituted products although at a sluggish rate. In the case of nitration reactions, additional complications arise from the fact that the normal nitrating reagents are strong acids; consequently amino substituents exist in the reaction medium as the salts. The resulting ammonium ion substituent is, of course, electron-withdrawing and therefore meta-directing. The nitration of amino aromatic and hydroxyl aromatic compounds is further complicated by the fact that both substituents are subject to oxidation.

Table II shows typical isomer distributions upon mononitration of a series of substituted benzenes. It will be noted that frequently all three isomers are formed during nitration. However, the ortho–para-directing substituents yield primarily the ortho and para products, while the primary product of meta-directing substituents is the meta product, with quantities of the other isomers being formed in some cases. It is the authors' opinion that much of the earlier work on isomer distribution (such as Holleman [104] and Lauer [105] should be reevaluated by modern instrumental techniques. Greater attention should be paid to the matter of the rate of formation of various isomers as well as the total yields.

As mentioned above, the direct nitration of aniline is difficult because of interference from the oxidative side reactions. As a matter of fact, the direct

TABLE II

Isomer Distribution Upon Mononitration of Substituted Benzenes[a]

Nitrating system	Initially present substituent	Percentages of isomers isolated			Reference
		Ortho	Meta	Para	
H_2SO_4–HNO_3	F	12.4	—	87.6	b, c
	Cl	30.1	—	69.9	b
		34.6	—	65.4	c
	Br	37.6	—	62.4	b
	I	41.1	—	58.7	b
	CH_3	56.4	4.8	38.4	c, d
	1,2-$(CH_3)_2$		55 of 3-nitro isomer		d
			45 of 4-nitro isomer		
	1,3-$(CH_3)_2$		14 of 2-nitro-isomer		d
			86 of 4-nitro isomer		
	CH_2OCH_3	28.6	18.1	53.3	e
	$C_2H_4OCH_3$	31.6	9.4	59.0	e
	NH_3^+	—[f]	—[f]	38	g
	$N(CH_3)_3^+$	—[f]	88	11	g
Ac_2O–HNO_3 or	F	8.7	—	91.3	b
$AcNO_3$–CH_3NO_2		6.2	—	93.8	c
	Cl	29.6	0.9	69.5	b
		19.5	—	80.5	c
	Br	36.5	1.2	62.4	b
	I	38.3	1.8	59.7	b
		61.4	1.6	37.0	b
	CH_3	56.1	2.5	41.4	c
	CH_3CH_2	45.9	3.3	50.8	b
	OCH_3	70.0	—	30.0	e
	CH_2OCH_3	51.3	6.8	41.9	e
	$C_2H_4OCH_3$	62.3	3.7	34.0	e

[a] Since some of these data were obtained at a variety of reaction conditions and temperatures and some are from competitive reactions, correlations may not always be valid.

[b] Olah et al. [106].

[c] Sparks [107].

[d] Olah et al. [108]; reactions carried out under heterogeneous conditions.

[e] Norman and Radda [109].

[f] Data not reported.

[g] Ridd and Utley [110].

interaction of aniline with concentrated or fuming nitric acid is considered extremely hazardous because of the possibility of ignition of the reaction mixture under certain circumstances. The acylation of aniline is generally preferable if the ortho- and para-nitrated products are desired. It is interesting to note that the tosylate of aniline affords a substantially higher ratio of para to ortho isomers than other mono-acylates [105].

The separation of ortho and para isomers is usually accomplished on the basis of the lower melting point and boiling point, the greater volatility with steam, and the greater solubility of the ortho isomers. While this approach may be perfectly satisfactory in other areas, this technique has led to incorrect identification of isomers and, therefore, it is recommended that more positive methods of proof of structure be considered, particularly in the synthesis of new compounds.

In the naphthalene series, mononitration usually takes place in the number 1 position. Nitronaphthalenes are normally produced by indirect procedures such as the replacement of a sulfonic acid group which is readily introduced into the number 2 position as discussed below. Further substitution of 1-nitronaphthalene leads to products with substitution in either the number 5 or the number 8 position.

In the dinitration of benzophenone with sulfuric acid–nitric acid mixtures, the concentration of sulfur trioxide in the reaction mixture was found to have a profound influence on the distribution of the isomers of the dinitrobenzophenone formed. In preparation 3-3, a procedure is given for the isolation of a product containing 93.5% m,m'-dinitrobenzophenone with only 6.5% m,p'-dinitrobenzophenone [111].

3-3. Preparation of 93.5% m,m'-Dinitrobenzophenone and 6.5% m,p'-Dinitrobenzophenone [111]

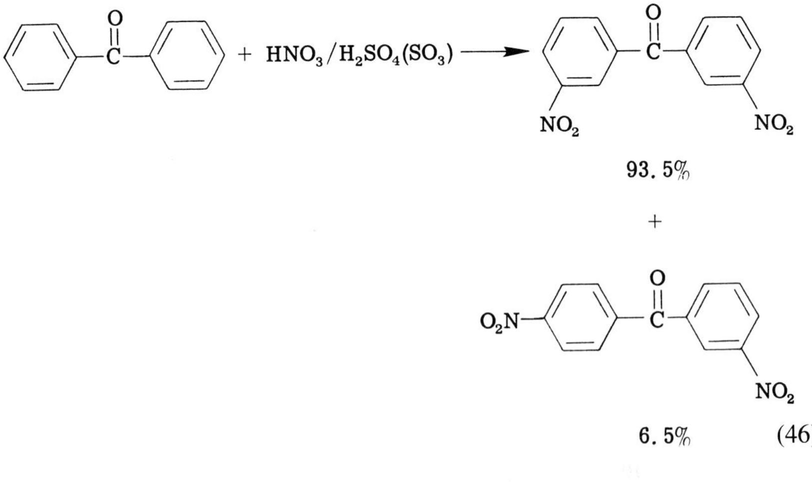

$$\text{C}_6\text{H}_5\text{COC}_6\text{H}_5 + \text{HNO}_3/\text{H}_2\text{SO}_4(\text{SO}_3) \longrightarrow \underset{93.5\%}{m,m'\text{-dinitrobenzophenone}} + \underset{6.5\%}{m,p'\text{-dinitrobenzophenone}} \quad (46)$$

§ 3. Aromatic Nitro Compounds

With suitable safety precautions, 200 gm (1.1 mole) of benzophenone is dissolved at a temperature not exceeding 20°C, with stirring, in 1260 gm of oleum (33% SO_3). Then, while still maintaining a temperature below 20°C, 165 gm of 90% nitric acid is added dropwise. The reaction mixture is then stirred at room temperature for ½ hr, followed by heating with stirring at 80°C for 2 hr.

The reaction mixture is then cooled, poured over 2 kg of a cracked ice–water mixture, and filtered. The solids are washed with 500 gm of water and digested in a Waring Blendor with 1 liter of 10% sodium hydroxide solution. The product is then collected on a filter, washed four times with 500-ml portions of water, and dried under reduced pressure to constant weight (at 100°C for approx. 16 hr). Yield: 232 gm (77%), containing (by HPLC) 93.5% of the m,m'-isomer and 6.5% of the m,p'-isomer. In this preparation, the prolonged heating step was included to degrade unwanted quantities of the o,m'-dinitro-isomer.

Among other mixed acid nitrating procedures are aromatic sulfonic acids–nitric acid systems on porous supports such as kieselguhr [112] and nitrations in the presence of a perfluorinated sulfonic acid resin (Nafion-H, a product of the Du Pont Company) and nitric acid [95]. The activity of the Nafion-H resin may be enhanced by promoting it with a mercury(II) compound like mercuric nitrate [96]. The isomer distribution upon mononitration of representative aromatic compounds is modified when mercury-promoted Nafion-H resin is used (cf. Table III).

b. NITRATION WITH ACYL NITRATES

While use of benzoyl nitrate as a nitrating reagent of aromatic compounds had been suggested some time ago [113] and it has been of considerable value in the elucidation of the mechanism of nitration [114], the use of this particular reagent in laboratory syntheses appears to be secondary. On the other hand, acetyl nitrates, or at least a nitrating composition consisting of nitric acid and acetic acid or nitric acid and acetic anhydride, sometimes in the presence of acetic acid, has found application. In this connection it is important to heed the warning of Powell and Johnson [115] that the addition of concentrated or fuming nitric acid to acetic acid–acetic anhydride solutions must be carried out well below 20°C, cooled by means of an ice–brine bath; otherwise the possibility of explosive reactions exists. [See also page 499 and Ref. 37.]

Nitration of a variety of aromatic compounds has been carried out in acetic anhydride. Thus, for example, mesitylene has been converted to 2-nitro-mesitylene [115].

The acetic anhydride medium for nitration has been used in the reaction of compounds such as anisole and anilides which are believed to be susceptible to oxidation or to hydrolytic attack by aqueous mixed acid [116].

In some cases, a change in the ratio of ortho- to para-substituted products is

TABLE III

Comparison of Isomer Distribution upon Nitration over a Nafion-H Catalyst with and without Mercury Promotion [96]

Substrate		Yield (%)	Isomer distribution		
			2-Nitro	3-Nitro	4-Nitro
Benzene					
(a)[a]		—	—	—	—
(b)[b]		71	—	—	—
Toluene					
(a)		—	56	4	40
(b)		67	33	7	60
Ethylbenzene HNO$_3$/H$_2$SO$_4$		—	44.7	2	53.0
(b)		66	38	5	44
			(and 13% acetophenone)	(and 13% acetophenone)	(and 13% acetophenone)
tert-Butylbenzene	(a)	—	18	11	71
	(b)	72	11	17	72
o-Xylene	(a)	—	—	45	55
	(b)	56	—	33	67
n-Xylene	(a)	—	—	—	—
	(b)	48	11	—	89
Chlorobenzene	(a)	—	38	1	61
	(b)	59	37	2	61
Bromobenzene	(a)	—	45	—	55
	(b)	76	44	—	56
			1-Nitro	2-Nitro	
Naphthalene	(a)	—	98%	2%	
	(b)	77	97%	3%	

[a] (a) Reaction on Nafion-H only.
[b] (b) Reaction on Nafion-H promoted with Hg(NO$_3$)$_2$.

observed when acetic anhydride–HNO$_3$ is used in place of the mixed acids [107,117]. This effect on the ortho- and para-substitution ratio is not universal since it is not observed in the case of toluene [107].

It has been found [117] that a small amount of sulfuric acid in a nitric acid–acetic anhydride composition increases the rate of nitration in the system, as compared to techniques which do not use the mineral acid.

The preparation of mixed chloronitrobenzenes is representative of the procedure.

3-4. Preparation of Mixed Chloronitrobenzenes [117]

$$\text{C}_6\text{H}_5\text{Cl} + \text{HNO}_3 \xrightarrow[\text{H}_2\text{SO}_4]{\text{Ac}_2\text{O}} \text{o-Cl-C}_6\text{H}_4\text{-NO}_2 + \text{p-Cl-C}_6\text{H}_4\text{-NO}_2 \qquad (47)$$

NOTE: See pages 499 and pages 503–507 on the hazards of acetyl nitrate.

In a hood, behind a shield, to 22.5 ml of reagent grade acetic anhydride cooled in an ice bath is slowly added 2.5 ml of reagent grade fuming nitric acid containing approximately 90% HNO_3. To this nitrating mixture is then added 1 drop (0.05 ml) of 96% sulfuric acid as a catalyst. To this nitrating solution are added dropwise with rapid stirring 2.25 gm (0.02 mole) of chlorobenzene, while maintaining the reaction temperature at 0°C. The reaction mixture is allowed to stir at 0°C for 20 min and is then added to a solution, maintained at 0°C, of 20 gm of sodium hydroxide in 175 ml of water. After standing at 0°C for at least 30 min to allow complete decomposition of the acetic anhydride, the crystalline product is filtered on a sintered glass funnel and washed thoroughly with ice water until free of alkali. It is then dried for several days and vacuum-desiccated over Drierite. By infrared analysis it was determined that the reaction product is approximately 90% p-chloronitrobenzene. Yield 2.8–3.0 gm (88–95%), mp 83°C.

Work involving the nitration of chlorobenzene with acetyl nitrate under only slightly different reaction conditions indicated that the product had the composition of 19.4% ortho, 0.9% meta, and 80.0% para isomers [118]. The differences in the isomer composition has been accounted for by analysis of the caustic wash solution which has a greater solvent action on the ortho isomer than on the para isomer [107].

Nitrations have also been carried out in glacial acetic acid without the use of acetic anhydride or catalytic sulfuric acid [119]. The preparation of 4-nitro-[2,2]-paracyclophane has been carried out by the nitration in glacial acetic acid [120].

c. Nitration with Nitronium Tetrafluoroborate and Related Systems

The nitrations of aromatic compounds in the presence of Friedel–Crafts-type reagents such as boron trifluoride, aluminum chloride, sulfuric acid, phosphoric acid, hydrogen fluoride, titanium tetrachloride, ferric chloride, and zirconium chloride have been studied extensively. Major contributions to this field and extensive reviews have been prepared by Olah and co-workers [27, 34, 106, 108, 118, 121–123].

The reaction using nitronium tetrafluoroborate as nitrating agent is generally quite rapid. Even with substituted aromatic compounds whose substituents deactivate the ring, high yields of the mononitro product are obtained. The reactions may be carried out with an excess of the aromatic reagent, thus reducing dinitration and permitting nitration of compounds which, under normal strongly acidic nitration conditions, undergo either hydrolysis or oxidation.

The preparation of 2-methyl-5-nitrobenzonitrile is an example of the technique using a starting material, *o*-toluonitrile, which would be subject to hydrolysis in ordinary nitration systems.

3-5. Preparation of 2-Methyl-5-nitrobenzonitrile [118]

$$\text{o-CH}_3\text{C}_6\text{H}_4\text{CN} + \text{NO}_2^+\text{BF}_4^- \longrightarrow \text{2-CH}_3\text{-5-NO}_2\text{-C}_6\text{H}_3\text{CN} \qquad (48)$$

Under anhydrous conditions, to a solution of 58.5 gm (0.5 mole) of *o*-toluonitrile in 100 ml of tetramethylene sulfone is added a solution of 20 gm (0.15 mole) of nitronium tetrafluoroborate in 300 ml of tetramethylene sulfone with stirring and external cooling at such a rate that the temperature is maintained between 15° and 25°C. After the addition has been completed, stirring is continued between 20° and 35°C for 30 min. The reaction mixture is then cautiously poured into 500 ml of cold water. The organic layer is separated, washed twice with water, and dried over calcium chloride. The excess aromatic substrate is distilled off under reduced pressure. Yield: 22 gm (90% based on NO_2BF_4 used).

The nitration with nitronium salts leads to significant differences in the nature and selectivity of the nitration of benzene and toluene as compared to the nitration of nitrobenzenes and nitrotoluenes. These differences have been attributed to differences in the nature of the transition states [124].

Benzene, alkylbenzenes, halobenzenes, and anisole have been nitrated with silver nitrate–boron trifluoride systems in acetonitrile [125].

N-nitropyridium salts [126] and *N*-nitropyrazole [127] have also been developed as nitrating agents in which the incipient nitronium ion is transferred from the nitrogen to which it is bound to the aromatic substrate.

d. Nitration with Oxides of Nitrogen

The various oxides of nitrogen have from time to time been suggested as nitrating agents for aromatic hydrocarbons under a variety of conditions.

This method may have some merit, particularly in the nitration of phenolic compounds. It must be pointed out, however, that the possibility exists that

§ 3. Aromatic Nitro Compounds

nitroso compounds will be formed as intermediates when oxides of nitrogen are used as "nitrating" agents. If nitroso compounds are generated, a further oxidation step is required to generate nitro compounds.

In one publication the use of a buffer solution in the nitration of phenolic compounds with sulfuric acid and sodium nitrite has been suggested [128]. An older publication mentions the generation of "nitrous fumes" in an aqueous layer in such a manner as to pass into a supernatent organic solution of a phenolic starting material. The preparation of 4,6-dinitroguaiacol in 55% yield is an example of the use of this method [129].

B. Indirect Nitration

a. DISPLACEMENT OF SULFONIC ACID GROUPS

Many aromatic compounds, particularly polynuclear hydroxyl compounds such as naphthols, are quite easily sulfonated. The sulfonic acids produced by such procedures readily react with nitric acid with the replacement of the sulfonic acid group by a nitro group.

The preparation of 2,4-dinitro-1-naphthol, taken from the famous Martius Yellow experiment, illustrates the simplicity of the procedure [130].

3-6. Preparation of 2,4-Dinitro-1-naphthol [130]

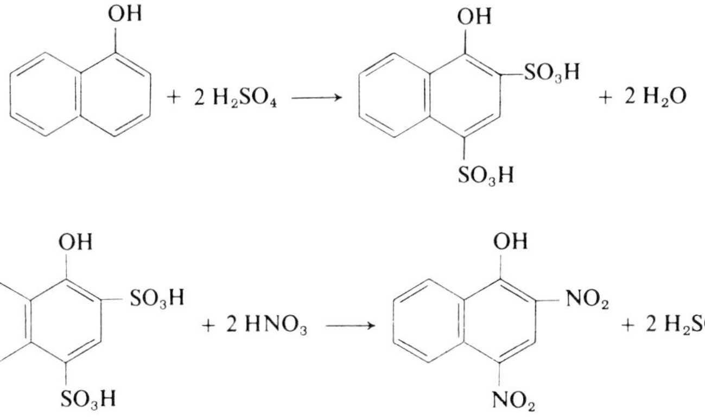

$$\text{naphthol-OH} + 2\,H_2SO_4 \longrightarrow \text{naphthol-OH-(SO_3H)_2} + 2\,H_2O \quad (49)$$

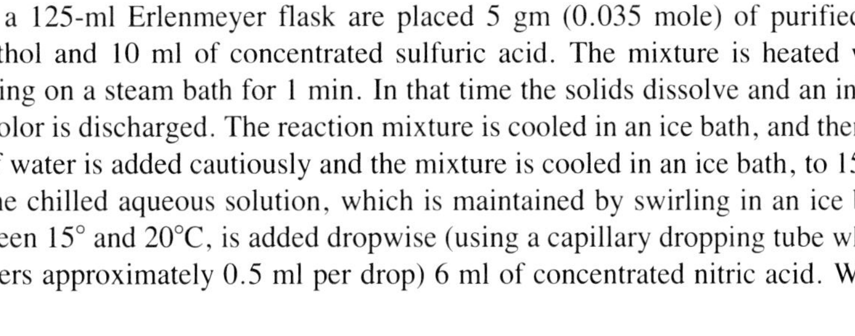

$$\text{naphthol-OH-(SO_3H)_2} + 2\,HNO_3 \longrightarrow \text{naphthol-OH-(NO_2)_2} + 2\,H_2SO_4 \quad (50)$$

In a 125-ml Erlenmeyer flask are placed 5 gm (0.035 mole) of purified α-naphthol and 10 ml of concentrated sulfuric acid. The mixture is heated with swirling on a steam bath for 1 min. In that time the solids dissolve and an initial red color is discharged. The reaction mixture is cooled in an ice bath, and then 25 ml of water is added cautiously and the mixture is cooled in an ice bath, to 15°C. To the chilled aqueous solution, which is maintained by swirling in an ice bath between 15° and 20°C, is added dropwise (using a capillary dropping tube which delivers approximately 0.5 ml per drop) 6 ml of concentrated nitric acid. When the addition is complete and the exothermic reaction has subsided (1–2 min), the reaction mixture is gently warmed to 50°C for 1 min. The nitration product separates as a stiff yellow paste. The full heat of the steam bath is applied for 1 min and the flask is filled with water. The paste is then stirred rapidly and the solid product is collected by filtration on a Büchner funnel. The solid is washed well with water. Contact of the skin with this yellow crude product and its orange ammonium salt, prepared below, must be avoided.

The purification of the product is carried out by conversion to the ammonium salt, followed by the generation of free dinitrophenol. This is carried out as follows.

The crude dinitro compound is washed into a 600-ml beaker with 100 ml of water. Then 150 ml of hot water and 5 ml of concentrated ammonia solution are added. The mixture is heated to the boiling point with stirring to dissolve the solid and then the hot solution is filtered with suction. Then 10 gm of ammonium chloride is added to the hot filtrate. On cooling, the ammonium salt, Martius Yellow, is precipitated out. The orange salt, after washing with a 2% solution of ammonium chloride in water, weighs 7.7 gm (88.5%).

To generate the purified 2,4-dinitro-1-naphthol, the moist salt may be used. It is dissolved in hot water. The solution is acidified with hydrochloric acid. The precipitated free 2,4-dinitro-1-naphthol is separated by filtration. The product may be crystallized after a charcoal treatment in methanol or ethanol. It forms yellow needles, mp 138°C.

b. Replacement of Diazonium Groups by Nitro Groups (Sandmeyer Reaction)

A number of nitro compounds have been prepared by diazotizing the corresponding amine and subjecting the diazonium salt to the Sandmeyer reaction in the presence of sodium nitrite and cupro-cupri sulfite. The method has limited application. It may be of special value in the preparation of specific position isomers, if starting materials can be prepared unequivocally.

The preparation of 2,5-dinitrotoluene from the corresponding diamine is an example of the preparation [94].

c. Nitration with Tetranitromethane

The nitration of phenols with tetranitromethane in pyridine is one process of nitration of aromatic compounds in a basic medium [131]. It is, therefore, of particular interest in the nitration of extremely sensitive materials such as those derived from biochemical sources.

Evidently the reaction is highyl specific, requiring careful control of the pH of the reaction. Its interest arises from the fact that it permits the nitration of tyrosine in the presence of other amino acids found in proteins. So far, cysteinyl residues are the only other groups found to react with tetranitromethane, probably because of oxidation by the reagent.

§ 3. Aromatic Nitro Compounds

The reaction is believed to be quantitative, although products are normally not isolated in the nitration of tyrosine, the interest being primarily in determining the absorption maximum of the nitrated derivative at 428 mμ. A product of the reaction is believed to be 3-nitrotyrosine, formed by the introduction of the nitro group ortho to the phenolic hydroxyl group [132].

C. Oxidation Methods

a. OXIDATION OF AROMATIC AMINES

The oxidation of aromatic amines to nitro compounds by way of nitroso intermediate with Caro's acids has been carried out successfully many times. Two related modern methods are believed to be simpler both from the standpoint of availability of reagents and the reaction procedures. The first oxidation involves the use of hydrogen peroxide in glacial acetic acid. The preparation of 3,5-dichloro-4-nitrobenzonitrile is a representative example of the procedure [133].

3-7. Preparation of 3,5-Dichloro-4-nitrobenzonitrile [133]

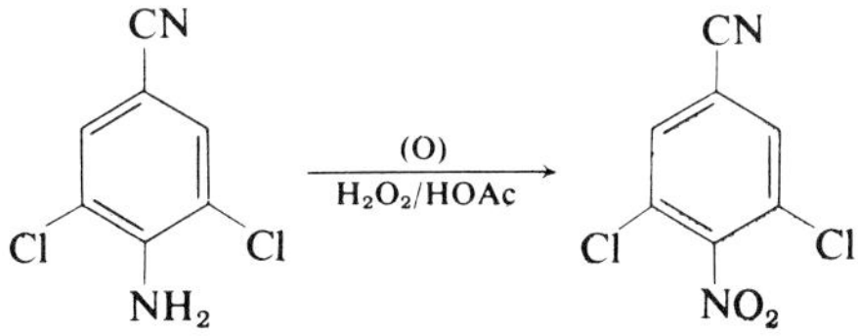

$$\text{3,5-dichloro-4-aminobenzonitrile} \xrightarrow[\text{H}_2\text{O}_2/\text{HOAc}]{(\text{O})} \text{3,5-dichloro-4-nitrobenzonitrile} \qquad (51)$$

On a steam bath, a mixture of 4.7 gm (0.025 mole) of 3,5-dichloro-4-aminobenzonitrile, 100 ml of glacial acetic acid, and 30 ml of 30% aqueous hydrogen peroxide, and 2 ml of concentrated sulfuric acid is heated. The reaction mixture is maintained at this temperature for 8½ hr. After the first ½ hr, the solution turns green and a precipitate of a white crystalline material appears. This, in all probability, is the dimer of the nitroso compound. After 3¼ hr, 15 ml more of the 30% hydrogen peroxide and 50 ml of glacial acetic acid are added, whereupon essentially all of the solid dissolves and the solution becomes light yellow. After the full 8½ hr, the reaction mixture is cooled and 200 ml of water is added. The solid which separates is filtered and recrystallized from 100 ml of hot methanol to yield 4.5 gm (83%) 3,5-dichloro-4-nitrobenzonitrile, mp 166°–167°C. The material may also be sublimed under reduced pressure for analytical samples.

The second more modern oxidation is based on the fact that, just as in the case of aliphatic amines [65], aromatic amines may be oxidized with peroxytrifluoroacetic acid. An example of this is the conversion of pentafluoroaniline to pentafluoronitrobenzene [134].

Recently polynitroarylamines have been converted to polynitroaromatic compounds by oxidation of the amine group with peroxydisulfuric acid in sulfuric acid [135].

b. OXIDATION OF NITROSO COMPOUNDS

As indicated above, the oxidation of amines evidently proceeds by way of the intermediate formation of nitroso compounds. The nitroso compounds themselves may, of course, be prepared by methods designed specifically for their preparation and these in turn may be oxidized to the corresponding nitro compounds. Thus, for example, 5-nitrosopyrimidines have been oxidized with aqueous 30% hydrogen peroxide in trifluoroacetic acid solution to the corresponding 5-nitropyramidines [136]. 4-Nitro-2,5-xylenol is prepared by the nitric acid oxidation of 4-nitroso-2,5-xylenol. In this particular oxidation, dilute nitric acids may be used. The 4-nitroso-2,5-xylenol is prepared by the reaction of 2,5-xylenol and acetic acid with sulfuric acid and sodium nitrite [137].

4. MISCELLANEOUS METHODS

(1) Reaction of nitrosyl chloride with unsaturated compounds [138].

$$CH_2=C(CH_3)-C(=O)-OCH_3 + NOCl \longrightarrow Cl-CH_2-C(Cl)(CH_3)-C(=O)-OCH_3 \quad (61\%)$$

$$\text{and} \quad O_2N-CH_2-C(Cl)(CH_3)-C(=O)-OCH_3 \quad (25\%) \tag{52}$$

(2) Displacement of an allylic bromine by a nitro group [139].

$$CH_3-CH(Br)-CH=CHCH_3 + N_2O_4 \xrightarrow{-20°C} CH_3-CH(NO_2)-CH=CH-CH_3 \quad (38.3\%) \tag{53}$$

(3) Reaction of sodium salts of secondary nitroalkanes with perchloryl fluoride [140].

$$[R_2C=NO_2^-]Na^+ + ClO_3F \longrightarrow R_2CFNO_2 + NaClO_3 + R_2C=O \text{ and}$$
$$R_2C(NO_2)-C(NO_2)R_2 \tag{54}$$

(4) Preparation of methyl and ethyl nitroacetate by treatment of dipotassium salt of nitroacetic acid in an alcoholic solution with sulfuric acid at Dry Ice temperatures [141].

§ 4. Miscellaneous Methods

$$2\ CH_3NO_2 + 2\ KOH \longrightarrow KO_2N{=}CHCO_2K + NH_3 + 2\ H_2O$$

$$KO_2N{=}CHCO_2K + CH_3OH \xrightarrow[-50°\text{ to }-60°\text{ C}]{2H_2SO_4} NO_2{-}CH_2CO_2CH_3 + 2\ KHSO_4 + H_2O \quad (55)$$

(5) Preparation of *gem*-dinitroalkanes by electrolytic procedures [142].

$$C_2H_5NO_2 \xrightarrow[\substack{\text{Electrolysis}\\ \text{at pH 11.5}\\ 60°\ C}]{NaNO_2} CH_3CH(NO_2)_2 \text{ and recovered nitroethane} \quad (56)$$

(6) Preparation of α-fluoronitroalkanes by reaction of olefins with nitric acid and anhydrous hydrogen fluoride at low temperatures [143].

(7) Cyanoethylation of nitroparaffins in the presence of a strongly basic ion-exchange resin [144].

$$3\ CH_2{=}CHCN + CH_3NO_2 \longrightarrow NC{-}CH_2CH_2{-}\underset{\underset{NO_2}{|}}{\overset{\overset{CH_2CH_2CN}{|}}{C}}{-}CH_2CH_2CN \quad (57)$$

(8) Nitroethylation of labile hydrogen compounds [145].

$$Ar{-}NH_2 + CH_2{=}CHNO_2 \longrightarrow ArNHCH_2CH_2NO_2 \quad (58)$$

(9) Nitroethylation of enamines [146].

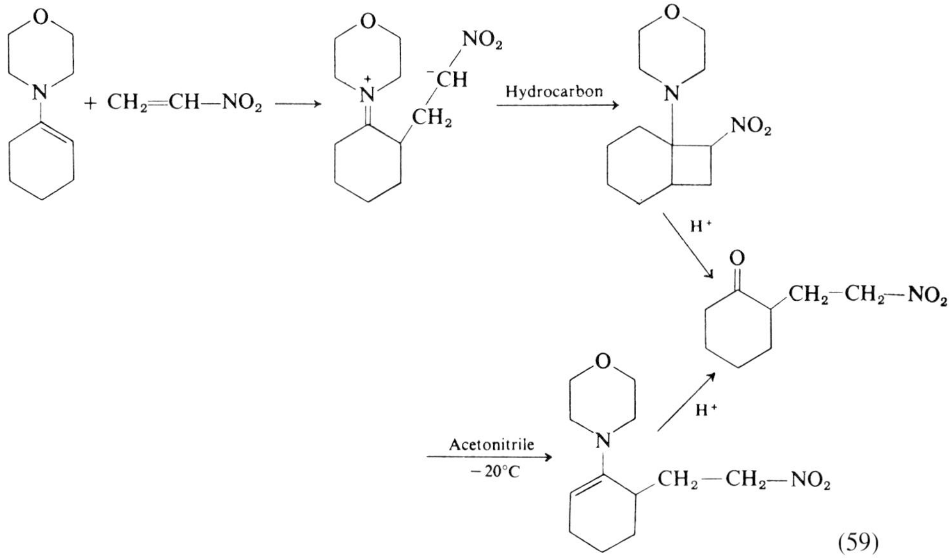

(10) Diels–Alder reaction with nitroolefins [147].

$$\text{(diene)} + \underset{\text{CH}_3\quad\text{NO}_2}{\overset{R\quad H}{\underset{\|}{\overset{|}{C}}\text{=}\overset{|}{C}}} \longrightarrow \underset{\text{NO}_2}{\overset{R}{\text{H}_3\text{C}}} \qquad R = H \text{ or } CH_3 \tag{60}$$

(See also Section 2,D,e).

(11) Nitration of aromatic hydrocarbons with nitric acid and with nitrates in the presence of anhydrous sulfur trioxide [148].

(12) Nitration of aromatic hydrocarbons with nitric acid and with nitrates in the presence of liquid sulfur dioxide [149].

(13) Nitration of 2,7-dimethylnaphthalene with dilute nitric acid [150].

(14) Nitration of aromatic hydrocarbons with mixed acid in tetramethylene sulfone [151].

(15) Nitration of toluene with alkyl nitrates and polyphosphoric acid [152].

(16) Preparation of β-*p*-nitrophenyl-γ-butyrolactone and γ-*p*-nitrophenyl-γ-butyrolactone from the corresponding phenyl-γ-butyrolactones with acetyl nitrate [153].

(17) Intramolecular cyclization of an α,β-unsaturated nitro compound [154].

(18) Synthesis of tetranitroethylene and Diels–Alder reactions thereof [155].

(19) Room temperature nitration of aromatic compounds with ammonium or metal nitrates in trifluoroacetic anhydride [156].

(20) Addition of Grignard reagents to nitroarenes as a means of alkylating aromatic compounds [157].

$$\text{Ar-NO}_2 + 2\text{RMgX} \longrightarrow \left[\text{intermediate} \right] \xrightarrow[\text{OH}^-]{\text{(O)} \atop \text{KMnO}_4} \text{Ar(R)-NO}_2 \tag{61}$$

References

1. H. B. Hass and E. F. Riley, *Chem. Rev.* **32**, 373 (1943).
2. J. Giber and C. Szantay, *Magy. Kem. Lapja* **11**, 147 (1956).
3. W. L. Reed, *Chem. Rev.* **64**, 19 (1964).
4. N. Kornblum, *Org. React.* **12**, 101 (1962).
5. V. V. Perekalin, "Unsaturated Nitro Compounds" (translated from the Russian). Israel Program for Scientific Translation, Jerusalem, 1963.
6. G. Schmidt, *J. Chem. Educ.* **27**, 557 (1950).
7. T. Slebodzinski, D. Bimer, and D. Salbut, *Przem. Chem.* **41**, No. 1, 18 (1962).
8. C. Matasa, "Nitro Compounds," Proc. Int. Symp., Warsaw, 1963, p. 95; J. H. Ridd, *ibid*, p. 43.
9. T. Kosuge, H. Ito, and E. Ota, *Shizuoka Yakka Daigaku Kaigaku 10-Shunen Kinen Rombunshu* p. 44 (1963).

§ References

10. W. R. Tomlinson, Jr., *Ind. Eng. Chem.* **49**, 1534 (1957); **50**, 1380 (1958); **51**, 1123 (1959); **52**, 545 (1960); **53**, 401 (1961).
11. J. Boileau, *Mem. poudres* **43**, 15 (1961).
12. A. I. Titov, *Usp. Khim.* **27**, 845 (1958); *Wiad. Chem.* **15**, 742 (1961); *Sb. Statei, Nauchn.-Issled. Inst. Org. Poluprod. Krasitelei* **2**, 46 (1961); *Zh. Obshch. Khim.* **16**, 1896 (1946); **18**, 733 (1948); **19**, 267, 517 (1949).
13. C. K. Ingold, *Proc. Chem. Soc.*, p. 480 (1948).
14. E. D. Hughes, C. K. Ingold, and co-workers, *J. Chem. Soc.* pp. 2400–2684 (1950).
15. F. H. Westheimer and M. S. Kharash, *J. Am. Chem. Soc.* **68**, 1871 (1946).
16. R. J. Gillespie and D. J. Millen, *Q. Rev. Chem. Soc.* **2**, 277 (1948).
17. D. V. Nightingale, *Chem. Rev.* **40**, 117 (1947).
18. R. C. Miller, D. S. Noyce, and T. Vermeulen, *Chem. Eng. News* **56**, 43 (1964).
19. C. Hanson, J. G. Marsland, and G. Wilson, *Chem. Ind. (London)* p. 675 (1966).
20. B. E. Berkman, "Industrial Synthesis of Aromatic Nitro Compounds and Amines." Khimiya, Moscow, 1964.
21. R. Nakajima, *Kagaku (Kyoto)* **18**, 821 (1963).
22. C. K. Ingold, "Structure and Mechanism in Organic Chemistry," 2nd ed. Cornell Univ. Press, Ithaca, New York, 1969.
23. J. Kovár and J. Novák, "Preparative Reactions in Organic Chemistry, III. Nitration, Nitrosation, and Sulfonation." Natl. Cesk. Akad. Ved., Prague, 1956.
24. A. U. Topchiev, "The Nitration of Hydrocarbons and Other Organic Compounds," 2nd ed. Akad. Nauk S.S.S.R., Moscow, 1956 (translation by C. Mathews, Pergamon, Oxford, 1959).
25. P. B. D. de la Mare and J. H. Ridd, "Aromatic Substitution—Nitration and Halogenation." Academic Press, New York, 1959.
26. G. A. Olah and S. J. Kuhn, "Friedel-Crafts and Related Reactions," Vol. 3, pp. 1393ff. Wiley (Interscience), New York, 1964.
27. T. Urbanski, "Nitro Compounds." Pergamon, Oxford, 1964.
28. J. H. Ridd, *Acc. Chem. Res.* **4**, 248 (1971).
29. H. Feuer, "The Chemistry of the Nitro and Nitroso Groups," Parts 1 and 2. Wiley (Interscience), New York, 1969, 1970.
30. H. G. Hoggett, R. B. Moodie, J. R. Penton, and K. Schofield, "Nitration and Aromatic Reactivity." Cambridge Univ. Press, London and New York, 1971.
31. J. Kochany, *Wiad. Chem.* **32**, 723 (1978).
32. D. Seebach, E. W. Colvin, F. Lehr, and T. Weller, *Chimia* **33**, 1 (1979).
33. G. A. Olah, *Aldrichimica Acta* **12**(3), 43 (1979).
34. R. H. Fisher and H. M. Weitz, *Synthesis* p. 261 (1980).
35. H. G. Adolph, R. E. Oesterling, and M. E. Sitzman, *J. Org. Chem.* **33**, 4296 (1968).
36. F. Cooke, *Chem. Eng. News* **59**, 3 (August 24, 1981), and Editor's note.
37. C. D. Bedford, *Chem. Eng. News* **58**, 33 (Sept. 1, 1980); E. E. Gilbert, *ibid.* p. 5 (Oct. 6, 1980); L. Bretherick, *ibid.* **59**, 59 (April 6, 1981).
38.a Anonymous, *Chem. Eng. News* **58**, 23 (Oct. 6, 1980).
38.b J. H. Bonfield, U.S. Patent 3,133,124 (1964).
38.c G. Gut, H. Reich, and A. Guyer, *Helv. Chim. Acta* **46**, 2526 (1963).
39. N. Levy and C. W. Scaife, *J. Chem. Soc.* pp. 1093, 1100 (1946); N. Levy, C. W. Scaife, and A. E. Wildersmith, *ibid.* p. 1096; p. 52 (1948); H. Baldock, N. Levy, and C. W. Scaife, *ibid.* P. 2627 (1949).
40. W. K. Seifert, *J. Org. Chem.* **28**, 125 (1963).
41. W. M. Cummings, Ger. Offen. 2,157,648 (1972); *Chem. Abstr.* **78**, 842848 (1973).
42. C. A. Burkhard and J. F. Brown, Jr., *J. Org. Chem.* **29**, 2235 (1964).
43. S. Yamada, H. Sasaki, and T. Tanaka, Japan Kokai 73/86,847 (1973); *Chem. Abstr.* **80**, p70440 (1974).

44. F. G. Bordwell and E. W. Garbisch, Jr., *J. Am. Chem. Soc.* **82**, 3588 (1960).
45. M. J. Haire and G. A. Boswell, Jr., *J. Org. Chem.* **42**, 4251 (1977).
46. D. I. Weisblat and D. A. Lyttle, *J. Am. Chem. Soc.* **71**, 3079 (1949).
47. F. D. Chattaway, *J. Chem. Soc., Trans.* **97**, Part II, 2099 (1910).
48. P. Liang, *Org. Synth. Collect. Vol.* **3**, 803 (1955).
49. W. Steinkopf and U. Kuhnel, *Ber. Dtsch. Chem. Ges. B* **751**, 1323 (1942).
50. G. B. Bachman and T. Hokama, *J. Am. Chem. Soc.* **81**, 4884 (1959); *J. Org. Chem.* **25**, 178(1960).
51. H. Shechter, F. Conrad, A. L. Daulton, and R. B. Kaplan, *J. Am. Chem. Soc.* **74**, 3052 (1952).
52. W. Wislicenus and A. Endress, *Ber. Dtsch. Chem. Ges.* **35**, 1755 (1902); **41**, 3334 (1908).
53. H. Feuer and P. M. Pivawer, *J. Org. Chem.* **31**, 3152 (1966).
54. H. Feuer, A. M. Hall, S. Golden, and R. L. Reitz, *J. Org. Chem.* **33**, 3622 (1968).
55. H. Özbal and W. W. Zajac, Jr., *J. Org. Chem.* **46**, 3082 (1981).
56. F. E. Elfehail and W. W. Zajac, Jr., *J. Org. Chem* **46**, 5151 (1981).
57. N. Kornblum, B. Taub, and H. E. Ungnade, *J. Am. Chem. Soc.* **76**, 3209 (1954).
58. N. Kornblum, H. O. Larson, R. K. Blackwood, D. D. Mooberry, E. P. Oliveto, and G. E. Graham, *J. Am. Chem. Soc.* **78**, 1497 (1956).
59. N. Kornblum and J. W. Powers, *J. Org. Chem.* **22**, 455 (1957).
60. N. Kornblum, R. K. Blackwood, and J. W. Powers, *J. Am. Chem. Soc.* **79**, 2507 (1957).
61. F. T. Williams, Jr., P. W. K. Flanagan, W. J. Taylor, and H. Shechter, *J. Org. Chem.* **30**, 2674 (1965).
62. T. Kametani, M. Tsubuki, and H. Nemoto, *J. Org. Chem.* **45**, 4391 (1980).
63. V. Auger, *Bull. Soc. Chim.* **23** (iii), 333 (1900); F. C. Whitmore and M. G. Whitmore, *Org. Synth. Collect. Vol.* **1**, 401 (1941).
64. E. J. Corey and H. Estreicher, *J. Am. Chem. Soc.* **100**, 6294 (1978).
65. W. D. Emmons and A. S. Pagano, *J. Am. Chem. Soc.* **77**, 4557 (1955).
66. D. C. Iffland, G. X. Criner, M. Koral, F. J. Lotspeich, Z. B. Papanastassiou, and S. M. White, Jr., *J. Am. Chem. Soc.* **75**, 4044 (1953); D. C. Iffland and G. X. Criner, *ibid.* p. 4047; D. C. Iffland and T.-F. Yen, *ibid.* **76**, 4083 (1954).
67. M. W. Barnes and J. M. Patterson, *J. Org. Chem.* **41**, 733 (1976).
68. W. D. Emmons, *J. Am. Chem. Soc.* **79**, 5528 (1957).
69. N. Kornblum, R. J. Clutter, and W. J. Jones, *J. Am. Chem. Soc.* **78**, 4003 (1956).
70. G. P. Sollott and E. E. Gilbert, *J. Org. Chem.* **45**, 5405 (1980).
71. H. R. Snyder and L. Katz, *J. Am. Chem. Soc.* **69**, 3140 (1947).
72. G. B. Bachman and T. Hokams, *J. Am. Chem. Soc.* **81**, 4882 (1959).
73. M. E. Kurz, P. Ngoviwatchai, and T. Tantrarant, *J. Org. Chem.* **46**, 4668 (1981).
74. M. E. Kurz and P. Ngoviwatchai, *J. Org. Chem.* **46**, 4672 (1981).
75. B. M. Vanderbilt and H. B. Hass, *Ind. Eng. Chem.* **32**, 34 (1940).
76. D. E. Worrall, *Org. Synth. Collect. Vol.* **1**, 413 (1948).
77. F. G. Borgardt, A. K. Seeler, and P. Noble, Jr., *J. Org. Chem.* **31**, 2806 (1966).
78. F. A. Ramirez and A. Burger, *J. Am. Chem. Soc.* **72**, 2781 (1950).
79. A. T. Shulgin, *J. Med. Chem.* **9**, 445 (1966).
80. J. W. Daly, J. Benigni, R. Minnis, Y. Kanaoka, and B. Witkop, *Biochemistry* **4**, 2513 (1965).
81. M. Bourillot and G. Descotes, *C. R. Hebd. Seances Acad. Sci.* **260**, 3107 (1965).
82. J. Strumza and S. Altschuler, *Isr. J. Chem.* **1**, 106 (1963).
83. J. Strumza and S. Altschuler, *Isr. J. Chem.* **1**, 25 (1963).
84. N. Kornblum and J. H. Eicher, *J. Am. Chem. Soc.* **78**, 1494 (1956).
85. J. M. Patterson, M. W. Barnes, and R. I. Johnson, *J. Org. Chem.* **31**, 3103 (1966).
86. S. Wakamatsu and K. Shimo, *J. Org. Chem.* **27**, 1609 (1962).

§ References

87. A. T. Blomquist and T. H. Shelley, Jr., *J. Am. Chem. Soc.* **70,** 147 (1948).
88. J. E. Fernandez, *Tetrahedron Lett.* p. 2889 (1964).
89. M. B. Winstead, R. G. Strachan, and H. W. Heine, *J. Org. Chem.* **26,** 4116 (1961).
90. D. Ranganathan, C. B. Rao, S. Ranganathan, A. K. Mehrotra, and R. Iyengar, *J. Org. Chem.* **45,** 1185 (1980).
91. D. A. Shirley, "Preparation of Organic Intermediates," p. 215. Wiley, New York, 1951; K. A. Kobe and H. Levin, *Ind. Eng. Chem.* **42,** 352 (1950).
92. Anonymous, *Chem. Eng. News* **45,** 41 (1967).
93. R. Adams and J. R. Johnson, "Laboratory Experiments in Organic Chemistry," 4th ed., pp. 299, 301, 303. Macmillan, New York, 1949.
94. H. H. Hodgson and F. Heyworth, *J. Chem. Soc.* p. 1624 (1949).
95. G. A. Olah, J. Kaspi, and J. Bukala, *J. Org. Chem.* **42,** 4187 (1977).
96. G. A. Olah, V. V. Krishnamarthy, and S. C. Narang, *J. Org. Chem.* **47,** 596 (1982).
97. A. Ault, *J. Chem. Educ.* **42,** 267 (1965).
98. L. H. Kent and J. H. R. Slade, *Biochem. J.* **77,** 12 (1960).
99. D. H. Hey, J. A. Leonard, and C. W. Rees, *J. Chem. Soc.* p. 4579 (1962).
100. A. R. Surrey and R. A. Cutler, *J. Am. Chem. Soc.* **73,** 2413 (1951).
101. W. E. Noland, L. R. Smith, and K. R. Rush, *J. Org. Chem.* **30,** 3457 (1965).
102. W. E. Noland, K. R. Rush, and L. R. Smith, *J. Org. Chem.* **31,** 65, 70 (1966).
103. G. Doddi, P. Mencarelli, A. Razzini, and F. Stegel, *J. Org. Chem.* **44,** 2321 (1179).
104. A. F. Holleman, *Chem. Rev.* **1,** 187 (1925).
105. K. Lauer, *J. Prakt. Chem.* [2] *137,* 175 (1933).
106. G. A. Olah, S. J. Kuhn, and S. H. Flood, *J. Am. Chem. Soc.* **83,** 4581 (1961).
107. A. K. Sparks, *J. Org. Chem.* **31,** 2299 (1966).
108. G. A. Olah, S. J. Kuhn, S. H. Flood, and J. C. Evans, *J. Am. Chem. Soc.* **84,** 3687 (1962).
109. R. O. C. Norman and G. K. Radda, *J. Chem. Soc.* p. 3030 (1961).
110. J. H. Ridd and J. H. P. Utley, *Proc. Chem. Soc., London* p. 24 (1964).
111. A. Onopchenko, E. T. Sabourin, and C. M. Selwitz, *J. Org. Chem.* **46,** 5014 (1981).
112. O. Manabe, T. Kameo, T. Ikeda, and T. Yamamoto, Japan Kokai 73/86,832 (1973); *Chem. Abstr.* **80,** P59662 (1974), O. Manabe, T. Kameo, T. Yamamoto, T. Ikeda, and S. Nishimura, Japan Kokai 74/18,833 (1974); *Chem. Abstr.* **80,** P133000 (1974).
113. F. E. Francis, *J. Chem. Soc.* **89,** 1 (1906).
114. V. Gold, E. D. Hughes, and C. K. Ingold, *J. Chem. Soc.* p. 2467 (1950).
115. G. Powell and F. R. Johnson, *Org. Synth. Collect. Vol.* **2,** 449 (1943); W. Kuhn, *ibid.* p. 447. 447.
116. M. A. Paul, *J. Am. Chem. Soc.* **80,** 5329 (1958).
117. M. A. Paul, *J. Am. Chem. Soc.* **80,** 5332 (1958).
118. S. J. Kuhn and G. A. Olah, *J. Am. Chem. Soc.* **83,** 4564 (1961).
119. D. Shapiro, *J. Am. Chem. Soc.* **72,** 2786 (1950).
120. B. E. Norcross, D. Becker, R. I. Cukier, and R. M. Schultz, *J. Org. Chem.* **32,** 320 (1967).
121. G. A. Olah, S. Kuhn, and A. Mlinko, *J. Chem. Soc.* p. 4257 (1956); G. A. Olah and S. J. Kuhn, U.S. Patent 3,162,675 (1964); G. A. Olah, J. A. Olah, and N. A. Overchuck, *J. Org. Chem.* **30,** 3373 (1965).
122. G. A. Olah, S. J. Kuhn, S. H. Flood, and J. C. Evans, *J. Am. Chem. Soc.* **84,** 3687 (1962).
123. D. Cook, S. J. Kuhn, and G. A. Olah, *J. Chem. Phys.* **33,** 1669 (1960).
124. G. A. Olah and H. C. Lin, *J. Org. Chem.* **46,** 549 (1981).
125. G. A. Olah, A. P. Fung, S. C. Narang, and J. A. Olah, *J. Org. Chem.* **46,** 3533 (1981).
126. C. A. Cupas and R. C. Pearson, *J. Am. Chem. Soc.* **90,** 4742 (1968).
127. G. A. Olah, S. C. Narang, and A. P. Fung, *J. Org. Chem.* **46,** 2706 (1981).
128. J. Gasparic, *Collect. Czech. Chem. Commun.* **29,** 6,1374 (1964).

129. W. Baker and R. Robinson, *J. Chem. Soc.* p. 152 (1929).
130. L. F. Fieser, "Experiments in Organic Chemistry," 3rd ed., p. 235. Heath, Boston, Massachusetts, 1955.
131. E. Schmidt and H. Fischer, *Chem. Ber.* **53,** 1529, 1537 (1920).
132. J. F. Riordan, M. Sokolovsky, and B. L. Vallee, *J. Am. Chem. Soc.* **88,** 4104 (1966).
133. R. R. Holmes and R. P. Bayer, *J. Am. Chem. Soc.* **82,** 3454 (1960).
134. G. M. Brooke, J. Burdon, and J. C. Tatlow, *J. Chem. Soc.* p. 802 (1961).
135. A. T. Nielsen, R. L. Atkins, W. P. Norris, C. L. Coon, and M. E. Sitzmann, *J. Org. Chem.* **45,** 2341 (1980).
136. E. C. Taylor and A. McKillop, *J. Org. Chem.* **30,** 3153 (1965).
137. G. R. Allen, Jr., J. P. Poletto, and M. J. Weiss, *J. Org. Chem.* **30,** 2897 (1965).
138. K. A. Oglobolin and V. P. Semenov, *Zh. Org. Khim.* **1,** 27 (1965).
139. Yu. V. Baskov and V. V. Perekalin, *Zh. Org. Khim.* **1,** 236 (1965).
140. H. Shechter and E. B. Roberson, Jr., *J. Org. Chem.* **25,** 175 (1960).
141. H. Feuer, H. H. Hass, and K. S. Warren, *J. Am. Chem. Soc.* **71,** 3078 (1949).
142. C. M. Wright, U.S. Patent 3,130,136 (1964).
143. I. L. Khunyants, L. S. German, and I. N. Rozhkov, *Izv. Akad. Nauk SSSR, Ser. Khim.* p. 1946 (1963).
144. N. I. Isagulyants and Z. Poredda, *Zh. Prikl. Khim. (Leningrad)* **37,** 1093 (1964).
145. H. Hopf and M. Capault, *Helv. Chim. Acta* **43,** 1898 (1960).
146. M. E. Kuehne and L. Foley, *J. Org. Chem.* **30,** 4280 (1965).
147. W. E. Noland and R. E. Bambury, *J. Am. Chem. Soc.* **77,** 6386 (1955).
148. W. Alama and K. Okon, *Biul. Wojskowej Akad. Tech.* **13** (6-142), 57 (1964).
149. W. Alama and K. Okon, *Biul. Wojskowej Akad. Tech.* **13** (6), 65 (1964).
150. J. B. Ashton, U.S. Patent 3,145,235 (1964).
151. A. K. Sparks, U.S. Patent 3,140,319 (1964).
152. S. M. Tsang, U.S. Patent 3,126,417 (1964).
153. S. J. Cristol and S. A. A. Osman, *J. Org. Chem.* **31,** 1654 (1966).
154. A. Berndt, *Angew. Chem., Int. Ed. Engl.* **7,** 627 (1968).
155. T. S. Griffin and K. Baum, *J. Org. Chem.* **45,** 2880 (1980).
156. J. V. Crivello, *J. Org. Chem.* **46,** 3056 (1981).
157. G. Bartoli, M. Bosco, and G. Baccolini, *J. Org. Chem.* **45,** 522 (1980).

CHAPTER 17 / NITRILES (CYANIDES)

1. Introduction	550
A. Hazards and Safe Handling Practices	551
2. Elimination Reactions	551
A. Dehydration of Amides	551
2-1. Preparation of cis-2,3-Dichloroacrylonitrile	552
2-2. Preparation of Benzonitrile	553
B. Conversion of Carbonyl Derivatives to Nitriles	553
a. Dehydration of Oximes	553
2-3. Preparation of 4-Methoxybenzonitrile	554
2-4. Preparation of Heptanonitrile	554
b. Reaction with O,N-Bis(trifluoroacetyl)hydroxylamine	555
2-5. Preparation of Heptanonitrile	555
c. Decomposition of N,N,N-Trimethylhydrazonium Salts	555
2-6. Preparation of Benzonitrile	556
d. Preparations from Other Carbonyl Derivatives	557
e. The Schmidt Reaction	557
2-7. Preparation of Bicyclo [3.3.1]non-6-ene-3-carbonitrile	558
f. A Novel Synthesis of Aromatic Nitriles from Aldehydes	559
2-8. Preparation of p-N,N-Dimethylaminobenzonitrile	559
3. Condensation Reactions	560
A. The Use of Cyanide Salts	560
a. Aliphatic Nitriles	560
3-1. Preparation of Pentanonitrile (Valeronitrile)	560
3-2. Preparation of 2-Methyloctanonitrile	562
3-3. Preparation of 1-Cyanooctane	563
3-4. Preparation of 1-Cyanododecane	564
b. Aromatic Nitriles	565
3-5. Preparation of 1,2-Dicyanoacenaphthylene	565
B. The Sandmeyer Reaction	566
3-6. Preparation of 1-Naphthonitrile	566
C. Nitriles via Cyanohydrin Formation	567
3-7. Preparation of 2-Chloroacetaldehyde Cyanohydrin and 2-Chloroacrylonitrile	568
D. Condensation of Existing Nitriles	570
a. Aldol Condensations	570
3-8. Preparation of 1-(4-Nitrophenyl)cinnamonitrile	570
E. Alkylation of Acetonitrile Derivatives	571
3-9. Preparation of Isopropyl-1-(2-morpholinoethyl)-1-phenylacetonitrile	571
F. Cyanoethylation	573
3-10. Preparation of 1,1,1-Tris(2-cyanoethyl)acetone	574
3-11. Preparation of 3-Trichlorosilylpropionitrile	575
4. Oxidation Reactions	575

549

4-1. Preparation of 4-Nitrobenzonitrile	575
5. Reduction Reactions	576
6. Miscellaneous Reactions	577
References	580

1. INTRODUCTION

Useful reviews of nitrile chemistry are references 1–3.

The best known reactions for the preparation of nitriles are the dehydration of amides and oximes, the condensation of halides with cyanides, and the Sandmeyer reaction. Other methods of preparation involve the addition of hydrogen cyanide to double bonds, oxidation and reduction reactions, and nitrile interchange reactions. A variety of condensation reactions such as the aldol reaction and alkylation of active methylene compounds are also used for the synthesis of nitriles. In these preparations the carbon skeleton associated with an existing nitrile-containing structure is changed in a controlled manner.

By the dehydration of amides, primary, secondary, and tertiary nitriles can be obtained in good yields.

The method is applicable to aliphatic and aromatic amides. For example, isobutyronitrile is formed in 86% yield by heating isobutyramide and phosphorus pentoxide in the absence of a solvent at 100°–200°C and distilling the product as it is formed. Table I illustrates a few of the products prepared by the most common methods.

The most common example of the addition of hydrogen cyanide to a double bond is the formation of cyanohydrins from aldehydes and ketones with this reagent. The addition of hydrogen cyanide to olefins and acetylenes is important industrially but requires special equipment and precautions.

TABLE I

REPRESENTATIVE CYANIDES PREPARED FROM THEIR CORRESPONDING HALIDE, AMIDE, OR OXIME

Compound	Yield (%)	Method	Reference
n-Butyl cyanide	93	NaCN–DMSO	4
tert-Butyl cyanide	80	Amide	5
tert-Butyl cyanide	73	Amide	6
n-Hexyl cyanide	72	NaCN–H_2O	7
Cyclohexyl cyanide	93	Amide	8
Benzyl cyanide	90	NaCN–H_2O	9
Benzyl cyanide	87	Amide	10
4-Cyanobiphenyl	50	Sandmeyer	11
o-Chlorophenylacetonitrile	64	Oxime	12

A. Hazards and Safe Handling Practices

It is of prime importance to recognize and become familiar with the extreme toxicity of cyanide salts and HCN gas.

The alkali cyanides are readily soluble in water and sparingly soluble in the lower alcohols. These solutions are extremely toxic if taken internally and can be absorbed through the skin. The alkali cyanides are alkaline and rapidly liberate HCN on contact with acids or slowly with moist air. Hydrogen cyanide is a volatile liquid, bp 25°–26°C, f.p. −13°C, flammable and explosive with air at 5.6°C at 40% concentration. It decomposes with great violence when exposed to bases and therefore is sold with an acid-stabilizer. When cyanides are being used, each person should work in a well-ventilated hood and wear rubber gloves, an apron, and a gas mask with an HCN canister good for HCN up to a 2% concentration. Antidotes such as sodium thiosulfate and amyl nitrite must be used promptly, but they can be administered only by a qualified physician. One should never work alone and the other party should also be familiar with the necessary precautions. Signs should be posted to alert people entering the area.

It should also be understood that the organic cyanides and nitriles are toxic and are easily absorbed through the skin.

In our own experience while working with sodium cyanide, the question has arisen as to whether exposure to a fatal dose has taken place. If it has, immediate and drastic action by a physician is mandatory; if it has not, administration of antidotes for prophylactic purposes is hazardous, since some of them may be quite toxic.

The safe disposal of excess cyanides and of waste by-products, as well as the proper clean-up of all equipment, is also a problem. Thorough soaking with potassium permanganate solutions is believed to be of some value in the decontamination of inorganic cyanides. Never dispose of cyanides by pouring down the drain. Pour solutions into special waste containers set aside for work with cyanides.

From the legal standpoint, in many, if not all, states in the United States, a Laboratory may *not* provide any first aid equipment or antidotes which must be administered orally, by injection, or by inhalation, unless this be done by a physician. Self-administration is done at the risk of the individual, administration of antidotes by a lab partner or other nonmedical individual might subject that person to the most serious legal difficulties under both civil and criminal law.

2. ELIMINATION REACTIONS

A. Dehydration of Amides

The dehydration of amides is probably the best known of the elimination reactions for the preparation of nitriles. One method, the dehydration with phosphorus pentoxide, has been used for many years although it is undoubtedly one of the most messy synthetic procedures available. The procedure normally involves a dry distillation of a mixture of a primary amide and phosphorus pentoxide. Under the reaction conditions, this composition usually melts, chars, and, because a gaseous product is formed, quickly foams up to fill the available space in the cooler parts of the reactor. The use of large-size equipment and oversize delivery tubes to a condenser and the heating of the reaction mixture from the top surface with a Bunsen burner often helps to overcome some of the difficulties. The reaction is quite general in applicability and excellent yields have been reported [cf. Refs. 5,6,8,10].

The preparation of cis-2,3-dichloroacrylonitrile is an example of this procedure.

2-1. Preparation of cis-2,3-Dichloroacrylonitrile [13]

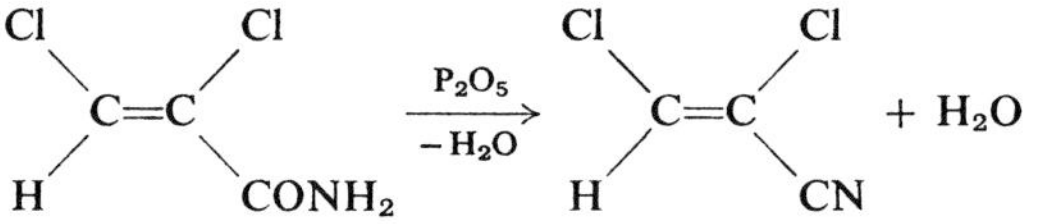

$$\underset{H}{\overset{Cl}{\diagdown}}C=C\underset{CONH_2}{\overset{Cl}{\diagup}} \xrightarrow[-H_2O]{P_2O_5} \underset{H}{\overset{Cl}{\diagdown}}C=C\underset{CN}{\overset{Cl}{\diagup}} + H_2O \qquad (1)$$

In a vacuum distillation apparatus leading to several cold traps, an intimate mixture of 8.0 gm (0.05 mole) of cis-2,3-dichloroacrylamide and 8.2 gm (0.058 mole) of phosphorus pentoxide is cautiously heated under a pressure of 225 mm of mercury at 200°C. Five and one-half grams of product is collected in the various receivers to give a 79% yield, bp 95°C (225 mm). The purity of the product was shown to be 99% by vapor phase chromatography.

A variety of other dehydrating agents has been suggested. Among these are p-toluenesulfonyl chloride as the dehydrating agent of a pyridine solution of an amide [14], a double salt of sodium chloride and aluminum chloride [15], thionyl chloride in a dimethylformamide (DMF) solution at a carefully controlled reaction temperature [16], [17], phosphorus oxychloride in DMF [18], polyphosphoric acid esters [19], and triphenylphosphine in carbon tetrachloride [20].

Aromatic amides, including N-substituted amides have been converted to nitriles by the use of a triphenylphosphine–rhodium chloride complex as indicated in Eq. 2 [21].

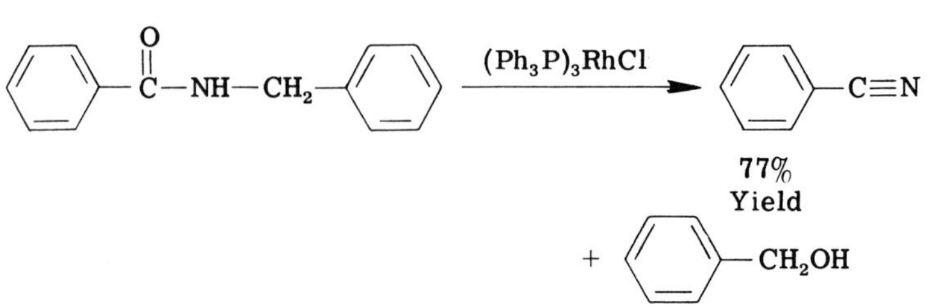

$$Ph-\overset{O}{\overset{\|}{C}}-NH-CH_2-Ph \xrightarrow{(Ph_3P)_3RhCl} Ph-C\equiv N$$
$$77\% \text{ Yield}$$
$$+ Ph-CH_2OH \qquad (2)$$

§ 2. Elimination Reactions

An intriguing new procedure for the dehydration of benzamide to benzonitrile in methylene dichloride solution was reported by Relles and Schluenz [22]. In this procedure, trisubstituted phosphine dichlorides supported on polystyrene acts as a dehydrating agent to afford a better than 78% yield. Unfortunately, we must warn against the use of the procedure of Relles and Schluenz. The preparation of their polymeric support involves the chloromethylation of cross-linked polystyrene with "chloromethyl methyl ether." Since the publication of this paper, chloromethyl methyl ether (monochloromethyl ether) has been implicated as the cause of extremely severe cases of cancer. Thus, the procedure of Ref. 22 is dangerous. On the other hand, monomeric chloromethylstyrene and poly(chloromethylstyrene) are available from commercial sources. Use of either of these materials under appropriate conditions should permit the development of a modification of the Relles and Schluenz procedure which is reasonably safe.

Aliphatic and aromatic amides, N-methyl aromatic amides, and benzaldoxime have been dehydrated with silazanes and related compounds [23]. Procedure 2-2 outlines the procedure which we believe to have wide applications.

2-2. Preparation of Benzonitrile [23]

$$3 \, \text{C}_6\text{H}_5-\overset{\overset{\text{O}}{\|}}{\text{C}}-\text{NH}_2 + \left(\begin{array}{c} \text{CH}_3 \\ | \\ \text{HNSi} \\ | \\ \text{CH}_3 \end{array} \right)_3 \longrightarrow 3 \, \text{C}_6\text{H}_5-\text{C}\equiv\text{N}$$

$$+ \, 3 \, \text{NH}_3 + \left(\begin{array}{c} \text{CH}_3 \\ | \\ \text{OSi} \\ | \\ \text{CH}_3 \end{array} \right)_x \quad (3)$$

In a reflux apparatus topped with a drying tube, equipped with a flask thermometer, and a stirrer, a mixture of 12.1 gm (0.1 mole) of benzamide and 7.8 gm (0.102 equivalents) of hexamethylcyclotrisilazane is heated for 20 hr at 220°C. When the evolution of ammonia has become negligible, the reaction mixture is distilled. The product is isolated at 70°C (approx. 10 mm). Yield: 9.6 gm (91%). By gas–liquid chromatography analysis this product was essentially pure.

B. Conversion of Carbonyl Derivatives to Nitriles

a. Dehydration of Oximes

The dehydration of oximes by such reagents as phosphorus pentoxide, thionyl chloride, acetic anhydride, acyl chlorides, and phosphorus pentachloride is well known [24]. In effect, this dehydration procedure permits the conversion of aldehydes to nitriles with the same number of carbon atoms. A modification applicable only to the aromatic series makes use of boiling acetic acid as a dehydrating agent [25]. With other dehydrating agents, aliphatic aldehydes also may be converted to nitriles. The preparation of 4-methoxybenzonitrile is representative of the new procedure.

2-3. Preparation of 4-Methoxybenzonitrile [25]

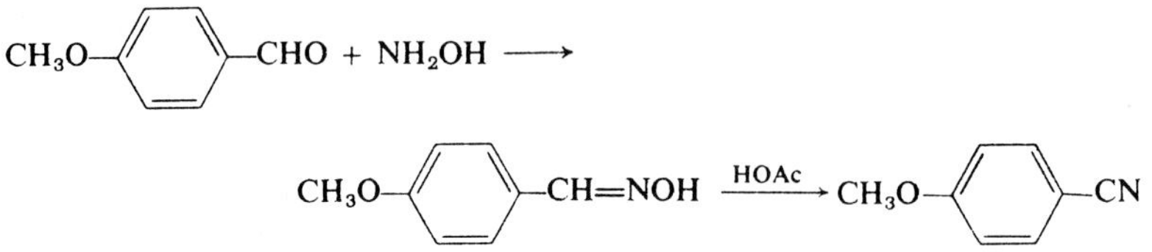

$$CH_3O\text{-}C_6H_4\text{-}CHO + NH_2OH \longrightarrow$$

$$CH_3O\text{-}C_6H_4\text{-}CH\text{=}NOH \xrightarrow{HOAc} CH_3O\text{-}C_6H_4\text{-}CN \quad (4)$$

To a solution of 4.3 gm (1.3 mole) of freshly fused sodium acetate and 20 ml of glacial acetic acid is added 3.1 gm (1.2 mole) of hydroxylamine hydrochloride and 5 gm (0.05 mole) of anisaldehyde. The reaction mixture is then heated for 16 hr under reflux. After cooling, the sodium chloride produced is filtered off and the acetic acid is removed by distillation. The residue is treated with water and extracted repeatedly with ether. The ether extracts are shaken several times with sodium carbonate solution and dried. Evaporation of the ether solution affords 3.3 gm (67%) of 4-methoxybenzonitrile (after recrystallization from petroleum ether), mp 59°–60°C.

Oximes have also been dehydrated with sodium in methanol and benzene [26] and with a chlorothiocarbonate,

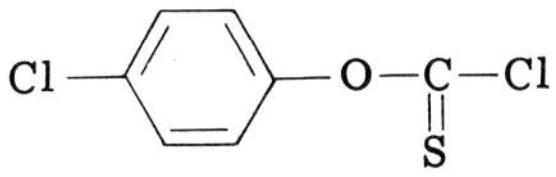

$$Cl\text{-}C_6H_4\text{-}O\text{-}\underset{\underset{S}{\|}}{C}\text{-}Cl$$

in pyridine [27].

Oximes may also be converted readily to nitriles by an acid-catalyzed reaction with ortho esters [28].

2-4. Preparation of Heptanonitrile [28]

$$CH_2CH_2CH_2CH_2CH_2CH_2CH\text{=}NOH + HC(OC_2H_5)_3$$

$$\downarrow H^+$$

$$CH_2CH_2CH_2CH_2CH_2CH_2C\text{≡}N + HCO_2C_2H_5 + 2\ C_2H_5OH \quad (5)$$

In a 100-ml flask attached to a short distillation column are placed 15.0 ml (0.1 mole) of n-heptaldehyde oxime and 20.0 ml (0.12 mole) of triethyl orthoformate containing one drop of methanesulfonic acid. The mixture is heated gently to distil out ethyl formate and ethanol as they form. After the coproducts have been removed, the residue is distilled under reduced pressure at 70°–72°C (10 mm). Yield: 95%.

b. REACTION WITH O,N-BIS(TRIFLUOROACETYL)HYDROXYLAMINE

A method of converting aldehydes to nitriles under relatively mild conditions has been reported to involve the reaction of an aldehyde with O,N-bis(trifluoroacetyl)hydroxylamine to yield nitriles in the presence of pyridine. This procedure appears to be applicable both to aliphatic and to aromatic nitriles and is illustrated here with the preparation of heptanonitrile [29]. It should be noted that this reaction is only of specialized interest since the reagent is not readily available. Further work will have to be done to establish its general nature.

2-5. *Preparation of Heptanonitrile* [29]

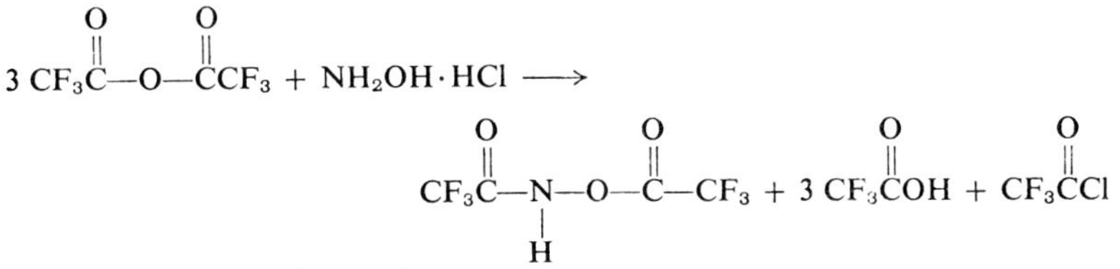

$$3\ CF_3\overset{O}{\overset{\|}{C}}-O-\overset{O}{\overset{\|}{C}}CF_3 + NH_2OH \cdot HCl \longrightarrow$$

$$CF_3\overset{O}{\overset{\|}{C}}-\underset{H}{N}-O-\overset{O}{\overset{\|}{C}}-CF_3 + 3\ CF_3\overset{O}{\overset{\|}{C}}OH + CF_3\overset{O}{\overset{\|}{C}}Cl \quad (6)$$

$$CH_3(CH_2)_5\overset{O}{\overset{\|}{\underset{H}{C}}} + CF_3\overset{O}{\overset{\|}{C}}-\underset{H}{N}O\overset{O}{\overset{\|}{C}}-CF_3 \xrightarrow{2\ C_5H_5N}$$

$$CH_3(CH_2)_5CN + 2\ C_5H_5N \cdot HO_2CCF_3 \quad (7)$$

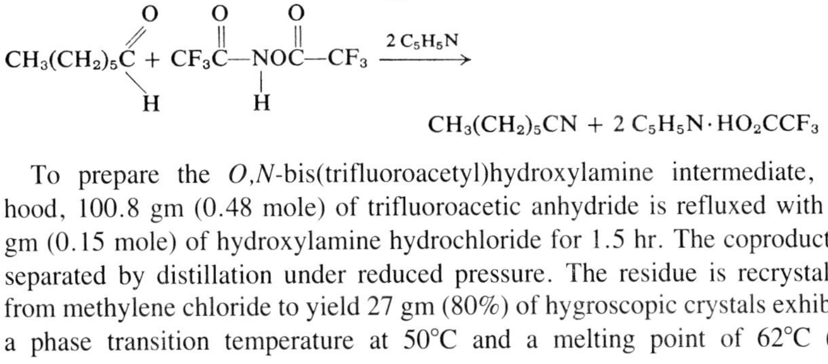

To prepare the O,N-bis(trifluoroacetyl)hydroxylamine intermediate, in a hood, 100.8 gm (0.48 mole) of trifluoroacetic anhydride is refluxed with 10.4 gm (0.15 mole) of hydroxylamine hydrochloride for 1.5 hr. The coproducts are separated by distillation under reduced pressure. The residue is recrystallized from methylene chloride to yield 27 gm (80%) of hygroscopic crystals exhibiting a phase transition temperature at 50°C and a melting point of 62°C (with sublimation).

In a solution of 11.4 gm (0.1 mole) of heptanal, 22.5 gm (0.1 mole) of O,N-bis(trifluoroacetyl)hydroxylamine and 15.8 gm (0.2 mole) of pyridine are refluxed in benzene for ½ hr. The pyridinium trifluoroacetate is separated by filtration and the resulting nitrile is isolated from the benzene solution by distillation to yield 7.9 gm (71.5%), bp 183°C, n_D^{20} 1.4945.

c. DECOMPOSITION OF N,N,N-TRIMETHYLHYDRAZONIUM SALTS

Another versatile method of converting both aliphatic and aromatic aldehydes to the corresponding nitriles involves a β-elimination reaction of the corresponding N,N,N-trimethylhydrazonium salts with bases [30], which is illustrated for the sake of simplicity by the following preparation of benzonitrile.

2-6. Preparation of Benzonitrile [30]

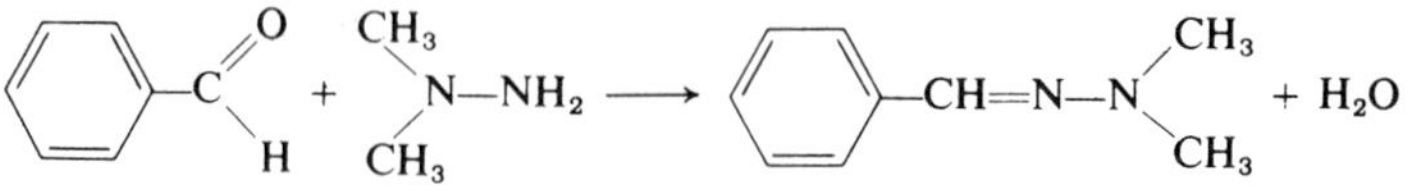

$$\text{PhCHO} + (CH_3)_2N-NH_2 \longrightarrow \text{Ph-CH=N-N}(CH_3)_2 + H_2O \quad (8)$$

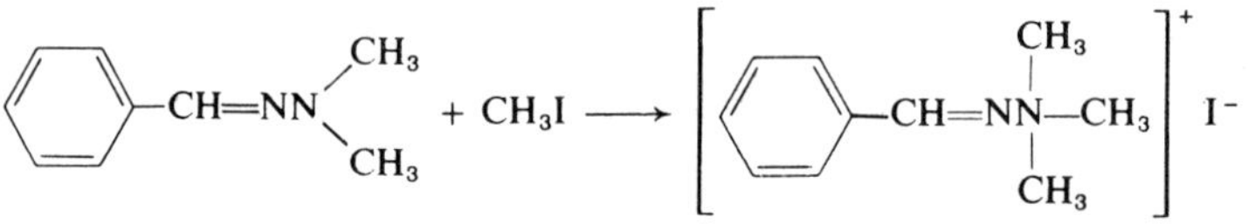

$$\text{Ph-CH=NN}(CH_3)_2 + CH_3I \longrightarrow [\text{Ph-CH=NN}(CH_3)_3]^+ I^- \quad (9)$$

$$[\text{Ph-CH=N-N}(CH_3)_3]^+ I^- + KOH \longrightarrow \text{Ph-CN} + N(CH_3)_3 + KI + H_2O \quad (10)$$

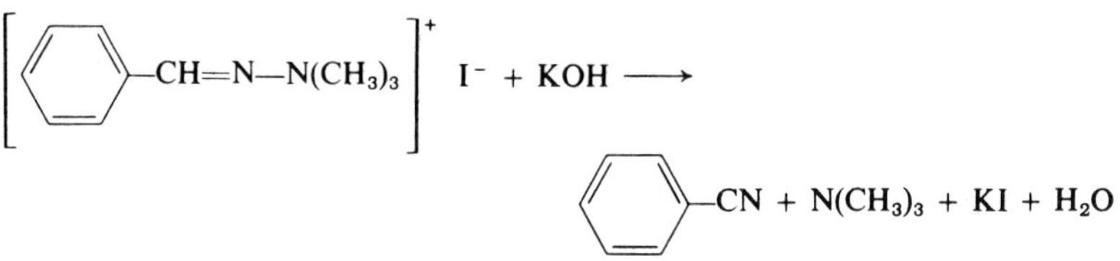

In a reflux apparatus equipped with a Dean–Stark water separator, a solution of 10.6 gm (0.1 mole) of benzaldehyde and 6.0 gm (0.1 mole) of N,N-dimethylhydrazine in 100 ml of dry benzene is refluxed until the theoretical amount of water has evolved. The solution is then concentrated to half volume, and, after cooling, 14.2 gm (0.1 mole) of methyl iodide is added. The solution is then refluxed for 2–4 hr, even if the hydrazonium salt does begin to separate promptly. The product is separated by the addition of an equal volume of anhydrous ether followed by filtration and drying in a vacuum desiccator. The trimethylhydrazonium salt may be recrystallized from methanol. Yield 26.6 gm (92%) mp 233°–235°C.

The following decomposition should be carried out in a hood. To a solution of 5.6 gm (0.1 mole) the potassium hydroxide in 150 ml of methanol, the dry quaternary salt prepared above is added. (Normally this reaction is carried out using sodium methoxide instead of potassium hydroxide.)

After the evolution of trimethylamine has ceased, the reaction mixture is refluxed until the amine odor is no longer detectable. The reaction mixture is then diluted with a large volume of water and the nitrile is isolated by extraction with ether.

The ether is separated from the product by distillation. An 85% yield of benzonitrile is obtained, bp 188°–190°C. If the nitrile formed in the reaction is a solid, it normally will separate from the reaction mixture upon dilution with water. In that case, the product is isolated by filtration.

§ 2. Elimination Reactions

d. PREPARATIONS FROM OTHER CARBONYL DERIVATIVES

Heating of p-tolylsulfonylhydrazones of ketones with hydrogen cyanide in decalin has resulted in the formation of nitriles [Eq. (11)] [31].

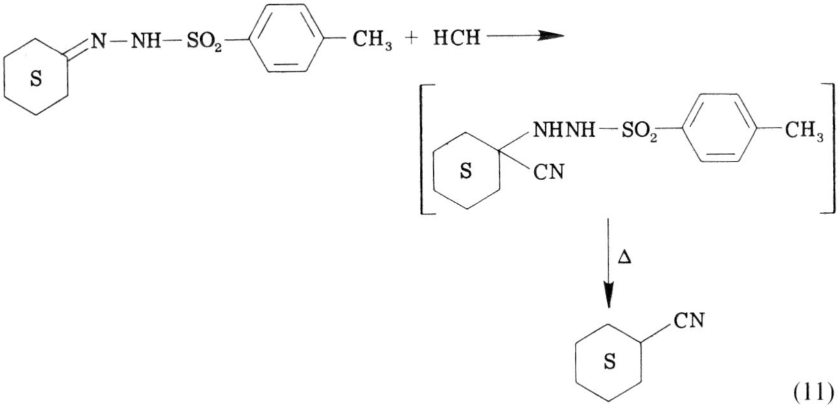

(11)

Koehler converted hydrazones of aldehydes to nitriles by treatment with potassium cyanide in alcohol [32].

Oximes of α-substituted cyclic ketone with substituents such as alkoxy, ethylthio, and alkylamino groups, on treatment with phosphorus pentachloride followed by hydrolysis, produced ω-cyanoaldehydes. This modification of the classical Beckmann rearrangement is particularly interesting since cyanoaldehydes are not readily produced by other methods [33].

A method of introducing an additional one-carbon unit into a ketone, i.e., $R_2C=O \rightarrow R_2CH\ CN$, at modest temperatures involves treatment of a ketone with tosylmethyl isocyanide ($CH_3-\langle\rangle-SO_2CH_2-N{=}C$) and a base in 1,2-dimethoxyethane. The reaction appears to be quite general [34].

e. THE SCHMIDT REACTION*

The Schmidt reaction is primarily considered as a method of converting carboxylic acids to amines by use of hydrazoic acid. However, other carbonyl compounds will react with hydrazoic acid also. Aldehydes may be reacted to produce nitriles and formamide derivatives, ketones may yield amides [35,36]. The mechanism of the reaction involving ketones is quite complex and in certain cases, nitriles have been isolated. A particularly interesting example of the reaction is the formation of bicyclo[3.3.1]non-6-ene-3-carbonitrile from adamantan-2-one. In this preparation, the ratio of methanesulfonic to acetic acids is important [37].

---

*A dicarbonyl diazide detonated during preparation [J. A. Landgrebe, *Chem. Eng. News* **59** (17), 47 (1981)]. Extreme precautions should be exercised in using this procedure.

2-7. Preparation of Bicyclo[3.3.1]non-6-ene-3-carbonitrile [37]

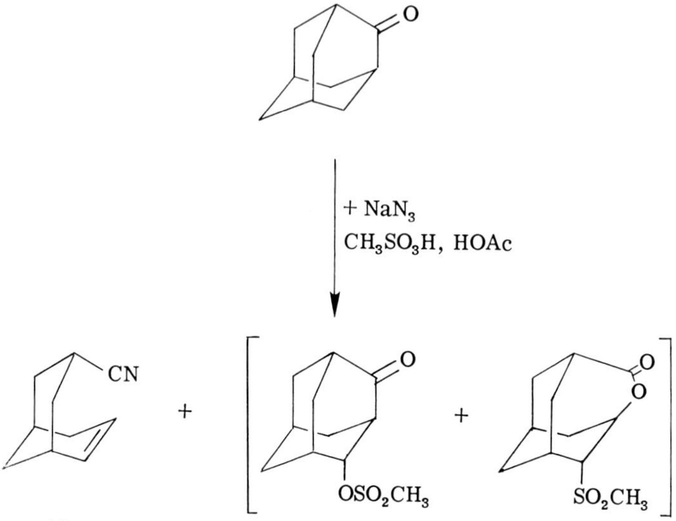

In a small Erlenmeyer flask equipped with a magnetic stirrer, to 1.0 gm (0.0067 mole) of adamantan-2-one dissolved in 3 ml of methanesulfonic acid and 4 ml glacial acetic acid is added in small portions 0.5 gm (0.008 mole) of sodium azide over a 45-min period at room temperature. Stirring is continued for an additional 20 min. Upon pouring the reaction mixture onto an ice–water mixture, a colorless solid precipitates. The solid is filtered off to give 0.6 gm of bicyclo[3.3.1]non-6-ene-3-carbonitrile(I) (61% yield); mp 160°–165°C. $\delta_{TMS}^{CCl_4}$ 6.20–5.70 (m, 2H vinyl protons), 2.95 (m, 1H–CHCN), 2.72–1.20 (m, 10 H, remaining protons); mass spectra m/e [rel. intensity 147 (100)].

The mother liquor upon neutralization with sodium bicarbonate affords 0.05 gm (3.0%) of 4(e)-methane-sulfonoxydamantan-2-one(II), mp 73°–75°, N_{max}^{KBr} 3000, 1720, 1340, 1190 and 1180 cm^{-1}). Extraction of the basic mother liquor with chloroform led to the isolation of 0.36 gm (33%) of 4-azatricyclo-[4.3.1.1.^{3,5}]-undecan-5-one(III) (mp > 300°, N_{max}^{KBr} 3200, 3080, and 1640 cm^{-1}).

Knittel and co-workers [38] observed that the thermal decomposition of α-azidoketones led to α-cyanoketones with one less carbon atom [Eq. (13)].

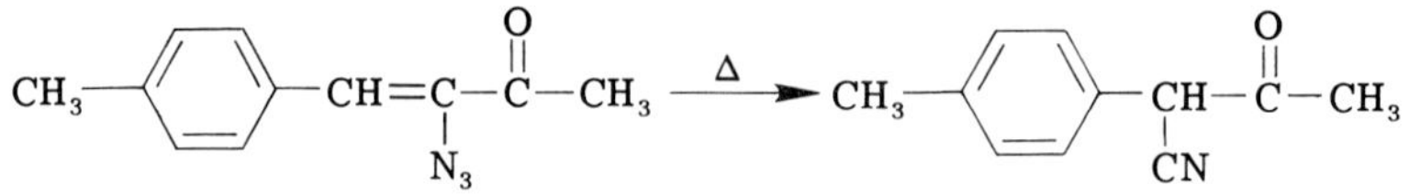

Yield 75% (13)

§ 2. Elimination Reactions

Other α-azide compounds such as azido-sulfides, -sulfoxides, -sulfones, and -nitriles upon base-catalyzed decomposition gave nitriles [39]. An example is the conversion of phenylacetonitrile to benzonitrile in a 90% yield. Equations (14) and (15) indicate the proposed reaction mechanism of the two-step procedure.

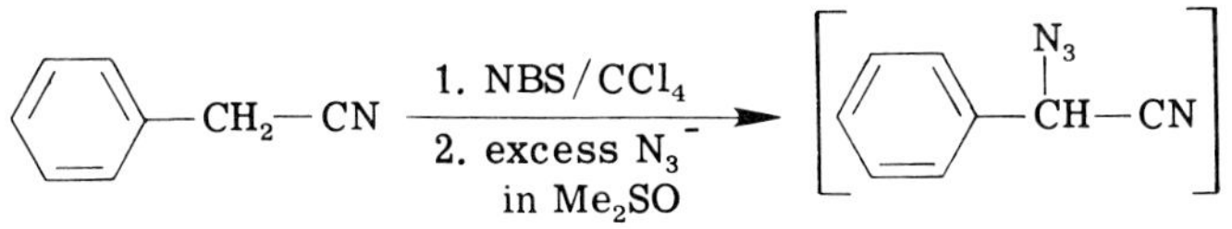

$$\text{Ph-CH}_2\text{-CN} \xrightarrow[\text{2. excess N}_3^-\text{ in Me}_2\text{SO}]{\text{1. NBS/CCl}_4} \left[\text{Ph-CH(N}_3\text{)-CN}\right] \quad (14)$$

$$\left[\text{Ph-CH(N}_3\text{)-CN}\right] \xrightarrow[-\text{HN}_3]{\text{N}_3^-} \left[\text{Ph-C(N}_3\text{)(-)-CN}\right] \xrightarrow{-\text{N}_2} \left[\text{Ph-C(=N}^-\text{)-CN}\right]$$

$$\downarrow \text{CN}^-$$

$$\text{Ph-CN} \quad (15)$$

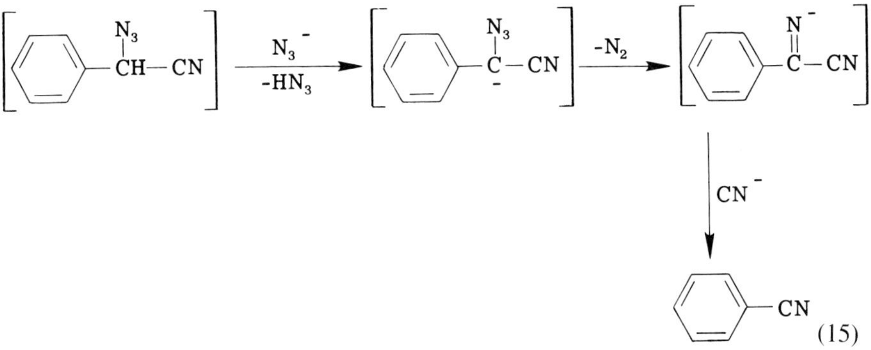

f. A Novel Synthesis of Aromatic Nitriles from Aldehydes

As diverse a group of aromatic nitriles as p-N,N-dimethylaminobenzonitrile, 3,4,5-trimethoxybenzonitrile, p-dichlorobenzonitrile, indole-3-carbonitrile, 5-bromoindole-3-carbonitrile, and 7-azoindole-3-carbonitrile has been prepared from the corresponding aldehydes upon treatment with ammonium dibasic phosphate and nitropropane in acetic acid. The mechanism of this reaction is not entirely clear [40,41]. Preparation 2-8 illustrates this process.

2-8. Preparation of p-N,N-Dimethylaminobenzonitrile [40]

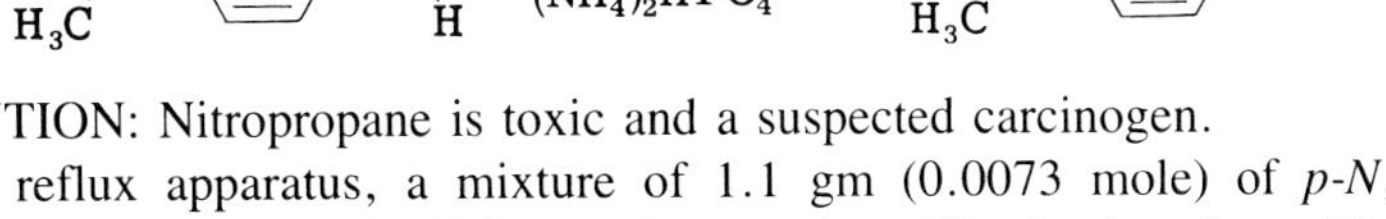

$$(\text{H}_3\text{C})_2\text{N-C}_6\text{H}_4\text{-CHO} \xrightarrow[(\text{NH}_4)_2\text{HPO}_4]{\text{C}_3\text{H}_7\text{NO}_2} (\text{H}_3\text{C})_2\text{N-C}_6\text{H}_4\text{-CN} \quad (16)$$

CAUTION: Nitropropane is toxic and a suspected carcinogen.

In a reflux apparatus, a mixture of 1.1 gm (0.0073 mole) of p-N,N-dimethylaminobenzaldehyde, 7.0 gm of ammonium dibasic phosphate, 30 ml of nitropropane, and 10 ml of glacial acetic acid is refluxed for 16 hr. During this time the colorless solution becomes, in turn, dark brown, light brown, and then yellow. The low boiling components of the reaction mixture are then removed under reduced pressure. To the residual oil is added, with stirring, 100 ml of water. After standing at room temperature for an hour, the oil crystallizes. The solid, after filtration weighs 0.95 gm, mp 70°–78°. After recrystallization from water–methanol 0.82 gm (77%) of pure product is isolated, mp 75°–77°C.

3. CONDENSATION REACTIONS

A. The Use of Cyanide Salts

a. ALIPHATIC NITRILES

The nucleophilic displacement reaction of both alkyl and aryl halides by cyanide ions in dipolar aprotic solvents has been studied extensively [4, 9, 42]. This reaction can be visualized as a condensation of NaCN with RX to give RCN. In the alkyl series, sodium cyanide is usually used with dimethyl sulfoxide or dimethylformamide as the solvent [43–51]. Some examples from Friedman and Shechter [4] are reproduced in Table II.

Perhaps the earliest example of a nucleophilic displacement reaction for the synthesis of aliphatic nitriles was the Pelouze reaction of dimethyl sulfates or potassium monoalkysulfates with potassium cyanide [52]. More recently alkyl tosylates [49,53] and sulfonates [44,–46] have been used. Since tosylates are readily prepared from alcohols, many nitriles become readily accessible, particularly when the reactions are carried out in aprotic solvents such as dimethylformamide [44,46,49] or dimethyl sulfoxide [49]. Hexamethylphosphoramide has also been suggested as a reaction solvent [49], but this solvent is believed to be carcinogenic.

The reaction is also applicable to both primary and secondary halides and to α-halo ethers [47]. Iodide ions catalyze the replacement of chlorine. Tertiary halides and alicyclic halides tend to form olefins or decomposition products. The preparation of pentanonitrile using dimethyl sulfoxide is representative of the procedure although dimethylformamide has also been used [48].

3-1. *Preparation of Pentanonitrile (Valeronitrile)* [4]

$$CH_3CH_2CH_2CH_2Cl + NaCN \xrightarrow{DMSO} CH_3CH_2CH_2CH_2CN + NaCl \quad (17)$$

CAUTION: See notes at the beginning of this chapter on the hazards in handling sodium cyanide. Dimethyl sulfoxide should also be handled with caution. This solvent is readily absorbed through the skin, carrying with it some impurities which may be present on the skin (e.g., residual soap), leading to physiological reactions. The solvent is also said to reduce the sensation of pain. Test animals have exhibited reactions in their eyes upon treatment with DMSO.

In a hood, with due precautions, to a rapidly stirred mixture of 53 gm (1.08 moles) of reagent grade sodium cyanide in 240 ml of dimethyl sulfoxide at 80°C is added 93 gm (1 mole) of 1-chlorobutane over a 15-min period. During this

TABLE II

REACTIONS OF ALKYL AND CYCLOALKYL HALIDES WITH SODIUM OR POTASSIUM CYANIDE IN DIMETHYL SULFOXIDE[a]

Halide	Cyanide[b]	Reaction temp (°C)	Reaction time (hr)[c]	Yield of nitrile (%)
1-Chlorobutane	NaCN	140°	0.25	93
1-Chlorobutane	KCN	120°–140°	10	69
1-Chloro-2-methylpropane	NaCN	140°[d]	0.5	88[e]
1-Chloro-3-methylbutane	NaCN	100°–140°	2	85[f]
1-Chloro-2-methyl-2-phenylpropane	NaCN	120°	24	26
Benzyl chloride	NaCN	35°–40°[g]	2.5	92[h]
p-Nitrobenzyl chloride	NaCN	35°–40°	1	0[i]
2-Chlorobutane	NaCN	120°–145°	3	64[j]
2-Chlorobutane	KCN	120°–138°[k]	24	42
Chlorocyclopentane	NaCN	125°–130°	3	70
Chlorocyclohexane[n]	NaCN	130°–80°	4	0[l]
2-Chloro-2-methylpropane	NaCN	130°–105°	4	0[m]
1-Bromobutane[o]	NaCN	60°–90°	0.6	92[o]
1-Bromo-2-methylpropane	NaCN	70°	2	62
2-Bromobutane	NaCN	70°	6	41[p]

[a]Reprinted from L. Friedman and H. Shechter, *J. Org. Chem.* **25**, 877 (1960). Copyright 1960 by the American Chemical Society. Reprinted by permission of the copyright owner.

[b]The ratio of halide (moles), cyanide (moles), and DMSO (ml) usually used was 1:1.2:250.

[c]The reaction time listed is the sum of that for addition of the halide and subsequent reaction at the given temperature.

[d]The halide was added in 10 min to the initial mixture at 80°C. The reaction is mildly exothermic and was completed by heating to 140°C until refluxing ceased.

[e]Bp 128°C, n_D^{20} 1.3926.

[f]Bp 151°–155°C, n_D^{20} 1.4047–1.4051; lit. b.p. 150°–155°C [H. Rupe and K. Glenz, *Helv. Chim. Acta* **5**, 939 (1922)].

[g]The reaction mixture was cooled externally.

[h]Bp 90.5°–91°C (5 mm), n_D^{20} 1.5237–1.5238; lit. bp 115°–120°C (10 mm), n_D^{20} 1.5242 [J. W. Bruhl, *Z. Phys. Chem.* **16**, 218 (1895)].

[i]4,4'-Dinitrostilbene is formed in 78% crude yield, m.p. 286°–288°C; lit. 288°C [P. Ruggli and F. Lang, *Helv. Chim. Acta* **21**, 42 (1938); R. Walden and A. Kernbaum, *Ber. Dtsch. Chem. Ges.* **23**, 1959 (1890)].

[j]Bp 123.5°–124°C (742 mm), n_D^{20} 1.3898–1.3900; lit. bp 125°C [M. Hanriot and L. Bouveault, *Bull. Soc. Chim. Fr.* [3] **1**, 172 (1889)].

[k]The halide was added dropwise in 6 hr to the mixture at 120°–125°C; the mixture was then heated for 18 hr until it reached 138°C.

[l]The reaction mixture became dark and gave cyclohexene, gases, and a black intractable product.

[m]Upon initiating reaction at 130°C, gases (2-methylpropene, hydrogen cyanide, and formaldehyde) were evolved, the temperature dropped to 105°C, and black resinous materials were formed. The desired product was not obtained at lower reaction temperatures.

[n]The bromide was added in 30 min to the cyanide mixture at 60°C while effecting cooling of the reaction; the mixture was then heated for 15 min at 90°C.

[o]Bp 138°–139°C (742 mm), n_D^{20} 1.3970.

[p]Upon addition of the 2-bromobutane, the temperature was maintained at 55°–60°C by intermittent cooling. In subsequent reaction gases were evolved and the mixture became progressively darker and malodorous.

period the reaction temperature rises rapidly and has to be controlled at 140° ± 5°C by water cooling. The reaction mixture cools rapidly after the addition is completed.

The reaction mixture is cooled, diluted with water to approximately 1 liter, and extracted with three 150-ml portions of ether.

In the hood, the ether extracts are washed in turn with 6 N hydrochloric acid (to hydrolyze a small amount of isocyanide) and with water, and are then dried over calcium chloride. The ether is evaporated and the residue is fractionally distilled from phosphorus pentoxide. The pentanonitrile passes over at 138°–139°C (747 mm). Yield 77 gm (93%).

The conversion of alcohols to nitriles has been accomplished by refluxing a solution of an alcohol, carbon tetrachloride, and triphenylphosphine successively with dimethyl sulfoxide and sodium cyanide under rigorously anhydrous conditions [50]. The optimum reaction conditions appear to occur when equimolar quantities of the phosphine, alcohol, and carbon tetrachloride are present. The general technique could be extended to the formation of nitrilo-derivatives of carbohydrates.

Procedure 3-2 is an example of the procedure.

3-2. Preparation of 2-Methyloctanonitrile [50]

$$CH_3\underset{OH}{CH}-CH_2CH_2CH_2CH_2CH_3 + NaCN \xrightarrow[\substack{(C_6H_5)_3P \\ DMSO}]{CCl_4} CH_3CH_2CH_2CH_2CH_2CH_2\underset{CN}{CH}CH_3 \quad (18)$$

With due safety precautions, in a stirred apparatus fitted with a total reflux–partial take-off distillation head, carefully protected from atmospheric moisture, are placed 14 ml of 2-octanol, 15 ml of carbon tetrachloride, and 20 gm of of triphenylphosphine. The mixture is heated on a steam bath for a few minutes. Then 80 ml of dimethyl sulfoxide is added and low boilers are distilled out until a temperature of approximately 90°C is achieved. While the solution is heated at reflux, 4.7 gm of sodium cyanide is added portionwise (use extreme caution). The reaction mixture is heated and stirred for an additional 2 hr under reflux. Then the solution is cooled and poured into an ice–ferrous sulfate solution. The mixture is allowed to warm to room temperature. Then it is extracted with three 100-ml portions of chloroform. The extract is dried and then fractionally distilled under reduced pressure. The product was obtained in a 70% yield, bp 85°–90°C (at the water pump), $n_D^{20.5}$ 1.4201.

Hydroxybenzyl alcohols have been converted directly to hydroxyphenylacetonitriles by simply refluxing the appropriate alcohol with sodium cyanide in dimethyl formamide [51].

The displacement reactions of alkyl halides with inorganic anions in water frequently is inhibited because the reaction can take place only at the interface between the aqueous and the organic phase. In many cases, cationic surfactants such as tetraalkylammonium or tetraalkylphosphonium salts effectively catalyze the process by bringing the anions into the organic layer for reaction. This process is now termed "phase-transfer catalysis" [54].

In Preparation 3-3 of 1-cyanooctane, refluxing of the indicated reaction mixture without the catalyst hexadecyltributylphosphonium bromide at 105°C for 3 hours did not result in the formation of any detectable 1-cyanooctane. When the phase-transfer catalyst was added, the reaction proceeded smoothly within less than 3 hours.

3-3. Preparation of 1-Cyanooctane [54]

$$CH_3CH_2CH_2CH_2CH_2CH_2CH_2CH_2Cl + NaCN \xrightarrow{(C_{16}H_{33})P^+(C_4H_9)_3Br^-} CH_3(CH_2)_6CH_2CN \quad (19)$$

With appropriate safety precautions, in a stirred reflux apparatus, 100 gm (0.67 mole) of 1-chlorooctane, 100 gm (2.0 mole) of sodium cyanide, 25 ml of water, and 25 ml decane are warmed to 50°C. Then 5 gm (0.01 mole) of hexadecyltributylphosphonium bromide is added. The mixture is rapidly heated to the reflux temperature and maintained at that temperature (105°C) for 145 min. The reaction mixture is cooled to room temperature and the organic phase is separated.

The organic phase is distilled in a short-residence time wiped-film evaporator at 140°/1 mm. Yield: 98.4 gm (95%). The product had a purity of 90.6% by gas–liquid chromatography.

An indication of the wide applicability of phase-transfer catalyzed reactions is that they made it possible to introduce a variety of functional groups in cross-linked polystyrene resins which had been chloromethylated. Reaction with sodium cyanide in the presence of a surfactant (Adogen 464) produced the expected cyanated polymer. Alkylation by the chloromethylated polystyrene of malononitrile produced a polymer with $-CH_2-CH(CN)_2$ functions [55].

High yields of nitriles have been obtained upon reaction of alkyl halides with alkaline cyanides in the presence of crown ethers (the so-called "naked" anion effect) [56]. In general, the reaction is more rapid in acetonitrile than in benzene. As a matter of fact hexamethylphosphoramide (HMPA) appears to be a particularly effective solvent for use with crown ether catalysis [57]. It is unfortunate that, as mentioned above, HMPA is suspected as being a serious cancer producing agent and can no longer be considered for routine syntheses; however, other aprotic solvents should be evaluated further.

In the reaction with crown ethers, it was observed that primary alkyl chlorides reacted much faster than the corresponding bromides. On the other hand, secondary alkyl bromides reacted more rapidly than secondary chlorides, although yield and low levels of alkenes were observed with the secondary chlorides. The reaction time and temperature and the simplicity of work-up procedures serve to recommend this approach over some of the older methods [56].

In recent years, Regen and co-workers [58–60] introduced an interesting variation to the concept of phase-transfer catalysis. The aqueous phase containing sodium cyanide and the organic phase containing the alkyl halide were contacted with a solid phase catalyst to produce alkyl cyanides. The solid phase catalyst was produced by reacting chloromethylated cross-linked polystyrene with butyldimethylamine [58] or with tributylphosphine [59] to provide quaternary ammonium salt groups in the former and quaternary phosphonium groups in the latter case. Another "triphase" catalyst was a neutral grade of alumina [60]. With alkyl halides the triphase reaction with cyanides was more effective than the reaction catalyzed by a crown ether such as 1,4,7,10,13-16-hexaoxacyclooctadecane ("18-crown-6"). In other displacement reactions, the situation was reversed.

Interestingly enough, another paper by Regen and co-workers published at the same time as Ref. [60] reports on a convenient synthesis of nitriles which we believe to have application to other displacement reactions. In this procedure, alumina is impregnated with sodium cyanide. This solid is stirred with a solution of an alkyl halide at 90°C. The insoluble reagent is filtered off and the filtrate is freed of solvent to leave a high purity residue of the nitrile in substantial yield [61]. While the use of impregnated ion exchange resins has been suggested before, this particular procedure seems particularly simple [cf Refs. 62 and 63]. Procedure 3-4 illustrates the newer technique.

3-4. *Preparation of 1-Cyanododecane* [61]

$$CH_3(CH_2)_{10}CH_2Br \xrightarrow{\text{NaCN on }Al_2O_3} CH_3(CH_2)_{10}CH_2CN \qquad (20)$$

(a) Preparation of alumina impregnated with sodium cyanide. In a 200-ml round-bottom flask is dissolved 8.0 gm (0.1232 mole) of sodium cyanide in 20 ml of distilled water. Then 16.0 gm of neutral alumina (Bio-Rad Laboratories, Alumina AG-7, 100–200 mesh) is added. The flask is then attached to a rotary evaporator and freed of water under reduced pressure while warming with a water bath maintained below 65°C. The impregnated alumina is then dried at 110°C. at 0.05 mm for 4 hr.

Synthesis of 1-cyanododecane. In a 100-ml round-bottom flask containing a Teflon-coated magnetic stirrer and equipped with a thermometer and a reflux condenser topped with a drying tube are placed 2.5 gm (0.01 mole) of 1-bromododecane, 15 gm of the impregnated alumina, and 30 ml of toluene. With stirring, this mixture is heated at 90°C for 45 hr.

At the end of this period, the reaction mixture is cooled to room temperature. The spent impregnated alumina is washed with 100 ml of toluene. The toluene extract is combined with the filtrate. The combined toluene solution is freed of solvent, under reduced pressure. The residue is a colorless liquid identical with an authentic sample by infrared and NMR spectra. Yield: 1.96 gm (100%).

b. AROMATIC NITRILES

Aryl halides are frequently converted to the corresponding nitriles with cuprous cyanide rather than with alkali cyanides in dimethylformamide [42] or in N-methylpyrrolidone [64]. Of these solvents, N-methylpyrrolidone has the advantage of being a solvent for cuprous cyanide. The work-up of reaction mixtures from the cuprous cyanide reaction is somewhat involved since cuprous halides form complexes with nitriles. Friedman and Shechter [42] recommend three approaches for the decomposition of these complexes. (1) Oxidation of the adduct with ferric chloride. Since the resultant cupric salts do not complex with nitrile, the product may be isolated readily as described below. (2) Formation of a more stable complex of cuprous and cupric ions with ethylenediamine (followed by separation of suspended copper derivatives with sodium cyanide). (3) Destruction of the complexes with an excess of sodium cyanide.

The first separation method is recommended for the relatively nonbasic nitriles (cyano acids, esters, ketones, aldehydes, etc.), the other two for more basic nitriles (such as cyanopyridines).

Besides the original work on this reaction due to Rosenmund [65] and von Braun [66], a number of examples of the reaction are cited here [16, 67–70].

The preparation of 1,2-dicyanoacenaphthylene [71] is a recent example of the method of Newman and Boden [64].

3-5. *Preparation of 1,2-Dicyanoacenaphthylene* [71]

$$\text{1,2-dibromoacenaphthylene} \xrightarrow[\text{N-methyl-pyrrolidone}]{(CuCN)_2} \text{1,2-dicyanoacenaphthylene} \quad (21)$$

59.5%

CAUTION: Although cuprous cyanide may be less toxic than sodium cyanide, the notes and precautions given at the beginning of this chapter should still be observed.

Under a nitrogen atmosphere, with stirring, 9.9 gm (0.112 mole) of cuprous cyanide, 55 ml of N-methyl-2-pyrrolidone, and 8.5 gm (0.274 mole) of 1,2-dibromoacenaphthylene are heated to 160°C for 2 hr. The reaction mixture is then cooled before pouring cautiously into a solution of 12 gm of sodium cyanide in water. The aqueous mixture is shaken repeatedly and then is extracted with three 350-ml portions of benzene. The organic layers are filtered and washed in turn with 200 ml of a 10% aqueous solution of sodium cyanide and 200 ml of water. The benzene solution is dried over anhydrous sodium sulfate. It is then evaporated under reduced pressure to leave a residue of 3.4 gm of a brown solid. This crude product is dissolved in methylene chloride, mixed with 30 gm of silica gel (Davison, grade 950; 60–200 mesh), and evaporated in a stream of dry nitrogen. The dried impregnated silica gel is placed on top of a 5 × 70 cm silica gel column (Davison, grade 950; 60–200 mesh) and eluted with benzene. The fractions collected are 250 ml in volume. Fraction 3 affords 40 mg of 1,2-dibromoacenaphthylene; fraction 5, 20 mg of 1-bromo-2-cyanoacenaphthylene; and fractions 9–14, 3.3 gm of 1,2-dicyanoacenaphthylene (59.5% yield, mp 240°C). Upon recrystallization from benzene, the mp is raised to 243°C, UV max (CH_2H_2) 345 (ϵ12,000), IR (KBr) 2240 cm^{-1}(CN); mass spectrum (68 eV), m/e (rel intensity) 202 (100), 175 (38), 174 (17), 100 (11), 76 (6).

Some aryl and vinyl halides have been converted to nitriles by use of sodium dicyanocuprate in DMF [72].

The use of sodium cyanide-impregnated alumina [cf. 61] was also applied to the preparation of aromatic nitriles. In this latter situation, however, tetrakis (triphenylphosphine)palladium (0) was added to improve the yield [73].

Aromatic thallium compounds have been used as intermediates in the formation of aromatic nitriles [74,75].

B. The Sandmeyer Reaction

The Sandmeyer reaction has been used for the conversion of aromatic amines via diazonium salts to nitriles for nearly a century [76]. Yields, however, are highly variable. While sodium or potassium cyanide solutions of cuprous cyanide have been used to furnish the nitrile group, improved yields are sometimes observed when a sodium or potassium cyanide solution of nickel cyanide is used [77].

The use of a neutral diazonium salt solution and its reaction with the cyanide complex at 70°–80°C appears to afford higher yields [78]. These observations are suggestive of improvements to be considered in carrying out Sandmeyer reactions. Unfortunately, the yield data are not sufficiently clear-cut in the report by Hodgson and Heyworth [78] to permit the drawing of definite conclusions because experiments were carried out with 2–5 gm of amines. Our example, therefore, is drawn from the somewhat older literature. This preparation should be considered illustrative of the general techniques. The reagent 1-naphthylamine cannot be used since it is considered carcinogenic.

3-6. Preparation of 1-Naphthonitrile [79a]

§ 3. Condensation Reactions

CAUTION: The cautions mentioned at the beginning of this chapter in regard to the handling of cyanides must be observed. 1-Naphthylamine is carcinogenic.

In a suitable open vessel, 28.6 gm (0.2 mole) of 1-naphthylamine is dissolved in a solution of 20 ml of concentrated hydrochloric acid in 600 ml of hot water.

With vigorous stirring, the solution is cooled to 0°C and maintained at that temperature during the addition in turn of 50 ml of concentrated hydrochloric acid and a solution of 14.4 gm of sodium nitrite in 60 ml of water. The slight excess of nitrous acid is then destroyed with a small quantity of urea. This diazonium salt solution is maintained at ice temperature, while the cyanide solution is being prepared.

In a hood, in a 5-liter flask, with suitable precautions, a solution of 72.7 gm of nickel nitrate in 100 ml of water is added to 250 ml of a solution containing 81.4 gm of potassium cyanide in 20 gm of sodium hydroxide. Then 150 ml of benzene and some crushed ice are added. The diazonium solution is then added over a ½ hr period with vigorous stirring while maintaining the temperature between 0° and 5°C. The mixture is then allowed to come to room temperature with stirring over a 2-hr period, heated to 50°C, cooled, and the aqueous layer is separated. Benzene and product are then separated from the reaction mixture by steam distillation (continued until 5 liters have been collected).

The benzene solution is washed with sodium hydroxide solution and dried. After the benzene has been distilled off, the residue is fractionally distilled under reduced pressure.

Redistillation of the fraction boiling between 120° and 200°C at 20 mm gave 17.8 gm (58%) of 1-naphthonitrile, bp 165°–170°C (20 mm). This product quickly solidifies.

By a similar technique 4-cyanobiphenyl has been produced (11). The more thermally stable diazonium tetrafluoroborate salts have been converted to nitriles using sodium tetracyanonickelate(II) [79b].

The reactions of diazonium salts with cuprous salts are usually called Sandmeyer reactions [76]. When metallic copper is used in conjunction with such reactions, it is usually termed a Gattermann reaction [80], while the reaction with cupric salts is sometimes referred to as a Körner–Contardi reaction [81].

C. Nitriles via Cyanohydrin Formation

Liquid hydrogen cyanide may be added readily to carbonyl compounds to afford cyanohydrins which may be converted to substituted acrylonitriles.

The reaction is sometimes called the Urech–Ultee method [82].

The reactions are usually carried out in the liquid phase with an amine or a cyanide as a basic catalyst. High yields can be obtained by carrying out the reaction at relatively low temperature and neutralizing the basic catalyst before isolating the product.

The preparation of acetone cyanohydrin is described in the literature. The procedure consists of adding sulfuric acid gradually to an aqueous solution of acetone and sodium cyanide at 10°–20°C. The cyanohydrin is recovered in 77–78% yield by decantation, extraction, and distillation [83].

The reaction is quite general. For example, C. F. H. Allen and R. K. Kimball made use of the 1,4-addition of hydrogen cyanide to a conjugated system as illustrated in Eq. (23) [84].

$$\text{Ph}-\text{CH}=\text{CH}-\overset{\text{O}}{\underset{\|}{\text{C}}}-\text{Ph} \xrightarrow[\text{Alcohol}]{\text{KCN, H}^+} \text{Ph}-\underset{\underset{\text{CN}}{|}}{\text{CH}}-\text{CH}_2-\overset{\text{O}}{\underset{\|}{\text{C}}}-\text{Ph} \qquad (23)$$

Other chalcones have reacted similarly [85].

Even acrylamide has been reacted with liquid hydrogen cyanide to produce 3-cyanopropionamide [86].

Other reactions involve the reaction of ketones with sodium cyanide and ammonium chloride [87].

In Preparation 3-7, the general features of the synthesis are illustrated. Care should be taken in the final distillation of the cyanohydrin. Overheating may result in explosive decomposition [88].

3-7. Preparation of 2-Chloroacetaldehyde Cyanohydrin and 2-Chloroacrylonitrile [88]

$$\text{Cl}-\text{CH}_2\overset{\overset{\text{O}}{\|}}{\underset{\text{H}}{\text{C}}} + \text{HCN} \longrightarrow \text{Cl}-\text{CH}_2-\underset{\underset{\text{H}}{|}}{\overset{\overset{\text{OH}}{|}}{\text{C}}}-\text{CN} \qquad (24)$$

$$\text{Cl}-\text{CH}_2-\underset{\underset{\text{H}}{|}}{\overset{\overset{\text{OH}}{|}}{\text{C}}}-\text{CN} + \text{CH}_3\overset{\text{O}}{\underset{\|}{\text{C}}}-\text{O}-\overset{\text{O}}{\underset{\|}{\text{C}}}\text{CH}_3 \longrightarrow \text{Cl}-\text{CH}_2-\underset{\underset{\text{H}}{|}}{\overset{\overset{\text{OCOCH}_3}{|}}{\text{C}}}-\text{CN} + \text{CH}_3\text{CO}_2\text{H} \qquad (25)$$

$$\text{Cl}-\text{CH}_2-\underset{\underset{\text{CN}}{|}}{\text{CHOCOCH}_3} \longrightarrow \text{Cl}-\text{CH}=\text{CHCN} + \text{CH}_3\text{CO}_2\text{H} \qquad (26)$$

CAUTION: The cautions mentioned at the beginning of this chapter in regard to handling cyanides must be observed. The distillation of the cyanohydrin must be carried out with great care—explosion hazard. Use rubber gloves and a gas mask suitable for HCN work throughout.

In a hood, to 8000 gm of a 50% aqueous solution of 2-chloroacetaldehyde

§ 3. Condensation Reactions

(whose pH has been adjusted to 7.5 ± 0.5 with solid sodium bicarbonate) are added 1580 gm of liquid hydrogen cyanide at a temperature of 0°–9°C over a 2-hr period. The solution is then allowed to stand overnight, treated with concentrated phosphoric acid till the solution has a pH of 2.5 or less, and freed of the excess of hydrogen cyanide under reduced pressure at room temperature. Most of the water is then removed by distillation at 50°C (15 mm). The resultant solution contains approximately 95% of 2-chloroacetaldehyde cyanohydrin.

A mixture of 1530 gm of acetic anhydride and 970 gm of the 95% aqueous solution of 2-chloroacetaldehyde cyanohydrin (1-hydroxy-2-chloropropionitrile) is warmed in turn to 40°C for 2 hr and 100°C for 3 hr. The excess acetic anhydride and acetic acid is separated by distillation at 5 mm pressure. The residual 1-acetoxy-2-chloropropionitrile is distilled at 70°C (215 mm). Yield 1200 gm (90%).

Into a glass tube packed with glass beads, heated to 535°C, 77 gm of 1-acetoxy-2-chloropropionitrile is fed over a 2-hr period (contact time, 10 sec). The liquid product is collected in a trap at room temperature. Crude yield 70 gm.

The crude product is poured into 100 ml of water to which sodium bicarbonate is added until the pH remains at 7.0. The mixture is extracted with ether and the ether layer is dried with sodium sulfate. Then the ether is removed by evaporation. The residue is distilled fractionally. The first fraction, bp 88°C, is 1-chloroacrylonitrile (13 gm). The higher boiling fraction is a 50:50 mixture of *cis*- and *trans*-2-chloroacrylonitrile. Redistillation of this mixture affords the *trans*-isomer, bp 118°C, mp 45°C and the *cis*-isomer, bp 145°–146°C. Both products are lachrymators and vesicants.

The flavoprotein D-oxynitrilase has been adsorbed on a cellulose-based ion exchange resin. To a column packed with this material, a mixture of very pure benzaldehyde and hydrogen cyanide in 50% methanol was added at a rate of 10–15 ml/min in a nitrogen atmosphere. From the elutade, D-(+)-mandelonitrile was isolated in 95%, mp 28°–29°C, $[\alpha]_D^{20} + 46°$ (c 5, $CHCl_3$). It has been stated that with milligram quantities of this enzyme, a continuous synthesis is possible which will yield kilogram quantities [89].

Nagatu and co-workers have developed new methods for the preparation of nitriles by the use of hydrogen cyanide and trialkylaluminum [90] or the use of diethylaluminum cyanide with α,β unsaturated ketones, aldehydes, and esters [91–93]. The technique for the preparation of 6-methoxy-3,4-dihydronaphthalene-1-carbonitrile from 6-methoxy-1-tetralone with a diethylaluminum cyanide solution is given in great detail in *Organic Syntheses* [93].

Related to the formation of cyanohydrins from carbonyl compounds is the hydrocyanation of olefinic bonds.

Lapworth [94] and Knoevenagel [95] have added HCN to double bonds conjugated with activating groups:

$$\text{Ph}-\text{CH}=\underset{\text{CN}}{\text{C}}-\text{Ph} + \text{HCN} \longrightarrow \text{Ph}-\underset{\text{CN}}{\text{CH}}-\underset{\text{CN}}{\text{CH}}-\text{Ph} \quad (27)$$

For detailed reviews of the addition of liquid HCN catalyzed by KCN, see Ref. [96].

Olefins, such as ethylene or propylene, have been cyanated in the presence of palladium(II) cyanide, nickel(II) cyanide, and copper(I) cyanide in solvents. Of these, only palladium(II) cyanide was an efficient reagent. Polar solvents played an important role in the reaction. The products of reaction were a complex mixture. For example, upon cyanation of propylene in benzonitrile, the identified products were methacrylonitrile, 3-butenonitrile, crotononitrile, isobutyronitrile, and butyronitrile along with oligomers (mainly hexene isomers) [97].

D. Condensation of Existing Nitriles

a. ALDOL CONDENSATIONS

In the aromatic and heterocyclic series, 2-hydroxy-1-nitriles have been produced by the condensation of the corresponding aromatic ketones with nitriles such as acetonitrile or propionitrile in the presence of sodium amide in ether [98]. Evidently reaction conditions were sufficiently mild in this procedure that the 1,2-cyanohydrin could be isolated rather than the olefinic nitrile expected from an aldol condensation.

The usual conditions for a Knoevenagel reaction have frequently been used to prepare a variety of nitriles. Aldehydes have been condensed with nitriles in the presence of piperidine [99], sodium methoxide, and other bases. In the example cited, a quaternary ammonium base is used as catalyst [100]. Zinc acetate has also been used as a condensation catalyst [101].

3-8. Preparation of 1-(4-Nitrophenyl)cinnamonitrile [100]

$$\text{NO}_2-\text{C}_6\text{H}_4-\text{CH}_2\text{CN} + \text{Ph}-\overset{\text{O}}{\underset{\|}{\text{C}}}-\text{H} \xrightarrow{\text{Base}} \text{NO}_2-\text{C}_6\text{H}_4-\underset{\text{CN}}{\text{C}}=\text{CH}-\text{Ph} \quad (28)$$

A mixture of 1.5 gm 1-(4-nitrophenyl)acetonitrile and 1 ml of benzaldehyde in 10 ml of ethanol is heated to dissolve the solid. Then five drops of a 40% aqueous solution of benzyltrimethylammonium hydroxide (available as Triton B from Aldrich Chemical Co.) is added. The solvent boils while drastic color changes take place, followed by precipitation of the final product. Yield: quantitative, mp 177°–178°C. The color of the product can be improved by washing with methanol, followed by recrystallization from propanol.

§ 3. Condensation Reactions

The preparation of vinylidene cyanide (1,1-dicyanoethene) is based on a Knoevenagel-type condensation of formaldehyde with malononitrile. The reaction proceeds through the formation of 1,1,3,3-tetracyanopropane which can be cracked to yield vinylidene cyanide according to Eq. (17) [102].

$$2 \begin{array}{c} CN \\ | \\ CH_2 \\ | \\ CN \end{array} + HCHO \longrightarrow H_2O + \begin{array}{c} CN \\ \diagdown \\ CH-CH_2-CH \\ \diagup \\ CN \end{array} \begin{array}{c} CN \\ \diagup \\ \diagdown \\ CN \end{array} \longrightarrow \begin{array}{c} CN \\ | \\ CH_2 \\ | \\ CN \end{array} + \begin{array}{c} CN \\ \diagup \\ CH_2{=}C \\ \diagdown \\ CN \end{array} \qquad (29)$$

Since vinylidene cyanide polymerizes rapidly in the presence of moisture and bases, great care must be taken to eliminate water from the reaction system. Dusting phosphorus pentoxide on glass wool which is suspended in the reaction flask, condenser, and receiver assists in preserving the monomer. Vinylidene cyanide has been found to be a very hazardous material. Inhalation of a small amount of its vapor hospitalized one member of our laboratory staff with symptoms resembling asthma. Even residual monomer in its polymer appeared to cause this difficulty.

E. Alkylation of Acetonitrile Derivatives

Methylene groups adjacent to nitrile functions are sufficiently activated to permit their alkylation very much like the methylene groups in acetoacetic esters or malonic esters. A variety of acetonitrile derivatives have been used as starting materials for such alkylations; among these are malononitrile, cyanoacetic esters, and substituted acetonitriles [103,104]. The reactions are usually carried out with sodamide as catalyst. Since this catalyst is now available commercially as a free-flowing powder, the process should be considerably simplified, although many workers still prefer to prepare sodamide *in situ* since some explosive hazard is said to have existed with certain batches of commercial sodamide [105].

3-9. Preparation of 1-Isopropyl-1-(2-morpholinoethyl)-1-phenylacetonitrile [106]

$$\underset{\substack{| \\ CH \\ / \diagdown \\ CH_3 \ CH_3}}{Ph{-}CHCN} + \underset{\substack{| \\ CH_2{-}CH_2Cl}}{\underset{H_2C \diagdown N \diagup CH_2}{\overset{H_2C \diagup O \diagdown CH_2}{}}} \xrightarrow{NaNH_2}$$

$$\underset{H_2C-CH_2}{\overset{H_2C-CH_2}{\underset{\diagdown}{O}}}\underset{\diagdown}{N}{-}CH_2CH_2{-}\underset{\substack{| \\ CH \\ / \diagdown \\ CH_3 \ CH_3}}{\overset{Ph}{\underset{|}{C}}}{-}CN + NaCl + NH_3 \qquad (30)$$

To a solution of 83 gm (0.52 mole) of 1-isopropyl-1-phenylacetonitrile in 1 liter of dry benzene, 20.4 gm (0.52 mole) of sodamide is added in small portions. The mixture is refluxed for 2 hr with stirring and 78 gm (0.52 mole) of 2-(N-morpholino)-1-chloroethane is added dropwise after a 1-hr period. Refluxing is continued for 6 hr. After cooling to room temperature, the excess of sodamide is cautiously decomposed by adding 400 ml of water (enough water is used to dissolve all the sodium chloride present).

The benzene layer is separated, and, since the product is a tertiary amine, extracted with 1.5 liter of 10% hydrochloric acid. The acid extract is washed with 400 ml of ether and then made alkaline with 10% sodium hydroxide to a phenolphthalein end point. The product separates as an oil which is diluted with 1 liter of ether. The ether solution is washed with water until neutral and dried with sodium sulfate. Distillation of the extract yields a solid which may be recrystallized from ligroin (bp 75°–120°C). Yield 118°8 gm (84%), mp 75.5°–77.5°C.

If the alkylation is to be carried out with simple alkyl halides, the final work would, naturally, involve washing the benzene solution of the product with dilute acid, followed by washing with water until neutral, drying the solution, and distilling the benzene solution. Other modifications involve evaporating the reaction mixture to dryness, neutralizing, and then extracting the product with ether. The ether extract is then dried and distilled [107].

Both the aldol condensates and the alkylated products of ethyl cyanoacetate must frequently be converted to decarboxylated products. The alternative procedures are (1) hydrolysis of the ester function to the free acid followed by decarboxylation, which depends on the selective hydrolysis of an ester in the presence of a nitrile; or (2) pyrolysis of the ester, which appears to be a selective reaction in which the ester portion of the molecule is converted to an ethylene and carbon dioxide [108,109].

By phase-transfer alkylation, α-phenylbutyronitrile was prepared from phenylacetonitrile and ethyl bromide in the presence of sodium hydroxide and a quaternary ammonium salt [110] and 2-phenyl-2-vinylbutyronitrile was produced from 2-phenylbutyronitrile and acetylene [111].

A recent example of alkylation of doubly activated nitriles with the new reagent 3-bromo-2-methoxy-1-butene in the presence of sodium hydride in tetrahydrofuran solution has shown that this compound is a useful three-carbon annelation reagent [112].

Organoboranes which may be obtained by hydroboration reactions react readily with a variety of chloronitriles such as dichloroacetonitrile, isobutylchloroacetonitrile, cyclopentylchloroacetonitrile in the presence of potassium 2,6-di-*tert*-butylphenoxide to afford alkylated nitriles. The reaction is also satisfactory for dialkylation reactions, [Eqs. (31) and (32)] [113].

§ 3. Condensation Reactions

$$R_3B + Cl_2CHCN \longrightarrow R-\underset{H}{\overset{Cl}{\underset{|}{C}}}-CN \qquad (31)$$

$$R_3'B + R\overset{Cl}{\underset{|}{C}}H_3CN \longrightarrow R-\underset{R'}{\overset{H}{\underset{|}{C}}}-CN \qquad (32)$$

F. Cyanoethylation

In the presence of an alkaline catalyst, a large variety of reactive hydrogen compounds add to olefinic nitriles such as acrylonitrile or 2,4-dicyano-1-butene. The reaction with acrylonitrile is, in effect, a chain-lengthening process adding three carbons to a molecule.

In his review of cyanoethylation, Bruson [114] gives 10 classes of active hydrogen compounds which undergo this reaction, usually in the presence of a basic catalyst. The following equations give examples of the types of starting compounds which may be used and the types of products which may be formed:

1. $CH_2{=}CHCN + RNH_2 \longrightarrow RNHCH_2CH_2CN$ (33)

    $CH_2{=}CHCN + RNHCH_2CH_2CN \longrightarrow RN(CH_2CH_2CN)_2$ (34)

2. $CH_2{=}CHCN + ROH \longrightarrow ROCH_2CH_2CN$ (35)

    $2\ CH_2{=}CHCN + C_6H_5AsH_2 \longrightarrow C_6H_5As(CH_2CH_2CN)_2$ (36)

3. $CH_2{=}CHCN + HCl \longrightarrow ClCH_2CH_2CN$ (37)

4. $CH_2{=}CHCN + CHCl_3 \longrightarrow Cl_3CCH_2CH_2CN$ (38)

5. $2\ CH_2{=}CHCN + C_6H_5SO_2CH_2C_6H_5 \longrightarrow C_6H_5SO_2\underset{\underset{C_6H_5}{|}}{C}(CH_2CH_2CN)_2$ (39)

6. $CH_2{=}CHCN + CH_3NO_2 \longrightarrow CN{-}CH_2CH_2CH_2NO_2$ (40)

7. $CH_2{=}CHCN + R-\overset{O}{\overset{\|}{C}}-CH_3 \longrightarrow R-\overset{O}{\overset{\|}{C}}-C(CH_2CH_2CN)_3$ (41)

8. $CH_2{=}CHCN + CH_2\!\!\begin{array}{c}\diagup CO_2R \\ \diagdown CO_2R\end{array} \longrightarrow CHCH_2CH_2CN\!\!\begin{array}{c}\diagup CO_2R \\ \diagdown CO_2R\end{array}$ (42)

9. $CH_2{=}CHCN + \underset{}{\bigcirc}\!\!-CH_2CN \longrightarrow \underset{}{\bigcirc}\!\!-\underset{\underset{CN}{|}}{CH}-CH_2CH_2CN$ (43)

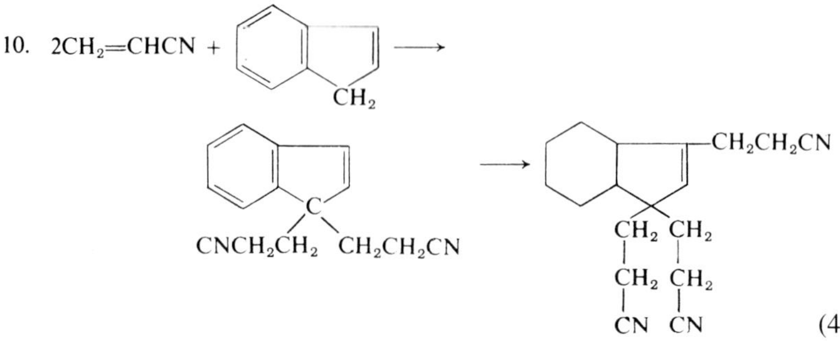

10. $2CH_2=CHCN +$ [indene] $\longrightarrow$

[structures showing cyanoethylated indene products] (44)

An example of the cyanoethylation reaction is the preparation of 1,1,1-*tris*-(2-cyanoethyl)acetone [115]. In this case, alcoholic potassium hydroxide is used as a catalyst, while benzyltrimethylammonium hydroxide is frequently used in other preparations.

3-10. Preparation of 1,1,1-Tris(2-cyanoethyl)acetone [115]

$$CH_3-\overset{O}{\underset{\|}{C}}-CH_3 + 3\ CH_2=CHCN \longrightarrow CH_3-\overset{O}{\underset{\|}{C}}-C(CH_2CH_2CN)_3 \quad (45)$$

To a stirred solution of 29 gm (0.5 mole) of acetone, 30 gm of *tert*-butyl alcohol, and 2.5 gm of 30% ethanolic potassium hydroxide solution maintained between 0° and 5°C, a solution of 80 gm (1.5 moles) of acrylonitrile in 37 gm of *tert*-butyl alcohol is added dropwise over a period of 1½ hr. Stirring is continued for 2 hr at 5°C. Then the product is filtered off. Yield 84 gm (79.5%), mp (after crystallization from water) 154°C.

An example of the cyanoethylation of an alcohol is given in Ref. [116]. Koelsch [117] cyanoethylated a diazonium salt:

$$ArN_2^+Cl^- + CH_2=CHCN \longrightarrow ArCH_2-\underset{\underset{Cl}{|}}{CH}-CN \quad (46)$$

An enamine was cyanoethylated in the preparation of *dl*-2-isopropyl-4-cyanobutyraldehyde [118].

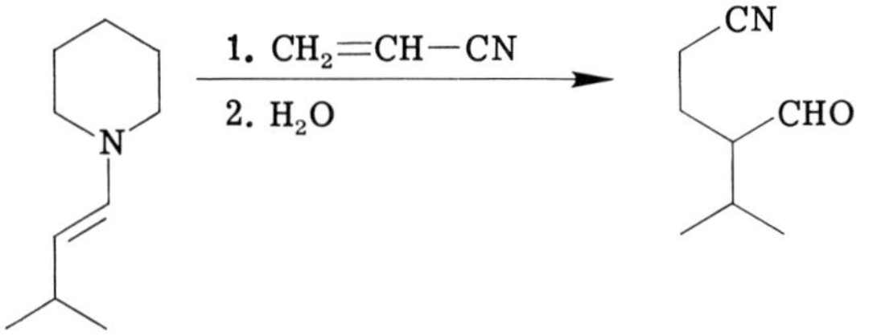

[enamine structure] $\xrightarrow{\text{1. } CH_2=CH-CN}_{\text{2. } H_2O}$ [product structure with CN and CHO] (47)

Trichlorosilane has also been cyanoethylated. The procedure is given in preparation 3-11 [119].

3-11. Preparation of 3-Trichlorosilylpropionitrile [119]

$$Cl-\underset{\underset{Cl}{|}}{\overset{\overset{Cl}{|}}{Si}}H \longrightarrow CH_2=CH-CN \xrightarrow{DMF} Cl-\underset{\underset{Cl}{|}}{\overset{\overset{Cl}{|}}{Si}}-CH_2CH_2CN \quad (48)$$

In a reflux apparatus protected from atmospheric moisture and equipped with an addition funnel 106 gm (2 moles) of acrylonitrile and 271 gm (2 moles) of trichlorosilane are refluxed with 7.2 gm (0.1 mole) of dimethylformamide for 24 hr. During this time the reaction temperature rose to 83°C. (*NOTE:* During an attempt to carry out the reaction in a sealed pressure bottle on a steam bath an explosion took place.)

Then an additional 135.5 gm (1 mole) of trichlorosilane is slowly added over a 48-hr period. The lower boiling materials are then distilled off. The residue is distilled under reduced pressure to give 3-trichlorosilylpropionitrile. Yield: 272 gm (72%), mp 32°–33°C, bp 78°–84°C at 7-8 mm.

4. OXIDATION REACTIONS

Another method of converting aldehydes to nitriles, which is particularly suitable in the aromatic series, involves the lead tetraacetate oxidation of aldimines which are formed *in situ* from the aldehydes during the reaction in the presence of ammonia [120].

The preparation of 4-nitrobenzonitrile is a typical example of the procedure.

4-1. Preparation of 4-Nitrobenzonitrile [120]

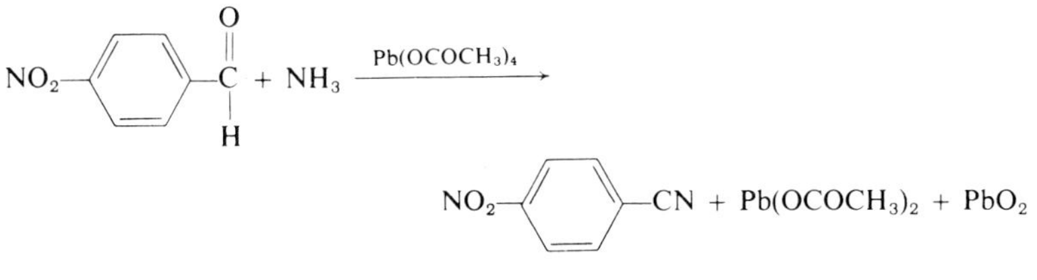

$$NO_2-C_6H_4-CHO + NH_3 \xrightarrow{Pb(OCOCH_3)_4} NO_2-C_6H_4-CN + Pb(OCOCH_3)_2 + PbO_2 \quad (49)$$

A solution of 2 millimoles of 4-nitrobenzaldehyde in carefully dried benzene is stirred under nitrogen, and dry ammonia gas is introduced at a rate of about two bubbles of ammonia per second. While the reaction is being cooled in an ice bath, 6 millimoles of lead tetraacetate in benzene is added in small portions over a 1-hr period. When the addition of the lead tetraacetate is completed, the addition of ammonia is also stopped. Stirring, however, is continued until the reaction is completed (approximately 16 hr). The reaction mixture is then diluted with ether and the precipitated mixed lead compounds are separated by filtration. The product solution is then treated in turn with 10% hydrochloric acid, water, and saturated sodium chloride solution. After drying the ether solution with sodium sulfate, the solvent is removed by distillation under reduced pressure, leaving a crude nitrile which may be purified by recrystallization. Yield: 81%, mp 136°–137°C.

The ammoxidation of aldehydes and olefins is well known. Large-scale industrial processes for the preparation of olefinic nitriles are based on the catalytic oxidation of an olefin in the presence of ammonia. Few processes have been published, and catalysts are subject to patent applications. The preparation of methacrylonitrile is illustrative of the process:

$$CH_3-\underset{\underset{CH_3}{|}}{C}=CH_2 + NH_3 + O_2 \xrightarrow[\text{Molybdenum oxide}]{\text{Heat}} CH_2=\underset{\underset{CH_3}{|}}{C}-CN \qquad (50)$$

Another patent, recommending a variety of other catalysts, involves reaction of propylene, ammonia, and oxygen in the presence of steam, and a bismuth/molybdenum catalyst with silica as a porous support (SOHIO process) [121–125].

N,N-Dimethylhydrazones of aldehydes have been oxidized with hydrogen peroxide [126].

Certain heavily arylated unsaturated hydrocarbons have been cyanated under oxidizing conditions in the presence of sodium cyanide in an aprotic solvent [127,128].

Oxidative dimerizations have been observed [Eq. (51)] [129].

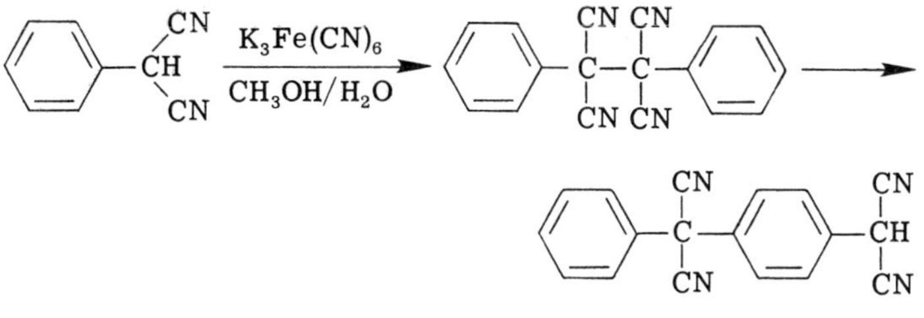

$$\text{Ph-CH(CN)}_2 \xrightarrow[\text{CH}_3\text{OH/H}_2\text{O}]{K_3\text{Fe(CN)}_6} \text{Ph-C(CN)}_2\text{-C(CN)}_2\text{-Ph} \longrightarrow \text{Ph-C(CN)}_2\text{-C}_6\text{H}_4\text{-CH(CN)}_2 \qquad (51)$$

5. REDUCTION REACTIONS

Unsaturated nitriles have been hydrogenated to saturated nitriles over palladium on charcoal [130]. Since the unsaturated nitriles are often prepared by cyanoacetic ester condensations, procedures combining the condensation and reduction have been devised [131]. Recently a reduction of α,β-unsaturated nitriles was published which gave high yields without reduction of the nitrile moiety. The reagent of choice was magnesium in methanol [132]. The method is said to be regioselective of a conjugated double bond in the presence of a nonconjugated double bond. Unfortunately, the published procedures are on a millimole scale. Procedures for handling fair quantities of material would be helpful.

6. MISCELLANEOUS REACTIONS

(1) By treating a carboxylic acid salt with an inorganic thiocyanate (Letts reaction) [133,134].

$$(RCO_2)_2Zn + Pb(SCN)_3 \longrightarrow 2\ RCN + PbS + ZnS + 2\ CO_2 \quad (52)$$

(2) Dehydrogenation of primary amines [135].

$$RCH_2NH_2 \xrightarrow{\text{heat}} RCN + 2H_2 \quad (53)$$

(3) Direct displacement of aminogroups of *para-* or *meta-*hydroxybenzylamines to phenolic acetonitriles with cyanide ions [136,137; see also 51].

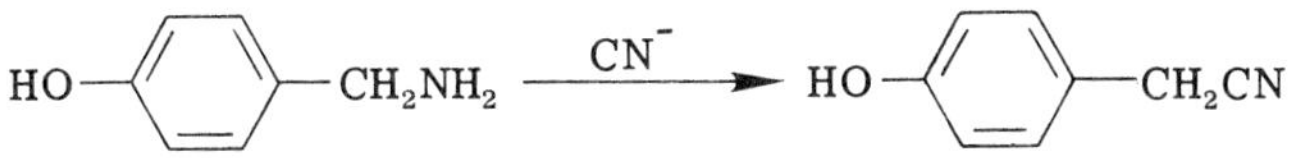

$$HO\text{-}C_6H_4\text{-}CH_2NH_2 \xrightarrow{CN^-} HO\text{-}C_6H_4\text{-}CH_2CN \quad (54)$$

(4) Decomposition of azomethines [138]. Aromatic or heterocyclic aldehydes and 4-amino-1,2,4-triazole from azomethines which may be decomposed to nitriles.

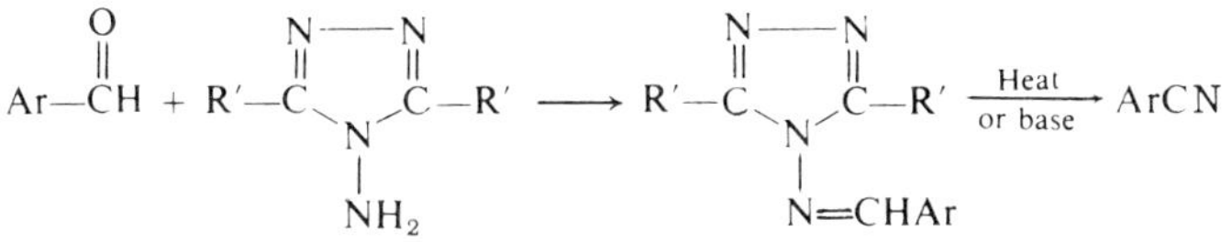

$$Ar\text{-}CHO + R'\text{-}C(=N\text{-}N=)C\text{-}R'(NH_2) \longrightarrow R'\text{-}C(=N\text{-}N=)C\text{-}R'(N=CHAr) \xrightarrow[\text{or base}]{\text{Heat}} ArCN \quad (55)$$

(5) Nitrile interchange [139]. Refluxing of adiponitrile with a fatty acid in the presence of *p*-toluenesulfonic acid is said to convert the fatty acid to the nitrile with the same number of carbon atoms. In a plasma discharge, acetonitrile and aromatic hydrocarbons form aromatic nitriles [140].

(6) From Grignard reagents [141].

$$RMgCl + ClCN \longrightarrow RCN + MgCl_2 \quad (56)$$

(7) From cyanogen chloride or bromide by Friedel–Crafts reaction [142].

$$ArH + ClCN \xrightarrow{AlCl_3} ArCN + HCl \quad (57)$$

(8) By halogen replacement of acyl bromides to form acyl cyanides [143,144].

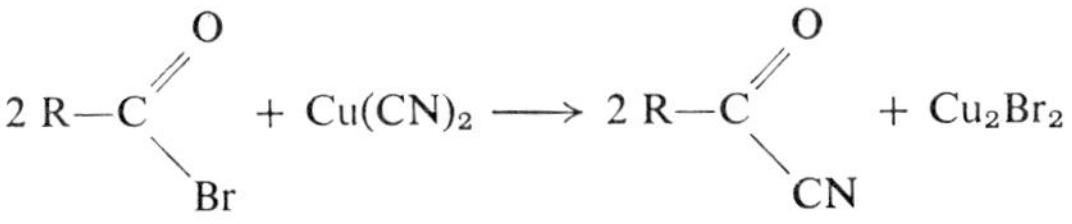

$$2\ R\text{-}C(=O)Br + Cu(CN)_2 \longrightarrow 2\ R\text{-}C(=O)CN + Cu_2Br_2 \quad (58)$$

(9) From Mannich bases [145].

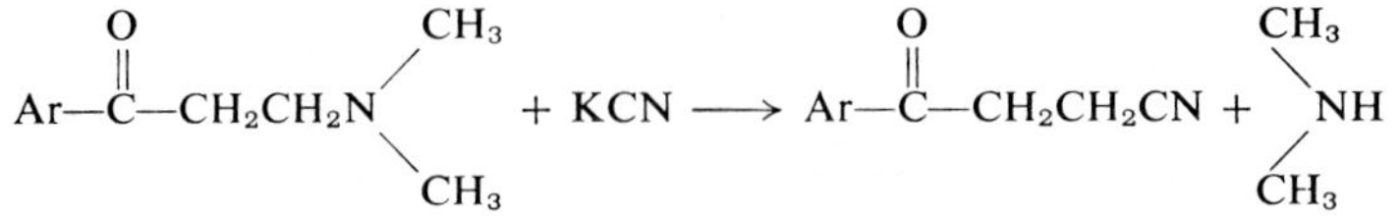

$$\text{Ar}-\overset{\text{O}}{\underset{\|}{\text{C}}}-\text{CH}_2\text{CH}_2\text{N}\begin{pmatrix}\text{CH}_3\\\text{CH}_3\end{pmatrix} + \text{KCN} \longrightarrow \text{Ar}-\overset{\text{O}}{\underset{\|}{\text{C}}}-\text{CH}_2\text{CH}_2\text{CN} + \begin{pmatrix}\text{CH}_3\\\text{CH}_3\end{pmatrix}\text{NH} \quad (59)$$

(10) Organoaminolysis of carbonyl compounds [146, 147].

(a)

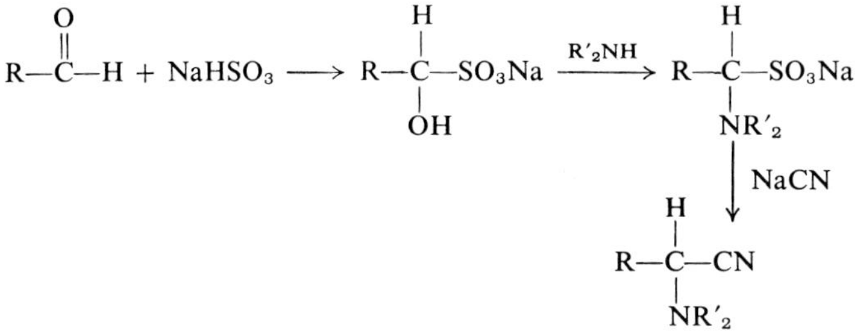

$$R-\overset{O}{\underset{\|}{C}}-H + NaHSO_3 \longrightarrow R-\overset{H}{\underset{OH}{\overset{|}{C}}}-SO_3Na \xrightarrow{R'_2NH} R-\overset{H}{\underset{NR'_2}{\overset{|}{C}}}-SO_3Na$$

$$\downarrow NaCN$$

$$R-\overset{H}{\underset{NR'_2}{\overset{|}{C}}}-CN \quad (60)$$

(b)

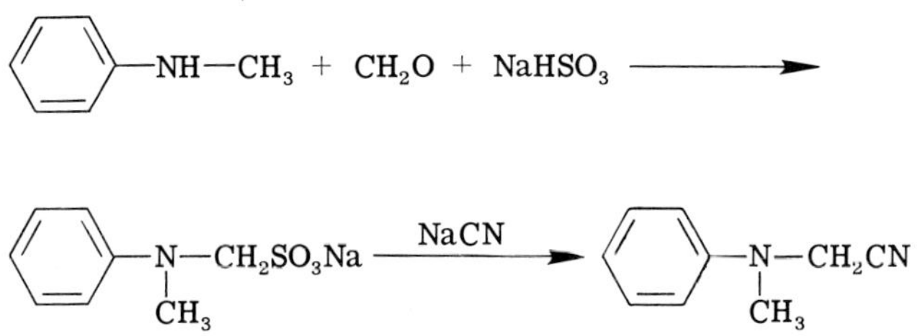

Ph—NH—CH₃ + CH₂O + NaHSO₃ ⟶

$$\text{Ph}-\underset{\text{CH}_3}{\overset{|}{\text{N}}}-\text{CH}_2\text{SO}_3\text{Na} \xrightarrow{\text{NaCN}} \text{Ph}-\underset{\text{CH}_3}{\overset{|}{\text{N}}}-\text{CH}_2\text{CN} \quad (61)$$

(10) Addition of hydrogen cyanide to epoxides [148,149].

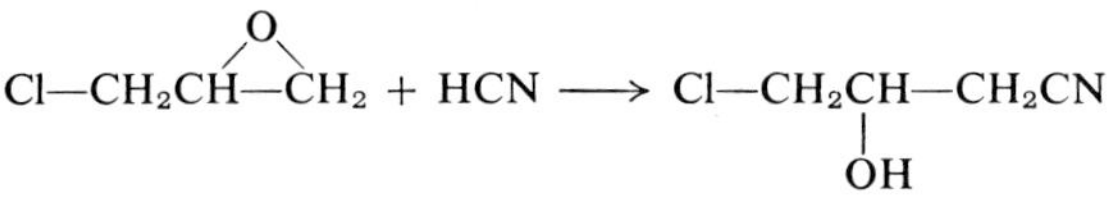

$$\text{Cl}-\text{CH}_2\overset{O}{\overset{\frown}{\text{CH}}}-\text{CH}_2 + \text{HCN} \longrightarrow \text{Cl}-\text{CH}_2\underset{\text{OH}}{\overset{|}{\text{CH}}}-\text{CH}_2\text{CN} \quad (62)$$

(11) Reaction of quaternary ammonium salts [150].

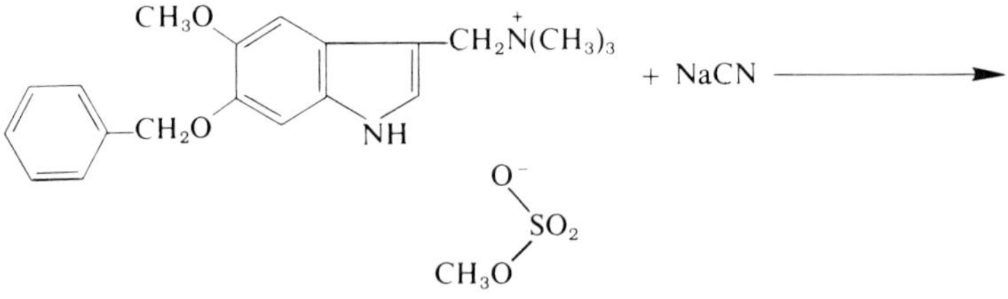

6-Benzyloxy-5-methoxygramine methosulfate + NaCN ⟶

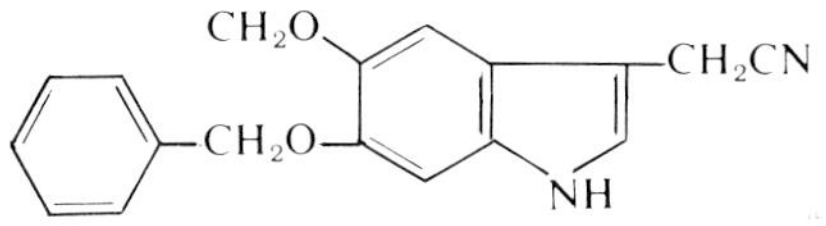

§ 6. Miscellaneous Reactions

$$\text{(structure: 5,6-bis(methoxymethyl)-2-phenylindole with } CH_2CN \text{ at 3-position)} \tag{63}$$

(12) Diels–Alder reaction with acrylonitrile [151].

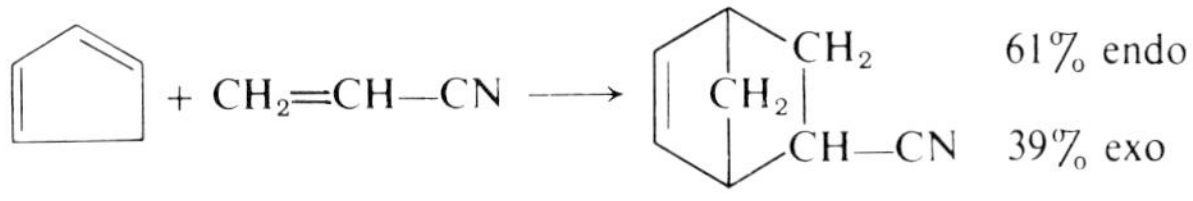

$$\text{cyclopentadiene} + CH_2=CH-CN \longrightarrow \text{norbornene-CN adduct} \quad \begin{array}{l} 61\% \text{ endo} \\ 39\% \text{ exo} \end{array} \tag{64}$$

(13) Ketonitrile formation from enamines and ClCN [152].

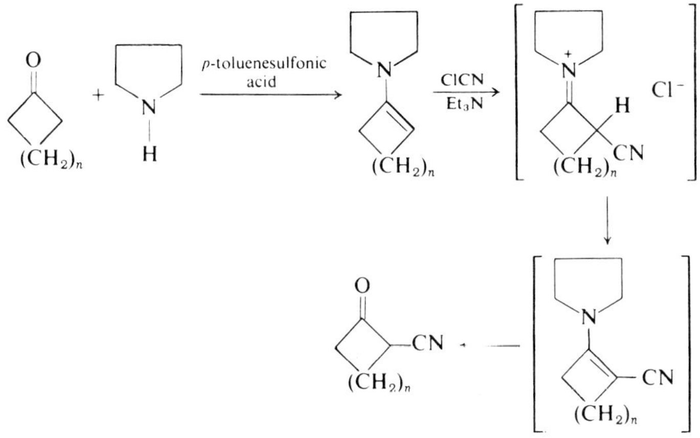

$$\text{cyclic ketone} + \text{pyrrolidine} \xrightarrow{p\text{-toluenesulfonic acid}} \text{enamine} \xrightarrow[\text{Et}_3N]{ClCN} \left[\text{iminium chloride intermediate} \right] \longrightarrow \text{ketonitrile} \leftarrow \left[\text{enamine-CN} \right] \tag{65}$$

(14) Pentenonitriles from N-(2-alkenyl)amides by rearrangement [153].

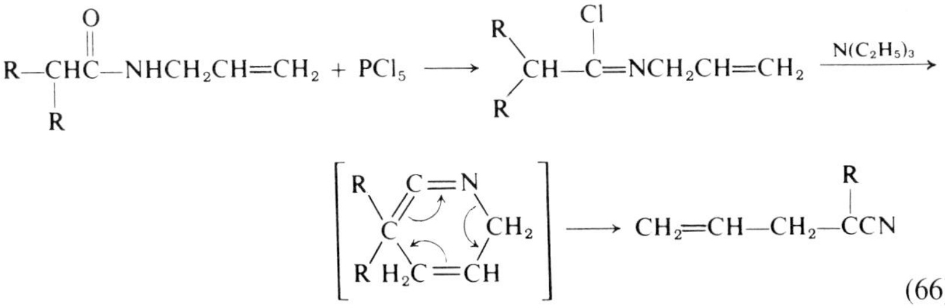

$$\underset{R}{R-CHC-NHCH_2CH=CH_2} + PCl_5 \longrightarrow \underset{R}{\overset{Cl}{\underset{|}{CH-C=NCH_2CH=CH_2}}} \xrightarrow{N(C_2H_5)_3}$$

$$\left[\text{cyclic intermediate} \right] \longrightarrow CH_2=CH-CH_2-\underset{R}{\overset{R}{\underset{|}{CCN}}} \tag{66}$$

(15) The chemistry of percyanocarbons had been of considerable synthetic interest during the 1960s. Tetracyanoethylene, for example, is a highly reactive reagent capable of undergoing Diels–Alder reactions at low temperatures in a few minutes; of reacting with active hydrogen compounds with loss of HCN; and of undergoing cyclizations with hydrogen sulfide. Refs. 154–161 are representative of the literature.

(16) Reissert Reaction [162].

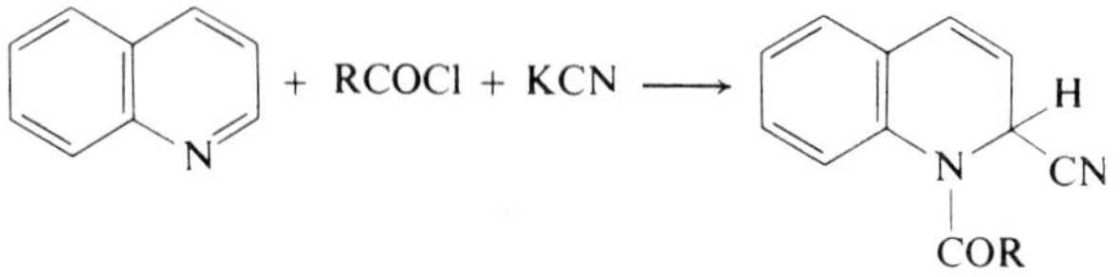

quinoline + RCOCl + KCN ⟶ 1,2-dihydroquinoline-2-carbonitrile with N—COR (67)

(17) Gas phase reaction of ammonia and alcohols [163,164].

$$C_4H_9OH + NH_3 \xrightarrow[320°C,\ 1\ atm.]{Cu/Al_2O_3} C_3H_7CN \qquad (68)$$

(18) Pyrolysis of trichloropropionitrile to provide chlorocyanoacetylene [165].
(19) Photochemical cyanation of acetophenone, by acetyl displacement [166].

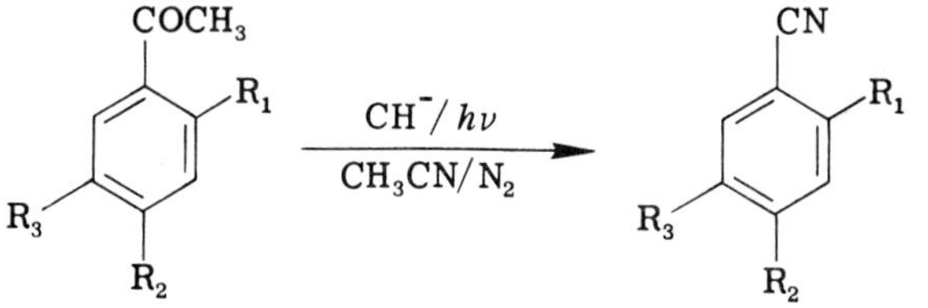

$$\text{Ar-COCH}_3 \xrightarrow[CH_3CN/N_2]{CH^-/h\nu} \text{Ar-CN} \qquad (69)$$

(with substituents R_1, R_2, R_3)

(20) Copper(II)-induced cleavage of *o*-benzoquinones to form mononitriles of muconic acids [167].

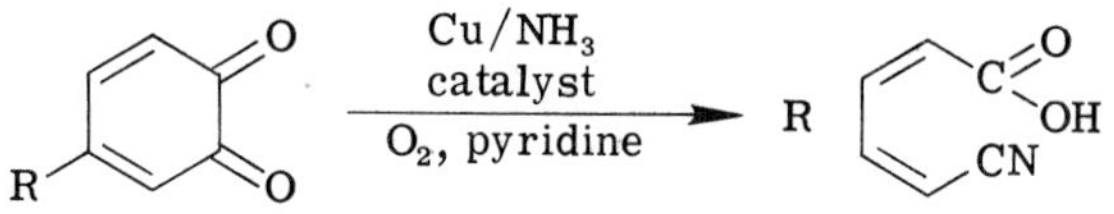

$$\text{R-o-benzoquinone} \xrightarrow[O_2,\ \text{pyridine}]{Cu/NH_3\ \text{catalyst}} \text{R-muconic acid mononitrile (—COOH, —CN)} \qquad (70)$$

(21) Preparation of aromatic nitrile oxides. Aromatic nitrile oxides have been prepared by dehydrogenation of oximate anions with alkali hypobromite [168].

$$R\text{—CH=NOH} \xrightarrow{NaOH} R\text{—CH=NO}^- \xrightarrow{OBr^-} R\text{—C}{\equiv}N{\to}O \qquad (71)$$

References

1. D. T. Mowry, *Chem. Rev.* **42**, 189 (1948).
2. Anonymous, "Cyanides in Organic Reactions, a Literature Review." Electrochem. Dept., Sodium Prod. Div., E. I. du Pont de Nemours and Co., Wilmington, Delaware (literature reviewed up to 1958).
3. Z. Rappoport, ed., "The Chemistry of the Cyano Group." Wiley (Interscience), New York, 1970.
4. L. Friedman and H. Schechter, *J. Org. Chem.* **25**, 877 (1960).
5. S. M. McElvain, R. L. Clarke, and G. D. Jones, *J. Am. Chem. Soc.* **64**, 1968 (1942).
6. F. C. Whitmore, C. I. Moll, and V. C. Meunier, *J. Am. Chem. Soc.* **61**, 683 (1939).
7. G. H. Jeffery and A. I. Vogel, *J. Chem. Soc.* p. 674 (1948).
8. C. H. Tilford, M. G. Van Campen, Jr., and R. S. Shelton, *J. Am. Chem. Soc.* **69**, 2902 (1947).
9. R. Adams and A. F. Thal, *Org. Synth. Collect. Vol.* **1**, 107 (1941).

10. J. A. Mitchell and E. E. Reid, *J. Am. Chem. Soc.* **53,** 321 (1931).
11. L. Bauer and J. Cymerman, *J. Chem. Soc.* p. 2078 (1950).
12. N. Campbell and J. E. McKail, *J. Chem. Soc.* p. 1251 (1948).
13. A. N. Kurtz, W. E. Billups, R. B. Greenlee, H. F. Hamil, and W. T. Pace, *J. Org. Chem.* **30,** 3141 (1965).
14. M. Pesson, S. Dupin, M. Antoine, D. Humbert, and M. Joannic, *Bull. Soc. Chim. Fr.* No. 8, p. 2262 (1965).
15. J. F. Norris and A. J. Klemka, *J. Am. Chem. Soc.* **62,** 1432 (1940).
16. E. A. Lawton and D. D. McRitchie, *J. Org. Chem.* **24,** 26 (1958).
17. V. V. Korshak, S. V. Vinogradova, and V. A. Pankratov, *Zh. Prikl. Khim. (Leningrad)* **45** (5), 1082 (1972); *Chem. Abstr.* **77,** 61483 (1972).
18. A. Albert, *J. Chem. Soc. C* p. 230 (1971).
19. Y. Kanaoka, T. Kuga, and K. Tanizawa, *Chem. Pharm. Bull.* **18,** 397 (1970).
20. E. Yamato and S. Sugasawa, *Tetrahedron Lett.*, p. 4384 (1970).
21. J. Blum and A. Fisher, *Tetrahedron Lett.*, p. 1963 (1970).
22. H. M. Relles and R. W. Schluenz, *J. Am. Chem. Soc.* **96,** 6469 (1974).
23. W. E. Dennis, *J. Org. Chem.* **35,** 3253 (1970).
24. Houben-Weyl, *in* "Methoden der organischen Chemie" (E. Mueller, ed.), 4th ed., Vol. VII, p. 325. Thieme, Stuttgart, 1952.
25. J. H. Hunt, *Chem. Ind. (London)* p. 1873 (1961).
26. T. Moriwake, *J. Med. Chem.* **9,** 163 (1966).
27. D. L. Clive, *Chem. Commun.* p. 1014 (1970).
28. M. M. Rogić, J. F. Van Peppen, K. P. Klein, and T. R. Demmin, *J. Org. Chem.* **39,** 3424 (1974).
29. J. H. Pomeroy and C. A. Craig, *J. Am. Chem. Soc.* **81,** 6340 (1959).
30. R. F. Smith and L. E. Walker, *J. Org. Chem.* **27,** 4372 (1962).
31. S. Cacchi, L. Caglioti, and G. Paolucci, *Chem. Ind. (London)* No. 5, p. 213 (1972).
32. W. Koehler, *Z. Chem.* **11**(9), 343 (1971).
33. M. Ohno, N. Naruse, S. Torimitsu, and I. Teresawa, *J. Am. Chem. Soc.* **88,** 3168 (1966).
34. O. H. Oldenziel, D. van Leusen, and A. M. van Leusen, *J. Org. Chem.* **42,** 3114 (1977).
35. R. F. Schmidt, *Ber. Dtsch. Chem. Ges.* **57,** 704 (1924).
36. H. Wolff, *Org. React.* **3,** 307 (1946).
37. T. Sasaki, S. Eguchi, and T. Toru, *J. Org. Chem.* **35,** 4109 (1970).
38. D. Knittel, H. Hemetsberger, R. Leipert, and H. Weidman, *Tetrahedron Lett.* p. 1459 (1970).
39. B. B. Jarvis and P. E. Nicholas, *J. Org. Chem.* **44,** 2951 (1979).
40. H. M. Blatter, H. Lukaszewski, and G. De Stevens, *J. Am. Chem. Soc.* **83,** 2203 (1961).
41. H. M. Blatter, H. Lukaszewski, and G. De Stevens, *Org. Synth. Collect. Vol.* **5,** 656 (1973).
42. L. Friedman and H. Schechter, *J. Org. Chem.* **26,** 2522 (1961).
43. R. A. Smiley and C. Arnold, *J. Org. Chem.* **25,** 257 (1960).
44. M. S. Newman and S. Otsuka, *J. Org. Chem.* **23,** 797 (1958).
45. M. Hirakura, M. Yanargita, and S. Inayama, *J. Org. Chem.* **26,** 3061 (1961).
46. A. J. Parker, *Q. Rev. Chem. Soc.* **16,** 163 (1962).
47. P. A. Argabright and D. W. Hall, *Chem. Ind. (London)* p. 1365 (1964).
48. Y. Oshiro, H. Tanisake, and S. Komori, *Yuki Gosei Kagaku Kyokaishi* **24,** 950 (1966).
49. J. J. Bloomfield and A. Mitra, *Chem. Ind. (London)* p. 2012 (1966).
50. D. Brett, I. M. Downie, and J. B. Lee, *J. Org. Chem.* **32,** 855 (1967).
51. M. A. Schwartz, M. Zoda, B. Vishnuvajjala, and I. Mami, *J. Org. Chem.* **41,** 2502 (1976).
52. J. Pelouze, *Justus Liebigs Ann. Chemie* **10,** 249 (1834).
53. V. C. Sekera and C. S. Marvel, *J. Am. Chem. Soc.* **55,** 345 (1933).
54. C. M. Starks, *J. Am. Chem. Soc.* **93,** 195 (1971).
55. J. M. J. Fréchet, M. D. de Smet, and M. I. Farrall, *J. Org. Chem.* **44,** 1774 (1979).

56. F. C. Cook, C. W. Bowers, and C. L. Liotta, *J. Org. Chem.* **39,** 3416 (1974).
57. J. E. Shaw, O. Y. Hsia, G. S. Parries, and T. K. Sawyer, *J. Org. Chem.* **43,** 1017 (1978).
58. S. L. Regen, *J. Am. Chem. Soc.* **97,** 5956 (1975).
59. S. L. Regen, J. C. Heh, and J. McLick, *J. Org. Chem.* **44,** 1961 (1979).
60. S. Quici and S. L. Regen, *J. Org. Chem.* **44,** 3436 (1979).
61. S. L. Regen, S. Quici, and S.-J. Liaw, *J. Org. Chem.* **44,** 2029 (1979).
62. M. Gordon and C. E. Griffin, *Chem. Ind. (London)* p. 1091 (1962).
63. Y. Urata, *Nippon Kagaku Zasshi* **83,** (10), 1105 (1962); *Chem. Abstr.* **59,** 11240c (1963).
64. M. S. Newman and H. Boden, *J. Org. Chem.* **26,** 1961 (1961).
65. K. W. Rosenmund and E. Struck, *Ber. Dtsch. Chem. Ges.* **52,** 1749 (1916).
66. J. von Braun and G. Manz, *Justus Liebigs Ann. Chem.* **488,** 111 (1931).
67. R. Calas and R. Lalande, *Bull. Soc. Chim. Fr.* p. 302 (1947).
68. J. E. Callen, L. A. Dornfeld, and G. H. Coleman, *Org. Synth. Collect. Vol.* **3,** 212 (1955).
69. M. S. Newman, *Org. Synth. Collect. Vol.* **3,** 631 (1955).
70. R. G. R. Bacon and H. A. O. Hill, *Q. Rev., Chem. Soc.* **19,** 121 (1965).
71. D. A. Herold and R. D. Rieke, *J. Org. Chem.* **44,** 1359 (1979).
72. H. O. House and W. F. Fischer, Jr., *J. Org. Chem.* **34,** 3626 (1969).
73. J. R. Dalton and S. L. Regen, *J. Org. Chem.* **44,** 4443 (1979).
74. E. C. Taylor, H. W. Altland, R. H. Danforth, G. McGillivray, and A. McKillop, *J. Am. Chem. Soc.* **92,** 3520 (1970).
75. S. Uemura, Y. Ikeda, and K. Ichikawa, *Tetrahedron* **28**(11), 3025 (1972).
76. T. Sandmeyer, *Ber. Dtsch. Chem. Ges.* **17,** 1633, 2654 (1884).
77. A. Korczyaski and B. Fandrich, *C. R Hebd. Seances Acad. Sci.*, **183,** 421 (1926).
78. H. H. Hodgson and F. Heyworth, *J. Chem. Soc.* p. 1131 (1949).
79a. J. A. McRae, *J. Am. Chem. Soc.*, **52,** 4550 (1930).
79b. F. Urano, K. Suzuki, and M. Sekiya, *Yuki Gosei Kagaku Kyokaishi* **30**(2), 154 (1972); *Chem. Abstr.* **76,** 112851 (1972).
80. L. Gattermann, *Ber. Dtsch. Chem. Ges.* **23,** 1218 (1890).
81. G. Körner and A. Contardi, *Atti Accad. Nazl. Lincei*, **23**(II), 464 (1914).
82. F. Urech, *Justus Liebigs Ann. Chem.*, **164,** 225 (1872); A. J. Ultee, *Recl. Trav. Chim. Pays-Bas* **28,** 1 (1909).
83. R. F. B. Cox and R. J. Stormont, *Org. Synth. Collect. Vol.* **2,** 7 (1943).
84. C. F. Allen and R. K. Kimball, *Org. Synth. Collect. Vol.* **2,** 498 (1943).
85. D. Lednicer, S. C. Lyster, B. D. Aspergren, and G. W. Duncan, *J. Med. Chem.* **9,** 172 (1966).
86. A. Kleeman and co-workers, *Angew. Chem., Int. Ed. Engl.* **19,** 627 (1980).
87. R. Steiger, *Org. Synth. Collect. Vol.* **3,** 6684 (1955).
88. F. Scotti and E. J. Frazza, *J. Org. Chem.* **29,** 1800 (1964).
89. W. Becker and E. Pfeil, *J. Am. Chem. Soc.* **88,** 4299 (1966).
90. W. Nagata, *Tetrahedron Lett.* p. 461 (1962).
91. W. Nagata, T. Okumura, and M. Yoshioka, *J. Chem. Soc. C* p. 2347 (1970).
92. W. Nagata, M. Yoshioka, and M. Murakami, *J. Chem. Soc. C* p. 2355 (1970).
93. W. Nagata, M. Yoshioka, and M. Murakami, *Org. Synth.* **52,** 96 (1972).
94. A. Lapworth, *J. Chem. Soc.* **83,** 1004 (1903).
95. E. Knoevenagel, *Ber. Dtsch. Chem. Ges.* **37,** 4065 (1904).
96. P. Kurtz, *Justus Liebigs Ann. Chem.* **572,** 28 (1951); H. F. Piepenbrink, *ibid.* p. 83.
97. Y. Odaira, T. Oishi, T. Yukawa, and S. Tsutsumi, *J. Am. Chem. Soc.* **88,** 4105 (1966).
98. C. Runti and L. Sindellari, *Boll. Chim. Farm.* **99,** 499 (1960); *Chem. Abstr.* **55,** 10468c (1961).
99. J. N. Walker, *J. Med. Chem.* **8,** 583 (1965).
100. C. F. H. Allen and G. P. Happ, *Can. J. Chem.* **42,** 641 (1964).

§ References 583

101. B. D. Halpern, J. Dickstein, and R.-M. Hoegerle, U.S. Patent 3,142,698 (1964).
102. M. E. Ardis, S. T. Averill, H. Gilbert, F. F. Miller, R. F. Smith, F. D. Stewart, and H. L. Trumbull, *J. Am. Chem. Soc.* **72,** 1385 (1950).
103. R. Robin, *Bull. Soc. Chim. Fr.* No. 8, p. 2275 (1965).
104. A. C. Cope, H. L. Holmes, and H. O. House, *Org. React.* **9,** 107 (1957).
105. C. R. Hauser, F. W. Swamer, and J. T. Adams, *Org. React.* **8,** 121 (1954).
106. S. Casadio, G. Pala, E. Crescenzi, T. Bruzzese, E. Marazzi-Uberti, and G. Coppi, *J. Med. Chem.* **8,** 589 (1965).
107. J.-P. Fleury and A. Bader, *Bull. Soc. Chim. Fr.* p. 951 (1965).
108. E. J. Corey, *J. Am. Chem. Soc.* **74,** 5897 (1952); **75,** 1163 (1953).
109. W. J. Bailey and J. J. Daly, Jr., *J. Am. Chem. Soc.* **81,** 5397 (1959).
110. M. Makosza and A. Jończyk, *Org. Synth.* **55,** 91 (1975).
111. M. Makosza, J. Czyzewski, and M. Jawdosiuk, *Org. Synth.* **55,** 99 (1975).
112. R. M. Jacobson, A. Abbaspour, and G. P. Lahm, *J. Org. Chem.* **43,** 4650 (1978).
113. H. Nambu and H. C. Brown, *J. Am. Chem. Soc.* **92,** 5790 (1970).
114. H. A. Bruson, *Org. React.* **5,** 79 (1949).
115. H. A. Bruson and T. W. Riener, *J. Am. Chem. Soc.* **64,** 2850 (1942).
116. B. A. Feit, J. Sinnreich, and A. Zilkha, *J. Org. Chem.* **28,** 3245 (1963).
117. C. F. Koelsch, *J. Am. Chem. Soc.* **65,** 57 (1943).
118. R. R. Johnson and J. A. Nicholson, *J. Org. Chem.* **30,** 2918 (1965).
119. J. G. Saam and J. L. Speir, *J. Org. Chem.* **24,** 427 (1959).
120. K. N. Parameswaran and O. M. Friedman, *Chem. Ind. (London)* p. 988 (1965).
121. W. F. Brill and J. H. Finley, *Ind. Eng. Chem., Prod. Res. Dev.* **3,** No. 2, 89 (1964).
122. R. Schonbeck, I. H. Konig, K. Krzemicki, and I. Kahofer, *Angew. Chem., Int. Ed. Engl.* **5,** No. 7, 642 (1966).
123. J. N. Cosby, U.S. Patent 2,481,826 (1949).
124. W. Brackman and P. J. Smit, *Recl. Trav. Chim. Pays-Bas* **82**(8), 757 (1963).
125. H. R. Sheely, Ger. Offen. 2,124,472 (1971).
126. R. F. Smith, J. A. Albright, and A. M. Waring, *J. Org. Chem.* **31,** 4100 (1966).
127. K. E. Witaker, B. E. Galbraith, and H. R. Snyder, *J. Org. Chem.* **34,** 1411 (1969).
128. R. B. Chapas, R. F. Nystrom, and H. R. Snyder, *J. Org. Chem.* **37,** 314 (1972).
129. H. D. Hartzler, *J. Org. Chem.* **31,** 2654 (1966).
130. A. C. Cope, C. M. Hoffman, C. Wyckoff, and E. Hardenbergh, *J. Am. Chem. Soc.* **63,** 3452 (1941).
131. A. C. Cope and E. R. Alexander, *J. Am. Chem. Soc.* **64,** 886 (1944).
132. J. A. Profitt, D. S. Watt, and E. J. Corey, *J. Org. Chem.* **40,** 127 (1975).
133. L. A. Letts, *Ber. Dtsch. Chem. Ges.* **5,** 669 (1872).
134. G. D. von Epps and E. E. Reid, *J. Am. Chem. Soc.* **38,** 2120 (1916).
135. L. M. Peters, K. E. Marple, T. W. Evans, S. H. McAllister, and R. C. Castner, *Ind. Eng. Chem.* **40,** 2046 (1948).
136. J. H. Short, D. A. Dannigan, and C. W. Ours, *Tetrahedron* **29,** 1931 (1973).
137. M. A. Schwartz and S. W. Scott, *J. Org. Chem.* **36,** 1827 (1971).
138. H. G. O. Becker and H. J. Timpe, *Z. Chem.* **4**(8), 304 (1964).
139. P. S. Pyryalova and E. N. Zil'berman, *Tr. Khim. Khim. Tekhnol.* No. 2, p. 353 (1963).
140. L. L. Miller and A. B. Szabo, *J. Org. Chem.* **44,** 1670 (1979).
141. V. Grignard and E. Courtot, *Bull. Soc. Chim. Fr.* **17,** No. 4, 228 (1915).
142. P. Karrer and E. Zeller, *Helv. Chim. Acta* **2,** 482 (1919); P. Karrer, A. Rebmann, and E. Zeller, *ibid.* **3,** 261 (1920).
143. C. D. Hurd, O. E. Edwards, and J. R. Road, *J. Am. Chem. Soc.* **66,** 2013 (1944).
144. V. F. Raaen, *J. Org. Chem.* **31,** 3310 (1966).

145. C. B. Knott, *J. Chem. Soc.* p. 1190 (1947).
146. C. F. H. Allen and J. A. Van Allen, *Org. Synth. Collect. Vol.* **3,** 275 (1955).
147. S. Hayao, H. J. Havera, W. G. Stryker, T. J. Leipzig, R. A. Kulp, and H. E. Hartzler, *J. Med. Chem.* **8,** 807 (1965).
148. R. Rambaud, *Bull. Soc. Chim. Fr.* [5] **3,** 138 (1936).
149. R. Legrand, *Bull. Acad. R. Belg.* [5] **29,** 256 (1943).
150. R. G. Tobarsky, P. Delvigs, and I. H. Page, *J. Med. Chem.* **8,** 855 (1965).
151. P. Wilder and D. B.Knight, *J. Org. Chem.* **30,** 3078 (1965).
152. M. E. Kuehne, *J. Am. Chem. Soc.* **81,** 5400 (1959).
153. K. C. Brannoch and R. O. Burpitt, *J. Org. Chem.* **30,** 2564 (1965).
154. Anonymous, *Chem. Eng. News* **35,** p. 28 (May 20, 1957).
155. B. C. McKusick and G. F. Biehn, *Chem. Eng. News* **38,** 114 (April 11, 1960).
156. B. C. McKusick, R. E. Heckert, T. L. Cairns, D. D. Coffman, and H. F. Mower, *J. Am. Chem. Soc.* **80,** 2806 (1958).
157. B. C. McKusick and L. A. Melby, *Org. Synth. Collect. Vol.* **4,** 953 (1963).
158. T. L. Cairns and B. C. McKusick, *Angew. Chem.* **73,** 520 (1961).
159. W. J. Linn, O. W. Webster, and R. E. Benson, *J. Am. Chem. Soc.* **87,** 3651, 3657, 3665 (1965).
160. O. W. Webster, M. Brown, and R. E. Benson, *J. Org. Chem.* **30,** 3223 (1965).
161. D. N. Dhar, *Chem. Rev.* **67,** 611 (1967).
162. W. E. McEwen and R. N. Hazlett, *J. Am. Chem. Soc.* **71,** 1949 (1949).
163. M. A. Popov and N. I. Shnikin, *Izv. Akad. Nauk SSSR Otd. Khim. Nauk* p. 1855 (1961).
164. R. J. Card and J. L. Schmitt, *J. Org. Chem.* **46,** 754 (1981).
165. N. Hashimoto, K. Matsumura, T. Saraie, Y. Kawano, and K. Morita, *J. Org. Chem.* **35,** 675, 828 (1970).
166. A. L. Colb, *J. Am. Chem. Soc.* **101,** 3416 (1979).
167. T. R. Demmin and M. M. Rogić, *J. Org. Chem.* **45,** 2737 (1980).
168. G. Grundmann and J. M. Dean, *J. Org. Chem.* **30,** 2809 (1965).

CHAPTER 18 / Mercaptans, Sulfides, and Disulfides

1. Introduction	586
2. Mercaptans (Thiols)	587
A. Condensation Reactions	587
a. Reaction of Metal Sulfides or Hydrosulfides with Alkyl Halides	587
2-1. Preparation of p-*Aminothiophenol*	587
2-2. Preparation of Triphenylmethyl Mercaptan	587
b. Reaction of Thiourea with Active Halides	588
2-3. General Procedure for Preparing Alkanethiols That Boil Below 130°C (760 mm)	589
2-4. General Procedure for the Preparation of Monothiols and Dithiols That Boil Above 130°C (760 mm)	589
2-5. Preparation of n-*Octyl Mercaptan*	590
c. The Reaction of Organometallics with Sulfur	591
d. Hydrolysis of Xanthates	591
2-6. Preparation of 3-Bromothiophenol	591
e. Addition of Hydrogen Sulfide to Olefins and Other Unsaturated Compounds	592
2-7. Preparation of n-*Butylmercaptan*	593
f. Miscellaneous Methods for Preparing Mercaptans	593
*2-8. Preparation of 1,3-Di-*p-*chlorophenylpropane-2,2-dithiol*	593
3. Sulfides	594
A. Condensation Reactions	594
a. Reaction of Sodium or Other Metal Mercaptides with Active Alkyl or Aryl Halides	594
3-1. Preparation of α-*Methylallyl Methyl Sulfide*	594
3-2. Preparation of α-*Methylallyl Phenyl Sulfide*	595
b. The Reaction of Metallic Sulfides with Halides	595
3-3. Preparation of Dibenzyl Sulfide	595
c. Mercaptylation of the Double Bond	596
3-4. Preparation of tert-*Butyl Sulfide*	596
3-5. Preparation of β-tert-Butylmercaptopropionitrile	596
d. Preparation of Episulfides (Thiiranes) from Epoxides	596
3-6. Preparation of 3-Chloropropylene Sulfide (Epichlorosulfide)	597
3-7. Preparation of Isobutylene Sulfide	597
e. Miscellaneous Methods for Preparing Sulfides	597
4. Disulfides	597
A. Condensation Reactions	597
*4-1. Preparation of Di-*n-*amyl Disulfide*	598
B. Oxidation	598
4-2. Oxidation of Benzenethiol (Thiophenol) to Diphenyl Disulfide Using Dimethyl Sulfoxide	598
C. Miscellaneous Methods for Preparing Disulfides	599
*4-3. Preparation of Di-*n-*butyl Disulfide*	599
References	599

18. Mercaptans, Sulfides, and Disulfides

1. INTRODUCTION

The reaction of sodium hydrosulfide with active aliphatic or aromatic halides yields mercaptans. If aromatic nitro substituents are present, they are reduced to amino groups.

$$RX + NaSH \longrightarrow RSH + NaX \qquad (1)$$

Dihalides yield dithiols. Tertiary aliphatic halides usually give olefins.

Two other methods which have found general applicability in the laboratory are the reaction of alkyl halides with thiourea with subsequent alkaline hydrolysis, and the reaction of free sulfur with aryl lithium or Grignard reagents as shown below.

$$(NH_2)_2CS + RX \longrightarrow NH_2-C(SR)=NH_2^+ + X^- \xrightarrow{NaOH} RSH + (NH_2CN)_x \qquad (2)$$

$$ArM + S \longrightarrow ArSM \xrightarrow{H_2O} ArSH + MOH \qquad (3)$$

$$M = MgX \text{ or } Li$$

A widely used laboratory method for the preparation of sulfides is the action of halides on metallic sulfides.

$$Na_2S + 2 RX \longrightarrow R_2S + 2 NaX \qquad (4)$$

The monohydrate of sodium sulfide is a satisfactory reagent to afford high yields of symmetrical sulfides.

The conversion of mercaptans to sulfides is effected by converting to the sodium mercaptide and reacting with active halides or dialkyl sulfates.

$$RSH \longrightarrow RSNa \xrightarrow[\text{or } R_2SO_4]{R'X} RS-R' \qquad (5)$$

Polar solvents such as amides favor this reaction.

The addition of hydrogen sulfide or mercaptans to olefins is another important method for the preparation of mercaptans and sulfides via ionic (6a) or free radical means (6b).

$$RCH=CH_2 + R'SH \xrightarrow{H^+} RCH-CH_3$$
$$\qquad\qquad\qquad\qquad\qquad\quad |$$
$$\qquad\qquad\qquad\qquad\quad\;\; SR'$$
$$\xrightarrow[\text{or } h\nu]{H_2O_2} RCH_2CH_2SR' \qquad (6)$$

$$R = H, \text{ alkyl, or aryl group}$$

The reaction of active aryl halides with sodium sulfide yields disulfides.

$$2 RX + Na_2S_2 \longrightarrow RSSR + 2 NaX \qquad (7)$$

§ 2. Mercaptans (Thiols)

The oxidation of mercaptans by hydrogen peroxide is the best method of preparing disulfides in good yields if the corresponding mercaptan is readily available. Recently it has been reported that dimethyl sulfoxide oxidizes thiols at 80°–90°C to give disulfides in good yields. Oxidation by free sulfur has also been reported.

2. MERCAPTANS (THIOLS)

A. Condensation Reactions

a. REACTION OF METAL SULFIDES OR HYDROSULFIDES WITH ALKYL HALIDES

Sodium of potassium hydrosulfide [1–4a] react with active halides to give mercaptans. Alkyl sulfates and primary or secondary halides act as alkylating agents. Since hydrosulfides are reducing agents some structural features of the alkylating agent may be reduced. Thus, for example, the nitro group in *p*-chloronitrobenzene is reduced to give *p*-amino thiophenol [4b]. Sodium sulfide reacts with 1 mole of an activated halide to give the sodium salt of the thiol. The addition of acid liberates the free thiol. In addition hydrogen sulfide may be capable of reacting with an intermediate carbonium ion to give a thiol.

Recently it has been reported that alkyl thiols and silyl-substituted alkyl thiols have been synthesized from the corresponding alkyl halides using a combination of hydrogen sulfide and ammonia or alkylamines [5].

2-1. Preparation of p-Aminothiophenol [5]

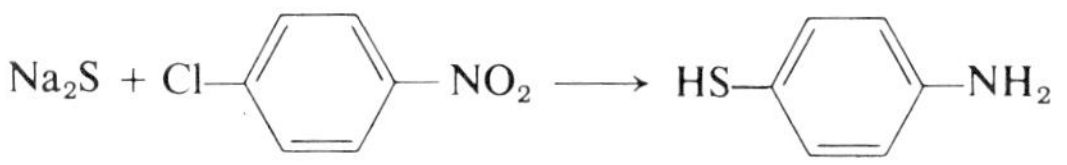

$$Na_2S + Cl{-}\!\!\bigcirc\!\!{-}NO_2 \longrightarrow HS{-}\!\!\bigcirc\!\!{-}NH_2 \qquad (8)$$

To a flask containing 480 gm (2 moles) of sodium sulfide monohydrate dissolved in 2 liters of water is added 128 gm (0.81 mole) of *p*-chloronitrobenzene and the mixture is refluxed for 8 hr. A small amount of an orange colored oil separates which is ether-extracted and discarded. The remaining aqueous layer is saturated with sodium chloride and 240 gm (4 moles) of glacial acetic acid is added. The liberated oil is extracted several times with ether and the ether extract is dried. Evaporation of the solvent leaves a residue which upon distillation under reduced pressure yields 70 gm (69%) of product, bp 143°–146°C (17 mm), mp 43°–45°C (literature 46°C) [6].

2-2. Preparation of Triphenylmethyl Mercaptan [1]

$$(C_6H_5)_3{-}C{-}Cl + H_2S \xrightarrow{Al_2O_3} (C_6H_5)_3{-}C{-}SH + HCl \qquad (9)$$

To 400 ml of dry dioxane (dried by refluxing with 4% of its weight of sodium for 4 hr) are added 100 gm of activated alumina (Alcoa F-20) and 100 gm of triphenylmethyl chloride. Dry hydrogen sulfide is then passed into the mixture below the level of the suspended alumina at such a rate as to agitate the solution and to keep the solution saturated. After 15 hr the alumina is filtered and washed with two 50-ml portions of dioxane. The filtrate and dioxane washings are poured into 2 liters of ice water and the contents are stirred until a granular product precipitates. The product is filtered, and then disolved in 500 ml of boiling isopropanol. Slow cooling yields 75–80 gm (75–80%) of pale yellow crystals of triphenyl mercaptan, mp 106°–107°C. The mother liquor yields no appreciable amount of product but a small amount of triphenylcarbinol is isolated.

b. Reaction of Thiourea with Active Halides

Thiourea reacts with active halides such as primary [7,8], secondary [9], tertiary [10], allyl [11], and benzyl [7,8] halides to give S-thiouronium salts. Hydrolysis with base yields the mercaptan in good yields and this method is excellent for laboratory scale preparations [12]. Isolation usually involves either a steam distillation or an ether extraction and sometimes both. An improved method [13a] involves (1) preparation of the isothiouronium salt in a high-boiling solvent, followed by (2) cleavage of the salt with a high-boiling amine. The thiol is directly distilled leaving the guanidine salt as a residue. This method is not applicable for the preparation of 1,2-ethanedithiol.

$$RX + H_2NCS-NH_2 \longrightarrow RS-\overset{\overset{NHHX}{\|}}{C}-NH_2 \xrightarrow{R'_2NH} RSH + R'_2N-\overset{\overset{NHHX}{\|}}{C}-NH_2 \quad (10)$$

The solvent and amine should be carefully chosen so that their boiling points are above that of the desired thiol. Triethylene glycol and tetraethylenepentamine

TABLE I[a]

Thiols RSH

R	Bp (°C)	n_D^{25}	Purity (%)	Yield (%)
C_2H_5-	35°	1.4269	99.7	68
$n-C_3H_7-$	65°	1.4345	98.5	79
$n-C_4H_9-$	96°–97°	1.4407	99.5	77
$n-C_5H_{11}-$	123°–124°	1.4439	100.0	75
$n-C_6H_{17}-$	91°–93°/24 mm	1.4518	97.7	84
$n-C_{10}H_{21}-$	94°/5 mm	1.4545	100.0	87

[a]Reprinted from B. C. Cossar, J. O. Fournier, D. L. Fields, and D. D. Reynolds, *J. Org. Chem.* **27,** 93 (1962). Copyright 1962 by the American Chemical Society. Reprinted by permission of the copyright owner.

TABLE II[a]

DITHIOLS HS(CH$_2$)$_n$SH

n	Bp (°C)	n_D^{25}	Purity (%)	Yield (%)
2	—	—	—	0
3	69°–69°/18 mm	1.5374	98.3	58
4	75°/18 mm	1.5280	97.7	78
5	103°/20 mm	1.5174	99.4	80

[a]Reprinted from B. C. Cossar, J. O. Fournier, D. L. Fields, and D. D. Reynolds, *J. Org. Chem.* **27**, 93 (1962). Copyright 1962 by the American Chemical Society. Reprinted by permission of the copyright owner.

are readily available at low cost and have been used [13a]. The time required for a 1-mole preparation is approximately 1 hr. Some of the products prepared by this method are summarized in Tables I and II.

2-3. *General Procedure for Preparing Alkanethiols That Boil Below 130°C (760 mm)* [13a]

Practical-grade triethylene glycol, practical-grade thiourea, Eastman White Label grade alkyl bromides, this and the technical-grade tetraethylenepentamine have been successfully used in the following procedures. The glycol and tetraethylenepentamine were heated under vacuum to remove materials which boiled below 150°C at 1.0 mm.

A mixture of 125 ml of triethylene glycol and 83.6 gm (1.1 moles) of thiourea is stirred in a 1-liter flask equipped with a magnetic stirrer, a thermometer, a dropping funnel, and a 14-inch, glass-helices-packed column topped with a variable reflux ratio still-head. The pot temperature is raised to 75°C and 1 mole of alkyl bromide is added through the dropping funnel. The reaction temperature is kept below 130°C. External cooling is applied when necessary. After the reaction mixture becomes homogeneous, the reaction is allowed to proceed for an additional 15 min, then 94.6 gm (1 mole) of tetraethylenepentamine is added via the dropping funnel. The resulting exothermic reaction causes the lower-boiling thiols to reflux. Heat is applied to the reaction flask and total reflux continued until the head temperature becomes constant. Distillation is then begun and the thiol collected.

2-4. *General Procedure for the Preparation of Monothiols and Dithiols That Boil Above 130°C (760 mm)* [13a]

In the apparatus described above, with provision for distillation under reduced pressure, a mixture of 250 ml of triethylene glycol and 167.4 gm (2.2 moles) of thiourea is stirred at 75°C. One mole of dibromoalkane (or 2 moles of alkyl bromide for the preparation of monothiols) is added. The mixture is stirred until it becomes homogeneous while the temperature is kept below 130°C. After an additional 15 min of stirring, a vacuum pump is attached to the system and the pressure of the flask reduced. Tetramethylenepentamine (180.3 gm; 1 mole) is added cautiously at a rate which prevents foaming or too rapid a reaction. Heat is applied and total reflux is continued until the head temperature becomes constant. Distillation is then begun and the product collected.

Procedures 2-3 and 2-4 are reprinted from B. C. Cossar, J. O. Fournier, D. L. Fields, and D. D. Reynolds, *J. Org. Chem.* **27**, 93 (1962). Copyright 1962 by the American Chemical Society. Reprinted by permission of the copyright owner.

Another interesting and synthetically useful modification involves the direct reaction of alcohols with thiourea in the presence of hydrobromic acid to give the *s*-isothiouronium group [8].

$$ROH + H_2NCSNH_2 + HBr \longrightarrow \left[H_2N-\overset{SR}{\underset{|}{C}}-NH_2 \right]^+ X^- \xrightarrow{NaOH} RSH \quad (11)$$

The advantage of this method is that one does not first have to convert the alcohols to the bromides as is true in the earlier methods. In fact, one may be able to substitute the sodium hydroxide with a high-boiling amine and the water with an organic solvent as described earlier [13a].

2-5. *Preparation of* n-*Octylmercaptan* [8]

$$CH_3(CH_2)_6-CH_2OH + H_2NCSNH_2 + HBr \longrightarrow$$

$$\left[H_2N-\overset{S-(CH_2)_7CH_3}{\underset{|}{C}}-NH_2 \right]^+ Br^- \xrightarrow{NaOH} CH_3(CH_2)_6CH_2SH \quad (12)$$

To a 1-liter three-necked flask equipped with a reflux condenser and stirrer are added 66 gm (0.5 mole) of *n*-octyl alcohol, 38 gm (0.5 mole) of thiourea, and 253 gm (1.5 mole) of hydrogen bromide as a 48% solution. The mixture is refluxed for 9 hr with stirring. A sodium hydroxide solution (60 gm/600 ml water) is added, nitrogen is passed over the liquid, and the mixture is refluxed for 2 hr without stirring. The organic layer is separated and the aqueous layer is acidified and extracted with three 50-ml portions of ether. The combined organic layer and ether extracts are dried over Drierite (CaSO$_4$) and fractionally distilled through a 1-ft helix-packed column to yield 54 gm (73%) of *n*-octylmercaptan, bp 86°C (15 mm), n_D^{20} 1.4540.

Under similar conditions, when hydrochloric acid is used, less than 5% of the mercaptan is isolated. Using sulfuric acid instead gave approximately 5% of the mercaptan.

A separate experiment in which the thiourea is left out of the reaction mixture indicates that *n*-octyl alcohol is converted to *n*-octyl bromide in the same period of time in 82% yield.

Yields of mercaptan by this method are as good as those obtained by starting with the bromides. The yields are best for primary and poorest for tertiary alcohols due to the latters' tendency to form olefins. The mechanism of this reaction is still not known. It may even be possible to carry this reaction out with olefins.

Recently it has been reported that *S*-methiodide derivatives of activated thioureas react with alcohols to give mercaptans in good yields (for nonsterically hindered alcohols [13b].

c. THE REACTION OF ORGANOMETALLICS WITH SULFUR

Phenyllithium [14, 15] and Grignard reagents [16,17] react with sulfur to give thiophenols or mercaptans.

d. HYDROLYSIS OF XANTHATES

Diazotization of aromatic amino groups and the reaction then with potassium ethyl xanthate, give a substituted xanthate which can be hydrolyzed to the thiophenol [18,19]. Reaction of carbon disulfide with alcoholic potassium hydroxide yields potassium ethyl xanthate [20].

$$ArN_2{}^+X + C_2H_5OCS_2K \longrightarrow C_2H_5OCS_2Ar \xrightarrow[H^+]{KOH} ArSH + COS + C_2H_5OH \quad (13)$$

2-6. *Preparation of 3-Bromothiophenol* [21]

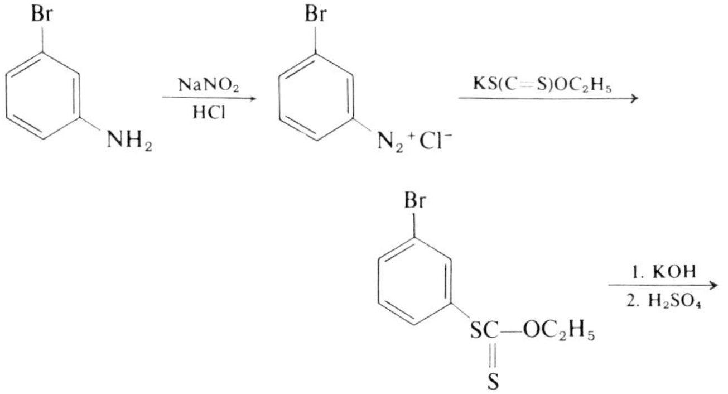

$$\text{3-Br-C}_6\text{H}_4\text{-NH}_2 \xrightarrow[\text{HCl}]{\text{NaNO}_2} \text{3-Br-C}_6\text{H}_4\text{-N}_2{}^+\text{Cl}^- \xrightarrow{KS(C=S)OC_2H_5} \text{3-Br-C}_6\text{H}_4\text{-SC(=S)-OC}_2\text{H}_5 \xrightarrow[\text{2. H}_2\text{SO}_4]{\text{1. KOH}} \text{3-Br-C}_6\text{H}_4\text{-SH} \quad (14)$$

A 1-liter flask equipped with a mechanical stirrer and low temperature thermometer is immersed in an ice bath. To the flask are added 50 ml of concentrated hydrochloric acid and 50 gm of crushed ice. While stirring, 51.5 gm (0.3 mole) of *m*-bromoaniline is slowly added. The mixture is cooled to 0°C and a cold solution of 22 gm (0.32 mole) of sodium nitrite in 50 ml of water is added at such a rate to keep the temperature below 4°C. The cold diazonium solution is transferred in 15-ml portions through a dropping funnel attached to another flask containing a solution of 70 gm of potassium ethyl xanthate in 90 ml of water at 40°–50°C until the entire diazonium solution has been added (1 hr). The mixture is then heated for an additional ½ hr, cooled, the organic layer separated, and the water layer is extracted twice with ether. The combined ether extracts and organic layer are washed with a 10% solution of sodium hydroxide and twice with water.

The ether solution is dried, the ether is removed using an aspirator, 300 ml of ethanol is added, and the solution heated to boiling. Potassium hydroxide pellets (70 gm, 1.2 mole) are then slowly added and the solution is refluxed for 7 hr. The solution is concentrated by distillation, water is added, and the alkaline solution is extracted twice with ether (100-ml portions). The ether layer is discarded and the solution is acidified with 6 N sulfuric acid. The acid solution is steam-distilled and the product is obtained from the distillate by ether extraction (two 100-ml portions). The combined extracts are dried and distilled to yield 28 gm (50%), bp 119°–121°C (20–22 mm), n_D^{25} 1.6310.

In order to prepare hindered aromatic thiols it has been found that the lithium aluminum hydride reduction of the xanthates gives higher yields than can be obtained with alkaline hydrolysis [22]. For example, o-thiocresol gives 39% yields by alkaline hydrolysis versus 89% using $LiAlH_4$ reduction of ethyl xanthate.

e. ADDITION OF HYDROGEN SULFIDE TO OLEFINS AND OTHER UNSATURATED COMPOUNDS [23–48]

Hydrogen sulfide adds to olefins and unsaturated compounds to give thiols in good to excellent yields. The reaction is usually very rapid and is catalyzed either by free radical initiators (peroxides or ultraviolet light) [23–28] or ionic types of catalysts such as transition metals, sulfuric acid, or even free sulfur [23–62]. The free radical initiated reaction addition gives anti-Markownikoff addition products [Eq. (15)] whereas the ionic catalyzed reaction gives Markownikoff addition products [Eq. (16)]. Both types of reactions work well in the laboratory and are also practiced commercially.

$$RCH{=}CH_2 + H_2S \xrightarrow[\text{or } h\nu]{ROOH} RCH_2{-}CH_2SH \qquad (15)$$

$$RCH{=}CH_2 + H_2S \xrightarrow[\substack{\text{or metals and} \\ N \text{ compounds}}]{H_2SO_4} \underset{SH}{RCH}{-}CH_3 \qquad (16)$$

2-7. Preparation of n-Butylmercaptan [23,31]

(a) Sealed Tube Experiment [23]. 1-Butene (0.044 mole) and hydrogen sulfide (0.088 mole) are sealed in a 10-mm I.D. quartz tube. The contents are cooled to 0°C and illuminated with a ultraviolet light source (quartz mercury arc having a wavelength below about 2900–3000 Ångstroms) for 4 min. The tube is cooled and opened. The unreacted (approx. 20%) butane and hydrogen sulfide are recovered and the product (3.8 ml) is distilled to give 85% yield of *n*-butylmercaptan (bp 97°–98°C) and 15% yield of di-*n*-butyl sulfide (bp 188°–189°C.)

(b) Pressure Autoclave [31]. A mixture of hydrogen sulfide (2.0 moles) and 1-butene (1.0 mole) is charged to a 3-gal autoclave equipped with a mechanical stirrer, cooling coil, and a quartz immersion well containing a 200-watt high pressure ultraviolet light. The pressure of the reaction vessel at the start is 200 psi and the reaction is maintained at 90°–130°F. The ultraviolet lamp is started and samples from the reaction are analyzed by gas chromatography. Olefin conversion is 7–22% per pass and mercaptan yield is about 90–97%. The unreacted olefin and hydrogen sulfide are recycled to the reaction mixture.

f. MISCELLANEOUS METHODS FOR PREPARING MERCAPTANS
   (1) The conversion of phenols to thiophenols via dialkylthiocarbamates [49].
   (2) Reduction of sulfonyl halides [50].
   (3) Reduction of disulfides [51] and polysulfides [52].
   (4) Addition of hydrogen sulfide to ethylenediamine [53].
   (5) Addition of amines to olefin sulfides [54].
   (6) Reaction of hydrogen sulfide with alcohols [55].
   (7) Reaction of aralkyl ketones with H_2S–HCl to give *gem*-dithiols [56].
   (8) A new synthesis of thiophenols using 2,4-dinitrophenylsulfenylchloride [58].
   (9) Addition of hydrogen sulfide to unsaturated amines [59].
   (10) Preparation of methyl mercaptan by reaction of a carbon oxide with hydrogen and hydrogen sulfide in the presence of a catalyst [60].
   (11) Mercaptans via the addition of thiolcarboxylic acids to olefin followed by hydrolysis [61].
   (12) Mercaptans by the catalytic cleavage of organic disulfides [62].

2-8. Preparation of 1,3-Di-p-chlorophenylpropane-2,2-dithiol [57]

$$Cl-C_6H_4-CH_2-\overset{O}{\underset{\|}{C}}-CH_2-C_6H_4-Cl + H_2S + HCl \longrightarrow$$
$$(Cl-C_6H_4-CH_2)_2-C(SH)_2 + H_2O \quad (17)$$

A solution of 8.5 gm (0.3 mole) of 1,3-di-*p*-chlorophenyl-2-propanone in 300 ml of absolute methanol is slowly added dropwise to a stirred solution of 200 ml of absolute methanol saturated with H_2S–HCl gas at 0°C. Hydrogen chloride and hydrogen sulfide are continuously bubbled through the solution during the addition. The temperature is kept at 0°–5°C and after 4½ hr the reaction flask is stoppered and placed in the refrigerator overnight. The resulting solid is filtered and dried in a vacuum desiccator over sodium hydroxide to yield 9.6 gm (95%), mp 90°–107°C. Recrystallization (eight times) from *n*-hexane-benzene yields 2.0 gm, mp 121°–124°C (dec) (white crystals).

3. SULFIDES

A. Condensation Reactions

a. Reaction of Sodium or Other Metal Mercaptides with Active Alkyl or Aryl Halides

Sodium mercaptides are prepared from the mercaptans and aqueous or alcoholic solutions of sodium hydroxide or alcoholic sodium ethoxide. The sodium mercaptide reacts with halides [63–65], chlorohydrins [66], esters of sulfonic acid [67], or alkyl sulfonates [68] to give sulfides in yields of 70% or more. A recent report [69] describes a general procedure for synthesizing aryl thioesters by a nucleophilic displacement of aryl halide with thiolate ion in amide solvents. No copper catalysis is necessary as in an Ullmann-type reaction [69].

$$ArSK + Ar'X \xrightarrow{DMF} ArSAr' + KX \tag{18}$$

CAUTION: Lithium *p*-nitrophenylthiolate (prepared from free thiol and *n*-butyllithium) has been reported to detonate fairly violently after drying and being exposed to the atmosphere [70].

3-1. *Preparation of α-Methylallyl Methyl Sulfide* [71]

$$CH_2{=}CH{-}\underset{\underset{CH_3}{|}}{CH}{-}Cl + CH_3SNa \longrightarrow CH_2{=}CH{-}\underset{\underset{CH_3}{|}}{CH}{-}S{-}CH_3 \tag{19}$$

To a flask containing 14.1 gm (0.156 mole) of a mixture of 90% α-methylallyl chloride and about 10% of the crotyl chloride (bp 64°–65°C, prepared from crotyl alcohol and thionyl chloride [72] is added 14.0 gm (0.175 mole) of sodium methyl sulfide (from methanethiol and sodium methoxide) in 100 ml of methanol. The methanolic mixture is refluxed for 1 hr. Fractional distillation through a 2-ft spinning band column yields 9.80 gm of α-methylallyl sulfide (62%), bp 103.5°–106.5°C and 1.9 gm bp 113°–124°C of a mixture of *cis*- and *trans*-crotyl methyl sulfides.

3-2. Preparation of α-Methylallyl Phenyl Sulfide [73]

$$H_2C=CH-\underset{\underset{CH_3}{|}}{CH}-Cl + C_6H_5SNa \longrightarrow CH_2=CH-\underset{\underset{CH_3}{|}}{CH}-S-C_6H_5 \quad (20)$$

To a flask containing 6.8 gm (0.1 moles) of sodium ethoxide in 30 ml of ethanol is added 11 gm (0.1 mole) of thiophenol followed by 10 gm (0.11 mole) of α-methylallyl chloride. The reaction is exothermic and sodium chloride precipitates. The reaction mixture is allowed to remain overnight. After this time it is no longer alkaline towards litmus. The alcohol and remaining α-methylallyl chloride are removed at reduced pressure and then 100 ml of water is added to dissolve the sodium chloride. An oil is separated, and the water layer extracted twice with 25-ml portions of ether. The combined organic and ether layers are dried over Drierite, the ether is removed, and the residue is distilled under reduced pressure to yield 18.2 gm (50%), bp 35°–36°C (0.05 mm), n_D^{23} 1.5564, λ_{max}^{EtOH} 256 mµ (ε 4350). Gas chromatography indicates approximately 6% of the γ-isomer.

Macrocyclic polyether sulfides have been synthesized by reaction of the appropriate dimercaptans and oligoethylene diepoxides using potassium or sodium hydroxide and ethanol as a solvent [74].

Cuprous oxide has been reported to facilitate the condensation of 2-thiophenethiol with 2-bromothiophene using dimethylformamide as a solvent in the presence of potassium hydroxide [75].

b. THE REACTION OF METALLIC SULFIDES WITH HALIDES

Sodium sulfide reacts with aqueous alcoholic solutions of the halides to give good yields of the symmetrical sulfide [76]. Halides containing β-carboxy [77], hydroxyl [78], ethoxyl [79], or diethylamino [80] groups are effective in this reaction. Long chain halides give cyclic suflides [81].

3-3. Preparation of Dibenzyl Sulfide [59]

$$2\ C_6H_5-CH_2Cl + Na_2S \longrightarrow [C_6H_5CH_2]_2S + 2\ NaCl \quad (21)$$

A flask containing 116 gm (0.56 mole) of benzyl chloride in 300 ml of 95% ethanol is heated in a steam bath while a solution of 35 gm (0.45 moles) of sodium sulfide in 125 ml of distilled water is slowly added. The stirred mixture is heated for 3 days on the steam bath, the alcohol is removed, and the residue is distilled under reduced pressure to remove water and unreacted benzyl chloride. The residue is placed in a refrigerator until it solidifies and then yields 48 gm (83%), mp 49°C.

c. Mercaptylation of the Double Bond

Mercaptans add to olefins in good yields according to Markovnikov's rule in the presence of sulfuric acid [82], boron trifluoride [83], or sulfur [84] and also in an anti-Markovnikov fashion in the presence of peroxides [85] or via photochemical means [86]. Vinyl chloride [85] and allyl alcohol [87,88] give lower yields than conjugated olefinic ketones [88,89], aldehydes [90], esters [91], and cyanides [92]. Cupric acetate is used as a catalyst for the reaction of methanethiol methyl mercaptan with acrolein to β-methylenecaptopropionaldehyde in 84% yield [93]. Allene reacts homolytically with methanethiol to give allyl sulfide and the 1,3- and 1,2-dimethylthiopropanes [94].

3-4. Preparation of tert-Butyl Sulfide [89]

$$(CH_3)_2C=CH_2 + (CH_3)_3C-SH \longrightarrow [(CH_3)_3C-]_2S \qquad (22)$$

To an ice cold mixture of 225 gm (2.3 mole) of concentrated sulfuric acid and 65 gm of water is added 50.4 gm (0.60 mole) of *tert*-butyl alcohol at such a rate to keep the temperature at 10°C. After the addition of the alcohol, 27 gm (0.30 mole) of *tert*-butylmercaptan is added over a 30-min period. The ice bath is removed and the mixture is warmed to room temperature. The mixture is poured into 500 gm of ice, extracted with ether, dried over magnesium sulfate, and concentrated. Distillation of the residue yields 27.9 gm (87% based on the reacted mercaptan), bp 148°–149°C.

3-5. Preparation of β-tert-Butylmercaptopropionitrile [95]

$$CH_2=CH-CN + CH_3-\underset{\underset{SH}{|}}{\overset{\overset{CH_3}{|}}{C}}-CH_3 \xrightarrow{NaOCH_3} (CH_3)_3C-S-CH_2-CH_2CN \qquad (23)$$

To a flask containing 19.4 gm (0.22 mole) of 2-methyl-2-propanethiol and 0.20 gm of sodium methoxide is added 38 gm (0.71 mole) of acrylonitrile over a 30-min period while some heat is applied to initiate the reaction. The temperature is not allowed to rise above 45°C. The contents are stirred at 25°C for 30 min, and the contents are decanted. The excess acrylonitrile is removed under reduced pressure. Upon vacuum distillation there is obtained 29 gm (95%), bp 113.5°–114°C (17 mm), N_D^{25} 1.4733.

d. Preparation of Episulfides (Thiiranes) From Epoxides

Epoxides react with either thiourea [96] or potassium and ammonium thiocyanate [97] at room temperature to give good yields of episulfides. Recently the use of silica gel either as a support for potassium thiocyanate or as a catalyst has been reported to give good yields of thiiranes with high stereospecificity [98]. This technique is also recommended in carrying out the preparations described below, and the original reference should be consulted for details.

3-6. Preparation of 3-Chloropropylene Sulfide (Epichlorosulfide) [62,96]

$$\underset{O}{CH_2{-}CH{-}CH_2Cl} + NH_2CSNH_2 \longrightarrow \underset{S}{CH_2{-}CH{-}CH_2Cl} + NH_2CONH_2 \quad (24)$$

To a 2-liter round-bottomed flask chilled to 0°C is added 210 gm (2.75 moles) of thiourea dissolved in 750 ml MeOH. Epichlorhydrin (232 gm, 2.51 mole) is added dropwise. The solution is stirred at about 0°C for 1 hr and then at 20°C for 3 hr. The solution, at room temperature, is poured into a separatory funnel containing 2 liters of water, shaken, and the bottom layer is collected (104 gm). The 104 gm of crude material is dried over Drierite and distilled under reduced pressure to yield 40 gm (14.7%), bp 79°–80°C (114 mm), n_D^{26} 1.5232.

3-7. Preparation of Isobutylene Sulfide [97]

$$\underset{O}{(CH_3)_2CH{-}CH_2} + KSCN \xrightarrow{H_2O} \underset{S}{(CH_3)_2{-}CH{-}CH_2} \quad (25)$$

To a vigorously stirred solution of 97 gm (1 mole) of potassium thiocyanate in 100 ml of water is slowly added 72 gm (1.29 mole) of isobutylene oxide over a period of 5 hr. The top layer is separated and then stirred with a fresh solution of 50 gm (5.16 mole) of potassium thiocyanate in 100 ml of water for an additional 5 hr while keeping the temperature below 40°C. The organic layer is dried and fractionally distilled to yield 64 gm (73%), bp 84°–86°C, n_D^{20} 1.4641.

e. MISCELLANEOUS METHODS FOR PREPARING SULFIDES

(1) Haloalkylation of mercaptans [99].
(2) Reaction of epoxides with H_2S or mercaptans [100].
(3) Reaction of mercaptans with lactones [101].
(4) Reaction of sodium sulfide on dithiocyanates to give episulfides [102].
(5) Reaction of mercaptans with sulfur to yield trisulfides [103].
(6) Reaction of carbon monoxide with thiols, sulfides, and disulfides [104].
(7) Reaction of aromatic thiocyanates with trialkyl phosphites [105].
(8) Condensation of formaldehyde with thiols [106].
(9) Preparation of polymeric sulfides [107].
(10) Conversion of disulfides to sulfides [108].

4. DISULFIDES

A. Condensation Reactions

The reaction of alkyl halides [109] or activated aryl halides [110] with sodium disulfide produces disulfides in good yields. 1,3-Dihalides [111] yield cyclic sulfides and polysulfides are produced with two to five equivalents of sodium sulfide. The hydroxyl [112] or nitro [113] groups do not interfere with the reaction.

Recently the preparation of unsymmetrical disulfides in fair to moderate yields has been reported to be feasible by the reaction of alkylthiosulfates with sodium alkylthiolates [114,115].

$$RB_2 + Na_2S_2O_3 \cdot 5H_2O \longrightarrow RS-SO_3Na + NaBr \qquad (26)$$

$$RS-SO_3Na + R'SNa \longrightarrow RS-SR' \qquad (27)$$

For somewhat related chemistry, see Preparation 4-3.

4-1. Preparation of Di-n-amyl Disulfide [68]

$$2\ CH_3(CH_2)_3CH_2Br + Na_2S_2 \longrightarrow [CH_3(CH_2)_3CH_2]_2S_2 \qquad (28)$$

To 180 gm (0.75 mole) of sodium sulfide dissolved in 750 ml of 95% ethanol are added while refluxing 24 gm (0.75 mole) of sulfur. The mixture is stirred until the sulfur dissolves. This hot solution is added to 151 gm (1.0 mole) of n-amyl bromide in 250 ml of 95% ethanol at such a rate as to maintain gentle refluxing (20 min). The mixture is refluxed for 3 hr and then allowed to stand overnight. One-third of the alcohol is removed under reduced pressure and the remaining solution is extracted with 500 ml of benzene. The extract is washed several times with water, dried, and distilled to yield 62 gm (69%), bp 90°–92°C (1 mm), n_D^{25} 1.4875.

B. Oxidation

Hydrogen peroxide oxidation of mercaptans yields disulfides [116–118]. Other useful mild oxidizing agents are iodine in ethanol [119] or alkaline solutions of iodine [120]. Amino or halo groups do not interfere with the reaction. Dimethyl sulfoxide has been reported to give good yields of disulfides by oxidizing thiols at 80°–90°C [121]. For example 1-butanethiol and benzenethiol give 86% and 100% yields, respectively [122]. More recently alkyl mercaptans have been oxidized to disulfide in 89–90% yields in benzene solution on treatment with basic alumina and exposed to air at room temperature [123].

4-2. Oxidation of Benzenethiol (Thiophenol) to Diphenyl Disulfide Using Dimethyl Sulfoxide [122]

$$2\ C_6H_5-SH + (CH_3)_2S=O \longrightarrow C_6H_5-S-S-C_6H_5 + H_2O + (CH_3)_2S \qquad (29)$$

To a 250-ml three-necked flask equipped with a magnetic stirrer, thermometer, a nitrogen gas inlet tube, and an outlet trap cooled in Dry Ice are added 11 gm (0.1 mole) of benzenethiol and 50 ml of dimethy sulfoxide. The flask is flushed continuously with a slow stream of nitrogen while the contents are heated with stirring at 80°C for 18 hr. The solution is poured into a tenfold volume of ice water and after standing 3 hr, the precipitated disulfide is filtered, washed three or four times with water, and dried under reduced pressure to yield 10.8 gm (100%), mp 61°–62°C.

C. Miscellaneous Methods for Preparing Disulfides

(1) Reaction of sulfonyl halides [124].
(2) Decomposition of alkyl thiosulfates [125,126].
(3) Disulfides via oxidation of thiols with potassium hexacyanoferrate [127].
(4) Sulfonic acids and sulfonyl derivatives are reduced to disulfides with iodide in the presence of boron halides [128].

4-3. Preparation of Di-n-butyl Disulfide [126]

$$C_4H_9Br + NaS\text{—}SO_2ONa \longrightarrow C_4H_9S\text{—}SO_2ONa \longrightarrow$$
$$(C_4H_9S)_2 + Na_2SO_4 + SO_2 \quad (30)$$

To 137 gm (1.0 mole) of n-butyl bromide in 500 ml of ethanol are added 300 gm of sodium thiosulfate dissolved in 400 ml of water. The mixture is refluxed for ½ hr and then 140 gm (2.5 mole) of potassium hydroxide in 300 ml of water are added. The mixture is heated for another ½ hr and the disulfide separates as an oil. The alkali sulfate and sulfite salts are precipitated and filtered off. After extraction with ether the combined organic layer is distilled to yield 42 gm (47%) of a slightly yellow liquid, bp 226°C (760 mm), n_D^{20} 1.4926.

REFERENCES

1. N. Kharasch and H. R. Williams, *J. Am. Chem. Soc.* **72**, 1843 (1950).
2. L. M. Ellis, Jr. and E. E. Reid. *J. Am. Chem. Soc.* **54**, 1674 (1932).
3. H. Gilman, M. A. Plunkett, L. Tolman, L. Fullhart, and H. S. Broadbent, *J. Am. Chem. Soc.* **67**, 1845 (1945).
4a. W. P. Hall and E. E. Reid, *J. Am. Chem. Soc.* **65**, 1466 (1943).
4b. H. Gilman and G. C. Gainer, *J. Am. Chem. Soc.* **71**, 1749 (1949).
5. J. E. Bittell and J. L. Speier, *J. Org. Chem.* **43**, 1687 (1978).
6. T. Zincke and P. Jorg, *Ber. Dtsch. Chem. Ges.* **42**, 3362 (1909).
7. A. I. Vogel, *J. Chem. Soc.* p. 1820 (1948).
8. R. Frank and P. V. Smith, *J. Am. Chem. Soc.* **68**, 2103 (1946).
9. J. A. King and F. M. McMillan, *J. Am. Chem. Soc.* **68**, 1369 (1946).
10. H. J. Backer, *Recl. Trav. Chim. Pays-Bas* **54**, 216 (1935).
11. H. J. Backer and J. Kramer, *Recl. Trav. Chim. Pays-Bas* **53**, 1102 (1934).
12. E. E. Reid, "Organic Chemistry of Bivalent Sulfur," Vol. 1, p. 32, Chem. Publ. Co., New York, 1950.
13a. B. C. Cossar, J. O. Fournier, D. L. Fields, and D. D. Reynolds, *J. Org. Chem.* **27**, 93 (1962).

13b. D. L. Klayman, R. J. Shine, and J. D. Bower, *J. Org. Chem.* **37**, 1532 (1972).
14. B. R. Backer, M. V. Query, S. Bernstein, S. R. Safir, and Y. Subbarow, *J. Org. Chem.* **12**, 171 (1947).
15. L. C. Cheney and J. R. Pieniny, *J. Am. Chem. Soc.* **67**, 733 (1945).
16. M. Seyhan, *Ber. Dtsch. Chem. Ges. B* **72**, 594 (1939).
17. Q. F. Soper, C. W. Whitehead, O. K. Behrens, J. J. Corse, and R. G. Jones, *J. Am. Chem. Soc.* **70**, 2849 (1948).
18. D. S. Tarbell and K. D. Fukushima, *Org. Synth.* **27**, 81 (1947).
19. C. M. Suter and H. L. Hansen, *J. Am. Chem. Soc.* **54**, 4102 (1932).
20. C. C. Price and G. W. Stacy, *Org. Synth.* **28**, 82 (1948).
21. H. F. Wilson and D. S. Tarbell, *J. Am. Chem. Soc.* **72**, 5203 (1950).
22. E. Campaigne and S. W. Osborn, *J. Org. Chem.* **22**, 561 (1957).
23. W. E. Vaughan and F. F. Rust, *J. Org. Chem.* **7**, 472 (1942).
24. F. F. Rust and W. E. Vaughan, U.S. Patent 2,392,294 (1946).
25. W. E. Vaughan and F. F. Rust, U.S. Patent 2,398,479 (1942).
26. W. E. Vaughan and F. F. Rust, U.S. Patent 2,398,480 (1946).
27. W. E. Vaughan and F. F. Rust, U.S. Patent 2,522,589 (1950).
28. W. E. Vaughan and F. F. Rust, U.S. Patent 2,522,590 (1950).
29. J. F. Harris, Jr. and F. W. Stacey, *J. Am. Chem. Soc.* **85**, 749 (1963).
30. P. F. Warner and J. W. Stanley, U.S. Patent 3,248,315 (1966).
31. J. R. Edwards, U.S. Patent 3,412,001 (1968).
32. J.-L. Seris, J. Suberlucq, H. T. Thuy, and C. Leirouici, French Patent 2,051,887 (1971); *Chem. Abstr.* **76**, 72008a (1972).
33. B. F. Dannels, U.S. Patent 4,052,283 (1977).
34. D. A. Dimnig, U.S. Patent 4,140,604 (1979).
35. J. Ollivier and G. Soulonmaic, European Patent Application 4,400 (1979).
36. G. N. Burkhardt, *Trans. Faraday Soc.* **30**, 18 (1934).
37. T. Kaneko, *J. Chem. Soc. Jpn.* **59**, 1139 (1938); *Chem. Abstr.* **33**, 2105 (1939).
38. M. S. Kharasch, A. T. Read, and E. R. Mayo, *Chem. Ind. (London)* **57**, 752 (1938).
39. F. R. Mayo and C. Walling, *Chem. Rev.* **27**, 351 (1940).
40. T. Posner, *Ber. Dtsch. Chem. Ges.* **38**, 646 (1905).
41. B. H. Nicolet, *J. Am. Chem. Soc.* **57**, 1098 (1935).
42. S. O. Jones and E. E. Reid, *J. Am. Chem. Soc.* **60**, 2452 (1938).
43. V. N. Ipatieff, H. Pines, and B. S. Friedman, *J. Am. Chem. Soc.* **60**, 2731 (1938).
44. R. P. Louthan, French Patent 1,378,539 (1964); *Chem. Abstr.* **62**, 14501 (1965).
45. T. Chiba, Y. Itsukaichi, K. Kawai, and K. Murata, Japanese Patent 72/49,600 (1972); *Chem. Abstr.* **78**, 110610s (1973).
46. V. E. Mazaev and M. A. Korshunov, *V.sb., Org. Soedin. Sery.* No. 1, p. 328 (1976); *Chem. Abstr.* **85**, 191810b(1976).
47. B. Buchholz, U.S. Patent 4,102,931 (7/25/78).
48. U. M. Dzhemiler, Y. U. Sangulov, S. R. Rafikov, G. A. Tolstikov, R. M. Massagutov, K. S. Minsker, A. I. Yudaev, S. M. Maksimov, and G. E. Ivonov, U.S.S.R. Patent 639,875 (1978); *Chem. Abstr.* **90**, 103400t (1979); R. W. Cleve, R. Lindenhahn, M. Helmut, and R. Scheibe, German (East) Patent 137,307 (1979); *Chem. Abstr.* **92**, 41337h (1980).
49. M. S. Newman and H. A. Karnes, *J. Org. Chem.* **31**, 3980 (1966).
50. R. Adams and C. S. Marvel, *Org. Synth. Collect. Vol.* **1**, 504 (1941).
51. C. F. H. Allen and D. D. MacKay, *Org. Synth. Collect. Vol.* **2**, 580 (1943).
52. E. L. Eliel, V. S. Rao, S. Smith, and R. O. Hutchins, *J. Org. Chem.* **40**, 524 (1975).
53. Nippon Shokubai Kagaku Kogyo Co. Ltd., Japan Kokai Tokkyo Kho 80/111,459 (1980).
54. H. R. Snyder, J. M. Stewart, and J. B. Ziegler, *J. Am. Chem. Soc.* **69**, 2672 (1947).

§ References

55. A. Binz and L. H. Pence, *J. Am. Chem. Soc.* **61**, 3134 (1939).
56. E. Campaigne and B. E. Edwards, *J. Org. Chem.* **27**, 3760 (1962).
57. J. Barrault, M. Guisnet, L. Jacques, and R. Makrel, Fr. Demande 2,333,784 (1977); *Chem. Abstr.* **88**, 74032a (1978).
58. N. Kharasch and R. Swidler, *J. Org. Chem.* **19**, 1704 (1954).
59. S. D. Turk, R. P. Louthan, R. L. Cobb, and C. R. Bresson, *J. Org. Chem.* **27**, 2846 (1962).
60. J. F. Olin, B. Buchholz, B. Loev, and R. H. Goshorn, U.S. Patent 3,070,632 (1962); B. Buchholz, Belgian Patent, 874,616 (1979); *Chem. Abstr.* **92**, 41331b (1980).
61. C. R. Morgan and R. W. Bush, U.S. Patent 4,117,017 (1978).
62. D. H. Kubicek, U.S. Patents 4,005,149, 4,059,636 (1977).
63. R. L. Shriner, H. C. Struck, and W. J. Jorison, *J. Am. Chem. Soc.* **52**, 2066 (1930).
64. W. R. Kirner and G. H. Richter, *J. Am. Chem. Soc.* **51**, 3131 (1929).
65. F. Kipnis and J. Ornfelt, *J. Am. Chem. Soc.* **71**, 3571 (1949).
66. Authors' Laboratory.
67. H. Gilman and N. J. Beaber, *J. Am. Chem. Soc.* **47**, 1449 (1925).
68. D. S. Tarbell and D. K. Fukushima, *J. Am. Chem. Soc.* **68**, 1458 (1946).
69. J. R. Campbell, *J. Org. Chem.* **29**, 1830 (1964).
70. M. Julia, *Chem. Eng. News* **59**,3 (April 6, 1981).
71. E. S. Heyser and R. M. Kellogg, *J. Org. Chem.* **30**, 2867 (1965).
72. F. F. Caserio, G. E. Dennis, R. H. DeWolfe, and W. G. Young, *J. Am. Chem. Soc.* **77**,4182 (1955).
73. W. E. Parham and S. H. Groen, *J. Org. Chem.* **30**, 730 (1965); C. D. Hurd and H. Greengard, *J. Am. Chem. Soc.* **52**, 3357 (1930).
74. J. S. Bradshaw, R. A. Reeder, M. D. Thompson, E. D. Flanders, R. L. Carruth, R. M. Izatt, and J. J. Christensen, *J. Org. Chem.* **41**, 134 (1976).
75. E. Jones and I. M. Moodie, *Org. Synth.* **50**, 75 (1978).
76. R. W. Bost and M. W. Conn, *Org. Synth. Collect. Vol.* **2**, 547 (1943).
77. G. M. Bennett and L. V. D. Scorah, *J. Chem. Soc.* p. 196 (1927).
78. E. N. Faber and G. E. Miller, *Org. Synth. Collect. Vol.* **2**, 576 (1943).
79. L. C. Swallen and C. E. Boord, *J. Am. Chem. Soc.* **52**, 657 (1930).
80. E. S. Cook and C. W. Kreke, *J. Am. Chem. Soc.* **61**, 2971 (1939).
81. R. F. Naylor, *J. Chem. Soc.* p. 1107 (1947).
82. V. N. Ipatieff and B. S. Friedman, *J. Am. Chem. Soc.* **61**, 71 (1939).
83. G. A. Dilbeck, L. Field, A. A. Gallo, and R. J. Gargiulo, *J. Org. Chem.* **43**, 4579 (1978).
84. S. O. Jones and E. E. Reid, *J. Am. Chem. Soc.* **60**, 2452 (1938).
85. N. A. LeBel, R. F. Czaja, and A. De Boer, *J. Org. Chem.* **34**, 3112 (1969).
86. J. E. Herweh and J. L. Work, U.S. Patent 4,035,337 (1977); *Chem. Abstr.* **87**, 118789p (1977).
87. R. C. Fuson, C. C. Price, and D. M. Burness, *J. Org. Chem.* **11**, 475 (1946).
88. J. L. Szabo and E. T. Stiller, *J. Am. Chem. Soc.* **70**, 3667 (1948).
89. E. A. Fehnel and M. Carmack, *J. Am. Chem. Soc.* **71**, 92 (1949).
90. R. H. Hall and B. K. Howe, *J. Chem. Soc.* p. 2723 (1949).
91. E. A. Fehnel and M. Carmack, *Org. Synth.* **30**, 65 (1950).
92. R. M. Ross, *J. Am. Chem. Soc.* **71**, 3458 (1949).
93. E. Pierson, M. Giella, and M. Tishler, *J. Am. Chem. Soc.* **70**, 1450 (1948).
94. K. Griesbaum, A. A. Oswald, E. R. Qutram, and W. Naegele, *J. Org. Chem.* **28**, 1952 (1963).
95. C. D. Hurd and I. I. Gershbeim, *J. Am. Chem. Soc.* **69**, 2330 (1947).
96. C. C. J. Culvenor, W. Davies, and K. H. Pausacker, *J. Chem. Soc.* p. 1050 (1946).
97. H. R. Snyder, J. M. Stewart, and J. R. Ziegler, *J. Am. Chem. Soc.* **69**, 2674 (1947).

98. M. O. Brimeyer, A. Mehrota, S. Quici, A. Nigram, and S. L. Regen, *J. Org. Chem.* **45,** 4254 (1980).
99. L. A. Walter, L. H. Goodson, and R. J. Fosbinder, *J. Am. Chem. Soc.* **67,** 655 (1945).
100. H. L. Gilman and L. Fullhart, *J. Am. Chem. Soc.* **71,** 1478 (1949).
101. H. Plieninger, *Chem. Ber.* **83,** 267 (1950).
102. M. A. Youtz and P. P. Perkins, *J. Am. Chem. Soc.* **51,** 3510 (1929).
103. B. D. Vineyard, *J. Org. Chem.* **31,** 601 (1966).
104. H. E. Holmquist and J. E. Carnahan, *J. Org. Chem.* **25,** 2240 (1960).
105. K. Pilgram and D. D. Phillips, *J. Org. Chem.* **30,** 2388 (1965).
106. G. F. Grillot and P. T. S. Lau, *J. Org. Chem.* **30,** 28 (1965).
107. G. A. Crosby, N. M. Weinshenker, and H.-S. Uh, *J. Am. Chem. Soc.* **97,** 2232 (1975).
108. D. N. Harpp and R. A. Smith, *Org. Synth.* **58,** 138 (1978).
109. E. Miller, F. S. Crossley, and M. L. Moore, *J. Am. Chem. Soc.* **64,** 2323 (1942).
110. M. T. Bogert and A. Stull, *Org. Synth. Collect. Vol.* **1,** 220 (1941).
111. H. J. Backer and N. Evenhuis, *Recl. Trav. Chim. Pays-Bas* **56,** 129 (1937).
112. B. Sjoberg, *Chem. Ber.* **75,** 26 (1942).
113. R. S. Schreiber and R. L. Shriner, *J. Am. Chem. Soc.* **56,** 115 (1934).
114. M. E. Alonso and H. Aragona, *Org. Synth.* **58,** 147 (1978).
115. K. C. Mattes, O. L. Chapman, and J. A. Klun, *J. Org. Chem.* **42,** 1814 (1977).
116. C. C. Price and G. W. Stacy, *Org. Synth.* **28,** 14 (1948).
117. J. Barnett, *J. Chem. Soc.* p. 5 (1944).
118. A. H. Nathan and M. T. Bogert, *J. Am. Chem. Soc.* **63,** 2363 (1941).
119. L. D. Small, J. H. Bailey, and C. J. Cavallito, *J. Am. Chem. Soc.* **69,** 1711 (1947).
120. E. Miller, F. S. Crossley, and M. L. Moore, *J. Am. Chem. Soc.* **64,** 2323 (1942).
121. C. N. Yiannois and J. V. Karabinos, *J. Org. Chem.* **28,** 3246 (1963).
122. W. W. Epstein and F. W. Sweat, *Chem. Rev.* **67,** 247 (1967).
123. K.-T. Liu and Y-C. Tong, *Synthesis* No. 9 p. 669 (1978).
124. N. E. Foss, J. J. Stehle, H. M. Shusett, and D. Hadburg, *J. Am. Chem. Soc.* **60,** 2729 (1938).
125. H. E. Westlake, Jr. and G. Dougherty, *J. Am. Chem. Soc.* **64,** 149 (1942).
126. R. E. Stutz and R. L. Shriner, *J. Am. Chem. Soc.* **55,** 1243 (1933).
127. F. S. Jargensen and J. P. Snyder, *J. Org. Chem.* **45,** 1015 (1980).
128. G. A. Olah, S. C. Narang, L. D. Field, and R. Karpeles, *J. Org. Chem.* **46,** 2408 (1981).

CHAPTER 19 / SULFOXIDES

1. Introduction	603
2. Oxidation Methods	603
2-1. *General Method of Oxidation of Sulfides to Sulfoxides Using Sodium Metaperiodate*	606
2-2. *Preparation of Tetramethylene Sulfoxide*	606
2-3. *Preparation of 1,4-Dithiane 1-Oxide*	607
3. Miscellaneous Methods	608
References	609

1. INTRODUCTION

The first reported synthesis of a sulfoxide was by Märcker [1a] in 1865. The methods [1b] generally involved the controlled oxidation of a sulfide by oxidizing agents such as hydrogen peroxide [2], ozone [3], peracids [4], hydroperoxides [5], manganese dioxide [6], selenium dioxide [7], nitric acid [8], chromic acid [9], dinitrogen tetroxide [10], iodosobenzene [11], and others. The chemistry of dimethyl sulfoxide for the period 1961–1965 has recently been reviewed [12].

$$R_2S \xrightarrow{[O]} R_2S{=}O \qquad (1)$$

2. OXIDATION METHODS

One reported method describes the convenient use of sodium metaperiodate as an oxidizing agent to form sulfoxides from sulfides free of sulfone contaminants [13]. The method finds use in preparing linear and cyclic aliphatic or aryl sulfoxides as shown in Table I [13].

The reaction is carried out by adding the sulfide in a methanol–water mixture to a slight excess of 0.5 M aqueous sodium metaperiodate at ice bath temperatures. Higher temperatures lead to sulfone formation. The reaction is complete in 3 to 12 hr, and yields of 90% or better of the sulfoxide are obtained. A more recent example of the use of sodium metaperiodate for this reaction has been reported [14].

$$R_2S + NaIO_4 \longrightarrow R_2SO + NaIO_3 \qquad (2)$$

TABLE I

Sulfoxides Produced by Sodium Metaperiodate Oxidation of the Corresponding Sulfides[a]

Name	Structure	Yield (%)	Mp (Bp) (°C) Found	Mp (Bp) (°C) Reptd.
1-Thiacyclooctan-5-one 1-oxide	$OC\overset{(CH_2)_3}{\underset{(CH_2)_3}{\diagup\diagdown}}SO$	91	91–92	—
1-Thiacyclohexan-4-one 1-oxide	$OC\overset{(CH_2)_2}{\underset{(CH_2)_2}{\diagup\diagdown}}SO$	97	109–110	113[c]
Methyl 4-ketopentyl sulfoxide	$CH_3SO(CH_2)_3COCH_3$	98	22.5–23.5 (99–101 / 0.12 mm)	—
Phenyl sulfoxide	$(C_6H_5)_2SO$	98	69–71	69–71[e,f]
Methyl phenyl sulfoxide	$CH_3SOC_6H_5$	99	29–30 (83–85 /0.1 mm)	29.5 (104.5/0.7 mm)[h,i]
Thian 1-oxide	$(CH_2)_5SO$	99	67–68.2	60–61.5[j]
1,4-Oxathian 4-oxide	$O\overset{(CH_2)_2}{\underset{(CH_2)_2}{\diagup\diagdown}}SO$	83[k]	46–47.2	44.5–45[l]
Bis(2-diethylaminoethyl) sulfoxide	$[(C_2H_5)_2N(CH_2)_2]_2SO$	85[m]	Dipricate 146–148	—
1-Benzylsulfinyl-2-propanone	$C_6H_5CH_2SOCH_2COCH_3$	89	126–126.5	125[n]
Acetoxymethyl methyl sulfoxide	$CH_3COOCH_2SOCH_3$	72	(85–90 /0.1 mm)	—
Phenylsulfinylacetic acid[o]	$C_6H_5SOCH_2COOH$	99	118–119.5	113–115[p]
Benzyl sulfoxide[r]	$(C_6H_5CH_2)_2SO$	96	135–136	132–133[f]
Ethyl sulfoxide	$(C_2H_5)_2SO$	65[s]	(45–47 /0.15 mm)	(88–89/15 mm)[t]

[a] Carbon tetrachloride solution. [b] Chloroform solution. [c] G. M. Bennett and W. B. Waddington [*J. Chem. Soc.* p. 2829 (1929)] reported mp 113°C, but were unable to repeat their preparation. [d] Repeated purification procedures did not improve analysis. [e] H. H. Szmant and R. L. Lapinski, *J. Am. Chem. Soc.* **78**, 458 (1956). [f] R. L. Shriner, H. C. Struck, and W. J. Jorison, *J. Am. Chem. Soc.* **52**, 2060 (1930). [g] $\lambda_{max}^{C_4H_5OH}$ 274 mμ (log of 3.3), 233 (4.2) [H. P. Koch, *J. Chem. Soc.* p. 2892 (1950)]. [h] L. Horner and F. Hübenett, *Justus Liebigs Ann. Chem.* **579**, 193 (1953). [i] C. C. Price and J. J. Hydock, *J. Am. Chem. Soc.* **74**, 1943 (1952). [j] M. Tamres and S. Searles, Jr., *J. Am. Chem. Soc.* **81**, 2100 (1959). [k] Yield based on technical thioxane.

TABLE I (Continued)

v_{max}S=O (cm^{-1})	Formula	C (%) Calcd.	C (%) Found	H (%) Calcd.	H (%) Found	Other identifying properties, Remarks
1049^a	C$_7$H$_{12}$O$_2$S	52.48	52.46	7.55	7.86	$v_{C=O}{}^a$ 1710 cm^{-1} extremely hygroscopic
1055^a	C$_5$H$_8$O$_2$S	45.43	44.72^d	6.10	6.19	$v_{C=O}{}^a$ 1725 cm^{-1}
1058^b	C$_6$H$_{12}$O$_2$S	48.66	48.86	8.17	8.25	$n_{D_1}^{25}$ 1.4873, $v_{C=O}{}^b$ 1718 cm^{-1}
1033^b	C$_{12}$H$_{10}$OS	—	—	—	—	λ_{max}^{C2H5OH} 274 mμ (log ϵ 3.2) 233 mμ (log ϵ 4.1)^g
1050^a	C$_7$H$_8$OS	59.90	59.75	5.71	6.18	Very hygroscopic
1045^a	C$_5$H$_{10}$OS	—	—	—	—	Liquifies immediately on exposure to the atmosphere
1026^a	C$_4$H$_8$O$_2$S	—	—	—	—	Hygroscopic
—	Dipricate C$_{24}$H$_{34}$N$_8$O$_{15}$S	40.80	40.98	4.86	5.16	—
1046^b	C$_{10}$H$_{12}$O$_2$S	61.17	61.33	6.16	6.22	$v_{C=O}{}^b$ 1705 cm^{-1}
1044^a	C$_4$H$_8$O$_3$S	35.28	35.17	5.92	5.80	n_D^{25} 1.4798; $v_{C=O}{}^a$ 1762 cm^{-1}
1015^q	C$_8$H$_8$O$_3$S	—	—	—	—	$v_{C=O}{}^q$ 1732 cm^{-1}
1025^b	C$_{14}$H$_{14}$OS	73.00	73.12	6.13	6.06	Analytically pure after single crystallization from ethanol
1047^a	C$_4$H$_{10}$OS	—	—	—	—	n_D^{25} 1.4676

^{l}French Patent 859,886 (1940), Chem. Abstr. **42**, 3783 (1948). ^{m}Crude yield. No formal purification of free base was made. Characterized as the dipricate. ^{n}C. Wahl, Ber. Dtsch. Chem. Ges. **55**, 1449 (1922). ^{o}Isolated by lyophilization of reaction mixture, followed by extraction with hot ethyl acetate. ^{p}A. Tananger, Ark. Kemi, Mineral. Geol. **24A**, No. 10 (1947). ^{q}Nujol mull. ^{r}Oxidation by 0.25 M sodium metaperiodate in 50% methanol. ^{s}Lower yield due to incomplete extraction. Ethyl sulfone (5%) was formed during heating used to concentrate reaction mixture prior to extraction. ^{t}R. Pummerer, Ber. Dtsch. Chem. Ges. **43**, 1401 (1910). ^{u}Reprinted from N. J. Leonard and C. R. Johnson, J. Org. Chem. **27**, 283 (1962). Copyright 1962 by the American Chemical Society. Reprinted by permission of the copyright owner.

CAUTION: Dimethyl sulfoxide has been reported to be easily absorbed through the skin and to then pass into the blood stream. Others have given some indication that skin irritation or burns and eye injuries may result from prolonged exposure to sulfoxides [15]. Therefore, great caution should be exercised in handling and preparing sulfoxides since these compounds possess this great tendency of skin penetration.

A laboratory explosion has been reported in the preparation of methyl sulfinyl carbanion with sodium hydride and dimethyl sulfoxide [16].

$$(CH_3)_2SO + NaH \longrightarrow CH_3SOCH_2^- Na^+ + H_2 \qquad (3)$$

Whether this occurs with other sulfoxides is unknown. Chapter 9 should be consulted for a further discussion of the possible hazards involved in the use of this reagent.

2-1. General Method of Oxidation of Sulfides to Sulfoxides Using Sodium Metaperiodate [17]

To 210 ml (0.105 mole) of a 0.5 M solution of sodium metaperiodate at 0°C is added 0.1 ml of sulfide. The mixture is stirred at ice bath temperature, usually overnight. The precipitated sodium iodate is removed by filtration, and the filtrate is extracted with chloroform. The extract is dried over anydrous magnesium sulfate, and the solvent is removed under reduced pressure. The sulfoxide is purified by distillation, crystallization, or sublimation.

A method of limited applicability is the reaction of dimethyl sulfoxide with straight chain aliphatic sulfides which leads to an oxygen exchange reaction yielding a new sulfoxide [18].

$$R_2S + (CH_3)_2SO \rightleftharpoons R_2SO + (CH_3)_2S \qquad (4)$$

This reaction has potential usefulness but a further investigation is required to extend its scope.

The reactants are heated at 160°–175°C for 8–12 hr and the dimethyl sulfide is removed as it is formed (bp 37°C). The product is separated from the black tarry mixture by reduced pressure distillation. Attempts to apply this reaction to di-*tert*-butyl, diphenyl, and pentamethylene sulfides and 3,3-dimethylthietane failed. However, yields of 55–58% of the sulfoxide were obtained from di-*n*-propyl, di-*n*-butyl, and tetramethylene sulfides.

2-2. Preparation of Tetramethylene Sulfoxide [19]

The preparation of tetramethylene sulfoxide by the hydrogen peroxide oxidation of tetramethylene sulfide is illustrative of a general method for the preparation of sulfoxides in the laboratory or on a commercial scale. For example, dimethyl sulfide and diphenyl sulfide give 50 and 68% yields, respectively, of the corresponding sulfoxides [19].

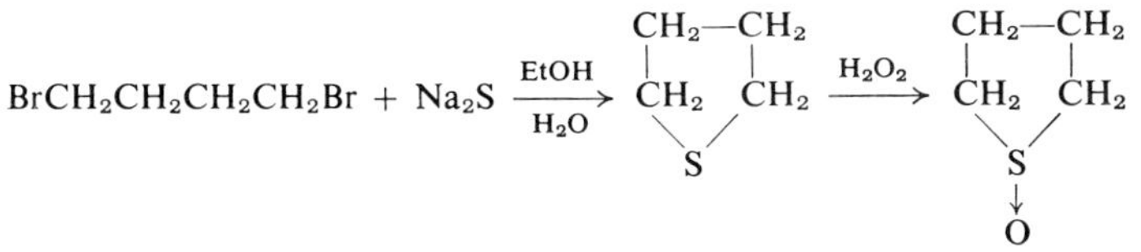

§ 2. Oxidation Methods

$$BrCH_2CH_2CH_2CH_2Br + Na_2S \xrightarrow[H_2O]{EtOH} \underset{S}{\underset{|}{CH_2}-\underset{|}{CH_2}}\! \overset{CH_2-CH_2}{} \xrightarrow{H_2O_2} \underset{\underset{O}{\underset{\downarrow}{S}}}{\underset{|}{CH_2}-\underset{|}{CH_2}}\! \overset{CH_2-CH_2}{} \quad (5)$$

To a round-bottomed flask containing 30 gm (0.34 mole) of tetramethylene sulfide and cooled with an ice bath is added dropwise with stirring 39 gm (0.36 mole) of 30% hydrogen peroxide. The mixture is stirred for 1 hr in the ice bath, at which point a homogeneous solution is obtained. After standing overnight, the water is stripped off under reduced pressure and the oil is distilled to yield 31.8 gm (90%), bp 105°–107°C (12 mm), n_D^{23} 1.5198.

A more recent application of this method is found in Procedure 2-3.

2-3. Preparation of 1,4-Dithiane 1-Oxide [20a]

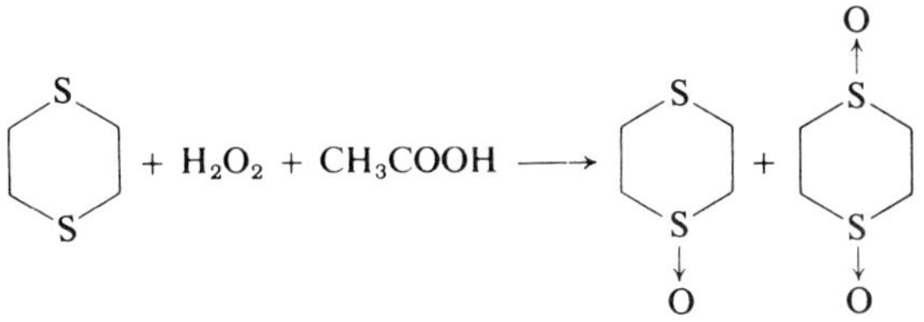

$$\text{(1,4-dithiane)} + H_2O_2 + CH_3COOH \longrightarrow \text{(1,4-dithiane 1-oxide)} + \text{(1,4-dithiane 1,4-dioxide)} \quad (6)$$

To a solution of 10.0 gm (0.083 mole) of 1,4-dithiane in 250 ml glacial acetic acid is added dropwise a solution of 4 ml (0.042 mole) of 30% hydrogen peroxide in 125 ml of glacial acetic acid at 23°–25°C. The mixture is stirred for 15 hr after the addition and is then distilled to remove the glacial acetic acid and unreacted 1,4-dithiane. The aqueous residue is evaporated, water is added, and the mixture is again evaporated. The remaining residue (5.2 gm) is added to 40 ml of boiling ethanol and the insoluble material (disulfoxide) is filtered off. Concentrating the ethanol solution yields 2.5 gm (22%) of the monosulfoxide, mp 125°C. The product is recrystallized from benzene.

Fluoroalkylsulfides are also oxidized to the sulfoxide using hydrogen peroxide and acetic acid [20b].

The chemistry of trimethylene sulfide (thietane) [21] is similar to that of tetramethylene sulfide, including hydrogen peroxide oxidation to the cyclic sulfoxide.

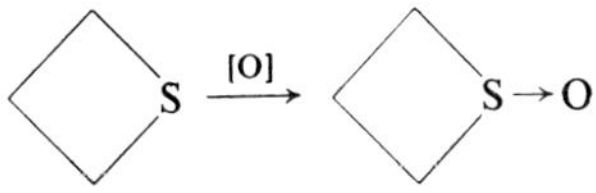

$$\text{(thietane)} \xrightarrow{[O]} \text{(thietane S-oxide)} \quad (7)$$

However, practically all thiiranes (ethylene episulfides) have been reported to undergo ring-opening to the sulfonic acid upon oxidization [22]. In some cases, polymers are formed [23].

$$\underset{H_2C-CH_2}{\overset{S}{\triangle}} + \text{Perbenzoic acid} \longrightarrow \text{Polymer} \quad (8)$$
$$\text{(polyethylene sulfone?)}$$

Most recently it has been reported that ethylene episulfide is oxidized with sodium metaperiodate in aqueous methanolic solution in 65% yields to ethylene episulfoxide [24a].

$$\underset{H_2C-CH_2}{\overset{S}{\triangle}} + NaIO_4 \xrightarrow[20°-25°C]{H_2O+CH_3OH} \underset{H_2C-CH_2}{\overset{S\uparrow O}{\triangle}} + NaIO_3 \quad (9)$$
$$\text{bp } 46°-48°C \text{ (2.0 mm)}, n_D^{25} \text{ 1.5210}$$

Using the same technique propylene episulfoxide, cyclohexene episulfoxide, and styrene episulfoxide were synthesized.

β-Hydroxy sulfoxides have been reported to be prepared by the cooxidation of α-olefins and arenethiols with oxygen catalyzed by irradiation with a black-light fluorescent lamp [24b].

3. MISCELLANEOUS METHODS

(1) The oxidation of sulfides to sulfoxides by dibenzyl selenoxide $(C_6H_5-CH_2)_2SeO$ [25].

(2) Vanadate catalysis of the hydrogen peroxide oxidation of sulfides to sulfoxides [26].

(3) The reaction of N-bromosuccinimide with acyl sulfides in aqueous media to give sulfoxides [27].

(4) Thermal rearrangement of sulfenates to sulfoxides [28].

(5) Preparation of 1-azulyl sulfoxides by electrophilic substitution using alkyl or arylsulfinyl chlorides [29].

(6) The preparation of symmetrical diaryl sulfoxides by the reaction of arenes with thionyl chloride and aluminum chloride [30].

(7) Disulfoxides via the transfer of oxygen to sulfide using dimethyl sulfoxide [31].

(8) Sulfide oxidation to sulfoxide with N-halosuccinimides [32].

(9) α-Chlorosulfoxide via reaction of diazo compounds with sulfinyl chlorides [33].

(10) α-Chlorosulfoxides via chlorination of disulfoxides [34].

(11) α,β-Unsaturated sulfoxides via α-phosphoryl sulfoxides [35].
(12) Sulfoxides via the alkylation of sulfoxides [36].

REFERENCES

1a. C. Märcker, *Justus Liebigs Ann. Chem.* **136,** 75 (1865).
1b. E. E. Reid, "Organic Chemistry of Bivalent Sulfur," Vol. 2, Chem. Publ. Co., New York, 1960.
2. D. Λ. Peak and T. I. Walkins, *J. Chem. Soc.* p. 445 (1950).
3. L. Horner, H. Schaefer, and W. Ludwig, *Chem. Ber.* **91,** 75 (1958); L. J. Hughes, T. D. McKennon, Jr., and J. C. Burleson, U.S. Patent 3,114,775 (1963).
4. C. G. Overberger and R. W. Cummins, *J. Am. Chem. Soc.* **75,** 4250 (1953).
5. D. Barnard, *J. Chem. Soc.* p. 489 (1956).
6. D. Edwards and J. B. Stenlake, *J. Chem. Soc.* p. 3272 (1954).
7. N. N. Mel'nikov, *Usp. Khim.* **5,** 443 (1936).
8. F. G. Bordwell and P. J. Bontan, *J. Am. Chem. Soc.* **79,** 717 (1957).
9. R. Knoll, *J. Prakt. Chem.* **[2] 113,** 40 (1926).
10. R. W. Whitaker and H. H. Sisler, *J. Org. Chem.* **25,** 1038 (1960).
11. W. H. Ford-Moore, *J. Chem. Soc.* p. 2126 (1949).
12. N. Kharasch and B. S. Thyagarajan, *Q. Rep. Sulfur Chem.* **1,** No. 1, 1 (1966).
13. N. J. Leonard and C. R. Johnson, *J. Org. Chem.* **27,** 282 (1962).
14. P. Friedman and P. Allen, Jr., *J. Org. Chem.* **30,** 780 (1965).
15. G. E. Hartzell and J. N. Paige, *J. Am. Chem. Soc.* **88,** 2617 (1966).
16. G. L. Olson, *Chem. Eng. News* **44,** No. 24, 7 (1966).
17. N. J. Leonard and C. R. Johnson, *J. Org. Chem.* **27,** 284 (1962).
18. S. Searles, Jr. and H. R. Hays, *J. Org. Chem.* **23,** 2028 (1958).
19. D. S. Tarbell and C. Weaver, *J. Am. Chem. Soc.* **63,** 2939 (1941).
20a. W. E. Parham and M. D. Bhovsac, *J. Org. Chem.* **28,** 2686 (1963).
20b. R. N. Hazeldine, R. R. Rigby, and A. E. Tipping, *J. Chem. Soc., Perkin Trans.* **1** No. 7, p. 676 (1978).
21. M. Sander, *Chem. Rev.* **66,** No. 3, 341 (1966).
22. C. C. J. Culvenor, W. Davies, and W. E. Savige, *J. Chem. Soc.* p. 282 (1949).
23. G. Hesse, E. Reichold, and S. Majmucher, *Chem. Ber.* **90,** 2106 (1957).
24a. G. E. Hartzell and J. N. Paige, *J. Am. Chem. Soc.* **88,** 2616 (1966).
24b. S. Irivchijima, K. Maniwa, T. Sakakibara, and G.-I. Tsuchihashi, *J. Org. Chem.* **39,** 1170 (1974).
25. D. Barnard and D. T. Woodbridge, *Chem. Ind. (London)* p. 1603 (1959).
26. H. S. Schultz, S. R. Buc, and H. B. Freyermuth, U.S. Patent 3,006,962 (1959).
27. W. Tagaki, K. K. Kukawa, K. Andi, and S. Oae, *Chem. Ind. (London)* p. 1624 (1964).
28. E. G. Miller, D. R. Rayner, and K. Mislow, *J. Am. Chem. Soc.* **88,** 3139 (1966).
29. L. L. Replogle and J. R. Maynard, *J. Org. Chem.* **32,** 1909 (1967).
30. H. H. Szmant, *Org. Sulfur Compd.* **1,** 158 (1961).
31. C. M. Hull and T. W. Barger, *J. Org. Chem.* **40,** 3152 (1975).
32. R. Harville and S. F. Reed, Jr., *J. Org. Chem.* **33,** 3976 (1968).
33. C. G. Venier, H.-H. Hsieh, and H. J. Barager, III, *J. Org. Chem.* **38,** 17 (1973).
34. B. B. Jarvis and H. E. Fried, *J. Org. Chem.* **40,** 1278 (1975).
35. M. Mikolajczyk, S. Grzejszczak, and A. Zatorski, *J. Org. Chem.* **40,** 1979 (1975).
36. P. O. Grieco and C. S. Pogonowski, *J. Org. Chem.* **39,** 732 (1974); P. A. Bartlett, *J. Am. Chem. Soc.* **98,** 3305 (1976).

CHAPTER 20 / **Sulfones**

1. Introduction	610
2. Oxidation Methods	611
2-1. Preparation of Tetramethylene Sulfone	611
2-2. Preparation of Sulfones by the Nitric Acid Oxidation of Sulfides and Sulfoxides	611
3. Condensation Methods	612
3-1. Preparation of Dibenzyl Sulfone	613
3-2. Preparation of Poly(p-xylylene) Sulfone	613
3-3. The Friedel–Crafts Preparation of Dimesityl Sulfone	613
3-4. Preparation of Phenyl Benzyl Sulfone	616
4. Miscellaneous Methods	617
References	617

1. INTRODUCTION

The oxidation of either sulfides or sulfoxides yields the corresponding sulfone. Some of the oxidizing agents that are described in the literature are hydrogen peroxide [1], peracids [2], oxygen, ozone, organic peroxides, potassium permanganate [3], potassium persulfate [4], sodium hypochlorite, hypochlorous acid, ruthenium tetroxide [5], oxides of nitrogen, nitric acid [6], and anodic oxidation.

Salts of sulfinic acids, especially benzene sulfinites, are easily alkylated by primary [7] and secondary [8] benzyl halides [9] and by alkyl sulfates [7] to sulfones.

$$ArSO_2Na + RX \longrightarrow ArSO_2R + NaX \tag{1}$$

Aryl halides also undergo this reaction, provided the halogen is activated by nitro groups in the ortho or para position [10].

The Friedel–Crafts condensation reaction of aromatic hydrocarbons with sulfonyl chlorides yields sulfones [11].

The reaction of Grignard reagents with sulfonyl chlorides also yields sulfones [12].

Sulfones are produced as by-products in the sulfonation of aromatic hydrocarbons, probably as a result of the condensation of the sulfonic acid with unreacted hydrocarbon [13]. A more recent modification of preparative value is the preparation of aromatic sulfones by the condensation of aromatic sulfonic acids and aromatic hydrocarbons in a polyphosphoric acid medium [14].

§ 2. Oxidation Methods

$$ArSO_3H + Ar'H \xrightarrow{-H_2O} ArSO_2Ar' \qquad (2)$$

2. OXIDATION METHODS

The most useful reagents for the laboratory preparation of sulfones are 30% hydrogen peroxide [1] or nitric acid. Other reagents have also been described [15].

More recently benzyl triethylammonium permanganate has been reported to be a mild, one-phase oxidizer of sulfides and sulfoxides to sulfones in 63–98% yield [16].

Oxidation of sulfoxides to sulfones with oxygen in the presence of soluble irridium and rhodium complexes has also been reported [17]. Metachloroperbenzoic acid has also been reported to be a good laboratory oxidizing agent for converting sulfides to sulfones [18].

2-1. Preparation of Tetramethylene Sulfone [1]

$$\underset{H_2C\diagdown_S\diagup CH_2}{\overset{H_2C-CH_2}{|\qquad|}} + H_2O_2 \longrightarrow \underset{H_2C\diagdown_{\underset{O_2}{S}}\diagup CH_2}{\overset{H_2C-CH_2}{|\qquad|}} \qquad (3)$$

To a flask containing 8.8 gm (0.1 mole) of tetramethylene sulfide is added in one portion 22.8 gm (0.2 mole) of 30% hydrogen peroxide. The reaction is exothermic and after 1 hr the solution becomes homogeneous. The solution is refluxed for 4 hr and then water is distilled off over a 1-hr period. The remaining solvent is stripped off under reduced pressure to leave 11.7 gm (87%) of colorless tetramethylene sulfone, mp 10°–10.5°C.

2-2. Preparation of Sulfones by the Nitric Acid Oxidation of Sulfides and Sulfoxides [6]

Nitric acid oxidizes sulfides to sulfoxides and then to sulfones at elevated temperatures (130°–180°C). The reaction appears to be favorable for the commercial preparation of sulfones from sulfides or sulfoxides since the nitrogen oxides can be recovered. Some of the results using this method are summarized in Table 1 [6,19].

$$3\,R_2S + 8\,HNO_3 \longrightarrow 3\,R_2SO_2 + 6\,NO_2 + 2\,NO + H_2O \qquad (4)$$

$$R_2SO + 2\,HNO_3 \longrightarrow R_2SO_2 + 2\,NO_2 + H_2O \qquad (5)$$

The sulfide and concentrated nitric acid are carefully mixed together at room temperature at a molar ratio of 1:2 to 1:6 and heated until the brown fumes cease to evolve. The dialkyl sulfoxides are heated in a similar manner. The results and experimental conditions are summarized in Table 1.

TABLE I

NITRIC ACID OXIDATION OF SULFIDES AND SULFOXIDES TO YIELD SULFONES[a]

Material oxidized	Quantity oxidized (mole)	Quantity nitric acid (moles)	Reaction temp (°C)	Reaction time (min)	Sulfone produced	Mp	Yield (%)
Dimethyl sulfide	1.00	3.0	122°–148°	85	Dimethyl	109°	85
Diethyl sulfide	0.45	0.96	140°	120	Diethyl	74°	77
Di-n-propyl sulfide	0.33	1.33	115–150°	45	Di-n-propyl	26°	97
Di-n-butyl sulfide	0.33	1.33	96°–120°	80	Di-n-butyl	43°	89
Di-n-octyl sulfide	0.33	1.33	126°–176°	12	Di-n-octyl	73°	83
Dimethyl sulfoxide	0.42	0.64	120°–150°	240	Dimethyl	109°	86

[a] Reprinted from D. W. Goheen and C. F. Bennett, *J. Org. Chem.* **26**, 1332 (1961). Copyright 1961 by the American Chemical Society. Reprinted by permission of the copyright owners.

3. CONDENSATION METHODS

Sulfones and polysulfones have been prepared from alkyl halides and sodium dithionite ($Na_2S_2O_4$) [20]. The by-products are SO_2 and sodium halide. Both reactants should be in solution and dimethylformamide (DMF) is a preferred solvent.

The condensation of sodium sulfinites with active halides as described earlier produces sulfones. For example, the condensation of sodium *p*-acetaminobenzenesulfinate and *p*-chloronitrobenzene yields 4-nitro-4'-acetylaminodiphenyl sulfone in 50–52% yields [21]. Reduction of the latter gives 4,4'-diaminodiphenyl sulfone in 74–77% yields [10]. Sodium benzenesulfinate condenses with benzyl chloride to give a 52% yield of phenyl benzyl sulfone.

The reaction of sulfur dioxide with olefins affords polysulfones and cyclic sulfones from dienes (sulfolene from butadiene) [22].

A novel single-step synthesis has been reported to take place by the condensation of olefins with electrophilic substituents (acid, amide, ester) and olefin dioxide in the presence of formic acid–tertiary amine adducts [23].

$$2\ R_1R_2C=CH_2 + SO_2 \xrightarrow{HCOOH} (R_1R_2CHCH_2)_2SO_2 \tag{6}$$

3. Condensation Methods

3-1. Preparation of Dibenzyl Sulfone [20]

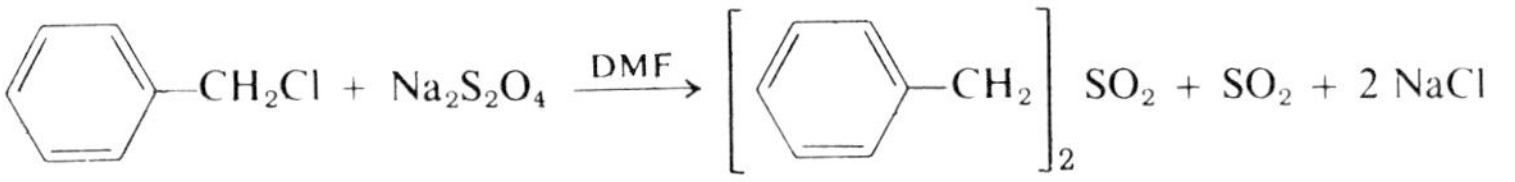

$$\text{C}_6\text{H}_5\text{-CH}_2\text{Cl} + \text{Na}_2\text{S}_2\text{O}_4 \xrightarrow{\text{DMF}} [\text{C}_6\text{H}_5\text{-CH}_2]_2\text{SO}_2 + \text{SO}_2 + 2\text{ NaCl} \quad (7)$$

To a round-bottomed flask are added 9.0 gm (0.05 mole) of sodium dithionite (sodium hydrosulfite), 12.6 gm (0.1 mole) of benzyl chloride, and 100 ml of dimethylformamide. The mixture is heated at 110°C with stirring for 9 hr and then poured into ice water in order to precipitate the product. Recrystallization from ethyl alcohol gives 2.0 gm (17%) of crystalline dibenzyl sulfone, mp 150°–151°C.

3-2. Preparation of Poly(p-xylylene) Sulfone [20]

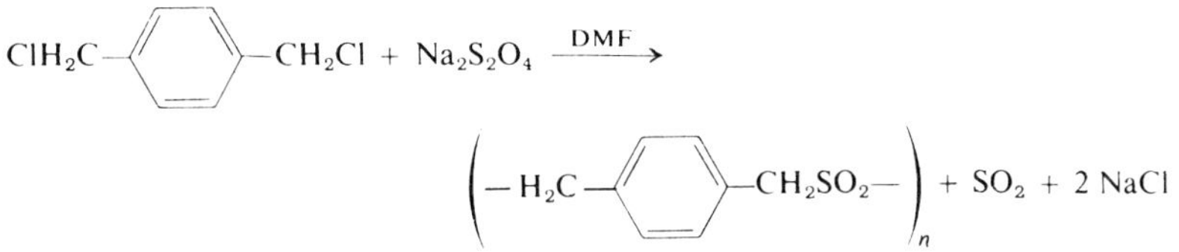

$$\text{ClH}_2\text{C-C}_6\text{H}_4\text{-CH}_2\text{Cl} + \text{Na}_2\text{S}_2\text{O}_4 \xrightarrow{\text{DMF}}$$

$$(-\text{H}_2\text{C-C}_6\text{H}_4\text{-CH}_2\text{SO}_2-)_n + \text{SO}_2 + 2\text{ NaCl} \quad (8)$$

To a stirred mixture of 18 gm (0.10 mole) of p-xylylene dichloride in dimethylformamide at 100°C is slowly added 18 gm (0.1 mole) of sodium dithionite. The reaction mixture is stirred for about 8 hr at which time the sulfur dioxide evolution ceases as indicated by a bubble counter. During the reaction a white dispersion is formed and then the mixture is poured into water to give a pale yellow solid. The solid is washed with hot ethanol and dried to give 10 gm (60%) of the polymer, mp > 360°C. The infrared spectrum is similar to that of dibenzyl sulfone, C—H, 3.4 μm (strong); C—H aromatic, 6.2, 6.6 μm (strong); C—H deformation, 6.75, 7.02 μm (very strong); SO_2, 7.6, 8.95 μm (very strong).

3-3. The Friedel–Crafts Preparation of Dimesityl Sulfone [11]

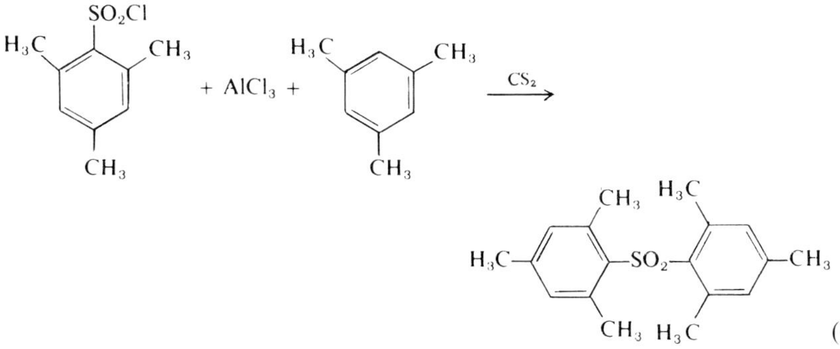

mesityl-SO$_2$Cl + AlCl$_3$ + mesitylene $\xrightarrow{\text{CS}_2}$ dimesityl sulfone (9)

TABLE II

Aromatic Sulfones Prepared in Polyphosphoric Acid at 80°[a]

Starting materials		Reaction time, hr	Sulfone product, RSO$_2$R'		Registry no.	% yield	Observed mp, °C	Lit. mp, °C
R in RSO$_3$H	Aromatic hydrocarbon		R	R'				
C$_6$H$_5$	Benzene	24	C$_6$H$_5$	C$_6$H$_5$	127-63-9	23	121	123
C$_6$H$_5$	Toluene	8	C$_6$H$_5$	p-CH$_3$C$_6$H$_4$	640-57-3	44	122	125[b]
C$_6$H$_5$	m-Xylene	8	C$_6$H$_5$	2,4-(CH$_3$)$_2$C$_6$H$_3$	4212-74-2	63	85	87[c]
C$_6$H$_5$	Mesitylene	7	C$_6$H$_5$	2,4,6-(CH$_3$)$_3$C$_6$H$_2$	3112-82-1	69	73	[e]
C$_6$H$_5$	Naphthalene	8	C$_6$H$_5$	α-C$_{10}$H$_7$	13249-96-2	49[d]	98	98
C$_6$H$_5$	Biphenyl	8	C$_6$H$_5$	p-C$_6$H$_5$-C$_6$H$_4$	1230-51-9	15[d]	147	148[b]
p-CH$_3$C$_6$H$_4$	Benzene	8	p-CH$_3$C$_6$H$_4$	C$_6$H$_5$	...	22	125	125[b]
p-CH$_3$C$_6$H$_4$	Chlorobenzene	8	...	...	...	0	...	...
p-CH$_3$C$_6$H$_4$	Toluene	8	p-CH$_3$C$_6$H$_4$	p-CH$_3$C$_6$H$_4$	599-66-6	47	156	158[e]
p-CH$_3$C$_6$H$_4$	m-Xylene	8	p-CH$_3$C$_6$H$_4$	2,4-(CH$_3$)$_2$C$_6$H$_3$	3249-97-3	77	49	51[f]
p-CH$_3$C$_6$H$_4$	Mesitylene	9	p-CH$_3$C$_6$H$_4$	2,4,6-(CH$_3$)$_3$C$_6$H$_2$	15184-64-5	94	117	119

p-$CH_3C_6H_4$	Naphthalene	8	p-$CH_3C_6H_4$	α-$C_{10}H_7$	13249-99-5	47[d]	119	121
2,4-$(CH_3)_2C_6H_3$ (Na salt)	m-Xylene	8	2,4-$(CH_3)_2C_6H_3$	2,4-$(CH_3)_2C_6H_3$	5184-75-8	82	120	121[g]
2,4-$(CH_3)_2C_6H_3$ (Na salt)	Mesitylene	8	2,4-$(CH_3)_2C_6H_3$	2,4,6-$(CH_3)_3C_6H_3$	13250-01-6	82	155[h]	156[i]
2,5-$(CH_3)_2C_6H_3$	p-Xylene	4	2,5-$(CH_3)_2C_6H_3$	2,5-$(CH_3)_2C_6H_3$	6632-44-6	50	143	142[g]
α-$C_{10}H_7$	Naphthalene	8	α-$C_{10}H_7$	α-$C_{10}H_7$	13250-03-8	55[d]	185[h]	187
β-$C_{10}H_7$	Naphthalene	8	β-$C_{10}H_7$	α-$C_{10}H_7$	13250-04-9	54[d]	119[h]	121
β-$C_{10}H_7$	m-Xylene	8	β-$C_{10}H_7$	2,4-$(CH_3)_2C_6H_3$	13250-05-0	72	129[h]	128
β-$C_{10}H_7$	Toluene	8	β-$C_{10}H_7$	p-$CH_3C_6H_4$	13250-06-1	50	154[h]	154
2,4,5-$(CH_3)_3C_6H_2$	Benzene	24	2,4,5-$(CH_3)_3C_6H_2$	C_6H_5	13250-07-2	10	159	160
2,4,6-$(CH_3)_3C_6H_2$	Mesitylene	8	2,4,6-$(CH_3)_3C_6H_2$	2,4,6-$(CH_3)_3C_6H_2$	3112-79-6	80	201	202[i]

[a]Reprinted from B. M. Graybill, *J. Org. Chem.* **32**, 2931 (1967). Copyright 1967 by the American Chemical Society. Reprinted by permission of the copyright owner. All reactions contained 0.02 mole each of sulfonic acid and aromatic hydrocarbon in 60 g of polyphosphoric acid. [b]C. A. Buehler and J. Masters, *J. Org. Chem.*, **4**, 262 (1939). [c]W. Steinkopf and R. Hubner, *J. Prakt. Chem.*, **141**, 193 (1934). [d]The product and unreacted aromatic hydrocarbon were separated by ether extraction. The sulfone was relatively insoluble in ether. [e]R. Otto, *Ber. Dtsch. Chem. Ges.* **12**, 1177 (1879). [f]H. Burton and P. F. G. Praill, *J. Chem. Soc.*, **887** (1955). [g]J. Pollok, *et al.*, *Monatsh. Chem.* **55**, 358 (1930). [h]Recrystallized from acetic acid. All other products were recrystallized from 95% alcohol. [i]W. Truce and O. Norman, *J. Am. Chem. Soc.*, **75**, 6023 (1953).

To a mixture of 147 gm (0.674 mole) of mesitylene sulfonyl chloride, 100 gm (0.83 mole) of mesitylene, and 1 liter of carbon disulfide in a flask equipped with a stirrer and a condenser with a drying tube is slowly added 100 gm (0.75 mole) of aluminum chloride. A mild reaction occurs while the addition takes place and then the mixture is refluxed for 11 hr. The carbon disulfide is removed by distillation and the contents of the flask are poured into ice. The product is filtered, dried, and is obtained in 75% yield (153 gm). The product is crystallized from glacial acetic acid, mp 202°–204°C (corrected).

3-4. Preparation of Phenyl Benzyl Sulfone [21,24]

$$C_6H_5SO_2Na + C_6H_5CH_2Cl \xrightarrow{C_2H_5OH} C_6H_5SO_2CH_2C_6H_5 + NaCl \quad (10)$$

To a flask containing 178 gm (1 mole) of sodium benzenesulfinate is added 127 gm (1 mole) of benzyl chloride in 500 ml of absolute alcohol. The mixture is refluxed for 8 hr and the hot mixture is poured into 1 liter of ice water. The crude product is filtered, dried, and recrystallized from ethanol to give 120 gm (52%), mp 146°–146.5°C. *Note:* The sulfinate does not completely dissolve during the reaction and sodium chloride is precipitated at the same time. Dimethylformamide may be a more useful solvent.

α-Nitroolefins can be converted to sulfones by the basic addition of hydrogen sulfide or thiols and subsequent oxidation with 30% hydrogen peroxide in glacial acetic acid [24].

$$CH_3C=CH_2 + CH_3SH \xrightarrow{CH_3ONa} CH_3-CH-CH_2SCH_3 \xrightarrow[CH_3COOH]{H_2O_2}$$
$$\underset{NO_2}{} \qquad\qquad \underset{NO_2}{}$$

$$CH_3-CH-CH_2-SO_2CH_3$$
$$\underset{NO_2}{}$$
$$\text{Mp } 69°–70°C \quad (11)$$

The synthesis of aryl sulfones is also possible by the condensation of aromatic sulfonic acids with aromatic hydrocarbons in the polyphosphoric acid [14]. This method appears to have the advantage that the starting materials are more easily obtained. The reaction is run at 80°C for about 8 hr in most cases. The isolation of the product is simple and gives a much purer product than obtained in the Friedel–Crafts method. For example, 3.8 gm (0.02 mole) of *p*-toluenesulfonic acid monohydrate. 2.1 gm (0.02 mole) of *m*-xylene, and 60 gm of polyphosphoric acid yield after 8 hr at 80°, 4.0 gm (77%) of *p*-tolyl-2,4-dimethylphenyl sulfone, mp 49° (recrystallized from ethanol). Some typical examples of the versatility and yields of this method of synthesis of sulfones is shown in Table II [14].

4. MISCELLANEOUS METHODS

(1) The preparation of thiosulfones, RS—SO_2R [25].
(2) Oxidation of sulfides with hydrogen peroxide catalyzed by sodium vanadates [26].
(3) Preparation of pyrimidyl sulfones [27].
(4) Reaction of formaldehyde and sulfinic acids [28].
(5) Reaction of 1,2-cyanoethylene and organic sulfinic acid salts [29].
(6) New catalysts for the oxidation of mercaptoethanols to sulfones [30].
(7) Sulfonic hydrogenation [31].
(8) Condensation of sulfonyl halides with Grignard reagents [32].
(9) Condensation of sulfinic acid with activated alkenes [33].
(10) Preparation of thirane dioxide [34].

REFERENCES

1. D. S. Tarbell and C. Weaver, *J. Am. Chem. Soc.* **63**, 2941 (1941); V. G. Kulkarni and G. V. Jadhov, *J. Indian Chem. Soc.* **34**, 245 (1957).
2. D. Swern, *Chem. Rev.* **45**, 33, 35 (1949).
3. H. Rheinboldt and E. Giesbrecht, *J. Am. Chem. Soc.* **68**, 973 1946).
4. E. Howard, Jr. and L. S. Levitt, *J. Am. Chem. Soc.* **75**, 6170 (1953).
5. C. Djerassi and R. R. Engle, *J. Am. Chem. Soc.* **75**, 3838 (1953).
6. D. W. Goheen and C. F. Bennett, *J. Org. Chem.* **26**, 1331 (1961).
7. A. T. Fuller, I. M. Tonkin, and J. Walker, *J. Chem. Soc.* p. 636 (1945).
8. W. A. Baldwin and R. Robinson, *J. Chem. Soc.* p. 1447 (1932).
9. R. L. Shriner, H. C. Struck, and W. J. Jorison, *J. Am. Chem. Soc.* **52**, 2060 (1930).
10. C. W. Ferry, J. S. Buck, and R. Baltzly, *Org. Synth.* **22**, 31 (1942).
11. M. E. Maclean and R. Adams, *J. Am. Chem. Soc.* **55**, 4685 (1933).
12. H. Gilman and R. E. Fothergill, *J. Am. Chem. Soc.* **51**, 3506 (1929).
13. H. Heyman and L. F. Fieser, *J. Am. Chem. Soc.* **67**, 1982 (1945).
14. B. M. Graybill, *J. Org. Chem.* **32**, 2931 (1967).
15. C. M. Suter, "The Organic Chemistry of Sulfur," pp. 660–667. Wiley, New York, 1944.
16. D. Scholz, *Monatsh. Chem.* **112** (2), 241 (1931); *Chem. Abstr.* **94**, 173724h (1981).
17. B. H. Henbert and J. Trocha-Grimshaw, *J. Chem. Soc., Perkins Trans. 1* no. 5, p. 607 (1974).
18. L. A. Carpino and L. V. McAdams, II, *Org. Synth.* **50**, 31 (1970).
19. D. W. Coheen and C. F. Bennett, *J. Org. Chem.* **26**, 1332 (1961).
20. E. Wellisch, E. Gripstein, and O. J. Sweeting, *J. Polym. Sci.* **B2**, No. 1, 35–37 (1964).
21. Authors' Laboratory, unpublished results.
22. S. R. Sandler and W. Karo, *Polymer Syntheses*, Vol. III, Academic Press, Inc., New York, 1980 pp. 1–41.
23. H. W. Gibson and D. A. McKenzie, *J. Org. Chem.* **39**, 2994 (1970).
24. R. L. Heath and A. Lambert, *J. Chem. Soc.* p. 1477 (1947).
25. G. Leandri and A. Tundo, *Ann. Chim. (Rome)* **47**, 575 (1957).
26. H. S. Schultz, S. R. Buc, and H. B. Freyermuth, U.S. Patent 3,006,962 (1959).
27. M. Ohta and R. Such, *J. Pharm. Soc. Jpn.* **71**, 511 (1951).
28. H. Bredereck and E. Baden, *Chem. Ber.* **87**, 129 (1954).
29. E. L. Martin, U.S. Patent 3,079,421 (1963).

30. H. S. Schultz, H. B. Freyermuth, and S. R. Buc, *J. Org. Chem.* **28,** 1140 (1963).
31. E. E. Huxley and M. E. Nash, U.S. Patent 4,275,218 (1981).
32. R. J. Koshar and R. A. Mitsch, *J. Org. Chem.* **38,** 3358 (1973).
33. P. Messinger, *Arch. Pharm. (Weinheim, Ger.)* **306** (6), 458 (1973); *Chem. Abstr.* **79,** 78294z (1973).
34. L. A. Carpino and J. R. Williams, *J. Org. Chem.* **39,** 2320 (1974).

CHAPTER 21 / SULFONIC ACIDS, SULFONIC ACID DERIVATIVES, AND SULFINIC ACIDS

1. Introduction		619
2. Sulfonic Acids		621
A. The Reaction of Sulfuric Acid and Its Derivatives with Aromatic Hydrocarbons		621
2-1.	Preparation of Sodium Benzenesulfonate	622
2-2.	Preparation of p-Toluenesulfonic Acid	623
2-3.	Preparation of Durenesulfonic Acid	624
2-4.	Preparation of Sodium 4-Phenoxybenzenesulfonate	624
2-5.	Preparation of Potassium 1-Methylnaphthalene-4-sulfonate	624
B. The Strecker Synthesis		625
2-6.	Preparation of β-Phenoxyethanesulfonic Acid	626
2-7.	Preparation of Sodium Isoamyl Sulfonate	626
C. The Addition of Bisulfites to Olefins		626
D. Reactions of Sulfur Trioxide to Yield Sulfonic Acids		627
2-8.	Preparation of α-Sulfopalmitic Acid	628
E. Oxidation Reactions		629
2-9.	Oxidation of Sodium α-Mercaptopalmitate to Sodium α-Sulfopalmitate	629
3. Derivatives of Sulfonic Acids		630
A. Sulfonyl Chlorides		630
3-1.	Preparation of p-Chlorobenzenesulfonyl Chloride	633
3-2.	Preparation of Phenoxybenzene-4,4'-disulfonyl Chloride	633
B. Sulfonic Esters		633
3-3.	General Procedure for the Preparation of Alkyl p-Toluenesulfonates	633
4. Sulfinic Acids		634
4-1.	Preparation of Sodium o-Chlorophenylsulfinate	634
4-2.	The Friedel–Crafts Method—Preparation of Sodium p-Fluorophenylsulfinate Dihydrate	635
4-3.	Preparation of p-Chlorobenzenesulfinic Acid	635
4-4.	Preparation of 1,4-Butanedisulfinic Acid	636
5. Miscellaneous Methods		637
References		637

1. INTRODUCTION

The literature on the sulfonation reactions is voluminous and several earlier reviews are worth consulting [1–5]. The industrial aspects of sulfonation with SO_3 to produce dodecylbenzenesulfonate salts and lauryl sulfate salts (sulfation reaction) have been reported in detail in an article also discussing plant design [6]. The use of sulfonation reaction to produce surfactants has been described [5].

The industrial chemist will find several references to the direct sulfonation of polymers to produce ion exchange resins and water-soluble materials. For example, chlorosulfonic acid is used to treat polyvinyl chloride to give an ion exchange resin [5]. In addition polystyrene and phenol formaldehyde resins are also sulfonated [5].

The most direct preparation of aromatic sulfonic acids is by the replacement of the hydrogen atom by one of the reagents shown below [7,8].

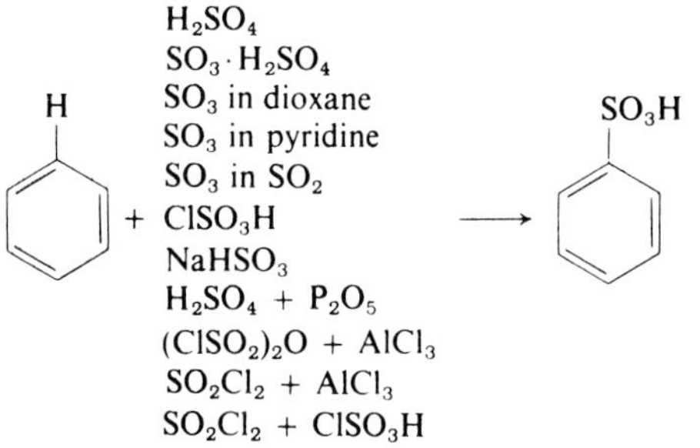

$$\text{C}_6\text{H}_6 + \begin{array}{l} \text{H}_2\text{SO}_4 \\ \text{SO}_3 \cdot \text{H}_2\text{SO}_4 \\ \text{SO}_3 \text{ in dioxane} \\ \text{SO}_3 \text{ in pyridine} \\ \text{SO}_3 \text{ in SO}_2 \\ \text{ClSO}_3\text{H} \\ \text{NaHSO}_3 \\ \text{H}_2\text{SO}_4 + \text{P}_2\text{O}_5 \\ (\text{ClSO}_2)_2\text{O} + \text{AlCl}_3 \\ \text{SO}_2\text{Cl}_2 + \text{AlCl}_3 \\ \text{SO}_2\text{Cl}_2 + \text{ClSO}_3\text{H} \end{array} \longrightarrow \text{C}_6\text{H}_5\text{SO}_3\text{H} \quad (1)$$

These sulfonations are more commonly used in the laboratory than the indirect methods which involve the oxidation of thiols, sulfinic acids, disulfides, or the conversion of the diazonium group into a sulfonic acid group.

Olefins have been found to react by a free radical addition with thiolacetic acid to form thiolacetates, which yield sulfonic acids upon oxidation by hydrogen peroxide–acetic acid [9].

$$\text{RCH}=\text{CH}_2 + \text{CH}_3\text{COSH} \longrightarrow \text{RCH}_2-\text{CH}_2\text{SCOCH}_3 \xrightarrow{\text{H}_2\text{O}_2-\text{CH}_3\text{COOH}} \text{RCH}_2\text{CH}_2\text{SO}_3\text{H} \quad (2)$$

where R represents straight and branched groups; the olefin can be terminal and internal.

In addition sodium bisulfite reacts with olefins by a free radical mechanism to give sodium sulfonates [10].

$$\text{RCH}=\text{CH}_2 + \text{NaHSO}_3 \longrightarrow \text{RCH}_2-\text{CH}_2\text{SO}_3\text{Na} \quad (3)$$

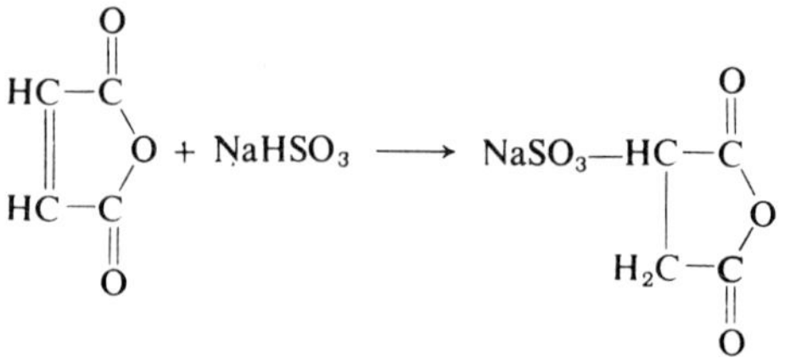

$$\text{maleic anhydride} + \text{NaHSO}_3 \longrightarrow \text{sulfosuccinate} \quad (4)$$

The Strecker synthesis involves the reaction of an active halogen compound with alkali or ammonium sulfites to give good yields of sulfonic acid salts [11].

$$RX + Na_2SO_3 \longrightarrow RSO_3Na + NaX \qquad (5)$$

The sulfoxidation and chlorosulfoxidation of hydrocarbons yields sulfonic acids and alkane sulfonyl chlorides, respectively.

The reaction of alkyl mercaptans and dialkyl disulfides with chlorine in an aqueous system yields alkane sulfonyl chlorides in good yield.

Sulfinic acids are related to sulfonic acids and are made either by the direct reaction of sulfur dioxide with diazonium salts, Grignard reagents, hydrocarbons and $AlCl_3$, or by reduction of the sulfonyl halide.

$$SO_2 + \begin{array}{c} ArN_2{}^+ \\ ArMgX \\ ArH + AlCl_3 \end{array} \longrightarrow ArSO_2H \qquad (6)$$

$$2\ RSO_2Cl \xrightarrow[\text{or } Na_2SO_3]{Zn,\ Na_2CO_3} 2\ RSO_2Na \qquad (7)$$

Sulfonamides are useful derivatives of sulfonic acids and are made by reacting sulfonyl halides with amines. The sulfonamides are solids and tables of the more common sulfonamides exist in the literature so that identification of an unknown is facilitated.

2. SULFONIC ACIDS

A. The Reaction of Sulfuric Acid and Its Derivatives with Aromatic Hydrocarbons

Sulfuric acid is satisfactory for the sulfonation of the more reactive aromatic hydrocarbons. However, a large excess of reagent is required to give a good yield since the reaction is reversible.

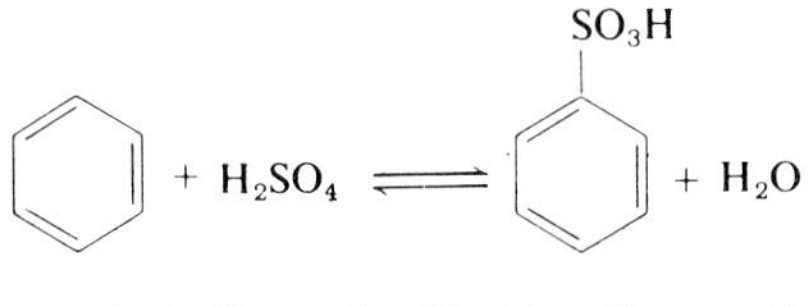

$$C_6H_6 + H_2SO_4 \rightleftharpoons C_6H_5SO_3H + H_2O \qquad (8)$$

Removal of the water as it is formed will drive the reaction to completion and will allow one to use the stiochiometric amount of sulfuric acid [12]. Aromatic sulfonic acids hydrolyze easily when heated in the presence of water and dilute acids.

Solvents are employed to moderate the sulfonation reaction as in the case of biphenyl [13], where chlorosulfonic acid in chloroform or tetrachloroethane is employed to give monosulfonation. A solvent also minimizes the formation of sulfonyl chlorides [14].

The sulfonic acids are usually isolated as their sodium salt and then hydrochloric acid is added to give the sulfonic acid [15]. The sulfonic acids are hygroscopic solids or liquids which are difficult to purify.

CAUTION: Some sulfonations are exothermic and may take place with explosive violence when the reaction is conducted at an elevated temperature.

Heating a solution of p-nitrotoluene and sulfuric acid at 160°C initiates an exothermic reaction which results in an explosion [16].

Benzene is monosulfonated at room temperature with the aid of sulfuric acid [17] and at 70°–90°C to m-benzenedisulfonic acid [18]. Sodium m-benzenedisulfonate is converted to 1,3,5-benzenetrisulfonic acid in 73% yield by heating at 275°C with 15% oleum and a mercury catalyst [19].

2-1. Preparation of Sodium Benzenesulfonate [20]

$$C_6H_6 + H_2SO_4 \longrightarrow C_6H_5SO_3H + H_2O \xrightarrow{NaCl} C_6H_5SO_3Na \qquad (9)$$

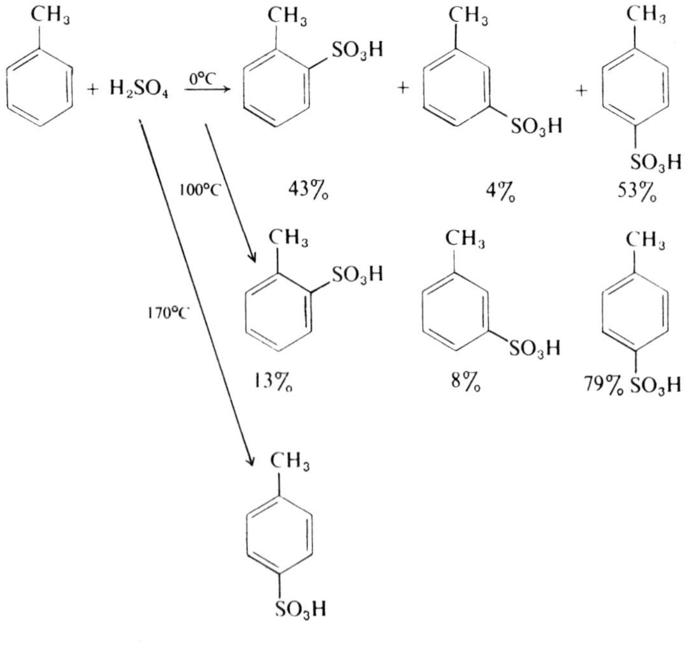

To an Erlenmeyer flask are added 39 gm (0.50 mole) of distilled benzene and 29 ml (53 gm, 0.52 mole) of concentrated sulfuric acid. The stoppered flask is shaken and put aside for 2 days. The upper benzene layer is separated. The sulfuric layer is added dropwise to 200 ml of a cold saturated (40 gm NaCl) sodium chloride solution. The yield of sodium benzenesulfonate after filtration and drying at 100°C is 13 gm ((14.5%) reported 52 gm) [17]. The saturated solution may be concentrated to about 100 ml to yield an additional 5 gm (5.5%) of sodium benzenesulfonate. The product is recrystallized from hot ethanol (95%) using 15-18 ml of ethanol for each gram of sodium benzenesulfonate.

The sulfonation reaction is reversible and the sulfonic acid group can migrate with changes in the temperature of the reaction (Jacobsen rearrangement) [21]. For example, the sulfonation of toluene is temperature-dependent, as is shown above [22]. It is estimated that in 1978 20 million pounds of the potassium and sodium salts of toluenesulfonic acid were produced in the United States.

The aromatic sulfonic acids are usually solids at room temperature and are very deliquescent. The anhydrous acids are difficult to prepare and this has been the source of reported varying physical properties. The sulfonation of naphthalene is similarly effected [23].

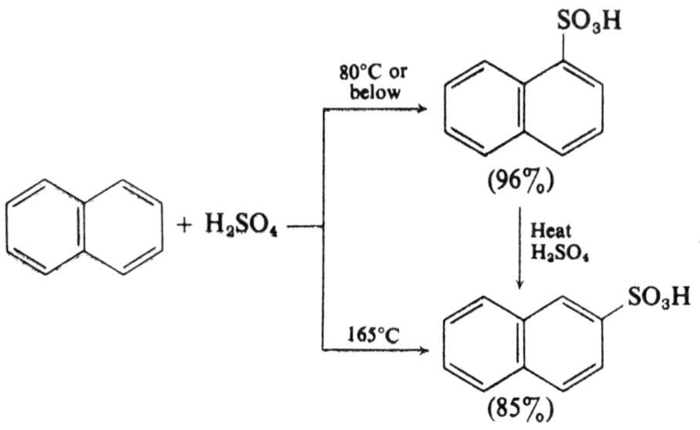

$$\text{naphthalene} + H_2SO_4 \xrightarrow{\text{80°C or below}} \text{1-naphthalenesulfonic acid} \quad (96\%) \quad (13)$$

$$\xrightarrow[\text{165°C}]{\text{Heat } H_2SO_4} \text{2-naphthalenesulfonic acid} \quad (85\%) \quad (14)$$

2-2. Preparation of p-Tolunesulfonic Acid [20]

$$C_6H_5CH_3 + H_2SO_4 \longrightarrow p\text{—}HSO_3\text{—}C_6H_4\text{—}CH_3 + H_2O + \text{some} \quad (15)$$
ortho and meta isomers

To an Erlenmeyer flask are added 53 ml (0.5 mole) of distilled toluene, a boiling chip, and 29 ml (52 gm, 0.52 mole) of concentrated sulfuric acid. The mixture is gently refluxed by heating in an oil bath. The flask is gently swirled so that the toluene and sulfuric acid can react. (A stirring hot plate would be more convenient.) After about 1 hr the toluene layer is almost gone and there is very little return of toluene. The flask is cooled and the crystals are filtered to give 79 gm (83%) of crude p-toluenesulfonic acid monohydrate. The acid is crystallized from chloroform and dried to give 71 gm, mp 104°–106°C. The product is predominantly the para isomer but contains small amounts of the ortho and meta isomers. Azeotropic removal of water has been reported to aid this reaction and increase the yield [24].

2-3. Preparation of Durenesulfonic Acid [25]

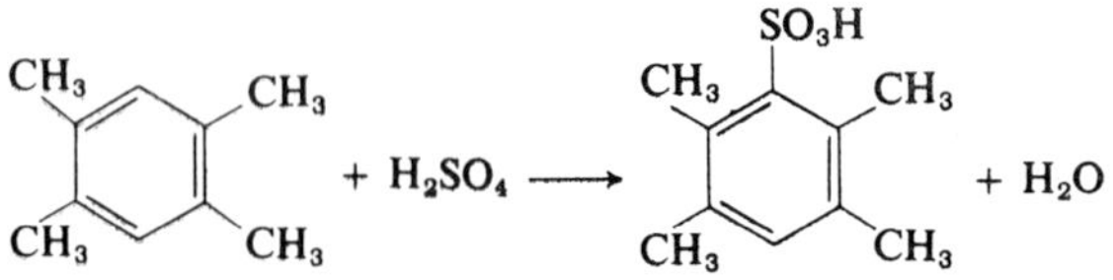

$$\text{durene} + H_2SO_4 \longrightarrow \text{durenesulfonic acid} + H_2O \qquad (16)$$

To an Erlenmeyer flask is added 10 gm (0.0785 mole) of durene and it is then covered with 27 ml (50 gm) of a solution of 1 part of 60% fuming sulfuric acid and 2 parts of concentrated sulfuric acid. The reaction mixture is stirred vigorously and any lumps of durene are broken up. The temperature rises about 20°C during the reaction and after 5–10 min it is poured with stirring into 250 gm of ice. The liquid is filtered immediately from the unmelted ice and unreacted durene. To the liquid filtrate at 0°–5°C is added concentrated sulfuric acid until it causes the material to solidify. The solid is filtered by suction and pressed to remove entrained liquids. The solid is dried on a porous plate and durenesulfonic acid is obtained in 94% yield (18.6 gm), mp 110°–112°C. The solid is recrystallized by dissolving in a minimum of 20% hydrochloric acid at 80°C, cooling to 0°C, filtering and drying to yield 14 gm (70%), mp 113°C.

2-4. Preparation of Sodium 4-Phenoxybenzenesulfonate [26]

$$C_6H_5O-C_6H_5 + H_2SO_4 \longrightarrow C_6H_5-O-C_6H_4-SO_3H \xrightarrow{NaOH} C_6H_5O-C_6H_4SO_3-Na \qquad (17)$$

To a flask containing 170 gm (1 mole) of diphenyl ether and 100 ml acetic anhydride is slowly added with agitation 69 ml (1.2 moles) of 95% sulfuric acid. The mixture is warmed on a steam bath for 1 hr and then poured into 1 liter of ice water. The unreacted diphenyl ether is filtered off and the sodium sulfonate of diphenyl ether is obtained by adding 80 gm (2 moles) of sodium hydroxide in 250 ml of water. The salt is filtered and dried and yields 212 gm (93%).

The sulfonation reaction is catalyzed by some metallic salts such as $HgSO_4$, $CaSO_4$, $Al_2(SO_4)_3$, $PbSO_4$, and $FeSO_4$ but not by $MnSO_4$ [27,28].

The sulfonate group has been found to be affected by the steric character of the aromatic groups. For example, *tert*-butylbenzene is sulfonated predominantly in the 4-position [29]. Other examples can be found in the literature [30].

2-5. Preparation of Potassium 1-Methylnaphthalene-4-sulfonate [31]

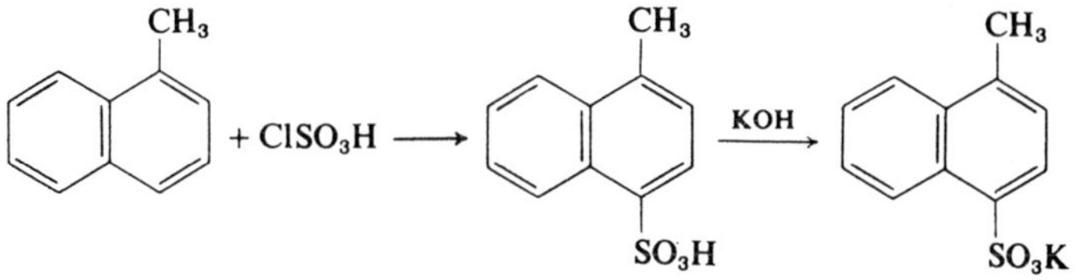

$$\text{1-methylnaphthalene} + ClSO_3H \longrightarrow \text{1-methylnaphthalene-4-SO}_3H \xrightarrow{KOH} \text{1-methylnaphthalene-4-SO}_3K \qquad (18)$$

§ 2. Sulfonic Acids

A flask containing 200 gm (1.41 moles) of α-methylnaphthalene in 425 ml of carbon tetrachloride is stirred with a tantalum stirrer and cooled to $-7°$ to $0°C$ while 169 gm (1.45 moles) of chlorosulfonic acid is added dropwise. As the chlorosulfonic acid is added, the product precipitates. The mixture soon becomes very thick and one must stir the contents manually. The supernatant solvent is decanted and to the residue is added 800 ml of water. Most of the solid dissolves and the solution is filtered. A layer of carbon tetrachloride separates and is rejected. The solution is neutralized with potassium hydroxide and the precipitated sulfonate is filtered and dried at $100°C$ (22 mm) to yield 322 gm (88%). The potassium salt can be recrystallized with 95% recovery from water.

Several functional groups may be present in the aromatic ring during the sulfonation reaction, such as hydroxyl [32], phenoxyl [21], carboxyl [33], and halo [32]. Amines react with sulfuric acid and are then rearranged to the amino sulfonic acid derivative by heating [34]. For example, see the preparation shown in Eq. (19) [34].

$$\underset{}{\text{o-toluidine}} + H_2SO_4 \longrightarrow \underset{}{\text{[}o\text{-CH}_3\text{C}_6\text{H}_4\text{NH}_3\bar{S}O_4H\text{]}} \xrightarrow{180°-195°C} \underset{SO_3H}{\text{2-amino-4-methylbenzenesulfonic acid}} \quad (79\text{–}83\%) \quad (19)$$

The sulfonation of *o*-toluidine with fuming sulfuric acid at $180°C$ gives the sulfonic acid directly [35].

The sulfonation of some heterocyclic nuclei is difficult. For example the yield of 3-pyridinesulfonic acid is only 13% by sulfonating at $390°C$ with oleum. However, the yield may be increased to 70% by the use of a catalyst such as mercuric sulfate [36–38].

It is easier to sulfonate heterocycles containing an aromatic ring. For example, 2-dibenzofuransulfonic acid is prepared in 75% yield by heating a mixture of dibenzofuran and concentrated sulfuric acid for 1 hr [39]. In addition, benzoguanamine is sulfonated in 96.5% yields with fuming sulfuric acid at $85°C$ [40].

B. The Strecker Synthesis

The halogen groups of halogen compounds, especially reactive halogens, is easily replaced by the $^-SO_3Na$ groups to give sodiumsulfonic acid salts [41]. Aliphatic compounds give high yields but branched chains give lower yields. *tert*-Butyl bromide yields only 23% of the sulfonate salt [42,43a]. Higher temperatures are required for high molecular weight halogen compounds. Phase transfer catalysts such as $R_4N^+Cl^-$ (R = Bu, Et) have been reported to facilitate this reaction [43b]. The Strecker method is not preferred in preparing sodium trifluoroethanesulfonate (10% yield) via the reaction of trifluoroethyl bromide with sodium sulfite [43c].

2-6. Preparation of β-Phenoxyethanesulfonic Acid [44]

$$C_6H_5OCH_2CH_2Cl + NaHSO_3 \longrightarrow C_6H_5OCH_2CH_2SO_3Na \xrightarrow{H^+} C_6H_5OCH_2CH_2SO_3H \quad (20)$$

To a flask containing 469.5 gm (3.0 moles) of β-chloroethyl phenyl ether is added a solution of 390 gm (3.0 moles) of sodium bisulfite in 1380 ml of water. The mixture is stirred vigorously and refluxed for 21 hr. Upon cooling to 20°C a crystalline precipitate of sodium β-phenoxyethanesulfonate appears. The product is washed with ether, and dried at 125°C to give 289 gm (43%).

An aqueous solution of 288 gm (1.29 moles) of the sodium salt is passed through an ion exchange column containing 454 gm of Dowex–50X, a cation exchange resin in the hydrogen form. Evaporation gives an almost quantitative recovery of β-phenoxyethanesulfonic acid which after drying over phosphorous pentoxide *in vacuo* melts below 100°C.

2-7. Preparation of Sodium Isoamyl Sulfonate [42]

$$CH_3-\underset{\underset{CH_3}{|}}{CH}-CH_2-CH_2-Br + NaHSO_3 \longrightarrow CH_3-\underset{\underset{CH_3}{|}}{CH}-CH_2-CH_2SO_3Na \quad (21)$$

To a flask containing 36.6 gm (0.243 mole) of isoamyl bromide is added 250 ml of a saturated solution of sodium bisulfite. The mixture is refluxed until the two layers disappear (approx. 24 hr). The sodium salt is obtained from the solution by evaporating to dryness and purified by repeated fractional crystallization from a 75% aqueous alcohol solution until the salt obtained is free of bromide. The yield is 40.4 gm (95.7%).

The free acid may be obtained by passing an aqueous solution of the sodium salt through a column with a suitable cation exchange resin (in hydrogen form).

C. The Addition of Bisulfites to Olefins

Bisulfites add to olefins by a free radical process in a contrary Markovnikov addition [45,46]. The reaction gives good yields with double bond compounds containing electron-withdrawing groups and with triple bonds. For example, Kharasch [47] found that the bisulfite addition to styrene is air-catalyzed and gives three products with 2-hydroxy-2-phenylethanesulfonic acid (III) predominating (65%).

§ 2. Sulfonic Acids

$$\underset{(I)}{\text{PhCH}_2\text{—CH}_2\text{SO}_3\text{H}} + \underset{(II)}{\text{PhCH CHSO}_3\text{H}} + \underset{(III)}{\text{PhCH(OH)—CH}_2\text{SO}_3\text{H}} \quad \text{from PhCH=CH}_2 + \text{HSO}_3^- \quad (22)$$

Recently the bisulfite addition has been shown to be greatly increased by using an organic perester and a water solution of an iron salt. For example, C_{10}–C_{20} primary olefins heated to 160°F in the presence of $NaHSO_3$ (0.016 mole Fe/mole olefin), *tert*-butyl perbenzylate, and 86% methanol gives 14.5% conversion after 2 hr but adding 0.001 mole Fe/mole olefin gives a 44.2% conversion after 1.5 hr [48a].

Sodium bisulfite has more recently been reported to add free radically to olefins in a mixed alcohol water solution [48b].

Acetylenes also add bisulfite to give disulfonic acids in good yields [46].

D. Reactions of Sulfur Trioxide to Yield Sulfonic Acids [49]

CAUTION: For all reactions involving sulfur trioxide one should wear rubber gloves and a face shield. The reaction should be carried out in a hood.

The reactions most often encountered with sulfur trioxide are [49]

$$SO_3 + \begin{cases} RH \longrightarrow RSO_3H & \text{Sulfonation [50]} \quad (23) \\ ROH \longrightarrow ROSO_3H & \text{Sulfation [50]} \quad (24) \\ R_2NH \longrightarrow R_2NSO_3H & \text{N-Sulfamation} \quad (25) \end{cases}$$

The reaction of sulfur trioxide (chloroform solution) with benzene is very fast at 0°–10°C and yields of 90% of benzenesulfonic acid are isolated [50,51]. In contrast, sulfuric acid and benzene (equal volumes) reach equilibrium when refluxed for 20–30 hr to give about 80% benzenesulfonic acid [52]. Sulfur trioxide is either used in chlorinated solvents such as chloroform, or ethylene chloride, or in sulfuric acid (oleum). Sulfur dioxide is also a good reaction medium for SO_3 sulfonations. Benzene is sulfonated by SO_3–SO_2 to give benzenesulfonic acid in yields greater than 95% [53,54].

The addition compounds of sulfur trioxide and bases such as pyridine or trimethylamine are weak sulfonation reagents. The less basic the agent forming the addition compound with SO_3, the more active it is as a sulfonation reagent.

Thus, SO_3—SO_3 (S_2O_6) > SO_3—$ClSO_3H$, SO_3—H_2SO_4 ($H_2S_2O_7$) > SO_3—dioxane [55] ≫ SO-pyridine [56] > SO_3—$(CH_3)_3N$.

The SO_3—$(CH_3)_3N$ complex has a low solubility in organic solvents but dissolves in water, 1.8 gm per 100 gm at 25°C. It is used for sulfating alcohols and phenols and for sulfamating aromatic amines and proteins [49].

Heterocycles react with SO_3-pyridine complex more cleanly than with sulfuric acid. There is very little reaction with SO_2—$(CH_3)_3N$. Furan, thiophene, pyrrole, and indole derivatives are described in detail using SO_3-pyridine in a recent review [49].

The reaction of $ClSO_3H$ with pyridine yields an immediate formation of SO_3-pyridine and a mole of pyridinium chloride.

$$2\,C_5H_5N + ClSO_3H \longrightarrow SO_3C_5H_5N + C_5H_5\overset{+}{N}HCl^- \qquad (26)$$

When the reaction is carried out in chloroform, the complex can be separated while the pyridinium hydrochloride is soluble in chloroform [57].

Olefins react with SO_3-dioxane to give unsaturated sulfonic acids in good yields. For example, styrene reacts with sulfur trioxide in ethylene dichloride to give 58–65% yields of sodium β-styrenesulfonate [58,59a]. The reaction of styrene with sodium bisulfite in the presence of oxygen gives 2-hydroxy-2-phenylethane sulfonic acid in 58% yields [47].

$$C_6H_5CH{=}CH_2 + SO_3 \cdot dioxane \longrightarrow C_6H_5CH{=}CHSO_3H \xrightarrow{NaOH} C_6H_5CH{=}CHSO_3Na \qquad (27)$$

$$C_6H_5CH{=}CH_2 \xrightarrow[\text{(Oxygen)}]{NaHSO_3} C_6H_5\underset{\underset{\displaystyle OH}{|}}{CH}{-}CH_2SO_3Na \qquad (28)$$

Recently the reaction of 1-hexadecene and 1-octene with sulfur trioxide has been reported to give mixtures of sulfones and sulfonic acids [59b]. The gas phase reaction of isobutylene and sulfur trioxide in the presence of an inert gas gives 75% mono and 19% disulfonic acids [59c].

The use of sulfur trioxide as a sulfonating agent has been known for a long time [60]. The technique for its application to the sulfonation of benzene has already been mentioned. In addition sulfur trioxide can be used to sulfonate α-hydrogen positions of acids [61].

2-8. *Preparation of* α-*Sulfopalmitic Acid* [62]

$$CH_3(CH_2)_{13}CH_2COOH + SO_3 \longrightarrow CH_3(CH_2)_{13}\underset{\underset{\displaystyle SO_3H}{|}}{CH}{-}COOH \qquad (29)$$

To 256 gm (1 mole) of palmitic acid in 1 liter of tetrachloroethylene is slowly added 128 gm (1.6 moles) sulfur trioxide. The reaction mixture is neutralized to give monosodium salt, extracted twice with hot acetone, and crystallized four times from water to give 218 gm (61%). Crude yields range in the order of 80–90%.

E. Oxidation Reactions

The most important oxidation method for the preparation of sulfonic acids involves the oxidation of mercaptans with $KMnO_4$ [63], CrO_3 [63], $Br_2 + H_2O$ [64], HNO_3 [65], and H_2O_2 [66a,b].

$$RSH \xrightarrow{[O]} RSO_3H \qquad (30)$$

Mercaptans are also oxidized by chlorine in aqueous hydrochloric acid to sulfonyl chlorides [66c]. Sulfonyl halides are also described in the next reaction. The sulfonyl chloride can be hydrolyzed to the sulfonic acid.

2-9. Oxidation of Sodium α-Mercaptopalmitate to Sodium α-Sulfopalmitate [62]

$$CH_3(CH_2)_{13}-\underset{SH}{CH}-COONa + NaOBr \xrightarrow[2.\ H_3O^+]{1.\ H_2O} CH_3(CH_2)_{13}-\underset{SO_3H}{CH}-COOH \qquad (31)$$

To a flask are added 5.4 gm (0.017 mole) of sodium α-mercaptopalmitate, 100 ml water, and 14 gm of sodium carbonate. Bromine is added dropwise with cooling. After the addition the solution is acidified and extracted with hot acetone. Concentration of the above yields a solid which upon recrystallization from water gives a 43% yield.

Recently it has been reported that sulfonic acids have been prepared in 56–91% yields from a series of 1-olefins and cycloolefins by the free radical addition of thiolacetic acid to form the thiolacetates followed by hydrogen peroxide–acetic acid oxidations [66c].

$$RCH=CH_2 \xrightarrow[2.\ H_2O_2-CH_3COOH]{1.\ CH_3COSH} RCH_2CH_2SO_3H \qquad (32)$$

Recently it has been reported that alkanes can be sulfoxidated to sulfonic acids by photochemical means [67a].

$$RH + SO_2 + O_2 \xrightarrow{h\nu} RSO_3H \qquad (33)$$

Disulfides are also oxidized to sulfonic acid in good yields [67b].

3. DERIVATIVES OF SULFONIC ACIDS

A. Sulfonyl Chlorides

The most common derivatives of sulfonic acids are the sulfonyl chlorides and the sulfonamides. Benzenesulfonyl chloride is made by reacting sodium benzenesulfonate with PCl_5 or $POCl_3$ at 180°C [67c]. Chlorosulfonic acid is also useful for the conversion of aromatic hydrocarbons to sulfonyl derivatives and can be used in the characterization of unknown hydrocarbons [68].

$$R\text{—}C_6H_5 + ClSO_3H \xrightarrow{CHCl_3} R\text{—}C_6H_4SO_2Cl \xrightarrow{RNH_2} R\text{—}C_6H_4SO_2NHR$$

$$R\text{—}C_6H_4SO_3Na + PCl_5 \text{ or } POCl_3 \qquad (34)$$

Some hydrocarbons which yield solid sulfonyl chloride derivatives are shown in Table I. The sulfonyl chloride group is introduced into the aromatic nucleus by treatment with excess chlorosulfonic acid in chloroform [69].

The sulfonyl chlorides can also be converted to sulfonamides by treatment with ammonium carbonate [69] and these yield solid derivatives, even from liquid sulfonyl chloride compounds [68]. A list of sulfonamide derivatives is given in Table II.

Sulfonyl chlorides have also been prepared by the reaction of Grignard reagents first with SO_2 and then with chlorine gas [70].

TABLE I

Solid Sulfonyl Chlorides from Alkylbenzenes[a]

Hydrocarbon, R–benzene R =	Sulfonyl chloride, R–benzenesulfonyl chloride R =	Yield (%)	Mp (°C, uncor.)
Me-	4-Me-	61–65	64°–66°
1,2-Di-Me-	3,4-Di-Me-	74–86	52°
1,3,5-Tri-Me-	2,4,6-Tri-Me-	65–72	50°–52°
tert-Bu	4-tert-Bu-	100	80°–82°
1,2,3,4-Tetra-Me-	2,3,4,5-Tetra-Me-	95	72°–73°
1,2,4,5-Tetra-Me	2,3,5,6-Tetra-Me	100	98°–99°
Penta-Me-	Penta-Me-	98	77°–78.5°
1,3-Di-Me-5-tert-Bu-	2,4-Di-Me-6-tert-Bu-(?)	97	66°–67°
Cyclohexyl-	4-Cyclohexyl-		51°–52.5°
1,2,4,6-Tetra-iso-Pr-	2,3,5,6-Tetra-iso-Pr-	77–86	141.5°–142°

[a] Reprinted from E. H. Huntress and F. H. Carten, *J. Am. Chem. Soc.* **62**, 513 (1940). Copyright 1940 by the American Chemical Society. Reprinted by permission of the copyright owner.

§ 3. Derivatives of Sulfonic Acids

TABLE II

SULFONAMIDES FROM ALKYLBENZENES[a]

R–Benzene (R)	R–Benzene-1-sulfonamide (R)	Yield (%)	Mp (°C, uncor.)
H	—	23	150.0°–150.5°
Me-	4-Me-	36–44	135.5°–136°[a]
Et-	4-Et-	73–95	109°–110°[b]
1,2-Di-Me	3,4-Di-Me-	67	143°–144°[c,d]
1,3-Di-Me-	2,4-Di-Me-	78	136.5°–137°[a,c,e]
1,4-Di-Me-	2,5-Di-Me-	84	145.5°–146.5°[d,e]
n-Pr-	4-n-Pr-	65–95	107°–108°[f]
iso-Pr-	4-iso-Pr-	82–88	104°5°–105.5°[f,g]
1,2,4-Tri-Me-	2,4,5-Tri-Me-	52	175°–176°
1,3,5-Tri-Me-	2,4,6-Tri-Me-	57	141.5°–142.5°
n-Bu-	4-n-Bu-	80	94.5°–95°[h,i,n]
sec-Bu-	4-sec-Bu-	63–72	81.0°–82.5°[j]
tert-Bu-	4-tert-Bu-	100	136°–137°[a]
iso-Bu	4-iso-Bu-	82	84°–85°[j]
iso-Me-4-i-Pr-	2-Me-5-iso-Pr-	84–87	114.5°–115.5°[j]
1,3-Di-Et-	2,4-Di-Et-(?)	57–58	98°–99°[h]
1,3-Di-Me-4-Et-	2,4-Di-Me-5-Et-(?)	80–85	147°–148°
1,2,3,4-Tetra-Me-	2,3,4-Tetra-Me-	85	183.5°–184.0°
1,2,3,5-Tetra-Me-	2,3,4,6-Tetra-Me-	82	141.5°–142°
1,2,4,5-Tetra-Me-	2,3,5,6-Tetra-Me-	——	153°–154°
n-Am-	4-n-Am-	80–100	85.5°–86.5°[h]
tert-Am-	4-tert-Am-	89–90	83°–84°[k,o]
1,3-Di-Me-4-n-Pr-	2,4-Di-Me-5-n-Pr-(?)	79–82	90°–93°
1,3-Di-Me-4-iso-Pr-	2,4-Di-Me-5-iso-Pr-(?)	68	155.5°–156°
1,3,5-Tri-Me-2-Et-	2,4,6-Tri-Me-3-Et-	64–71	131°–132°
Penta Me-	Penta Me-	90–92	182°–183°
n-Hexyl-	4-n-Hexyl-	81	85°–85.5°[p]
1,3-Di-Me-4-tert-Bu-	—	81	128°–130°[l,q]
1,3-Di-Me-5-tert-Bu	2,4-Di-Me-6-tert-Bu-(?)	86	132°–133°[l]
1,3,5-Tri-Et-	2,4,6-Tri-Et-	94	118°–118.5°
1,4-Di-tert-Bu-	2,5-Di-tert-Bu	—	135.5°–136.5°[a,r]
n-Nonyl-	4-n-Nonyl-	54–67	94.5°–95°[h,i,s]
n-Undecyl-	4-n-Undecyl-	—	95.7°–96.2°[h]
Cyclohexyl	4-Cyclohexyl-	85–87	180°–180.5°[l]
1,2,4,5-Tetra-iso-Pr-	2,3,5,6-Tetra-iso-Pr-	80–85	154.5°–155°[m]

[a]The mpm (melting point of a mixture) of the sulfonamides (mp 135.5–136.5°C uncor.) from 1,4-di-tert-Bu-benzene and that (mp 136°–137°C uncor.) from tert-Bu-benzene was not depressed, i.e., mp 135.5°–136.5°C uncor. However, each of these compounds when mixed with p-toluenesulfonamide (mp 136°C uncor.) or with m-xylenesulfonamide-4 (mp 136.5°–137°C uncor.) showed substantial depression to 95°–115°C.

[b]The mpm of this product with the corresponding sulfone (mp 97.5°–98°C) was 75°–95°.

[c]The mpm of the sulfonamides from o-xylene and m-xylene was 112°–115°C.

[d]The mpm of the sulfonamides from o-xylene and p-xylene was 114°–120°C.

$$\text{RMgCl} + \text{SO}_2 \longrightarrow \text{RSO}_2\text{MgCl} \xrightarrow{\text{Cl}_2} \text{RSO}_2\text{Cl} + \text{MgCl}_2 \tag{35}$$

Aliphatic hydrocarbons can also be chlorosulfonated to sulfonyl chloride derivatives [71a]. For example 2,3-dimethylbutane yields 20% 2,3-dimethylbutane-1-sulfonyl chloride by such a procedure [71a].

As described in Section 2,E above, mercaptans (or disulfides) on reaction with chlorine in aqueous hydrochloric acid can be converted to the sulfonyl chloride [66c].

$$\text{RSH} + 3\,\text{Cl}_2 + 2\text{H}_2\text{O} \xrightarrow{\text{conc. HCl}} \text{RSO}_2\text{Cl} + 5\text{HCl} \tag{36}$$

$$\text{RSSR} + 5\,\text{Cl}_2 + 4\text{H}_2\text{O} \xrightarrow{\text{aq. HCl}} 2\,\text{RSO}_2\text{Cl} + 8\text{HCl} \tag{37}$$

More recently, the photolytic chlorosulfonylation of alkanes has been reported to give alkanesulfonyl halides [71b].

TABLE II (*Continued*)

[e] The mpm of the sulfonamides from *m*-xylene and *p*-xylene was 108°–112°C.

[f] The mpm of the sulfonamides from *n*-Pr-benzene and iso-Pr-benzene was 83°–93°C.

[g] The mpm of this sulfonamide and the corresponding sulfone (mp 105°–107°C uncor.) was 75°–86°C.

[h] The mpm of the sulfonamide of *n*-undecylbenzenes with that from *n*-nonylbenzene was 85°–91°C; with that from *n*-butylbenzene 60°–84°C; with that from *m*-diethylbenzene 70°–80°C.

[i] The mpm of the sulfonamides from *n*-butylbenzene and *n*-nonylbenzene was 86°–94°C.

[j] The mpm of the sulfonamides from *sec*-butyl- and iso-butylbenzenes was 80°–84°C.

[k] The mpm of the sulfonamides from *n*-amyl- and *tert*-amylbenzenes was 55°–75°C.

[l] The mpm of the sulfonamides from 1,3-di-Me-4-*tert*-Bu-benzene and 1,3-di-Me-5-*tert*-Bu-benzene varied in the range between the values of the two individuals.

[m] This product could not be obtained from the sulfonyl chloride via the ammonium carbonate method but only by long treatment of the dry ligroin solution with gaseous ammonia. Recrystallization of the sulfonyl chloride from the dilute methanol yielded the methyl ester, m.p. 126°–126.5°C uncor., or from dilute ethanol the ethyl ester, mp 99°–99.5°C uncor. The analogous behavior of pentaethylbenzenesulfonyl chloride was reported during the progress of this work by L. I. Smith and C. O. Guss [*J. Am. Chem. Soc.* **62**, 2634 (1940)].

		Analyses	
			Nitrogen
	Formula	Calcd.	Found
[n]	$C_{10}H_{15}O_2NS$	6.57	6.62 6.78
[o]	$C_{11}H_{17}O_2NS$	6.16	6.45 6.34
[p]	$C_{12}N_{10}O_2NS$	5.80	6.05 6.02
[q]	$C_{12}H_{10}O_2NS$	5.80	5.59 5.63
[r]	$C_{14}H_{23}O_2NS$	5.20	5.51 5.62
[s]	$C_{15}H_{25}O_2NS$	4.94	4.90 4.91
[t]	$C_{12}H_{17}O_2NS$	5.85	5.85 5.93

[u] Reprinted from E. H. Huntress and J. S. Autenrieth, *J. Am. Chem. Soc.* **63**, 3447 (1941). Copyright 1941 by the American Chemical Society. Reprinted by permission of the copyright owner.

§ 3. Derivatives of Sulfonic Acids

A process for preparing alkanesulfonyl fluorides from mercaptans, hydrogen fluoride, and nitrogen dioxide has been reported [71c].

3-1. Preparation of p-Chlorobenzenesulfonyl Chloride [72]

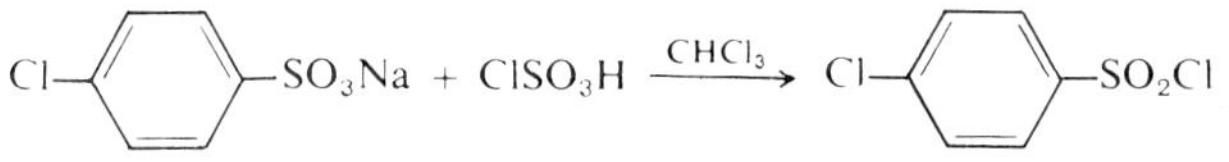

$$\text{Cl}-\text{C}_6\text{H}_4-\text{SO}_3\text{Na} + \text{ClSO}_3\text{H} \xrightarrow{\text{CHCl}_3} \text{Cl}-\text{C}_6\text{H}_4-\text{SO}_2\text{Cl} \quad (38)$$

To a flask containing 335 gm (1.56 moles) of dry sodium p-chlorobenzenesulfonate in 700 ml of chloroform is added dropwise with stirring 370 gm (3.18 moles) of chlorosulfonic acid at such a rate as to keep the temperature below 60°C. The addition takes about 15 min. The resulting thick reaction mixture is heated at 55°–60°C for 6 hr, cooked, and poured into ice water. The organic layer is separated, washed three times with cold water, dried with $CaCl_2$, and the solvent removed. Distillation of the residue yields 293 gm (89%), bp 140°C (12 mm), mp 52°–53°C.

3-2. Preparation of Phenoxybenzene-4,4'-disulfonyl Chloride [20,22]

$$C_6H_5-O-C_6H_5 + 2\ ClSO_3H \longrightarrow ClSO_2-C_6H_4-O-C_6H_4-SO_2Cl + 2\ H_2O \quad (39)$$

To 85 gm (0.5 mole) of diphenyl ether is added dropwise 200 ml of chlorosulfonic acid. After stirring for 2 hr the mixture is poured into cold water, and the crude product is filtered. The yield obtained is 160 gm (88%), mp 118°–120°C (crude), mp 128°–129°C after recrystallizing from petroleum ether (bp 90°/120°C).

B. Sulfonic Esters

Aliphatic and aromatic sulfonyl halides react with alcohols to give high yields of sulfonic esters. A basic medium is required and can be obtained using sodium hydroxide. The sodium or potassium salts of the alcohol reacting in pyridine is also a satisfactory procedure [73–75].

3-3. General Procedure for the Preparation of Alkyl p-Toluenesulfonates [76a]

$$\text{ROH} + p\text{-CH}_3-C_6H_4SO_2Cl \xrightarrow{\text{Pyridine}} p\text{-CH}_3-C_6H_4SO_2OR \quad (40)$$

Esters prepared from alcohols with more than 10 carbon atoms are solids.
To a solution of 1 mole of the alcohol and 4 moles of pyridine cooled below −20°C is added portionwise with stirring 1.1 moles of p-toluenesulfonyl chloride. The reaction mixture is stirred for 2 hr after the addition while cooling to keep the temperature below 20°C. The reaction mixture is then treated with 300 ml of concentrated hydrochloric acid in 1 liter of ice for each 0.5 mole of alcohol

starting material. The ester which separates is filtered with suction, dried, and recrystallized from alcohol or petroleum ether.

For example, lauryl alcohol yields 75% ester, mp 30°C, and stearyl alcohol yields 57% ester, mp 56°C.

Methanesulfonyl chloride (Pennwalt) reacts with alcohols in the presence of pyridine at $-5°$ to $0°C$ to give the ester [76b].

4. SULFINIC ACIDS

Sulfinic acids are either made by the reduction of sulfonic acid chlorides [72] or by the reaction of SO_2 with hydrocarbons ($AlCl_3$-catalyzed) [76a], organometallics [77], or the diazonium salt [78].

The reaction of SO_2 with the diazonium salt usually gives excellent yields. The replacement of the diazonium group by SO_2 is catalyzed by such metals as copper, copper bronze [78], or zinc dust–$CuSO_4$ compositions.

The preparation of aliphatic sulfinic acids has been reviewed [79].

4-1. Preparation of Sodium o-Chlorophenylsulfinate [72]

$$\underset{NH_2,\,Cl}{\bigcirc} + NaNO_2 + H_2SO_4 \longrightarrow \underset{\overset{+}{N_2} + H\bar{S}O_4,\,Cl}{\bigcirc} \xrightarrow[\text{Cu-Bronze}]{SO_2} \xrightarrow{NaOH} \underset{SO_2Na,\,Cl}{\bigcirc}$$

(86%) (41)

To a flask containing 25 gm (0.196 mole) of o-chloroaniline is added 600 gm of 30% sulfuric acid. The solution is cooled to 0°C and a 20% sodium nitrite solution is added dropwise until about 700 ml have been added. As the diazotization proceeds, the amine sulfate goes into solution. To the latter solution is added an ice cold solution of 100 gm of concentrated sulfuric acid and 80 gm of water. Sulfur dioxide is bubbled into the solution until there is a net gain in weight of the solution of 15 gm/100 ml solution. At this point copper bronze is slowly added to the solution at 0°–5°C while sulfur dioxide is being bubbled into it. When the nitrogen evolution ceases, the addition of copper bronze is terminated. The solution is filtered, and the precipitate is stirred into 400 ml of 10% sodium carbonate solution. The reaction mixture is filtered, acidified, and the crude sulfinic acid is redissolved in the calculated amount of sodium hydroxide. The solution is evaporated to yield 38 gm (86%) of sodium o-chlorophenylsulfinate.

4-2. The Friedel–Crafts Method—Preparation of Sodium p-Fluorophenylsulfinate Dihydrate [80]

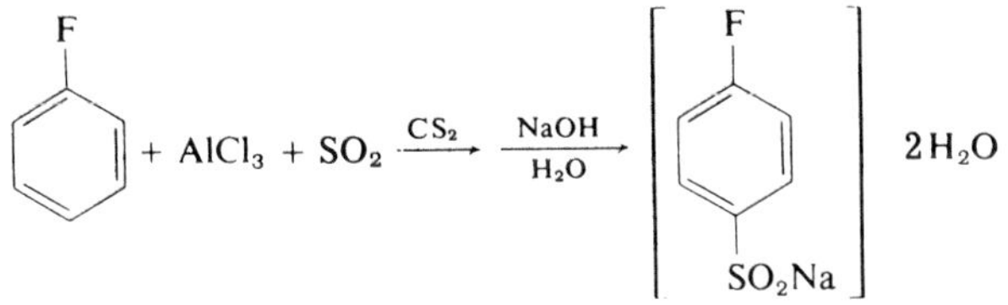

$$\text{F-C}_6\text{H}_5 + \text{AlCl}_3 + \text{SO}_2 \xrightarrow[\text{H}_2\text{O}]{\text{CS}_2, \text{NaOH}} \left[\text{F-C}_6\text{H}_4\text{-SO}_2\text{Na} \right] \cdot 2\text{H}_2\text{O} \quad (42)$$

To a flask at 0°C containing 15 gm (0.113 mole) of anhydrous aluminum chloride and 10 gm of fluorobenzene in 25 ml of carbon disulfide is added dry hydrogen chloride gas in order to saturate the mixture. Dry sulfur dioxide is bubbled in until the aluminum chloride turns to a green oily layer. On standing overnight, the latter crystallizes at room temperature. The mixture is decomposed by pouring into 200 ml of ice water containing 70 ml of 20% sodium hydroxide and digesting on the steam bath for 1 hr. The mixture is filtered and the soluble aluminum salts are precipitated by passing in carbon dioxide. The mixture is filtered and concentrated to 50 ml, at which point 12.7 gm of pure salt separates. Upon further concentration an additional 17.0 gm precipitates to give a total yield of 75%. The product is recrystallized from hot water to give glistening diamond-shaped crystals of the dihydrate.

Reducing sulfonyl chlorides with zinc and hot water yields the zinc salts of sulfinic acid [81,82]. Good results are also obtained by reducing sulfonyl halides with sodium sulfite solution [83a]. Methanesulfonyl chloride has been reported to be reduced to sodium methanesulfinate by means of aqueous sodium sulfite [83b].

4-3. Preparation of p-Chlorobenzenesulfinic Acid [72]

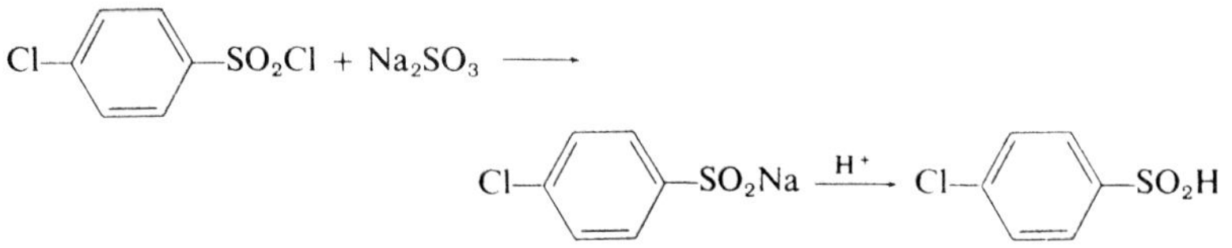

$$\text{Cl-C}_6\text{H}_4\text{-SO}_2\text{Cl} + \text{Na}_2\text{SO}_3 \longrightarrow$$

$$\text{Cl-C}_6\text{H}_4\text{-SO}_2\text{Na} \xrightarrow{\text{H}^+} \text{Cl-C}_6\text{H}_4\text{-SO}_2\text{H} \quad (43)$$

To a flask containing 630 gm (5.0 moles) sodium sulfite in 2 liters of water at 70°C is slowly added with vigorous stirring 194.5 gm (1.0 mole) p-chlorobenzenesulfonyl chloride. The reaction mixture is kept at 55°–60°C for 5 hr, acidified with concentrated HCl, cooled, and filtered to yield 141.0 gm (80%) of p-chlorobenzene sulfinic acid in three successive crops.

Alkane sulfinic acids have been prepared by reacting the alkylmagnesium halides with sulfur dioxide at −5°C [77]. The reaction mixture is hydrolyzed to

TABLE III

ALKANESULFINIC ACIDS RSO_2H[a]

Decane	Yield of acid from mg salt (%)	Mp (°C)	Analyses (%)					
			Carbon		Hydrogen		Sulfur	
			Calcd.	Found	Calcd.	Found	Calcd.	Found
1-Tetra-	65.5	48 –48.4	64.12	64.46	11.43	11.51	12.2	12.06
1-Hexa-	58	54 –55	66.21	65.90	11.71	11.58	11.05	11.15
1-Octa-	69.2	60 –60.5	67.92	67.68	11.94	12.14	10.05	10.20

[a]Reprinted from C. S. Marvel and N. A. Meinhardt, *J. Am. Chem. Soc.* **73**, 859 (1951). Copyright 1951 by the American Chemical Society. Reprinted by permission of the copyright owner.

the free sulfinic acid with sulfuric acid and then immediately extracted with ether.

Several C_{14}–C_{18}, alkanesulfinic acids were prepared by this method and it was observed that the free sulfinic acids are unstable and decompose in a few days [84]. Some results are given in Table III.

Recently it has been reported that 1,4-butanedisulfinic acid can be obtained in a stable cyrstalline form [85]. As mentioned earlier the lower aliphatic monosulfinic acids were found to be unstable [84]. Surprisingly it was found that 1,3-propanedisulfinic acid, 1,5-pentanedisulfinic acid, and 1,10-decanedisulfinic acid are unstable and disproportionate on heating [85].

4-4. Preparation of 1,4-Butanedisulfinic Acid [85]

$ClSO_2CH_2-CH_2-CH_2-CH_2SO_2Cl + 2\ Na_2SO_3 \longrightarrow$

$NaSO_2-CH_2(CH_2)_2-CH_2SO_2Na \xrightarrow{HCl} HSO_2CH_2(CH_2)_2CH_2SO_2H$ (44)

To a flask containing a solution of 26.5 gm (0.21 mole) of sodium sulfite and 36.1 gm (0.43 mole) of sodium bicarbonate in 100 ml of water is slowly added over a 1-hr period 25.5 gm (0.1 mole) of 1,4-butanedisulfonyl chloride [86] while stirring at 45°–50°C. Afterwards the solution is stored at 70°–80°C for 2 hr, cooled to 50°C, and filtered. The filtrate is cooled to 5°C, filtered again to remove inorganic salts, and acidified with 19 ml of concentrated hydrochloric acid. A colorless precipitate forms which is filtered and stirred with 30 ml of water at room temperature. The crystals are filtered and dried under reduced pressure to yield 11.25 gm (60.3%), mp 124°–125°C. The infrared absorption spectrum shows the characteristic absorption band at 1047 cm^{-1} and none at 1176 cm^{-1} indicative of the sulfonic acid group.

The disulfonyl chloride is prepared by oxidative chlorination of the corresponding diisothiuronium salt [86], using the method of Sprague and Johnson [87].

5. MISCELLANEOUS METHODS

(1) Preparation of methanesulfonyl fluoride [88].
(2) Addition of methanesulfonic acid to olefins [89].
(3) The cleavage of sulfides and sulfones to give sulfinic acid [90].
(4) Alkene sulfonic acid via alcohol–H_2SO_4 or from oxiranes and $NaHSO_3$ [91].
(5) The reaction of cyclohexene with dioxane–SO_2 complex to give cyclohexenesulfonic acid [92].

REFERENCES

1. E. E. Gilbert, "Sulfonation and Related Reactions." Wiley (Interscience), New York, 1965.
2. G. F. Lisk, *Ind. Eng. Chem.* **42,** 1746 (1950).
3. E. E. Gilbert and E. P. Jones, *Ind. Eng. Chem.* **50,** 1406 (1958).
4. E. E. Gilbert and E. P. Jones, *Ind. Eng. Chem.* **51,** 1148 (1959).
5. E. E. Gilbert and E. P. Jones, *Ind. Eng. Chem.* **52,** 629 (1960).
6. E. J. Carson, G. Flint, E. E. Gilbert, and H. R.Nychka, *Ind. Eng. Chem.* **50,** 276 (1958).
7. C. M. Suter and A. W. Weston, *Org. React.* **3,** 141 (1946).
8. C. M. Suter, "Organic Chemistry of Sulfur," pp. 94, 195. Wiley, New York, 1944.
9. J. S. Showell, J. R. Russell, and D. Swern, *J. Org. Chem.* **27,** 2853 (1962).
10. M. S. Kharasch, E. M. May, and F. R. Mayo, *J. Org. Chem.* **3,** 175 (1938).
11. F. M. Beringer and R. A. Falk, *J. Am. Chem. Soc.* **81,** 2997 (1959).
12. H. Meyer, *Justus Liebigs Ann. Chem.* **433,** 327 (1923).
13. E. Gebauer-Fulnegg, E. Riess, and S. Ilse, *Monatsh. Chem.* **49,** 41 (1928).
14. R. Pschorr, *Ber. Dtsch. Chem. Ges.* **34,** 3998 (1901).
15. O. N. Witt, *Ber. Dtsch. Chem. Ges.* **48,** 743 (1915).
16. J. K. Hunt, *Chem. Eng. News* **27,** 2504 (1949).
17. I. Tanasescu and M. Macarovici, *Bull. Soc. Chim. Fr.* [5] **5,** 1126 (1938).
18. H. E. Fierz-David and G. Stammn, *Helv. Chim. Acta* **25,** 3681 (1942).
19. C. M. Suter and G. A. Harrington, *J. Am. Chem. Soc.* **59,** 2577 (1937).
20. Authors' Laboratory.
21. O. Jacobsen, *Chem. Ber.* **15,** 1853 (1882); see Chapter 1 of this text.
22. A. F. Holleman and P. Caland, *Ber. Dtsch. Chem. Ges.* **44,** 2504 (1911).
23. D. C. J. Euwes, *Recl. Trav. Chim. Pays-Bas* **28,** 298 (1909).
24. Chemetron Corp., British Patent 968,874 (1964).
25. L. I. Smith and O. W. Cass, *J. Am. Chem. Soc.* **54,** 1612 (1932).
26. C. M. Suter, *J. Am. Chem. Soc.* **53,** 1114 (1931).
27. R. Behrend and M. Mertelsmann, *Justus Liebigs Ann. Chem.* **378,** 352 (1911).
28. G. Mohrmann, *Justus Liebigs Ann. Chem,* **410,** 373 (1915).
29. M. Senkowski, *Ber. Dtsch. Chem. Ges.* **23,** 2412 (1890).
30. M. Phillips, *J. Am. Chem. Soc.* **46,** 686 (1924).
31. L. F. Fieser and D. M. Bowen, *J. Am. Chem. Soc.* **62,** 2105 (1940).

32. J. H. Crowell and L. C. Raiford, *J. Prakt. Chem.* [2] **42,** 145 (1920).
33. K. Lauer, *J. Prakt. Chem.* [2] **138,** 81 (1933).
34. C. H. F. Allen and J. A. Van Allan, *Org. Synth. Collect. Vol.* **3,** 824 (1955).
35. M. N. Schultz and H. J. Lucas, *J. Am. Chem. Soc.* **49,** 299 (1927).
36. S. M. McElvain and M. A. Goese, *J. Am. Chem. Soc.* **65,** 2233 (1943).
37. J. L. Webb and A. H. Corwin, *J. Am. Chem. Soc.* **66,** 1456 (1944).
38. Pyridium Corp., British Patent 602,822 (1948).
39. R. T. Wendland, C. H. Smith, and R. Muracu, *J. Am. Chem. Soc.* **71,** 1593 (1949).
40. Rohm and Haas Bulletin on Benzoguanamine, SP-142, Rohm and Haas Co., Philadelphia, Pennsylvania, 1960.
41. R. M. Reed and H. V. Tarter, *J. Am. Chem. Soc.* **57,** 571 (1935).
42. S. Zuffanti, *J. Am. Chem. Soc.* **62,** 1044 (1940).
43a. F. M. McElvain, A. Jelinek, and K. Rorig, *J. Am. Chem. Soc.* **67,** 1578 (1945).
43b. R. Lantzsch, A. Marhold, and K. Fikehment, Ger. Offen. 2,545,644 (1977); *Chem. Abstr.* **87,** 38874 (1977).
43c. C. Bunyagidj, H. Piotrowska, and M. A. Aldridge, *J. Org. Chem.* **46,** 3335 (1981).
44. R. M. Beringer and R. A. Falk, *J. Am. Chem. Soc.* **81,** 2997 (1959).
45. W. M. Lauer and C. M. Langkammerer, *J. Am. Chem. Soc.* **56,** 1628 (1934).
46. H. J. Backer and A. E. Bente, *Recl. Trav. Chim. Pays-Bas* **54,** 601, 621 (1935).
47. M. S. Kharasch, R. T. E. Schenek, and F. R. Mayo, *J. Am. Chem. Soc.* **61,** 3093 (1939).
48a. E. E. Johnson and R. T. Adams, U.S. Patent 3,150,169 (1964).
48b. R. Tokosh, J. Barillo, and W. Urban, U.S. Patent 4,275,013 (1981).
49. E. E. Gilbert, *Chem. Rev.* **62,** 549 (1962).
50. R. J. Brooks and B. Brooks, Belgian Patent 636,074 (1963).
51. C. Courtot and J. Bonnet, *C. R. Hebd. Seances Acad. Sci.* **182,** 855 (1926).
52. A. Michael and A. Adair, *Ber. Dtsch. Chem. Ges.* **10,** 583 (1877).
53. L. Leiserson, R. W. Bost, and R. LeBaron, *Ind. Eng. Chem.* **40,** 508 (1948).
54. W. H. C. Rueggeberg, T. W. Sands, and S. L. Norwood, *J. Org. Chem.* **20,** 455 (1955).
55. C. M. Suter, P. B. Evans, and J. M. Kiefer, *J. Am. Chem. Soc.* **60,** 538 (1938).
56. P. Baumgarten, *Ber. Dtsch. Chem. Ges.* **59,** 1976 (1926).
57. H. H. Sisler and L. F. Audrieth, *Inorg. Synth.* **2,** 173 (1946).
58. C. S. Rondesvedt, Jr. and F. G. Bordwell, *Org. Synth. Collect. Vol.* **4,** 846 (1963).
59a. F. G. Bordwell and C. S. Rondestvedt, Jr., *J. Am. Chem. Soc.* **70,** 2429 (1948).
59b. A. E. Straus, W. A. Sweeney, R. House, and S. H. Sharman, U.S. Patent 3,721,707 (1973); *Chem. Abstr.* **79,** 5002t (1973).
59c. O. Okumura, K. Yaguchi, and M. Nagayama, Japan Kokai 73/62,722 (1973); *Chem. Abstr.* **79,** 145985r (1973).
60. W. F. Luder and S. Zuffanti, *Chem. Rev.* **34,** 349 (1944).
61. J. K. Weil, R. G. Bistline, Jr., and A. J. Stirton, *Org. Synth. Collect. Vol.* **4,** 862 (1963).
62. J. K. Weil, L. P. Witnauer, and A. J. Stirton, *J. Am. Chem. Soc.* **75,** 2526 (1952); P. A. Levine, T. Mori, and L. A. Mikeska, *J. Biol. Chem.* **75,** 354 (1927).
63. G. Collin, T. P. Hilditch, P. Marsh, and A. F. McLeod, *J. Soc. Chem. Ind., London* **52,** 272 (1933).
64. H. A. Young, *J. Am. Chem. Soc.* **59,** 811 (1937).
65. R. C. Murray, *J. Chem. Soc.* p. 739 (1933).
66a. H. Becker, *Recl. Trav. Chim. Pays-Bas* **54,** 205 (1935).
66b. G. Schroyer, E. F. Geiger, and J. Hensel, Ger. Offen. 2,504,235 (1976); *Chem. Abstr.* **86,** 4934f (1977).
66c. R. Hueter, U.S. Patent 2,277,325 (1942); S. L. Griolito and H. O. Hofmann, U.S. Patent 3,600,136 (1971); *Chem. Abstr.* **75,** 129353v (1971). R. M. Guertin, Ger. Offen. 1,811,768

(1969); F. Hubenett, U.S. Patent 4,280,966 (1981); J. S. Showell, J. R. Russell, and D. Swern, *J. Org. Chem.* **27**, 2853 (1962).
67a. U. Szalaiko and K. Lewandowski, *Przem. Chem.* **58**(7), 348 (1979); *Chem. Abstr.* **91**, 157202u (1979).
67b. H. L. Diamond, V. J. Pascarella, and A. C. Whitaker, U.S. Patent 3,392,895 (1968); C. E. Johnson and W. F. Wolff, U.S. Patent 2,697,722 (1954).
67c. R. Adams and C. S. Marvel, *Org. Synth. Collect. Vol.* **1**, 84 (1941).
68. E. H. Huntress and J. S. Autenrieth, *J. Am. Chem. Soc.* **63**, 3446 (1941).
69. E. H. Huntress and F. H. Carten, *J. Am. Chem. Soc.* **62**, 511 (1940).
70. R. B. Scott, Jr., J. B. Gayle, M. S. Heller, and R. E. Lutz, *J. Org. Chem.* **20**, 1165 (1955).
71a. R. B. Scott, Jr. and M. S. Heller, *J. Org. Chem.* **20**, 1149 (1955).
71b. G. F. Lisk, *Ind. Eng. Chem.* **40**, 1671 (1948); S. R. Detrick, W. H. Lockwood, and N. Whitman, U.S. Patent 2,462,730 (1949); H. Berthold, P. Jodl, H. Leue, G. Lipfert, E. Ukirauch, and H. Winter, German (East) Patent 117,875 (1976); *Chem. Abstr.* **85**, 62655f (1976); C. Bierge, M. Gellato, J. L. Seris, and J. Suberlucq, Ger. Offen. 2,447,133 (1975); *Chem. Abstr.* **83**, 186294r (1975).
71c. E. Plattner and C. Comninellis. Ger. Offen. 2,442,105 (1975); *Chem. Abstr.* **82**, 155334d (1975).
72. M. Kulka, *J. Am. Chem. Soc.* **72**, 1215 (1950).
73. A. T. Roos, H. Gilman, and N. J. Beaber, *Org. Synth. Collect. Vol.* **1**, 145 (1941).
74. C. S. Marvel and V. C. Sekera, *Org. Synth. Collect. Vol.* **3**, 366 (1955).
75. V. C. Sekera and C. S. Marvel, *J. Am. Chem. Soc.* **55**, 345 (1933).
76a. W. M. Ziegler and R. Connor, *J. Am. Chem. Soc.* **62**, 2596 (1940).
76b. J. M. Photis and L. A. Paquette, *Org. Synth.* **57**, 53 (1977).
77. H. G. Houlton and H. V. Tartar, *J. Am. Chem. Soc.* **60**, 544 (1938); A. V. Kuchin, L. I. Akhnetov, V. P. Yur'ev, and G. A. Tolstikov, *Zh. Obshch. Khim.* **49(26), 401 (1979)**; *Chem. Abstr.* **91**, 4686a (1979).
78. H. R. Todd and R. L. Shriner, *J. Am. Chem. Soc.* **56**, 1382 (1934).
79. W. E. Truce and A. M. Murphy, *Chem. Rev.* **48**, 69 (1951).
80. R. M. Hann, *J. Am. Chem. Soc.* **57**, 2166 (1935).
81. F. C. Whitmore and F. H. Hamilton, *Org. Synth. Collect. Vol.* **1**, 492 (1941).
82. H. Gilman, E. W. Smith, and H. J. Oatfield, *J. Am. Chem. Soc.* **56**, 1413 (1934).
83a. S. Krishna and H. Singh, *J. Am. Chem. Soc.* **60**, 794 (1928).
83b. M. S. A. Vrijland, *Org. Synth.* **59**, 88 (1977).
84. C. S. Marvel and N. A. Meinhardt, *J. Am. Chem. Soc.* **73**, 859 (1951).
85. M. T. Beachem, J. T. Shaw, G. D. Sargent, R. B. Fortenbaugh, and J. M. Salsburg, *J. Am. Chem. Soc.* **81**, 5430 (1959).
86. C. H. Grogan, L. M. Rice, and M. X. Sullivan, *J. Org. Chem.* **18**, 728 (1953).
87. J. M. Sprague and T. B. Johnson, *J. Am. Chem. Soc.* **59**, 2440 (1937).
88. L. Martin, U.S. Patent 3,920,738 (1975).
89. W. J. Roberts, C. L. Smart, and J. DiPietro, U.S. Patent 3,337,615 (1967).
90. W. E. Truce, D. P. Tate, and D. N. Burdge, *J. Am. Chem. Soc.* **83**, 2872 (1960).
91. A. Lambert and J. D. Rose, *J. Chem. Soc.* p. 46 (1949).
92. R. Sperling, *J. Chem. Soc.* p. 1925 (1949).

Index

A

Acetic anhydride, nitration of, 507–508
Acetonitrile derivatives, alkylation of, 571–573
Acetophenone, preparation of, 215, 224–226
3-β-Acetoxy-16-β-diazoacetylisopregn-5-en-20-one, preparation of, 486–487
Acetylacetone, 2-heptanone preparation from, 222
N-Acetyl-L-cysteine, preparation of, 331
Acetylenes
  condensation reactions and, 62–64, 86–91
  elimination reactions, 83–86
  miscellaneous methods, 94
  oxidation reactions, 91–93, 185–186
  rearrangement reactions, 93
Acetylnitrate, nitration by, 503–505
9-Acetylphenanthrene, preparation of, 214–215
Acid(s)
  halogenation of, 167–169
  mixed, nitration and, 529–535
Acid chlorides, halides and, 155–158
Acid derivatives
  hydrolysis
    amides, 271
    of esters, acylhalides, anhydrides and trihalides, 271–273
    nitriles, 270
    of 1,1,1-trihalomethyl derivatives, 273–274
  reduction of, 189–191
Acrolein, preparation of, 182–183
Acylation, nitrocompounds and, 524
Acylaziridine, reduction by lithium aluminum hydride, 190–191
Acylhalides
  condensation with amides, 323–330
  hydrolysis, carboxylic acids and, 271–273
  reaction with hydroxy compounds, 293–295
Acylnitrates, nitration with, 535–537
Addition reactions
  of amides, 347–348
  for thiocyanates, 369–370
Adipaldehyde, preparation using lead tetraacetate, 184–185
β-Alanine, preparation of, 392–393

Alcohols
  condensation reactions for, 105–111
  condensation with aldehydes, olefins, acetylenes, alkyl sulfates and oxides, 133–136
  conversion to alkyl halides, 149–150
  dehydration, olefins and, 41, 46–49
  hydrochlorination, general procedure for, 152–154
  hydrolysis reactions and, 102–105
  miscellaneous methods, 122–123
  oxidation, 115–120
    carboxylic acids and, 239–241
  primary, oxidation of, 181–184, 304–305
  reaction
    with carboxylic acids, 291–293
    with lactones, 303
  rearrangement reactions for
    aryl esters to phenolic ketones, 121
    aryl hydroxylamines, 120–121
    epoxides to allylic alcohols, 121
    ethers, 120
    sigmatropic, of olefinic alcohols, 121
  reduction reactions and, 12–13, 111–114
Aldehydes
  condensation reactions and, 64–65, 191–195
    miscellaneous methods, 195–196
  elimination reactions and, 196–198
    miscellaneous reactions, 198–199
  halogenation of, 167–169
  miscellaneous reactions, 199–200
  novel synthesis of aromatic nitriles from, 559–560
  oxidation reactions and
    of alkylgroups, 186–187
    carboxylic acids and, 239–241
    direct, 306
    of glycols, 184–185
    of olefins and acetylenes, 185–186
    of primary alcohols, 181–184
  reduction reactions and
    of nitriles, 187–189
    of other acid derivatives, 189–191
    miscellaneous reducing agents, 191
Aldol condensation
  alcohols and, 108–109
  of nitriles, 570–571

nitro compounds and, 524–526
olefins and, 65
Alkanethiols, boiling below 130°C, general procedure for preparation of, 589
Alkylation
  of acetonitrile derivatives, 571–573
  for esters, 300–302
  of hydrazine derivatives, 439–445
  nitro compounds and, 323–324
Alkylation and acylation, of amides, 342–345
Alkyl groups, oxidation of, 186–187
Alkyl halides, *see also* Halides
  reaction with metal sulfides or hydrosulfides, 587–588
N-Alkyl-N-nitrosoacyl amides, decomposition of, 472–473
1-Alkyl-1-nitroso-3-nitroguanidine, decomposition of, 475–476
N-Alkyl-N-nitroso-p-toluenesulfonamides, decomposition of, 469–471
Alkyl side chains, oxidation, carboxylic acids and, 238–239
Alkyl p-toluenesulfonates, preparation, general procedures for, 633–634
N-Allyacetamide, preparation of, 332–333
Allyl alcohol
  oxidation with manganese dioxide, 182–183
  reaction with carbon tetrachloride, 164–165
Allyldiethylamine, preparation of, 385
Allyl iodide, preparation of, 170
Allyl isocyanate, preparation of, 366
2-Allylphenol, preparation of, 72
Allylphenyl ether, preparation of, 72, 133
Amides
  addition reactions of, 347–348
  condensation reactions and
    with acid anhydrides, 330–335
    with acyl halides, 323–330
    with esters, 336–340
    Grignard reactions, 347
    N- and C-alkylation and N-acylation of, 342–347
    transamidation reactions, 340–342
  dehydration of, 551–553
  dehydration of ammonium salts and, 318–323
  hydration reactions with nitriles, 348–351
  hydrolysis
    amines and, 397–398
    carboxylic acids and, 271
    miscellaneous methods, 354–355
    oxidation reactions and, 351

rearrangement reactions
  Beckman rearrangement, 352
  Schmidt and Curtius rearrangements, 352–353
  Wolff rearrangement, 353
  reduction reactions and, 351–352, 402–403
Amine(s)
  addition to double bonds, 398–399
  aromatic, oxidation of, 541–542
  condensation reactions and
    Hofmann alkylation reactions, 380–390
    miscellaneous reactions, 390–401
  miscellaneous preparations, 420–429
  oxidation, nitro compounds and, 522–523
  primary aliphatic, diazotization of, 477–478
  rearrangement and related reactions for
    benzidine reaction, 414–415
    Curtius reaction, 418–420
    Hofmann reaction, 415–417
    Schmidt reaction, 418
  reduction reactions for, 401–402
    of amides, 402–403
    Leuckart reaction, 412–414
    of nitriles, 403–405
    of nitro compounds, 405–411
  secondary, preparation of, 383–384
  tertiary, preparation of, 384–385
  tertiary cyclic, preparation of, 387
2-Amino-2-methyl-1-propanol, preparation of, 407
p-Aminothiophenol, preparation of, 587
Ammonium salts, dehydration of, 318–323
Amyl acetate, preparation of, 292–293
Amyl 8-nitrooctanoate, preparation of, 511–512
Anhydrides
  condensation of amides with, 330–335
  hydrolysis, carboxylic acids and, 271–273
  reaction with hydroxy compounds, 295–296
Aniline, preparation of, 408–409
Anisoin, synthesis of, 109–111
Anisyl alcohol, preparation of, 119
9-Anthroic acid, preparation of, 241, 257–258
1-Anthroquinonediazonium chloride, preparation of, 490–491
Arndt–Eistert rearrangement, for carboxylic acids, 265–266
Aromatic compounds, reduction of, 9
Aromatic nucleus, carboxylation of, 255–259
Aryldiazonium fluoroborate, preparation of, 490
Azo compounds, reduction of, hydrazines and, 453–454

INDEX

B

Beckman rearrangement, amides and, 352
Benzaldehyde, preparation of, 197–198
Benzaldehyde p-phenylhydrazone, preparation of, 455–456
1,2-Benzanthracene, synthesis of, 29
Benzene, alkylation of, 16
Benzenediazonium chloride, preparation of, 489
Benzenethiol, oxidation to diphenyl disulfide, 598–599
Benzidine rearrangement, for amines, 414–415
Benzilic acid, rearrangement of, 252
Benzoic acid, preparation of, 249
Benzonitrile, preparation of, 553, 556
Benzophenone, reduction to 1,1-diphenylethane, 11–12
β-Benzoylacrylic acid, preparation of, 257
N-Benzyloctamide, preparation of, 321–322
6-Benzyloxy-5-methoxygramine, preparation of, 401
Biphenyl-2-acetic acid, preparation of, 265–266
Birch reduction, modification, olefins and, 70–71
N,N-Bis(2-chloroethyl)-2-(dichloroacetamido)acetamide, preparation of, 322–323
Bis(chloromethyl)durene, preparation of, 155
Bis(N-methyl-N-nitroso)terephthalamide, decomposition of, 471–472
O,N-Bis(trifluoroacetyl)hydroxylamine, nitriles and, 555
Bisulfites, addition to olefins, 626–627
Boord method, preparation of olefins and, 54–57
Bromobenzene, preparation of, 159–160
α-Bromobutyric acid, preparation of, 169
1-Bromohexane, preparation of, 173
6-Bromo-2-naphthyl-β-D-glucuronide, preparation of, 242
N-(9-Bromononyl)phthalimide, preparation of, 343–345
1-Bromooctane, reduction of, 14
1-Bromo-1-propene, propyne preparation and, 84
3-Bromothiophenol, preparation of, 591–592
1-Bromoundecanoamide, preparation of, 353
1,4-Butanedisulfinic acid, preparation of, 636–637
tert-Butyl acetate, preparation of, 294, 299–300
tert-Butylacetic acid, preparation of, 248–249
tert-Butyl acetoacetate, 3-isobutyl-2-heptanone preparation from, 222–224
tert-Butyl acrylate, preparation of, 300
n-Butylamine, preparation of, 382
N-tert-Butylbenzylamine, preparation, 384
n-Butyl bromide, preparation of, 150–151
n-Butyl n-butyrate, preparation of, 305–306
3-tert-Butyl-1-cyclohexene, preparation of, 51–52
N-tert-Butyl-N-cyclohexyl-2-chloroacetamide, preparation of, 329
n-Butyl isothiocyanate, preparation of, 374
n-Butylmercaptan, preparation of, 593
β-tert-Butylmercaptopropionitrile, preparation of, 596
N-tert-Butyl-N-phenylhydrazine, preparation of, 449–450
tert-Butyl sulfide, preparation of, 596
1-Butyne, preparation of, 84
2-Butyne, condensation reaction and, 63–64
Butyramide, preparation of, 318–320

C

Cannizzaro reaction
　for carboxylic acids, 251–252
　crossed, for alcohols, 119
Caproic acid, preparation of, 264–265
Carbon monoxide, carboxylic acids and, 278–280
Carbon tetrachloride, reaction with allyl alcohol, 164–165
Carbonyl compounds
　conversion to nitriles, 553–560
　reduction of
　　Clemmensen method, 10–11
　　thioketals or thioacetals, 11–12
　　Wolff-Kishner method, 9–10
Carboxylic acids
　bimolecular oxidation-reduction reactions for, 251–252
　carbonation of organometallic compounds and, 253–255
　carboxylation of aromatic nuclei and, 255–259
　condensation reactions for, 280–282
　　of active methylene compounds with chloral, 266–267
　　Diels–Alder reaction, 269–270
　　ethyl acetoacetate ester synthesis, 264–266

Knoevenagel condensation, 260–261
  other reactions, 261–263
  Perkin reaction, 259–260
  Reformatskii reaction, 268–269
  Strecker amino acid synthesis, 266
hydrolysis and elimination reactions, 275–277
hydrolysis of acid derivates and
  of amides, 271
  of esters, acyl halides, anhydrides and trihalides, 271–273
  of nitriles, 270
  of 1,1,1-trihalomethyl derivatives, 273–274
miscellaneous methods, 274–282
by oxidation of ketones and quinones, 246–248
  haloform reactions, 248–249
  Willgerodt reaction, 249–251
oxidation reactions, 274–275
  of alcohols and aldehydes, 239–241
  of alkyl side chains, 238–239
  catalytic oxidation with oxygen, 241–242
  of olefins, 242–246
reaction
  with alcohols, 291–293
  with olefins, 299–300
salts of, reaction with halides, 296–297
Cellosolve acrylate, preparation of, 297–298
Chugaev method, preparation of olefins and, 51–52
Chloral, condensation with active methylene compounds, 266–267
2-Chloroacetaldehyde cyanohydrin, preparation of, 568–569
cis-β-Chloroacrylamide, preparation of, 348–349
2-Chloroacrylonitrile, preparation of, 568–569
2-Chloroaniline, preparation of, 406
Chlorobenzene, reduction of, 15
p-Chlorobenzene sulfinic acid, preparation of, 635–636
p-Chlorobenzene sulfonyl chloride, preparation of, 633
p-Chlorocinnamic acid, decarboxylation of, 52–53
Chloroethers, reaction with olefins and organometallic reagents, 136
2-Chloro-3-hydroxycyclohexene, preparation of, 57
Chloromethylation reaction, halides and, 154–155
Chloromethyl ether, preparation of, 133–134

3-Chloro-4-methyl-α-methyl styrene, preparation of, 53–64
3-Chloro-3-methylstyrene, preparation of, 53–54
Chloronitrobenzenes, mixed, preparation of, 537
m-Chlorophenylacetic acid, preparation of, 247–248
p-Chlorophenyl isothiocyanate, preparation of, 372–373
4-Chlorophenylmethylcarbinol, preparation of, 114
2-Chloro-2-phenylpropane, preparation of, 163–164
3-Chloropropylene sulfide, preparation of, 597
p-Chlorostyrene, preparation of, 52–53
Δ^4-3-Cholestenone, preparation of, 213
Cinnamic acid, preparation of, 261–262
Citraconic acid, preparation of, 273
Claisen condensation, for carboxylic acids, 261–262
Claisen rearrangement
  of ethers, 120
  for olefins, 72
Cleavage reactions, halides and, 169–170
  Hunsdiecker reaction, 172–173
  Sandmeyer and Schiemann reaction, 170–172
Clemmensen method, for reduction of carbonyl compounds, 10–11
Condensation reactions
  for acetylenes, 86–91
  for alcohols and phenols, 105–111
  aliphatic diazo compounds and, 486–488
  of amides
    with acid anhydrides, 330–335
    with acyl halides, 323–330
    with esters, 336–340
    Grignard reactions, 347
    N- and C-alkylation and N-acylation of, 342–347
    transamidation reactions, 340–342
  for amines
    Hofmann alkylation of ammonia and amines, 380–390
    miscellaneous reactions, 390–401
  for carboxylic acids, 280–282
    of active methylene compounds with chloral, 266–267
    Diels–Alder reaction, 269–270
    ethyl acetoacetate ester synthesis, 264–266
    Knoevenagel condensation, 260–261

other reactions, 261-263
Perkin reaction, 259-260
Reformatskii reaction, 268-269
Strecker amino acid synthesis, 266
for disulfides, 597-598
for esters
  of acyl halides and hydroxy compounds, 293-295
  of alcohols and carboxylic acids, 291-293
  alkylation reactions, 300-302
  of anhydrides with hydroxy compounds, 295-296
  of carboxylic acids with olefins, 299-300
  ester interchange, 297-299
  of halides with salts of carboxylic acids, 296
  of lactones with hydroxy compounds, 303
  miscellaneous reactions, 303-304
  preparation of lactones, 302-303
for ethers and oxides, 131-132
  of alcohols with aldehydes, olefins, acetylenes, alkyl sulfates and oxides, 132-136
  of chloroethers with olefins and organometallic reagents, 136
  Darzens glycidic ester synthesis, 138-139
  of oxiranes, 136-138
  Williamson synthesis, 132-133
of existing nitriles, 570-571
halides and
  acid chlorides, 155-158
  chloromethylation reaction, 154-155
  conversion of alcohols to alkylhalides, 149-154
  halogenation of aldehydes, ketones and acids, 167-169
  halogenation reactions, 158-161
  reactions of olefins with halogens and halogen derivatives, 161-167
for hydrocarbons
  coupling reactions, 23-27
  Diels-Alder reaction, 22-23
  Friedel-Crafts reaction, 15-19
  hydrocarbon polymers, 19-20
  small ring syntheses, 20-22
for isocyanates, 364-366
for isothiocyanates, 371-374
of ketones, 213-224
for mercaptans
  addition of hydrogen sulfide to olefins, 592-593
  hydrolysis of xanthates, 591-592
  miscellaneous methods, 593-594

reaction of metal sulfides with alkyl halides, 587-588
reaction of organometallics with sulfur, 591
reaction of thiourea with active halides, 588-591
for nitriles
  alkylation of acetonitrile derivatives, 571-573
  condensation of existing nitriles, 570-571
  cyanoethylation, 573-575
  cyanohydrin formation, 567-570
  Sandmeyer reaction, 566-567
  using cyanide salts, 570-571
for olefins, 73-74
  acetylenes and, 62-64
  of aldehydes and ketones, 64-65
  Diels-Alder reaction, 67-68
  Grignard reaction, 65-67
  Wittig synthesis, 59-62
for sulfides
  mercaptylation of double bonds, 596
  miscellaneous methods, 597
  preparation of episulfides from epoxides, 596-597
  reaction of metallic sulfides with halides, 595
  reaction of sodium or other metal mercaptides with active halides, 594-595
for sulfones, 612-616
for thiocyanates, 369
Coupling reactions, of hydrocarbons, 23-27
Crossed Cannizzaro reaction, for alcohols, 119
Curtius rearrangement
  amides and, 352-353
  for amines, 418-420
Cyanates, preparation of, 362-363
Cyanide(s), see also Nitriles
Cyanide salts, preparation of nitriles with, 560-566
  in aliphatic nitrile preparation, 560-565
  in aromatic nitrile preparation, 565-566
1-Cyanododecane, preparation of, 564-565
Cyanoethylation, nitriles by, 573-575
Cyanohydrin, formation, nitriles via, 567-570
1-Cyanooctane, preparation of, 563-564
p-Cyanophenylmethylcarbinol acetate, pyrolysis of, 51
p-Cyanostyrene, preparation of, 51
Cyclic enol acetates, nitration of, 512-514
Cyclic enolates, nitration of, 512-514

Cyclodecyne, preparation of, 92–93
Cyclohexene
  bromination of, 162–163
  preparation of, 46
cis-4-Cyclohexene-1,2-dicarboxylic anhydride, preparation of, 67–68
Cyclohexylcarboxamide, preparation of, 350–351
Cyclohexylhydrazine hydrogen sulfate, preparation of, 445–446
Cyclohexyl isocyanate, preparation of, 366–367
(±)-trans-2-Cyclohexyloxycyclopropylamine, preparation of, 419–420
Cyclopentene, from 2-chloro-3-hydroxycyclohexene, 57
Cyclopropanecarboxaldehyde, preparation of, 190–191

D

Darzens glycidic ester synthesis, 138–139
Darzens reaction, for aldehydes, 197
Decomposition reactions
  for isocyanates, 366–367
  for isothiocyanates, 374
Dehydrogenations, hydrocarbons and, 29–30
Dehydrohalogenation reaction, preparation of olefins and, 53–58
Delépine reaction, amines and, 391–392
Diallyl isophthalate, preparation of, 67
Diamines, monoalkyl derivatives, preparation of, 387–390
$\alpha_3\delta$-Diaminoadipic acid, preparation of, 393–394
Di(2-amino-4-chlorophenyl) sulfone, preparation of, 407–408
1,8-Diamino-3,6-dioxaoctane, preparation of, 394–396
2,4-Diamino-1-naphthol dihydrochloride, preparation of, 409–410
Di-n-amyl disulfide, preparation of, 598
β,β-Dianisylacrylic acid, preparation
  using oxalyl chloride, 257-259
  using phosgene, 259
Diazo compounds, aliphatic, 467–469
  condensation reactions, 486–488
  by decomposition of N-nitroso compounds by alkalies, 469–477
  by diazotization of primary amines, 477–478
  by diazo transfer reactions, 483–488

miscellaneous preparations, 492–493
  by oxidation of hydrazones, 479–483
  by reaction of dichlorocarbene with hydrazine, 478–479
Diazoethane, preparation of, 473, 474
Diazomethane, preparation of, 469–470, 472, 479
Diazonium groups, replacement by nitro groups, 540
Diazonium salts
  aromatic, preparation of, 488–492
  preparation of solutions in aqueous media, 488–489
  stabilized, 489–492
Diazo-n-pentane, preparation of, 476
1-Diazopropane, preparation of, 475
2-Diazopropane, preparation of, 480–481
Diazo transfer reactions, aliphatic diazo compounds and, 483–485
2,4-Dibenzyloxy-5-methoxy-β-nitrostyrene, preparation of, 526
Dibenzyl sulfide, preparation of, 595
Dibenzyl sulfone, preparation of, 613
1,3-Dibromoacetone, preparation of, 167–169
3,3'-Dibromobenzidine, preparation of, 415
2-Dibromobutane, 1-butyne preparation from, 84
1,2-Dibromocyclohexane, preparation of, 162–163
1,1-Dibromo-2,2-diphenylcyclopropane, preparation of, 165–167
2,6-Dibromo-4-nitrophenol, preparation of, 160–161
1,2-Di(2-bromophenyl)hydrazine, preparation of, 454
Di-n-butyl disulfide, preparation of, 599
cis-2,3-Dichloroacrylonitrile, preparation of, 552
3,5-Dichloro-4-nitrobenzonitrile, preparation of, 541
2,5-Dichloroterphthaloyl chloride, preparation of, 156–157
1,2-Dicyanoacenaphthylene, preparation of, 565–566
N,N-Dicyclohexylbenzamide-O-diazonium fluoroborate, preparation of, 492
Diels–Alder reaction
  for carboxylic acids, 269–270
  hydrocarbons and, 22–23
  with nitriles, 579
  nitro compounds and, 529
  olefins and, 67–68

2,2'-Diethylbiphenyl, synthesis of, 25
Diethyl sec-butylmalonate, preparation of, 301–302
1,1-Diethylhydrazine, preparation of, 448–449
N,N-Diethyl-1-hydroxy-4-methylcyclohexaneacetamide, preparation of, 346
Diethylmaleate, hydrogenation of, 5–6
Diethyl nitromalonate, preparation of, 506–507
Diethyl succinate, from diethyl maleate, 5–6
Diglycidyl isophthalate, preparation of, 294–295
Dihalides, dehalogenation of, 58–59
1,1-Dihalocyclopropanes
  insertion of a carbon atom in, 54–57
  preparation, general procedure for, 165
4,4'-Dihydrazinooctafluorobiphenyl, preparation of, 441–442
2,5-Dihydroethylbenzene, preparation of, 70–71
N-(1,1-Dihydroperfluorobutyl)acrylamide, preparation of, 327–328
Dihydroxyoctadecane, preparation from oleyl alcohol, 105
1,1-Diisobutylhydrazine, preparation of, 450
N,N-Diisopropylfumaramide, preparation of, 351
Dimesityl sulfone, preparation of, 613, 616
Dimethylacetophenones, preparation of, 215–217
N,N-Dimethylalkylamines, preparation of, by reduction of quaternary ammonium salts, 385–387
p-N,N-Dimethylaminobenzonitrile, preparation of, 559–560
2(N,N-Dimethylamino)-5-methylbenzaldehyde, preparation of, 183–184
3,4-Dimethylaniline, preparation of, 418
3,5-Dimethyl-2-fluoro-1-bromobenzene, preparation of, 171–172
2,3-Dimethylheptanoic acid, preparation of, 243–244
2,2-Dimethyl-1-hexanol, preparation of, 106
α,α-Dimethyl-β-hydroxypropionaldehyde, preparation of, 108–109
2,3-Dimethylnaphthoquinone, preparation of, 210–211
N,N-Dimethyl-(+)-neomenthylamine, preparation of, 386–387
2,2-Dimethyl-3-pentanone, preparation of, 226–227
3,3-Dimethyl-2-pentanone, preparation of, 226–227

Dimethylphenylethanol, synthesis of, 112
2(2,4-Dimethylphenyl)ethanol, preparation of, 108
Dimethylstyrenes, preparation of, 46–49
Dimethyl sulfoxide, oxidation of heptyltosylate and, 186–187
5,6-Dimethyltetralin, dehydrogenation of, 18
m,m'-Dinitrobenzophenone, preparation of, 534–535
m,p'-Dinitrobenzophenone, preparation of, 534–535
2,4-Dinitrobromobenzene, preparation of, 531
Dinitrogen tetroxide, nitration by, 501–503
2,4-Dinitro-1-naphthol, preparation of, 539–540
Di-p-nitrophenylacetylene, preparation of, 91
2,4-Dinitrophenylhydrazine, preparation of, 441
Diphenic acid, preparation of, 247
Diphenylacetylene, preparation of, 86
1,1-Diphenyl-2-bromo-3-acetoxy-1-propene, preparation of, 57–58
Diphenyl disulfide, preparation of, 598–599
1,1-Diphenylethane, preparation of, 12–13
1,2-Diphenylhydrazine, preparation of, 453–454
Diphenylketene
  preparation of, 217–219
  reaction with vinyl acetate, 219
1,2-Distyrylbenzene, preparation of, 60–61
Disulfides
  condensation reactions for, 597–598
  miscellaneous methods for, 599
  oxidation reactions for, 598–599
1,4-Dithiane-1-oxide, preparation of, 607
Dithiols, boiling above 130°C, general procedure for preparation of, 589–590
N,N'-Di(trifluoro)-meso-2,6-diaminopimelic acid, preparation of, 334–335
Duresulfonic acid, preparation of, 624

E

Elbs reaction, hydrocarbons and, 29
Elimination reactions
  of acetylenes, 83–86
  for aldehydes, 196–199
  for epoxides, 140
  for ethers, 139
  halides and, 169–173
  of hydrocarbons, 27–29

ketones and, 224–226
for nitriles
  conversion of carbonyl compounds, 553–560
  dehydration of amides, 551–553
for preparation of olefins, 72–73
  dehalogenation of dihalides, 58–59
  dehydration of alcohols, 41–49
  dehydrohalogenation, 53–58
  pyrolysis, 49–53
Enamines, cycloadducts with ketones, 220–221
Enzyme reaction, for carboxylic acids, 282
Episulfides, preparation from epoxides, 596–597
Epoxides, preparation of episulfides from, 596–597
2,3-Epoxy-*trans*-decalin, preparation of, 142
Eschweiler–Clarke modification, 413–414
Esters
  condensation reactions for, 290–291
    of acyl halides and hydroxy compounds, 293–295
    of alcohols and carboxylic acids, 291–293
    alkylation reactions, 300–302
    with amides, 336–340
    of anhydrides with hydroxy compounds, 295–296
    of carboxylic acids with olefins, 299–300
    ester interchange, 297–299
    of halides with salts of carboxylic acids, 296
    of lactones with hydroxy compounds, 303
    miscellaneous reactions, 303–304
    preparation of lactones, 302–303
  hydrolysis, carboxylic acids and, 271–273
  interchange of, 297–299
  miscellaneous methods, 308–309
  oxidation of, 306
  oxidation reactions for
    direct, of aldehydes and ketones, 306
    of ethers, 306
    of primary alcohols, 304–305
    Tischtschenko–Cannizzaro reaction of aldehydes, 305–306
  rearrangement of
    n-alkyl phenyl ethers, 120
    aryl alkyl ethers to alcohols, 120
    Claisen rearrangement, 120
  rearrangement reactions for
    Arndt–Eistert reaction, 307
    Favorskii reaction, 307–308
  reduction reactions for, 307

Ethers (oxides)
  condensation reactions
    of alcohols with aldehydes, olefins, acetylenes, alkyl sulfates and oxides, 132–136
    of chloroethers with olefins and organometallic reagents, 136
    Darzens glycidic ester synthesis, 138–139
    of oxiranes, 136–138
    Williamson synthesis, 132–133
  elimination reactions and, 139–140
  miscellaneous methods, 143
  oxidation reactions for, 140–143
1-Ethoxy-2-propanol, preparation of, 135–136
Ethyl acetoacetic ester synthesis, carboxylic acids and, 264–266
2-Ethylaniline, preparation of, 410–411
p-Ethylbenzyl acetate, preparation of, 297
Ethyl *n*-butylmalonate, preparation of, 262–263
α-Ethylbutyrolactone, preparation of, 302–303
α-Ethylcinnamaldehyde, preparation of, 65
Ethyl cinnamate, preparation of, 261–262
2-Ethylcyclohexanone, preparation of, 212
Ethyl 2-diazo-3-hydroxybutyrate, preparation of, 487–488
N-Ethyl-1,1-dihydroheptafluorobutylamine, preparation of, 402–403
Ethylenediamine, 1-ethynyl-1-cyclohexanol and, 87–89
Ethyl 4-ethyl-3-hydroxy-2-octanoate, preparation of, 268–269
α-Ethylglutarimide, preparation of, 335
Ethylhydrazine, preparation of, 440
1-Ethyl-2-isopropylhydrazine, preparation of, 451–452
Ethyl β-methylaminopropionate, preparation of, 398–399
Ethyl α-nitro-α-carbethoxy-β-(3-indole) propionate, preparation of, 523–524
Ethyl α-nitrovalerate, preparation of, 518–519
Ethyl oleate, reduction to ethyl stearate, 6–8
N-Ethylperfluorobutyramide, preparation of, 336–338
Ethyl stearate
  formation from ethyl oleate, 6–8
  reduction to 1-octadecanol, 112–114
1-Ethynyl-1-cyclohexanol, preparation of, 87–89
Exchange reactions, for isocyanates, 367–368

INDEX 649

F

Flavoprotein D-oxynitrilase
  in nitrile preparation, 569
Fluorene-9-carboxylic acid, preparation of, 254
[2-(2-Fluoro-3,4-dimethoxyphenyl)ethyl]-amine, preparation of, 404–405
Formyl fluoride, preparation of tolualdehyde and, 193–194
3-Formylindole, preparation of, 194–195
Friedel–Crafts reaction
  for alcohols, 108
  alkylation of hydrocarbons and, 15–19, 28
  for carboxylic acids, 256–259
  dimethylacetophenones and, 215–217
  for sodium p-fluorophenylsulfinate dihydrate, 635
  for sulfones, 613, 616
Furoic acid, preparation of, 241, 251–252
N-(2-Furyl)benzamide, preparation of, 347

G

Gabriel condensation, Ing-Manske modification, 392–396
Gattermann synthesis, for aldehydes, 192–193
Glycidyl benzoate, preparation of, 136–137
Glycols, oxidation of, 184–185
Grignard reaction
  acetophenone preparation and, 215
  for alcohols, 106–107
  amides and, 347
  carboxylic acids and, 253–254
  hydrocarbon coupling reactions and, 23, 24–25
  ketone preparation and, 214–215
  olefins and, 65–67

H

Halides
  active, reaction with thiourea, 588–591
  condensation reactions for
    acid chlorides, 155–158
    chloromethylation reaction, 154–155
    conversion of alcohols to alkyl halides, 149–154
    halogenation of aldehydes, ketones and acids, 167–169
    halogenation reactions, 158–161
    reactions of olefins with halogens and halogen derivatives, 161–167
  elimination and cleavage reactions and, 169–170
  Hunsdiecker reaction, 172–173
  Sandmeyer and Schiemann reactions, 170–172
  miscellaneous methods and, 173–175
  reaction of metallic sulfides with, 595
  reaction of sodium or other metal mercaptides with, 594–595
  reaction with salts of carboxylic acids, 296–297
  treatment of amines with, 380–382
Haloform reactions, for carboxylic acids, 248–249
Halogenation reaction, for halides, 158–161
Halogens, reaction with olefins, 161–167
Hell–Volhard–Zelinsky reaction, halides and, 169
Heptaldehyde, preparation of, 186–187
n-Heptane-γ-carboxylic acid, preparation of, 240
Heptanoic acid, preparation of 1-bromohexane from, 173
2-Heptanone, preparation of, 222
Heptanonitrile, preparation of, 554–555
Heptyl tosylate, oxidation by dimethyl sulfoxide, 186–187
1,4-Hexadiene, preparation of, 54–57
2,4-Hexadiyne-1,6-diol, preparation of, 91–92
Hexamethylbenzene, formation of, 26–27
Hexamethylbicyclo[2-2.0]-2,5-hexadiene, 63–64
Hexamethyl–Dewar benzene, preparation of, 63–64
1-Hexene, conversion to n-hexane by hydroboration method, 4–5
1-Hexene oxide, preparation of, 142
2-Hexenoic acid, preparation of, 260–261
n-Hexylcyclopropane, preparation of, 21–22
n-Hexyl isocyanate, preparation of, 367–368
1-Hexyne, hydroboration to 1-hexene, 69–70
Hofmann alkylation reaction, for amines, 380–390
Hofmann rearrangement, for amines, 415–417
Homophthalic acid, preparation of, 245–246
Huang–Minlon method, for preparation of 2-(n-octyl)naphthalene, 10
Hunsdiecker reaction, halides and, 172–173
Hydrazides, preparation of, 457–459, 461–463
Hydrazines
  preparation of, 438, 460–461
  alkylation of, 439–445

reductive methods, 448–455
  syntheses involving formation of
    nitrogen-nitrogen bonds, 445–448
  reaction of dichlorocarbene with,
    478–479
Hydrazones
  oxidation of, 479–483
  preparation of, 455–457, 461
  reduction of, hydrazines and, 451–453
Hydroboration, amines and, 390–391
Hydrocarbons
  condensation reactions
    coupling reactions, 23–27
    Diels–Alder reaction, 22–23
    Friedel–Crafts alkylation, 15–19
    polymerization, 19–20
    small ring syntheses, 20–22
  by dehydrogenations, 29–30
  elimination reactions of, 27–29
  miscellaneous methods, 31
  reduction reactions
    of alcohols, 12–13
    of aromatic compounds, 9
    of carbonyl compounds, 9–12
    of halides, 13–15
    of unsaturated compounds (olefins),
      3–8
Hydrogen sulfide, addition to olefins and other
  unsaturated compounds, 592–593
Hydrolysis reactions, for alcohols and phenols,
  102–105
Hydrotropaldehyde, preparation of, 197
2-exo-Hydroxy-4-azahomoadamantan-5-one,
  preparation of, 349–350
Hydroxy compounds
  reaction with acyl halides, 293–295
  reaction with anhydrides, 295–296
N-(2-Hydroxyethyl)acetamide, preparation of,
  340
β-2-Hydroxyethyl naphthyl ether, preparation
  of, 107–108
N-Hydroxymethylphenylacetamide, preparation
  of, 342

I

2-Iodoformanilide, preparation of,
  340–341
Isobutylene sulfide, preparation of, 597
Isobutyl ethyl ether, preparation of, 135
3-Isobutyl-2-heptanone, preparation of,
  222–224

Isocyanates, preparation of
  condensation reactions, 364–366
  decomposition reactions, 366–367
  exchange reactions, 367–368
  miscellaneous reactions, 368–369
  oxidation reactions, 368
  pyrolysis reaction, 368
  rearrangement reactions, 368
Isomerization reactions, for olefins, 71–72,
  74–75
Isophorone oxide, preparation of, 142–143
Isophthalic acid, preparation of, 251
1-Isopropyl-1-(2-morpholinoethyl)-
  1-phenylacetonitrile, preparation of,
  571–572
p-Isopropylphenol, preparation
  from diazonium compound, 103
  from sodium cumenesulfonate, 104
Isothiocyanates, preparation of
  condensation reactions, 371–374
  decomposition reactions, 374
  miscellaneous reactions, 374–375

J

Jacobsen reaction, of 6,7-dimethyltetralin, 18

K

Ketenes, cyclo adducts with enamines,
  220–221
Ketones
  condensation reactions and, 64–65,
    213–224
  conversion to alcohols and reduction to
    alkanes, 12–13
  direct oxidation of, 306
  elimination reactions and, 224–226
  halogenation of, 167–169
  miscellaneous methods, 227–229
  oxidation, 207–213
    carboxylic acids and, 246–248
      haloform reactions, 248–249
      Willgerodt reaction, 249–251
  rearrangement reactions and, 226–227
β-Keto sulfoxides
  acetophenone preparation from, 224–226
Knoevenagel condensation, for carboxylic
  acids, 260–261
Kolbe–Schmitt reaction, for carboxylic acids,
  256

INDEX

L

Lactones
  preparation of, 302–303
  reaction with alcohols, 303
Lead tetraacetate, preparation of adipaldehyde and, 184–185
Letts reaction, 577
Leuckart reaction, for amines
  classical, 412–413
  Eschweiler-Clarke modification, 413–414
Lithium acetylide, 1-ethynyl-1-cyclohexanol and, 87–89
Lithium aluminum hydride
  preparation of aldehydes and, 190–191
  reduction of halides and, 13, 14
Lithium reagents, carboxylic acids and, 254
Lithium triethoxyaluminum hydride, reduction of nitriles and, 189

M

McFadyen–Stevens reaction, for aldehydes, 197–198
Magnesium, reduction of halides and, 13, 15
Malonic ester synthesis, carboxylic acids and, 262–263
Manganese dioxide, oxidation of allyl alcohol and, 182–183
Mannich reaction
  amines and, 400–401
  nitro compounds and, 527–528
Markovnikov hydration, of olefins, general procedure for, 116–117
Meerwein reduction reaction, for alcohols, 114
1-Menthone, preparation of, 209–210
Mercaptans, condensation reactions for
  addition of hydrogen sulfide to olefins, 592–593
  hydrolysis of xanthates, 591–592
  miscellaneous methods, 593–594
  reaction of metal sulfides with alkyl halides, 587–588
  reaction of organometallics with sulfur, 591
  reaction of thiourea with active halides, 588–591
Mercaptides, of sodium or other metal, reaction with active alkyl or aryl halides, 594–595
Mercaptylation, of double bonds, 596
Metal sulfides, reaction with alkyl halides and, 587–588

4-Methoxybenzonitrile, preparation of, 554
p-Methoxydiphenylacetylene, preparation of, 90
α-Methoxyisobutyric acid, preparation of, 273–274
6-Methoxy-2-nitrobenzamide, preparation of, 349
N-(2-Methoxy-4-nitrophenyl)-benzoylacetamide, preparation of, 338–339
p-Methoxyphenylacetic acid, preparation of, 250–251
2-(4-Methoxyphenyl)-N-methylacetamide, preparation of, 342–343
Methyl acetate, preparation of, 292
N-Methylacrylamide, preparation of, 326–327
α-Methylallyl methyl sulfide, preparation of, 594
α-Methylallyl phenyl sulfide, preparation of, 595
α-Methylbutyric acid, preparation of, 253–254
Methyl chloride, reduction of, 14–15
Methyl 2-chloro-3-nitropropionate, preparation of, 509–510
1-Methylcyclobutylamine, preparation of, 396–397
2-Methylcyclohexanone, preparation of, 211–212
4-Methylcyclohexanone, preparation of, 208
trans-2-Methylcyclohexylamine, preparation of, 391
trans-2-Methylcyclopentanol, preparation of, 117–119
Methyl-α-diazo-α-[(methylthio)carbonyl]acetate, preparation of, 485
N-Methyldihydropyran-2-methylamine, preparation of, 383
Methylene compounds
  active
    condensation reactions with nitro compounds, 523–529
    condensation reactions for olefins and, 64–65
    condensation with chloral, 266–267
    nitration and, 506–507, 510–512
Methylenecyclohexane, preparation of, 60
5-Methylhexanoic acid, preparation of, 244
Methyl hydrogen phthalate, preparation of, 295–296
Methyl 1-methylcyclohexane carboxylate, preparation of, 308
Methyl neopentyl ketone, preparation of, 211

2-Methyl-5-nitrobenzonitrile, preparation of, 538
2-Methyl-3-nitrocinnamic acid, preparation of, 259–260
2-Methyl-*endo*-norborneol, hydrochlorination of, 154
2-Methyloctanonitrile, preparation of, 562
1-(Methyl-3-phenylpropyl)piperidine, preparation of, 412–413
α-Methylstyrene, hydrochlorination of, 163–164
1-Methyl-2-(*p*-tolyl)piperidine, preparation of, 413–414
Michael condensation, nitro compounds and, 526–527
Monoacetylethylene diamine, preparation of, 339–340
Monomethylethylenediamine, preparation of
 method 1, 388–389
 method 2, 389–390
Monothiols, boiling above 130°C, general procedure for preparation of, 589–590

N

1-Naphthonitrile, preparation of, 566–567
N-(α-Naphthyl)acetamide, preparation of, 341
Neohexane, formation of, 23, 24–25
Nitration, direct
 of aliphatic compounds, 500–514
 of aromatic compounds, 529–539
Nitric acid, nitration by, 529–535
Nitric acid–sodium nitrite, nitration by, 505–506
Nitriles
 aliphatic, preparation by condensation reactions, 560–565
 aromatic, preparation by condensation reactions, 565–566
 by dehydration reactions, 551
 hazards and safe handling practices, 551
 hydration reactions, amides and, 348–351
 hydrolysis, carboxylic acids and, 270
 preparation of
  condensation reactions, 560–575
  elimination reactions, 551–560
  miscellaneous reactions, 577–580
  oxidation reactions, 575–576
  reduction reactions, 576–577
 reduction of, 187–189, 403–405
m-Nitrobenzoic acid, preparation of, 272–273
4-Nitrobenzonitrile, preparation of, 575–576

p-Nitrobromobenzene, preparation of, 173, 530–531
3-Nitro-2-butyl acetate, preparation of, 505
N-(2-Nitrobutyl)diethylamine, preparation of, 528
Nitro-*tert*-butyl methyl ether, preparation of, 134–135
p-Nitrochlorobenzene, preparation of, 171
6-Nitrocholesteryl acetate, preparation of, 506
Nitro compounds
 aliphatic
  condensation of active methylene compounds, 523–529
  direct nitration, 500–514
  indirect nitration, 514–520
  miscellaneous methods, 542–543
  oxidation of oximes and amines, 520–523
 aromatic
  direct nitration, 529–539
  indirect nitration, 539–541
  miscellaneous methods, 543–544
  oxidation methods, 541–542
 bimolecular reduction of, hydrazines and, 454
 reduction of using
  activated iron and water, 405–406
  catalytic hydrogenation, 411
  hydrazine hydrate, 407–408
  iron and ferrous sulfate, 406–407
  sodium hydrosulfite, 409–410
  sodium sulfide and related compounds, 410–411
  tervalent phosphorus reagents, 411
  tin and hydrochloric acid, 408–409
1-Nitrocyclohexene, preparation of, 519–520
2-Nitrocyclooctanone, preparation of, 511–512
1-Nitrocyclooctene, preparation of, 502–503
Nitrogen, oxides of, nitration with, 538–538
4-Nitroheptane, preparation of, 520–522
2-Nitro-4-methylcyclohexanone, preparation of, 512
3-Nitro-3(and 4)-methylcyclopentanone, preparation of, 513–514
4-Nitro-4-methylheptadiamide, preparation of, 527
4-Nitro-4-methylpimeldiamide, preparation of, 347–348
Nitronium tetrafluoroborate, nitration with, 537–539
2-*endo*-Nitronorbornene, preparation of, 529
1-Nitrooctane, preparation of, 514, 516–517
2-Nitrooctane, preparation of, 517–518

INDEX

p-Nitrophenol, bromination of, 160–161
1-(4-Nitrophenyl)cinnamonitrile, preparation of, 570
p-Nitrophenyl isocyanate, preparation of, 364–366
p-Nitrophenyl isothiocyanate, preparation of, 373–374
N-Nitrosamines, reduction of, hydrazines and, 448–450
N-Nitroso-β-alkylaminoisobutyl ketones, decomposition of, 476–477
N-Nitroso-N-alkylurethanes, decomposition of, 473–475
N-Nitroso compounds
  decomposition with alkalies, 469–477
  oxidation, nitro compounds and, 542
2-Nitro-2,4,4-trimethylpentane, preparation of, 522–523
Nitryl chloride, nitration with, 509–510
Nomenclature, of cyanates, isocyanates, thiocyanates and isothiocyanates, 361
[3,3,1]Non-6-ene-3-carbonitrile, preparation of, 558
Nonylamine, preparation of, 416–417
3-Nonyne, preparation of, 87
Nylon 6/10, 329–330

O

1-Octadecanol, synthesis of, 112–114
n-Octaldehyde, preparation of, 185–186, 188
1-Octanol, preparation of, 107
n-Octyl mercaptan, preparation of, 590–591
2-(n-Octyl)naphthalene, preparation by
  Clemmensen method, 11
  Huang-Minlon method, 10
1-Octyne, oxidation of, 185–186
Oleamide, preparation of, 341–342
Olefins
  addition
    of bisulfite to, 626–627
    of hydrogen sulfide to, 592–593
  chloronitration with nitryl chloride, 509–510
  condensation reactions for
    acetylenes and, 62–64
    aldehydes and ketones and other active methylene compounds, 64–65
    coupling and Grignard reactions, 65–67
    Diels–Alder reaction, 67–68
    Wittig synthesis, 59–62
  elimination reactions and
    dehalogenation of dihalides, 58–59

  dehydration of alcohols, 41–49
  dehydrohalogenation, 53–58
  pyrolysis, 49–53
Markovnikov hydration of, general procedure for, 116–117
miscellaneous methods, 72–75
nitration using
  acetyl nitrate, 503–505
  dinitrogen tetroxide, 501–503
  nitric acid–sodium nitrite, 505–506
oxidation, 185–186
  carboxylic acids and, 242–246
  peroxidation to give oxiranes, 140–141
preparation of, 39–41
  computer-assisted analysis, 41, 42–45
  reaction with carboxylic acids, 299–300
  reaction with halogens or halogen derivatives, 161–167
Olemorpholide, preparation of, 320–321
Oleyl alcohol, dihydroxyoctadecanes from, 105
Oppenauer oxidation
  for aldehydes, 183–184
  ketones and, 212
Organometallic compounds
  carbonation of, carboxylic acids and, 253–255, 277–278
  reaction with chloroethers, 136
  reaction with sulfur, 591
Oxalyl chloride, carboxylic acids and, 257–259
Oxidation reactions
  for acetylenes, 91–93
  for alcohols and phenols, 115–120
  for aldehydes
    of alkyl groups, 186–187
    of glycols, 184–185
    of olefins and acetylenes, 185–186
    of primary alcohols, 181–184
  amides and, 351
  for aromatic nitro compounds, 541–542
  for carboxylic acids
    of alcohols and aldehydes, 239–241
    of alkyl side chains, 238–239
    catalytic oxidation with oxygen, 241–242
  for disulfides, 598–599
  for esters
    direct, of aldehydes and ketones, 306
    of ethers, 306
    of primary alcohols, 304–305
    Tischtschenko–Cannizzaro reaction of aldehydes, 305–306

654 INDEX

for ethers and oxides, 140–143
hydrazines and, 454–455
for isocyanates, 368
for ketones, 207–213
for nitriles, 575–576
for olefins, 74
for sulfones, 611–612
for sulfonic acids, 629
for sulfoxides, 603–608
Oxidation–reduction reactions, bimolecular, for carboxylic acids, 251–252
Oxides, see Ethers
Oximes
 dehydration of, 553–555
 oxidation, nitro compounds and, 520–522
Oxirane compounds, condensation to give substituted oxiranes (epoxides), 136–138
3-Oxo-2,2-diphenylcyclobutyl acetate, preparation of, 219
Oxygen, catalytic oxidation with, carboxylic acids and, 241–242

P

Pelargonic acid, preparation of, 263
2,3-Pentanediol, preparation of, 104–105
Pentanonitrile, preparation of, 560–562
2-Pentylhydrazine, preparation of, 446–447
1-Pentyn-3-ol, preparation of, 89–90
Percyanocarbons, 579
Perkin reaction, for carboxylic acids, 259–260
Phenanthridone, preparation of, 352
9-Phenanthroic acid, preparation of, 270
Phenol(s), see also Alcohols
Phenol, by oxidation of phenylmagnesium bromide, 115–116
Phenoxybenzene-4,4'-disulfonyl chloride, preparation of, 633
β-Phenoxyethanesulfonic acid, preparation of, 626
Phenyl benzyl sulfone, preparation of, 616
α-Phenylbutyric acid, preparation of, 271
Phenyl cyanate, preparation of, 363
2-Phenylcyclohexanone, preparation of, 208–209
Phenyldiazomethane, preparation of, 470–471
1-Phenyldiazopropane, preparation of, 483
2-Phenyl-3-(2'-furyl)propionamide, preparation of, 333–334
Phenylmethylglycidic ester, preparation of, 139
Phosgene, carboxylic acids and, 259

β-Pinene, hydrogenation of, 6
Pivalaldehyde, preparation of, 189
Polyacrylic hydrazide, preparation of, 458–459
Poly(hexamethylenesebacamide) (Nylon 6/10), preparation of, 329–330
Polystyrene, synthesis of, 19–20
Poly(styrenediazonium chloride), preparation of, 491–492
Poly(p-xylene)sulfone, preparation of, 613
Potassium 1-methylnaphthalene-4-sulfonate, preparation of, 624–625
Propionaldehyde, preparation of, 182
Propyne, preparation of, 84
3-(2-Pyridyl)acrylic acid, preparation of, 267
Pyrolysis reactions
 for isocyanates, 368
 preparation of olefins and, 49–53
Pyromellitoyl chloride, preparation of, 156

Q

Quaternary ammonium salts, reduction of, 385–387
Quinones, oxidation
 carboxylic acids and, 246–248
 haloform reactions, 248–249
 Willgerodt reaction, 249–251

R

Raschig synthesis, modified, hydrazines and, 445–448
Rearrangement reactions
 for acetylenes, 93
 alcohols and phenols and
  aryl esters to phenolic ketones, 121
  aryl hydroxylamines, 120–121
  epoxides to allylic alcohols, 121
  ethers, 120
  sigmatropic, of olefinic alcohols, 121
 amides and
  Beckman rearrangement, 252
  Schmidt and Curtius rearrangements, 352–353
  Wolff rearrangement, 353
 for amines
  benzidine reaction, 414–415
  Curtius reaction, 418–420
  Hofmann reaction, 415–418
  Schmidt reaction, 418

INDEX

for esters
  Arndt–Eistert reaction, 307
  Favorskii reaction, 307–308
for isocyanates, 368
for ketones, 226–227
Reduction reactions
  for alcohols and phenols, 111–114
  for aldehydes
    miscellaneous reducing agents, 191
    nitriles and, 187–189
    other acid derivatives, 189–191
  amides and, 351–352
  for amines, 401–402
    amides and, 402–403
    Leuckart reaction, 412–414
    nitriles, 403–405
    nitro compounds, 405–411
  for esters, 307
  for hydrazines, 448–455
  hydrocarbons and
    alcohols, 12–13
    aromatic compounds, 9
    carbonyl compounds, 9–12
    halides, 13–15
    unsaturated compounds, 3–8
  for nitriles, 576–577
  for olefins, 69–71, 74
Reformatskii reaction, for carboxylic acids, 268–269
Reissert reaction, 580
Resorcylaldehyde, preparation of, 192–193
β-Resorcylic acid, preparation of, 256
Ritter reaction, amines and, 396–397
Rope trick, 329–330

S

Sandmeyer reaction
  halides and, 171
  for nitriles, 566–567
Schiemann reaction, halides and, 171–172
Schmidt reaction
  for amines, 418
  nitriles and, 557–559
Schmidt rearrangement, amides and, 352–353
Silver nitrate condensations, indirect nitration and, 514–516
Sodium benzenesulfonate, preparation of, 622
Sodium borohydride, reduction of methyl chloride and, 14–15

Sodium o-chlorophenylsulfinate, preparation of, 634
Sodium cumenesulfonate, p-isopropylphenol from, 104
Sodium p-fluorophenylsulfinate dihydrate, preparation of, 635
Sodium isoamyl sulfonate, preparation of, 626
Sodium α-mercaptopalmitate, oxidation of, 629
Sodium phenoxybenzenesulfonate, preparation of, 624
Sodium reagents, carboxylic acids and, 254–255
Sodium α-sulfopalmitate, preparation of, 629
Sommelet reaction, for aldehydes, 187
N-Stearoylsuccinimide, preparation of, 345–346
Stephen method, for aldehydes, 188
$trans$-Stilbene, preparation with phosphonate carbanions, 61–62
Strecker amino acid synthesis, carboxylic acids and, 266
Strecker synthesis, for sulfonic acids, 625–626
Styrene, polymerization of
  emulsion and, 20
  thermally, 19
Sulfides
  condensation reactions for
    mercaptylation of double bonds, 596
    miscellaneous methods for, 597
    preparation of episulfides from epoxides, 596–597
    reaction of metallic sulfides with halides, 595
    reaction of sodium or other metal mercaptides with active halides, 594–595
  metallic, reaction with halides, 595
  oxidation to sulfones, 611–612
  oxidation to sulfoxides, general method, 606
Sulfinic acids, preparation of, 634–637
Sulfones
  condensation methods for, 612–616
  miscellaneous methods, 617
  oxidation methods, 611–612
Sulfonic acid groups, displacement, nitration and, 539–540
Sulfonic acids
  addition of bisulfites to olefins, 626–627
  derivatives
    sulfonic esters, 633–634
    sulfonyl chlorides, 630–633

oxidation reactions, 629
reaction of sulfuric acid and derivatives with aromatic hydrocarbons, 621–625
reactions of sulfur trioxide, 627–629
Strecker synthesis, 625–626
Sulfonic esters, preparation of, 633–634
Sulfonyl chlorides, preparation of, 630–633
α-Sulfopalmitic acid, preparation of, 628–629
Sulfoxides
  miscellaneous methods for, 603–608
  oxidation methods for, 608–609
  oxidation to sulfones, 611–612
Sulfur, reaction with organometallics, 591
Sulfuric acid, and derivatives, reaction with aromatic hydrocarbons, 621–625
Sulfur trioxide, sulfonic acids and, 627–629

T

Terephthalic dihydrazide, preparation of, 457–458
2,2′,6,6′-Tetrabromobisphenol, diglycidyl ether, preparation of, 137–138
Tetrachlorobutanol, preparation of, 164–165
1,1,2,2-Tetra(4-fluorophenyl)hydrazine, preparation of, 455
Tetrahydrophthalic anhydride, 67
  preparation of, 269–270
Tetramethylene sulfone, preparation of, 611
Tetramethylene sulfoxide, preparation of, 606–607
Tetranitromethane
  nitration with, 540–541
  preparation of, 508
Thiiranes, see Episulfides
Thioacetals, reduction of, 11–12
Thiocyanates, preparation of
  addition reactions, 369–370
  condensation reactions, 369
  miscellaneous reactions, 370–371
cis and trans-3-Thiocyanoacrylamide, preparation of, 370
Thioketals, reduction of, 11–12
Thiols, see Mercaptans
3-Thiophenealdehyde, preparation of, 187
Thiourea, reaction with active halides, 588–591
Tischtschenko-Cannizzaro reactions, of aldehydes, 305–306
Tolualdehyde, preparation of, 193–194
Toluene, formylation of, 193–194

p-Toluenesulfonic acid, preparation of, 623
α-Toluic acid, preparation of, 238–239
N-(o-Tolyl)-2′,6′-acetoxylidide, preparation of, 345
p-Tolylacetylene, preparation of, 84–86
1-p-Tolyl-1-chloroethylene, p-tolylacetylene and, 84–86
p-Tolyldiazonium chloride, conversion to toluene, 28
p-Tolylsulfonyl hydrazones, of ketones, nitriles and, 557
Tosylates, reaction with amines, 382–383
Transamidation reactions, of amides, 340–342
3-Trichlorosilylpropionitrile, preparation of, 575
4,4′,4″-Trichlorotribenzamide, preparation of, 328–329
1,3,5-Triethylbenzene, preparation by Friedel-Crafts reaction, 17–18
1,1,1-Trifluoro-2-diazopropane, preparation of, 477–478
Trihalides, hydrolysis, carboxylic acids and, 271–273
1,1,1-Trihalomethyl derivatives, hydrolysis, carboxylic acids and, 273–274
Trimellitic anhydride acid chloride, preparation of, 157
Trimellitoyl chloride, preparation of, 158
Trimesoyl chloride, preparation of, 157
Trimethylacetic acid, preparation of, 246
Trimethylacetonitrile, reduction with lithium triethoxyaluminum hydride, 189
2,2,3-Trimethyl-1-butanol, hydrogenolysis of, 12
N,N,N-Trimethylhydrazonium salts, decomposition of, 555–556
7,7,10-Trimethyl-$\Delta^{1(9)}$-octalin, preparation of, 11–12
2,2,2-Trinitroethanol, preparation of, 525–526
Triphenylmethyl ethyl ether, preparation of, 132–133
Triphenylmethyl mercaptan, preparation of, 587–588
1,1,1-Tris(2-cyanoethyl)acetone, preparation of, 574
1,1,1-Tris(hydroxymethyl)-2-methylpropane, preparation of, 119–120

U

Ullmann synthesis, of 2,2′-diethylbiphenyl, 25

INDEX

Undecyl thiocyanate, preparation of, 369
Unsaturated compounds, addition of hydrogen sulfide to, 592–593

V

Valeronitrile, preparation of, 560–562
Vilsmeier method, for aldehydes, 194–195
Vinyl acetate, reaction with diphenylketene, 219
Vinyl caproate, preparation of, 298–299
Vinyl chloroacetate, preparation of, 63

W

Willgerodt reaction, for carboxylic acids, 249–251
Williamson synthesis, for ethers, 132–133
Wittig synthesis, of olefins, 59–62
Wolff–Kishner method, for reduction of carbonyl compounds, 9–10
Wolff rearrangement, amides and, 353

X

Xanthates, hydrolysis of, 591–592

ORGANIC CHEMISTRY
A SERIES OF MONOGRAPHS

EDITOR

HARRY H. WASSERMAN

Department of Chemistry
Yale University
New Haven, Connecticut

1. Wolfgang Kirmse. CARBENE CHEMISTRY, 1964; 2nd Edition, 1971
2. Brandes H. Smith. BRIDGED AROMATIC COMPOUNDS, 1964
3. Michael Hanack. CONFORMATION THEORY, 1965
4. Donald J. Cram. FUNDAMENTALS OF CARBANION CHEMISTRY, 1965
5. Kenneth B. Wiberg (Editor). OXIDATION IN ORGANIC CHEMISTRY, PART A, 1965; Walter S. Trahanovsky (Editor). OXIDATION IN ORGANIC CHEMISTRY, PART B, 1973; PART C, 1978; PART D, 1982
6. R. F. Hudson. STRUCTURE AND MECHANISM IN ORGANO-PHOSPHORUS CHEMISTRY, 1965
7. A. William Johnson. YLID CHEMISTRY, 1966
8. Jan Hamer (Editor). 1,4-CYCLOADDITION REACTIONS, 1967
9. Henri Ulrich. CYCLOADDITION REACTIONS OF HETEROCUMULENES, 1967
10. M. P. Cava and M. J. Mitchell. CYCLOBUTADIENE AND RELATED COMPOUNDS, 1967
11. Reinhard W. Hoffmann. DEHYDROBENZENE AND CYCLOALKYNES, 1967
12. Stanley R. Sandler and Wolf Karo. ORGANIC FUNCTIONAL GROUP PREPARATIONS, VOLUME I, 1968; 2ND EDITION, 1983; VOLUME II, 1971; VOLUME III, 1972
13. Robert J. Cotter and Markus Matzner. RING-FORMING POLYMERIZATIONS, PART A, 1969; PART B, 1; B, 2, 1972
14. R. H. DeWolfe, CARBOXYLIC ORTHO ACID DERIVATIVES, 1970
15. R. Foster. ORGANIC CHARGE-TRANSFER COMPLEXES, 1969
16. James P. Snyder (Editor). NONBENZENOID AROMATICS, VOLUME I, 1969; VOLUME II, 1971
17. C. H. Rochester. ACIDITY FUNCTIONS, 1970
18. Richard J. Sundberg. THE CHEMISTRY OF INDOLES, 1970
19. A. R. Katritzky and J. M. Lagowski. CHEMISTRY OF THE HETEROCYCLIC N-OXIDES, 1970
20. Ivar Ugi (Editor). ISONITRILE CHEMISTRY, 1971
21. G. Chiurdoglu (Editor). CONFORMATIONAL ANALYSIS, 1971
22. Gottfried Schill. CATENANES, ROTAXANES, AND KNOTS, 1971
23. M. Liler. REACTION MECHANISMS IN SULPHURIC ACID AND OTHER STRONG ACID SOLUTIONS, 1971
24. J. B. Stothers. CARBON-13 NMR SPECTROSCOPY, 1972

25. Maurice Shamma. THE ISOQUINOLINE ALKALOIDS: CHEMISTRY AND PHARMACOLOGY, 1972

26. Samuel P. McManus (Editor). ORGANIC REACTIVE INTERMEDIATES, 1973

27. H. C. Van der Plas. RING TRANSFORMATIONS OF HETEROCYCLES, VOLUMES 1 AND 2, 1973

28. Paul N. Rylander. ORGANIC SYNTHESES WITH NOBLE CATALYSTS, 1973

29. Stanley R. Sandler and Wolf Karo. POLYMER SYNTHESES, VOLUME I, 1974; VOLUME II, 1977; VOLUME III, 1980

30. Robert T. Blickenstaff, Anil C. Ghosh, and Gordon C. Wolf. TOTAL SYNTHESIS OF STEROIDS, 1974

31. Barry M. Trost and Lawrence S. Melvin, Jr. SULFUR YLIDES: EMERGING SYNTHETIC INTERMEDIATES, 1975

32. Sidney D. Ross, Manuel Finkelstein, and Eric J. Rudd. ANODIC OXIDATION, 1975

33. Howard Alper (Editor). TRANSITION METAL ORGANOMETALLICS IN ORGANIC SYNTHESIS, VOLUME I, 1976; VOLUME II, 1978

34. R. A. Jones and G. P. Bean. THE CHEMISTRY OF PYRROLES, 1976

35. Alan P. Marchand and Roland E. Lehr (Editors). PERICYCLIC REACTIONS, VOLUME I, 1977; VOLUME II, 1977

36. Pierre Crabbé (Editor). PROSTAGLANDIN RESEARCH, 1977

37. Eric Block. REACTIONS OF ORGANOSULFUR COMPOUNDS, 1978

38. Arthur Greenberg and Joel F. Liebman, STRAINED ORGANIC MOLECULES, 1978

39. Philip S. Bailey. OZONATION IN ORGANIC CHEMISTRY, VOLUME I, 1978; VOLUME II, 1982

40. Harry H. Wasserman and Robert W. Murray (Editors). SINGLET OXYGEN, 1979

41. Roger F. C. Brown. PYROLYTIC METHODS IN ORGANIC CHEMISTRY: APPLICATIONS OF FLOW AND FLASH VACUUM PYROLYTIC TECHNIQUES, 1980

42. Paul de Mayo (Editor). REARRANGEMENTS IN GROUND AND EXCITED STATES, VOLUME I, 1980; VOLUME II, 1980; VOLUME III, 1980

43. Elliot N. Marvell. THERMAL ELECTROCYCLIC REACTIONS, 1980

44. Joseph J. Gajewski. HYDROCARBON THERMAL ISOMERIZATIONS, 1981

45. Philip M. Keehn and Stuart M. Rosenfeld (Editors). CYCLOPHANES VOLUME I, 1983; VOLUME II, 1983